# Triangles and Angles

AF870498

***Right Triangle***

Triangle has one 90° (right) angle.

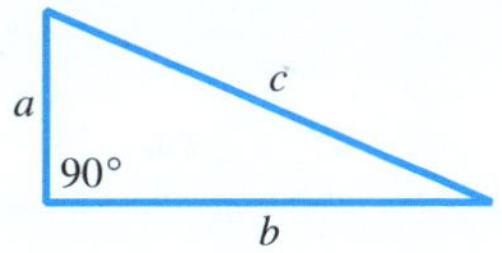


***Pythagorean Formula (for right triangles)***

$c^2 = a^2 + b^2$

***Right Angle***

Measure is 90°.

***Isosceles Triangle***

Two sides are equal.

$AB = BC$

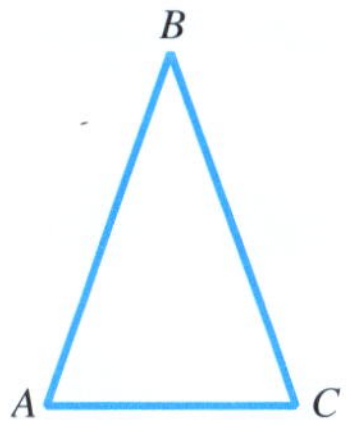


***Straight Angle***

Measure is 180°.

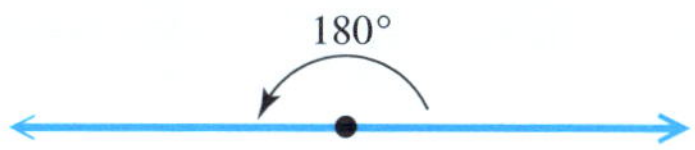


***Equilateral Triangle***

All sides are equal.

$AB = BC = CA$

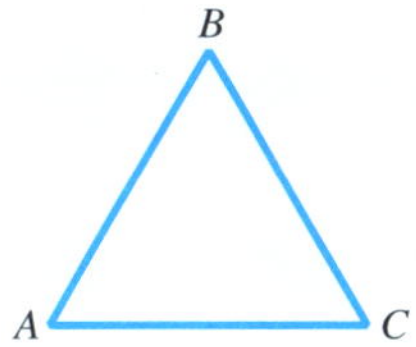


***Complementary Angles***

The sum of the measures of two complementary angles is 90°.

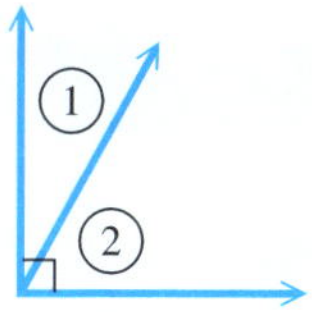


Angles (1) and (2) are complementary.

***Sum of the Angles of Any Triangle***

$A + B + C = 180°$

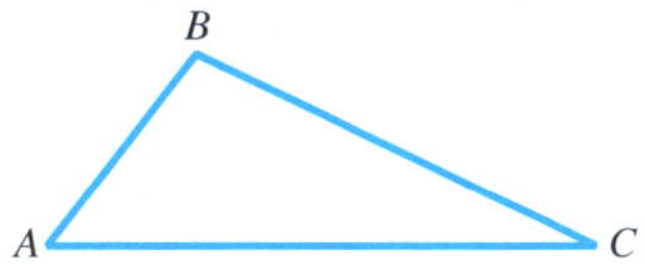


***Supplementary Angles***

The sum of the measures of two supplementary angles is 180°.

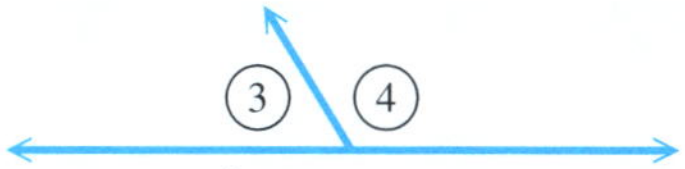


Angles (3) and (4) are supplementary.

***Similar Triangles***

Corresponding angles are equal; corresponding sides are proportional.

$A = D, B = E, C = F$

$$\frac{AB}{DE} = \frac{AC}{DF} = \frac{BC}{EF}$$

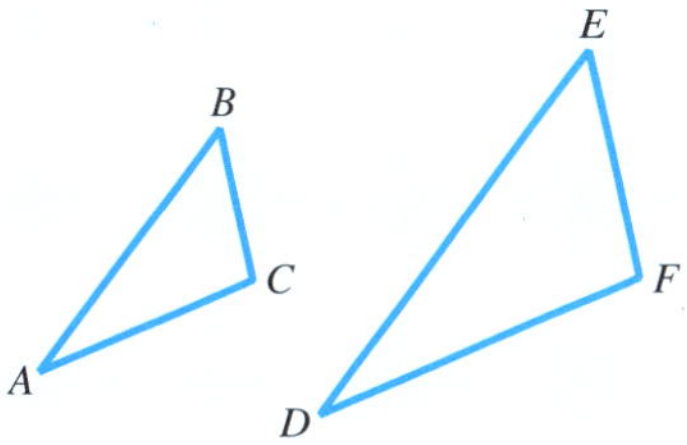


***Vertical Angles***

Vertical angles have equal measures.

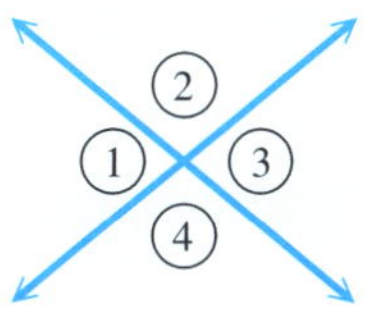


Angle (1) = Angle (3)

Angle (2) = Angle (4)

Taken from:

*Beginning and Intermediate Algebra*, Third Edition
by Margaret L. Lial, John Hornsby, and Terry McGinnis
Copyright © 2004 by Pearson Education, Inc.
Published by Addison-Wesley
Boston, Massachusetts 02116

Copyright © 2005 by Pearson Custom Publishing
All rights reserved.

Permission in writing must be obtained from the publisher before any part of this work may be reproduced or transmitted in any form or by any means, electronic or mechanical, including photocopying and recording, or by any information storage or retrieval system.

Printed in the United States of America

10 9 8 7

ISBN 0-536-84775-4

2004360449

LR

Please visit our web site at *www.pearsoncustom.com*

PEARSON CUSTOM PUBLISHING
75 Arlington Street, Suite 300, Boston, MA 02116
A Pearson Education Company

# Contents

# Math 101 Formulas

## Course Formulas

Formulas that students are required to memorize in this course:

### Chapter 1

- Multiplying and Dividing Fractions

$$\frac{a}{b}\cdot\frac{c}{d}=\frac{a\cdot c}{b\cdot d} \quad (b\neq 0,\ d\neq 0)$$

$$\frac{a}{b}\div\frac{c}{d}=\frac{a}{b}\cdot\frac{d}{c} \quad (b\neq 0,\ c\neq 0,\ d\neq 0)$$

- Adding and Subtracting Fractions

$$\frac{a}{b}+\frac{c}{b}=\frac{a+c}{b} \quad (b\neq 0)$$

$$\frac{a}{b}-\frac{c}{b}=\frac{a-c}{b} \quad (b\neq 0)$$

- Double Negative Rule $-(-x) = x$

- Absolute Value of $x$ $\quad |x|=\begin{cases} x & \text{if } x\geq 0 \\ -x & \text{if } x<0 \end{cases}$

- Multiplying and Dividing Real Numbers

$x\cdot 0 = 0 \qquad x(-y) = -(xy) \qquad (-x)y = -(xy) \qquad -x(-y) = xy$

$\frac{0}{a} = 0$ *for* $a \neq 0 \qquad \frac{a}{0}$ is undefined for $a\neq 0 \qquad \frac{0}{0}$ is indeterminate

$$\frac{-x}{y}=\frac{x}{-y}=-\frac{x}{y} \quad \text{and} \quad \frac{-x}{-y}=\frac{x}{y} \quad \text{for } y\neq 0$$

- Properties of Addition & Multiplication

Commutative Properties $\quad a+b=b+a \quad \& \quad ab=ba$

Associative Properties $\quad (a+b)+c=a+(b+c) \quad \& \quad (ab)c=a(bc)$

Distributive Properties $\quad a(b+c)=ab+ac \quad \& \quad a(b-c)=ab-ac$

Chapter 2

- Sum of the Angles in a Triangle $A + B + C = 180°$

- Geometric Formulas

| | Area | Perimeter |
|---|---|---|
| Square | $A = s^2$ | $P = 4s$ |
| Rectangle | $A = LW$ | $P = 2L + 2W$ |
| Triangle | $A = \frac{1}{2}(bh)$ | $P = a + b + c$ |
| Circle | $A = \pi r^2$ | $C = 2\pi r$ |

- Interest $I = prt$

Chapter 4

- Properties of Exponents

Zero Exponent $a^0 = 1, \quad a \neq 0$

Negative Exponent $a^{-n} = \frac{1}{a^n}, \quad a \neq 0$

Product Rule $a^m \cdot a^n = a^{m+n}$

Quotient Rule $\frac{a^m}{a^n} = a^{m-n}, \quad a \neq 0$

Power Rules

$(a^m)^n = a^{mn}$

$(ab)^m = a^m b^m$

$\left(\frac{a}{b}\right)^m = \frac{a^m}{b^m}, \quad b \neq 0$

$\frac{a^{-m}}{b^{-n}} = \frac{b^n}{a^m}, \quad a \neq 0, b \neq 0$

$\left(\frac{a}{b}\right)^{-m} = \left(\frac{b}{a}\right)^m, \quad a \neq 0, b \neq 0$

- Square of a Binomial: $(x+y)^2 = x^2 + 2xy + y^2$

$(x-y)^2 = x^2 - 2xy + y^2$

- Product of the Sum and Difference of Two Terms $(x+y)(x-y) = x^2 - y^2$

- Dividing a Polynomial by a Monomial $\frac{a+b}{c} = \frac{a}{c} + \frac{b}{c} \quad (c \neq 0)$

## Chapter 5

- Difference of Squares $x^2 - y^2 = (x + y)(x - y)$

- Sums and Differences of Cubes $x^3 + y^3 = (x + y)(x^2 - xy + y^2)$
  $x^3 - y^3 = (x - y)(x^2 + xy + y^2)$

- Pythagorean Theorem $a^2 + b^2 = c^2$

## Chapter 6

- Multiplying and Dividing Rational Expressions

$$\frac{P}{Q} \cdot \frac{R}{S} = \frac{PR}{QS} \quad (Q \neq 0, S \neq 0) \qquad \frac{P}{Q} \div \frac{R}{S} = \frac{P}{Q} \cdot \frac{S}{R} = \frac{PS}{QR} \quad (Q \neq 0, R \neq 0, S \neq 0)$$

- Adding and Subtracting Rational Expressions

$$\frac{P}{Q} + \frac{R}{Q} = \frac{P+R}{Q} \quad \text{and} \quad \frac{P}{Q} - \frac{R}{Q} = \frac{P-R}{Q} \quad (Q \neq 0)$$

## Suggestions to Help You Succeed in a Math Course

1. **READ YOUR TEXTBOOK.** Always read the material for a topic before attempting the homework. Read carefully, with pencil and paper at hand *working problems through* as you go. You may find that you have to reread something many times to understand. If possible, read the material for a topic *before* it is covered in class.

2. **ATTEND ALL CLASSES AND TAKE DETAILED NOTES**. The instructor will present many theorems, definitions and problems in class. Many examples will be very helpful when you do the homework assigned. Keep a notebook with daily notes and handouts for the class together.

3. **STUDY IN THE SAME PLACE EVERY DAY.** Studies have shown that you will learn better if you are consistent with study time and location. A general guideline is to study about 2 hours for every 50 minutes spent in class.

4. **COPY OVER AND REORGANIZE CLASS NOTES.** Between every two consecutive classes, copy over your notes and organize them in a way that is most easy for you to understand. This can be time consuming, but it really works!

5. **ATTEMPT ALL HOMEWORK PROBLEMS.** Do not give up too easily on a problem. If you get frustrated or stumped on a problem, leave it alone for a while and then come back to it. Sometimes all you need is a fresh perspective to solve the problem. It is generally of little help to find the answer in the back of the book and try to work backwards to the answer. Use the solutions in the back of the book to check all work, and only as a very last resort if you have attempted a problem multiple times without success.

6. **ORGANIZE A STUDY GROUP.** Work in groups with other students in class. If you can't get together on campus or off campus, exchange phone numbers with several classmates to call at night with homework questions, and when studying for tests.

7. **REVIEW DAILY.** Each time you sit down to study, go back to previous sections and work a few problems. Keep definitions, theorems, methods and review problems on index cards, and randomly pull out a few each night to review.

8. **REDO INCORRECT HOMEWORK, QUIZ OR TEST PROBLEMS**. When graded work is returned, spend time figuring out what went wrong. Determine a good way to complete the problems. Find similar problems in the text and work them until you master the concept.

9. **ASK MANY QUESTIONS.** Do not be shy when you are not following something in class. Ask the instructor to stop and clarify. The instructor cannot read your mind, so speak up when something confuses you. Chances are, many people will have the same questions that you do.

10. **USE ALL RESOURCES AVAILABLE TO YOU.** Make use of the Math Help Room, the Math Lab, Tutorial Services and the instructor's office hours whenever you need to.

# OPPORTUNITIES AVAILABLE TO MATH STUDENTS

## Extra Help

- Contact your **instructor** during office hours, or make an appointment.

- **Math Lab, Room 1315**: Computer tutorials and video tape lessons for almost all topics covered in math classes. Check to the right of the door for hours of operation.

- **Math Help Room, Room 1343**: Quiet study area with faculty and student tutors present for individual help. Staffing schedule and hours are posted on the door.

- **Tutorial Services, Room 10-444A**; Individual tutors are available for 100 level classes. Call 512-2792 to make arrangements.

- **Tutorial Websites**:
  MyMathLab accompanies the Mat 101/102 text. Your individual access code comes packaged with the software. Check with your instructor for the Course Compass ID.

  The Math Lab lists current websites tied to individual courses on cards inside the Lab.

- **Computerized Practice Exams** for Math 101 and 102 in the Math Lab

- **Retention Program**: Must be recommended by instructor to Retention Specialist. Eligible students are those:
  1. Who have satisfied the prerequisite for the course
  2. Are attending class regularly
  3. Attempting all required assignments
  4. Scored a high F or D on the first test, and are in "striking distance" of succeeding in the class.

- **Test Review Sessions** for MAT 101 and MAT 102 ( schedule is posted in the Math Lab, Room 1315)

- **Workshops** on Math Anxiety, Test Taking Tips, Study Skills for Math Classes (schedule is posted in the Math Lab, Room 1315)

Note: The Retention Program, Test Review Sessions, and special workshops are not available during the summer months.

# The Real Number System 1

The table indicates the figures for several stocks listed on the NASDAQ stock exchange for Wednesday, May 29, 2002. The final column represents the change in the price of the stock from the previous day. These changes are indicated by signed numbers. A positive number, such as +.75, means that the price of one share of the stock rose 75 cents from the previous day's closing; a negative number, such as −2.24, means that the price fell \$2.24. In the exercises for Sections 1.1 and 1.4, we apply the concepts of this chapter to investigate situations involving stock prices.

| *Stock* | *Vol.* | *Close* | *Chg.* |
|---|---|---|---|
| AAONs | 14 | 31.85 | +.75 |
| ABC Bcp | 7 | 14.55 | |
| AC Moore | 58 | 40.35 | −2.24 |
| ACT Tele | 400 | 3.30 | +.14 |
| ADC Tel | 4752 | 3.49 | −.12 |
| AFC Ent | 225 | 30.40 | −.96 |
| AHL Sev | 79 | 2.03 | −.21 |

*Source: Times Picayune.*

# 1.1 Fractions

**OBJECTIVES**

1. Learn the definition of *factor.*
2. Write fractions in lowest terms.
3. Multiply and divide fractions.
4. Add and subtract fractions.
5. Solve applied problems that involve fractions.
6. Interpret data in a circle graph.

As preparation for the study of algebra, we begin this section with a brief review of arithmetic. In everyday life, the numbers seen most often are the **natural numbers,**

$$1, 2, 3, 4, \ldots,$$

the **whole numbers,**

$$0, 1, 2, 3, 4, \ldots,$$

and **fractions,** such as

$$\frac{1}{2}, \quad \frac{2}{3}, \quad \text{and} \quad \frac{15}{7}.$$

The parts of a fraction are named as follows.

Fraction bar → $\dfrac{4}{7}$ ← Numerator (4) ← Denominator (7)

As we will see later, the fraction bar represents division $\left(\frac{a}{b} = a \div b\right)$ and also serves as a grouping symbol.

If the numerator of a fraction is less than the denominator, we call it a **proper fraction.** A proper fraction has a value less than 1. If the numerator is greater than or equal to the denominator, the fraction is an **improper fraction.** An improper fraction that has a value greater than 1 is often written as a **mixed number.** For example, $\frac{12}{5}$ may be written as $2\frac{2}{5}$. In algebra, we prefer to use the improper form because it is easier to work with. In applications, we usually convert answers to mixed form, which is more meaningful.

**OBJECTIVE 1 Learn the definition of *factor.*** In the statement $2 \times 9 = 18$, the numbers 2 and 9 are called **factors** of 18. Other factors of 18 include 1, 3, 6, and 18. The result of the multiplication, 18, is called the **product.**

There are several ways of representing the product of two numbers. For example,

$$6 \times 3, \quad 6 \cdot 3, \quad (6)(3), \quad 6(3), \quad \text{and} \quad (6)3$$

all represent the product of 6 and 3.

The number 18 is **factored** by writing it as the product of two or more numbers. For example, 18 can be factored as $6 \cdot 3$, $18 \cdot 1$, $9 \cdot 2$, or $3 \cdot 3 \cdot 2$. (In algebra, a raised dot $\cdot$ is often used instead of the $\times$ symbol to indicate multiplication.)

A natural number (except 1) is **prime** if it has only itself and 1 as factors. "Factors" are understood here to mean natural number factors. (By agreement, the number 1 is not a prime number.) The first dozen primes are

$$2, 3, 5, 7, 11, 13, 17, 19, 23, 29, 31, 37.$$

A natural number (except 1) that is not prime, such as 4, 6, 8, 9, and 12, is a **composite** number.

It is often useful to find all **prime factors** of a number—those factors that are prime numbers. For example, the only prime factors of 18 are 2 and 3.

**EXAMPLE 1 Factoring Numbers**

Write each number as the product of prime factors.

**(a)** 35

Write 35 as the product of the prime fractors 5 and 7, or as

$$35 = 5 \cdot 7.$$

**(b)** 24

One way to begin is to divide by the smallest prime, 2, to get

$$24 = 2 \cdot 12.$$

Now divide 12 by 2 to find factors of 12.

$$24 = 2 \cdot 2 \cdot 6$$

Since 6 can be written as $2 \cdot 3$,

$$24 = 2 \cdot 2 \cdot 2 \cdot 3,$$

where all factors are prime.

**Now Try Exercises 9 and 19.**

**NOTE** It is not necessary to start with the smallest prime factor, as shown in Example 1(b). In fact, no matter which prime factor we start with, we will *always* obtain the same prime factorization.

**OBJECTIVE 2 Write fractions in lowest terms.** We use prime numbers to write fractions in *lowest terms.* A fraction is in **lowest terms** when the numerator and denominator have no factors in common (other than 1). We use the **basic principle of fractions** to write a fraction in lowest terms.

**Basic Principle of Fractions**

If the numerator and denominator of a fraction are multiplied or divided by the same nonzero number, the value of the fraction is not changed.

For example, $\frac{12}{16}$ can be written in lowest terms as follows:

$$\frac{12}{16} = \frac{3 \cdot 4}{4 \cdot 4} = \frac{3}{4} \cdot \frac{4}{4} = \frac{3}{4} \cdot 1 = \frac{3}{4}.$$

This procedure uses the rule for multiplying fractions (covered in the next objective) and the multiplication property of 1 (covered in Section 1.7).

To write a fraction in lowest terms, use these steps.

**Writing a Fraction in Lowest Terms**

*Step 1* Write the numerator and the denominator as the product of prime factors.

*Step 2* Divide the numerator and the denominator by the **greatest common factor,** the product of all factors common to both.

**EXAMPLE 2 Writing Fractions in Lowest Terms**

Write each fraction in lowest terms.

**(a)** $\frac{10}{15} = \frac{2 \cdot 5}{3 \cdot 5} = \frac{2 \cdot 1}{3 \cdot 1} = \frac{2}{3}$

Since 5 is the greatest common factor of 10 and 15, dividing both numerator and denominator by 5 gives the fraction in lowest terms.

**(b)** $\frac{15}{45} = \frac{3 \cdot 5}{3 \cdot 3 \cdot 5} = \frac{1 \cdot 1}{3 \cdot 1 \cdot 1} = \frac{1}{3}$

The factored form shows that 3 and 5 are the common factors of both 15 and 45. Dividing both 15 and 45 by $3 \cdot 5 = 15$ gives $\frac{15}{45}$ in lowest terms as $\frac{1}{3}$.

**Now Try Exercises 25 and 29.**

We can simplify this process by finding the greatest common factor in the numerator and denominator by inspection. For instance, in Example 2(b), we can use 15 rather than $3 \cdot 5$.

$$\frac{15}{45} = \frac{15}{3 \cdot 15} = \frac{1}{3 \cdot 1} = \frac{1}{3}$$

**CAUTION** Errors may occur when writing fractions in lowest terms if the factor 1 is not included. To see this, refer to Example 2(b). In the equation

$$\frac{3 \cdot 5}{3 \cdot 3 \cdot 5} = \frac{?}{3},$$

if 1 is not written in the numerator when dividing out common factors, you may make an error. The ? should be replaced by 1.

**OBJECTIVE 3 Multiply and divide fractions.** We multiply two fractions by first multiplying their numerators and then multiplying their denominators. This rule is written in symbols as follows.

**Multiplying Fractions**

If $\frac{a}{b}$ and $\frac{c}{d}$ are fractions, then $\frac{a}{b} \cdot \frac{c}{d} = \frac{a \cdot c}{b \cdot d}$.

**EXAMPLE 3 Multiplying Fractions**

Find each product, and write it in lowest terms.

**(a)** $\frac{3}{8} \cdot \frac{4}{9} = \frac{3 \cdot 4}{8 \cdot 9}$ Multiply numerators. Multiply denominators.

$= \frac{3 \cdot 4}{2 \cdot 4 \cdot 3 \cdot 3}$ Factor.

$= \frac{1}{2 \cdot 3} = \frac{1}{6}$ Write in lowest terms.

**(b)** $2\frac{1}{3} \cdot 5\frac{1}{2} = \frac{7}{3} \cdot \frac{11}{2}$ Write as improper fractions.

$= \frac{77}{6}$ or $12\frac{5}{6}$ Multiply numerators and denominators.

**Now Try Exercises 35 and 39.**

Two fractions are **reciprocals** of each other if their product is 1. For example, $\frac{3}{4}$ and $\frac{4}{3}$ are reciprocals since

$$\frac{3}{4} \cdot \frac{4}{3} = \frac{12}{12} = 1.$$

Also, $\frac{7}{11}$ and $\frac{11}{7}$ are reciprocals of each other, as are $\frac{1}{6}$ and 6. Because division is the opposite (or inverse) of multiplication, we use reciprocals to divide fractions. To divide two fractions, multiply the first fraction by the reciprocal of the second fraction.

**Dividing Fractions**

For the fractions $\frac{a}{b}$ and $\frac{c}{d}$, $\quad \frac{a}{b} \div \frac{c}{d} = \frac{a}{b} \cdot \frac{d}{c}$.

That is, to divide by a fraction, multiply by its reciprocal.

We will explain why this method works in Chapter 6. However, as an example, we know that $20 \div 10 = 2$ and $20 \cdot \frac{1}{10} = 2$. The answer to a division problem is called a **quotient.** For example, the quotient of 20 and 10 is 2.

## EXAMPLE 4 Dividing Fractions

Find each quotient, and write it in lowest terms.

**(a)** $\frac{3}{4} \div \frac{8}{5} = \frac{3}{4} \cdot \frac{5}{8} = \frac{3 \cdot 5}{4 \cdot 8} = \frac{15}{32}$

Multiply by the reciprocal of the second fraction.

**(b)** $\frac{3}{4} \div \frac{5}{8} = \frac{3}{4} \cdot \frac{8}{5} = \frac{3 \cdot 8}{4 \cdot 5} = \frac{3 \cdot 4 \cdot 2}{4 \cdot 5} = \frac{6}{5}$ or $1\frac{1}{5}$

**(c)** $\frac{5}{8} \div 10 = \frac{5}{8} \div \frac{10}{1} = \frac{5}{8} \cdot \frac{1}{10} = \frac{1}{16}$

Write 10 as $\frac{10}{1}$.

**(d)** $1\frac{2}{3} \div 4\frac{1}{2} = \frac{5}{3} \div \frac{9}{2}$ Write as improper fractions.

$= \frac{5}{3} \cdot \frac{2}{9}$ Multiply by the reciprocal of the second fraction.

$= \frac{10}{27}$

**Now Try Exercises 43, 45, and 47.**

**OBJECTIVE 4 Add and subtract fractions.** The result of adding two numbers is called the **sum** of the numbers. For example, $2 + 3 = 5$, so 5 is the sum of 2 and 3. To find the sum of two fractions having the same denominator, add the numerators and keep the same denominator.

**Adding Fractions**

If $\frac{a}{b}$ and $\frac{c}{b}$ are fractions, then $\frac{a}{b} + \frac{c}{b} = \frac{a+c}{b}$.

**EXAMPLE 5 Adding Fractions with the Same Denominator**

Add.

**(a)** $\frac{3}{7} + \frac{2}{7} = \frac{3+2}{7} = \frac{5}{7}$ Add numerators; keep the same denominator.

**(b)** $\frac{2}{10} + \frac{3}{10} = \frac{2+3}{10} = \frac{5}{10} = \frac{1}{2}$ Write in lowest terms.

**Now Try Exercise 51.**

If the fractions to be added do not have the same denominators, we can still use the procedure above, but only *after* we rewrite the fractions with a common denominator. For example, to rewrite $\frac{3}{4}$ as a fraction with denominator 32,

$$\frac{3}{4} = \frac{?}{32},$$

we find the number that can be multiplied by 4 to give 32. Since $4 \cdot 8 = 32$, we use the number 8. By the basic principle of fractions, we multiply the numerator and the denominator by 8.

$$\frac{3}{4} = \frac{3 \cdot 8}{4 \cdot 8} = \frac{24}{32}$$

**Finding the Least Common Denominator**

To add or subtract fractions with different denominators, find the **least common denominator (LCD)** as follows.

*Step 1* Factor both denominators.

*Step 2* For the LCD, use every factor that appears in any factored form. If a factor is repeated, use the largest number of repeats in the LCD.

The next example shows this procedure.

**EXAMPLE 6 Adding Fractions with Different Denominators**

Add. Write the sums in lowest terms.

**(a)** $\frac{4}{15} + \frac{5}{9}$

To find the least common denominator, first factor both denominators.

$$15 = 5 \cdot 3 \qquad \text{and} \qquad 9 = 3 \cdot 3$$

Since 5 and 3 appear as factors, and 3 is a factor of 9 twice, the LCD is

$$\overbrace{5 \cdot 3}^{15} \cdot 3 \quad (\text{with } 3 \cdot 3 = 9) \qquad \text{or} \qquad 45.$$

Write each fraction with 45 as denominator.

$$\frac{4}{15} = \frac{4 \cdot 3}{15 \cdot 3} = \frac{12}{45} \qquad \text{and} \qquad \frac{5}{9} = \frac{5 \cdot 5}{9 \cdot 5} = \frac{25}{45}$$

Now add the two equivalent fractions.

$$\frac{4}{15} + \frac{5}{9} = \frac{12}{45} + \frac{25}{45} = \frac{37}{45}$$

**(b)** $3\frac{1}{2} + 2\frac{3}{4}$

We add mixed numbers using either of two methods.

*Method 1*

Rewrite both numbers as improper fractions.

$$3\frac{1}{2} = 3 + \frac{1}{2} = \frac{3}{1} + \frac{1}{2} = \frac{6}{2} + \frac{1}{2} = \frac{6+1}{2} = \frac{7}{2}$$

$$2\frac{3}{4} = 2 + \frac{3}{4} = \frac{8}{4} + \frac{3}{4} = \frac{8+3}{4} = \frac{11}{4}$$

Now add. The common denominator is 4.

$$3\frac{1}{2} + 2\frac{3}{4} = \frac{7}{2} + \frac{11}{4} = \frac{14}{4} + \frac{11}{4} = \frac{25}{4} \quad \text{or} \quad 6\frac{1}{4}$$

*Method 2*

Write $3\frac{1}{2}$ as $3\frac{2}{4}$. Then add vertically.

$$\begin{array}{r} 3\frac{1}{2} \\ +\ 2\frac{3}{4} \\ \hline \end{array} \quad \rightarrow \quad \begin{array}{r} 3\frac{2}{4} \\ +\ 2\frac{3}{4} \\ \hline 5\frac{5}{4} \end{array}$$

Since $\frac{5}{4} = 1\frac{1}{4}$,

$$5\frac{5}{4} = 5 + 1\frac{1}{4} = 6\frac{1}{4} \quad \text{or} \quad \frac{25}{4}.$$

**Now Try Exercises 53 and 55.**

The **difference** between two numbers is found by subtracting the numbers. For example, $9 - 5 = 4$, so the difference between 9 and 5 is 4. Subtraction of fractions is similar to addition. Just subtract the numerators instead of adding them; again, keep the same denominator.

**Subtracting Fractions**

If $\frac{a}{b}$ and $\frac{c}{b}$ are fractions, then $\frac{a}{b} - \frac{c}{b} = \frac{a-c}{b}$.

**EXAMPLE 7 Subtracting Fractions**

Subtract. Write the differences in lowest terms.

**(a)** $\frac{15}{8} - \frac{3}{8} = \frac{15-3}{8}$ Subtract numerators; keep the same denominator.

$= \frac{12}{8} = \frac{3}{2}$ or $1\frac{1}{2}$ Lowest terms

**(b)** $\frac{7}{18} - \frac{4}{15}$

Here, $18 = 2 \cdot 3 \cdot 3$ and $15 = 3 \cdot 5$, so the LCD is $2 \cdot 3 \cdot 3 \cdot 5 = 90$.

$$\frac{7}{18} - \frac{4}{15} = \frac{7 \cdot 5}{2 \cdot 3 \cdot 3 \cdot 5} - \frac{4 \cdot 2 \cdot 3}{2 \cdot 3 \cdot 3 \cdot 5} = \frac{35}{90} - \frac{24}{90} = \frac{11}{90}$$

**(c)** $\frac{15}{32} - \frac{11}{45}$

Since $32 = 2 \cdot 2 \cdot 2 \cdot 2 \cdot 2$ and $45 = 3 \cdot 3 \cdot 5$, there are no common factors, and the LCD is $32 \cdot 45 = 1440$.

$\frac{15}{32} - \frac{11}{45} = \frac{15 \cdot 45}{32 \cdot 45} - \frac{11 \cdot 32}{45 \cdot 32}$ Get a common denominator.

$= \frac{675}{1440} - \frac{352}{1440}$

$= \frac{323}{1440}$ Subtract.

**Now Try Exercises 57 and 59.**

**OBJECTIVE 5 Solve applied problems that involve fractions.** Applied problems often require work with fractions. For example, when a carpenter reads diagrams and plans, he or she often must work with fractions whose denominators are 2, 4, 8, 16, or 32, as shown in the next example.

**EXAMPLE 8 Adding Fractions to Solve an Applied Problem**

The diagram in Figure 1 appears in the book *Woodworker's 39 Sure-Fire Projects.* It is the front view of a corner bookcase/desk. Add the fractions shown in the diagram to find the height of the bookcase/desk to the top of the writing surface.

We must add the following measures (″ means inches):

$$\frac{3}{4},\quad 4\frac{1}{2},\quad 9\frac{1}{2},\quad \frac{3}{4},\quad 9\frac{1}{2},\quad \frac{3}{4},\quad 4\frac{1}{2}.$$

We begin by changing $4\frac{1}{2}$ to $4\frac{2}{4}$ and $9\frac{1}{2}$ to $9\frac{2}{4}$, since the common denominator is 4. Then we use Method 2 from Example 6(b).

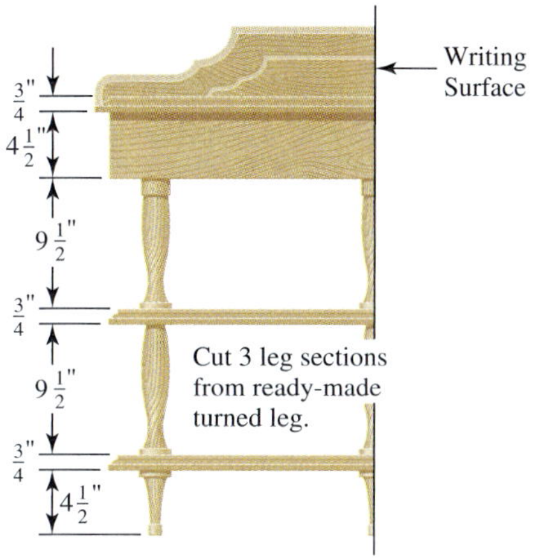


$$\begin{array}{rcr} \frac{3}{4} & & \frac{3}{4} \\ 4\frac{1}{2} & & 4\frac{2}{4} \\ 9\frac{1}{2} & & 9\frac{2}{4} \\ \frac{3}{4} & \rightarrow & \frac{3}{4} \\ 9\frac{1}{2} & & 9\frac{2}{4} \\ \frac{3}{4} & & \frac{3}{4} \\ +\ 4\frac{1}{2} & & 4\frac{2}{4} \\ \hline & & 26\frac{17}{4} \end{array}$$

FIGURE 1

Since $\frac{17}{4} = 4\frac{1}{4}$, $26\frac{17}{4} = 26 + 4\frac{1}{4} = 30\frac{1}{4}$. The height is $30\frac{1}{4}$ in. It is best to give answers as mixed numbers in applications like this.

**Now Try Exercise 69.**

**OBJECTIVE 6 Interpret data in a circle graph.** A **circle graph** or **pie chart** is often used to give a pictorial representation of data. A circle is used to indicate the total of all the categories represented. The circle is divided into sectors, or wedges (like pieces of pie), whose sizes show the relative magnitudes of the categories. The sum of all the fractional parts must be 1 (for 1 whole circle).

### EXAMPLE 9 Using a Circle Graph (Pie Chart) to Interpret Information

The 1999 market share for satellite-TV home subscribers is shown in the circle graph in Figure 2. The number of subscribers reached 12 million in August 1999.

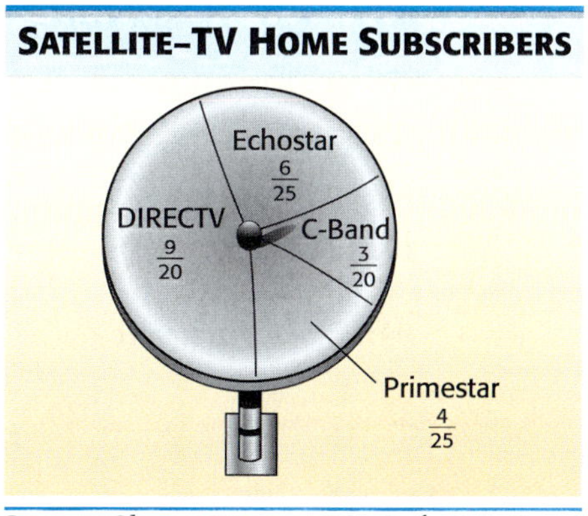


***Source:*** Skyreport.com; *USA Today.*

FIGURE 2

**(a)** Which provider had the largest share of the home subscriber market in August 1999? What was that share?

In the circle graph, the sector for DIRECTV is the largest, so DIRECTV had the largest market share, $\frac{9}{20}$.

**(b)** Estimate the number of home subscribers to DIRECTV in August 1999.

A market share of $\frac{9}{20}$ can be rounded to $\frac{10}{20}$, or $\frac{1}{2}$. We multiply $\frac{1}{2}$ by the total number of subscribers, 12 million. A good estimate for the number of DIRECTV subscribers would be

$$\frac{1}{2}(12) = 6 \text{ million.}$$

**(c)** How many actual home subscribers to DIRECTV were there?

To find the answer, we multiply the actual fraction from the graph for DIRECTV, $\frac{9}{20}$, by the number of subscribers, 12 million:

$$\frac{9}{20}(12) = \frac{9}{20} \cdot \frac{12}{1} = \frac{27}{5} = 5\frac{2}{5}.$$

Thus, $5\frac{2}{5}$ million, or 5,400,000, homes subscribed to DIRECTV. This is reasonable given our estimate in part (b).

**Now Try Exercises 75 and 77.**

## 1.1 EXERCISES

**For Extra Help**

- Student's Solutions Manual
- MyMathLab
- InterAct Math Tutorial Software
- AW Math Tutor Center
- MathXL
- Digital Video Tutor CD 1/Videotape 1

*Decide whether each statement is* true *or* false. *If it is false, say why.*

**1.** In the fraction $\frac{3}{7}$, 3 is the numerator and 7 is the denominator.

**2.** The mixed number equivalent of $\frac{41}{5}$ is $8\frac{1}{5}$.

**3.** The fraction $\frac{17}{51}$ is in lowest terms.

**4.** The reciprocal of $\frac{8}{2}$ is $\frac{4}{1}$.

**5.** The product of 8 and 2 is 10.

**6.** The difference between 12 and 2 is 6.

*Identify each number as* prime, composite, *or* neither. *If the number is composite, write it as the product of prime factors. See Example 1.*

| | | | |
|---|---|---|---|
| **7.** 19 | **8.** 31 | **9.** 64 | **10.** 99 |
| **11.** 3458 | **12.** 1025 | **13.** 1 | **14.** 0 |
| **15.** 30 | **16.** 40 | **17.** 500 | **18.** 700 |
| **19.** 124 | **20.** 120 | **21.** 29 | **22.** 83 |

*Write each fraction in lowest terms. See Example 2.*

| | | | |
|---|---|---|---|
| **23.** $\frac{8}{16}$ | **24.** $\frac{4}{12}$ | **25.** $\frac{15}{18}$ | **26.** $\frac{16}{20}$ |
| **27.** $\frac{18}{60}$ | **28.** $\frac{16}{64}$ | **29.** $\frac{144}{120}$ | **30.** $\frac{132}{77}$ |

**31.** One of the following is the correct way to write $\frac{16}{24}$ in lowest terms. Which one is it?

**A.** $\dfrac{16}{24} = \dfrac{8+8}{8+16} = \dfrac{8}{16} = \dfrac{1}{2}$ **B.** $\dfrac{16}{24} = \dfrac{4 \cdot 4}{4 \cdot 6} = \dfrac{4}{6}$

**C.** $\dfrac{16}{24} = \dfrac{8 \cdot 2}{8 \cdot 3} = \dfrac{2}{3}$ **D.** $\dfrac{16}{24} = \dfrac{14+2}{21+3} = \dfrac{2}{3}$

**32.** For the fractions $\frac{p}{q}$ and $\frac{r}{s}$, which one of the following can serve as a common denominator?

**A.** $q \cdot s$ **B.** $q + s$ **C.** $p \cdot r$ **D.** $p + r$

*Find each product or quotient, and write it in lowest terms. See Examples 3 and 4.*

**33.** $\dfrac{4}{5} \cdot \dfrac{6}{7}$ **34.** $\dfrac{5}{9} \cdot \dfrac{10}{7}$ **35.** $\dfrac{1}{10} \cdot \dfrac{12}{5}$ **36.** $\dfrac{6}{11} \cdot \dfrac{2}{3}$

**37.** $\dfrac{15}{4} \cdot \dfrac{8}{25}$ **38.** $\dfrac{4}{7} \cdot \dfrac{21}{8}$ **39.** $2\dfrac{2}{3} \cdot 5\dfrac{4}{5}$ **40.** $3\dfrac{3}{5} \cdot 7\dfrac{1}{6}$

**41.** $\dfrac{5}{4} \div \dfrac{3}{8}$ **42.** $\dfrac{7}{6} \div \dfrac{9}{10}$ **43.** $\dfrac{32}{5} \div \dfrac{8}{15}$ **44.** $\dfrac{24}{7} \div \dfrac{6}{21}$

**45.** $\dfrac{3}{4} \div 12$ **46.** $\dfrac{2}{5} \div 30$ **47.** $2\dfrac{5}{8} \div 1\dfrac{15}{32}$ **48.** $2\dfrac{3}{10} \div 7\dfrac{4}{5}$

**49.** Write a summary explaining how to multiply and divide two fractions. Give examples.

**50.** Write a summary explaining how to add and subtract two fractions. Give examples.

*Find each sum or difference, and write it in lowest terms. See Examples 5–7.*

**51.** $\dfrac{7}{12} + \dfrac{1}{12}$ **52.** $\dfrac{3}{16} + \dfrac{5}{16}$ **53.** $\dfrac{5}{9} + \dfrac{1}{3}$ **54.** $\dfrac{4}{15} + \dfrac{1}{5}$

**55.** $3\dfrac{1}{8} + 2\dfrac{1}{4}$ **56.** $5\dfrac{3}{4} + 3\dfrac{1}{3}$ **57.** $\dfrac{13}{15} - \dfrac{3}{15}$ **58.** $\dfrac{11}{12} - \dfrac{3}{12}$

**59.** $\dfrac{7}{12} - \dfrac{1}{9}$ **60.** $\dfrac{11}{16} - \dfrac{1}{12}$ **61.** $6\dfrac{1}{4} - 5\dfrac{1}{3}$ **62.** $8\dfrac{4}{5} - 7\dfrac{4}{9}$

*Solve each applied problem. See Example 8.*

*Use the table, which appears on a package of Quaker Quick Grits, to answer Exercises 63 and 64.*

| | Microwave | Stove Top | | |
|---|---|---|---|---|
| *Servings* | 1 | 1 | 4 | 6 |
| Water | $\frac{3}{4}$ cup | 1 cup | 3 cups | 4 cups |
| Grits | 3 Tbsp | 3 Tbsp | $\frac{3}{4}$ cup | 1 cup |
| Salt (optional) | Dash | Dash | $\frac{1}{4}$ tsp | $\frac{1}{2}$ tsp |

**63.** How many cups of water would be needed for eight microwave servings?

**64.** How many tsp of salt would be needed for five stove top servings? (*Hint:* 5 is halfway between 4 and 6.)

*Stock prices are now reported using decimal numerals. However, prior to April 9, 2001, fractions were used for stock prices. With this in mind, answer Exercises 65 and 66.*

**65.** On Tuesday, February 10, 1998, Earthlink stock on the NASDAQ exchange closed the day at $4\frac{5}{8}$ (dollars) ahead of where it had opened. It closed at $38\frac{5}{8}$ (dollars). What was its opening price?

**66.** A report in *USA Today* on February 10, 1998, stated that Teva Pharmaceutical skidded $9\frac{9}{16}$ (dollars) to $37\frac{1}{2}$ (dollars) after the Israeli drugmaker said fourth-quarter net income was likely to be below expectations. What was its price before the skid?

**67.** A hardware store sells a 40-piece socket wrench set. The measure of the largest socket is $\frac{3}{4}$ in., while the measure of the smallest socket is $\frac{3}{16}$ in. What is the difference between these measures?

**68.** Two sockets in a socket wrench set have measures of $\frac{9}{16}$ in. and $\frac{3}{8}$ in. What is the difference between these two measures?

**69.** A motel owner has decided to expand his business by buying a piece of property next to the motel. The property has an irregular shape, with five sides as shown in the figure. Find the total distance around the piece of property. (This is called the *perimeter* of the figure.)

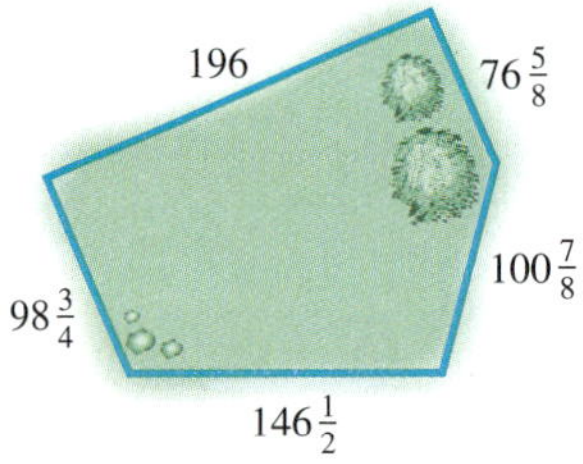


Measurements in feet

**70.** Find the perimeter of the triangle in the figure.

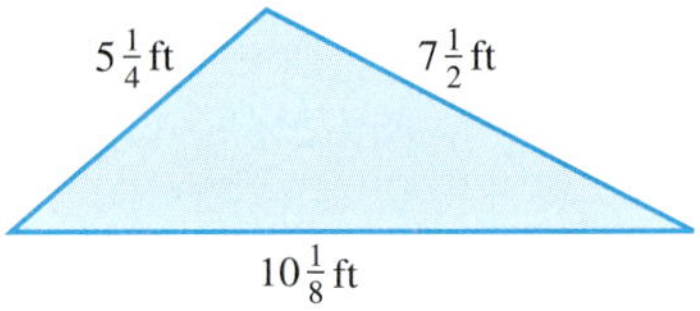


**71.** A piece of board is $15\frac{5}{8}$ in. long. If it must be divided into 3 pieces of equal length, how long must each piece be?

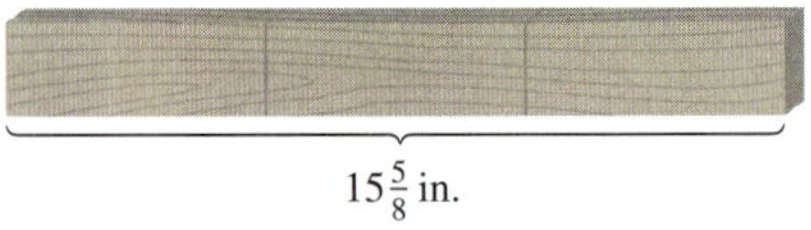


**72.** Under existing standards, most of the holes in Swiss cheese must have diameters between $\frac{11}{16}$ and $\frac{13}{16}$ in. To accommodate new high-speed slicing machines, the USDA wants to reduce the minimum size to $\frac{3}{8}$ in. How much smaller is $\frac{3}{8}$ in. than $\frac{11}{16}$ in.? (*Source:* U.S. Department of Agriculture.)

**73.** Tex's favorite recipe for barbecue sauce calls for $2\frac{1}{3}$ cups of tomato sauce. The recipe makes enough barbecue sauce to serve 7 people. How much tomato sauce is needed for 1 serving?

**74.** A cake recipe calls for $1\frac{3}{4}$ cups of sugar. A caterer has $15\frac{1}{2}$ cups of sugar on hand. How many cakes can he make?

*More than 8 million immigrants were admitted to the United States between 1990 and 1997. The circle graph gives the fractional number from each region of birth for these immigrants. Use the graph to answer the following questions. See Example 9.*

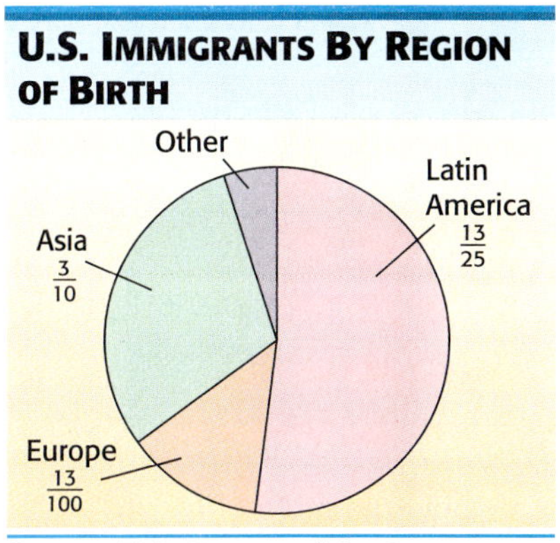


***Source:*** U.S. Bureau of the Census.

**75.** What fractional part of the immigrants were from Other regions?

**76.** What fractional part of the immigrants were from Latin America or Asia?

**77.** How many (in millions) were from Europe?

**78.** At the conclusion of the Addison Wesley softball league season, batting statistics for five players were as follows.

| *Player* | *At-Bats* | *Hits* | *Home Runs* |
|---|---|---|---|
| Alison Romike | 40 | 9 | 2 |
| Jennifer Crum | 36 | 12 | 3 |
| Jason Jordan | 11 | 5 | 1 |
| Greg Tobin | 16 | 8 | 0 |
| Greg Erb | 20 | 10 | 2 |

Use the table to answer each question, using estimation skills as necessary.

**(a)** Which player got a hit in exactly $\frac{1}{3}$ of his or her at-bats?

**(b)** Which player got a hit in just less than $\frac{1}{2}$ of his or her at-bats?

**(c)** Which player got a home run in just less than $\frac{1}{10}$ of his or her at-bats?

**(d)** Which player got a hit in just less than $\frac{1}{4}$ of his or her at-bats?

**(e)** Which two players got hits in exactly the same fractional parts of their at-bats? What was the fractional part, expressed in lowest terms?

**79.** For each description, write a fraction in lowest terms that represents the region described.

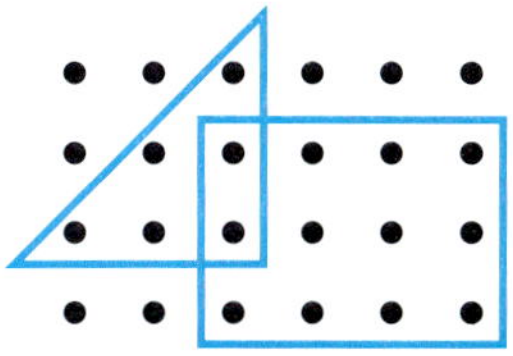

**(a)** The dots in the rectangle as a part of the dots in the entire figure

**(b)** The dots in the triangle as a part of the dots in the entire figure

**(c)** The dots in the overlapping region of the triangle and the rectangle as a part of the dots in the triangle alone

**(d)** The dots in the overlapping region of the triangle and the rectangle as a part of the dots in the rectangle alone

**80.** Estimate the best approximation for the following sum:

$$\frac{14}{26} + \frac{98}{99} + \frac{100}{51} + \frac{90}{31} + \frac{13}{27}.$$

**A.** 6 **B.** 7 **C.** 5 **D.** 8

# 1.2 Exponents, Order of Operations, and Inequality

**OBJECTIVES**

1. Use exponents.
2. Use the order of operations rules.
3. Use more than one grouping symbol.
4. Know the meanings of ≠, <, >, ≤, and ≥.
5. Translate word statements to symbols.
6. Write statements that change the direction of inequality symbols.
7. Interpret data in a bar graph.

**OBJECTIVE 1 Use exponents.** In the prime factored form of 81, written

$$81 = 3 \cdot 3 \cdot 3 \cdot 3,$$

the factor 3 appears four times. In algebra, repeated factors are written with an *exponent*. For example, in $3 \cdot 3 \cdot 3 \cdot 3$, the number 3 appears as a factor four times, so the product is written as $3^4$, and is read "3 to the fourth power."

$$\underbrace{3 \cdot 3 \cdot 3 \cdot 3}_{\text{4 factors of 3}} = 3^4$$

Exponent (4); Base (3)

The number 4 is the **exponent** or **power** and 3 is the **base** in the **exponential expression** $3^4$. A natural number exponent, then, tells how many times the base is used as a factor. A number raised to the first power is simply that number. For example, $5^1 = 5$ and $\left(\frac{1}{2}\right)^1 = \frac{1}{2}$.

**EXAMPLE 1 Evaluating Exponential Expressions**

Find the value of each exponential expression.

**(a)** $5^2 = 5 \cdot 5 = 25$ 5 is used as a factor 2 times.

Read $5^2$ as "5 squared" or "the square of 5."

**(b)** $6^3 = 6 \cdot 6 \cdot 6 = 216$ 6 is used as a factor 3 times.

Read $6^3$ as "6 cubed" or "the cube of 6."

**(c)** $2^5 = 2 \cdot 2 \cdot 2 \cdot 2 \cdot 2 = 32$ 2 is used as a factor 5 times.

Read $2^5$ as "2 to the fifth power."

**(d)** $\left(\frac{2}{3}\right)^3 = \frac{2}{3} \cdot \frac{2}{3} \cdot \frac{2}{3} = \frac{8}{27}$ $\frac{2}{3}$ is used as a factor 3 times.

**Now Try Exercises 5 and 17.**

**CAUTION** Squaring, or raising a number to the second power, is *not* the same as doubling the number. For example,

$$\mathbf{3^2 \text{ means } 3 \cdot 3,} \quad \textit{not} \quad 2 \cdot 3.$$

Thus $3^2 = 9$, not 6. Similarly, cubing, or raising a number to the third power, does *not* mean tripling the number.

**OBJECTIVE 2 Use the order of operations rules.** Many problems involve more than one operation. To indicate the order in which the operations should be performed, we often use **grouping symbols.** If no grouping symbols are used, we apply the order of operations rules discussed below.

Consider the expression $5 + 2 \cdot 3$. To show that the multiplication should be performed before the addition, we can use parentheses to write

$$5 + (2 \cdot 3) = 5 + 6 = 11.$$

If addition is to be performed first, the parentheses should group $5 + 2$ as follows.

$$(5 + 2) \cdot 3 = 7 \cdot 3 = 21$$

Other grouping symbols used in more complicated expressions are brackets [ ], braces { }, and fraction bars. (For example, in $\frac{8-2}{3}$, the expression $8 - 2$ is considered to be grouped in the numerator.)

To work problems with more than one operation, we use the following **order of operations.** This order is used by most calculators and computers.

**Order of Operations**

*If grouping symbols are present,* simplify within them, innermost first (and above and below fraction bars separately), in the following order.

*Step 1* Apply all **exponents.**

*Step 2* Do any **multiplications** or **divisions** in the order in which they occur, working from left to right.

*Step 3* Do any **additions** or **subtractions** in the order in which they occur, working from left to right.

*If no grouping symbols are present,* start with Step 1.

When no operation is indicated, as in $3(7)$ or $(-5)(-4)$, multiplication is understood.

**EXAMPLE 2 Using the Order of Operations**

Find the value of each expression.

**(a)** $9(6 + 11)$

Using the order of operations given above, work first inside the parentheses.

$$\begin{aligned} 9(6 + 11) &= 9(17) && \text{Work inside parentheses.} \\ &= 153 && \text{Multiply.} \end{aligned}$$

**(b)** $6 \cdot 8 + 5 \cdot 2$

Do the multiplications, working from left to right, and then add.

$$\begin{aligned} 6 \cdot 8 + 5 \cdot 2 &= 48 + 10 && \text{Multiply.} \\ &= 58 && \text{Add.} \end{aligned}$$

**(c)** 
$$\begin{aligned} 2(5 + 6) + 7 \cdot 3 &= 2(11) + 7 \cdot 3 && \text{Work inside parentheses.} \\ &= 22 + 21 && \text{Multiply.} \\ &= 43 && \text{Add.} \end{aligned}$$

**(d)** $9 - 2^3 + 5$

Find $2^3$ first.

$$\begin{aligned} 9 - 2^3 + 5 &= 9 - 2 \cdot 2 \cdot 2 + 5 && \text{Use the exponent.} \\ &= 9 - 8 + 5 && \text{Multiply.} \\ &= 1 + 5 && \text{Subtract.} \\ &= 6 && \text{Add.} \end{aligned}$$

**(e)**
$$\begin{aligned} 72 \div 2 \cdot 3 + 4 \cdot 2^3 - 3^3 &= 72 \div 2 \cdot 3 + 4 \cdot 8 - 27 && \text{Use the exponents.} \\ &= 36 \cdot 3 + 4 \cdot 8 - 27 && \text{Divide.} \\ &= 108 + 32 - 27 && \text{Multiply.} \\ &= 140 - 27 && \text{Add.} \\ &= 113 && \text{Subtract.} \end{aligned}$$

Notice that the multiplications and divisions are performed from left to right *as they appear;* then the additions and subtractions should be done from left to right, *as they appear.*

**Now Try Exercises 23, 27, and 29.**

**OBJECTIVE 3 Use more than one grouping symbol.** An expression with double (or *nested*) parentheses, such as $2(8 + 3(6 + 5))$, can be confusing. For clarity, square brackets, [ ], often are used in place of one pair of parentheses. Fraction bars also act as grouping symbols.

## EXAMPLE 3 Using Brackets and Fraction Bars as Grouping Symbols

Simplify each expression.

**(a)** $2[8 + 3(6 + 5)]$

Work first within the parentheses, and then simplify inside the brackets until a single number remains.

$$\begin{aligned} 2[8 + 3(6 + 5)] &= 2[8 + 3(11)] \\ &= 2[8 + 33] \\ &= 2[41] \\ &= 82 \end{aligned}$$

**(b)** $\dfrac{4(5 + 3) + 3}{2(3) - 1}$

The expression can be written as the quotient

$$[4(5 + 3) + 3] \div [2(3) - 1],$$

which shows that the fraction bar groups the numerator and denominator separately. Simplify both numerator and denominator, then divide, if possible.

$$\begin{aligned} \frac{4(5 + 3) + 3}{2(3) - 1} &= \frac{4(8) + 3}{2(3) - 1} && \text{Work inside parentheses.} \\ &= \frac{32 + 3}{6 - 1} && \text{Multiply.} \end{aligned}$$

$$= \frac{35}{5} \qquad \text{Add and subtract.}$$

$$= 7 \qquad \text{Divide.}$$

**Now Try Exercises 31 and 35.**

**NOTE** Parentheses and fraction bars are used as grouping symbols to indicate an expression that represents a single number. That is why we must first simplify within parentheses and above and below fraction bars.

**OBJECTIVE 4 Know the meanings of $\neq$, $<$, $>$, $\leq$, and $\geq$.** So far, we have used the symbols for the operations of arithmetic and the symbol for equality ($=$). The symbols $\neq$, $<$, $>$, $\leq$, and $\geq$ are used to express an **inequality,** a statement that two expressions are not equal. The equality symbol with a slash through it, $\neq$, means "is not equal to." For example,

$$7 \neq 8$$

indicates that 7 is not equal to 8.

If two numbers are not equal, then one of the numbers must be less than the other. The symbol $<$ represents "is less than," so that "7 is less than 8" is written

$$7 < 8.$$

The word *is* in the phrase "is less than" is a verb, which actually makes $7 < 8$ a sentence. This is not the case for expressions, like $7 + 8$, $7 - 8$, and so on.

The symbol $>$ means "is greater than." Write "8 is greater than 2" as

$$8 > 2.$$

To keep the meanings of the symbols $<$ and $>$ clear, remember that the symbol always *points to the lesser number.* For example, write "8 is less than 15" by pointing the symbol toward the 8:

Lesser number $\rightarrow$ $8 < 15.$

Two other symbols, $\leq$ and $\geq$, also represent the idea of inequality. The symbol $\leq$ means "is less than or equal to," so

$$5 \leq 9$$

means "5 is less than or equal to 9." This statement is true, since $5 < 9$ is true. ***If either the < part or the = part is true, then the inequality ≤ is true.***

The symbol $\geq$ means "is greater than or equal to." Again,

$$9 \geq 5$$

is true because $9 > 5$ is true. Also, $8 \leq 8$ is true since $8 = 8$ is true. But it is not true that $13 \leq 9$ because neither $13 < 9$ nor $13 = 9$ is true.

## EXAMPLE 4 Using Inequality Symbols

Determine whether each statement is true or false.

**(a)** $6 \neq 5 + 1$

The statement is false because 6 *is equal to* $5 + 1$.

**(b)** $5 + 3 < 19$
Since $5 + 3$ represents a number (8) that is less than 19, this statement is true.

**(c)** $15 \leq 20 \cdot 2$
The statement $15 \leq 20 \cdot 2$ is true, since $15 < 40$.

**(d)** $25 \geq 30$
Both $25 > 30$ and $25 = 30$ are false. Because of this, $25 \geq 30$ is false.

**(e)** $12 \geq 12$
Since $12 = 12$, this statement is true.

**(f)** $9 < 9$
Since $9 = 9$, this statement is false.

**Now Try Exercises 41 and 43.**

**OBJECTIVE 5 Translate word statements to symbols.** An important part of algebra deals with translating words into algebraic notation.

**PROBLEM SOLVING**

As we will see throughout this book, the ability to solve problems using mathematics is based on translating the words of the problem into symbols. The next example is the first of many that illustrate such translations.

### EXAMPLE 5 Translating from Words to Symbols

Write each word statement in symbols.

**(a)** Twelve equals ten plus two.
$12 = 10 + 2$

**(b)** Nine is less than ten.
$9 < 10$

**(c)** Fifteen is not equal to eighteen.
$15 \neq 18$

**(d)** Seven is greater than four.
$7 > 4$

**(e)** Thirteen is less than or equal to forty.
$13 \leq 40$

**(f)** Eleven is greater than or equal to eleven.
$11 \geq 11$

**Now Try Exercises 55 and 57.**

**OBJECTIVE 6 Write statements that change the direction of inequality symbols.** Any statement with $<$ can be converted to one with $>$, and any statement with $>$ can be converted to one with $<$. We do this by reversing the order of the numbers and the direction of the symbol. For example, the statement $6 < 10$ can be written with $>$ as $10 > 6$. Similarly, the statement $4 \leq 10$ can be changed to $10 \geq 4$.

### EXAMPLE 6 Converting between Inequality Symbols

The following examples show the same statement written in two equally correct ways.

**(a)** $9 < 16$ $16 > 9$

**(b)** $5 > 2$ $2 < 5$

**(c)** $3 \leq 8$ $8 \geq 3$

**(d)** $12 \geq 5$ $5 \leq 12$

Note that in each pair of inequalities, the inequality symbol points toward the smaller number.

**Now Try Exercise 71.**

Here is a summary of the symbols discussed in this section.

**Symbols of Equality and Inequality**

| | | | |
|---|---|---|---|
| $=$ | is equal to | $\neq$ | is not equal to |
| $<$ | is less than | $>$ | is greater than |
| $\leq$ | is less than or equal to | $\geq$ | is greater than or equal to |

**CAUTION** The equality and inequality symbols are used to write mathematical *sentences* that describe the relationship between two numbers. On the other hand, the symbols for operators ($+$, $-$, $\times$, $\div$) are used to write mathematical *expressions* that represent a single number. For example, compare the sentence $4 < 10$ with the expression $4 + 10$, which represents the number 14.

**OBJECTIVE 7 Interpret data in a bar graph.** **Bar graphs** are often used to summarize data in a concise manner.

### EXAMPLE 7 Interpreting Inequality Concepts Using a Bar Graph

The bar graph in Figure 3 shows the federal budget outlays for national defense during the 1990s.

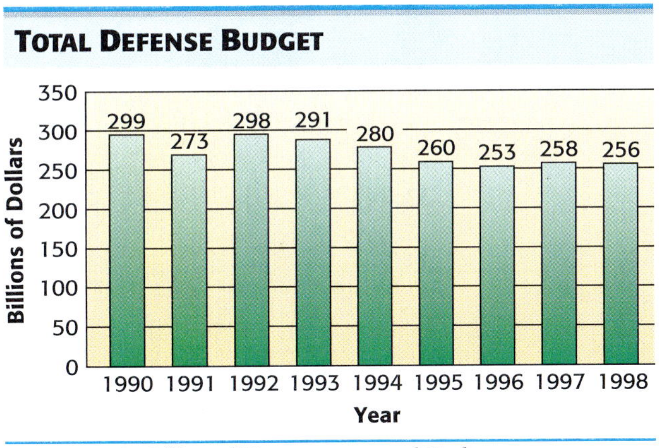


*Source:* U.S. Office of Management and Budget.

FIGURE 3

**(a)** Which years had outlays greater than 280 billion dollars?

Look for numbers greater than 280 at the tops of the columns. In 1990, 1992, and 1993, the outlays were 299 billion, 298 billion, and 291 billion dollars, respectively.

**(b)** Which years had outlays greater than 300 billion dollars?

Since the tops of all the bars are below the line representing 300 billion, there were no such years. This is reinforced by the fact that all of the numbers at the tops of the bars are less than 300 (billion).

**Now Try Exercise 79.**

## 1.2 EXERCISES

**For Extra Help**

Student's Solutions Manual

MyMathLab

InterAct Math Tutorial Software

AW Math Tutor Center

MathXL

Digital Video Tutor CD 1/Videotape 1

*Decide whether each statement is* true *or* false. *If it is false, explain why.*

**1.** When evaluated, $4 + 3(8 - 2)$ is equal to 42.

**2.** $3^3 = 9$

**3.** The statement "4 is 12 less than 16" is interpreted $4 = 12 - 16$.

**4.** The statement "6 is 4 less than 10" is interpreted $6 < 10 - 4$.

*Find the value of each exponential expression. See Example 1.*

**5.** $7^2$ **6.** $4^2$ **7.** $12^2$ **8.** $14^2$

**9.** $4^3$ **10.** $5^3$ **11.** $10^3$ **12.** $11^3$

**13.** $3^4$ **14.** $6^4$ **15.** $4^5$ **16.** $3^5$

**17.** $\left(\frac{2}{3}\right)^4$ **18.** $\left(\frac{3}{4}\right)^3$ **19.** $(.04)^3$ **20.** $(.05)^4$

**21.** Explain in your own words how to evaluate a power of a number, such as $6^3$.

**22.** Explain why any power of 1 must be equal to 1.

*Find the value of each expression. See Examples 2 and 3.*

**23.** $64 \div 4 \times 2$ **24.** $250 \div 5 \times 2$ **25.** $\frac{1}{4} \cdot \frac{2}{3} + \frac{2}{5} \cdot \frac{11}{3}$

**26.** $\frac{9}{4} \cdot \frac{2}{3} + \frac{4}{5} \cdot \frac{5}{3}$ **27.** $9 \cdot 4 - 8 \cdot 3$ **28.** $11 \cdot 4 + 10 \cdot 3$

**29.** $3(4 + 2) + 8 \cdot 3$ **30.** $9(1 + 7) + 2 \cdot 5$ **31.** $5[3 + 4(2^2)]$

**32.** $6[2 + 8(3^3)]$ **33.** $3^2[(11 + 3) - 4]$ **34.** $4^2[(13 + 4) - 8]$

**35.** $\dfrac{6(3^2 - 1) + 8}{8 - 2^2}$ **36.** $\dfrac{2(8^2 - 4) + 8}{29 - 3^3}$

**37.** $\dfrac{4(6 + 2) + 8(8 - 3)}{6(4 - 2) - 2^2}$ **38.** $\dfrac{6(5 + 1) - 9(1 + 1)}{5(8 - 6) - 2^3}$

**39.** Write an explanation of how you would use the order of operations to simplify $4 + 3(2^2 - 1)^3$.

**40.** When evaluating $(4^2 + 3^3)^4$, what is the *last* exponent that would be applied?

*First simplify both sides of each inequality. Then tell whether the given statement is* true *or* false. *See Examples 2–4.*

**41.** $9 \cdot 3 - 11 \leq 16$ **42.** $6 \cdot 5 - 12 \leq 18$

**43.** $5 \cdot 11 + 2 \cdot 3 \leq 60$

**44.** $9 \cdot 3 + 4 \cdot 5 \geq 48$

**45.** $0 \geq 12 \cdot 3 - 6 \cdot 6$

**46.** $10 \leq 13 \cdot 2 - 15 \cdot 1$

**47.** $45 \geq 2[2 + 3(2 + 5)]$

**48.** $55 \geq 3[4 + 3(4 + 1)]$

**49.** $[3 \cdot 4 + 5(2)] \cdot 3 > 72$

**50.** $2 \cdot [7 \cdot 5 - 3(2)] \leq 58$

**51.** $\dfrac{3 + 5(4 - 1)}{2 \cdot 4 + 1} \geq 3$

**52.** $\dfrac{7(3 + 1) - 2}{3 + 5 \cdot 2} \leq 2$

**53.** $3 \geq \dfrac{2(5 + 1) - 3(1 + 1)}{5(8 - 6) - 4 \cdot 2}$

**54.** $7 \leq \dfrac{3(8 - 3) + 2(4 - 1)}{9(6 - 2) - 11(5 - 2)}$

*Write each word statement in symbols. See Example 5.*

**55.** Fifteen is equal to five plus ten.

**56.** Twelve is equal to twenty minus eight.

**57.** Nine is greater than five minus four.

**58.** Ten is greater than six plus one.

**59.** Sixteen is not equal to nineteen.

**60.** Three is not equal to four.

**61.** Two is less than or equal to three.

**62.** Five is less than or equal to nine.

*Write each statement in words and decide whether it is* true *or* false.

**63.** $7 < 19$

**64.** $9 < 10$

**65.** $3 \neq 6$

**66.** $9 \neq 13$

**67.** $8 \geq 11$

**68.** $4 \leq 2$

**69.** Construct a true statement that involves an addition on the left side, the symbol $\geq$, and a multiplication on the right side.

**70.** Construct a false statement that involves subtraction on the left side, the symbol $\leq$, and a division on the right side. Then tell why the statement is false and how it could be changed to become true.

*Write each statement with the inequality symbol reversed while keeping the same meaning. See Example 6.*

**71.** $5 < 30$

**72.** $8 > 4$

**73.** $12 \geq 3$

**74.** $25 \leq 41$

**75.** What English-language phrase is used to express the fact that one person's age *is less than* another person's age?

**76.** What English-language phrase is used to express the fact that one person's height *is greater than* another person's height?

**77.** $12 \geq 12$ is a true statement. Suppose that someone tells you the following: "$12 \geq 12$ is false, because even though 12 is equal to 12, 12 is not greater than 12." How would you respond to this?

**78.** The table shows results of a science literacy survey by Jon Miller of the International Center for the Advancement of Science Literacy in Chicago.

| *Country* | *Science Literacy Index* |
|---|---|
| United States | 56 |
| Netherlands | 52 |
| France | 50 |
| Canada | 45 |
| Greece | 38 |
| Japan | 36 |

**(a)** Which countries scored more than 50?
**(b)** Which countries scored at most 40?
**(c)** For which countries were scores not less than 50?

*The bar graph shows world coal production by year for the years 1993 through 1999. Use the graph to answer Exercises 79 and 80. See Example 7.*

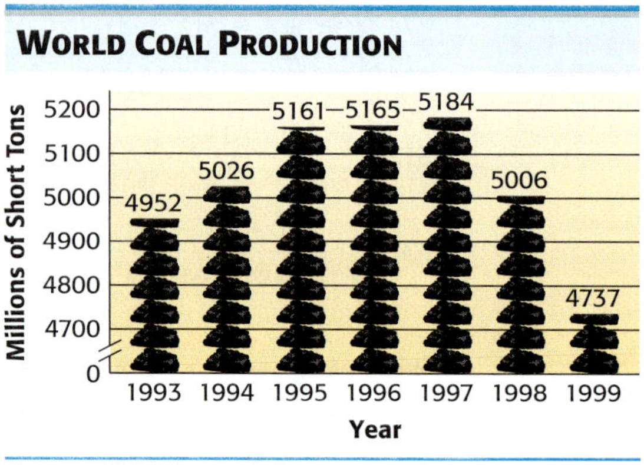


***Source:*** U.S. Energy Information Administration, *International Energy Review*, 1999.

**79.** In two of the years represented, coal production was less than production in the previous year. What were these two years?

**80.** What was the first year in which production was greater than 5100 millions of short tons?

**81.** According to an article in the April 1, 2002 issue of *USA Today,* the stock for Clear Channel, the largest radio station owner in the country, closed at \$36.80 per share on September 20, 2001. A year prior to that date, the same stock had closed at \$53.76.

**(a)** By how much had the stock price decreased?

**(b)** To find the percent of decrease, divide the amount from part (a) by 53.76 and convert to a percent. What was the percent of decrease? (Round to the nearest tenth of a percent.)

**82.** In March 2001, farmers were paid \$15.00 per hundred pounds of lettuce. Due to a cold snap in Arizona and California, one year later the price had soared to \$86.50. (*Source:* "Prices Soar as Cold Snap Shreds Iceberg Lettuce Supply," *USA Today,* April 1, 2001.)

**(a)** By how much had the price increased?

**(b)** What was the percent of increase? (Round to the nearest percent.) (*Hint:* See Exercise 81(b).)

*Insert one pair of parentheses to make the left side of each equation equal to the right side.*

**83.** $3 \cdot 6 + 4 \cdot 2 = 60$

**84.** $2 \cdot 8 - 1 \cdot 3 = 42$

**85.** $10 - 7 - 3 = 6$

**86.** $15 - 10 - 2 = 7$

**87.** $8 + 2^2 = 100$

**88.** $4 + 2^2 = 36$

# 1.3 Variables, Expressions, and Equations

**OBJECTIVES**

1. Define *variable*, and find the value of an algebraic expression, given the values of the variables.
2. Convert phrases from words to algebraic expressions.
3. Identify solutions of equations.
4. Identify solutions of equations from a set of numbers.
5. Distinguish between an *expression* and an *equation*.

**OBJECTIVE 1 Define *variable*, and find the value of an algebraic expression, given the values of the variables.** A **variable** is a symbol, usually a letter such as $x$, $y$, or $z$, used to represent any unknown number. An **algebraic expression** is a collection of numbers, variables, operation symbols, and grouping symbols (such as parentheses). For example,

$$6(x+5), \qquad 2m-9, \qquad \text{and} \qquad 8p^2+6p+2$$

are all algebraic expressions. In the algebraic expression $2m - 9$, the expression $2m$ indicates the product of 2 and $m$, just as $8p^2$ shows the product of 8 and $p^2$. Also, $6(x + 5)$ means the product of 6 and $x + 5$. An algebraic expression has different numerical values for different values of the variable.

### EXAMPLE 1 Evaluating Expressions

Find the numerical value of each algebraic expression when $m = 5$.

**(a)** $8m$

Replace $m$ with 5, to get

$$8m = 8 \cdot 5 = 40.$$

**(b)** $3m^2$

For $m = 5$,

$$3m^2 = 3 \cdot 5^2 = 3 \cdot 25 = 75.$$

**Now Try Exercises 15 and 17.**

**CAUTION** In Example 1(b), notice that $3m^2$ means $3 \cdot m^2$; it *does not* mean $3m \cdot 3m$. The product $3m \cdot 3m$ is indicated by $(3m)^2$.

### EXAMPLE 2 Evaluating Expressions

Find the value of each expression when $x = 5$ and $y = 3$.

**(a)** $2x + 7y$

Replace $x$ with 5 and $y$ with 3. Follow the order of operations: Multiply first, then add.

$$\begin{aligned} 2x + 7y &= 2 \cdot 5 + 7 \cdot 3 && \text{Let } x = 5 \text{ and } y = 3. \\ &= 10 + 21 && \text{Multiply.} \\ &= 31 && \text{Add.} \end{aligned}$$

**(b)**

$$\begin{aligned} \frac{9x - 8y}{2x - y} &= \frac{9 \cdot 5 - 8 \cdot 3}{2 \cdot 5 - 3} && \text{Let } x = 5 \text{ and } y = 3. \\ &= \frac{45 - 24}{10 - 3} && \text{Multiply.} \\ &= \frac{21}{7} && \text{Subtract.} \\ &= 3 && \text{Divide.} \end{aligned}$$

(c) $x^2 - 2y^2 = 5^2 - 2 \cdot 3^2$ Let $x = 5$ and $y = 3$.

$= 25 - 2 \cdot 9$ Use the exponents.

$= 25 - 18$ Multiply.

$= 7$ Subtract.

**Now Try Exercises 27, 35, and 37.**

**OBJECTIVE 2 Convert phrases from words to algebraic expressions.**

**PROBLEM SOLVING**

Sometimes variables must be used to change word phrases into algebraic expressions. Such translations are used in problem solving.

### EXAMPLE 3 Using Variables to Change Word Phrases to Algebraic Expressions

Change each word phrase to an algebraic expression. Use $x$ as the variable.

**(a)** The sum of a number and 9

"Sum" is the answer to an addition problem. This phrase translates as

$$x + 9 \quad \text{or} \quad 9 + x.$$

**(b)** 7 minus a number

"Minus" indicates subtraction, so the answer is $7 - x$.

**(c)** 7 less than a number

Write 7 less than a number as $x - 7$. In this case $7 - x$ would not be correct, because "less than" means "subtracted from."

**(d)** The product of 11 and a number

$$11 \cdot x \quad \text{or} \quad 11x$$

As mentioned earlier, $11x$ means 11 times $x$. No symbol is needed to indicate the product of a number and a variable.

**(e)** 5 divided by a number

This translates as $\frac{5}{x}$. The expression $\frac{x}{5}$ would *not* be correct here.

**(f)** The product of 2, and the sum of a number and 8

We are multiplying 2 times another number. This number is the sum of $x$ and 8, written $x + 8$. Using parentheses for this sum, the final expression is

$$2(x + 8).$$

**Now Try Exercises 43, 49, and 53.**

**CAUTION** In Example 3(b), the response $x - 7$ would *not* be correct; this statement translates as "a number minus 7," not "7 minus a number." The expressions $7 - x$ and $x - 7$ are rarely equal. For example, if $x = 10$, $10 - 7 \neq 7 - 10$. ($7 - 10$ is a *negative number,* discussed in Section 1.4.)

**OBJECTIVE 3 Identify solutions of equations.** An **equation** is a statement that two algebraic expressions are equal. Therefore, an equation *always* includes the equality symbol, =. Examples of equations are

$$x + 4 = 11, \qquad 2y = 16, \qquad \text{and} \qquad 4p + 1 = 25 - p.$$

To **solve** an equation means to find the values of the variable that make the equation true. Such values of the variable are called the **solutions** of the equation.

### EXAMPLE 4 Deciding Whether a Number Is a Solution of an Equation

Decide whether the given number is a solution of the equation.

**(a)** $5p + 1 = 36; \quad 7$

Replace $p$ with 7.

$$\begin{aligned} 5p + 1 &= 36 & & \\ 5 \cdot 7 + 1 &= 36 \quad ? & & \text{Let } p = 7. \\ 35 + 1 &= 36 \quad ? & & \\ 36 &= 36 & & \text{True} \end{aligned}$$

The number 7 is a solution of the equation.

**(b)** $9m - 6 = 32; \quad 4$

$$\begin{aligned} 9m - 6 &= 32 & & \\ 9 \cdot 4 - 6 &= 32 \quad ? & & \text{Let } m = 4. \\ 36 - 6 &= 32 \quad ? & & \\ 30 &= 32 & & \text{False} \end{aligned}$$

The number 4 is not a solution of the equation.

**Now Try Exercise 61.**

**OBJECTIVE 4 Identify solutions of equations from a set of numbers.** A **set** is a collection of objects. In mathematics, these objects are most often numbers. The objects that belong to the set, called **elements** of the set, are written between **set braces.** For example, the set containing the numbers 1, 2, 3, 4, and 5 is written as

$$\{1, 2, 3, 4, 5\}.$$

For more information about sets, see Appendix C at the back of this book.

In some cases, the set of numbers from which the solutions of an equation must be chosen is specifically stated. One way of determining solutions is direct substitution of all possible replacements. The ones that lead to a true statement are solutions.

### EXAMPLE 5 Finding a Solution from a Given Set

Change each word statement to an equation. Use $x$ as the variable. Then find all solutions of the equation from the set

$$\{0, 2, 4, 6, 8, 10\}.$$

**(a)** The sum of a number and four is six.

The word *is* suggests "equals." If $x$ represents the unknown number, then translate as follows.

$$\underbrace{\text{The sum of a number and four}}_{x+4}\ \underbrace{\text{is}}_{=}\ \underbrace{\text{six.}}_{6}$$

Try each number from the given set {0, 2, 4, 6, 8, 10}, in turn, to see that 2 is the only solution of $x + 4 = 6$.

**(b)** 9 more than five times a number is 49.

Use $x$ to represent the unknown number. Start with $5x$ and then add 9 to it. The word *is* translates as =.

$$5x + 9 = 49$$

Try each number from {0, 2, 4, 6, 8, 10}. The solution is 8, since $5 \cdot 8 + 9 = 49$.

**(c)** The sum of a number and 12 is equal to four times the number.

If $x$ represents the number, "the sum of a number and 12," is represented by $x + 12$. The translation is

$$x + 12 = 4x.$$

Trying each replacement leads to a true statement when $x = 4$, since $4 + 12 = 4(4) = 16$.

**Now Try Exercise 69.**

**OBJECTIVE 5 Distinguish between an *expression* and an *equation*.** Students often have trouble distinguishing between equations and expressions. Remember that an equation is a sentence; an expression is a phrase.

$$4x + 5 = 9 \qquad\qquad 4x + 5$$

Equation (to solve) — Expression (to simplify or evaluate)

### EXAMPLE 6 Distinguishing between Equations and Expressions

Decide whether each of the following is an equation or an expression.

**(a)** $2x - 5y$

There is no equals sign, so this is an expression.

**(b)** $2x = 5y$

Because an equals sign is present, this is an equation.

**Now Try Exercises 77 and 81.**

# 1.3 EXERCISES

**For Extra Help**

- Student's Solutions Manual
- MyMathLab
- InterAct Math Tutorial Software
- AW Math Tutor Center
- MathXL
- Digital Video Tutor CD 1/Videotape 1

*Fill in each blank with the correct response.*

**1.** If $x = 3$, then the value of $x + 7$ is _____.

**2.** If $x = 1$ and $y = 2$, then the value of $4xy$ is _____.

**3.** The sum of 12 and $x$ is represented by the expression _____. If $x = 9$, the value of that expression is _____.

**4.** If $x$ can be chosen from the set $\{0, 1, 2, 3, 4, 5\}$, the only solution of $x + 5 = 9$ is _____.

**5.** Will the equation $x = x + 4$ ever have a solution? _____.

**6.** $2x + 3$ is an ______________ (equation/expression), while $2x + 3 = 8$ is an ______________ (equation/expression).

*Exercises 7–12 cover some of the concepts introduced in this section. Give a short explanation for each.*

**7.** Explain why $2x^3$ is not the same as $2x \cdot 2x \cdot 2x$.

**8.** Why are "5 less than a number" and "5 is less than a number" translated differently?

**9.** Explain why, when evaluating the expression $4x^2$ for $x = 3$, 3 must be squared *before* multiplying by 4.

**10.** What value of $x$ would cause the expression $2x + 3$ to equal 9? Explain your reasoning.

**11.** There are many pairs of values of $x$ and $y$ for which $2x + y$ will equal 6. Name two such pairs and describe how you determined them.

**12.** Suppose that for the equation $3x - y = 9$, the value of $x$ is given as 4. What would be the corresponding value of $y$? How do you know this?

*Find the numerical value if **(a)** $x = 4$ and **(b)** $x = 6$. See Example 1.*

**13.** $x + 9$ **14.** $x - 1$ **15.** $5x$ **16.** $7x$ **17.** $4x^2$

**18.** $5x^2$ **19.** $\frac{x + 1}{3}$ **20.** $\frac{x - 2}{5}$ **21.** $\frac{3x - 5}{2x}$ **22.** $\frac{4x - 1}{3x}$

**23.** $3x^2 + x$ **24.** $2x + x^2$ **25.** $6.459x$ **26.** $.74x^2$

*Find the numerical value if **(a)** $x = 2$ and $y = 1$ and **(b)** $x = 1$ and $y = 5$. See Example 2.*

**27.** $8x + 3y + 5$ **28.** $4x + 2y + 7$ **29.** $3(x + 2y)$ **30.** $2(2x + y)$

**31.** $x + \frac{4}{y}$ **32.** $y + \frac{8}{x}$ **33.** $\frac{x}{2} + \frac{y}{3}$ **34.** $\frac{x}{5} + \frac{y}{4}$

**35.** $\frac{2x + 4y - 6}{5y + 2}$ **36.** $\frac{4x + 3y - 1}{x}$ **37.** $2y^2 + 5x$ **38.** $6x^2 + 4y$

**39.** $\frac{3x + y^2}{2x + 3y}$ **40.** $\frac{x^2 + 1}{4x + 5y}$ **41.** $.841x^2 + .32y^2$ **42.** $.941x^2 + .2y^2$

*Change each word phrase to an algebraic expression. Use x as the variable to represent the number. See Example 3.*

**43.** Twelve times a number

**44.** Nine times a number

**45.** Seven added to a number

**46.** Thirteen added to a number

**47.** Two subtracted from a number

**48.** Eight subtracted from a number

**49.** A number subtracted from seven

**50.** A number subtracted from fourteen

**51.** The difference between a number and 6

**52.** The difference between 6 and a number

**53.** 12 divided by a number

**54.** A number divided by 12

**55.** The product of 6, and four less than a number.

**56.** The product of 9, and five more than a number.

**57.** In the phrase "Four more than the product of a number and 6," does the word *and* signify the operation of addition? Explain.

**58.** Suppose that the directions on a test read "Solve the following expressions." How would you politely correct the person who wrote these directions? What alternative directions might you suggest?

*Decide whether the given number is a solution of the equation. See Example 4.*

**59.** $5m + 2 = 7;\quad 1$

**60.** $3r + 5 = 8;\quad 1$

**61.** $2y + 3(y - 2) = 14;\quad 3$

**62.** $6a + 2(a + 3) = 14;\quad 2$

**63.** $6p + 4p + 9 = 11;\quad \frac{1}{5}$

**64.** $2x + 3x + 8 = 20;\quad \frac{12}{5}$

**65.** $3r^2 - 2 = 46;\quad 4$

**66.** $2x^2 + 1 = 19;\quad 3$

**67.** $\frac{z + 4}{2 - z} = \frac{13}{5};\quad \frac{1}{3}$

**68.** $\frac{x + 6}{x - 2} = \frac{37}{5};\quad \frac{13}{4}$

*Change each word statement to an equation. Use x as the variable. Find all solutions from the set* {2, 4, 6, 8, 10}. *See Example 5.*

**69.** The sum of a number and 8 is 18.

**70.** A number minus three equals 1.

**71.** Sixteen minus three-fourths of a number is 13.

**72.** The sum of six-fifths of a number and 2 is 14.

**73.** One more than twice a number is 5.

**74.** The product of a number and 3 is 6.

**75.** Three times a number is equal to 8 more than twice the number.

**76.** Twelve divided by a number equals $\frac{1}{3}$ times that number.

*Identify each as an* expression *or an* equation. *See Example 6.*

**77.** $3x + 2(x - 4)$

**78.** $5y - (3y + 6)$

**79.** $7t + 2(t + 1) = 4$

**80.** $9r + 3(r - 4) = 2$

**81.** $x + y = 3$

**82.** $x + y - 3$

*A* **mathematical model** *is an equation that describes the relationship between two quantities. For example, based on data from the United States Olympic Committee, the winning distances for the men's discus throw can be approximated by the equation*

$$y = 1.2304x - 2224.5,$$

*where x is the year (between 1896 and 1996) and y is in feet. Use this model to approximate the winning distances for the following years.*

**83.** 1912

**84.** 1936

**85.** 1960

**86.** 1992

# 1.4 Real Numbers and the Number Line

**OBJECTIVES**

1. Classify numbers and graph them on number lines.
2. Tell which of two different real numbers is smaller.
3. Find additive inverses and absolute values of real numbers.
4. Interpret the meanings of real numbers from a table of data.

**OBJECTIVE 1 Classify numbers and graph them on number lines.** In Section 1.1 we introduced two important sets of numbers, the *natural numbers* and the *whole numbers.*

**Natural Numbers***

{1, 2, 3, 4, . . .} is the set of **natural numbers.**

**Whole Numbers**

{0, 1, 2, 3, 4, . . .} is the set of **whole numbers.**

**NOTE** The three dots show that the list of numbers continues in the same way indefinitely.

These numbers, along with many others, can be represented on **number lines** like the one in Figure 4. The **graph** of a number is the point on the number line associated with that number. The number is called the **coordinate** of the point. We draw a number line by choosing any point on the line and labeling it 0. Then we choose any point to the right of 0 and label it 1. The distance between 0 and 1 gives a unit of measure used to locate other points, as shown in Figure 4. The points labeled in Figure 4 correspond to the first few whole numbers.

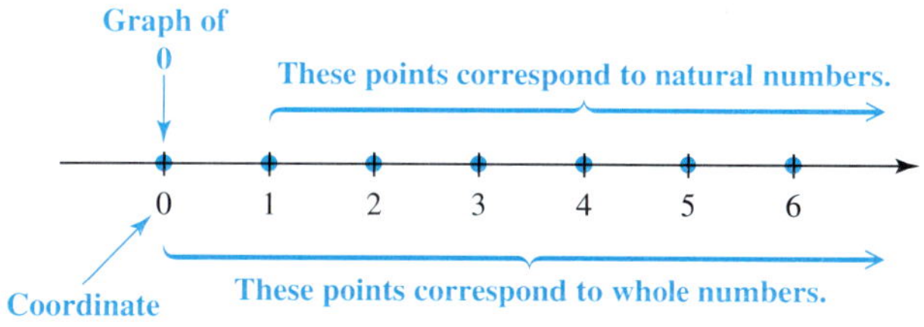


**FIGURE 4**

The natural numbers are located to the right of 0 on the number line. But numbers may also be placed to the left of 0. For each natural number we can place a corresponding number to the left of 0. These numbers, written −1, −2, −3, −4, and so on, are shown in Figure 5 on the next page. Each is the **opposite** or **negative** of a natural number. The natural numbers, their opposites, and 0 form a new set of numbers called the *integers.*

**Integers**

{. . . , −3, −2, −1, 0, 1, 2, 3, . . .} is the set of **integers.**

*The symbols { and } are braces used in conjunction with sets. See Appendix C.

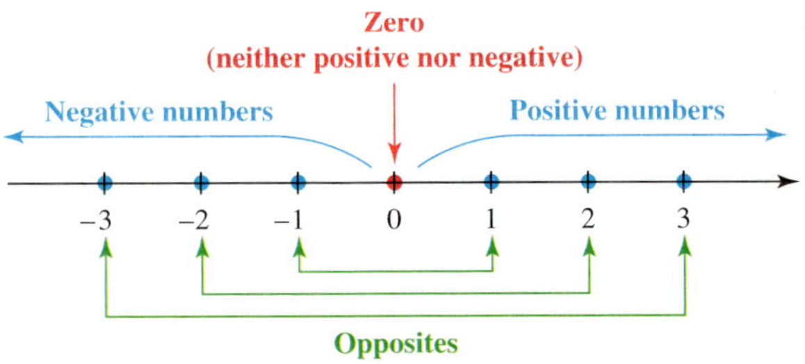

The points correspond to integers.

**FIGURE 5**

Positive numbers and negative numbers are called **signed numbers.** There are many practical applications of negative numbers. For example, a Fahrenheit temperature on a cold January day might be $-10°$, and a business that spends more than it takes in has a negative "profit."

### EXAMPLE 1 Using Negative Numbers in Applications

Use an integer to express the number in italics in each application.

**(a)** The lowest Fahrenheit temperature ever recorded in meteorological records was *129°* below zero at Vostok, Antarctica, on July 21, 1983. (*Source: World Almanac and Book of Facts,* 2000).
Use $-129°$ because "below zero" indicates a negative number.

**(b)** The shore surrounding the Dead Sea is *1340* ft below sea level. (*Source: Microsoft Encarta Encyclopedia 2000*).
Again, "below sea level" indicates a negative number, $-1340$.

**Now Try Exercises 3 and 5.**

Fractions, introduced in Section 1.1, are examples of *rational numbers.*

**Rational Numbers**

$\{x \mid x$ is a quotient of two integers, with denominator not $0\}$ is the set of **rational numbers.**

(Read the part in the braces as "the set of all numbers $x$ such that $x$ is a quotient of two integers, with denominator not 0.")

**NOTE** The set symbolism used in the definition of rational numbers,

$$\{x \mid x \text{ has a certain property}\},$$

is called **set-builder notation.** This notation is convenient to use when it is not possible to list all the elements of a set.

Since any integer can be written as the quotient of itself and 1, all integers also are rational numbers. Figure 6 shows a number line with the graphs of several rational numbers.

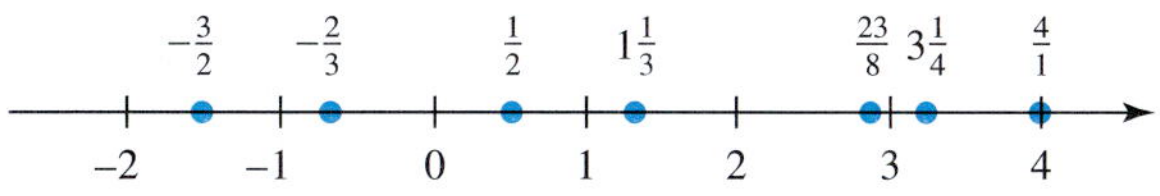


**FIGURE 6**

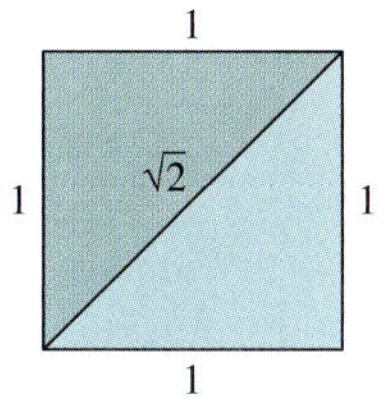


**FIGURE 7**

Although many numbers are rational, not all are. For example, a square that measures one unit on a side has a diagonal whose length is the square root of 2, written $\sqrt{2}$. See Figure 7. It can be shown that $\sqrt{2}$ cannot be written as a quotient of integers. Because of this, $\sqrt{2}$ is not rational; it is *irrational*. Other examples of **irrational numbers** are $\sqrt{3}$, $\sqrt{7}$, $-\sqrt{10}$, and $\pi$ (the ratio of the *circumference* of a circle to its diameter).

Both rational and irrational numbers can be represented by points on the number line and together form the set of *real numbers.*

## Real Numbers

$\{x \mid x$ is a rational or an irrational number$\}$ is the set of **real numbers.**

Real numbers can be written as decimals. Any rational number will have a decimal that either comes to an end (terminates) or repeats in a fixed "block" of digits. For example, $\frac{2}{5} = .4$ and $\frac{27}{100} = .27$ are rational numbers with terminating decimals; $\frac{1}{3} = .333\ldots$ and $\frac{3}{11} = .27272727\ldots$ are repeating decimals, often written as $.\overline{3}$ and $.\overline{27}$ with a bar over the repeating digit(s). The decimal representation of an irrational number will neither terminate nor repeat.

An example of a number that is not a real number is the square root of a negative number like $\sqrt{-5}$. These numbers are discussed in Chapter 10.

Two ways to represent the relationships among the various types of numbers are shown in Figure 8. Part (a) also gives some examples. Notice that every real number is either a rational number or an irrational number.

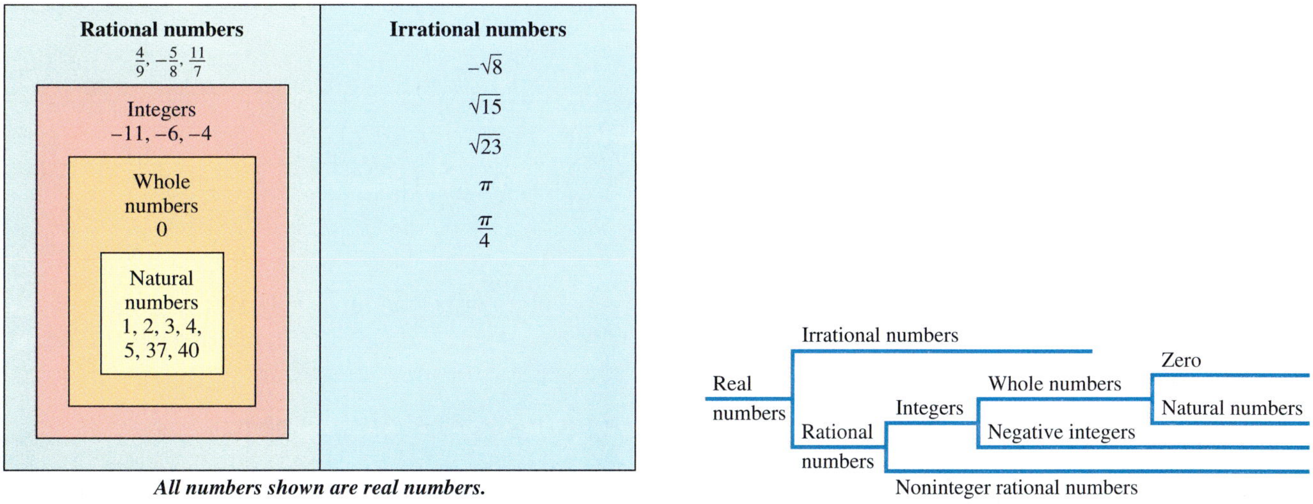


**(a)** **(b)**

**FIGURE 8**

**EXAMPLE 2 Determining Whether a Number Belongs to a Set**

List the numbers in the set

$$\left\{-5, -\frac{2}{3}, 0, \sqrt{2}, 3\frac{1}{4}, 5, 5.8\right\}$$

that belong to each set of numbers.

**(a)** Natural numbers

The only natural number in the set is 5.

**(b)** Whole numbers

The whole numbers consist of the natural numbers and 0. So the elements of the set that are whole numbers are 0 and 5.

**(c)** Integers

The integers in the set are $-5$, 0, and 5.

**(d)** Rational numbers

The rational numbers are $-5, -\frac{2}{3}, 0, 3\frac{1}{4}, 5, 5.8$, since each of these numbers *can* be written as the quotient of two integers. For example, $3\frac{1}{4} = \frac{13}{4}$ and $5.8 = \frac{58}{10}$.

**(e)** Irrational numbers

The only irrational number in the set is $\sqrt{2}$.

**(f)** Real numbers

All the numbers in the set are real numbers.

**Now Try Exercise 19.**

**OBJECTIVE 2 Tell which of two different real numbers is smaller.** Given any two whole numbers, you probably can tell which number is smaller. But what happens with negative numbers, as in the set of integers? Positive numbers decrease as the corresponding points on the number line go to the left. For example, $8 < 12$ because 8 is to the left of 12 on the number line. This ordering is extended to all real numbers by definition.

**Ordering of Real Numbers**

For any two real numbers $a$ and $b$, ***a* is less than *b*** if $a$ is to the left of $b$ on the number line.

This means that any negative number is smaller than 0, and any negative number is smaller than any positive number. Also, 0 is smaller than any positive number.

**EXAMPLE 3 Determining the Order of Real Numbers**

Is it true that $-3 < -1$?

To decide whether the statement is true, locate both numbers, $-3$ and $-1$, on a number line, as shown in Figure 9. Since $-3$ is to the left of $-1$ on the number line, $-3$ is smaller than $-1$. The statement $-3 < -1$ is true.

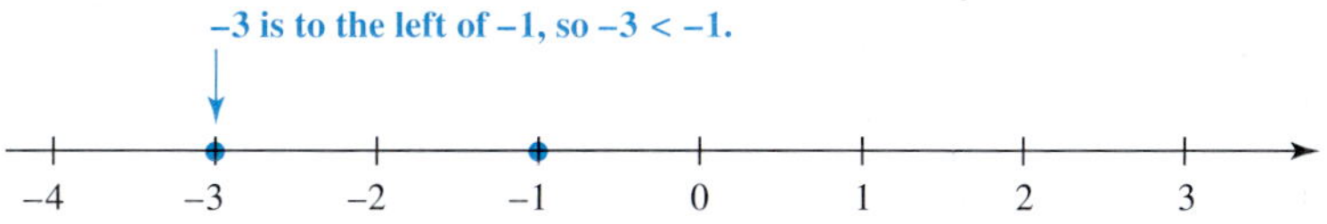


FIGURE 9

Now Try Exercise 53.

**NOTE** In Section 1.2 we saw how it is possible to rewrite a statement involving $<$ as an equivalent statement involving $>$. The question in Example 2 can also be worded as follows: Is it true that $-1 > -3$? This is, or course, also a true statement.

We can also say that for any two real numbers $a$ and $b$, ***a* is greater than *b*** if $a$ is to the right of $b$ on the number line.

**OBJECTIVE 3 Find additive inverses and absolute values of real numbers.** By a property of the real numbers, for any real number $x$ (except 0), there is exactly one number on the number line the same distance from 0 as $x$ but on the opposite side of 0. For example, Figure 10 shows that the numbers 1 and $-1$ are each the same distance from 0 but are on opposite sides of 0. The numbers 1 and $-1$ are called *additive inverses,* or *opposites,* of each other.

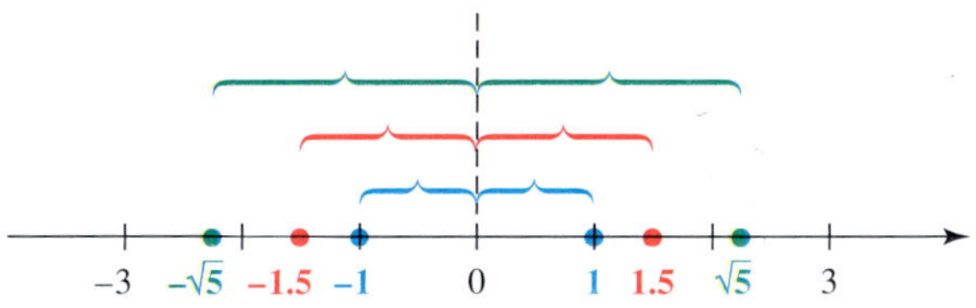


FIGURE 10

**Additive Inverse**

The **additive inverse** of a number $x$ is the number that is the same distance from 0 on the number line as $x$, but on the opposite side of 0.

The additive inverse of a number can be indicated by writing the symbol $-$ in front of the number. With this symbol, the additive inverse of 7 is written $-7$. The additive inverse of $-3$ is written $-(-3)$, and can be read "the opposite of $-3$" or "the negative of $-3$." Figure 10 suggests that 3 is the additive inverse of $-3$. A number can have only one additive inverse, and so the symbols 3 and $-(-3)$ must represent the same number, which means that

$$-(-3) = 3.$$

This idea can be generalized as follows.

**Double Negative Rule***

For any real number $x$, $\quad -(-x) = x.$

| Number | Additive Inverse |
|---|---|
| $-4$ | $-(-4)$ or 4 |
| 0 | 0 |
| 19 | $-19$ |
| $-\frac{2}{3}$ | $\frac{2}{3}$ |
| .52 | $-.52$ |

The table in the margin shows several numbers and their additive inverses. It suggests that the additive inverse of a number is found by changing the sign of the number. An important property of additive inverses will be studied later in this chapter: $a + (-a) = (-a) + a = 0$ for all real numbers $a$.

As previously mentioned, additive inverses are numbers that are the same distance from 0 on the number line. See Figure 10. We can also express this idea by saying that a number and its additive inverse have the same absolute value. The **absolute value** of a real number can be defined as the distance between 0 and the number on the number line. The symbol for the absolute value of the number $x$ is $|x|$, read "the absolute value of $x$." For example, the distance between 2 and 0 on the number line is 2 units, so

$$|2| = 2.$$

Because the distance between $-2$ and 0 on the number is also 2 units,

$$|-2| = 2.$$

Since distance is a physical measurement, which is never negative, ***the absolute value of a number is never negative.*** For example, $|12| = 12$ and $|-12| = 12$, since both 12 and $-12$ lie at a distance of 12 units from 0 on the number line. Also, since 0 is a distance of 0 units from 0, $|0| = 0$.

In symbols, the absolute value of $x$ is defined as follows.

**Absolute Value**

$$|x| = \begin{cases} x & \text{if } x \geq 0 \\ -x & \text{if } x < 0 \end{cases}$$

By this definition, if $x$ is a positive number or 0, then its absolute value is $x$ itself. For example, since 8 is a positive number, $|8| = 8$. However, if $x$ is a negative number, then its absolute value is the additive inverse of $x$. This means that if $x = -9$, then $|-9| = -(-9) = 9$, since the additive inverse of $-9$ is 9.

**CAUTION** The definition of absolute value can be confusing if it is not read carefully. The "$-x$" in the second part of the definition *does not* represent a negative number. Since $x$ is negative in the second part, $-x$ represents the opposite of a negative number, that is, a positive number. *The absolute value of a number is never negative.*

*This rule is justified by interpreting $-(-x)$ as $-1 \cdot (-x) = [(-1)(-1)]x = 1 \cdot x = x$. This requires concepts covered in later sections of this chapter.

**EXAMPLE 4 Finding Absolute Value**

Simplify by finding the absolute value.

(a) $|5| = 5$

(b) $|-5| = -(-5) = 5$

(c) $-|5| = -(5) = -5$

(d) $-|-14| = -(14) = -14$

(e) $|8 - 2| = |6| = 6$

(f) $-|8 - 2| = -|6| = -6$

**Now Try Exercises 35, 37, and 39.**

Parts (e) and (f) of Example 4 show that absolute value bars are also grouping symbols. You must perform any operations that appear inside absolute value symbols before finding the absolute value.

**OBJECTIVE 4 Interpret the meanings of real numbers from a table of data.** The Producer Price Index is the oldest continuous statistical series published by the Bureau of Labor Statistics. It measures the average changes in prices received by producers of all commodities produced in the United States. The next example shows how signed numbers can be used to interpret such data.

**EXAMPLE 5 Interpreting Data**

The table shows the percent change in the Producer Price Index for selected commodities from 1998 to 1999 and from 1999 to 2000. Use the table to answer each question.

| *Commodity* | *Change from 1998 to 1999* | *Change from 1999 to 2000* |
|---|---|---|
| Fresh fruits and melons | 13.1 | −12.8 |
| Gasoline | 11.3 | 30.1 |
| X-ray and electromedical equipment | −2.4 | −2.5 |
| Prepared animal feeds | −9.6 | 4.6 |
| Office and store machines and equipment | 0 | .6 |

*Source:* U.S. Bureau of Labor Statistics.

**(a)** What commodity in which year represents the greatest percent decrease?

We must find the negative number with the greatest absolute value. The number that satisfies this condition is $-12.8$; the greatest percent decrease was shown by fresh fruits and melons from 1999 to 2000.

**(b)** Which commodity in which year represents the least change?

In this case, we must find the number (either positive, negative, or zero) with the least absolute value. From 1998 to 1999, office and store machines and equipment showed no change, as represented by 0.

**Now Try Exercises 65 and 67.**

## 1.4 EXERCISES

**For Extra Help**

- Student's Solutions Manual
- MyMathLab
- InterAct Math Tutorial Software
- AW Math Tutor Center
- MathXL
- Digital Video Tutor CD 1/Videotape 1

*In the applications in Exercises 1–6, use an integer to express each number in italics representing a change. In Exercises 7 and 8, use a rational number. See Example 1.*

**1.** Between 1995 and 2000, the number of U.S. citizens 65 years of age or older increased by *1,198,000.* (*Source:* U.S. Bureau of the Census.)

**2.** Between 1990 and 2000, the mean SAT verbal score for Florida residents increased by *3,* while the mathematics score increased by *7.* (*Source:* The College Board.)

**3.** From 1990 to 2000, the number of cable TV systems in the United States went from 9575 to 10,500, representing an increase of *925.* (*Source: Television and Cable Factbook.*)

**4.** From 1990 to 1998, the population of Glen Ellyn, Illinois, went from 25,956 to 24,919, representing a decrease of *1037* people. (*Source:* U.S. Bureau of the Census.)

**5.** In 1935, there were 15,295 banks in the United States. By 1999, this number was 10,221, representing a decrease of *5074* banks. (*Source:* Federal Deposit Insurance Corporation.)

**6.** Death Valley lies *282* feet below sea level. (Sea level is considered 0 feet.) (*Source:* U.S. Geological Survey, Department of the Interior.)

**7.** On Thursday, May 30, 2002, the Dow Jones Industrial Average closed at 9911.69. On the previous day it had closed at 9923.04. Thus on Thursday it closed down *11.35.* (*Source: Times Picayune.*)

**8.** On Thursday, May 30, 2002, the NASDAQ closed at 1631.92. On the previous day it had closed at 1624.39. Thus on Thursday it closed up *7.53.* (*Source: Times Picayune.*)

*In Exercises 9–14, give a number that satisfies the given condition.*

**9.** An integer between 3.5 and 4.5

**10.** A rational number between 3.8 and 3.9

**11.** A whole number that is not positive and is less than 1

**12.** A whole number greater than 4.5

**13.** An irrational number that is between $\sqrt{11}$ and $\sqrt{13}$

**14.** A real number that is neither negative nor positive

*In Exercises 15–18, decide whether each statement is* true *or* false.

**15.** Every natural number is positive.

**16.** Every whole number is positive.

**17.** Every integer is a rational number.

**18.** Every rational number is a real number.

*For Exercises 19 and 20, see Example 2.*

**19.** List all numbers from the set

$$\left\{-9, -\sqrt{7}, -1\frac{1}{4}, -\frac{3}{5}, 0, \sqrt{5}, 3, 5.9, 7\right\}$$

that are

**(a)** natural numbers; **(b)** whole numbers; **(c)** integers;
**(d)** rational numbers; **(e)** irrational numbers; **(f)** real numbers.

**20.** List all numbers from the set

$$\left\{-5.3, -5, -\sqrt{3}, -1, -\frac{1}{9}, 0, 1.2, 1.8, 3, \sqrt{11}\right\}$$

that are

**(a)** natural numbers; **(b)** whole numbers; **(c)** integers;
**(d)** rational numbers; **(e)** irrational numbers; **(f)** real numbers.

**21.** Explain in your own words the different sets of numbers introduced in this section, and give an example of each kind.

**22.** What two possible situations exist for the decimal representation of a rational number?

*Graph each group of numbers on a number line. See Figures 5 and 6.*

**23.** 0, 3, −5, −6

**24.** 2, 6, −2, −1

**25.** −2, −6, −4, 3, 4

**26.** −5, −3, −2, 0, 4

**27.** $\frac{1}{4}, 2\frac{1}{2}, -3\frac{4}{5}, -4, -1\frac{5}{8}$

**28.** $5\frac{1}{4}, 4\frac{5}{9}, -2\frac{1}{3}, 0, -3\frac{2}{5}$

**29.** Match each expression in Column I with its value in Column II. Choices in Column II may be used once, more than once, or not at all.

| I | II |
|---|---|
| **(a)** $\lvert -7 \rvert$ | **A.** 7 |
| **(b)** $-(-7)$ | **B.** −7 |
| **(c)** $-\lvert -7 \rvert$ | **C.** Neither A nor B |
| **(d)** $-\lvert -(-7) \rvert$ | **D.** Both A and B |

**30.** Fill in the blanks with the correct values: The opposite of −2 is _____, while the absolute value of −2 is _____. The additive inverse of −2 is _____, while the additive inverse of the absolute value of −2 is _____.

*Find **(a)** the opposite (or additive inverse) of each number and **(b)** the absolute value of each number.*

**31.** −2 **32.** −8 **33.** 6 **34.** 11

*Simplify by finding the absolute value. See Example 4.*

**35.** $\lvert -6 \rvert$ **36.** $\lvert -12 \rvert$ **37.** $-\lvert -12 \rvert$

**38.** $-\lvert -6 \rvert$ **39.** $\lvert 6 - 3 \rvert$ **40.** $-\lvert 6 - 3 \rvert$

*Select the smaller of the two given numbers. See Examples 3 and 4.*

**41.** −12, −4 **42.** −9, −14 **43.** −8, −1

**44.** −15, −16 **45.** $3, \lvert -4 \rvert$ **46.** $5, \lvert -2 \rvert$

**47.** $\lvert -3 \rvert, \lvert -4 \rvert$ **48.** $\lvert -8 \rvert, \lvert -9 \rvert$ **49.** $-\lvert -6 \rvert, -\lvert -4 \rvert$

**50.** $-\lvert -2 \rvert, -\lvert -3 \rvert$ **51.** $\lvert 5 - 3 \rvert, \lvert 6 - 2 \rvert$ **52.** $\lvert 7 - 2 \rvert, \lvert 8 - 1 \rvert$

*Decide whether each statement is* true *or* false. *See Examples 3 and 4.*

**53.** $-5 < -2$
**54.** $-8 > -2$
**55.** $-4 \leq -(-5)$
**56.** $-6 \leq -(-3)$
**57.** $|-6| < |-9|$
**58.** $|-12| < |-20|$
**59.** $-|8| > |-9|$
**60.** $-|12| > |-15|$
**61.** $-|-5| \geq -|-9|$
**62.** $-|-12| \leq -|-15|$
**63.** $|6 - 5| \geq |6 - 2|$
**64.** $|13 - 8| \leq |7 - 4|$

*The table shows the percent change in the Producer Price Index for selected construction commodities from 1998 to 1999 and from 1999 to 2000.* Use the table to answer Exercises 65–68. See Example 5.*

| Commodity | Change from 1998 to 1999 | Change from 1999 to 2000 |
|---|---|---|
| Softwood plywood | 32.1 | −33.2 |
| Outdoor lighting equipment | −.5 | 2.4 |
| Steel pipe and tubes | −6.9 | 4.3 |
| Gypsum products | 30.4 | −6.6 |
| Paving mixtures and blocks | .4 | 17.4 |

*2000 data are preliminary.
*Source:* U.S. Bureau of Labor Statistics.

**65.** Which commodity in which year represents the greatest percentage increase?

**66.** Which commodity in which year represents the greatest percentage decrease?

**67.** Which commodity in which year represents the least change?

**68.** Which commodity represents an increase for both years?

*Give three numbers between* $-6$ *and* 6 *that satisfy each given condition.*

**69.** Positive real numbers but not integers

**70.** Real numbers but not positive numbers

**71.** Real numbers but not whole numbers

**72.** Rational numbers but not integers

**73.** Real numbers but not rational numbers

**74.** Rational numbers but not negative numbers

**75.** Students often say "Absolute value is always positive." Is this true? Explain.

**76.** True or false: If $a$ is negative, then $|a| = -a$.

# 1.5 Adding and Subtracting Real Numbers

**OBJECTIVES**

1. Add two numbers with the same sign.
2. Add positive and negative numbers.

In this and the next section, we extend the rules for operations with positive numbers to the negative numbers.

**OBJECTIVE 1 Add two numbers with the same sign.** A number line can be used to illustrate adding real numbers.

3 Use the definition of subtraction.

4 Use the order of operations with real numbers.

5 Interpret words and phrases involving addition and subtraction.

6 Use signed numbers to interpret data.

**EXAMPLE 1 Adding Numbers on a Number Line**

**(a)** Use a number line to find the sum $2 + 3$.

Add the positive numbers 2 and 3 on the number line by starting at 0 and drawing an arrow 2 units to the *right,* as shown in Figure 11. This arrow represents the number 2 in the sum $2 + 3$. Then, from the right end of this arrow draw another arrow 3 units to the right. The number below the end of this second arrow is 5, so $2 + 3 = 5$.

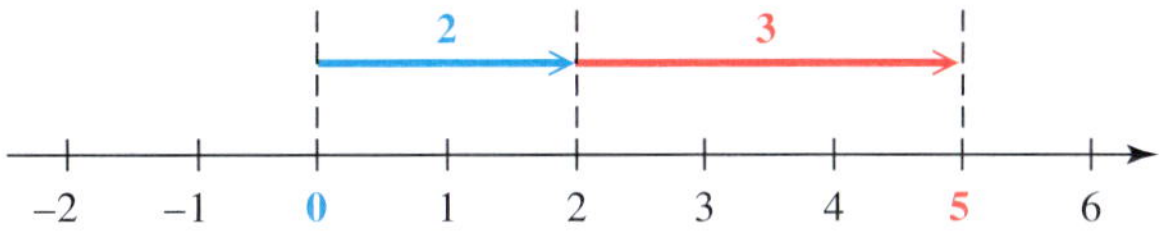


FIGURE 11

**(b)** Use a number line to find the sum $-2 + (-4)$. (Parentheses are placed around the $-4$ to avoid the confusing use of $+$ and $-$ next to each other.)

Add the negative numbers $-2$ and $-4$ on the number line by starting at 0 and drawing an arrow 2 units to the *left,* as shown in Figure 12. The arrow is drawn to the left to represent the addition of a *negative* number. From the left end of the first arrow, draw a second arrow 4 units to the left. The number below the end of this second arrow is $-6$, so $-2 + (-4) = -6$.

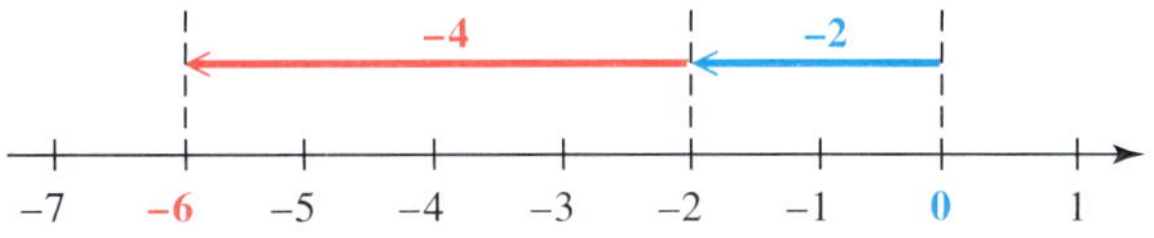


FIGURE 12

**Now Try Exercise 1.**

In Example 1, the sum of the two negative numbers $-2$ and $-4$ is a negative number whose distance from 0 is the sum of the distance of $-2$ from 0 and the distance of $-4$ from 0. That is, *the sum of two negative numbers is the negative of the sum of their absolute values.*

$$-2 + (-4) = -(|-2| + |-4|) = -(2 + 4) = -6$$

**Adding Numbers with the Same Sign**

To add two numbers with the *same* sign, add the absolute values of the numbers. The sum has the same sign as the numbers being added.

**EXAMPLE 2 Using the Rule to Add Two Negative Numbers**

Find each sum.

**(a)** $-2 + (-9) = -(|-2| + |-9|) = -(2 + 9) = -11$

**(b)** $-8 + (-12) = -20$ **(c)** $-15 + (-3) = -18$

**Now Try Exercise 7.**

**OBJECTIVE 2 Add positive and negative numbers.** We can use a number line to illustrate the sum of a positive number and a negative number.

### EXAMPLE 3 Adding Numbers with Different Signs

Use a number line to find the sum $-2 + 5$.

Find the sum $-2 + 5$ on the number line by starting at 0 and drawing an arrow 2 units to the left. From the left end of this arrow, draw a second arrow 5 units to the right, as shown in Figure 13. The number below the end of the second arrow is 3, so $-2 + 5 = 3$.

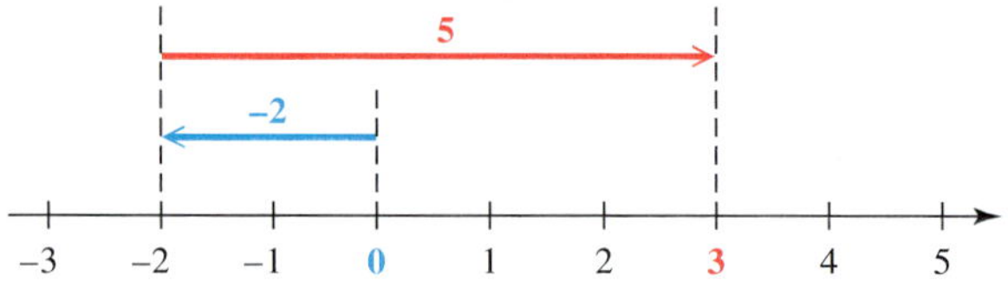


**FIGURE 13**

**Now Try Exercise 3.**

**Adding Numbers with Different Signs**

To add two numbers with *different* signs, subtract the smaller absolute value from the larger absolute value. The answer has the sign of the number with the larger absolute value.

For example, to add $-12$ and 5, find their absolute values: $|-12| = 12$ and $|5| = 5$. Then find the difference between these absolute values: $12 - 5 = 7$. Since $|-12| > |5|$, the sum will be negative, so the final answer is $-12 + 5 = -7$.

While a number line is useful in showing the rules for addition, it is important to be able to find sums mentally.

### EXAMPLE 4 Adding Mentally

Check each answer, trying to work the addition mentally. If you have trouble, use a number line.

**(a)** $7 + (-4) = 3$

**(b)** $-8 + 12 = 4$

**(c)** $-\frac{1}{2} + \frac{1}{8} = -\frac{4}{8} + \frac{1}{8} = -\frac{3}{8}$ Remember to find a common denominator first.

**(d)** $\frac{5}{6} + \left(-\frac{4}{3}\right) = -\frac{1}{2}$

**(e)** $-4.6 + 8.1 = 3.5$

**(f)** $-16 + 16 = 0$

**(g)** $42 + (-42) = 0$

**Now Try Exercises 15 and 27.**

Parts (f) and (g) in Example 4 suggest that the sum of a number and its additive inverse is 0. This is always true, and this property is discussed further in Section 1.7.

The rules for adding signed numbers are summarized as follows.

**Adding Signed Numbers**

*Same sign* Add the absolute values of the numbers. The sum has the same sign as the given numbers.

*Different signs* Find the difference between the larger absolute value and the smaller. The sum has the sign of the number with the larger absolute value.

**OBJECTIVE 3 Use the definition of subtraction.** We can illustrate subtraction of 4 from 7, written $7 - 4$, using a number line. As seen in Figure 14, we begin at 0 and draw an arrow 7 units to the right. From the right end of this arrow, we draw an arrow 4 units to the left. The number at the end of the second arrow shows that $7 - 4 = 3$.

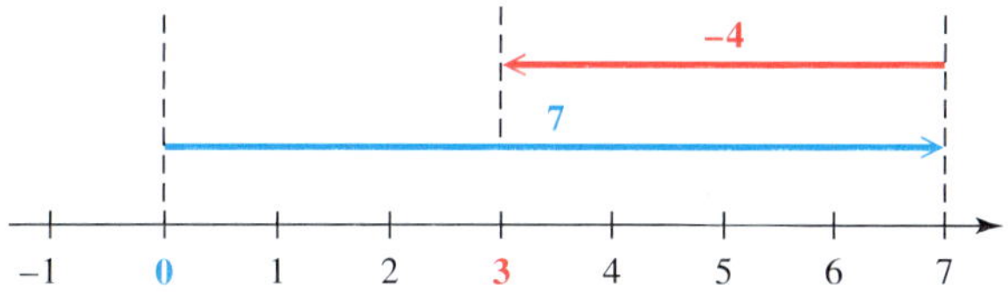


**FIGURE 14**

The procedure used to find $7 - 4$ is exactly the same procedure that would be used to find $7 + (-4)$, so

$$7 - 4 = 7 + (-4).$$

This suggests that *subtracting* a positive number from a larger positive number is the same as *adding* the additive inverse of the smaller number to the larger. This result leads to the definition of subtraction for all real numbers.

**Subtraction**

For any real numbers $x$ and $y$,

$$x - y = x + (-y).$$

That is, to *subtract* $y$ from $x$, *add the additive inverse* (or opposite) of $y$ to $x$.

The definition gives the following procedure for subtracting signed numbers.

**Subtracting Signed Numbers**

*Step 1* Change the subtraction symbol to the addition symbol.
*Step 2* Change the sign of the number being subtracted.
*Step 3* Add.

### EXAMPLE 5 Using the Definition of Subtraction

Subtract.

Change − to +.
No change — Additive inverse of 3

**(a)** $12 - 3 = 12 + (-3) = 9$

**(b)** $5 - 7 = 5 + (-7) = -2$

Change − to +.
No change — Additive inverse of −5

**(c)** $-3 - (-5) = -3 + (5) = 2$

**(d)** $-6 - 9 = -6 + (-9) = -15$

**(e)** $\frac{4}{3} - \left(-\frac{1}{2}\right) = \frac{4}{3} + \frac{1}{2} = \frac{8}{6} + \frac{3}{6} = \frac{11}{6}$ or $1\frac{5}{6}$

**Now Try Exercises 43, 51, and 55.**

We have now used the symbol − for three purposes:

1. to represent subtraction, as in $9 - 5 = 4$;
2. to represent negative numbers, such as $-10$, $-2$, and $-3$;
3. to represent the opposite (or negative) of a number, as in "the opposite (or negative) of 8 is $-8$."

We may see more than one use of − in the same problem, such as $-6 - (-9)$, where $-9$ is subtracted from $-6$. The meaning of the − symbol depends on its position in the algebraic expression.

**OBJECTIVE 4 Use the order of operations with real numbers.** As before, with problems that have grouping symbols, first do any operations inside the parentheses and brackets. Work within the innermost set of grouping symbols first, and then work outward.

### EXAMPLE 6 Adding and Subtracting with Grouping Symbols

Perform each indicated operation.

**(a)**
$$\begin{aligned} -6 - [2 - (8 + 3)] &= -6 - [2 - 11] && \text{Add.} \\ &= -6 - [2 + (-11)] && \text{Definition of subtraction} \\ &= -6 - (-9) && \text{Add.} \\ &= -6 + (9) && \text{Definition of subtraction} \\ &= 3 && \text{Add.} \end{aligned}$$

**(b)** $5 + [(-3 - 2) - (4 - 1)] = 5 + [(-3 + (-2)) - 3]$
$= 5 + [(-5) - 3]$
$= 5 + [(-5) + (-3)]$
$= 5 + (-8)$
$= -3$

**(c)** $\frac{2}{3} - \left[\frac{1}{12} - \left(-\frac{1}{4}\right)\right] = \frac{8}{12} - \left[\frac{1}{12} - \left(-\frac{3}{12}\right)\right]$ Find a common denominator.
$= \frac{8}{12} - \left[\frac{1}{12} + \frac{3}{12}\right]$ Definition of subtraction
$= \frac{8}{12} - \frac{4}{12}$ Add.
$= \frac{4}{12}$ Subtract.
$= \frac{1}{3}$ Lowest terms

**(d)** $|4 - 7| + 2|6 - 3| = |-3| + 2|3|$ Work within absolute value bars.
$= 3 + 2 \cdot 3$ Evaluate absolute values.
$= 3 + 6$ Multiply.
$= 9$ Add.

**Now Try Exercises 65, 75, and 79.**

**OBJECTIVE 5 Interpret words and phrases involving addition and subtraction.**

### PROBLEM SOLVING

As we mentioned earlier, problem solving often requires translating words and phrases into symbols. The word *sum* indicates addition. The table lists some of the words and phrases that also signify addition.

| *Word or Phrase* | *Example* | *Numerical Expression and Simplification* |
|---|---|---|
| Sum of | The *sum of* −3 and 4 | $-3 + 4 = 1$ |
| Added to | 5 *added to* −8 | $-8 + 5 = -3$ |
| More than | 12 *more than* −5 | $-5 + 12 = 7$ |
| Increased by | −6 *increased by* 13 | $-6 + 13 = 7$ |
| Plus | 3 *plus* 14 | $3 + 14 = 17$ |

**EXAMPLE 7 Interpreting Words and Phrases Involving Addition**

Write a numerical expression for each phrase and simplify the expression.

**(a)** The sum of $-8$ and 4 and 6

$$-8 + 4 + 6 = -4 + 6 = 2 \quad \text{Add in order from left to right.}$$

**(b)** 3 more than $-5$, increased by 12

$$-5 + 3 + 12 = -2 + 12 = 10$$

**Now Try Exercise 87.**

**PROBLEM SOLVING**

To solve problems that involve subtraction, we must be able to interpret key words and phrases that indicate subtraction. *Difference* is one of them. Some of these are given in the table.

| *Word, Phrase, or Sentence* | *Example* | *Numerical Expression and Simplification* |
|---|---|---|
| Difference between | The *difference between* $-3$ and $-8$ | $-3 - (-8) = -3 + 8 = 5$ |
| Subtracted from | 12 *subtracted from* 18 | $18 - 12 = 6$ |
| From . . . , subtract. . . . | From 12, subtract 8. | $12 - 8 = 12 + (-8) = 4$ |
| Less | 6 *less* 5 | $6 - 5 = 1$ |
| Less than | 6 *less than* 5 | $5 - 6 = 5 + (-6) = -1$ |
| Decreased by | 9 *decreased by* $-4$ | $9 - (-4) = 9 + 4 = 13$ |
| Minus | 8 *minus* 5 | $8 - 5 = 3$ |

**CAUTION** When you are subtracting two numbers, it is important that you write them in the correct order, because, in general, $a - b \neq b - a$. For example, $5 - 3 \neq 3 - 5$. For this reason, *think carefully before interpreting an expression involving subtraction.* (This difficulty did not arise for addition.)

**EXAMPLE 8 Interpreting Words and Phrases Involving Subtraction**

Write a numerical expression for each phrase and simplify the expression.

**(a)** The difference between $-8$ and 5

In this book, we will write the numbers in the order they are given when "difference between" is used.*

$$-8 - 5 = -8 + (-5) = -13$$

**(b)** 4 subtracted from the sum of 8 and $-3$

Here addition is also used, as indicated by the word *sum*. First, add 8 and $-3$. Next, subtract 4 *from* this sum.

$$[8 + (-3)] - 4 = 5 - 4 = 1$$

*In some cases, people interpret "the difference between" (at least for two positive numbers) to represent the larger minus the smaller. However, we will not do so in this book.

**(c)** 4 less than $-6$

Be careful with order. Here, 4 must be taken *from* $-6$.

$$-6 - 4 = -6 + (-4) = -10$$

Notice that "4 less than $-6$" differs from "4 *is less than* $-6$." The second of these is symbolized $4 < -6$ (which is a false statement).

**(d)** 8, decreased by 5 less than 12

First, write "5 less than 12" as $12 - 5$. Next, subtract $12 - 5$ from 8.

$$8 - (12 - 5) = 8 - 7 = 1$$

**Now Try Exercises 91 and 97.**

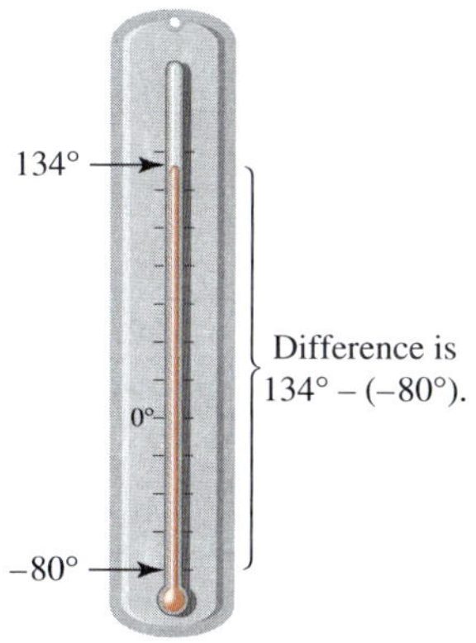


**FIGURE 15**

### EXAMPLE 9 Solving a Problem Involving Subtraction

The record high temperature in the United States was 134° Fahrenheit, recorded at Death Valley, California in 1913. The record low was $-80$°F, at Prospect Creek, Alaska, in 1971. See Figure 15. What is the difference between these highest and lowest temperatures? (*Source: World Almanac and Book of Facts,* 2000.)

We must subtract the lowest temperature from the highest temperature.

$$134 - (-80) = 134 + 80 \quad \text{Definition of subtraction}$$
$$= 214 \quad \text{Add.}$$

The difference between the two temperatures is 214°F.

**Now Try Exercise 113.**

**OBJECTIVE 6** **Use signed numbers to interpret data.**

### EXAMPLE 10 Using a Signed Number in Data Interpretation

The bar graph in Figure 16 gives the Producer Price Index (PPI) for crude materials between 1994 and 1999.

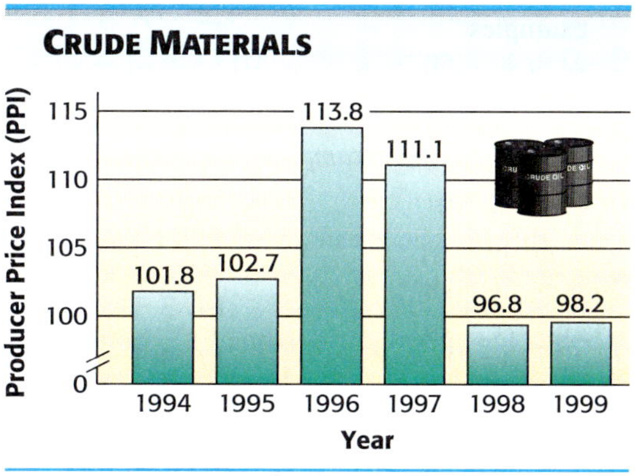


*Source:* U.S. Bureau of Labor Statistics, Producer Price Indexes, monthly and annual.

**FIGURE 16**

**(a)** Use a signed number to represent the change in the PPI from 1995 to 1996.

**95.** The sum of 9 and $-4$, decreased by 7

**96.** The sum of 12 and $-7$, decreased by 14

**97.** 12 less than the difference between 8 and $-5$

**98.** 19 less than the difference between 9 and $-2$

*The bar graph shows federal budget outlays for the U.S. Treasury Department for the years 1998 through 2001. Use a signed number to represent the change in outlay for each time period. See Example 10.*

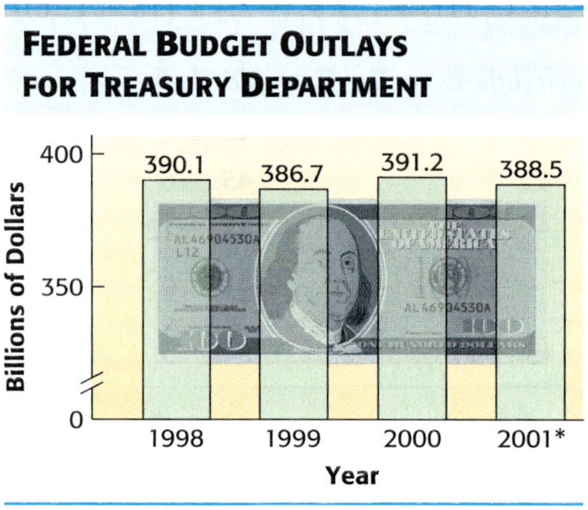


*Estimated

***Source:*** U.S. Office of Management and Budget.

**99.** 1998 to 1999

**100.** 1999 to 2000

**101.** 2000 to 2001

**102.** 1998 to 2001

*The two tables show the heights of some selected mountains and the depths of some selected trenches. Use the information given to answer Exercises 103–108.*

| *Mountain* | *Height (in feet)* |
|---|---|
| Foraker | 17,400 |
| Wilson | 14,246 |
| Pikes Peak | 14,110 |

| *Trench* | *Depth (in feet, as a negative number)* |
|---|---|
| Philippine | $-32,995$ |
| Cayman | $-24,721$ |
| Java | $-23,376$ |

*Source: World Almanac and Book of Facts,* 2000.

**103.** What is the difference between the height of Mt. Foraker and the depth of the Philippine Trench?

**104.** What is the difference between the height of Pikes Peak and the depth of the Java Trench?

**105.** How much deeper is the Cayman Trench than the Java Trench?

**106.** How much deeper is the Philippine Trench than the Cayman Trench?

**Pikes Peak**

**107.** How much higher is Mt. Wilson than Pikes Peak?

**108.** If Mt. Wilson and Pikes Peak were stacked one on top of the other, how much higher would they be than Mt. Foraker?

*Solve each problem. See Example 9.*

**109.** Based on census population projections for 2020, New York will lose 5 seats in the U.S. House of Representatives, Pennsylvania will lose 4 seats, and Ohio will lose 3. Write a signed number that represents the total number of seats these three states are projected to lose. (*Source:* Population Reference Bureau.)

**110.** Michigan is projected to lose 3 seats in the U.S. House of Representatives and Illinois 2 in 2020. The states projected to gain the most seats are California with 9, Texas with 5, Florida with 3, Georgia with 2, and Arizona with 2. Write a signed number that represents the algebraic sum of these changes. (*Source:* Population Reference Bureau.)

**111.** On January 23, 1943, the temperature rose 49°F in two minutes in Spearfish, South Dakota. The starting temperature was −4°F. What was the temperature two minutes later? (*Source: Guinness Book of World Records,* 2002.)

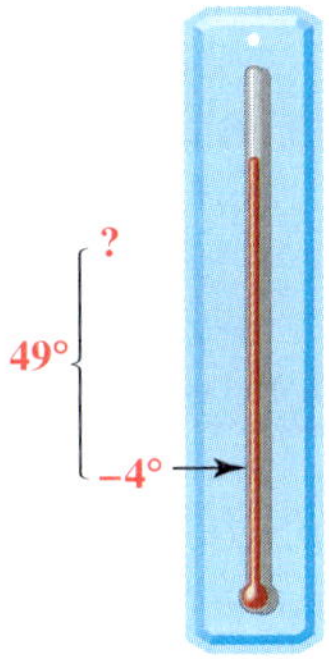


**112.** The lowest temperature ever recorded in Arkansas was −29°F. The highest temperature ever recorded there was 149°F more than the lowest. What was this highest temperature? (*Source: World Almanac and Book of Facts,* 2000.)

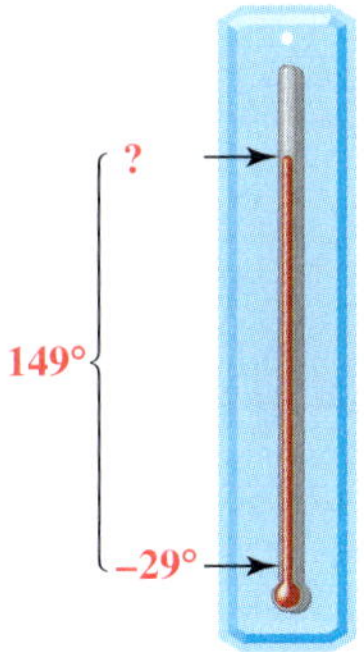


**113.** The coldest temperature recorded in Chicago, Illinois, was −35°F in 1996. The record low in South Dakota was set in 1936 and was 23°F lower than Chicago's record low. What was the record low in South Dakota? (*Source: World Almanac and Book of Facts,* 2000.)

**114.** The top of Mt. Whitney, visible from Death Valley, has an altitude of 14,494 ft above sea level. The bottom of Death Valley is 282 ft below sea level. Using 0 as sea level, find the difference between these two elevations. (*Source: World Almanac and Book of Facts,* 2000.)

**115.** No one knows just why humpback whales heave their 45-ton bodies out of the water, but leap they do. (This activity is called *breaching.*) Mark and Debbie, two researchers based on the island of Maui, noticed that one of their favorite whales, "Pineapple," leaped 15 ft above the surface of the ocean while her mate cruised 12 ft below the surface. What is the difference between these two distances?

**116.** The surface, or rim, of a canyon is at altitude 0. On a hike down into the canyon, a party of hikers stops for a rest at 130 m below the surface. They then descend another 54 m. Write the new altitude as a signed number.

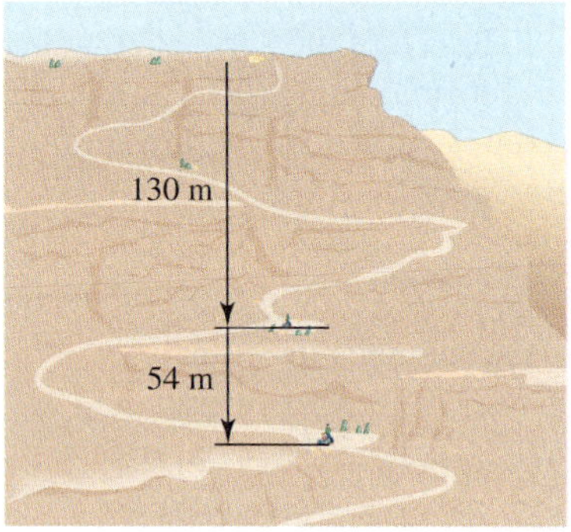


**117.** A pilot announces to the passengers that the current altitude of their plane is 34,000 ft. Because of turbulence, the pilot is forced to descend 2100 ft. Write the new altitude as a signed number.

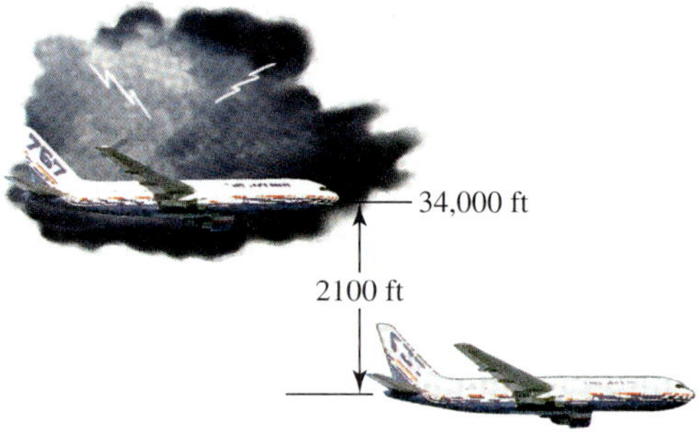


**118.** A welder working with stainless steel must use precise measurements. Suppose a welder attaches two pieces of steel that are each 3.60 in. long, and then attaches an additional three pieces that are each 9.10 in. long. She finally cuts off a piece that is 7.60 in. long. Find the length of the welded piece of steel.

**119.** Jennifer owes $153 to a credit card company. She makes a $14 purchase with the card and then pays $60 on the account. What is her current balance as a signed number?

**120.** A female polar bear weighed 660 lb when she entered her winter den. She lost 45 lb during each of the first two months of hibernation, and another 205 lb before leaving the den with her two cubs in March. How much did she weigh when she left the den?

## 1.6 Multiplying and Dividing Real Numbers

**OBJECTIVES**

1. Find the product of a positive number and a negative number.
2. Find the product of two negative numbers.
3. Identify factors of integers.

In this section we learn how to multiply with positive and negative numbers. We already know how to multiply positive numbers and that the product of two positive numbers is positive. We also know that the product of 0 and any positive number is 0, so we extend that property to all real numbers.

**Multiplication by Zero**

For any real number $x$, $\quad \mathbf{x \cdot 0 = 0.}$

4 Use the reciprocal of a number to apply the definition of division.

5 Use the order of operations when multiplying and dividing signed numbers.

6 Evaluate expressions involving variables.

7 Interpret words and phrases involving multiplication and division.

8 Translate simple sentences into equations.

**OBJECTIVE 1 Find the product of a positive number and a negative number.** To define the product of a positive and a negative number so that the result is consistent with the multiplication of two positive numbers, look at the following pattern.

$$\begin{aligned} 3 \cdot 5 &= 15 \\ 3 \cdot 4 &= 12 \\ 3 \cdot 3 &= 9 \\ 3 \cdot 2 &= 6 \\ 3 \cdot 1 &= 3 \\ 3 \cdot 0 &= 0 \\ 3 \cdot (-1) &= ? \end{aligned}$$

The products decrease by 3.

What should $3(-1)$ equal? The product $3(-1)$ represents the sum

$$-1 + (-1) + (-1) = -3,$$

so the product should be $-3$. Also,

$$3(-2) = -2 + (-2) + (-2) = -6$$

and

$$3(-3) = -3 + (-3) + (-3) = -9.$$

These results maintain the pattern in the list, which suggests the following rule.

**Multiplying Numbers with Different Signs**

For any positive real numbers $x$ and $y$,

$$x(-y) = -(xy) \quad \text{and} \quad (-x)y = -(xy).$$

That is, the product of two numbers with opposite signs is negative.

**EXAMPLE 1 Multiplying a Positive Number and a Negative Number**

Find each product using the multiplication rule given in the box.

**(a)** $8(-5) = -(8 \cdot 5) = -40$ **(b)** $5(-4) = -(5 \cdot 4) = -20$

**(c)** $-7(2) = -(7 \cdot 2) = -14$ **(d)** $-9(3) = -(9 \cdot 3) = -27$

**Now Try Exercise 13.**

**OBJECTIVE 2 Find the product of two negative numbers.** The product of two positive numbers is positive, and the product of a positive and a negative number is negative. What about the product of two negative numbers? Look at another pattern.

$$\begin{aligned} -5(4) &= -20 \\ -5(3) &= -15 \\ -5(2) &= -10 \\ -5(1) &= -5 \\ -5(0) &= 0 \\ -5(-1) &= ? \end{aligned}$$

The products increase by 5.

The numbers on the left of the equals sign (in color) decrease by 1 for each step down the list. The products on the right increase by 5 for each step down the list. To maintain this pattern, $-5(-1)$ should be 5 more than $-5(0)$, or 5 more than 0, so

$$-5(-1) = 5.$$

The pattern continues with

$$-5(-2) = 10$$
$$-5(-3) = 15$$
$$-5(-4) = 20$$
$$-5(-5) = 25,$$

and so on, which suggests the next rule.

**Multiplying Two Negative Numbers**

For any positive real numbers $x$ and $y$,

$$-x(-y) = xy.$$

That is, the product of two negative numbers is positive.

**EXAMPLE 2 Multiplying Two Negative Numbers**

Find each product using the multiplication rule given in the box.

**(a)** $-9(-2) = 9 \cdot 2 = 18$ **(b)** $-6(-12) = 6 \cdot 12 = 72$

**(c)** $-8(-1) = 8 \cdot 1 = 8$ **(d)** $-15(-2) = 15 \cdot 2 = 30$

**Now Try Exercise 11.**

A summary of multiplying signed numbers is given here.

**Multiplying Signed Numbers**

The product of two numbers having the *same* sign is *positive,* and the product of two numbers having *different* signs is *negative.*

**OBJECTIVE 3 Identify factors of integers.** In Section 1.1 the definition of a *factor* was given for whole numbers. (For example, since $9 \cdot 5 = 45$, both 9 and 5 are factors of 45.) The definition can now be extended to integers.

If the product of two integers is a third integer, then each of the two integers is a *factor* of the third. For example, $(-3)(-4) = 12$, so $-3$ and $-4$ are both factors of 12. The integer factors of 12 are $-12, -6, -4, -3, -2, -1, 1, 2, 3, 4, 6,$ and 12.

The table on the next page shows several integers and the factors of those integers.

| Integer | 18 | 20 | 15 | 7 | 1 |
|---|---|---|---|---|---|
| Pairs of factors | 1, 18 | 1, 20 | 1, 15 | 1, 7 | 1, 1 |
| | 2, 9 | 2, 10 | 3, 5 | −1, −7 | −1, −1 |
| | 3, 6 | 4, 5 | −1, −15 | | |
| | −1, −18 | −1, −20 | −3, −5 | | |
| | −2, −9 | −2, −10 | | | |
| | −3, −6 | −4, −5 | | | |

Now Try Exercise 29.

**OBJECTIVE 4 Use the reciprocal of a number to apply the definition of division.** In Section 1.5 we saw that the difference between two numbers is found by adding the additive inverse of the second number to the first. Similarly, the *quotient* of two numbers is found by *multiplying* by the *reciprocal,* or *multiplicative inverse.* By definition, since

$$8 \cdot \frac{1}{8} = \frac{8}{8} = 1 \qquad \text{and} \qquad \frac{5}{4} \cdot \frac{4}{5} = \frac{20}{20} = 1,$$

the reciprocal or multiplicative inverse of 8 is $\frac{1}{8}$, and of $\frac{5}{4}$ is $\frac{4}{5}$.

### Reciprocal or Multiplicative Inverse

Pairs of numbers whose product is 1 are **reciprocals**, or **multiplicative inverses,** of each other.

| Number | *Multiplicative Inverse (Reciprocal)* |
|---|---|
| 4 | $\frac{1}{4}$ |
| $.3 = \frac{3}{10}$ | $\frac{10}{3}$ |
| −5 | $\frac{1}{-5}$ or $-\frac{1}{5}$ |
| $-\frac{5}{8}$ | $-\frac{8}{5}$ |
| 0 | None |
| 1 | 1 |
| −1 | −1 |

The table in the margin shows several numbers and their multiplicative inverses. Why is there no multiplicative inverse for the number 0? Suppose that $k$ is to be the multiplicative inverse of 0. Then $k \cdot 0$ should equal 1. But $k \cdot 0 = 0$ for any number $k$. Since there is no value of $k$ that is a solution of the equation $k \cdot 0 = 1$, the following statement can be made.

0 has no multiplicative inverse.

By definition, the quotient of $x$ and $y$ is the product of $x$ and the multiplicative inverse of $y$.

### Division

For any real numbers $x$ and $y$, with $y \neq 0$, $\quad \frac{x}{y} = x \cdot \frac{1}{y}.$

The definition of division indicates that $y$, the number to divide by, cannot be 0. The reason is that 0 has no multiplicative inverse, so that $\frac{1}{0}$ is not a number. **Because 0 has no multiplicative inverse, *division by 0 is undefined.* If a division problem turns out to involve division by 0, write "undefined."**

**NOTE** While division *by* 0 $\left(\frac{a}{0}\right)$ is undefined, we may divide 0 by any nonzero number. In fact, if $a \neq 0$,

$$\frac{0}{a} = 0.$$

Since division is defined in terms of multiplication, all the rules for multiplying signed numbers also apply to dividing them.

### EXAMPLE 3 Using the Definition of Division

Find each quotient using the definition of division.

**(a)** $\frac{12}{3} = 12 \cdot \frac{1}{3} = 4$

**(b)** $\frac{-10}{2} = -10 \cdot \frac{1}{2} = -5$

**(c)** $\frac{-14}{-7} = -14\left(\frac{1}{-7}\right) = 2$

**(d)** $-\frac{2}{3} \div \left(-\frac{4}{5}\right) = -\frac{2}{3} \cdot \left(-\frac{5}{4}\right) = \frac{5}{6}$

**(e)** $\frac{0}{13} = 0\left(\frac{1}{13}\right) = 0$  $\frac{0}{a} = 0 \quad (a \neq 0)$

**(f)** $\frac{-10}{0}$ Undefined

Now Try Exercises 33, 39, and 45.

When dividing fractions, multiplying by the reciprocal works well. However, using the definition of division directly with integers is awkward. It is easier to divide in the usual way, then determine the sign of the answer.

**Dividing Signed Numbers**

The quotient of two numbers having the *same* sign is *positive;* the quotient of two numbers having *different* signs is *negative.*

Note that these are the same as the rules for multiplication.

### EXAMPLE 4 Dividing Signed Numbers

Find each quotient.

**(a)** $\frac{8}{-2} = -4$

**(b)** $\frac{-4.5}{-.09} = 50$

**(c)** $-\frac{1}{8} \div \left(-\frac{3}{4}\right) = -\frac{1}{8} \cdot \left(-\frac{4}{3}\right) = \frac{1}{6}$

Now Try Exercises 35, 41, and 43.

From the definitions of multiplication and division of real numbers,

$$\frac{-40}{8} = -40 \cdot \frac{1}{8} = -5, \quad \text{and} \quad \frac{40}{-8} = 40\left(\frac{1}{-8}\right) = -5, \quad \text{so}$$

$$\frac{-40}{8} = \frac{40}{-8}.$$

Based on this example, the quotient of a positive and a negative number can be expressed in any of the following three forms.

For any positive real numbers $x$ and $y$, $\frac{-x}{y} = \frac{x}{-y} = -\frac{x}{y}$.

Similarly, the quotient of two negative numbers can be expressed as a quotient of two positive numbers.

For any positive real numbers $x$ and $y$, $\frac{-x}{-y} = \frac{x}{y}$.

**OBJECTIVE 5 Use the order of operations when multiplying and dividing signed numbers.**

### EXAMPLE 5 Using the Order of Operations

Perform each indicated operation.

**(a)** $-9(2) - (-3)(2)$

First find all products, working from left to right.

$$\begin{aligned} -9(2) - (-3)(2) &= -18 - (-6) \\ &= -18 + 6 \\ &= -12 \end{aligned}$$

**(b)** $\begin{aligned}[t] -6(-2) - 3(-4) &= 12 - (-12) \\ &= 12 + 12 \\ &= 24 \end{aligned}$

**(c)** $-5(-2 - 3) = -5(-5) = 25$

**(d)** $\frac{5(-2) - 3(4)}{2(1 - 6)}$

Simplify the numerator and denominator separately. Then write in lowest terms.

$$\frac{5(-2) - 3(4)}{2(1 - 6)} = \frac{-10 - 12}{2(-5)} = \frac{-22}{-10} = \frac{11}{5}$$

**Now Try Exercises 53 and 67.**

We summarize the rules for operations with signed numbers on the next page.

**Operations with Signed Numbers**

*ADDITION*

*Same sign* Add the absolute values of the numbers. The sum has the same sign as the numbers.

$$-4 + (-6) = -10$$

*Different signs* Subtract the number with the smaller absolute value from the one with the larger. Give the sum the sign of the number having the larger absolute value.

$$4 + (-6) = -(6 - 4) = -2$$

*SUBTRACTION*

Add the opposite of the second number to the first number.

$$8 - (-3) = 8 + 3 = 11$$

*MULTIPLICATION AND DIVISION*

*Same sign* The product or quotient of two numbers with the same sign is positive.

$$-5(-6) = 30 \quad \text{and} \quad \frac{-36}{-12} = 3$$

*Different signs* The product or quotient of two numbers with different signs is negative.

$$-5(6) = -30 \quad \text{and} \quad \frac{18}{-6} = -3$$

Division by 0 is undefined.

**OBJECTIVE 6 Evaluate expressions involving variables.** The next examples show numbers substituted for variables where the rules for multiplying and dividing signed numbers must be used.

### EXAMPLE 6 Evaluating Expressions for Numerical Values

Evaluate each expression, given that $x = -1$, $y = -2$, and $m = -3$.

**(a)** $(3x + 4y)(-2m)$

First substitute the given values for the variables. Then use the order of operations to find the value of the expression.

$(3x + 4y)(-2m) = [3(-1) + 4(-2)][-2(-3)]$ Put parentheses around the value for each variable.

$= [-3 + (-8)][6]$ Multiply.

$= (-11)(6)$ Add inside the parentheses.

$= -66$ Multiply.

**(b)** $2x^2 - 3y^2$

Use parentheses as shown.

$$\begin{aligned} 2(-1)^2 - 3(-2)^2 &= 2(1) - 3(4) && \text{Apply the exponents.} \\ &= 2 - 12 && \text{Multiply.} \\ &= -10 && \text{Subtract.} \end{aligned}$$

**(c)** $\dfrac{4y^2 + x}{m}$

$$\begin{aligned} \frac{4(-2)^2 + (-1)}{-3} &= \frac{4(4) + (-1)}{-3} && \text{Apply the exponent.} \\ &= \frac{16 + (-1)}{-3} && \text{Multiply.} \\ &= \frac{15}{-3} && \text{Add.} \\ &= -5 && \text{Divide.} \end{aligned}$$

Notice how the fraction bar was used as a grouping symbol.

**Now Try Exercises 77 and 85.**

**OBJECTIVE 7 Interpret words and phrases involving multiplication and division.** Just as there are words and phrases that indicate addition and subtraction, certain words also indicate multiplication and division.

### PROBLEM SOLVING

The word *product* refers to multiplication. The table gives other key words and phrases that indicate multiplication.

| *Word or Phrase* | *Example* | *Numerical Expression and Simplification* |
|---|---|---|
| Product of | The *product of* $-5$ and $-2$ | $-5(-2) = 10$ |
| Times | 13 *times* $-4$ | $13(-4) = -52$ |
| Twice (meaning "2 times") | *Twice* 6 | $2(6) = 12$ |
| Of (used with fractions) | $\frac{1}{2}$ *of* 10 | $\frac{1}{2}(10) = 5$ |
| Percent of | 12*% of* $-16$ | $.12(-16) = -1.92$ |
| As much as | $\frac{2}{3}$ *as much as* 30 | $\frac{2}{3}(30) = 20$ |

### EXAMPLE 7 Interpreting Words and Phrases Involving Multiplication

Write a numerical expression for each phrase and simplify. Use the order of operations.

**(a)** The product of 12 and the sum of 3 and $-6$

Here, 12 is multiplied by "the sum of 3 and $-6$."

$$12[3 + (-6)] = 12(-3) = -36$$

**(b)** Twice the difference between 8 and $-4$

$$2[8 - (-4)] = 2[8 + 4] = 2(12) = 24$$

**(c)** Two-thirds of the sum of $-5$ and $-3$

$$\frac{2}{3}[-5 + (-3)] = \frac{2}{3}[-8] = -\frac{16}{3}$$

**(d)** 15% of the difference between 14 and $-2$

Remember that 15% = .15.

$$.15[14 - (-2)] = .15[14 + 2] = .15(16) = 2.4$$

**(e)** Double the product of 3 and 4

$$2 \cdot (3 \cdot 4) = 2 \cdot 12 = 24$$

**Now Try Exercises 89, 93, and 101.**

**PROBLEM SOLVING**

The word *quotient* refers to the answer in a division problem. In algebra, quotients are usually represented with a fraction bar; the symbol ÷ is seldom used. When translating applied problems involving division, use a fraction bar. The table gives some key phrases associated with division.

| *Phrase* | *Example* | *Numerical Expression and Simplification* |
|---|---|---|
| Quotient of | The *quotient of* $-24$ and 3 | $\frac{-24}{3} = -8$ |
| Divided by | $-16$ *divided by* $-4$ | $\frac{-16}{-4} = 4$ |
| Ratio of | The *ratio of* 2 to 3 | $\frac{2}{3}$ |

It is customary to write the first number named as the numerator and the second as the denominator when interpreting a phrase involving division, as shown in the next example.

### EXAMPLE 8 Interpreting Words and Phrases Involving Division

Write a numerical expression for each phrase and simplify the expression.

**(a)** The quotient of 14 and the sum of $-9$ and 2

"Quotient" indicates division. The number 14 is the numerator and "the sum of $-9$ and 2" is the denominator.

$$\frac{14}{-9 + 2} = \frac{14}{-7} = -2$$

**(b)** The product of 5 and $-6$, divided by the difference between $-7$ and 8

The numerator of the fraction representing the division is obtained by multiplying 5 and $-6$. The denominator is found by subtracting $-7$ and 8.

$$\frac{5(-6)}{-7 - 8} = \frac{-30}{-15} = 2$$

**Now Try Exercise 95.**

**OBJECTIVE 8 Translate simple sentences into equations.** In this section and the previous one, important words and phrases involving the four operations of arithmetic have been introduced. We can use these words and phrases to interpret sentences that translate into equations.

### EXAMPLE 9 Translating Words into Equations

Write each sentence in symbols, using $x$ as the variable, and guess or use trial and error to find the solution. All solutions come from the list of integers between $-12$ and 12, inclusive.

**(a)** Three times a number is $-18$.

The word *times* indicates multiplication, and the word *is* translates as the equals sign (=).

$$3x = -18$$

Since the integer between $-12$ and 12 inclusive that makes this statement true is $-6$, the solution of the equation is $-6$.

**(b)** The sum of a number and 9 is 12.

$$x + 9 = 12$$

Since $3 + 9 = 12$, the solution of this equation is 3.

**(c)** The difference between a number and 5 is 0.

$$x - 5 = 0$$

Since $5 - 5 = 0$, the solution of this equation is 5.

**(d)** The quotient of 24 and a number is $-2$.

$$\frac{24}{x} = -2$$

Here, $x$ must be a negative number, since the numerator is positive and the quotient is negative. Since $\frac{24}{-12} = -2$, the solution is $-12$.

**Now Try Exercises 103 and 107.**

**CAUTION** It is important to recognize the distinction between the types of problems found in Examples 7 and 8 and Example 9. In Examples 7 and 8, the *phrases* translate as *expressions,* while in Example 9, the *sentences* translate as *equations.* Remember that an equation is a sentence with an = sign, while an expression is a phrase.

$$\frac{5(-6)}{-7-8} \qquad 3x = -18$$

Expression (points to $\frac{5(-6)}{-7-8}$) — Equation (points to the = sign in $3x = -18$)

## 1.6 EXERCISES

**For Extra Help**

- Student's Solutions Manual
- MyMathLab
- InterAct Math Tutorial Software
- AW Math Tutor Center
- MathXL
- Digital Video Tutor CD 2/Videotape 2

*Fill in each blank with one of the following:* greater than 0, less than 0, equal to 0.

**1.** A positive number is ____________.

**2.** A negative number is ____________.

**3.** The product or the quotient of two numbers with the same sign is ____________.

**4.** The product or the quotient of two numbers with different signs is ____________.

**5.** If three negative numbers are multiplied together, the product is ____________.

**6.** If two negative numbers are multiplied and then their product is divided by a negative number, the result is ____________.

**7.** If a negative number is squared and the result is added to a positive number, the final answer is ____________.

**8.** The reciprocal of a negative number is ____________.

**9.** If three positive numbers, five negative numbers, and zero are multiplied, the product is ____________.

**10.** The fifth power of a negative number is ____________.

*Find each product. See Examples 1 and 2.*

**11.** $-3(-4)$ **12.** $-3(4)$ **13.** $3(-4)$ **14.** $-2(-8)$

**15.** $-10(-12)$ **16.** $9(-5)$ **17.** $3(-11)$ **18.** $3(-15)$

**19.** $15(-11)$ **20.** $-9(-4)$ **21.** $-\frac{3}{8} \cdot \left(-\frac{10}{9}\right)$ **22.** $-\frac{5}{4} \cdot \left(-\frac{5}{8}\right)$

**23.** $\left(-1\frac{1}{4}\right)\left(\frac{2}{15}\right)$ **24.** $\left(\frac{3}{7}\right)\left(-1\frac{5}{9}\right)$ **25.** $(-8)\left(-\frac{3}{4}\right)$ **26.** $(-6)\left(-\frac{5}{3}\right)$

*Find all integer factors of each number.*

**27.** 32 **28.** 36 **29.** 40 **30.** 50 **31.** 31 **32.** 17

*Find each quotient. See Examples 3 and 4.*

**33.** $\frac{15}{5}$ **34.** $\frac{25}{5}$ **35.** $\frac{-30}{6}$ **36.** $\frac{-28}{14}$

**37.** $\frac{-28}{-4}$ **38.** $\frac{-35}{-7}$ **39.** $\frac{96}{-16}$ **40.** $\frac{38}{-19}$

**41.** $-\frac{4}{3} \div \left(-\frac{1}{8}\right)$ **42.** $-\frac{5}{6} \div \left(-\frac{15}{7}\right)$ **43.** $\frac{-8.8}{2.2}$ **44.** $\frac{-4.6}{-.23}$

**45.** $\frac{0}{-2}$ **46.** $\frac{0}{-8}$ **47.** $\frac{12}{0}$ **48.** $\frac{6}{0}$

*Perform each indicated operation. See Examples 5(a), (b), and (c).*

**49.** $7 - 3 \cdot 6$ **50.** $8 - 2 \cdot 5$ **51.** $-10 - (-4)(2)$

**52.** $-11 - (-3)(6)$ **53.** $-7(3 - 8)$ **54.** $-5(4 - 7)$

**55.** $(12 - 14)(1 - 4)$ **56.** $(8 - 9)(4 - 12)$ **57.** $(7 - 10)(10 - 4)$

**58.** $(5 - 12)(19 - 4)$ **59.** $(-2 - 8)(-6) + 7$ **60.** $(-9 - 4)(-2) + 10$

**61.** $3(-5) + |3 - 10|$ **62.** $4(-8) + |4 - 15|$

*Perform each indicated operation. See Example 5(d).*

**63.** $\dfrac{-5(-6)}{9 - (-1)}$ **64.** $\dfrac{-12(-5)}{7 - (-5)}$ **65.** $\dfrac{-21(3)}{-3 - 6}$

**66.** $\dfrac{-40(3)}{-2 - 3}$ **67.** $\dfrac{-10(2) + 6(2)}{-3 - (-1)}$ **68.** $\dfrac{8(-1) - |(-4)(-3)|}{-6 - (-1)}$

**69.** $\dfrac{-27(-2) - |6 \cdot 4|}{-2(3) - 2(2)}$ **70.** $\dfrac{-13(-4) - (-8)(-2)}{(-10)(2) - 4(-2)}$ **71.** $\dfrac{-5(2) + [3(-2) - 4]}{-3 - (-1)}$

**72.** Explain the method you would use to evaluate $3x + 2y$ when $x = -3$ and $y = 4$.

**73.** If $x$ and $y$ are both replaced by negative numbers, is the value of $4x + 8y$ positive or negative?

**74.** Repeat Exercise 73, but replace the first word *negative* with *positive.*

*Evaluate each expression if $x = 6$, $y = -4$, and $a = 3$. See Example 6.*

**75.** $5x - 2y + 3a$ **76.** $6x - 5y + 4a$

**77.** $(2x + y)(3a)$ **78.** $(5x - 2y)(-2a)$

**79.** $\left(\frac{1}{3}x - \frac{4}{5}y\right)\left(-\frac{1}{5}a\right)$ **80.** $\left(\frac{5}{6}x + \frac{3}{2}y\right)\left(-\frac{1}{3}a\right)$

**81.** $(-5 + x)(-3 + y)(3 - a)$ **82.** $(6 - x)(5 + y)(3 + a)$

**83.** $-2y^2 + 3a$ **84.** $5x - 4a^2$

**85.** $\dfrac{2y^2 - x}{a + 10}$ **86.** $\dfrac{xy + 8a}{x - y}$

*Write a numerical expression for each phrase and simplify. See Examples 7 and 8.*

**87.** The product of $-9$ and 2, added to 9

**88.** The product of 4 and $-7$, added to $-12$

**89.** Twice the product of $-1$ and 6, subtracted from $-4$

**90.** Twice the product of $-8$ and 2, subtracted from $-1$

**91.** Nine subtracted from the product of 1.5 and $-3.2$

**92.** Three subtracted from the product of 4.2 and $-8.5$

**93.** The product of 12 and the difference between 9 and $-8$

**94.** The product of $-3$ and the difference between 3 and $-7$

**95.** The quotient of $-12$ and the sum of $-5$ and $-1$

**96.** The quotient of $-20$ and the sum of $-8$ and $-2$

**97.** The sum of 15 and $-3$, divided by the product of 4 and $-3$

**98.** The sum of $-18$ and $-6$, divided by the product of 2 and $-4$

**99.** Twice the sum of 8 and 9

**100.** Three-fourths of the sum of $-8$ and 12

**101.** 20% of the product of $-5$ and 6

**102.** $\frac{2}{3}$ as much as the difference between 8 and $-1$

*Write each statement in symbols, using x as the variable, and find the solution by guessing or by using trial and error. All solutions come from the set of integers between* −12 *and* 12, *inclusive. See Example 9.*

**103.** The quotient of a number and 3 is −3.

**104.** The quotient of a number and 4 is −1.

**105.** 6 less than a number is 4.

**106.** 7 less than a number is 2.

**107.** When 5 is added to a number, the result is −5.

**108.** When 6 is added to a number, the result is −3.

*To find the* **average** *of a group of numbers, we add the numbers and then divide the sum by the number of terms added. For example, to find the average of* 14, 8, 3, 9, *and* 1, *we add them and then divide by* 5:

$$\frac{14 + 8 + 3 + 9 + 1}{5} = \frac{35}{5} = 7.$$

*The average of these numbers is* 7.

*Exercises 109–114 involve finding the average of a group of numbers.*

**109.** Find the average of 23, 18, 13, −4, and −8.

**110.** Find the average of 18, 12, 0, −4, and −10.

**111.** What is the average of all integers between −10 and 14, inclusive of both?

**112.** What is the average of all even integers between −18 and 4, inclusive of both?

**113.** If the average of a group of numbers is 0, what is the sum of all the numbers?

**114.** Suppose there is a group of numbers with some positive and some negative. Under what conditions will the average be a positive number? Under what conditions will the average be negative?

*The operation of division is used in* divisibility *tests. A divisibility test allows us to determine whether a given number is divisible (without remainder) by another number. For example, a number is divisible by* 2 *if its last digit is divisible by* 2, *and not otherwise.*

**115.** Tell why **(a)** 3,473,986 is divisible by 2 and **(b)** 4,336,879 is not divisible by 2.

**116.** An integer is divisible by 3 if the sum of its digits is divisible by 3, and not otherwise. Show that **(a)** 4,799,232 is divisible by 3 and **(b)** 2,443,871 is not divisible by 3.

**117.** An integer is divisible by 4 if its last two digits form a number divisible by 4, and not otherwise. Show that **(a)** 6,221,464 is divisible by 4 and **(b)** 2,876,335 is not divisible by 4.

**118.** An integer is divisible by 5 if its last digit is divisible by 5, and not otherwise. Show that **(a)** 3,774,595 is divisible by 5 and **(b)** 9,332,123 is not divisible by 5.

**119.** An integer is divisible by 6 if it is divisible by both 2 and 3, and not otherwise. Show that **(a)** 1,524,822 is divisible by 6 and **(b)** 2,873,590 is not divisible by 6.

**120.** An integer is divisible by 8 if its last three digits form a number divisible by 8, and not otherwise. Show that **(a)** 2,923,296 is divisible by 8 and **(b)** 7,291,623 is not divisible by 8.

**121.** An integer is divisible by 9 if the sum of its digits is divisible by 9, and not otherwise. Show that **(a)** 4,114,107 is divisible by 9 and **(b)** 2,287,321 is not divisible by 9.

**122.** An integer is divisible by 12 if it is divisible by both 3 and 4, and not otherwise. Show that **(a)** 4,253,520 is divisible by 12 and **(b)** 4,249,474 is not divisible by 12.

## SUMMARY EXERCISES ON OPERATIONS WITH REAL NUMBERS

### Addition

*Same sign* Add the absolute values of the numbers. The sum has the same sign as the numbers.

*Different signs* Subtract the number with the smaller absolute value from the one with the larger. Give the sum the sign of the number having the larger absolute value.

### Subtraction

Add the opposite of the second number to the first number.

### Multiplication and Division

*Same sign* The product or quotient of two numbers with the same sign is positive.

*Different signs* The product or quotient of two numbers with different signs is negative.

Division by 0 is undefined.

*Perform each indicated operation.*

**1.** $14 - 3 \cdot 10$

**2.** $-3(8) - 4(-7)$

**3.** $(3 - 8)(-2) - 10$

**4.** $-6(7 - 3)$

**5.** $7 - (-3)(2 - 10)$

**6.** $-4[(-2)(6) - 7]$

**7.** $(-4)(7) - (-5)(2)$

**8.** $-5[-4 - (-2)(-7)]$

**9.** $40 - (-2)[8 - 9]$

**10.** $\dfrac{5(-4)}{-7 - (-2)}$

**11.** $\dfrac{-3 - (-9 + 1)}{-7 - (-6)}$

**12.** $\dfrac{5(-8 + 3)}{13(-2) + (-7)(-3)}$

**13.** $\dfrac{6^2 - 8}{-2(2) + 4(-1)}$

**14.** $\dfrac{16(-8 + 5)}{15(-3) + (-7 - 4)(-3)}$

**15.** $\dfrac{9(-6) - 3(8)}{4(-7) + (-2)(-11)}$

**16.** $\dfrac{2^2 + 4^2}{5^2 - 3^2}$

**17.** $\dfrac{(2 + 4)^2}{(5 - 3)^2}$

**18.** $\dfrac{4^3 - 3^3}{-5(-4 + 2)}$

**19.** $\dfrac{-9(-6) + (-2)(27)}{3(8 - 9)}$

**20.** $|-4(9)| - |-11|$

**21.** $\dfrac{6(-10 + 3)}{15(-2) - 3(-9)}$

**22.** $\dfrac{(-9)^2 - 9^2}{3^2 - 5^2}$

**23.** $\dfrac{(-10)^2 + 10^2}{-10(5)}$

**24.** $-\dfrac{3}{4} \div \left(-\dfrac{5}{8}\right)$

**25.** $\dfrac{1}{2} \div \left(-\dfrac{1}{2}\right)$

**26.** $\dfrac{8^2 - 12}{(-5)^2 + 2(6)}$

**27.** $\left[\dfrac{5}{8} - \left(-\dfrac{1}{16}\right)\right] + \dfrac{3}{8}$

**28.** $\left(\dfrac{1}{2} - \dfrac{1}{3}\right) - \dfrac{5}{6}$

**29.** $-.9(-3.7)$

**30.** $-5.1(-.2)$

**31.** $-3^2 - 2^2$

**32.** $|-2(3) + 4| - |-2|$

**33.** $40 - (-2)[-5 - 3]$

*Evaluate each expression if $x = -2$, $y = 3$, and $a = 4$.*

**34.** $-x + y - 3a$

**35.** $(x + 6)^3 - y^3$

**36.** $(x - y) - (a - 2y)$

**37.** $\left(\dfrac{1}{2}x + \dfrac{2}{3}y\right)\left(-\dfrac{1}{4}a\right)$

**38.** $\dfrac{2x + 3y}{a - xy}$

**39.** $\dfrac{x^2 - y^2}{x^2 + y^2}$

**40.** $-x^2 + 3y$

# 1.7 Properties of Real Numbers

**OBJECTIVES**

1 Use the commutative properties.

2 Use the associative properties.

3 Use the identity properties.

4 Use the inverse properties.

5 Use the distributive property.

If you were asked to find the sum $3 + 89 + 97$, you might mentally add $3 + 97$ to get 100, and then add $100 + 89$ to get 189. While the rule for order of operations says to add from left to right, we may change the order of the terms and group them in any way we choose without affecting the sum. These are examples of shortcuts that we use in everyday mathematics. These shortcuts are justified by the basic properties of addition and multiplication, discussed in this section. In these properties, $a$, $b$, and $c$ represent real numbers.

**OBJECTIVE 1 Use the commutative properties.** The word *commute* means to go back and forth. Many people commute to work or to school. If you travel from home to work and follow the same route from work to home, you travel the same distance each time. The **commutative properties** say that if two numbers are added or multiplied in any order, the result is the same.

**Commutative Properties**

$$a + b = b + a$$
$$ab = ba$$

**EXAMPLE 1 Using the Commutative Properties**

Use a commutative property to complete each statement.

**(a)** $-8 + 5 = 5 + \_\_\_\_$

By the commutative property for addition, the missing number is $-8$, since $-8 + 5 = 5 + (-8)$.

**(b)** $(-2)7 = \_\_\_\_(-2)$

By the commutative property for multiplication, the missing number is 7, since $(-2)7 = 7(-2)$.

**Now Try Exercises 1 and 3.**

**OBJECTIVE 2 Use the associative properties.** When we *associate* one object with another, we tend to think of those objects as being grouped together. The **associative properties** say that when we add or multiply three numbers, we can group the first two together or the last two together and get the same answer.

**Associative Properties**

$$(a + b) + c = a + (b + c)$$
$$(ab)c = a(bc)$$

**EXAMPLE 2 Using the Associative Properties**

Use an associative property to complete each statement.

**(a)** $8 + (-1 + 4) = (8 + \_\_\_\_) + 4$

The missing number is $-1$.

**(b)** $[2 \cdot (-7)] \cdot 6 = 2 \cdot \_\_\_\_$

The completed expression on the right should be $2 \cdot [(-7) \cdot 6]$.

**Now Try Exercises 5 and 7.**

By the associative property of addition, the sum of three numbers will be the same no matter how the numbers are "associated" in groups. For this reason, parentheses can be left out in many addition problems. For example, both

$$(-1 + 2) + 3 \qquad \text{and} \qquad -1 + (2 + 3)$$

can be written as

$$-1 + 2 + 3.$$

In the same way, parentheses also can be left out of many multiplication problems.

**EXAMPLE 3 Distinguishing between the Associative and Commutative Properties**

**(a)** Is $(2 + 4) + 5 = 2 + (4 + 5)$ an example of the associative property or the commutative property?

The order of the three numbers is the same on both sides of the equals sign. The only change is in the *grouping,* or association, of the numbers. Therefore, this is an example of the associative property.

**(b)** Is $6(3 \cdot 10) = 6(10 \cdot 3)$ an example of the associative property or the commutative property?

The same numbers, 3 and 10, are grouped on each side. On the left, the 3 appears first, but on the right, the 10 appears first. Since the only change involves the *order* of the numbers, this statement is an example of the commutative property.

**(c)** Is $(8 + 1) + 7 = 8 + (7 + 1)$ an example of the associative property or the commutative property?

In the statement, both the order and the grouping are changed. On the left the order of the three numbers is 8, 1, and 7. On the right it is 8, 7, and 1. On the left the 8 and 1 are grouped, and on the right the 7 and 1 are grouped. Therefore, both the associative and the commutative properties are used.

**Now Try Exercises 15 and 23.**

**EXAMPLE 4 Using the Commutative and Associative Properties**

Find the sum $23 + 41 + 2 + 9 + 25$.

The commutative and associative properties make it possible to choose pairs of numbers whose sums are easy to add.

$$\begin{aligned} 23 + 41 + 2 + 9 + 25 &= (41 + 9) + (23 + 2) + 25 \\ &= 50 + 25 + 25 \\ &= 100 \end{aligned}$$

**Now Try Exercise 37.**

**OBJECTIVE 3 Use the identity properties.** If a child wears a costume on Halloween, the child's appearance is changed, but his or her *identity* is unchanged. The identity of a real number is left unchanged when identity properties are applied. The **identity properties** say that the sum of 0 and any number equals that number, and the product of 1 and any number equals that number.

**Identity Properties**

$$a + 0 = a \quad \text{and} \quad 0 + a = a$$
$$a \cdot 1 = a \quad \text{and} \quad 1 \cdot a = a$$

The number 0 leaves the identity, or value, of any real number unchanged by addition. For this reason, 0 is called the **identity element for addition,** or the **additive identity.** Since multiplication by 1 leaves any real number unchanged, 1 is the **identity element for multiplication,** or the **multiplicative identity.**

**EXAMPLE 5 Using the Identity Properties**

These statements are examples of the identity properties.

**(a)** $-3 + 0 = -3$ **(b)** $1 \cdot \frac{1}{2} = \frac{1}{2}$

**Now Try Exercise 31.**

We use the identity property for multiplication to write fractions in lowest terms and to find common denominators.

**EXAMPLE 6 Using the Identity Property to Simplify Expressions**

Simplify each expression.

**(a)**
$$\begin{aligned} \frac{49}{35} &= \frac{7 \cdot 7}{5 \cdot 7} && \text{Factor.} \\ &= \frac{7}{5} \cdot \frac{7}{7} && \text{Write as a product.} \\ &= \frac{7}{5} \cdot 1 && \text{Divide.} \\ &= \frac{7}{5} && \text{Identity property} \end{aligned}$$

**(b)**
$$\begin{aligned} \frac{3}{4} + \frac{5}{24} &= \frac{3}{4} \cdot 1 + \frac{5}{24} && \text{Identity property} \\ &= \frac{3}{4} \cdot \frac{6}{6} + \frac{5}{24} && \text{Find a common denominator.} \\ &= \frac{18}{24} + \frac{5}{24} && \text{Multiply.} \\ &= \frac{23}{24} && \text{Add.} \end{aligned}$$

**Now Try Exercise 27.**

**OBJECTIVE 4 Use the inverse properties.** Each day before you go to work or school, you probably put on your shoes before you leave. Before you go to sleep at night, you probably take them off, and this leads to the same situation that existed before you put them on. These operations from everyday life are examples of *inverse* operations. The **inverse properties** of addition and multiplication lead to the additive and multiplicative identities, respectively. Recall that $-a$ is the **additive inverse,** or **opposite,** of $a$ and $\frac{1}{a}$ is the **multiplicative inverse,** or **reciprocal,** of the nonzero number $a$. The sum of the numbers $a$ and $-a$ is 0, and the product of the nonzero numbers $a$ and $\frac{1}{a}$ is 1.

**Inverse Properties**

$$a + (-a) = 0 \quad \text{and} \quad -a + a = 0$$

$$a \cdot \frac{1}{a} = 1 \quad \text{and} \quad \frac{1}{a} \cdot a = 1 \quad (a \neq 0)$$

### EXAMPLE 7 Using the Inverse Properties

The following statements are examples of the inverse properties.

**(a)** $\frac{2}{3} \cdot \frac{3}{2} = 1$  **(b)** $-5\left(-\frac{1}{5}\right) = 1$

**(c)** $-\frac{1}{2} + \frac{1}{2} = 0$  **(d)** $4 + (-4) = 0$

**Now Try Exercise 19.**

In the next example, we show how the various properties are used to simplify an expression.

### EXAMPLE 8 Using Properties to Simplify an Expression

Simplify $-2x + 10 + 2x$, using the properties discussed in this section.

$$\begin{aligned} -2x + 10 + 2x &= (-2x + 10) + 2x && \text{Order of operations} \\ &= [10 + (-2x)] + 2x && \text{Commutative property} \\ &= 10 + [(-2x) + 2x] && \text{Associative property} \\ &= 10 + 0 && \text{Inverse property} \\ &= 10 && \text{Identity property} \end{aligned}$$

Note that for *any* value of $x$, $-2x$ and $2x$ are additive inverses; that is why we can use the inverse property in this simplification.

**Now Try Exercise 43.**

**NOTE** The detailed procedure shown in Example 8 is seldom, if ever, used in practice. We include the example to show how the properties of this section apply, even though steps may be skipped when actually doing the simplification.

**OBJECTIVE 5 Use the distributive property.** The everyday meaning of the word *distribute* is "to give out from one to several." An important property of real number operations involves this idea.

Look at the value of the following expressions.

$$2(5+8) = 2(13) = 26$$
$$2(5) + 2(8) = 10 + 16 = 26$$

Since both expressions equal 26,

$$2(5+8) = 2(5) + 2(8).$$

This result is an example of the *distributive property,* the only property involving *both* addition and multiplication. With this property, a product can be changed to a sum or difference. This idea is illustrated by the divided rectangle in Figure 17.

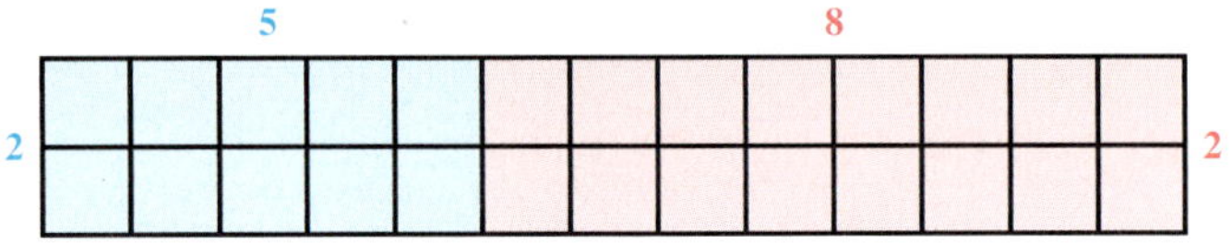


The area of the left part is 2(5) = 10.
The area of the right part is 2(8) = 16.
The total area is 2(5 + 8) = 2(13) = 26 or the total area is 2(5) + 2(8) = 10 + 16 = 26.
Thus, 2(5 + 8) = 2(5) + 2(8).

**FIGURE 17**

The **distributive property** says that multiplying a number $a$ by a sum of numbers $b + c$ gives the same result as multiplying $a$ by $b$ and $a$ by $c$ and then adding the two products.

**Distributive Property**

$$a(b+c) = ab + ac \quad \text{and} \quad (b+c)a = ba + ca$$

As the arrows show, the $a$ outside the parentheses is "distributed" over the $b$ and $c$ inside. Another form of the distributive property is valid for subtraction.

$$a(b-c) = ab - ac \quad \text{and} \quad (b-c)a = ba - ca$$

The distributive property also can be extended to more than two numbers.

$$a(b+c+d) = ab + ac + ad$$

The distributive property can be used "in reverse." For example, we can write

$$ac + bc = (a+b)c.$$

**EXAMPLE 9 Using the Distributive Property**

Use the distributive property to rewrite each expression.

**(a)** $5(9+6) = 5 \cdot 9 + 5 \cdot 6$ Distributive property

$= 45 + 30$ Multiply.

$= 75$ Add.

**(b)** $4(x + 5 + y) = 4x + 4 \cdot 5 + 4y$ Distributive property
$= 4x + 20 + 4y$ Multiply.

**(c)** $-2(x + 3) = -2x + (-2)(3)$ Distributive property
$= -2x - 6$ Multiply.

**(d)** $3(k - 9) = 3k - 3 \cdot 9$ Distributive property
$= 3k - 27$ Multiply.

**(e)** $8(3r + 11t + 5z) = 8(3r) + 8(11t) + 8(5z)$ Distributive property
$= (8 \cdot 3)r + (8 \cdot 11)t + (8 \cdot 5)z$ Associative property
$= 24r + 88t + 40z$ Multiply.

**(f)** $6 \cdot 8 + 6 \cdot 2 = 6(8 + 2)$ Distributive property
$= 6(10) = 60$ Add, then multiply.

**(g)** $4x - 4m = 4(x - m)$ Distributive property

**(h)** $6x - 12 = 6 \cdot x - 6 \cdot 2 = 6(x - 2)$ Distributive property

**Now Try Exercises 57, 63, 65, and 69.**

The symbol $-a$ may be interpreted as $-1 \cdot a$. Similarly, when a negative sign precedes an expression within parentheses, it may also be interpreted as a factor of $-1$. Thus, we can use the distributive property to remove parentheses from expressions such as $-(2y + 3)$. We do this by first writing $-(2y + 3)$ as $-1 \cdot (2y + 3)$.

$-(2y + 3) = -1 \cdot (2y + 3)$
$= -1 \cdot 2y + (-1) \cdot 3$ Distributive property
$= -2y - 3$ Multiply.

### EXAMPLE 10 Using the Distributive Property to Remove Parentheses

Write without parentheses.

**(a)** $-(7r - 8) = -1(7r - 8)$
$= -1(7r) + (-1)(-8)$ Distributive property
$= -7r + 8$ Multiply.

**(b)** $-(-9w + 2) = -1(-9w + 2)$
$= 9w - 2$

**Now Try Exercise 75.**

The properties discussed here are the basic properties that justify how we do algebra. Here is a summary of these properties.

**Properties of Addition and Multiplication**

For any real numbers $a$, $b$, and $c$, the following properties hold.

*Commutative Properties* $\quad \mathbf{a + b = b + a} \quad \mathbf{ab = ba}$

*Associative Properties* $\quad \mathbf{(a + b) + c = a + (b + c)}$
$\mathbf{(ab)c = a(bc)}$

(continued)

| | |
|---|---|
| *Identity Properties* | There is a real number 0 such that $\mathbf{a + 0 = a}$ and $\mathbf{0 + a = a}$. There is a real number 1 such that $\mathbf{a \cdot 1 = a}$ and $\mathbf{1 \cdot a = a}$. |
| *Inverse Properties* | For each real number $a$, there is a single real number $-a$ such that $\mathbf{a + (-a) = 0}$ and $\mathbf{(-a) + a = 0}$. For each nonzero real number $a$, there is a single real number $\frac{1}{a}$ such that $\mathbf{a \cdot \frac{1}{a} = 1}$ and $\mathbf{\frac{1}{a} \cdot a = 1}$. |
| *Distributive Properties* | $\mathbf{a(b + c) = ab + ac}$ $\mathbf{(b + c)a = ba + ca}$ |

## 1.7 EXERCISES

**For Extra Help**

- Student's Solutions Manual
- MyMathLab
- InterAct Math Tutorial Software
- AW Math Tutor Center
- MathXL
- Digital Video Tutor CD 2/Videotape 2

*Use the commutative or the associative property to complete each statement. State which property is used. See Examples 1 and 2.*

**1.** $-12 + 6 = 6 +$ _____

**2.** $8 + (-4) = -4 +$ _____

**3.** $-6 \cdot 3 =$ _____ $\cdot (-6)$

**4.** $-12 \cdot 6 = 6 \cdot$ _____

**5.** $(4 + 7) + 8 = 4 + ($_____ $+ 8)$

**6.** $(-2 + 3) + 6 = -2 + ($_____ $+ 6)$

**7.** $8 \cdot (3 \cdot 6) = ($_____ $\cdot 3) \cdot 6$

**8.** $6 \cdot (4 \cdot 2) = (6 \cdot$ _____$) \cdot 2$

**9.** Match each item in Column I with the correct choice(s) from Column II. Choices may be used once, more than once, or not at all.

| I | II |
|---|---|
| **(a)** Identity element for addition | **A.** $(5 \cdot 4) \cdot 3 = 5 \cdot (4 \cdot 3)$ |
| **(b)** Identity element for multiplication | **B.** 0 |
| **(c)** Additive inverse of $a$ | **C.** $-a$ |
| **(d)** Multiplicative inverse, or reciprocal, of the nonzero number $a$ | **D.** $-1$ |
| **(e)** The number that is its own additive inverse | **E.** $5 \cdot 4 \cdot 3 = 60$ |
| **(f)** The two numbers that are their own multiplicative inverses | **F.** 1 |
| **(g)** The only number that has no multiplicative inverse | **G.** $(5 \cdot 4) \cdot 3 = 3 \cdot (5 \cdot 4)$ |
| **(h)** An example of the associative property | **H.** $5(4 + 3) = 5 \cdot 4 + 5 \cdot 3$ |
| **(i)** An example of the commutative property | **I.** $\frac{1}{a}$ |
| **(j)** An example of the distributive property | |

**10.** Explain the difference between the commutative and associative properties.

*Decide whether each statement is an example of the* commutative, associative, identity, inverse, *or* distributive property. *See Examples 1, 2, 3, 5, 6, 7, and 9.*

**11.** $7 + 18 = 18 + 7$

**12.** $13 + 12 = 12 + 13$

**13.** $5 \cdot (13 \cdot 7) = (5 \cdot 13) \cdot 7$

**14.** $-4 \cdot (2 \cdot 6) = (-4 \cdot 2) \cdot 6$

**15.** $-6 + (12 + 7) = (-6 + 12) + 7$

**16.** $(-8 + 13) + 2 = -8 + (13 + 2)$

**17.** $-6 + 6 = 0$

**18.** $12 + (-12) = 0$

**19.** $\frac{2}{3}\left(\frac{3}{2}\right) = 1$

**20.** $\frac{5}{8}\left(\frac{8}{5}\right) = 1$

**21.** $2.34 + 0 = 2.34$

**22.** $-8.456 + 0 = -8.456$

**23.** $(4 + 17) + 3 = 3 + (4 + 17)$

**24.** $(-8 + 4) + 12 = 12 + (-8 + 4)$

**25.** $6(x + y) = 6x + 6y$

**26.** $14(t + s) = 14t + 14s$

**27.** $-\frac{5}{9} = -\frac{5}{9} \cdot \frac{3}{3} = -\frac{15}{27}$

**28.** $\frac{13}{12} = \frac{13}{12} \cdot \frac{7}{7} = \frac{91}{84}$

**29.** $5(2x) + 5(3y) = 5(2x + 3y)$

**30.** $3(5t) - 3(7r) = 3(5t - 7r)$

**31.** The following conversation actually took place between one of the authors of this book and his son, Jack, when Jack was 4 years old:

DADDY: "Jack, what is 3 + 0?"
JACK: "3."
DADDY: "Jack, what is 4 + 0?"
JACK: "4. And Daddy, *string* plus zero equals *string*!"

What property of addition did Jack recognize?

**32.** The distributive property holds for multiplication with respect to addition. Is there a distributive property for addition with respect to multiplication? If not, give an example to show why.

**33.** Write a paragraph explaining in your own words the identity and inverse properties of addition and multiplication.

**34.** Write a paragraph explaining in your own words the distributive property of multiplication with respect to addition. Give examples.

*Find each sum. Use the commutative and associative properties to make your work easier. See Example 4.*

**35.** $97 + 13 + 3 + 37$

**36.** $49 + 199 + 1 + 1$

**37.** $1999 + 2 + 1 + 8$

**38.** $2998 + 3 + 2 + 17$

**39.** $159 + 12 + 141 + 88$

**40.** $106 + 8 + (-6) + (-8)$

**41.** $843 + 627 + (-43) + (-27)$

**42.** $1846 + 1293 + (-46) + (-93)$

*Use the properties of this section to simplify each expression. See Examples 7 and 8.*

**43.** $6t + 8 - 6t + 3$

**44.** $9r + 12 - 9r + 1$

**45.** $\frac{2}{3}x - 11 + 11 - \frac{2}{3}x$

**46.** $\frac{1}{5}y + 4 - 4 - \frac{1}{5}y$

**47.** $\left(\frac{9}{7}\right)(-.38)\left(\frac{7}{9}\right)$

**48.** $\left(\frac{4}{5}\right)(-.73)\left(\frac{5}{4}\right)$

**49.** $t + (-t) + \frac{1}{2}(2)$

**50.** $w + (-w) + \frac{1}{4}(4)$

**51.** Evaluate $25 - (6 - 2)$ and evaluate $(25 - 6) - 2$. Do you think subtraction is associative?

**52.** Evaluate $180 \div (15 \div 3)$ and evaluate $(180 \div 15) \div 3$. Do you think division is associative?

**53.** Suppose that a student shows you the following work.

$$-3(4 - 6) = -3(4) - 3(6) = -12 - 18 = -30$$

The student has made a very common error. Explain the student's mistake, and work the problem correctly.

54. Explain how the procedure of changing $\frac{3}{4}$ to $\frac{9}{12}$ requires the use of the multiplicative identity element, 1.

*Use the distributive property to rewrite each expression. Simplify if possible. See Example 9.*

55. $5(9 + 8)$
56. $6(11 + 8)$
57. $4(t + 3)$
58. $5(w + 4)$
59. $-8(r + 3)$
60. $-11(x + 4)$
61. $-5(y - 4)$
62. $-9(g - 4)$
63. $-\frac{4}{3}(12y + 15z)$
64. $-\frac{2}{5}(10b + 20a)$
65. $8z + 8w$
66. $4s + 4r$
67. $7(2v) + 7(5r)$
68. $13(5w) + 13(4p)$
69. $8(3r + 4s - 5y)$
70. $2(5u - 3v + 7w)$
71. $-3(8x + 3y + 4z)$
72. $-5(2x - 5y + 6z)$
73. $5x + 15$
74. $9p + 18$

*Use the distributive property to write each expression without parentheses. See Example 10.*

75. $-(4t + 3m)$
76. $-(9x + 12y)$
77. $-(-5c - 4d)$
78. $-(-13x - 15y)$
79. $-(-3q + 5r - 8s)$
80. $-(-4z + 5w - 9y)$

81. The operations of "getting out of bed" and "taking a shower" are not commutative. Give an example of another pair of everyday operations that are not commutative.

82. The phrase "dog biting man" has two different meanings, depending on how the words are associated:

(dog biting) man     dog (biting man)

Give another example of a three-word phrase that has different meanings depending on how the words are associated.

## RELATING CONCEPTS (EXERCISES 83–86)

**For Individual or Group Work**

In Section 1.6 we used a pattern to see that the product of two negative numbers is a positive number. In the group of exercises that follows, we show another justification for determining the sign of the product of two negative numbers. ***Work Exercises 83–86 in order.***

83. Evaluate the expression $-3[5 + (-5)]$ by using the order of operations.

84. Write the expression in Exercise 83 using the distributive property. Do not simplify the products.

85. The product $-3(5)$ should be part of the answer you wrote for Exercise 84. Based on the results in Section 1.6, what is this product?

86. In Exercise 83, you should have obtained 0 as an answer. Now, consider the following, using the results of Exercises 83 and 85.

$$-3[5 + (-5)] = -3(5) + (-3)(-5)$$
$$0 = -15 + ?$$

The question mark represents the product $(-3)(-5)$. When added to $-15$, it must give a sum of 0. Therefore, how must we interpret $(-3)(-5)$?

**87.** Consider the statement

$$-2(5 \cdot 7) = (-2 \cdot 5) \cdot (-2 \cdot 7).$$

**(a)** Is this statement true?
**(b)** If it is not true, tell which property you think was erroneously applied.

**88.** Consider the statement

$$4 + (3 \cdot 5) = (4 + 3) \cdot (4 + 5).$$

**(a)** Is this statement true?
**(b)** If it is not true, tell which property you think was erroneously applied.

# 1.8 Simplifying Expressions

**OBJECTIVES**

1 Simplify expressions.
2 Identify terms and numerical coefficients.
3 Identify like terms.
4 Combine like terms.
5 Simplify expressions from word phrases.

**OBJECTIVE 1 Simplify expressions.** In this section we show how to simplify expressions using the properties of addition and multiplication introduced in the previous section.

### EXAMPLE 1 Simplifying Expressions

Simplify each expression.

**(a)** $4x + 8 + 9$

Since $8 + 9 = 17$,

$$4x + 8 + 9 = 4x + 17.$$

**(b)** $4(3m - 2n)$

Use the distributive property first.

$$\begin{aligned} 4(3m - 2n) &= 4(3m) - 4(2n) && \text{Arrows denote distributive property.} \\ &= (4 \cdot 3)m - (4 \cdot 2)n && \text{Associative property} \\ &= 12m - 8n \end{aligned}$$

**(c)**
$$\begin{aligned} 6 + 3(4k + 5) &= 6 + 3(4k) + 3(5) && \text{Distributive property} \\ &= 6 + (3 \cdot 4)k + 3(5) && \text{Associative property} \\ &= 6 + 12k + 15 \\ &= 6 + 15 + 12k && \text{Commutative property} \\ &= 21 + 12k \end{aligned}$$

**(d)**
$$\begin{aligned} 5 - (2y - 8) &= 5 - 1 \cdot (2y - 8) && -a = -1 \cdot a \\ &= 5 - 1(2y) - 1(-8) && \text{Distributive property} \\ &= 5 - 2y + 8 \\ &= 5 + 8 - 2y && \text{Commutative property} \\ &= 13 - 2y \end{aligned}$$

**Now Try Exercises 3 and 5.**

**NOTE** In Examples 1(c) and 1(d), a different use of the commutative property would have resulted in answers of $12k + 21$ and $-2y + 13$. These answers also would be acceptable.

The steps using the commutative and associative properties will not be shown in the rest of the examples, but you should be aware that they are usually involved.

**OBJECTIVE 2 Identify terms and numerical coefficients.** A **term** is a number, a variable, or a product or quotient of numbers and variables raised to powers.* Examples of terms include

$$-9x^2, \quad 15y, \quad -3, \quad 8m^2n, \quad \frac{2}{p}, \quad \text{and} \quad k.$$

The **numerical coefficient** of the term $9m$ is 9, the numerical coefficient of $-15x^3y^2$ is $-15$, the numerical coefficient of $x$ is 1, and the numerical coefficient of 8 is 8. In the expression $\frac{x}{3}$, the numerical coefficient of $x$ is $\frac{1}{3}$ since $\frac{x}{3} = \frac{1x}{3} = \frac{1}{3}x$.

**Now Try Exercises 9 and 15.**

| Term | Numerical Coefficient |
|---|---|
| $-7y$ | $-7$ |
| $34r^3$ | 34 |
| $-26x^5yz^4$ | $-26$ |
| $-k$ | $-1$ |
| $r$ | 1 |
| $\frac{3x}{8} = \frac{3}{8}x$ | $\frac{3}{8}$ |

**NOTE** It is important to be able to distinguish between *terms* and *factors.* For example, in the expression $8x^3 + 12x^2$, there are two *terms,* $8x^3$ and $12x^2$. On the other hand, in the one-term expression $(8x^3)(12x^2)$, $8x^3$ and $12x^2$ are *factors.*

Examples of terms and their numerical coefficients are shown in the table in the margin.

**OBJECTIVE 3 Identify like terms.** Terms with exactly the same variables that have the same exponents are **like terms.** For example, $9m$ and $4m$ have the same variable and are like terms. Also, $6x^3$ and $-5x^3$ are like terms. The terms $-4y^3$ and $4y^2$ have different exponents and are **unlike terms.**

Here are some additional examples.

| | | |
|---|---|---|
| $5x$ and $-12x$ | $3x^2y$ and $5x^2y$ | Like terms |
| $4xy^2$ and $5xy$ | $-7w^3z^3$ and $2xz^3$ | Unlike terms |

**Now Try Exercises 19 and 25.**

**OBJECTIVE 4 Combine like terms.** Recall the distributive property.

$$x(y + z) = xy + xz.$$

As seen in the previous section, this statement can also be written "backward" as

$$xy + xz = x(y + z).$$

This form of the distributive property may be used to find the sum or difference of like terms. For example,

$$3x + 5x = (3 + 5)x = 8x.$$

This process is called **combining like terms.**

**NOTE** Remember that *only like terms may be combined.*

*Another name for certain terms, **monomial**, is introduced in Chapter 4.

### EXAMPLE 2 Combining Like Terms

Combine like terms in each expression.

**(a)** $9m + 5m$

Use the distributive property as given above.

$$9m + 5m = (9 + 5)m = 14m$$

**(b)** $6r + 3r + 2r = (6 + 3 + 2)r = 11r$ Distributive property

**(c)** $4x + x = 4x + 1x = (4 + 1)x = 5x$ (Note: $x = 1x$.)

**(d)** $16y^2 - 9y^2 = (16 - 9)y^2 = 7y^2$

**(e)** $32y + 10y^2$ cannot be combined because $32y$ and $10y^2$ are unlike terms. We cannot use the distributive property here to combine coefficients.

**Now Try Exercises 31, 35, and 45.**

When an expression involves parentheses, the distributive property is used both "forward" and "backward" to combine like terms, as shown in the following example.

### EXAMPLE 3 Simplifying Expressions Involving Like Terms

Combine like terms in each expression.

**(a)**
$$\begin{aligned} 14y + 2(6 + 3y) &= 14y + 2(6) + 2(3y) && \text{Distributive property} \\ &= 14y + 12 + 6y && \text{Multiply.} \\ &= 20y + 12 && \text{Combine like terms.} \end{aligned}$$

**(b)**
$$\begin{aligned} 9k - 6 - 3(2 - 5k) &= 9k - 6 - 3(2) - 3(-5k) && \text{Distributive property} \\ &= 9k - 6 - 6 + 15k && \text{Multiply.} \\ &= 24k - 12 && \text{Combine like terms.} \end{aligned}$$

**(c)**
$$\begin{aligned} -(2 - r) + 10r &= -1(2 - r) + 10r && -(2 - r) = -1(2 - r) \\ &= -1(2) - 1(-r) + 10r && \text{Distributive property} \\ &= -2 + 1r + 10r && \text{Multiply.} \\ &= -2 + 11r && \text{Combine like terms.} \end{aligned}$$

**(d)**
$$\begin{aligned} 100[.03(x + 4)] &= [(100)(.03)](x + 4) && \text{Associative property} \\ &= 3(x + 4) && \text{Multiply.} \\ &= 3x + 12 && \text{Distributive property} \end{aligned}$$

**(e)**
$$\begin{aligned} 5(2a - 6) - 3(4a - 9) &= 10a - 30 - 12a + 27 && \text{Distributive property} \\ &= -2a - 3 && \text{Combine like terms.} \end{aligned}$$

**Now Try Exercises 49, 51, 53, and 55.**

Example 3(e) suggests that the commutative property can be used with subtraction by treating the subtracted terms as the addition of their additive inverses.

**NOTE** Examples 2 and 3 suggest that like terms may be combined by adding or subtracting the coefficients of the terms and keeping the same variable factors.

**OBJECTIVE 5 Simplify expressions from word phrases.** Earlier we saw how to translate words, phrases, and statements into expressions and equations. Now we can simplify translated expressions by combining like terms.

### EXAMPLE 4 Translating Words to a Mathematical Expression

Translate to a mathematical expression, and simplify: The sum of 9, five times a number, four times the number, and six times the number.

The word "sum" indicates that the terms should be added. Use $x$ to represent the number. Then the phrase translates as

$$9 + 5x + 4x + 6x,$$ Write with symbols.

which simplifies to

$$9 + 15x.$$ Combine like terms.

**Now Try Exercise 67.**

**CAUTION** In Example 4, we are dealing with an expression to be simplified, *not* an equation to be solved.

## 1.8 EXERCISES

**For Extra Help**

- Student's Solutions Manual
- MyMathLab
- InterAct Math Tutorial Software
- AW Math Tutor Center
- MathXL
- Digital Video Tutor CD 2/Videotape 2

*Simplify each expression. See Example 1.*

**1.** $4r + 19 - 8$ **2.** $7t + 18 - 4$ **3.** $5 + 2(x - 3y)$

**4.** $8 + 3(s - 6t)$ **5.** $-2 - (5 - 3p)$ **6.** $-10 - (7 - 14r)$

**7.** $6 + (4 - 3x) - 8$ **8.** $-12 + (7 - 8x) + 6$

*Give the numerical coefficient of each term.*

**9.** $-12k$ **10.** $-23y$ **11.** $5m^2$ **12.** $-3n^6$ **13.** $xw$

**14.** $pq$ **15.** $-x$ **16.** $-t$ **17.** $74$ **18.** $98$

*Identify each group of terms as* like *or* unlike.

**19.** $8r, -13r$ **20.** $-7a, 12a$ **21.** $5z^4, 9z^3$ **22.** $8x^5, -10x^3$

**23.** $4, 9, -24$ **24.** $7, 17, -83$ **25.** $x, y$ **26.** $t, s$

*Simplify each expression by combining like terms. See Examples 1–3.*

**27.** $9y + 8y$ **28.** $15m + 12m$

**29.** $-4a - 2a$ **30.** $-3z - 9z$

**31.** $12b + b$ **32.** $30x + x$

**33.** $2k + 9 + 5k + 6$ **34.** $2 + 17z + 1 + 2z$

**35.** $-5y + 3 - 1 + 5 + y - 7$ **36.** $2k - 7 - 5k + 7k - 3 - k$

**37.** $-2x + 3 + 4x - 17 + 20$ **38.** $r - 6 - 12r - 4 + 6r$

**39.** $16 - 5m - 4m - 2 + 2m$ **40.** $6 - 3z - 2z - 5 + z - 3z$

41. $-10 + x + 4x - 7 - 4x$

42. $-p + 10p - 3p - 4 - 5p$

43. $1 + 7x + 11x - 1 + 5x$

44. $-r + 2 - 5r + 3 + 4r$

45. $6y^2 + 11y^2 - 8y^2$

46. $-9m^3 + 3m^3 - 7m^3$

47. $2p^2 + 3p^2 - 8p^3 - 6p^3$

48. $5y^3 + 6y^3 - 3y^2 - 4y^2$

49. $2(4x + 6) + 3$

50. $4(6y - 9) + 7$

51. $100[.05(x + 3)]$

52. $100[.06(x + 5)]$

53. $-4(y - 7) - 6$

54. $-5(t - 13) - 4$

55. $-5(5y - 9) + 3(3y + 6)$

56. $-3(2t + 4) + 8(2t - 4)$

57. $-3(2r - 3) + 2(5r + 3)$

58. $-4(5y - 7) + 3(2y - 5)$

59. $8(2k - 1) - (4k - 3)$

60. $6(3p - 2) - (5p + 1)$

61. $-2(-3k + 2) - (5k - 6) - 3k - 5$

62. $-2(3r - 4) - (6 - r) + 2r - 5$

63. $-4(-3k + 3) - (6k - 4) - 2k + 1$

64. $-5(8j + 2) - (5j - 3) - 3j + 17$

65. $-7.5(2y + 4) - 2.9(3y - 6)$

66. $8.4(6t - 6) + 2.4(9 - 3t)$

*Translate each phrase into a mathematical expression. Use x as the variable. Combine like terms when possible. See Example 4.*

67. Five times a number, added to the sum of the number and three

68. Six times a number, added to the sum of the number and six

69. A number multiplied by $-7$, subtracted from the sum of 13 and six times the number

70. A number multiplied by 5, subtracted from the sum of 14 and eight times the number

71. Six times a number added to $-4$, subtracted from twice the sum of three times the number and 4 (*Hint: Twice* means two times.)

72. Nine times a number added to 6, subtracted from triple the sum of 12 and 8 times the number (*Hint: Triple* means three times.)

## RELATING CONCEPTS (EXERCISES 73–80)

**For Individual or Group Work**

***Work Exercises 73–80 in order.*** *They will help prepare you for graphing later in the text.*

73. Evaluate the expression $x + 2$ for the values of $x$ shown in the table.

| $x$ | $x + 2$ |
|---|---|
| 0 | |
| 1 | |
| 2 | |
| 3 | |

74. Based on your results from Exercise 73, complete the following statement: For every increase of 1 unit for $x$, the value of $x + 2$ increases by ____ unit(s).

75. Repeat Exercise 73 for these expressions:

**(a)** $x + 1$ **(b)** $x + 3$ **(c)** $x + 4$

(continued)

**76.** Based on your results from Exercises 73 and 75, make a conjecture (an educated guess) about what happens to the value of an expression of the form $x + b$ for any value of $b$, as $x$ increases by 1 unit.

**77.** Repeat Exercise 73 for these expressions:

**(a)** $2x + 2$ **(b)** $3x + 2$ **(c)** $4x + 2$

**78.** Based on your results from Exercise 77, complete the following statement: For every increase of 1 unit for $x$, the value of $mx + 2$ increases by _____ units.

**79.** Repeat Exercise 73 and compare your results to those in Exercise 77 for these expressions:

**(a)** $2x + 7$ **(b)** $3x + 5$ **(c)** $4x + 1$

**80.** Based on your results from Exercises 73–79, complete the following statement: For every increase of 1 unit for $x$, the value of $mx + b$ increases by _____ units.

**81.** There is an old saying, "You can't add apples and oranges." Explain how this saying can be applied to the goal of Objective 4 in this section.

**82.** Explain how the distributive property is used in combining $6t + 5t$ to get $11t$.

**83.** Write the expression $9x - (x + 2)$ using words, as in Exercises 67–72.

**84.** Write the expression $2(3x + 5) - 2(x + 4)$ using words, as in Exercises 67–72.

## Chapter 1 Group Activity

### Comparing Floor Plans of Houses

**OBJECTIVE** Use arithmetic skills to make comparisons.

This activity explores perimeters and areas of different shaped homes. Floor plans for three different homes are given below.

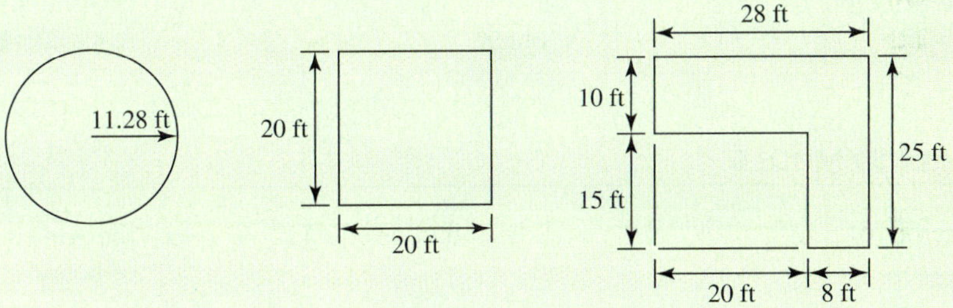


Plan one
$A = \pi r^2$
$C = 2\pi r$

Plan two
$A = s^2$

Plan three
$A = L \cdot W$

**A.** As a group, look at the dimensions of the given floor plans. Considering only the dimensions, which plan do you think has the greatest area?

**B.** Now have each student in your group pick one floor plan. For each plan, find the following. (Round all answers to the nearest whole number. In Plan one, let $\pi = 3.14$.)

**1.** The area of the plan

**2.** The perimeter or circumference (the distance around the outside) of the plan

**C.** Share your findings with the group and answer the following questions.

**1.** What did you determine about the areas of the three floor plans?

**2.** Which plan has the smallest perimeter? Which has the largest perimeter?

**3.** Why do you think houses with round floor plans might be more energy efficient?

**4.** What advantages do you think houses with square or rectangular floor plans have? Why do you think floor plans with these shapes are most common for homes today?

# CHAPTER 1 SUMMARY

## KEY TERMS

**1.1** natural numbers
whole numbers
fractions
numerator
denominator
proper fraction
improper fraction
mixed number
factor
product
factored
prime number
composite number
greatest common factor
lowest terms
basic principle of fractions
reciprocal
quotient
sum
least common denominator (LCD)
difference

**1.2** exponent (power)
base
exponential expression
grouping symbols
order of operations
inequality

**1.3** variable
algebraic expression
equation
solution
set
element

**1.4** number line
graph
coordinate
negative numbers
positive numbers
signed numbers
integers
rational numbers
set-builder notation
irrational numbers
real numbers
additive inverse (opposite)
absolute value

**1.6** multiplicative inverse (reciprocal)

**1.7** commutative property
associative property
identity property
identity element for addition (additive identity)
identity element for multiplication (multiplicative identity)
inverse property
distributive property

**1.8** term
numerical coefficient
like terms
unlike terms
combining like terms

## NEW SYMBOLS

| | |
|---|---|
| $a^n$ | $n$ factors of $a$ |
| [ ] | square brackets |
| $=$ | is equal to |
| $\neq$ | is not equal to |
| $<$ | is less than |
| $>$ | is greater than |
| $\leq$ | is less than or equal to |
| $\geq$ | is greater than or equal to |
| { } | set braces |
| $\{x \mid x$ **has a certain property**$\}$ | set-builder notation |
| $-x$ | the additive inverse, or opposite, of $x$ |
| $\lvert x \rvert$ | absolute value of $x$ |
| $\frac{1}{x}$ | the multiplicative inverse, or reciprocal, of the nonzero number $x$ |
| $a(b)$, $(a)b$, $(a)(b)$, $a \cdot b$, or $ab$ | $a$ times $b$ |
| $\frac{a}{b}$ or $a/b$ | $a$ divided by $b$ |

## TEST YOUR WORD POWER

*See how well you have learned the vocabulary in this chapter. Answers, with examples, follow the Quick Review.*

1. A **factor** is
   **A.** the answer in an addition problem
   **B.** the answer in a multiplication problem
   **C.** one of two or more numbers that are added to get another number
   **D.** one of two or more numbers that are multiplied to get another number.

2. A number is **prime** if
   **A.** it cannot be factored
   **B.** it has just one factor
   **C.** it has only itself and 1 as factors
   **D.** it has at least two different factors.

3. An **exponent** is
   **A.** a symbol that tells how many numbers are being multiplied
   **B.** a number raised to a power
   **C.** a number that tells how many times a factor is repeated
   **D.** one of two or more numbers that are multiplied.

4. A **variable** is
   **A.** a symbol used to represent an unknown number
   **B.** a value that makes an equation true
   **C.** a solution of an equation
   **D.** the answer in a division problem.

5. An **integer** is
   **A.** a positive or negative number
   **B.** a natural number, its opposite, or zero

C. any number that can be graphed on a number line
D. the quotient of two numbers.

6. A **coordinate** is
A. the number that corresponds to a point on a number line
B. the graph of a number
C. any point on a number line
D. the distance from 0 on a number line.

7. The **absolute value** of a number is
A. the graph of the number
B. the reciprocal of the number
C. the opposite of the number
D. the distance between 0 and the number on a number line.

8. A **term** is
A. a numerical factor
B. a number or a product or quotient of numbers and variables raised to powers
C. one of several variables with the same exponents
D. a sum of numbers and variables raised to powers.

9. A **numerical coefficient** is
A. the numerical factor in a term
B. the number of terms in an expression
C. a variable raised to a power
D. the variable factor in a term.

## QUICK REVIEW

| CONCEPTS | EXAMPLES |
|---|---|

### 1.1 FRACTIONS

**Operations with Fractions**

*Addition/Subtraction:*

1. To add/subtract fractions with the same denominator, add/subtract the numerators and keep the same denominator.
2. To add/subtract fractions with different denominators, find the LCD and write each fraction with this LCD. Then follow the procedure above.

*Multiplication:* Multiply numerators and multiply denominators.

*Division:* Multiply the first fraction by the reciprocal of the second fraction.

Perform each operation.

$$\frac{2}{5}+\frac{7}{5}=\frac{2+7}{5}=\frac{9}{5}$$

$$\frac{2}{3}-\frac{1}{2}=\frac{4}{6}-\frac{3}{6}$$ 6 is the LCD.

$$=\frac{4-3}{6}=\frac{1}{6}$$

$$\frac{4}{3}\cdot\frac{5}{6}=\frac{20}{18}=\frac{10}{9}$$

$$\frac{6}{5}\div\frac{1}{4}=\frac{6}{5}\cdot\frac{4}{1}=\frac{24}{5}$$

### 1.2 EXPONENTS, ORDER OF OPERATIONS, AND INEQUALITY

**Order of Operations**

Simplify within parentheses or above and below fraction bars first, in the following order.

*Step 1* Apply all exponents.

*Step 2* Do any multiplications or divisions from left to right.

*Step 3* Do any additions or subtractions from left to right.

If no grouping symbols are present, start with Step 1.

Simplify $36 - 4(2^2 + 3)$.

$$\begin{aligned}36 - 4(2^2 + 3) &= 36 - 4(4 + 3)\\ &= 36 - 4(7)\\ &= 36 - 28\\ &= 8\end{aligned}$$

(continued)

| CONCEPTS | EXAMPLES |
|---|---|

## 1.3 VARIABLES, EXPRESSIONS, AND EQUATIONS

Evaluate an expression with a variable by substituting a given number for the variable.

Evaluate $2x + y^2$ if $x = 3$ and $y = -4$.

$$\begin{aligned} 2x + y^2 &= 2(3) + (-4)^2 \\ &= 6 + 16 \\ &= 22 \end{aligned}$$

Values of a variable that make an equation true are solutions of the equation.

Is 2 a solution of $5x + 3 = 18$?

$$\begin{aligned} 5(2) + 3 &= 18 \quad ? \\ 13 &= 18 \quad \text{False} \end{aligned}$$

2 is not a solution.

## 1.4 REAL NUMBERS AND THE NUMBER LINE

**The Ordering of Real Numbers**
*a* is less than *b* if *a* is to the left of *b* on the number line.

Graph $-2$, 0, and 3.

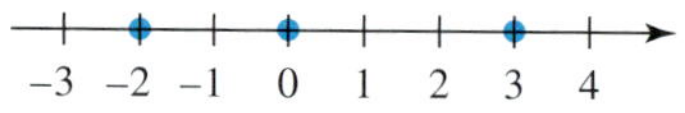


$-2 < 3 \qquad 3 > 0 \qquad 0 < 3$

The additive inverse of $x$ is $-x$.

$-(5) = -5 \qquad -(-7) = 7 \qquad -0 = 0$

The absolute value of $x$, $|x|$, is the distance between $x$ and 0 on the number line.

$|13| = 13 \qquad |0| = 0 \qquad |-5| = 5$

## 1.5 ADDING AND SUBTRACTING REAL NUMBERS

**Rules for Adding Signed Numbers**
To add two numbers with the same sign, add their absolute values. The sum has that same sign.

To add two numbers with different signs, subtract their absolute values. The sum has the sign of the number with larger absolute value.

Add.

$$\begin{aligned} 9 + 4 &= 13 \\ -8 + (-5) &= -13 \\ 7 + (-12) &= -5 \\ -5 + 13 &= 8 \end{aligned}$$

**Definition of Subtraction**

$$x - y = x + (-y)$$

**Rules for Subtracting Signed Numbers**
1. Change the subtraction symbol to the addition symbol.
2. Change the sign of the number being subtracted.
3. Add, using the rules for addition.

Subtract.

$$5 - (-2) = 5 + 2 = 7$$

$$-3 - 4 = -3 + (-4) = -7$$

$$-2 - (-6) = -2 + 6 = 4$$

$$13 - (-8) = 13 + 8 = 21$$

| CONCEPTS | EXAMPLES |
|---|---|

## 1.6 MULTIPLYING AND DIVIDING REAL NUMBERS

**Rules for Multiplying and Dividing Signed Numbers**

The product (or quotient) of two numbers having the *same sign* is *positive;* the product (or quotient) of two numbers having *different signs* is *negative.*

Multiply or divide.

$$6 \cdot 5 = 30 \qquad -7(-8) = 56 \qquad \frac{20}{4} = 5$$

$$\frac{-24}{-6} = 4 \qquad -6(5) = -30 \qquad 6(-5) = -30$$

$$\frac{-18}{9} = -2 \qquad \frac{49}{-7} = -7$$

**Definition of Division**

$$\frac{x}{y} = x \cdot \frac{1}{y}, \quad y \neq 0$$

$$\frac{10}{2} = 10 \cdot \frac{1}{2} = 5$$

Division by 0 is undefined.

$\frac{5}{0}$ is undefined.

0 divided by a nonzero number equals 0.

$$\frac{0}{5} = 0$$

## 1.7 PROPERTIES OF REAL NUMBERS

**Commutative Properties**

$$a + b = b + a$$
$$ab = ba$$

$$7 + (-1) = -1 + 7$$
$$5(-3) = (-3)5$$

**Associative Properties**

$$(a + b) + c = a + (b + c)$$
$$(ab)c = a(bc)$$

$$(3 + 4) + 8 = 3 + (4 + 8)$$
$$[-2(6)]4 = -2[(6)4]$$

**Identity Properties**

$$a + 0 = a \qquad 0 + a = a$$
$$a \cdot 1 = a \qquad 1 \cdot a = a$$

$$-7 + 0 = -7 \qquad 0 + (-7) = -7$$
$$9 \cdot 1 = 9 \qquad 1 \cdot 9 = 9$$

**Inverse Properties**

$$a + (-a) = 0 \qquad -a + a = 0$$
$$a \cdot \frac{1}{a} = 1 \qquad \frac{1}{a} \cdot a = 1 \quad (a \neq 0)$$

$$7 + (-7) = 0 \qquad -7 + 7 = 0$$
$$-2\left(-\frac{1}{2}\right) = 1 \qquad -\frac{1}{2}(-2) = 1$$

**Distributive Properties**

$$a(b + c) = ab + ac$$
$$(b + c)a = ba + ca$$
$$a(b - c) = ab - ac$$

$$5(4 + 2) = 5(4) + 5(2)$$
$$(4 + 2)5 = 4(5) + 2(5)$$
$$9(5 - 4) = 9(5) - 9(4)$$

(continued)

| CONCEPTS | EXAMPLES |
|---|---|
| **1.8 SIMPLIFYING EXPRESSIONS** | |
| Only like terms may be combined. | $-3y^2 + 6y^2 + 14y^2 = 17y^2$<br>$4(3 + 2x) - 6(5 - x)$<br>$= 12 + 8x - 30 + 6x$ Distributive property<br>$= 14x - 18$ |

**Answers to Test Your Word Power**

**1.** D; *Example:* Since $2 \times 5 = 10$, the numbers 2 and 5 are factors of 10; other factors of 10 are $-10, -5, -2, -1, 1$, and 10. **2.** C; *Examples:* 2, 3, 11, 41, 53 **3.** C; *Example:* In $2^3$, the number 3 is the exponent (or power), so 2 is a factor three times; $2^3 = 2 \cdot 2 \cdot 2 = 8$. **4.** A; *Examples:* $a, b, c$ **5.** B; *Examples:* $-9, 0, 6$ **6.** A; *Example:* The point graphed 3 units to the right of 0 on a number line has coordinate 3. **7.** D; *Examples:* $|2| = 2$ and $|-2| = 2$ **8.** B; *Examples:* $6, \frac{x}{2}, -4ab^2$ **9.** A; *Examples:* The term 3 has numerical coefficient 3, $8z$ has numerical coefficient 8, and $-10x^4y$ has numerical coefficient $-10$.

# CHAPTER 1 REVIEW EXERCISES

**[1.1]** *Perform each indicated operation.**

**1.** $\frac{8}{5} \div \frac{32}{15}$

**2.** $\frac{3}{8} + 3\frac{1}{2} - \frac{3}{16}$

**3.** The circle graph illustrates how 400 people responded to a survey that asked "Do you think that the CEOs of corporations guilty of accounting fraud should go to jail?" What fractional part of the group did not have an opinion?

**4.** Based on the graph in Exercise 3, how many people responded "yes"?

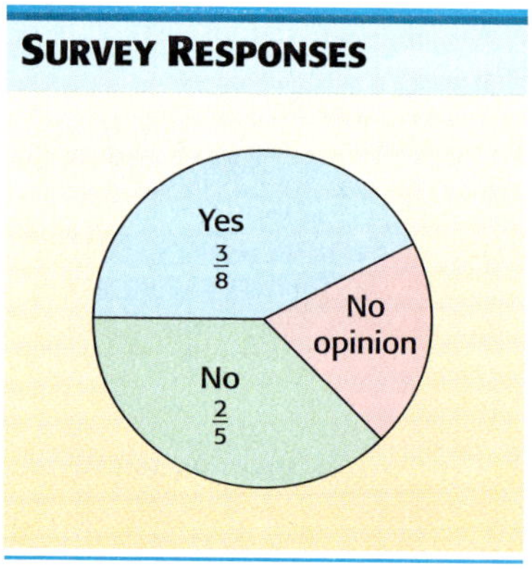


**[1.2]** *Find the value of each exponential expression.*

**5.** $5^4$

**6.** $\left(\frac{3}{5}\right)^3$

**7.** $(.02)^5$

**8.** $(.001)^3$

*For help with the Review Exercises in this text, refer to the appropriate section given in brackets.

*Find the value of each expression.*

**9.** $8 \cdot 5 - 13$

**10.** $7[3 + 6(3^2)]$

**11.** $\frac{9(4^2 - 3)}{4 \cdot 5 - 17}$

**12.** $\frac{6(5 - 4) + 2(4 - 2)}{3^2 - (4 + 3)}$

*Tell whether each statement is* true *or* false.

**13.** $12 \cdot 3 - 6 \cdot 6 \leq 0$

**14.** $3[5(2) - 3] > 20$

**15.** $9 \leq 4^2 - 8$

*Write each word statement in symbols.*

**16.** Thirteen is less than seventeen.

**17.** Five plus two is not equal to ten.

**18.** The bar graph shows the number of worldwide airline fatalities for the years 1990 to 1999.

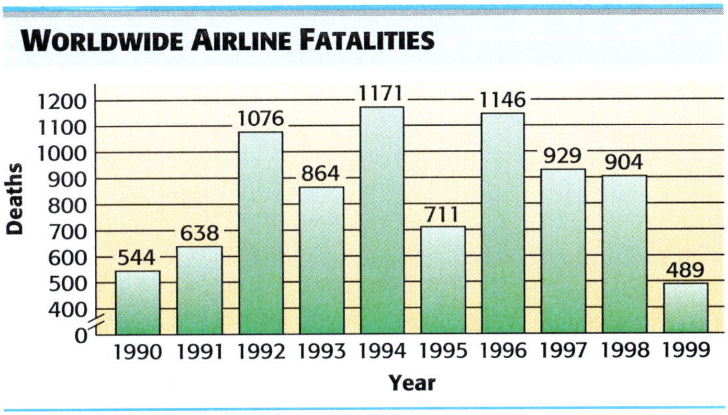


*Source:* International Civil Aviation Organization.

**(a)** In which years were *fewer than* 700 people killed?
**(b)** In which years were *at least* 905 people killed?
**(c)** How many people were killed in the 5 yr having the largest numbers of deaths?

[1.3] *Find the numerical value of each expression if $x = 6$ and $y = 3$.*

**19.** $2x + 6y$

**20.** $4(3x - y)$

**21.** $\frac{x}{3} + 4y$

**22.** $\frac{x^2 + 3}{3y - x}$

*Change each word phrase to an algebraic expression. Use x as the variable to represent the number.*

**23.** Six added to a number

**24.** A number subtracted from eight

**25.** Nine subtracted from six times a number

**26.** Three-fifths of a number added to 12

*Decide whether the given number is a solution of the given equation.*

**27.** $5x + 3(x + 2) = 22$; 2

**28.** $\frac{t + 5}{3t} = 1$; 6

*Change each word statement to an equation. Use x as the variable. Then find the solution from the set* $\{0, 2, 4, 6, 8, 10\}$.

**29.** Six less than twice a number is 10.

**30.** The product of a number and 4 is 8.

[1.4] *Graph each group of numbers on a number line.*

**31.** $-4, -\frac{1}{2}, 0, 2.5, 5$

**32.** $-2, |-3|, -3, |-1|$

*Classify each number, using the sets* natural numbers, whole numbers, integers, rational numbers, irrational numbers, real numbers.

**33.** $\frac{4}{3}$

**34.** $\sqrt{6}$

*Select the smaller number in each pair.*

**35.** $-10, 5$

**36.** $-8, -9$

**37.** $-\frac{2}{3}, -\frac{3}{4}$

**38.** $0, -|23|$

*Decide whether each statement is* true *or* false.

**39.** $12 > -13$

**40.** $0 > -5$

**41.** $-9 < -7$

**42.** $-13 \geq -13$

*For each number,* **(a)** *find the opposite of the number and* **(b)** *find the absolute value of the number.*

**43.** $-9$

**44.** $0$

**45.** $6$

**46.** $-\frac{5}{7}$

*Simplify each number by removing absolute value symbols.*

**47.** $|-12|$

**48.** $-|3|$

**49.** $-|-19|$

**50.** $-|9 - 2|$

[1.5] *Perform each indicated operation.*

**51.** $-10 + 4$

**52.** $14 + (-18)$

**53.** $-8 + (-9)$

**54.** $\frac{4}{9} + \left(-\frac{5}{4}\right)$

**55.** $-13.5 + (-8.3)$

**56.** $(-10 + 7) + (-11)$

**57.** $[-6 + (-8) + 8] + [9 + (-13)]$

**58.** $(-4 + 7) + (-11 + 3) + (-15 + 1)$

**59.** $-7 - 4$

**60.** $-12 - (-11)$

**61.** $5 - (-2)$

**62.** $-\frac{3}{7} - \frac{4}{5}$

**63.** $2.56 - (-7.75)$

**64.** $(-10 - 4) - (-2)$

**65.** $(-3 + 4) - (-1)$

**66.** $-(-5 + 6) - 2$

*Write a numerical expression for each phrase and simplify the expression.*

**67.** 19 added to the sum of $-31$ and 12

**68.** 13 more than the sum of $-4$ and $-8$

**69.** The difference between $-4$ and $-6$

**70.** Five less than the sum of 4 and $-8$

*Find the solution of each equation from the set* $\{-3, -2, -1, 0, 1, 2, 3\}$ *by guessing or by trial and error.*

**71.** $x + (-2) = -4$

**72.** $12 + x = 11$

*Solve each problem.*

**73.** Like many people, Otis Taylor neglects to keep up his checkbook balance. When he finally balanced his account, he found the balance was $-\$23.75$, so he deposited \$50.00. What is his new balance?

**74.** The low temperature in Yellowknife, in the Canadian Northwest Territories, one January day was $-26°F$. It rose $16°$ that day. What was the high temperature?

**75.** Eric owed his brother \$28. He repaid \$13 but then borrowed another \$14. What positive or negative amount represents his present financial status?

**76.** If the temperature drops $7°$ below its previous level of $-3°$, what is the new temperature?

**77.** A football team gained 3 yd on the first play from scrimmage, lost 12 yd on the second play, and then gained 13 yd on the third play. How many yards did the team gain or lose altogether?

**78.** On Wednesday, May 29, 2002, the Dow Jones Industrial Average closed at 9923.04, down 58.54 from the previous day. What was the closing value the previous day? (*Source: Times Picayune.*)

[1.6] *Perform each indicated operation.*

**79.** $(-12)(-3)$ **80.** $15(-7)$ **81.** $-\frac{4}{3}\left(-\frac{3}{8}\right)$ **82.** $(-4.8)(-2.1)$

**83.** $5(8 - 12)$ **84.** $(5 - 7)(8 - 3)$ **85.** $2(-6) - (-4)(-3)$

**86.** $3(-10) - 5$ **87.** $\frac{-36}{-9}$ **88.** $\frac{220}{-11}$

**89.** $-\frac{1}{2} \div \frac{2}{3}$ **90.** $-33.9 \div (-3)$ **91.** $\frac{-5(3) - 1}{8 - 4(-2)}$

**92.** $\frac{5(-2) - 3(4)}{-2[3 - (-2)] - 1}$ **93.** $\frac{10^2 - 5^2}{8^2 + 3^2 - (-2)}$ **94.** $\frac{(.6)^2 + (.8)^2}{(-1.2)^2 - (-.56)}$

*Evaluate each expression if $x = -5$, $y = 4$, and $z = -3$.*

**95.** $6x - 4z$ **96.** $5x + y - z$ **97.** $5x^2$ **98.** $z^2(3x - 8y)$

*Write a numerical expression for each phrase and simplify the expression.*

**99.** Nine less than the product of $-4$ and 5

**100.** Five-sixths of the sum of 12 and $-6$

**101.** The quotient of 12 and the sum of 8 and $-4$

**102.** The product of $-20$ and 12, divided by the difference between 15 and $-15$

*Write each sentence in symbols, using x as the variable, and find the solution by guessing or by trial and error. All solutions come from the list of integers between $-12$ and 12.*

**103.** 8 times a number is $-24$.

**104.** The quotient of a number and 3 is $-2$.

*Find the average of each group of numbers.*

**105.** 26, 38, 40, 20, 4, 14, 96, 18

**106.** −12, 28, −36, 0, 12, −10

**[1.7]** *Decide whether each statement is an example of the* commutative, associative, identity, inverse, *or* distributive *property.*

**107.** $6 + 0 = 6$

**108.** $5 \cdot 1 = 5$

**109.** $-\frac{2}{3}\left(-\frac{3}{2}\right) = 1$

**110.** $17 + (-17) = 0$

**111.** $5 + (-9 + 2) = [5 + (-9)] + 2$

**112.** $w(xy) = (wx)y$

**113.** $3x + 3y = 3(x + y)$

**114.** $(1 + 2) + 3 = 3 + (1 + 2)$

*Use the distributive property to rewrite each expression. Simplify if possible.*

**115.** $7y + 14$

**116.** $-12(4 - t)$

**117.** $3(2s) + 3(5y)$

**118.** $-(-4r + 5s)$

**119.** Evaluate $25 - (5 - 2)$ and $(25 - 5) - 2$. Use this example to explain why subtraction is not associative.

**120.** Evaluate $180 \div (15 \div 5)$ and $(180 \div 15) \div 5$. Use this example to explain why division is not associative.

**[1.8]** *Combine like terms whenever possible.*

**121.** $2m + 9m$

**122.** $15p^2 - 7p^2 + 8p^2$

**123.** $5p^2 - 4p + 6p + 11p^2$

**124.** $-2(3k - 5) + 2(k + 1)$

**125.** $7(2m + 3) - 2(8m - 4)$

**126.** $-(2k + 8) - (3k - 7)$

*In Exercises 127–130, choose the letter of the correct response.*

**127.** Which one of the following is true for all real numbers $x$?

**A.** $6 + 2x = 8x$ **B.** $6 - 2x = 4x$
**C.** $6x - 2x = 4x$ **D.** $3 + 8(4x - 6) = 11(4x - 6)$

**128.** Which one of the following is an example of a pair of like terms?

**A.** $6t, 6w$ **B.** $-8x^2y, 9xy^2$ **C.** $5ry, 6yr$ **D.** $-5x^2, 2x^3$

**129.** Which one of the following is an example of a term with numerical coefficient 5?

**A.** $5x^3y^7$ **B.** $x^5$ **C.** $\frac{x}{5}$ **D.** $5^2xy^3$

**130.** Which one of the following is a correct translation for "six times a number, subtracted from the product of eleven and the number" (if $x$ represents the number)?

**A.** $6x - 11x$ **B.** $11x - 6x$ **C.** $(11 + x) - 6x$ **D.** $6x - (11 + x)$

## MIXED REVIEW EXERCISES*

*Perform each indicated operation.*

**131.** $[(-2) + 7 - (-5)] + [-4 - (-10)]$

**132.** $\left(-\frac{5}{6}\right)^2$

*The order of exercises in this final group does not correspond to the order in which topics occur in the chapter. This random ordering should help you prepare for the chapter test in yet another way.

**133.** $\dfrac{6(-4) + 2(-12)}{5(-3) + (-3)}$

**134.** $\dfrac{3}{8} - \dfrac{5}{12}$

**135.** $\dfrac{8^2 + 6^2}{7^2 + 1^2}$

**136.** $-16(-3.5) - 7.2(-3)$

**137.** $2\dfrac{5}{6} - 4\dfrac{1}{3}$

**138.** $-8 + [(-4 + 17) - (-3 - 3)]$

**139.** $-\dfrac{12}{5} \div \dfrac{9}{7}$

**140.** $(-8 - 3) - 5(2 - 9)$

**141.** $5x^2 - 12y^2 + 3x^2 - 9y^2$

**142.** $-4(2t + 1) - 8(-3t + 4)$

**143.** Write a sentence or two explaining the special considerations involving 0 when dividing.

**144.** "Two negatives give a positive" is often heard from students. Is this correct? Use more precise language in explaining what this means.

**145.** Use $x$ as the variable and write an expression for "the product of 5, and the sum of a number and 7." Then use the distributive property to rewrite the expression.

**146.** The highest temperature ever recorded in Albany, New York, was 99°F, while the lowest was 112° less than the highest. What was the lowest temperature ever recorded in Albany? (*Source: World Almanac and Book of Facts,* 2000.)

# CHAPTER 1 TEST

**1.** Write $\frac{63}{99}$ in lowest terms.

**2.** Add: $\dfrac{5}{8} + \dfrac{11}{12} + \dfrac{7}{15}$.

**3.** Divide: $\dfrac{19}{15} \div \dfrac{6}{5}$.

**4.** The circle graph indicates the market share of different means of intercity transportation in a recent year, based on 1230 million passengers carried.

**(a)** How many of these passengers used air travel?

**(b)** How many of these passengers did not use the bus?

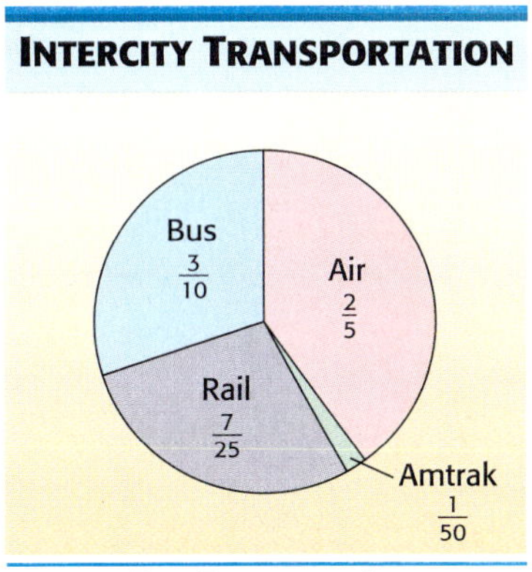


***Source:*** Eno Transportation Foundation, Inc.

**5.** Decide whether $4[-20 + 7(-2)] \leq 135$ is true or false.

**6.** Graph the group of numbers $-1, -3, |-4|, |-1|$ on a number line.

**7.** To which of the following sets does $-\frac{2}{3}$ belong: natural numbers, whole numbers, integers, rational numbers, irrational numbers, real numbers?

**8.** Explain how a number line can be used to show that $-8$ is less than $-1$.

**9.** Write in symbols: The quotient of $-6$ and the sum of 2 and $-8$. Simplify the expression.

*Perform each indicated operation.*

**10.** $-2 - (5 - 17) + (-6)$

**11.** $-5\frac{1}{2} + 2\frac{2}{3}$

**12.** $-6 - [-7 + (2 - 3)]$

**13.** $4^2 + (-8) - (2^3 - 6)$

**14.** $(-5)(-12) + 4(-4) + (-8)^2$

**15.** $\frac{-7 - (-6 + 2)}{-5 - (-4)}$

**16.** $\frac{30(-1 - 2)}{-9[3 - (-2)] - 12(-2)}$

*Find the solution for each equation from the set* $\{-6, -4, -2, 0, 2, 4, 6\}$ *by guessing or by trial and error.*

**17.** $-x + 3 = -3$

**18.** $-3x = -12$

*Evaluate each expression, given* $x = -2$ *and* $y = 4$.

**19.** $3x - 4y^2$

**20.** $\frac{5x + 7y}{3(x + y)}$

*Solve each problem.*

**21.** The highest elevation in Argentina is Mt. Aconcagua, which is 6960 m above sea level. The lowest point in Argentina is the Valdes Peninsula, 40 m below sea level. Find the difference between the highest and lowest elevations.

**22.** For a certain system of rating major league baseball relief pitchers, 3 points are awarded for a save, 3 points are awarded for a win, 2 points are subtracted for a loss, and 2 points are subtracted for a blown save. If Mariano Rivera of the New York Yankees has 4 saves, 3 wins, 2 losses, and 1 blown save, how many points does he have?

**23.** The bar graph shows the federal budget outlays for national defense for the years 1990 through 1998. Use a signed number to represent the change in outlays for each time period. For example, the change from 1995 to 1996 was $253.3 − $259.6 = −$6.3 billion.

**(a)** 1991–1992
**(b)** 1993–1994
**(c)** 1996–1997
**(d)** 1997–1998

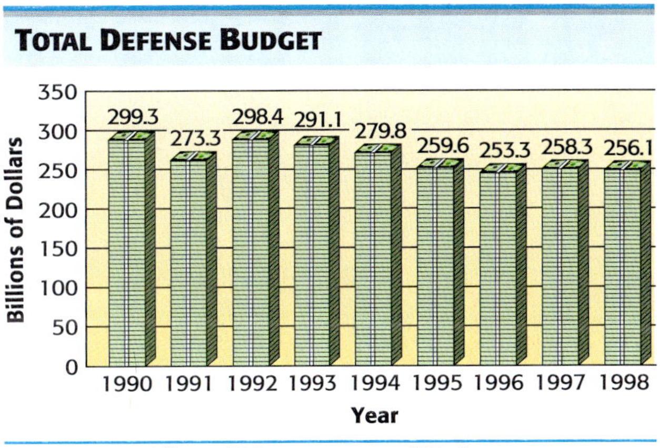


***Source:*** U.S. Office of Management and Budget.

*Match each property in Column I with the example of it in Column II.*

| I | II |
|---|---|
| **24.** Commutative property | **A.** $3x + 0 = 3x$ |
| **25.** Associative property | **B.** $(5 + 2) + 8 = 8 + (5 + 2)$ |
| **26.** Inverse property | **C.** $-3(x + y) = -3x + (-3y)$ |
| **27.** Identity property | **D.** $-5 + (3 + 2) = (-5 + 3) + 2$ |
| **28.** Distributive property | **E.** $-\frac{5}{3}\left(-\frac{3}{5}\right) = 1$ |

**29.** What property is used to show that $3(x + 1) = 3x + 3$?

**30.** Consider the expression $-6[5 + (-2)]$.
**(a)** Evaluate it by first working within the brackets.
**(b)** Evaluate it by using the distributive property.
**(c)** Why must the answers in items (a) and (b) be the same?

*Simplify by combining like terms.*

**31.** $8x + 4x - 6x + x + 14x$

**32.** $5(2x - 1) - (x - 12) + 2(3x - 5)$

# Linear Equations and Inequalities in One Variable; Applications

# 2

From a gathering of 13 nations in Athens in 1896, the Olympic Games have truly become a worldwide event. The XIX Olympic Winter Games in Salt Lake City in 2002 attracted athletes from 82 countries and entertained 3.5 billion viewers around the world.

One ceremonial aspect of the games is the flying of the Olympic flag with its five interlocking rings of different colors on a white background. First introduced at the 1920 Games in Antwerp, Belgium, the five rings on the flag symbolize unity among the nations of Africa, the Americas, Asia, Australia, and Europe. (*Source: Microsoft Encarta Encyclopedia 2002;* www.olympic-usa.org) Throughout this chapter we use linear equations to solve applications about the Olympics.

# 2.1 The Addition Property of Equality

**OBJECTIVES**

1. Identify linear equations.
2. Use the addition property of equality.
3. Simplify equations, and then use the addition property of equality.

Recall from Section 1.3 that an *equation* is a statement that two algebraic expressions are equal. The simplest type of equation is a *linear equation.*

**OBJECTIVE 1 Identify linear equations.**

**Linear Equation in One Variable**

A **linear equation in one variable** can be written in the form

$$Ax + B = C$$

for real numbers $A$, $B$, and $C$, with $A \neq 0$.

For example,

$$4x + 9 = 0, \qquad 2x - 3 = 5, \qquad \text{and} \qquad x = 7$$

are linear equations in one variable ($x$). The final two can be written in the specified form using properties developed in this chapter. However,

$$x^2 + 2x = 5, \qquad \frac{1}{x} = 6, \qquad \text{and} \qquad |2x + 6| = 0$$

are *not* linear equations.

As we saw in Section 1.3, a *solution* of an equation is a number that makes the equation true when it replaces the variable. An equation is solved by finding its **solution set,** the set of all solutions. Equations that have exactly the same solution sets are **equivalent equations.** Linear equations are solved by using a series of steps to produce a simpler equivalent equation of the form

$$\mathbf{x = a\ number} \qquad \text{or} \qquad \mathbf{a\ number = x}.$$

**OBJECTIVE 2 Use the addition property of equality.** In the equation $x - 5 = 2$, both $x - 5$ and 2 represent the same number because this is the meaning of the equals sign. To solve the equation, we change the left side from $x - 5$ to just $x$. We do this by adding 5 to $x - 5$. We use 5 because 5 is the opposite (additive inverse) of $-5$, and $-5 + 5 = 0$. To keep the two sides equal, we must also add 5 to the right side.

| | |
|---|---|
| $x - 5 = 2$ | Given equation |
| $x - 5 + 5 = 2 + 5$ | Add 5 to each side. |
| $x + 0 = 7$ | Additive inverse property |
| $x = 7$ | Additive identity property |

The solution of the given equation is 7. We check by replacing $x$ with 7 in the original equation.

*Check:*

| | | |
|---|---|---|
| $x - 5 = 2$ | | Original equation |
| $7 - 5 = 2$ | ? | Let $x = 7$. |
| $2 = 2$ | | True |

Since the final equation is true, 7 checks as the solution and $\{7\}$ is the solution set.

To solve the equation, we added the same number to each side. The **addition property of equality** justifies this step.

**Addition Property of Equality**

If $A$, $B$, and $C$ are real numbers, then the equations

$$A = B \quad \text{and} \quad A + C = B + C$$

are equivalent equations.

That is, we can add the same number to each side of an equation without changing the solution.

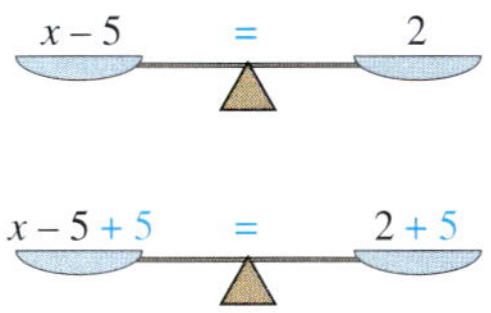


**FIGURE 1**

In the addition property, $C$ represents a real number. This means that any quantity that represents a real number can be added to each side of an equation to change it to an equivalent equation.

**NOTE** Equations can be thought of in terms of a balance. Thus, adding the same quantity to each side does not affect the balance. See Figure 1.

### EXAMPLE 1 Using the Addition Property of Equality

Solve $x - 16 = 7$.

Our goal is to get an equivalent equation of the form $x =$ a number. To do this, we use the addition property of equality and add 16 to each side.

$$x - 16 = 7$$

$$x - 16 + 16 = 7 + 16 \qquad \text{Add 16 to each side.}$$

$$x = 23 \qquad \text{Combine like terms.}$$

Note that we combined the steps that change $x - 16 + 16$ to $x + 0$ and $x + 0$ to $x$. We will combine these steps from now on. We check by substituting 23 for $x$ in the *original* equation.

*Check:*

$$x - 16 = 7 \qquad \text{Original equation}$$

$$23 - 16 = 7 \quad ? \qquad \text{Let } x = 23.$$

$$7 = 7 \qquad \text{True}$$

Since a true statement results, $\{23\}$ is the solution set.

**Now Try Exercise 7.**

### EXAMPLE 2 Using the Addition Property of Equality

Solve $x - 2.9 = -6.4$.

To get $x$ alone on the left side, we must eliminate the $-2.9$. We use the addition property of equality and add 2.9 to each side.

$$x - 2.9 = -6.4$$

$$x - 2.9 + 2.9 = -6.4 + 2.9 \qquad \text{Add 2.9 to each side.}$$

$$x = -3.5$$

**EXAMPLE 6 Using the Distributive Property to Simplify an Equation**

Solve $3(2 + 5x) - (1 + 14x) = 6$.

$$\begin{aligned}
3(2 + 5x) - (1 + 14x) &= 6 \\
3(2 + 5x) - 1(1 + 14x) &= 6 && -(1 + 14x) = -1(1 + 14x) \\
3(2) + 3(5x) - 1(1) - 1(14x) &= 6 && \text{Distributive property} \\
6 + 15x - 1 - 14x &= 6 && \text{Multiply.} \\
x + 5 &= 6 && \text{Combine like terms.} \\
x + 5 - 5 &= 6 - 5 && \text{Subtract 5 from each side.} \\
x &= 1 && \text{Combine like terms.}
\end{aligned}$$

Check by substituting 1 for $x$ in the original equation. The solution set is $\{1\}$.

**Now Try Exercise 49.**

**CAUTION** Be careful to apply the distributive property correctly in a problem like that in Example 6, or a sign error may result.

## 2.1 EXERCISES

**For Extra Help**

- Student's Solutions Manual
- MyMathLab
- InterAct Math Tutorial Software
- AW Math Tutor Center
- MathXL
- Digital Video Tutor CD 3/Videotape 3

**1.** Which pairs of equations are equivalent equations?

**A.** $x + 2 = 6$ and $x = 4$ **B.** $10 - x = 5$ and $x = -5$
**C.** $x + 3 = 9$ and $x = 6$ **D.** $4 + x = 8$ and $x = -4$

**2.** Decide whether each of the following is an expression or an equation. If it is an expression, simplify it. If it is an equation, solve it.

**(a)** $5x + 8 - 4x + 7$ **(b)** $-6y + 12 + 7y - 5$
**(c)** $5x + 8 - 4x = 7$ **(d)** $-6y + 12 + 7y = -5$

**3.** In your own words, state the addition property of equality. Give an example.

**4.** Explain how to check a solution of an equation.

*Solve each equation, and check your solution. See Examples 1–4.*

| | | |
|---|---|---|
| **5.** $x - 4 = 8$ | **6.** $x - 8 = 9$ | **7.** $y - 12 = 19$ |
| **8.** $t - 15 = 25$ | **9.** $x - 5 = -8$ | **10.** $x - 7 = -9$ |
| **11.** $r + 9 = 13$ | **12.** $y + 6 = 10$ | **13.** $x + 26 = 17$ |
| **14.** $x + 45 = 24$ | **15.** $7 + r = -3$ | **16.** $8 + k = -4$ |
| **17.** $2 = p + 15$ | **18.** $3 = z + 17$ | **19.** $-2 = x - 12$ |
| **20.** $-6 = x - 21$ | **21.** $x - 8.4 = -2.1$ | **22.** $y - 15.5 = -5.1$ |
| **23.** $t + 12.3 = -4.6$ | **24.** $x + 21.5 = -13.4$ | **25.** $\frac{2}{5}w - 6 = \frac{7}{5}w$ |
| **26.** $-\frac{2}{7}z + 2 = \frac{5}{7}z$ | **27.** $5.6x + 2 = 4.6x$ | **28.** $9.1x - 5 = 8.1x$ |

29. $3p + 6 = 10 + 2p$

30. $8b - 4 = -6 + 7b$

31. $1.2y - 4 = .2y - 4$

32. $7.7r + 6 = 6.7r + 6$

33. $\frac{1}{2}x + 2 = -\frac{1}{2}x$

34. $\frac{1}{5}x - 7 = -\frac{4}{5}x$

35. $3x + 7 - 2x = 0$

36. $5x + 4 - 4x = 0$

37. Which of the following are *not* linear equations in one variable?

**A.** $x^2 - 5x + 6 = 0$ **B.** $x^3 = x$
**C.** $3x - 4 = 0$ **D.** $7x - 6x = 3 + 9x$

Define a linear equation in one variable in words.

38. Refer to the definition of linear equation in one variable given in this section. Why is the restriction $A \neq 0$ necessary?

*Solve each equation, and check your solution. See Examples 5 and 6.*

39. $5t + 3 + 2t - 6t = 4 + 12$

40. $4x + 3x - 6 - 6x = 10 + 3$

41. $6x + 5 + 7x + 3 = 12x + 4$

42. $4x - 3 - 8x + 1 = -5x + 9$

43. $5.2q - 4.6 - 7.1q = -.9q - 4.6$

44. $-4.0x + 2.7 - 1.6x = -4.6x + 2.7$

45. $\frac{5}{7}x + \frac{1}{3} = \frac{2}{5} - \frac{2}{7}x + \frac{2}{5}$

46. $\frac{6}{7}s - \frac{3}{4} = \frac{4}{5} - \frac{1}{7}s + \frac{1}{6}$

47. $(5y + 6) - (3 + 4y) = 10$

48. $(8r - 3) - (7r + 1) = -6$

49. $2(p + 5) - (9 + p) = -3$

50. $4(k - 6) - (3k + 2) = -5$

51. $-6(2b + 1) + (13b - 7) = 0$

52. $-5(3w - 3) + (1 + 16w) = 0$

53. $10(-2x + 1) = -19(x + 1)$

54. $2(2 - 3r) = -5(r - 3)$

55. $-2(8p + 2) - 3(2 - 7p) - 2(4 + 2p) = 0$

56. $-5(1 - 2z) + 4(3 - z) - 7(3 + z) = 0$

57. $4(7x - 1) + 3(2 - 5x) - 4(3x + 5) = -6$

58. $9(2m - 3) - 4(5 + 3m) - 5(4 + m) = -3$

59. Write an equation that requires the use of the addition property of equality, where 6 must be added to each side to solve the equation, and the solution is a negative number.

60. Write an equation that requires the use of the addition property of equality, where $\frac{1}{2}$ must be subtracted from each side, and the solution is a positive number.

*Write an equation using the information given in the problem. Use x as the variable. Then solve the equation.*

61. Three times a number is 17 more than twice the number. Find the number.

62. One, added to three times a number, is three less than four times the number. Find the number.

63. If six times a number is subtracted from seven times the number, the result is $-9$. Find the number.

64. If five times a number is added to three times the number, the result is the sum of seven times the number and 9. Find the number.

# 2.2 The Multiplication Property of Equality

**OBJECTIVES**

1. Use the multiplication property of equality.
2. Simplify equations, and then use the multiplication property of equality.

**OBJECTIVE 1 Use the multiplication property of equality.** The addition property of equality alone is not enough to solve some equations, such as $3x + 2 = 17$.

$$3x + 2 = 17$$
$$3x + 2 - 2 = 17 - 2 \qquad \text{Subtract 2 from each side.}$$
$$3x = 15 \qquad \text{Combine like terms.}$$

Notice that the coefficient of $x$ on the left side is 3, not 1 as desired. Another property is needed to change $3x = 15$ to an equation of the form

$$x = \text{a number.}$$

If $3x = 15$, then $3x$ and 15 both represent the same number. Multiplying both $3x$ and 15 by the same number will also result in an equality. The **multiplication property of equality** states that we can multiply each side of an equation by the same nonzero number without changing the solution.

**Multiplication Property of Equality**

If $A$, $B$, and $C$ ($C \neq 0$) represent real numbers, then the equations

$$\boldsymbol{A = B} \qquad \text{and} \qquad \boldsymbol{AC = BC}$$

are equivalent equations.

That is, we can multiply each side of an equation by the same nonzero number without changing the solution.

This property can be used to solve $3x = 15$. The $3x$ on the left must be changed to $1x$, or $x$, instead of $3x$. To isolate $x$, we multiply each side of the equation by $\frac{1}{3}$. We use $\frac{1}{3}$ because $\frac{1}{3}$ is the reciprocal of 3, and $\frac{1}{3} \cdot 3 = \frac{3}{3} = 1$.

$$3x = 15$$
$$\frac{1}{3}(3x) = \frac{1}{3} \cdot 15 \qquad \text{Multiply each side by } \tfrac{1}{3}.$$
$$\left(\frac{1}{3} \cdot 3\right)x = \frac{1}{3} \cdot 15 \qquad \text{Associative property}$$
$$1x = 5 \qquad \text{Multiplicative inverse property}$$
$$x = 5 \qquad \text{Multiplicative identity property}$$

We check by substituting 5 for $x$ in the original equation. The solution set of the equation is $\{5\}$. From now on, we will combine the last two steps shown in the preceding example.

Just as the addition property of equality permits *subtracting* the same number from each side of an equation, the multiplication property of equality permits *dividing* each side of an equation by the same nonzero number. For example, the equation $3x = 15$, which we just solved by multiplying each side by $\frac{1}{3}$, could also be solved by dividing each side by 3.

$$3x = 15$$

$$\frac{3x}{3} = \frac{15}{3} \qquad \text{Divide each side by 3.}$$

$$x = 5$$

We can divide each side of an equation by the same nonzero number without changing the solution. Do not, however, divide each side by a variable, as that may result in losing a valid solution.

**NOTE** In practice, it is usually easier to multiply on each side if the coefficient of the variable is a fraction, and divide on each side if the coefficient is an integer. For example, to solve

$$-\frac{3}{4}x = 12,$$

it is easier to multiply by $-\frac{4}{3}$, the reciprocal of $-\frac{3}{4}$, than to divide by $-\frac{3}{4}$. On the other hand, to solve

$$-5x = -20,$$

it is easier to divide by $-5$ than to multiply by $-\frac{1}{5}$.

### EXAMPLE 1 Dividing Each Side of an Equation by a Nonzero Number

Solve $25p = 30$.

Transform the equation so that $p$ (instead of $25p$) is on the left by using the multiplication property of equality. Divide each side of the equation by 25, the coefficient of $p$.

$$25p = 30$$

$$\frac{25p}{25} = \frac{30}{25} \qquad \text{Divide each side by 25.}$$

$$p = \frac{30}{25} = \frac{6}{5} \qquad \text{Write in lowest terms.}$$

To check, substitute $\frac{6}{5}$ for $p$ in the original equation.

*Check:*

$$25p = 30$$

$$\frac{25}{1}\left(\frac{6}{5}\right) = 30 \quad ? \qquad \text{Let } p = \tfrac{6}{5}.$$

$$30 = 30 \qquad \text{True}$$

The solution set is $\left\{\frac{6}{5}\right\}$.

**Now Try Exercise 25.**

**EXAMPLE 2 Solving an Equation with Decimals**

Solve $-2.1x = 6.09$.

$$-2.1x = 6.09$$

$$\frac{-2.1x}{-2.1} = \frac{6.09}{-2.1} \qquad \text{Divide each side by } -2.1.$$

A calculator will simplify the work at this point.

$$x = -2.9 \qquad \text{Divide.}$$

*Check:*

$$-2.1x = 6.09 \qquad \text{Original equation}$$

$$-2.1(-2.9) \stackrel{?}{=} 6.09 \qquad \text{Let } x = -2.9.$$

$$6.09 = 6.09 \qquad \text{True}$$

The solution set is $\{-2.9\}$.

**Now Try Exercise 39.**

In the next two examples, multiplication produces the solution more quickly than division would.

**EXAMPLE 3 Using the Multiplication Property of Equality**

Solve $\frac{a}{4} = 3$.

Replace $\frac{a}{4}$ by $\frac{1}{4}a$, since dividing by 4 is the same as multiplying by $\frac{1}{4}$. To get $a$ alone on the left, multiply each side by 4, the reciprocal of the coefficient of $a$.

$$\frac{a}{4} = 3$$

$$\frac{1}{4}a = 3 \qquad \text{Change } \tfrac{a}{4} \text{ to } \tfrac{1}{4}a.$$

$$4 \cdot \frac{1}{4}a = 4 \cdot 3 \qquad \text{Multiply by 4.}$$

$$a = 12 \qquad \text{Multiplicative inverse property; multiplicative identity property}$$

Check that 12 is the solution.

*Check:*

$$\frac{a}{4} = 3 \qquad \text{Original equation}$$

$$\frac{12}{4} \stackrel{?}{=} 3 \qquad \text{Let } a = 12.$$

$$3 = 3 \qquad \text{True}$$

The solution set is $\{12\}$.

**Now Try Exercise 43.**

**EXAMPLE 4 Using the Multiplication Property of Equality**

Solve $\frac{3}{4}h = 6.$

To transform the equation so that $h$ is alone on the left, multiply each side of the equation by $\frac{4}{3}$. Use $\frac{4}{3}$ because $\frac{4}{3} \cdot \frac{3}{4}h = 1 \cdot h = h.$

$$\frac{3}{4}h = 6$$

$$\frac{4}{3}\left(\frac{3}{4}h\right) = \frac{4}{3} \cdot 6 \qquad \text{Multiply by } \tfrac{4}{3}.$$

$$1 \cdot h = \frac{4}{3} \cdot \frac{6}{1} \qquad \text{Multiplicative inverse property}$$

$$h = 8 \qquad \text{Multiplicative identity property; multiply fractions.}$$

Check by substituting 8 for $h$ in the original equation. The solution set is $\{8\}$.

**Now Try Exercise 47.**

In Section 2.1, we obtained the equation $-k = -17$ in our alternate solution to Example 4. We reasoned that since this equation says that the additive inverse (or opposite) of $k$ is $-17$, then $k$ must equal 17. We can also use the multiplication property of equality to obtain the same result, as shown in the next example.

**EXAMPLE 5 Using the Multiplication Property of Equality When the Coefficient of the Variable Is −1**

Solve $-k = -17.$

On the left side, change $-k$ to $k$ by first writing $-k$ as $-1 \cdot k$.

$$-k = -17$$

$$-1 \cdot k = -17 \qquad -k = -1 \cdot k$$

$$-1(-1 \cdot k) = -1(-17) \qquad \text{Multiply by } -1 \text{, since } -1(-1) = 1.$$

$$[-1(-1)] \cdot k = 17 \qquad \text{Associative property; multiply.}$$

$$1 \cdot k = 17 \qquad \text{Multiplicative inverse property}$$

$$k = 17 \qquad \text{Multiplicative identity property}$$

*Check:*

$$-k = -17 \qquad \text{Original equation}$$

$$-(17) = -17 \quad ? \qquad \text{Let } k = 17.$$

$$-17 = -17 \qquad \text{True}$$

The solution, 17, checks, so $\{17\}$ is the solution set.

**Now Try Exercise 53.**

From Example 5, we see that the following is true.

$$\text{If} \quad -x = a, \quad \text{then} \quad x = -a.$$

**OBJECTIVE 2 Simplify equations, and then use the multiplication property of equality.** In the next example, it is necessary to simplify the equation before using the multiplication property of equality.

### EXAMPLE 6 Simplifying an Equation

Solve $5m + 6m = 33$.

$$5m + 6m = 33$$

$$11m = 33 \quad \text{Combine like terms.}$$

$$\frac{11m}{11} = \frac{33}{11} \quad \text{Divide by 11.}$$

$$1m = 3 \quad \text{Divide.}$$

$$m = 3 \quad \text{Multiplicative identity property}$$

Check this proposed solution. The solution set is $\{3\}$.

**Now Try Exercise 57.**

#### CONNECTIONS

The use of algebra to solve equations and applied problems is very old. The 3600-year-old Rhind Papyrus includes the following "word problem." "Aha, its whole, its seventh, it makes 19." This brief sentence describes the equation

$$x + \frac{x}{7} = 19.$$

The solution of this equation is $16\frac{5}{8}$. The word *algebra* is from the work *Hisab al-jabr m'al muquabalah,* written in the ninth century by Muhammed ibn Musa Al-Khowarizmi. The title means "the science of transposition and cancellation." From Latin versions of Khowarizmi's text, "al-jabr" became the broad term covering the art of equation solving.

## 2.2 EXERCISES

**For Extra Help**

- Student's Solutions Manual
- MyMathLab
- InterAct Math Tutorial Software
- AW Math Tutor Center
- MathXL
- Digital Video Tutor CD 3/Videotape 3

**1.** In your own words, state the multiplication property of equality. Give an example.

**2.** In the statement of the multiplication property of equality in this section, there is a restriction that $C \neq 0$. What would happen if you multiplied each side of an equation by 0?

**3.** Which equation does *not* require the use of the multiplication property of equality?

**A.** $3x - 5x = 6$ **B.** $-\frac{1}{4}x = 12$ **C.** $5x - 4x = 7$ **D.** $\frac{x}{3} = -2$

**4.** Tell whether you would use the addition or multiplication property of equality to solve each equation. Explain your answer.

**(a)** $3x = 12$ **(b)** $3 + x = 12$

5. A student tried to solve the equation $4x = 8$ by dividing each side by 8. Why is this wrong?

6. State how you would find the solution of a linear equation if your next-to-last step reads "$-x = 5$."

*By what number is it necessary to multiply both sides of each equation to get just x on the left side? Do not actually solve these equations.*

7. $\frac{2}{3}x = 8$
8. $\frac{4}{5}x = 6$
9. $\frac{x}{10} = 3$
10. $\frac{x}{100} = 8$
11. $-\frac{9}{2}x = -4$
12. $-\frac{8}{3}x = -11$
13. $-x = .36$
14. $-x = .29$

*By what number is it necessary to divide both sides of each equation to get just x on the left side? Do not actually solve these equations.*

15. $6x = 5$
16. $7x = 10$
17. $-4x = 13$
18. $-13x = 6$
19. $.12x = 48$
20. $.21x = 63$
21. $-x = 23$
22. $-x = 49$

*Solve each equation, and check your solution. See Examples 1–5.*

23. $5x = 30$
24. $7x = 56$
25. $2m = 15$
26. $3m = 10$
27. $3a = -15$
28. $5k = -70$
29. $-3x = 12$
30. $-4x = 36$
31. $10t = -36$
32. $4s = -34$
33. $-6x = -72$
34. $-8x = -64$
35. $2r = 0$
36. $5x = 0$
37. $.2t = 8$
38. $.9x = 18$
39. $-2.1m = 25.62$
40. $-3.9a = 31.2$
41. $\frac{1}{4}y = -12$
42. $\frac{1}{5}p = -3$
43. $\frac{z}{6} = 12$
44. $\frac{x}{5} = 15$
45. $\frac{x}{7} = -5$
46. $\frac{k}{8} = -3$
47. $\frac{2}{7}p = 4$
48. $\frac{3}{8}y = 9$
49. $-\frac{5}{6}t = -15$
50. $-\frac{3}{4}k = -21$
51. $-\frac{7}{9}c = \frac{3}{5}$
52. $-\frac{5}{6}d = \frac{4}{9}$
53. $-y = 12$
54. $-t = 14$
55. $-x = -\frac{3}{4}$
56. $-y = -\frac{1}{2}$

*Solve each equation, and check your solution. See Example 6.*

57. $4x + 3x = 21$
58. $9x + 2x = 121$
59. $3r - 5r = 10$
60. $9p - 13p = 24$
61. $5m + 6m - 2m = 63$
62. $11r - 5r + 6r = 168$
63. $-6x + 4x - 7x = 0$
64. $-5x + 4x - 8x = 0$
65. $9w - 5w + w = -3$
66. $10y - 6y + 3y = -4$

67. Write an equation that requires the use of the multiplication property of equality, where each side must be multiplied by $\frac{2}{3}$, and the solution is a negative number.

68. Write an equation that requires the use of the multiplication property of equality, where each side must be divided by 100, and the solution is not an integer.

*Write an equation using the information given in the problem. Use x as the variable. Then solve the equation.*

**69.** When a number is multiplied by 4, the result is 6. Find the number.

**70.** When a number is multiplied by $-4$, the result is 10. Find the number.

**71.** When a number is divided by $-5$, the result is 2. Find the number.

**72.** If twice a number is divided by 5, the result is 4. Find the number.

# 2.3 More on Solving Linear Equations

**OBJECTIVES**

1. Learn and use the four steps for solving a linear equation.
2. Solve equations with fractions or decimals as coefficients.
3. Solve equations with no solution or infinitely many solutions.
4. Write expressions for two related unknown quantities.

**OBJECTIVE 1 Learn and use the four steps for solving a linear equation.** In this section, we use the addition and multiplication properties together to solve more complicated equations. We use the following four-step method.

**Solving a Linear Equation**

*Step 1* **Simplify each side separately.** Clear parentheses using the distributive property, if needed, and combine like terms.

*Step 2* **Isolate the variable term on one side.** Use the addition property if necessary so that the variable term is on one side of the equation and a number is on the other.

*Step 3* **Isolate the variable.** Use the multiplication property if necessary to get the equation in the form $x =$ a number.

*Step 4* **Check.** Substitute the proposed solution into the *original* equation to see if a true statement results.

## EXAMPLE 1 Using the Four Steps to Solve an Equation

Solve $3r + 4 - 2r - 7 = 4r + 3$.

| | | |
|---|---|---|
| *Step 1* | $3r + 4 - 2r - 7 = 4r + 3$ | |
| | $r - 3 = 4r + 3$ | Combine like terms. |
| *Step 2* | $r - 3 + 3 = 4r + 3 + 3$ | Add 3. |
| | $r = 4r + 6$ | Combine like terms. |
| | $r - 4r = 4r + 6 - 4r$ | Subtract $4r$. |
| | $-3r = 6$ | Combine like terms. |
| *Step 3* | $\frac{-3r}{-3} = \frac{6}{-3}$ | Divide by $-3$. |
| | $r = -2$ | $\frac{-3}{-3} = 1$; $1r = r$ |

*Step 4* Substitute $-2$ for $r$ in the original equation to check.

$$3r + 4 - 2r - 7 = 4r + 3 \quad \text{Original equation}$$
$$3(-2) + 4 - 2(-2) - 7 \stackrel{?}{=} 4(-2) + 3 \quad \text{Let } r = -2.$$
$$-6 + 4 + 4 - 7 \stackrel{?}{=} -8 + 3 \quad \text{Multiply.}$$
$$-5 = -5 \quad \text{True}$$

The solution set of the equation is $\{-2\}$.

**Now Try Exercise 7.**

**NOTE** In Step 2 of Example 1, we added and subtracted the terms in such a way that the variable term ended up on the left side of the equation. Choosing differently would have put the variable term on the right side of the equation. Either way, the same solution results.

**EXAMPLE 2 Using the Four Steps to Solve an Equation**

Solve $4(k - 3) - k = k - 6$.

*Step 1* Before combining like terms, use the distributive property to simplify $4(k - 3)$.

$$4(k - 3) - k = k - 6$$
$$4(k) + 4(-3) - k = k - 6 \quad \text{Distributive property}$$
$$4k - 12 - k = k - 6$$
$$3k - 12 = k - 6 \quad \text{Combine like terms.}$$

*Step 2*
$$3k - 12 - k = k - 6 - k \quad \text{Subtract } k.$$
$$2k - 12 = -6 \quad \text{Combine like terms.}$$
$$2k - 12 + 12 = -6 + 12 \quad \text{Add 12.}$$
$$2k = 6 \quad \text{Combine like terms.}$$

*Step 3*
$$\frac{2k}{2} = \frac{6}{2} \quad \text{Divide by 2.}$$
$$k = 3$$

*Step 4* Check by substituting 3 for $k$ in the original equation. Work inside the parentheses first.

$$4(k - 3) - k = k - 6 \quad \text{Original equation}$$
$$4(3 - 3) - 3 \stackrel{?}{=} 3 - 6 \quad \text{Let } k = 3.$$
$$4(0) - 3 \stackrel{?}{=} 3 - 6$$
$$0 - 3 \stackrel{?}{=} 3 - 6$$
$$-3 = -3 \quad \text{True}$$

The solution set of the equation is $\{3\}$.

**Now Try Exercise 11.**

### EXAMPLE 3 Using the Four Steps to Solve an Equation

Solve $8a - (3 + 2a) = 3a + 1$.

*Step 1* Clear parentheses using the distributive property.

$$8a - (3 + 2a) = 3a + 1$$

$$8a - 1(3 + 2a) = 3a + 1 \quad \text{Multiplicative identity property}$$

$$8a - 3 - 2a = 3a + 1 \quad \text{Distributive property}$$

$$6a - 3 = 3a + 1$$

*Step 2*

$$6a - 3 - 3a = 3a + 1 - 3a \quad \text{Subtract } 3a.$$

$$3a - 3 = 1$$

$$3a - 3 + 3 = 1 + 3 \quad \text{Add 3.}$$

$$3a = 4$$

*Step 3*

$$\frac{3a}{3} = \frac{4}{3} \quad \text{Divide by 3.}$$

$$a = \frac{4}{3}$$

*Step 4* Check this solution in the original equation.

$$8a - (3 + 2a) = 3a + 1 \quad \text{Original equation}$$

$$8\left(\frac{4}{3}\right) - \left[3 + 2\left(\frac{4}{3}\right)\right] \stackrel{?}{=} 3\left(\frac{4}{3}\right) + 1 \quad \text{Let } a = \tfrac{4}{3}.$$

$$\frac{32}{3} - \left[3 + \frac{8}{3}\right] \stackrel{?}{=} 4 + 1$$

$$\frac{32}{3} - \left[\frac{9}{3} + \frac{8}{3}\right] \stackrel{?}{=} 5$$

$$\frac{32}{3} - \frac{17}{3} \stackrel{?}{=} 5$$

$$5 = 5 \quad \text{True}$$

The check shows that $\{\frac{4}{3}\}$ is the solution set.

**Now Try Exercise 9.**

**CAUTION** Be very careful with signs when solving an equation like the one in Example 3. When clearing parentheses in the expression

$$8 - (3 + 2a),$$

remember that the $-$ sign acts like a factor of $-1$ and affects the sign of *every* term within the parentheses. Thus,

$$8 - (3 + 2a) = 8 + (-1)(3 + 2a)$$

$$= 8 - 3 - 2a.$$

Change to $-$ in *both* terms.

### EXAMPLE 4 Using the Four Steps to Solve an Equation

Solve $4(8 - 3t) = 32 - 8(t + 2)$.

*Step 1* $4(8 - 3t) = 32 - 8(t + 2)$

$32 - 12t = 32 - 8t - 16$ Distributive property

$32 - 12t = 16 - 8t$

*Step 2* $32 - 12t + 12t = 16 - 8t + 12t$ Add $12t$.

$32 = 16 + 4t$

$32 - 16 = 16 + 4t - 16$ Subtract 16.

$16 = 4t$

*Step 3* $\frac{16}{4} = \frac{4t}{4}$ Divide by 4.

$4 = t$ or $t = 4$

*Step 4* *Check:*

$4(8 - 3t) = 32 - 8(t + 2)$ Original equation

$4(8 - 3 \cdot 4) \stackrel{?}{=} 32 - 8(4 + 2)$ Let $t = 4$.

$4(8 - 12) \stackrel{?}{=} 32 - 8(6)$

$4(-4) \stackrel{?}{=} 32 - 48$

$-16 = -16$ True

Since a true statement results, the solution set is $\{4\}$.

**Now Try Exercise 13.**

**OBJECTIVE 2 Solve equations with fractions or decimals as coefficients.** We can clear an equation of fractions by multiplying each side by the least common denominator (LCD) of all the fractions in the equation. It is a good idea to do this *before* starting the four-step method to avoid working with fractions.

### EXAMPLE 5 Solving an Equation with Fractions as Coefficients

Solve $\frac{2}{3}x - \frac{1}{2}x = -\frac{1}{6}x - 2$.

The LCD of all the fractions in the equation is 6, so multiply each side by 6.

$$\frac{2}{3}x - \frac{1}{2}x = -\frac{1}{6}x - 2$$

$$6\left(\frac{2}{3}x - \frac{1}{2}x\right) = 6\left(-\frac{1}{6}x - 2\right) \quad \text{Multiply by 6.}$$

$$6\left(\frac{2}{3}x\right) + 6\left(-\frac{1}{2}x\right) = 6\left(-\frac{1}{6}x\right) + 6(-2) \quad \text{Distributive property}$$

$$4x - 3x = -x - 12$$

Now use the four steps to solve this equivalent equation.

*Step 1* $x = -x - 12$ Combine like terms.

*Step 2* $x + x = -x - 12 + x$ Add $x$.

$2x = -12$ Combine like terms.

*Step 3* $\frac{2x}{2} = \frac{-12}{2}$ Divide by 2.

$x = -6$

*Step 4* *Check:*

$$\frac{2}{3}x - \frac{1}{2}x = -\frac{1}{6}x - 2 \quad \text{Original equation}$$

$$\frac{2}{3}(-6) - \frac{1}{2}(-6) = -\frac{1}{6}(-6) - 2 \quad ? \quad \text{Let } x = -6.$$

$$-4 + 3 = 1 - 2 \quad ?$$

$$-1 = 1 \quad \text{True}$$

The solution set of the equation is $\{-6\}$.

**Now Try Exercise 23.**

**CAUTION** When clearing an equation of fractions, be sure to multiply *every* term on each side of the equation by the LCD.

The multiplication property is also used to clear an equation of decimals.

## EXAMPLE 6 Solving an Equation with Decimals as Coefficients

Solve $.1t + .05(20 - t) = .09(20)$.

The decimals are expressed as tenths and hundredths. Choose the smallest exponent on 10 needed to eliminate the decimals; in this case, use $10^2 = 100$. A number can be multiplied by 100 by moving the decimal point two places to the right.

$$.10t + .05(20 - t) = .09(20) \quad .1 = .10$$

$$10t + 5(20 - t) = 9(20) \quad \text{Multiply by 100.}$$

Now use the four steps.

*Step 1* $10t + 5(20) + 5(-t) = 180$ Distributive property

$10t + 100 - 5t = 180$

$5t + 100 = 180$ Combine like terms.

*Step 2* $5t + 100 - 100 = 180 - 100$ Subtract 100.

$5t = 80$ Combine like terms.

*Step 3* $\frac{5t}{5} = \frac{80}{5}$ Divide by 5.

$t = 16$

*Step 4* Check that $\{16\}$ is the solution set by substituting 16 for $t$ in the original equation.

**Now Try Exercise 29.**

**OBJECTIVE 3 Solve equations with no solution or infinitely many solutions.** Each equation that we have solved so far has had exactly one solution. An equation with exactly one solution is a **conditional equation** because it is only true under certain conditions. As the next examples show, linear equations may also have no solution or infinitely many solutions. (The four steps are not identified in these examples. See if you can identify them.)

### EXAMPLE 7 Solving an Equation That Has Infinitely Many Solutions

Solve $5x - 15 = 5(x - 3)$.

$$5x - 15 = 5(x - 3)$$
$$5x - 15 = 5x - 15 \quad \text{Distributive property}$$
$$5x - 15 - 5x = 5x - 15 - 5x \quad \text{Subtract } 5x.$$
$$-15 = -15$$
$$-15 + 15 = -15 + 15 \quad \text{Add 15.}$$
$$0 = 0 \quad \text{True}$$

The variable has "disappeared." Since the last statement $(0 = 0)$ is true, *any* real number is a solution. (We could have predicted this from the line in the solution that says $5x - 15 = 5x - 15$, which is certainly true for *any* value of $x$.) An equation with both sides exactly the same, like $0 = 0$, is called an **identity.** An identity is true for all replacements of the variables. We indicate this by writing the solution set as {all real numbers}.

**Now Try Exercise 17.**

**CAUTION** When solving an equation like the one in Example 7, do not write {0} as the solution set. Although 0 is a solution, there are infinitely many other solutions.

### EXAMPLE 8 Solving an Equation That Has No Solution

Solve $2x + 3(x + 1) = 5x + 4$.

$$2x + 3(x + 1) = 5x + 4$$
$$2x + 3x + 3 = 5x + 4 \quad \text{Distributive property}$$
$$5x + 3 = 5x + 4$$
$$5x + 3 - 5x = 5x + 4 - 5x \quad \text{Subtract } 5x.$$
$$3 = 4 \quad \text{False}$$

Again, the variable has disappeared, but this time a false statement $(3 = 4)$ results. Whenever this happens in solving an equation, it is a signal that the equation has no solution. An equation with no solution is called a **contradiction.** Its solution set is the **empty set,** or **null set,** symbolized $\emptyset$.

**Now Try Exercise 21.**

The table on the next page summarizes the solution sets of the three types of linear equations.

| Type of Linear Equation | Final Equation in Solution | Number of Solutions | Solution Set |
|---|---|---|---|
| Conditional | $x =$ a number | One | {a number} |
| Identity | A true statement with no variable, such as $0 = 0$ | Infinite | {all real numbers} |
| Contradiction | A false statement with no variable, such as $3 = 4$ | None | $\emptyset$ |

**OBJECTIVE 4 Write expressions for two related unknown quantities.** Next, we continue our work with translating from words to symbols.

**PROBLEM SOLVING**

Often we are given a problem in which the sum of two quantities is a particular number, and we are asked to find the values of the two quantities. Example 9 shows how to express the unknown quantities in terms of a single variable.

### EXAMPLE 9 Translating a Phrase into an Algebraic Expression

Two numbers have a sum of 23. If one of the numbers is represented by $k$, find an expression for the other number.

First, suppose that the sum of two numbers is 23, and one of the numbers is 10. How would you find the other number? You would subtract 10 from 23 to get 13; $23 - 10 = 13$. So instead of using 10 as one of the numbers, use $k$ as stated in the problem. The other number would be obtained in the same way. You must subtract $k$ from 23. Therefore, an expression for the other number is $23 - k$.

**Now Try Exercise 47.**

**NOTE** The approach used in Example 9, first writing an expression with a trial number, is also useful when translating applied problems to equations.

## 2.3 EXERCISES

**For Extra Help**

- Student's Solutions Manual
- MyMathLab
- InterAct Math Tutorial Software
- AW Math Tutor Center
- MathXL
- Digital Video Tutor CD 3/Videotape 3

**1.** In your own words, give the four steps used to solve a linear equation. Give an example to demonstrate the steps.

**2.** After working correctly through several steps of the solution of a linear equation, a student obtains the equation $7x = 3x$. Then the student divides each side by $x$ to get $7 = 3$, and gives $\emptyset$ as the answer. Is this correct? If not, explain why.

**3.** Which one of the following linear equations does *not* have {all real numbers} as its solution set?

**A.** $5x = 4x + x$ **B.** $2(x + 6) = 2x + 12$ **C.** $\frac{1}{2}x = .5x$ **D.** $3x = 2x$

**4.** Based on the discussion in this section, if an equation has decimals or fractions as coefficients, what additional step will make the work easier?

*Solve each equation, and check your solution. See Examples 1–4, 7, and 8.*

**5.** $3x + 8 = 5x + 10$

**6.** $10p + 6 = 12p - 4$

**7.** $12h - 5 = 11h + 5 - h$

**8.** $-4x - 1 = -5x + 1 + 3x$

**9.** $-2p + 7 = 3 - (5p + 1)$

**10.** $4x + 9 = 3 - (x - 2)$

**11.** $3(4x + 2) + 5x = 30 - x$

**12.** $5(2m + 3) - 4m = 8m + 27$

**13.** $6(3w + 5) = 2(10w + 10)$

**14.** $4(2x - 1) = -6(x + 3)$

**15.** $6(4x - 1) = 12(2x + 3)$

**16.** $6(2x + 8) = 4(3x - 6)$

**17.** $3(2x - 4) = 6(x - 2)$

**18.** $3(6 - 4x) = 2(-6x + 9)$

**19.** $7r - 5r + 2 = 5r - r$

**20.** $9p - 4p + 6 = 7p - 3p$

**21.** $11x - 5(x + 2) = 6x + 5$

**22.** $6x - 4(x + 1) = 2x + 4$

*Solve each equation by clearing fractions or decimals. Check your solution. See Examples 5 and 6.*

**23.** $\frac{3}{5}t - \frac{1}{10}t = t - \frac{5}{2}$

**24.** $-\frac{2}{7}r + 2r = \frac{1}{2}r + \frac{17}{2}$

**25.** $-\frac{1}{4}(x - 12) + \frac{1}{2}(x + 2) = x + 4$

**26.** $\frac{1}{9}(y + 18) + \frac{1}{3}(2y + 3) = y + 3$

**27.** $\frac{2}{3}k - \left(k + \frac{1}{4}\right) = \frac{1}{12}(k + 4)$

**28.** $-\frac{5}{6}q - \left(q - \frac{1}{2}\right) = \frac{1}{4}(q + 1)$

**29.** $.20(60) + .05x = .10(60 + x)$

**30.** $.30(30) + .15x = .20(30 + x)$

**31.** $1.00x + .05(12 - x) = .10(63)$

**32.** $.92x + .98(12 - x) = .96(12)$

**33.** $.06(10{,}000) + .08x = .072(10{,}000 + x)$

**34.** $.02(5000) + .03x = .025(5000 + x)$

*Solve each equation, and check your solution. See Examples 1–8.*

**35.** $10(2x - 1) = 8(2x + 1) + 14$

**36.** $9(3k - 5) = 12(3k - 1) - 51$

**37.** $-(4y + 2) - (-3y - 5) = 3$

**38.** $-(6k - 5) - (-5k + 8) = -3$

**39.** $\frac{1}{2}(x + 2) + \frac{3}{4}(x + 4) = x + 5$

**40.** $\frac{1}{3}(x + 3) + \frac{1}{6}(x - 6) = x + 3$

**41.** $.10(x + 80) + .20x = 14$

**42.** $.30(x + 15) + .40(x + 25) = 25$

**43.** $4(x + 8) = 2(2x + 6) + 20$

**44.** $4(x + 3) = 2(2x + 8) - 4$

**45.** $9(v + 1) - 3v = 2(3v + 1) - 8$

**46.** $8(t - 3) + 4t = 6(2t + 1) - 10$

*Write the answer to each problem as an algebraic expression. See Example 9.*

**47.** Two numbers have a sum of 11. One of the numbers is $q$. Find the other number.

**48.** The product of two numbers is 9. One of the numbers is $k$. What is the other number?

**49.** A football player gained $x$ yd rushing. On the next down he gained 7 yd. How many yards did he gain altogether?

**50.** A baseball player got 65 hits one season. He got $h$ of the hits in one game. How many hits did he get in the rest of the games?

**51.** Mary is $a$ yr old. How old will she be in 12 yr? How old was she 5 yr ago?

**52.** Tom has $r$ quarters. Find the value of the quarters in cents.

**53.** A bank teller has $t$ dollars, all in \$5 bills. How many \$5 bills does the teller have?

**54.** A plane ticket costs $b$ dollars for an adult and $d$ dollars for a child. Find the total cost for 3 adults and 2 children.

## SUMMARY EXERCISES ON SOLVING LINEAR EQUATIONS

This section of miscellaneous linear equations provides practice in solving all the types introduced in Sections 2.1–2.3. Refer to the examples in these sections to review the various solution methods.

*Solve each equation, and check your solution.*

**1.** $a + 2 = -3$

**2.** $2m + 8 = 16$

**3.** $12.5k = -63.75$

**4.** $-x = -12$

**5.** $\frac{4}{5}x = -20$

**6.** $7m - 5m = -12$

**7.** $5x - 9 = 4(x - 3)$

**8.** $\frac{a}{-2} = 8$

**9.** $-3(m - 4) + 2(5 + 2m) = 29$

**10.** $\frac{2}{3}y + 8 = \frac{1}{4}y$

**11.** $.08x + .06(x + 9) = 1.24$

**12.** $x - 16.2 = 7.5$

**13.** $4x + 2(3 - 2x) = 6$

**14.** $-.3x + 2.1(x - 4) = -6.6$

**15.** $-x = 6$

**16.** $3(m + 5) - 1 + 2m = 5(m + 2)$

**17.** $7m - (2m - 9) = 39$

**18.** $7(p - 2) + p = 2(p + 2)$

**19.** $-2t + 5t - 9 = 3(t - 4) - 5$

**20.** $-6z = -14$

**21.** $.02(50) + .08r = .04(50 + r)$

**22.** $2.3x + 13.7 = 1.3x + 2.9$

**23.** $2(3 + 7x) - (1 + 15x) = 2$

**24.** $6q - 9 = 12 + 3q$

**25.** $2(4 + 3r) = 3(r + 1) + 11$

**26.** $r + 9 + 7r = 4(3 + 2r) - 3$

**27.** $\frac{1}{4}x - 4 = \frac{3}{2}x + \frac{3}{4}x$

**28.** $.06(100 - x) + .04x = .05(92)$

**29.** $\frac{3}{4}(a - 2) - \frac{1}{3}(5 - 2a) = -2$

**30.** $2 - (m + 4) = 3m + 8$

# 2.4 An Introduction to Applications of Linear Equations

**OBJECTIVES**

1. Learn the six steps for solving applied problems.
2. Solve problems involving unknown numbers.

**OBJECTIVE 1 Learn the six steps for solving applied problems.** We now look at how algebra is used to solve applied problems. Since many meaningful applications of mathematics require concepts that are beyond the level of this book, some of the problems you encounter will seem contrived, and to some extent they are. But the skills you develop in solving simple problems will help you solve more realistic problems in chemistry, physics, biology, business, and other fields. While there is not one specific method that enables you to solve all kinds of applied problems, the following six-step method is suggested.

3 Solve problems involving sums of quantities.

4 Solve problems involving supplementary and complementary angles.

5 Solve problems involving consecutive integers.

**Solving an Applied Problem**

*Step 1* **Read** the problem carefully until you understand what is given and what is to be found.

*Step 2* **Assign a variable** to represent the unknown value, using diagrams or tables as needed. Write down what the variable represents. If necessary, express any other unknown values in terms of the variable.

*Step 3* **Write an equation** using the variable expression(s).

*Step 4* **Solve** the equation.

*Step 5* **State the answer.** Does it seem reasonable?

*Step 6* **Check** the answer in the words of the *original* problem.

The third step in solving an applied problem is often the hardest. To translate the problem into an equation, write the given phrases as mathematical expressions. Replace any words that mean *equal* or *same* with an = sign. Other forms of the verb "to be," such as *is, are, was,* and *were,* also translate this way. The = sign leads to an equation to be solved.

**OBJECTIVE 2 Solve problems involving unknown numbers.** Some of the simplest applied problems involve unknown numbers.

### EXAMPLE 1 Finding the Value of an Unknown Number

The product of 4, and a number decreased by 7, is 100. What is the number?

*Step 1* **Read** the problem carefully. Decide what you are being asked to find.

*Step 2* **Assign a variable** to represent the unknown quantity. In this problem, we are asked to find a number, so we write

$$\text{Let } x = \text{the number.}$$

There are no other unknown quantities to find.

*Step 3* **Write an equation.**

| The product of 4, | and | a number | decreased by | 7, | is | 100. |
|---|---|---|---|---|---|---|
| ↓ | | ↓ | ↓ | ↓ | ↓ | ↓ |
| $4\cdot$ | | $(x$ | $-$ | $7)$ | $=$ | $100$ |

Because of the commas in the given problem, writing the equation as $4x - 7 = 100$ is incorrect. The equation $4x - 7 = 100$ corresponds to the statement "The product of 4 and a number, decreased by 7, is 100."

*Step 4* **Solve** the equation.

$$4(x - 7) = 100$$

$$4x - 28 = 100 \qquad \text{Distributive property}$$

$$4x - 28 + 28 = 100 + 28 \qquad \text{Add 28.}$$

$$4x = 128 \qquad \text{Combine like terms.}$$

$$\frac{4x}{4} = \frac{128}{4} \qquad \text{Divide by 4.}$$

$$x = 32$$

*Step 5* **State the answer.** The number is 32.

*Step 6* **Check.** When 32 is decreased by 7, we get $32 - 7 = 25$. If 4 is multiplied by 25, we get 100, as required. The answer, 32, is correct.

**Now Try Exercise 5.**

**OBJECTIVE 3 Solve problems involving sums of quantities.** A common type of problem in elementary algebra involves finding two quantities when the sum of the quantities is known. In Example 9 of the previous section, we prepared for this type of problem by writing mathematical expressions for two related unknown quantities.

**PROBLEM SOLVING**

In general, to solve problems involving sums of quantities, choose a variable to represent one of the unknowns. Then represent the other quantity in terms of the *same* variable, using information from the problem. Write an equation based on the words of the problem.

### EXAMPLE 2 Finding Numbers of Olympic Medals

In the 2002 Winter Olympics in Salt Lake City, the United States won 10 more medals than Norway. The two countries won a total of 58 medals. How many medals did each country win? (*Source:* U.S. Olympic Committee.)

*Step 1* **Read** the problem. We are given information about the total number of medals and asked to find the number each country won.

*Step 2* **Assign a variable.**

Let $x$ = the number of medals Norway won.

Then $x + 10$ = the number of medals the U.S. won.

*Step 3* **Write an equation.**

| The total | is | the number of medals Norway won | plus | the number of medals the U.S. won. |
|---|---|---|---|---|
| ↓ | ↓ | ↓ | ↓ | ↓ |
| $58$ | $=$ | $x$ | $+$ | $(x + 10)$ |

*Step 4* **Solve** the equation.

$$58 = 2x + 10 \quad \text{Combine like terms.}$$

$$58 - 10 = 2x + 10 - 10 \quad \text{Subtract 10.}$$

$$48 = 2x \quad \text{Combine like terms.}$$

$$\frac{48}{2} = \frac{2x}{2} \quad \text{Divide by 2.}$$

$$24 = x \quad \text{or} \quad x = 24$$

*Step 5* **State the answer.** The variable $x$ represents the number of medals Norway won, so Norway won 24 medals. Then the number of medals the United States won is $x + 10 = 24 + 10 = 34$.

*Step 6* **Check.** Since the United States won 34 medals and Norway won 24, the total number of medals was $34 + 24 = 58$. Because $34 - 24 = 10$, the United States won 10 more medals than Norway. This information agrees with what is given in the problem, so the answer checks.

**Now Try Exercise 11.**

**NOTE** The problem in Example 2 could also be solved by letting $x$ represent the number of medals the United States won. Then $x - 10$ would represent the number of medals Norway won. The equation would be

$$58 = x + (x - 10).$$

The solution of this equation is 34, which is the number of U.S. medals. The number of Norwegian medals would be $34 - 10 = 24$. The answers are the same, whichever approach is used.

### EXAMPLE 3 Finding the Number of Orders for Tea

The owner of P. J.'s Coffeehouse found that on one day the number of orders for tea was $\frac{1}{3}$ the number of orders for coffee. If the total number of orders for the two drinks was 76, how many orders were placed for tea?

*Step 1* **Read** the problem. It asks for the number of orders for tea.

*Step 2* **Assign a variable.** Because of the way the problem is stated, let the variable represent the number of orders for coffee.

Let $x =$ the number of orders for coffee.

Then $\frac{1}{3}x =$ the number of orders for tea.

*Step 3* **Write an equation.** Use the fact that the total number of orders was 76.

| The total | is | orders for coffee | plus | orders for tea. |
|---|---|---|---|---|
| ↓ | ↓ | ↓ | ↓ | ↓ |
| $76$ | $=$ | $x$ | $+$ | $\frac{1}{3}x$ |

*Step 4* **Solve** the equation.

$$76 = \frac{4}{3}x \qquad x = 1x = \tfrac{3}{3}x;\ \text{Combine like terms.}$$

$$\frac{3}{4}(76) = \frac{3}{4}\left(\frac{4}{3}x\right) \qquad \text{Multiply by } \tfrac{3}{4}.$$

$$57 = x$$

*Step 5* **State the answer.** In this problem, *$x$ does not represent the quantity that we are asked to find.* The number of orders for tea was $\frac{1}{3}x$. So $\frac{1}{3}(57) = 19$ is the number of orders for tea.

*Step 6* **Check.** The number of coffee orders ($x$) was 57 and the number of tea orders was 19; 19 is one-third of 57, and $19 + 57 = 76$. Since this agrees with the information given in the problem, the answer is correct.

**Now Try Exercise 23.**

**PROBLEM SOLVING**

In Example 3, it was easier to let the variable represent the quantity that was *not* asked for. This required extra work in Step 5 to find the number of orders for tea. In some cases, this approach is easier than letting the variable represent the quantity that we are asked to find. Experience in solving problems will indicate when this approach is useful, and experience comes only from solving many problems.

## EXAMPLE 4 Analyzing a Gasoline/Oil Mixture

A lawn trimmer uses a mixture of gasoline and oil. The mixture contains 16 oz of gasoline for each ounce of oil. If the tank holds 68 oz of the mixture, how many ounces of oil and how many ounces of gasoline does it require when it is full?

*Step 1* **Read** the problem. We must find how many ounces of oil and gasoline are needed to fill the tank.

*Step 2* **Assign a variable.**

Let $x =$ the number of ounces of oil required.

Then $16x =$ the number of ounces of gasoline required.

*Step 3* **Write an equation.**

| Amount of gasoline | plus | amount of oil | is | total amount in tank. |
|---|---|---|---|---|
| ↓ | ↓ | ↓ | ↓ | ↓ |
| $16x$ | $+$ | $x$ | $=$ | $68$ |

*Step 4* **Solve.**

$$17x = 68 \quad \text{Combine like terms.}$$

$$\frac{17x}{17} = \frac{68}{17} \quad \text{Divide by 17.}$$

$$x = 4$$

*Step 5* **State the answer.** The lawn trimmer requires 4 oz of oil and $16(4) = 64$ oz of gasoline when full.

*Step 6* **Check.** Since $4 + 64 = 68$, and 64 is 16 times 4, the answer checks.

**Now Try Exercise 25.**

**PROBLEM SOLVING**

Sometimes it is necessary to find three unknown quantities in an applied problem. Frequently the three unknowns are compared in *pairs.* When this happens, it is usually easiest to let the variable represent the unknown found in both pairs.

## EXAMPLE 5 Dividing a Board into Pieces

The instructions for a woodworking project call for three pieces of wood. The longest piece must be twice the length of the middle-sized piece, and the shortest piece must be 10 in. shorter than the middle-sized piece. Maria Gonzales has a board 70 in. long that she wishes to use. How long can each piece be?

*Step 1* **Read** the problem. There will be three answers.

*Step 2* **Assign a variable.** Since the middle-sized piece appears in both pairs of comparisons, let $x$ represent the length, in inches, of the middle-sized piece. We have

$$x = \text{the length of the middle-sized piece,}$$
$$2x = \text{the length of the longest piece, and}$$
$$x - 10 = \text{the length of the shortest piece.}$$

A sketch is helpful here. See Figure 2.

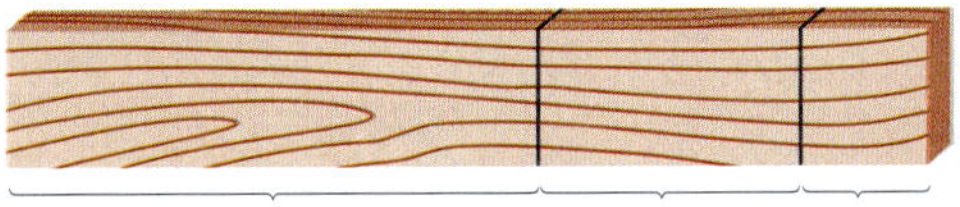


FIGURE 2

*Step 3* **Write an equation.**

| Longest | plus | middle-sized | plus | shortest | is | total length. |
|---|---|---|---|---|---|---|
| ↓ | ↓ | ↓ | ↓ | ↓ | ↓ | ↓ |
| $2x$ | $+$ | $x$ | $+$ | $(x - 10)$ | $=$ | $70$ |

*Step 4* **Solve.**

$$4x - 10 = 70 \quad \text{Combine like terms.}$$
$$4x - 10 + 10 = 70 + 10 \quad \text{Add 10.}$$
$$4x = 80 \quad \text{Combine like terms.}$$
$$\frac{4x}{4} = \frac{80}{4} \quad \text{Divide by 4.}$$
$$x = 20$$

*Step 5* **State the answer.** The middle-sized piece is 20 in. long, the longest piece is $2(20) = 40$ in. long, and the shortest piece is $20 - 10 = 10$ in. long.

*Step 6* **Check.** The sum of the lengths is 70 in. All conditions of the problem are satisfied.

**Now Try Exercise 29.**

**OBJECTIVE 4 Solve problems involving supplementary and complementary angles.** An angle can be measured by a unit called the **degree** (°), which is $\frac{1}{360}$ of a complete rotation. Two angles whose sum is 90° are said to be **complementary,** or complements of each other. An angle that measures 90° is a **right angle.** Two angles whose sum is 180° are said to be **supplementary,** or supplements of each other. One angle *supplements* the other to form a **straight angle** of 180°. See Figure 3. If $x$ represents the degree measure of an angle, then

$90 - x$ represents the degree measure of its complement,

and $180 - x$ represents the degree measure of its supplement.

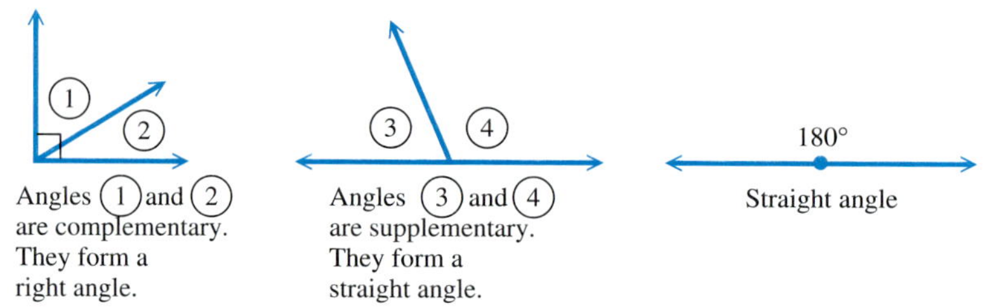


**FIGURE 3**

### EXAMPLE 6 Finding the Measure of an Angle

Find the measure of an angle whose supplement is 10° more than twice its complement.

*Step 1* **Read** the problem. We are to find the measure of an angle, given information about its complement and its supplement.

*Step 2* **Assign a variable.**

Let $x =$ the degree measure of the angle.

Then $90 - x =$ the degree measure of its complement,

and $180 - x =$ the degree measure of its supplement.

*Step 3* **Write an equation.**

| Supplement | is | 10 | more than | twice | its complement. |
|---|---|---|---|---|---|
| ↓ | ↓ | ↓ | ↓ | ↓ | ↓ |
| $180 - x$ | $=$ | $10$ | $+$ | $2 \cdot$ | $(90 - x)$ |

*Step 4* **Solve.**

$$180 - x = 10 + 180 - 2x \quad \text{Distributive property}$$
$$180 - x = 190 - 2x \quad \text{Combine like terms.}$$
$$180 - x + 2x = 190 - 2x + 2x \quad \text{Add } 2x.$$
$$180 + x = 190 \quad \text{Combine like terms.}$$
$$180 + x - 180 = 190 - 180 \quad \text{Subtract 180.}$$
$$x = 10$$

*Step 5* **State the answer.** The measure of the angle is 10°.

*Step 6* **Check.** The complement of 10° is 80° and the supplement of 10° is 170°. 170° is equal to 10° more than twice 80° ($170 = 10 + 2(80)$ is true); therefore, the answer is correct.

**Now Try Exercise 43.**

**OBJECTIVE 5 Solve problems involving consecutive integers.** Two integers that differ by 1 are called **consecutive integers.** For example, 3 and 4, 6 and 7, and −2 and −1 are pairs of consecutive integers. In general, if $x$ represents an integer, $x + 1$ represents the next larger consecutive integer.

Consecutive *even* integers, such as 8 and 10, differ by 2. Similarly, consecutive *odd* integers, such as 9 and 11, also differ by 2. In general, if $x$ represents an even integer, $x + 2$ represents the next larger consecutive even integer. The same holds true for odd integers; that is, if $x$ is an odd integer, $x + 2$ is the next larger odd integer.

**PROBLEM SOLVING**

When solving consecutive integer problems, if $x =$ the first integer, then for any

two consecutive integers, use $\quad$ $x,\quad x + 1;$

two consecutive *even* integers, use $\quad$ $x,\quad x + 2;$

two consecutive *odd* integers, use $\quad$ $x,\quad x + 2.$

**EXAMPLE 7 Finding Consecutive Integers**

Two pages that face each other in this book have 569 as the sum of their page numbers. What are the page numbers?

*Step 1* **Read** the problem. Because the two pages face each other, they must have page numbers that are consecutive integers.

*Step 2* **Assign a variable.**

Let $\quad x =$ the smaller page number.

Then $\quad x + 1 =$ the larger page number.

*Step 3* **Write an equation.** Because the sum of the page numbers is 569, the equation is

$$x + (x + 1) = 569.$$

*Step 4* **Solve.**

$$2x + 1 = 569 \quad \text{Combine like terms.}$$
$$2x = 568 \quad \text{Subtract 1.}$$
$$x = 284 \quad \text{Divide by 2.}$$

*Step 5* **State the answer.** The smaller page number is 284, and the larger page number is $284 + 1 = 285$.

*Step 6* **Check.** The sum of 284 and 285 is 569. The answer is correct.

**Now Try Exercise 47.**

In the final example, we do not number the steps. See if you can identify them.

**EXAMPLE 8 Finding Consecutive Odd Integers**

If the smaller of two consecutive odd integers is doubled, the result is 7 more than the larger of the two integers. Find the two integers.

Let $x$ be the smaller integer. Since the two numbers are consecutive *odd* integers, then $x + 2$ is the larger. Now we write an equation.

| If the smaller is doubled, | the result is | 7 | more than | the larger. |
|---|---|---|---|---|
| ↓ | ↓ | ↓ | ↓ | ↓ |
| $2x$ | $=$ | $7$ | $+$ | $(x + 2)$ |

$$2x = 9 + x \quad \text{Combine like terms.}$$

$$x = 9 \quad \text{Subtract } x.$$

The first integer is 9 and the second is $9 + 2 = 11$. To check, we see that when 9 is doubled, we get 18, which is 7 more than the larger odd integer, 11. The answers are correct.

**Now Try Exercise 49.**

**CONNECTIONS**

George Polya (1888–1985), a native of Budapest, Hungary, wrote the modern classic *How to Solve It.* In this book he proposed a four-step process for problem solving:

1. Understand the problem.
2. Devise a plan.
3. Carry out the plan.
4. Look back and check.

**For Discussion or Writing**

Compare Polya's four-step process with the six steps given in this section. Identify which of the steps in our list match Polya's four steps. Trial and error is also a useful problem-solving tool. Where does this method fit into Polya's steps?

## 2.4 EXERCISES

**For Extra Help**

- Student's Solutions Manual
- MyMathLab
- InterAct Math Tutorial Software
- AW Math Tutor Center
- MathXL
- Digital Video Tutor CD 3/Videotape 3

**1.** In your own words, write the general procedure for solving applications as outlined in this section.

**2.** List some of the words that translate as "=" when writing an equation to solve an applied problem.

**3.** Suppose that a problem requires you to find the number of cars on a dealer's lot. Which one of the following would *not* be a reasonable answer? Justify your answer.

**A.** 0 **B.** 45 **C.** 1 **D.** $6\frac{1}{2}$

**4.** Suppose that a problem requires you to find the number of hours a light bulb is on during a day. Which one of the following would *not* be a reasonable answer? Justify your answer.

**A.** 0 **B.** 4.5 **C.** 13 **D.** 25

*Solve each problem. See Example 1.*

**5.** If 2 is added to five times a number, the result is equal to 5 more than four times the number. Find the number.

**6.** If four times a number is added to 8, the result is three times the number added to 5. Find the number.

**7.** If 2 is subtracted from a number and this difference is tripled, the result is 6 more than the number. Find the number.

**8.** If 3 is added to a number and this sum is doubled, the result is 2 more than the number. Find the number.

**9.** The sum of three times a number and 7 more than the number is the same as the difference between $-11$ and twice the number. What is the number?

**10.** If 4 is added to twice a number and this sum is multiplied by 2, the result is the same as if the number is multiplied by 3 and 4 is added to the product. What is the number?

*Solve each problem. See Example 2.*

**11.** The number of drive-in movie screens has declined steadily in the United States since the 1960s. California and New York were two of the states with the most remaining drive-in movie screens in 2001. California had 11 more screens than New York, and there were 107 screens total in the two states. How many drive-in movie screens remained in each state? (*Source:* National Association of Theatre Owners.)

**12.** Thursday is the most-watched night for the major broadcast TV networks (ABC, CBS, NBC, and Fox), with 20 million more viewers than Saturday, the least-watched night. The total for the two nights is 102 million viewers. How many viewers of the major networks are there on each of these nights? (*Source:* Nielsen Media Research.)

**13.** The U.S. Senate has 100 members. During the 106th session (1999–2001), there were 10 more Republicans than Democrats. How many Democrats and Republicans were there in the Senate? (*Source: World Almanac and Book of Facts,* 2000.)

**14.** The total number of Democrats and Republicans in the U.S. House of Representatives during the 106th session was 434. There were 12 more Republicans than Democrats. How many members of each party were there? (*Source: World Almanac and Book of Facts,* 2000.)

**15.** The rock band U2 generated top revenue on the concert circuit in 2001. U2 and second-place 'N Sync together took in $196.5 million from ticket sales. If 'N Sync took in $22.9 million less than U2, how much revenue did each generate? (*Source:* Pollstar.)

**16.** The Toyota Camry was the top-selling passenger car in the United States in 2000, followed by the Honda Accord. Honda Accord sales were 18 thousand less than Toyota Camry sales, and 828 thousand of these two cars were sold. How many of each make of car were sold? (*Source:* Ward's Communications.)

**17.** In the 2001–2002 NBA regular season, the Sacramento Kings won two less than three times as many games as they lost. The Kings played 82 games. How many wins and losses did the team have? (*Source:* nba.com)

**18.** In the 2000–2001 regular baseball season, the Oakland Athletics won 18 less than twice as many games as they lost. They played 162 regular season games. How many wins and losses did the team have? (*Source: World Almanac and Book of Facts,* 2002.)

*Solve each problem. See Examples 3 and 4.*

**19.** A September 2000 issue of *Coin World* listed the value of a "Mint State-65" (uncirculated) 1950 Jefferson nickel minted at Denver at $\frac{7}{6}$ the value of a similar condition 1945 nickel minted at Philadelphia. Together the total value of the two coins is $26.00. What is the value of each coin?

**20.** The largest sheep ranch in the world is located in Australia. The number of sheep on the ranch is $\frac{8}{3}$ the number of uninvited kangaroos grazing on the pastureland. Together, herds of these two animals number 88,000. How many sheep and how many kangaroos roam the ranch? (*Source: Guinness Book of Records.*)

**21.** In 1988, a dairy in Alberta, Canada, created a sundae with approximately 1 lb of topping for every 83.2 lb of ice cream. The total of the two ingredients weighed approximately 45,225 lb. To the nearest tenth of a pound, how many pounds of ice cream and how many pounds of topping were there? (*Source: Guinness Book of Records.*)

**22.** A husky running the Iditarod (a thousand-mile race between Anchorage and Nome, Alaska) burns $5\frac{3}{8}$ calories in exertion for every 1 calorie burned in thermoregulation in extreme cold. According to one scientific study, a husky in top condition burns an amazing total of 11,200 calories per day. How many calories are burned for exertion, and how many are burned for regulation of body temperature? Round answers to the nearest whole number.

**23.** A pharmacist found that at the end of the day she had $\frac{4}{3}$ as many prescriptions for antibiotics as she did for tranquilizers. She had 42 prescriptions altogether for these two types of drugs. How many did she have for tranquilizers?

**24.** In a mixture of concrete, there are 3 lb of cement mix for every 1 lb of gravel. If the mixture contains a total of 140 lb of these two ingredients, how many pounds of gravel are there?

**25.** A mixture of nuts contains only peanuts and cashews. For every 1 oz of cashews there are 5 oz of peanuts. If the mixture contains a total of 27 oz, how many ounces of each type of nut does the mixture contain?

**26.** An insecticide contains 95 cg of inert ingredient for every 1 cg of active ingredient. If a quantity of the insecticide weighs 336 cg, how much of each type of ingredient does it contain?

*Solve each problem. See Example 5.*

**27.** In one day, Akilah Cadet received 13 packages. Federal Express delivered three times as many as Airborne Express, while United Parcel Service delivered 2 fewer than Airborne Express. How many packages did each service deliver to Akilah?

**28.** In his job at the post office, Eddie Thibodeaux works a 6.5-hr day. He sorts mail, sells stamps, and does supervisory work. One day he sold stamps twice as long as he sorted mail, and he supervised .5 hr longer than he sorted mail. How many hours did he spend at each task?

**29.** The United States earned an all-time high of 34 medals at the 2002 Winter Olympics. The number of gold medals earned was 1 less than the number of bronze medals. The number of silver medals earned was 2 more than the number of bronze medals. How many of each kind of medal did the United States earn? (*Source:* U.S. Olympic Committee.)

**30.** Nagaraj Nanjappa has a party-length submarine sandwich 59 in. long. He wants to cut it into three pieces so that the middle piece is 5 in. longer than the shortest piece and the shortest piece is 9 in. shorter than the longest piece. How long should the three pieces be?

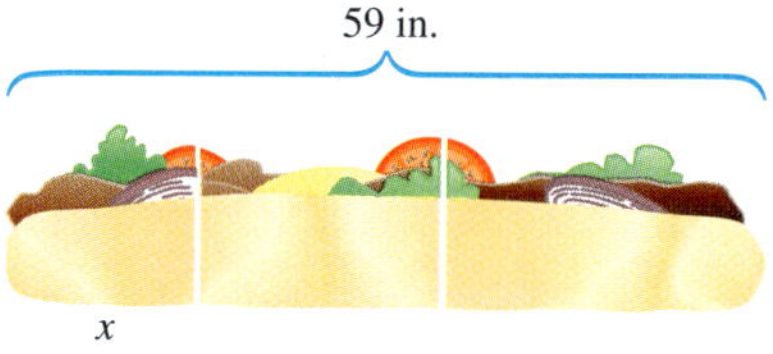


**31.** Venus is 31.2 million mi farther from the sun than Mercury, while Earth is 57 million mi farther from the sun than Mercury. If the total of the distances from these three planets to the sun is 196.2 million mi, how far away from the sun is Mercury? (All distances given here are *mean* (*average*) distances.) (*Source: Universal Almanac,* 1997.)

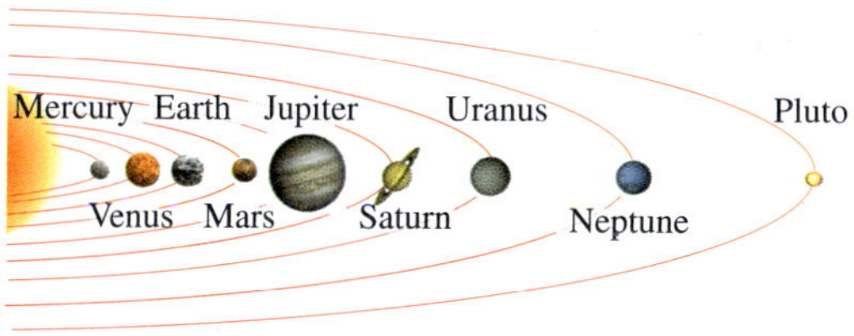

**32.** Together, Saturn, Jupiter, and Mars have a total of 36 known satellites (moons). Jupiter has 2 fewer satellites than Saturn, and Mars has 16 fewer satellites than Saturn. How many known satellites does Mars have? (*Source: World Almanac and Book of Facts,* 2000.)

**33.** The sum of the measures of the angles of any triangle is 180°. In triangle *ABC*, angles *A* and *B* have the same measure, while the measure of angle *C* is 60° larger than each of *A* and *B*. What are the measures of the three angles?

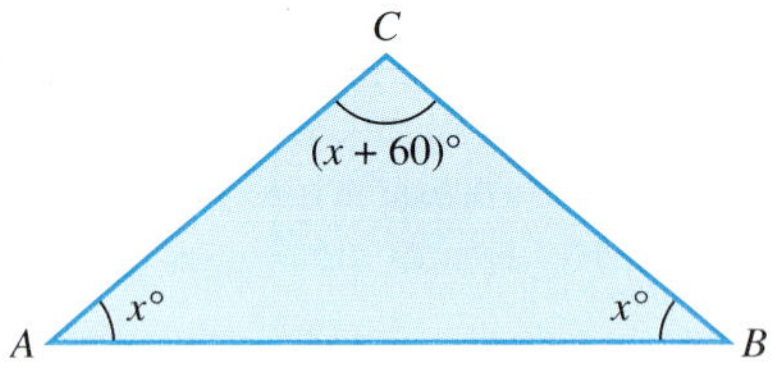


**34.** In triangle *ABC*, the measure of angle *A* is 141° more than the measure of angle *B*. The measure of angle *B* is the same as the measure of angle *C*. Find the measure of each angle. (*Hint:* See Exercise 33.)

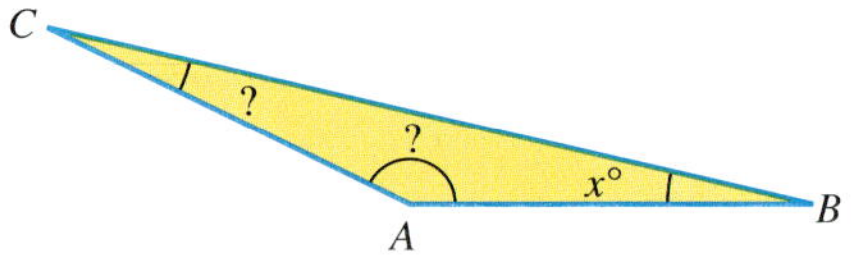


*Use the concepts of this section to answer each question.*

**35.** If the sum of two numbers is $k$, and one of the numbers is $m$, how can you express the other number?

**36.** If the product of two numbers is $r$, and one of the numbers is $s$ ($s \neq 0$), how can you express the other number?

**37.** Is there an angle whose supplement is equal to its complement? If so, what is the measure of the angle?

**38.** Is there an angle that is equal to its supplement? Is there an angle that is equal to its complement? If the answer is yes to either question, give the measure of the angle.

**39.** If $x$ represents an integer, how can you express the next smaller consecutive integer in terms of $x$?

**40.** If $x$ represents an integer, how can you express the next smaller even integer in terms of $x$?

*Solve each problem. See Example 6.*

**41.** Find the measure an an angle whose complement is four times its measure.

**42.** Find the measure of an angle whose supplement is three times its measure.

**43.** Find the measure of an angle whose supplement measures 39° more than twice its complement.

**44.** Find the measure of an angle whose supplement measures 38° less than three times its complement.

**45.** Find the measure of an angle such that the difference between the measures of its supplement and three times its complement is 10°.

**46.** Find the measure of an angle such that the sum of the measures of its complement and its supplement is 160°.

*Solve each problem. See Examples 7 and 8.*

**47.** The numbers on two consecutively numbered gym lockers have a sum of 137. What are the locker numbers?

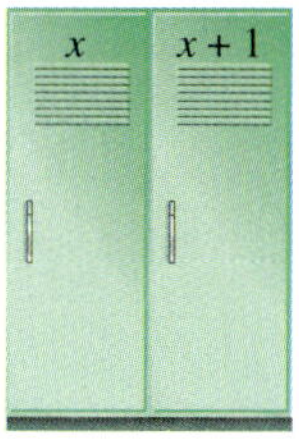


**48.** The sum of two consecutive checkbook check numbers is 357. Find the numbers.

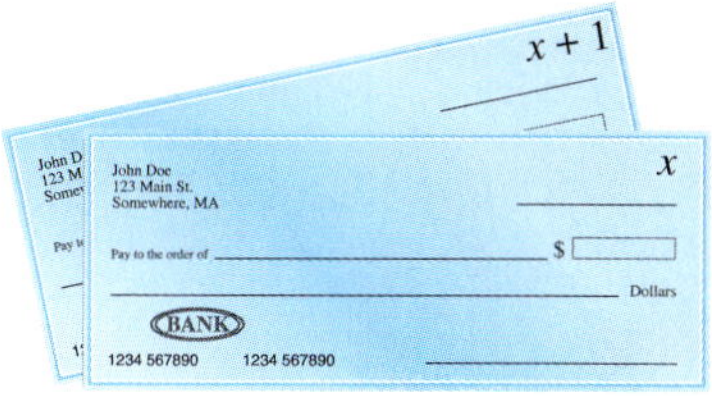


**49.** Find two consecutive even integers such that the smaller added to three times the larger gives a sum of 46.

**50.** Find two consecutive odd integers such that twice the larger is 17 more than the smaller.

**51.** Two pages that are back-to-back in this book have 203 as the sum of their page numbers. What are the page numbers?

**52.** Two houses on the same side of the street have house numbers that are consecutive even integers. The sum of the integers is 58. What are the two house numbers?

**53.** When the smaller of two consecutive integers is added to three times the larger, the result is 43. Find the integers.

**54.** If five times the smaller of two consecutive integers is added to three times the larger, the result is 59. Find the integers.

**55.** If the sum of three consecutive even integers is 60, what is the smallest even integer? (*Hint:* If $x$ and $x + 2$ represent the first two consecutive even integers, how would you represent the third consecutive even integer?)

**56.** If the sum of three consecutive odd integers is 69, what is the largest odd integer?

**57.** If 6 is subtracted from the largest of three consecutive odd integers, with this result multiplied by 2, the answer is 23 less than the sum of the first and twice the second of the integers. Find the integers.

**58.** If the first and third of three consecutive even integers are added, the result is 22 less than three times the second integer. Find the integers.

*Apply the ideas of this section to solve Exercises 59 and 60, which are based on the graphs.*

**59.** In a recent year, the funding for Head Start programs increased by .55 billion dollars from the funding in the previous year. The following year, the increase was .20 billion dollars more. For this 3-year period, the total funding was 9.64 billion dollars. How much was funded in each of these years? (*Source:* U.S. Department of Health and Human Services.)

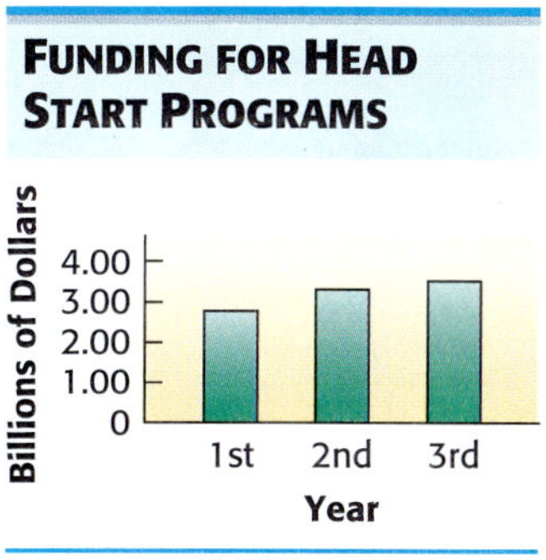

**60.** According to data provided by the National Safety Council for a recent year, the number of serious injuries per 100,000 participants in football, bicycling, and golf is illustrated in the graph. There were 800 more in bicycling than in golf, and there were 1267 more in football than in bicycling. Altogether there were 3179 serious injuries per 100,000 participants. How many such serious injuries were there in each sport?

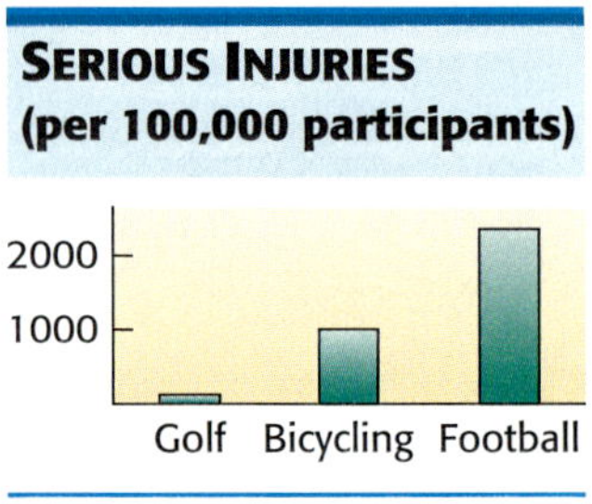


# 2.5 Formulas and Applications from Geometry

**OBJECTIVES**

1. Solve a formula for one variable, given the values of the other variables.
2. Use a formula to solve an applied problem.
3. Solve problems involving vertical angles and straight angles.
4. Solve a formula for a specified variable.

Many applied problems can be solved with formulas. For example, formulas exist for geometric figures such as squares and circles, for distance, for money earned on bank savings, and for converting English measurements to metric measurements. The formulas used in this book are given on the inside covers.

**OBJECTIVE 1 Solve a formula for one variable, given the values of the other variables.** Given the values of all but one of the variables in a formula, we can find the value of the remaining variable by using the methods introduced in this chapter.

In Example 1, we use the idea of *area*. The **area** of a plane (two-dimensional) geometric figure is a measure of the surface covered by the figure.

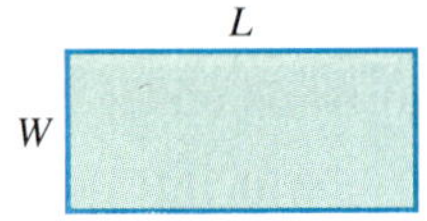


Rectangle
$A = LW$

**FIGURE 4**

**EXAMPLE 1 Using Formulas to Evaluate Variables**

Find the value of the remaining variable in each formula.

**(a)** $A = LW$; $A = 64$, $L = 10$

As shown in Figure 4, this formula gives the area of a rectangle with length $L$ and width $W$. Substitute the given values into the formula and then solve for $W$.

$$A = LW$$

$$64 = 10W \qquad \text{Let } A = 64 \text{ and } L = 10.$$

$$\frac{64}{10} = \frac{10W}{10} \qquad \text{Divide by 10.}$$

$$6.4 = W$$

The width of the rectangle is 6.4. Since $10(6.4) = 64$, the given area, the answer checks.

**(b)** $A = \frac{1}{2}h(b + B)$; $A = 210$, $B = 27$, $h = 10$

This formula gives the area of a trapezoid with parallel sides of lengths $b$ and $B$ and distance $h$ between the parallel sides. See Figure 5. Again, begin by substituting the given values into the formula.

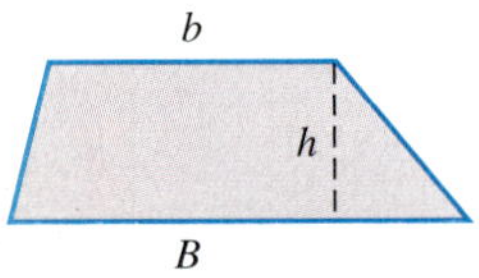


Trapezoid
$A = \frac{1}{2}h(b + B)$

**FIGURE 5**

$$A = \frac{1}{2}h(b + B)$$

$$210 = \frac{1}{2}(10)(b + 27) \qquad A = 210,\ h = 10,\ B = 27$$

Now solve for $b$.

$$210 = 5(b + 27) \quad \text{Multiply.}$$
$$210 = 5b + 135 \quad \text{Distributive property}$$
$$210 - 135 = 5b + 135 - 135 \quad \text{Subtract 135.}$$
$$75 = 5b \quad \text{Combine like terms.}$$
$$\frac{75}{5} = \frac{5b}{5} \quad \text{Divide by 5.}$$
$$15 = b$$

Check that the length of the shorter parallel side, $b$, is 15.

**Now Try Exercises 19 and 23.**

**OBJECTIVE 2 Use a formula to solve an applied problem.** As the next examples show, formulas are often used to solve applied problems. *It is a good idea to draw a sketch when a geometric figure is involved.* Example 2 uses the idea of perimeter. The **perimeter** of a plane (two-dimensional) geometric figure is the distance around the figure, that is, the sum of the lengths of its sides. We use the six steps introduced in the previous section.

### EXAMPLE 2 Finding the Width of a Rectangular Lot

A rectangular lot has perimeter 80 m and length 25 m. Find the width of the lot.

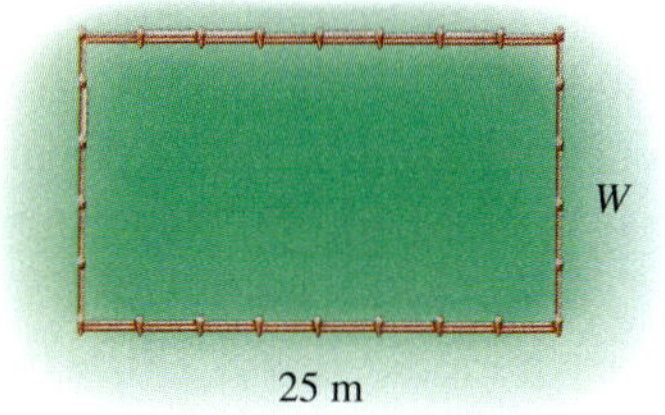


**FIGURE 6**

*Step 1* **Read.** We are told to find the width of the lot.

*Step 2* **Assign a variable.** Let $W =$ the width of the lot in meters. See Figure 6.

*Step 3* **Write an equation.** The formula for the perimeter of a rectangle is

$$P = 2L + 2W.$$

We find the width by substituting 80 for $P$ and 25 for $L$ in the formula.

$$80 = 2(25) + 2W \quad P = 80,\ L = 25$$

*Step 4* **Solve** the equation.

$$80 = 50 + 2W \quad \text{Multiply.}$$
$$80 - 50 = 50 + 2W - 50 \quad \text{Subtract 50.}$$
$$30 = 2W \quad \text{Combine like terms.}$$
$$\frac{30}{2} = \frac{2W}{2} \quad \text{Divide by 2.}$$
$$15 = W$$

*Step 5* **State the answer.** The width is 15 m.

*Step 6* **Check.** If the width is 15 m and the length is 25 m, the distance around the rectangular lot (perimeter) is $2(25) + 2(15) = 50 + 30 = 80$ m, as required.

**Now Try Exercise 37.**

FIGURE 7

## EXAMPLE 3 Finding the Height of a Triangular Sail

The area of a triangular sail of a sailboat is 126 $ft^2$. (Recall that $ft^2$ means "square feet.") The base of the sail is 12 ft. Find the height of the sail.

*Step 1* **Read.** We must find the height of the triangular sail.

*Step 2* **Assign a variable.** Let $h =$ the height of the sail in feet. See Figure 7.

*Step 3* **Write an equation.** The formula for the area of a triangle is $A = \frac{1}{2}bh$, where $A$ is the area, $b$ is the base, and $h$ is the height. Using the information given in the problem, we substitute 126 for $A$ and 12 for $b$ in the formula.

$$A = \frac{1}{2}bh$$

$$126 = \frac{1}{2}(12)h \qquad A = 126,\ b = 12$$

*Step 4* **Solve.**

$$126 = 6h \qquad \text{Multiply.}$$

$$21 = h \qquad \text{Divide by 6.}$$

*Step 5* **State the answer.** The height of the sail is 21 ft.

*Step 6* **Check** to see that the values $A = 126$, $b = 12$, and $h = 21$ satisfy the formula for the area of a triangle.

**Now Try Exercise 39.**

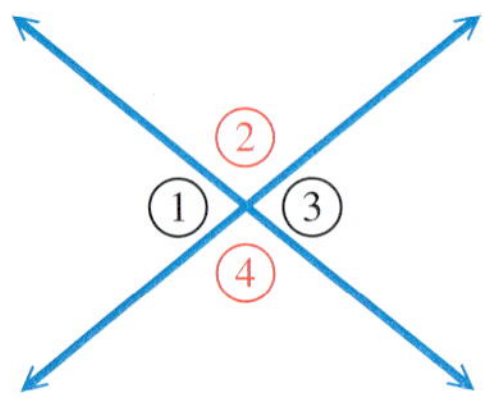


FIGURE 8

**OBJECTIVE 3 Solve problems involving vertical angles and straight angles.** Figure 8 shows two intersecting lines forming angles that are numbered ①, ②, ③, and ④. Angles ① and ③ lie "opposite" each other. They are called **vertical angles.** Another pair of vertical angles is ② and ④. In geometry, it is shown that vertical angles have equal measures.

Now look at angles ① and ②. When their measures are added, we get 180°, the measure of a straight angle. There are three other such pairs of angles: ② and ③, ③ and ④, and ① and ④.

The next example uses these ideas.

## EXAMPLE 4 Finding Angle Measures

Refer to the appropriate figure in each part.

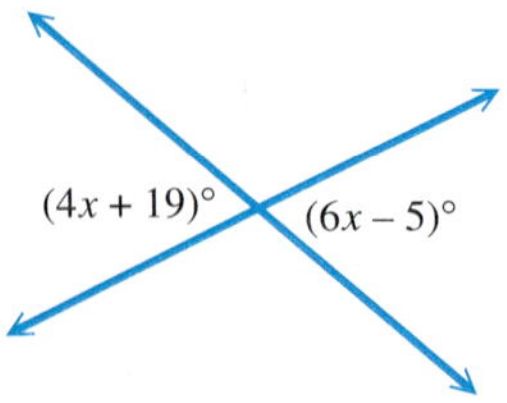


FIGURE 9

**(a)** Find the measure of each marked angle in Figure 9.

Since the marked angles are vertical angles, they have equal measures. Set $4x + 19$ equal to $6x - 5$ and solve.

$$4x + 19 = 6x - 5$$

$$19 = 2x - 5 \qquad \text{Subtract } 4x.$$

$$24 = 2x \qquad \text{Add 5.}$$

$$12 = x \qquad \text{Divide by 2.}$$

Since $x = 12$, one angle has measure $4(12) + 19 = 67$ degrees. The other has the same measure, since $6(12) - 5 = 67$ as required. Each angle measures 67°.

**(b)** Find the measure of each marked angle in Figure 10.

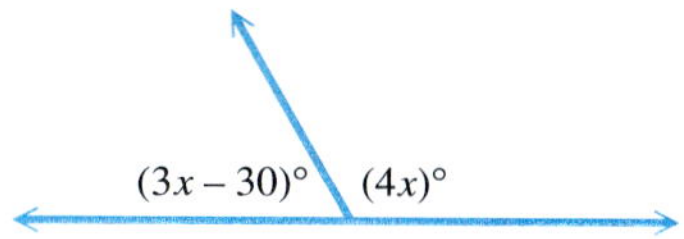


FIGURE 10

The measures of the marked angles must add to 180° because together they form a straight angle. The equation to solve is

$$(3x - 30) + 4x = 180.$$
$$7x - 30 = 180 \quad \text{Combine like terms.}$$
$$7x = 210 \quad \text{Add 30.}$$
$$x = 30 \quad \text{Divide by 7.}$$

To find the measures of the angles, replace $x$ with 30 in the two expressions.

$$3x - 30 = 3(30) - 30 = 90 - 30 = 60$$
$$4x = 4(30) = 120$$

The two angle measures are 60° and 120°.

**Now Try Exercises 45 and 47.**

**CAUTION** In Example 4, the answer is *not* the value of $x$. Remember to substitute the value of the variable into the expression given for each angle.

**OBJECTIVE 4 Solve a formula for a specified variable.** Sometimes it is necessary to solve a number of problems that use the same formula. For example, a surveying class might need to solve several problems that involve the formula for the area of a rectangle, $A = LW$. Suppose that in each problem the area ($A$) and the length ($L$) of a rectangle are given, and the width ($W$) must be found. Rather than solving for $W$ each time the formula is used, it would be simpler to *rewrite the formula* so that it is solved for $W$. This process is called **solving for a specified variable.**

In solving a formula for a specified variable, we treat the specified variable as if it were the *only* variable in the equation, and treat the other variables as if they were numbers. We use the same steps to solve the equation for the specified variable that we use to solve equations with just one variable.

## EXAMPLE 5 Solving for a Specified Variable

Solve $A = LW$ for $W$.

Think of undoing what has been done to $W$. Since $W$ is multiplied by $L$, undo the multiplication by dividing each side of $A = LW$ by $L$.

$$A = LW$$
$$\frac{A}{L} = \frac{LW}{L} \quad \text{Divide by } L.$$
$$\frac{A}{L} = W \quad \text{or} \quad W = \frac{A}{L} \quad \frac{L}{L} = 1;\ 1W = W$$

**Now Try Exercise 51.**

### EXAMPLE 6 Solving for a Specified Variable

Solve $P = 2L + 2W$ for $L$.

We want to get $L$ alone on one side of the equation. We begin by subtracting $2W$ from each side.

$$P = 2L + 2W$$

$$P - 2W = 2L + 2W - 2W \quad \text{Subtract } 2W.$$

$$P - 2W = 2L \quad \text{Combine like terms.}$$

$$\frac{P - 2W}{2} = \frac{2L}{2} \quad \text{Divide by 2.}$$

$$\frac{P - 2W}{2} = L \quad \tfrac{2}{2} = 1;\ 1L = L$$

or

$$L = \frac{P - 2W}{2}$$

The last step gives the formula solved for $L$, as required.

**Now Try Exercise 67.**

### EXAMPLE 7 Solving for a Specified Variable

Solve $F = \frac{9}{5}C + 32$ for $C$. (This is the formula for converting from Celsius to Fahrenheit.)

We need to isolate $C$ on one side of the equation. First we undo the addition of 32 to $\frac{9}{5}C$ by subtracting 32 from each side.

$$F = \frac{9}{5}C + 32$$

$$F - 32 = \frac{9}{5}C + 32 - 32 \quad \text{Subtract 32.}$$

$$F - 32 = \frac{9}{5}C$$

Now we multiply each side by $\frac{5}{9}$, using parentheses on the left.

$$\frac{5}{9}(F - 32) = \frac{5}{9} \cdot \frac{9}{5}C \quad \text{Multiply by } \tfrac{5}{9}.$$

$$\frac{5}{9}(F - 32) = C \quad \text{or} \quad C = \frac{5}{9}(F - 32)$$

This last result is the formula for converting temperatures from Fahrenheit to Celsius.

**Now Try Exercise 69.**

### EXAMPLE 8 Solving for a Specified Variable

Solve $A = \frac{1}{2}(b + B)h$ for $B$.

To get $B$ alone, begin by multiplying each side by 2 to clear the fraction.

$$A = \frac{1}{2}(b + B)h$$

$$2A = 2 \cdot \frac{1}{2}(b + B)h$$ Multiply by 2. (*Do not distribute* the 2 on the right side.)

$$2A = (b + B)h$$ $2 \cdot \frac{1}{2} = \frac{2}{2} = 1$

$$2A = bh + Bh$$ Distributive property

$$2A - bh = bh + Bh - bh$$ Subtract $bh$.

$$2A - bh = Bh$$ Combine like terms.

$$\frac{2A - bh}{h} = \frac{Bh}{h}$$ Divide by $h$.

$$\frac{2A - bh}{h} = B \quad \text{or} \quad B = \frac{2A - bh}{h}$$

**Now Try Exercise 73.**

**NOTE** The result in Example 8 can be written in a different form as follows.

$$B = \frac{2A - bh}{h} = \frac{2A}{h} - \frac{bh}{h} = \frac{2A}{h} - b$$

Either form is correct.

## 2.5 EXERCISES

**For Extra Help**

- Student's Solutions Manual
- MyMathLab
- InterAct Math Tutorial Software
- AW Math Tutor Center
- MathXL
- Digital Video Tutor CD 4/Videotape 4

**1.** In your own words, explain what is meant by each term.

**(a)** Perimeter of a plane geometric figure
**(b)** Area of a plane geometric figure

**2.** Perimeter is to a polygon as ______________ is to a circle.

**3.** If a formula has exactly five variables, how many values would you need to be given in order to find the value of any one variable?

**4.** The formula for changing Celsius to Fahrenheit is given in Example 7 as $F = \frac{9}{5}C + 32$. Sometimes it is seen as $F = \frac{9C}{5} + 32$. These are both correct. Why is it true that $\frac{9}{5}C$ is equal to $\frac{9C}{5}$?

*Decide whether perimeter or area would be used to solve a problem concerning the measure of the quantity.*

**5.** Sod for a lawn

**6.** Carpeting for a bedroom

**7.** Baseboards for a living room

**8.** Fencing for a yard

**9.** Fertilizer for a garden

**10.** Tile for a bathroom

**11.** Determining the cost of planting rye grass in a lawn for the winter

**12.** Determining the cost of replacing a linoleum floor with a wood floor

*In the following exercises a formula is given, along with the values of all but one of the variables in the formula. Find the value of the variable that is not given. (When necessary, use 3.14 as an approximation for $\pi$.) See Example 1.*

**13.** $P = 2L + 2W$ (perimeter of a rectangle); $L = 8$, $W = 5$

**14.** $P = 2L + 2W$; $L = 6$, $W = 4$

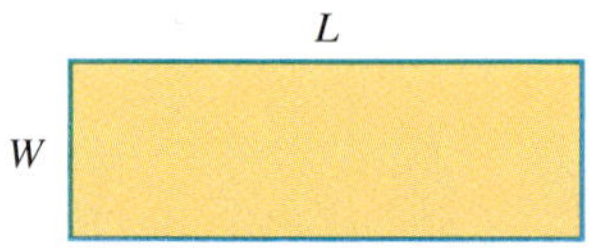


**15.** $A = \frac{1}{2}bh$ (area of a triangle); $b = 8$, $h = 16$

**16.** $A = \frac{1}{2}bh$; $b = 10$, $h = 14$

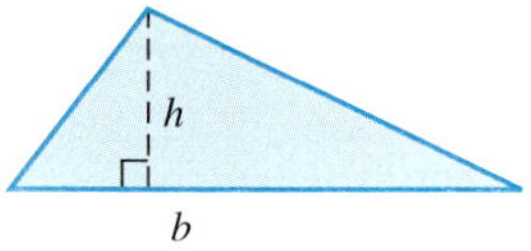


**17.** $P = a + b + c$ (perimeter of a triangle); $P = 12$, $a = 3$, $c = 5$

**18.** $P = a + b + c$; $P = 15$, $a = 3$, $b = 7$

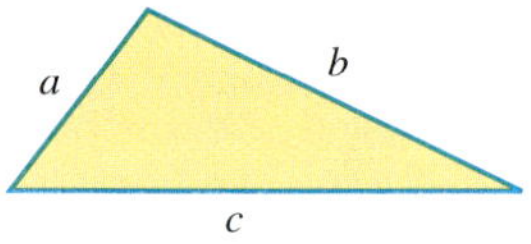


**19.** $d = rt$ (distance formula); $d = 252$, $r = 45$

**20.** $d = rt$; $d = 100$, $t = 2.5$

**21.** $I = prt$ (simple interest); $p = 7500$, $r = .035$, $t = 6$

**22.** $I = prt$; $p = 5000$, $r = .025$, $t = 7$

**23.** $A = \frac{1}{2}h(b + B)$ (area of a trapezoid); $A = 91$, $h = 7$, $b = 12$

**24.** $A = \frac{1}{2}h(b + B)$; $A = 75$, $b = 19$, $B = 31$

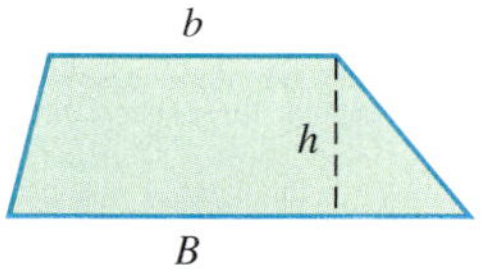


**25.** $C = 2\pi r$ (circumference of a circle); $C = 16.328$

**26.** $C = 2\pi r$; $C = 8.164$

**27.** $A = \pi r^2$ (area of a circle); $r = 4$

**28.** $A = \pi r^2$; $r = 12$

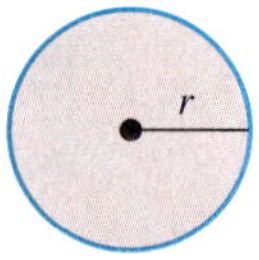


*The **volume** of a three-dimensional object is a measure of the space occupied by the object. For example, we would need to know the volume of a gasoline tank in order to know how many gallons of gasoline it would take to completely fill the tank. In the following exercises, a formula for the volume (V) of a three-dimensional object is given, along with values for the other variables. Evaluate V. (Use 3.14 as an approximation for $\pi$.) See Example 1.*

**29.** $V = LWH$ (volume of a rectangular box); $L = 10$, $W = 5$, $H = 3$

**30.** $V = LWH$; $L = 12$, $W = 8$, $H = 4$

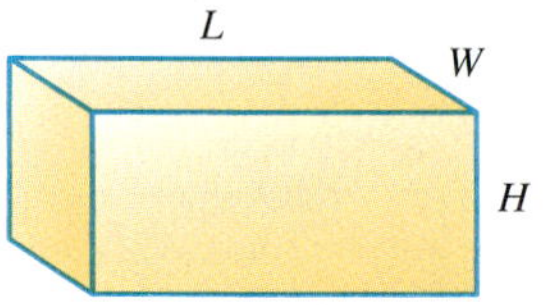


**31.** $V = \frac{1}{3}Bh$ (volume of a pyramid); $B = 12$, $h = 13$

**32.** $V = \frac{1}{3}Bh$; $B = 36$, $h = 4$

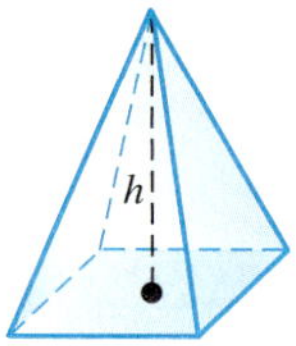


**33.** $V = \frac{4}{3}\pi r^3$ (volume of a sphere); $r = 12$

**34.** $V = \frac{4}{3}\pi r^3$; $r = 6$

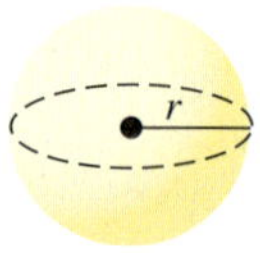

*Use a formula to write an equation for each application, and then use the problem-solving method of Section 2.4 to solve. (Use* 3.14 *as an approximation for* $\pi$.*) Formulas are found on the inside covers of this book. See Examples 2 and 3.*

**35.** Recently, a prehistoric ceremonial site dating to about 3000 B.C. was discovered at Stanton Drew in southwestern England. The site, which is larger than Stonehenge, is a nearly perfect circle, consisting of nine concentric rings that probably held upright wooden posts. Around this timber temple is a wide, encircling ditch enclosing an area with a diameter of 443 ft. Find this enclosed area to the nearest thousand square feet. (*Hint:* Find the radius. Then use $A = \pi r^2$.) (*Source: Archaeology,* vol. 51, no. 1, Jan./Feb. 1998.)

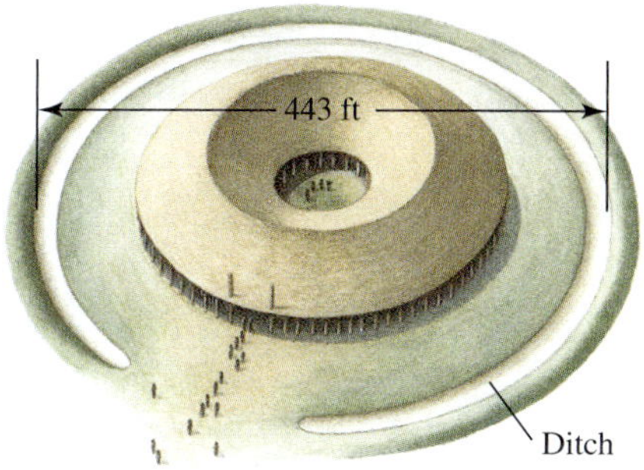


**36.** The Skydome in Toronto, Canada, is the first stadium with a hard-shell, retractable roof. The steel dome is 630 ft in diameter. To the nearest foot, what is the circumference of this dome? (*Source:* www.4ballparks.com)

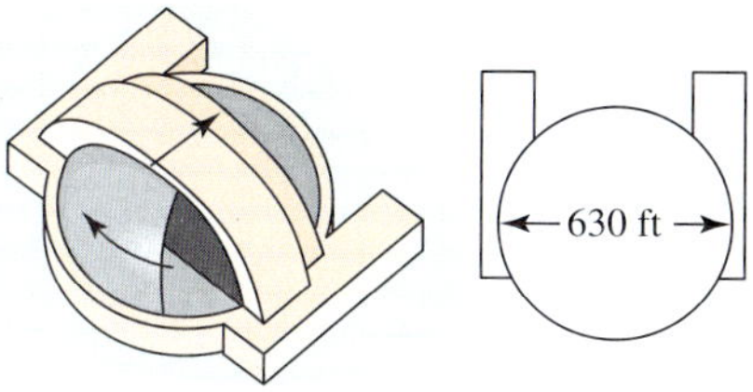


**37.** The *Daily Banner,* published in Roseberg, Oregon, in the 19th century, had page size 3 in. by 3.5 in. What was the perimeter? What was the area? (*Source: Guinness Book of Records.*)

**38.** The newspaper *The Constellation,* printed in 1859 in New York City as part of the Fourth of July celebration, had length 51 in. and width 35 in. What was the perimeter? What was the area? (*Source: Guinness Book of Records.*)

**39.** The largest drum ever constructed was played at the Royal Festival Hall in London in 1987. It had a diameter of 13 ft. What was the area of the circular face of the drum? (*Hint:* $A = \pi r^2$.) (*Source: Guinness Book of Records.*)

**40.** What was the circumference of the drum described in Exercise 39? (*Hint:* Use $C = 2\pi r$.)

**41.** The survey plat depicted here shows two lots that form a trapezoid. The measures of the parallel sides are 115.80 ft and 171.00 ft. The height of the trapezoid is 165.97 ft. Find the combined area of the two lots. Round your answer to the nearest hundredth of a square foot.

**42.** Lot A in the figure is in the shape of a trapezoid. The parallel sides measure 26.84 ft and 82.05 ft. The height of the trapezoid is 165.97 ft. Find the area of Lot A. Round your answer to the nearest hundredth of a square foot.

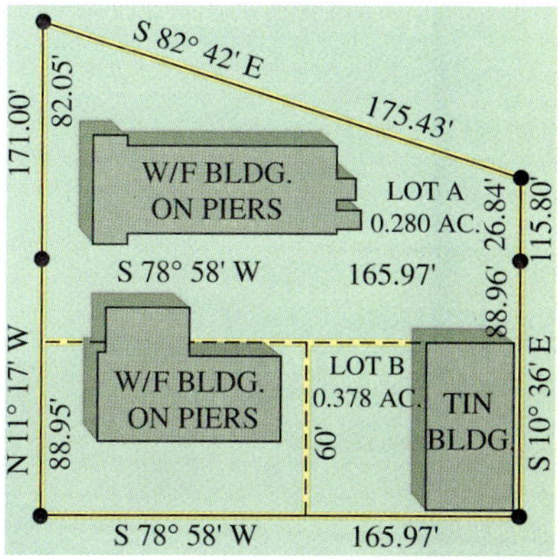


*Source:* Property survey in New Roads, Louisiana.

**43.** The U.S. Postal Service requires that any box sent through the mail have length plus girth (distance around) totaling no more than 108 in. The maximum volume that meets this condition is contained by a box with a square end 18 in. on each side. What is the length of the box? What is the maximum volume?

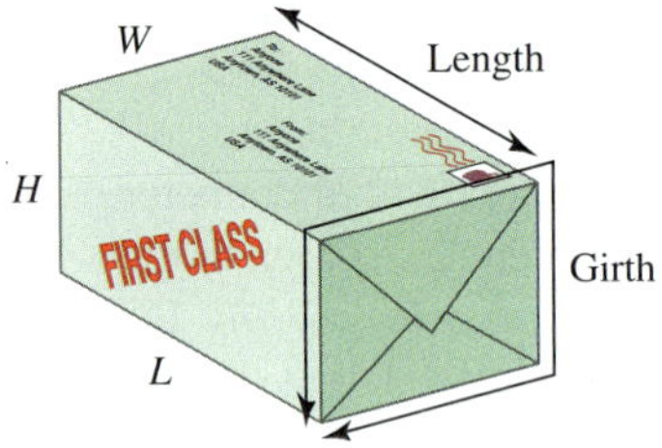


**44.** The largest box of popcorn was filled by students in Jacksonville, Florida. The box was approximately 40 ft long, $20\frac{2}{3}$ ft wide, and 8 ft high. To the nearest cubic foot, what was the volume of the box? (*Source: Guinness Book of Records.*)

*Find the measure of each marked angle. See Example 4.*

**45.**

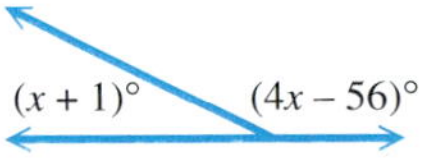


**46.**

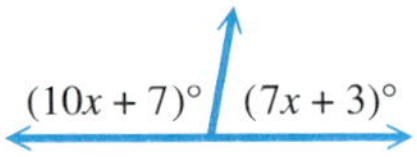


**47.**

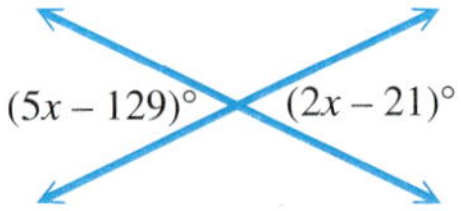


**48.**

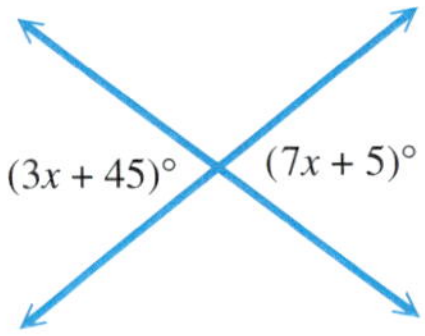


**49.**

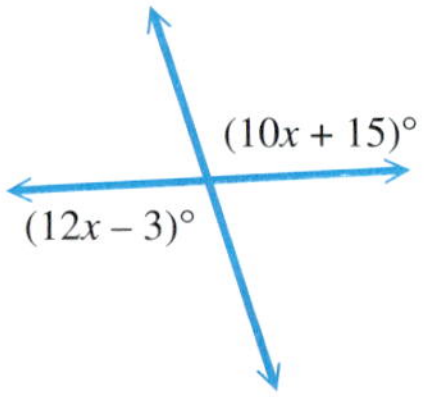


**50.**

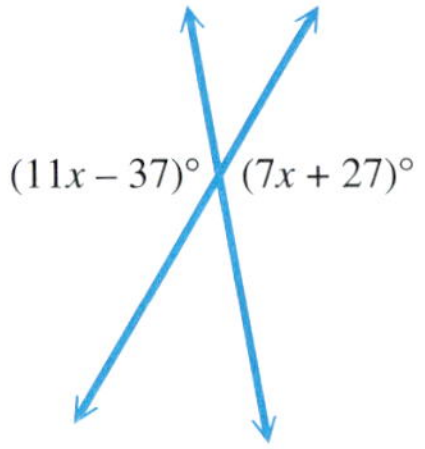


*Solve each formula for the specified variable. See Examples 5–8.*

**51.** $d = rt$ for $t$

**52.** $d = rt$ for $r$

**53.** $A = bh$ for $b$

**54.** $A = LW$ for $L$

**55.** $C = \pi d$ for $d$

**56.** $P = 4s$ for $s$

**57.** $V = LWH$ for $H$

**58.** $V = LWH$ for $W$

**59.** $I = prt$ for $r$

**60.** $I = prt$ for $p$

**61.** $A = \frac{1}{2}bh$ for $h$

**62.** $A = \frac{1}{2}bh$ for $b$

**63.** $V = \frac{1}{3}\pi r^2 h$ for $h$

**64.** $V = \pi r^2 h$ for $h$

**65.** $P = a + b + c$ for $b$

**66.** $P = a + b + c$ for $a$

**67.** $P = 2L + 2W$ for $W$

**68.** $A = p + prt$ for $r$

**69.** $y = mx + b$ for $m$

**70.** $y = mx + b$ for $x$

**71.** $Ax + By = C$ for $y$

**72.** $Ax + By = C$ for $x$

**73.** $M = C(1 + r)$ for $r$

**74.** $C = \frac{5}{9}(F - 32)$ for $F$

# 2.6 Ratios and Proportions

**OBJECTIVES**

1 Write ratios.
2 Solve proportions.
3 Solve applied problems using proportions.

**OBJECTIVE 1 Write ratios.** A **ratio** is a comparison of two quantities using a quotient.

**Ratio**

The ratio of the number $a$ to the number $b$ ($b \neq 0$) is written

$$\boldsymbol{a} \textbf{ to } \boldsymbol{b}, \qquad \boldsymbol{a:b}, \qquad \text{or} \qquad \frac{\boldsymbol{a}}{\boldsymbol{b}}.$$

The last way of writing a ratio is most common in algebra.

Percents are ratios where the second number is always 100. For example, 50% represents the ratio of 50 to 100, 27% represents the ratio of 27 to 100, and so on.

**EXAMPLE 1 Writing Word Phrases as Ratios**

Write a ratio for each word phrase.

**(a)** The ratio of 5 hr to 3 hr is

$$\frac{5 \text{ hr}}{3 \text{ hr}} = \frac{5}{3}.$$

**(b)** To find the ratio of 6 hr to 3 days, first convert 3 days to hours.

$$3 \text{ days} = 3 \cdot 24$$
$$= 72 \text{ hr}$$

The ratio of 6 hr to 3 days is thus

$$\frac{6 \text{ hr}}{3 \text{ days}} = \frac{6 \text{ hr}}{72 \text{ hr}} = \frac{6}{72} = \frac{1}{12}.$$

**Now Try Exercises 3 and 7.**

An application of ratios is in unit pricing, to see which size of an item offered in different sizes produces the best price per unit. To do this, set up the ratio of the price of the item to the number of units on the label. Then divide to obtain the price per unit.

**EXAMPLE 2 Finding Price per Unit**

The Winn-Dixie supermarket in Mandeville, Louisiana, charges the prices shown in the table for a box of trash bags. Which size is the best buy? That is, which size has the lowest unit price?

| *Size* | *Price* |
|---|---|
| 10-count | \$1.28 |
| 20-count | \$2.68 |
| 30-count | \$3.88 |

To find the best buy, write ratios comparing the price for each box size to the number of units (bags) per box. The results in the table on the next page are rounded to the nearest thousandth.

| *Size* | *Unit Cost (dollars per bag)* |
|---|---|
| 10-count | $\frac{\$1.28}{10} = \$.128$ ← The best buy |
| 20-count | $\frac{\$2.68}{20} = \$.134$ |
| 30-count | $\frac{\$3.88}{30} = \$.129$ |

Because the 10-count size produces the lowest unit cost, it is the best buy. This example shows that buying the largest size does not always provide the best buy, although this is often true.

**Now Try Exercise 15.**

**OBJECTIVE 2 Solve proportions.** A ratio is used to compare two numbers or amounts. A **proportion** says that two ratios are equal, so it is a special type of equation. For example,

$$\frac{3}{4} = \frac{15}{20}$$

is a proportion that says that the ratios $\frac{3}{4}$ and $\frac{15}{20}$ are equal. In the proportion

$$\frac{a}{b} = \frac{c}{d} \quad (b, d \neq 0),$$

$a$, $b$, $c$, and $d$ are the **terms** of the proportion. The $a$ and $d$ terms are called the **extremes,** and the $b$ and $c$ terms are called the **means.** We read the proportion $\frac{a}{b} = \frac{c}{d}$ as "$a$ is to $b$ as $c$ is to $d$." Beginning with this proportion and multiplying each side by the common denominator, $bd$, gives

$$\frac{a}{b} = \frac{c}{d}$$

$$bd \cdot \frac{a}{b} = bd \cdot \frac{c}{d}$$

$$\frac{b}{b}(d \cdot a) = \frac{d}{d}(b \cdot c) \qquad \text{Associative and commutative properties}$$

$$ad = bc. \qquad \text{Commutative and identity properties}$$

We can also find the products $ad$ and $bc$ by multiplying diagonally.

$$\frac{a}{b} = \frac{c}{d} \quad \begin{matrix} bc \\ ad \end{matrix}$$

For this reason, $ad$ and $bc$ are called **cross products.**

**Cross Products**

If $\frac{a}{b} = \frac{c}{d}$, then the cross products $ad$ and $bc$ are equal.

Also, if $ad = bc$, then $\frac{a}{b} = \frac{c}{d}$ $(b, d \neq 0)$.

From this rule, if $\frac{a}{b} = \frac{c}{d}$ then $ad = bc$; that is, the product of the extremes equals the product of the means.

**NOTE** If $\frac{a}{c} = \frac{b}{d}$, then $ad = cb$, or $ad = bc$. This means that the two proportions are equivalent, and

$$\text{the proportion } \frac{a}{b} = \frac{c}{d} \text{ can also be written as } \frac{a}{c} = \frac{b}{d} \quad (c \neq 0).$$

Sometimes one form is more convenient to work with than the other.

### EXAMPLE 3 Deciding Whether Proportions Are True

Decide whether each proportion is true or false.

**(a)** $\frac{3}{4} = \frac{15}{20}$

Check to see whether the cross products are equal.

$$4 \cdot 15 = 60$$

$$\frac{3}{4} = \frac{15}{20}$$

$$3 \cdot 20 = 60$$

The cross products are equal, so the proportion is true.

**(b)** $\frac{6}{7} = \frac{30}{32}$

The cross products are $6 \cdot 32 = 192$ and $7 \cdot 30 = 210$. The cross products are not equal, so the proportion is false.

**Now Try Exercises 23 and 25.**

Four numbers are used in a proportion. If any three of these numbers are known, the fourth can be found.

### EXAMPLE 4 Finding an Unknown in a Proportion

Solve the proportion

$$\frac{5}{9} = \frac{x}{63}.$$

The cross products must be equal.

$$5 \cdot 63 = 9 \cdot x \quad \text{Cross products}$$

$$315 = 9x \quad \text{Multiply.}$$

$$35 = x \quad \text{Divide by 9.}$$

Check by substituting 35 for $x$ in the proportion. The solution set is $\{35\}$.

**Now Try Exercise 29.**

**CAUTION** The cross product method cannot be used directly if there is more than one term on either side of the equals sign.

**EXAMPLE 5 Solving an Equation Using Cross Products**

Solve the equation

$$\frac{m-2}{5}=\frac{m+1}{3}.$$

Find the cross products.

$$\begin{aligned} 3(m-2) &= 5(m+1) && \text{Be sure to use parentheses.}\\ 3m-6 &= 5m+5 && \text{Distributive property}\\ 3m &= 5m+11 && \text{Add 6.}\\ -2m &= 11 && \text{Subtract } 5m.\\ m &= -\frac{11}{2} && \text{Divide by } -2. \end{aligned}$$

The solution set is $\left\{-\frac{11}{2}\right\}$.

**Now Try Exercise 37.**

**NOTE** When you set cross products equal to each other, you are really multiplying each ratio in the proportion by a common denominator.

**OBJECTIVE 3 Solve applied problems using proportions.** Proportions are useful in many practical applications. We continue to use the six-step method, although the steps are not numbered here.

**EXAMPLE 6 Applying Proportions**

After Lee Ann Spahr pumped 5.0 gal of gasoline, the display showing the price read \$7.90. When she finished pumping the gasoline, the price display read \$21.33. How many gallons did she pump?

To solve this problem, set up a proportion, with prices in the numerators and gallons in the denominators. Make sure that the corresponding numbers appear together.

Let $x =$ the number of gallons she pumped. Then

$$\begin{aligned} \text{Price} \rightarrow \frac{\$7.90}{5.0} &= \frac{\$21.33}{x} \leftarrow \text{Price} \\ \text{Gallons} \rightarrow \quad &\quad \leftarrow \text{Gallons}\\ 7.90x &= 5.0(21.33) && \text{Cross products}\\ 7.90x &= 106.65 && \text{Multiply.}\\ x &= 13.5. && \text{Divide by 7.90.} \end{aligned}$$

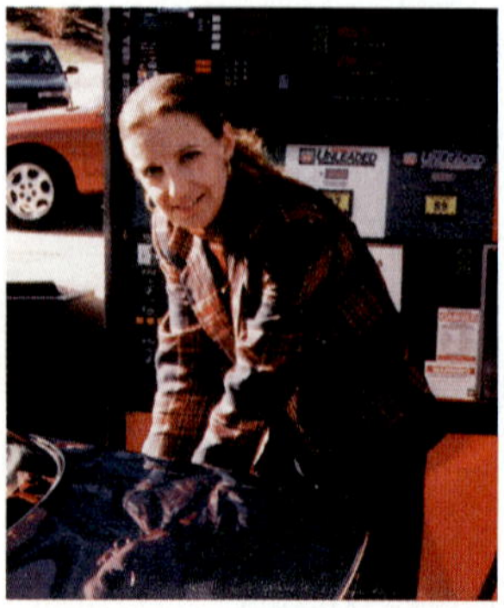

She pumped 13.5 gal. Check this answer. Using a calculator to perform the arithmetic reduces the possibility of errors. Notice that the way the proportion was set up uses the fact that the unit price is the same, no matter how many gallons are purchased.

**Now Try Exercise 43.**

## 2.6 EXERCISES

**For Extra Help**

- Student's Solutions Manual
- MyMathLab
- InterAct Math Tutorial Software
- AW Math Tutor Center
- MathXL
- Digital Video Tutor CD 4/Videotape 4

**1.** Match each ratio in Column I with the ratio equivalent to it in Column II.

| I | II |
|---|---|
| **(a)** 75 to 100 | **A.** 80 to 100 |
| **(b)** 5 to 4 | **B.** 50 to 100 |
| **(c)** $\frac{1}{2}$ | **C.** 3 to 4 |
| **(d)** 4 to 5 | **D.** 15 to 12 |

**2.** Give three different, equivalent forms of the ratio $\frac{4}{3}$.

*Write a ratio for each word phrase. In Exercises 7–12, first write the amounts with the same units. Write fractions in lowest terms. See Example 1.*

**3.** 60 ft to 70 ft

**4.** 40 mi to 30 mi

**5.** 72 dollars to 220 dollars

**6.** 120 people to 90 people

**7.** 30 in. to 8 ft

**8.** 20 yd to 8 ft

**9.** 16 min to 1 hr

**10.** 24 min to 2 hr

**11.** 5 days to 40 hr

**12.** 60 in. to 2 yd

*A supermarket was surveyed to find the prices charged for items in various sizes. Find the best buy (based on price per unit) for each item. See Example 2.*

**13.** Seasoning mix

8-oz size: $1.75
17-oz size: $2.88

**14.** Red beans

1-lb package: $.89
2-lb package: $1.79

**15.** Prune juice

32-oz can: $1.95
48-oz can: $2.89
64-oz can: $3.29

**16.** Corn oil

24-oz bottle: $2.08
64-oz bottle: $3.94
128-oz bottle: $7.65

**17.** Artificial sweetener packets

50-count: $1.19
100-count: $1.85
250-count: $3.79
500-count: $6.38

**18.** Chili (no beans)

7.5-oz can: $1.19
10.5-oz can: $1.29
15-oz can: $1.78
25-oz can: $2.59

**19.** Extra crunchy peanut butter

12-oz size: $1.49
28-oz size: $1.99
40-oz size: $3.99

**20.** Tomato ketchup

14-oz size: $.93
32-oz size: $1.19
44-oz size: $2.19

**21.** Explain the distinction between *ratio* and *proportion.* Give examples.

**22.** Suppose that someone told you to use cross products to multiply fractions. How would you explain to the person what is wrong with his or her thinking?

*Decide whether each proportion is* true *or* false. *See Example 3.*

**23.** $\frac{5}{35} = \frac{8}{56}$

**24.** $\frac{4}{12} = \frac{7}{21}$

**25.** $\frac{120}{82} = \frac{7}{10}$

**26.** $\dfrac{27}{160} = \dfrac{18}{110}$ **27.** $\dfrac{\frac{1}{2}}{5} = \dfrac{1}{10}$ **28.** $\dfrac{\frac{1}{3}}{6} = \dfrac{1}{18}$

*Solve each equation. See Examples 4 and 5.*

**29.** $\dfrac{k}{4} = \dfrac{175}{20}$ **30.** $\dfrac{x}{6} = \dfrac{18}{4}$ **31.** $\dfrac{49}{56} = \dfrac{z}{8}$

**32.** $\dfrac{20}{100} = \dfrac{z}{80}$ **33.** $\dfrac{a}{24} = \dfrac{15}{16}$ **34.** $\dfrac{x}{4} = \dfrac{12}{30}$

**35.** $\dfrac{z}{2} = \dfrac{z+1}{3}$ **36.** $\dfrac{m}{5} = \dfrac{m-2}{2}$ **37.** $\dfrac{3y-2}{5} = \dfrac{6y-5}{11}$

**38.** $\dfrac{2r+8}{4} = \dfrac{3r-9}{3}$ **39.** $\dfrac{5k+1}{6} = \dfrac{3k-2}{3}$ **40.** $\dfrac{x+4}{6} = \dfrac{x+10}{8}$

**41.** $\dfrac{2p+7}{3} = \dfrac{p-1}{4}$ **42.** $\dfrac{3m-2}{5} = \dfrac{4-m}{3}$

*Solve each problem. See Example 6.*

**43.** If nine pairs of jeans cost \$121.50, find the cost of five pairs. (Assume all are equally priced.)

**44.** If 7 shirts cost \$87.50, find the cost of 11 shirts. (Assume all are equally priced.)

**45.** If 6 gal of premium unleaded gasoline cost \$11.34, how much would it cost to completely fill a 15-gal tank?

**46.** If sales tax on a \$16.00 compact disc is \$1.32, how much would the sales tax be on a \$120.00 compact disc player?

**47.** The distance between Kansas City, Missouri, and Denver is 600 mi. On a certain wall map, this is represented by a length of 2.4 ft. On the map, how many feet would there be between Memphis and Philadelphia, two cities that are actually 1000 mi apart?

**48.** The distance between Singapore and Tokyo is 3300 mi. On a certain wall map, this distance is represented by 11 in. The actual distance between Mexico City and Cairo is 7700 mi. How far apart are they on the same map?

**49.** A chain saw requires a mixture of 2-cycle engine oil and gasoline. According to the directions on a bottle of Oregon 2-cycle Engine Oil, for a 50 to 1 ratio requirement, approximately 2.5 fluid oz of oil are required for 1 gal of gasoline. For 2.75 gal, how many fluid ounces of oil are required?

**50.** The directions on the bottle mentioned in Exercise 49 indicate that if the ratio requirement is 24 to 1, approximately 5.5 oz of oil are required for 1 gal of gasoline. If gasoline is to be mixed with 22 oz of oil, how much gasoline is to be used?

**51.** In a recent year, the average exchange rate between British pounds and U.S. dollars was 1 pound to \$1.6762. Margaret went to London and exchanged her U.S. currency for British pounds, and received 400 pounds. How much in U.S. dollars did Margaret exchange? (*Note:* Great Britain and Switzerland (see Exercise 52) are among the European nations not using euros.)

**52.** If 3 U.S. dollars can be exchanged for 4.5204 Swiss francs, how many Swiss francs can be obtained for \$49.20? (Round to the nearest hundredth.)

**53.** Biologists tagged 500 fish in Willow Lake on October 5. At a later date they found 7 tagged fish in a sample of 700. Estimate the total number of fish in Willow Lake to the nearest hundred.

**54.** On May 13 researchers at Argyle Lake tagged 840 fish. When they returned a few weeks later, their sample of 1000 fish contained 18 that were tagged. Give an approximation of the fish population in Argyle Lake to the nearest hundred.

*The International Olympic Committee has come to rely more and more on television rights and major corporate sponsors to finance the games. The circle graphs show funding plans for the first Olympics in Athens and the 1996 Olympics in Atlanta 100 yr later. Use proportions and the graphs to work Exercises 55 and 56.*

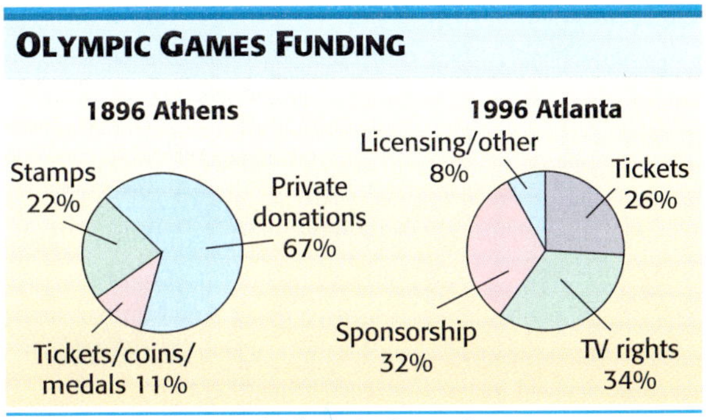


***Source:*** International Olympic Committee.

**55.** In the 1996 Olympics, total revenue of \$350 million was raised. There were 10 major sponsors.

**(a)** Write a proportion to find the amount of revenue provided by tickets. Solve it.

**(b)** What amount was provided by sponsors? Assuming the sponsors contributed equally, how much was provided per sponsor?

**(c)** What amount was raised by TV rights?

**56.** Suppose the amount of revenue raised in the 1896 Olympics was equivalent to the \$350 million in 1996.

**(a)** Write a proportion for the amount of revenue provided by stamps and solve it.

**(b)** What amount (in dollars) would have been provided by private donations?

Two triangles are **similar** if they have the same shape (but not necessarily the same size). Similar triangles have sides that are proportional. The figure shows two similar triangles. Notice that the ratios of the corresponding sides all equal $\frac{3}{2}$:

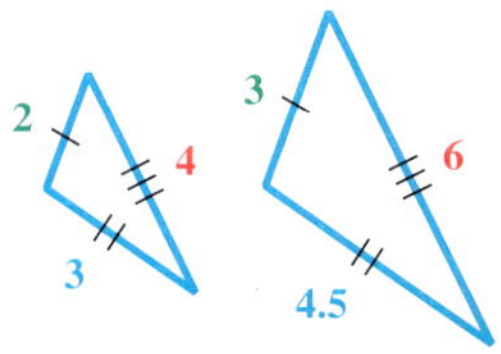


$$\frac{3}{2} = \frac{3}{2} \qquad \frac{4.5}{3} = \frac{3}{2} \qquad \frac{6}{4} = \frac{3}{2}.$$

If we know that two triangles are similar, we can set up a proportion to solve for the length of an unknown side. (continued)

*Use a proportion to find the length x, given that each pair of triangles is similar.*

**57.**

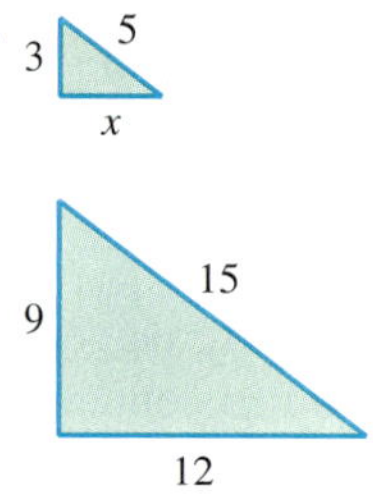


**58.**

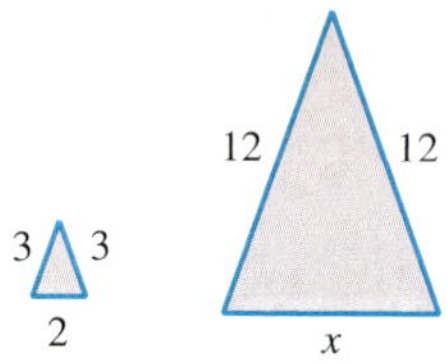


**59.**

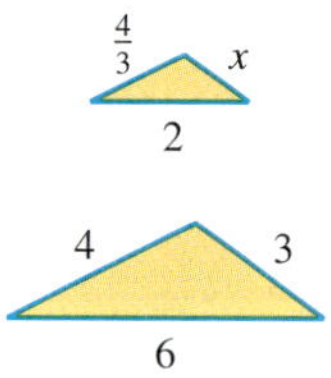


**60.**

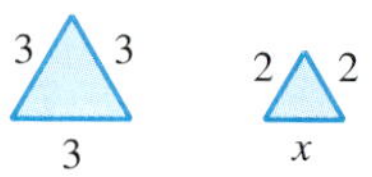


*For Exercises 61 and 62,* ***(a)*** *draw a sketch consisting of two right triangles, depicting the situation described, and* ***(b)*** *solve the problem. (Source: Guinness Book of Records.)*

**61.** An enlarged version of the chair used by George Washington at the Constitutional Convention casts a shadow 18 ft long at the same time a vertical pole 12 ft high casts a shadow 4 ft long. How tall is the chair?

**62.** One of the tallest candles ever constructed was exhibited at the 1897 Stockholm Exhibition. If it cast a shadow 5 ft long at the same time a vertical pole 32 ft high cast a shadow 2 ft long, how tall was the candle?

The Consumer Price Index provides a means of determining the purchasing power of the U.S. dollar from one year to the next. Using the period from 1982 to 1984 as a measure of 100.0, the Consumer Price Index for selected years from 1990 through 2000 is shown in the table. To use the Consumer Price Index to predict a price in a particular year, we set up a proportion and compare it with a known price in another year, as follows:

$$\frac{\text{price in year } A}{\text{index in year } A} = \frac{\text{price in year } B}{\text{index in year } B}.$$

| Year | Consumer Price Index |
|---|---|
| 1990 | 130.7 |
| 1992 | 140.3 |
| 1994 | 148.2 |
| 1996 | 156.9 |
| 1998 | 163.0 |
| 1999 | 166.6 |
| 2000 | 172.2 |

*Source:* U.S. Bureau of Labor Statistics.

*Use the Consumer Price Index figures in the table to find the amount that would be charged for the use of the same amount of electricity that cost \$225 in 1990. Give your answer to the nearest dollar.*

**63.** in 1996 **64.** in 1998 **65.** in 1999 **66.** in 2000

## RELATING CONCEPTS (EXERCISES 67–70)

**For Individual or Group Work**

*In Section 2.3 we solved equations with fractions by first multiplying each side of the equation by the common denominator. A proportion with a variable is this kind of equation.* ***Work Exercises 67–70 in order.*** *The steps justify the method of solving a proportion by cross products.*

**67.** What is the LCD of the fractions in the equation $\frac{x}{6} = \frac{2}{5}$?

**68.** Solve the equation in Exercise 67 as follows.

**(a)** Multiply each side by the LCD. What equation do you get?
**(b)** Solve the equation from part (a) by dividing each side by the coefficient of $x$.

**69.** Solve the equation in Exercise 67 using cross products.

**70.** Compare your solutions from Exercises 68 and 69. What do you notice?

# 2.7 More about Problem Solving

**OBJECTIVES**

1. Use percent in problems involving rates.
2. Solve problems involving mixtures.
3. Solve problems involving simple interest.
4. Solve problems involving denominations of money.
5. Solve problems involving distance, rate, and time.

**OBJECTIVE 1 Use percent in problems involving rates.** Recall that percent means "per hundred." Thus, percents are ratios where the second number is always 100. For example, 50% represents the ratio of 50 to 100 and 27% represents the ratio of 27 to 100.

**PROBLEM SOLVING**

Percents are often used in problems involving mixing different concentrations of a substance or different interest rates. In each case, to get the amount of pure substance or the interest, we multiply.

| Mixture Problems | Interest Problems (annual) |
|---|---|
| **base × rate (%) = percentage** | **principal × rate (%) = interest** |
| $b \times r = p$ | $p \times r = I$ |

In an equation, *percent is always written as a decimal.* For example, 35% is written .35, *not* 35, and 7% is written .07, *not* 7.

**EXAMPLE 1 Using Percents to Find Percentages**

**(a)** If a chemist has 40 L of a 35% acid solution, then the amount of pure acid in the solution is

$$40 \times .35 = 14 \text{ L.}$$

40 ↑ Amount of solution; .35 ↑ Rate of concentration; 14 L ↑ Amount of pure acid

**(b)** If \$1300 is invested for one year at 7% simple interest, the amount of interest earned in the year is

$$\$1300 \times .07 = \$91.$$

\$1300 ↑ Principal; .07 ↑ Interest rate; \$91 ↑ Interest earned

**Now Try Exercises 1 and 3.**

**PROBLEM SOLVING**

In the examples that follow, we use tables to organize the information in the problems. A table enables us to more easily set up an equation, which is usually the most difficult step.

**OBJECTIVE 2 Solve problems involving mixtures.** In the next example, we use percent to solve a mixture problem.

## EXAMPLE 2 Solving a Mixture Problem

A chemist needs to mix 20 L of 40% acid solution with some 70% acid solution to get a mixture that is 50% acid. How many liters of the 70% acid solution should be used?

*Step 1* **Read** the problem. Note the percent of each solution and of the mixture.

*Step 2* **Assign a variable.**

Let $x$ = the number of liters of 70% acid solution needed.

Recall from Example 1(a) that the amount of pure acid in this solution will be given by the product of the percent of strength and the number of liters of solution, or

$$\text{liters of pure acid in } x \text{ L of 70\% solution} = .70x.$$

The amount of pure acid in the 20 L of 40% solution is

$$\text{liters of pure acid in the 40\% solution} = .40(20) = 8.$$

The new solution will contain $(x + 20)$ L of 50% solution. The amount of pure acid in this solution is

$$\text{liters of pure acid in the 50\% solution} = .50(x + 20).$$

Figure 11 illustrates this information, which is summarized in the table.

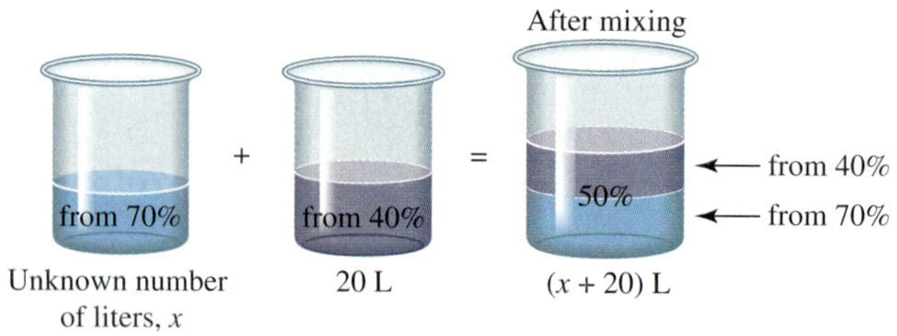


**FIGURE 11**

| Liters of Solution | Rate (as a decimal) | Liters of Pure Acid |
|---|---|---|
| $x$ | .70 | $.70x$ |
| 20 | .40 | $.40(20) = 8$ |
| $x + 20$ | .50 | $.50(x + 20)$ |

Sum must equal

*Step 3* **Write an equation.** The number of liters of pure acid in the 70% solution added to the number of liters of pure acid in the 40% solution will equal the number of liters of pure acid in the final mixture, so the equation is

| Pure acid in 70% | plus | pure acid in 40% | is | pure acid in 50%. |
|---|---|---|---|---|
| ↓ | ↓ | ↓ | ↓ | ↓ |
| $.70x$ | $+$ | $.40(20)$ | $=$ | $.50(x + 20)$. |

*Step 4* **Solve** the equation. Clear parentheses on the right side, then multiply by 100 to clear decimals.

$$.70x + .40(20) = .50x + .50(20) \quad \text{Distributive property}$$
$$70x + 40(20) = 50x + 50(20) \quad \text{Multiply by 100.}$$
$$70x + 800 = 50x + 1000$$
$$20x + 800 = 1000 \quad \text{Subtract } 50x.$$
$$20x = 200 \quad \text{Subtract 800.}$$
$$x = 10 \quad \text{Divide by 20.}$$

*Step 5* **State the answer.** The chemist needs to use 10 L of 70% solution.

*Step 6* **Check.** Since

$$.70(10) + .40(20) = 7 + 8 = 15$$

and $$.50(10 + 20) = .50(30) = 15,$$

the answer checks.

**Now Try Exercise 15.**

**NOTE** In a problem such as Example 2, the concentration of the final mixture must be *between* the concentrations of the two solutions making up the mixture.

**OBJECTIVE 3 Solve problems involving simple interest.** The next example uses the formula for simple interest, $I = prt$. Remember that when $t = 1$, the formula becomes $I = pr$, as shown in the Problem-Solving box at the beginning of this section. Once again, the idea of multiplying the total amount (principal) by the rate (rate of interest) gives the percentage (amount of interest).

### EXAMPLE 3 Solving a Simple Interest Problem

Elizabeth Suco receives an inheritance. She plans to invest part of it at 9% and $2000 more than this amount at 10%. To earn $1150 per year in interest, how much should she invest at each rate?

*Step 1* **Read** the problem again.

*Step 2* **Assign a variable.**

Let $x$ = the amount invested at 9% (in dollars).

Then $x + 2000$ = the amount invested at 10% (in dollars).

Use a table to arrange the information given in this problem.

| *Amount Invested in Dollars* | *Rate of Interest* | *Interest for One Year* |
|---|---|---|
| $x$ | .09 | $.09x$ |
| $x + 2000$ | .10 | $.10(x + 2000)$ |

*Step 3* **Write an equation.** Multiply amount by rate to get the interest earned. Since the total interest is to be \$1150, the equation is

Interest at 9% plus interest at 10% is total interest.

$$.09x + .10(x + 2000) = 1150.$$

*Step 4* **Solve** the equation. Clear parentheses; then clear decimals.

$$.09x + .10x + .10(2000) = 1150 \quad \text{Distributive property}$$
$$9x + 10x + 10(2000) = 115{,}000 \quad \text{Multiply by 100.}$$
$$9x + 10x + 20{,}000 = 115{,}000$$
$$19x + 20{,}000 = 115{,}000 \quad \text{Combine like terms.}$$
$$19x = 95{,}000 \quad \text{Subtract 20,000.}$$
$$x = 5000 \quad \text{Divide by 19.}$$

*Step 5* **State the answer.** She should invest \$5000 at 9% and \$5000 + \$2000 = \$7000 at 10%.

*Step 6* **Check.** Investing \$5000 at 9% and \$7000 at 10% gives total interest of .09(\$5000) + .10(\$7000) = \$450 + \$700 = \$1150, as required in the original problem.

**Now Try Exercise 25.**

**OBJECTIVE 4 Solve problems involving denominations of money.**

**PROBLEM SOLVING**

Problems that involve different denominations of money or items with different monetary values are very similar to mixture and interest problems. To get the total value, we multiply.

**Money Problems**

**number × value of one item = total value**

For example, 30 dimes have a monetary value of 30(\$.10) = \$3. Fifteen \$5 bills have a value of 15(\$5) = \$75. A table is helpful for these problems, too.

**EXAMPLE 4 Solving a Money Problem**

A bank teller has 25 more \$5 bills than \$10 bills. The total value of the money is \$200. How many of each denomination of bill does she have?

*Step 1* **Read** the problem. We must find the number of each denomination of bill that the teller has.

*Step 2* **Assign a variable.**

Let $x =$ the number of \$10 bills.

Then $x + 25 =$ the number of \$5 bills.

Organize the given information in a table.

| Number of Bills | Denomination | Total Value |
|---|---|---|
| $x$ | 10 | $10x$ |
| $x + 25$ | 5 | $5(x + 25)$ |

*Step 3* **Write an equation.** Multiplying the number of bills by the denomination gives the monetary value. The value of the tens added to the value of the fives must be \$200.

Value of tens ↓ $10x$ plus ↓ $+$ value of fives ↓ $5(x + 25)$ is ↓ $=$ \$200. ↓ $200$

$$10x + 5(x + 25) = 200$$

*Step 4* **Solve.**

$$10x + 5x + 125 = 200 \quad \text{Distributive property}$$
$$15x + 125 = 200 \quad \text{Combine like terms.}$$
$$15x = 75 \quad \text{Subtract 125.}$$
$$x = 5 \quad \text{Divide by 15.}$$

*Step 5* **State the answer.** The teller has 5 tens and $5 + 25 = 30$ fives.

*Step 6* **Check.** The teller has $30 - 5 = 25$ more fives, and the value of the money is $5(\$10) + 30(\$5) = \$200$, as required.

**Now Try Exercise 29.**

**OBJECTIVE 5 Solve problems involving distance, rate, and time.** If your car travels at an average rate of 50 mph for 2 hr, then it travels $50 \times 2 = 100$ mi. This is an example of the basic relationship between distance, rate, and time,

$$\text{distance} = \text{rate} \times \text{time},$$

given by the formula $d = rt$. By solving, in turn, for $r$ and $t$ in the formula, we obtain two other equivalent forms of the formula. The three forms are given here.

**Distance, Rate, and Time Relationship**

$$d = rt \qquad r = \frac{d}{t} \qquad t = \frac{d}{r}$$

The next examples illustrate the uses of these formulas.

### EXAMPLE 5 Finding Distance, Rate, or Time

**(a)** The speed of sound is 1088 ft per sec at sea level at 32°F. In 5 sec under these conditions, sound travels

$$\underset{\text{Rate}}{1088} \times \underset{\text{Time}}{5} = \underset{\text{Distance}}{5440 \text{ ft.}}$$

Here, we found distance given rate and time, using $d = rt$.

**(b)** The winner of the first Indianapolis 500 race (in 1911) was Ray Harroun, driving a Marmon Wasp at an average speed of 74.59 mph. (*Source: Universal Almanac,* 1997.) To complete the 500 mi, it took him

$$\frac{500 \leftarrow \text{Distance}}{74.59 \leftarrow \text{Rate}} = 6.70 \text{ hr} \quad \text{(rounded).} \leftarrow \text{Time}$$

Here, we found time given rate and distance using $t = \frac{d}{r}$. To convert .70 hr to minutes, we multiply by 60 to get $.70(60) = 42$. It took Harroun about 6 hr, 42 min to complete the race.

**(c)** At the 2000 Olympic Games in Sydney, Australia, Dutch swimmer Inge de Bruijn set a world record in the women's 50-m freestyle swimming event of 24.13 sec. (*Source: World Almanac and Book of Facts,* 2002.) Her rate was

$$\text{Rate} = \frac{\text{Distance}}{\text{Time}} \rightarrow \frac{50}{24.13} = 2.07 \text{ m per sec (rounded).}$$

**Now Try Exercises 37, 39, and 41.**

### EXAMPLE 6 Solving a Motion Problem

Two cars leave Baton Rouge, Louisiana, at the same time and travel east on Interstate 10. One travels at a constant speed of 55 mph, and the other travels at a constant speed of 63 mph. In how many hours will the distance between them be 24 mi?

*Step 1* **Read** the problem. We must find the time it will take for the distance between the cars to be 24 mi.

*Step 2* **Assign a variable.** Since we are looking for time, we let $t$ = the number of hours until the distance between them is 24 mi. The sketch in Figure 12 shows what is happening in the problem.

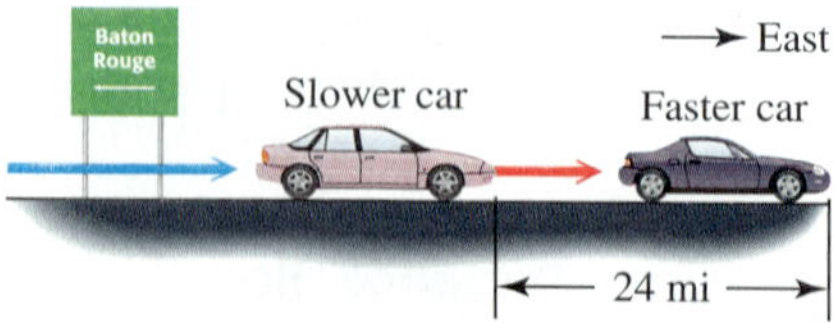


FIGURE 12

Now we construct a table using the information given in the problem and $t$ for the time traveled by each car. We multiply rate by time to get the expressions for distances traveled.

| | Rate × | Time = | Distance |
|---|---|---|---|
| Faster Car | 63 | $t$ | $63t$ |
| Slower Car | 55 | $t$ | $55t$ |

Difference is 24 mi.

The quantities $63t$ and $55t$ represent the two distances. Refer to Figure 12, and notice that the *difference* between the larger distance and the smaller distance is 24 mi.

*Step 3* **Write an equation.**

$$63t - 55t = 24$$

*Step 4* **Solve.**

$$8t = 24 \quad \text{Combine like terms.}$$

$$t = 3 \quad \text{Divide by 8.}$$

*Step 5* **State the answer.** It will take the cars 3 hr to be 24 mi apart.

*Step 6* **Check.** After 3 hr the faster car will have traveled $63 \times 3 = 189$ mi, and the slower car will have traveled $55 \times 3 = 165$ mi. Since $189 - 165 = 24$, the conditions of the problem are satisfied.

**Now Try Exercise 45.**

**NOTE** In motion problems like the one in Example 6, once you have filled in two pieces of information in each row of the table, you should automatically fill in the third piece of information, using the appropriate form of the formula relating distance, rate, and time. Set up the equation based on your sketch and the information in the table.

## EXAMPLE 7 Solving a Motion Problem

Two planes leave Los Angeles at the same time. One heads south to San Diego; the other heads north to San Francisco. The San Francisco plane flies 50 mph faster. In $\frac{1}{2}$ hr, the planes are 275 mi apart. What are their speeds?

*Step 1* **Read** the problem carefully.

*Step 2* **Assign a variable.**

Let $r =$ the speed of the slower plane.

Then $r + 50 =$ the speed of the faster plane.

Fill in a table.

| | Rate | Time | Distance |
|---|---|---|---|
| Slower Plane | $r$ | $\frac{1}{2}$ | $\frac{1}{2}r$ |
| Faster Plane | $r + 50$ | $\frac{1}{2}$ | $\frac{1}{2}(r + 50)$ |

Sum is 275 mi.

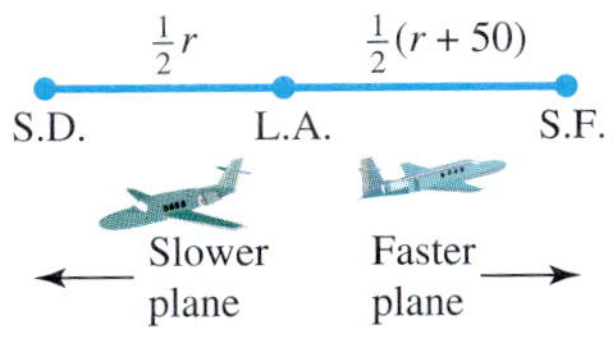

FIGURE 13

*Step 3* **Write an equation.** As Figure 13 shows, the planes are headed in *opposite* directions. The *sum* of their distances equals 275 mi, so

$$\frac{1}{2}r + \frac{1}{2}(r + 50) = 275.$$

*Step 4* **Solve.** $\frac{1}{2}r + \frac{1}{2}(r + 50) = 275$

$$r + (r + 50) = 550 \quad \text{Multiply by 2.}$$
$$2r + 50 = 550 \quad \text{Combine like terms.}$$
$$2r = 500 \quad \text{Subtract 50.}$$
$$r = 250 \quad \text{Divide by 2.}$$

*Step 5* **State the answer.** The slower plane (headed south) has a speed of 250 mph. The speed of the faster plane is $250 + 50 = 300$ mph.

*Step 6* **Check.** Verify that $\frac{1}{2}(250) + \frac{1}{2}(300) = 275$ mi.

**Now Try Exercise 51.**

Another way to solve the problems in this section is given in Chapter 8.

## 2.7 EXERCISES

**For Extra Help**

- Student's Solutions Manual
- MyMathLab
- InterAct Math Tutorial Software
- AW Math Tutor Center
- MathXL
- Digital Video Tutor CD 4/Videotape 4

*Use the concepts of this section to answer each question. See Example 1 and the Problem-Solving box before Example 4.*

**1.** How much pure acid is in 250 mL of a 14% acid solution?

**2.** How much pure alcohol is in 150 L of a 30% alcohol solution?

**3.** If \$10,000 is invested for 1 yr at 3.5% simple interest, how much interest is earned?

**4.** If \$25,000 is invested at 3% simple interest for 2 yr, how much interest is earned?

**5.** What is the monetary amount of 283 nickels?

**6.** What is the monetary amount of 35 half-dollars?

*Solve each percent problem. Remember that base $\times$ rate $=$ percentage.*

**7.** The 2000 U.S. Census showed that the population of Alabama was 4,447,000, with 26.0% represented by African-Americans. What is the best estimate of the African-American population in Alabama? (*Source:* U.S. Bureau of the Census.)

**A.** 500,000 **B.** 750,000 **C.** 1,100,000 **D.** 1,500,000

**8.** The 2000 U.S. Census showed that the population of New Mexico was 1,819,000, with 42.1% being Hispanic. What is the best estimate of the Hispanic population in New Mexico? (*Source:* U.S. Bureau of the Census.)

**A.** 720,000 **B.** 72,000 **C.** 650,000
**D.** 36,000

**9.** The graph shows the breakdown, by approximate percents, of colors chosen for new compact/sports cars in the 2000 model year. If about 2.5 million compact/sports cars were sold in 2000, about how many were each color? (*Source:* Ward's Communications.)

**(a)** Red **(b)** White **(c)** Black

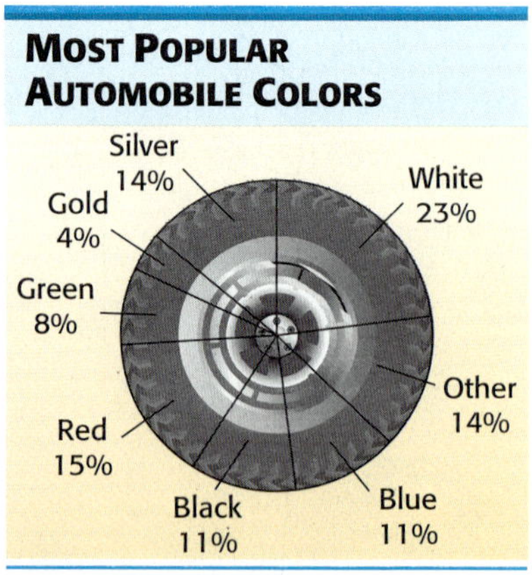

*Source:* DuPont Automotive Products.

**10.** An average middle-income family will spend $160,140 to raise a child born in 1999 from birth to age 17. The graph shows the breakdown, by approximate percents, for various expense categories. To the nearest dollar, about how much will be spent to provide the following?

**(a)** Housing **(b)** Food
**(c)** Health care

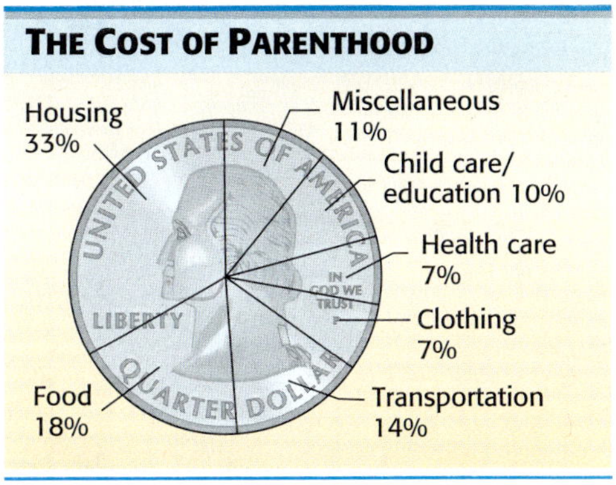


***Source:*** U.S. Department of Agriculture.

**11.** In 1998, the U.S. civilian labor force consisted of 137,673,000 persons. Of this total, 6,210,000 were unemployed. To the nearest tenth, what was the percent of unemployment? (*Source:* U.S. Bureau of Labor Statistics.)

**12.** In 1998, the U.S. labor force (excluding agricultural employees, self-employed persons, and the unemployed) consisted of 116,730,000 persons. Of this total, 16,211,000 were union members. To the nearest tenth, what percent of this labor force belonged to unions? (*Source:* U.S. Bureau of Labor Statistics.)

*Use the concepts of this section to answer each question.*

**13.** Suppose that a chemist is mixing two acid solutions, one of 20% concentration and the other of 30% concentration. Which one of the following concentrations could *not* be obtained?

**A.** 22% **B.** 24% **C.** 28% **D.** 32%

**14.** Suppose that pure alcohol is added to a 24% alcohol mixture. Which one of the following concentrations could *not* be obtained?

**A.** 22% **B.** 26% **C.** 28% **D.** 30%

*Work each mixture problem. See Example 2.*

**15.** How many gallons of 50% antifreeze must be mixed with 80 gal of 20% antifreeze to get a mixture that is 40% antifreeze?

| Gallons of Mixture | Rate | Gallons of Antifreeze |
|---|---|---|
| $x$ | .50 | $.50x$ |
| 80 | .20 | $.20(80)$ |
| $x + 80$ | .40 | $.40(x + 80)$ |

**16.** How many liters of 25% acid solution must be added to 80 L of 40% solution to get a solution that is 30% acid?

| Liters of Solution | Rate | Liters of Acid |
|---|---|---|
| $x$ | .25 | $.25x$ |
| 80 | .40 | $.40(80)$ |
| $x + 80$ | .30 | $.30(x + 80)$ |

**17.** A certain metal is 20% tin. How many kilograms of this metal must be mixed with 80 kg of a metal that is 70% tin to get a metal that is 50% tin?

| Kilograms of Metal | Rate | Kilograms of Pure Tin |
|---|---|---|
| $x$ | .20 | |
| | .70 | |
| | .50 | |

**18.** A pharmacist has 20 L of a 10% drug solution. How many liters of 5% solution must be added to get a mixture that is 8%?

| Liters of Solution | Rate | Liters of Pure Drug |
|---|---|---|
| 20 | | 20(.10) |
| | .05 | |
| | .08 | |

**19.** In a chemistry class, 12 L of a 12% alcohol solution must be mixed with a 20% solution to get a 14% solution. How many liters of the 20% solution are needed?

**20.** How many liters of a 60% acid solution must be mixed with a 75% acid solution to get 20 L of a 72% solution?

**21.** How many liters of a 10% alcohol solution must be mixed with 40 L of a 50% solution to get a 40% solution?

**22.** How many gallons of a 12% indicator solution must be mixed with a 20% indicator solution to get 10 gal of a 14% solution?

**23.** Minoxidil is a drug that has recently proven to be effective in treating male pattern baldness. A pharmacist wishes to mix a solution that is 2% minoxidil. She has on hand 50 mL of a 1% solution, and she wishes to add some 4% solution to it to obtain the desired 2% solution. How much 4% solution should she add?

**24.** Water must be added to 20 mL of a 4% minoxidil solution to dilute it to a 2% solution. How many milliliters of water should be used? (*Hint:* Water is 0% minoxidil.)

*Work each investment problem using simple interest. See Example 3.*

**25.** Li Nguyen invested some money at 3% and $4000 less than that amount at 5%. The two investments produced a total of $200 interest in 1 yr. How much was invested at each rate?

**26.** LaShondra Williams inherited some money from her uncle. She deposited part of the money in a savings account paying 2%, and $3000 more than that amount in a different account paying 3%. Her annual interest income was $690. How much did she deposit at each rate?

**27.** With income earned by selling the rights to his life story, an actor invests some of the money at 3% and $30,000 more than twice as much at 4%. The total annual interest earned from the investments is $5600. How much is invested at each rate?

**28.** An artist invests her earnings in two ways. Some goes into a tax-free bond paying 6%, and $6000 more than three times as much goes into mutual funds paying 5%. Her total annual interest income from the investments is $825. How much does she invest at each rate?

*Work each problem involving monetary values. See Example 4.*

**29.** A bank teller has some $5 bills and some $20 bills. The teller has 5 more twenties than fives. The total value of the money is $725. Find the number of $5 bills that the teller has.

| Number of Bills | Denomination | Total Value |
|---|---|---|
| $x$ | 5 | |
| $x + 5$ | 20 | |

**30.** A coin collector has \$1.70 in dimes and nickels. She has 2 more dimes than nickels. How many nickels does she have?

| Number of Coins | Denomination | Total Value |
|---|---|---|
| $x$ | .05 | $.05x$ |
| | .10 | |

**31.** A cashier has a total of 126 bills, made up of fives and tens. The total value of the money is \$840. How many of each kind does he have?

**32.** A convention manager finds that she has \$1290, made up of twenties and fifties. She has a total of 42 bills. How many of each kind does she have?

**33.** A merchant wishes to mix candy worth \$5 per lb with 40 lb of candy worth \$2 per lb to get a mixture that can be sold for \$3 per lb. How many pounds of \$5 candy should be used?

**34.** At Vern's Grill, hamburgers cost \$2.70 each, and a bag of french fries costs \$1.20. How many hamburgers and how many bags of french fries can a customer buy with \$26.40 if he wants twice as many hamburgers as bags of french fries?

*Solve each problem involving distance, rate, and time. See Example 5.*

**35.** Which choice is the best estimate for the average speed of a trip of 405 mi that lasted 8.2 hr?

**A.** 50 mph **B.** 30 mph **C.** 60 mph **D.** 40 mph

**36.** Suppose that an automobile averages 45 mph, and travels for 30 min. Is the distance traveled $45 \times 30 = 1350$ mi? If not, explain why not, and give the correct distance.

**37.** A driver averaged 53 mph and took 10 hr to travel from Memphis to Chicago. What is the distance between Memphis and Chicago?

**38.** A small plane traveled from Warsaw to Rome, averaging 164 mph. The trip took 2 hr. What is the distance from Warsaw to Rome?

**39.** The winner of the 1998 Charlotte 500 (mile) race was Mark Martin, who drove his Ford to victory with a rate of 123.188 mph. What was his time? (*Source: Sports Illustrated 2000 Sports Almanac.*)

**40.** In 1998, Jeff Gordon drove his Chevrolet to victory in the North Carolina 400 (mile) race. His rate was 128.423 mph. What was his time? (*Source: Sports Illustrated 2000 Sports Almanac.*)

*In Exercises 41–44, find the rate based on the information provided. Use a calculator and round your answers to the nearest hundredth. All events were at the 2000 Summer Olympics.* (*Source:* http://espn.go.com/oly/summer00)

| | Event | Participant | Distance | Time |
|---|---|---|---|---|
| **41.** | 100-m hurdles, Women | Olga Shishigina, Kazakhstan | 100 m | 12.65 sec |
| **42.** | 400-m hurdles, Women | Irina Privalova, Russia | 400 m | 53.02 sec |
| **43.** | 400-m hurdles, Men | Angelo Taylor, USA | 400 m | 47.50 sec |
| **44.** | 400-m dash, Men | Michael Johnson, USA | 400 m | 43.84 sec |

*Solve each motion problem. See Examples 6 and 7.*

**45.** St. Louis and Portland are 2060 mi apart. A small plane leaves Portland, traveling toward St. Louis at an average speed of 90 mph. Another plane leaves St. Louis at the same time, traveling toward Portland, averaging 116 mph. How long will it take them to meet?

| | $r$ | $t$ | $d$ |
|---|---|---|---|
| *Plane Leaving Portland* | 90 | $t$ | $90t$ |
| *Plane Leaving St. Louis* | 116 | $t$ | $116t$ |

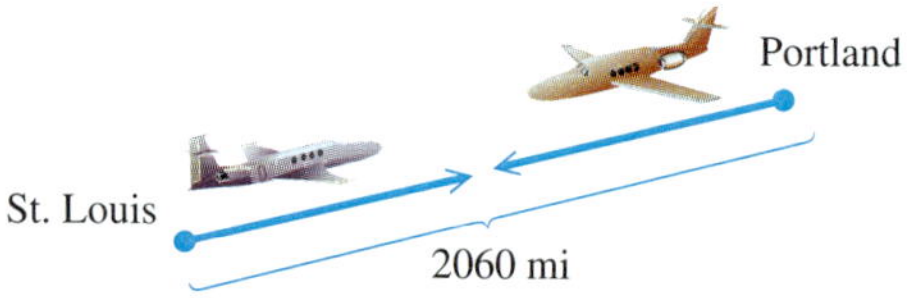


**46.** Atlanta and Cincinnati are 440 mi apart. John leaves Cincinnati, driving toward Atlanta at an average speed of 60 mph. Pat leaves Atlanta at the same time, driving toward Cincinnati in her antique auto, averaging 28 mph. How long will it take them to meet?

| | $r$ | $t$ | $d$ |
|---|---|---|---|
| *John* | 60 | $t$ | $60t$ |
| *Pat* | 28 | $t$ | $28t$ |

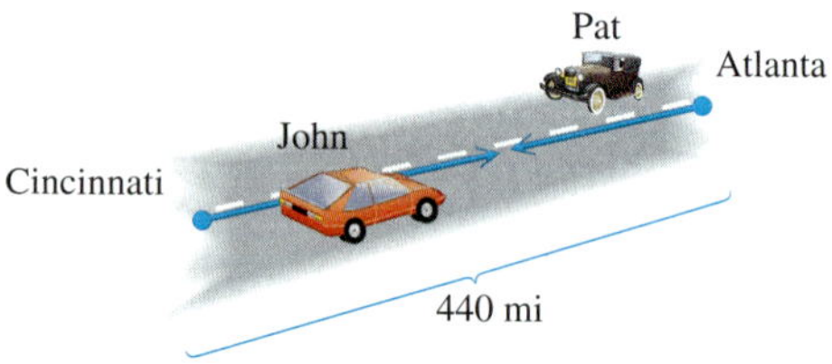


**47.** Two steamers leave a port on a river at the same time, traveling in opposite directions. Each is traveling 22 mph. How long will it take for them to be 110 mi apart?

**48.** A train leaves Kansas City, Kansas, and travels north at 85 km per hr. Another train leaves at the same time and travels south at 95 km per hour. How long will it take before they are 315 km apart?

**49.** At a given hour two steamboats leave a city in the same direction on a straight canal. One travels at 18 mph and the other travels at 25 mph. In how many hours will the boats be 35 mi apart?

**50.** From a point on a straight road, Lupe and Maria ride bicycles in the same direction. Lupe rides 10 mph and Maria rides 12 mph. In how many hours will they be 5 mi apart?

**51.** Two trains leave a city at the same time. One travels north and the other travels south at 20 mph faster. In 2 hr the trains are 280 mi apart. Find their speeds.

| | $r$ | $t$ | $d$ |
|---|---|---|---|
| *Northbound* | $x$ | 2 | |
| *Southbound* | $x + 20$ | 2 | |

# SUPPLEMENT TO SECTION 2.7

## EXAMPLE

The price of a car was reduced to a sale price of \$13,559. This was a 9% reduction. What was the original price?

Solution:

1. Let x be the original price of the car

2. Write an equation:

   Original price - reduction = sale price

   x - .09x = \$13,559

3. Solve the equation

   $x - .09x = 13{,}559$

   $.91x = 13{,}559$

   $x = 13{,}559/.91$

   $x = 14{,}900$

4. Answer with a sentence

   The original price of the car was \$ 14,900

   1. Check:

   9% x \$14, 900 = .09 x 14,900 = \$1341

   \$14, 900 - \$1341 = \$13, 559.

## EXAMPLE

A receipt from Village Stationers shows that the total paid (including tax) for a box of stationery was \$31.50. If the sales tax is 5%, find the cost of the merchandise before the tax is added.

Solution:

1. Define a variable

   Let x be the cost of the stationery before the tax is added

2. Write an equation

   Price before tax + 5% of price before tax = \$ 31.50

   $$x + .05x = 31.50$$

3. Solve the equation

   $$x + .05x = 31.50$$

   $$1.05x = 31.50$$

   $$x = 31.50/1.05$$

   $$x = 30$$

4. Answer with a sentence

   The price of the stationery before the tax was \$30.00

5. Check:

   Tax = \$30 x .05 = \$1.50

   Cost of stationery = \$30 + \$1.50 = \$31.50

## PRACTICE EXERCISES

1. After a 40% reduction, a shirt is on sale for $9.60. What was the original price of the shirt ( that is, the price before the reduction )?

2. After a 34% reduction, a blouse is on sale for $9. 24. What was the original price?

Answers:

1. The original price was $16.00

2. The original price was $14.00

## PRACTICE EXERCISES

1. Doreen paid $920.20 for some building materials. How much did the materials cost if the sales tax was 7%?

2. Jim wrote a check at Office Depot for office supplies that totaled $38.69. If the sales tax was 6% , what was the cost of the supplies?

Answers:

1. The cost of the building materials was $ 860.

2. The cost of the office supplies was $36.50.

**52.** Two planes leave an airport at the same time, one flying east, the other flying west. The eastbound plane travels 150 mph slower. They are 2250 mi apart after 3 hr. Find the speed of each plane.

| | *r* | *t* | *d* |
|---|---|---|---|
| *Eastbound* | $x - 150$ | 3 | |
| *Westbound* | $x$ | 3 | |

**53.** Two cars leave towns 230 km apart at the same time, traveling directly toward one another. One car travels 15 km per hr slower than the other. They pass one another 2 hr later. What are their speeds?

**54.** Two cars start from towns 400 mi apart and travel toward each other. They meet after 4 hr. Find the speed of each car if one travels 20 mph faster than the other.

# 2.8 Solving Linear Inequalities

**OBJECTIVES**

1 Graph intervals on a number line.
2 Use the addition property of inequality.
3 Use the multiplication property of inequality.
4 Solve linear inequalities using both properties of inequality.
5 Solve linear inequalities with three parts.
6 Solve applied problems using inequalities.

Inequalities are algebraic expressions related by

$<$ "is less than," $\quad$ $\leq$ "is less than or equal to,"

$>$ "is greater than," $\quad$ $\geq$ "is greater than or equal to."

We solve an inequality by finding all real number solutions for it. For example, the solution set of $x \leq 2$ includes *all real numbers* that are less than or equal to 2, not just the *integers* less than or equal to 2.

**OBJECTIVE 1 Graph intervals on a number line.** A good way to show the solution set of an inequality is by graphing. We graph all the real numbers satisfying $x \leq 2$ by placing a square bracket at 2 on a number line and drawing an arrow extending from the bracket to the left (to represent the fact that all numbers less than 2 are also part of the graph). The graph is shown in Figure 14.

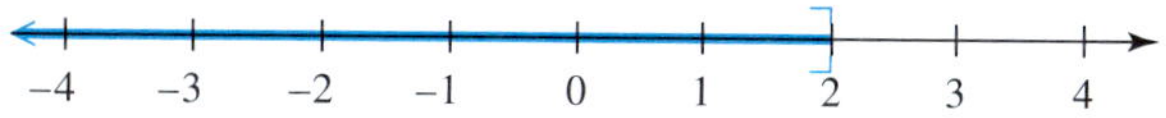


**FIGURE 14**

The set of numbers less than or equal to 2 is an example of an **interval** on the number line. To write intervals, we use **interval notation.** For example, using this notation, the interval of all numbers less than or equal to 2 is written as $(-\infty, 2]$. The **negative infinity** symbol $-\infty$ does not indicate a number. It is used to show that the interval includes all real numbers less than 2. As on the number line, the square bracket indicates that 2 is part of the solution. A parenthesis is always used next to the infinity symbol. The set of real numbers is written in interval notation as $(-\infty, \infty)$.

**EXAMPLE 1 Graphing Intervals Written in Interval Notation on a Number Line**

Write each inequality in interval notation and graph the interval.

**(a)** $x > -5$

The statement $x > -5$ says that $x$ can represent any number greater than $-5$ but cannot equal $-5$. The interval is written $(-5, \infty)$. We show this on a graph by placing a parenthesis at $-5$ and drawing an arrow to the right, as shown in Figure 15. The parenthesis at $-5$ indicates that $-5$ is not part of the graph.

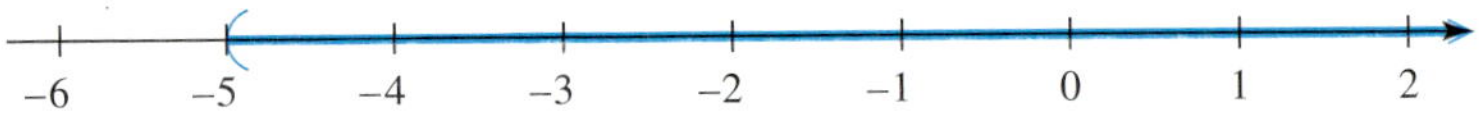


FIGURE 15

**(b)** $-1 \leq x < 3$

The statement is read "$-1$ is less than or equal to $x$ *and* $x$ is less than 3." Thus, we want the set of numbers that are *between* $-1$ and 3, with $-1$ included and 3 excluded. In interval notation, we write $[-1, 3)$, using a square bracket at $-1$ because $-1$ is part of the graph, and a parenthesis at 3 because 3 is not part of the graph. The graph is shown in Figure 16.

FIGURE 16

**Now Try Exercises 7 and 15.**

We summarize the various types of intervals here.

**Interval Notation**

| Type of Interval | Set | Interval Notation | Graph |
|---|---|---|---|
| Open interval | $\{x \mid a < x\}$ | $(a, \infty)$ |  |
| | $\{x \mid a < x < b\}$ | $(a, b)$ |  |
| | $\{x \mid x < b\}$ | $(-\infty, b)$ |  |
| | $\{x \mid x \text{ is a real number}\}$ | $(-\infty, \infty)$ |  |
| Half-open interval | $\{x \mid a \leq x\}$ | $[a, \infty)$ |  |
| | $\{x \mid a < x \leq b\}$ | $(a, b]$ |  |
| | $\{x \mid a \leq x < b\}$ | $[a, b)$ |  |
| | $\{x \mid x \leq b\}$ | $(-\infty, b]$ |  |
| Closed interval | $\{x \mid a \leq x \leq b\}$ | $[a, b]$ |  |

**OBJECTIVE 2 Use the addition property of inequality.** Solving inequalities is similar to solving equations.

**Linear Inequality in One Variable**

A **linear inequality in one variable** can be written in the form

$$Ax + B < C,$$

where $A$, $B$, and $C$ are real numbers, with $A \neq 0$.

(Throughout this section we give definitions and rules only for $<$, but they are also valid for $>$, $\leq$, and $\geq$.)

Examples of linear inequalities in one variable include

$$x + 5 < 2, \qquad y - 3 \geq 5, \qquad \text{and} \qquad 2k + 5 \leq 10.$$

Consider the inequality $2 < 5$. If 4 is added to each side of this inequality, the result is

$$2 + 4 < 5 + 4$$
$$6 < 9,$$

a true sentence. Now subtract 8 from each side:

$$2 - 8 < 5 - 8$$
$$-6 < -3.$$

The result is again a true sentence. These examples suggest the **addition property of inequality,** which states that the same real number can be added to each side of an inequality without changing the solutions.

**Addition Property of Inequality**

For any real numbers $A$, $B$, and $C$, the inequalities

$$A < B \qquad \text{and} \qquad A + C < B + C$$

have exactly the same solutions.

That is, the same number may be added to each side of an inequality without changing the solutions.

As with the addition property of equality, the same number may also be *subtracted* from each side of an inequality.

**EXAMPLE 2 Using the Addition Property of Inequality**

Solve $7 + 3k > 2k - 5$.

Use the addition property of inequality twice, once to get the terms containing $k$ alone on one side of the inequality and a second time to get the integers together on the other side. (These steps can be done in either order.)

$$7 + 3k > 2k - 5$$

$$7 + 3k - 2k > 2k - 5 - 2k \qquad \text{Subtract } 2k.$$

$$7 + k > -5 \qquad \text{Combine like terms.}$$

$$7 + k - 7 > -5 - 7 \qquad \text{Subtract 7.}$$

$$k > -12 \qquad \text{Combine like terms.}$$

The solution set is $(-12, \infty)$. Its graph is shown in Figure 17.

**FIGURE 17**

**Now Try Exercise 21.**

**OBJECTIVE 3 Use the multiplication property of inequality.** The addition property of inequality cannot be used to solve inequalities such as $4y \geq 28$. These inequalities require the *multiplication property of inequality.* To see how this property works, we look at some examples.

First, start with the inequality $3 < 7$ and multiply each side by the positive number 2.

$$3 < 7$$

$$2(3) < 2(7) \qquad \text{Multiply each side by 2.}$$

$$6 < 14 \qquad \text{True}$$

Now multiply each side of $3 < 7$ by the negative number $-5$.

$$3 < 7$$

$$-5(3) < -5(7) \qquad \text{Multiply each side by } -5.$$

$$-15 < -35 \qquad \text{False}$$

To get a true statement when multiplying each side by $-5$, we must reverse the direction of the inequality symbol.

$$3 < 7$$

$$-5(3) > -5(7) \qquad \text{Multiply by } -5\text{; reverse the symbol.}$$

$$-15 > -35 \qquad \text{True}$$

Take the inequality $-6 < 2$ as another example. Multiply each side by the positive number 4.

$$-6 < 2$$

$$4(-6) < 4(2) \qquad \text{Multiply by 4.}$$

$$-24 < 8 \qquad \text{True}$$

Multiplying each side of $-6 < 2$ by $-5$ *and at the same time reversing the direction of the inequality symbol* gives

$$-6 < 2$$

$$-5(-6) > -5(2) \qquad \text{Multiply by } -5\text{; reverse the symbol.}$$

$$30 > -10. \qquad \text{True}$$

In summary, the **multiplication property of inequality** has two parts.

**Multiplication Property of Inequality**

For any real numbers $A$, $B$, and $C$, with $C \neq 0$,

1. if $C$ is *positive,* then the inequalities

$$A < B \quad \text{and} \quad AC < BC$$

have exactly the same solutions;

2. if $C$ is *negative,* then the inequalities

$$A < B \quad \text{and} \quad AC > BC$$

have exactly the same solutions.

That is, each side of an inequality may be multiplied by the same positive number without changing the solutions. If the multiplier is negative, we must reverse the direction of the inequality symbol.

The multiplication property of inequality also permits *division* of each side of an inequality by the same nonzero number.

It is important to remember the differences in the multiplication property for positive and negative numbers.

1. When each side of an inequality is multiplied or divided by a positive number, the direction of the inequality symbol *does not change.* (Adding or subtracting terms on each side also does not change the symbol.)
2. When each side of an inequality is multiplied or divided by a negative number, the direction of the symbol *does change.* ***Reverse the direction of the inequality symbol only when multiplying or dividing each side by a negative number.***

### EXAMPLE 3 Using the Multiplication Property of Inequality

Solve each inequality and graph the solution set.

**(a)** $3r < -18$

Using the multiplication property of inequality, we divide each side by 3. Since 3 is a positive number, the direction of the inequality symbol *does not* change.

$$3r < -18$$

$$\frac{3r}{3} < \frac{-18}{3} \qquad \text{Divide by 3.}$$

$$r < -6$$

The solution set is $(-\infty, -6)$. The graph is shown in Figure 18.

**FIGURE 18**

**(b)** $-4t \geq 8$

Here each side of the inequality must be divided by $-4$, a negative number, which *does* change the direction of the inequality symbol.

$$-4t \geq 8$$

$$\frac{-4t}{-4} \leq \frac{8}{-4} \quad \text{Divide by } -4\text{; reverse the symbol.}$$

$$t \leq -2$$

The solution set $(-\infty, -2]$ is graphed in Figure 19.

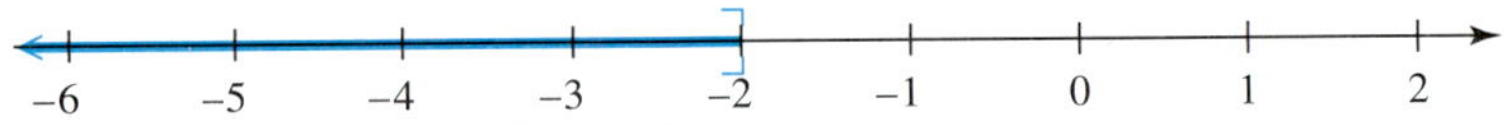


FIGURE 19

**Now Try Exercises 29 and 31.**

**CAUTION** Even though the number on the right side of the inequality in Example 3(a) is negative $(-18)$, *do not reverse the direction of the inequality symbol.* Reverse the symbol only when multiplying or dividing by a negative number, as shown in Example 3(b).

**OBJECTIVE 4 Solve linear inequalities using both properties of inequality.** To solve a linear inequality, follow these steps.

### Solving a Linear Inequality

*Step 1* **Simplify each side separately.** Use the distributive property to clear parentheses and combine like terms on each side as needed.

*Step 2* **Isolate the variable terms on one side.** Use the addition property of inequality to get all terms with variables on one side of the inequality and all numbers on the other side.

*Step 3* **Isolate the variable.** Use the multiplication property of inequality to change the inequality to the form $x < k$ or $x > k$.

*Remember:* Reverse the direction of the inequality symbol ***only when multiplying or dividing each side of an inequality by a negative number.***

### EXAMPLE 4 Solving a Linear Inequality

Solve $3z + 2 - 5 > -z + 7 + 2z$ and graph the solution set.

*Step 1* Combine like terms and simplify.

$$3z + 2 - 5 > -z + 7 + 2z$$

$$3z - 3 > z + 7$$

*Step 2* Use the addition property of inequality.

$$\begin{aligned} 3z - 3 + 3 &> z + 7 + 3 && \text{Add 3.} \\ 3z &> z + 10 \\ 3z - z &> z + 10 - z && \text{Subtract } z. \\ 2z &> 10 \end{aligned}$$

*Step 3* Use the multiplication property of inequality.

$$\begin{aligned} \frac{2z}{2} &> \frac{10}{2} && \text{Divide by 2.} \\ z &> 5 \end{aligned}$$

Since 2 is positive, the direction of the inequality symbol is not changed in Step 3. The solution set is $(5, \infty)$. Its graph is shown in Figure 20.

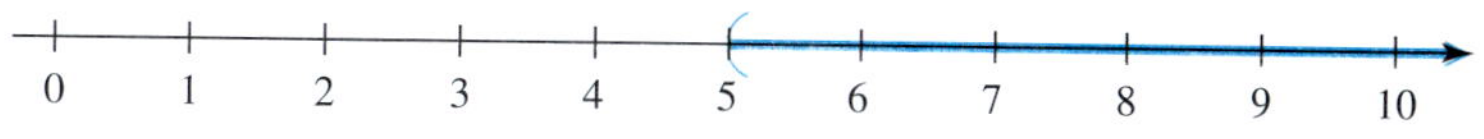

**FIGURE 20**

**Now Try Exercise 41.**

## EXAMPLE 5 Solving a Linear Inequality

Solve $5(k - 3) - 7k \geq 4(k - 3) + 9$ and graph the solution set.

*Step 1* Simplify and combine like terms.

$$\begin{aligned} 5(k - 3) - 7k &\geq 4(k - 3) + 9 \\ 5k - 15 - 7k &\geq 4k - 12 + 9 && \text{Distributive property} \\ -2k - 15 &\geq 4k - 3 && \text{Combine like terms.} \end{aligned}$$

*Step 2* Use the addition property.

$$\begin{aligned} -2k - 15 - 4k &\geq 4k - 3 - 4k && \text{Subtract } 4k. \\ -6k - 15 &\geq -3 \\ -6k - 15 + 15 &\geq -3 + 15 && \text{Add 15.} \\ -6k &\geq 12 \end{aligned}$$

*Step 3* Divide each side by $-6$, a negative number. Change the direction of the inequality symbol.

$$\begin{aligned} \frac{-6k}{-6} &\leq \frac{12}{-6} && \text{Divide by } -6\text{; reverse the symbol.} \\ k &\leq -2 \end{aligned}$$

The solution set is $(-\infty, -2]$. Its graph is shown in Figure 21.

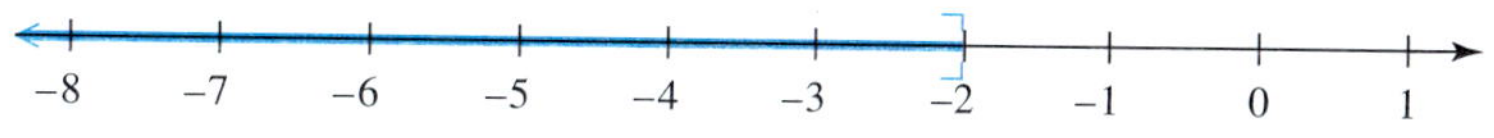

**FIGURE 21**

**Now Try Exercise 45.**

## EXAMPLE 6 Solving a Linear Inequality with Fractions

Solve $-\frac{2}{3}(r-3)-\frac{1}{2}<\frac{1}{2}(5-r)$ and graph the solution set.

To clear fractions, multiply each side by the least common denominator, 6.

$$6\left[-\frac{2}{3}(r-3)-\frac{1}{2}\right]<6\left[\frac{1}{2}(5-r)\right] \qquad \text{Multiply by 6.}$$

$$6\left[-\frac{2}{3}(r-3)\right]-6\left(\frac{1}{2}\right)<6\left[\frac{1}{2}(5-r)\right] \qquad \text{Distributive property}$$

$$-4(r-3)-3<3(5-r)$$

*Step 1*
$$-4r+12-3<15-3r \qquad \text{Distributive property}$$
$$-4r+9<15-3r$$

*Step 2*
$$-4r+9+3r<15-3r+3r \qquad \text{Add } 3r.$$
$$-r+9<15$$
$$-r+9-9<15-9 \qquad \text{Subtract 9.}$$
$$-r<6$$

*Step 3* To solve for $r$, multiply each side of the inequality by $-1$. Since $-1$ is negative, change the direction of the inequality symbol.

$$-1(-r)>-1(6) \qquad \text{Multiply by } -1\text{, change } < \text{ to } >.$$
$$r>-6$$

The solution set is $(-6, \infty)$. See the graph in Figure 22.

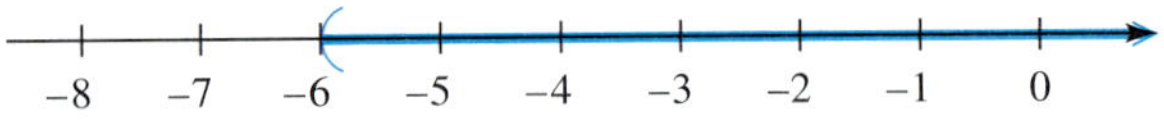


FIGURE 22

**Now Try Exercise 47.**

**OBJECTIVE 5 Solve linear inequalities with three parts.** Inequalities that say that one number is *between* two other numbers are **three-part inequalities.** For example,

$$-3<5<7$$

says that 5 is between $-3$ and 7. For some applications, it is necessary to work with an inequality such as

$$3<x+2<8,$$

where $x+2$ is between 3 and 8. To solve this inequality, we subtract 2 from each of the three parts of the inequality, giving

$$3-2<x+2-2<8-2$$
$$1<x<6.$$

The idea is to get the inequality in the form

$$\text{a number} < x < \text{another number},$$

using "is less than." The solution set (in this case the interval $(1, 6)$) can then easily be graphed.

**CAUTION** When inequalities have three parts, the order of the parts is important. It would be *wrong* to write an inequality as $8 < x + 2 < 3$, since this would imply that $8 < 3$, a false statement. In general, three-part inequalities are written so that the symbols point in the same direction, and both point toward the smaller number.

## EXAMPLE 7 Solving Three-Part Inequalities

Solve each inequality and graph the solution set.

**(a)** $$4 \le 3x - 5 < 6$$

$$4 + 5 \le 3x - 5 + 5 < 6 + 5 \quad \text{Add 5 to each part.}$$

$$9 \le 3x < 11$$

$$\frac{9}{3} \le \frac{3x}{3} < \frac{11}{3} \quad \text{Divide each part by 3.}$$

$$3 \le x < \frac{11}{3}$$

The solution set is $\left[3, \frac{11}{3}\right)$. Its graph is shown in Figure 23.

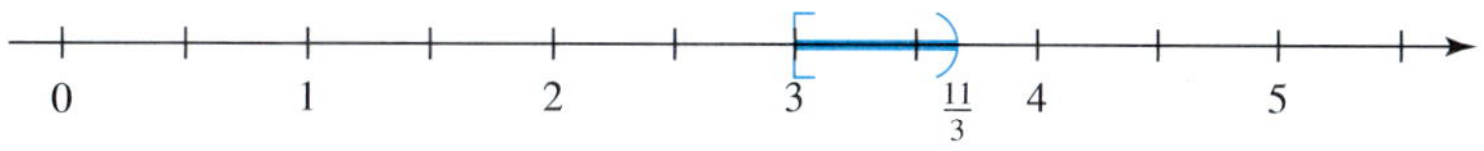


**FIGURE 23**

**(b)** $-4 \le \frac{2}{3}m - 1 < 8$

To clear the fraction, multiply each part by 3.

$$3(-4) \le 3\left(\frac{2}{3}m - 1\right) < 3(8) \quad \text{Multiply each part by 3.}$$

$$-12 \le 2m - 3 < 24 \quad \text{Distributive property}$$

$$-12 + 3 \le 2m - 3 + 3 < 24 + 3 \quad \text{Add 3 to each part.}$$

$$-9 \le 2m < 27$$

$$\frac{-9}{2} \le \frac{2m}{2} < \frac{27}{2} \quad \text{Divide each part by 2.}$$

$$-\frac{9}{2} \le m < \frac{27}{2}$$

The solution set is $\left[-\frac{9}{2}, \frac{27}{2}\right)$. Its graph is shown in Figure 24.

**FIGURE 24**

**Now Try Exercises 59 and 65.**

**NOTE** The inequality in Example 7(b) can also be solved by first adding 1 to each part, and then multiplying each part by $\frac{3}{2}$. Do this and confirm that the same solution set results.

Examples of the types of solution sets to be expected from solving linear equations or linear inequalities are shown below.

**Solution Sets of Linear Equations and Inequalities**

| *Equation or Inequality* | *Typical Solution Set* | *Graph of Solution Set* |
|---|---|---|
| Linear equation $5x + 4 = 14$ | $\{2\}$ | 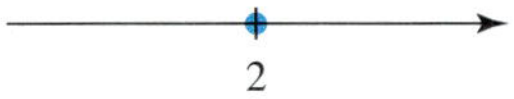 |
| Linear inequality $5x + 4 < 14$ | $(-\infty, 2)$ | 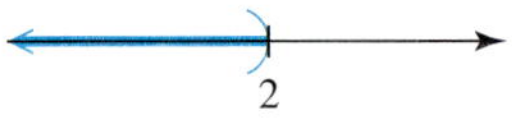 |
| Linear inequality $5x + 4 > 14$ | $(2, \infty)$ | 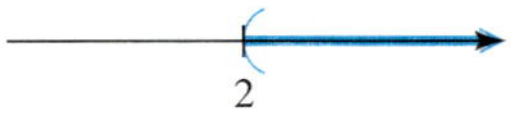 |
| Three-part inequality $-1 \leq 5x + 4 \leq 14$ | $[-1, 2]$ |  |

**OBJECTIVE 6 Solve applied problems using inequalities.** Until now, the applied problems that we have studied have all led to equations.

**PROBLEM SOLVING**

Inequalities can be used to solve applied problems involving phrases that suggest inequality. The table gives some of the more common such phrases along with examples and translations.

| *Phrase* | *Example* | *Inequality* |
|---|---|---|
| Is more than | A number *is more than* 4. | $x > 4$ |
| Is less than | A number *is less than* $-12$. | $x < -12$ |
| Is at least | A number *is at least* 6. | $x \geq 6$ |
| Is at most | A number *is at most* 8. | $x \leq 8$ |

We use the same six problem-solving steps from Section 2.4, changing Step 3 to "Write an inequality" instead of "Write an equation."

**CAUTION** Do not confuse statements like "5 is more than a number" with the phrase "5 more than a number." The first of these is expressed as $5 > x$ while the second is expressed as $x + 5$ or $5 + x$.

The next example shows an application of algebra that is important to anyone who has ever asked, "What score can I make on my next test and have a (particular grade) in this course?" It uses the idea of finding the average of a number of grades. In general, to find the average of $n$ numbers, add the numbers, then divide by $n$.

### EXAMPLE 8 Finding an Average Test Score

Brent has test grades of 86, 88, and 78 on his first three tests in geometry. If he wants an average of at least 80 after his fourth test, what are the possible scores he can make on his fourth test?

*Step 1* **Read** the problem again.

*Step 2* **Assign a variable.** Let $x$ = Brent's score on his fourth test.

*Step 3* **Write an inequality.** To find his average after 4 tests, add the test scores and divide by 4.

Average — is at least — 80.

$$\frac{86 + 88 + 78 + x}{4} \geq 80$$

*Step 4* **Solve.**

$$\frac{252 + x}{4} \geq 80 \qquad \text{Add the known scores.}$$

$$4\left(\frac{252 + x}{4}\right) \geq 4(80) \qquad \text{Multiply by 4.}$$

$$252 + x \geq 320$$

$$252 + x - 252 \geq 320 - 252 \qquad \text{Subtract 252.}$$

$$x \geq 68 \qquad \text{Combine like terms.}$$

*Step 5* **State the answer.** He must score 68 or more on the fourth test to have an average of *at least* 80.

*Step 6* **Check.**

$$\frac{86 + 88 + 78 + 68}{4} = \frac{320}{4} = 80$$

**Now Try Exercise 77.**

**CAUTION** Errors often occur when the phrases "at least" and "at most" appear in applied problems. Remember that

at least translates as greater than or equal to

and

at most translates as less than or equal to.

### EXAMPLE 9 Using a Linear Inequality to Solve a Rental Problem

A rental company charges \$15 to rent a chain saw, plus \$2 per hr. Al Ghandi can spend no more than \$35 to clear some logs from his yard. What is the *maximum* amount of time he can use the rented saw?

*Step 1* **Read** the problem again.

*Step 2* **Assign a variable.** Let $h$ = the number of hours he can rent the saw.

*Step 3* **Write an inequality.** He must pay \$15, plus \$2$h$, to rent the saw for $h$ hours, and this amount must be *no more than* \$35.

| Cost of renting | is no more than | 35 dollars. |
|---|---|---|
| $15 + 2h$ | $\leq$ | $35$ |

*Step 4* **Solve.** $2h \leq 20$ Subtract 15.

$h \leq 10$ Divide by 2.

*Step 5* **State the answer.** He can use the saw for a maximum of 10 hr. (Of course, he may use it for less time, as indicated by the inequality $h \leq 10$.)

*Step 6* **Check.** If Al uses the saw for 10 hr, he will spend $15 + 2(10) = 35$ dollars, the maximum amount.

**Now Try Exercise 87.**

## CONNECTIONS

Many applications from economics involve inequalities rather than equations. For example, a company that produces videocassettes has found that revenue from the sales of the cassettes is \$5 per cassette less sales costs of \$100. Production costs are \$125 plus \$4 per cassette. Profit ($P$) is given by revenue ($R$) less cost ($C$), so the company must find the production level $x$ that makes

$$P = R - C > 0.$$

### For Discussion or Writing

Write an expression for revenue, letting $x$ represent the production level (number of cassettes to be produced). Write an expression for production costs using $x$. Write an expression for profit and solve the inequality shown above. Describe the solution in terms of the problem.

## 2.8 EXERCISES

**For Extra Help**

- Student's Solutions Manual
- MyMathLab
- InterAct Math Tutorial Software
- AW Math Tutor Center
- MathXL
- Digital Video Tutor CD 4/Videotape 4

**1.** Explain how to determine whether to use a parenthesis or a square bracket at the endpoint when graphing an inequality on a number line.

**2.** How does the graph of $t \geq -7$ differ from the graph of $t > -7$?

*Write an inequality involving the variable x that describes each set of numbers graphed. See Example 1.*

**3.**

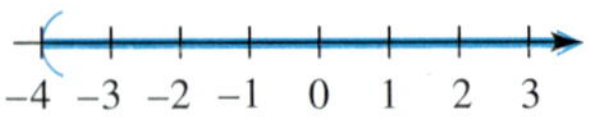


**4.**

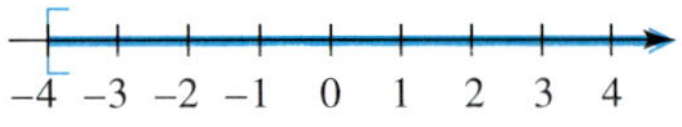


**5.**

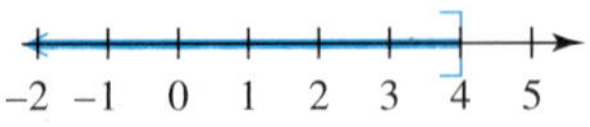


**6.**

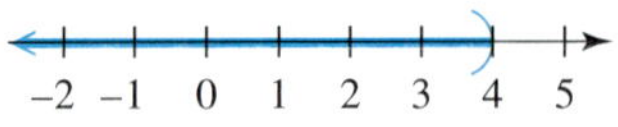

*Write each inequality in interval notation and graph the interval. See Example 1.*

**7.** $k \le 4$ **8.** $r \le -11$ **9.** $x < -3$ **10.** $y < 3$

**11.** $t > 4$ **12.** $m > 5$ **13.** $8 \le x \le 10$ **14.** $3 \le x \le 5$

**15.** $0 < y \le 10$ **16.** $-3 \le x < 5$

**17.** Why is it wrong to write $3 < x < -2$ to indicate that $x$ is between $-2$ and 3?

**18.** If $p < q$ and $r < 0$, which one of the following statements is false?

**A.** $pr < qr$ **B.** $pr > qr$ **C.** $p + r < q + r$ **D.** $p - r < q - r$

*Solve each inequality. Write the solution set in interval notation and graph it. See Example 2.*

**19.** $z - 8 \ge -7$ **20.** $p - 3 \ge -11$ **21.** $2k + 3 \ge k + 8$

**22.** $3x + 7 \ge 2x + 11$ **23.** $3n + 5 < 2n - 6$ **24.** $5x - 2 < 4x - 5$

**25.** Under what conditions must the inequality symbol be reversed when solving an inequality?

**26.** Your friend tells you that when solving the inequality $6x < -42$ he reversed the direction of the inequality symbol because of the $-42$. How would you respond?

*Solve each inequality. Write the solution set in interval notation and graph it. See Example 3.*

**27.** $3x < 18$ **28.** $5x < 35$ **29.** $2y \ge -20$ **30.** $6m \ge -24$

**31.** $-8t > 24$ **32.** $-7x > 49$ **33.** $-x \ge 0$ **34.** $-k < 0$

**35.** $-\frac{3}{4}r < -15$ **36.** $-\frac{7}{8}t < -14$ **37.** $-.02x \le .06$ **38.** $-.03v \ge -.12$

*Solve each inequality. Write the solution set in interval notation and graph it. See Examples 4–6.*

**39.** $5r + 1 \ge 3r - 9$ **40.** $6t + 3 < 3t + 12$

**41.** $6x + 3 + x < 2 + 4x + 4$ **42.** $-4w + 12 + 9w \ge w + 9 + w$

**43.** $-x + 4 + 7x \le -2 + 3x + 6$ **44.** $14y - 6 + 7y > 4 + 10y - 10$

**45.** $5(x + 3) - 6x \le 3(2x + 1) - 4x$ **46.** $2(x - 5) + 3x < 4(x - 6) + 1$

**47.** $\frac{2}{3}(p + 3) > \frac{5}{6}(p - 4)$ **48.** $\frac{7}{9}(y - 4) \le \frac{4}{3}(y + 5)$

**49.** $-\frac{1}{4}(p + 6) + \frac{3}{2}(2p - 5) < 10$ **50.** $\frac{3}{5}(k - 2) - \frac{1}{4}(2k - 7) \le 3$

**51.** $4x - (6x + 1) \le 8x + 2(x - 3)$ **52.** $2y - (4y + 3) > 6y + 3(y + 4)$

**53.** $5(2k + 3) - 2(k - 8) > 3(2k + 4) + k - 2$

**54.** $2(3z - 5) + 4(z + 6) \ge 2(3z + 2) + 3z - 15$

*Write a three-part inequality involving the variable x that describes each set of numbers graphed. See Example 1(b).*

**55.** 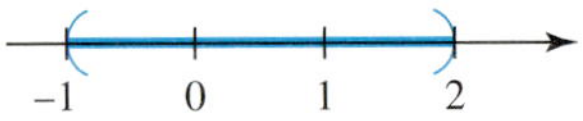


**56.** 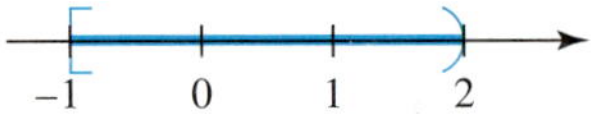


**57.** 

**58.** 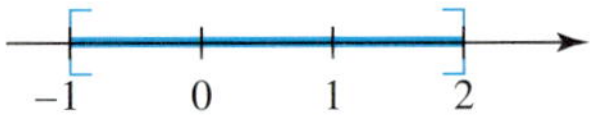


*Solve each inequality. Write the solution set in interval notation and graph it. See Example 7.*

**59.** $-5 \leq 2x - 3 \leq 9$

**60.** $-7 \leq 3x - 4 \leq 8$

**61.** $5 < 1 - 6m < 12$

**62.** $-1 \leq 1 - 5q \leq 16$

**63.** $10 < 7p + 3 < 24$

**64.** $-8 \leq 3r - 1 \leq -1$

**65.** $-12 \leq \frac{1}{2}z + 1 \leq 4$

**66.** $-6 \leq 3 + \frac{1}{3}a \leq 5$

**67.** $1 \leq 3 + \frac{2}{3}p \leq 7$

**68.** $2 < 6 + \frac{3}{4}y < 12$

**69.** $-7 \leq \frac{5}{4}r - 1 \leq -1$

**70.** $-12 \leq \frac{3}{7}a + 2 \leq -4$

## RELATING CONCEPTS (EXERCISES 71–76)

**For Individual or Group Work**

*The methods for solving linear equations and linear inequalities are quite similar.* ***Work Exercises 71–76 in order,*** *to see how the solutions of an inequality are closely connected to the solution of the corresponding equation.*

**71.** Solve the equation $3x + 2 = 14$ and graph the solution set as a single point on the number line.

**72.** Solve the inequality $3x + 2 > 14$ and graph the solution set as an interval on the number line.

**73.** Solve the inequality $3x + 2 < 14$ and graph the solution set as an interval on the number line.

**74.** If you were to graph all the solution sets from Exercises 71–73 on the same number line, what would the graph be? (This is called the *union* of all the solution sets.)

**75.** Based on your results from Exercises 71–74, if you were to graph the union of the solution sets of

$$-4x + 3 = -1, \qquad -4x + 3 > -1, \qquad \text{and} \qquad -4x + 3 < -1,$$

what do you think the graph would be?

**76.** Comment on the following statement: *Equality* is the boundary between *less than* and *greater than.*

*Solve each problem. See Examples 8 and 9.*

**77.** Inkie Landry has scores of 76 and 81 on her first two algebra tests. If she wants an average of at least 80 after her third test, what possible scores can she make on her third test?

**78.** Mabimi Pampo has scores of 96 and 86 on his first two geometry tests. What possible scores can he make on his third test so that his average is at least 90?

**79.** When 2 is added to the difference between six times a number and 5, the result is greater than 13 added to five times the number. Find all such numbers.

**80.** When 8 is subtracted from the sum of three times a number and 6, the result is less than 4 more than the number. Find all such numbers.

**81.** The formula for converting Fahrenheit temperature to Celsius is

$$C = \frac{5}{9}(F - 32).$$

If the Celsius temperature on a certain summer day in Toledo is never more than 30°, how would you describe the corresponding Fahrenheit temperatures?

**82.** The formula for converting Celsius temperature to Fahrenheit is

$$F = \frac{9}{5}C + 32.$$

The Fahrenheit temperature of Key West, Florida, has never exceeded 95°. How would you describe this using Celsius temperature?

**83.** For what values of $x$ would the rectangle have perimeter of at least 400?

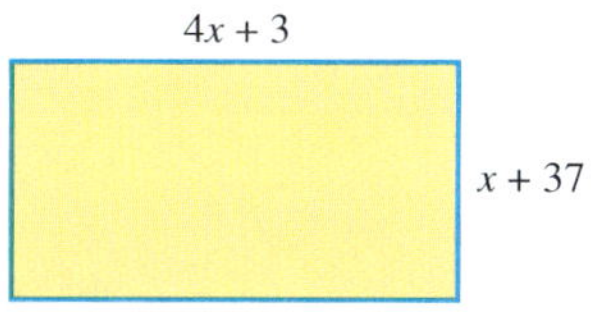


**84.** For what values of $x$ would the triangle have perimeter of at least 72?

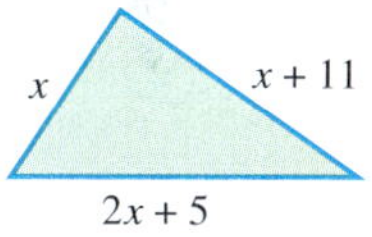


**85.** A long-distance phone call costs \$2.00 for the first three minutes plus \$.30 per minute for each minute or fractional part of a minute after the first three minutes. If $x$ represents the number of minutes of the length of the call after the first three minutes, then $2 + .30x$ represents the cost of the call. If Jorge has \$5.60 to spend on a call, what is the maximum total time he can use the phone?

**86.** If the call described in Exercise 85 costs between \$5.60 and \$6.50, what are the possible total time lengths for the call?

**87.** At the Speedy Gas'n Go, a car wash costs \$3.00, and gasoline is selling for \$1.50 per gal. Lee Ann Spahr has \$17.25 to spend, and her car is so dirty that she must have it washed. What is the maximum number of gallons of gasoline that she can purchase?

**88.** A taxicab driver charges \$1.50 for the first $\frac{1}{5}$ mi and \$.25 for each additional $\frac{1}{5}$ mi. Dantrell Jackson has only \$3.75. What is the maximum distance he can travel (not including a tip for the cabbie)?

**89.** A product will break even or produce a profit if the revenue $R$ from selling the product is at least equal to the cost $C$ of producing it. Suppose that the cost $C$ (in dollars) to produce $x$ units of bicycle helmets is $C = 50x + 5000$, while the revenue $R$ (in dollars) collected from the sale of $x$ units is $R = 60x$. For what values of $x$ does the product break even or produce a profit?

**90.** (See Exercise 89.) If the cost to produce $x$ units of baseball cards is $C = 100x + 6000$ (in dollars), and the revenue collected from selling $x$ units is $R = 500x$ (in dollars), for what values of $x$ does the product break even or produce a profit?

**91.** A BMI (body mass index) between 19 and 25 is considered healthy. Use the formula

$$\text{BMI} = \frac{704 \times (\text{weight in pounds})}{(\text{height in inches})^2}$$

to find the weight range $w$, to the nearest pound, that gives a healthy BMI for each height. (*Source: Washington Post.*)

**(a)** 72 in. **(b)** Your height in inches

**92.** To achieve the maximum benefit from exercising, the heart rate in beats per minute should be in the target heart rate zone (THR). For a person aged $A$, the formula is

$$.7(220 - A) \leq \text{THR} \leq .85(220 - A).$$

Find the THR to the nearest whole number for each age. (*Source:* Hockey, Robert V., *Physical Fitness: The Pathway to Healthful Living,* Times Mirror/Mosby College Publishing, 1989.)

**(a)** 35 **(b)** Your age

*The July 14th weather forecast by time of day for the 2000 U.S. Olympic Track and Field Trials, held July 14–23, 2000, in Sacramento, California, is shown in the figure. Use this graph to work Exercises 93–96.*

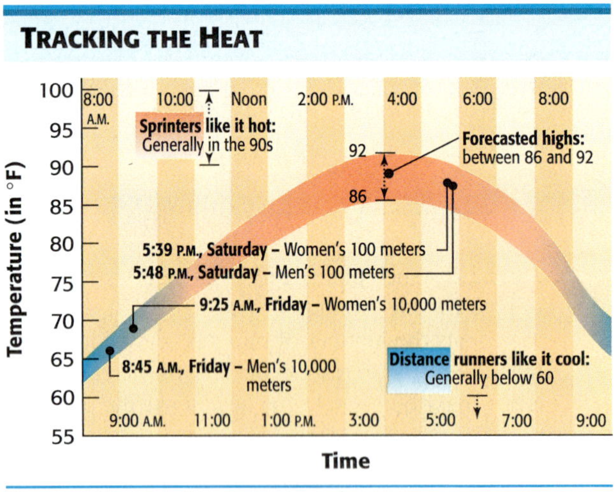


*Source:* Accuweather, Bee research.

**93.** Sprinters prefer Fahrenheit temperatures in the 90s. Using the upper boundary of the forecast, in what time period is the temperature expected to be at least 90°F?

**94.** Distance runners prefer cool temperatures. During what time period are temperatures predicted to be no more than 70°F? Use the lower forecast boundary.

**95.** What range of temperatures is predicted for the Women's 100-m event?

**96.** What range of temperatures is forecast for the Men's 10,000-m event?

## Chapter 2 Group Activity

### Are You a Race-Walker?

**OBJECTIVE** Use proportions to calculate walking speeds.

**MATERIALS** Students will need a large area for walking. Each group will need a stopwatch.

Race-walking at speeds exceeding 8 mph is a high fitness, long-distance competitive sport. The table contains gold medal winners in Olympic race-walking competition. Complete the table by applying the proportion given below to find the race-walker's steps per minute.

**Use 10 km ≈ 6.21 mi.** (Round all answers except those for steps per minute to the nearest thousandth. Round steps per minute to the nearest whole number.)

$$\frac{70 \text{ steps per min}}{2 \text{ mph}} = \frac{x \text{ steps per min}}{y \text{ mph}}$$

| *Event* | *Gold Medal Winner* | *Country* | *Time in Hours: Minutes: Seconds* | *Time in Minutes* | *Time in Hours* | *y Miles per Hour* | *x Steps per Minute* |
|---|---|---|---|---|---|---|---|
| 10-km Walk, Women | Yelena Nikolayeva | Russia | 0:41:49 | | | | |
| 20-km Walk, Men | Jefferson Perez | Ecuador | 1:20:07 | | | | |
| 50-km Walk, Men | Robert Korzeniowski | Poland | 3:43:30 | | | | |

*Source: World Almanac and Book of Facts,* 1999.

**A.** Using a stopwatch, take turns counting how many steps each member of the group takes in one minute while walking at a normal pace. Record the results in the following table. Then do it again at a fast pace. Record these results.

| | *Normal Pace* | | *Fast Pace* | |
|---|---|---|---|---|
| *Name* | *x Steps per Minute* | *y Miles per Hour* | *x Steps per Minute* | *y Miles per Hour* |
| | | | | |
| | | | | |
| | | | | |
| | | | | |
| | | | | |

**B.** Use the given proportion to convert the numbers from part A to miles per hour and complete the table.

1. Find the average speed for the group at a normal pace and at a fast pace.
2. What is the minimum number of steps per minute you would have to take to be a race-walker?
3. At a fast pace did anyone in the group walk fast enough to be a race-walker? Explain how you decided.

# CHAPTER 2 SUMMARY

## KEY TERMS

**2.1** linear equation in one variable
solution set
equivalent equations

**2.3** conditional equation
identity
contradiction
empty (null) set

**2.4** degree
complementary angles
right angle
supplementary angles
straight angle
consecutive integers

**2.5** area
perimeter
vertical angles
volume

**2.6** ratio
proportion
terms of a proportion
extremes
means
cross products

**2.8** interval on the number line
interval notation
linear inequality in one variable
three-part inequality

## NEW SYMBOLS

$\emptyset$ empty set

$1^\circ$ one degree

$a$ **to** $b$, $a{:}b$, **or** $\frac{a}{b}$ the ratio of $a$ to $b$

$(a, b)$ interval notation for $a < x < b$

$[a, b]$ interval notation for $a \leq x \leq b$

$\infty$ infinity

$-\infty$ negative infinity

$(-\infty, \infty)$ set of all real numbers

## TEST YOUR WORD POWER

*See how well you have learned the vocabulary in this chapter. Answers, with examples, follow the Quick Review.*

**1.** A **solution set** is the set of numbers that
- **A.** make an expression undefined
- **B.** make an equation false
- **C.** make an equation true
- **D.** make an expression equal to 0.

**2.** The **empty set** is a set
- **A.** with 0 as its only element
- **B.** with an infinite number of elements
- **C.** with no elements
- **D.** of ideas.

**3.** **Complementary angles** are angles
- **A.** formed by two parallel lines
- **B.** whose sum is 90°
- **C.** whose sum is 180°
- **D.** formed by perpendicular lines.

**4.** **Supplementary angles** are angles
- **A.** formed by two parallel lines
- **B.** whose sum is 90°
- **C.** whose sum is 180°
- **D.** formed by perpendicular lines.

**5.** A **ratio**
- **A.** compares two quantities using a quotient
- **B.** says that two quotients are equal
- **C.** is a product of two quantities
- **D.** is a difference between two quantities.

**6.** A **proportion**
- **A.** compares two quantities using a quotient
- **B.** says that two quotients are equal
- **C.** is a product of two quantities
- **D.** is a difference between two quantities.

**7.** An **inequality** is
- **A.** a statement that two algebraic expressions are equal
- **B.** a point on a number line
- **C.** an equation with no solutions
- **D.** a statement with algebraic expressions related by $<$, $\leq$, $>$, or $\geq$.

**8.** **Interval notation** is
- **A.** a portion of a number line
- **B.** a special notation for describing a point on a number line
- **C.** a way to use symbols to describe an interval on a number line
- **D.** a notation to describe unequal quantities.

## QUICK REVIEW

| CONCEPTS | EXAMPLES |
|---|---|
| **2.1 THE ADDITION PROPERTY OF EQUALITY** | |
| The same number may be added to (or subtracted from) each side of an equation without changing the solution. | Solve. $x - 6 = 12$<br>$x - 6 + 6 = 12 + 6$ Add 6.<br>$x = 18$ Combine like terms.<br>Solution set: $\{18\}$ |
| **2.2 THE MULTIPLICATION PROPERTY OF EQUALITY** | |
| Each side of an equation may be multiplied (or divided) by the same nonzero number without changing the solution. | Solve. $\frac{3}{4}x = -9$<br>$\frac{4}{3} \cdot \frac{3}{4}x = \frac{4}{3}(-9)$ Multiply by $\frac{4}{3}$.<br>$x = -12$<br>Solution set: $\{-12\}$ |
| **2.3 MORE ON SOLVING LINEAR EQUATIONS** | |
| | Solve. |
| *Step 1* Simplify each side separately. | $2x + 2(x + 1) = 14 + x$<br>$2x + 2x + 2 = 14 + x$ Distributive property<br>$4x + 2 = 14 + x$ Combine like terms. |
| *Step 2* Isolate the variable term on one side. | $4x + 2 - x - 2 = 14 + x - x - 2$ Subtract $x$; subtract 2.<br>$3x = 12$ Combine like terms. |
| *Step 3* Isolate the variable. | $\frac{3x}{3} = \frac{12}{3}$ Divide by 3.<br>$x = 4$ |
| *Step 4* Check. | *Check:*<br>$2(4) + 2(4 + 1) \stackrel{?}{=} 14 + 4$ Let $x = 4$.<br>$18 = 18$ True<br>Solution set: $\{4\}$ |
| **2.4 AN INTRODUCTION TO APPLICATIONS OF LINEAR EQUATIONS** | |
| | One number is 5 more than another. Their sum is 21. What are the numbers? |
| *Step 1* Read. | We are looking for two numbers. |
| *Step 2* Assign a variable. | Let $x$ represent the smaller number. Then $x + 5$ represents the larger number. |

(continued)

| CONCEPTS | EXAMPLES |
| --- | --- |
| *Step 3* Write an equation. | $x + (x + 5) = 21$ |
| *Step 4* Solve the equation. | $2x + 5 = 21$ Combine like terms.<br>$2x + 5 - 5 = 21 - 5$ Subtract 5.<br>$2x = 16$ Combine like terms.<br>$\frac{2x}{2} = \frac{16}{2}$ Divide by 2.<br>$x = 8$ |
| *Step 5* State the answer.<br>*Step 6* Check. | The numbers are 8 and 13.<br>13 is 5 more than 8, and $8 + 13 = 21$. It checks. |

**2.5 FORMULAS AND APPLICATIONS FROM GEOMETRY**

| CONCEPTS | EXAMPLES |
| --- | --- |
| To find the value of one of the variables in a formula, given values for the others, substitute the known values into the formula. | Find $L$ if $A = LW$, given that $A = 24$ and $W = 3$.<br>$24 = L \cdot 3$ $A = 24, W = 3$<br>$\frac{24}{3} = \frac{L \cdot 3}{3}$ Divide by 3.<br>$8 = L$ |
| To solve a formula for one of the variables, isolate that variable by treating the other variables as numbers and using the steps for solving equations. | Solve $P = 2L + 2W$ for $W$.<br>$P - 2L = 2L + 2W - 2L$ Subtract $2L$.<br>$P - 2L = 2W$ Combine like terms.<br>$\frac{P - 2L}{2} = \frac{2W}{2}$ Divide by 2.<br>$\frac{P - 2L}{2} = W$ or $W = \frac{P - 2L}{2}$ |

**2.6 RATIOS AND PROPORTIONS**

| CONCEPTS | EXAMPLES |
| --- | --- |
| To write a ratio, express quantities in the same units. | 4 ft to 8 in. = 48 in. to 8 in. $= \frac{48}{8} = \frac{6}{1}$ |
| To solve a proportion, use the method of cross products. | Solve $\frac{x}{12} = \frac{35}{60}$.<br>$60x = 12 \cdot 35$ Cross products<br>$60x = 420$ Multiply.<br>$\frac{60x}{60} = \frac{420}{60}$ Divide by 60.<br>$x = 7$<br>Solution set: $\{7\}$ |

| CONCEPTS | EXAMPLES |
|---|---|

## 2.7 MORE ABOUT PROBLEM SOLVING

*Step 1* Read.

A sum of money is invested at simple interest in two ways. Part is invested at 12%, and \$20,000 less than that amount is invested at 10%. If the total interest for 1 yr is \$9000, find the amount invested at each rate.

*Step 2* Assign a variable. Make a table to help solve the problems in this section.

Let $x$ = amount invested at 12%.

Then $x - 20,000$ = amount invested at 10%.

| Dollars Invested | Rate of Interest | Interest for One Year |
|---|---|---|
| x | .12 | .12x |
| x − 20,000 | .10 | .10(x − 20,000) |

*Step 3* Write an equation.

*Step 4* Solve the equation.

$$.12x + .10(x - 20,000) = 9000$$

$$.12x + .10x + .10(-20,000) = 9000 \quad \text{Distributive property}$$

$$12x + 10x + 10(-20,000) = 900,000 \quad \text{Multiply by 100.}$$

$$12x + 10x - 200,000 = 900,000$$

$$22x - 200,000 = 900,000$$

$$22x = 1,100,000 \quad \text{Add 200,000.}$$

$$x = 50,000 \quad \text{Divide by 22.}$$

*Steps 5 and 6* State the answer and check the solution.

\$50,000 is invested at 12% and \$30,000 is invested at 10%.

The three forms of the formula relating distance, rate, and time are

$$d = rt, \quad r = \frac{d}{t}, \quad \text{and} \quad t = \frac{d}{r}.$$

Two cars leave from the same point, traveling in opposite directions. One travels at 45 mph and the other at 60 mph. How long will it take them to be 210 mi apart?

Let $t$ = time it takes for them to be 210 mi apart.

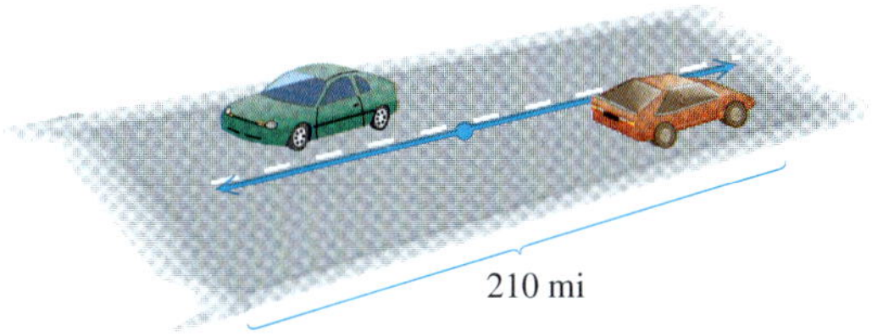


The table gives the information from the problem.

| | Rate | Time | Distance |
|---|---|---|---|
| One Car | 45 | t | 45t |
| Other Car | 60 | t | 60t |

The sum of the distances, $45t$ and $60t$, must be 210 mi.

$$45t + 60t = 210$$

$$105t = 210 \quad \text{Combine like terms.}$$

$$t = 2 \quad \text{Divide by 105.}$$

It will take them 2 hr to be 210 mi apart. (continued)

| CONCEPTS | EXAMPLES |
|---|---|

**2.8 SOLVING LINEAR INEQUALITIES**

*Step 1* Simplify each side separately.

*Step 2* Isolate the variable term on one side.

*Step 3* Isolate the variable.
***Be sure to reverse the direction of the inequality symbol when multiplying or dividing by a negative number.***

Solve and graph the solution set.

$$3(1 - x) + 5 - 2x > 9 - 6$$
$$3 - 3x + 5 - 2x > 9 - 6 \quad \text{Clear parentheses.}$$
$$8 - 5x > 3 \quad \text{Combine like terms.}$$
$$8 - 5x - 8 > 3 - 8 \quad \text{Subtract 8.}$$
$$-5x > -5 \quad \text{Combine like terms.}$$
$$\frac{-5x}{-5} < \frac{-5}{-5} \quad \text{Divide by } -5\text{; change} > \text{to} <.$$
$$x < 1$$

Solution set: $(-\infty, 1)$

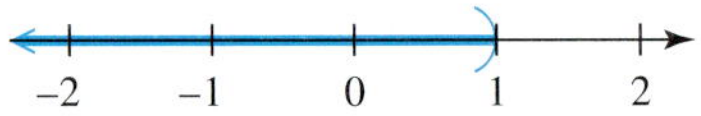

To solve an inequality such as

$$4 < 2x + 6 < 8$$

work with all three expressions at the same time.

Solve.

$$4 < 2x + 6 < 8$$
$$4 - 6 < 2x + 6 - 6 < 8 - 6 \quad \text{Subtract 6.}$$
$$-2 < 2x < 2 \quad \text{Combine like terms.}$$
$$\frac{-2}{2} < \frac{2x}{2} < \frac{2}{2} \quad \text{Divide by 2.}$$
$$-1 < x < 1$$

Solution set: $(-1, 1)$

**Answers to Test Your Word Power**

**1.** C; *Example:* $\{8\}$ is the solution set of $2x + 5 = 21$. **2.** C; *Example:* The empty set $\emptyset$ is the solution set of $5x + 3 = 5x + 4$. **3.** B; *Example:* Angles with measures 35° and 55° are complementary angles. **4.** C; *Example:* Angles with measures 112° and 68° are supplementary angles. **5.** A; *Example:* $\frac{7 \text{ in.}}{12 \text{ in.}}$ or $\frac{7}{12}$ **6.** B; *Example:* $\frac{2}{3} = \frac{8}{12}$ **7.** D; *Examples:* $x < 5$, $7 + 2y \geq 11$, $-5 < 2z - 1 \leq 3$ **8.** C; *Examples:* $(-\infty, 5]$, $(1, \infty)$, $[-3, 3)$

# CHAPTER 2 REVIEW EXERCISES

**[2.1–2.3]** *Solve each equation.*

**1.** $m - 5 = 1$ **2.** $y + 8 = -4$ **3.** $3k + 1 = 2k + 8$

**4.** $5k = 4k + \frac{2}{3}$ **5.** $(4r - 2) - (3r + 1) = 8$ **6.** $3(2y - 5) = 2 + 5y$

**7.** $7k = 35$ **8.** $12r = -48$ **9.** $2p - 7p + 8p = 15$

**10.** $\frac{m}{12} = -1$ **11.** $\frac{5}{8}k = 8$ **12.** $12m + 11 = 59$

**13.** $3(2x + 6) - 5(x + 8) = x - 22$ **14.** $5x + 9 - (2x - 3) = 2x - 7$

**15.** $\frac{1}{2}r - \frac{r}{3} = \frac{r}{6}$

**16.** $.10(x + 80) + .20x = 14$

**17.** $3x - (-2x + 6) = 4(x - 4) + x$

**18.** $2(y - 3) - 4(y + 12) = -2(y + 27)$

[2.4] *Use the six-step method to solve each problem.*

**19.** In a recent year, the state of Florida had a total of 120 members in its House of Representatives, consisting of only Democrats and Republicans. There were 30 more Democrats than Republicans. How many representatives from each party were there?

**20.** The land area of Hawaii is 5213 $\text{mi}^2$ greater than the area of Rhode Island. Together, the areas total 7637 $\text{mi}^2$. What is the area of each of the two states?

**21.** The height of Seven Falls in Colorado is $\frac{5}{2}$ the height of Twin Falls in Idaho. The sum of the heights is 420 ft. Find the height of each. (*Source: World Almanac and Book of Facts.*)

**22.** The supplement of an angle measures 10 times the measure of its complement. What is the measure of the angle?

**23.** Find two consecutive odd integers such that when the smaller is added to twice the larger, the result is 24 more than the larger integer.

[2.5] *A formula is given along with the values for all but one of the variables. Find the value of the variable that is not given.*

**24.** $A = \frac{1}{2}bh$; $A = 44$, $b = 8$

**25.** $A = \frac{1}{2}h(b + B)$; $b = 3$, $B = 4$, $h = 8$

**26.** $C = 2\pi r$; $C = 29.83$, $\pi = 3.14$

**27.** $V = \frac{4}{3}\pi r^3$; $r = 6$, $\pi = 3.14$

*Solve each formula for the specified variable.*

**28.** $A = bh$ for $h$

**29.** $A = \frac{1}{2}h(b + B)$ for $h$

*Find the measure of each marked angle.*

**30.**

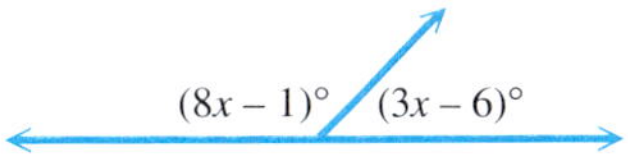


**31.**

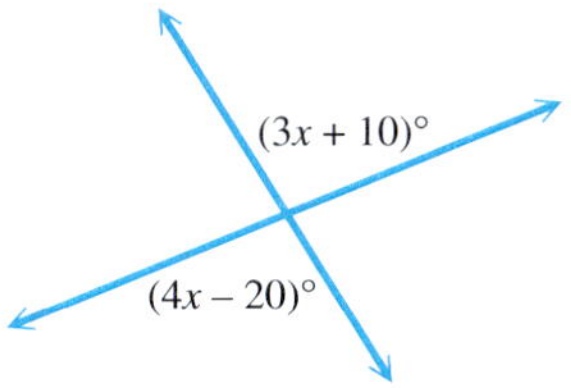


*Solve each problem.*

**32.** The perimeter of a certain rectangle is 16 times the width. The length is 12 cm more than the width. Find the width of the rectangle.

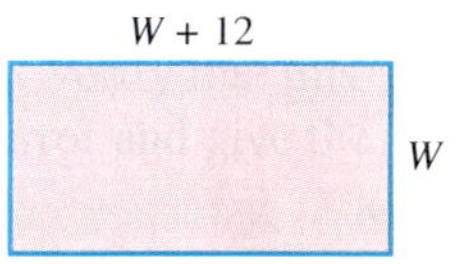


**33.** The Ziegfield Room in Reno, Nevada, has a circular turntable on which its showgirls dance. The circumference of the table is 62.5 ft. What is the diameter? What is the radius? What is the area? (Use $\pi = 3.14$.) (*Source: Guinness Book of Records.*)

(d) Describe the pattern indicated by the points on the scatter diagram. What is happening to graduation rates for 4-yr college students within 5 yr?

**72.** The maximum benefit for the heart from exercising occurs if the heart rate is in the target heart rate zone. The lower limit of this target zone can be approximated by the linear equation

$$y = -.7x + 154,$$

where $x$ represents age and $y$ represents heartbeats per minute. (*Source:* Hockey, R. V., *Physical Fitness: The Pathway to Healthy Living,* Times Mirror/Mosby College Publishing, 1989.)

| Age | Heartbeats (per minute) |
|---|---|
| 20 | |
| 40 | |
| 60 | |
| 80 | |

(a) Complete the table of values for this linear equation.
(b) Write the data from the table of values as ordered pairs.
(c) Make a scatter diagram of the data. Do the points lie in an approximately linear pattern?

**73.** (See Exercise 72.) The upper limit of the target heart rate zone can be approximated by the linear equation

$$y = -.8x + 186,$$

where $x$ represents age and $y$ represents heartbeats per minute. (*Source:* Hockey, R. V., *Physical Fitness: The Pathway to Healthy Living,* Times Mirror/Mosby College Publishing, 1989.)

| Age | Heartbeats (per minute) |
|---|---|
| 20 | |
| 40 | |
| 60 | |
| 80 | |

(a) Complete the table of values for this linear equation.
(b) Write the data from the table of values as ordered pairs.
(c) Make a scatter diagram of the data. Describe the pattern indicated by the data.

**74.** Refer to Exercises 72 and 73. What is the target heart rate zone for age 20? age 40?

# 3.2 Graphing Linear Equations in Two Variables

**OBJECTIVES**

1. Graph linear equations.
2. Find intercepts.
3. Graph linear equations of the form $Ax + By = 0$.
4. Graph linear equations of the form $y = b$ or $x = a$.
5. Use a linear equation to model data.

In this section we use a few ordered pairs that satisfy a linear equation to graph the equation.

**OBJECTIVE 1 Graph linear equations.** We know that infinitely many ordered pairs satisfy a linear equation. Some ordered pairs that are solutions of $x + 2y = 7$ are graphed in Figure 8. Notice that the points plotted in this figure all appear to lie on a straight line as shown in Figure 9. In fact,

**every point on the line represents a solution of the equation $x + 2y = 7$, and each solution of the equation corresponds to a point on the line.**

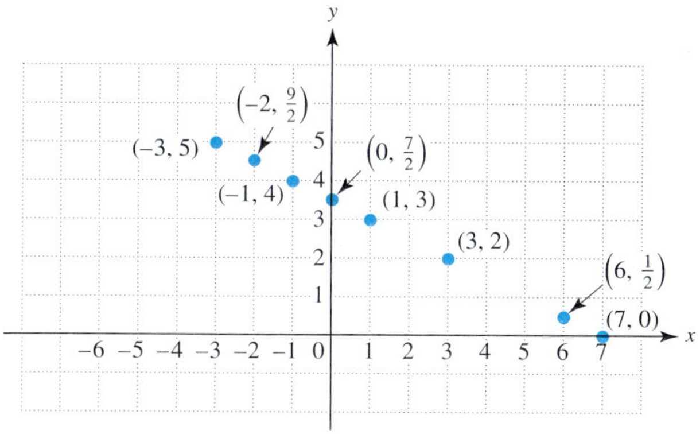


**FIGURE 8**

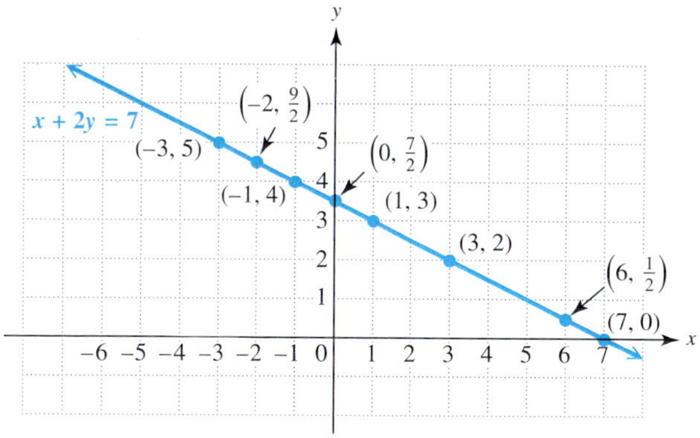


**FIGURE 9**

The line is a "picture" of all the solutions of the equation $x + 2y = 7$. Only a portion of the line is shown here, but it extends indefinitely in both directions, as suggested by the arrowhead on each end of the line. The line is called the **graph** of the equation, and the process of plotting the ordered pairs and drawing the line through the corresponding points is called **graphing.**

The preceding discussion can be generalized.

**Graph of a Linear Equation**

The graph of any linear equation in two variables is a straight line.

Notice that the word *line* appears in the name "*line*ar equation."

Since two distinct points determine a line, we can graph a straight line by finding any two different points on the line. However, it is a good idea to plot a third point as a check.

### EXAMPLE 1 Graphing a Linear Equation

Graph the linear equation $2y = -3x + 6$.

Although this equation is not in the form $Ax + By = C$, it *could* be put in that form, and so is a linear equation. For most linear equations, we can find two different points on the graph by first letting $x = 0$, and then letting $y = 0$. Doing this gives the ordered pairs $(0, 3)$ and $(2, 0)$. We get a third ordered pair (as a check) by letting $x$ or $y$ equal some other number. For example, if $x = -2$, we find that $y = 6$, giving the ordered pair $(-2, 6)$. These three ordered pairs are shown in the table of values with Figure 10. The table of values is an alternative way of listing the ordered pairs (0, 3), (2, 0), and (−2, 6).

| $x$ | $y$ |
|---|---|
| 0 | 3 |
| 2 | 0 |
| −2 | 6 |

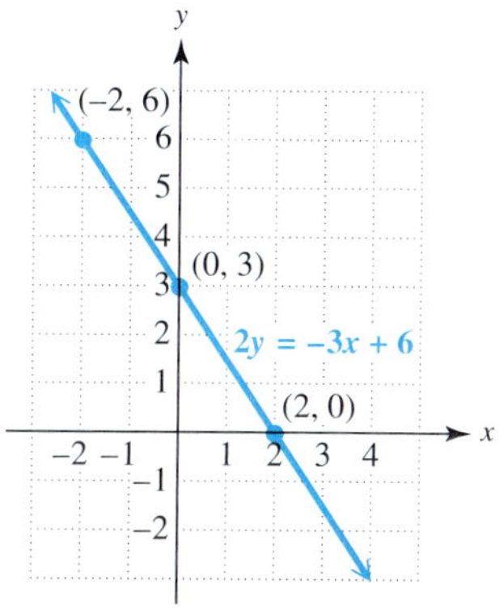


**FIGURE 10**

We plot the corresponding points, then draw a line through them. This line, shown in Figure 10, is the graph of $2y = -3x + 6$.

**Now Try Exercise 5.**

**OBJECTIVE 2 Find intercepts.** In Figure 10 the graph intersects (crosses) the $y$-axis at $(0, 3)$ and the $x$-axis at $(2, 0)$. For this reason $(0, 3)$ is called the ***y*-intercept** and $(2, 0)$ is called the ***x*-intercept** of the graph. The intercepts are particularly useful for graphing linear equations.

**Finding Intercepts**

To find the $x$-intercept, let $y = 0$ in the given equation and solve for $x$. Then $(x, 0)$ is the $x$-intercept.

To find the $y$-intercept, let $x = 0$ in the given equation and solve for $y$. Then $(0, y)$ is the $y$-intercept.

## EXAMPLE 2 Finding Intercepts

Find the intercepts for the graph of $2x + y = 4$. Draw the graph.

Find the $y$-intercept by letting $x = 0$; find the $x$-intercept by letting $y = 0$.

| | | | |
|---|---|---|---|
| $2x + y = 4$ | | $2x + y = 4$ | |
| $2(0) + y = 4$ | Let $x = 0$. | $2x + 0 = 4$ | Let $y = 0$. |
| $0 + y = 4$ | | $2x = 4$ | |
| $y = 4$ | | $x = 2$ | |

The $y$-intercept is $(0, 4)$. The $x$-intercept is $(2, 0)$. The graph, with the two intercepts shown in red, is given in Figure 11. Get a third point as a check. For example, choosing $x = 4$ gives $y = -4$. These three ordered pairs are shown in the table with Figure 11. Plot $(0, 4)$, $(2, 0)$, and $(4, -4)$ and draw a line through them. This line, shown in Figure 11, is the graph of $2x + y = 4$.

| $x$ | $y$ |
|---|---|
| 0 | 4 |
| 2 | 0 |
| 4 | −4 |

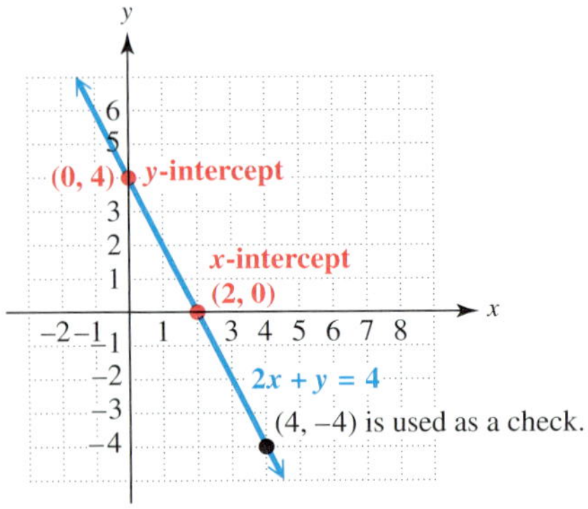


FIGURE 11

**Now Try Exercise 13.**

**CAUTION** When choosing $x$- or $y$-values to find ordered pairs to plot, be careful to choose so that the resulting points are not too close together. For example, using $(-1, -1)$, $(0, 0)$, and $(1, 1)$ may result in an inaccurate line. It is better to choose points where the $x$-values differ by at least 2.

**OBJECTIVE 3 Graph linear equations of the form $Ax + By = 0$.** In earlier examples, the $x$- and $y$-intercepts were used to help draw the graphs. This is not always possible. Example 3 shows what to do when the $x$- and $y$-intercepts are the same point.

## EXAMPLE 3 Graphing an Equation of the Form $Ax + By = 0$

Graph $x - 3y = 0$.

If we let $x = 0$, then $y = 0$, giving the ordered pair $(0, 0)$. Letting $y = 0$ also gives $(0, 0)$. This is the same ordered pair, so we choose two *other* values for $x$ or $y$. Choosing 2 for $y$ gives $x - 3 \cdot 2 = 0$, or $x = 6$, giving the ordered pair $(6, 2)$. For a check point, we choose $-6$ for $x$ getting $-2$ for $y$. We use the ordered pairs $(-6, -2)$, $(0, 0)$, and $(6, 2)$ to get the graph shown in Figure 12.

| $x$ | $y$ |
|---|---|
| 0 | 0 |
| 6 | 2 |
| $-6$ | $-2$ |

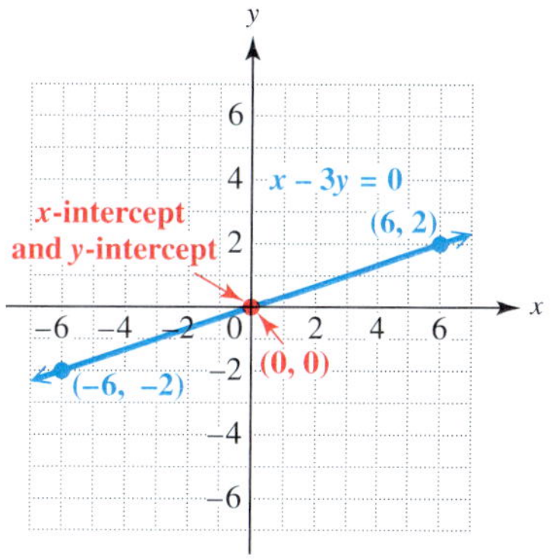


**FIGURE 12**

**Now Try Exercise 31.**

Example 3 can be generalized as follows.

**Line through the Origin**

If $A$ and $B$ are nonzero real numbers, the graph of a linear equation of the form

$$Ax + By = 0$$

goes through the origin $(0, 0)$.

**OBJECTIVE 4 Graph linear equations of the form $y = b$ or $x = a$.** The equation $y = -4$ is a linear equation in which the coefficient of $x$ is 0. (Write $y = -4$ as $0x + y = -4$ to see this.) Also, $x = 3$ is a linear equation in which the coefficient of $y$ is 0. These equations lead to horizontal or vertical straight lines.

### EXAMPLE 4 Graphing an Equation of the Form $y = b$

Graph $y = -4$.

As the equation states, for any value of $x$, $y$ is always equal to $-4$. To get ordered pairs that are solutions of this equation, we choose any numbers for $x$, always using $-4$ for $y$. Three ordered pairs that satisfy the equation are shown in the table of values with Figure 13. Drawing a line through these points gives the horizontal line shown in Figure 13. The $y$-intercept is $(0, -4)$; there is no $x$-intercept.

| $x$ | $y$ |
|---|---|
| $-2$ | $-4$ |
| 0 | $-4$ |
| 3 | $-4$ |

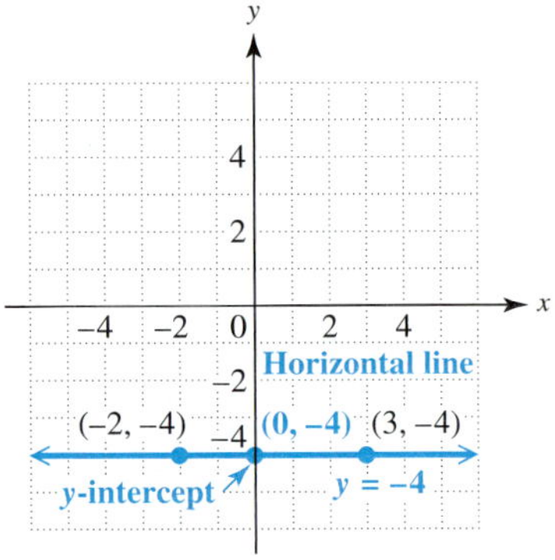


FIGURE 13

**Now Try Exercise 35.**

**Horizontal Line**

The graph of a linear equation $\boldsymbol{y = b}$, where $b$ is a real number, is the horizontal line with $y$-intercept $(0, b)$ and no $x$-intercept.

### EXAMPLE 5 Graphing an Equation of the Form $x = a$

Graph $x - 3 = 0$.

First we add 3 to each side of the equation $x - 3 = 0$ to get $x = 3$. All the ordered pairs that are solutions of this equation have an $x$-value of 3. Any number can be used for $y$. We show three ordered pairs that satisfy the equation in the table of values with Figure 14. Drawing a line through these points gives the vertical line shown in Figure 14. The $x$-intercept is $(3, 0)$; there is no $y$-intercept.

| $x$ | $y$ |
|---|---|
| 3 | 3 |
| 3 | 0 |
| 3 | $-2$ |

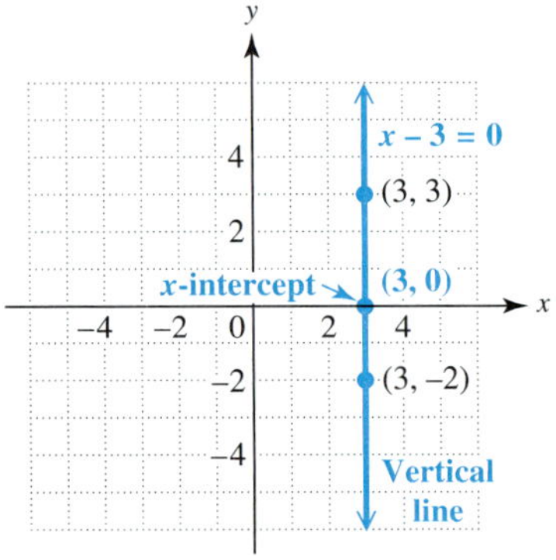


FIGURE 14

**Now Try Exercise 37.**

**Vertical Line**

The graph of a linear equation $\boldsymbol{x = a}$, where $a$ is a real number, is the vertical line with $x$-intercept $(a, 0)$ and no $y$-intercept.

In particular, notice that the horizontal line $y = 0$ is the $x$-axis and the vertical line $x = 0$ is the $y$-axis.

**CAUTION** The equations of horizontal and vertical lines are often confused with each other. Remember that the graph of $y = b$ is parallel to the $x$-axis and that of $x = a$ is parallel to the $y$-axis.

The different forms of straight-line equations and the methods of graphing them are given in the following summary.

**Graphing Straight Lines**

| *Equation* | *To Graph* | *Example* |
|---|---|---|
| $y = b$ | Draw a horizontal line, through $(0, b)$. | 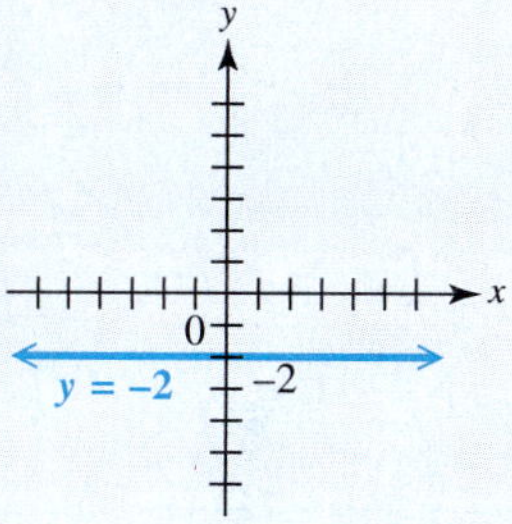 |
| $x = a$ | Draw a vertical line, through $(a, 0)$. | 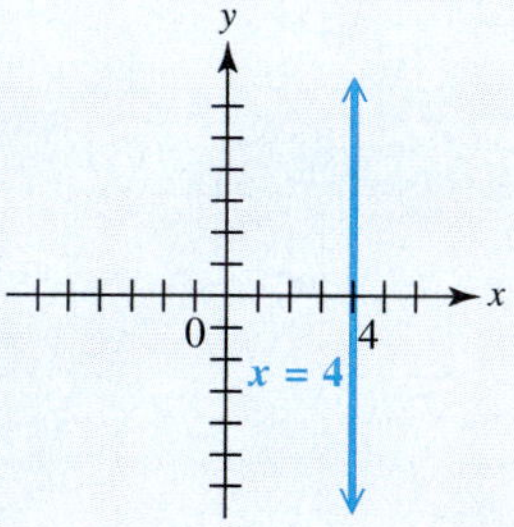 |
| $Ax + By = 0$ | Graph goes through $(0, 0)$. Get additional points that lie on the graph by choosing any value of $x$ or $y$, except 0. | 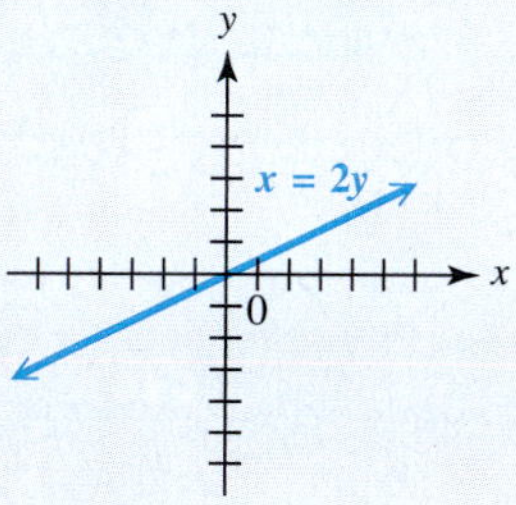 |

(continued)

| | | |
|---|---|---|
| **$Ax + By = C$ but not of the types above** | Find any two points the line goes through. A good choice is to find the intercepts: let $x = 0$, and find the corresponding value of $y$; then let $y = 0$, and find $x$. As a check, get a third point by choosing a value of $x$ or $y$ that has not yet been used. | 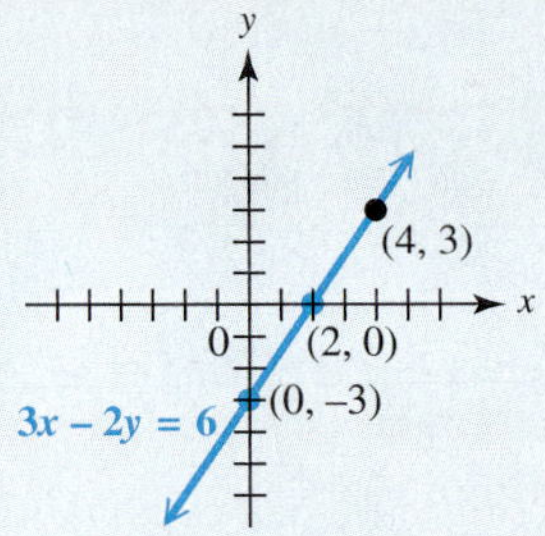  |

## CONNECTIONS

Beginning in this chapter we include information on the basic features of graphing calculators. The most obvious feature is their ability to graph equations. We must solve the equation for $y$ in order to enter it into the calculator. Also, we must select an appropriate "window" for the graph. The window is determined by the minimum and maximum values of $x$ and $y$. Graphing calculators have a standard window, often from $x = -10$ to $x = 10$ and from $y = -10$ to $y = 10$. Sometimes this is written $[-10, 10]$, $[-10, 10]$, with the $x$-interval shown first.

For example, to graph the equation $2x + y = 4$, discussed in Example 2, we first solve for $y$.

$$2x + y = 4$$

$$y = -2x + 4 \qquad \text{Subtract } 2x.$$

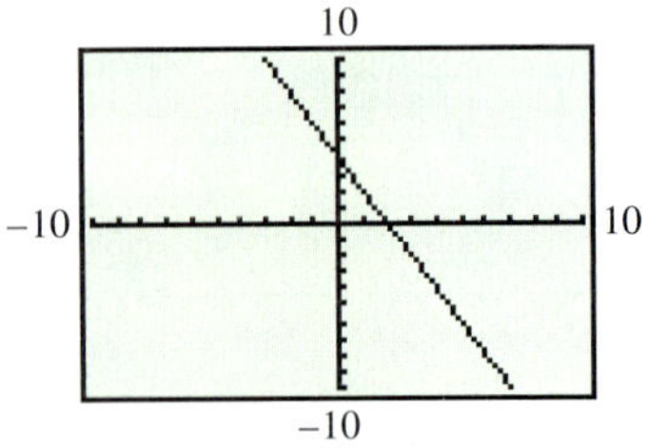


If we enter this equation as $y = -2x + 4$ and choose the standard window, the calculator displays the graph shown in the margin.

The $x$-value of the point on the graph where $y = 0$ (the $x$-intercept) gives the solution of the equation

$$y = 0,$$

or $$-2x + 4 = 0. \qquad \text{Substitute } -2x + 4 \text{ for } y.$$

Since each tick mark on the $x$-axis represents 1, the graph shows that $(2, 0)$ is the $x$-intercept.

### For Discussion or Writing

How would you rewrite these equations, with one side equal to 0, to enter them into a graphing calculator for solution? (It is not necessary to clear parentheses or combine terms.)

**1.** $3x + 4 - 2x - 7 = 4x + 3$ **2.** $5x - 15 = 3(x - 2)$

**OBJECTIVE 5** **Use a linear equation to model data.**

### EXAMPLE 6 Using a Linear Equation to Model Credit Card Debt

Credit card debt in the United States increased steadily from 1992 through 1999. The amount of debt $y$ in billions of dollars can be modeled by the linear equation

$$y = 47.3x + 281,$$

where $x = 0$ represents 1992, $x = 1$ represents 1993, and so on. (*Source:* Board of Governors of the Federal Reserve System.)

**(a)** Use the equation to approximate credit card debt in the years 1992, 1993, and 1999.

For 1992: $y = 47.3(0) + 281$ Replace $x$ with 0.

$y = 281$ billion dollars

For 1993: $y = 47.3(1) + 281$ Replace $x$ with 1.

$y = 328.3$ billion dollars

For 1999: $y = 47.3(7) + 281$ $1999 - 1992 = 7$; replace $x$ with 7.

$y = 612.1$ billion dollars

**(b)** Write the information from part (a) as three ordered pairs, and use them to graph the given linear equation.

Since $x$ represents the year and $y$ represents the debt in billions of dollars, the ordered pairs are (0, 281), (1, 328.3), and (7, 612.1). Figure 15 shows a graph of these ordered pairs and the line through them. (Note that arrowheads are not included with the graphed line since the data are for the years 1992 to 1999 only, that is, from $x = 0$ to $x = 7$.)

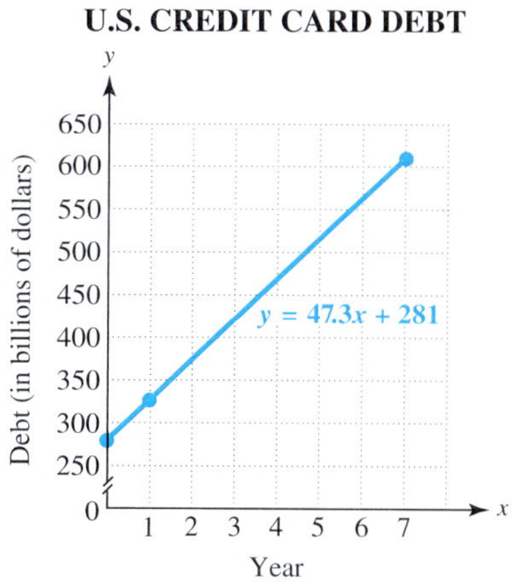


**FIGURE 15**

**(c)** Use the graph and then the equation to approximate credit card debt in 1996.

For 1996, $x = 4$. On the graph, find 4 on the horizontal axis and move up to the graphed line, then across to the vertical axis. It appears that credit card debt in 1996 was about 470 billion dollars.

To use the equation, substitute 4 for $x$.

$y = 47.3(4) + 281$ Let $x = 4$.

$y = 470.2$ billion dollars

This result is quite similar to our estimate using the graph. ■

**Now Try Exercise 45.**

## 3.2 EXERCISES

**For Extra Help**

- Student's Solutions Manual
- MyMathLab
- InterAct Math Tutorial Software
- AW Math Tutor Center
- MathXL
- Digital Video Tutor CD 5/Videotape 5

*Complete the given ordered pairs using the given equation. Then graph each equation by plotting the points and drawing a line through them. See Example 1.*

**1.** $y = -x + 5$
$(0, \quad), (\quad, 0), (2, \quad)$

**2.** $y = x - 2$
$(0, \quad), (\quad, 0), (5, \quad)$

**3.** $y = \frac{2}{3}x + 1$
$(0, \quad), (3, \quad), (-3, \quad)$

**4.** $y = -\frac{3}{4}x + 2$
$(0, \quad), (4, \quad), (-4, \quad)$

**5.** $3x = -y - 6$
$(0, \quad), (\quad, 0), \left(-\frac{1}{3}, \quad\right)$

**6.** $x = 2y + 3$
$(\quad, 0), (0, \quad), \left(\quad, \frac{1}{2}\right)$

*In Exercises 7–12, match the information about each graph in Column I with the correct linear equation in Column II.*

**I**

**7.** The graph of the equation has $x$-intercept $(4, 0)$.

**8.** The graph of the equation has $y$-intercept $(0, -4)$.

**9.** The graph of the equation goes through the origin.

**10.** The graph of the equation is a vertical line.

**11.** The graph of the equation is a horizontal line.

**12.** The graph of the equation goes through $(9, 2)$.

**II**

**A.** $x = 5$

**B.** $y = -3$

**C.** $2x - 5y = 8$

**D.** $x + 4y = 0$

**E.** $3x + y = -4$

*Find the x-intercept and the y-intercept for the graph of each equation. See Examples 2, 4, and 5.*

**13.** $2x - 3y = 24$ **14.** $-3x + 8y = 48$ **15.** $x + 6y = 0$ **16.** $3x - y = 0$

**17.** $5x - 2y = 20$ **18.** $-3x + 2y = 12$ **19.** $x - 4 = 0$ **20.** $y = 5$

**21.** What is the equation of the $x$-axis? What is the equation of the $y$-axis?

**22.** A student attempted to graph $4x + 5y = 0$ by finding intercepts. She first let $x = 0$ and found $y$; then she let $y = 0$ and found $x$. In both cases, the resulting point was $(0, 0)$. She knew that she needed at least two different points to graph the line, but was unsure what to do next since finding intercepts gave her only one point. How would you explain to her what to do next?

*Graph each linear equation. See Examples 1–5.*

**23.** $x = y + 2$ **24.** $x = -y + 6$ **25.** $x - y = 4$ **26.** $x - y = 5$

**27.** $2x + y = 6$ **28.** $-3x + y = -6$ **29.** $3x + 7y = 14$ **30.** $6x - 5y = 18$

**31.** $y - 2x = 0$ **32.** $y + 3x = 0$ **33.** $y = -6x$ **34.** $y = 4x$

**35.** $y + 1 = 0$ **36.** $y - 3 = 0$ **37.** $x = -2$ **38.** $x = 4$

**39.** Write a paragraph summarizing how to graph a linear equation in two variables.

## TECHNOLOGY INSIGHTS (EXERCISES 40–44)

*In Exercises 40–43, we give the graph of a linear equation in two variables solved for y and a corresponding linear equation in one variable, with y replaced by 0. Solve the equation in one variable using the methods of Section 2.3. Is the solution you get the same as the x-intercept (labeled "Zero") on the calculator screen?*

**40.** $8 - 2(3x - 4) - 2x = 0$

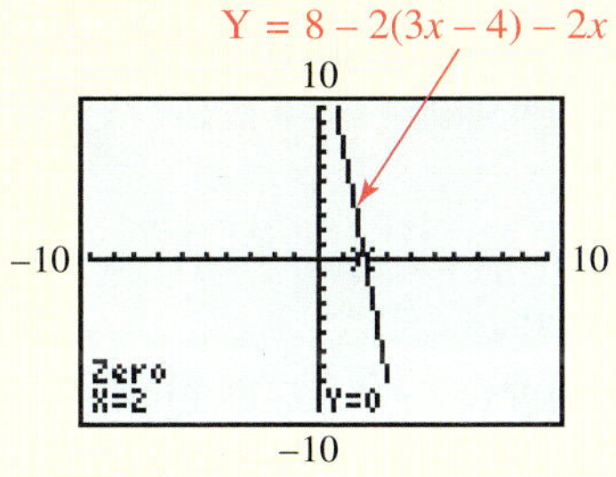


**41.** $5(2x - 1) - 4(2x + 1) - 7 = 0$

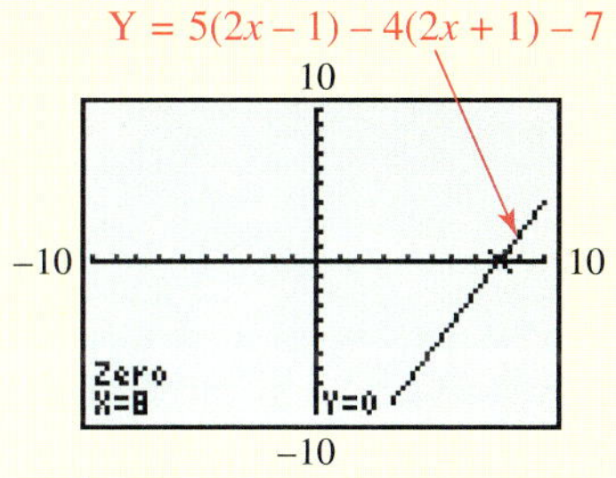


**42.** $.6x - .1x - x + 2.5 = 0$

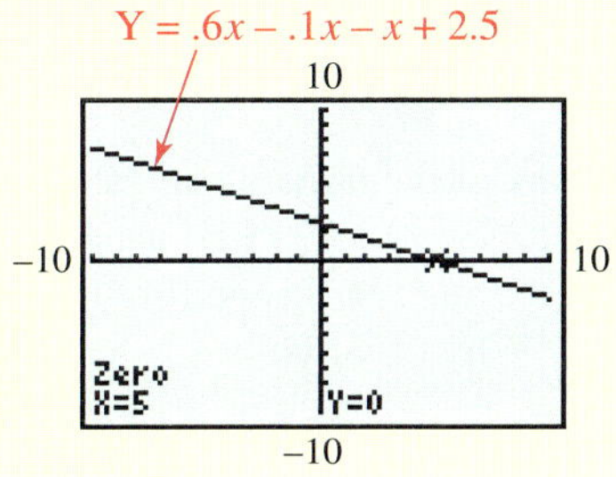


**43.** $-\frac{2}{7}x + 2x - \frac{1}{2}x - \frac{17}{2} = 0$

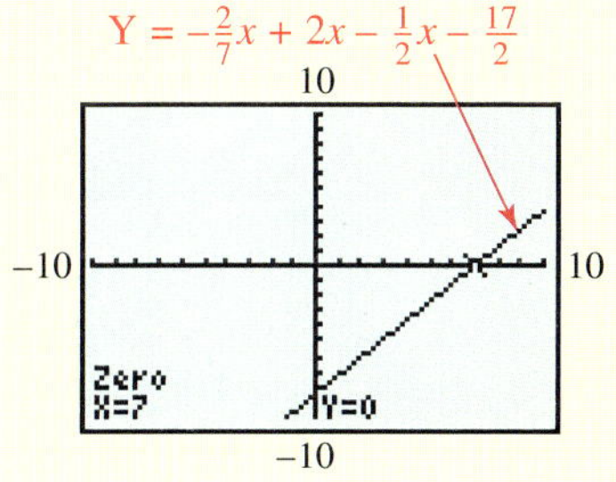


**44.** From the results in Exercises 40–43, what can you conclude?

*Solve each problem. See Example 6.*

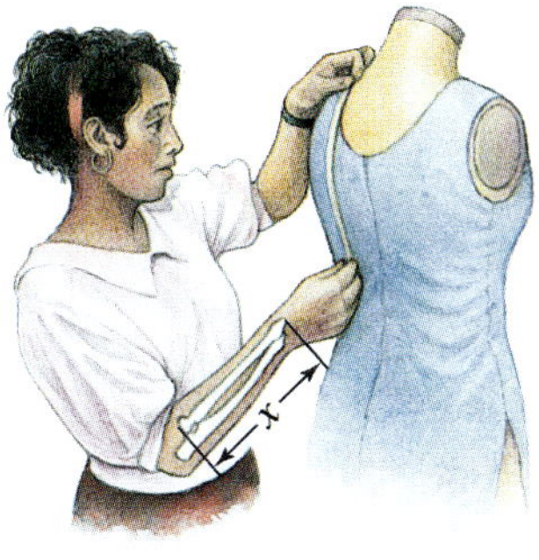


**45.** The height $y$ (in centimeters) of a woman is related to the length of her radius bone $x$ (from the wrist to the elbow) and is approximated by the linear equation

$$y = 3.9x + 73.5.$$

**(a)** Use the equation to find the approximate heights of women with radius bones of lengths 20 cm, 26 cm, and 22 cm.

**(b)** Graph the equation using the data from part (a).

**(c)** Use the graph to estimate the length of the radius bone in a woman who is 167 cm tall. Then use the equation to find the length of this radius bone to the nearest centimeter. (*Hint:* Substitute for $y$ in the equation.)

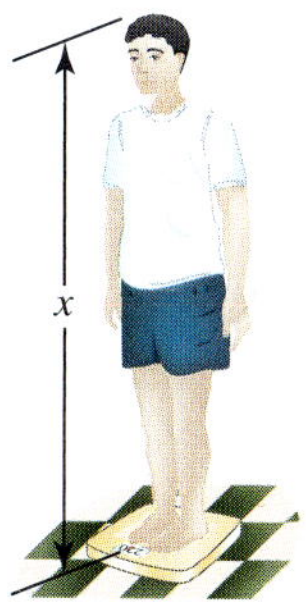


**46.** The weight $y$ (in pounds) of a man taller than 60 in. can be roughly approximated by the linear equation

$$y = 5.5x - 220,$$

where $x$ is the height of the man in inches.

**(a)** Use the equation to approximate the weights of men whose heights are 62 in., 66 in., and 72 in.

**(b)** Graph the equation using the data from part (a).

**(c)** Use the graph to estimate the height of a man who weighs 155 lb. Then use the equation to find the height of this man to the nearest inch. (*Hint:* Substitute for $y$ in the equation.)

**47.** Refer to Section 3.1 Exercise 72. Draw a line through the points you plotted in the scatter diagram there.

**(a)** Use the graph to estimate the lower limit of the target heart rate zone for age 30.

**(b)** Use the linear equation given there to approximate the lower limit for age 30.

**(c)** How does the approximation using the equation compare to the estimate from the graph?

**48.** Refer to Section 3.1 Exercise 73. Draw a line through the points you plotted in the scatter diagram there.

**(a)** Use the graph to estimate the upper limit of the target heart rate zone for age 30.

**(b)** Use the linear equation given there to approximate the upper limit for age 30.

**(c)** How does the approximation using the equation compare to the estimate from the graph?

**49.** Use the results of Exercises 47(b) and 48(b) to determine the target heart rate zone for age 30.

**50.** Should the graphs of the target heart rate zone in Section 3.1 Exercises 72 and 73 be used to estimate the target heart rate zone for ages below 20 or above 80? Why or why not?

**51.** Per capita consumption of carbonated soft drinks increased for the years 1992 through 1997 as shown in the graph. If $x = 0$ represents 1992, $x = 1$ represents 1993, and so on, per capita consumption can be modeled by the linear equation

$$y = .8x + 49,$$

where $y$ is in gallons.

**(a)** Use the equation to approximate consumption in 1993, 1995, and 1997.

**(b)** Use the graph to estimate consumption for the same years.

**(c)** How do the approximations using the equation compare to the estimates from the graph?

**SOFT DRINK CONSUMPTION**

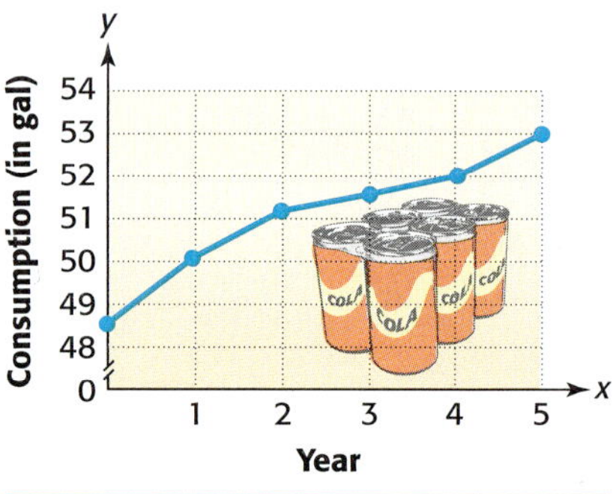


*Source:* U.S. Department of Agriculture.

**52.** Sporting goods sales $y$ (in billions of dollars) from 1995 through 2000 are modeled by the linear equation

$$y = 3.606x + 41.86,$$

where $x = 0$ corresponds to 1990, $x = 5$ corresponds to 1995, and so on.

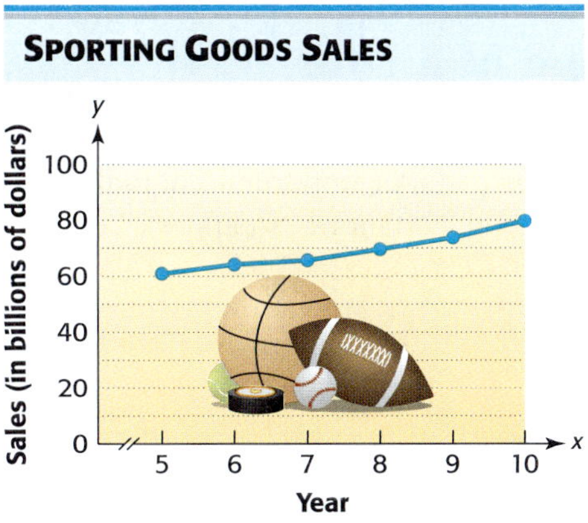


***Source:*** U.S. Bureau of the Census.

**(a)** Use the equation to approximate sporting goods sales in 1995, 1997, and 2000. Round your answers to the nearest billion dollars.

**(b)** Use the graph to estimate sales for the same years.

**(c)** How do the approximations using the equation compare to the estimates using the graph?

**53.** The graph shows the value of a certain sport utility vehicle over the first 5 yr of ownership.

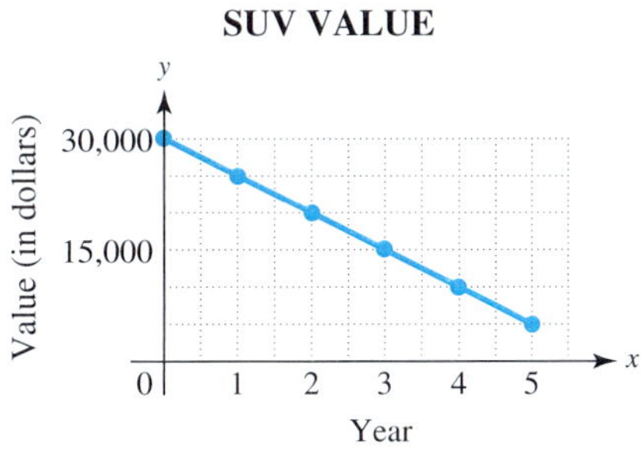


Use the graph to do the following.

**(a)** Determine the initial value of the SUV.

**(b)** Find the *depreciation* (loss in value) from the original value after the first 3 yr.

**(c)** What is the annual or yearly depreciation in each of the first 5 yr?

**(d)** What does the ordered pair $(5, 5000)$ mean in the context of this problem?

**54.** Demand for an item is often closely related to its price. As price increases, demand decreases, and as price decreases, demand increases. Suppose demand for a video game is 2000 units when the price is \$40, and demand is 2500 units when the price is \$30.

**(a)** Let $x$ be the price and $y$ be the demand for the game. Graph the two given pairs of prices and demands.

**(b)** Assume the relationship is linear. Draw a line through the two points from part (a). From your graph, estimate the demand if the price drops to \$20.

**(c)** Use the graph to estimate the price if the demand is 3500 units.

# 3.3 The Slope of a Line

**OBJECTIVES**

1. Find the slope of a line given two points.
2. Find the slope from the equation of a line.
3. Use slope to determine whether two lines are parallel, perpendicular, or neither.

An important characteristic of the lines we graphed in the previous section is their slant or "steepness." See Figure 16.

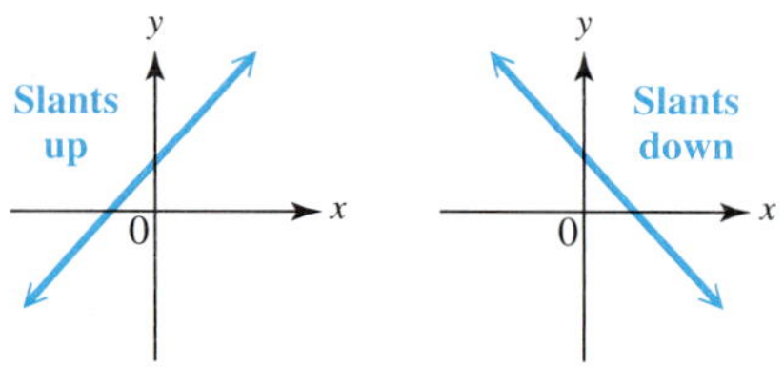


FIGURE 16

One way to measure the steepness of a line is to compare the vertical change in the line to the horizontal change while moving along the line from one fixed point to another. This measure of steepness is called the *slope* of the line.

**OBJECTIVE 1 Find the slope of a line given two points.** Figure 17 shows a line through two nonspecific points $(x_1, y_1)$ and $(x_2, y_2)$. (This notation is called **subscript notation.** Read $x_1$ as "$x$-sub-one" and $x_2$ as "$x$-sub-two.")

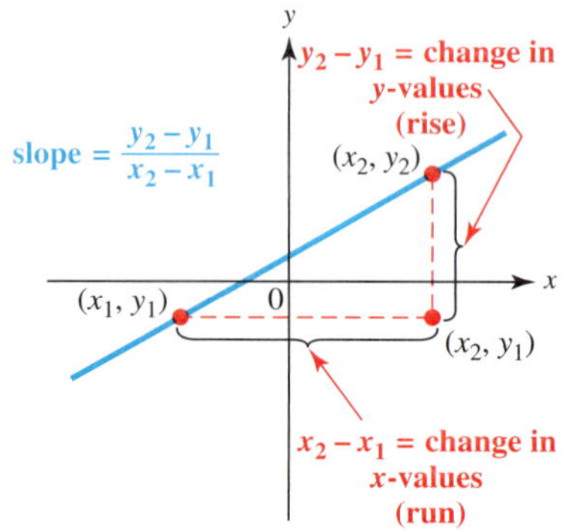


FIGURE 17

Moving along the line from the point $(x_1, y_1)$ to the point $(x_2, y_2)$ causes $y$ to change by $y_2 - y_1$ units. This is the vertical change or **rise.** Similarly, $x$ changes by $x_2 - x_1$ units, which is the horizontal change or **run.** (In both cases, the change is expressed as a *difference.*) Remember from Section 2.6 that one way to compare two numbers is by using a ratio. **Slope** is the ratio of the vertical change in $y$ to the horizontal change in $x$.

## EXAMPLE 1 Comparing Rise to Run

Figure 18 shows two lines. Find the slope ratio of the rise to the run for each line.

We use the two points shown on each line. For the line in Figure 18(a), we find the rise from point $Q$ to point $P$ by determining the vertical change from $-4$ to 4, the

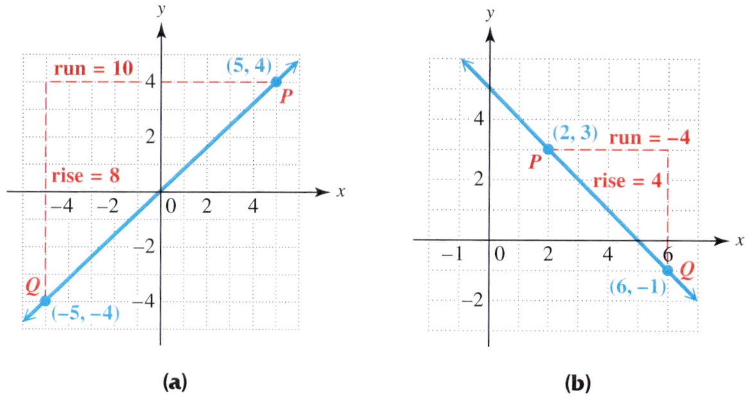


FIGURE 18

difference $4 - (-4) = 8$. Similarly, we find the run by determining the horizontal change from $Q$ to $P$, $5 - (-5) = 10$. Slope is the ratio

$$\frac{\text{rise}}{\text{run}} = \frac{8}{10} \text{ or } \frac{4}{5}.$$

From $Q$ to $P$, the line in Figure 18(b) has rise $3 - (-1) = 4$ and run $2 - 6 = -4$, so the slope is the ratio

$$\frac{\text{rise}}{\text{run}} = \frac{4}{-4} \text{ or } -1.$$

To confirm our slope ratios, count grid squares from one point on the line to another. For example, starting at the origin in Figure 18(a), count up 4 squares $\left(\text{the rise in the slope ratio } \frac{4}{5}\right)$ and then 5 squares to the right (the run) to arrive at the point $(5, 4)$ on the line.

**Now Try Exercise 3.**

The slopes we found for the two lines in Figure 18 suggest that a line with positive slope slants upward from left to right, and a line with negative slope slants downward. These facts can be generalized.

**Positive and Negative Slopes**

A line with positive slope rises from left to right.

A line with negative slope falls from left to right.

Hill

The idea of slope is used in many everyday situations. For example, because $10\% = \frac{1}{10}$, a highway with a 10% grade (or slope) rises 1 meter for every 10 horizontal meters. A highway sign is used to warn of a downgrade ahead that may be long or steep. Architects specify the pitch of a roof using slope; a $\frac{5}{12}$ roof means that the roof rises 5 ft for every 12 ft in the horizontal direction. See the figure on the next page. The slope of a stairwell also indicates the ratio of the vertical rise to the horizontal run. The slope of the stairs in the figure is $\frac{8}{14}$.

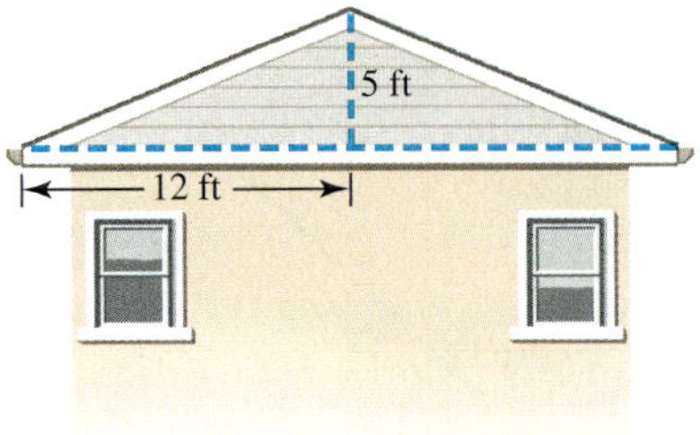


$\frac{5}{12}$ roof pitch

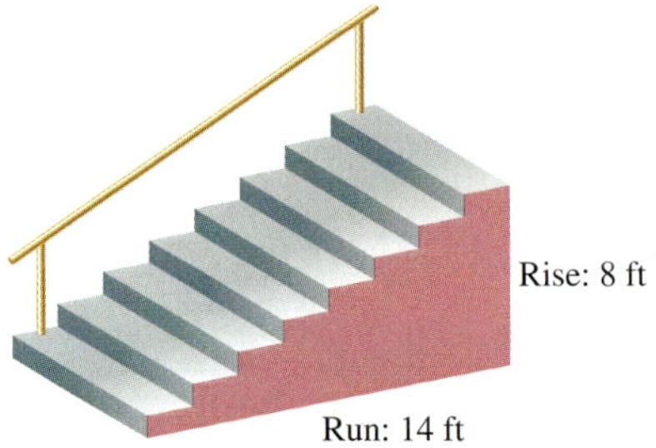


Slope of a stairwell

Traditionally, the letter $m$ represents slope. The slope $m$ of a line is defined as follows.

**Slope Formula**

The **slope** of the line through the points $(x_1, y_1)$ and $(x_2, y_2)$ is

$$m = \frac{\textbf{change in } y}{\textbf{change in } x} = \frac{y_2 - y_1}{x_2 - x_1}, \quad \text{if } x_1 \neq x_2.$$

The slope of a line tells how fast $y$ changes for each unit of change in $x$; that is, the slope gives the rate of change in $y$ for each unit of change in $x$.

## EXAMPLE 2 Finding Slopes of Lines

Find the slope of each line.

**(a)** The line through $(1, -2)$ and $(-4, 7)$

Use the slope formula. Let $(-4, 7) = (x_2, y_2)$ and $(1, -2) = (x_1, y_1)$. Then

$$\text{slope } m = \frac{\text{change in } y}{\text{change in } x} = \frac{y_2 - y_1}{x_2 - x_1} = \frac{7 - (-2)}{-4 - 1} = \frac{9}{-5} = -\frac{9}{5}.$$

See Figure 19.

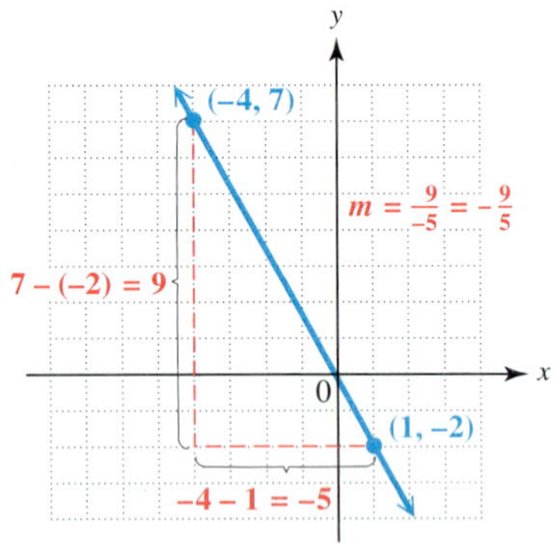


FIGURE 19

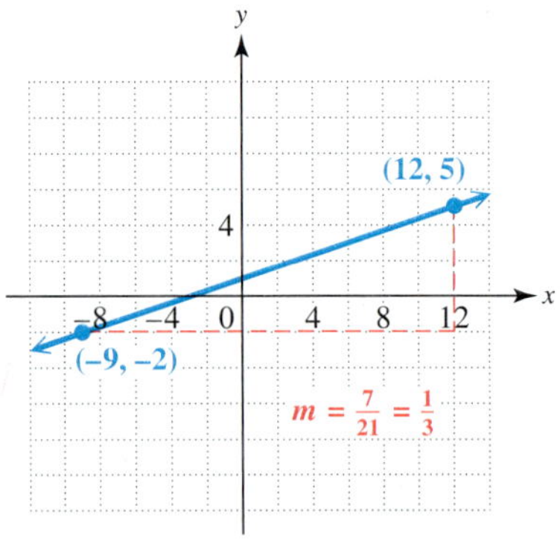


FIGURE 20

**(b)** The line through $(-9, -2)$ and $(12, 5)$

$$m = \frac{y_2 - y_1}{x_2 - x_1} = \frac{5 - (-2)}{12 - (-9)} = \frac{7}{21} = \frac{1}{3}$$

See Figure 20. Note that the same slope is obtained by subtracting in reverse order.

$$m = \frac{-2 - 5}{-9 - 12} = \frac{-7}{-21} = \frac{1}{3}$$

**Now Try Exercises 17 and 21.**

**CAUTION** It makes no difference which point is $(x_1, y_1)$ or $(x_2, y_2)$; however, be consistent. Start with the $x$- and $y$-values of one point (either one) and subtract the corresponding values of the other point. Also, the slope of a line is the same for *any* two points on the line.

## EXAMPLE 3 Finding the Slope of a Horizontal Line

Find the slope of the line through $(-8, 4)$ and $(2, 4)$.

Use the slope formula.

$$m = \frac{4 - 4}{-8 - 2} = \frac{0}{-10} = 0 \quad \text{Zero slope}$$

As shown in Figure 21, the line through these two points is horizontal, with equation $y = 4$. *All horizontal lines have slope 0,* since the difference in $y$-values is always 0.

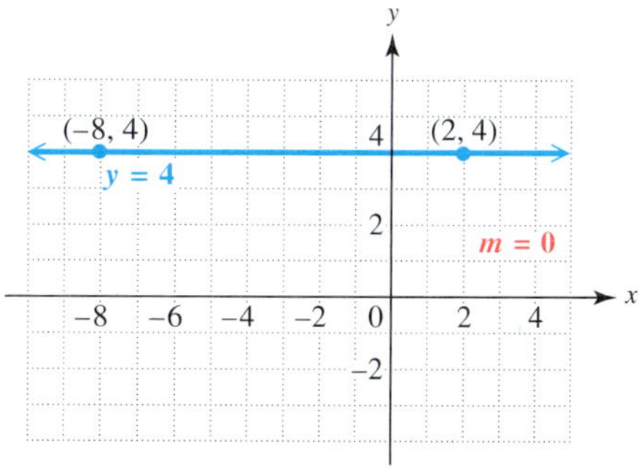


FIGURE 21

**Now Try Exercise 23.**

## EXAMPLE 4 Finding the Slope of a Vertical Line

Find the slope of the line through $(6, 2)$ and $(6, -9)$.

$$m = \frac{2 - (-9)}{6 - 6} = \frac{11}{0} \quad \text{Undefined slope}$$

Since division by 0 is undefined, the slope is undefined. The graph in Figure 22 shows that the line through these two points is vertical with equation $x = 6$. All points on a vertical line have the same $x$-value, so *the slope of any vertical line is undefined.*

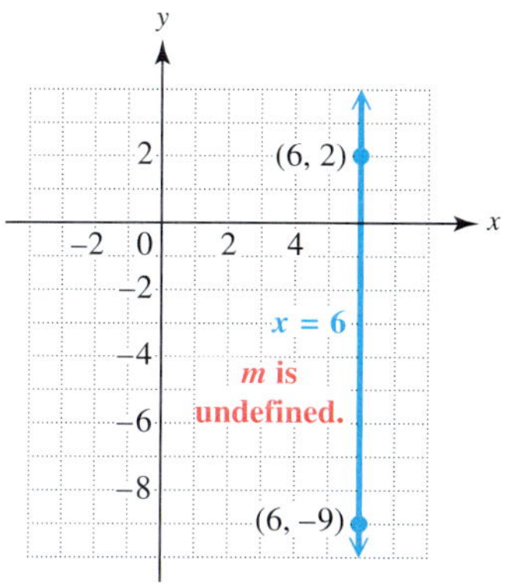


FIGURE 22

**Now Try Exercise 25.**

**Slopes of Horizontal and Vertical Lines**

**Horizontal lines,** with equations of the form $y = b$, have **slope 0.**

**Vertical lines,** with equations of the form $x = a$, have **undefined slope.**

**OBJECTIVE 2 Find the slope from the equation of a line.** The slope of a line can be found directly from its equation. For example, the slope of the line

$$y = -3x + 5$$

is found using any two points on the line. We get these two points by first choosing two different values of $x$ and then finding the corresponding values of $y$. We choose $x = -2$ and $x = 4$.

| | | | |
|---|---|---|---|
| $y = -3x + 5$ | | $y = -3x + 5$ | |
| $y = -3(-2) + 5$ | Let $x = -2$. | $y = -3(4) + 5$ | Let $x = 4$. |
| $y = 6 + 5$ | | $y = -12 + 5$ | |
| $y = 11$ | | $y = -7$ | |

The ordered pairs are $(-2, 11)$ and $(4, -7)$. Now we use the slope formula.

$$m = \frac{11 - (-7)}{-2 - 4} = \frac{18}{-6} = -3$$

The slope, $-3$, is the same number as the coefficient of $x$ in the equation $y = -3x + 5$. It can be shown that this always happens, *as long as the equation is solved for y.* This fact is used to find the slope of a line from its equation.

**Finding the Slope of a Line from Its Equation**

*Step 1* Solve the equation for $y$.

*Step 2* The slope is given by the coefficient of $x$.

## EXAMPLE 5 Finding Slopes from Equations

Find the slope of each line.

**(a)** $2x - 5y = 4$

Solve the equation for $y$.

$$2x - 5y = 4$$

$$-5y = -2x + 4 \quad \text{Subtract } 2x \text{ from each side.}$$

$$y = \frac{2}{5}x - \frac{4}{5} \quad \text{Divide by } -5.$$

The slope is given by the coefficient of $x$, so the slope is $\frac{2}{5}$.

**(b)** $8x + 4y = 1$

Solve the equation for $y$.

$$8x + 4y = 1$$

$$4y = -8x + 1 \quad \text{Subtract } 8x.$$

$$y = -2x + \frac{1}{4} \quad \text{Divide by 4.}$$

The slope of this line is given by the coefficient of $x$, $-2$.

**Now Try Exercise 33.**

**OBJECTIVE 3 Use slope to determine whether two lines are parallel, perpendicular, or neither.** Two lines in a plane that never intersect are **parallel.** We use slopes to tell whether two lines are parallel. For example, Figure 23 shows the graphs of $x + 2y = 4$ and $x + 2y = -6$. These lines appear to be parallel. Solve for $y$ to find that both $x + 2y = 4$ and $x + 2y = -6$ have slope $-\frac{1}{2}$. Nonvertical parallel lines always have equal slopes.

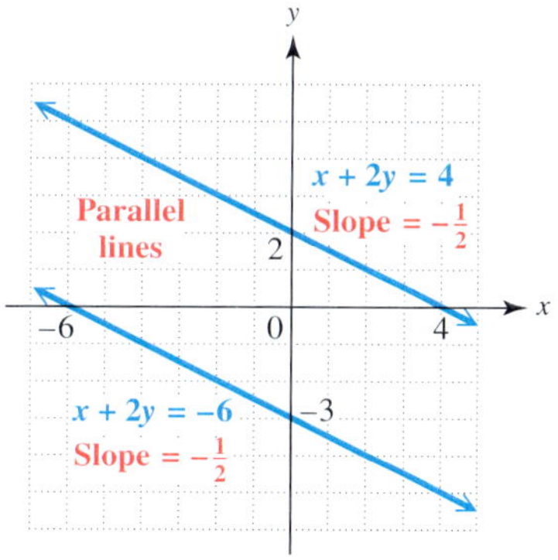


FIGURE 23

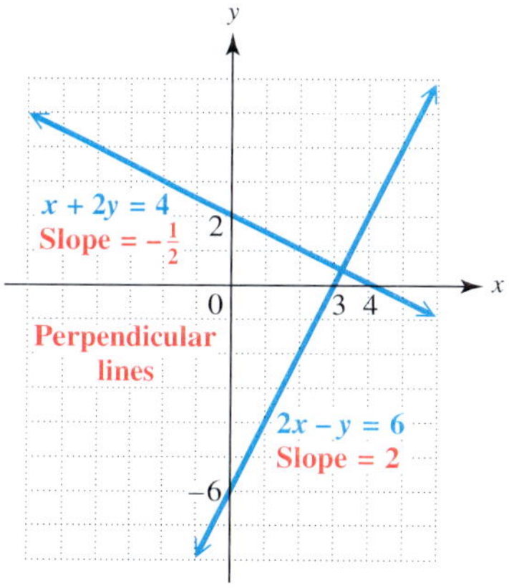


FIGURE 24

Figure 24 shows the graphs of $x + 2y = 4$ and $2x - y = 6$. These lines appear to be **perpendicular** (that is, they intersect at a 90° angle). Solving for $y$ shows that the slope of $x + 2y = 4$ is $-\frac{1}{2}$, while the slope of $2x - y = 6$ is 2. The product of $-\frac{1}{2}$ and 2 is

$$-\frac{1}{2}(2) = -1.$$

This is true in general; the product of the slopes of two perpendicular lines, neither of which is vertical, is always $-1$. This means that the slopes of perpendicular lines are negative reciprocals; if one slope is the nonzero number $a$, the other is $-\frac{1}{a}$.

**Slopes of Parallel and Perpendicular Lines**

Two lines with the same slope are parallel.

Two lines whose slopes have a product of $-1$ are perpendicular.

**EXAMPLE 6 Deciding Whether Two Lines Are Parallel or Perpendicular**

Decide whether each pair of lines is *parallel, perpendicular,* or *neither.*

**(a)** $x + 2y = 7$

$-2x + y = 3$

Find the slope of each line by first solving each equation for $y$.

| | |
|---|---|
| $x + 2y = 7$ | $-2x + y = 3$ |
| $2y = -x + 7$ | $y = 2x + 3$ |
| $y = -\frac{1}{2}x + \frac{7}{2}$ | |
| Slope: $-\frac{1}{2}$ | Slope: 2 |

Since the slopes are not equal, the lines are not parallel. Check the product of the slopes: $-\frac{1}{2}(2) = -1$. The two lines are perpendicular because the product of their slopes is $-1$, indicating that the slopes are negative reciprocals.

**(b)** $3x - y = 4$ → Solve for $y$. → $y = 3x - 4$

$6x - 2y = -12$ → $y = 3x + 6$

Both lines have slope 3, so the lines are parallel.

**(c)** $4x + 3y = 6$ → Solve for $y$. → $y = -\frac{4}{3}x + 2$

$2x - y = 5$ → $y = 2x - 5$

Here the slopes are $-\frac{4}{3}$ and 2. These lines are neither parallel nor perpendicular because $-\frac{4}{3} \neq 2$ and $-\frac{4}{3} \cdot 2 \neq -1$.

**(d)** $5x - y = 1$ → Solve for $y$. → $y = 5x - 1$

$x - 5y = -10$ → $y = \frac{1}{5}x + 2$

The slopes are 5 and $\frac{1}{5}$. The lines are not parallel, nor are they perpendicular. (*Be careful!* $5\left(\frac{1}{5}\right) = 1$, *not* $-1$.)

**Now Try Exercises 51, 55, and 57.**

## CONNECTIONS

Because the viewing window of a graphing calculator is a rectangle, the graphs of perpendicular lines will not appear perpendicular unless appropriate intervals are used for $x$ and $y$. Graphing calculators usually have a key to select a "square" window automatically. In a square window, the $x$-interval is about 1.5 times the $y$-interval because the screen is about 1.5 times as wide as it is high. The equations from Figure 24 are graphed with the standard (nonsquare) window and then with a square window on the next page.

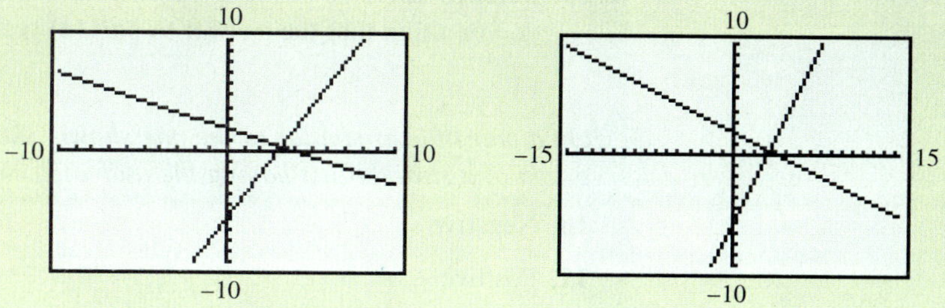


A standard (nonsquare) window
Lines do not appear perpendicular.

A square window
Lines appear perpendicular.

## 3.3 EXERCISES

**For Extra Help**

- Student's Solutions Manual
- MyMathLab
- InterAct Math Tutorial Software
- AW Math Tutor Center
- MathXL
- Digital Video Tutor CD 5/Videotape 5

**1.** In the context of the graph of a straight line, what is meant by "rise"? What is meant by "run"?

*Use the coordinates of the indicated points to find the ratio of rise to run for each line. See Example 1.*

**2.**

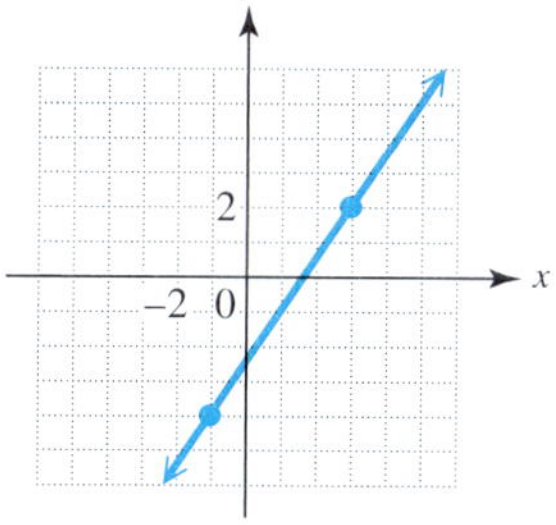


**3.**

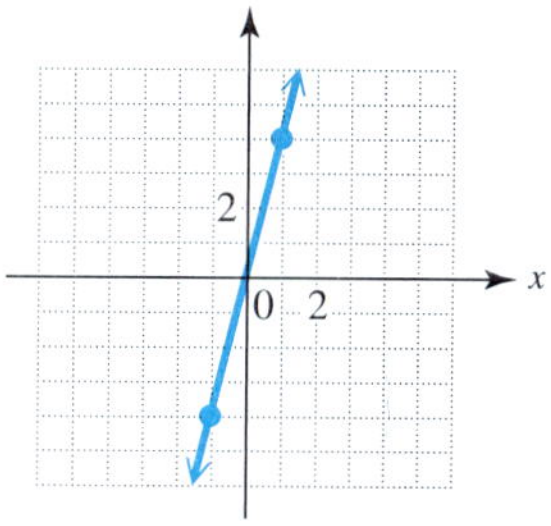


**4.**

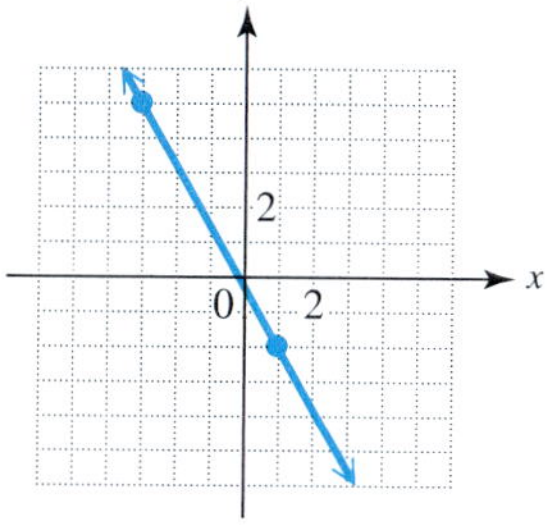


**5.**

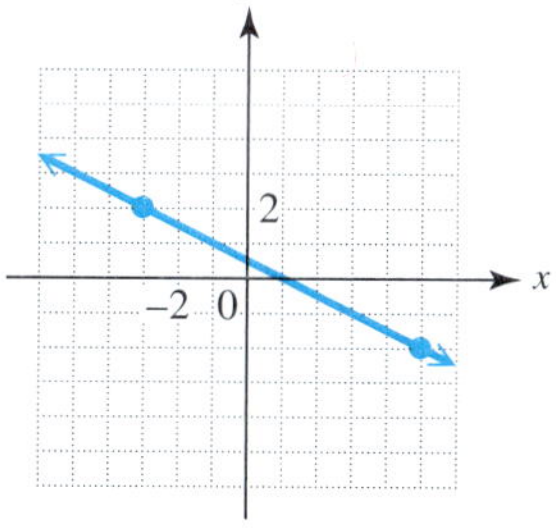


**6.**

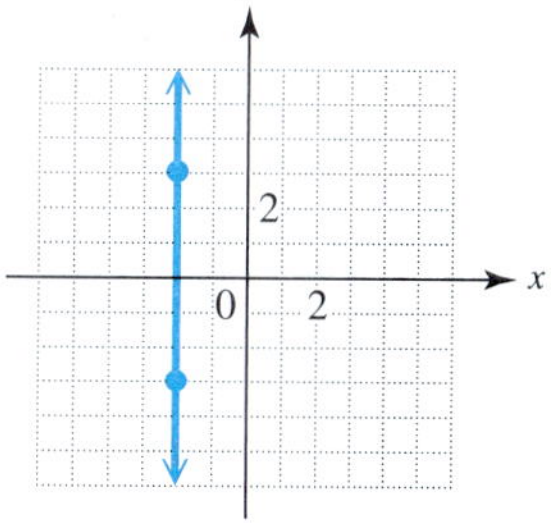


**7.**

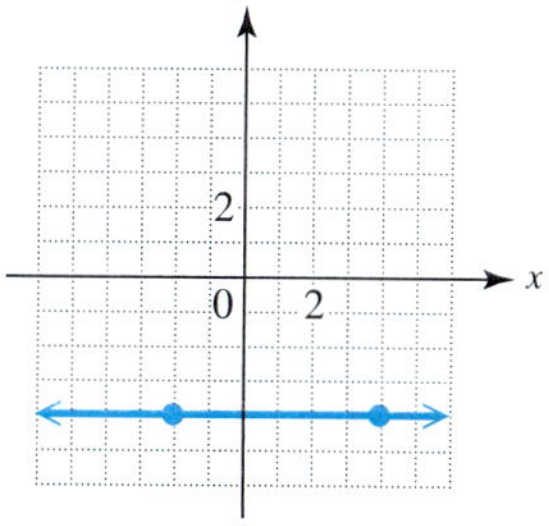


**8.** Look at the graph in Exercise 2 and answer the following.

**(a)** Start at the point $(-1, -4)$ and count vertically up to the horizontal line that goes through the other plotted point. What is this vertical change? (Remember: "up" means positive, "down" means negative.)

**(b)** From this new position, count horizontally to the other plotted point. What is this horizontal change? (Remember: "right" means positive, "left" means negative.)

**(c)** What is the quotient of the numbers found in parts (a) and (b)? What do we call this number?

9. Refer to Exercise 8. If we were to *start* at the point $(3, 2)$ and *end* at the point $(-1, -4)$, do you think that the answer to part (c) would be the same? Explain why or why not.

*On a pair of axes similar to the one shown, sketch the graph of a straight line having the indicated slope.*

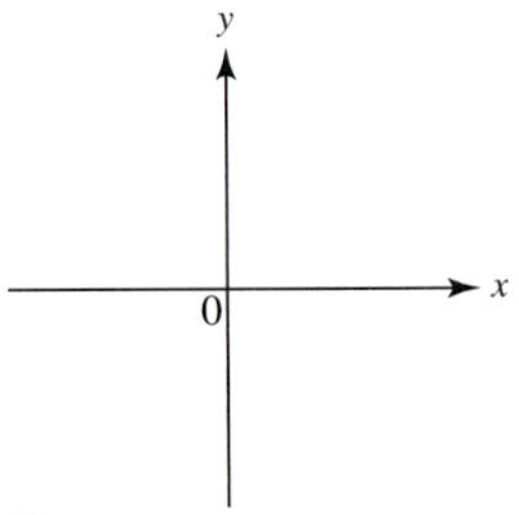


10. Negative

11. Positive

12. Undefined

13. Zero

14. Explain in your own words what is meant by *slope* of a line.

15. A student was asked to find the slope of the line through the points $(2, 5)$ and $(-1, 3)$. His answer, $-\frac{2}{3}$, was incorrect. He showed his work as

$$\frac{3-5}{2-(-1)} = \frac{-2}{3} = -\frac{2}{3}.$$

What was his error? Give the correct slope.

*Find the slope of the line through each pair of points. See Examples 2–4.*

16. $(4, -1)$ and $(-2, -8)$
17. $(1, -2)$ and $(-3, -7)$
18. $(-8, 0)$ and $(0, -5)$
19. $(0, 3)$ and $(-2, 0)$
20. $(-4, -5)$ and $(-5, -8)$
21. $(-2, 4)$ and $(-3, 7)$
22. $(6, -5)$ and $(-12, -5)$
23. $(4, 3)$ and $(-6, 3)$
24. $(-8, 6)$ and $(-8, -1)$
25. $(-12, 3)$ and $(-12, -7)$
26. $(3.1, 2.6)$ and $(1.6, 2.1)$
27. $\left(-\frac{7}{5}, \frac{3}{10}\right)$ and $\left(\frac{1}{5}, -\frac{1}{2}\right)$

*Find the slope of each line. See Example 5.*

28. $y = 2x - 3$
29. $y = 5x + 12$
30. $2y = -x + 4$
31. $4y = x + 1$
32. $-6x + 4y = 4$
33. $3x - 2y = 3$
34. $2x + 4y = 5$
35. $-3x + 2y = 5$
36. $x = -2$
37. $y = -5$
38. $y = 4$
39. $x = 6$
40. What is the slope of a line whose graph is parallel to the graph of $-5x + y = -3$? Perpendicular to the graph of $-5x + y = -3$?
41. What is the slope of a line whose graph is parallel to the graph of $3x + y = 7$? Perpendicular to the graph of $3x + y = 7$?

*The figure shows a line that has a positive slope (because it rises from left to right) and a positive y-value for the y-intercept (because it intersects the y-axis above the origin).*

*For each line in Exercises 42–47, decide whether* **(a)** *the slope is* positive, negative, *or* zero *and* **(b)** *the y-value of the y-intercept is* positive, negative, *or* zero.

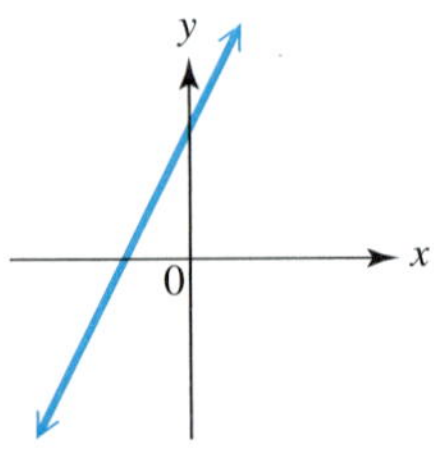

**42.**

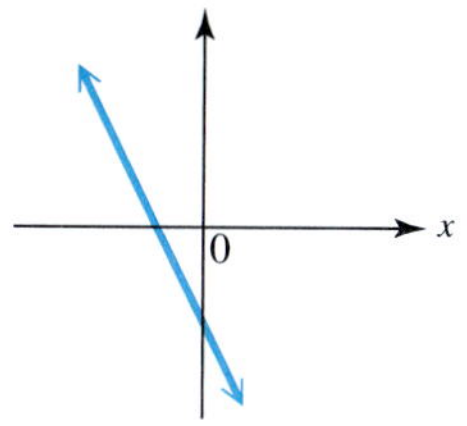


**43.**

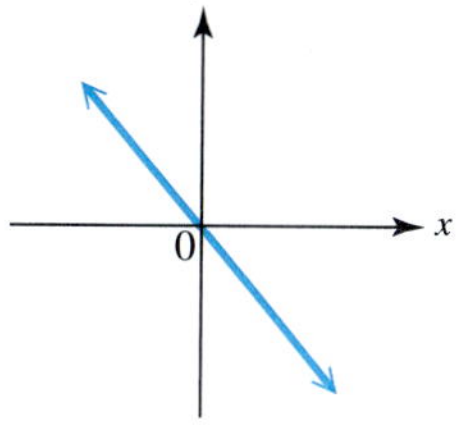


**44.**

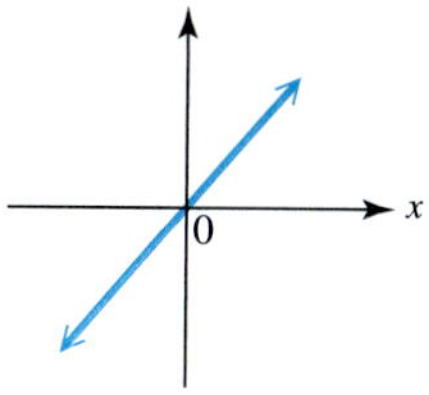


**45.**

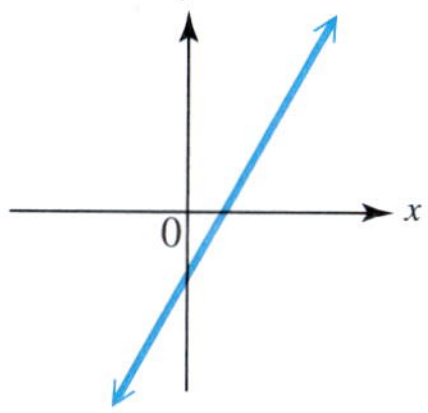


**46.**

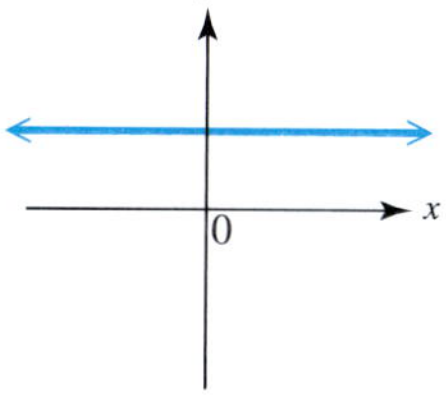


**47.**

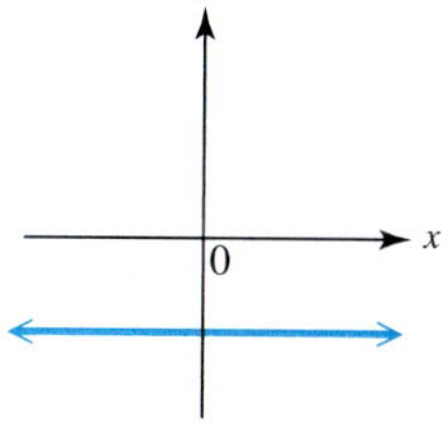


**48.** If two nonvertical lines are parallel, what do we know about their slopes? If two lines are perpendicular and neither is parallel to an axis, what do we know about their slopes? Why must the lines be nonvertical?

**49.** If two lines are both vertical or both horizontal, which of the following are they?

**A.** Parallel **B.** Perpendicular **C.** Neither parallel nor perpendicular

**50.** If a line is vertical, what is true of any line that is perpendicular to it?

*For each pair of equations, give the slopes of the lines and then determine whether the two lines are* parallel, perpendicular, *or* neither parallel nor perpendicular. *See Example 6.*

**51.** $2x + 5y = 4$
$4x + 10y = 1$

**52.** $-4x + 3y = 4$
$-8x + 6y = 0$

**53.** $8x - 9y = 6$
$8x + 6y = -5$

**54.** $5x - 3y = -2$
$3x - 5y = -8$

**55.** $3x - 2y = 6$
$2x + 3y = 3$

**56.** $3x - 5y = -1$
$5x + 3y = 2$

**57.** $5x - y = 1$
$x - 5y = -10$

**58.** $3x - 4y = 12$
$4x + 3y = 12$

**59.** What is the slope (or pitch) of this roof?

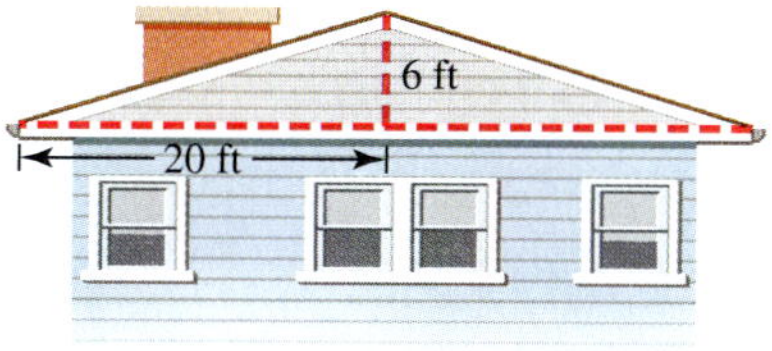


**60.** What is the slope (or grade) of this hill?

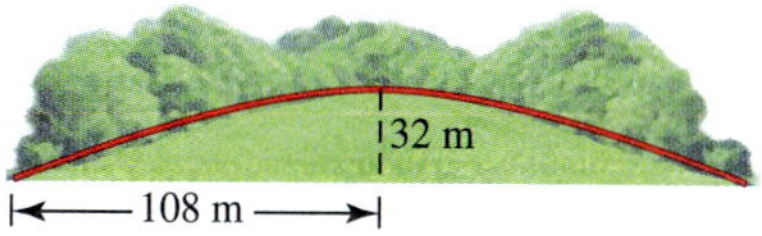

## RELATING CONCEPTS (EXERCISES 61–66)

**For Individual or Group Work**

*Figure A gives public school enrollment (in thousands) in grades 9–12 in the United States. Figure B gives the (average) number of public school students per computer.*

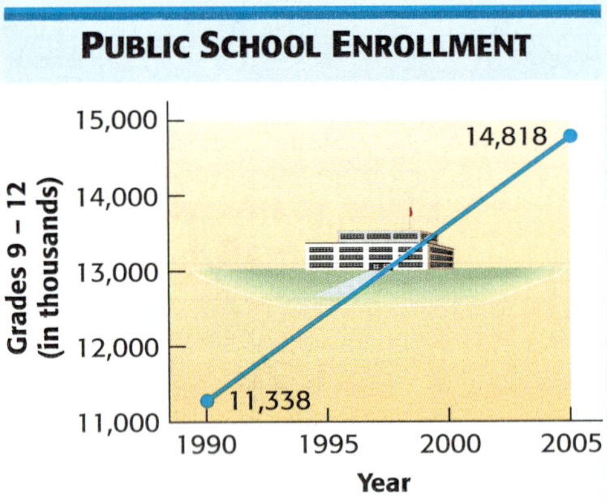


***Source:*** *Digest of Educational Statistics*, annual, and *Projections of Educational Statistics*, annual.

**FIGURE A**

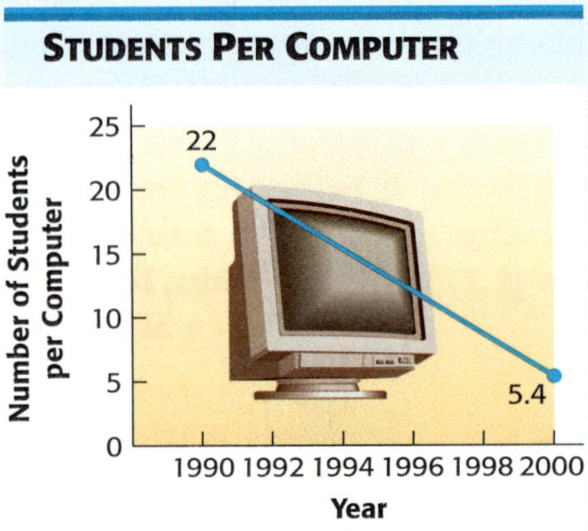


***Source:*** *Quality Education Data.*

**FIGURE B**

***Work Exercises 61–66 in order.***

**61.** Use the ordered pairs (1990, 11,338) and (2005, 14,818) to find the slope of the line in Figure A.

**62.** The slope of the line in Figure A is ______________ (positive/negative). This means that during the period represented, enrollment ______________ (increased/decreased).

**63.** The slope of a line represents its *rate of change.* Based on Figure A, what was the increase in students *per year* during the period shown?

**64.** Use the given information to find the slope of the line in Figure B.

**65.** The slope of the line in Figure B is ____________________ (positive/negative). This means that during the period represented, the number of students per computer ____________________ (increased/decreased).

**66.** Based on Figure B, what was the decrease in students per computer *per year* during the period shown?

**67.** The growth in retail square footage, in billions, is shown in the line graph. This graph looks like a straight line. If the change in square footage each year is the same, then it is a straight line. Find the change in square footage for the years shown in the graph. (*Hint:* To find the change in square footage from 1991 to 1992, subtract the $y$-value for 1991 from the $y$-value for 1992.) Is the graph a straight line?

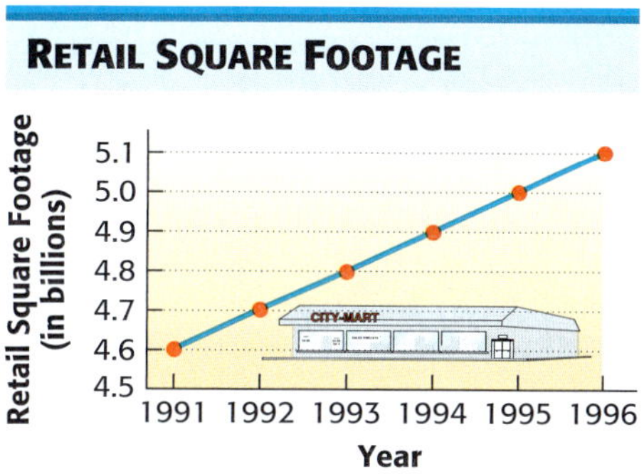


*Source*: International Council of Shopping Centers.

**68.** Find the slope of the line in Exercise 67 by using any two of the points shown on the line. How does the slope compare with the yearly change in square footage?

## TECHNOLOGY INSIGHTS (EXERCISES 69–72)

*Some graphing calculators have the capability of displaying a table of points for a graph. The table shown here gives several points that lie on a line designated* $Y_1$.

| X | $Y_1$ |
|---|---|
| -12 | -.8 |
| -10 | 0 |
| -8 | .8 |
| -6 | 1.6 |
| -4 | 2.4 |
| -2 | 3.2 |
| 0 | 4 |

X= -12

**69.** Use any two of the displayed ordered pairs to find the slope of the line.

**70.** What is the $x$-intercept of the line?

**71.** What is the $y$-intercept of the line?

**72.** Which one of the two lines shown is the graph of $Y_1$?

**A.**

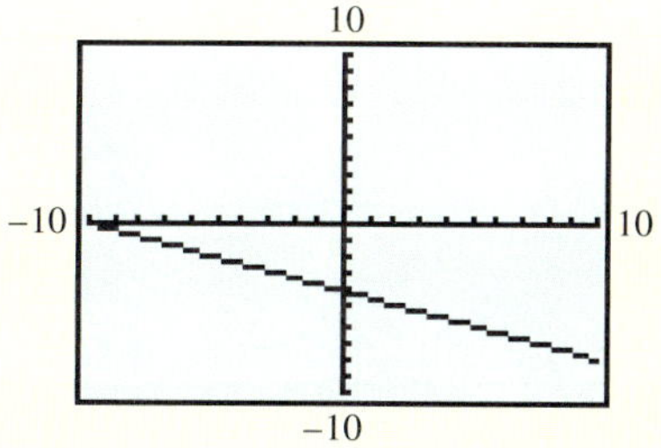


**B.**

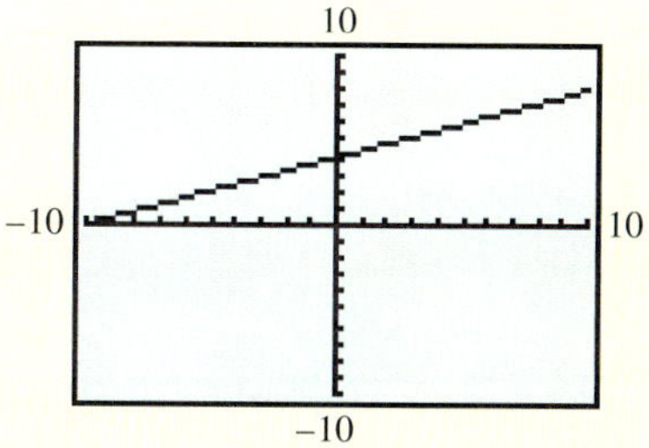

# 3.4 Equations of a Line

**OBJECTIVES**

1. Write an equation of a line given its slope and $y$-intercept.
2. Graph a line given its slope and a point on the line.
3. Write an equation of a line given its slope and any point on the line.
4. Write an equation of a line given two points on the line.

In the previous section we found the slope (steepness) of a line from the equation of the line by solving the equation for $y$. In that form, the slope is the coefficient of $x$. For example, the slope of the line with equation $y = 2x + 3$ is 2, the coefficient of $x$. What does the number 3 represent? If $x = 0$, the equation becomes

$$y = 2(0) + 3 = 0 + 3 = 3.$$

Since $y = 3$ corresponds to $x = 0$, $(0, 3)$ is the $y$-intercept of the graph of $y = 2x + 3$. An equation like $y = 2x + 3$ that is solved for $y$ is said to be in *slope-intercept form* because both the slope and the $y$-intercept of the line can be read directly from the equation.

**Slope-Intercept Form**

The **slope-intercept form** of the equation of a line with slope $m$ and $y$-intercept $(0, b)$ is

$$\boldsymbol{y = mx + b.}$$

Remember that the intercept in the slope-intercept form is the *y-intercept*.

**NOTE** The slope-intercept form is the most useful form for a linear equation because of the information we can determine from it. It is also the form used by graphing calculators and the one that describes a *linear function,* an important concept in mathematics.

**OBJECTIVE 1 Write an equation of a line given its slope and $y$-intercept.** Given the slope and $y$-intercept of a line, we can use the slope-intercept form to find an equation of the line.

**EXAMPLE 1 Finding an Equation of a Line**

Find an equation of the line with slope $\frac{2}{3}$ and $y$-intercept $(0, -1)$.

Here $m = \frac{2}{3}$ and $b = -1$, so the equation is

Slope ↓ ↓ $y$-intercept

$$y = mx + b$$

$$y = \frac{2}{3}x - 1.$$

**Now Try Exercise 11.**

**OBJECTIVE 2 Graph a line given its slope and a point on the line.** We can use the slope and $y$-intercept to graph a line. For example, to graph $y = \frac{2}{3}x - 1$, we first locate the $y$-intercept, $(0, -1)$, on the $y$-axis. From the definition of slope and the fact that the slope of the line is $\frac{2}{3}$,

$$m = \frac{\text{rise}}{\text{run}} = \frac{2}{3}.$$

Another point $P$ on the graph of the line can be found by counting from the $y$-intercept 2 units up and then counting 3 units to the right. We then draw the line through point $P$ and the $y$-intercept, as shown in Figure 25.

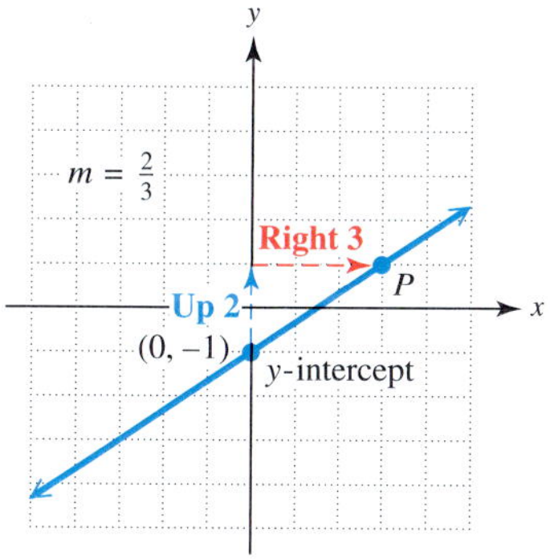


**FIGURE 25**

This method can be extended to graph a line given its slope and any point on the line, not just the $y$-intercept.

### EXAMPLE 2 Graphing a Line Given a Point and the Slope

Graph the line through $(-2, 3)$ with slope $-4$.

First, locate the point $(-2, 3)$. Write the slope as

$$m = \frac{\text{rise}}{\text{run}} = -4 = \frac{-4}{1}.$$

Locate another point on the line by counting 4 units down (because of the negative sign) and then 1 unit to the right. Finally, draw the line through this new point $P$ and the given point $(-2, 3)$. See Figure 26.

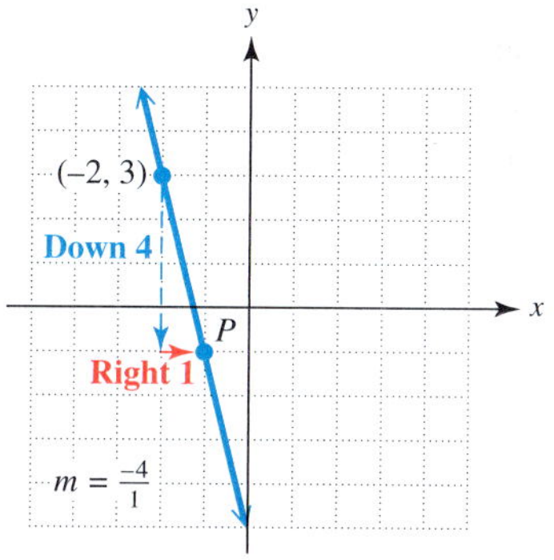


**FIGURE 26**

**Now Try Exercise 19.**

**NOTE** In Example 2, we could have written the slope as $\frac{4}{-1}$ instead. In this case, we would move 4 units up from $(-2, 3)$ and then 1 unit to the left (because of the negative sign). Verify that this produces the same line.

**OBJECTIVE 3 Write an equation of a line given its slope and any point on the line.** Let $m$ represent the slope of a line and let $(x_1, y_1)$ represent a given point on the line. Let $(x, y)$ represent any other point on the line. Then by the definition of slope,

$$\frac{y - y_1}{x - x_1} = m \quad \text{or} \quad y - y_1 = m(x - x_1).$$

This result is the *point-slope form* of the equation of a line.

**Point-Slope Form**

The **point-slope form** of the equation of a line with slope $m$ going through $(x_1, y_1)$ is

$$y - y_1 = m(x - x_1).$$

**EXAMPLE 3 Using the Point-Slope Form to Write Equations**

Find an equation of each line. Write the equation in slope-intercept form.

**(a)** Through $(-2, 4)$, with slope $-3$

The given point is $(-2, 4)$ so $x_1 = -2$ and $y_1 = 4$. Also, $m = -3$. First substitute these values into the point-slope form to get an equation of the line. Then solve for $y$ to write the equation in slope-intercept form.

$$\begin{aligned} y - y_1 &= m(x - x_1) && \text{Point-slope form} \\ y - 4 &= -3[x - (-2)] && \text{Let } y_1 = 4, m = -3, x_1 = -2. \\ y - 4 &= -3(x + 2) && \\ y - 4 &= -3x - 6 && \text{Distributive property} \\ y &= -3x - 2 && \text{Add 4.} \end{aligned}$$

The last equation is in slope-intercept form.

**(b)** Through $(4, 2)$, with slope $\frac{3}{5}$

$$\begin{aligned} y - y_1 &= m(x - x_1) && \\ y - 2 &= \frac{3}{5}(x - 4) && \text{Let } y_1 = 2, m = \tfrac{3}{5}, x_1 = 4. \\ y - 2 &= \frac{3}{5}x - \frac{12}{5} && \text{Distributive property} \\ y &= \frac{3}{5}x - \frac{12}{5} + \frac{10}{5} && \text{Add } 2 = \tfrac{10}{5} \text{ to both sides.} \\ y &= \frac{3}{5}x - \frac{2}{5} && \text{Combine terms.} \end{aligned}$$

We did not clear fractions after the substitution step because we want the equation in slope-intercept form; that is, solved for $y$.

**Now Try Exercises 29 and 31.**

**OBJECTIVE 4 Write an equation of a line given two points on the line.** We can also use the point-slope form to find an equation of a line when two points on the line are known.

**EXAMPLE 4 Finding the Equation of a Line Given Two Points**

Find an equation of the line through the points $(-2, 5)$ and $(3, 4)$. Write the equation in slope-intercept form.

First, find the slope of the line, using the slope formula.

$$\text{slope } m = \frac{y_2 - y_1}{x_2 - x_1} = \frac{5 - 4}{-2 - 3} = \frac{1}{-5} = -\frac{1}{5}$$

Now use either $(-2, 5)$ or $(3, 4)$ and the point-slope form. We choose $(3, 4)$.

$$y - y_1 = m(x - x_1)$$

$$y - 4 = -\frac{1}{5}(x - 3) \qquad \text{Let } y_1 = 4,\ m = -\tfrac{1}{5},\ x_1 = 3.$$

$$y - 4 = -\frac{1}{5}x + \frac{3}{5} \qquad \text{Distributive property}$$

$$y = -\frac{1}{5}x + \frac{3}{5} + \frac{20}{5} \qquad \text{Add } 4 = \tfrac{20}{5} \text{ to both sides.}$$

$$y = -\frac{1}{5}x + \frac{23}{5} \qquad \text{Combine terms.}$$

The same result would be found by using $(-2, 5)$ for $(x_1, y_1)$.

**Now Try Exercise 39.**

Many of the linear equations in Sections 3.1 through 3.3 were given in the form $Ax + By = C$, which is called *standard form.*

**Standard Form**

A linear equation is in **standard form** if it is written as

$$Ax + By = C,$$

where $A$, $B$, and $C$ are integers and $A > 0$, $B \neq 0$.

**NOTE** The definition of standard form is not the same in all texts. A linear equation can be written in this form in many different, equally correct, ways. For example, $3x + 4y = 12$, $6x + 8y = 24$, and $9x + 12y = 36$ represent the same set of ordered pairs. Let us agree that $3x + 4y = 12$ is preferable to the other forms because the greatest common factor of 3, 4, and 12 is 1.

A summary of the forms of linear equations follows.

**Linear Equations**

| | |
|---|---|
| $x = a$ | **Vertical line**<br>Slope is undefined; $x$-intercept is $(a, 0)$. |
| $y = b$ | **Horizontal line**<br>Slope is 0; $y$-intercept is $(0, b)$. |
| $y = mx + b$ | **Slope-intercept form**<br>Slope is $m$; $y$-intercept is $(0, b)$. |
| $y - y_1 = m(x - x_1)$ | **Point-slope form**<br>Slope is $m$; line goes through $(x_1, y_1)$. |
| $Ax + By = C$ | **Standard form**<br>Slope is $-\frac{A}{B}$; $x$-intercept is $\left(\frac{C}{A}, 0\right)$; $y$-intercept is $\left(0, \frac{C}{B}\right)$. |

## 3.4 EXERCISES

**For Extra Help**

- Student's Solutions Manual
- MyMathLab
- InterAct Math Tutorial Software
- AW Math Tutor Center
- MathXL
- Digital Video Tutor CD 5/Videotape 5

*Match the correct equation in Column II with the description in Column I.*

| I | II |
|---|---|
| **1.** Slope $= -2$, through the point $(4, 1)$ | **A.** $y = 4x$ |
| **2.** Slope $= -2$, $y$-intercept $(0, 1)$ | **B.** $y = \frac{1}{4}x$ |
| **3.** Passing through the points $(0, 0)$ and $(4, 1)$ | **C.** $y = -2x + 1$ |
| **4.** Passing through the points $(0, 0)$ and $(1, 4)$ | **D.** $y - 1 = -2(x - 4)$ |

**5.** Explain why the equation of a vertical line cannot be written in the form $y = mx + b$.

**6.** Match each equation with the graph that would most closely resemble its graph.

**(a)** $y = x + 3$

**(b)** $y = -x + 3$

**(c)** $y = x - 3$

**(d)** $y = -x - 3$

**A.**

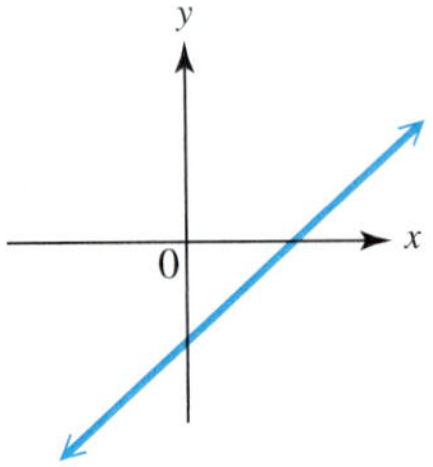


**B.**

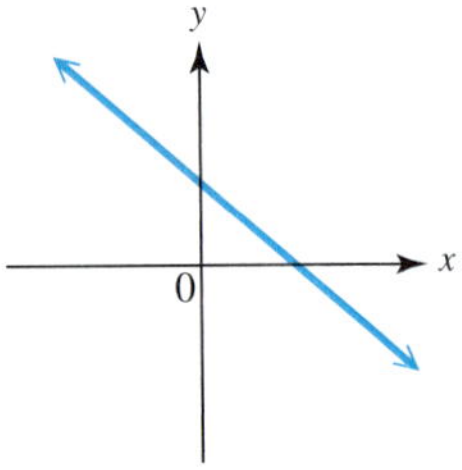


**C.**

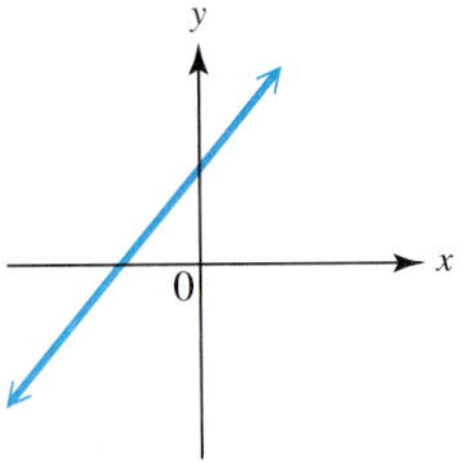


**D.**

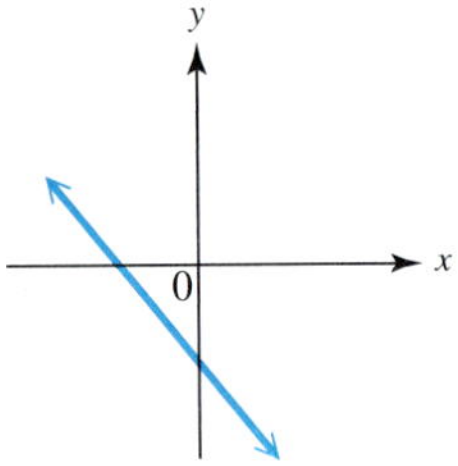


*Use the geometric interpretation of slope* (rise divided by run, *from Section 3.3) to find the slope of each line. Then, by identifying the y-intercept from the graph, write the slope-intercept form of the equation of the line.*

**7.**

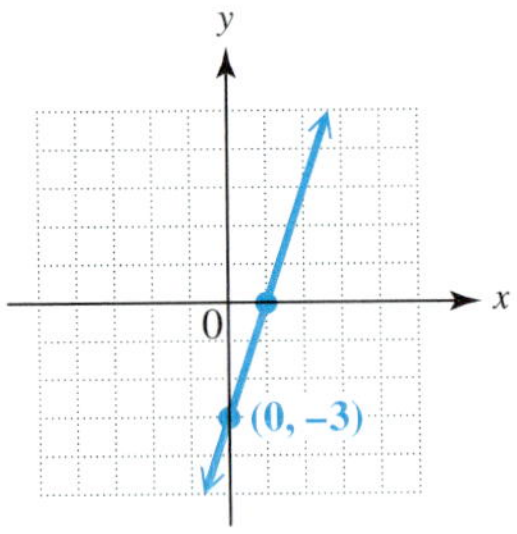


**8.**

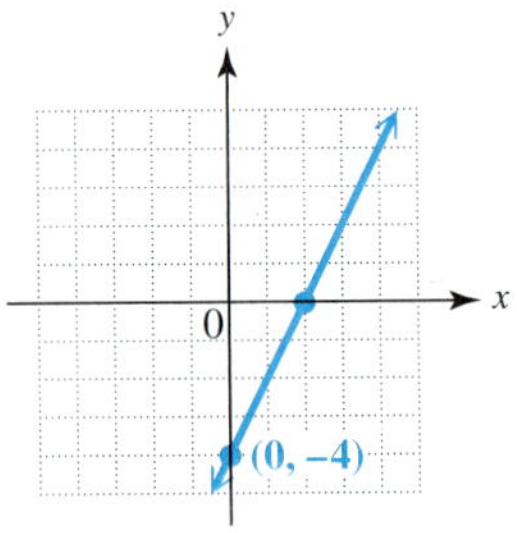


**9.**

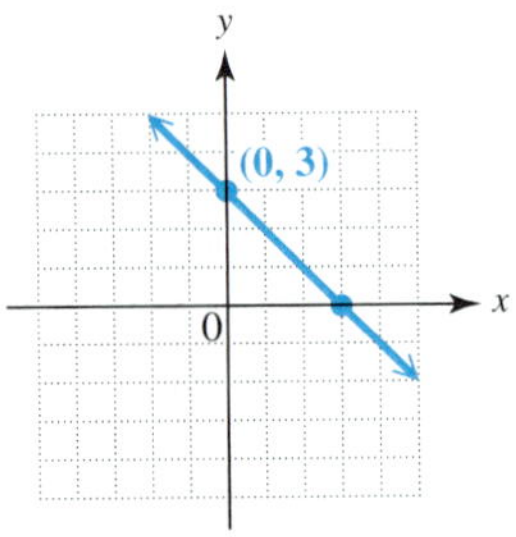


**10.**

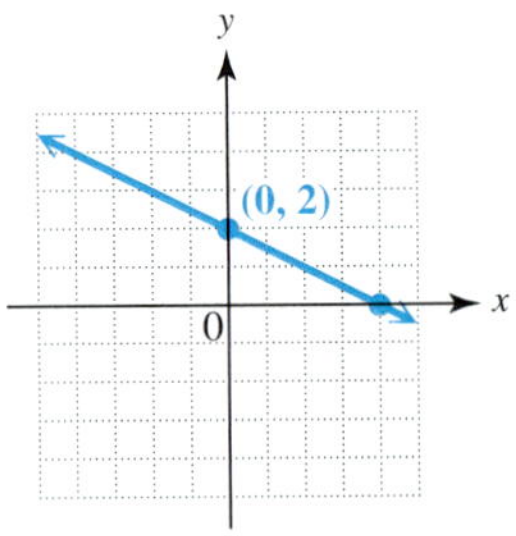

*Write the equation of each line with the given slope and y-intercept. See Example 1.*

**11.** $m = 4, (0, -3)$

**12.** $m = -5, (0, 6)$

**13.** $m = 0, (0, 3)$

**14.** $m = 0, (0, -4)$

**15.** Undefined slope, $(0, -2)$

**16.** Undefined slope, $(0, 5)$

*Graph each line passing through the given point and having the given slope. (In Exercises 22–25, recall the types of lines having slope 0 and undefined slope.) See Example 2.*

**17.** $(0, 2), m = 3$

**18.** $(0, -5), m = -2$

**19.** $(1, -5), m = -\frac{2}{5}$

**20.** $(2, -1), m = -\frac{1}{3}$

**21.** $(-1, 4), m = \frac{2}{5}$

**22.** $(3, 2), m = 0$

**23.** $(-2, 3), m = 0$

**24.** $(3, -2)$, undefined slope

**25.** $(2, 4)$, undefined slope

**26.** What is the common name given to a vertical line whose $x$-intercept is the origin?

**27.** What is the common name given to a line with slope 0 whose $y$-intercept is the origin?

*Write an equation for each line passing through the given point and having the given slope. Write the equation in slope-intercept form. See Example 3.*

**28.** $(4, 1), m = 2$

**29.** $(-1, 3), m = -4$

**30.** $(2, 7), m = 3$

**31.** $(-2, 5), m = \frac{2}{3}$

**32.** $(4, 2), m = -\frac{1}{3}$

**33.** $(-4, 1), m = \frac{3}{4}$

**34.** $(6, -3), m = -\frac{4}{5}$

**35.** $(2, 1), m = \frac{5}{2}$

**36.** $(7, -2), m = -\frac{7}{2}$

**37.** If a line passes through the origin and a second point whose $x$- and $y$-coordinates are equal, what is an equation of the line?

*Write an equation for the line passing through the given pair of points. Write each equation in slope-intercept form. See Example 4.*

**38.** $(8, 5)$ and $(9, 6)$

**39.** $(4, 10)$ and $(6, 12)$

**40.** $(-1, -7)$ and $(-8, -2)$

**41.** $(-2, -1)$ and $(3, -4)$

**42.** $(0, -2)$ and $(-3, 0)$

**43.** $(-4, 0)$ and $(0, 2)$

**44.** $\left(\frac{1}{2}, \frac{3}{2}\right)$ and $\left(-\frac{1}{4}, \frac{5}{4}\right)$

**45.** $\left(-\frac{2}{3}, \frac{8}{3}\right)$ and $\left(\frac{1}{3}, \frac{7}{3}\right)$

**46.** Describe in your own words the slope-intercept and point-slope forms of the equation of a line. Tell what information must be given to use each form to write an equation. Include examples.

## RELATING CONCEPTS (EXERCISES 47–54)

**For Individual or Group Work**

*If we think of ordered pairs of the form $(C, F)$, then the two most common methods of measuring temperature, Celsius and Fahrenheit, can be related as follows: When $C = 0$, $F = 32$, and when $C = 100$, $F = 212$.* ***Work Exercises 47–54 in order.***

**47.** Write two ordered pairs relating these two temperature scales.

**48.** Find the slope of the line through the two points.

**49.** Use the point-slope form to find an equation of the line. (Your variables should be $C$ and $F$ rather than $x$ and $y$.)

**50.** Write an equation for $F$ in terms of $C$.

**51.** Use the equation from Exercise 50 to write an equation for $C$ in terms of $F$.

**52.** Use the equation from Exercise 50 to find the Fahrenheit temperature when $C = 30$.

**53.** Use the equation from Exercise 51 to find the Celsius temperature when $F = 50$.

**54.** For what temperature is $F = C$?

*Write an equation of the line satisfying the given conditions. Write the equation in slope-intercept form.*

**55.** Through $(2, -3)$, parallel to $3x = 4y + 5$

**56.** Through $(-1, 4)$, perpendicular to $2x + 3y = 8$

**57.** Perpendicular to $x - 2y = 7$, $y$-intercept $(0, -3)$

**58.** Parallel to $5x = 2y + 10$, $y$-intercept $(0, 4)$

*The cost to produce $x$ items is, in some cases, expressed as $y = mx + b$. The number $b$ gives the* fixed cost *(the cost that is the same no matter how many items are produced), and the number $m$ is the* variable cost *(the cost to produce an additional item). Use this information to work Exercises 59 and 60.*

**59.** It costs \$400 to start up a business selling snow cones. Each snow cone costs \$.25 to produce.

**(a)** What is the fixed cost?
**(b)** What is the variable cost?
**(c)** Write the cost equation.
**(d)** What will be the cost to produce 100 snow cones, based on the cost equation?
**(e)** How many snow cones will be produced if total cost is \$775?

**60.** It costs \$2000 to purchase a copier, and each copy costs \$.02 to make.

**(a)** What is the fixed cost?
**(b)** What is the variable cost?
**(c)** Write the cost equation.
**(d)** What will be the cost to produce 10,000 copies, based on the cost equation?
**(e)** How many copies will be produced if total cost is \$2600?

## TECHNOLOGY INSIGHTS (EXERCISES 61–64)

*In Exercises 61 and 62, two graphing calculator views of the same line are shown. Use the displays at the bottom of the screen to find an equation of the form $y = mx + b$ for each line.*

**61.**

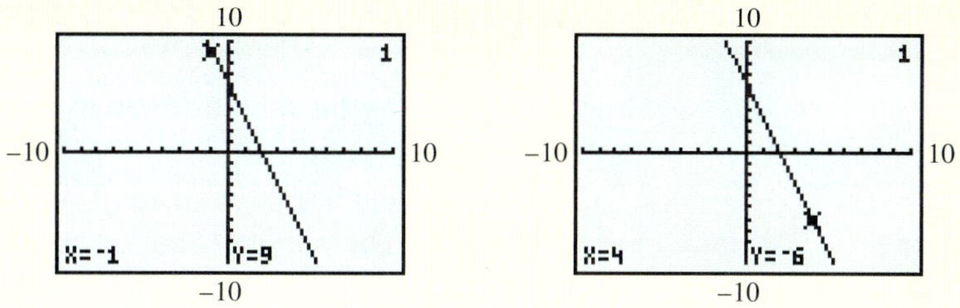


**62.**

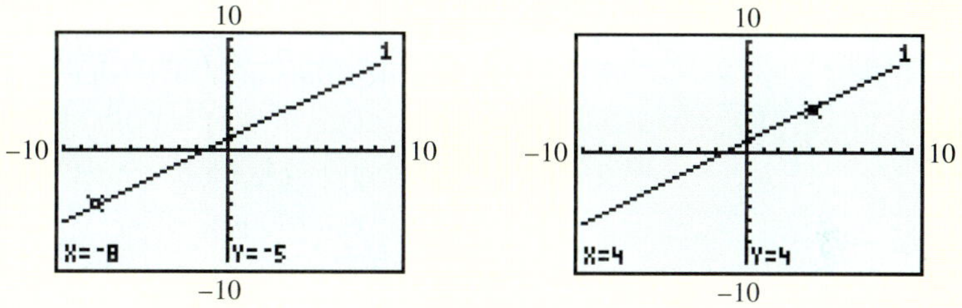


*In Exercises 63 and 64, a table of points, generated by a graphing calculator, is shown for a line $Y_1$.* Use any two points to find the equation of each line. Write the equation in slope-intercept form.*

**63.**

| X | $Y_1$ |
|---|---|
| -2 | -.5 |
| -1 | .25 |
| 0 | 1 |
| 1 | 1.75 |
| 2 | 2.5 |
| 3 | 3.25 |
| 4 | 4 |

X= -2

**64.**

| X | $Y_1$ |
|---|---|
| -4 | 14 |
| -3 | 10 |
| -2 | 6 |
| -1 | 2 |
| 0 | -2 |
| 1 | -6 |
| 2 | -10 |

X= -4

*With graphing calculators, we use capital $Y_1$ and X like the calculator does.

## Chapter 3 Group Activity

### Determining Business Profit

**OBJECTIVE** Use equations, tables, and graphs to make business decisions.

Graphs are used by businesses to analyze and make decisions. This activity will explore such a process.

The Parent Teacher Organization of a school decides to sell bags of popcorn as a fund-raiser. The parents want to estimate their profit. Costs include \$14 for popcorn (enough for 680 bags) and a total of \$7 for bags to hold the popcorn.

**A.** Develop a profit formula.

1. From the information above, determine the total costs. (Assume the organization is buying enough supplies for 680 bags of popcorn.)
2. If the filled bags sell for \$.25 each, write an expression for the total revenue from $n$ bags of popcorn.
3. Since Profit = Revenue − Cost, write an equation that represents the profit $P$ for $n$ bags of popcorn sold.
4. Work in pairs. One person should use the profit equation from part 3 to complete the table below. The other person should decide on an appropriate scale and graph the profit equation.

| $n$ | $P$ |
|---|---|
| 0 | |
| | 0 |
| 100 | |

**B.** Choose a different price for a bag of popcorn (between \$.20 and \$.75).

1. Write the profit equation for this cost.
2. Switch roles, that is, if you drew the graph in part A, now make a table of values for this equation. Have your partner graph the profit equation on the same coordinate system you used in part A.

**C.** Compare your findings and answer the following.

1. What is the break-even point (that is, when profits are 0 or revenue equals costs) for each equation?
2. What does it mean if you end up with a negative value for $P$?
3. If you sell all 680 bags, what will your profits be for the two different prices? Explain how you would estimate this from your graph.
4. What are the advantages and/or disadvantages of charging a higher price?
5. Together, decide on the price you would charge for a bag of popcorn. Explain why you chose this price.

# CHAPTER 3 SUMMARY

## KEY TERMS

**3.1** linear equation in two variables
ordered pair
table of values
$x$-axis
$y$-axis
rectangular (Cartesian) coordinate system
quadrant
origin
plane
coordinates
plot
scatter diagram

**3.2** graph, graphing
$y$-intercept
$x$-intercept

**3.3** subscript notation
rise
run
slope
parallel lines
perpendicular lines

## NEW SYMBOLS

$(a, b)$ an ordered pair

$(x_1, y_1)$ $x$-sub-one, $y$-sub-one

$m$ slope

## TEST YOUR WORD POWER

*See how well you have learned the vocabulary in this chapter. Answers, with examples, follow the Quick Review.*

1. An **ordered pair** is a pair of numbers written
   **A.** in numerical order between brackets
   **B.** between parentheses or brackets
   **C.** between parentheses in which order is important
   **D.** between parentheses in which order does not matter.

2. An **intercept** is
   **A.** the point where the $x$-axis and $y$-axis intersect
   **B.** a pair of numbers written in parentheses in which order matters
   **C.** one of the four regions determined by a rectangular coordinate system
   **D.** the point where a graph intersects the $x$-axis or the $y$-axis.

3. The **slope** of a line is
   **A.** the measure of the run over the rise of the line
   **B.** the distance between two points on the line
   **C.** the ratio of the change in $y$ to the change in $x$ along the line
   **D.** the horizontal change compared to the vertical change of two points on the line.

4. Two lines in a plane are **parallel** if
   **A.** they represent the same line
   **B.** they never intersect
   **C.** they intersect at a 90° angle
   **D.** one has a positive slope and one has a negative slope.

5. Two lines in a plane are **perpendicular** if
   **A.** they represent the same line
   **B.** they never intersect
   **C.** they intersect at a 90° angle
   **D.** one has a positive slope and one has a negative slope.

## QUICK REVIEW

**CONCEPTS** | **EXAMPLES**

### 3.1 READING GRAPHS; LINEAR EQUATIONS IN TWO VARIABLES

An ordered pair is a solution of an equation if it satisfies the equation.

Is $(2, -5)$ or $(0, -6)$ a solution of $4x - 3y = 18$?

$4(2) - 3(-5) = 23 \neq 18$
$(2, -5)$ is not a solution.

$4(0) - 3(-6) = 18$
$(0, -6)$ is a solution.

If a value of either variable in an equation is given, the other variable can be found by substitution.

Complete the ordered pair $(0, \quad)$ for $3x = y + 4$.

$$3(0) = y + 4$$
$$0 = y + 4$$
$$-4 = y$$

The ordered pair is $(0, -4)$.

Plot the ordered pair $(-3, 4)$ by starting at the origin, going 3 units to the left, then going 4 units up.

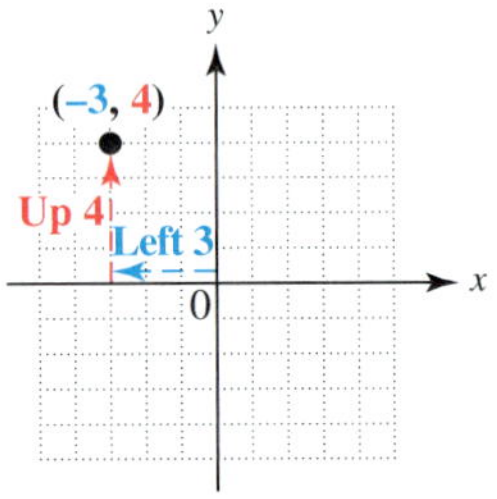


### 3.2 GRAPHING LINEAR EQUATIONS IN TWO VARIABLES

The graph of $\boldsymbol{y = b}$ is a horizontal line through $(0, b)$.

The graph of $\boldsymbol{x = a}$ is a vertical line through $(a, 0)$.

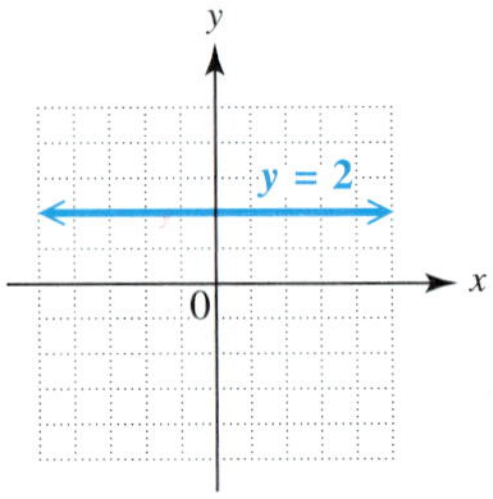


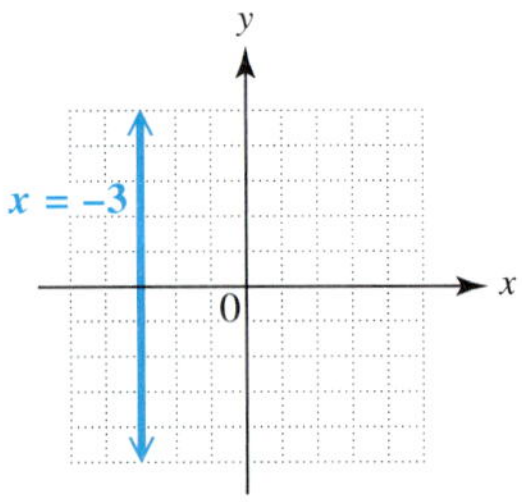


The graph of $\boldsymbol{Ax + By = 0}$ goes through the origin. Find and plot another point that satisfies the equation. Then draw the line through the two points.

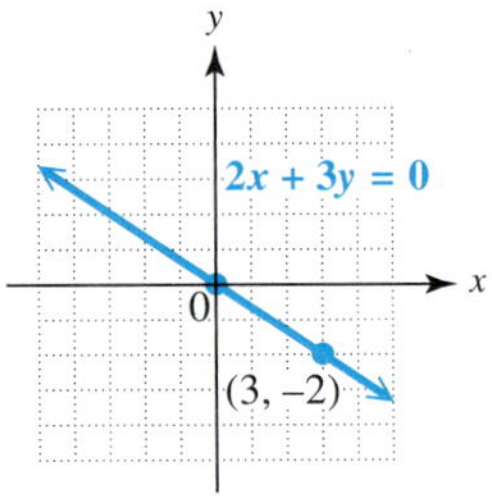


To graph a linear equation:

*Step 1* Find at least two ordered pairs that satisfy the equation.

*Step 2* Plot the corresponding points.

*Step 3* Draw a straight line through the points.

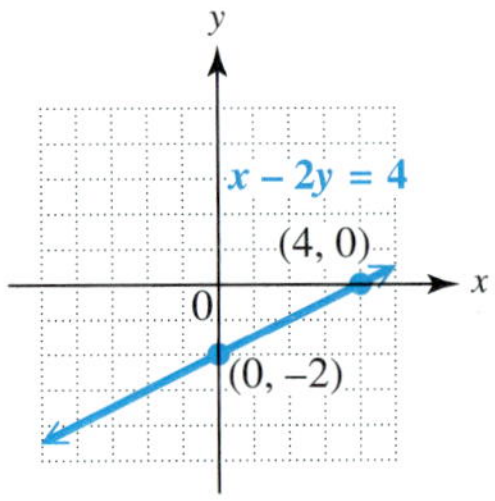

| CONCEPTS | EXAMPLES |
|---|---|
| **3.3 THE SLOPE OF A LINE** | |
| The slope of the line through $(x_1, y_1)$ and $(x_2, y_2)$ is $$m = \frac{y_2 - y_1}{x_2 - x_1} \quad (x_1 \neq x_2).$$ | The line through $(-2, 3)$ and $(4, -5)$ has slope $$m = \frac{-5 - 3}{4 - (-2)} = \frac{-8}{6} = -\frac{4}{3}.$$ |
| Horizontal lines have slope 0. | The line $y = -2$ has slope 0. |
| Vertical lines have undefined slope. | The line $x = 4$ has undefined slope. |
| To find the slope of a line from its equation, solve for $y$. The slope is the coefficient of $x$. | Find the slope of $3x - 4y = 12$. $$-4y = -3x + 12$$ $$y = \frac{3}{4}x - 3$$ The slope is $\frac{3}{4}$. |
| Parallel lines have the same slope. | The lines $y = 3x - 1$ and $y = 3x + 4$ are parallel because both have slope 3. |
| The slopes of perpendicular lines are negative reciprocals (that is, their product is $-1$). | The lines $y = -3x - 1$ and $y = \frac{1}{3}x + 4$ are perpendicular because their slopes are $-3$ and $\frac{1}{3}$, and $-3\left(\frac{1}{3}\right) = -1$. |
| **3.4 EQUATIONS OF A LINE** | |
| **Slope-Intercept Form**<br>$y = mx + b$<br>$m$ is the slope.<br>$(0, b)$ is the $y$-intercept. | Find an equation of the line with slope 2 and $y$-intercept $(0, -5)$.<br>The equation is $y = 2x - 5$. |
| **Point-Slope Form**<br>$y - y_1 = m(x - x_1)$<br>$m$ is the slope.<br>$(x_1, y_1)$ is a point on the line. | Find an equation of the line with slope $-\frac{1}{2}$ through $(-4, 5)$. $$y - 5 = -\frac{1}{2}[x - (-4)]$$ $$y - 5 = -\frac{1}{2}(x + 4)$$ $$y - 5 = -\frac{1}{2}x - 2$$ $$y = -\frac{1}{2}x + 3$$ |
| **Standard Form**<br>$Ax + By = C$<br>$A$, $B$, and $C$ are integers and $A > 0$, $B \neq 0$. | This equation is written in standard form as $$x + 2y = 6,$$ with $A = 1$, $B = 2$, and $C = 6$. |

**Answers to Test Your Word Power**

**1.** C; *Examples:* (0, 3), (−3, 8), (4, 0) **2.** D; *Example:* The graph of the equation $4x - 3y = 12$ has $x$-intercept (3, 0) and $y$-intercept (0, −4). **3.** C; *Example:* The line through (3, 6) and (5, 4) has slope $\frac{4-6}{5-3} = \frac{-2}{2} = -1$. **4.** B; *Example:* See Figure 23 in Section 3.3. **5.** C; *Example:* See Figure 24 in Section 3.3.

# CHAPTER 3 REVIEW EXERCISES

**[3.1]**

**1.** The line graph shows average prices for a gallon of gasoline in Cedar Rapids, Iowa, from June 1999 through June 2000. Use the graph to work parts (a)–(d).

**(a)** About how much did a gallon of gas cost in June 1999? in June 2000?

**(b)** How much did the price of a gallon of gas increase over this 1-yr period? What percent increase is this?

**(c)** During which months did the biggest increase in the price of a gallon of gas occur? About how much did the price increase during this time?

**(d)** Between which months did the price of a gallon of gas decrease?

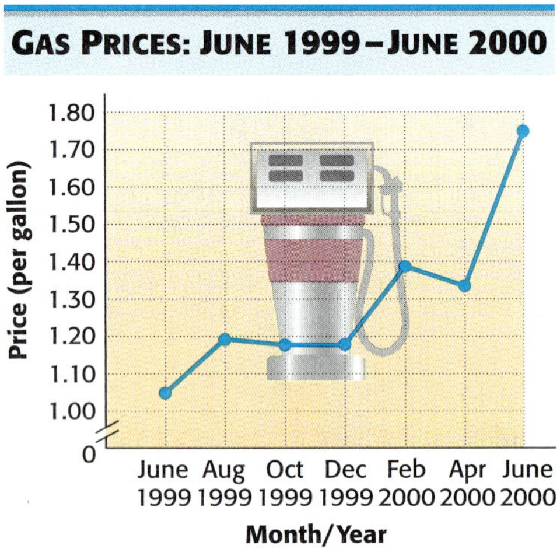


*Source:* Iowa DNR.

*Complete the given ordered pairs for each equation.*

**2.** $y = 3x + 2$; $(-1,\ \ ), (0,\ \ ), (\ \ , 5)$

**3.** $4x + 3y = 6$; $(0,\ \ ), (\ \ , 0), (-2,\ \ )$

**4.** $x = 3y$; $(0,\ \ ), (8,\ \ ), (\ \ , -3)$

**5.** $x - 7 = 0$; $(\ \ , -3), (\ \ , 0), (\ \ , 5)$

*Determine whether the given ordered pair is a solution of the given equation.*

**6.** $x + y = 7$; $(2, 5)$

**7.** $2x + y = 5$; $(-1, 3)$

**8.** $3x - y = 4$; $\left(\frac{1}{3}, -3\right)$

*Name the quadrant in which each pair lies. Then plot each ordered pair in a rectangular coordinate system.*

**9.** $(2, 3)$

**10.** $(-4, 2)$

**11.** $(3, 0)$

**12.** $(0, -6)$

**13.** If $xy > 0$, in what quadrant or quadrants must $(x, y)$ lie?

**[3.2]** *Find the x- and y-intercepts for the line that is the graph of each equation, and graph the line.*

**14.** $y = 2x + 5$

**15.** $3x + 2y = 8$

**16.** $x + 2y = -4$

**[3.3]** *Find the slope of each line.*

**17.** Through $(2, 3)$ and $(-4, 6)$

**18.** Through $(2, 5)$ and $(2, 8)$

**19.** $y = 3x - 4$

**20.** $y = 5$

**21.**

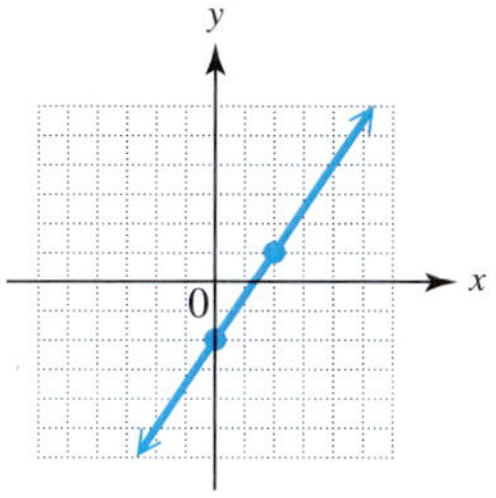


**22.**

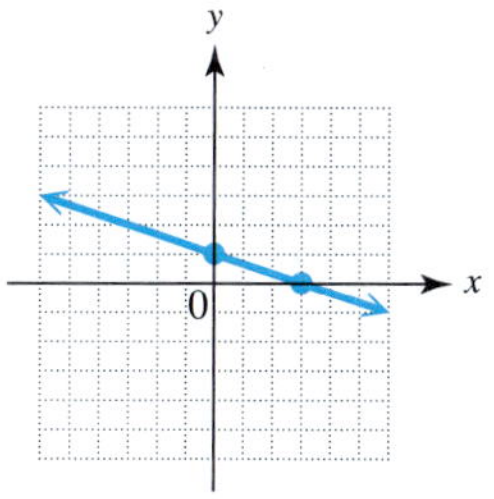


**23.** The line having these points

| $x$ | $y$ |
|---|---|
| 0 | 1 |
| 2 | 4 |
| 6 | 10 |

**24.** **(a)** A line parallel to the graph of $y = 2x + 3$

**(b)** A line perpendicular to the graph of $y = -3x + 3$

*Decide whether each pair of lines is* parallel, perpendicular, *or* neither.

**25.** $3x + 2y = 6$
$6x + 4y = 8$

**26.** $x - 3y = 1$
$3x + y = 4$

**27.** $x - 2y = 8$
$x + 2y = 8$

[3.4] *Write an equation for each line in the form $y = mx + b$, if possible.*

**28.** $m = -1, b = \frac{2}{3}$

**29.** Through $(2, 3)$ and $(-4, 6)$

**30.** Through $(4, -3)$, $m = 1$

**31.** Through $(-1, 4)$, $m = \frac{2}{3}$

**32.** Through $(1, -1)$, $m = -\frac{3}{4}$

**33.** $m = -\frac{1}{4}, b = \frac{3}{2}$

**34.** Slope 0, through $(-4, 1)$

**35.** Through $\left(\frac{1}{3}, -\frac{5}{4}\right)$ with undefined slope

## MIXED REVIEW EXERCISES

*In Exercises 36–41, match each statement to the appropriate graph or graphs in A–D. Graphs may be used more than once.*

**A.**

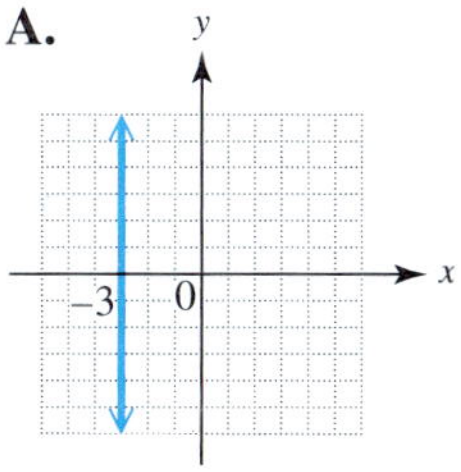


**B.**

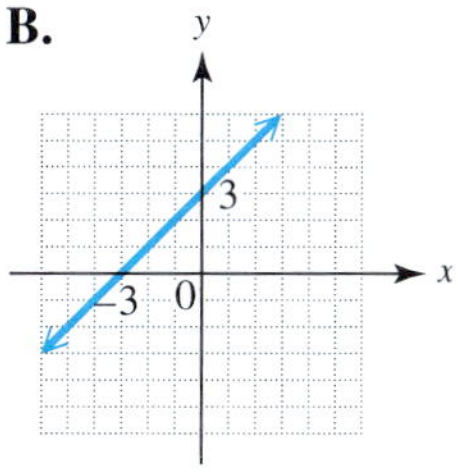


**C.**

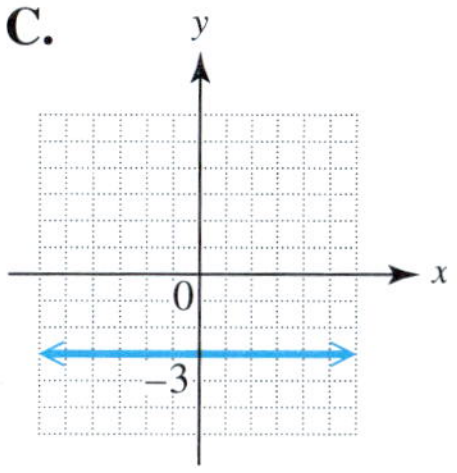


**D.**

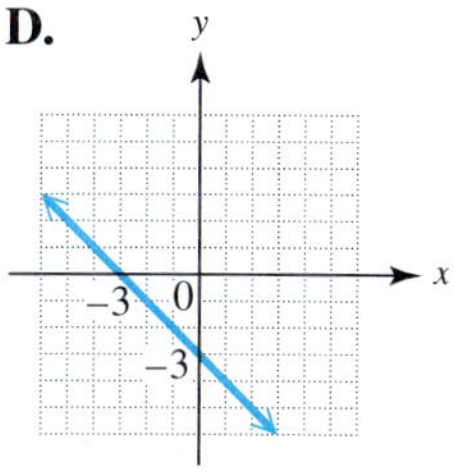


**36.** The line shown in the graph has undefined slope.

**37.** The graph of the equation has $y$-intercept $(0, -3)$.

**38.** The graph of the equation has $x$-intercept $(-3, 0)$.

**39.** The line shown in the graph has negative slope.

**40.** The graph is that of the equation $y = -3$.

**41.** The line shown in the graph has slope 1.

*Find the intercepts and the slope of each line. Then graph the line.*

**42.** $y = -2x - 5$

**43.** $x + 3y = 0$

**44.** $y - 5 = 0$

*Write an equation for each line. Write the equation in slope-intercept form.*

**45.** $m = -\frac{1}{4}, b = -\frac{5}{4}$

**46.** Through (8, 6), $m = -3$

**47.** Through $(3, -5)$ and $(-4, -1)$

## RELATING CONCEPTS (EXERCISES 48–54)

**For Individual or Group Work**

*The total amount spent (in billions of dollars) on video rentals in the United States from 1996 through 2000 is shown in the graph. Use the graph to **work Exercises 48–54 in order.***

**VIDEO RENTALS**

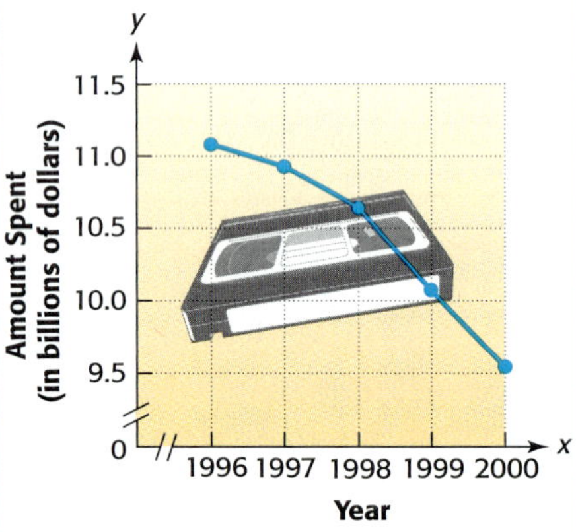


*Source:* Blockbuster.

**48.** About how much did the amount spent on video rentals decrease during the years shown in the graph?

**49.** Since the points of the graph lie approximately in a linear pattern, a straight line can be used to model the data. Will this line have positive or negative slope? Explain.

**50.** The table gives the actual amounts spent on video rentals in 1996 and 2000.

**(a)** Write two ordered pairs for the data.

**(b)** Use the ordered pairs to find the slope of the line through them.

| Year x | Amount y (in billions of dollars) |
|---|---|
| 1996 | 11.1 |
| 2000 | 9.6 |

**51.** The amount spent on video rentals $y$ in billions of dollars from 1996 through 2000 can be modeled by the linear equation

$$y = -.375x + 759.6,$$

where $x$ represents the year. Based on this equation, what is the slope of the line? Does it agree with your answers in Exercises 49 and 50?

**52.** Use the equation from Exercise 51 to approximate the amount spent on video rentals from 1997 through 1999, and complete the table. Round your answers to the nearest tenth.

| x | y |
|---|---|
| 1996 | 11.1 |
| 1997 | |
| 1998 | |
| 1999 | |
| 2000 | 9.6 |

**53.** The actual amounts spent on video rentals are given in the following ordered pairs.

(1996, 11.1), (1997, 10.9), (1998, 10.6), (1999, 10.1), (2000, 9.6)

How do the actual amounts compare to those you found in Exercise 52 using the linear equation?

**54.** Since the equation in Exercise 51 models the data fairly well, use it to estimate the amount spent on video rentals in 2002.

# CHAPTER 3 TEST

1. Complete these ordered pairs for the equation $3x + 5y = -30$: $(0, \quad)$, $(\quad, 0)$, $(\quad, -3)$.
2. Is $(4, -1)$ a solution of $4x - 7y = 9$?
3. How do you find the $x$-intercept of the graph of a linear equation in two variables? How do you find the $y$-intercept?

*Graph each linear equation. Give the x- and y-intercepts.*

4. $3x + y = 6$
5. $y - 2x = 0$
6. $x + 3 = 0$
7. $y = 1$
8. $x - y = 4$

*Find the slope of each line.*

9. Through $(-4, 6)$ and $(-1, -2)$
10. $2x + y = 10$
11. $x + 12 = 0$
12. 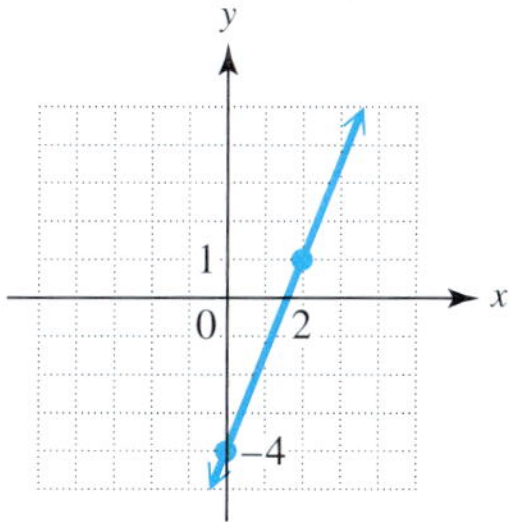

13. A line parallel to the graph of $y - 4 = 6$

*Write an equation for each line. Write the equation in slope-intercept form.*

14. Through $(-1, 4)$; $m = 2$
15. The line in Exercise 12
16. Through $(2, -6)$ and $(1, 3)$

*The graph shows total food and drink sales at U.S. restaurants from 1970 through 2000, where 1970 corresponds to $x = 0$. Use the graph to work Exercises 17–20 on the next page.*

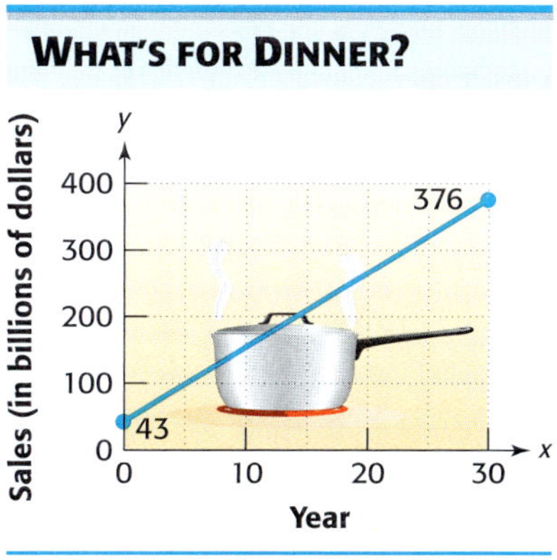

*Source:* National Restaurant Association.

17. Is the slope of the line in the graph positive or negative? Explain.

18. Write two ordered pairs for the data points shown in the graph. Use them to find the slope of the line.

19. The linear equation

$$y = 11.1x + 43$$

approximates food and drink sales $y$ in billions of dollars, where $x = 0$ again represents 1970. Use the equation to approximate food and drink sales for 1990 and 1995.

20. What does the ordered pair (30, 376) mean in the context of this situation?

## CUMULATIVE REVIEW EXERCISES CHAPTERS 1–3

*Perform each indicated operation.*

1. $10\frac{5}{8} - 3\frac{1}{10}$

2. $\frac{3}{4} \div \frac{1}{8}$

3. $5 - (-4) + (-2)$

4. $\frac{(-3)^2 - (-4)(2^4)}{5(2) - (-2)^3}$

5. True or false? $\frac{4(3 - 9)}{2 - 6} \geq 6$

6. Find the value of $xz^3 - 5y^2$ when $x = -2$, $y = -3$, and $z = -1$.

7. What property does $3(-2 + x) = -6 + 3x$ illustrate?

8. Simplify $-4p - 6 + 3p + 8$ by combining terms.

*Solve.*

9. $V = \frac{1}{3}\pi r^2 h$ for $h$

10. $6 - 3(1 + a) = 2(a + 5) - 2$

11. $-(m - 3) = 5 - 2m$

12. $\frac{x - 2}{3} = \frac{2x + 1}{5}$

*Solve each inequality, and graph the solutions.*

13. $-2.5x < 6.5$

14. $4(x + 3) - 5x < 12$

15. $\frac{2}{3}x - \frac{1}{6}x \leq -2$

*Solve each problem.*

16. The gap in average annual earnings by level of education continues to increase. Based on the most recent statistics available, a person with a bachelor's degree can expect to earn \$17,583 more each year than someone with a high school diploma. Together the individuals would earn \$63,373. How much can a person at each level of education expect to earn? (*Source:* U.S. Bureau of the Census.)

**17.** Mount Mayon in the Philippines is the most perfectly shaped conical volcano in the world. Its base is a perfect circle with circumference 80 mi, and it has a height of about 8200 ft. (One mile is 5280 ft.) Find the radius of the circular base to the nearest mile. (*Hint:* This problem has some unneeded information.) (*Source: Microsoft Encarta Encyclopedia 2000.*)

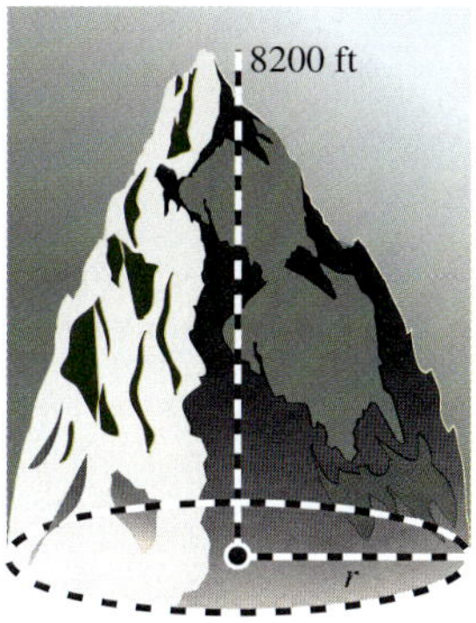

Circumference = 80 mi

**18.** The winning times in seconds for the women's 1000-m speed skating event in the Winter Olympics for the years 1960 through 1998 can be closely approximated by the linear equation

$$y = -.4685x + 95.07,$$

where $x$ is the number of years since 1960. That is, $x = 4$ represents 1964, $x = 8$ represents 1968, and so on. (*Source: The Universal Almanac,* 1998.)

**(a)** Use this equation to complete the table of values. Round times to the nearest hundredth of a second.

**(b)** What does the ordered pair $(20, 85.7)$ mean in the context of the problem?

| $x$ | $y$ |
|---|---|
| 12 | |
| 28 | |
| 36 | |

**19.** Baby boomers are expected to inherit \$10.4 trillion from their parents over the next 45 yr, an average of \$50,000 each. The circle graph shows how they plan to spend their inheritances.

**(a)** How much of the \$50,000 is expected to go toward home purchase?

**(b)** How much is expected to go toward retirement?

**(c)** Use the answer from part (b) to estimate the amount expected to go toward paying off debts or funding children's education.

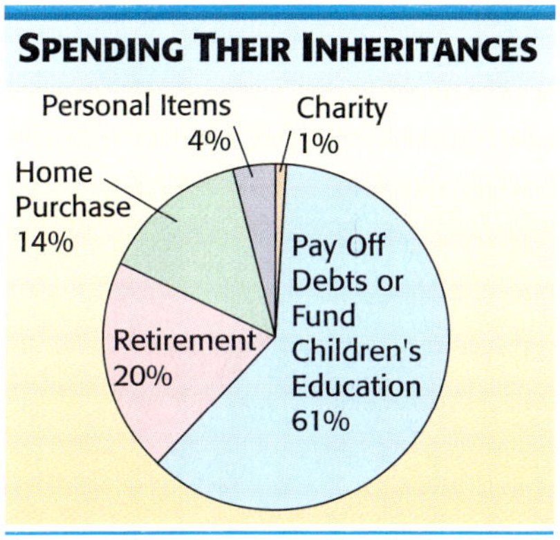

***Source:*** First Interstate Bank Trust and Private Banking Group.

*Consider the linear equation* $-3x + 4y = 12$. *Find the following.*

**20.** The $x$- and $y$-intercepts **21.** The slope **22.** The graph

**23.** Are the lines with equations $x + 5y = -6$ and $y = 5x - 8$ *parallel, perpendicular,* or *neither*?

*Write an equation for each line. Write the equation in slope-intercept form, if possible.*

**24.** Through $(2, -5)$ with slope 3 **25.** Through $(0, 4)$ and $(2, 4)$

# Exponents and Polynomials

# 4

The expression $100x - 13x^2$ gives the distance in feet a car going approximately 68 mph will skid in $x$ sec. This expression in $x$ is an example of a *polynomial,* one topic of this chapter. Accident investigators use polynomials like this to determine the length of a skid or the elapsed time during a skid. In Exercises 95 and 96 of Section 4.4, we use this polynomial to approximate skidding distance.

# 4.1 The Product Rule and Power Rules for Exponents

**OBJECTIVES**

1. Identify bases and exponents.
2. Use the product rule for exponents.
3. Use the rule $(a^m)^n = a^{mn}$.
4. Use the rule $(ab)^m = a^m b^m$.
5. Use the rule $\left(\frac{a}{b}\right)^m = \frac{a^m}{b^m}$.
6. Use combinations of rules.

**OBJECTIVE 1 Identify bases and exponents.** Recall from Section 1.2 that in the expression $5^2$, the number 5 is the *base* and 2 is the *exponent* or *power*. The expression $5^2$ is called an *exponential expression*. Usually we do not write the exponent when it is 1; however, sometimes it is convenient to do so. In general, for any quantity $a$, $a^1 = a$.

**EXAMPLE 1 Determining the Base and Exponent in Exponential Expressions**

Evaluate each exponential expression. Name the base and the exponent.

| | Base | Exponent |
|---|---|---|
| **(a)** $5^4 = 5 \cdot 5 \cdot 5 \cdot 5 = 625$ | 5 | 4 |
| **(b)** $-5^4 = -1 \cdot 5^4 = -1 \cdot (5 \cdot 5 \cdot 5 \cdot 5) = -625$ | 5 | 4 |
| **(c)** $(-5)^4 = (-5)(-5)(-5)(-5) = 625$ | −5 | 4 |

**Now Try Exercises 15 and 17.**

**CAUTION** Note the differences between parts (b) and (c) of Example 1. In $-5^4$ the lack of parentheses shows that the exponent 4 refers only to the base 5, not $-5$; in $(-5)^4$ the parentheses show that the exponent 4 refers to the base $-5$. In summary, $-a^n$ and $(-a)^n$ are not necessarily the same.

| Expression | Base | Exponent | Example |
|---|---|---|---|
| $-a^n$ | $a$ | $n$ | $-3^2 = -(3 \cdot 3) = -9$ |
| $(-a)^n$ | $-a$ | $n$ | $(-3)^2 = (-3)(-3) = 9$ |

**OBJECTIVE 2 Use the product rule for exponents.** By the definition of exponents,

$$\begin{aligned} 2^4 \cdot 2^3 &= \underbrace{(2 \cdot 2 \cdot 2 \cdot 2)}_{\text{4 factors}}\underbrace{(2 \cdot 2 \cdot 2)}_{\text{3 factors}} \\ &= \underbrace{2 \cdot 2 \cdot 2 \cdot 2 \cdot 2 \cdot 2 \cdot 2}_{4 + 3 = 7 \text{ factors}} \\ &= 2^7. \end{aligned}$$

Also,

$$\begin{aligned} 6^2 \cdot 6^3 &= (6 \cdot 6)(6 \cdot 6 \cdot 6) \\ &= 6 \cdot 6 \cdot 6 \cdot 6 \cdot 6 \\ &= 6^5. \end{aligned}$$

Generalizing from these examples, $2^4 \cdot 2^3 = 2^{4+3} = 2^7$ and $6^2 \cdot 6^3 = 6^{2+3} = 6^5$, suggests the **product rule for exponents.**

**Product Rule for Exponents**

For any positive integers $m$ and $n$, $\quad a^m \cdot a^n = a^{m+n}$.
(Keep the same base; add the exponents.)

*Example:* $6^2 \cdot 6^5 = 6^{2+5} = 6^7$

**CAUTION** Avoid the common error of multiplying the bases when using the product rule:

$$6^2 \cdot 6^5 \neq 36^7.$$

**Keep the same base and add the exponents.**

### EXAMPLE 2 Using the Product Rule

Use the product rule for exponents to find each result when possible.

**(a)** $6^3 \cdot 6^5 = 6^{3+5} = 6^8$

**(b)** $(-4)^7(-4)^2 = (-4)^{7+2} = (-4)^9$

**(c)** $x^2 \cdot x = x^2 \cdot x^1 = x^{2+1} = x^3$

**(d)** $m^4m^3m^5 = m^{4+3+5} = m^{12}$

**(e)** $2^3 \cdot 3^2$

The product rule does not apply to the product $2^3 \cdot 3^2$, since the bases are different.

$$2^3 \cdot 3^2 = 8 \cdot 9 = 72$$

**(f)** $2^3 + 2^4$

The product rule does not apply to $2^3 + 2^4$, since this is a *sum,* not a *product.*

$$2^3 + 2^4 = 8 + 16 = 24$$

**(g)** $(2x^3)(3x^7)$

Since $2x^3$ means $2 \cdot x^3$ and $3x^7$ means $3 \cdot x^7$, we use the associative and commutative properties to get

$$(2x^3)(3x^7) = (2 \cdot 3) \cdot (x^3 \cdot x^7) = 6x^{10}.$$

**Now Try Exercises 25, 29, 31, 35, and 39.**

**CAUTION** Be sure you understand the difference between *adding* and *multiplying* exponential expressions. For example,

$$8x^3 + 5x^3 = (8 + 5)x^3 = 13x^3,$$

but $$(8x^3)(5x^3) = (8 \cdot 5)x^{3+3} = 40x^6.$$

**OBJECTIVE 3 Use the rule $(a^m)^n = a^{mn}$.** We simplify an expression such as $(8^3)^2$ with the product rule for exponents.

$$(8^3)^2 = (8^3)(8^3) = 8^{3+3} = 8^6$$

The exponents in $(8^3)^2$ are multiplied to give the exponent in $8^6$. As another example,

$$\begin{aligned}(5^2)^4 &= 5^2 \cdot 5^2 \cdot 5^2 \cdot 5^2 && \text{Definition of exponent}\\ &= 5^{2+2+2+2} && \text{Product rule}\\ &= 5^8\end{aligned}$$

and $2 \cdot 4 = 8$. These examples suggest **power rule (a) for exponents.**

**Power Rule (a) for Exponents**

For any positive integers $m$ and $n$, $\quad (a^m)^n = a^{mn}$.
(Raise a power to a power by multiplying exponents.)
*Example:* $(3^2)^4 = 3^{2\cdot 4} = 3^8$

**EXAMPLE 3 Using Power Rule (a)**

Use power rule (a) for exponents to simplify each expression.

**(a)** $(2^5)^3 = 2^{5\cdot 3} = 2^{15}$ **(b)** $(5^7)^2 = 5^{7(2)} = 5^{14}$ **(c)** $(x^2)^5 = x^{2(5)} = x^{10}$

**Now Try Exercises 43 and 45.**

**OBJECTIVE 4 Use the rule $(ab)^m = a^m b^m$.** We can use the properties studied in Chapter 1 to develop two more rules for exponents. Using the definition of an exponential expression and the commutative and associative properties, we can rewrite the expression $(4x)^3$ as follows.

$$(4x)^3 = (4x)(4x)(4x) \quad \text{Definition of exponent}$$
$$= (4 \cdot 4 \cdot 4)(x \cdot x \cdot x) \quad \text{Commutative and associative properties}$$
$$= 4^3 \cdot x^3 \quad \text{Definition of exponent}$$

This example suggests **power rule (b) for exponents.**

**Power Rule (b) for Exponents**

For any positive integer $m$, $\quad (ab)^m = a^m b^m$.
(Raise a product to a power by raising each factor to the power.)
*Example:* $(2p)^5 = 2^5 p^5$

**EXAMPLE 4 Using Power Rule (b)**

Use power rule (b) for exponents to simplify each expression.

**(a)** $(3xy)^2 = 3^2x^2y^2 = 9x^2y^2$ Power rule (a)

**(b)** $5(pq)^2 = 5(p^2q^2)$ Power rule (b)
$= 5p^2q^2$ Multiply.

**(c)** $3(2m^2p^3)^4 = 3[2^4(m^2)^4(p^3)^4]$ Power rule (b)
$= 3 \cdot 2^4 m^8 p^{12}$ Power rule (a)
$= 48m^8p^{12}$ Multiply.

**(d)** $(-5^6)^3 = (-1 \cdot 5^6)^3$
$= (-1)^3 \cdot (5^6)^3$
$= -1 \cdot 5^{18}$
$= -5^{18}$

**Now Try Exercises 49 and 53.**

**CAUTION** Power rule (b) *does not* apply to a *sum*.

$$(x + 4)^2 \neq x^2 + 4^2$$

**OBJECTIVE 5 Use the rule** $\left(\frac{a}{b}\right)^m = \frac{a^m}{b^m}$**.** Since the quotient $\frac{a}{b}$ can be written as $a\left(\frac{1}{b}\right)$, we use power rule (b), together with some of the properties of real numbers, to get **power rule (c) for exponents.**

**Power Rule (c) for Exponents**

For any positive integer $m$,

$$\left(\frac{a}{b}\right)^m = \frac{a^m}{b^m} \quad (b \neq 0).$$

(Raise a quotient to a power by raising both numerator and denominator to the power.)

*Example:* $\left(\frac{5}{3}\right)^2 = \frac{5^2}{3^2}$

**EXAMPLE 5 Using Power Rule (c)**

Use power rule (c) for exponents to simplify each expression.

**(a)** $\left(\frac{2}{3}\right)^5 = \frac{2^5}{3^5}$ **(b)** $\left(\frac{m}{n}\right)^3 = \frac{m^3}{n^3} \quad (n \neq 0)$

**Now Try Exercises 59 and 61.**

In the following box, we list the rules for exponents discussed in this section. These rules are basic to the study of algebra.

**Rules for Exponents**

For positive integers $m$ and $n$:

| | | **Examples** |
|---|---|---|
| **Product rule** | $a^m \cdot a^n = a^{m+n}$ | $6^2 \cdot 6^5 = 6^{2+5} = 6^7$ |
| **Power rules (a)** | $(a^m)^n = a^{mn}$ | $(3^2)^4 = 3^{2 \cdot 4} = 3^8$ |
| **(b)** | $(ab)^m = a^m b^m$ | $(2p)^5 = 2^5 p^5$ |
| **(c)** | $\left(\frac{a}{b}\right)^m = \frac{a^m}{b^m} \quad (b \neq 0)$ | $\left(\frac{5}{3}\right)^2 = \frac{5^2}{3^2}$ |

**OBJECTIVE 6 Use combinations of rules.** As shown in the next example, more than one rule may be needed to simplify an expression with exponents.

**EXAMPLE 6 Using Combinations of Rules**

Use the rules for exponents to simplify each expression.

**(a)** $\left(\frac{2}{3}\right)^2 \cdot 2^3 = \frac{2^2}{3^2} \cdot \frac{2^3}{1}$ Power rule (c)

$= \frac{2^2 \cdot 2^3}{3^2 \cdot 1}$ Multiply fractions.

$= \frac{2^5}{3^2}$ Product rule

**(b)** $(5x)^3(5x)^4 = (5x)^7$ Product rule

$= 5^7x^7$ Power rule (b)

**(c)** $(2x^2y^3)^4(3xy^2)^3 = 2^4(x^2)^4(y^3)^4 \cdot 3^3x^3(y^2)^3$ Power rule (b)

$= 2^4x^8y^{12} \cdot 3^3x^3y^6$ Power rule (a)

$= 2^4 \cdot 3^3x^8x^3y^{12}y^6$ Commutative and associative properties

$= 16 \cdot 27x^{11}y^{18}$ Product rule

$= 432x^{11}y^{18}$

**(d)** $(-x^3y)^2(-x^5y^4)^3 = (-1 \cdot x^3y)^2(-1 \cdot x^5y^4)^3$

$= (-1)^2(x^3)^2y^2 \cdot (-1)^3(x^5)^3(y^4)^3$ Power rule (b)

$= (-1)^2(x^6)(y^2)(-1)^3(x^{15})(y^{12})$ Power rule (a)

$= (-1)^5(x^{21})(y^{14})$ Product rule

$= -x^{21}y^{14}$

**Now Try Exercises 63, 67, 75, and 77.**

**CAUTION** Refer to Example 6(c). Notice that

$$(2x^2y^3)^4 = 2^4x^{2\cdot 4}y^{3\cdot 4}, \quad \text{not} \quad (2 \cdot 4)x^{2\cdot 4}y^{3\cdot 4}.$$

Do not multiply the coefficient 2 and the exponent 4.

## 4.1 EXERCISES

**For Extra Help**

- Student's Solutions Manual
- MyMathLab
- InterAct Math Tutorial Software
- AW Math Tutor Center
- MathXL
- Digital Video Tutor CD 6/Videotape 6

*Decide whether each statement is* true *or* false.

**1.** $3^3 = 9$

**2.** $(-2)^4 = 2^4$

**3.** $(a^2)^3 = a^5$

**4.** $\left(\frac{1}{4}\right)^2 = \frac{1}{4^2}$

*Write each expression using exponents.*

**5.** $w \cdot w \cdot w \cdot w \cdot w \cdot w$

**6.** $t \cdot t \cdot t \cdot t \cdot t \cdot t \cdot t$

**7.** $\frac{1}{4 \cdot 4 \cdot 4 \cdot 4}$

**8.** $\frac{1}{3 \cdot 3 \cdot 3}$

**9.** $(-7x)(-7x)(-7x)(-7x)$

**10.** $(-8p)(-8p)$

**11.** $\left(\frac{1}{2}\right)\left(\frac{1}{2}\right)\left(\frac{1}{2}\right)\left(\frac{1}{2}\right)\left(\frac{1}{2}\right)\left(\frac{1}{2}\right)$

**12.** $\left(-\frac{1}{4}\right)\left(-\frac{1}{4}\right)\left(-\frac{1}{4}\right)\left(-\frac{1}{4}\right)\left(-\frac{1}{4}\right)$

13. Explain how the expressions $(-3)^4$ and $-3^4$ are different.

14. Explain how the expressions $(5x)^3$ and $5x^3$ are different.

*Identify the base and the exponent for each exponential expression. In Exercises 15–18, also evaluate each expression. See Example 1.*

15. $3^5$
16. $2^7$
17. $(-3)^5$
18. $(-2)^7$
19. $(-6x)^4$
20. $(-8x)^4$
21. $-6x^4$
22. $-8x^4$

23. Explain why the product rule does not apply to the expression $5^2 + 5^3$. Then evaluate the expression by finding the individual powers and adding the results.

24. Repeat Exercise 23 for the expression $(-4)^3 + (-4)^4$.

*Use the product rule, if possible, to simplify each expression. Write each answer in exponential form. See Example 2.*

25. $5^2 \cdot 5^6$
26. $3^6 \cdot 3^7$
27. $4^2 \cdot 4^7 \cdot 4^3$
28. $5^3 \cdot 5^8 \cdot 5^2$
29. $(-7)^3(-7)^6$
30. $(-9)^8(-9)^5$
31. $t^3 \cdot t^8 \cdot t^{13}$
32. $n^5 \cdot n^6 \cdot n^9$
33. $(-8r^4)(7r^3)$
34. $(10a^7)(-4a^3)$
35. $(-6p^5)(-7p^5)$
36. $(-5w^8)(-9w^8)$
37. $(5x^2)(-2x^3)(3x^4)$
38. $(12y^3)(4y)(-3y^5)$
39. $3^8 + 3^9$
40. $4^{12} + 4^5$
41. $5^8 \cdot 3^9$
42. $6^3 \cdot 8^9$

*Use the power rules for exponents to simplify each expression. Write each answer in exponential form. See Examples 3–5.*

43. $(4^3)^2$
44. $(8^3)^6$
45. $(t^4)^5$
46. $(y^6)^5$
47. $(7r)^3$
48. $(11x)^4$
49. $(5xy)^5$
50. $(9pq)^6$
51. $(-5^2)^6$
52. $(-9^4)^8$
53. $(-8^3)^5$
54. $(-7^5)^7$
55. $8(qr)^3$
56. $4(vw)^5$
57. $\left(\frac{1}{2}\right)^3$
58. $\left(\frac{1}{3}\right)^5$
59. $\left(\frac{a}{b}\right)^3 \quad (b \neq 0)$
60. $\left(\frac{r}{t}\right)^4 \quad (t \neq 0)$
61. $\left(\frac{9}{5}\right)^8$
62. $\left(\frac{12}{7}\right)^3$

*Use a combination of the rules of exponents to simplify each expression. See Example 6.*

63. $\left(\frac{5}{2}\right)^3 \cdot \left(\frac{5}{2}\right)^2$
64. $\left(\frac{3}{4}\right)^5 \cdot \left(\frac{3}{4}\right)^6$
65. $\left(\frac{9}{8}\right)^3 \cdot 9^2$
66. $\left(\frac{8}{5}\right)^4 \cdot 8^3$
67. $(2x)^9(2x)^3$
68. $(6y)^5(6y)^8$
69. $(-6p)^4(-6p)$
70. $(-13q)^3(-13q)$
71. $(6x^2y^3)^5$
72. $(5r^5t^6)^7$
73. $(x^2)^3(x^3)^5$
74. $(y^4)^5(y^3)^5$
75. $(2w^2x^3y)^2(x^4y)^5$
76. $(3x^4y^2z)^3(yz^4)^5$
77. $(-r^4s)^2(-r^2s^3)^5$

**78.** $(-ts^6)^4(-t^3s^5)^3$ **79.** $\left(\dfrac{5a^2b^5}{c^6}\right)^3 \quad (c \neq 0)$ **80.** $\left(\dfrac{6x^3y^9}{z^5}\right)^4 \quad (z \neq 0)$

**81.** A student tried to simplify $(10^2)^3$ as $1000^6$. Is this correct? If not, how is it simplified using the product rule for exponents?

**82.** Explain why $(3x^2y^3)^4$ is *not* equivalent to $(3 \cdot 4)x^8y^{12}$.

*Find the area of each figure. Use the formulas found on the inside covers. (The small squares in the figures indicate 90° right angles.)*

**83.**

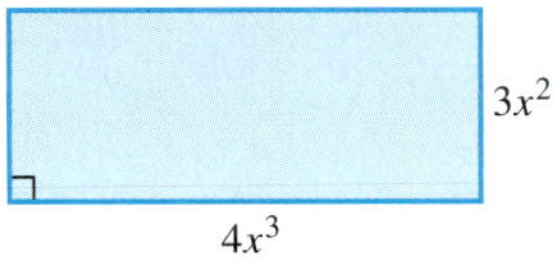


**84.**

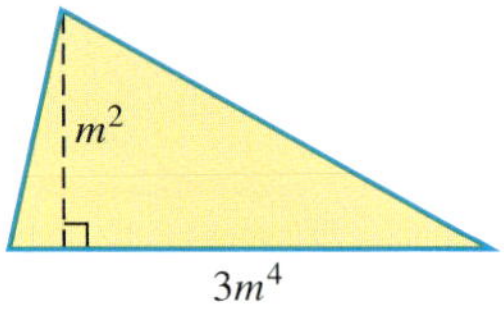


**85.**

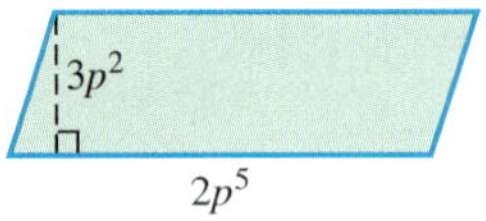


**86.**

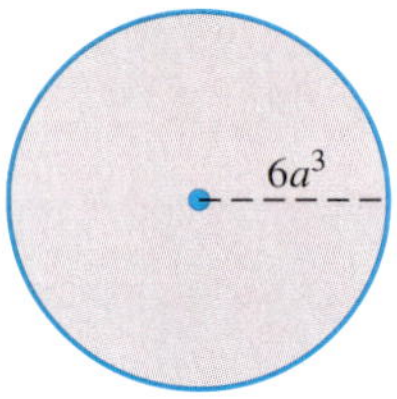


*Find the volume of each figure. Use the formulas found on the inside covers.*

**87.**

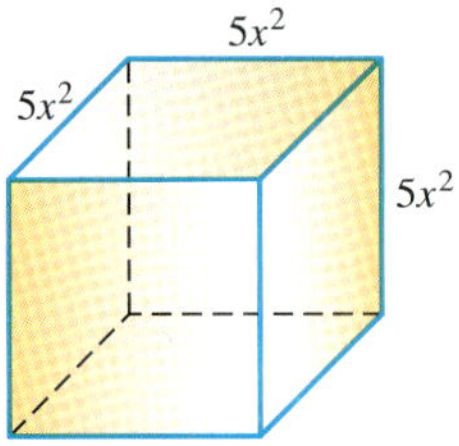


**88.**

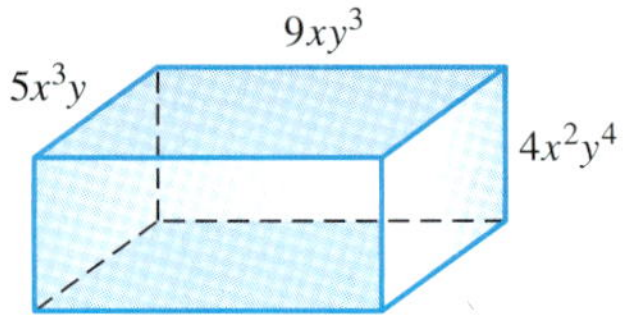


**89.** Assume $a$ is a positive number greater than 1. Arrange the following terms in order from smallest to largest: $-(-a)^3$, $-a^3$, $(-a)^4$, $-a^4$. Explain how you decided on the order.

**90.** Devise a rule to tell whether an exponential expression with a negative base is positive or negative.

In Chapter 2 we used the formula for simple interest, $I = prt$, which deals with interest paid only on the principal. With **compound interest,** interest is paid on the principal and the interest earned earlier. The formula for compound interest, which involves an exponential expression, is

$$A = P(1 + r)^n.$$

Here $A$ is the amount accumulated from a principal of $P$ dollars left untouched for $n$ years with an annual interest rate $r$ (expressed as a decimal).

*In Exercises 91–94, use this formula and a calculator to find A to the nearest cent.*

**91.** $P = \$250$, $r = .04$, $n = 5$

**92.** $P = \$400$, $r = .04$, $n = 3$

**93.** $P = \$1500$, $r = .035$, $n = 6$

**94.** $P = \$2000$, $r = .025$, $n = 4$

# 4.2 Integer Exponents and the Quotient Rule

**OBJECTIVES**

1. Use 0 as an exponent.
2. Use negative numbers as exponents.
3. Use the quotient rule for exponents.
4. Use combinations of rules.

In Section 4.1 we studied the product rule for exponents. In all our earlier work, exponents were positive integers. Now we want to develop meaning for exponents that are not positive integers.

Consider the following list of exponential expressions.

$$2^4 = 16$$
$$2^3 = 8$$
$$2^2 = 4$$

Do you see the pattern in the values? Each time we reduce the exponent by 1, the value is divided by 2 (the base). Using this pattern, we can continue the list to smaller and smaller integer exponents.

$$2^1 = 2$$
$$2^0 = 1$$
$$2^{-1} = \frac{1}{2}$$
$$2^{-2} = \frac{1}{4}$$
$$2^{-3} = \frac{1}{8}$$

From the preceding list, it appears that we should define $2^0$ as 1 and negative exponents as reciprocals.

**OBJECTIVE 1 Use 0 as an exponent.** We want the definitions of 0 and negative exponents to satisfy the rules for exponents from Section 4.1. For example, if $6^0 = 1$,

$$6^0 \cdot 6^2 = 1 \cdot 6^2 = 6^2 \quad \text{and} \quad 6^0 \cdot 6^2 = 6^{0+2} = 6^2,$$

so the product rule is satisfied. Check that the power rules are also valid for a 0 exponent. Thus, we define a 0 exponent as follows.

**Zero Exponent**

For any nonzero real number $a$, $\boldsymbol{a^0 = 1}$.

*Example:* $17^0 = 1$

### EXAMPLE 1 Using Zero Exponents

Evaluate each exponential expression.

**(a)** $60^0 = 1$

**(b)** $(-60)^0 = 1$

**(c)** $-60^0 = -(1) = -1$

**(d)** $y^0 = 1 \quad (y \neq 0)$

**(e)** $6y^0 = 6(1) = 6 \quad (y \neq 0)$

**(f)** $(6y)^0 = 1 \quad (y \neq 0)$

**Now Try Exercises 1 and 5.**

**CAUTION** Notice the difference between parts (b) and (c) of Example 1. In $(-60)^0$, the base is $-60$ and the exponent is 0. Any nonzero base raised to the 0 exponent is 1. But in $-60^0$, the base is 60. Then $60^0 = 1$, and $-60^0 = -1$.

**OBJECTIVE 2 Use negative numbers as exponents.** From the lists at the beginning of this section, since $2^{-2} = \frac{1}{4}$ and $2^{-3} = \frac{1}{8}$, we can deduce that $2^{-n}$ should equal $\frac{1}{2^n}$. Is the product rule valid in such cases? For example, if we multiply $6^{-2}$ by $6^2$, we get

$$6^{-2} \cdot 6^2 = 6^{-2+2} = 6^0 = 1.$$

The expression $6^{-2}$ behaves as if it were the reciprocal of $6^2$ because their product is 1. The reciprocal of $6^2$ is also $\frac{1}{6^2}$, leading us to define $6^{-2}$ as $\frac{1}{6^2}$. This is a particular case of the definition of negative exponents.

**Negative Exponents**

For any nonzero real number $a$ and any integer $n$, $\quad \boldsymbol{a^{-n} = \frac{1}{a^n}}$.

*Example:* $3^{-2} = \frac{1}{3^2}$

By definition, $a^{-n}$ and $a^n$ are reciprocals, since

$$a^n \cdot a^{-n} = a^n \cdot \frac{1}{a^n} = 1.$$

Since $1^n = 1$, the definition of $a^{-n}$ can also be written

$$a^{-n} = \frac{1}{a^n} = \frac{1^n}{a^n} = \left(\frac{1}{a}\right)^n.$$

For example,

$$6^{-3} = \left(\frac{1}{6}\right)^3 \quad \text{and} \quad \left(\frac{1}{3}\right)^{-2} = 3^2.$$

**EXAMPLE 2 Using Negative Exponents**

Simplify by writing each expression with positive exponents.

**(a)** $3^{-2} = \frac{1}{3^2} = \frac{1}{9}$

**(b)** $5^{-3} = \frac{1}{5^3} = \frac{1}{125}$

**(c)** $\left(\frac{1}{2}\right)^{-3} = 2^3 = 8$ $\quad$ $\frac{1}{2}$ and 2 are reciprocals.

Notice that we can change the base to its reciprocal if we also change the sign of the exponent.

**(d)** $\left(\frac{2}{5}\right)^{-4} = \left(\frac{5}{2}\right)^4$ $\quad$ $\frac{2}{5}$ and $\frac{5}{2}$ are reciprocals.

**(e)** $\left(\frac{4}{3}\right)^{-5} = \left(\frac{3}{4}\right)^5$

**(f)** $4^{-1} - 2^{-1} = \frac{1}{4} - \frac{1}{2} = \frac{1}{4} - \frac{2}{4} = -\frac{1}{4}$

We apply the exponents first, then subtract.

**(g)** $p^{-2} = \frac{1}{p^2} \quad (p \neq 0)$

**(h)** $\frac{1}{x^{-4}} \quad (x \neq 0)$

$$\frac{1}{x^{-4}} = \frac{1^{-4}}{x^{-4}} \qquad 1^{-4} = 1$$

$$= \left(\frac{1}{x}\right)^{-4} \qquad \text{Power rule (c)}$$

$$= x^4 \qquad \frac{1}{x} \text{ and } x \text{ are reciprocals.}$$

**Now Try Exercises 19, 21, 23, and 27.**

**CAUTION** A negative exponent does not indicate a negative number; negative exponents lead to reciprocals.

| Expression | Example | |
|---|---|---|
| $a^{-n}$ | $3^{-2} = \frac{1}{3^2} = \frac{1}{9}$ | Not negative |
| $-a^{-n}$ | $-3^{-2} = -\frac{1}{3^2} = -\frac{1}{9}$ | Negative |

The definition of negative exponents allows us to move factors across a fraction bar if we also change the signs of the exponents. For example,

$$\frac{2^{-3}}{3^{-4}} = \frac{\frac{1}{2^3}}{\frac{1}{3^4}} = \frac{1}{2^3} \cdot \frac{3^4}{1} = \frac{3^4}{2^3},$$

so

$$\frac{2^{-3}}{3^{-4}} = \frac{3^4}{2^3}.$$

**Changing from Negative to Positive Exponents**

For any nonzero numbers $a$ and $b$, and any integers $m$ and $n$,

$$\frac{a^{-m}}{b^{-n}} = \frac{b^n}{a^m} \quad \text{and} \quad \left(\frac{a}{b}\right)^{-m} = \left(\frac{b}{a}\right)^m.$$

*Examples:* $\frac{3^{-5}}{2^{-4}} = \frac{2^4}{3^5}$ and $\left(\frac{4}{5}\right)^{-3} = \left(\frac{5}{4}\right)^3$

## EXAMPLE 3 Changing from Negative to Positive Exponents

Write with only positive exponents. Assume all variables represent nonzero real numbers.

**(a)** $\frac{4^{-2}}{5^{-3}} = \frac{5^3}{4^2}$ or $\frac{125}{16}$

**(b)** $\frac{m^{-5}}{p^{-1}} = \frac{p^1}{m^5} = \frac{p}{m^5}$

**(c)** $\frac{a^{-2}b}{3d^{-3}} = \frac{bd^3}{3a^2}$

Notice that $b$ in the numerator and 3 in the denominator were not affected.

**(d)** $x^3y^{-4} = \frac{x^3y^{-4}}{1} = \frac{x^3}{y^4}$

**(e)** $\left(\frac{x}{2y}\right)^{-4} = \left(\frac{2y}{x}\right)^4 = \frac{2^4y^4}{x^4}$

**Now Try Exercises 31, 35, and 47.**

**CAUTION** Be careful. We cannot change negative exponents to positive exponents using this rule if the exponents occur in a sum of terms. For example,

$$\frac{5^{-2} + 3^{-1}}{7 - 2^{-3}}$$

cannot be written with positive exponents using the rule given here. We would have to use the definition of a negative exponent to rewrite this expression with positive exponents, as

$$\frac{\frac{1}{5^2} + \frac{1}{3}}{7 - \frac{1}{2^3}}.$$

**OBJECTIVE 3 Use the quotient rule for exponents.** How should we handle the quotient of two exponential expressions with the same base? We know that

$$\frac{6^5}{6^3} = \frac{6 \cdot 6 \cdot 6 \cdot 6 \cdot 6}{6 \cdot 6 \cdot 6} = 6^2.$$

Notice that the difference between the exponents, $5 - 3 = 2$, is the exponent in the quotient. Also,

$$\frac{6^2}{6^4} = \frac{6 \cdot 6}{6 \cdot 6 \cdot 6 \cdot 6} = \frac{1}{6^2} = 6^{-2}.$$

Here, $2 - 4 = -2$. These examples suggest the **quotient rule for exponents.**

### Quotient Rule for Exponents

For any nonzero real number $a$ and any integers $m$ and $n$,

$$\frac{a^m}{a^n} = a^{m-n}.$$

(Keep the base and subtract the exponents.)

*Example:* $\frac{5^8}{5^4} = 5^{8-4} = 5^4$

**CAUTION** A common **error** is to write $\frac{5^8}{5^4} = 1^{8-4} = 1^4$. Notice that by the quotient rule, the quotient should have the *same base,* 5. That is,

$$\frac{5^8}{5^4} = 5^{8-4} = 5^4.$$

If you are not sure, use the definition of an exponent to write out the factors:

$$5^8 = 5 \cdot 5 \cdot 5 \cdot 5 \cdot 5 \cdot 5 \cdot 5 \cdot 5 \quad \text{and} \quad 5^4 = 5 \cdot 5 \cdot 5 \cdot 5.$$

Then it is clear that the quotient is $5^4$.

### EXAMPLE 4 Using the Quotient Rule

Simplify, using the quotient rule for exponents. Write answers with positive exponents.

**(a)** $\frac{5^8}{5^6} = 5^{8-6} = 5^2$ or 25

**(b)** $\frac{4^2}{4^9} = 4^{2-9} = 4^{-7} = \frac{1}{4^7}$

**(c)** $\frac{5^{-3}}{5^{-7}} = 5^{-3-(-7)} = 5^4$ or 625

**(d)** $\frac{q^5}{q^{-3}} = q^{5-(-3)} = q^8 \quad (q \neq 0)$

**(e)** $\frac{3^2x^5}{3^4x^3} = \frac{3^2}{3^4} \cdot \frac{x^5}{x^3} = 3^{2-4} \cdot x^{5-3} = 3^{-2}x^2 = \frac{x^2}{3^2} \quad (x \neq 0)$

**(f)** $\frac{(m+n)^{-2}}{(m+n)^{-4}} = (m+n)^{-2-(-4)} = (m+n)^{-2+4} = (m+n)^2 \quad (m \neq -n)$

**(g)** $\frac{7x^{-3}y^2}{2^{-1}x^2y^{-5}} = \frac{7 \cdot 2^1y^2y^5}{x^2x^3} = \frac{14y^7}{x^5} \quad (x, y \neq 0)$

**Now Try Exercises 29, 41, 45, and 49.**

The definitions and rules for exponents given in this section and Section 4.1 are summarized here.

**Definitions and Rules for Exponents**

For any integers $m$ and $n$:

| | | Examples |
|---|---|---|
| **Product rule** | $a^m \cdot a^n = a^{m+n}$ | $7^4 \cdot 7^5 = 7^9$ |
| **Zero exponent** | $a^0 = 1 \quad (a \neq 0)$ | $(-3)^0 = 1$ |
| **Negative exponent** | $a^{-n} = \frac{1}{a^n} \quad (a \neq 0)$ | $5^{-3} = \frac{1}{5^3}$ |
| **Quotient rule** | $\frac{a^m}{a^n} = a^{m-n} \quad (a \neq 0)$ | $\frac{2^2}{2^5} = 2^{2-5} = 2^{-3} = \frac{1}{2^3}$ |
| **Power rules (a)** | $(a^m)^n = a^{mn}$ | $(4^2)^3 = 4^6$ |
| **(b)** | $(ab)^m = a^mb^m$ | $(3k)^4 = 3^4k^4$ |
| **(c)** | $\left(\frac{a}{b}\right)^m = \frac{a^m}{b^m} \quad (b \neq 0)$ | $\left(\frac{2}{3}\right)^2 = \frac{2^2}{3^2}$ |

(continued)

| | | |
|---|---|---|
| **Negative to positive rules** | $\dfrac{a^{-m}}{b^{-n}} = \dfrac{b^n}{a^m}$ $(a \neq 0, b \neq 0)$ | $\dfrac{2^{-4}}{5^{-3}} = \dfrac{5^3}{2^4}$ |
| | $\left(\dfrac{a}{b}\right)^{-m} = \left(\dfrac{b}{a}\right)^m$ | $\left(\dfrac{4}{7}\right)^{-2} = \left(\dfrac{7}{4}\right)^2$ |

**OBJECTIVE 4 Use combinations of rules.** We may sometimes need to use more than one rule to simplify an expression.

### EXAMPLE 5 Using Combinations of Rules

Use the rules for exponents to simplify each expression. Assume all variables represent nonzero real numbers.

**(a)** $\dfrac{(4^2)^3}{4^5} = \dfrac{4^6}{4^5}$ Power rule (a)

$= 4^{6-5}$ Quotient rule

$= 4^1 = 4$

**(b)** $(2x)^3(2x)^2 = (2x)^5$ Product rule

$= 2^5x^5$ or $32x^5$ Power rule (b)

**(c)** $\left(\dfrac{2x^3}{5}\right)^{-4} = \left(\dfrac{5}{2x^3}\right)^4$ Negative to positive rule

$= \dfrac{5^4}{2^4x^{12}}$ or $\dfrac{625}{16x^{12}}$ Power rules (a)–(c)

**(d)** $\left(\dfrac{3x^{-2}}{4^{-1}y^3}\right)^{-3} = \dfrac{3^{-3}x^6}{4^3y^{-9}}$ Power rules (a)–(c)

$= \dfrac{x^6y^9}{4^3 \cdot 3^3}$ or $\dfrac{x^6y^9}{1728}$ Negative to positive rule

**(e)** $\dfrac{(4m)^{-3}}{(3m)^{-4}} = \dfrac{4^{-3}m^{-3}}{3^{-4}m^{-4}}$ Power rule (b)

$= \dfrac{3^4m^4}{4^3m^3}$ Negative to positive rule

$= \dfrac{3^4m^{4-3}}{4^3}$ Quotient rule

$= \dfrac{3^4m}{4^3}$ or $\dfrac{81m}{64}$

**Now Try Exercises 57, 63, 65, and 67.**

**NOTE** Since the steps can be done in several different orders, there are many equally correct ways to simplify expressions like those in Examples 5(d) and 5(e).

## 4.2 EXERCISES

**For Extra Help**

- Student's Solutions Manual
- MyMathLab
- InterAct Math Tutorial Software
- AW Math Tutor Center
- MathXL
- Digital Video Tutor CD 6/Videotape 6

*Each expression is either equal to* 0, 1, *or* −1. *Decide which is correct. See Example 1.*

**1.** $9^0$ **2.** $5^0$ **3.** $(-4)^0$ **4.** $(-10)^0$

**5.** $-9^0$ **6.** $-5^0$ **7.** $(-2)^0 - 2^0$ **8.** $(-8)^0 - 8^0$

**9.** $\frac{0^{10}}{10^0}$ **10.** $\frac{0^5}{5^0}$

*Match each expression in Column I with the equivalent expression in Column II. Choices in Column II may be used once, more than once, or not at all.*

| I | II |
|---|---|
| **11.** $-2^{-4}$ | **A.** $8$ |
| **12.** $(-2)^{-4}$ | **B.** $16$ |
| **13.** $2^{-4}$ | **C.** $-\frac{1}{16}$ |
| **14.** $\frac{1}{2^{-4}}$ | **D.** $-8$ |
| **15.** $\frac{1}{-2^{-4}}$ | **E.** $-16$ |
| **16.** $\frac{1}{(-2)^{-4}}$ | **F.** $\frac{1}{16}$ |

*Evaluate each expression. See Examples 1 and 2.*

**17.** $7^0 + 9^0$ **18.** $8^0 + 6^0$ **19.** $4^{-3}$ **20.** $5^{-4}$

**21.** $\left(\frac{1}{2}\right)^{-4}$ **22.** $\left(\frac{1}{3}\right)^{-3}$ **23.** $\left(\frac{6}{7}\right)^{-2}$ **24.** $\left(\frac{2}{3}\right)^{-3}$

**25.** $(-3)^{-4}$ **26.** $(-4)^{-3}$ **27.** $5^{-1} + 3^{-1}$ **28.** $6^{-1} + 2^{-1}$

*Use the quotient rule to simplify each expression. Write the expression with positive exponents. Assume that all variables represent nonzero real numbers. See Examples 2–4.*

**29.** $\frac{5^8}{5^5}$ **30.** $\frac{11^6}{11^3}$ **31.** $\frac{3^{-2}}{5^{-3}}$ **32.** $\frac{4^{-3}}{3^{-2}}$ **33.** $\frac{5}{5^{-1}}$

**34.** $\frac{6}{6^{-2}}$ **35.** $\frac{x^{12}}{x^{-3}}$ **36.** $\frac{y^4}{y^{-6}}$ **37.** $\frac{1}{6^{-3}}$ **38.** $\frac{1}{5^{-2}}$

**39.** $\frac{2}{r^{-4}}$ **40.** $\frac{3}{s^{-8}}$ **41.** $\frac{4^{-3}}{5^{-2}}$ **42.** $\frac{6^{-2}}{5^{-4}}$ **43.** $p^5q^{-8}$

**44.** $x^{-8}y^4$ **45.** $\frac{r^5}{r^{-4}}$ **46.** $\frac{a^6}{a^{-4}}$ **47.** $\frac{x^{-3}y}{4z^{-2}}$ **48.** $\frac{p^{-5}q^{-4}}{9r^{-3}}$

**49.** $\frac{(a+b)^{-3}}{(a+b)^{-4}}$ **50.** $\frac{(x+y)^{-8}}{(x+y)^{-9}}$ **51.** $\frac{(x+2y)^{-3}}{(x+2y)^{-5}}$ **52.** $\frac{(p-3q)^{-2}}{(p-3q)^{-4}}$

## RELATING CONCEPTS (EXERCISES 53–56)

**For Individual or Group Work**

*In Objective 1, we showed how $6^0$ acts as 1 when it is applied to the product rule, thus motivating the definition for 0 as an exponent. We can also use the quotient rule to motivate this definition.* ***Work Exercises 53–56 in order.***

**53.** Consider the expression $\frac{25}{25}$. What is its simplest form?

**54.** Because $25 = 5^2$, the expression $\frac{25}{25}$ can be written as the quotient of powers of 5. Write the expression in this way.

**55.** Apply the quotient rule for exponents to the expression you wrote in Exercise 54. Give the answer as a power of 5.

**56.** Your answers in Exercises 53 and 55 must be equal because they both represent $\frac{25}{25}$. Write this equality. What definition does this result support?

*Use a combination of the rules for exponents to simplify each expression. Write answers with only positive exponents. Assume that all variables represent nonzero real numbers. See Example 5.*

**57.** $\frac{(7^4)^3}{7^9}$

**58.** $\frac{(5^3)^2}{5^2}$

**59.** $x^{-3} \cdot x^5 \cdot x^{-4}$

**60.** $y^{-8} \cdot y^5 \cdot y^{-2}$

**61.** $\frac{(3x)^{-2}}{(4x)^{-3}}$

**62.** $\frac{(2y)^{-3}}{(5y)^{-4}}$

**63.** $\left(\frac{x^{-1}y}{z^2}\right)^{-2}$

**64.** $\left(\frac{p^{-4}q}{r^{-3}}\right)^{-3}$

**65.** $(6x)^4(6x)^{-3}$

**66.** $(10y)^9(10y)^{-8}$

**67.** $\frac{(m^7n)^{-2}}{m^{-4}n^3}$

**68.** $\frac{(m^8n^{-4})^2}{m^{-2}n^5}$

**69.** $\frac{(x^{-1}y^2z)^{-2}}{(x^{-3}y^3z)^{-1}}$

**70.** $\frac{(a^{-2}b^{-3}c^{-4})^{-5}}{(a^2b^3c^4)^5}$

**71.** $\left(\frac{xy^{-2}}{x^2y}\right)^{-3}$

**72.** $\left(\frac{wz^{-5}}{w^{-3}z}\right)^{-2}$

**73.** $\frac{(4a^2b^3)^{-2}(2ab^{-1})^3}{(a^3b)^{-4}}$

**74.** $\frac{(m^6n)^{-2}(m^2n^{-2})^3}{m^{-1}n^{-2}}$

**75.** $\frac{(2y^{-1}z^2)^2(3y^{-2}z^{-3})^3}{(y^3z^2)^{-1}}$

**76.** $\frac{(3p^{-2}q^3)^2(5p^{-1}q^{-4})^{-1}}{(p^2q^{-2})^{-3}}$

**77.** $\frac{(9^{-1}z^{-2}x)^{-1}(4z^2x^4)^{-2}}{(5z^{-2}x^{-3})^2}$

**78.** $\frac{(4^{-1}a^{-1}b^{-2})^{-2}(5a^{-3}b^4)^{-2}}{(3a^{-3}b^{-5})^2}$

**79.** Consider the following typical student **error:**

$$\frac{16^3}{2^2} = \left(\frac{16}{2}\right)^{3-2} = 8^1 = 8.$$

Explain what the student did incorrectly, and then give the correct answer.

**80.** Consider the following typical student **error:**

$$-5^4 = (-5)^4 = 625.$$

Explain what the student did incorrectly, and then give the correct answer.

## SUMMARY EXERCISES ON THE RULES FOR EXPONENTS

*Use the rules for exponents to simplify each expression. Use only positive exponents in your answers. Assume all variable expressions represent positive numbers.*

**1.** $\left(\frac{6x^2}{5}\right)^{12}$

**2.** $\left(\frac{rs^2t^3}{3t^4}\right)^6$

**3.** $(10x^2y^4)^2(10xy^2)^3$

**4.** $(-2ab^3c)^4(-2a^2b)^3$

**5.** $\left(\frac{9wx^3}{y^4}\right)^3$

**6.** $(4x^{-2}y^{-3})^{-2}$

**7.** $\frac{c^{11}(c^2)^4}{(c^3)^3(c^2)^{-6}}$

**8.** $\left(\frac{k^4t^2}{k^2t^{-4}}\right)^{-2}$

**9.** $5^{-1} + 6^{-1}$

**10.** $\frac{(3y^{-1}z^3)^{-1}(3y^2)}{(y^3z^2)^{-3}}$

**11.** $\frac{(2xy^{-1})^3}{2^3x^{-3}y^2}$

**12.** $-8^0 + (-8)^0$

**13.** $(z^4)^{-3}(z^{-2})^{-5}$

**14.** $\left(\frac{r^2st^5}{3r}\right)^{-2}$

**15.** $\frac{(3^{-1}x^{-3}y)^{-1}(2x^2y^{-3})^2}{(5x^{-2}y^2)^{-2}}$

**16.** $\left(\frac{5x^2}{3x^{-4}}\right)^{-1}$

**17.** $\left(\frac{-2x^{-2}}{2x^2}\right)^{-2}$

**18.** $\frac{(x^{-4}y^2)^3(x^2y)^{-1}}{(xy^2)^{-3}}$

**19.** $\frac{(a^{-2}b^3)^{-4}}{(a^{-3}b^2)^{-2}(ab)^{-4}}$

**20.** $(2a^{-30}b^{-29})(3a^{31}b^{30})$

**21.** $5^{-2} + 6^{-2}$

**22.** $\left[\frac{(x^{47}y^{23})^2}{x^{-26}y^{-42}}\right]^0$

**23.** $\left(\frac{7a^2b^3}{2}\right)^3$

**24.** $-(-12^0)$

**25.** $-(-12)^0$

**26.** $\frac{0^{12}}{12^0}$

**27.** $\frac{(2xy^{-3})^{-2}}{(3x^{-2}y^4)^{-3}}$

**28.** $\left(\frac{a^2b^3c^4}{a^{-2}b^{-3}c^{-4}}\right)^{-2}$

**29.** $(6x^{-5}z^3)^{-3}$

**30.** $(2p^{-2}qr^{-3})(2p)^{-4}$

**31.** $\frac{(xy)^{-3}(xy)^5}{(xy)^{-4}}$

**32.** $42^0 - (-12)^0$

**33.** $\frac{(7^{-1}x^{-3})^{-2}(x^4)^{-6}}{7^{-1}x^{-3}}$

**34.** $\left(\frac{3^{-4}x^{-3}}{3^{-3}x^{-6}}\right)^{-2}$

**35.** $(5p^{-2}q)^{-3}(5pq^3)^4$

**36.** $8^{-1} + 6^{-1}$

**37.** $\left[\frac{4r^{-6}s^{-2}t}{2r^8s^{-4}t^2}\right]^{-1}$

**38.** $(13x^{-6}y)(13x^{-6}y)^{-1}$

**39.** $\frac{(8pq^{-2})^4}{(8p^{-2}q^{-3})^3}$

**40.** $\left(\frac{mn^{-2}p}{m^2np^4}\right)^{-2}\left(\frac{mn^{-2}p}{m^2np^4}\right)^3$

**41.** $-(-3^0)^0$

**42.** $5^{-1} - 8^{-1}$

# 4.3 An Application of Exponents: Scientific Notation

**OBJECTIVES**

1. Express numbers in scientific notation.
2. Convert numbers in scientific notation to numbers without exponents.
3. Use scientific notation in calculations.

**OBJECTIVE 1 Express numbers in scientific notation.** One example of the use of exponents comes from science. The numbers occurring in science are often extremely large (such as the distance from Earth to the sun, 93,000,000 mi) or extremely small (the wavelength of yellow-green light, approximately .0000006 m). Because of the difficulty of working with many zeros, scientists often express such numbers with exponents. Each number is written as $a \times 10^n$, where $1 \leq |a| < 10$ and $n$ is an integer. This form is called **scientific notation.** There is always one nonzero digit before the decimal point. This is shown in the following examples. (In work with scientific notation, the times symbol, $\times$, is commonly used.)

$3.19 \times 10^1 = 3.19 \times 10 = 31.9$ Decimal point moves 1 place to the right.

$3.19 \times 10^2 = 3.19 \times 100 = 319.$ Decimal point moves 2 places to the right.

$3.19 \times 10^3 = 3.19 \times 1000 = 3190.$ Decimal point moves 3 places to the right.

$3.19 \times 10^{-1} = 3.19 \times .1 = .319$ Decimal point moves 1 place to the left.

$3.19 \times 10^{-2} = 3.19 \times .01 = .0319$ Decimal point moves 2 places to the left.

$3.19 \times 10^{-3} = 3.19 \times .001 = .00319$ Decimal point moves 3 places to the left.

A number in scientific notation is always written with the decimal point after the first nonzero digit and then multiplied by the appropriate power of 10. For example, 35 is written $3.5 \times 10^1$, or $3.5 \times 10$; 56,200 is written $5.62 \times 10^4$, since

$$56{,}200 = 5.62 \times 10{,}000 = 5.62 \times 10^4.$$

To write a number in scientific notation, follow these steps.

**Writing a Number in Scientific Notation**

*Step 1* Move the decimal point to the right of the first nonzero digit.

*Step 2* Count the number of places you moved the decimal point.

*Step 3* The number of places in Step 2 is the absolute value of the exponent on 10.

*Step 4* The exponent on 10 is positive if the original number is larger than the number in Step 1; the exponent is negative if the original number is smaller than the number in Step 1. If the decimal point is not moved, the exponent is 0.

**EXAMPLE 1 Using Scientific Notation**

Write each number in scientific notation.

**(a)** 93,000,000

The number will be written in scientific notation as $9.3 \times 10^n$. To find the value of $n$, first compare the original number, 93,000,000, with 9.3. Here 93,000,000 is *larger* than 9.3. Therefore, multiply by a *positive* power of 10 so the product $9.3 \times 10^n$ will equal the larger number.

Move the decimal point to follow the first nonzero digit (the 9). Count the number of places the decimal point was moved.

$$93{,}000{,}000 \qquad \text{7 places}$$

Since the decimal point was moved 7 places, and since $n$ is positive,

$$93{,}000{,}000 = 9.3 \times 10^7.$$

**(b)** $63{,}200{,}000{,}000 = 6.3200000000 = 6.32 \times 10^{10}$

10 places

**(c)** .00462

Move the decimal point to the right of the first nonzero digit, and count the number of places the decimal point was moved.

$$.00462 \qquad \text{3 places}$$

Since .00462 is *smaller* than 4.62, the exponent must be *negative*.

$$.00462 = 4.62 \times 10^{-3}$$

**(d)** $.0000762 = 7.62 \times 10^{-5}$

**Now Try Exercises 15 and 19.**

**NOTE** To choose the exponent when writing a number in scientific notation, think: If the original number is "large," like 93,000,000, use a *positive* exponent on 10, since positive is larger than negative. However, if the original number is "small," like .00462, use a *negative* exponent on 10, since negative is smaller than positive.

**OBJECTIVE 2 Convert numbers in scientific notation to numbers without exponents.** To convert a number written in scientific notation to a number without exponents, work in reverse. Multiplying by a positive power of 10 will make the number larger; multiplying by a negative power of 10 will make the number smaller.

### EXAMPLE 2 Writing Numbers without Exponents

Write each number without exponents.

**(a)** $6.2 \times 10^3$

Since the exponent is positive, make 6.2 larger by moving the decimal point 3 places to the right, inserting zeros as needed.

$$6.2 \times 10^3 = 6.200 = 6200$$

**(b)** $4.283 \times 10^5 = 4.28300 = 428{,}300$ Move 5 places to the right.

**(c)** $7.04 \times 10^{-3} = .00704$ Move 3 places to the left.

The exponent tells the number of places and the direction that the decimal point is moved.

**Now Try Exercises 23 and 29.**

**OBJECTIVE 3 Use scientific notation in calculations.** The next example uses scientific notation with products and quotients.

### EXAMPLE 3 Multiplying and Dividing with Scientific Notation

Write each product or quotient without exponents.

**(a)** $(6 \times 10^3)(5 \times 10^{-4}) = (6 \times 5)(10^3 \times 10^{-4})$ Commutative and associative properties

$= 30 \times 10^{-1}$ Product rule

$= 3$ Write without exponents.

**(b)** $\dfrac{4 \times 10^{-5}}{2 \times 10^3} = \dfrac{4}{2} \times \dfrac{10^{-5}}{10^3} = 2 \times 10^{-8} = .00000002$

**Now Try Exercises 33 and 43.**

**NOTE** Multiplying or dividing numbers written in scientific notation may produce an answer in the form $a \times 10^0$. Since $10^0 = 1$, $a \times 10^0 = a$. For example,

$$(8 \times 10^{-4})(5 \times 10^4) = 40 \times 10^0 = 40.$$

### EXAMPLE 4 Calculating Using Scientific Notation

Convert to scientific notation, perform each computation, then give the result without scientific notation.

**(a)** To determine helium usage at Kennedy Space Center, scientists must calculate 70,000(.0283)(1000). (*Source: NASA-AMATYC-NSF Mathematics Explorations II,* Capital Community College, 2000.)

$$\begin{aligned} 70{,}000(.0283)(1000) &= (7 \times 10^4)(2.83 \times 10^{-2})(1 \times 10^3) \\ &= (7 \times 2.83 \times 1)(10^{4-2+3}) \\ &= 19.81 \times 10^5 \\ &= 1{,}981{,}000 \end{aligned}$$

**(b)** The ratio of the tidal force exerted by the moon compared to that exerted by the sun is given by

$$\frac{73.5 \times 10^{21} \times (1.5 \times 10^8)^3}{1.99 \times 10^{30} \times (3.84 \times 10^5)^3}.$$

(*Source:* Kastner, Bernice, *Space Mathematics,* NASA.)

$$\begin{aligned} \frac{7.35 \times 10^1 \times 10^{21} \times 1.5^3 \times 10^{24}}{1.99 \times 10^{30} \times 3.84^3 \times 10^{15}} &\approx .22 \times 10^{1+21+24-30-15} \\ &= .22 \times 10^1 \\ &= 2.2 \end{aligned}$$

**Now Try Exercises 61 and 65.**

## CONNECTIONS

Charles F. Richter devised a scale in 1935 to compare the intensities, or relative power, of earthquakes. The *intensity* of an earthquake is measured relative to the intensity of a standard *zero-level* earthquake of intensity $I_0$. The relationship is equivalent to $I = I_0 \times 10^R$, where $R$ is the *Richter scale* measure. For example, if an earthquake has magnitude 5.0 on the Richter scale, then its intensity is calculated as $I = I_0 \times 10^{5.0} = I_0 \times 100{,}000$, which is 100,000 times as intense as a zero-level earthquake. The following diagram illustrates the intensities of earthquakes and their Richter scale magnitudes.

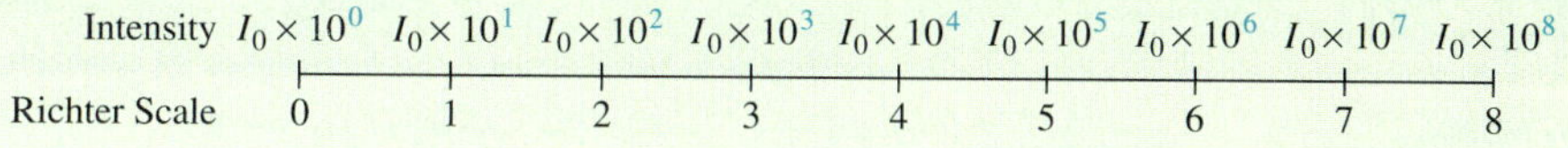


To compare two earthquakes to each other, a ratio of the intensities is calculated. For example, to compare an earthquake that measures 8.0 on the Richter scale to one that measures 5.0, simply find the ratio of the intensities:

$$\frac{\text{intensity } 8.0}{\text{intensity } 5.0} = \frac{I_0 \times 10^{8.0}}{I_0 \times 10^{5.0}} = \frac{10^8}{10^5} = 10^{8-5} = 10^3 = 1000.$$

Therefore an earthquake that measures 8.0 on the Richter Scale is 1000 times as intense as one that measures 5.0.

### For Discussion or Writing

The table gives Richter scale measurements for several earthquakes.

| Earthquake | | Richter Scale Measurement |
|---|---|---|
| 1960 | Concepción, Chile | 9.5 |
| 1906 | San Francisco, California | 8.3 |
| 1939 | Erzincan, Turkey | 8.0 |
| 1998 | Sumatra, Indonesia | 7.0 |
| 1998 | Adana, Turkey | 6.3 |

*Source: World Almanac and Books of Facts,* 2000.

1. Compare the intensity of the 1939 Erzincan earthquake to the 1998 Sumatra earthquake.
2. Compare the intensity of the 1998 Adana earthquake to the 1906 San Francisco earthquake.
3. Compare the intensity of the 1939 Erzincan earthquake to the 1998 Adana earthquake.
4. Suppose an earthquake measures 7.2 on the Richter scale. How would the intensity of a second earthquake compare if its Richter scale measure differed by +3.0? By −1.0?

## 4.3 EXERCISES

**For Extra Help**

Student's Solutions Manual

MyMathLab

InterAct Math Tutorial Software

AW Math Tutor Center

MathXL

Digital Video Tutor CD 6/Videotape 6

*Match each number written in scientific notation in Column I with the correct choice from Column II.*

| I | II |
|---|---|
| **1.** $4.6 \times 10^{-4}$ | **A.** .00046 |
| **2.** $4.6 \times 10^{4}$ | **B.** 46,000 |
| **3.** $4.6 \times 10^{5}$ | **C.** 460,000 |
| **4.** $4.6 \times 10^{-5}$ | **D.** .000046 |

*Determine whether or not each number is written in scientific notation as defined in Objective 1. If it is not, write it as such.*

**5.** $4.56 \times 10^{3}$ **6.** $7.34 \times 10^{5}$ **7.** 5,600,000 **8.** 34,000

**9.** $.8 \times 10^{2}$ **10.** $.9 \times 10^{3}$ **11.** .004 **12.** .0007

**13.** Explain in your own words what it means for a number to be written in scientific notation. Give examples.

**14.** Explain how to multiply a number by a positive power of ten. Then explain how to multiply a number by a negative power of ten.

*Write each number in scientific notation. See Example 1.*

**15.** 5,876,000,000 **16.** 9,994,000,000 **17.** 82,350 **18.** 78,330

**19.** .000007 **20.** .0000004 **21.** .00203 **22.** .0000578

*Write each number without exponents. See Example 2.*

**23.** $7.5 \times 10^{5}$ **24.** $8.8 \times 10^{6}$ **25.** $5.677 \times 10^{12}$ **26.** $8.766 \times 10^{9}$

**27.** $6.21 \times 10^{0}$ **28.** $8.56 \times 10^{0}$ **29.** $7.8 \times 10^{-4}$ **30.** $8.9 \times 10^{-5}$

**31.** $5.134 \times 10^{-9}$ **32.** $7.123 \times 10^{-10}$

*Use properties and rules for exponents to perform the indicated operations, and write each answer without exponents. See Example 3.*

**33.** $(2 \times 10^{8}) \times (3 \times 10^{3})$ **34.** $(4 \times 10^{7}) \times (3 \times 10^{3})$

**35.** $(5 \times 10^{4}) \times (3 \times 10^{2})$ **36.** $(8 \times 10^{5}) \times (2 \times 10^{3})$

**37.** $(3 \times 10^{-4}) \times (2 \times 10^{8})$ **38.** $(4 \times 10^{-3}) \times (2 \times 10^{7})$

**39.** $\dfrac{9 \times 10^{-5}}{3 \times 10^{-1}}$ **40.** $\dfrac{12 \times 10^{-4}}{4 \times 10^{-3}}$ **41.** $\dfrac{8 \times 10^{3}}{2 \times 10^{2}}$

**42.** $\dfrac{5 \times 10^{4}}{1 \times 10^{3}}$ **43.** $\dfrac{2.6 \times 10^{-3}}{2 \times 10^{2}}$ **44.** $\dfrac{9.5 \times 10^{-1}}{5 \times 10^{3}}$

### TECHNOLOGY INSIGHTS (EXERCISES 45–50)

*Graphing calculators such as the TI-83 Plus can display numbers in scientific notation (when in scientific mode), using the format shown in the screen at the top left on the next page. For example, the calculator displays* 5.4E3 *to represent* $5.4 \times 10^{3}$, *the scientific notation form for* 5400. *The display* 5.4E−4 *means* $5.4 \times 10^{-4}$. *It will also perform operations with numbers entered in scientific notation, as shown in the screen on the right. Notice how the rules for exponents are applied.*

```
5400
            5.4E3
5.4*10^(-4)
           5.4E-4
```

```
(2E5)*(4E-2)
              8E3
(2E5)/(4E-2)
              5E6
(2E5)³
             8E15
```

*Predict the display the calculator would give for the expression shown in each screen.*

45.
```
.00000047
```

46.
```
.000021
```

47.
```
(8E5)/(4E-2)
```

48.
```
(9E-4)/(3E3)
```

49.
```
(2E6)*(2E-3)/(4E
2)
```

50.
```
(5E-3)*(1E9)/(5E
3)
```

*Each statement comes from* Astronomy! A Brief Edition *by James B. Kaler (Addison-Wesley, 1997). If the number in italics is in scientific notation, write it without exponents. If the number is written without exponents, write it in scientific notation.*

**51.** Multiplying this view over the whole sky yields a galaxy count of more than *10 billion*. (page 496)

**52.** The circumference of the solar orbit is . . . about *4.7 million* km . . . . (in reference to the orbit of Jupiter, page 395)

**53.** The solar luminosity requires that $2 \times 10^9$ kg of mass be converted into energy every second. (page 327)

**54.** At maximum, a cosmic ray particle—a mere atomic nucleus of only $10^{-13}$ cm across—can carry the energy of a professionally pitched baseball. (page 445)

Either the horizontal or the vertical method may be used to add and subtract polynomials.

We add and subtract polynomials in more than one variable by combining like terms, just as with single variable polynomials.

**EXAMPLE 9 Adding and Subtracting Polynomials with More Than One Variable**

Add or subtract as indicated.

**(a)** $(4a + 2ab - b) + (3a - ab + b) = 4a + 2ab - b + 3a - ab + b$

$= 7a + ab$ Combine like terms.

**(b)** $(2x^2y + 3xy + y^2) - (3x^2y - xy - 2y^2)$

$= 2x^2y + 3xy + y^2 - 3x^2y + xy + 2y^2$

$= -x^2y + 4xy + 3y^2$

**Now Try Exercises 73 and 75.**

**OBJECTIVE 6 Graph equations defined by polynomials of degree 2.** In Chapter 3 we introduced graphs of straight lines. These graphs were defined by linear equations (which are actually polynomial equations of degree 1). By selective point-plotting, we can graph polynomial equations of degree 2.

**EXAMPLE 10 Graphing Equations Defined by Polynomials with Degree 2**

Graph each equation.

**(a)** $y = x^2$

Select several values for $x$; then find the corresponding $y$-values. For example, selecting $x = 2$ gives

$$y = 2^2 = 4,$$

and so the point (2, 4) is on the graph of $y = x^2$. (Recall that in an ordered pair such as (2, 4), the $x$-value comes first and the $y$-value second.) We show some ordered pairs that satisfy $y = x^2$ in the table next to Figure 1. If we plot the ordered pairs from the table on a coordinate system and draw a smooth curve through them, we obtain the graph shown in Figure 1.

| $x$ | $y$ |
|---|---|
| 3 | 9 |
| 2 | 4 |
| 1 | 1 |
| 0 | 0 |
| −1 | 1 |
| −2 | 4 |
| −3 | 9 |

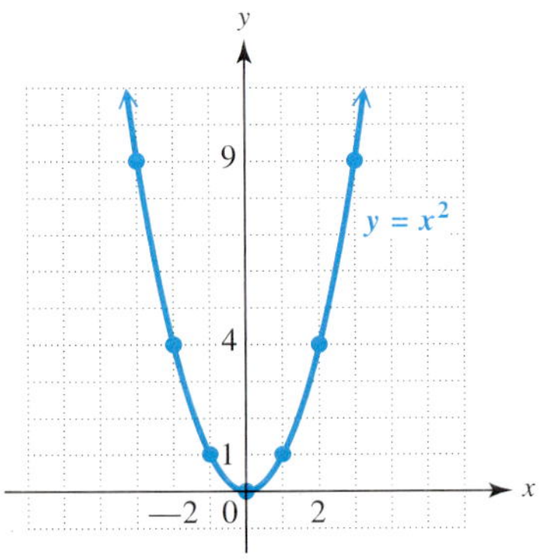


FIGURE 1

The graph of $y = x^2$ is the graph of a function, since each input $x$ is related to just one output $y$. The curve in Figure 1 is called a **parabola.** The point (0, 0), the lowest point on this graph, is called the **vertex** of the parabola. The vertical line through the vertex (the $y$-axis here) is called the **axis** of the parabola. The axis of a parabola is a **line of symmetry** for the graph. If the graph is folded on this line, the two halves will match.

**(b)** $y = -x^2 + 3$

Once again plot points to obtain the graph. For example, if $x = -2$,

$$y = -(-2)^2 + 3 = -4 + 3 = -1.$$

This point and several others are shown in the table that accompanies the graph in Figure 2. The vertex of this parabola is (0, 3). This time the vertex is the *highest* point on the graph. The graph opens downward because $x^2$ has a negative coefficient.

| $x$ | $y$ |
|---|---|
| −2 | −1 |
| −1 | 2 |
| 0 | 3 |
| 1 | 2 |
| 2 | −1 |

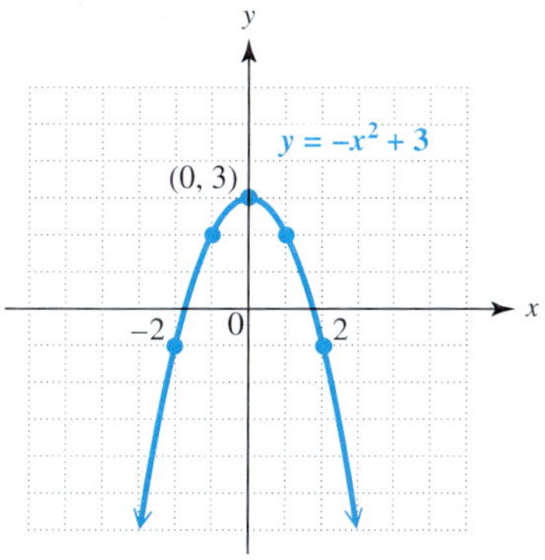


**FIGURE 2**

**Now Try Exercises 87 and 91.**

**NOTE** All polynomials of degree 2 have parabolas as their graphs. When graphing by plotting points, it is necessary to continue finding points until the vertex and points on either side of it are located. (In this section, all parabolas have their vertices on the $x$-axis or the $y$-axis.)

## CONNECTIONS

In Section 3.1 we introduced the idea of a *function:* for every input $x$, there is one output $y$. For example, according to the U.S. National Aeronautics and Space Administration (NASA), the budget in millions of dollars for space station research for 1996–2001 can be approximated by the polynomial equation

$$y = -10.25x^2 - 126.04x + 5730.21,$$

where $x = 0$ represents 1996, $x = 1$ represents 1997, and so on, up to $x = 5$ representing 2001. The actual budget for 1998 was 5327 million dollars; an input of $x = 2$ (for 1998) in the equation gives approximately $y = 5437$. Considering the magnitude of the numbers, this is a good approximation.

(continued)

**For Discussion or Writing**

Use the given polynomial equation to approximate the budget in other years between 1996 and 2001. Compare to the actual figures given here.

| Year | Budget (in millions of dollars) |
|---|---|
| 1996 | 5710 |
| 1997 | 5675 |
| 1998 | 5327 |
| 1999 | 5306 |
| 2000 | 5077 |
| 2001 | 4832 |

## 4.4 EXERCISES

**For Extra Help**

- Student's Solutions Manual
- MyMathLab
- InterAct Math Tutorial Software
- AW Math Tutor Center
- MathXL
- Digital Video Tutor CD 6/Videotape 6

*Fill in each blank with the correct response.*

**1.** In the term $7x^5$, the coefficient is _____ and the exponent is _____.

**2.** The expression $5x^3 - 4x^2$ has __________ term(s). (how many?)

**3.** The degree of the term $-4x^8$ is _____.

**4.** The polynomial $4x^2 - y^2$ __________ an example of a trinomial. (is/is not)

**5.** When $x^2 + 10$ is evaluated for $x = 4$, the result is _____.

**6.** $5x^{\_\_} + 3x^3 - 7x$ is a trinomial of degree 4.

**7.** $3xy + 2xy - 5xy =$ _____.

**8.** _____ is an example of a monomial with coefficient 5, in the variable $x$, having degree 9.

*For each polynomial, determine the number of terms and name the coefficients of the terms. See Example 1.*

**9.** $6x^4$ **10.** $-9y^5$ **11.** $t^4$ **12.** $s^7$

**13.** $-19r^2 - r$ **14.** $2y^3 - y$ **15.** $x + 8x^2 + 5x^3$ **16.** $v - 2v^3 - v^7$

*In each polynomial, add like terms whenever possible. Write the result in descending powers of the variable. See Example 2.*

**17.** $-3m^5 + 5m^5$ **18.** $-4y^3 + 3y^3$

**19.** $2r^5 + (-3r^5)$ **20.** $-19y^2 + 9y^2$

**21.** $.2m^5 - .5m^2$ **22.** $-.9y + .9y^2$

**23.** $-3x^5 + 2x^5 - 4x^5$ **24.** $6x^3 - 8x^3 + 9x^3$

**25.** $-4p^7 + 8p^7 + 5p^9$

**26.** $-3a^8 + 4a^8 - 3a^2$

**27.** $-4xy^2 + 3xy^2 - 2xy^2 + xy^2$

**28.** $3pr^5 - 8pr^5 + pr^5 + 2pr^5$

*For each polynomial, first simplify, if possible, and write it in descending powers of the variable. Then give the degree of the resulting polynomial and tell whether it is a* monomial, binomial, trinomial, *or* none of these. *See Example 3.*

**29.** $6x^4 - 9x$

**30.** $7t^3 - 3t$

**31.** $5m^4 - 3m^2 + 6m^4 - 7m^3$

**32.** $6p^5 + 4p^3 - 8p^5 + 10p^2$

**33.** $\frac{5}{3}x^4 - \frac{2}{3}x^4$

**34.** $\frac{4}{5}r^6 + \frac{1}{5}r^6$

**35.** $.8x^4 - .3x^4 - .5x^4 + 7$

**36.** $1.2t^3 - .9t^3 - .3t^3 + 9$

*Find the value of each polynomial when **(a)** $x = 2$ and when **(b)** $x = -1$. See Example 4.*

**37.** $2x^5 - 4x^4 + 5x^3 - x^2$

**38.** $2x^2 + 5x + 1$

**39.** $-3x^2 + 14x - 2$

**40.** $-2x^2 + 3$

**41.** $2x^2 - 3x - 5$

**42.** $x^2 + 5x - 10$

*Add. See Example 5.*

**43.** $$\begin{array}{r} 2x^2 - 4x \\ 3x^2 + 2x \\ \hline \end{array}$$

**44.** $$\begin{array}{r} -5y^3 + 3y \\ 8y^3 - 4y \\ \hline \end{array}$$

**45.** $$\begin{array}{r} 3m^2 + 5m + 6 \\ 2m^2 - 2m - 4 \\ \hline \end{array}$$

**46.** $$\begin{array}{r} 4a^3 - 4a^2 - 4 \\ 6a^3 + 5a^2 - 8 \\ \hline \end{array}$$

**47.** $$\begin{array}{r} \frac{2}{3}x^2 + \frac{1}{5}x + \frac{1}{6} \\ \frac{1}{2}x^2 - \frac{1}{3}x + \frac{2}{3} \\ \hline \end{array}$$

**48.** $$\begin{array}{r} \frac{4}{7}y^2 - \frac{1}{5}y + \frac{7}{9} \\ \frac{1}{3}y^2 - \frac{1}{3}y + \frac{2}{5} \\ \hline \end{array}$$

**49.** $$\begin{array}{r} 9m^3 - 5m^2 + 4m - 8 \\ -3m^3 + 6m^2 + 8m - 6 \\ \hline \end{array}$$

**50.** $$\begin{array}{r} 12r^5 + 11r^4 - 7r^3 - 2r^2 \\ -8r^5 + 10r^4 + 3r^3 + 2r^2 \\ \hline \end{array}$$

*Subtract. See Example 8.*

**51.** $$\begin{array}{r} 5y^3 - 3y^2 \\ 2y^3 + 8y^2 \\ \hline \end{array}$$

**52.** $$\begin{array}{r} -6t^3 + 4t^2 \\ 8t^3 - 6t^2 \\ \hline \end{array}$$

**53.** $$\begin{array}{r} 12x^4 - x^2 + x \\ 8x^4 + 3x^2 - 3x \\ \hline \end{array}$$

**54.** $$\begin{array}{r} 13y^5 - y^3 - 8y^2 \\ 7y^5 + 5y^3 + y^2 \\ \hline \end{array}$$

**55.** $$\begin{array}{r} 12m^3 - 8m^2 + 6m + 7 \\ -3m^3 + 5m^2 - 2m - 4 \\ \hline \end{array}$$

**56.** $$\begin{array}{r} 5a^4 - 3a^3 + 2a^2 - a + 6 \\ -6a^4 + a^3 - a^2 + a - 1 \\ \hline \end{array}$$

**57.** After reading Examples 5–8, do you have a preference regarding horizontal or vertical addition and subtraction of polynomials? Explain your answer.

**58.** Write a paragraph explaining how to add and subtract polynomials. Give an example using addition.

*Perform each indicated operation. See Examples 6 and 7.*

**59.** $(8m^2 - 7m) - (3m^2 + 7m - 6)$

**60.** $(x^2 + x) - (3x^2 + 2x - 1)$

**61.** $(16x^3 - x^2 + 3x) + (-12x^3 + 3x^2 + 2x)$

**62.** $(-2b^6 + 3b^4 - b^2) + (b^6 + 2b^4 + 2b^2)$

**63.** $(7y^4 + 3y^2 + 2y) - (18y^4 - 5y^2 + y)$

**64.** $(8t^5 + 3t^3 + 5t) - (19t^5 - 6t^3 + t)$

**65.** $(9a^4 - 3a^2 + 2) + (4a^4 - 4a^2 + 2) + (-12a^4 + 6a^2 - 3)$

**66.** $(4m^2 - 3m + 2) + (5m^2 + 13m - 4) - (16m^2 + 4m - 3)$

**67.** $[(8m^2 + 4m - 7) - (2m^2 - 5m + 2)] - (m^2 + m + 1)$

**68.** $[(9b^3 - 4b^2 + 3b + 2) - (-2b^3 - 3b^2 + b)] - (8b^3 + 6b + 4)$

**69.** $[(3x^2 - 2x + 7) - (4x^2 + 2x - 3)] - [(9x^2 + 4x - 6) + (-4x^2 + 4x + 4)]$

**70.** $[(6t^2 - 3t + 1) - (12t^2 + 2t - 6)] - [(4t^2 - 3t - 8) + (-6t^2 + 10t - 12)]$

**71.** Without actually performing the operations, determine mentally the coefficient of $x^2$ in the simplified form of $(-4x^2 + 2x - 3) - (-2x^2 + x - 1) + (-8x^2 + 3x - 4)$.

**72.** Without actually performing the operations, determine mentally the coefficient of $x$ in the simplified form of $(-8x^2 - 3x + 2) - (4x^2 - 3x + 8) - (-2x^2 - x + 7)$.

*Add or subtract as indicated. See Example 9.*

**73.** $(6b + 3c) + (-2b - 8c)$

**74.** $(-5t + 13s) + (8t - 3s)$

**75.** $(4x + 2xy - 3) - (-2x + 3xy + 4)$

**76.** $(8ab + 2a - 3b) - (6ab - 2a + 3b)$

**77.** $(5x^2y - 2xy + 9xy^2) - (8x^2y + 13xy + 12xy^2)$

**78.** $(16t^3s^2 + 8t^2s^3 + 9ts^4) - (-24t^3s^2 + 3t^2s^3 - 18ts^4)$

*For Exercises 79–82, use the formulas found on the inside covers. Find the perimeter of each rectangle.*

**79.**

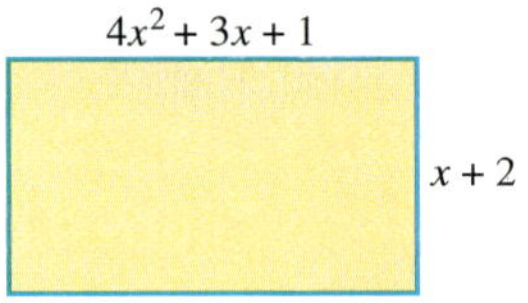


**80.**

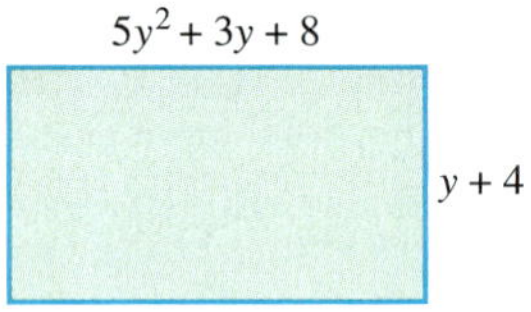


*Find **(a)** a polynomial representing the perimeter of each triangle and **(b)** the measures of the angles of the triangle.*

**81.**

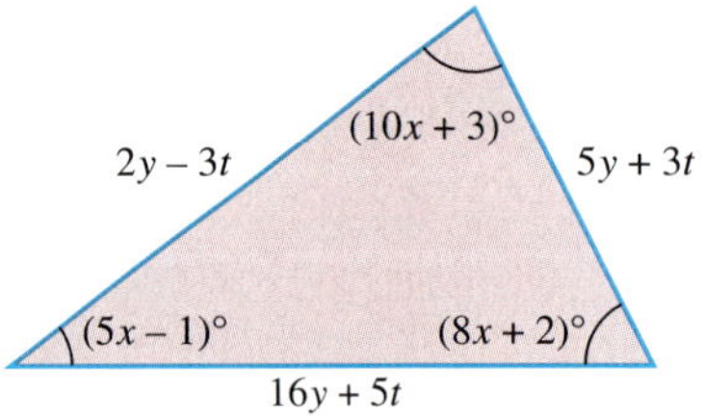


**82.**

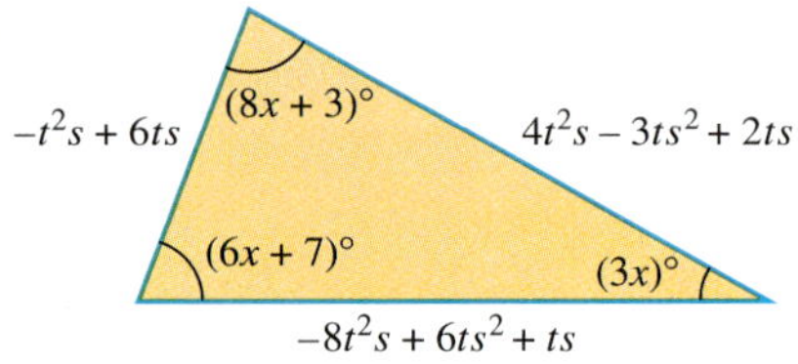


*The concepts required to work Exercises 83–86 have been covered, but the usual wording of the problem has been changed. Perform the indicated operations.*

**83.** Subtract $9x^2 - 6x + 5$ from $3x^2 - 2$.

**84.** Find the difference when $9x^4 + 3x^2 + 5$ is subtracted from $8x^4 - 2x^3 + x - 1$.

**85.** Find the difference between the sum of $5x^2 + 2x - 3$ and $x^2 - 8x + 2$ and the sum of $7x^2 - 3x + 6$ and $-x^2 + 4x - 6$.

**86.** Subtract the sum of $9t^3 - 3t + 8$ and $t^2 - 8t + 4$ from the sum of $12t + 8$ and $t^2 - 10t + 3$.

*Graph each equation by completing the table of values. See Example 10.*

**87.** $y = x^2 - 4$

| $x$ | $y$ |
|---|---|
| $-2$ | |
| $-1$ | |
| 0 | |
| 1 | |
| 2 | |

**88.** $y = x^2 - 6$

| $x$ | $y$ |
|---|---|
| $-2$ | |
| $-1$ | |
| 0 | |
| 1 | |
| 2 | |

**89.** $y = 2x^2 - 1$

| $x$ | $y$ |
|---|---|
| $-2$ | |
| $-1$ | |
| 0 | |
| 1 | |
| 2 | |

**90.** $y = 2x^2 + 2$

| $x$ | $y$ |
|---|---|
| $-2$ | |
| $-1$ | |
| 0 | |
| 1 | |
| 2 | |

**91.** $y = -x^2 + 4$

| $x$ | $y$ |
|---|---|
| $-2$ | |
| $-1$ | |
| 0 | |
| 1 | |
| 2 | |

**92.** $y = -x^2 + 2$

| $x$ | $y$ |
|---|---|
| $-2$ | |
| $-1$ | |
| 0 | |
| 1 | |
| 2 | |

**93.** $y = (x + 3)^2$

| $x$ | $-5$ | $-4$ | $-3$ | $-2$ | $-1$ |
|---|---|---|---|---|---|
| $y$ | | | | | |

**94.** $y = (x - 4)^2$

| $x$ | 2 | 3 | 4 | 5 | 6 |
|---|---|---|---|---|---|
| $y$ | | | | | |

*In the introduction to this chapter, we gave a polynomial that models the distance in feet that a car going approximately 68 mph will skid in x seconds. If we let D represent this distance, then*

$$D = 100x - 13x^2.$$

*Each time we evaluate this polynomial for a value of x, we get one and only one output value D. This idea is basic to the concept of a function, introduced in Section 3.1. Exercises 95 and 96 illustrate the idea with this polynomial.*

**95.** Use the given polynomial equation to approximate the skidding distance in feet if $x = 5$ sec.

**96.** Use the given polynomial equation to find the distance the car will skid in 1 sec. Write an ordered pair of the form $(x, D)$.

## RELATING CONCEPTS (EXERCISES 97–100)

**For Individual or Group Work**

*As explained earlier in this section, the polynomial equation*

$$y = -10.25x^2 - 126.04x + 5730.21$$

*gives a good approximation of NASA's budget for space station research, in millions of dollars, for 1996 through 2001, where $x = 0$ represents 1996, $x = 1$ represents*

(continued)

*1997, and so on. If we evaluate the polynomial for a specific input value x, we will get one and only one output value y as a result. This idea is basic to the study of functions, one of the most important concepts in mathematics.* ***Work Exercises 97–100 in order.***

**97.** If gasoline costs \$1.25 per gallon, then the monomial $1.25x$ gives the cost of $x$ gallons. Evaluate this monomial for 4, and then use the result to fill in the blanks: If _____ gallons are purchased, then the cost is _____.

**98.** If it costs \$15 to rent a chain saw plus \$2 per day, the binomial $2x + 15$ gives the cost to rent the chain saw for $x$ days. Evaluate this polynomial for 6 and then use the result to fill in the blanks: If the saw is rented for _____ days, then the cost is _____.

**99.** If an object is thrown upward under certain conditions, its height in feet is given by the trinomial $-16x^2 + 60x + 80$, where $x$ is in seconds. Evaluate this polynomial for 2.5 and then use the result to fill in the blanks: If _____ seconds have elapsed, then the height of the object is _____ feet.

**100.** The polynomial $2.69x^2 + 4.75x + 452.43$ gives a good approximation for the number of revenue passenger miles, in billions, for the U.S. airline industry during the period from 1990 to 1995, where $x = 0$ represents 1990. Use this polynomial to approximate the number of revenue passenger miles in 1991. (*Hint:* Any power of 1 is equal to 1, so simply add the coefficients and the constant.) (*Source:* Air Transportation Association of America.)

# 4.5 Multiplying Polynomials

**OBJECTIVES**

1. Multiply a monomial and a polynomial.
2. Multiply two polynomials.
3. Multiply binomials by the FOIL method.

**OBJECTIVE 1 Multiply a monomial and a polynomial.** As shown earlier, we find the product of two monomials by using the rules for exponents and the commutative and associative properties. For example,

$$(-8m^6)(-9n^6) = (-8)(-9)(m^6)(n^6) = 72m^6n^6.$$

**CAUTION** Do not confuse *addition* of terms with *multiplication* of terms. For example,

$$7q^5 + 2q^5 = 9q^5, \quad \text{but} \quad (7q^5)(2q^5) = 7 \cdot 2q^{5+5} = 14q^{10}.$$

To find the product of a monomial and a polynomial with more than one term, we use the distributive property and multiplication of monomials.

**EXAMPLE 1 Multiplying Monomials and Polynomials**

Use the distributive property to find each product.

**(a)** $4x^2(3x + 5)$

$$4x^2(3x + 5) = 4x^2(3x) + 4x^2(5) \quad \text{Distributive property}$$
$$= 12x^3 + 20x^2 \quad \text{Multiply monomials.}$$

**(b)** $-8m^3(4m^3 + 3m^2 + 2m - 1)$

$$= -8m^3(4m^3) + (-8m^3)(3m^2)$$
$$+ (-8m^3)(2m) + (-8m^3)(-1) \quad \text{Distributive property}$$
$$= -32m^6 - 24m^5 - 16m^4 + 8m^3 \quad \text{Multiply monomials.}$$

**Now Try Exercises 11 and 19.**

**OBJECTIVE 2 Multiply two polynomials.** We use the distributive property repeatedly to find the product of any two polynomials. For example, to find the product of the polynomials $x^2 + 3x + 5$ and $x - 4$, think of $x - 4$ as a single quantity and use the distributive property as follows.

$$(x^2 + 3x + 5)(x - 4) = x^2(x - 4) + 3x(x - 4) + 5(x - 4)$$

Now use the distributive property three times to find $x^2(x - 4)$, $3x(x - 4)$, and $5(x - 4)$.

$$x^2(x - 4) + 3x(x - 4) + 5(x - 4)$$
$$= x^2(x) + x^2(-4) + 3x(x) + 3x(-4) + 5(x) + 5(-4)$$
$$= x^3 - 4x^2 + 3x^2 - 12x + 5x - 20 \quad \text{Multiply monomials.}$$
$$= x^3 - x^2 - 7x - 20 \quad \text{Combine like terms.}$$

This example suggests the following rule.

**Multiplying Polynomials**

To multiply two polynomials, multiply each term of the second polynomial by each term of the first polynomial and add the products.

**EXAMPLE 2 Multiplying Two Polynomials**

Multiply $(m^2 + 5)(4m^3 - 2m^2 + 4m)$.

Multiply each term of the second polynomial by each term of the first.

$$(m^2 + 5)(4m^3 - 2m^2 + 4m)$$
$$= m^2(4m^3) + m^2(-2m^2) + m^2(4m) + 5(4m^3) + 5(-2m^2) + 5(4m)$$
$$= 4m^5 - 2m^4 + 4m^3 + 20m^3 - 10m^2 + 20m$$
$$= 4m^5 - 2m^4 + 24m^3 - 10m^2 + 20m \quad \text{Combine like terms.}$$

**Now Try Exercise 25.**

When at least one of the factors in a product of polynomials has three or more terms, the multiplication can be simplified by writing one polynomial above the other vertically.

## EXAMPLE 3 Multiplying Polynomials Vertically

Multiply $(x^3 + 2x^2 + 4x + 1)(3x + 5)$ using the vertical method.

Write the polynomials as follows.

$$\begin{array}{r} x^3 + 2x^2 + 4x + 1 \\ 3x + 5 \\ \hline \end{array}$$

It is not necessary to line up terms in columns, because any terms may be multiplied (not just like terms). Begin by multiplying each of the terms in the top row by 5.

$$\begin{array}{r} x^3 + 2x^2 + 4x + 1 \\ 3x + 5 \\ \hline 5x^3 + 10x^2 + 20x + 5 \end{array} \qquad 5(x^3 + 2x^2 + 4x + 1)$$

Notice how this process is similar to multiplication of whole numbers. Now multiply each term in the top row by $3x$. Be careful to place like terms in columns, since the final step will involve addition (as in multiplying two whole numbers).

$$\begin{array}{r} x^3 + 2x^2 + 4x + 1 \\ 3x + 5 \\ \hline 5x^3 + 10x^2 + 20x + 5 \\ 3x^4 + 6x^3 + 12x^2 + 3x \phantom{{}+5} \end{array} \qquad 3x(x^3 + 2x^2 + 4x + 1)$$

Add like terms.

$$\begin{array}{r} x^3 + 2x^2 + 4x + 1 \\ 3x + 5 \\ \hline 5x^3 + 10x^2 + 20x + 5 \\ 3x^4 + 6x^3 + 12x^2 + 3x \phantom{{}+5} \\ \hline 3x^4 + 11x^3 + 22x^2 + 23x + 5 \end{array}$$

The product is $3x^4 + 11x^3 + 22x^2 + 23x + 5$.

**Now Try Exercise 29.**

## EXAMPLE 4 Multiplying Polynomials with Fractional Coefficients Vertically

Find the product of $4m^3 - 2m^2 + 4m$ and $\frac{1}{2}m^2 + \frac{5}{2}$.

$$\begin{array}{r} 4m^3 - 2m^2 + 4m \\ \frac{1}{2}m^2 + \frac{5}{2} \\ \hline 10m^3 - 5m^2 + 10m \\ 2m^5 - m^4 + 2m^3 \phantom{{} - 5m^2 + 10m} \\ \hline 2m^5 - m^4 + 12m^3 - 5m^2 + 10m \end{array}$$

Terms of top row multiplied by $\frac{5}{2}$

Terms of top row multiplied by $\frac{1}{2}m^2$

Add.

**Now Try Exercise 35.**

We can use a rectangle to model polynomial multiplication. For example, to find the product

$$(2x + 1)(3x + 2),$$

label a rectangle with each term as shown here.

| | $3x$ | $2$ |
|---|---|---|
| $2x$ | | |
| $1$ | | |

Now put the product of each pair of monomials in the appropriate box.

| | $3x$ | $2$ |
|---|---|---|
| $2x$ | $6x^2$ | $4x$ |
| $1$ | $3x$ | $2$ |

The product of the original binomials is the sum of these four monomial products.

$$\begin{aligned}(2x + 1)(3x + 2) &= 6x^2 + 4x + 3x + 2 \\ &= 6x^2 + 7x + 2\end{aligned}$$

This approach can be extended to polynomials with any number of terms.

**OBJECTIVE 3 Multiply binomials by the FOIL method.** In algebra, many of the polynomials to be multiplied are both binomials (with just two terms). For these products, the **FOIL method** reduces the rectangle method to a systematic approach without the rectangle. To develop the FOIL method, we use the distributive property to find $(x + 3)(x + 5)$.

$$\begin{aligned}(x + 3)(x + 5) &= (x + 3)x + (x + 3)5 \\ &= x(x) + 3(x) + x(5) + 3(5) \\ &= x^2 + 3x + 5x + 15 \\ &= x^2 + 8x + 15\end{aligned}$$

Here is where the letters of the word FOIL originate.

$(x + 3)(x + 5)$ Multiply the **First** terms: $x(x)$. **F**

$(x + 3)(x + 5)$ Multiply the **Outer** terms: $x(5)$. **O**
This is the **outer product.**

$(x + 3)(x + 5)$ Multiply the **Inner** terms: $3(x)$. **I**
This is the **inner product.**

$(x + 3)(x + 5)$ Multiply the **Last** terms: $3(5)$. **L**

The inner product and the outer product should be added mentally so that the three terms of the answer can be written without extra steps as

$$(x + 3)(x + 5) = x^2 + 8x + 15.$$

A summary of the steps in the FOIL method follows.

### Multiplying Binomials by the FOIL Method

*Step 1* Multiply the two **F**irst terms of the binomials to get the first term of the answer.

*Step 2* Find the **O**uter product and the **I**nner product and add them (when possible) to get the middle term of the answer.

*Step 3* Multiply the two **L**ast terms of the binomials to get the last term of the answer.

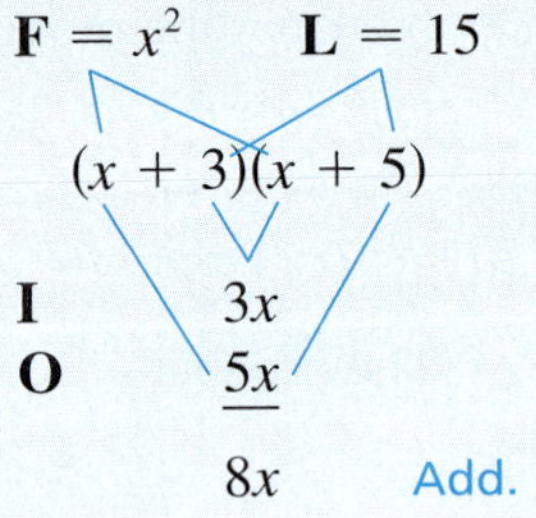


## EXAMPLE 5 Using the FOIL Method

Use the FOIL method to find the product $(x + 8)(x - 6)$.

*Step 1* **F** Multiply the first terms.

$$x(x) = x^2$$

*Step 2* **O** Find the outer product.

$$x(-6) = -6x$$

**I** Find the inner product.

$$8(x) = 8x$$

Add the outer and inner products mentally.

$$-6x + 8x = 2x$$

*Step 3* **L** Multiply the last terms.

$$8(-6) = -48$$

The product of $x + 8$ and $x - 6$ is the sum of the four terms found in three steps above, so

$$(x + 8)(x - 6) = x^2 + 2x - 48.$$

As a shortcut, this product can be found in the following manner.

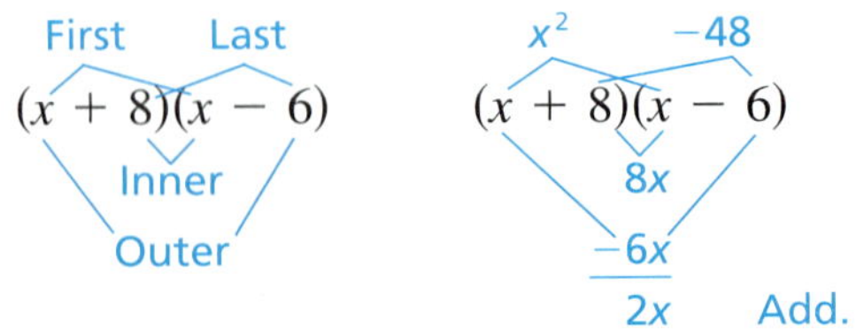


**Now Try Exercise 37.**

It is not possible to add the inner and outer products of the FOIL method if unlike terms result, as shown in the next example.

### EXAMPLE 6 Using the FOIL Method

Multiply $(9x - 2)(3y + 1)$.

| | | | |
|---|---|---|---|
| *First* | $(9x - 2)(3y + 1)$ | $27xy$ | |
| *Outer* | $(9x - 2)(3y + 1)$ | $9x$ | Unlike terms |
| *Inner* | $(9x - 2)(3y + 1)$ | $-6y$ | Unlike terms |
| *Last* | $(9x - 2)(3y + 1)$ | $-2$ | |

$$(9x - 2)(3y + 1) = \overset{F}{27xy} + \overset{O}{9x} - \overset{I}{6y} - \overset{L}{2}$$

**Now Try Exercise 51.**

### EXAMPLE 7 Using the FOIL Method

Find each product.

**(a)** $(2k + 5y)(k + 3y) = \overset{F}{2k(k)} + \overset{O}{2k(3y)} + \overset{I}{5y(k)} + \overset{L}{5y(3y)}$

$= 2k^2 + 6ky + 5ky + 15y^2$

$= 2k^2 + 11ky + 15y^2$

**(b)** $(7p + 2q)(3p - q) = 21p^2 - pq - 2q^2$ FOIL

**(c)** $2x^2(x - 3)(3x + 4) = 2x^2(3x^2 - 5x - 12)$ FOIL

$= 6x^4 - 10x^3 - 24x^2$ Distributive property

**Now Try Exercises 53 and 55.**

> **NOTE** Example 7(c) showed one way to multiply three polynomials. We could have multiplied $2x^2$ and $x - 3$ first, then multiplied that product and $3x + 4$ as follows.
>
> $2x^2(x - 3)(3x + 4) = (2x^3 - 6x^2)(3x + 4)$
>
> $= 6x^4 - 10x^3 - 24x^2$ FOIL

## 4.5 EXERCISES

**For Extra Help**

- Student's Solutions Manual
- MyMathLab
- InterAct Math Tutorial Software
- AW Math Tutor Center
- MathXL
- Digital Video Tutor CD 7/Videotape 7

**1.** Match each product in Column I with the correct monomial in Column II.

| I | II |
|---|---|
| **(a)** $(5x^3)(6x^5)$ | **A.** $125x^{15}$ |
| **(b)** $(-5x^5)(6x^3)$ | **B.** $30x^8$ |
| **(c)** $(5x^5)^3$ | **C.** $-216x^9$ |
| **(d)** $(-6x^3)^3$ | **D.** $-30x^8$ |

**2.** Match each product in Column I with the correct polynomial in Column II.

| I | II |
|---|---|
| **(a)** $(x - 5)(x + 3)$ | **A.** $x^2 + 8x + 15$ |
| **(b)** $(x + 5)(x + 3)$ | **B.** $x^2 - 8x + 15$ |
| **(c)** $(x - 5)(x - 3)$ | **C.** $x^2 - 2x - 15$ |
| **(d)** $(x + 5)(x - 3)$ | **D.** $x^2 + 2x - 15$ |

*Find each product. Use the rules for exponents discussed earlier in the chapter.*

**3.** $(5y^4)(3y^7)$
**4.** $(10p^2)(5p^3)$
**5.** $(-15a^4)(-2a^5)$
**6.** $(-3m^6)(-5m^4)$
**7.** $(5p)(3q^2)$
**8.** $(4a^3)(3b^2)$
**9.** $(-6m^3)(3n^2)$
**10.** $(9r^3)(-2s^2)$

*Find each product. See Example 1.*

**11.** $2m(3m + 2)$
**12.** $4x(5x + 3)$
**13.** $3p(-2p^3 + 4p^2)$
**14.** $4x(3 + 2x + 5x^3)$
**15.** $-8z(2z + 3z^2 + 3z^3)$
**16.** $-7y(3 + 5y^2 - 2y^3)$
**17.** $2y^3(3 + 2y + 5y^4)$
**18.** $2m^4(3m^2 + 5m + 6)$
**19.** $-4r^3(-7r^2 + 8r - 9)$
**20.** $-9a^5(-3a^6 - 2a^4 + 8a^2)$
**21.** $3a^2(2a^2 - 4ab + 5b^2)$
**22.** $4z^3(8z^2 + 5zy - 3y^2)$
**23.** $7m^3n^2(3m^2 + 2mn - n^3)$
**24.** $2p^2q(3p^2q^2 - 5p + 2q^2)$

*Find each product. See Examples 2–4.*

**25.** $(6x + 1)(2x^2 + 4x + 1)$
**26.** $(9y - 2)(8y^2 - 6y + 1)$
**27.** $(9a + 2)(9a^2 + a + 1)$
**28.** $(2r - 1)(3r^2 + 4r - 4)$
**29.** $(4m + 3)(5m^3 - 4m^2 + m - 5)$
**30.** $(y + 4)(3y^4 - 2y^2 + 1)$
**31.** $(2x - 1)(3x^5 - 2x^3 + x^2 - 2x + 3)$
**32.** $(2a + 3)(a^4 - a^3 + a^2 - a + 1)$
**33.** $(5x^2 + 2x + 1)(x^2 - 3x + 5)$
**34.** $(2m^2 + m - 3)(m^2 - 4m + 5)$
**35.** $(6x^4 - 4x^2 + 8x)\left(\frac{1}{2}x + 3\right)$
**36.** $(8y^6 + 4y^4 - 12y^2)\left(\frac{3}{4}y^2 + 2\right)$

*Find each product. Use the FOIL method. See Examples 5–7.*

**37.** $(m + 7)(m + 5)$
**38.** $(n - 1)(n + 4)$
**39.** $(x + 5)(x - 5)$
**40.** $(y + 8)(y - 8)$
**41.** $(2x + 3)(6x - 4)$
**42.** $(4m + 3)(4m + 3)$
**43.** $(3x - 2)(3x - 2)$
**44.** $(b + 8)(6b - 2)$
**45.** $(5a + 1)(2a + 7)$
**46.** $(8 - 3a)(2 + a)$
**47.** $(6 - 5m)(2 + 3m)$
**48.** $(-4 + k)(2 - k)$
**49.** $(5 - 3x)(4 + x)$
**50.** $(2m - 3n)(m + 5n)$
**51.** $(4x + 3)(2y - 1)$
**52.** $(5x + 7)(3y - 8)$
**53.** $(3x + 2y)(5x - 3y)$
**54.** $x(2x - 5)(x + 3)$
**55.** $3y^3(2y + 3)(y - 5)$
**56.** $5t^4(t + 3)(3t - 1)$
**57.** $-8r^3(5r^2 + 2)(5r^2 - 2)$

**58.** Find a polynomial that represents the area of this square.

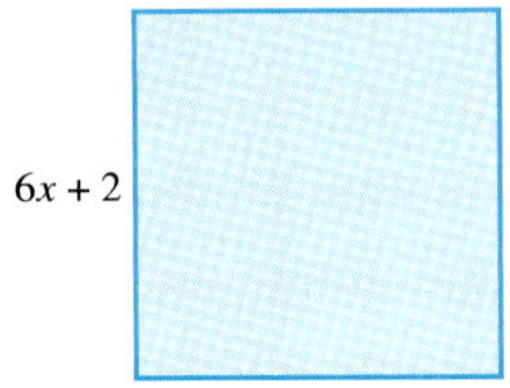


**59.** Find a polynomial that represents the area of this rectangle.

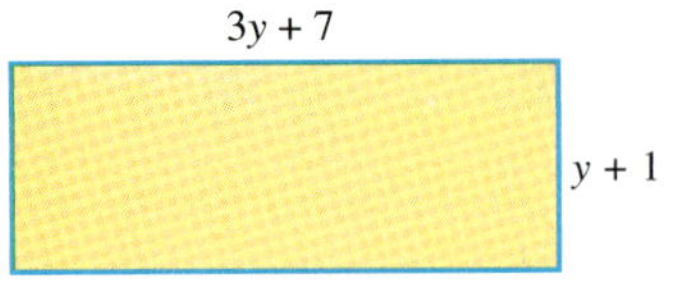

**60.** Perform the indicated multiplications, and then describe the pattern that you observe in the products.

**(a)** $(x + 4)(x - 4)$; $(y + 2)(y - 2)$; $(r + 7)(r - 7)$
**(b)** $(x + 4)(x + 4)$; $(y - 2)(y - 2)$; $(r + 7)(r + 7)$

*Find each product. In Exercises 71–74 and 78–80, apply the meaning of exponents.*

**61.** $\left(3p + \frac{5}{4}q\right)\left(2p - \frac{5}{3}q\right)$

**62.** $\left(-x + \frac{2}{3}y\right)\left(3x - \frac{3}{4}y\right)$

**63.** $(x + 7)^2$

**64.** $(m + 6)^2$

**65.** $(a - 4)(a + 4)$

**66.** $(b - 10)(b + 10)$

**67.** $(2p - 5)^2$

**68.** $(3m + 1)^2$

**69.** $(5k + 3q)^2$

**70.** $(8m - 3n)^2$

**71.** $(m - 5)^3$

**72.** $(p + 3)^3$

**73.** $(2a + 1)^3$

**74.** $(3m - 1)^3$

**75.** $7(4m - 3)(2m + 1)$

**76.** $-4r(3r + 2)(2r - 5)$

**77.** $-3a(3a + 1)(a - 4)$

**78.** $(k + 1)^4$

**79.** $(3r - 2s)^4$

**80.** $(2z + 5y)^4$

**81.** $3p^3(2p^2 + 5p)(p^3 + 2p + 1)$

**82.** $5k^2(k^2 - k + 4)(k^3 - 3)$

**83.** $-2x^5(3x^2 + 2x - 5)(4x + 2)$

**84.** $-4x^3(3x^4 + 2x^2 - x)(-2x + 1)$

*Find a polynomial that represents the area of each shaded region. In Exercises 87 and 88 leave $\pi$ in your answer. Use the formulas found on the inside covers.*

**85.**

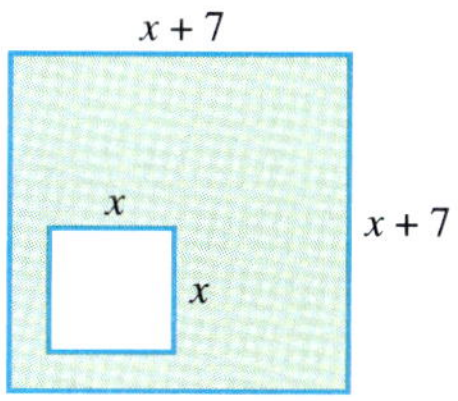


**86.**

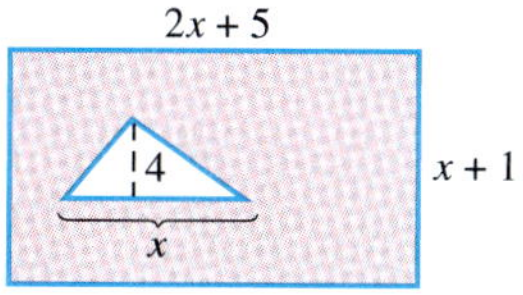


**87.**

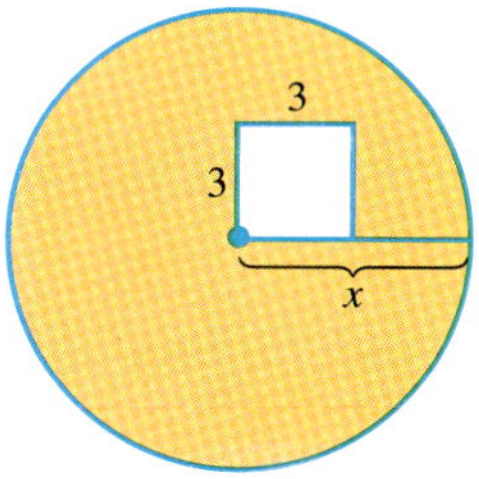


**88.**

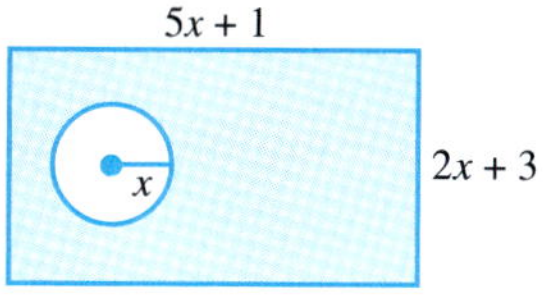


## RELATING CONCEPTS (EXERCISES 89–96)

**For Individual or Group Work**

***Work Exercises 89–96 in order.*** *Refer to the figure as necessary.*

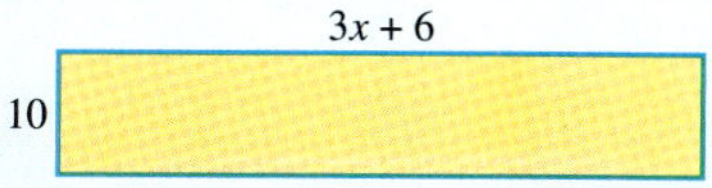


**89.** Find a polynomial that represents the area of the rectangle. (continued)

**90.** Suppose you know that the area of the rectangle is 600 $yd^2$. Use this information and the polynomial from Exercise 89 to write an equation that allows you to solve for $x$.

**91.** Solve for $x$.

**92.** What are the dimensions of the rectangle (assume units are all in yards)?

**93.** Suppose the rectangle represents a lawn and it costs \$3.50 per $yd^2$ to lay sod on the lawn. How much will it cost to sod the entire lawn?

**94.** Use the result of Exercise 92 to find the perimeter of the lawn.

**95.** Again, suppose the rectangle represents a lawn and it costs \$9.00 per yd to fence the lawn. How much will it cost to fence the lawn?

**96.** **(a)** Suppose that it costs $k$ dollars per $yd^2$ to sod the lawn. Determine a polynomial in the variables $x$ and $k$ that represents the cost to sod the entire lawn.

**(b)** Suppose that it costs $r$ dollars per yd to fence the lawn. Determine a polynomial in the variables $x$ and $r$ that represents the cost to fence the lawn.

**97.** Explain the FOIL method for multiplying two binomials. Give an example.

**98.** Why does the FOIL method not apply to the product of a binomial and a trinomial? Give an example.

# 4.6 Special Products

**OBJECTIVES**

1 Square binomials.
2 Find the product of the sum and difference of two terms.
3 Find higher powers of binomials.

In this section, we develop shortcuts to find certain binomial products that occur frequently.

**OBJECTIVE 1 Square binomials.** The square of a binomial can be found quickly by using the method suggested by Example 1.

### EXAMPLE 1 Squaring a Binomial

Find $(m + 3)^2$.

Squaring $m + 3$ by the FOIL method gives

$$(m + 3)(m + 3) = m^2 + 3m + 3m + 9$$
$$= m^2 + 6m + 9.$$

**Now Try Exercise 1.**

The result above has the squares of the first and the last terms of the binomial:

$$m^2 = m^2 \quad \text{and} \quad 3^2 = 9.$$

The middle term is twice the product of the two terms of the binomial, since the outer and inner products are $m(3)$ and $3(m)$, and

$$m(3) + 3(m) = 2(m)(3) = 6m.$$

This example suggests the following rules.

**Square of a Binomial**

The square of a binomial is a trinomial consisting of the square of the first term of the binomial, plus twice the product of the two terms, plus the square of the last term of the binomial. For $x$ and $y$,

$$(x + y)^2 = x^2 + 2xy + y^2.$$

Also,

$$(x - y)^2 = x^2 - 2xy + y^2.$$

**EXAMPLE 2 Squaring Binomials**

Use the rules to square each binomial.

$$(x - y)^2 = x^2 - 2 \cdot x \cdot y + y^2$$

**(a)** $(5z - 1)^2 = (5z)^2 - 2(5z)(1) + (1)^2$

$= 25z^2 - 10z + 1$ $\quad (5z)^2 = 5^2z^2 = 25z^2$

**(b)** $(3b + 5r)^2 = (3b)^2 + 2(3b)(5r) + (5r)^2$

$= 9b^2 + 30br + 25r^2$

**(c)** $(2a - 9x)^2 = 4a^2 - 36ax + 81x^2$

**(d)** $\left(4m + \frac{1}{2}\right)^2 = (4m)^2 + 2(4m)\left(\frac{1}{2}\right) + \left(\frac{1}{2}\right)^2$

$= 16m^2 + 4m + \frac{1}{4}$

**Now Try Exercises 5, 7, 9, and 15.**

Notice that in the square of a sum all of the terms are positive, as in Examples 2(b) and (d). In the square of a difference, the middle term is negative, as in Examples 2(a) and (c).

**CAUTION** A common error when squaring a binomial is to forget the middle term of the product. In general,

$$(x + y)^2 \neq x^2 + y^2.$$

**OBJECTIVE 2 Find the product of the sum and difference of two terms.** Binomial products of the form $(x + y)(x - y)$ also occur frequently. In these products, one binomial is the sum of two terms, and the other is the difference of the same two terms. For example, the product of $x + 2$ and $x - 2$ is

$$(x + 2)(x - 2) = x^2 - 2x + 2x - 4$$
$$= x^2 - 4.$$

As this example suggests, the product of $x + y$ and $x - y$ is the difference between two squares.

**Product of the Sum and Difference of Two Terms**

$$(x + y)(x - y) = x^2 - y^2$$

# CHAPTER 4 SUMMARY

## KEY TERMS

4.3 scientific notation
4.4 polynomial
descending powers
degree of a term
degree of a polynomial
trinomial
binomial
monomial
parabola
vertex
axis
line of symmetry
4.5 outer product
inner product
FOIL

## NEW SYMBOLS

$x^{-n}$ $x$ to the negative $n$ power

## TEST YOUR WORD POWER

*See how well you have learned the vocabulary in this chapter. Answers, with examples, follow the Quick Review.*

1. A **polynomial** is an algebraic expression made up of
   **A.** a term or a finite product of terms with positive coefficients and exponents
   **B.** a term or a finite sum of terms with real coefficients and whole number exponents
   **C.** the product of two or more terms with positive exponents
   **D.** the sum of two or more terms with whole number coefficients and exponents.

2. The **degree of a term** is
   **A.** the number of variables in the term
   **B.** the product of the exponents on the variables
   **C.** the smallest exponent on the variables
   **D.** the sum of the exponents on the variables.

3. A **trinomial** is a polynomial with
   **A.** only one term
   **B.** exactly two terms
   **C.** exactly three terms
   **D.** more than three terms.

4. A **binomial** is a polynomial with
   **A.** only one term
   **B.** exactly two terms
   **C.** exactly three terms
   **D.** more than three terms.

5. A **monomial** is a polynomial with
   **A.** only one term
   **B.** exactly two terms
   **C.** exactly three terms
   **D.** more than three terms.

6. **FOIL** is a method for
   **A.** adding two binomials
   **B.** adding two trinomials
   **C.** multiplying two binomials
   **D.** multiplying two trinomials.

## QUICK REVIEW

| CONCEPTS | EXAMPLES |
|---|---|
| **4.1 THE PRODUCT RULE AND POWER RULES FOR EXPONENTS** | |
| For any integers $m$ and $n$ with no denominators zero: | Perform the operations by using rules for exponents. |
| **Product Rule** $a^m \cdot a^n = a^{m+n}$ | $2^4 \cdot 2^5 = 2^9$ |
| **Power Rules** (a) $(a^m)^n = a^{mn}$ | $(3^4)^2 = 3^8$ |
| (b) $(ab)^m = a^m b^m$ | $(6a)^5 = 6^5 a^5$ |
| (c) $\left(\frac{a}{b}\right)^m = \frac{a^m}{b^m}$ $(b \neq 0)$ | $\left(\frac{2}{3}\right)^4 = \frac{2^4}{3^4}$ |

| CONCEPTS | EXAMPLES |
|---|---|

## 4.2 INTEGER EXPONENTS AND THE QUOTIENT RULE

If $a \neq 0$, for integers $m$ and $n$:

**Zero Exponent** $a^0 = 1$

**Negative Exponent** $a^{-n} = \dfrac{1}{a^n}$

**Quotient Rule** $\dfrac{a^m}{a^n} = a^{m-n}$

**Negative to Positive Rules**

$$\frac{a^{-m}}{b^{-n}} = \frac{b^n}{a^m} \quad (b \neq 0)$$

$$\left(\frac{a}{b}\right)^{-m} = \left(\frac{b}{a}\right)^m \quad (b \neq 0)$$

Simplify by using the rules for exponents.

$$15^0 = 1$$

$$5^{-2} = \frac{1}{5^2} = \frac{1}{25}$$

$$\frac{4^8}{4^3} = 4^5$$

$$\frac{4^{-2}}{3^{-5}} = \frac{3^5}{4^2}$$

$$\left(\frac{6}{5}\right)^{-3} = \left(\frac{5}{6}\right)^3$$

## 4.3 AN APPLICATION OF EXPONENTS: SCIENTIFIC NOTATION

To write a number in scientific notation (as $a \times 10^n$), move the decimal point to follow the first nonzero digit. If moving the decimal point makes the number smaller, $n$ is positive. If it makes the number larger, $n$ is negative. If the decimal point is not moved, $n$ is 0.

Write in scientific notation.

$$247 = 2.47 \times 10^2$$

$$.0051 = 5.1 \times 10^{-3}$$

$$4.8 = 4.8 \times 10^0$$

Write without exponents.

$$3.25 \times 10^5 = 325{,}000$$

$$8.44 \times 10^{-6} = .00000844$$

## 4.4 ADDING AND SUBTRACTING POLYNOMIALS; GRAPHING SIMPLE POLYNOMIALS

**Adding Polynomials**
Add like terms.

Add.

$$\begin{array}{r} 2x^2 + 5x - 3 \\ \underline{5x^2 - 2x + 7} \\ 7x^2 + 3x + 4 \end{array}$$

**Subtracting Polynomials**
Change the signs of the terms in the second polynomial and add to the first polynomial.

Subtract.

$$\begin{aligned} &(2x^2 + 5x - 3) - (5x^2 - 2x + 7) \\ &= (2x^2 + 5x - 3) + (-5x^2 + 2x - 7) \\ &= -3x^2 + 7x - 10 \end{aligned}$$

**Graphing Simple Polynomials**
To graph a simple polynomial equation such as $y = x^2 - 2$, plot points near the vertex. (In this chapter, all parabolas have a vertex on the $x$-axis or the $y$-axis.)

Graph $y = x^2 - 2$.

| $x$ | $y$ |
|---|---|
| $-2$ | 2 |
| $-1$ | $-1$ |
| 0 | $-2$ |
| 1 | $-1$ |
| 2 | 2 |

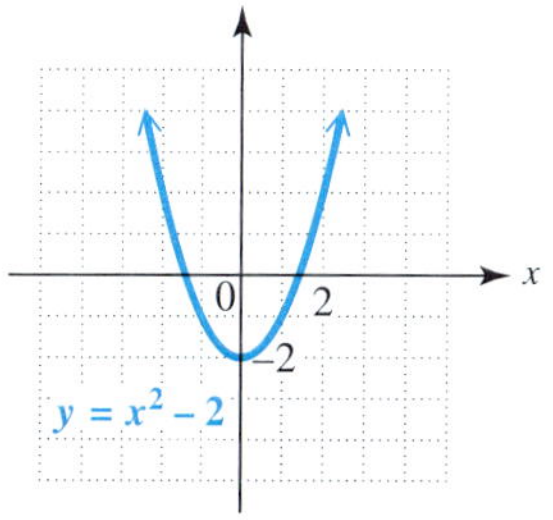


(continued)

| CONCEPTS | EXAMPLES |
|---|---|

## 4.5 MULTIPLYING POLYNOMIALS

**General Method for Multiplying Polynomials**

Multiply each term of the first polynomial by each term of the second polynomial. Then add like terms.

Multiply.

$$\begin{array}{rrrrr} & 3x^3 & -\ 4x^2 & +\ 2x & -\ 7 \\ & & & 4x & +\ 3 \\ \hline & 9x^3 & -\ 12x^2 & +\ 6x & -\ 21 \\ 12x^4 & -\ 16x^3 & +\ 8x^2 & -\ 28x & \\ \hline 12x^4 & -\ 7x^3 & -\ 4x^2 & -\ 22x & -\ 21 \end{array}$$

**FOIL Method for Multiplying Binomials**

*Step 1* Multiply the two first terms to get the first term of the answer.

*Step 2* Find the outer product and the inner product and mentally add them, when possible, to get the middle term of the answer.

*Step 3* Multiply the two last terms to get the last term of the answer.

Multiply. $(2x + 3)(5x - 4)$

$$2x(5x) = 10x^2$$

$$2x(-4) + 3(5x) = 7x$$

$$3(-4) = -12$$

The product of $(2x + 3)$ and $(5x - 4)$ is $10x^2 + 7x - 12$.

## 4.6 SPECIAL PRODUCTS

**Square of a Binomial**

$$(x + y)^2 = x^2 + 2xy + y^2$$
$$(x - y)^2 = x^2 - 2xy + y^2$$

Multiply.

$$(3x + 1)^2 = 9x^2 + 6x + 1$$
$$(2m - 5n)^2 = 4m^2 - 20mn + 25n^2$$

**Product of the Sum and Difference of Two Terms**

$$(x + y)(x - y) = x^2 - y^2$$

$$(4a + 3)(4a - 3) = 16a^2 - 9$$

## 4.7 DIVIDING POLYNOMIALS

**Dividing a Polynomial by a Monomial**

Divide each term of the polynomial by the monomial.

$$\frac{a + b}{c} = \frac{a}{c} + \frac{b}{c}$$

Divide.

$$\frac{4x^3 - 2x^2 + 6x - 8}{2x} = 2x^2 - x + 3 - \frac{4}{x}$$

**Dividing a Polynomial by a Polynomial**

Use "long division."

Divide.

$$\begin{array}{r} 2x - 5 \\ 3x + 4 \overline{)6x^2 - 7x - 21} \\ \underline{6x^2 + 8x} \phantom{- 21} \\ -15x - 21 \\ \underline{-15x - 20} \\ -1 \end{array} \leftarrow \text{Remainder}$$

The final answer is $2x - 5 + \dfrac{-1}{3x + 4}$.

**Answers to Test Your Word Power**

1. B; *Example:* $5x^3 + 2x^2 - 7$ 2. D; *Examples:* The term 6 has degree 0, $3x$ has degree 1, $-2x^8$ has degree 8, and $5x^2y^4$ has degree 6.
3. C; *Example:* $2a^2 - 3ab + b^2$ 4. B; *Example:* $3t^3 + 5t$ 5. A; *Examples:* $-5$ and $4xy^5$
6. C; *Example:* $(m + 4)(m - 3) = \overset{F}{m(m)} \overset{O}{- 3m} \overset{I}{+ 4m} \overset{L}{+ 4(-3)} = m^2 + m - 12$

# CHAPTER 4 REVIEW EXERCISES

[4.1] *Use the product rule, power rules, or both to simplify each expression. Write the answers in exponential form.*

**1.** $4^3 \cdot 4^8$

**2.** $(-5)^6(-5)^5$

**3.** $(-8x^4)(9x^3)$

**4.** $(2x^2)(5x^3)(x^9)$

**5.** $(19x)^5$

**6.** $(-4y)^7$

**7.** $5(pt)^4$

**8.** $\left(\frac{7}{5}\right)^6$

**9.** $(6x^2z^4)^2(x^3yz^2)^4$

**10.** $\left(\frac{2m^3n}{p^2}\right)^3$

**11.** Why does the product rule for exponents not apply to the expression $7^2 + 7^4$?

[4.2] *Evaluate each expression.*

**12.** $6^0 + (-6)^0$

**13.** $(-23)^0 - (-23)^0$

**14.** $-10^0$

*Write each expression using only positive exponents.*

**15.** $-7^{-2}$

**16.** $\left(\frac{5}{8}\right)^{-2}$

**17.** $(5^{-2})^{-4}$

**18.** $9^3 \cdot 9^{-5}$

**19.** $2^{-1} + 4^{-1}$

**20.** $\frac{6^{-5}}{6^{-3}}$

*Simplify. Write each answer with only positive exponents. Assume that all variables represent nonzero real numbers.*

**21.** $\frac{x^{-7}}{x^{-9}}$

**22.** $\frac{y^4 \cdot y^{-2}}{y^{-5}}$

**23.** $(3r^{-2})^{-4}$

**24.** $(3p)^4(3p^{-7})$

**25.** $\frac{ab^{-3}}{a^4b^2}$

**26.** $\frac{(6r^{-1})^2(2r^{-4})}{r^{-5}(r^2)^{-3}}$

[4.3] *Write each number in scientific notation.*

**27.** 48,000,000

**28.** 28,988,000,000

**29.** .0000000824

*Write each number without exponents.*

**30.** $2.4 \times 10^4$

**31.** $7.83 \times 10^7$

**32.** $8.97 \times 10^{-7}$

*Perform each indicated operation and write the answer without exponents.*

**33.** $(2 \times 10^{-3}) \times (4 \times 10^5)$

**34.** $\frac{8 \times 10^4}{2 \times 10^{-2}}$

**35.** $\frac{12 \times 10^{-5} \times 5 \times 10^4}{4 \times 10^3 \times 6 \times 10^{-2}}$

*Each quote is taken from the source cited. Write each number that is in scientific notation in the quote without exponents.*

**36.** The muon, a close relative of the electron produced by the bombardment of cosmic rays against the upper atmosphere, has a half-life of 2 millionths of a second ($2 \times 10^{-6}$ s). (Excerpt from *Conceptual Physics,* 6th edition, by Paul G. Hewitt. Copyright © by Paul G. Hewitt. Published by HarperCollins College Publishers.)

**37.** There are 13 red balls and 39 black balls in a box. Mix them up and draw 13 out one at a time without returning any ball . . . the probability that the 13 drawings each will produce a red ball is . . . $1.6 \times 10^{-12}$. (Weaver, Warren, *Lady Luck,* Doubleday, 1963, pp. 298–299.)

*Each quote is taken from the source cited. Write each number in scientific notation.*

**38.** An electron and a positron attract each other in two ways: the electromagnetic attraction of their opposite electric charges, and the gravitational attraction of their two masses. The electromagnetic attraction is

4,200,000,000,000,000,000,000,000,000,000,000,000,000,000,000

times as strong as the gravitational. (Asimov, Isaac, *Isaac Asimov's Book of Facts,* Bell Publishing Company, 1981, p. 106.)

**39.** The aircraft carrier USS John Stennis is a 97,000-ton nuclear powered floating city with a crew of 5000. (*Source:* Seelye, Katharine Q., "Staunch Allies Hard to Beat: Defense Dept., Hollywood," *New York Times,* in *Plain Dealer,* June 10, 2002.)

**40.** A googol is

10,000,000,000,000,000,000,000,000,000,000,000,000,000,000,000,000,
000,000,000,000,000,000,000,000,000,000,000,000,000,000,000,000,000.

The Web search engine Google is named after a googol. Sergey Brin, president and co-founder of Google, Inc., was a math major. He chose the name Google to describe the vast reach of this search engine. (*Source: The Gazette,* March 2, 2001.)

**41.** According to Campbell, Mitchell, and Reece in *Biology Concepts and Connections* (Benjamin Cummings, 1994, p. 230), "The amount of DNA in a human cell is about 1000 times greater than the DNA in *E. coli.* Does this mean humans have 1000 times as many genes as the 2000 in *E. coli*? The answer is probably no; the human genome is thought to carry between 50,000 and 100,000 genes, which code for various proteins (as well as for tRNA and rRNA)."

Write each number from this quote using scientific notation.

**(a)** 1000 **(b)** 2000 **(c)** 50,000 **(d)** 100,000

[4.4] *Combine like terms where possible in each polynomial. Write the answer in descending powers of the variable. Give the degree of the answer. Identify the polynomial as a* monomial, binomial, trinomial, *or* none of these.

**42.** $9m^2 + 11m^2 + 2m^2$

**43.** $-4p + p^3 - p^2 + 8p + 2$

**44.** $12a^5 - 9a^4 + 8a^3 + 2a^2 - a + 3$

**45.** $-7y^5 - 8y^4 - y^5 + y^4 + 9y$

**46.** $(12r^4 - 7r^3 + 2r^2) - (5r^4 - 3r^3 + 2r^2 - 1)$

**47.** Simplify $(5x^3y^2 - 3xy^5 + 12x^2) - (-9x^2 - 8x^3y^2 + 2xy^5)$.

*Add or subtract as indicated.*

**48.** Add.

$$\begin{array}{r} -2a^3 + 5a^2 \\ \underline{3a^3 - \phantom{5}a^2} \end{array}$$

**49.** Subtract.

$$\begin{array}{r} 6y^2 - 8y + 2 \\ \underline{5y^2 + 2y - 7} \end{array}$$

**50.** Subtract.

$$\begin{array}{r} -12k^4 - 8k^2 + \phantom{1}7k \\ \underline{k^4 + 7k^2 - 11k} \end{array}$$

*Graph each equation by completing the table of values.*

**51.** $y = -x^2 + 5$

| $x$ | $-2$ | $-1$ | $0$ | $1$ | $2$ |
|---|---|---|---|---|---|
| $y$ | | | | | |

**52.** $y = 3x^2 - 2$

| $x$ | $-2$ | $-1$ | $0$ | $1$ | $2$ |
|---|---|---|---|---|---|
| $y$ | | | | | |

**[4.5]** *Find each product.*

**53.** $(a + 2)(a^2 - 4a + 1)$

**54.** $(3r - 2)(2r^2 + 4r - 3)$

**55.** $(5p^2 + 3p)(p^3 - p^2 + 5)$

**56.** $(m - 9)(m + 2)$

**57.** $(3k - 6)(2k + 1)$

**58.** $(a + 3b)(2a - b)$

**59.** $(6k + 5q)(2k - 7q)$

**60.** $(s - 1)^3$

**61.** Find a polynomial that represents the area of the rectangle shown.

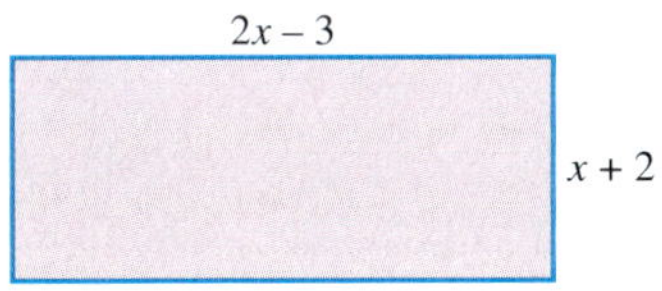


**62.** If the side of a square has a measure represented by $5x^4 + 2x^2$, what polynomial represents its area?

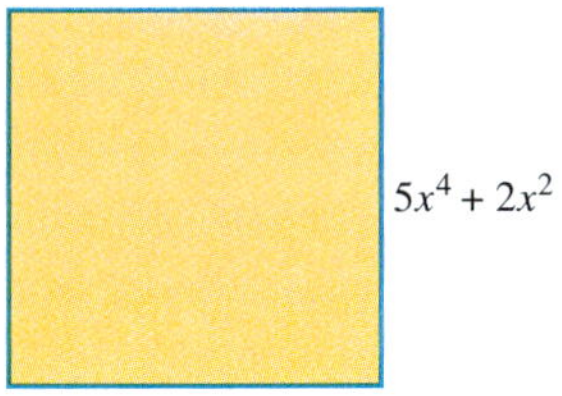


**[4.6]** *Find each product.*

**63.** $(a + 4)^2$

**64.** $(2r + 5t)^2$

**65.** $(6m - 5)(6m + 5)$

**66.** $(5a + 6b)(5a - 6b)$

**67.** $(r + 2)^3$

**68.** $t(5t - 3)^2$

**69.** Choose values for $x$ and $y$ to show that, in general, the following hold true.

**(a)** $(x + y)^2 \neq x^2 + y^2$ **(b)** $(x + y)^3 \neq x^3 + y^3$

**70.** Write an explanation on how to raise a binomial to the third power. Give an example.

**71.** Refer to Exercise 69. Suppose that you happened to let $x = 0$ and $y = 1$. Would your results be sufficient to illustrate the truth, in general, of the inequalities shown? If not, what would you need to do as your next step in working the exercise?

**72.** What is the volume of a cube with one side having length $x^2 + 2$ cm?

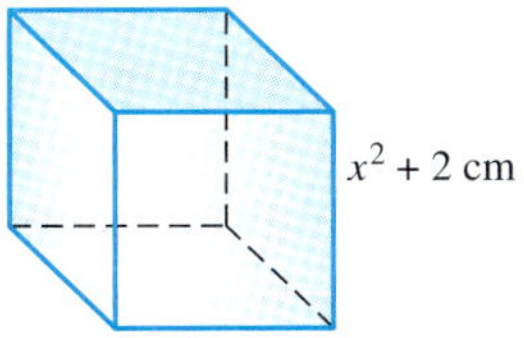


A cube

**73.** What is the volume of a sphere with radius $x + 1$ in.?

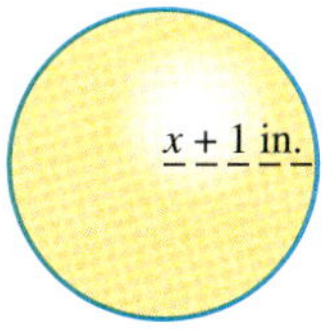


A sphere

[4.7] *Perform each division.*

**74.** $\dfrac{-15y^4}{9y^2}$

**75.** $\dfrac{6y^4 - 12y^2 + 18y}{6y}$

**76.** $(-10m^4n^2 + 5m^3n^2 + 6m^2n^4) \div (5m^2n)$

**77.** What polynomial, when multiplied by $6m^2n$, gives the product $12m^3n^2 + 18m^6n^3 - 24m^2n^2$?

**78.** One of your friends in class simplified $\dfrac{6x^2 - 12x}{6}$ as $x^2 - 12x$. Is this correct? If not, what error did your friend make and how would you explain the correct method of performing the division?

*Perform each division.*

**79.** $\dfrac{2r^2 + 3r - 14}{r - 2}$

**80.** $\dfrac{10a^3 + 9a^2 - 14a + 9}{5a - 3}$

**81.** $\dfrac{x^4 - 5x^2 + 3x^3 - 3x + 4}{x^2 - 1}$

**82.** $\dfrac{m^4 + 4m^3 - 12m - 5m^2 + 6}{m^2 - 3}$

**83.** $\dfrac{16x^2 - 25}{4x + 5}$

**84.** $\dfrac{25y^2 - 100}{5y + 10}$

**85.** $\dfrac{y^3 - 8}{y - 2}$

**86.** $\dfrac{1000x^6 + 1}{10x^2 + 1}$

**87.** $\dfrac{6y^4 - 15y^3 + 14y^2 - 5y - 1}{3y^2 + 1}$

**88.** $\dfrac{4x^5 - 8x^4 - 3x^3 + 22x^2 - 15}{4x^2 - 3}$

## MIXED REVIEW EXERCISES

*Perform each indicated operation. Write with positive exponents only. Assume all variables represent nonzero real numbers.*

**89.** $5^0 + 7^0$

**90.** $\left(\dfrac{6r^2p}{5}\right)^3$

**91.** $(12a + 1)(12a - 1)$

**92.** $2^{-4}$

**93.** $(8^{-3})^4$

**94.** $\dfrac{2p^3 - 6p^2 + 5p}{2p^2}$

**95.** $\dfrac{(2m^{-5})(3m^2)^{-1}}{m^{-2}(m^{-1})^2}$

**96.** $(3k - 6)(2k^2 + 4k + 1)$

**97.** $\dfrac{r^9 \cdot r^{-5}}{r^{-2} \cdot r^{-7}}$

**98.** $(2r + 5s)^2$

**99.** $(-5y^2 + 3y - 11) + (4y^2 - 7y + 15)$

**100.** $(2r + 5)(5r - 2)$

**101.** $\dfrac{2y^3 + 17y^2 + 37y + 7}{2y + 7}$

**102.** $(25x^2y^3 - 8xy^2 + 15x^3y) \div (10x^2y^3)$

**103.** $(6p^2 - p - 8) - (-4p^2 + 2p - 3)$

**104.** $\dfrac{5^8}{5^{19}}$

**105.** $(-7 + 2k)^2$

**106.** $\left(\dfrac{x}{y^{-3}}\right)^{-4}$

# CHAPTER 4 TEST

*Evaluate each expression.*

1. $5^{-4}$
2. $(-3)^0 + 4^0$
3. $4^{-1} + 3^{-1}$
4. Use the rules for exponents to simplify $\frac{(3x^2y)^2(xy^3)^2}{(xy)^3}$. Assume $x$ and $y$ are nonzero.

*Simplify, and write the answer using only positive exponents. Assume that variables represent nonzero numbers.*

5. $\frac{8^{-1} \cdot 8^4}{8^{-2}}$
6. $\frac{(x^{-3})^{-2}(x^{-1}y)^2}{(xy^{-2})^2}$
7. Determine whether each expression represents a number that is *positive, negative,* or *zero.*

   **(a)** $3^{-4}$ **(b)** $(-3)^4$ **(c)** $-3^4$ **(d)** $3^0$ **(e)** $(-3)^0 - 3^0$ **(f)** $(-3)^{-3}$

8. **(a)** Write 45,000,000,000 using scientific notation.
   **(b)** Write $3.6 \times 10^{-6}$ without using exponents.
   **(c)** Write the quotient without using exponents: $\frac{9.5 \times 10^{-1}}{5 \times 10^3}$.

9. A satellite galaxy of our own Milky Way, known as the Large Magellanic Cloud, is 1000 light-years across. A *light-year* is equal to 5,890,000,000,000 mi. (*Source:* "Images of Brightest Nebula Unveiled," *USA Today,* June 12, 2002.)

   **(a)** Write these two numbers using scientific notation.
   **(b)** How many miles across is the Large Magellanic Cloud?

*For each polynomial, combine like terms when possible and write the polynomial in descending powers of the variable. Give the degree of the simplified polynomial. Decide whether the simplified polynomial is a* monomial, binomial, trinomial, *or* none of these.

10. $5x^2 + 8x - 12x^2$
11. $13n^3 - n^2 + n^4 + 3n^4 - 9n^2$
12. Use the table to complete a set of ordered pairs that lie on the graph of $y = 2x^2 - 4$. Then graph the equation.

| $x$ | $-2$ | $-1$ | $0$ | $1$ | $2$ |
|---|---|---|---|---|---|
| $y$ | | | | | |

*Perform each indicated operation.*

13. $(2y^2 - 8y + 8) + (-3y^2 + 2y + 3) - (y^2 + 3y - 6)$
14. $(-9a^3b^2 + 13ab^5 + 5a^2b^2) - (6ab^5 + 12a^3b^2 + 10a^2b^2)$
15. Subtract.

$$\begin{array}{r} 9t^3 - 4t^2 + 2t + 2 \\ \underline{9t^3 + 8t^2 - 3t - 6} \end{array}$$

16. $3x^2(-9x^3 + 6x^2 - 2x + 1)$

**17.** $(t - 8)(t + 3)$

**18.** $(4x + 3y)(2x - y)$

**19.** $(5x - 2y)^2$

**20.** $(10v + 3w)(10v - 3w)$

**21.** $(2r - 3)(r^2 + 2r - 5)$

**22.** What polynomial expression represents the area of this square?

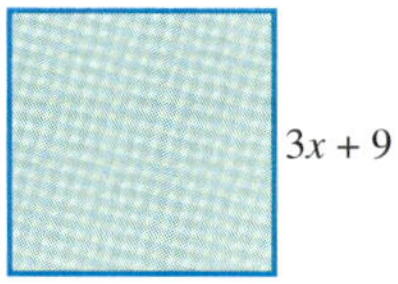


*Perform each division.*

**23.** $\dfrac{8y^3 - 6y^2 + 4y + 10}{2y}$

**24.** $(-9x^2y^3 + 6x^4y^3 + 12xy^3) \div (3xy)$

**25.** $(3x^3 - x + 4) \div (x - 2)$

## CUMULATIVE REVIEW EXERCISES CHAPTERS 1–4

*Write each fraction in lowest terms.*

**1.** $\dfrac{28}{16}$

**2.** $\dfrac{55}{11}$

*Perform each operation.*

**3.** $\dfrac{2}{3} + \dfrac{1}{8}$

**4.** $\dfrac{7}{4} - \dfrac{9}{5}$

**5.** A contractor installs toolsheds. Each requires $1\frac{1}{4}$ $yd^3$ of concrete. How much concrete would be needed for 25 sheds?

**6.** A retailer has \$34,000 invested in her business. She finds that last year she earned 5.4% on this investment. How much did she earn?

**7.** List all positive integer factors of 45.

**8.** Find the value of

$$\frac{4x - 2y}{x + y}$$

if $x = -2$ and $y = 4$.

*Perform each indicated operation.*

**9.** $\dfrac{(-13 + 15) - (3 + 2)}{6 - 12}$

**10.** $-7 - 3[2 + (5 - 8)]$

*Decide which property justifies each statement.*

**11.** $(9 + 2) + 3 = 9 + (2 + 3)$

**12.** $6(4 + 2) = 6(4) + 6(2)$

**13.** Simplify the expression $-3(2x^2 - 8x + 9) - (4x^2 + 3x + 2)$.

*Solve each equation.*

**14.** $2 - 3(t - 5) = 4 + t$

**15.** $2(5h + 1) = 10h + 4$

**16.** $d = rt$ for $r$

**17.** $\frac{x}{5} = \frac{x - 2}{7}$

**18.** $\frac{1}{3}p - \frac{1}{6}p = -2$

**19.** $.05x + .15(50 - x) = 5.50$

**20.** $4 - (3x + 12) = (2x - 9) - (5x - 1)$

*Solve each problem.*

**21.** A 1-oz mouse takes about 16 times as many breaths as does a 3-ton elephant. If the two animals take a combined total of 170 breaths per minute, how many breaths does each take during that time period? (*Source:* McGowan, Christopher, *Dinosaurs, Spitfires, and Sea Dragons,* Harvard University Press, 1991.)

**22.** If a number is subtracted from 8 and this difference is tripled, the result is 3 times the number. Find this number, and you will learn how many times a dolphin rests during a 24-hr period.

**23.** One side of a triangle is twice as long as a second side. The third side of the triangle is 17 ft long. The perimeter of the triangle cannot be more than 50 ft. Find the longest possible values for the other two sides of the triangle.

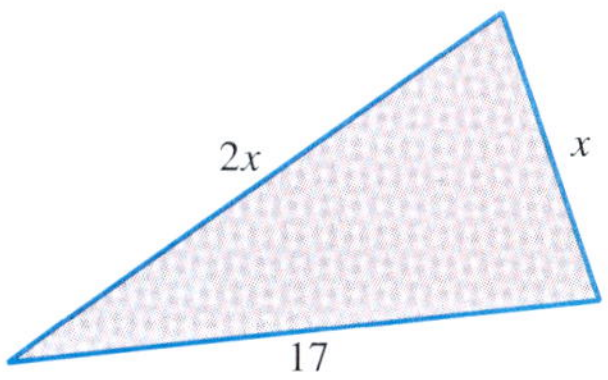


*Solve each inequality.*

**24.** $-8x \leq -80$

**25.** $-2(x + 4) > 3x + 6$

**26.** $-3 \leq 2x + 5 < 9$

**27.** Graph $y = -3x + 6$.

**28.** Consider the two points $(-1, 5)$ and $(2, 8)$.

**(a)** Find the slope of the line joining them.
**(b)** Find the equation of the line joining them.

*Evaluate each expression.*

**29.** $4^{-1} + 3^0$

**30.** $2^{-4} \cdot 2^5$

**31.** $\frac{8^{-5} \cdot 8^7}{8^2}$

**32.** Write with positive exponents only: $\frac{(a^{-3}b^2)^2}{(2a^{-4}b^{-3})^{-1}}$.

**33.** Write in scientific notation: 34,500.

**34.** Write without exponents: $5.36 \times 10^{-7}$.

**35.** It takes about $3.6 \times 10^1$ sec at a speed of $3.0 \times 10^5$ km per sec for light from the sun to reach Venus. How far is Venus from the sun? (*Source: World Almanac and Book of Facts,* 2000.)

**36.** Graph $y = (x + 4)^2$, using the $x$-values $-6, -5, -4, -3$, and $-2$ to obtain a set of points.

*Perform each indicated operation.*

**37.** $(7x^3 - 12x^2 - 3x + 8) + (6x^2 + 4) - (-4x^3 + 8x^2 - 2x - 2)$

**38.** $6x^5(3x^2 - 9x + 10)$

**39.** $(7x + 4)(9x + 3)$

**40.** $(5x + 8)^2$

**41.** $\dfrac{14x^3 - 21x^2 + 7x}{7x}$

**42.** $\dfrac{y^3 - 3y^2 + 8y - 6}{y - 1}$

# Factoring and Applications

# 5

Galileo Galilei, born near Pisa, Italy, in 1564, became a professor of mathematics at the University of Pisa at age 25. He and his students conducted experiments involving the famous Leaning Tower to investigate the relationship between an object's speed of fall and its weight. (*Source: Microsoft Encarta Encyclopedia 2002.*) We use the concepts of this chapter and the formula Galileo developed from his experiments in Section 5.6.

# 5.1 The Greatest Common Factor; Factoring by Grouping

**OBJECTIVES**

1. Find the greatest common factor of a list of terms.
2. Factor out the greatest common factor.
3. Factor by grouping.

Recall from Chapter 1 that to **factor** means to write a quantity as a product. That is, factoring is the opposite of multiplying. For example,

| *Multiplying* | *Factoring* |
|---|---|
| $6 \cdot 2 = 12,$ | $12 = 6 \cdot 2.$ |
| ↑ ↑ ↑ | ↑ ↑ ↑ |
| Factors Product | Product Factors |

Other **factored forms** of 12 are

$$-6(-2), \quad 3 \cdot 4, \quad -3(-4), \quad 12 \cdot 1, \quad \text{and} \quad -12(-1).$$

More than two factors may be used, so another factored form of 12 is $2 \cdot 2 \cdot 3$. The positive integer factors of 12 are

$$1, 2, 3, 4, 6, 12.$$

**OBJECTIVE 1 Find the greatest common factor of a list of terms.** An integer that is a factor of two or more integers is called a **common factor** of those integers. For example, 6 is a common factor of 18 and 24 since 6 is a factor of both 18 and 24. Other common factors of 18 and 24 are 1, 2, and 3. The **greatest common factor (GCF)** of a list of integers is the largest common factor of those integers. Thus, 6 is the greatest common factor of 18 and 24, since it is the largest of their common factors.

**NOTE** Factors of a number are also divisors of the number. The greatest common factor is actually the same as the greatest common divisor. There are many rules for deciding what numbers divide into a given number. Here are some especially useful divisibility rules for small numbers. It is surprising how many people do not know them.

| *A Whole Number Divisible by:* | *Must Have the Following Property:* |
|---|---|
| 2 | Ends in 0, 2, 4, 6, or 8 |
| 3 | Sum of its digits is divisible by 3. |
| 4 | Last two digits form a number divisible by 4 |
| 5 | Ends in 0 or 5 |
| 6 | Divisible by both 2 and 3 |
| 8 | Last three digits form a number divisible by 8 |
| 9 | Sum of its digits is divisible by 9. |
| 10 | Ends in 0 |

Recall from Chapter 1 that a prime number has only itself and 1 as factors. In Section 1.1 we factored numbers into prime factors. This is the first step in finding the greatest common factor of a list of numbers. We find the greatest common factor (GCF) of a list of numbers as follows.

**Finding the Greatest Common Factor (GCF)**

*Step 1* **Factor.** Write each number in prime factored form.

*Step 2* **List common factors.** List each prime number that is a factor of every number in the list. (If a prime does not appear in one of the prime factored forms, it cannot appear in the greatest common factor.)

*Step 3* **Choose smallest exponents.** Use as exponents on the common prime factors the *smallest* exponent from the prime factored forms.

*Step 4* **Multiply.** Multiply the primes from Step 3. If there are no primes left after Step 3, the greatest common factor is 1.

**EXAMPLE 1 Finding the Greatest Common Factor for Numbers**

Find the greatest common factor for each list of numbers.

**(a)** 30, 45

First write each number in prime factored form.

$$30 = 2 \cdot 3 \cdot 5$$
$$45 = 3 \cdot 3 \cdot 5$$

Use each prime the *least* number of times it appears in *all* the factored forms. There is no 2 in the prime factored form of 45, so there will be no 2 in the greatest common factor. The least number of times 3 appears in all the factored forms is 1, and the least number of times 5 appears is also 1. From this, the

$$\text{GCF} = 3^1 \cdot 5^1 = 15.$$

**(b)** 72, 120, 432

Find the prime factored form of each number.

$$72 = 2 \cdot 2 \cdot 2 \cdot 3 \cdot 3$$
$$120 = 2 \cdot 2 \cdot 2 \cdot 3 \cdot 5$$
$$432 = 2 \cdot 2 \cdot 2 \cdot 2 \cdot 3 \cdot 3 \cdot 3$$

The least number of times 2 appears in all the factored forms is 3, and the least number of times 3 appears is 1. There is no 5 in the prime factored form of either 72 or 432, so the

$$\text{GCF} = 2^3 \cdot 3^1 = 24.$$

**(c)** 10, 11, 14

Write the prime factored form of each number.

$$10 = 2 \cdot 5$$
$$11 = 11$$
$$14 = 2 \cdot 7$$

There are no primes common to all three numbers, so the GCF is 1.

**Now Try Exercise 1.**

The greatest common factor can also be found for a list of variable terms. For example, the terms $x^4$, $x^5$, $x^6$, and $x^7$ have $x^4$ as the greatest common factor because each of these terms can be written with $x^4$ as a factor.

$$x^4 = 1 \cdot x^4, \quad x^5 = x \cdot x^4, \quad x^6 = x^2 \cdot x^4, \quad x^7 = x^3 \cdot x^4$$

**NOTE** The exponent on a variable in the GCF is the *smallest* exponent that appears in *all* the common factors.

### EXAMPLE 2 Finding the Greatest Common Factor for Variable Terms

Find the greatest common factor for each list of terms.

**(a)** $21m^7$, $-18m^6$, $45m^8$, $-24m^5$

$$\begin{aligned} 21m^7 &= 3 \cdot 7 \cdot m^7 \\ -18m^6 &= -1 \cdot 2 \cdot 3^2 \cdot m^6 \\ 45m^8 &= 3^2 \cdot 5 \cdot m^8 \\ -24m^5 &= -1 \cdot 2^3 \cdot 3 \cdot m^5 \end{aligned}$$

First, 3 is the greatest common factor of the coefficients 21, $-18$, 45, and $-24$. The smallest exponent on $m$ is 5, so the GCF of the terms is $3m^5$.

**(b)** $x^4y^2$, $x^7y^5$, $x^3y^7$, $y^{15}$

$$\begin{aligned} x^4y^2 &= x^4 \cdot y^2 \\ x^7y^5 &= x^7 \cdot y^5 \\ x^3y^7 &= x^3 \cdot y^7 \\ y^{15} &= y^{15} \end{aligned}$$

There is no $x$ in the last term, $y^{15}$, so $x$ will not appear in the greatest common factor. There is a $y$ in each term, however, and 2 is the smallest exponent on $y$. The GCF is $y^2$.

**(c)** $-a^2b$, $-ab^2$

$$\begin{aligned} -a^2b &= -1a^2b = -1 \cdot 1 \cdot a^2b \\ -ab^2 &= -1ab^2 = -1 \cdot 1 \cdot ab^2 \end{aligned}$$

The factors of $-1$ are $-1$ and 1. Since $1 > -1$, the GCF is $1ab$ or $ab$.

**Now Try Exercises 9 and 13.**

**NOTE** In a list of negative terms, sometimes a negative common factor is preferable (even though it is not the greatest common factor). In Example 2(c), for instance, we might prefer $-ab$ as the common factor. In factoring exercises, either answer will be acceptable.

**OBJECTIVE 2 Factor out the greatest common factor.** Writing a polynomial (a sum) in factored form as a product is called **factoring.** For example, the polynomial

$$3m + 12$$

has two terms, $3m$ and 12. The greatest common factor for these two terms is 3. We can write $3m + 12$ so that each term is a product with 3 as one factor.

$$3m + 12 = 3 \cdot m + 3 \cdot 4$$

Now we use the distributive property.

$$3m + 12 = 3 \cdot m + 3 \cdot 4 = 3(m + 4)$$

The factored form of $3m + 12$ is $3(m + 4)$. This process is called **factoring out the greatest common factor.**

**CAUTION** The polynomial $3m + 12$ is *not* in factored form when written as

$$3 \cdot m + 3 \cdot 4.$$

The *terms* are factored, but the polynomial is not. The factored form of $3m + 12$ is the *product*

$$3(m + 4).$$

## EXAMPLE 3 Factoring Out the Greatest Common Factor

Factor out the greatest common factor.

**(a)** $5y^2 + 10y = 5y(y) + 5y(2)$ GCF $= 5y$

$= 5y(y + 2)$ Distributive property

To check, multiply out the factored form: $5y(y + 2) = 5y^2 + 10y$, which is the original polynomial.

**(b)** $20m^5 + 10m^4 + 15m^3$

The GCF for the terms of this polynomial is $5m^3$.

$$20m^5 + 10m^4 + 15m^3 = 5m^3(4m^2) + 5m^3(2m) + 5m^3(3)$$
$$= 5m^3(4m^2 + 2m + 3)$$

*Check:* $5m^3(4m^2 + 2m + 3) = 20m^5 + 10m^4 + 15m^3$, which is the original polynomial.

**(c)** $x^5 + x^3 = x^3(x^2) + x^3(1) = x^3(x^2 + 1)$ Don't forget the 1.

**(d)** $20m^7p^2 - 36m^3p^4 = 4m^3p^2(5m^4) - 4m^3p^2(9p^2)$ GCF $= 4m^3p^2$

$= 4m^3p^2(5m^4 - 9p^2)$

**(e)** $\frac{1}{6}n^2 + \frac{5}{6}n = \frac{1}{6}n(n) + \frac{1}{6}n(5) = \frac{1}{6}n(n + 5)$ GCF $= \frac{1}{6}n$

**Now Try Exercises 37, 41, and 47.**

**CAUTION** Be sure to include the 1 in a problem like Example 3(c). *Always* check that the factored form can be multiplied out to give the original polynomial.

### EXAMPLE 4 Factoring Out the Greatest Common Factor

Factor out the greatest common factor.

**(a)** $a(a + 3) + 4(a + 3)$

The binomial $a + 3$ is the greatest common factor here.

Same

$$a(a + 3) + 4(a + 3) = (a + 3)(a + 4)$$

**(b)** $x^2(x + 1) - 5(x + 1) = (x + 1)(x^2 - 5)$ Factor out $x + 1$.

**Now Try Exercise 55.**

**OBJECTIVE 3 Factor by grouping.** When a polynomial has four terms, common factors can sometimes be used to **factor by grouping.**

### EXAMPLE 5 Factoring by Grouping

Factor by grouping.

**(a)** $2x + 6 + ax + 3a$

The first two terms have a common factor of 2, and the last two terms have a common factor of $a$.

$$2x + 6 + ax + 3a = 2(x + 3) + a(x + 3)$$

The expression is still not in factored form because it is the *sum* of two terms. Now, however, $x + 3$ is a common factor and can be factored out.

$$\begin{aligned} 2x + 6 + ax + 3a &= 2(x + 3) + a(x + 3) \\ &= (x + 3)(2 + a) \end{aligned}$$

The final result is in factored form because it is a *product.* Note that the goal in factoring by grouping is to get a common factor, $x + 3$ here, so that the last step is possible.

*Check:* $(x + 3)(2 + a) = 2x + 6 + ax + 3a$, which is the original polynomial.

**(b)**
$$\begin{aligned} 2x^2 - 10x + 3xy - 15y &= (2x^2 - 10x) + (3xy - 15y) && \text{Group terms.} \\ &= 2x(x - 5) + 3y(x - 5) && \text{Factor each group.} \\ &= (x - 5)(2x + 3y) && \text{Factor out the common factor, } x - 5. \end{aligned}$$

*Check:*
$$\begin{aligned} (x - 5)(2x + 3y) &= 2x^2 + 3xy - 10x - 15y && \text{FOIL} \\ &= 2x^2 - 10x + 3xy - 15y && \text{Original polynomial} \end{aligned}$$

**(c)**
$$\begin{aligned} t^3 + 2t^2 - 3t - 6 &= (t^3 + 2t^2) + (-3t - 6) && \text{Group terms.} \\ &= t^2(t + 2) - 3(t + 2) && \text{Factor out } -3 \text{ so there is a common factor, } t + 2;\ -3(t + 2) = -3t - 6. \\ &= (t + 2)(t^2 - 3) && \text{Factor out } t + 2. \end{aligned}$$

Check by multiplying.

**Now Try Exercises 69, 73, and 77.**

**CAUTION** Use negative signs carefully when grouping, as in Example 5(c). Otherwise, sign errors may result. Always check by multiplying.

Use these steps when factoring four terms by grouping.

**Factoring by Grouping**

*Step 1* **Group terms.** Collect the terms into two groups so that each group has a common factor.

*Step 2* **Factor within groups.** Factor out the greatest common factor from each group.

*Step 3* **Factor the entire polynomial.** Factor a common binomial factor from the results of Step 2.

*Step 4* **If necessary, rearrange terms.** If Step 2 does not result in a common binomial factor, try a different grouping.

**EXAMPLE 6 Rearranging Terms Before Factoring by Grouping**

Factor by grouping.

**(a)** $10x^2 - 12y + 15x - 8xy$

Factoring out the common factor of 2 from the first two terms and the common factor of $x$ from the last two terms gives

$$10x^2 - 12y + 15x - 8xy = 2(5x^2 - 6y) + x(15 - 8y).$$

This did not lead to a common factor, so we try rearranging the terms. There is usually more than one way to do this. We try

$$10x^2 - 8xy - 12y + 15x,$$

and group the first two terms and the last two terms as follows.

$$\begin{aligned} 10x^2 - 8xy - 12y + 15x &= 2x(5x - 4y) + 3(-4y + 5x) \\ &= 2x(5x - 4y) + 3(5x - 4y) \\ &= (5x - 4y)(2x + 3) \end{aligned}$$

*Check:* $(5x - 4y)(2x + 3) = 10x^2 + 15x - 8xy - 12y$ FOIL

$= 10x^2 - 12y + 15x - 8xy$ Original polynomial

**(b)** $2xy + 3 - 3y - 2x$

We need to rearrange these terms to get two groups that each have a common factor. Trial and error suggests the following grouping.

$2xy + 3 - 3y - 2x = (2xy - 3y) + (-2x + 3)$ Group terms.

$= y(2x - 3) - 1(2x - 3)$ Factor each group.

$= (2x - 3)(y - 1)$ Factor out the common binomial factor.

Since the quantities in parentheses in the second step must be the same, we factored out $-1$ rather than 1. Check by multiplying.

**Now Try Exercise 81.**

## 5.1 EXERCISES

**For Extra Help**

- Student's Solutions Manual
- MyMathLab
- InterAct Math Tutorial Software
- AW Math Tutor Center
- MathXL
- Digital Video Tutor CD 8/Videotape 8

*Find the greatest common factor for each list of numbers. See Example 1.*

**1.** 40, 20, 4 **2.** 50, 30, 5 **3.** 18, 24, 36, 48 **4.** 15, 30, 45, 75

**5.** How can you check your answer when you factor a polynomial?

**6.** Explain how to find the greatest common factor of a list of terms. Use examples.

*Find the greatest common factor for each list of terms. See Examples 1 and 2.*

**7.** $16y$, $24$

**8.** $18w$, $27$

**9.** $30x^3$, $40x^6$, $50x^7$

**10.** $60z^4$, $70z^8$, $90z^9$

**11.** $12m^3n^2$, $18m^5n^4$, $36m^8n^3$

**12.** $25p^5r^7$, $30p^7r^8$, $50p^5r^3$

**13.** $-x^4y^3$, $-xy^2$

**14.** $-a^4b^5$, $-a^3b$

**15.** $42ab^3$, $-36a$, $90b$, $-48ab$

**16.** $45c^3d$, $75c$, $90d$, $-105cd$

*An expression is factored when it is written as a product, not a sum. Which of the following are not factored?*

**17.** $2k^2(5k)$

**18.** $2k^2(5k + 1)$

**19.** $2k^2 + (5k + 1)$

**20.** $(2k^2 + 1)(5k + 1)$

**21.** Is $-xy$ a common factor of $-x^4y^3$ and $-xy^2$? If so, what is the other factor that when multiplied by $-xy$ gives $-x^4y^3$?

**22.** Is $-a^5b^2$ a common factor of $-a^4b^5$ and $-a^3b$?

*Complete each factoring.*

**23.** $9m^4 = 3m^2(\qquad)$

**24.** $12p^5 = 6p^3(\qquad)$

**25.** $-8z^9 = -4z^5(\qquad)$

**26.** $-15k^{11} = -5k^8(\qquad)$

**27.** $6m^4n^5 = 3m^3n(\qquad)$

**28.** $27a^3b^2 = 9a^2b(\qquad)$

**29.** $12y + 24 = 12(\qquad)$

**30.** $18p + 36 = 18(\qquad)$

**31.** $10a^2 - 20a = 10a(\qquad)$

**32.** $15x^2 - 30x = 15x(\qquad)$

**33.** $8x^2y + 12x^3y^2 = 4x^2y(\qquad)$

**34.** $18s^3t^2 + 10st = 2st(\qquad)$

*Factor out the greatest common factor. See Examples 3 and 4.*

**35.** $27m^3 - 9m$

**36.** $36p^3 + 24p$

**37.** $16z^4 + 24z^2$

**38.** $25k^4 - 15k^2$

**39.** $\frac{1}{4}d^2 - \frac{3}{4}d$

**40.** $\frac{1}{5}z^2 + \frac{3}{5}z$

**41.** $12x^3 + 6x^2$

**42.** $21b^3 - 7b^2$

**43.** $65y^{10} + 35y^6$

**44.** $100a^5 + 16a^3$

**45.** $11w^3 - 100$

**46.** $13z^5 - 80$

**47.** $8m^2n^3 + 24m^2n^2$

**48.** $19p^2y - 38p^2y^3$

**49.** $13y^8 + 26y^4 - 39y^2$

**50.** $5x^5 + 25x^4 - 20x^3$

**51.** $45q^4p^5 + 36qp^6 + 81q^2p^3$

**52.** $125a^3z^5 + 60a^4z^4 - 85a^5z^2$

**53.** $a^5 + 2a^3b^2 - 3a^5b^2 + 4a^4b^3$

**54.** $x^6 + 5x^4y^3 - 6xy^4 + 10xy$

**55.** $c(x + 2) - d(x + 2)$

**56.** $r(5 - x) + t(5 - x)$

**57.** $m(m + 2n) + n(m + 2n)$

**58.** $3p(1 - 4p) - 2q(1 - 4p)$

Students often have difficulty when factoring by grouping because they are not able to tell when the polynomial is completely factored. For example,

$$5y(2x - 3) + 8t(2x - 3)$$

is not in factored form, because it is the *sum* of two terms, $5y(2x - 3)$ and $8t(2x - 3)$. However, because $2x - 3$ is a common factor of these two terms, the expression can now be factored as

$$(2x - 3)(5y + 8t).$$

The factored form is a *product* of two factors, $2x - 3$ and $5y + 8t$.

*Determine whether each expression is in factored form or is not in factored form. If it is not in factored form, factor it if possible.*

**59.** $8(7t + 4) + x(7t + 4)$

**60.** $3r(5x - 1) + 7(5x - 1)$

**61.** $(8 + x)(7t + 4)$

**62.** $(3r + 7)(5x - 1)$

**63.** $18x^2(y + 4) + 7(y + 4)$

**64.** $12k^3(s - 3) + 7(s + 3)$

**65.** Tell why it is not possible to factor the expression in Exercise 64.

**66.** Summarize the method of factoring a polynomial with four terms by grouping. Give an example.

*Factor by grouping. See Examples 5 and 6.*

**67.** $p^2 + 4p + pq + 4q$

**68.** $m^2 + 2m + mn + 2n$

**69.** $a^2 - 2a + ab - 2b$

**70.** $y^2 - 6y + yw - 6w$

**71.** $7z^2 + 14z - az - 2a$

**72.** $5m^2 + 15mp - 2mr - 6pr$

**73.** $18r^2 + 12ry - 3xr - 2xy$

**74.** $8s^2 - 4st + 6sy - 3yt$

**75.** $3a^3 + 3ab^2 + 2a^2b + 2b^3$

**76.** $4x^3 + 3x^2y + 4xy^2 + 3y^3$

**77.** $1 - a + ab - b$

**78.** $6 - 3x - 2y + xy$

**79.** $16m^3 - 4m^2p^2 - 4mp + p^3$

**80.** $10t^3 - 2t^2s^2 - 5ts + s^3$

**81.** $5m - 6p - 2mp + 15$

**82.** $y^2 + 3x - 3y - xy$

**83.** $18r^2 - 2ty + 12ry - 3rt$

**84.** $2b^3 + 3a^3 + 3ab^2 + 2a^2b$

**85.** $a^5 - 3 + 2a^5b - 6b$

**86.** $4b^3 + a^2b - 4a - ab^4$

## RELATING CONCEPTS (EXERCISES 87–90)

**For Individual or Group Work**

*In many cases, the choice of which pairs of terms to group when factoring by grouping can be made in different ways. To see this for Example 6(b),* ***work Exercises 87–90 in order.***

**87.** Start with the polynomial from Example 6(b), $2xy + 3 - 3y - 2x$, and rearrange the terms as follows: $2xy - 2x - 3y + 3$. What property from Section 1.7 allows this?

**88.** Group the first two terms and the last two terms of the rearranged polynomial in Exercise 87. Then factor each group.

**89.** Is your result from Exercise 88 in factored form? Explain your answer.

**90.** If your answer to Exercise 89 is *no*, factor the polynomial. Is the result the same as the one shown for Example 6(b)?

**91.** Refer to Exercise 77. The answer given in the back of the book is $(1 - a)(1 - b)$. A student factored this same polynomial and got the result $(a - 1)(b - 1)$.

**(a)** Is the student's answer correct?

**(b)** If your answer to part (a) is *yes*, explain why these two seemingly different answers are both acceptable.

**92.** A student factored $18x^3y^2 + 9xy$ as $9xy(2x^2y)$. Is this correct? If not, explain the error and factor correctly.

## 5.2 Factoring Trinomials

**OBJECTIVES**

1 Factor trinomials with a coefficient of 1 for the squared term.

2 Factor such trinomials after factoring out the greatest common factor.

Using FOIL, the product of the binomials $k - 3$ and $k + 1$ is

$$(k - 3)(k + 1) = k^2 - 2k - 3. \qquad \text{Multiplying}$$

Suppose instead that we are given the polynomial $k^2 - 2k - 3$ and want to rewrite it as the product $(k - 3)(k + 1)$. That is,

$$k^2 - 2k - 3 = (k - 3)(k + 1). \qquad \text{Factoring}$$

Recall from the previous section that this process is called factoring the polynomial. Factoring reverses or "undoes" multiplying.

**OBJECTIVE 1 Factor trinomials with a coefficient of 1 for the squared term.** When factoring polynomials with integer coefficients, we use only integers in the factors. For example, we can factor $x^2 + 5x + 6$ by finding integers $m$ and $n$ such that

$$x^2 + 5x + 6 = (x + m)(x + n).$$

To find these integers $m$ and $n$, we first use FOIL to multiply the two binomials on the right side of the equation:

$$(x + m)(x + n) = x^2 + nx + mx + mn.$$

By the distributive property,

$$x^2 + nx + mx + mn = x^2 + (n + m)x + mn.$$

Comparing this result with $x^2 + 5x + 6$ shows that we must find integers $m$ and $n$ having a sum of 5 and a product of 6.

Product of $m$ and $n$ is 6.

$$x^2 + 5x + 6 = x^2 + (n + m)x + mn$$

Sum of $m$ and $n$ is 5.

Since many pairs of integers have a sum of 5, it is best to begin by listing those pairs of integers whose product is 6. Both 5 and 6 are positive, so we consider only pairs in which both integers are positive.

| *Pair* | *Product* | *Sum* | |
|---|---|---|---|
| 1, 6 | $1 \cdot 6 = 6$ | $1 + 6 = 7$ | |
| 2, 3 | $2 \cdot 3 = 6$ | $2 + 3 = 5$ | Sum is 5. |

Both pairs have a product of 6, but only the pair 2 and 3 has a sum of 5. So 2 and 3 are the required integers, and

$$x^2 + 5x + 6 = (x + 2)(x + 3).$$

Check by multiplying the binomials using FOIL. Make sure that the sum of the outer and inner products produces the correct middle term.

$$(x + 2)(x + 3) = x^2 + 5x + 6$$

$2x$
$3x$
$5x$ Add.

We can use this method of factoring only for trinomials that have 1 as the coefficient of the squared term. We give methods for factoring other trinomials in the next section.

### EXAMPLE 1 Factoring a Trinomial with All Positive Terms

Factor $m^2 + 9m + 14$.

Look for two integers whose product is 14 and whose sum is 9. List the pairs of integers whose product is 14. Then examine the sums. Again, only positive integers are needed because all signs in $m^2 + 9m + 14$ are positive.

| *Factors of 14* | *Sums of Factors* | |
|---|---|---|
| 14, 1 | 14 + 1 = 15 | |
| 7, 2 | 7 + 2 = 9 | Sum is 9. |

From the list, 7 and 2 are the required integers, since $7 \cdot 2 = 14$ and $7 + 2 = 9$. Thus

$$m^2 + 9m + 14 = (m + 7)(m + 2).$$

Check by multiplying on the right side to be sure the original trinomial results.

**Now Try Exercise 25.**

**NOTE** In Example 1, the answer also could have been written $(m + 2)(m + 7)$. Because of the commutative property of multiplication, the order of the factors does not matter. Always check by multiplying.

### EXAMPLE 2 Factoring a Trinomial with a Negative Middle Term

Factor $x^2 - 9x + 20$.

We must find two integers whose product is 20 and whose sum is $-9$. Since the numbers we are looking for have a positive product and a negative sum, we consider only pairs of negative integers.

| *Factors of 20* | *Sums of Factors* | |
|---|---|---|
| −20, −1 | −20 + (−1) = −21 | |
| −10, −2 | −10 + (−2) = −12 | |
| −5, −4 | −5 + (−4) = −9 | Sum is −9. |

The required integers are −5 and −4, so

$$x^2 - 9x + 20 = (x - 5)(x - 4).$$

*Check:* $(x - 5)(x - 4) = x^2 - 4x - 5x + 20$
$= x^2 - 9x + 20$

**Now Try Exercise 29.**

### EXAMPLE 3 Factoring a Trinomial with Two Negative Terms

Factor $p^2 - 2p - 15$.

Find two integers whose product is −15 and whose sum is −2. If these numbers do not come to mind right away, find them (if they exist) by listing all the pairs of integers whose product is −15. Because the last term, −15, is negative, list pairs of integers with different signs.

| *Factors of −15* | *Sums of Factors* | |
|---|---|---|
| 15, −1 | 15 + (−1) = 14 | |
| −15, 1 | −15 + 1 = −14 | |
| 5, −3 | 5 + (−3) = 2 | |
| −5, 3 | −5 + 3 = −2 | Sum is −2. |

The required integers are −5 and 3, so

$$p^2 - 2p - 15 = (p - 5)(p + 3).$$

*Check:* Multiply $(p - 5)(p + 3)$ to get $p^2 - 2p - 15$.

**Now Try Exercise 35.**

**NOTE** In Examples 1–3, notice that we listed factors in descending order (disregarding sign) when we were looking for the required pair of integers. This helps avoid skipping the correct combination.

As shown in the next example, some trinomials cannot be factored using only integers. We call such trinomials **prime polynomials.**

### EXAMPLE 4 Deciding Whether Polynomials Are Prime

Factor each trinomial.

**(a)** $x^2 - 5x + 12$

As in Example 2, both factors must be negative to give a positive product and a negative sum. First, list all pairs of negative integers whose product is 12. Then examine the sums.

| *Factors of 12* | *Sums of Factors* |
|---|---|
| −12, −1 | −12 + (−1) = −13 |
| −6, −2 | −6 + (−2) = −8 |
| −4, −3 | −4 + (−3) = −7 |

None of the pairs of integers has a sum of $-5$. Therefore, the trinomial $x^2 - 5x + 12$ *cannot be factored using only integers;* it is a *prime polynomial.*

**(b)** $k^2 - 8k + 11$

There is no pair of integers whose product is 11 and whose sum is $-8$, so $k^2 - 8k + 11$ is a prime polynomial.

**Now Try Exercise 31.**

The procedure for factoring a trinomial of the form $x^2 + bx + c$ is summarized here.

**Factoring $x^2 + bx + c$**

Find two integers whose product is $c$ and whose sum is $b$.

1. Both integers must be positive if $b$ and $c$ are positive.
2. Both integers must be negative if $c$ is positive and $b$ is negative.
3. One integer must be positive and one must be negative if $c$ is negative.

### EXAMPLE 5 Factoring a Trinomial with Two Variables

Factor $z^2 - 2bz - 3b^2$.

The coefficient of $z$ in the middle term is $-2b$, so we need to find two expressions whose product is $-3b^2$ and whose sum is $-2b$. The expressions are $-3b$ and $b$, so

$$z^2 - 2bz - 3b^2 = (z - 3b)(z + b).$$

Check by multiplying.

**Now Try Exercise 45.**

**OBJECTIVE 2 Factor such trinomials after factoring out the greatest common factor.** The trinomial in the next example does not have a coefficient of 1 for the squared term. (In fact, there is no squared term.) However, there may be a common factor.

### EXAMPLE 6 Factoring a Trinomial with a Common Factor

Factor $4x^5 - 28x^4 + 40x^3$.

First, factor out the greatest common factor, $4x^3$.

$$4x^5 - 28x^4 + 40x^3 = 4x^3(x^2 - 7x + 10)$$

Now factor $x^2 - 7x + 10$. The integers $-5$ and $-2$ have a product of 10 and a sum of $-7$. The complete factored form is

$$4x^5 - 28x^4 + 40x^3 = 4x^3(x - 5)(x - 2).$$

Remember to include $4x^3$.

*Check:* $4x^3(x - 5)(x - 2) = 4x^3(x^2 - 7x + 10)$
$= 4x^5 - 28x^4 + 40x^3$

**Now Try Exercise 53.**

**CAUTION** When factoring, always look for a common factor first. Remember to include the common factor as part of the answer. As a check, multiplying out the factored form should always give the original polynomial.

## 5.2 EXERCISES

**For Extra Help**

- Student's Solutions Manual
- MyMathLab
- InterAct Math Tutorial Software
- AW Math Tutor Center
- MathXL
- Digital Video Tutor CD 8/Videotape 8

*In Exercises 1–4, list all pairs of integers with the given product. Then find the pair whose sum is given.*

**1.** Product: 48; Sum: $-19$

**2.** Product: 48; Sum: 14

**3.** Product: $-24$; Sum: $-5$

**4.** Product: $-36$; Sum: $-16$

**5.** In factoring a trinomial in $x$ as $(x + a)(x + b)$, what must be true of $a$ and $b$ if the coefficient of the last term of the trinomial is negative?

**6.** In Exercise 5, what must be true of $a$ and $b$ if the coefficient of the last term is positive?

**7.** What is meant by a *prime polynomial*?

**8.** How can you check your work when factoring a trinomial? Does the check ensure that the trinomial is completely factored?

**9.** Which one of the following is the correct factored form of $x^2 - 12x + 32$?

**A.** $(x - 8)(x + 4)$ **B.** $(x + 8)(x - 4)$
**C.** $(x - 8)(x - 4)$ **D.** $(x + 8)(x + 4)$

**10.** What would be the first step in factoring $2x^3 + 8x^2 - 10x$?

*Complete each factoring. See Examples 1–4.*

**11.** $p^2 + 11p + 30 = (p + 5)(\qquad)$

**12.** $x^2 + 10x + 21 = (x + 7)(\qquad)$

**13.** $x^2 + 15x + 44 = (x + 4)(\qquad)$

**14.** $r^2 + 15r + 56 = (r + 7)(\qquad)$

**15.** $x^2 - 9x + 8 = (x - 1)(\qquad)$

**16.** $t^2 - 14t + 24 = (t - 2)(\qquad)$

**17.** $y^2 - 2y - 15 = (y + 3)(\qquad)$

**18.** $t^2 - t - 42 = (t + 6)(\qquad)$

**19.** $x^2 + 9x - 22 = (x - 2)(\qquad)$

**20.** $x^2 + 6x - 27 = (x - 3)(\qquad)$

**21.** $y^2 - 7y - 18 = (y + 2)(\qquad)$

**22.** $y^2 - 2y - 24 = (y + 4)(\qquad)$

*Factor completely. If the polynomial cannot be factored, write* prime. *See Examples 1–4.*

**23.** $y^2 + 9y + 8$

**24.** $a^2 + 9a + 20$

**25.** $b^2 + 8b + 15$

**26.** $x^2 + 6x + 8$

**27.** $m^2 + m - 20$

**28.** $p^2 + 4p - 5$

**29.** $y^2 - 8y + 15$

**30.** $y^2 - 6y + 8$

**31.** $x^2 + 4x + 5$

**32.** $t^2 + 11t + 12$

**33.** $z^2 - 15z + 56$

**34.** $x^2 - 13x + 36$

**35.** $r^2 - r - 30$

**36.** $q^2 - q - 42$

**37.** $a^2 - 8a - 48$

**38.** $m^2 - 10m - 25$

**39.** $x^2 + 3x - 39$

**40.** $d^2 + 4d - 45$

**41.** Explain how you would factor $8 + 6x + x^2$.

**42.** Use your answer to Exercise 41 to factor $5 - 4x - x^2$.

*Factor completely. See Examples 5 and 6.*

**43.** $r^2 + 3ra + 2a^2$
**44.** $x^2 + 5xa + 4a^2$
**45.** $t^2 - tz - 6z^2$
**46.** $a^2 - ab - 12b^2$
**47.** $x^2 + 4xy + 3y^2$
**48.** $p^2 + 9pq + 8q^2$
**49.** $v^2 - 11vw + 30w^2$
**50.** $v^2 - 11vx + 24x^2$
**51.** $4x^2 + 12x - 40$
**52.** $5y^2 - 5y - 30$
**53.** $2t^3 + 8t^2 + 6t$
**54.** $3t^3 + 27t^2 + 24t$
**55.** $2x^6 + 8x^5 - 42x^4$
**56.** $4y^5 + 12y^4 - 40y^3$
**57.** $5m^5 + 25m^4 - 40m^2$
**58.** $12k^5 - 6k^3 + 10k^2$
**59.** $m^3n - 10m^2n^2 + 24mn^3$
**60.** $y^3z + 3y^2z^2 - 54yz^3$

**61.** Use the FOIL method from Section 4.5 to show that $(2x + 4)(x - 3) = 2x^2 - 2x - 12$. If you are asked to completely factor $2x^2 - 2x - 12$, why would it be incorrect to give $(2x + 4)(x - 3)$ as your answer?

**62.** If you are asked to completely factor the polynomial $3x^2 + 9x - 12$, why would it be incorrect to give $(x - 1)(3x + 12)$ as your answer?

*Use a combination of the factoring methods discussed in this section to factor each polynomial.*

**63.** $a^5 + 3a^4b - 4a^3b^2$
**64.** $m^3n - 2m^2n^2 - 3mn^3$
**65.** $y^3z + y^2z^2 - 6yz^3$
**66.** $k^7 - 2k^6m - 15k^5m^2$
**67.** $z^{10} - 4z^9y - 21z^8y^2$
**68.** $x^9 + 5x^8w - 24x^7w^2$
**69.** $(a + b)x^2 + (a + b)x - 12(a + b)$
**70.** $(x + y)n^2 + (x + y)n + 16(x + y)$
**71.** $(2p + q)r^2 - 12(2p + q)r + 27(2p + q)$
**72.** $(3m - n)k^2 - 13(3m - n)k + 40(3m - n)$
**73.** What polynomial can be factored as $(a + 9)(a + 4)$?
**74.** What polynomial can be factored as $(y - 7)(y + 3)$?

# 5.3 More on Factoring Trinomials

**OBJECTIVES**

1. Factor trinomials by grouping when the coefficient of the squared term is not 1.
2. Factor trinomials using FOIL.

Trinomials like $2x^2 + 7x + 6$, in which the coefficient of the squared term is *not* 1, are factored with extensions of the methods from the previous sections. One such method uses factoring by grouping from Section 5.1.

**OBJECTIVE 1 Factor trinomials by grouping when the coefficient of the squared term is not 1.** Recall that a trinomial such as $m^2 + 3m + 2$ is factored by finding two numbers whose product is 2 and whose sum is 3. To factor $2x^2 + 7x + 6$, we look for two integers whose product is $2 \cdot 6 = 12$ and whose sum is 7.

Sum is 7.

$$2x^2 + 7x + 6$$

Product is $2 \cdot 6 = 12$.

By considering pairs of positive integers whose product is 12, we find the necessary integers to be 3 and 4. We use these integers to write the middle term, $7x$, as $7x = 3x + 4x$. The trinomial $2x^2 + 7x + 6$ becomes

$$2x^2 + 7x + 6 = 2x^2 + \underbrace{3x + 4x}_{7x} + 6.$$

$$= (2x^2 + 3x) + (4x + 6) \qquad \text{Group terms.}$$

$$= x(2x + 3) + 2(2x + 3) \qquad \text{Factor each group.}$$

Must be the same factor

$$2x^2 + 7x + 6 = (2x + 3)(x + 2) \qquad \text{Factor out } 2x + 3.$$

*Check:* $(2x + 3)(x + 2) = 2x^2 + 7x + 6$

In this example, we could have written $7x$ as $4x + 3x$. Factoring by grouping this way would give the same answer.

## EXAMPLE 1 Factoring Trinomials by Grouping

Factor each trinomial.

**(a)** $6r^2 + r - 1$

We must find two integers with a product of $6(-1) = -6$ and a sum of 1.

Sum is 1.

$$6r^2 + r - 1 = 6r^2 + 1r - 1$$

Product is $6(-1) = -6$.

The integers are $-2$ and 3. We write the middle term, $r$, as $-2r + 3r$.

$$6r^2 + r - 1 = 6r^2 - 2r + 3r - 1 \qquad r = -2r + 3r$$

$$= (6r^2 - 2r) + (3r - 1) \qquad \text{Group terms.}$$

$$= 2r(3r - 1) + 1(3r - 1) \qquad \text{The binomials must be the same.}$$

$$= (3r - 1)(2r + 1) \qquad \text{Factor out } 3r - 1.$$

*Check:* $(3r - 1)(2r + 1) = 6r^2 + r - 1$

**(b)** $12z^2 - 5z - 2$

Look for two integers whose product is $12(-2) = -24$ and whose sum is $-5$. The required integers are 3 and $-8$, so

$$12z^2 - 5z - 2 = 12z^2 + 3z - 8z - 2 \qquad -5z = 3z - 8z$$

$$= (12z^2 + 3z) + (-8z - 2) \qquad \text{Group terms.}$$

$$= 3z(4z + 1) - 2(4z + 1) \qquad \text{Factor each group; be careful with signs.}$$

$$= (4z + 1)(3z - 2). \qquad \text{Factor out } 4z + 1.$$

*Check:* $(4z + 1)(3z - 2) = 12z^2 - 5z - 2$

**(c)** $10m^2 + mn - 3n^2$

Two integers whose product is $10(-3) = -30$ and whose sum is 1 are $-5$ and 6. Rewrite the trinomial with four terms.

$$10m^2 + mn - 3n^2 = 10m^2 - 5mn + 6mn - 3n^2 \qquad mn = -5mn + 6mn$$
$$= 5m(2m - n) + 3n(2m - n) \qquad \text{Group terms; factor each group.}$$
$$= (2m - n)(5m + 3n) \qquad \text{Factor out } 2m - n.$$

Check by multiplying.

**Now Try Exercises 21, 27, and 47.**

### EXAMPLE 2 Factoring a Trinomial with a Common Factor by Grouping

Factor $28x^5 - 58x^4 - 30x^3$.

First factor out the greatest common factor, $2x^3$.

$$28x^5 - 58x^4 - 30x^3 = 2x^3(14x^2 - 29x - 15)$$

To factor $14x^2 - 29x - 15$, find two integers whose product is $14(-15) = -210$ and whose sum is $-29$. Factoring 210 into prime factors gives

$$210 = 2 \cdot 3 \cdot 5 \cdot 7.$$

Combine these prime factors in pairs in different ways, using one positive and one negative (to get $-210$). The factors 6 and $-35$ have the correct sum. Now rewrite the given trinomial and factor it.

$$28x^5 - 58x^4 - 30x^3 = 2x^3(14x^2 + 6x - 35x - 15)$$
$$= 2x^3[(14x^2 + 6x) + (-35x - 15)]$$
$$= 2x^3[2x(7x + 3) - 5(7x + 3)]$$
$$= 2x^3[(7x + 3)(2x - 5)]$$
$$= 2x^3(7x + 3)(2x - 5)$$

Check by multiplying.

**Now Try Exercise 43.**

**CAUTION** Remember to include the common factor in the final result.

**OBJECTIVE 2 Factor trinomials using FOIL.** In the rest of this section we show an alternative method of factoring trinomials in which the coefficient of the squared term is not 1. This method generalizes the factoring procedure explained in Section 5.2.

To factor $2x^2 + 7x + 6$ (the trinomial factored at the beginning of this section) by the method in Section 5.2, use FOIL backwards. We want to write $2x^2 + 7x + 6$ as the product of two binomials.

$$2x^2 + 7x + 6 = (\qquad)(\qquad)$$

The product of the two first terms of the binomials is $2x^2$. The possible factors of $2x^2$ are $2x$ and $x$ or $-2x$ and $-x$. Since all terms of the trinomial are positive, we consider only positive factors. Thus, we have

$$2x^2 + 7x + 6 = (2x \qquad)(x \qquad).$$

The product of the two last terms, 6, can be factored as $1 \cdot 6$, $6 \cdot 1$, $2 \cdot 3$, or $3 \cdot 2$. Try each pair to find the pair that gives the correct middle term, $7x$.

$(2x + 1)(x + 6)$ Incorrect

$x$
$12x$
$13x$ Add.

$(2x + 6)(x + 1)$ Incorrect

$6x$
$2x$
$8x$ Add.

Since $2x + 6 = 2(x + 3)$, the binomial $2x + 6$ has a common factor of 2, while $2x^2 + 7x + 6$ has no common factor other than 1. The product $(2x + 6)(x + 1)$ cannot be correct.

**NOTE** If the original polynomial has no common factor, then none of its binomial factors will either.

Now try the numbers 2 and 3 as factors of 6. Because of the common factor of 2 in $2x + 2$, $(2x + 2)(x + 3)$ will not work, so we try $(2x + 3)(x + 2)$.

$(2x + 3)(x + 2) = 2x^2 + 7x + 6$ Correct

$3x$
$4x$
$7x$ Add.

Thus, $2x^2 + 7x + 6$ factors as

$$2x^2 + 7x + 6 = (2x + 3)(x + 2).$$

Check by multiplying $2x + 3$ and $x + 2$.

### EXAMPLE 3 Factoring a Trinomial with All Positive Terms Using FOIL

Factor $8p^2 + 14p + 5$.

The number 8 has several possible pairs of factors, but 5 has only 1 and 5 or $-1$ and $-5$. For this reason, it is easier to begin by considering the factors of 5. Ignore the negative factors since all coefficients in the trinomial are positive. If $8p^2 + 14p + 5$ can be factored, the factors will have the form

$$(\quad + 5)(\quad + 1).$$

The possible pairs of factors of $8p^2$ are $8p$ and $p$, or $4p$ and $2p$. Try various combinations, checking in each case to see if the middle term is $14p$.

$(8p + 5)(p + 1)$ Incorrect

$5p$
$8p$
$13p$ Add.

$(p + 5)(8p + 1)$ Incorrect

$40p$
$p$
$41p$ Add.

$(4p + 5)(2p + 1)$ Correct

$10p$
$4p$
$14p$ Add.

Since $14p$ is the correct middle term,

$$8p^2 + 14p + 5 = (4p + 5)(2p + 1).$$

*Check:* $(4p + 5)(2p + 1) = 8p^2 + 14p + 5$

**Now Try Exercise 23.**

### EXAMPLE 4 Factoring a Trinomial with a Negative Middle Term Using FOIL

Factor $6x^2 - 11x + 3$.

Since 3 has only 1 and 3 or $-1$ and $-3$ as factors, it is better here to begin by factoring 3. The last term of the trinomial $6x^2 - 11x + 3$ is positive and the middle term has a negative coefficient, so we consider only negative factors. We need two negative factors because the *product* of two negative factors is positive and their *sum* is negative, as required. Use $-3$ and $-1$ as factors of 3:

$$(\quad - 3)(\quad - 1).$$

The factors of $6x^2$ may be either $6x$ and $x$, or $2x$ and $3x$. We try $2x$ and $3x$.

$(2x - 3)(3x - 1)$ Correct

$-9x$
$-2x$
$-11x$ Add.

These factors give the correct middle term, so

$$6x^2 - 11x + 3 = (2x - 3)(3x - 1).$$

Check by multiplying.

**Now Try Exercise 29.**

### EXAMPLE 5 Factoring a Trinomial with a Negative Last Term Using FOIL

Factor $8x^2 + 6x - 9$.

The integer 8 has several possible pairs of factors, as does $-9$. Since the last term is negative, one positive factor and one negative factor of $-9$ are needed. Since the coefficient of the middle term is small, it is wise to avoid large factors such as 8 or 9. We try $4x$ and $2x$ as factors of $8x^2$, and 3 and $-3$ as factors of $-9$, and check the middle term.

$(4x + 3)(2x - 3)$ Incorrect

$6x$
$-12x$
$-6x$ Add.

Now, we try interchanging 3 and $-3$, since only the sign of the middle term is incorrect.

$(4x - 3)(2x + 3)$ Correct

$-6x$
$12x$
$6x$ Add.

This combination produces the correct middle term, so

$$8x^2 + 6x - 9 = (4x - 3)(2x + 3).$$

**Now Try Exercise 33.**

### EXAMPLE 6 Factoring a Trinomial with Two Variables

Factor $12a^2 - ab - 20b^2$.

There are several pairs of factors of $12a^2$, including $12a$ and $a$, $6a$ and $2a$, and $3a$ and $4a$, just as there are many pairs of factors of $-20b^2$, including $20b$ and $-b$,

$-20b$ and $b$, $10b$ and $-2b$, $-10b$ and $2b$, $4b$ and $-5b$, and $-4b$ and $5b$. Once again, since the desired middle term is small, avoid the larger factors. Try the factors $6a$ and $2a$, and $4b$ and $-5b$.

$$(6a + 4b)(2a - 5b)$$

This cannot be correct, as mentioned before, since $6a + 4b$ has a common factor while the given trinomial has none. Try $3a$ and $4a$ with $4b$ and $-5b$.

$$(3a + 4b)(4a - 5b) = 12a^2 + ab - 20b^2 \quad \text{Incorrect}$$

Here the middle term has the wrong sign, so we interchange the signs in the factors.

$$(3a - 4b)(4a + 5b) = 12a^2 - ab - 20b^2 \quad \text{Correct}$$

**Now Try Exercise 41.**

## EXAMPLE 7 Factoring Trinomials with Common Factors

Factor each trinomial.

**(a)** $15y^3 + 55y^2 + 30y$

First factor out the greatest common factor, $5y$.

$$15y^3 + 55y^2 + 30y = 5y(3y^2 + 11y + 6)$$

Now factor $3y^2 + 11y + 6$. Try $3y$ and $y$ as factors of $3y^2$, and 2 and 3 as factors of 6.

$$(3y + 2)(y + 3) = 3y^2 + 11y + 6 \quad \text{Correct}$$

The complete factored form of $15y^3 + 55y^2 + 30y$ is

$$15y^3 + 55y^2 + 30y = 5y(3y + 2)(y + 3).$$

Check by multiplying.

**(b)** $-24a^3 - 42a^2 + 45a$

The common factor could be $3a$ or $-3a$. If we factor out $-3a$, the first term of the trinomial will be positive, which makes it easier to factor.

$$\begin{aligned} -24a^3 - 42a^2 + 45a &= -3a(8a^2 + 14a - 15) && \text{Factor out } -3a. \\ &= -3a(4a - 3)(2a + 5) && \text{Use FOIL.} \end{aligned}$$

Check by multiplying.

**Now Try Exercise 45.**

**CAUTION** This caution bears repeating: Remember to include the common factor in the final factored form.

The two methods for factoring trinomials are summarized here.

### Factoring Trinomials of the Form $ax^2 + bx + c$

Write the middle term of the trinomial as the sum of two terms to get a polynomial with four terms. Use factoring by grouping as shown in Section 5.1.

Factor the trinomial using FOIL backwards as explained in Section 5.2 and this section.

## 5.3 EXERCISES

**For Extra Help**

- Student's Solutions Manual
- MyMathLab
- InterAct Math Tutorial Software
- AW Math Tutor Center
- MathXL
- Digital Video Tutor CD 8/Videotape 8

*Factor each polynomial by grouping. (The middle term of an equivalent trinomial has already been rewritten.) See Example 1.*

**1.** $10t^2 + 5t + 4t + 2$

**2.** $6x^2 + 9x + 4x + 6$

**3.** $15z^2 - 10z - 9z + 6$

**4.** $12p^2 - 9p - 8p + 6$

**5.** $8s^2 - 4st + 6st - 3t^2$

**6.** $3x^2 - 7xy + 6xy - 14y^2$

**7.** Which pair of integers would be used to rewrite the middle term when factoring $12y^2 + 5y - 2$ by grouping?

**A.** $-8, 3$ **B.** $8, -3$
**C.** $-6, 4$ **D.** $6, -4$

**8.** Which pair of integers would be used to rewrite the middle term when factoring $20b^2 - 13b + 2$ by grouping?

**A.** $10, 3$ **B.** $-10, -3$
**C.** $8, 5$ **D.** $-8, -5$

*Complete the steps to factor each trinomial by grouping.*

**9.** $2m^2 + 11m + 12$

**(a)** Find two integers whose product is ________ · ________ = ________ and whose sum is ________.
**(b)** The required integers are ________ and ________.
**(c)** Write the middle term $11m$ as ________ + ________.
**(d)** Rewrite the given trinomial as ____________________.
**(e)** Factor the polynomial in part (d) by grouping.
**(f)** Check by multiplying.

**10.** $6y^2 - 19y + 10$

**(a)** Find two integers whose product is ________ · ________ = ________ and whose sum is ________.
**(b)** The required integers are ________ and ________.
**(c)** Write the middle term $-19y$ as ________ + ________.
**(d)** Rewrite the given trinomial as ____________________.
**(e)** Factor the polynomial in part (d) by grouping.
**(f)** Check by multiplying.

*Decide which is the correct factored form of the given polynomial.*

**11.** $2x^2 - x - 1$

**A.** $(2x - 1)(x + 1)$
**B.** $(2x + 1)(x - 1)$

**12.** $3a^2 - 5a - 2$

**A.** $(3a + 1)(a - 2)$
**B.** $(3a - 1)(a + 2)$

**13.** $4y^2 + 17y - 15$

**A.** $(y + 5)(4y - 3)$
**B.** $(2y - 5)(2y + 3)$

**14.** $12c^2 - 7c - 12$

**A.** $(6c - 2)(2c + 6)$
**B.** $(4c + 3)(3c - 4)$

*Complete each factoring.*

**15.** $6a^2 + 7ab - 20b^2 = (3a - 4b)(\qquad)$

**16.** $9m^2 - 3mn - 2n^2 = (3m + n)(\qquad)$

**17.** $2x^2 + 6x - 8 = 2(\qquad\qquad)$
$= 2(\qquad)(\qquad)$

**18.** $3x^2 - 9x - 30 = 3(\qquad\qquad)$
$= 3(\qquad)(\qquad)$

**19.** For the polynomial $12x^2 + 7x - 12$, 2 is not a common factor. Explain why the binomial $2x - 6$, then, cannot be a factor of the polynomial.

**20.** Factor $4k^2 + 7k - 15$ twice, using the two methods discussed in the text. Do your answers agree? Which method do you prefer?

*Factor each trinomial completely. See Examples 1–7.*

**21.** $3a^2 + 10a + 7$
**22.** $7r^2 + 8r + 1$
**23.** $2y^2 + 7y + 6$
**24.** $5z^2 + 12z + 4$
**25.** $15m^2 + m - 2$
**26.** $6x^2 + x - 1$
**27.** $12s^2 + 11s - 5$
**28.** $20x^2 + 11x - 3$
**29.** $10m^2 - 23m + 12$
**30.** $6x^2 - 17x + 12$
**31.** $8w^2 - 14w + 3$
**32.** $9p^2 - 18p + 8$
**33.** $20y^2 - 39y - 11$
**34.** $10x^2 - 11x - 6$
**35.** $3x^2 - 15x + 16$
**36.** $2t^2 + 13t - 18$
**37.** $20x^2 + 22x + 6$
**38.** $36y^2 + 81y + 45$
**39.** $24x^2 - 42x + 9$
**40.** $48b^2 - 74b - 10$
**41.** $40m^2q + mq - 6q$
**42.** $15a^2b + 22ab + 8b$
**43.** $15n^4 - 39n^3 + 18n^2$
**44.** $24a^4 + 10a^3 - 4a^2$
**45.** $15x^2y^2 - 7xy^2 - 4y^2$
**46.** $14a^2b^3 + 15ab^3 - 9b^3$
**47.** $5a^2 - 7ab - 6b^2$
**48.** $6x^2 - 5xy - y^2$
**49.** $12s^2 + 11st - 5t^2$
**50.** $25a^2 + 25ab + 6b^2$
**51.** $6m^6n + 7m^5n^2 + 2m^4n^3$
**52.** $12k^3q^4 - 4k^2q^5 - kq^6$
**53.** $5 - 6x + x^2$
**54.** $7 + 8x + x^2$
**55.** $16 + 16x + 3x^2$
**56.** $18 + 65x + 7x^2$
**57.** $-10x^3 + 5x^2 + 140x$
**58.** $-18k^3 - 48k^2 + 66k$

**59.** On a quiz, a student factored $16x^2 - 24x + 5$ by grouping as follows.

$$\begin{aligned} 16x^2 - 24x + 5 &= 16x^2 - 4x - 20x + 5 \\ &= 4x(4x - 1) - 5(4x - 1) \qquad \text{His answer} \end{aligned}$$

He thought his answer was correct since it checked by multiplying. Why was the answer marked wrong? What is the correct factored form?

**60.** On the same quiz, another student factored $3k^3 - 12k^2 - 15k$ by first factoring out the common factor $3k$ to get $3k(k^2 - 4k - 5)$. Then she wrote

$$\begin{aligned} k^2 - 4k - 5 &= k^2 - 5k + k - 5 \\ &= k(k - 5) + 1(k - 5) \\ &= (k - 5)(k + 1). \qquad \text{Her answer} \end{aligned}$$

Why was the answer marked wrong? What is the correct factored form?

If a trinomial has a negative coefficient for the squared term, such as $-2x^2 + 11x - 12$, it is usually easier to factor by first factoring out the common factor $-1$:

$$\begin{aligned} -2x^2 + 11x - 12 &= -1(2x^2 - 11x + 12) \\ &= -1(2x - 3)(x - 4). \end{aligned}$$

*Use this method to factor each trinomial. See Example 7(b).*

**61.** $-x^2 - 4x + 21$
**62.** $-x^2 + x + 72$
**63.** $-3x^2 - x + 4$
**64.** $-5x^2 + 2x + 16$
**65.** $-2a^2 - 5ab - 2b^2$
**66.** $-3p^2 + 13pq - 4q^2$

**67.** The answer given in the back of the book for Exercise 61 is $-1(x + 7)(x - 3)$. Is $(x + 7)(3 - x)$ also a correct answer? Explain.

**68.** One answer for Exercise 62 is $-1(x + 8)(x - 9)$. Is $(-x - 8)(-x + 9)$ also a correct answer? Explain.

*Factor each polynomial. Remember to factor out the greatest common factor as the first step.*

**69.** $25q^2(m+1)^3 - 5q(m+1)^3 - 2(m+1)^3$

**70.** $18x^2(y-3)^2 - 21x(y-3)^2 - 4(y-3)^2$

**71.** $15x^2(r+3)^3 - 34xy(r+3)^3 - 16y^2(r+3)^3$

**72.** $4t^2(k+9)^7 + 20ts(k+9)^7 + 25s^2(k+9)^7$

*Find all integers k so that the trinomial can be factored using the methods of this section.* (Hint: *Try all possible factored forms with the given first and last terms. The coefficient of the variable will give the values of k.*)

**73.** $5x^2 + kx - 1$

**74.** $2c^2 + kc - 3$

**75.** $2m^2 + km + 5$

**76.** $3y^2 + ky + 3$

# 5.4 Special Factoring Rules

**OBJECTIVES**

1 Factor a difference of squares.
2 Factor a perfect square trinomial.
3 Factor a difference of cubes.
4 Factor a sum of cubes.

By reversing the rules for multiplication of binomials from the last chapter, we get rules for factoring polynomials in certain forms.

**OBJECTIVE 1 Factor a difference of squares.** The formula for the product of the sum and difference of the same two terms is

$$(x+y)(x-y) = x^2 - y^2.$$

Reversing this rule leads to the following special factoring rule.

**Factoring a Difference of Squares**

$$x^2 - y^2 = (x+y)(x-y)$$

For example,

$$m^2 - 16 = m^2 - 4^2 = (m+4)(m-4).$$

As the next examples show, the following conditions must be true for a binomial to be a difference of squares.

**1.** Both terms of the binomial must be squares, such as

$$x^2, \quad 9y^2, \quad 25, \quad 1, \quad m^4.$$

**2.** The second terms of the binomials must have different signs (one positive and one negative).

### EXAMPLE 1 Factoring Differences of Squares

Factor each binomial, if possible.

$$\begin{array}{l} \quad\; x^2 - y^2 = (x+y)(x-y) \\ \quad\;\;\downarrow\quad\;\downarrow\qquad\downarrow\;\;\;\downarrow\;\downarrow\quad\downarrow \\ \textbf{(a)}\; a^2 - 49 = a^2 - 7^2 = (a+7)(a-7) \end{array}$$

**(b)** $y^2 - m^2 = (y + m)(y - m)$

**(c)** $z^2 - \frac{9}{16} = z^2 - \left(\frac{3}{4}\right)^2 = \left(z + \frac{3}{4}\right)\left(z - \frac{3}{4}\right)$

**(d)** $x^2 - 8$

Because 8 is not the square of an integer, this binomial is not a difference of squares. It is a prime polynomial.

**(e)** $p^2 + 16$

Since $p^2 + 16$ is a *sum* of squares, it is not equal to $(p + 4)(p - 4)$. Also, using FOIL,

$$(p - 4)(p - 4) = p^2 - 8p + 16 \neq p^2 + 16$$

and

$$(p + 4)(p + 4) = p^2 + 8p + 16 \neq p^2 + 16,$$

so $p^2 + 16$ is a prime polynomial.

**Now Try Exercises 7, 9, and 11.**

**CAUTION** As Example 1(e) suggests, after any common factor is removed, a *sum* of squares cannot be factored.

## EXAMPLE 2 Factoring Differences of Squares

Factor each difference of squares.

$$x^2 - y^2 = (x + y)(x - y)$$

**(a)** $25m^2 - 16 = (5m)^2 - 4^2 = (5m + 4)(5m - 4)$

**(b)** $49z^2 - 64 = (7z)^2 - 8^2 = (7z + 8)(7z - 8)$

**Now Try Exercise 13.**

**NOTE** As in previous sections, you should always check a factored form by multiplying.

## EXAMPLE 3 Factoring More Complex Differences of Squares

Factor completely.

**(a)** $81y^2 - 36$

First factor out the common factor, 9.

$$\begin{aligned} 81y^2 - 36 &= 9(9y^2 - 4) && \text{Factor out 9.} \\ &= 9[(3y)^2 - 2^2] \\ &= 9(3y + 2)(3y - 2) && \text{Difference of squares} \end{aligned}$$

**(b)** $9x^2 - 4z^2 = (3x)^2 - (2z)^2 = (3x + 2z)(3x - 2z)$

**(c)** $p^4 - 36 = (p^2)^2 - 6^2 = (p^2 + 6)(p^2 - 6)$

Neither $p^2 + 6$ nor $p^2 - 6$ can be factored further.

**(d)** $m^4 - 16 = (m^2)^2 - 4^2$

$= (m^2 + 4)(m^2 - 4)$ Difference of squares

$= (m^2 + 4)(m + 2)(m - 2)$ Difference of squares again

**Now Try Exercises 17, 21, and 25.**

**CAUTION** Remember to factor again when any of the factors is a difference of squares, as in Example 3(d). Check by multiplying.

**OBJECTIVE 2 Factor a perfect square trinomial.** The expressions 144, $4x^2$, and $81m^6$ are called *perfect squares* because

$$144 = 12^2, \qquad 4x^2 = (2x)^2, \qquad \text{and} \qquad 81m^6 = (9m^3)^2.$$

A **perfect square trinomial** is a trinomial that is the square of a binomial. For example, $x^2 + 8x + 16$ is a perfect square trinomial because it is the square of the binomial $x + 4$:

$$x^2 + 8x + 16 = (x + 4)(x + 4) = (x + 4)^2.$$

For a trinomial to be a perfect square, *two of its terms must be perfect squares.* For this reason, $16x^2 + 4x + 15$ is not a perfect square trinomial because only the term $16x^2$ is a perfect square.

On the other hand, even if two of the terms are perfect squares, the trinomial may not be a perfect square trinomial. For example, $x^2 + 6x + 36$ has two perfect square terms, $x^2$ and 36, but it is not a perfect square trinomial. (Try to find a binomial that can be squared to give $x^2 + 6x + 36$.)

We can multiply to see that the square of a binomial gives one of the following perfect square trinomials.

**Factoring Perfect Square Trinomials**

$$x^2 + 2xy + y^2 = (x + y)^2$$
$$x^2 - 2xy + y^2 = (x - y)^2$$

The middle term of a perfect square trinomial is always twice the product of the two terms in the squared binomial (as shown in Section 4.6). Use this rule to check any attempt to factor a trinomial that appears to be a perfect square.

## EXAMPLE 4 Factoring a Perfect Square Trinomial

Factor $x^2 + 10x + 25$.

The term $x^2$ is a perfect square, and so is 25. Try to factor the trinomial as

$$x^2 + 10x + 25 = (x + 5)^2.$$

To check, take twice the product of the two terms in the squared binomial.

$$2 \cdot x \cdot 5 = 10x$$

Twice — First term of binomial — Last term of binomial

Since $10x$ is the middle term of the trinomial, the trinomial is a perfect square and can be factored as $(x + 5)^2$. Thus,

$$x^2 + 10x + 25 = (x + 5)^2.$$

**Now Try Exercise 33.**

### EXAMPLE 5 Factoring Perfect Square Trinomials

Factor each trinomial.

**(a)** $x^2 - 22x + 121$

The first and last terms are perfect squares ($121 = 11^2$ or $(-11)^2$). Check to see whether the middle term of $x^2 - 22x + 121$ is twice the product of the first and last terms of the binomial $x - 11$.

$$2 \cdot x \cdot (-11) = -22x$$

Twice — First term — Last term

Since twice the product of the first and last terms of the binomial is the middle term, $x^2 - 22x + 121$ is a perfect square trinomial and

$$x^2 - 22x + 121 = (x - 11)^2.$$

Notice that the sign of the second term in the squared binomial is the same as the sign of the middle term in the trinomial.

**(b)** $9m^2 - 24m + 16 = (3m)^2 + 2(3m)(-4) + (-4)^2 = (3m - 4)^2$

Twice — First term — Last term

**(c)** $25y^2 + 20y + 16$

The first and last terms are perfect squares.

$$25y^2 = (5y)^2 \qquad \text{and} \qquad 16 = 4^2$$

Twice the product of the first and last terms of the binomial $5y + 4$ is

$$2 \cdot 5y \cdot 4 = 40y,$$

which is not the middle term of $25y^2 + 20y + 16$. This trinomial is not a perfect square. In fact, the trinomial cannot be factored even with the methods of the previous sections; it is a prime polynomial.

**(d)** $12z^3 + 60z^2 + 75z$

Factor out the common factor, $3z$, first.

$$\begin{aligned} 12z^3 + 60z^2 + 75z &= 3z(4z^2 + 20z + 25) \\ &= 3z[(2z)^2 + 2(2z)(5) + 5^2] \\ &= 3z(2z + 5)^2 \end{aligned}$$

**Now Try Exercises 35, 43, and 51.**

**NOTE** As mentioned in Example 5(a), the sign of the second term in the squared binomial is always the same as the sign of the middle term in the trinomial. Also, the first and last terms of a perfect square trinomial must be *positive,* because they are squares. For example, the polynomial $x^2 - 2x - 1$ cannot be a perfect square because the last term is negative.

Perfect square trinomials can also be factored using grouping or FOIL, although using the method of this section is often easier.

**OBJECTIVE 3 Factor a difference of cubes.** Just as we factored the difference of squares in Objective 1, we can also factor the **difference of cubes** using the following pattern.

**Factoring a Difference of Cubes**

$$x^3 - y^3 = (x - y)(x^2 + xy + y^2)$$

This pattern *should be memorized.* Multiply on the right to see that the pattern gives the correct factors.

$$\begin{array}{rrrr} & x^2 & + xy & + y^2 \\ & & x & - y \\ \hline & - x^2y & - xy^2 & - y^3 \\ x^3 & + x^2y & + xy^2 & \\ \hline x^3 & & & - y^3 \end{array}$$

Notice the pattern of the terms in the factored form of $x^3 - y^3$.

- $x^3 - y^3 =$ (a binomial factor)(a trinomial factor)
- The binomial factor has the difference of the cube roots of the given terms.
- The terms in the trinomial factor are all positive.
- What you write in the binomial factor determines the trinomial factor:

First term squared + positive product of the terms + second term squared

$$x^3 - y^3 = (x - y)(\;x^2 \;+\; xy \;+\; y^2\;).$$

**CAUTION** The polynomial $x^3 - y^3$ is not equivalent to $(x - y)^3$, because $(x - y)^3$ can also be written as

$$\begin{aligned} (x - y)^3 &= (x - y)(x - y)(x - y) \\ &= (x - y)(x^2 - 2xy + y^2) \end{aligned}$$

but

$$x^3 - y^3 = (x - y)(x^2 + xy + y^2).$$

### EXAMPLE 6 Factoring Differences of Cubes

Factor the following.

**(a)** $m^3 - 125$

Let $x = m$ and $y = 5$ in the pattern for the difference of cubes.

$$x^3 - y^3 = (x - y)(x^2 + xy + y^2)$$

$$m^3 - 125 = m^3 - 5^3 = (m - 5)(m^2 + 5m + 5^2) \quad \text{Let } x = m, y = 5.$$
$$= (m - 5)(m^2 + 5m + 25)$$

**(b)** $8p^3 - 27$

Since $8p^3 = (2p)^3$ and $27 = 3^3$,

$$8p^3 - 27 = (2p)^3 - 3^3$$
$$= (2p - 3)[(2p)^2 + (2p)3 + 3^2]$$
$$= (2p - 3)(4p^2 + 6p + 9).$$

**(c)** $4m^3 - 32 = 4(m^3 - 8)$

$$= 4(m^3 - 2^3)$$
$$= 4(m - 2)(m^2 + 2m + 4)$$

**(d)** $125t^3 - 216s^6 = (5t)^3 - (6s^2)^3$

$$= (5t - 6s^2)[(5t)^2 + 5t(6s^2) + (6s^2)^2]$$
$$= (5t - 6s^2)(25t^2 + 30ts^2 + 36s^4)$$

**Now Try Exercises 59, 63, and 69.**

**CAUTION** A common error in factoring a difference of cubes, such as $x^3 - y^3 = (x - y)(x^2 + xy + y^2)$, is to try to factor $x^2 + xy + y^2$. It is easy to confuse this factor with a perfect square trinomial, $x^2 + 2xy + y^2$. Because there is no 2 in $x^2 + xy + y^2$, it is very unusual to be able to further factor an expression of the form $x^2 + xy + y^2$.

**OBJECTIVE 4 Factor a sum of cubes.** A sum of squares, such as $m^2 + 25$, cannot be factored using real numbers, but a **sum of cubes** can be factored by the following pattern, *which should be memorized.*

**Factoring a Sum of Cubes**

$$x^3 + y^3 = (x + y)(x^2 - xy + y^2)$$

Compare the pattern for the *sum* of cubes with the pattern for the *difference* of cubes. The only difference between them is the positive and negative signs.

Positive

$$x^3 - y^3 = (x - y)(x^2 + xy + y^2)$$ Difference of cubes

Same sign — Opposite sign

Positive

$$x^3 + y^3 = (x + y)(x^2 - xy + y^2)$$ Sum of cubes

Same sign — Opposite sign

Observing these relationships should help you to remember these patterns.

**EXAMPLE 7 Factoring Sums of Cubes**

Factor.

**(a)** $$\begin{aligned} k^3 + 27 &= k^3 + 3^3 \\ &= (k + 3)(k^2 - 3k + 3^2) \\ &= (k + 3)(k^2 - 3k + 9) \end{aligned}$$

**(b)** $$\begin{aligned} 8m^3 + 125 &= (2m)^3 + 5^3 \\ &= (2m + 5)[(2m)^2 - 2m(5) + 5^2] \\ &= (2m + 5)(4m^2 - 10m + 25) \end{aligned}$$

**(c)** $$\begin{aligned} 1000a^6 + 27b^3 &= (10a^2)^3 + (3b)^3 \\ &= (10a^2 + 3b)[(10a^2)^2 - (10a^2)(3b) + (3b)^2] \\ &= (10a^2 + 3b)(100a^4 - 30a^2b + 9b^2) \end{aligned}$$

**Now Try Exercises 61 and 71.**

The methods of factoring discussed in this section are summarized here.

**Special Factorizations**

| | |
|---|---|
| **Difference of squares** | $x^2 - y^2 = (x + y)(x - y)$ |
| **Perfect square trinomials** | $x^2 + 2xy + y^2 = (x + y)^2$ |
| | $x^2 - 2xy + y^2 = (x - y)^2$ |
| **Difference of cubes** | $x^3 - y^3 = (x - y)(x^2 + xy + y^2)$ |
| **Sum of cubes** | $x^3 + y^3 = (x + y)(x^2 - xy + y^2)$ |

Remember the *sum* of *squares* can be factored only if the terms have a common factor.

## 5.4 EXERCISES

**For Extra Help**

Student's Solutions Manual

MyMathLab

InterAct Math Tutorial Software

AW Math Tutor Center

MathXL

Digital Video Tutor CD 8/Videotape 8

**1.** To help you factor the difference of squares, complete the following list of squares.

| | | | | |
|---|---|---|---|---|
| $1^2 =$ ____ | $2^2 =$ ____ | $3^2 =$ ____ | $4^2 =$ ____ | $5^2 =$ ____ |
| $6^2 =$ ____ | $7^2 =$ ____ | $8^2 =$ ____ | $9^2 =$ ____ | $10^2 =$ ____ |
| $11^2 =$ ____ | $12^2 =$ ____ | $13^2 =$ ____ | $14^2 =$ ____ | $15^2 =$ ____ |
| $16^2 =$ ____ | $17^2 =$ ____ | $18^2 =$ ____ | $19^2 =$ ____ | $20^2 =$ ____ |

**2.** The following powers of $x$ are all perfect squares: $x^2$, $x^4$, $x^6$, $x^8$, $x^{10}$. Based on this observation, we may make a conjecture (an educated guess) that if the power of a variable is divisible by ____ (with 0 remainder), then we have a perfect square.

**3.** To help you factor the sum or difference of cubes, complete the following list of cubes.

| | | | | |
|---|---|---|---|---|
| $1^3 =$ ____ | $2^3 =$ ____ | $3^3 =$ ____ | $4^3 =$ ____ | $5^3 =$ ____ |
| $6^3 =$ ____ | $7^3 =$ ____ | $8^3 =$ ____ | $9^3 =$ ____ | $10^3 =$ ____ |

**4.** The following powers of $x$ are all perfect cubes: $x^3$, $x^6$, $x^9$, $x^{12}$, $x^{15}$. Based on this observation, we may make a conjecture that if the power of a variable is divisible by ____ (with 0 remainder), then we have a perfect cube.

**5.** Identify each monomial as a perfect square, a perfect cube, both of these, or neither of these.

**(a)** $64x^6y^{12}$ **(b)** $125t^6$ **(c)** $49x^{12}$ **(d)** $81r^{10}$

**6.** What must be true for $x^n$ to be both a perfect square and a perfect cube?

*Factor each binomial completely. Use your answers from Exercises 1 and 2 as necessary. See Examples 1–3.*

**7.** $y^2 - 25$ **8.** $t^2 - 16$ **9.** $p^2 - \frac{1}{9}$ **10.** $q^2 - \frac{1}{4}$

**11.** $m^2 + 64$ **12.** $k^2 + 49$ **13.** $9r^2 - 4$ **14.** $4x^2 - 9$

**15.** $36m^2 - \frac{16}{25}$ **16.** $100b^2 - \frac{4}{49}$ **17.** $36x^2 - 16$ **18.** $32a^2 - 8$

**19.** $196p^2 - 225$ **20.** $361q^2 - 400$ **21.** $16r^2 - 25a^2$ **22.** $49m^2 - 100p^2$

**23.** $100x^2 + 49$ **24.** $81w^2 + 16$ **25.** $p^4 - 49$ **26.** $r^4 - 25$

**27.** $x^4 - 1$ **28.** $y^4 - 16$ **29.** $p^4 - 256$ **30.** $16k^4 - 1$

**31.** When a student was directed to factor $x^4 - 81$ completely, his teacher did not give him full credit for the answer $(x^2 + 9)(x^2 - 9)$. The student argued that since his answer does indeed give $x^4 - 81$ when multiplied out, he should be given full credit. Was the teacher justified in her grading of this item? Why or why not?

**32.** The binomial $4x^2 + 16$ is a sum of squares that *can* be factored. How is this binomial factored? When can the sum of squares be factored?

*Factor each trinomial completely. It may be necessary to factor out the greatest common factor first. See Examples 4 and 5.*

**33.** $w^2 + 2w + 1$ **34.** $p^2 + 4p + 4$ **35.** $x^2 - 8x + 16$

**36.** $x^2 - 10x + 25$ **37.** $t^2 + t + \frac{1}{4}$ **38.** $m^2 + \frac{2}{3}m + \frac{1}{9}$

**39.** $x^2 - 1.0x + .25$

**40.** $y^2 - 1.4y + .49$

**41.** $2x^2 + 24x + 72$

**42.** $3y^2 - 48y + 192$

**43.** $16x^2 - 40x + 25$

**44.** $36y^2 - 60y + 25$

**45.** $49x^2 - 28xy + 4y^2$

**46.** $4z^2 - 12zw + 9w^2$

**47.** $64x^2 + 48xy + 9y^2$

**48.** $9t^2 + 24tr + 16r^2$

**49.** $-50h^2 + 40hy - 8y^2$

**50.** $-18x^2 - 48xy - 32y^2$

**51.** $4k^3 - 4k^2 + 9k$

**52.** $9r^3 + 6r^2 + 16r$

**53.** $25z^4 + 5z^3 + z^2$

**54.** In the polynomial $9y^2 + 14y + 25$, the first and last terms are perfect squares. Can the polynomial be factored? If it can, factor it. If it cannot, explain why it is not a perfect square trinomial.

*Find the value of the indicated variable.* (Hint: *Perform the squaring on the right side, then solve the equation using the steps given in Chapter 2.)*

**55.** Find a value of $b$ so that $x^2 + bx + 25 = (x + 5)^2$.

**56.** For what value of $c$ is $4m^2 - 12m + c = (2m - 3)^2$?

**57.** Find $a$ so that $ay^2 - 12y + 4 = (3y - 2)^2$.

**58.** Find $b$ so that $100a^2 + ba + 9 = (10a + 3)^2$.

*Factor each binomial completely. Use your answers from Exercises 3 and 4 as necessary. See Examples 6 and 7.*

**59.** $a^3 - 1$

**60.** $m^3 - 8$

**61.** $m^3 + 8$

**62.** $b^3 + 1$

**63.** $27x^3 - 64$

**64.** $64y^3 - 27$

**65.** $6p^3 + 6$

**66.** $81x^3 + 3$

**67.** $5x^3 + 40$

**68.** $128y^3 - 54$

**69.** $2y^3 - 16x^3$

**70.** $27w^3 - 216z^3$

**71.** $8p^3 + 729q^3$

**72.** $64x^3 + 125y^3$

**73.** $27a^3 + 64b^3$

**74.** $125m^3 - 8p^3$

**75.** $125t^3 + 8s^3$

**76.** $27r^3 + 1000s^3$

**77.** State and give the name of each of the five special factoring rules. For each rule, give the numbers of three exercises from this exercise set that are examples.

## RELATING CONCEPTS (EXERCISES 78–81)

**For Individual or Group Work**

*We have seen that multiplication and factoring are reverse processes. We know that multiplication and division are also related: To check a division problem, we multiply the quotient by the divisor to get the dividend. To see how factoring and division are related,* ***work Exercises 78–81 in order.***

**78.** Factor $10x^2 + 11x - 6$.

**79.** Use long division to divide $10x^2 + 11x - 6$ by $2x + 3$.

**80.** Could we have predicted the result in Exercise 79 from the result in Exercise 78? Explain.

**81.** Divide $x^3 - 1$ by $x - 1$. Use your answer to factor $x^3 - 1$.

*Extend the methods of factoring presented so far in this chapter to factor each polynomial completely.*

**82.** $(m + n)^2 - (m - n)^2$

**83.** $(a - b)^3 - (a + b)^3$

**84.** $m^2 - p^2 + 2m + 2p$

**85.** $3r - 3k + 3r^2 - 3k^2$

## SUMMARY EXERCISES ON FACTORING

These mixed exercises are included to give you practice in selecting an appropriate method for factoring a particular polynomial. As you factor a polynomial, ask yourself these questions to decide on a suitable factoring technique.

**Factoring a Polynomial**

1. Is there a common factor? If so, factor it out.
2. How many terms are in the polynomial?

   *Two terms:* Check to see whether it is a difference of squares or the sum or difference of cubes. If so, factor as in Section 5.4.

   *Three terms:* Is it a perfect square trinomial? If the trinomial is not a perfect square, check to see whether the coefficient of the squared term is 1. If so, use the method of Section 5.2. If the coefficient of the squared term of the trinomial is not 1, use the general factoring methods of Section 5.3.

   *Four terms:* Try to factor the polynomial by grouping as in Section 5.1.
3. Can any factors be factored further? If so, factor them.

*Factor each polynomial completely.*

**1.** $a^2 - 4a - 12$

**2.** $a^2 + 17a + 72$

**3.** $6y^2 - 6y - 12$

**4.** $7y^6 + 14y^5 - 168y^4$

**5.** $6a + 12b + 18c$

**6.** $m^2 - 3mn - 4n^2$

**7.** $p^2 - 17p + 66$

**8.** $z^2 - 6z + 7z - 42$

**9.** $10z^2 - 7z - 6$

**10.** $2m^2 - 10m - 48$

**11.** $m^2 - n^2 + 5m - 5n$

**12.** $15y + 5$

**13.** $8a^5 - 8a^4 - 48a^3$

**14.** $8k^2 - 10k - 3$

**15.** $z^2 - 3za - 10a^2$

**16.** $50z^2 - 100$

**17.** $x^2 - 4x - 5x + 20$

**18.** $100n^2r^2 + 30nr^3 - 50n^2r$

**19.** $6n^2 - 19n + 10$

**20.** $9y^2 + 12y - 5$

**21.** $16x + 20$

**22.** $m^2 + 2m - 15$

**23.** $6y^2 - 5y - 4$

**24.** $m^2 - 81$

**25.** $6z^2 + 31z + 5$

**26.** $5z^2 + 24z - 5 + 3z + 15$

**27.** $4k^2 - 12k + 9$

**28.** $8p^2 + 23p - 3$

**29.** $54m^2 - 24z^2$

**30.** $8m^2 - 2m - 3$

**31.** $3k^2 + 4k - 4$

**32.** $45a^3b^5 - 60a^4b^2 + 75a^6b^4$

**33.** $14k^3 + 7k^2 - 70k$

**34.** $5 + r - 5s - rs$

**35.** $y^4 - 16$

**36.** $20y^5 - 30y^4$

**37.** $8m - 16m^2$

**38.** $k^2 - 16$

**39.** $z^3 - 8$

**40.** $y^2 - y - 56$

**41.** $k^2 + 9$

**42.** $27p^{10} - 45p^9 - 252p^8$

**43.** $32m^9 + 16m^5 + 24m^3$

**44.** $8m^3 + 125$

**45.** $16r^2 + 24rm + 9m^2$

**46.** $z^2 - 12z + 36$

**47.** $15h^2 + 11hg - 14g^2$

**48.** $5z^3 - 45z^2 + 70z$

**49.** $k^2 - 11k + 30$

**50.** $64p^2 - 100m^2$

**51.** $3k^3 - 12k^2 - 15k$

**52.** $y^2 - 4yk - 12k^2$

**53.** $1000p^3 + 27$

**54.** $64r^3 - 343$

**55.** $6 + 3m + 2p + mp$

**56.** $2m^2 + 7mn - 15n^2$

**57.** $16z^2 - 8z + 1$

**58.** $125m^4 - 400m^3n + 195m^2n^2$

**59.** $108m^2 - 36m + 3$

**60.** $100a^2 - 81y^2$

**61.** $64m^2 - 40mn + 25n^2$

**62.** $4y^2 - 25$

**63.** $32z^3 + 56z^2 - 16z$

**64.** $10m^2 + 25m - 60$

**65.** $20 + 5m + 12n + 3mn$

**66.** $4 - 2q - 6p + 3pq$

**67.** $6a^2 + 10a - 4$

**68.** $36y^6 - 42y^5 - 120y^4$

**69.** $a^3 - b^3 + 2a - 2b$

**70.** $16k^2 - 48k + 36$

**71.** $64m^2 - 80mn + 25n^2$

**72.** $72y^3z^2 + 12y^2 - 24y^4z^2$

**73.** $8k^2 - 2kh - 3h^2$

**74.** $2a^2 - 7a - 30$

**75.** $(m + 1)^3 + 1$

**76.** $8a^3 - 27$

**77.** $10y^2 - 7yz - 6z^2$

**78.** $m^2 - 4m + 4$

**79.** $8a^2 + 23ab - 3b^2$

**80.** $a^4 - 625$

## RELATING CONCEPTS (EXERCISES 81–88)

**For Individual or Group Work**

A binomial may be *both* a difference of squares *and* a difference of cubes. One example of such a binomial is $x^6 - 1$. Using the techniques of Section 5.4, one factoring method will give the complete factored form, while the other will not. ***Work Exercises 81–88 in order*** *to determine the method to use if you have to make such a decision.*

**81.** Factor $x^6 - 1$ as the difference of squares.

**82.** The factored form obtained in Exercise 81 consists of a difference of cubes multiplied by a sum of cubes. Factor each binomial further.

**83.** Now start over and factor $x^6 - 1$ as the difference of cubes.

**84.** The factored form obtained in Exercise 83 consists of a binomial which is a difference of squares and a trinomial. Factor the binomial further.

**85.** Compare your results in Exercises 82 and 84. Which one of these is the completely factored form?

**86.** Verify that the trinomial in the factored form in Exercise 84 is the product of the two trinomials in the factored form in Exercise 82.

**87.** Use the results of Exercises 81–86 to complete the following statement: In general, if I must choose between factoring first using the method for difference of squares or the method for difference of cubes, I should choose the __________ method to eventually obtain the complete factored form.

**88.** Find the *complete* factored form of $x^6 - 729$ using the knowledge you have gained in Exercises 81–87.

# 5.5 Solving Quadratic Equations by Factoring

**OBJECTIVES**

1. Solve quadratic equations by factoring.
2. Solve other equations by factoring.

Galileo Galilei (1564–1642) developed theories to explain physical phenomena and set up experiments to test his ideas. According to legend, Galileo dropped objects of different weights from the Leaning Tower of Pisa to disprove the belief that heavier objects fall faster than lighter objects. He developed a formula for freely falling objects described by

$$d = 16t^2,$$

where $d$ is the distance in feet that an object falls (disregarding air resistance) in $t$ seconds, regardless of weight.

The equation $d = 16t^2$ is a *quadratic equation,* the subject of this section. A quadratic equation contains a squared term and no terms of higher degree.

### Quadratic Equation

A **quadratic equation** is an equation that can be written in the form

$$\boldsymbol{ax^2 + bx + c = 0},$$

where $a$, $b$, and $c$ are real numbers, with $a \neq 0$.

The form $ax^2 + bx + c = 0$ is the **standard form** of a quadratic equation. For example,

$$x^2 + 5x + 6 = 0, \qquad 2a^2 - 5a = 3, \qquad \text{and} \qquad y^2 = 4$$

are all quadratic equations, but only $x^2 + 5x + 6 = 0$ is in standard form.

Up to now, we have factored *expressions,* including many quadratic expressions of the form $ax^2 + bx + c$. In this section, we see how we can use factored quadratic expressions to solve quadratic *equations.*

**OBJECTIVE 1 Solve quadratic equations by factoring.** We use the **zero-factor property** to solve a quadratic equation by factoring.

### Zero-Factor Property

If $a$ and $b$ are real numbers and if $ab = 0$, then $a = 0$ or $b = 0$.

That is, if the product of two numbers is 0, then at least one of the numbers must be 0. One number *must* be 0, but both *may* be 0.

### EXAMPLE 1 Using the Zero-Factor Property

Solve each equation.

**(a)** $(x + 3)(2x - 1) = 0$

The product $(x + 3)(2x - 1)$ is equal to 0. By the zero-factor property, the only way that the product of these two factors can be 0 is if at least one of the factors equals 0. Therefore, either $x + 3 = 0$ or $2x - 1 = 0$. Solve each of these two linear equations as in Chapter 2.

$$x + 3 = 0 \quad \text{or} \quad 2x - 1 = 0 \qquad \text{Zero-factor property}$$
$$x = -3 \quad \text{or} \quad 2x = 1 \qquad \text{Add 1 to each side.}$$
$$x = \frac{1}{2} \qquad \text{Divide each side by 2.}$$

The given equation, $(x + 3)(2x - 1) = 0$, has two solutions, $-3$ and $\frac{1}{2}$. Check these solutions by substituting $-3$ for $x$ in the original equation, $(x + 3)(2x - 1) = 0$. Then start over and substitute $\frac{1}{2}$ for $x$.

If $x = -3$, then

$$(x + 3)(2x - 1) = 0$$
$$(-3 + 3)[2(-3) - 1] = 0 \quad ?$$
$$0(-7) = 0. \quad \text{True}$$

If $x = \frac{1}{2}$, then

$$(x + 3)(2x - 1) = 0$$
$$\left(\frac{1}{2} + 3\right)\left(2 \cdot \frac{1}{2} - 1\right) = 0 \quad ?$$
$$\frac{7}{2}(1 - 1) = 0 \quad ?$$
$$\frac{7}{2} \cdot 0 = 0. \quad \text{True}$$

Both $-3$ and $\frac{1}{2}$ result in true equations, so the solution set is $\left\{-3, \frac{1}{2}\right\}$.

**(b)** $y(3y - 4) = 0$

$$y(3y - 4) = 0$$
$$y = 0 \quad \text{or} \quad 3y - 4 = 0 \qquad \text{Zero-factor property}$$
$$3y = 4$$
$$y = \frac{4}{3}$$

Check these solutions by substituting each one in the original equation. The solution set is $\left\{0, \frac{4}{3}\right\}$.

**Now Try Exercises 3 and 5.**

**NOTE** The word *or* as used in Example 1 means "one or the other or both."

In Example 1, each equation to be solved was given with the polynomial in factored form. If the polynomial in an equation is not already factored, first make sure that the equation is in standard form. Then factor.

### EXAMPLE 2 Solving Quadratic Equations

Solve each equation.

**(a)** $x^2 - 5x = -6$

First, rewrite the equation in standard form by adding 6 to each side.

$$x^2 - 5x = -6$$
$$x^2 - 5x + 6 = 0 \qquad \text{Add 6.}$$

Now factor $x^2 - 5x + 6$. Find two numbers whose product is 6 and whose sum is $-5$. These two numbers are $-2$ and $-3$, so the equation becomes

$$(x-2)(x-3) = 0. \quad \text{Factor.}$$
$$x-2=0 \quad \text{or} \quad x-3=0 \quad \text{Zero-factor property}$$
$$x=2 \quad \text{or} \quad x=3 \quad \text{Solve each equation.}$$

*Check:* If $x = 2$, then

$$x^2 - 5x = -6$$
$$2^2 - 5(2) = -6 \quad ?$$
$$4 - 10 = -6 \quad ?$$
$$-6 = -6. \quad \text{True}$$

If $x = 3$, then

$$x^2 - 5x = -6$$
$$3^2 - 5(3) = -6 \quad ?$$
$$9 - 15 = -6 \quad ?$$
$$-6 = -6. \quad \text{True}$$

Both solutions check, so the solution set is $\{2, 3\}$.

**(b)** $y^2 = y + 20$

Rewrite the equation in standard form.

$$y^2 = y + 20$$
$$y^2 - y - 20 = 0 \quad \text{Subtract } y \text{ and } 20.$$
$$(y-5)(y+4) = 0 \quad \text{Factor.}$$
$$y-5=0 \quad \text{or} \quad y+4=0 \quad \text{Zero-factor property}$$
$$y=5 \quad \text{or} \quad y=-4 \quad \text{Solve each equation.}$$

Check these solutions by substituting each one in the original equation. The solution set is $\{-4, 5\}$.

**Now Try Exercise 21.**

In summary, follow these steps to solve quadratic equations by factoring.

**Solving a Quadratic Equation by Factoring**

*Step 1* **Write the equation in standard form,** that is, with all terms on one side of the equals sign in descending powers of the variable and 0 on the other side.

*Step 2* **Factor** completely.

*Step 3* **Use the zero-factor property** to set each factor with a variable equal to 0, and solve the resulting equations.

*Step 4* **Check** each solution in the original equation.

**NOTE** Not all quadratic equations can be solved by factoring. A more general method for solving such equations is given in Chapter 11.

## EXAMPLE 3 Solving a Quadratic Equation with a Common Factor

Solve $4p^2 + 40 = 26p$.

Subtract $26p$ from each side and write the equation in standard form to get

$$4p^2 - 26p + 40 = 0.$$

$$2(2p^2 - 13p + 20) = 0 \quad \text{Factor out 2.}$$
$$2p^2 - 13p + 20 = 0 \quad \text{Divide each side by 2.}$$
$$(2p - 5)(p - 4) = 0 \quad \text{Factor.}$$
$$2p - 5 = 0 \quad \text{or} \quad p - 4 = 0 \quad \text{Zero-factor property}$$
$$2p = 5 \quad \text{or} \quad p = 4$$
$$p = \frac{5}{2}$$

Check that the solution set is $\left\{\frac{5}{2}, 4\right\}$ by substituting each solution in the original equation.

**Now Try Exercise 31.**

## EXAMPLE 4 Solving Quadratic Equations

Solve each equation.

**(a)** $16m^2 - 25 = 0$

Factor the left side of the equation as the difference of squares.

$$(4m + 5)(4m - 5) = 0$$
$$4m + 5 = 0 \quad \text{or} \quad 4m - 5 = 0 \quad \text{Zero-factor property}$$
$$4m = -5 \quad \text{or} \quad 4m = 5 \quad \text{Solve each equation.}$$
$$m = -\frac{5}{4} \quad \text{or} \quad m = \frac{5}{4}$$

Check the two solutions, $-\frac{5}{4}$ and $\frac{5}{4}$, in the original equation. The solution set is $\left\{-\frac{5}{4}, \frac{5}{4}\right\}$.

**(b)** $y^2 = 2y$

First write the equation in standard form.

$$y^2 - 2y = 0 \quad \text{Standard form}$$
$$y(y - 2) = 0 \quad \text{Factor.}$$
$$y = 0 \quad \text{or} \quad y - 2 = 0 \quad \text{Zero-factor property}$$
$$y = 2 \quad \text{Solve.}$$

The solution set is $\{0, 2\}$.

**(c)** $k(2k + 1) = 3$

Write the equation in standard form.

$$k(2k + 1) = 3$$
$$2k^2 + k = 3 \quad \text{Distributive property}$$
$$2k^2 + k - 3 = 0 \quad \text{Subtract 3.}$$
$$(k - 1)(2k + 3) = 0 \quad \text{Factor.}$$
$$k - 1 = 0 \quad \text{or} \quad 2k + 3 = 0 \quad \text{Zero-factor property}$$
$$k = 1 \quad \text{or} \quad 2k = -3$$
$$k = -\frac{3}{2}$$

The solution set is $\left\{1, -\frac{3}{2}\right\}$.

**Now Try Exercises 37, 41, and 45.**

**CAUTION** In Example 4(b) it is tempting to begin by dividing both sides of the equation $y^2 = 2y$ by $y$ to get $y = 2$. Note that we do not get the other solution, 0, if we divide by a variable. (We may divide each side of an equation by a *nonzero* real number, however. For instance, in Example 3 we divided each side by 2.)

In Example 4(c) we could not use the zero-factor property to solve the equation $k(2k + 1) = 3$ in its given form because of the 3 on the right. Remember that the zero-factor property applies only to a product that equals 0.

**OBJECTIVE 2 Solve other equations by factoring.** We can also use the zero-factor property to solve equations that involve more than two factors with variables, as shown in Example 5. (These equations are *not* quadratic equations. Why not?)

### EXAMPLE 5 Solving Equations with More Than Two Variable Factors

Solve each equation.

**(a)**
$$6z^3 - 6z = 0$$
$$6z(z^2 - 1) = 0 \quad \text{Factor out } 6z.$$
$$6z(z + 1)(z - 1) = 0 \quad \text{Factor } z^2 - 1.$$

By an extension of the zero-factor property, this product can equal 0 only if at least one of the factors is 0. Write and solve three equations, one for each factor with a variable.

$$6z = 0 \quad \text{or} \quad z + 1 = 0 \quad \text{or} \quad z - 1 = 0$$
$$z = 0 \quad \text{or} \quad z = -1 \quad \text{or} \quad z = 1$$

Check by substituting, in turn, 0, $-1$, and 1 in the original equation. The solution set is $\{-1, 0, 1\}$.

**(b)**
$$(3x - 1)(x^2 - 9x + 20) = 0$$
$$(3x - 1)(x - 5)(x - 4) = 0 \quad \text{Factor } x^2 - 9x + 20.$$
$$3x - 1 = 0 \quad \text{or} \quad x - 5 = 0 \quad \text{or} \quad x - 4 = 0 \quad \text{Zero-factor property}$$
$$x = \frac{1}{3} \quad \text{or} \quad x = 5 \quad \text{or} \quad x = 4$$

The solutions of the original equation are $\frac{1}{3}$, 4, and 5. Check each solution to verify that the solution set is $\{\frac{1}{3}, 4, 5\}$.

**Now Try Exercises 51 and 55.**

**CAUTION** In Example 5(b), it would be unproductive to begin by multiplying the two factors together. Keep in mind that the zero-factor property requires the *product* of two or more factors; this product must equal 0. Always consider first whether an equation is given in an appropriate form for the zero-factor property.

### EXAMPLE 6 Solving an Equation Requiring Multiplication Before Factoring

Solve $(3x + 1)x = (x + 1)^2 + 5$.

The zero-factor property requires the *product* of two or more factors to equal 0. To write this equation in the required form, we must first multiply on both sides and collect terms on one side.

$$(3x + 1)x = (x + 1)^2 + 5$$

$$3x^2 + x = x^2 + 2x + 1 + 5 \qquad \text{Multiply.}$$

$$3x^2 + x = x^2 + 2x + 6 \qquad \text{Combine like terms.}$$

$$2x^2 - x - 6 = 0 \qquad \text{Standard form}$$

$$(2x + 3)(x - 2) = 0 \qquad \text{Factor.}$$

$$2x + 3 = 0 \quad \text{or} \quad x - 2 = 0 \qquad \text{Zero-factor property}$$

$$x = -\frac{3}{2} \quad \text{or} \quad x = 2$$

Check that the solution set is $\left\{-\frac{3}{2}, 2\right\}$.

**Now Try Exercise 65.**

## 5.5 EXERCISES

**For Extra Help**

Student's Solutions Manual

MyMathLab

InterAct Math Tutorial Software

AW Math Tutor Center

MathXL

Digital Video Tutor CD 8/Videotape 8

*Solve each equation and check your solutions. See Example 1.*

**1.** $(x + 5)(x - 2) = 0$ **2.** $(x - 1)(x + 8) = 0$ **3.** $(2m - 7)(m - 3) = 0$

**4.** $(6k + 5)(k + 4) = 0$ **5.** $t(6t + 5) = 0$ **6.** $w(4w + 1) = 0$

**7.** $2x(3x - 4) = 0$ **8.** $6y(4y + 9) = 0$

**9.** $\left(x + \frac{1}{2}\right)\left(2x - \frac{1}{3}\right) = 0$ **10.** $\left(a + \frac{2}{3}\right)\left(5a - \frac{1}{2}\right) = 0$

**11.** $(.5z - 1)(2.5z + 2) = 0$ **12.** $(.25x + 1)(x - .5) = 0$

**13.** $(x - 9)(x - 9) = 0$ **14.** $(2y + 1)(2y + 1) = 0$

**15.** What is wrong with this "solution"?

$$2x(3x - 4) = 0$$

$$x = 2 \quad \text{or} \quad x = 0 \quad \text{or} \quad 3x - 4 = 0$$

$$x = \frac{4}{3}$$

The solution set is $\left\{2, 0, \frac{4}{3}\right\}$.

**16.** What is wrong with this "solution"?

$$x(7x - 1) = 0$$

$$7x - 1 = 0 \qquad \text{Zero-factor property}$$

$$x = \frac{1}{7}$$

The solution set is $\left\{\frac{1}{7}\right\}$.

*Solve each equation and check your solutions. See Examples 2–4.*

**17.** $y^2 + 3y + 2 = 0$ **18.** $p^2 + 8p + 7 = 0$ **19.** $y^2 - 3y + 2 = 0$

**20.** $r^2 - 4r + 3 = 0$ **21.** $x^2 = 24 - 5x$ **22.** $t^2 = 2t + 15$

**23.** $x^2 = 3 + 2x$ **24.** $m^2 = 4 + 3m$ **25.** $z^2 + 3z = -2$

**26.** $p^2 - 2p = 3$ **27.** $m^2 + 8m + 16 = 0$ **28.** $b^2 - 6b + 9 = 0$

**29.** $3x^2 + 5x - 2 = 0$ **30.** $6r^2 - r - 2 = 0$ **31.** $12p^2 = 8 - 10p$

**32.** $18x^2 = 12 + 15x$ **33.** $9s^2 + 12s = -4$ **34.** $36x^2 + 60x = -25$

**35.** $y^2 - 9 = 0$ **36.** $m^2 - 100 = 0$ **37.** $16k^2 - 49 = 0$

**38.** $4w^2 - 9 = 0$ **39.** $n^2 = 121$ **40.** $x^2 = 400$

*Solve each equation and check your solutions. See Examples 4–6.*

**41.** $x^2 = 7x$ **42.** $t^2 = 9t$ **43.** $6r^2 = 3r$

**44.** $10y^2 = -5y$ **45.** $g(g - 7) = -10$ **46.** $r(r - 5) = -6$

**47.** $3z(2z + 7) = 12$ **48.** $4b(2b + 3) = 36$

**49.** $2(y^2 - 66) = -13y$ **50.** $3(t^2 + 4) = 20t$

**51.** $(2r + 5)(3r^2 - 16r + 5) = 0$ **52.** $(3m + 4)(6m^2 + m - 2) = 0$

**53.** $(2x + 7)(x^2 + 2x - 3) = 0$ **54.** $(x + 1)(6x^2 + x - 12) = 0$

**55.** $9y^3 - 49y = 0$ **56.** $16r^3 - 9r = 0$

**57.** $r^3 - 2r^2 - 8r = 0$ **58.** $x^3 - x^2 - 6x = 0$

**59.** $a^3 + a^2 - 20a = 0$ **60.** $y^3 - 6y^2 + 8y = 0$

**61.** $r^4 = 2r^3 + 15r^2$ **62.** $x^3 = 3x + 2x^2$

**63.** $3x(x + 1) = (2x + 3)(x + 1)$ **64.** $2k(k + 3) = (3k + 1)(k + 3)$

**65.** $x^2 + (x + 1)^2 = (x + 2)^2$ **66.** $(x - 7)^2 + x^2 = (x + 1)^2$

**67.** $(2x)^2 = (2x + 4)^2 - (x + 5)^2$ **68.** $5 - (x - 1)^2 = (x - 2)^2$

**69.** $6p^2(p + 1) = 4(p + 1) - 5p(p + 1)$

**70.** $6x^2(2x + 3) - 5x(2x + 3) = 4(2x + 3)$

**71.** $(k + 3)^2 - (2k - 1)^2 = 0$

**72.** $(4y - 3)^3 - 9(4y - 3) = 0$

**73.** Galileo's formula for freely falling objects, $d = 16t^2$, was given at the beginning of this section. The distance $d$ in feet an object falls depends on the time elapsed $t$ in seconds. (This is an example of an important mathematical concept, the *function*.)

**(a)** Use Galileo's formula and complete the following table. (*Hint:* Substitute each given value into the formula and solve for the unknown value.)

| $t$ in seconds | 0 | 1 | 2 | 3 | | |
|---|---|---|---|---|---|---|
| $d$ in feet | 0 | 16 | | | 256 | 576 |

**(b)** When $t = 0$, $d = 0$. Explain this in the context of the problem.

**(c)** When you substituted 256 for $d$ and solved for $t$, you should have found two solutions, 4 and $-4$. Why doesn't $-4$ make sense as an answer?

## TECHNOLOGY INSIGHTS (EXERCISES 74–77)

*In Section 3.2 we showed how an equation in one variable can be solved with a graphing calculator by getting 0 on one side, then replacing 0 with y to get a corresponding equation in two variables. The x-values of the x-intercepts of the graph of the two-variable equation then give the solutions of the original equation.*

*Use the calculator screens to determine the solution set of each quadratic equation. Verify your answers by substitution.*

**74.** $x^2 + .4x - .05 = 0$

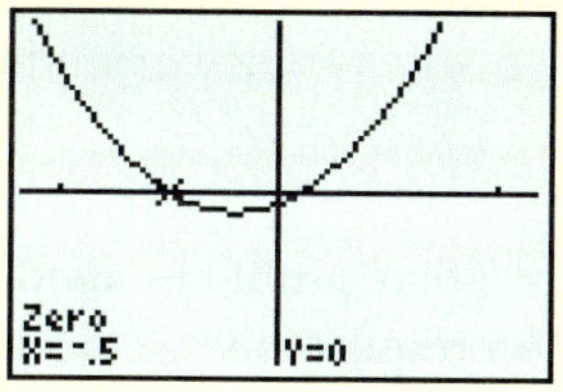


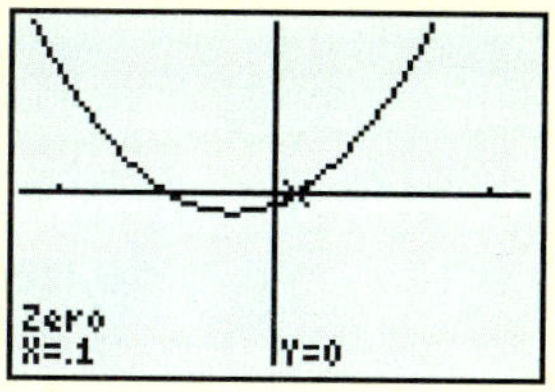


**75.** $2x^2 - 7.2x + 6.3 = 0$

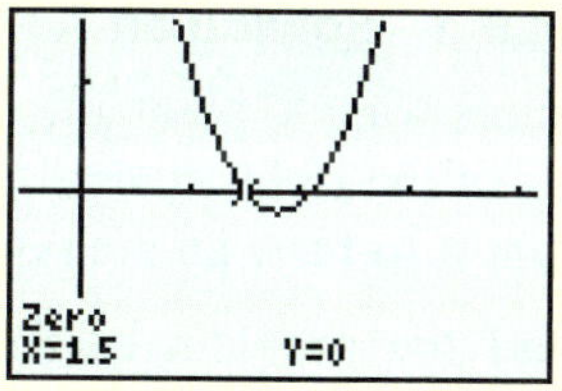


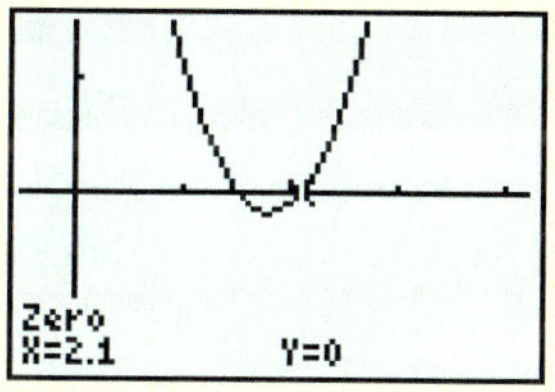


**76.** $2x^2 + 7.2x + 5.5 = 0$

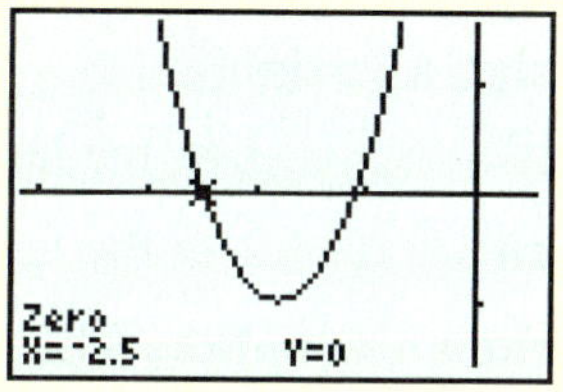


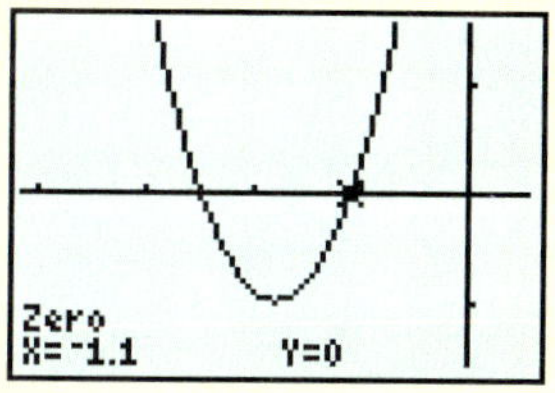


**77.** $4x^2 - x - 33 = 0$

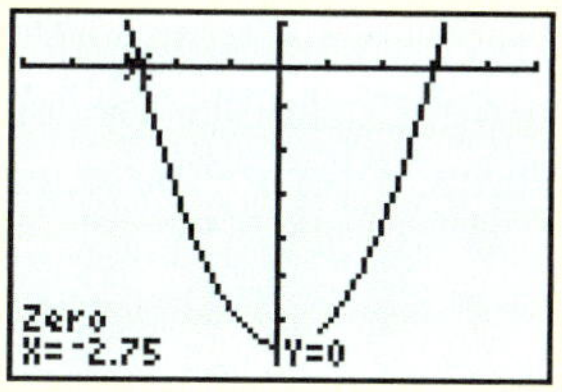


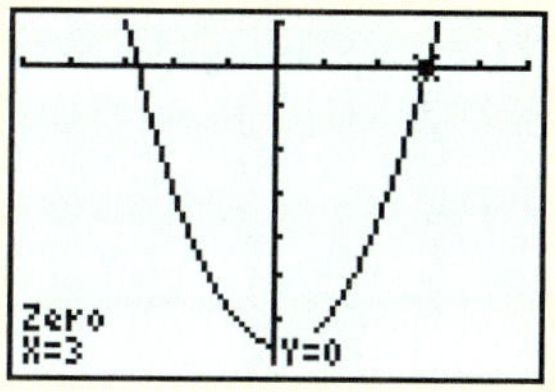


# 5.6 Applications of Quadratic Equations

**OBJECTIVES**

1. Solve problems about geometric figures.
2. Solve problems about consecutive integers.
3. Solve problems using the Pythagorean formula.
4. Solve problems using given quadratic models.

We can now use factoring to solve quadratic equations that arise in application problems. We follow the same six problem-solving steps given in Section 2.4.

**Solving an Applied Problem**

*Step 1* **Read** the problem carefully until you understand what is given and what is to be found.

*Step 2* **Assign a variable** to represent the unknown value, using diagrams or tables as needed. Write down what the variable represents. If necessary, express any other unknown values in terms of the variable.

(continued)

*Step 3* **Write an equation** using the variable expression(s).

*Step 4* **Solve** the equation.

*Step 5* **State the answer.** Does it seem reasonable?

*Step 6* **Check** the answer in the words of the original problem.

**OBJECTIVE 1 Solve problems about geometric figures.** Some of the applied problems in this section require one of the formulas given on the inside covers of the text.

### EXAMPLE 1 Solving an Area Problem

The O'Connors want to plant a flower bed in a triangular area in a corner of their garden. One leg of the right-triangular flower bed will be 2 m shorter than the other leg, and they want it to have an area of 24 $m^2$. See Figure 1. Find the lengths of the legs.

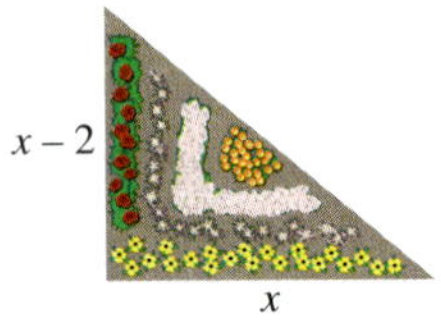


FIGURE 1

*Step 1* **Read** the problem carefully. We need to find the lengths of the legs of a right triangle with area 24 $m^2$.

*Step 2* **Assign a variable.**

Let $x =$ the length of one leg.

Then $x - 2 =$ the length of the other leg.

*Step 3* **Write an equation.** The area of a right triangle is given by the formula

$$\text{area} = \frac{1}{2} \times \text{base} \times \text{height} = \frac{1}{2}bh.$$

In a right triangle, the legs are the base and height, so we substitute 24 for the area, $x$ for the base, and $x - 2$ for the height in the formula.

$$A = \frac{1}{2}bh$$

$$24 = \frac{1}{2}x(x - 2) \qquad \text{Let } A = 24,\ b = x,\ h = x - 2.$$

*Step 4* **Solve.**

$$48 = x(x - 2) \qquad \text{Multiply by 2.}$$

$$48 = x^2 - 2x \qquad \text{Distributive property}$$

$$x^2 - 2x - 48 = 0 \qquad \text{Standard form}$$

$$(x + 6)(x - 8) = 0 \qquad \text{Factor.}$$

$$x + 6 = 0 \quad \text{or} \quad x - 8 = 0 \qquad \text{Zero-factor property}$$

$$x = -6 \quad \text{or} \quad x = 8$$

*Step 5* **State the answer.** The solutions are $-6$ and 8. Because a triangle cannot have a side of negative length, we discard the solution $-6$. Then the lengths of the legs will be 8 m and $8 - 2 = 6$ m.

*Step 6* **Check.** The length of one leg is 2 m less than the length of the other leg, and the area is $\frac{1}{2}(8)(6) = 24$ $m^2$, as required.

**Now Try Exercise 7.**

**CAUTION** When solving applied problems, *always* check solutions against physical facts and discard any answers that are not appropriate.

**OBJECTIVE 2 Solve problems about consecutive integers.** Recall from our work in Section 2.4 that consecutive integers are integers that are next to each other on a number line, such as 5 and 6, or $-11$ and $-10$. Consecutive odd integers are *odd* integers that are next to each other, such as 5 and 7, or $-13$ and $-11$. Consecutive even integers are defined similarly; for example, 4 and 6 are consecutive even integers, as are $-10$ and $-8$. The following list may be helpful.

**Consecutive Integers**

Let $x$ represent the first of the integers.

| | |
|---|---|
| Two consecutive integers | $x, x + 1$ |
| Three consecutive integers | $x, x + 1, x + 2$ |
| Two consecutive even or odd integers | $x, x + 2$ |
| Three consecutive even or odd integers | $x, x + 2, x + 4$ |

**EXAMPLE 2 Solving a Consecutive Integer Problem**

The product of the second and third of three consecutive integers is 2 more than 7 times the first integer. Find the integers.

*Step 1* **Read** the problem. Note that the integers are consecutive.

*Step 2* **Assign a variable.**

Let $x =$ the first integer.

Then $x + 1 =$ the second integer,

and $x + 2 =$ the third integer.

*Step 3* **Write an equation.**

The product of the second and third is 2 more than 7 times the first.

$$(x + 1)(x + 2) = 7x + 2$$

*Step 4* **Solve.**

$$x^2 + 3x + 2 = 7x + 2 \quad \text{Multiply.}$$
$$x^2 - 4x = 0 \quad \text{Standard form}$$
$$x(x - 4) = 0 \quad \text{Factor.}$$
$$x = 0 \quad \text{or} \quad x = 4 \quad \text{Zero-factor property}$$

*Step 5* **State the answer.** The solutions 0 and 4 each lead to a correct answer:

0, 1, 2 or 4, 5, 6.

*Step 6* **Check.** Since $1 \cdot 2 = 7 \cdot 0 + 2$ and $5 \cdot 6 = 7 \cdot 4 + 2$, both sets of consecutive integers satisfy the statement of the problem.

**Now Try Exercise 17.**

**OBJECTIVE 3 Solve problems using the Pythagorean formula.** The next example requires the Pythagorean formula from geometry.

**Pythagorean Formula**

If a right triangle (a triangle with a 90° angle) has longest side of length $c$ and two other sides of lengths $a$ and $b$, then

$$a^2 + b^2 = c^2.$$

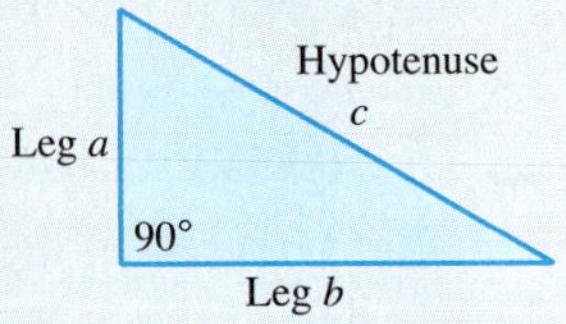


The longest side, the **hypotenuse,** is opposite the right angle. The two shorter sides are the **legs** of the triangle.

### EXAMPLE 3 Using the Pythagorean Formula

Ed and Mark leave their office, with Ed traveling north and Mark traveling east. When Mark is 1 mi farther than Ed from the office, the distance between them is 2 mi more than Ed's distance from the office. Find their distances from the office and the distance between them.

*Step 1* **Read** the problem again. There will be three answers to this problem.

*Step 2* **Assign a variable.** Let $x$ represent Ed's distance from the office, $x + 1$ represent Mark's distance from the office, and $x + 2$ represent the distance between them. Place these on a right triangle, as in Figure 2.

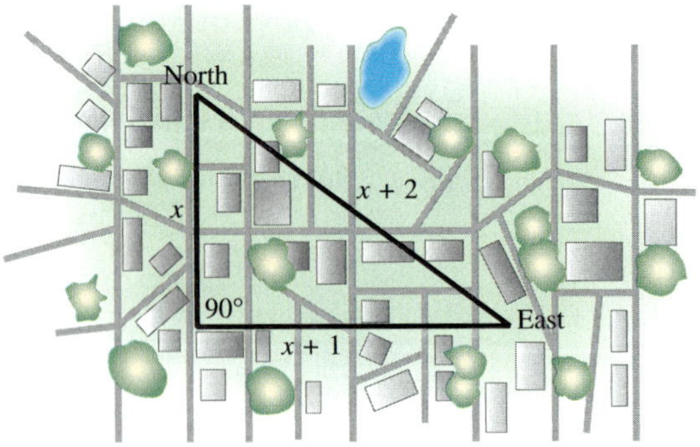


**FIGURE 2**

*Step 3* **Write an equation.** Substitute into the Pythagorean formula.

$$a^2 + b^2 = c^2$$
$$x^2 + (x + 1)^2 = (x + 2)^2$$

*Step 4* **Solve.**

$$x^2 + x^2 + 2x + 1 = x^2 + 4x + 4$$
$$x^2 - 2x - 3 = 0 \quad \text{Standard form}$$
$$(x - 3)(x + 1) = 0 \quad \text{Factor.}$$
$$x - 3 = 0 \quad \text{or} \quad x + 1 = 0 \quad \text{Zero-factor property}$$
$$x = 3 \quad \text{or} \quad x = -1$$

*Step 5* **State the answer.** Since $-1$ cannot represent a distance, 3 is the only possible answer. Ed's distance is 3 mi, Mark's distance is $3 + 1 = 4$ mi, and the distance between them is $3 + 2 = 5$ mi.

*Step 6* **Check.** Since $3^2 + 4^2 = 5^2$, the answers are correct.

**Now Try Exercise 25.**

**CAUTION** When solving a problem involving the Pythagorean formula, be sure that the expressions for the sides are properly placed.

$$\text{leg}^2 + \text{leg}^2 = \text{hypotenuse}^2$$

**OBJECTIVE 4 Solve problems using given quadratic models.** In Examples 1–3, we wrote quadratic equations to model, or mathematically describe, various situations and then solved the equations. In the final examples, we are given the quadratic models and must use them to determine data.

## EXAMPLE 4 Finding the Height of a Ball

A tennis player's serve travels 180 ft per sec (125 mph). If she hits the ball directly upward, the height $h$ of the ball in feet at time $t$ in seconds is modeled by the quadratic equation

$$h = -16t^2 + 180t + 6.$$

How long will it take for the ball to reach a height of 206 ft?

A height of 206 ft means $h = 206$, so we substitute 206 for $h$ in the equation.

$$206 = -16t^2 + 180t + 6 \quad \text{Let } h = 206.$$

To solve the equation, we first write it in standard form. For convenience, we reverse the sides of the equation.

$$-16t^2 + 180t + 6 = 206$$
$$-16t^2 + 180t - 200 = 0 \quad \text{Standard form}$$
$$4t^2 - 45t + 50 = 0 \quad \text{Divide by } -4.$$
$$(4t - 5)(t - 10) = 0 \quad \text{Factor.}$$
$$4t - 5 = 0 \quad \text{or} \quad t - 10 = 0 \quad \text{Zero-factor property}$$
$$t = \frac{5}{4} \quad \text{or} \quad t = 10$$

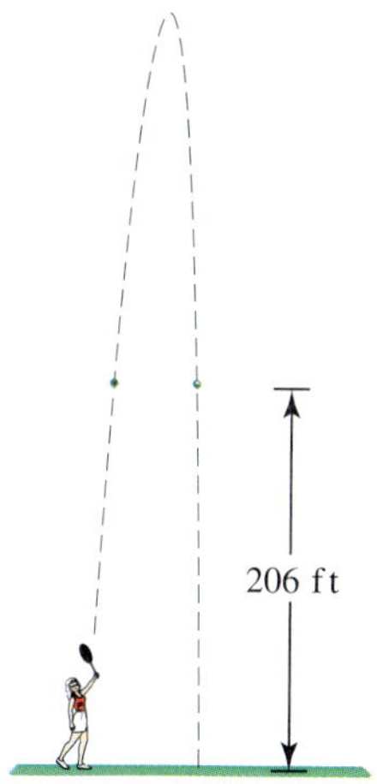


FIGURE 3

Since we found two acceptable answers, the ball will be 206 ft above the ground twice (once on its way up and once on its way down)—at $\frac{5}{4}$ sec and at 10 sec. See Figure 3.

**Now Try Exercise 29.**

## EXAMPLE 5 Modeling Increases in Drug Prices

The annual percent increase $y$ in the amount pharmacies paid wholesalers for drugs in the years 1990–1999 can be modeled by the quadratic equation

$$y = .23x^2 - 2.6x + 9,$$

where $x = 0$ represents 1990, $x = 1$ represents 1991, and so on. (*Source: IMS Health,* Retail and Provider Perspective.)

**(a)** Use the model to find the annual percent increase to the nearest tenth in 1997.
In 1997, $x = 1997 - 1990 = 7$. Substitute 7 for $x$ in the equation.

$$y = .23(7)^2 - 2.6(7) + 9 \quad \text{Let } x = 7.$$
$$y = 2.07 \quad \text{Use a calculator.}$$

To the nearest tenth, pharmacies paid about 2.1% more for drugs in 1997.

**(b)** Repeat part (a) for 1999.
For 1999, $x = 9$.

$$y = .23(9)^2 - 2.6(9) + 9 \quad \text{Let } x = 1999 - 1990 = 9.$$
$$y = 4.23$$

In 1999, pharmacies paid about 4.2% more for drugs.

**(c)** The model used in parts (a) and (b) was developed using the data shown in the table. How do the results in parts (a) and (b) compare to the actual data from the table?

| Year | Percent Increase |
|---|---|
| 1990 | 8.4 |
| 1991 | 7.2 |
| 1992 | 5.5 |
| 1993 | 3.0 |
| 1994 | 1.7 |
| 1995 | 1.9 |
| 1996 | 1.6 |
| 1997 | 2.5 |
| 1998 | 3.2 |
| 1999 | 4.2 |

From the table, the actual data for 1997 is 2.5%. Our answer, 2.1%, is a little low. For 1999, the actual data is 4.2%, which is the same as our answer in part (b).

**Now Try Exercise 31.**

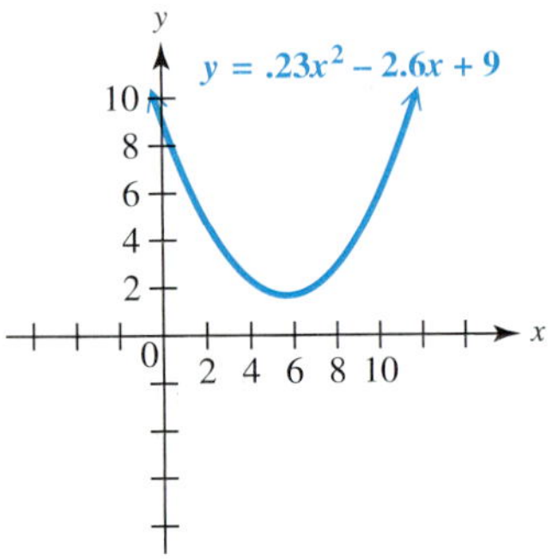


FIGURE 4

**NOTE** A graph of the quadratic equation from Example 5 is shown in Figure 4. Notice the basic shape of this graph, which follows the general pattern of the data in the table—it decreases from 1990 to 1996 (with the exception of the data for 1995) and then increases from 1997 to 1999. We consider such graphs of quadratic equations, called *parabolas,* in more detail in Chapter 11.

## 5.6 EXERCISES

**For Extra Help**

- Student's Solutions Manual
- MyMathLab
- InterAct Math Tutorial Software
- AW Math Tutor Center
- MathXL
- Digital Video Tutor CD 8/Videotape 8

**1.** To review the six problem-solving steps first introduced in Section 2.4, complete each statement.

*Step 1:* ________ the problem carefully until you understand what is given and what must be found.

*Step 2:* Assign a ________ to represent the unknown value.

*Step 3:* Write a(n) ________ using the variable expression(s).

*Step 4:* ________ the equation.

*Step 5:* State the ________.

*Step 6:* ________ the answer in the words of the ________ problem.

**2.** A student solves an applied problem and gets 6 or $-3$ for the length of the side of a square. Which of these answers is reasonable? Explain.

*In Exercises 3–6, a figure and a corresponding geometric formula are given. Using x as the variable, complete Steps 3–6 for each problem. (Refer to the steps in Exercise 1 as needed.)*

**3.**

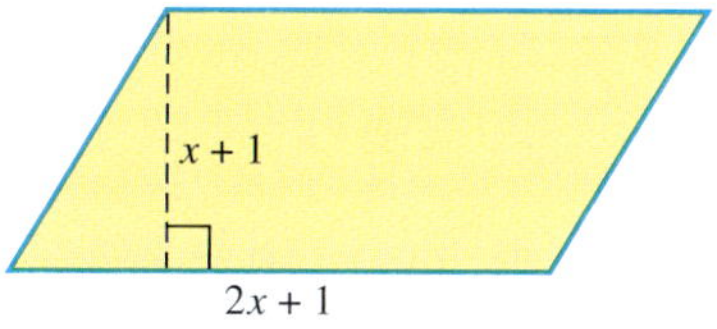


Area of a parallelogram: $A = bh$

The area of this parallelogram is 45 sq. units. Find its base and height.

**4.**

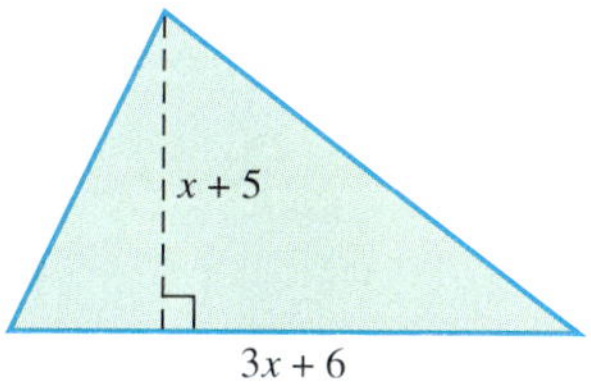


Area of a triangle: $A = \frac{1}{2}bh$

The area of this triangle is 60 sq. units. Find its base and height.

**5.**

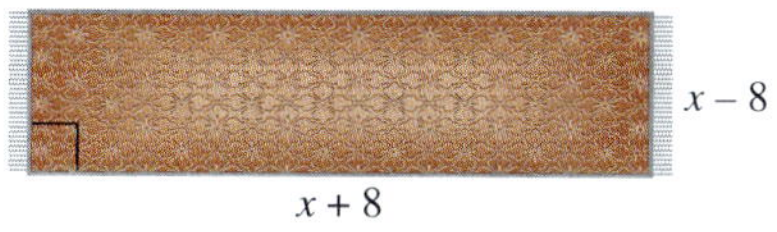


Area of a rectangular rug: $A = LW$

The area of this rug is 80 sq. units. Find its length and width.

**6.**

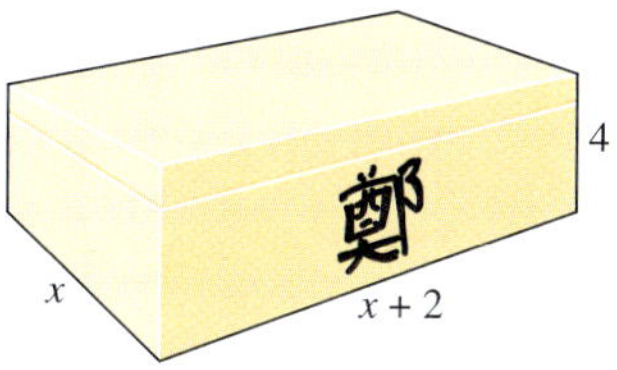


Volume of a rectangular Chinese box: $V = LWH$

The volume of this box is 192 cu. units. Find its length and width.

*Solve each problem. Check your answers to be sure they are reasonable. Refer to the formulas on the inside covers. See Example 1.*

**7.** The length of a VHS videocassette shell is 3 in. more than its width. The area of the rectangular top side of the shell is 28 in.$^2$. Find the length and width of the videocassette shell.

**8.** A plastic box that holds a standard audiocassette has length 4 cm longer than its width. The area of the rectangular top of the box is 77 cm$^2$. Find the length and width of the box.

**9.** The dimensions of a Gateway EV700 computer monitor screen are such that its length is 3 in. more than its width. If the length is increased by 1 in. while the width remains the same, the area is increased by 10 in.$^2$. What are the dimensions of the screen? (*Source:* Author's computer.)

**10.** The keyboard of the computer in Exercise 9 is 11 in. longer than it is wide. If both its length and width are increased by 2 in., the area of the top of the keyboard is increased by 54 in.$^2$. Find the length and width of the keyboard. (*Source:* Author's computer.)

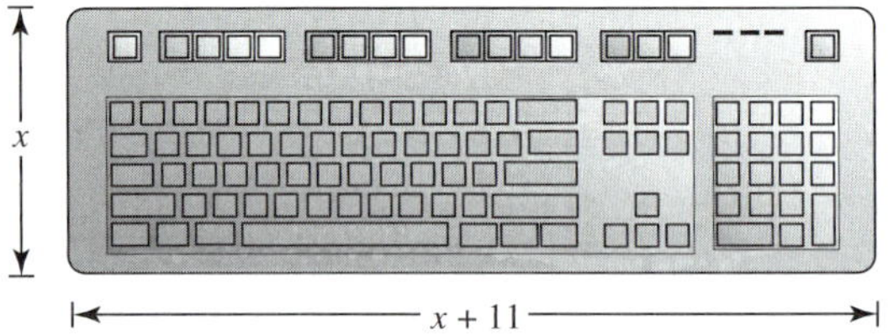


**11.** The area of a triangle is 30 in.$^2$. The base of the triangle measures 2 in. more than twice the height of the triangle. Find the measures of the base and the height.

**12.** A certain triangle has its base equal in measure to its height. The area of the triangle is 72 m$^2$. Find the equal base and height measure.

**13.** A 10-gal aquarium is 3 in. higher than it is wide. Its length is 21 in., and its volume is 2730 in.$^3$. What are the height and width of the aquarium?

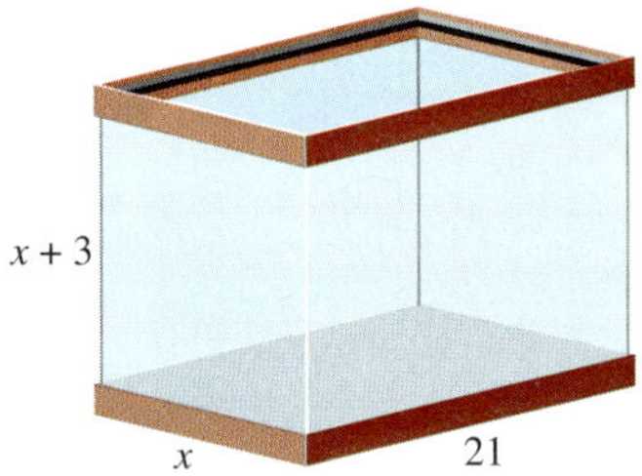


**14.** A toolbox is 2 ft high, and its width is 3 ft less than its length. If its volume is 80 ft$^3$, find the length and width of the box.

**15.** A square mirror has sides measuring 2 ft less than the sides of a square painting. If the difference between their areas is 32 ft$^2$, find the lengths of the sides of the mirror and the painting.

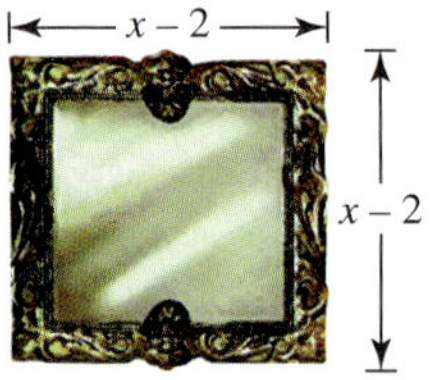

16. The sides of one square have length 3 m more than the sides of a second square. If the area of the larger square is subtracted from 4 times the area of the smaller square, the result is 36 m$^2$. What are the lengths of the sides of each square?

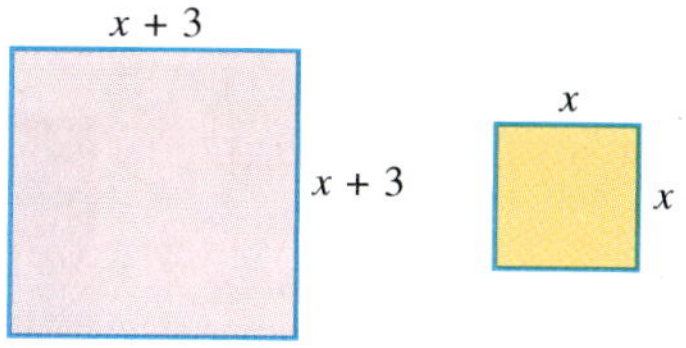


*Solve each problem about consecutive integers. See Example 2.*

17. The product of two consecutive integers is 11 more than their sum. Find the integers.

18. The product of two consecutive integers is 4 less than 4 times their sum. Find the integers.

19. Find three consecutive odd integers such that 3 times the sum of all three is 18 more than the product of the smaller two.

20. Find three consecutive odd integers such that the sum of all three is 42 less than the product of the larger two.

21. Find three consecutive even integers such that the sum of the squares of the smaller two is equal to the square of the largest.

22. Find three consecutive even integers such that the square of the sum of the smaller two is equal to twice the largest.

*Use the Pythagorean formula to solve each problem. See Example 3.*

23. The hypotenuse of a right triangle is 1 cm longer than the longer leg. The shorter leg is 7 cm shorter than the longer leg. Find the length of the longer leg of the triangle.

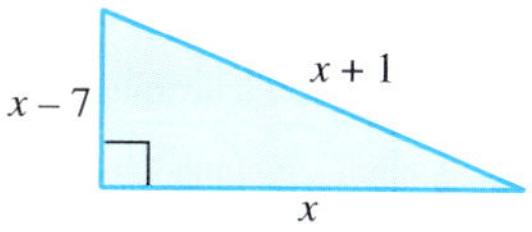


24. The longer leg of a right triangle is 1 m longer than the shorter leg. The hypotenuse is 1 m shorter than twice the shorter leg. Find the length of the shorter leg of the triangle.

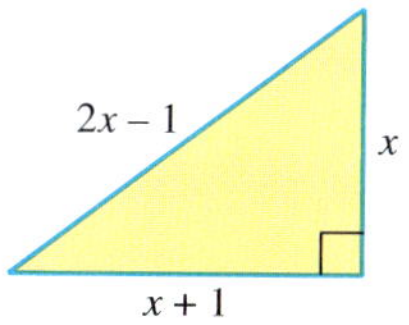


25. Tram works due north of home. Her husband Alan works due east. They leave for work at the same time. By the time Tram is 5 mi from home, the distance between them is 1 mi more than Alan's distance from home. How far from home is Alan?

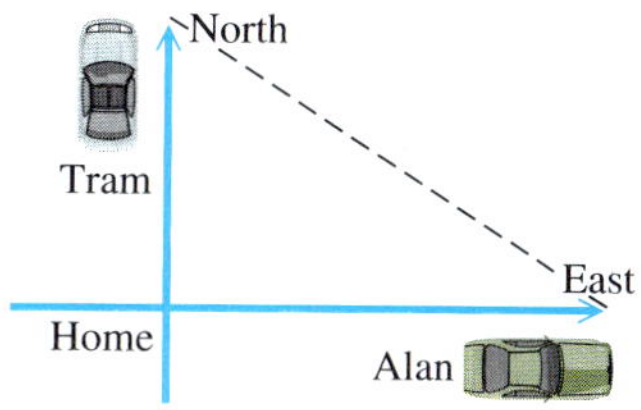


26. Two cars left an intersection at the same time. One traveled north. The other traveled 14 mi farther, but to the east. How far apart were they then, if the distance between them was 4 mi more than the distance traveled east?

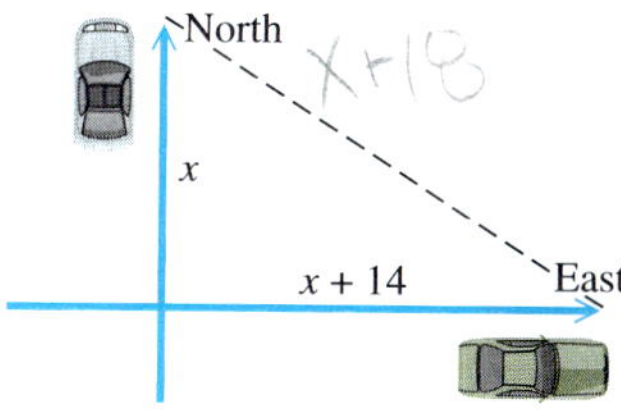

| CONCEPTS | EXAMPLES |
|---|---|

## 5.5 SOLVING QUADRATIC EQUATIONS BY FACTORING

**Zero-Factor Property**

If $a$ and $b$ are real numbers and if $ab = 0$, then $a = 0$ or $b = 0$.

If $(x - 2)(x + 3) = 0$, then $x - 2 = 0$ or $x + 3 = 0$.

**Solving a Quadratic Equation by Factoring**

*Step 1* Write in standard form.

*Step 2* Factor.

*Step 3* Use the zero-factor property.

*Step 4* Check.

Solve $2x^2 = 7x + 15$.

$$2x^2 - 7x - 15 = 0$$

$$(2x + 3)(x - 5) = 0$$

$$2x + 3 = 0 \quad \text{or} \quad x - 5 = 0$$

$$2x = -3 \qquad x = 5$$

$$x = -\frac{3}{2}$$

Both solutions satisfy the original equation; the solution set is $\left\{-\frac{3}{2}, 5\right\}$.

## 5.6 APPLICATIONS OF QUADRATIC EQUATIONS

**Pythagorean Formula**

In a right triangle, the square of the hypotenuse equals the sum of the squares of the legs.

$$a^2 + b^2 = c^2$$

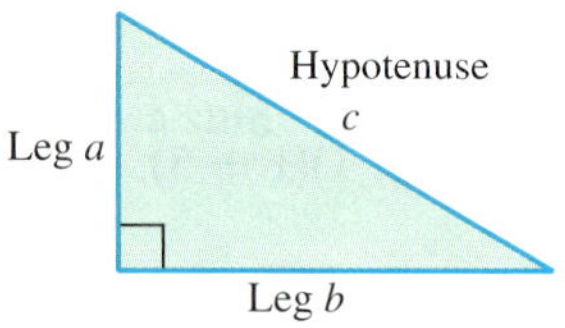


In a right triangle, one leg measures 2 ft longer than the other. The hypotenuse measures 4 ft longer than the shorter leg. Find the lengths of the three sides of the triangle.

Let $x =$ the length of the shorter leg. Then

$$x^2 + (x + 2)^2 = (x + 4)^2.$$

Verify that the solutions of this equation are $-2$ and 6. Discard $-2$ as a solution. Check that the sides are 6, $6 + 2 = 8$, and $6 + 4 = 10$ ft in length.

**Answers to Test Your Word Power**

**1.** B; *Example:* $x^2 - 5x - 14 = (x - 7)(x + 2)$ **2.** D; *Example:* The factored form of $x^2 - 5x - 14$ is $(x - 7)(x + 2)$. **3.** A; *Example:* $a^2 + 2a + 1$ is a perfect square trinomial; its factored form is $(a + 1)^2$. **4.** B; *Examples:* $y^2 - 3y + 2 = 0$, $x^2 - 9 = 0$, $2m^2 = 6m + 8$ **5.** C; *Example:* In the triangle above, the hypotenuse is the side labeled $c$.

# CHAPTER 5 REVIEW EXERCISES

[5.1] *Factor out the greatest common factor or factor by grouping.*

**1.** $7t + 14$

**2.** $60z^3 + 30z$

**3.** $2xy - 8y + 3x - 12$

**4.** $6y^2 + 9y + 4xy + 6x$

[5.2] *Factor completely.*

**5.** $x^2 + 5x + 6$

**6.** $y^2 - 13y + 40$

**7.** $q^2 + 6q - 27$

**8.** $r^2 - r - 56$

**9.** $r^2 - 4rs - 96s^2$

**10.** $p^2 + 2pq - 120q^2$

**11.** $8p^3 - 24p^2 - 80p$

**12.** $3x^4 + 30x^3 + 48x^2$

**13.** $p^7 - p^6q - 2p^5q^2$

**14.** $3r^5 - 6r^4s - 45r^3s^2$

[5.3]

**15.** To begin factoring $6r^2 - 5r - 6$, what are the possible first terms of the two binomial factors if we consider only positive integer coefficients?

**16.** What is the first step you would use to factor $2z^3 + 9z^2 - 5z$?

*Factor completely.*

**17.** $2k^2 - 5k + 2$

**18.** $3r^2 + 11r - 4$

**19.** $6r^2 - 5r - 6$

**20.** $10z^2 - 3z - 1$

**21.** $8v^2 + 17v - 21$

**22.** $24x^5 - 20x^4 + 4x^3$

**23.** $-6x^2 + 3x + 30$

**24.** $10r^3s + 17r^2s^2 + 6rs^3$

[5.4]

**25.** Which one of the following is the difference of squares?

**A.** $32x^2 - 1$ **B.** $4x^2y^2 - 25z^2$ **C.** $x^2 + 36$ **D.** $25y^3 - 1$

**26.** Which one of the following is a perfect square trinomial?

**A.** $x^2 + x + 1$ **B.** $y^2 - 4y + 9$ **C.** $4x^2 + 10x + 25$ **D.** $x^2 - 20x + 100$

*Factor completely.*

**27.** $n^2 - 49$

**28.** $25b^2 - 121$

**29.** $49y^2 - 25w^2$

**30.** $144p^2 - 36q^2$

**31.** $x^2 + 100$

**32.** $r^2 - 12r + 36$

**33.** $9t^2 - 42t + 49$

**34.** $m^3 + 1000$

**35.** $125k^3 + 64x^3$

**36.** $343x^3 - 64$

[5.5] *Solve each equation and check your solutions.*

**37.** $(4t + 3)(t - 1) = 0$

**38.** $(x + 7)(x - 4)(x + 3) = 0$

**39.** $x(2x - 5) = 0$

**40.** $z^2 + 4z + 3 = 0$

**41.** $m^2 - 5m + 4 = 0$

**42.** $x^2 = -15 + 8x$

**43.** $3z^2 - 11z - 20 = 0$

**44.** $81t^2 - 64 = 0$

**45.** $y^2 = 8y$

**46.** $n(n - 5) = 6$

**47.** $t^2 - 14t + 49 = 0$

**48.** $t^2 = 12(t - 3)$

**49.** $(5z + 2)(z^2 + 3z + 2) = 0$

**50.** $x^2 = 9$

[5.6] *Solve each problem.*

**51.** The length of a rug is 6 ft more than the width. The area is 40 ft$^2$. Find the length and width of the rug.

**52.** The surface area $S$ of a box is given by

$$S = 2WH + 2WL + 2LH.$$

A treasure chest from a sunken galleon has dimensions as shown in the figure. Its surface area is 650 ft$^2$. Find its width.

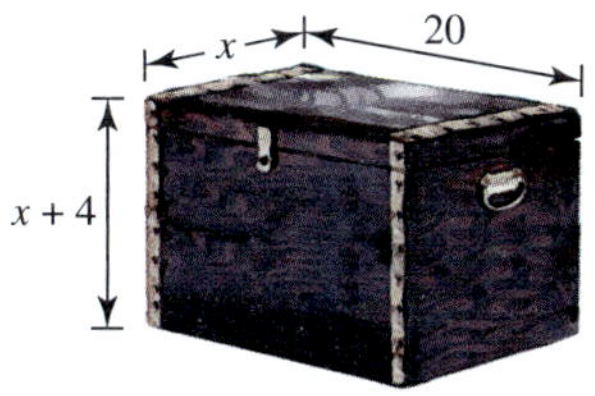

**53.** The length of a rectangle is three times the width. If the width were increased by 3 m while the length remained the same, the new rectangle would have an area of 30 $m^2$. Find the length and width of the original rectangle.

**54.** The volume of a rectangular box is 120 $m^3$. The width of the box is 4 m, and the height is 1 m less than the length. Find the length and height of the box.

**55.** The product of two consecutive integers is 29 more than their sum. What are the integers?

**56.** Two cars left an intersection at the same time. One traveled west, and the other traveled 14 mi less, but to the south. How far apart were they then, if the distance between them was 16 mi more than the distance traveled south?

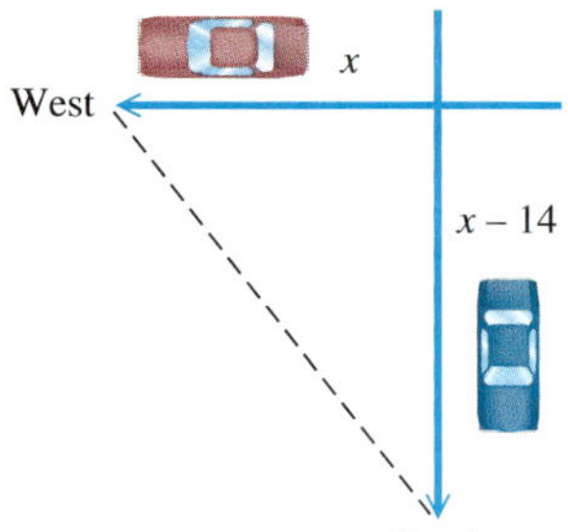


*If an object is propelled upward with an initial velocity of* 128 *ft per sec, its height h after t sec is*

$$h = 128t - 16t^2.$$

*Find the height of the object after each period of time.*

**57.** 1 sec

**58.** 2 sec

**59.** 4 sec

**60.** For the object described above, when does it return to the ground?

**61.** Annual revenue in millions of dollars for eBay is shown in the table.

| Year | Annual Revenue (in millions of dollars) |
|---|---|
| 1997 | 5.1 |
| 1998 | 47.4 |
| 1999 | 224.7 |

*Source:* eBay.

Using the data, we developed the quadratic equation

$$y = 67.5x^2 - 25.2x + 5.1$$

to model eBay revenues $y$ in year $x$, where $x = 0$ represents 1997, $x = 1$ represents 1998, and so on. Because only three years of data were used to determine the model, we must be careful about using it to predict revenue for years beyond 1999.

**(a)** Use the model to predict annual revenue for eBay in 2000.

**(b)** The revenue for eBay through the first half of 2000 was $183.2 million. Given this information, do you think your prediction in part (a) is reliable? Explain.

## MIXED REVIEW EXERCISES

**62.** Which of the following is *not* factored completely?

**A.** $3(7t)$ **B.** $3x(7t + 4)$ **C.** $(3 + x)(7t + 4)$ **D.** $3(7t + 4) + x(7t + 4)$

**63.** Although $(2x + 8)(3x - 4) = 6x^2 + 16x - 32$ is a true statement, the polynomial is not factored completely. Explain why and give the complete factored form.

*Factor completely.*

**64.** $z^2 - 11zx + 10x^2$

**65.** $3k^2 + 11k + 10$

**66.** $15m^2 + 20m - 12mp - 16p$

**67.** $y^4 - 625$

**68.** $6m^3 - 21m^2 - 45m$

**69.** $24ab^3c^2 - 56a^2bc^3 + 72a^2b^2c$

**70.** $25a^2 + 15ab + 9b^2$

**71.** $12x^2yz^3 + 12xy^2z - 30x^3y^2z^4$

**72.** $2a^5 - 8a^4 - 24a^3$

**73.** $12r^2 + 18rq - 10r - 15q$

**74.** $1000a^3 + 27$

**75.** $49t^2 + 56t + 16$

*Solve.*

**76.** $t(t - 7) = 0$

**77.** $x^2 + 3x = 10$

**78.** $25x^2 + 20x + 4 = 0$

**79.** The numbers of alternative-fueled vehicles, in thousands, in use for the years 1998–2001 are given in the table. Using statistical methods, we constructed the quadratic equation

$$y = .25x^2 - 25.65x + 496.6$$

to model the number of vehicles $y$ in year $x$. Here we used $x = 98$ for 1998, $x = 99$ for 1999, and so on. Because only four years of data were used to determine the model, we must be particularly careful about using it to estimate for years before 1998 or after 2001.

**ALTERNATIVE-FUELED VEHICLES**

| Year | Number (in thousands) |
|---|---|
| 1998 | 384 |
| 1999 | 407 |
| 2000 | 432 |
| 2001 | 456 |

*Source:* Energy Information Administration, Alternatives to Traditional Fuels, 2001.

**(a)** What prediction for 2002 is given by the equation?

**(b)** Why might the prediction for 2002 be unreliable?

**80.** A lot is shaped like a right triangle. The hypotenuse is 3 m longer than the longer leg. The longer leg is 6 m longer than twice the length of the shorter leg. Find the lengths of the sides of the lot.

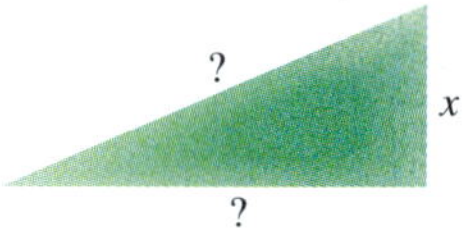


**81.** A pyramid has a rectangular base with a length that is 2 m more than the width. The height of the pyramid is 6 m, and its volume is 48 $m^3$. Find the length and width of the base.

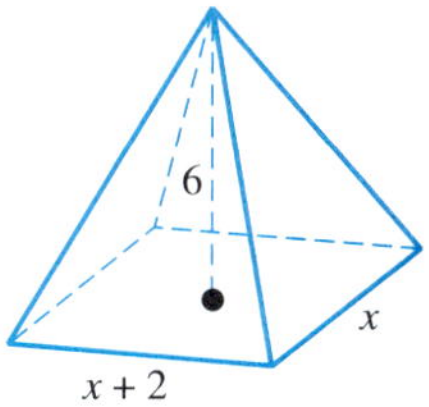


**82.** The product of the smaller two of three consecutive integers is equal to 23 plus the largest. Find the integers.

**83.** If an object is dropped, the distance $d$ in feet it falls in $t$ sec (disregarding air resistance) is given by the quadratic equation

$$d = 16t^2.$$

Find the distance an object would fall in the following times.

**(a)** 4 sec **(b)** 8 sec

**84.** The floor plan for a house is a rectangle with length 7 m more than its width. The area is 170 $m^2$. Find the width and length of the house.

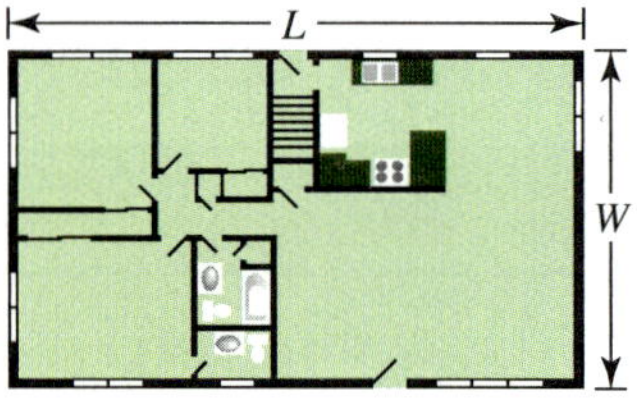


**85.** The triangular sail of a schooner has an area of 30 $m^2$. The height of the sail is 4 m more than the base. Find the base of the sail.

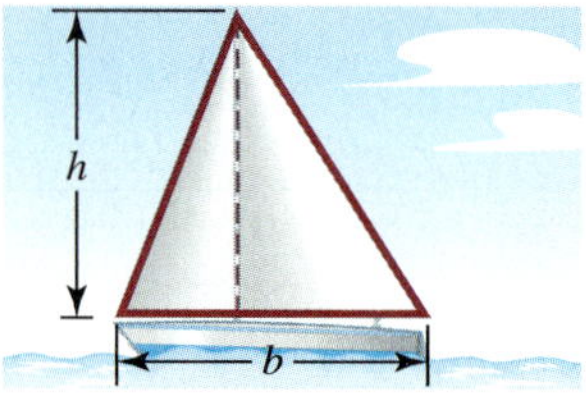


# CHAPTER 5 TEST

**1.** Which one of the following is the correct, completely factored form of $2x^2 - 2x - 24$?

**A.** $(2x + 6)(x - 4)$ **B.** $(x + 3)(2x - 8)$
**C.** $2(x + 4)(x - 3)$ **D.** $2(x + 3)(x - 4)$

*Factor each polynomial completely.*

**2.** $12x^2 - 30x$
**3.** $2m^3n^2 + 3m^3n - 5m^2n^2$
**4.** $2ax - 2bx + ay - by$
**5.** $x^2 - 5x - 24$
**6.** $2x^2 + x - 3$
**7.** $10z^2 - 17z + 3$
**8.** $t^2 + 2t + 3$
**9.** $x^2 + 36$
**10.** $12 - 6a + 2b - ab$
**11.** $9y^2 - 64$
**12.** $4x^2 - 28xy + 49y^2$
**13.** $-2x^2 - 4x - 2$
**14.** $6t^4 + 3t^3 - 108t^2$
**15.** $r^3 - 125$
**16.** $8k^3 + 64$
**17.** $x^4 - 81$

**18.** Why is $(p + 3)(p + 3)$ not the correct factored form of $p^2 + 9$?

*Solve each equation.*

**19.** $2r^2 - 13r + 6 = 0$
**20.** $25x^2 - 4 = 0$
**21.** $x(x - 20) = -100$
**22.** $t^3 = 9t$

*Solve each problem.*

**23.** The length of a rectangular flower bed is 3 ft less than twice its width. The area of the bed is 54 $ft^2$. Find the dimensions of the flower bed.

**24.** Find two consecutive integers such that the square of the sum of the two integers is 11 more than the smaller integer.

**25.** A carpenter needs to cut a brace to support a wall stud, as shown in the figure. The brace should be 7 ft less than three times the length of the stud. If the brace will be anchored on the floor 15 ft away from the stud, how long should the brace be?

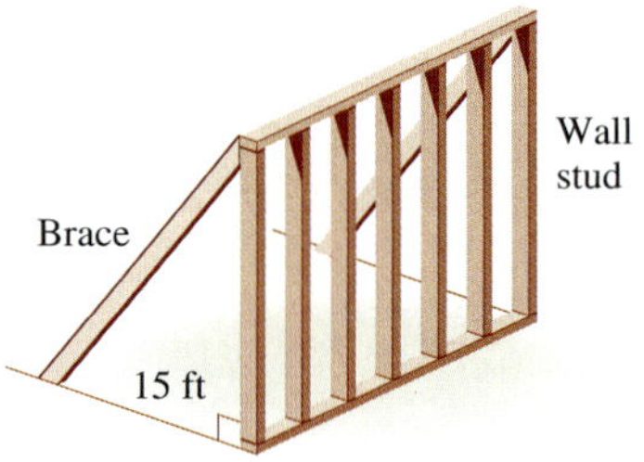

**26.** TV viewers have more choices than ever. The number of cable TV channels $y$ from 1984 through 1999 can be approximated by the quadratic equation

$$y = .57x^2 + .31x + 48,$$

where $x = 0$ represents 1984, $x = 1$ represents 1985, and so on. (*Source:* National Cable Television Association.) Use the model to estimate the number of cable TV channels in 1999. Round your answer to the nearest whole number.

# CUMULATIVE REVIEW EXERCISES CHAPTERS 1–5

*Solve each equation.*

**1.** $3x + 2(x - 4) = 4(x - 2)$

**2.** $.3x + .9x = .06$

**3.** $\frac{2}{3}y - \frac{1}{2}(y - 4) = 3$

**4.** Solve for $P$: $A = P + Prt$.

**5.** From a list of "everyday items" often taken for granted, adults were recently surveyed as to those items they wouldn't want to live without. Complete the results shown in the table if 500 adults were surveyed.

| *Item* | *Percent That Wouldn't Want to Live Without* | *Number That Wouldn't Want to Live Without* |
|---|---|---|
| Toilet paper | 69% | |
| Zipper | 42% | |
| Frozen foods | | 190 |
| Self-stick note pads | | 75 |

(Other items included tape, hairspray, pantyhose, paper clips, and Velcro.)
*Source:* Market Facts for Kleenex Cottonelle.

*Solve each problem.*

**6.** At the 1998 Winter Olympics in Nagano, Japan, the top medal winner was Germany with 29. Germany won 1 more silver medal than bronze and 3 more gold medals than silver. Find the number of each type of medal won. (*Source: World Almanac and Book of Facts,* 2000.)

**7.** In July 2000, roughly 144 million people surfed the Web from home. This was a 35% increase from the same month the previous year. How many people, to the nearest million, surfed the Web from home in July 1999? (*Source: The Gazette,* September 3, 2000.)

**8.** Find the measures of the marked angles.

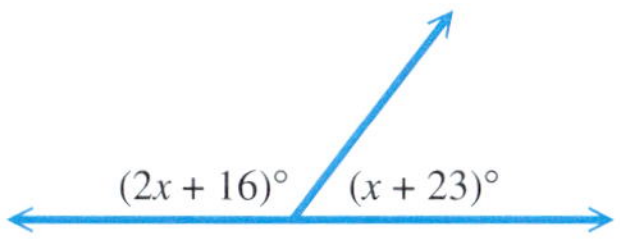


**9.** Fill in each blank with *positive* or *negative.* The point with coordinates $(a, b)$ is in

**(a)** quadrant II if $a$ is ________ and $b$ is ________.
**(b)** quadrant III if $a$ is ________ and $b$ is ________.

*Consider the equation* $y = 12x + 3$. *Find the following.*

**10.** The $x$- and $y$-intercepts

**11.** The slope

**12.** The graph

**13.** The points on the graph show the number of U.S. radio stations in the years 1993 through 1999, along with the graph of a linear equation that models the data. Use the ordered pairs shown on the graph to find the slope of the line to the nearest whole number. Interpret the slope.

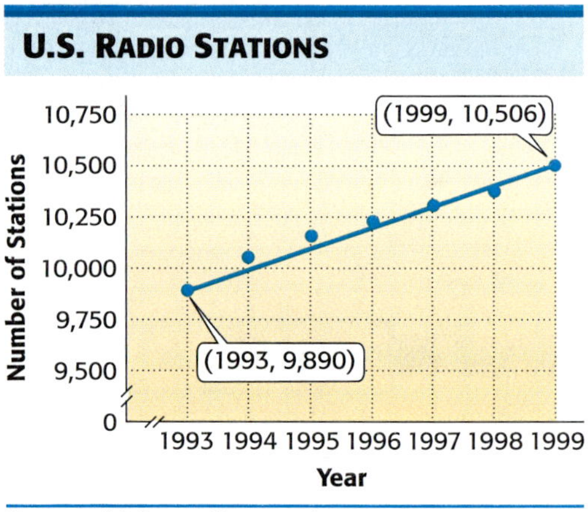


*Source:* M Street Corporation.

*Write an equation for each line. Write the equation in slope-intercept form if possible.*

**14.** Slope $-\frac{1}{2}$, $y$-intercept $(0, 5)$

**15.** Through $\left(\frac{1}{4}, -\frac{2}{3}\right)$, slope 0

*Evaluate each expression.*

**16.** $2^{-3} \cdot 2^5$

**17.** $\left(\frac{3}{4}\right)^{-2}$

**18.** $\left(\frac{4^{-3} \cdot 4^4}{4^5}\right)^{-1}$

*Simplify each expression and write the answer using only positive exponents. Assume no denominators are* 0.

**19.** $\frac{(p^2)^3p^{-4}}{(p^{-3})^{-1}p}$

**20.** $\frac{(m^{-2})^3m}{m^5m^{-4}}$

*Perform each indicated operation.*

**21.** $(2k^2 + 4k) - (5k^2 - 2) - (k^2 + 8k - 6)$

**22.** $(9x + 6)(5x - 3)$

**23.** $(3p + 2)^2$

**24.** $\frac{8x^4 + 12x^3 - 6x^2 + 20x}{2x}$

**25.** To make a pound of honey, bees may travel 55,000 mi and visit more than 2,000,000 flowers. (*Source: Home & Garden.*) Write the two given numbers in scientific notation.

*Factor completely.*

**26.** $2a^2 + 7a - 4$

**27.** $10m^2 + 19m + 6$

**28.** $8t^2 + 10tv + 3v^2$

**29.** $4p^2 - 12p + 9$

**30.** $25r^2 - 81t^2$

**31.** $2pq + 6p^3q + 8p^2q$

*Solve each equation.*

**32.** $6m^2 + m - 2 = 0$

**33.** $8x^2 = 64x$

**34.** The length of the hypotenuse of a right triangle is twice the length of the shorter leg, plus 3 m. The longer leg is 7 m longer than the shorter leg. Find the lengths of the sides.

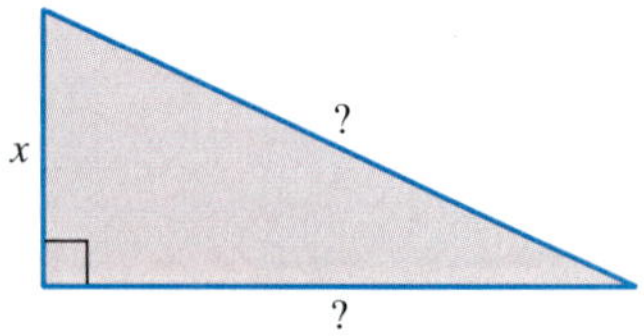

# Rational Expressions and Applications

# 6

Americans have been car crazy ever since the first automobiles hit the road early in the twentieth century. Today there are about 213.5 million vehicles in the United States driving on 3.4 million mi of paved roadways. There is even a new museum devoted exclusively to our four-wheeled passion and its influence on our lives and culture. The Museum of Automobile History in Syracuse, N.Y., features some 200 years of automobile memorabilia, including rare advertising pieces, designer drawings, and Hollywood movie posters. (*Source: Home and Away,* May/June 2002.) In Exercises 75 and 76 of Section 6.4, we use a rational expression to determine the cost of restoring a vintage automobile.

# 6.1 The Fundamental Property of Rational Expressions

**OBJECTIVES**

1. Find the numerical value of a rational expression.
2. Find the values of the variable for which a rational expression is undefined.
3. Write rational expressions in lowest terms.
4. Recognize equivalent forms of rational expressions.

The quotient of two integers (with denominator not 0), such as $\frac{2}{3}$ or $-\frac{3}{4}$, is called a *rational number.* In the same way, the quotient of two polynomials with denominator not equal to 0 is called a *rational expression.*

**Rational Expression**

A **rational expression** is an expression of the form

$$\frac{P}{Q},$$

where $P$ and $Q$ are polynomials, with $Q \neq 0$.

Examples of rational expressions include

$$\frac{-6x}{x^3 + 8}, \quad \frac{9x}{y + 3}, \quad \text{and} \quad \frac{2m^3}{8}.$$

Our work with rational expressions requires much of what we learned in Chapters 4 and 5 on polynomials and factoring, as well as the rules for fractions from Section 1.1.

**OBJECTIVE 1 Find the numerical value of a rational expression.** We use substitution to evaluate a rational expression for a given value of the variable.

**EXAMPLE 1 Evaluating Rational Expressions**

Find the numerical value of $\frac{3x + 6}{2x - 4}$ for each value of $x$.

**(a)** $x = 1$

$$\frac{3x + 6}{2x - 4} = \frac{3(1) + 6}{2(1) - 4} \qquad \text{Let } x = 1.$$

$$= \frac{9}{-2} \quad \text{or} \quad -\frac{9}{2}$$

**(b)** $x = -2$

$$\frac{3x + 6}{2x - 4} = \frac{3(-2) + 6}{2(-2) - 4} \qquad \text{Let } x = -2.$$

$$= \frac{0}{-8} \quad \text{or} \quad 0$$

**Now Try Exercise 3.**

**OBJECTIVE 2 Find the values of the variable for which a rational expression is undefined.** In the definition of a rational expression $\frac{P}{Q}$, $Q$ cannot equal 0. The denominator of a rational expression cannot equal 0 because division by 0 is undefined.

For instance, in the rational expression

$$\frac{3x + 6}{2x - 4} \leftarrow \text{Denominator cannot equal 0.}$$

from Example 1, the variable $x$ can take on any real number value except 2. If $x$ is 2, then the denominator becomes $2(2) - 4 = 0$, making the expression undefined. Thus, $x$ cannot equal 2. We indicate this restriction by writing $x \neq 2$.

**NOTE** The *numerator* of a rational expression may be *any* real number. If the numerator equals 0 and the denominator does not equal 0, then the rational expression equals 0. See Example 1(b).

To determine the values for which a rational expression is undefined, use the following procedure.

**Determining When a Rational Expression is Undefined**

*Step 1* Set the denominator of the rational expression equal to 0.

*Step 2* Solve this equation.

*Step 3* The solutions of the equation are the values that make the rational expression undefined.

## EXAMPLE 2 Finding Values That Make Rational Expressions Undefined

Find any values of the variable for which each rational expression is undefined.

**(a)** $\dfrac{p + 5}{3p + 2}$

Remember that the *numerator* may be any real number; you must find any value of $p$ that makes the *denominator* equal to 0 since division by 0 is undefined.

*Step 1* Set the denominator equal to 0.

$$3p + 2 = 0$$

*Step 2* Solve this equation.

$$3p = -2$$

$$p = -\frac{2}{3}$$

*Step 3* The given expression is undefined for $-\frac{2}{3}$, so $p \neq -\frac{2}{3}$.

**(b)** $\dfrac{8x^2 + 1}{x - 3}$

The denominator $x - 3 = 0$ when $x$ is 3. The given expression is undefined for 3, so $x \neq 3$.

**(c)** $\dfrac{9m^2}{m^2 - 5m + 6}$

Set the denominator equal to 0, and then find the solutions of the equation $m^2 - 5m + 6 = 0$.

$$(m - 2)(m - 3) = 0 \qquad \text{Factor.}$$
$$m - 2 = 0 \quad \text{or} \quad m - 3 = 0 \qquad \text{Zero-factor property}$$
$$m = 2 \quad \text{or} \quad m = 3$$

The given expression is undefined for 2 and 3, so $m \neq 2$ and $m \neq 3$.

**(d)** $\dfrac{2r}{r^2 + 1}$

This denominator will not equal 0 for any value of $r$ because $r^2$ is always greater than or equal to 0, and adding 1 makes the sum greater than 0. Thus, there are no values for which this rational expression is undefined.

**Now Try Exercises 17 and 21.**

**OBJECTIVE 3 Write rational expressions in lowest terms.** A fraction such as $\frac{2}{3}$ is said to be in *lowest terms.* How can "lowest terms" be defined? We use the idea of greatest common factor for this definition, which applies to all rational expressions.

**Lowest Terms**

A rational expression $\frac{P}{Q}$ $(Q \neq 0)$ is in **lowest terms** if the greatest common factor of its numerator and denominator is 1.

The properties of rational numbers also apply to rational expressions. We use the **fundamental property of rational expressions** to write a rational expression in lowest terms.

**Fundamental Property of Rational Expressions**

If $\frac{P}{Q}$ $(Q \neq 0)$ is a rational expression and if $K$ represents any polynomial, where $K \neq 0$, then

$$\frac{PK}{QK} = \frac{P}{Q}.$$

This property is based on the identity property of multiplication, since

$$\frac{PK}{QK} = \frac{P}{Q} \cdot \frac{K}{K} = \frac{P}{Q} \cdot 1 = \frac{P}{Q}.$$

The next example shows how to write both a rational number and a rational expression in lowest terms. Notice the similarity in the procedures. In both cases, we factor and then divide out the greatest common factor.

## EXAMPLE 3 Writing in Lowest Terms

Write each expression in lowest terms.

**(a)** $\dfrac{30}{72}$

Begin by factoring.

$$\frac{30}{72} = \frac{2 \cdot 3 \cdot 5}{2 \cdot 2 \cdot 2 \cdot 3 \cdot 3}$$

**(b)** $\dfrac{14k^2}{2k^3}$

Write $k^2$ as $k \cdot k$ and $k^3$ as $k \cdot k \cdot k$.

$$\frac{14k^2}{2k^3} = \frac{2 \cdot 7 \cdot k \cdot k}{2 \cdot k \cdot k \cdot k}$$

Group any factors common to the numerator and denominator.

$$\frac{30}{72} = \frac{5 \cdot (2 \cdot 3)}{2 \cdot 2 \cdot 3 \cdot (2 \cdot 3)} \qquad \frac{14k^2}{2k^3} = \frac{7(2 \cdot k \cdot k)}{k(2 \cdot k \cdot k)}$$

Use the fundamental property.

$$\frac{30}{72} = \frac{5}{2 \cdot 2 \cdot 3} = \frac{5}{12} \qquad \frac{14k^2}{2k^3} = \frac{7}{k}$$

**Now Try Exercise 27.**

To write a rational expression in lowest terms, follow these steps.

**Writing a Rational Expression in Lowest Terms**

*Step 1* **Factor** the numerator and denominator completely.

*Step 2* **Use the fundamental property** to divide out any common factors.

## EXAMPLE 4 Writing in Lowest Terms

Write each rational expression in lowest terms.

**(a)** $\dfrac{3x - 12}{5x - 20}$

*Step 1* Factor both numerator and denominator.

$$\frac{3x - 12}{5x - 20} = \frac{3(x - 4)}{5(x - 4)}$$

*Step 2* Use the fundamental property.

$$= \frac{3}{5} \qquad \text{Divide out the common factor.}$$

The given expression is equal to $\frac{3}{5}$ for all values of $x$, where $x \neq 4$ (since the denominator of the original rational expression is 0 when $x$ is 4).

**(b)** $\dfrac{2y^2 - 8}{2y + 4} = \dfrac{2(y^2 - 4)}{2(y + 2)}$ Factor. (Step 1)

$$= \frac{2(y + 2)(y - 2)}{2(y + 2)} \qquad \text{Be sure to factor completely.}$$

$$= y - 2 \qquad \text{Fundamental property (Step 2)}$$

Here, $y \neq -2$ since the denominator of the original expression is 0 for this value.

(c) $\dfrac{m^2 + 2m - 8}{2m^2 - m - 6} = \dfrac{(m + 4)(m - 2)}{(2m + 3)(m - 2)}$ Factor.

$= \dfrac{m + 4}{2m + 3}$ Fundamental property

Here, $m \neq -\frac{3}{2}$ and $m \neq 2$, since the denominator of the original expression is 0 for these values of $m$.

**Now Try Exercises 33, 41, and 49.**

From now on, we write statements of equality of rational expressions with the understanding that they apply only to those real numbers that make neither denominator equal to 0.

**CAUTION** ***Rational expressions cannot be written in lowest terms until after the numerator and denominator have been factored. Only common factors can be divided out,*** not common terms. For example,

$$\frac{6x + 9}{4x + 6} = \frac{3(2x + 3)}{2(2x + 3)} = \frac{3}{2} \qquad \Bigg| \qquad \frac{6 + x}{4x} \leftarrow \text{Numerator cannot be factored.}$$

↑ Divide out the common factor. — Already in lowest terms

## EXAMPLE 5 Writing in Lowest Terms (Factors Are Opposites)

Write $\dfrac{x - y}{y - x}$ in lowest terms.

At first glance, there does not seem to be any way in which $x - y$ and $y - x$ can be factored to get a common factor. However, the denominator $y - x$ can be factored as

$$y - x = -1(-y + x) = -1(x - y).$$

Now, use the fundamental property to simplify.

$$\frac{x - y}{y - x} = \frac{1(x - y)}{-1(x - y)} = \frac{1}{-1} = -1$$

**Now Try Exercise 63.**

**NOTE** Either the numerator or the denominator could have been factored in the first step in Example 5. Factor $-1$ from the numerator and confirm that the result is the same.

In Example 5, notice that $y - x$ is the opposite of $x - y$. A general rule for this situation follows.

If the numerator and the denominator of a rational expression are opposites, as in $\frac{x - y}{y - x}$, then the rational expression is equal to $-1$.

**EXAMPLE 6 Writing in Lowest Terms (Factors Are Opposites)**

Write each rational expression in lowest terms.

**(a)** $\dfrac{2-m}{m-2}$

Each term in the numerator differs only in sign from the same term in the denominator, so we can factor $-1$ from either the numerator or the denominator.

$$\frac{2-m}{m-2}=\frac{-1(m-2)}{m-2}=-1$$

**(b)** $\dfrac{4x^2-9}{6-4x}$

Factor the numerator and denominator.

$$\frac{4x^2-9}{6-4x}=\frac{(2x+3)(2x-3)}{2(3-2x)}$$

$$=\frac{(2x+3)(2x-3)}{2(-1)(2x-3)} \qquad \text{Write } 3-2x \text{ as } -1(2x-3).$$

$$=\frac{2x+3}{2(-1)}$$

$$=\frac{2x+3}{-2} \quad \text{or} \quad -\frac{2x+3}{2} \qquad \frac{a}{-b}=-\frac{a}{b}$$

**(c)** $\dfrac{3+r}{3-r}$

The quantity $3-r$ is *not* the opposite of $3+r$ because the 3 is positive in both cases. This rational expression is already in lowest terms.

**Now Try Exercises 65 and 69.**

**OBJECTIVE 4 Recognize equivalent forms of rational expressions.** When working with rational expressions, it is important to be able to recognize equivalent forms of an expression. For example, the common fraction $-\frac{5}{6}$ can also be written $\frac{-5}{6}$ and $\frac{5}{-6}$. Consider also the rational expression

$$-\frac{2x+3}{2}.$$

The $-$ sign representing the $-1$ factor is in front of the expression. The $-1$ factor may instead be placed in the numerator or in the denominator. Thus, some other equivalent forms of this rational expression are

Use parentheses.

$$\frac{-(2x+3)}{2} \quad \text{and} \quad \frac{2x+3}{-2}.$$

By the distributive property,

$$\frac{-(2x+3)}{2} \quad \text{can also be written} \quad \frac{-2x-3}{2}.$$

**CAUTION** $\frac{-2x + 3}{2}$ is *not* an equivalent form of $\frac{-(2x + 3)}{2}$. The sign preceding 3 in the numerator of $\frac{-2x + 3}{2}$ should be $-$ rather than $+$. Be careful to apply the distributive property correctly.

### EXAMPLE 7 Writing Equivalent Forms of a Rational Expression

Write four equivalent forms of the rational expression

$$-\frac{3x + 2}{x - 6}.$$

If we apply the negative sign to the numerator, we obtain the equivalent forms

$$\frac{-(3x + 2)}{x - 6} \quad \text{or, by the distributive property,} \quad \frac{-3x - 2}{x - 6}.$$

If we apply the negative sign to the denominator, we obtain

$$\frac{3x + 2}{-(x - 6)} \quad \text{or, distributing once again,} \quad \frac{3x + 2}{-x + 6}.$$

**Now Try Exercise 73.**

**CAUTION** Recall that $-\frac{5}{6} \neq \frac{-5}{-6}$. Thus, in Example 7, it would be incorrect to distribute the negative sign to *both* the numerator *and* the denominator. (Doing this would actually lead to the *opposite* of the original expression.)

**CONNECTIONS**

In Chapter 4 we used long division to find the quotient of two polynomials. For example, we found $(2x^2 + 5x - 12) \div (2x - 3)$ as follows:

$$\begin{array}{r} x + \phantom{0}4 \\ 2x - 3\overline{)2x^2 + 5x - 12} \\ \underline{2x^2 - 3x\phantom{ - 12}} \\ 8x - 12 \\ \underline{8x - 12} \\ 0 \end{array}$$

The quotient is $x + 4$. We also get the same quotient by expressing the division problem as a rational expression (fraction) and writing this rational expression in lowest terms.

$$\frac{2x^2 + 5x - 12}{2x - 3} = \frac{(2x - 3)(x + 4)}{2x - 3}$$
$$= x + 4$$

### For Discussion or Writing

What kind of division problem has a quotient that cannot be found by writing a fraction in lowest terms? Try using rational expressions to solve each division problem. Then use long division and compare.

**1.** $(3x^2 + 11x + 8) \div (x + 2)$ **2.** $(x^3 - 8) \div (x^2 + 2x + 4)$

## 6.1 EXERCISES

### For Extra Help

- Student's Solutions Manual
- MyMathLab
- InterAct Math Tutorial Software
- AW Math Tutor Center
- MathXL
- Digital Video Tutor CD 9/Videotape 9

**1.** Define *rational expression* in your own words, and give an example.

*Find the numerical value of each rational expression when* **(a)** $x = 2$ *and* **(b)** $x = -3$. *See Example 1.*

**2.** $\dfrac{5x - 2}{4x}$ **3.** $\dfrac{3x + 1}{5x}$ **4.** $\dfrac{2x^2 - 4x}{3x - 1}$

**5.** $\dfrac{x^2 - 4}{2x + 1}$ **6.** $\dfrac{(-3x)^2}{4x + 12}$ **7.** $\dfrac{(-2x)^3}{3x + 9}$

**8.** $\dfrac{5x + 2}{2x^2 + 11x + 12}$ **9.** $\dfrac{7 - 3x}{3x^2 - 7x + 2}$

**10.** Fill in each blank with the correct response: The rational expression $\frac{x+5}{x-3}$ is undefined when $x$ is ______, so $x \neq$ ______. This rational expression is equal to 0 when $x =$ ______.

**11.** Why can't the denominator of a rational expression equal 0?

**12.** If 2 is substituted for $x$ in the rational expression $\dfrac{x - 2}{x^2 - 4}$, the result is $\dfrac{0}{0}$. An often-heard statement is "Any number divided by itself is 1." Does this mean that this expression is equal to 1 for $x = 2$? If not, explain.

*Find any values for which each rational expression is undefined. See Example 2.*

**13.** $\dfrac{12}{5y}$ **14.** $\dfrac{-7}{3z}$ **15.** $\dfrac{x + 1}{x - 6}$ **16.** $\dfrac{m - 2}{m - 5}$

**17.** $\dfrac{4x^2}{3x + 5}$ **18.** $\dfrac{2x^3}{3x + 4}$ **19.** $\dfrac{5m + 2}{m^2 + m - 6}$ **20.** $\dfrac{2r - 5}{r^2 - 5r + 4}$

**21.** $\dfrac{x^2 + 3x}{4}$ **22.** $\dfrac{x^2 - 4x}{6}$ **23.** $\dfrac{3x - 1}{x^2 + 2}$ **24.** $\dfrac{4q + 2}{q^2 + 9}$

**25.** **(a)** Identify the two *terms* in the numerator and the two *terms* in the denominator of the rational expression $\dfrac{x^2 + 4x}{x + 4}$.

**(b)** Describe the steps you would use to write this rational expression in lowest terms. (*Hint:* It simplifies to $x$.)

**26.** Only one of the following rational expressions can be simplified. Which one is it?

**A.** $\dfrac{x^2 + 2}{x^2}$ **B.** $\dfrac{x^2 + 2}{2}$ **C.** $\dfrac{x^2 + y^2}{y^2}$ **D.** $\dfrac{x^2 - 5x}{x}$

*Write each rational expression in lowest terms. See Examples 3 and 4.*

**27.** $\frac{18r^3}{6r}$

**28.** $\frac{27p^2}{3p}$

**29.** $\frac{4(y-2)}{10(y-2)}$

**30.** $\frac{15(m-1)}{9(m-1)}$

**31.** $\frac{(x+1)(x-1)}{(x+1)^2}$

**32.** $\frac{(t+5)(t-3)}{(t-1)(t+5)}$

**33.** $\frac{7m+14}{5m+10}$

**34.** $\frac{8z-24}{4z-12}$

**35.** $\frac{6m-18}{7m-21}$

**36.** $\frac{5r+20}{3r+12}$

**37.** $\frac{m^2-n^2}{m+n}$

**38.** $\frac{a^2-b^2}{a-b}$

**39.** $\frac{2t+6}{t^2-9}$

**40.** $\frac{5s-25}{s^2-25}$

**41.** $\frac{12m^2-3}{8m-4}$

**42.** $\frac{20p^2-45}{6p-9}$

**43.** $\frac{3m^2-3m}{5m-5}$

**44.** $\frac{6t^2-6t}{2t-2}$

**45.** $\frac{9r^2-4s^2}{9r+6s}$

**46.** $\frac{16x^2-9y^2}{12x-9y}$

**47.** $\frac{5k^2-13k-6}{5k+2}$

**48.** $\frac{7t^2-31t-20}{7t+4}$

**49.** $\frac{x^2+2x-15}{x^2+6x+5}$

**50.** $\frac{y^2-5y-14}{y^2+y-2}$

**51.** $\frac{2x^2-3x-5}{2x^2-7x+5}$

**52.** $\frac{3x^2+8x+4}{3x^2-4x-4}$

*These exercises involve factoring by grouping (Section 5.1) and factoring sums and differences of cubes (Section 5.4). Write each rational expression in lowest terms.*

**53.** $\frac{zw+4z-3w-12}{zw+4z+5w+20}$

**54.** $\frac{km+4k+4m+16}{km+4k+5m+20}$

**55.** $\frac{m^2-n^2-4m-4n}{2m-2n-8}$

**56.** $\frac{x^2y+y+x^2z+z}{xy+xz}$

**57.** $\frac{b^3-a^3}{a^2-b^2}$

**58.** $\frac{k^3+8}{k^2-4}$

**59.** $\frac{z^3+27}{z^3-3z^2+9z}$

**60.** $\frac{1-8r^3}{8r^2+4r+2}$

**61.** Which two of the following rational expressions equal $-1$?

**A.** $\frac{2x+3}{2x-3}$ **B.** $\frac{2x-3}{3-2x}$ **C.** $\frac{2x+3}{3+2x}$ **D.** $\frac{2x+3}{-2x-3}$

**62.** Make the correct choice for the blank: $\frac{4-r^2}{4+r^2}$ ____________ equal to $-1$.
(is/is not)

*Write each rational expression in lowest terms. See Examples 5 and 6.*

**63.** $\frac{6-t}{t-6}$

**64.** $\frac{2-k}{k-2}$

**65.** $\frac{m^2-1}{1-m}$

**66.** $\frac{a^2-b^2}{b-a}$

**67.** $\frac{q^2-4q}{4q-q^2}$

**68.** $\frac{z^2-5z}{5z-z^2}$

**69.** $\frac{p+6}{p-6}$

**70.** $\frac{5-x}{5+x}$

**71.** Only one of the following rational expressions is *not* equivalent to $\frac{x-3}{4-x}$. Which one is it?

**A.** $\dfrac{3-x}{x-4}$ **B.** $\dfrac{x+3}{4+x}$ **C.** $-\dfrac{3-x}{4-x}$ **D.** $-\dfrac{x-3}{x-4}$

**72.** Make the correct choice for the blank: $\frac{5+2x}{3-x}$ and $\frac{-5-2x}{x-3}$ __________ equivalent rational expressions. (are/are not)

*Write four equivalent forms for each rational expression. See Example 7.*

**73.** $-\dfrac{x+4}{x-3}$ **74.** $-\dfrac{x+6}{x-1}$ **75.** $-\dfrac{2x-3}{x+3}$

**76.** $-\dfrac{5x-6}{x+4}$ **77.** $\dfrac{-3x+1}{5x-6}$ **78.** $\dfrac{-2x-9}{3x+1}$

**79.** The area of the rectangle is represented by

$$x^4 + 10x^2 + 21.$$

What is the width? $\left(\textit{Hint: } \text{Use } W = \frac{A}{L}.\right)$

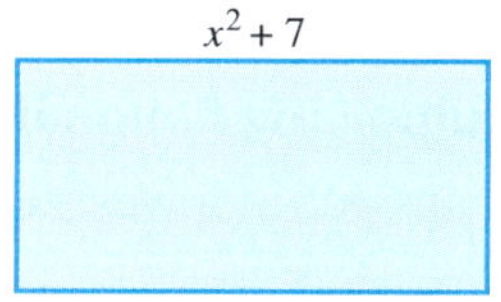


**80.** The volume of the box is represented by

$$(x^2 + 8x + 15)(x + 4).$$

Find the polynomial that represents the area of the bottom of the box.

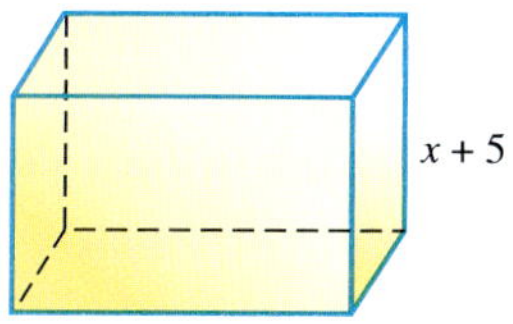


*Solve each problem.*

**81.** The average number of vehicles waiting in line to enter a sports arena parking area is approximated by the rational expression

$$\frac{x^2}{2(1-x)},$$

where $x$ is a quantity between 0 and 1 known as the *traffic intensity.* (*Source:* Mannering, F. and W. Kilareski, *Principles of Highway Engineering and Traffic Control,* John Wiley and Sons, 1990.) To the nearest tenth, find the average number of vehicles waiting if the traffic intensity is

**(a)** .1 **(b)** .8 **(c)** .9.

**(d)** What happens to waiting time as traffic intensity increases?

**82.** The percent of deaths caused by smoking is modeled by the rational expression

$$\frac{x-1}{x},$$

where $x$ is the number of times a smoker is more likely to die of lung cancer than a nonsmoker. This is called the *incidence rate.* (*Source:* Walker, A., *Observation and Inference: An Introduction to the Methods of Epidemiology,* Epidemiology Resources Inc., 1991.) For example, $x = 10$ means that a smoker is 10 times more likely than a nonsmoker to die of lung cancer. Find the percent of deaths if the incidence rate is

**(a)** 5 **(b)** 10 **(c)** 20.

**(d)** Can the incidence rate equal 0? Explain.

# 6.2 Multiplying and Dividing Rational Expressions

**OBJECTIVES**

1. Multiply rational expressions.
2. Divide rational expressions.

**OBJECTIVE 1 Multiply rational expressions.** The product of two fractions is found by multiplying the numerators and multiplying the denominators. Rational expressions are multiplied in the same way.

**Multiplying Rational Expressions**

The product of the rational expressions $\frac{P}{Q}$ and $\frac{R}{S}$ is

$$\frac{P}{Q} \cdot \frac{R}{S} = \frac{PR}{QS}.$$

That is, to multiply rational expressions, multiply the numerators and multiply the denominators.

In the following example, the parallel discussion with rational numbers and rational expressions lets you compare the steps.

**EXAMPLE 1 Multiplying Rational Expressions**

Multiply. Write each answer in lowest terms.

**(a)** $\frac{3}{10} \cdot \frac{5}{9}$ **(b)** $\frac{6}{x} \cdot \frac{x^2}{12}$

Indicate the product of the numerators and the product of the denominators.

$$\frac{3}{10} \cdot \frac{5}{9} = \frac{3 \cdot 5}{10 \cdot 9} \qquad \frac{6}{x} \cdot \frac{x^2}{12} = \frac{6 \cdot x^2}{x \cdot 12}$$

Leave the products in factored form because common factors are needed to write the product in lowest terms. Factor the numerator and denominator to further identify any common factors. Then use the fundamental property to write each product in lowest terms.

$$\frac{3}{10} \cdot \frac{5}{9} = \frac{3 \cdot 5}{2 \cdot 5 \cdot 3 \cdot 3} = \frac{1}{6} \qquad \frac{6}{x} \cdot \frac{x^2}{12} = \frac{6 \cdot x \cdot x}{2 \cdot 6 \cdot x} = \frac{x}{2}$$

**Now Try Exercise 3.**

**NOTE** It is also possible to divide out common factors in the numerator and denominator *before* multiplying the rational expressions. For example,

$$\frac{6}{5} \cdot \frac{35}{22} = \frac{2 \cdot 3}{5} \cdot \frac{5 \cdot 7}{2 \cdot 11} \quad \text{Identify common factors.}$$

$$= \frac{3 \cdot 7}{11} \quad \text{Lowest terms}$$

$$= \frac{21}{11}. \quad \text{Multiply in numerator.}$$

**EXAMPLE 2 Multiplying Rational Expressions**

Multiply. Write the answer in lowest terms.

$$\frac{x+y}{2x} \cdot \frac{x^2}{(x+y)^2} = \frac{(x+y)x^2}{2x(x+y)^2}$$ Multiply numerators. Multiply denominators.

$$= \frac{(x+y)x \cdot x}{2x(x+y)(x+y)}$$ Factor; identify common factors.

$$= \frac{x}{2(x+y)}$$ Lowest terms

Notice the quotient of factors $\frac{(x+y)x}{x(x+y)}$ in the second line of the solution. Since it is equal to 1, the final product is $\frac{x}{2(x+y)}$.

**Now Try Exercise 9.**

**EXAMPLE 3 Multiplying Rational Expressions**

Multiply. Write the answer in lowest terms.

$$\frac{x^2+3x}{x^2-3x-4} \cdot \frac{x^2-5x+4}{x^2+2x-3} = \frac{(x^2+3x)(x^2-5x+4)}{(x^2-3x-4)(x^2+2x-3)}$$ Definition of multiplication

$$= \frac{x(x+3)(x-4)(x-1)}{(x-4)(x+1)(x+3)(x-1)}$$ Factor.

$$= \frac{x}{x+1}$$ Lowest terms

The quotients $\frac{x+3}{x+3}$, $\frac{x-4}{x-4}$, and $\frac{x-1}{x-1}$ all equal 1, justifying the final product $\frac{x}{x+1}$.

**Now Try Exercise 43.**

**OBJECTIVE 2 Divide rational expressions.** To develop a method for dividing rational numbers and rational expressions, consider the following problem. Suppose you have $\frac{7}{8}$ gal of milk and want to find how many quarts you have. Since 1 qt is $\frac{1}{4}$ gal, you ask yourself, "How many $\frac{1}{4}$s are there in $\frac{7}{8}$?" This would be interpreted as

$$\frac{7}{8} \div \frac{1}{4} \quad \text{or} \quad \frac{\frac{7}{8}}{\frac{1}{4}}$$

since the fraction bar means division.

The fundamental property of rational expressions discussed earlier can be applied to rational number values of $P$, $Q$, and $K$. With $P = \frac{7}{8}$, $Q = \frac{1}{4}$, and $K = 4$ $\left(\text{since 4 is the reciprocal of } Q = \frac{1}{4}\right)$,

$$\frac{P}{Q} = \frac{P \cdot K}{Q \cdot K} = \frac{\frac{7}{8} \cdot 4}{\frac{1}{4} \cdot 4} = \frac{\frac{7}{8} \cdot 4}{1} = \frac{7}{8} \cdot \frac{4}{1}.$$

So, to divide $\frac{7}{8}$ by $\frac{1}{4}$, we multiply $\frac{7}{8}$ by the reciprocal of $\frac{1}{4}$, namely 4. Since $\frac{7}{8}(4) = \frac{7}{2}$, there are $\frac{7}{2}$ or $3\frac{1}{2}$ qt in $\frac{7}{8}$ gal.

The preceding discussion illustrates the rule for dividing common fractions. To divide $\frac{a}{b}$ by $\frac{c}{d}$, multiply $\frac{a}{b}$ by the reciprocal of $\frac{c}{d}$. Division of rational expressions is defined in the same way.

**Dividing Rational Expressions**

If $\frac{P}{Q}$ and $\frac{R}{S}$ are any two rational expressions, with $\frac{R}{S} \neq 0$, then

$$\frac{P}{Q} \div \frac{R}{S} = \frac{P}{Q} \cdot \frac{S}{R} = \frac{PS}{QR}.$$

That is, to divide one rational expression by another rational expression, multiply the first rational expression by the reciprocal of the second rational expression.

The next example shows the division of two rational numbers and the division of two rational expressions.

### EXAMPLE 4 Dividing Rational Expressions

Divide. Write each answer in lowest terms.

**(a)** $\frac{5}{8} \div \frac{7}{16}$

**(b)** $\frac{y}{y+3} \div \frac{4y}{y+5}$

Multiply the first expression by the reciprocal of the second.

**(a)**

$$\frac{5}{8} \div \frac{7}{16} = \frac{5}{8} \cdot \frac{16}{7}$$ Reciprocal of $\frac{7}{16}$

$$= \frac{5 \cdot 16}{8 \cdot 7}$$

$$= \frac{5 \cdot 8 \cdot 2}{8 \cdot 7}$$

$$= \frac{10}{7}$$

**(b)**

$$\frac{y}{y+3} \div \frac{4y}{y+5}$$

$$= \frac{y}{y+3} \cdot \frac{y+5}{4y}$$ Reciprocal of $\frac{4y}{y+5}$

$$= \frac{y(y+5)}{(y+3)(4y)}$$

$$= \frac{y+5}{4(y+3)}$$

**Now Try Exercise 19.**

### EXAMPLE 5 Dividing Rational Expressions

Divide. Write the answer in lowest terms.

$$\frac{(3m)^2}{(2p)^3} \div \frac{6m^3}{16p^2} = \frac{(3m)^2}{(2p)^3} \cdot \frac{16p^2}{6m^3}$$ Multiply by the reciprocal.

$$= \frac{9m^2}{8p^3} \cdot \frac{16p^2}{6m^3}$$ Power rule for exponents

$$= \frac{9 \cdot 16m^2p^2}{8 \cdot 6p^3m^3}$$ Multiply numerators. Multiply denominators.

$$= \frac{3}{mp}$$ Lowest terms

Now Try Exercise 17.

## EXAMPLE 6 Dividing Rational Expressions

Divide. Write the answer in lowest terms.

$$\frac{x^2 - 4}{(x + 3)(x - 2)} \div \frac{(x + 2)(x + 3)}{-2x}$$

$$= \frac{x^2 - 4}{(x + 3)(x - 2)} \cdot \frac{-2x}{(x + 2)(x + 3)}$$ Multiply by the reciprocal.

$$= \frac{-2x(x^2 - 4)}{(x + 3)(x - 2)(x + 2)(x + 3)}$$ Multiply numerators. Multiply denominators.

$$= \frac{-2x(x + 2)(x - 2)}{(x + 3)(x - 2)(x + 2)(x + 3)}$$ Factor.

$$= \frac{-2x}{(x + 3)^2}$$ Lowest terms

$$= -\frac{2x}{(x + 3)^2}$$ $\frac{-a}{b} = -\frac{a}{b}$

Now Try Exercise 35.

## EXAMPLE 7 Dividing Rational Expressions (Factors Are Opposites)

Divide. Write the answer in lowest terms.

$$\frac{m^2 - 4}{m^2 - 1} \div \frac{2m^2 + 4m}{1 - m}$$

$$= \frac{m^2 - 4}{m^2 - 1} \cdot \frac{1 - m}{2m^2 + 4m}$$ Multiply by the reciprocal.

$$= \frac{(m^2 - 4)(1 - m)}{(m^2 - 1)(2m^2 + 4m)}$$ Multiply numerators. Multiply denominators.

$$= \frac{(m + 2)(m - 2)(1 - m)}{(m + 1)(m - 1)(2m)(m + 2)}$$ Factor; $1 - m$ and $m - 1$ differ only in sign.

$$= \frac{-1(m - 2)}{2m(m + 1)}$$ From Section 6.1, $\frac{1 - m}{m - 1} = -1$.

$$= \frac{-m + 2}{2m(m + 1)} \quad \text{or} \quad \frac{2 - m}{2m(m + 1)}$$ Distribute the negative sign in the numerator.

Now Try Exercise 37.

In summary, follow these steps to multiply or divide rational expressions.

**Multiplying or Dividing Rational Expressions**

*Step 1* **Note the operation.** If the operation is division, use the definition of division to rewrite as multiplication.

*Step 2* **Multiply** numerators and multiply denominators.

*Step 3* **Factor** all numerators and denominators completely.

*Step 4* **Write in lowest terms** using the fundamental property.

*Note:* Steps 2 and 3 may be interchanged based on personal preference.

## 6.2 EXERCISES

**For Extra Help**

- Student's Solutions Manual
- MyMathLab
- InterAct Math Tutorial Software
- AW Math Tutor Center
- MathXL
- Digital Video Tutor CD 9/Videotape 9

**1.** Match each multiplication problem in Column I with the correct product in Column II.

| I | II |
|---|---|
| (a) $\frac{5x^3}{10x^4} \cdot \frac{10x^7}{2x}$ | A. $\frac{2}{5x^5}$ |
| (b) $\frac{10x^4}{5x^3} \cdot \frac{10x^7}{2x}$ | B. $\frac{5x^5}{2}$ |
| (c) $\frac{5x^3}{10x^4} \cdot \frac{2x}{10x^7}$ | C. $\frac{1}{10x^7}$ |
| (d) $\frac{10x^4}{5x^3} \cdot \frac{2x}{10x^7}$ | D. $10x^7$ |

**2.** Match each division problem in Column I with the correct quotient in Column II.

| I | II |
|---|---|
| (a) $\frac{5x^3}{10x^4} \div \frac{10x^7}{2x}$ | A. $\frac{5x^5}{2}$ |
| (b) $\frac{10x^4}{5x^3} \div \frac{10x^7}{2x}$ | B. $10x^7$ |
| (c) $\frac{5x^3}{10x^4} \div \frac{2x}{10x^7}$ | C. $\frac{2}{5x^5}$ |
| (d) $\frac{10x^4}{5x^3} \div \frac{2x}{10x^7}$ | D. $\frac{1}{10x^7}$ |

*Multiply. Write each answer in lowest terms. See Examples 1 and 2.*

**3.** $\frac{15a^2}{14} \cdot \frac{7}{5a}$

**4.** $\frac{27k^3}{9k} \cdot \frac{24}{9k^2}$

**5.** $\frac{12x^4}{18x^3} \cdot \frac{-8x^5}{4x^2}$

**6.** $\frac{12m^5}{-2m^2} \cdot \frac{6m^6}{28m^3}$

**7.** $\frac{2(c+d)}{3} \cdot \frac{18}{6(c+d)^2}$

**8.** $\frac{4(y-2)}{x} \cdot \frac{3x}{6(y-2)^2}$

**9.** $\frac{(x-y)^2}{2} \cdot \frac{24}{3(x-y)}$

**10.** $\frac{(a+b)^2}{5} \cdot \frac{30}{2(a+b)}$

**11.** $\frac{t-4}{8} \cdot \frac{4t^2}{t-4}$

**12.** $\frac{z+9}{12} \cdot \frac{3z^2}{z+9}$

**13.** $\frac{3x}{x+3} \cdot \frac{(x+3)^2}{6x^2}$

**14.** $\frac{(t-2)^2}{4t^2} \cdot \frac{2t}{t-2}$

*Divide. Write each answer in lowest terms. See Examples 4 and 5.*

**15.** $\frac{9z^4}{3z^5} \div \frac{3z^2}{5z^3}$

**16.** $\frac{35q^8}{9q^5} \div \frac{25q^6}{10q^5}$

**17.** $\frac{4t^4}{2t^5} \div \frac{(2t)^3}{-6}$

**18.** $\frac{-12a^6}{3a^2} \div \frac{(2a)^3}{27a}$

**19.** $\frac{3}{2y-6} \div \frac{6}{y-3}$

**20.** $\frac{4m+16}{10} \div \frac{3m+12}{18}$

21. $\frac{7t+7}{-6} \div \frac{4t+4}{15}$

22. $\frac{8z-16}{-20} \div \frac{3z-6}{40}$

23. $\frac{2x}{x-1} \div \frac{x^2}{x+2}$

24. $\frac{y^2}{y+1} \div \frac{3y}{y-3}$

25. $\frac{(x-3)^2}{6x} \div \frac{x-3}{x^2}$

26. $\frac{2a}{a+4} \div \frac{a^2}{(a+4)^2}$

27. Use an example to explain how to multiply rational expressions.

28. Use an example to explain how to divide rational expressions.

*Multiply or divide. Write each answer in lowest terms. See Examples 3, 6, and 7.*

29. $\frac{5x-15}{3x+9} \cdot \frac{4x+12}{6x-18}$

30. $\frac{8r+16}{24r-24} \cdot \frac{6r-6}{3r+6}$

31. $\frac{2-t}{8} \div \frac{t-2}{6}$

32. $\frac{4}{m-2} \div \frac{16}{2-m}$

33. $\frac{27-3z}{4} \cdot \frac{12}{2z-18}$

34. $\frac{5-x}{5+x} \cdot \frac{x+5}{x-5}$

35. $\frac{p^2+4p-5}{p^2+7p+10} \div \frac{p-1}{p+4}$

36. $\frac{z^2-3z+2}{z^2+4z+3} \div \frac{z-1}{z+1}$

37. $\frac{m^2-4}{16-8m} \div \frac{m+2}{8}$

38. $\frac{2}{3-x} \div \frac{2x+6}{x^2-9}$

39. $\frac{2x^2-7x+3}{x-3} \cdot \frac{x+2}{x-1}$

40. $\frac{3x^2-5x-2}{x-2} \cdot \frac{x-3}{x+1}$

41. $\frac{2k^2-k-1}{2k^2+5k+3} \div \frac{4k^2-1}{2k^2+k-3}$

42. $\frac{2m^2-5m-12}{m^2+m-20} \div \frac{4m^2-9}{m^2+4m-5}$

43. $\frac{2k^2+3k-2}{6k^2-7k+2} \cdot \frac{4k^2-5k+1}{k^2+k-2}$

44. $\frac{2m^2-5m-12}{m^2-10m+24} \div \frac{4m^2-9}{m^2-9m+18}$

45. $\frac{m^2+2mp-3p^2}{m^2-3mp+2p^2} \div \frac{m^2+4mp+3p^2}{m^2+2mp-8p^2}$

46. $\frac{r^2+rs-12s^2}{r^2-rs-20s^2} \div \frac{r^2-2rs-3s^2}{r^2+rs-30s^2}$

47. $\frac{m^2+3m+2}{m^2+5m+4} \cdot \frac{m^2+10m+24}{m^2+5m+6}$

48. $\frac{z^2-z-6}{z^2-2z-8} \cdot \frac{z^2+7z+12}{z^2-9}$

49. $\frac{y^2+y-2}{y^2+3y-4} \div \frac{y+2}{y+3}$

50. $\frac{r^2+r-6}{r^2+4r-12} \div \frac{r+3}{r-1}$

51. $\frac{2m^2+7m+3}{m^2-9} \cdot \frac{m^2-3m}{2m^2+11m+5}$

52. $\frac{m^2+2mp-3p^2}{m^2-3mp+2p^2} \div \frac{m^2+4mp+3p^2}{m^2+2mp-8p^2}$

53. $\frac{r^2+rs-12s^2}{r^2-rs-20s^2} \div \frac{r^2-2rs-3s^2}{r^2+rs-30s^2}$

54. $\frac{(x+1)^3(x+4)}{x^2+5x+4} \div \frac{x^2+2x+1}{x^2+3x+2}$

55. $\frac{(q-3)^4(q+2)}{q^2+3q+2} \div \frac{q^2-6q+9}{q^2+4q+4}$

56. $\frac{(x+4)^3(x-3)}{x^2-9} \div \frac{x^2+8x+16}{x^2+6x+9}$

*These exercises involve grouping symbols (Section 1.2), factoring by grouping (Section 5.1), and factoring sums and differences of cubes (Section 5.4). Multiply or divide as indicated. Write each answer in lowest terms.*

57. $\frac{x+5}{x+10} \div \left(\frac{x^2+10x+25}{x^2+10x} \cdot \frac{10x}{x^2+15x+50}\right)$

58. $\frac{m-8}{m-4} \div \left(\frac{m^2-12m+32}{8m} \cdot \frac{m^2-8m}{m^2-8m+16}\right)$

59. $\frac{3a-3b-a^2+b^2}{4a^2-4ab+b^2} \cdot \frac{4a^2-b^2}{2a^2-ab-b^2}$

*Find the LCD for the fractions in each list. See Examples 1–3.*

**5.** $\frac{7}{15}, \frac{21}{20}$

**6.** $\frac{9}{10}, \frac{12}{25}$

**7.** $\frac{17}{100}, \frac{23}{120}, \frac{43}{180}$

**8.** $\frac{17}{250}, \frac{-21}{300}, \frac{127}{360}$

**9.** $\frac{9}{x^2}, \frac{8}{x^5}$

**10.** $\frac{12}{m^7}, \frac{13}{m^8}$

**11.** $\frac{-2}{5p}, \frac{15}{6p}$

**12.** $\frac{14}{15k}, \frac{9}{4k}$

**13.** $\frac{17}{15y^2}, \frac{55}{36y^4}$

**14.** $\frac{4}{25m^3}, \frac{-9}{10m^4}$

**15.** $\frac{6}{21r^3}, \frac{8}{12r^5}$

**16.** $\frac{9}{35t^2}, \frac{5}{49t^6}$

**17.** $\frac{13}{5a^2b^3}, \frac{29}{15a^5b}$

**18.** $\frac{-7}{3r^4s^5}, \frac{-22}{9r^6s^8}$

**19.** $\frac{7}{6p}, \frac{15}{4p-8}$

**20.** $\frac{7}{8k}, \frac{-23}{12k-24}$

**21.** $\frac{9}{28m^2}, \frac{3}{12m-20}$

**22.** $\frac{15}{27a^3}, \frac{8}{9a-45}$

**23.** $\frac{7}{5b-10}, \frac{11}{6b-12}$

**24.** $\frac{3}{7x^2+21x}, \frac{1}{5x^2+15x}$

## RELATING CONCEPTS (EXERCISES 25–28)

**For Individual or Group Work**

*Suppose we want to find the LCD for the two common fractions*

$$\frac{1}{24} \quad \text{and} \quad \frac{1}{20}.$$

*In their prime factored forms, the denominators are*

$$24 = 2^3 \cdot 3 \quad \text{and} \quad 20 = 2^2 \cdot 5.$$

*Refer to this information as necessary, and* ***work Exercises 25–28 in order.***

**25.** What is the prime factored form of the LCD of the two fractions?

**26.** Suppose that two algebraic fractions have denominators $(t+4)^3(t-3)$ and $(t+4)^2(t+8)$. What is the factored form of the LCD of these?

**27.** What is the similarity between your answers in Exercises 25 and 26?

**28.** Comment on the following statement: The method for finding the LCD for two algebraic fractions is the same as the method for finding the LCD for two common fractions.

*Find the LCD for the fractions in each list. See Examples 1–3.*

**29.** $\frac{37}{6r-12}, \frac{25}{9r-18}$

**30.** $\frac{-14}{5p-30}, \frac{5}{6p-36}$

**31.** $\frac{5}{12p + 60}, \frac{17}{p^2 + 5p}, \frac{16}{p^2 + 10p + 25}$

**32.** $\frac{13}{r^2 + 7r}, \frac{-3}{5r + 35}, \frac{-7}{r^2 + 14r + 49}$

**33.** $\frac{3}{8y + 16}, \frac{22}{y^2 + 3y + 2}$

**34.** $\frac{-2}{9m - 18}, \frac{-9}{m^2 - 7m + 10}$

**35.** $\frac{5}{c - d}, \frac{8}{d - c}$

**36.** $\frac{4}{y - x}, \frac{7}{x - y}$

**37.** $\frac{12}{m - 3}, \frac{-4}{3 - m}$

**38.** $\frac{-17}{8 - a}, \frac{2}{a - 8}$

**39.** $\frac{29}{p - q}, \frac{18}{q - p}$

**40.** $\frac{16}{z - x}, \frac{8}{x - z}$

**41.** $\frac{3}{k^2 + 5k}, \frac{2}{k^2 + 3k - 10}$

**42.** $\frac{1}{z^2 - 4z}, \frac{4}{z^2 - 3z - 4}$

**43.** $\frac{6}{a^2 + 6a}, \frac{-5}{a^2 + 3a - 18}$

**44.** $\frac{8}{y^2 - 5y}, \frac{-2}{y^2 - 2y - 15}$

**45.** $\frac{5}{p^2 + 8p + 15}, \frac{3}{p^2 - 3p - 18}, \frac{2}{p^2 - p - 30}$

**46.** $\frac{10}{y^2 - 10y + 21}, \frac{2}{y^2 - 2y - 3}, \frac{5}{y^2 - 6y - 7}$

**47.** $\frac{-5}{k^2 + 2k - 35}, \frac{-8}{k^2 + 3k - 40}, \frac{9}{k^2 - 2k - 15}$

**48.** $\frac{19}{z^2 + 4z - 12}, \frac{16}{z^2 + z - 30}, \frac{6}{z^2 + 2z - 24}$

**49.** Suppose that $(2x - 5)^2$ is the LCD for two fractions. Is $(5 - 2x)^2$ also acceptable as an LCD? Why or why not?

**50.** Suppose that $(4t - 3)(5t - 6)$ is the LCD for two fractions. Is $(3 - 4t)(6 - 5t)$ also acceptable as an LCD? Why or why not?

## RELATING CONCEPTS (EXERCISES 51–56)

**For Individual or Group Work**

***Work Exercises 51–56 in order.***

**51.** Suppose that you want to write $\frac{3}{4}$ as an equivalent fraction with denominator 28. By what number must you multiply both the numerator and the denominator?

**52.** If you write $\frac{3}{4}$ as an equivalent fraction with denominator 28, by what number are you actually multiplying the fraction?

**53.** What property of multiplication is being used when we write a common fraction as an equivalent one with a larger denominator? (See Section 1.7.)

**54.** Suppose that you want to write $\frac{2x + 5}{x - 4}$ as an equivalent fraction with denominator $7x - 28$. By what number must you multiply both the numerator and the denominator?

**55.** If you write $\frac{2x + 5}{x - 4}$ as an equivalent fraction with denominator $7x - 28$, by what number are you actually multiplying the fraction?

**56.** Repeat Exercise 53, changing "a common" to "an algebraic."

*Rewrite each rational expression with the indicated denominator. See Examples 4 and 5.*

57. $\frac{4}{11} = \frac{}{55}$

58. $\frac{6}{7} = \frac{}{42}$

59. $\frac{-5}{k} = \frac{}{9k}$

60. $\frac{-3}{q} = \frac{}{6q}$

61. $\frac{15m^2}{8k} = \frac{}{32k^4}$

62. $\frac{5t^2}{3y} = \frac{}{9y^2}$

63. $\frac{19z}{2z-6} = \frac{}{6z-18}$

64. $\frac{2r}{5r-5} = \frac{}{15r-15}$

65. $\frac{-2a}{9a-18} = \frac{}{18a-36}$

66. $\frac{-5y}{6y+18} = \frac{}{24y+72}$

67. $\frac{6}{k^2-4k} = \frac{}{k(k-4)(k+1)}$

68. $\frac{15}{m^2-9m} = \frac{}{m(m-9)(m+8)}$

69. $\frac{36r}{r^2-r-6} = \frac{}{(r-3)(r+2)(r+1)}$

70. $\frac{4m}{m^2-8m+15} = \frac{}{(m-5)(m-3)(m+2)}$

71. $\frac{a+2b}{2a^2+ab-b^2} = \frac{}{2a^3b+a^2b^2-ab^3}$

72. $\frac{m-4}{6m^2+7m-3} = \frac{}{12m^3+14m^2-6m}$

73. $\frac{4r-t}{r^2+rt+t^2} = \frac{}{t^3-r^3}$

74. $\frac{3x-1}{x^2+2x+4} = \frac{}{x^3-8}$

75. $\frac{2(z-y)}{y^2+yz+z^2} = \frac{}{y^4-z^3y}$

76. $\frac{2p+3q}{p^2+2pq+q^2} = \frac{}{(p+q)(p^3+q^3)}$

77. Write an explanation of how to find the least common denominator for a group of denominators. Give an example.

78. Write an explanation of how to write a rational expression as an equivalent rational expression with a given denominator. Give an example.

# 6.4 Adding and Subtracting Rational Expressions

**OBJECTIVES**

1. Add rational expressions having the same denominator.
2. Add rational expressions having different denominators.

To add and subtract rational expressions, we need the skills developed in the previous section to find least common denominators and to write equivalent fractions with the LCD.

**OBJECTIVE 1 Add rational expressions having the same denominator.** We find the sum of two rational expressions with the same procedure that we used for adding two fractions in Section 1.1.

**3** Subtract rational expressions.

**Adding Rational Expressions**

If $\frac{P}{Q}$ and $\frac{R}{Q}$ $(Q \neq 0)$ are rational expressions, then

$$\frac{P}{Q} + \frac{R}{Q} = \frac{P+R}{Q}.$$

That is, to add rational expressions with the same denominator, add the numerators and keep the same denominator.

The first example shows how addition of rational expressions compares with that of rational numbers.

**EXAMPLE 1 Adding Rational Expressions with the Same Denominator**

Add. Write each answer in lowest terms.

**(a)** $\frac{4}{9} + \frac{2}{9}$

**(b)** $\frac{3x}{x+1} + \frac{3}{x+1}$

The denominators are the same, so the sum is found by adding the two numerators and keeping the same (common) denominator.

$$\frac{4}{9} + \frac{2}{9} = \frac{4+2}{9} = \frac{6}{9} = \frac{2}{3}$$

$$\frac{3x}{x+1} + \frac{3}{x+1} = \frac{3x+3}{x+1} = \frac{3(x+1)}{x+1} = 3$$

**Now Try Exercises 9 and 19.**

**OBJECTIVE 2 Add rational expressions having different denominators.** We use the following steps to add two rational expressions with different denominators. These are the same steps we used to add fractions with different denominators in Section 1.1.

**Adding with Different Denominators**

*Step 1* **Find the least common denominator (LCD).**

*Step 2* **Rewrite each rational expression** as an equivalent rational expression with the LCD as the denominator.

*Step 3* **Add** the numerators to get the numerator of the sum. The LCD is the denominator of the sum.

*Step 4* **Write in lowest terms** using the fundamental property.

**EXAMPLE 2 Adding Rational Expressions with Different Denominators**

Add. Write each answer in lowest terms.

**(a)** $\frac{1}{12}+\frac{7}{15}$

**(b)** $\frac{2}{3y}+\frac{1}{4y}$

*Step 1* First find the LCD using the methods of the previous section.

$\text{LCD}=2^2\cdot 3\cdot 5=60$ | $\text{LCD}=2^2\cdot 3\cdot y=12y$

*Step 2* Now rewrite each rational expression as a fraction with the LCD (either 60 or 12$y$) as the denominator.

$$\frac{1}{12}+\frac{7}{15}=\frac{1(5)}{12(5)}+\frac{7(4)}{15(4)}=\frac{5}{60}+\frac{28}{60}$$

$$\frac{2}{3y}+\frac{1}{4y}=\frac{2(4)}{3y(4)}+\frac{1(3)}{4y(3)}=\frac{8}{12y}+\frac{3}{12y}$$

*Step 3* Since the fractions now have common denominators, add the numerators.

*Step 4* Write in lowest terms if necessary.

$$\frac{5}{60}+\frac{28}{60}=\frac{5+28}{60}=\frac{33}{60}=\frac{11}{20}$$

$$\frac{8}{12y}+\frac{3}{12y}=\frac{8+3}{12y}=\frac{11}{12y}$$

**Now Try Exercises 25 and 31.**

**EXAMPLE 3 Adding Rational Expressions**

Add. Write the answer in lowest terms.

$$\frac{2x}{x^2-1}+\frac{-1}{x+1}$$

*Step 1* Since the denominators are different, find the LCD.

$$x^2-1=(x+1)(x-1)$$

$x+1$ is prime.

The LCD is $(x+1)(x-1)$.

*Step 2* Rewrite each rational expression as a fraction with common denominator $(x+1)(x-1)$.

$$\frac{2x}{x^2-1}+\frac{-1}{x+1}=\frac{2x}{(x+1)(x-1)}+\frac{-1(x-1)}{(x+1)(x-1)}$$ Multiply the second fraction by $\frac{x-1}{x-1}$.

$$=\frac{2x}{(x+1)(x-1)}+\frac{-x+1}{(x+1)(x-1)}$$ Distributive property

*Step 3* $$=\frac{2x-x+1}{(x+1)(x-1)}$$ Add numerators; keep the same denominator.

$$=\frac{x+1}{(x+1)(x-1)}$$ Combine like terms.

*Step 4* $= \dfrac{1(x+1)}{(x+1)(x-1)}$ Identity property of multiplication

$= \dfrac{1}{x-1}$ Fundamental property

**Now Try Exercise 41.**

## EXAMPLE 4 Adding Rational Expressions

Add. Write the answer in lowest terms.

$$\frac{2x}{x^2+5x+6} + \frac{x+1}{x^2+2x-3}$$

$$= \frac{2x}{(x+2)(x+3)} + \frac{x+1}{(x+3)(x-1)} \quad \text{Factor the denominators.}$$

The LCD is $(x+2)(x+3)(x-1)$. Use the fundamental property.

$$= \frac{2x(x-1)}{(x+2)(x+3)(x-1)} + \frac{(x+1)(x+2)}{(x+2)(x+3)(x-1)}$$

$$= \frac{2x(x-1)+(x+1)(x+2)}{(x+2)(x+3)(x-1)} \quad \text{Add numerators; keep the same denominator.}$$

$$= \frac{2x^2-2x+x^2+3x+2}{(x+2)(x+3)(x-1)} \quad \text{Multiply.}$$

$$= \frac{3x^2+x+2}{(x+2)(x+3)(x-1)} \quad \text{Combine like terms.}$$

It is usually more convenient to leave the denominator in factored form. The numerator cannot be factored here, so the expression is in lowest terms.

**Now Try Exercise 43.**

Rational expressions to be added or subtracted may have denominators that are opposites of each other.

## EXAMPLE 5 Adding Rational Expressions with Denominators That Are Opposites

Add. Write the answer in lowest terms.

$$\frac{y}{y-2} + \frac{8}{2-y}$$

One way to get a common denominator is to multiply the second expression by $-1$ in both the numerator and the denominator, giving $y-2$ as a common denominator.

$$\frac{y}{y-2} + \frac{8}{2-y} = \frac{y}{y-2} + \frac{8(-1)}{(2-y)(-1)} \quad \text{Fundamental property}$$

$$= \frac{y}{y-2} + \frac{-8}{y-2} \quad \text{Distributive property}$$

$$= \frac{y-8}{y-2} \quad \text{Add numerators; keep the same denominator.}$$

If we had chosen to use $2 - y$ as the common denominator, the final answer would be $\frac{8-y}{2-y}$, which is equivalent to $\frac{y-8}{y-2}$.

**Now Try Exercise 51.**

**OBJECTIVE 3 Subtract rational expressions.** To subtract rational expressions, use the following rule.

**Subtracting Rational Expressions**

If $\frac{P}{Q}$ and $\frac{R}{Q}$ $(Q \neq 0)$ are rational expressions, then

$$\frac{P}{Q} - \frac{R}{Q} = \frac{P - R}{Q}.$$

That is, to subtract rational expressions with the same denominator, subtract the numerators and keep the same denominator.

**EXAMPLE 6 Subtracting Rational Expressions with the Same Denominator**

Subtract. Write the answer in lowest terms.

$$\frac{2m}{m-1} - \frac{m+3}{m-1} = \frac{2m - (m+3)}{m-1}$$

Use parentheses around the quantity being subtracted.

Subtract numerators; keep the same denominator.

$$= \frac{2m - m - 3}{m-1}$$

Distributive property

$$= \frac{m-3}{m-1}$$

Combine like terms.

**Now Try Exercise 15.**

**CAUTION** Sign errors often occur in subtraction problems like the one in Example 6. Remember that the numerator of the fraction being subtracted must be treated as a single quantity. Be sure to use parentheses after the subtraction sign.

**EXAMPLE 7 Subtracting Rational Expressions with Different Denominators**

Subtract. Write the answer in lowest terms.

$$\frac{9}{x-2} - \frac{3}{x} = \frac{9x}{x(x-2)} - \frac{3(x-2)}{x(x-2)}$$

The LCD is $x(x-2)$.

$$= \frac{9x - 3(x-2)}{x(x-2)}$$

Subtract numerators; keep the same denominator.

$$= \frac{9x - 3x + 6}{x(x - 2)} \qquad \text{Distributive property; Be careful with signs.}$$

$$= \frac{6x + 6}{x(x - 2)} \qquad \text{Combine like terms.}$$

$$= \frac{6(x + 1)}{x(x - 2)} \qquad \text{Factor the numerator.}$$

**Now Try Exercise 39.**

**NOTE** We factored the final numerator in Example 7 to get $\frac{6(x + 1)}{x(x - 2)}$; however, the fundamental property does not apply since there are no common factors that allow us to write the answer in lower terms.

## EXAMPLE 8 Subtracting Rational Expressions with Denominators That Are Opposites

Subtract. Write the answer in lowest terms.

$$\frac{3x}{x - 5} - \frac{2x - 25}{5 - x}$$

The denominators are opposites, so either may be used as the common denominator. We choose $x - 5$.

$$\frac{3x}{x - 5} - \frac{2x - 25}{5 - x} = \frac{3x}{x - 5} - \frac{(2x - 25)(-1)}{(5 - x)(-1)} \qquad \text{Fundamental property}$$

$$= \frac{3x}{x - 5} - \frac{-2x + 25}{x - 5} \qquad \text{Multiply.}$$

$$= \frac{3x - (-2x + 25)}{x - 5} \qquad \text{Subtract numerators; use parentheses.}$$

$$= \frac{3x + 2x - 25}{x - 5} \qquad \text{Distributive property; Be careful with signs.}$$

$$= \frac{5x - 25}{x - 5} \qquad \text{Combine like terms.}$$

$$= \frac{5(x - 5)}{x - 5} \qquad \text{Factor.}$$

$$= 5 \qquad \text{Lowest terms}$$

**Now Try Exercise 53.**

## EXAMPLE 9 Subtracting Rational Expressions

Subtract. Write the answer in lowest terms.

$$\frac{6x}{x^2 - 2x + 1} - \frac{1}{x^2 - 1}$$

We begin by factoring the denominators.

$$x^2 - 2x + 1 = (x-1)(x-1) \qquad \text{and} \qquad x^2 - 1 = (x-1)(x+1)$$

From the factored denominators, we identify the LCD, $(x-1)(x-1)(x+1)$. We use the factor $x - 1$ twice, since it appears twice in the first denominator.

$$\frac{6x}{(x-1)(x-1)} - \frac{1}{(x-1)(x+1)}$$

$$= \frac{6x(x+1)}{(x-1)(x-1)(x+1)} - \frac{1(x-1)}{(x-1)(x-1)(x+1)}$$ Fundamental property

$$= \frac{6x(x+1) - 1(x-1)}{(x-1)(x-1)(x+1)}$$ Subtract numerators.

$$= \frac{6x^2 + 6x - x + 1}{(x-1)(x-1)(x+1)}$$ Distributive property

$$= \frac{6x^2 + 5x + 1}{(x-1)(x-1)(x+1)}$$ Combine like terms.

$$= \frac{(2x+1)(3x+1)}{(x-1)^2(x+1)}$$ Factor the numerator.

Verify that the final expression is in lowest terms.

**Now Try Exercise 63.**

## 6.4 EXERCISES

**For Extra Help**

- Student's Solutions Manual
- MyMathLab
- InterAct Math Tutorial Software
- AW Math Tutor Center
- MathXL
- Digital Video Tutor CD 9/Videotape 9

*Match the problem in Column I with the correct sum or difference in Column II.*

| I | II |
|---|---|
| **1.** $\frac{x}{x+6} + \frac{6}{x+6}$ | **A.** $2$ |
| **2.** $\frac{2x}{x-6} - \frac{12}{x-6}$ | **B.** $\frac{x-6}{x+6}$ |
| **3.** $\frac{6}{x-6} - \frac{x}{x-6}$ | **C.** $-1$ |
| **4.** $\frac{6}{x+6} - \frac{x}{x+6}$ | **D.** $\frac{6+x}{6x}$ |
| **5.** $\frac{x}{x+6} - \frac{6}{x+6}$ | **E.** $1$ |
| **6.** $\frac{1}{x} + \frac{1}{6}$ | **F.** $0$ |
| **7.** $\frac{1}{6} - \frac{1}{x}$ | **G.** $\frac{x-6}{6x}$ |
| **8.** $\frac{1}{6x} - \frac{1}{6x}$ | **H.** $\frac{6-x}{x+6}$ |

*Note: When adding and subtracting rational expressions, several different equivalent forms of the answer often exist. If your answer does not look exactly like the one given in the back of the book, check to see whether you have written an equivalent form. (See Example 7 of Section 6.1.)*

*Add or subtract. Write each answer in lowest terms. See Examples 1 and 6.*

**9.** $\frac{4}{m} + \frac{7}{m}$ **10.** $\frac{5}{p} + \frac{11}{p}$ **11.** $\frac{5}{y + 4} - \frac{1}{y + 4}$

**12.** $\frac{4}{y + 3} - \frac{1}{y + 3}$ **13.** $\frac{x}{x + y} + \frac{y}{x + y}$ **14.** $\frac{a}{a + b} + \frac{b}{a + b}$

**15.** $\frac{5m}{m + 1} - \frac{1 + 4m}{m + 1}$ **16.** $\frac{4x}{x + 2} - \frac{2 + 3x}{x + 2}$ **17.** $\frac{a + b}{2} - \frac{a - b}{2}$

**18.** $\frac{x - y}{2} - \frac{x + y}{2}$ **19.** $\frac{x^2}{x + 5} + \frac{5x}{x + 5}$ **20.** $\frac{t^2}{t - 3} + \frac{-3t}{t - 3}$

**21.** $\frac{y^2 - 3y}{y + 3} + \frac{-18}{y + 3}$ **22.** $\frac{r^2 - 8r}{r - 5} + \frac{15}{r - 5}$

**23.** Explain with an example how to add or subtract rational expressions with the same denominator.

**24.** Explain with an example how to add or subtract rational expressions with different denominators.

*Add or subtract. Write each answer in lowest terms. See Examples 2, 3, 4, and 7.*

**25.** $\frac{z}{5} + \frac{1}{3}$ **26.** $\frac{p}{8} + \frac{3}{5}$ **27.** $\frac{5}{7} - \frac{r}{2}$

**28.** $\frac{10}{9} - \frac{z}{3}$ **29.** $-\frac{3}{4} - \frac{1}{2x}$ **30.** $-\frac{5}{8} - \frac{3}{2a}$

**31.** $\frac{6}{5x} + \frac{9}{2x}$ **32.** $\frac{3}{2x} + \frac{4}{7x}$ **33.** $\frac{x + 1}{6} + \frac{3x + 3}{9}$

**34.** $\frac{2x - 6}{4} + \frac{x + 5}{6}$ **35.** $\frac{x + 3}{3x} + \frac{2x + 2}{4x}$ **36.** $\frac{x + 2}{5x} + \frac{6x + 3}{3x}$

**37.** $\frac{7}{3p^2} - \frac{2}{p}$ **38.** $\frac{12}{5m^2} - \frac{5}{m}$ **39.** $\frac{1}{k + 4} - \frac{2}{k}$

**40.** $\frac{3}{m + 1} - \frac{4}{m}$ **41.** $\frac{x}{x - 2} + \frac{-8}{x^2 - 4}$ **42.** $\frac{2x}{x - 1} + \frac{-4}{x^2 - 1}$

**43.** $\frac{4m}{m^2 + 3m + 2} + \frac{2m - 1}{m^2 + 6m + 5}$ **44.** $\frac{a}{a^2 + 3a - 4} + \frac{4a}{a^2 + 7a + 12}$

**45.** $\frac{4y}{y^2 - 1} - \frac{5}{y^2 + 2y + 1}$ **46.** $\frac{2x}{x^2 - 16} - \frac{3}{x^2 + 8x + 16}$

**47.** $\frac{t}{t + 2} + \frac{5 - t}{t} - \frac{4}{t^2 + 2t}$ **48.** $\frac{2p}{p - 3} + \frac{2 + p}{p} - \frac{-6}{p^2 - 3p}$

**49.** What are the two possible LCDs that could be used for the sum $\frac{10}{m - 2} + \frac{5}{2 - m}$?

**50.** If one form of the correct answer to a sum or difference of rational expressions is $\frac{4}{k - 3}$, what would an alternative form of the answer be if the denominator is $3 - k$?

*Add or subtract. Write each answer in lowest terms. See Examples 5 and 8.*

**51.** $\dfrac{4}{x-5}+\dfrac{6}{5-x}$

**52.** $\dfrac{10}{m-2}+\dfrac{5}{2-m}$

**53.** $\dfrac{-1}{1-y}-\dfrac{4y-3}{y-1}$

**54.** $\dfrac{-4}{p-3}-\dfrac{p+1}{3-p}$

**55.** $\dfrac{2}{x-y^2}+\dfrac{7}{y^2-x}$

**56.** $\dfrac{-8}{p-q^2}+\dfrac{3}{q^2-p}$

**57.** $\dfrac{x}{5x-3y}-\dfrac{y}{3y-5x}$

**58.** $\dfrac{t}{8t-9s}-\dfrac{s}{9s-8t}$

**59.** $\dfrac{3}{4p-5}+\dfrac{9}{5-4p}$

**60.** $\dfrac{8}{3-7y}-\dfrac{2}{7y-3}$

*In these subtraction problems, the rational expression that follows the subtraction sign has a numerator with more than one term. Be very careful with signs and find each difference. See Example 9.*

**61.** $\dfrac{2m}{m-n}-\dfrac{5m+n}{2m-2n}$

**62.** $\dfrac{5p}{p-q}-\dfrac{3p+1}{4p-4q}$

**63.** $\dfrac{5}{x^2-9}-\dfrac{x+2}{x^2+4x+3}$

**64.** $\dfrac{1}{a^2-1}-\dfrac{a-1}{a^2+3a-4}$

**65.** $\dfrac{2q+1}{3q^2+10q-8}-\dfrac{3q+5}{2q^2+5q-12}$

**66.** $\dfrac{4y-1}{2y^2+5y-3}-\dfrac{y+3}{6y^2+y-2}$

*Perform each indicated operation. See Examples 1–9.*

**67.** $\dfrac{4}{r^2-r}+\dfrac{6}{r^2+2r}-\dfrac{1}{r^2+r-2}$

**68.** $\dfrac{6}{k^2+3k}-\dfrac{1}{k^2-k}+\dfrac{2}{k^2+2k-3}$

**69.** $\dfrac{x+3y}{x^2+2xy+y^2}+\dfrac{x-y}{x^2+4xy+3y^2}$

**70.** $\dfrac{m}{m^2-1}+\dfrac{m-1}{m^2+2m+1}$

**71.** $\dfrac{r+y}{18r^2+12ry-3ry-2y^2}+\dfrac{3r-y}{36r^2-y^2}$

**72.** $\dfrac{2x-z}{2x^2-4xz+5xz-10z^2}-\dfrac{x+z}{x^2-4z^2}$

**73.** Refer to the rectangle in the figure.

**(a)** Find an expression that represents its perimeter. Give the simplified form.

**(b)** Find an expression that represents its area. Give the simplified form.

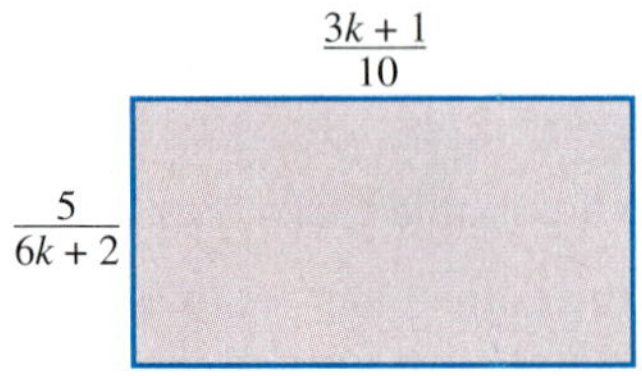


**74.** Refer to the triangle in the figure. Find an expression that represents its perimeter.

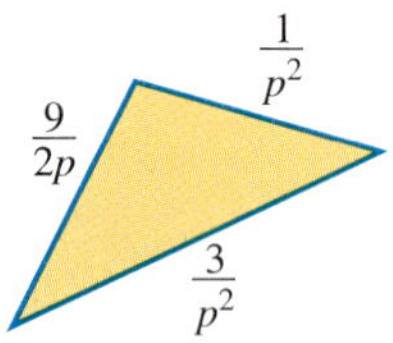

*A* concours d'elegance *is a competition in which a maximum of* 100 *points is awarded to a car based on its general attractiveness. The rational expression*

$$\frac{1010}{49(101 - x)} - \frac{10}{49}$$

*approximates the cost, in thousands of dollars, of restoring a car so that it will win x points.*

*Use this information to work Exercises 75 and 76.*

**75.** Simplify the expression by performing the indicated subtraction.

**76.** Use the simplified expression from Exercise 75 to determine how much it would cost to win 95 points.

# 6.5 Complex Fractions

**OBJECTIVES**

1. Simplify a complex fraction by writing it as a division problem (Method 1).
2. Simplify a complex fraction by multiplying by the least common denominator (Method 2).

The quotient of two mixed numbers in arithmetic, such as $2\frac{1}{2} \div 3\frac{1}{4}$, can be written as a fraction:

$$2\frac{1}{2} \div 3\frac{1}{4} = \frac{2\frac{1}{2}}{3\frac{1}{4}} = \frac{2 + \frac{1}{2}}{3 + \frac{1}{4}}.$$

The last expression is the quotient of expressions that involve fractions. In algebra, some rational expressions also have fractions in the numerator, or denominator, or both.

**Complex Fraction**

A rational expression with one or more fractions in the numerator, denominator, or both, is called a **complex fraction.**

Examples of complex fractions include

$$\frac{2 + \frac{1}{2}}{3 + \frac{1}{4}}, \quad \frac{\frac{3x^2 - 5x}{6x^2}}{2x - \frac{1}{x}}, \quad \text{and} \quad \frac{3 + x}{5 - \frac{2}{x}}.$$

The parts of a complex fraction are named as follows.

$$\frac{\frac{2}{p} - \frac{1}{q}}{\frac{3}{p} + \frac{5}{q}}$$

← Numerator of complex fraction
← Main fraction bar
← Denominator of complex fraction

**OBJECTIVE 1 Simplify a complex fraction by writing it as a division problem (Method 1).**
Since the main fraction bar represents division in a complex fraction, one method of simplifying a complex fraction involves division.

**Method 1**

To simplify a complex fraction:

*Step 1* Write both the numerator and denominator as single fractions.

*Step 2* Change the complex fraction to a division problem.

*Step 3* Perform the indicated division.

Once again, in this section the first example shows complex fractions from both arithmetic and algebra.

## EXAMPLE 1 Simplifying Complex Fractions (Method 1)

Simplify each complex fraction.

**(a)** $\dfrac{\frac{2}{3}+\frac{5}{9}}{\frac{1}{4}+\frac{1}{12}}$ **(b)** $\dfrac{6+\frac{3}{x}}{\frac{x}{4}+\frac{1}{8}}$

*Step 1* First, write each numerator as a single fraction.

$$\frac{2}{3}+\frac{5}{9}=\frac{2(3)}{3(3)}+\frac{5}{9}=\frac{6}{9}+\frac{5}{9}=\frac{11}{9}$$

$$6+\frac{3}{x}=\frac{6}{1}+\frac{3}{x}=\frac{6x}{x}+\frac{3}{x}=\frac{6x+3}{x}$$

Do the same thing with each denominator.

$$\frac{1}{4}+\frac{1}{12}=\frac{1(3)}{4(3)}+\frac{1}{12}=\frac{3}{12}+\frac{1}{12}=\frac{4}{12}$$

$$\frac{x}{4}+\frac{1}{8}=\frac{x(2)}{4(2)}+\frac{1}{8}=\frac{2x}{8}+\frac{1}{8}=\frac{2x+1}{8}$$

*Step 2* The original complex fraction can now be written as follows.

$$\frac{\frac{11}{9}}{\frac{4}{12}}=\frac{11}{9}\div\frac{4}{12}$$

$$\frac{\frac{6x+3}{x}}{\frac{2x+1}{8}}=\frac{6x+3}{x}\div\frac{2x+1}{8}$$

*Step 3* Now use the rule for division and the fundamental property.

Multiply by the reciprocal.

$$\frac{11}{9}\div\frac{4}{12}=\frac{11}{9}\cdot\frac{12}{4}$$

Multiply by the reciprocal.

$$\frac{6x+3}{x}\div\frac{2x+1}{8}=\frac{6x+3}{x}\cdot\frac{8}{2x+1}$$

$$= \frac{11 \cdot 3 \cdot 4}{3 \cdot 3 \cdot 4}$$

$$= \frac{11}{3}$$

$$= \frac{3(2x + 1)}{x} \cdot \frac{8}{2x + 1}$$

$$= \frac{24}{x}$$

**Now Try Exercises 1 and 15.**

**EXAMPLE 2 Simplifying a Complex Fraction (Method 1)**

Simplify the complex fraction.

$$\frac{\frac{xp}{q^3}}{\frac{p^2}{qx^2}} = \frac{xp}{q^3} \div \frac{p^2}{qx^2} = \frac{xp}{q^3} \cdot \frac{qx^2}{p^2} = \frac{x^3}{q^2 p}$$

**Now Try Exercise 9.**

**EXAMPLE 3 Simplifying a Complex Fraction (Method 1)**

Simplify the complex fraction.

$$\frac{\frac{3}{x+2} - 4}{\frac{2}{x+2} + 1} = \frac{\frac{3}{x+2} - \frac{4(x+2)}{x+2}}{\frac{2}{x+2} + \frac{1(x+2)}{x+2}}$$ Write both second terms with a denominator of $x + 2$.

$$= \frac{\frac{3 - 4(x+2)}{x+2}}{\frac{2 + 1(x+2)}{x+2}}$$ Subtract in the numerator. Add in the denominator.

$$= \frac{\frac{3 - 4x - 8}{x+2}}{\frac{2 + x + 2}{x+2}}$$ Distributive property

$$= \frac{\frac{-5 - 4x}{x+2}}{\frac{4 + x}{x+2}}$$ Combine like terms.

$$= \frac{-5 - 4x}{x+2} \cdot \frac{x+2}{4+x}$$ Multiply by the reciprocal.

$$= \frac{-5 - 4x}{4 + x}$$ Lowest terms

**Now Try Exercise 33.**

**OBJECTIVE 2 Simplify a complex fraction by multiplying by the least common denominator (Method 2).** Since any expression can be multiplied by a form of 1 to get an equivalent expression, we can multiply both the numerator and the denominator of a complex fraction by the same nonzero expression to get an equivalent complex fraction. If we choose the expression to be the LCD of all the fractions within the complex fraction, the complex fraction will be simplified. This is Method 2.

**Method 2**

To simplify a complex fraction:

*Step 1* Find the LCD of all fractions within the complex fraction.

*Step 2* Multiply both the numerator and the denominator of the complex fraction by this LCD using the distributive property as necessary. Write in lowest terms.

In the next example, Method 2 is used to simplify the complex fractions from Example 1.

**EXAMPLE 4 Simplifying Complex Fractions (Method 2)**

Simplify each complex fraction.

**(a)** $\dfrac{\frac{2}{3}+\frac{5}{9}}{\frac{1}{4}+\frac{1}{12}}$ **(b)** $\dfrac{6+\frac{3}{x}}{\frac{x}{4}+\frac{1}{8}}$

*Step 1* Find the LCD for all denominators in the complex fraction.

The LCD for 3, 9, 4, and 12 is 36. | The LCD for $x$, 4, and 8 is $8x$.

*Step 2* Multiply numerator and denominator of the complex fraction by the LCD.

(a)

$$\frac{\frac{2}{3}+\frac{5}{9}}{\frac{1}{4}+\frac{1}{12}} = \frac{36\left(\frac{2}{3}+\frac{5}{9}\right)}{36\left(\frac{1}{4}+\frac{1}{12}\right)}$$

$$= \frac{36\left(\frac{2}{3}\right)+36\left(\frac{5}{9}\right)}{36\left(\frac{1}{4}\right)+36\left(\frac{1}{12}\right)}$$

$$= \frac{24+20}{9+3}$$

$$= \frac{44}{12} = \frac{4\cdot 11}{4\cdot 3}$$

$$= \frac{11}{3}$$

(b)

$$\frac{6+\frac{3}{x}}{\frac{x}{4}+\frac{1}{8}} = \frac{8x\left(6+\frac{3}{x}\right)}{8x\left(\frac{x}{4}+\frac{1}{8}\right)}$$

$$= \frac{8x(6)+8x\left(\frac{3}{x}\right)}{8x\left(\frac{x}{4}\right)+8x\left(\frac{1}{8}\right)} \quad \text{Distributive property}$$

$$= \frac{48x+24}{2x^2+x}$$

$$= \frac{24(2x+1)}{x(2x+1)} \quad \text{Factor.}$$

$$= \frac{24}{x} \quad \text{Lowest terms}$$

**Now Try Exercises 2 and 19.**

**EXAMPLE 5 Simplifying a Complex Fraction (Method 2)**

Simplify the complex fraction.

$$\frac{\frac{3}{5m}-\frac{2}{m^2}}{\frac{9}{2m}+\frac{3}{4m^2}}=\frac{20m^2\left(\frac{3}{5m}-\frac{2}{m^2}\right)}{20m^2\left(\frac{9}{2m}+\frac{3}{4m^2}\right)}$$

The LCD for $5m$, $m^2$, $2m$, and $4m^2$ is $20m^2$.

$$=\frac{20m^2\left(\frac{3}{5m}\right)-20m^2\left(\frac{2}{m^2}\right)}{20m^2\left(\frac{9}{2m}\right)+20m^2\left(\frac{3}{4m^2}\right)}$$

Distributive property

$$=\frac{12m-40}{90m+15}$$

**Now Try Exercise 25.**

Either of the two methods can be used to simplify a complex fraction. You may want to choose one method and stick with it to eliminate confusion. However, some students prefer to use Method 1 for problems like Example 2, which is the quotient of two fractions. They prefer Method 2 for problems like Examples 1, 3, 4, and 5, which have sums or differences in the numerators or denominators or both.

**EXAMPLE 6 Deciding on a Method and Simplifying Complex Fractions**

Simplify each complex fraction.

**(a)** $$\frac{\frac{1}{y}+\frac{2}{y+2}}{\frac{4}{y}-\frac{3}{y+2}}$$

Although either method will work, we use Method 2 here since there are sums and differences in the numerator and denominator. The LCD is $y(y+2)$. Multiply the numerator and denominator by the LCD.

$$\frac{\frac{1}{y}+\frac{2}{y+2}}{\frac{4}{y}-\frac{3}{y+2}}\cdot\frac{y(y+2)}{y(y+2)}=\frac{1(y+2)+2y}{4(y+2)-3y}$$

Fundamental property

$$=\frac{y+2+2y}{4y+8-3y}$$

Distributive property

$$=\frac{3y+2}{y+8}$$

Combine like terms.

**(b)** $\dfrac{\dfrac{x+2}{x-3}}{\dfrac{x^2-4}{x^2-9}}$

Since this is simply a quotient of two rational expressions, we use Method 1.

$$\frac{\dfrac{x+2}{x-3}}{\dfrac{x^2-4}{x^2-9}} = \frac{x+2}{x-3} \div \frac{x^2-4}{x^2-9}$$

$$= \frac{x+2}{x-3} \cdot \frac{x^2-9}{x^2-4} \qquad \text{Definition of division}$$

$$= \frac{x+2}{x-3} \cdot \frac{(x+3)(x-3)}{(x+2)(x-2)} \qquad \text{Factor.}$$

$$= \frac{x+3}{x-2} \qquad \text{Lowest terms}$$

**Now Try Exercises 29 and 31.**

## 6.5 EXERCISES

**For Extra Help**

- Student's Solutions Manual
- MyMathLab
- InterAct Math Tutorial Software
- AW Math Tutor Center
- MathXL
- Digital Video Tutor CD 10/Videotape 10

*Note: In many problems involving complex fractions, several different equivalent forms of the answer exist. If your answer does not look exactly like the one given in the back of the book, check to see whether you have written an equivalent form.*

**1.** Consider the complex fraction $\dfrac{\frac{1}{2}-\frac{1}{3}}{\frac{5}{6}-\frac{1}{12}}$. Answer each part, outlining Method 1 for simplifying this complex fraction.

**(a)** To combine the terms in the numerator, we must find the LCD of $\frac{1}{2}$ and $\frac{1}{3}$. What is this LCD? Determine the simplified form of the numerator of the complex fraction.

**(b)** To combine the terms in the denominator, we must find the LCD of $\frac{5}{6}$ and $\frac{1}{12}$. What is this LCD? Determine the simplified form of the denominator of the complex fraction.

**(c)** Now use the results from parts (a) and (b) to write the complex fraction as a division problem using the symbol $\div$.

**(d)** Perform the operation from part (c) to obtain the final simplification.

**2.** Consider the same complex fraction given in Exercise 1, $\dfrac{\frac{1}{2}-\frac{1}{3}}{\frac{5}{6}-\frac{1}{12}}$. Answer each part, outlining Method 2 for simplifying this complex fraction.

**(a)** We must determine the LCD of all the fractions within the complex fraction. What is this LCD?

**(b)** Multiply every term in the complex fraction by the LCD found in part (a), but do not combine the terms in the numerator and the denominator yet.

**(c)** Combine the terms from part (b) to obtain the simplified form of the complex fraction.

3. Which complex fraction is equivalent to $\dfrac{3-\frac{1}{2}}{2-\frac{1}{4}}$? Answer this question without showing any work, and explain your reasoning.

A. $\dfrac{3+\frac{1}{2}}{2+\frac{1}{4}}$ B. $\dfrac{-3+\frac{1}{2}}{2-\frac{1}{4}}$ C. $\dfrac{-3-\frac{1}{2}}{-2-\frac{1}{4}}$ D. $\dfrac{-3+\frac{1}{2}}{-2+\frac{1}{4}}$

4. Only one of these choices is equal to $\dfrac{\frac{1}{2}+\frac{1}{4}}{\frac{1}{3}+\frac{1}{12}}$. Which one is it? Answer this question without showing any work, and explain your reasoning.

A. $\dfrac{9}{5}$ B. $-\dfrac{9}{5}$ C. $-\dfrac{5}{9}$ D. $-12$

5. Describe Method 1 for simplifying complex fractions. Illustrate with the example $\dfrac{\frac{1}{2}}{\frac{2}{3}}$.

6. Describe Method 2 for simplifying complex fractions. Illustrate with the example $\dfrac{\frac{1}{2}}{\frac{2}{3}}$.

*Simplify each complex fraction. Use either method. See Examples 1–6.*

7. $\dfrac{-\frac{4}{3}}{\frac{2}{9}}$ 8. $\dfrac{-\frac{5}{6}}{\frac{5}{4}}$ 9. $\dfrac{\frac{x}{y^2}}{\frac{x^2}{y}}$ 10. $\dfrac{\frac{p^4}{r}}{\frac{p^2}{r^2}}$

11. $\dfrac{\frac{4a^4b^3}{3a}}{\frac{2ab^4}{b^2}}$ 12. $\dfrac{\frac{2r^4t^2}{3t}}{\frac{5r^2t^5}{3r}}$ 13. $\dfrac{\frac{m+2}{3}}{\frac{m-4}{m}}$ 14. $\dfrac{\frac{q-5}{q}}{\frac{q+5}{3}}$

15. $\dfrac{\frac{2}{x}-3}{\frac{2-3x}{2}}$ 16. $\dfrac{6+\frac{2}{r}}{\frac{3r+1}{4}}$ 17. $\dfrac{\frac{1}{x}+x}{\frac{x^2+1}{8}}$ 18. $\dfrac{\frac{3}{m}-m}{\frac{3-m^2}{4}}$

19. $\dfrac{a-\frac{5}{a}}{a+\frac{1}{a}}$ 20. $\dfrac{q+\frac{1}{q}}{q+\frac{4}{q}}$ 21. $\dfrac{\frac{5}{8}+\frac{2}{3}}{\frac{7}{3}-\frac{1}{4}}$ 22. $\dfrac{\frac{6}{5}-\frac{1}{9}}{\frac{2}{5}+\frac{5}{3}}$

23. $\dfrac{\frac{1}{x^2}+\frac{1}{y^2}}{\frac{1}{x}-\frac{1}{y}}$ 24. $\dfrac{\frac{1}{a^2}-\frac{1}{b^2}}{\frac{1}{a}-\frac{1}{b}}$ 25. $\dfrac{\frac{2}{p^2}-\frac{3}{5p}}{\frac{4}{p}+\frac{1}{4p}}$ 26. $\dfrac{\frac{2}{m^2}-\frac{3}{m}}{\frac{2}{5m^2}+\frac{1}{3m}}$

27. $\dfrac{\frac{5}{x^2y}-\frac{2}{xy^2}}{\frac{3}{x^2y^2}+\frac{4}{xy}}$ 28. $\dfrac{\frac{1}{m^3p}+\frac{2}{mp^2}}{\frac{4}{mp}+\frac{1}{m^2p}}$ 29. $\dfrac{\frac{1}{4}-\frac{1}{a^2}}{\frac{1}{2}+\frac{1}{a}}$ 30. $\dfrac{\frac{1}{9}-\frac{1}{m^2}}{\frac{1}{3}+\frac{1}{m}}$

31. $\dfrac{\dfrac{1}{z+5}}{\dfrac{4}{z^2-25}}$

32. $\dfrac{\dfrac{1}{a+1}}{\dfrac{2}{a^2-1}}$

33. $\dfrac{\dfrac{1}{m+1}-1}{\dfrac{1}{m+1}+1}$

34. $\dfrac{\dfrac{2}{x-1}+2}{\dfrac{2}{x-1}-2}$

35. $\dfrac{\dfrac{1}{m-1}+\dfrac{2}{m+2}}{\dfrac{2}{m+2}-\dfrac{1}{m-3}}$

36. $\dfrac{\dfrac{5}{r+3}-\dfrac{1}{r-1}}{\dfrac{2}{r+2}+\dfrac{3}{r+3}}$

37. In a fraction, what operation does the fraction bar represent?

38. What property of real numbers justifies Method 2 of simplifying complex fractions?

**RELATING CONCEPTS** (EXERCISES 39–42)

**For Individual or Group Work**

*To find the average of two numbers, we add them and divide by 2. Suppose that we wish to find the average of $\frac{3}{8}$ and $\frac{5}{6}$.* ***Work Exercises 39–42 in order,*** *to see how a complex fraction occurs in a problem like this.*

39. Write in symbols: the sum of $\frac{3}{8}$ and $\frac{5}{6}$, divided by 2. Your result should be a complex fraction.

40. Simplify the complex fraction from Exercise 39 using Method 1.

41. Simplify the complex fraction from Exercise 39 using Method 2.

42. Your answers in Exercises 40 and 41 should be the same. Which method did you prefer? Why?

## 6.6 Solving Equations with Rational Expressions

**OBJECTIVES**

1 Distinguish between operations with rational expressions and equations with terms that are rational expressions.

2 Solve equations with rational expressions.

3 Solve a formula for a specified variable.

In Section 2.3 we solved equations with fractions as coefficients. By using the multiplication property of equality, we cleared the fractions by multiplying by the LCD. We continue this work here.

**OBJECTIVE 1 Distinguish between operations with rational expressions and equations with terms that are rational expressions.** Before solving equations with rational expressions, you must understand the difference between *sums* and *differences* of terms with rational coefficients, and *equations* with terms that are rational expressions. ***Sums and differences are operations to perform, while equations are solved.***

**EXAMPLE 1 Distinguishing between Operations and Equations**

Identify each of the following as an operation or an equation. Then perform the operation or solve the equation.

**(a)** $\dfrac{3}{4}x - \dfrac{2}{3}x$

This is a difference of two terms, so it is an operation. (There is no equals sign.) Find the LCD, write each coefficient with this LCD, and combine like terms.

$$\frac{3}{4}x - \frac{2}{3}x = \frac{9}{12}x - \frac{8}{12}x \qquad \text{Get a common denominator.}$$

$$= \frac{1}{12}x \qquad \text{Combine like terms.}$$

**(b)** $\frac{3}{4}x - \frac{2}{3}x = \frac{1}{2}$

Because of the equals sign, this is an equation to be solved. Proceed as in Section 2.3, using the multiplication property of equality to clear fractions. The LCD is 12.

$$\frac{3}{4}x - \frac{2}{3}x = \frac{1}{2}$$

$$12\left(\frac{3}{4}x - \frac{2}{3}x\right) = 12\left(\frac{1}{2}\right) \qquad \text{Multiply by 12.}$$

$$12\left(\frac{3}{4}x\right) - 12\left(\frac{2}{3}x\right) = 12\left(\frac{1}{2}\right) \qquad \text{Distributive property; Multiply } \textit{each} \text{ term by 12.}$$

$$9x - 8x = 6 \qquad \text{Multiply.}$$

$$x = 6 \qquad \text{Combine like terms.}$$

*Check:*

$$\frac{3}{4}x - \frac{2}{3}x = \frac{1}{2} \qquad \text{Original equation}$$

$$\frac{3}{4}(6) - \frac{2}{3}(6) = \frac{1}{2} \quad ? \qquad \text{Let } x = 6.$$

$$\frac{9}{2} - 4 = \frac{1}{2} \quad ? \qquad \text{Multiply.}$$

$$\frac{1}{2} = \frac{1}{2} \qquad \text{True}$$

Since a true statement results, $\{6\}$ is the solution set of the equation.

**Now Try Exercises 1 and 3.**

The ideas of Example 1 can be summarized as follows.

> When adding or subtracting, the LCD must be kept throughout the simplification.
>
> When solving an equation, the LCD is used to multiply each side so that denominators are eliminated.

**OBJECTIVE 2 Solve equations with rational expressions.** When an equation involves fractions as in Example 1(b), we use the multiplication property of equality to clear the fractions. Choose as multiplier the LCD of all denominators in the fractions of the equation.

**EXAMPLE 2 Solving an Equation with Rational Expressions**

Solve $\frac{p}{2} - \frac{p-1}{3} = 1$.

$$6\left(\frac{p}{2} - \frac{p-1}{3}\right) = 6(1) \quad \text{Multiply by the LCD, 6.}$$

$$6\left(\frac{p}{2}\right) - 6\left(\frac{p-1}{3}\right) = 6 \quad \text{Distributive property}$$

$$3p - 2(p - 1) = 6 \quad \text{Multiply; use parentheses around } p - 1.$$

$$3p - 2(p) - 2(-1) = 6 \quad \text{Distributive property}$$

$$3p - 2p + 2 = 6 \quad \text{Be careful with signs.}$$

$$p + 2 = 6 \quad \text{Combine like terms.}$$

$$p = 4 \quad \text{Subtract 2.}$$

Check to see that {4} is the solution set by replacing $p$ with 4 in the original equation.

**Now Try Exercise 37.**

**CAUTION** Note that the use of the LCD here is different from its use in the previous section. Here, we use the multiplication property of equality to multiply each side of an *equation* by the LCD. Earlier, we used the fundamental property to multiply a *fraction* by another fraction that had the LCD as both its numerator and denominator. Be careful not to confuse these two methods.

Recall from Section 6.1 that the denominator of a rational expression cannot equal 0 since division by 0 is undefined. Therefore, when solving an equation with rational expressions that have variables in the denominator, *the solution cannot be a number that makes the denominator equal 0.* If it does, the proposed solution must be rejected.

**EXAMPLE 3 Solving an Equation with Rational Expressions**

Solve $\frac{x}{x-2} = \frac{2}{x-2} + 2$. Check the proposed solution.

Note that $x \neq 2$ here since both denominators in the given expression equal 0 if $x$ is 2. The common denominator is $x - 2$. Solve the equation by multiplying each side by $x - 2$.

$$(x-2)\left(\frac{x}{x-2}\right) = (x-2)\left(\frac{2}{x-2} + 2\right)$$

$$(x-2)\left(\frac{x}{x-2}\right) = (x-2)\left(\frac{2}{x-2}\right) + (x-2)(2) \quad \text{Distributive property}$$

$$x = 2 + 2x - 4$$

$$x = -2 + 2x \quad \text{Combine like terms.}$$

$$-x = -2 \quad \text{Subtract } 2x.$$

$$x = 2 \quad \text{Multiply by } -1.$$

As noted, $x$ cannot equal 2 since replacing $x$ with 2 in the original equation causes the denominators to equal 0.

*Check:*
$$\frac{2}{2-2} \stackrel{?}{=} \frac{2}{2-2} + 2 \qquad \text{Let } x = 2 \text{ in the original equation.}$$
$$\frac{2}{0} \stackrel{?}{=} \frac{2}{0} + 2$$

Thus, 2 must be rejected as a solution, and the solution set is $\emptyset$.

**Now Try Exercise 31.**

While it is always a good idea to check solutions to guard against arithmetic and algebraic errors, it is *essential* to check proposed solutions when variables appear in denominators in the original equation. Some students like to determine which numbers cannot be solutions *before* solving the equation, as we did in Example 3.

The steps used to solve an equation with rational expressions follow.

**Solving an Equation with Rational Expressions**

*Step 1* **Multiply each side of the equation by the LCD** to clear the equation of fractions.

*Step 2* **Solve** the resulting equation.

*Step 3* **Check** each proposed solution by substituting it in the original equation. Reject any that cause a denominator to equal 0.

## EXAMPLE 4 Solving an Equation with Rational Expressions

Solve $\frac{2}{x^2 - x} = \frac{1}{x^2 - 1}$. Check the proposed solution.

*Step 1* Factor the denominators to find the LCD.

$$\frac{2}{x(x-1)} = \frac{1}{(x+1)(x-1)}$$

The LCD is $x(x + 1)(x - 1)$. Notice that $x \neq 0, -1$, or 1; otherwise a denominator will equal 0. Thus, 0, $-1$, and 1 cannot be solutions of this equation. Multiply each side of the equation by $x(x + 1)(x - 1)$.

$$x(x+1)(x-1)\frac{2}{x(x-1)} = x(x+1)(x-1)\frac{1}{(x+1)(x-1)}$$

*Step 2*
$$2(x+1) = x$$
$$2x + 2 = x \qquad \text{Distributive property}$$
$$2 = -x \qquad \text{Subtract } 2x.$$
$$x = -2 \qquad \text{Multiply by } -1.$$

*Step 3* The proposed solution is $-2$, which does not make any denominator equal 0.

*Check:*

$$\frac{2}{x^2 - x} = \frac{1}{x^2 - 1} \qquad \text{Original equation}$$

$$\frac{2}{(-2)^2 - (-2)} \stackrel{?}{=} \frac{1}{(-2)^2 - 1} \qquad \text{Let } x = -2.$$

$$\frac{2}{4 + 2} \stackrel{?}{=} \frac{1}{4 - 1}$$

$$\frac{1}{3} = \frac{1}{3} \qquad \text{True}$$

Thus, the solution set is $\{-2\}$.

**Now Try Exercise 43.**

### EXAMPLE 5 Solving an Equation with Rational Expressions

Solve $\dfrac{2m}{m^2 - 4} + \dfrac{1}{m - 2} = \dfrac{2}{m + 2}$.

Factor the first denominator on the left to find the LCD.

$$\frac{2m}{(m + 2)(m - 2)} + \frac{1}{m - 2} = \frac{2}{m + 2}$$

The LCD is $(m + 2)(m - 2)$. Here $m \neq -2$ or 2, so these numbers cannot be solutions of this equation. Multiply each side by the LCD.

$$(m + 2)(m - 2)\left(\frac{2m}{(m + 2)(m - 2)} + \frac{1}{m - 2}\right) = (m + 2)(m - 2)\frac{2}{m + 2}$$

$$(m + 2)(m - 2)\frac{2m}{(m + 2)(m - 2)} + (m + 2)(m - 2)\frac{1}{m - 2} = (m + 2)(m - 2)\frac{2}{m + 2}$$

$$2m + m + 2 = 2(m - 2)$$

$$3m + 2 = 2m - 4 \qquad \text{Combine like terms; distributive property}$$

$$m + 2 = -4 \qquad \text{Subtract } 2m.$$

$$m = -6 \qquad \text{Subtract 2.}$$

A check verifies that $\{-6\}$ is the solution set.

**Now Try Exercise 53.**

### EXAMPLE 6 Solving an Equation with Rational Expressions

Solve $\dfrac{1}{x - 1} + \dfrac{1}{2} = \dfrac{2}{x^2 - 1}$.

Factor the denominator on the right.

$$\frac{1}{x - 1} + \frac{1}{2} = \frac{2}{(x + 1)(x - 1)}$$

Notice that $x \neq 1$ or $-1$. Multiply each side of the equation by the LCD, $2(x+1)(x-1)$.

$$2(x+1)(x-1)\left(\frac{1}{x-1}+\frac{1}{2}\right)=2(x+1)(x-1)\frac{2}{(x+1)(x-1)}$$

$$2(x+1)(x-1)\frac{1}{x-1}+2(x+1)(x-1)\frac{1}{2}=2(x+1)(x-1)\frac{2}{(x+1)(x-1)}$$

$$2(x+1)+(x+1)(x-1)=4$$

$$2x+2+x^2-1=4 \quad \text{Distributive property}$$

$$x^2+2x+1=4 \quad \text{Combine like terms.}$$

$$x^2+2x-3=0 \quad \text{Subtract 4.}$$

$$(x+3)(x-1)=0 \quad \text{Factor.}$$

$$x+3=0 \quad \text{or} \quad x-1=0 \quad \text{Zero-factor property}$$

$$x=-3 \quad \text{or} \quad x=1$$

$-3$ and 1 are proposed solutions. However, as noted, 1 makes an original denominator equal 0, so 1 is not a solution. Check that $-3$ is a solution.

*Check:*

$$\frac{1}{x-1}+\frac{1}{2}=\frac{2}{x^2-1} \quad \text{Original equation}$$

$$\frac{1}{-3-1}+\frac{1}{2}\stackrel{?}{=}\frac{2}{(-3)^2-1} \quad \text{Let } x=-3.$$

$$\frac{1}{-4}+\frac{1}{2}\stackrel{?}{=}\frac{2}{9-1} \quad \text{Simplify.}$$

$$\frac{1}{4}=\frac{1}{4} \quad \text{True}$$

The solution set is $\{-3\}$.

**Now Try Exercise 63.**

### EXAMPLE 7 Solving an Equation with Rational Expressions

Solve $\dfrac{1}{k^2+4k+3}+\dfrac{1}{2k+2}=\dfrac{3}{4k+12}$.

Factoring each denominator gives the equation

$$\frac{1}{(k+1)(k+3)}+\frac{1}{2(k+1)}=\frac{3}{4(k+3)}.$$

The LCD is $4(k+1)(k+3)$, indicating that $k \neq -1$ or $-3$. Multiply each side by this LCD.

$$4(k+1)(k+3)\left(\frac{1}{(k+1)(k+3)}+\frac{1}{2(k+1)}\right)$$
$$=4(k+1)(k+3)\frac{3}{4(k+3)}$$

$$4(k+1)(k+3)\frac{1}{(k+1)(k+3)}+2\cdot 2(k+1)(k+3)\frac{1}{2(k+1)}$$
$$=4(k+1)(k+3)\frac{3}{4(k+3)}$$

$$4 + 2(k + 3) = 3(k + 1)$$
$$4 + 2k + 6 = 3k + 3 \quad \text{Distributive property}$$
$$2k + 10 = 3k + 3 \quad \text{Combine like terms.}$$
$$7 = k \quad \text{Subtract } 2k \text{ and } 3.$$

The proposed solution, 7, does not make an original denominator equal 0. A check shows that the algebra is correct, so {7} is the solution set.

**Now Try Exercise 65.**

**OBJECTIVE 3 Solve a formula for a specified variable.** Solving a formula for a specified variable was first discussed in Chapter 2. Remember to treat the variable for which you are solving as if it were the only variable, and all others as if they were constants.

### EXAMPLE 8 Solving for a Specified Variable

Solve each formula for the specified variable.

**(a)** $F = \frac{k}{d - D}$ for $d$

We want to get $d$ alone on one side of the equation. We begin by multiplying each side by $d - D$ to clear the fraction.

$$F = \frac{k}{d - D} \quad \text{Given equation}$$
$$F(d - D) = \frac{k}{d - D}(d - D) \quad \text{Clear the fraction.}$$
$$F(d - D) = k \quad \text{Simplify.}$$
$$Fd - FD = k \quad \text{Distributive property}$$
$$Fd = k + FD \quad \text{Add } FD.$$
$$d = \frac{k + FD}{F} \quad \text{or} \quad d = \frac{k}{F} + D \quad \text{Divide by } F.$$

**(b)** $\frac{1}{a} = \frac{1}{b} + \frac{1}{c}$ for $c$

The LCD of all the fractions in the equation is $abc$, so we multiply each side by $abc$.

$$abc\left(\frac{1}{a}\right) = abc\left(\frac{1}{b} + \frac{1}{c}\right)$$
$$abc\left(\frac{1}{a}\right) = abc\left(\frac{1}{b}\right) + abc\left(\frac{1}{c}\right) \quad \text{Distributive property}$$
$$bc = ac + ab$$

Since we are solving for $c$, we need to get all terms with $c$ on one side of the equation. We do this by subtracting $ac$ from each side.

$$bc - ac = ab \quad \text{Subtract } ac.$$
$$c(b - a) = ab \quad \text{Factor out } c.$$

Finally, we divide each side by the coefficient of $c$, which is $b - a$.

$$c = \frac{ab}{b - a}$$

**Now Try Exercises 77 and 83.**

**CAUTION** Students often have trouble in the step that involves factoring out the variable for which they are solving. In Example 8(b), we had to factor out $c$ on the left side so that we could divide both sides by $b - a$.

When solving an equation for a specified variable, *be sure that the specified variable appears alone on only one side of the equals sign in the final equation.*

## 6.6 EXERCISES

**For Extra Help**

- Student's Solutions Manual
- MyMathLab
- InterAct Math Tutorial Software
- AW Math Tutor Center
- MathXL
- Digital Video Tutor CD 10/Videotape 10

*Identify as an* expression *or an* equation. *Then perform the operation or solve the equation. See Example 1.*

**1.** $\frac{7}{8}x + \frac{1}{5}x$

**2.** $\frac{4}{7}x + \frac{3}{5}x$

**3.** $\frac{7}{8}x + \frac{1}{5}x = 1$

**4.** $\frac{4}{7}x + \frac{3}{5}x = 1$

**5.** $\frac{3}{5}x - \frac{7}{10}x$

**6.** $\frac{2}{3}x - \frac{7}{4}x$

**7.** $\frac{3}{5}x - \frac{7}{10}x = 1$

**8.** $\frac{2}{3}x - \frac{7}{4}x = -13$

*When solving an equation with variables in denominators, we must determine the values that cause these denominators to equal 0, so we can reject these values if they appear as possible solutions. Find all values for which at least one denominator is equal to 0. Do not solve. See Examples 3–7.*

**9.** $\frac{3}{x + 2} - \frac{5}{x} = 1$

**10.** $\frac{7}{x} + \frac{9}{x - 3} = 5$

**11.** $\frac{-1}{(x + 3)(x - 4)} = \frac{1}{2x + 1}$

**12.** $\frac{8}{(x - 8)(x + 2)} = \frac{7}{3x - 10}$

**13.** $\frac{4}{x^2 + 8x - 9} + \frac{1}{x^2 - 4} = 0$

**14.** $\frac{-3}{x^2 + 9x - 10} - \frac{12}{x^2 - 16} = 0$

**15.** Explain how the LCD is used in a different way when adding and subtracting rational expressions compared to solving equations with rational expressions.

**16.** If we multiply each side of the equation $\frac{6}{x + 5} = \frac{6}{x + 5}$ by $x + 5$, we get $6 = 6$. Are all real numbers solutions of this equation? Explain.

*Solve each equation, and check your answers. See Examples 1(b), 2, and 3.*

**17.** $\frac{5}{m} - \frac{3}{m} = 8$

**18.** $\frac{4}{y} + \frac{1}{y} = 2$

**19.** $\frac{5}{y} + 4 = \frac{2}{y}$

**20.** $\frac{11}{q} = 3 - \frac{1}{q}$

**21.** $\frac{3x}{5} - 6 = x$

**22.** $\frac{5t}{4} + t = 9$

23. $\frac{4m}{7} + m = 11$

24. $a - \frac{3a}{2} = 1$

25. $\frac{z-1}{4} = \frac{z+3}{3}$

26. $\frac{r-5}{2} = \frac{r+2}{3}$

27. $\frac{3p+6}{8} = \frac{3p-3}{16}$

28. $\frac{2z+1}{5} = \frac{7z+5}{15}$

29. $\frac{2x+3}{x} = \frac{3}{2}$

30. $\frac{5-2x}{x} = \frac{1}{4}$

31. $\frac{k}{k-4} - 5 = \frac{4}{k-4}$

32. $\frac{-5}{a+5} = \frac{a}{a+5} + 2$

33. $\frac{q+2}{3} + \frac{q-5}{5} = \frac{7}{3}$

34. $\frac{t}{6} + \frac{4}{3} = \frac{t-2}{3}$

35. $\frac{x}{2} = \frac{5}{4} + \frac{x-1}{4}$

36. $\frac{8p}{5} = \frac{3p-4}{2} + \frac{5}{2}$

*Solve each equation, and check your answers. Be careful with signs. See Example 2.*

37. $\frac{a+7}{8} - \frac{a-2}{3} = \frac{4}{3}$

38. $\frac{x+3}{7} - \frac{x+2}{6} = \frac{1}{6}$

39. $\frac{p}{2} - \frac{p-1}{4} = \frac{5}{4}$

40. $\frac{r}{6} - \frac{r-2}{3} = -\frac{4}{3}$

41. $\frac{3x}{5} - \frac{x-5}{7} = 3$

42. $\frac{8k}{5} - \frac{3k-4}{2} = \frac{5}{2}$

*Solve each equation, and check your answers. See Examples 3–7.*

43. $\frac{4}{x^2-3x} = \frac{1}{x^2-9}$

44. $\frac{2}{t^2-4} = \frac{3}{t^2-2t}$

45. $\frac{2}{m} = \frac{m}{5m+12}$

46. $\frac{x}{4-x} = \frac{2}{x}$

47. $\frac{-2}{z+5} + \frac{3}{z-5} = \frac{20}{z^2-25}$

48. $\frac{3}{r+3} - \frac{2}{r-3} = \frac{-12}{r^2-9}$

49. $\frac{3}{x-1} + \frac{2}{4x-4} = \frac{7}{4}$

50. $\frac{2}{p+3} + \frac{3}{8} = \frac{5}{4p+12}$

51. $\frac{x}{3x+3} = \frac{2x-3}{x+1} - \frac{2x}{3x+3}$

52. $\frac{2k+3}{k+1} - \frac{3k}{2k+2} = \frac{-2k}{2k+2}$

53. $\frac{2p}{p^2-1} = \frac{2}{p+1} - \frac{1}{p-1}$

54. $\frac{2x}{x^2-16} - \frac{2}{x-4} = \frac{4}{x+4}$

55. $\frac{5x}{14x+3} = \frac{1}{x}$

56. $\frac{m}{8m+3} = \frac{1}{3m}$

57. $\frac{2}{x-1} - \frac{2}{3} = \frac{-1}{x+1}$

58. $\frac{5}{p-2} = 7 - \frac{10}{p+2}$

59. $\frac{x}{2x+2} = \frac{-2x}{4x+4} + \frac{2x-3}{x+1}$

60. $\frac{5t+1}{3t+3} = \frac{5t-5}{5t+5} + \frac{3t-1}{t+1}$

61. $\frac{8x+3}{x} = 3x$

62. $\frac{2}{x} = \frac{x}{5x-12}$

63. $\frac{1}{x+4} + \frac{x}{x-4} = \frac{-8}{x^2-16}$

64. $\frac{x}{x-3} + \frac{4}{x+3} = \frac{18}{x^2-9}$

**65.** $\frac{4}{3x+6} - \frac{3}{x+3} = \frac{8}{x^2+5x+6}$

**66.** $\frac{-13}{t^2+6t+8} + \frac{4}{t+2} = \frac{3}{2t+8}$

**67.** $\frac{3x}{x^2+5x+6} = \frac{5x}{x^2+2x-3} - \frac{2}{x^2+x-2}$

**68.** $\frac{m}{m^2+m-2} + \frac{m}{m^2-1} = \frac{m}{m^2+3m+2}$

**69.** $\frac{x+4}{x^2-3x+2} - \frac{5}{x^2-4x+3} = \frac{x-4}{x^2-5x+6}$

**70.** $\frac{3}{r^2+r-2} - \frac{1}{r^2-1} = \frac{7}{2(r^2+3r+2)}$

**71.** If you are solving a formula for the letter $k$, and your steps lead to the equation $kr - mr = km$, what would be your next step?

**72.** If you are solving a formula for the letter $k$, and your steps lead to the equation $kr - km = mr$, what would be your next step?

*Solve each formula for the specified variable. See Example 8.*

**73.** $m = \frac{kF}{a}$ for $F$

**74.** $I = \frac{kE}{R}$ for $E$

**75.** $m = \frac{kF}{a}$ for $a$

**76.** $I = \frac{kE}{R}$ for $R$

**77.** $I = \frac{E}{R+r}$ for $R$

**78.** $I = \frac{E}{R+r}$ for $r$

**79.** $h = \frac{2A}{B+b}$ for $A$

**80.** $d = \frac{2S}{n(a+L)}$ for $S$

**81.** $d = \frac{2S}{n(a+L)}$ for $a$

**82.** $h = \frac{2A}{B+b}$ for $B$

**83.** $\frac{1}{x} = \frac{1}{y} - \frac{1}{z}$ for $y$

**84.** $\frac{3}{k} = \frac{1}{p} + \frac{1}{q}$ for $q$

**85.** $9x + \frac{3}{z} = \frac{5}{y}$ for $z$

**86.** $\frac{1}{a} = \frac{1}{b} + \frac{1}{c}$ for $a$

## SUMMARY EXERCISES ON OPERATIONS AND EQUATIONS WITH RATIONAL EXPRESSIONS

We have performed the four operations of arithmetic with rational expressions and solved equations with rational expressions. The exercises in this summary include a mixed variety of problems of these types. Since students often confuse *operations* on rational expressions with *solving equations* with rational expressions, we review the four operations using the rational expressions $\frac{1}{x}$ and $\frac{1}{x-2}$ as follows.

Add:

$$\frac{1}{x} + \frac{1}{x-2} = \frac{1(x-2)}{x(x-2)} + \frac{x(1)}{x(x-2)}$$ Write with a common denominator.

$$= \frac{x-2+x}{x(x-2)}$$ Add numerators; keep the same denominator.

$$= \frac{2x-2}{x(x-2)}$$ Combine like terms.

Subtract:

$$\frac{1}{x} - \frac{1}{x-2} = \frac{1(x-2)}{x(x-2)} - \frac{x(1)}{x(x-2)}$$ Write with a common denominator.

$$= \frac{x-2-x}{x(x-2)}$$ Subtract numerators; keep the same denominator.

$$= \frac{-2}{x(x-2)}$$ Combine like terms.

Multiply:

$$\frac{1}{x} \cdot \frac{1}{x-2} = \frac{1}{x(x-2)}$$ Multiply numerators and multiply denominators.

Divide:

$$\frac{1}{x} \div \frac{1}{x-2} = \frac{1}{x} \cdot \frac{x-2}{1} = \frac{x-2}{x}$$ Multiply by the reciprocal of the second fraction.

On the other hand, consider the *equation*

$$\frac{1}{x} + \frac{1}{x-2} = \frac{3}{4}.$$

Neither 0 nor 2 can be a solution of this equation, since each will cause a denominator to equal 0. We use the multiplication property of equality and multiply each side by the LCD, $4x(x-2)$, to clear fractions.

$$4x(x-2)\frac{1}{x} + 4x(x-2)\frac{1}{x-2} = 4x(x-2)\frac{3}{4}$$

$$4(x-2) + 4x = 3x(x-2)$$

$$4x - 8 + 4x = 3x^2 - 6x$$ Distributive property

$$0 = 3x^2 - 14x + 8$$ Get 0 on one side.

$$0 = (3x-2)(x-4)$$ Factor.

$$3x - 2 = 0 \quad \text{or} \quad x - 4 = 0$$ Zero-factor property

$$x = \frac{2}{3} \quad \text{or} \quad x = 4$$

Both $\frac{2}{3}$ and 4 are solutions since neither makes a denominator equal 0; the solution set is $\left\{\frac{2}{3}, 4\right\}$.

In conclusion, remember the following points when working exercises involving rational expressions.

**Points to Remember When Working with Rational Expressions**

1. The fundamental property is applied only after numerators and denominators have been *factored.*
2. When adding and subtracting rational expressions, the common denominator must be kept throughout the problem and in the final result.
3. Always look to see if the answer is in lowest terms; if it is not, use the fundamental property.

4. When solving equations, the LCD is used to clear the equation of fractions. Multiply each side by the LCD. (Notice how this differs from the use of the LCD in Point 2.)
5. When solving equations with rational expressions, reject any proposed solution that causes an original denominator to equal 0.

*For each exercise, indicate* "operation" *if an operation is to be performed or* "equation" *if an equation is to be solved. Then perform the operation or solve the equation.*

**1.** $\frac{4}{p} + \frac{6}{p}$

**2.** $\frac{x^3y^2}{x^2y^4} \cdot \frac{y^5}{x^4}$

**3.** $\frac{1}{x^2 + x - 2} \div \frac{4x^2}{2x - 2}$

**4.** $\frac{8}{m - 5} = 2$

**5.** $\frac{2y^2 + y - 6}{2y^2 - 9y + 9} \cdot \frac{y^2 - 2y - 3}{y^2 - 1}$

**6.** $\frac{2}{k^2 - 4k} + \frac{3}{k^2 - 16}$

**7.** $\frac{x - 4}{5} = \frac{x + 3}{6}$

**8.** $\frac{3t^2 - t}{6t^2 + 15t} \div \frac{6t^2 + t - 1}{2t^2 - 5t - 25}$

**9.** $\frac{4}{p + 2} + \frac{1}{3p + 6}$

**10.** $\frac{1}{x} + \frac{1}{x - 3} = -\frac{5}{4}$

**11.** $\frac{3}{t - 1} + \frac{1}{t} = \frac{7}{2}$

**12.** $\frac{6}{y} - \frac{2}{3y}$

**13.** $\frac{5}{4z} - \frac{2}{3z}$

**14.** $\frac{k + 2}{3} = \frac{2k - 1}{5}$

**15.** $\frac{1}{m^2 + 5m + 6} + \frac{2}{m^2 + 4m + 3}$

**16.** $\frac{2k^2 - 3k}{20k^2 - 5k} \div \frac{2k^2 - 5k + 3}{4k^2 + 11k - 3}$

**17.** $\frac{2}{x + 1} + \frac{5}{x - 1} = \frac{10}{x^2 - 1}$

**18.** $\frac{3}{x + 3} + \frac{4}{x + 6} = \frac{9}{x^2 + 9x + 18}$

**19.** $\frac{4t^2 - t}{6t^2 + 10t} \div \frac{8t^2 + 2t - 1}{3t^2 + 11t + 10}$

**20.** $\frac{x}{x - 2} + \frac{3}{x + 2} = \frac{8}{x^2 - 4}$

# 6.7 Applications of Rational Expressions

**OBJECTIVES**

1 Solve problems about numbers.
2 Solve problems about distance, rate, and time.
3 Solve problems about work.

When we learn how to solve a new type of equation, we are able to apply our knowledge to solving new types of applications. In Section 6.6 we solved equations with rational expressions; now we can solve applications that involve this type of equation. The six-step problem solving method of Chapter 2 still applies.

**OBJECTIVE 1 Solve problems about numbers.** We begin with an example about an unknown number.

### EXAMPLE 1 Solving a Problem about an Unknown Number

If the same number is added to both the numerator and the denominator of the fraction $\frac{2}{5}$, the result is equivalent to $\frac{2}{3}$. Find the number.

*Step 1* **Read** the problem carefully. We are trying to find a number.

*Step 2* **Assign a variable.** Here, we let $x =$ the number added to the numerator and the denominator.

*Step 3* **Write an equation.** The fraction

$$\frac{2+x}{5+x}$$

represents the result of adding the same number to both the numerator and the denominator. Since this result is equivalent to $\frac{2}{3}$, the equation is

$$\frac{2+x}{5+x}=\frac{2}{3}.$$

*Step 4* **Solve** this equation by multiplying each side by the LCD, $3(5+x)$.

$$3(5+x)\frac{2+x}{5+x}=3(5+x)\frac{2}{3}$$

$$3(2+x)=2(5+x)$$

$$6+3x=10+2x \qquad \text{Distributive property}$$

$$x=4 \qquad \text{Subtract } 2x\text{; subtract 6.}$$

*Step 5* **State the answer.** The number is 4.

*Step 6* **Check** the solution in the words of the original problem. If 4 is added to both the numerator and the denominator of $\frac{2}{5}$, the result is $\frac{6}{9}=\frac{2}{3}$, as required.

**Now Try Exercise 3.**

**OBJECTIVE 2 Solve problems about distance, rate, and time.** Recall from Chapter 2 the following formulas relating distance, rate, and time.

**Distance, Rate, and Time Relationship**

$$d = rt \qquad r = \frac{d}{t} \qquad t = \frac{d}{r}$$

You may wish to refer to Example 5 in Section 2.7 to review the basic use of these formulas. We continue our work with motion problems here.

## EXAMPLE 2 Solving a Problem about Distance, Rate, and Time

The Tickfaw River has a current of 3 mph. A motorboat takes as long to go 12 mi downstream as to go 8 mi upstream. What is the speed of the boat in still water?

*Step 1* **Read** the problem again. We are looking for the speed of the boat in still water.

*Step 2* **Assign a variable.** Let $x$ = the speed of the boat in still water. Because the current pushes the boat when the boat is going downstream, the speed of the boat downstream will be the sum of the speed of the boat and the speed of the current, $x + 3$ mph. Because the current slows down the boat when the boat is going upstream, the boat's speed going upstream is given by the difference between the speed of the boat and the speed of the current, $x - 3$ mph. See Figure 1.

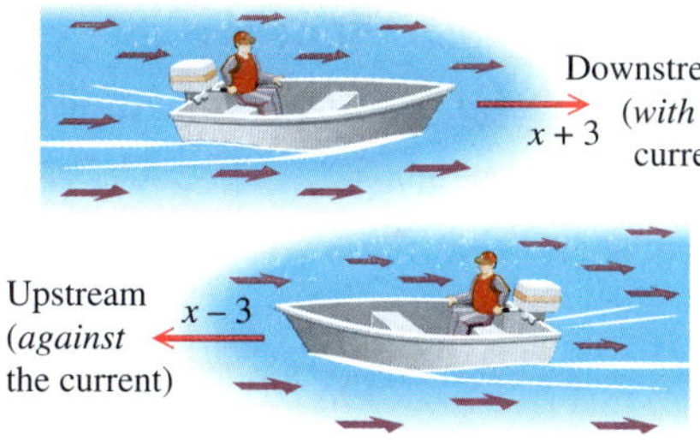


FIGURE 1

This information is summarized in the table.

| | d | r | t |
|---|---|---|---|
| Downstream | 12 | $x + 3$ | |
| Upstream | 8 | $x - 3$ | |

We fill in the column representing time by using the formula $t = \frac{d}{r}$. Then the time downstream is the distance divided by the rate, or

$$t = \frac{d}{r} = \frac{12}{x + 3},$$

and the time upstream is also the distance divided by the rate, or

$$t = \frac{d}{r} = \frac{8}{x - 3}.$$

The completed table follows.

| | $d$ | $r$ | $t$ |
|---|---|---|---|
| *Downstream* | 12 | $x + 3$ | $\frac{12}{x+3}$ |
| *Upstream* | 8 | $x - 3$ | $\frac{8}{x-3}$ |

Times are equal.

*Step 3* **Write an equation.** According to the original problem, the time downstream equals the time upstream. The two times from the table must therefore be equal, giving the equation

$$\frac{12}{x+3} = \frac{8}{x-3}.$$

*Step 4* **Solve.** We begin by multiplying each side by the LCD, $(x + 3)(x - 3)$.

$$(x+3)(x-3)\frac{12}{x+3} = (x+3)(x-3)\frac{8}{x-3}$$

$$12(x-3) = 8(x+3)$$

$$12x - 36 = 8x + 24 \quad \text{Distributive property}$$

$$4x = 60 \quad \text{Subtract } 8x\text{; add 36.}$$

$$x = 15 \quad \text{Divide by 4.}$$

*Step 5* **State the answer.** The speed of the boat in still water is 15 mph.

*Step 6* **Check.** First we find the speed of the boat going downstream, which is $15 + 3 = 18$ mph. Traveling 12 mi would take

$$t = \frac{d}{r} = \frac{12}{18} = \frac{2}{3} \text{ hr.}$$

On the other hand, the speed of the boat going upstream is $15 - 3 = 12$ mph, and traveling 8 mi would take

$$t = \frac{d}{r} = \frac{8}{12} = \frac{2}{3} \text{ hr.}$$

The time upstream equals the time downstream, as required.

**Now Try Exercise 21.**

**OBJECTIVE 3 Solve problems about work.** Suppose that you can mow your lawn in 4 hr. Then after 1 hr, you will have mowed $\frac{1}{4}$ of the lawn. After 2 hr, you will have mowed $\frac{2}{4}$ or $\frac{1}{2}$ of the lawn, and so on. This idea is generalized as follows.

**Rate of Work**

If a job can be completed in $t$ units of time, then the rate of work is

$$\frac{1}{t} \textbf{ job per unit of time.}$$

**PROBLEM SOLVING**

The relationship between problems involving work and problems involving distance is a close one. Recall that the formula $d = rt$ says that distance traveled is equal to rate of travel multiplied by time traveled. Similarly, the fractional part of a job accomplished is equal to the rate of work multiplied by the time worked. In the lawn mowing example, after 3 hr, the fractional part of the job done is

$$\underbrace{\frac{1}{4}}_{\text{Rate of work}} \cdot \underbrace{3}_{\text{Time worked}} = \underbrace{\frac{3}{4}}_{\text{Fractional part of job done}}.$$

After 4 hr, $\frac{1}{4}(4) = 1$ whole job has been done.

**EXAMPLE 3 Solving a Problem about Work Rates**

With spraying equipment, Mateo can paint the woodwork in a small house in 8 hr. His assistant, Chet, needs 14 hr to complete the same job painting by hand. If Mateo and Chet work together, how long will it take them to paint the woodwork?

*Step 1* **Read** the problem again. We are looking for time working together.

*Step 2* **Assign a variable.** Let $x =$ the number of hours it will take for Mateo and Chet to paint the woodwork, working together.

Certainly, $x$ will be less than 8, since Mateo alone can complete the job in 8 hr. We begin by making a table as shown. Remember that based on the previous discussion, Mateo's rate alone is $\frac{1}{8}$ job per hour, and Chet's rate is $\frac{1}{14}$ job per hour.

| | *Rate* | *Time Working Together* | *Fractional Part of the Job Done When Working Together* |
|---|---|---|---|
| *Mateo* | $\frac{1}{8}$ | $x$ | $\frac{1}{8}x$ |
| *Chet* | $\frac{1}{14}$ | $x$ | $\frac{1}{14}x$ |

Sum is 1 whole job.

*Step 3* **Write an equation.** Since together Mateo and Chet complete 1 whole job, we must add their individual fractional parts and set the sum equal to 1.

$$\underbrace{\frac{1}{8}x}_{\text{Fractional part done by Mateo}} + \underbrace{\frac{1}{14}x}_{\text{Fractional part done by Chet}} = \underbrace{1}_{\text{1 whole job.}}$$

*Step 4* **Solve.**

$$56\left(\frac{1}{8}x + \frac{1}{14}x\right) = 56(1) \qquad \text{Multiply by the LCD, 56.}$$

$$56\left(\frac{1}{8}x\right) + 56\left(\frac{1}{14}x\right) = 56(1) \qquad \text{Distributive property}$$
$$7x + 4x = 56$$
$$11x = 56 \qquad \text{Combine like terms.}$$
$$x = \frac{56}{11} \qquad \text{Divide by 11.}$$

*Step 5* **State the answer.** Working together, Mateo and Chet can paint the woodwork in $\frac{56}{11}$ hr, or $5\frac{1}{11}$ hr.

*Step 6* **Check** to be sure the answer is correct.

**Now Try Exercise 33.**

**NOTE** An alternative approach in work problems is to consider the part of the job that can be done in 1 hr. For instance, in Example 3 Mateo can do the entire job in 8 hr, and Chet can do it in 14 hr. Thus, their work rates, as we saw in Example 3, are $\frac{1}{8}$ and $\frac{1}{14}$, respectively. Since it takes them $x$ hr to complete the job when working together, in 1 hr they can paint $\frac{1}{x}$ of the woodwork. The amount painted by Mateo in 1 hr plus the amount painted by Chet in 1 hr must equal the amount they can do together. This leads to the equation

Amount by Chet ↓

Amount by Mateo → $\frac{1}{8} + \frac{1}{14} = \frac{1}{x}$. ← Amount together

Compare this with the equation in Example 3. Multiplying each side by $56x$ leads to

$$7x + 4x = 56,$$

the same equation found in the third line of Step 4 in the example. The same solution results.

## 6.7 EXERCISES

**For Extra Help**

- Student's Solutions Manual
- MyMathLab
- InterAct Math Tutorial Software
- AW Math Tutor Center
- MathXL
- Digital Video Tutor CD 10/Videotape 10

*Use Steps 2 and 3 of the six-step method to set up the equation you would use to solve each problem. (Remember that Step 1 is to read the problem carefully.) Do not actually solve the equation. See Example 1.*

**1.** The numerator of the fraction $\frac{5}{6}$ is increased by an amount so that the value of the resulting fraction is equivalent to $\frac{13}{3}$. By what amount was the numerator increased?

**(a)** Let $x =$ ______________. (*Step 2*)

**(b)** Write an expression for "the numerator of the fraction $\frac{5}{6}$ is increased by an amount."

**(c)** Set up an equation to solve the problem. (*Step 3*)

**2.** If the same number is added to the numerator and subtracted from the denominator of $\frac{23}{12}$, the resulting fraction is equivalent to $\frac{3}{2}$. What is the number?

**(a)** Let $x =$ ______________. (*Step 2*)

**(b)** Write an expression for "a number is added to the numerator of $\frac{23}{12}$." Then write an expression for "the same number is subtracted from the denominator of $\frac{23}{12}$."

**(c)** Set up an equation to solve the problem. (*Step 3*)

*Use the six-step method to solve each problem. See Example 1.*

**3.** In a certain fraction, the denominator is 6 more than the numerator. If 3 is added to both the numerator and the denominator, the resulting fraction is equivalent to $\frac{5}{7}$. What was the original fraction?

**4.** In a certain fraction, the denominator is 4 less than the numerator. If 3 is added to both the numerator and the denominator, the resulting fraction is equivalent to $\frac{3}{2}$. What was the original fraction?

**5.** The numerator of a certain fraction is four times the denominator. If 6 is added to both the numerator and the denominator, the resulting fraction is equivalent to 2. What was the original fraction?

**6.** The denominator of a certain fraction is three times the numerator. If 2 is added to the numerator and subtracted from the denominator, the resulting fraction is equivalent to 1. What was the original fraction?

**7.** One-third of a number is 2 more than one-sixth of the same number. What is the number?

**8.** One-sixth of a number is 5 more than the same number. What is the number?

**9.** A quantity, its $\frac{2}{3}$, its $\frac{1}{2}$, and its $\frac{1}{7}$, added together, become 33. What is the quantity? (*Source:* Rhind Mathematical Papyrus.)

**10.** A quantity, its $\frac{3}{4}$, its $\frac{1}{2}$, and its $\frac{1}{3}$, added together, become 93. What is the quantity? (*Source:* Rhind Mathematical Papyrus.)

*Solve each problem. See Example 5 in Section 2.7.*

**11.** In the 2002 Winter Olympics, Catriona LeMay Doan of Canada won the 500-m speed skating event for women. Her rate was 6.6889 m per sec. What was her time (to the nearest hundredth of a second)? (*Source:* www.espn.com)

**12.** In the 1998 World Championships, Amy Van Dyken of the United States won the 50-m freestyle swimming event for women. Her rate was 1.9881 m per sec. What was her time (to the nearest hundredth of a second)? (*Source: Sports Illustrated 2000 Sports Almanac.*)

**13.** The winner of the women's 1500-m race in the 2000 Olympics was Nouria Merah-Benida of Algeria with a time of 4.085 min. What was her rate (to three decimal places)? (*Source: World Almanac and Book of Facts,* 2002.)

**14.** Gabriela Szabo of Romania won the women's 5000-m race in the 2000 Olympics with a time of 14.680 min. What was her rate (to three decimal places)? (*Source: World Almanac and Book of Facts,* 2002.)

**15.** The winner of the 2001 Daytona 500 (mile) race was Michael Waltrip, who drove his Chevrolet to victory with a rate of 161.794 mph. What was his time (to the nearest thousandth of an hour)? (*Source: World Almanac and Book of Facts,* 2002.)

**16.** In 2001, Helio Castroneves drove his Reynard-Honda to victory in the Indianapolis 500 (mile) race. His rate was 131.294 mph. What was his time (to the nearest thousandth of an hour)? (*Source: World Almanac and Book of Facts,* 2002.)

*Use the concepts of this section to solve each problem.*

**17.** Suppose Stephanie walks $D$ mi at $R$ mph in the same time that Wally walks $d$ mi at $r$ mph. Give an equation relating $D$, $R$, $d$, and $r$.

**18.** If a migrating hawk travels $m$ mph in still air, what is its rate when it flies into a steady headwind of 5 mph? What is its rate with a tailwind of 5 mph?

*Set up the equation you would use to solve each problem. Do not actually solve the equation. See Example 2.*

**19.** Julio flew his airplane 500 mi against the wind in the same time it took him to fly 600 mi with the wind. If the speed of the wind was 10 mph, what was the average speed of his plane? (Let $x$ = speed of the plane in still air.)

| | $d$ | $r$ | $t$ |
|---|---|---|---|
| *Against the Wind* | 500 | $x - 10$ | |
| *With the Wind* | 600 | $x + 10$ | |

**20.** Luvenia can row 4 mph in still water. She takes as long to row 8 mi upstream as 24 mi downstream. How fast is the current? (Let $x$ = speed of the current.)

| | $d$ | $r$ | $t$ |
|---|---|---|---|
| *Upstream* | 8 | $4 - x$ | |
| *Downstream* | 24 | $4 + x$ | |

*Solve each problem. See Example 2.*

**21.** A boat can go 20 mi against a current in the same time that it can go 60 mi with the current. The current is 4 mph. Find the speed of the boat in still water.

**22.** Lucinda can fly her plane 200 mi against the wind in the same time it takes her to fly 300 mi with the wind. The wind blows at 30 mph. Find the speed of her plane in still air.

**23.** An airplane, maintaining a constant airspeed, takes as long to go 450 mi with the wind as it does to go 375 mi against the wind. If the wind is blowing at 15 mph, what is the speed of the plane?

**24.** A river has a current of 4 km per hr. Find the speed of Lynn McTernan's boat in still water if it goes 40 km downstream in the same time that it takes to go 24 km upstream.

**25.** Kellen's boat goes 12 mph. Find the rate of the current of the river if she can go 6 mi upstream in the same amount of time she can go 10 mi downstream.

**26.** Kasey can travel 8 mi upstream in the same time it takes her to go 12 mi downstream. Her boat goes 15 mph in still water. What is the rate of the current?

**27.** The distance from Seattle, Washington, to Victoria, British Columbia, is about 148 mi by ferry. It takes about 4 hr less to travel by the same ferry from Victoria to Vancouver, British Columbia, a distance of about 74 mi. What is the average speed of the ferry?

**28.** Driving from Tulsa to Detroit, Jeff averaged 50 mph. He figured that if he had averaged 60 mph, his driving time would have decreased 3 hr. How far is it from Tulsa to Detroit?

*Use the concepts of this section to solve each problem.*

**29.** If it takes Elayn 10 hr to do a job, what is her rate?

**30.** If it takes Clay 12 hr to do a job, how much of the job does he do in 8 hr?

*In Exercises 31 and 32, set up the equation you would use to solve each problem. Do not actually solve the equation. See Example 3.*

**31.** Working alone, Jorge can paint a room in 8 hr. Caterina can paint the same room working alone in 6 hr. How long will it take them if they work together? (Let $x$ represent the time working together.)

| | $r$ | $t$ | $w$ |
|---|---|---|---|
| *Jorge* | | $x$ | |
| *Caterina* | | $x$ | |

**32.** Edwin Bedford can tune up his Chevy in 2 hr working alone. His son, Beau, can do the job in 3 hr working alone. How long would it take them if they worked together? (Let $t$ represent the time working together.)

| | $r$ | $t$ | $w$ |
|---|---|---|---|
| *Edwin* | | $t$ | |
| *Beau* | | $t$ | |

*Solve each problem. See Example 3.*

**33.** Geraldo and Luisa Hernandez operate a small laundry. Luisa, working alone, can clean a day's laundry in 9 hr. Geraldo can clean a day's laundry in 8 hr. How long would it take them if they work together?

**34.** Lea can groom the horses in her boarding stable in 5 hr, while Tran needs 4 hr. How long will it take them to groom the horses if they work together?

**35.** A pump can pump the water out of a flooded basement in 10 hr. A smaller pump takes 12 hr. How long would it take to pump the water from the basement using both pumps?

**36.** Doug Todd's copier can do a printing job in 7 hr. Scott's copier can do the same job in 12 hr. How long would it take to do the job using both copiers?

**37.** An experienced employee can enter tax data into a computer twice as fast as a new employee. Working together, it takes the employees 2 hr. How long would it take the experienced employee working alone?

**38.** One roofer can put a new roof on a house three times faster than another. Working together they can roof a house in 4 days. How long would it take the faster roofer working alone?

**39.** One pipe can fill a swimming pool in 6 hr, and another pipe can do it in 9 hr. How long will it take the two pipes working together to fill the pool $\frac{3}{4}$ full?

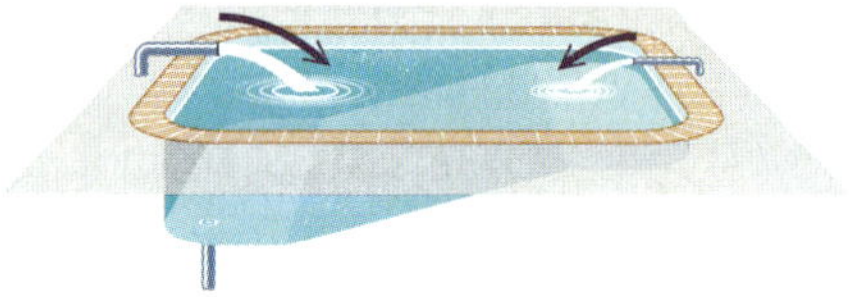

**40.** An inlet pipe can fill a swimming pool in 9 hr, and an outlet pipe can empty the pool in 12 hr. Through an error, both pipes are left open. How long will it take to fill the pool?

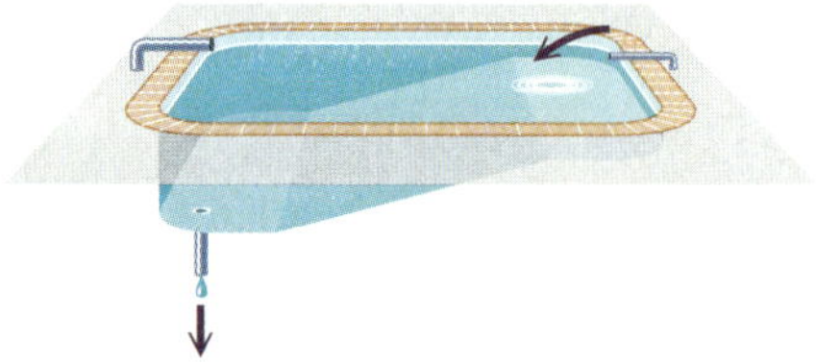

**41.** A cold water faucet can fill a sink in 12 min, and a hot water faucet can fill it in 15 min. The drain can empty the sink in 25 min. If both faucets are on and the drain is open, how long will it take to fill the sink?

**42.** Refer to Exercise 40. Assume the error was discovered after both pipes had been running for 3 hr, and the outlet pipe was then closed. How much more time would then be required to fill the pool? (*Hint:* Consider how much of the job had been done when the error was discovered.)

## Chapter 6 Group Activity

### Buying a Car

**OBJECTIVE** Use a complex fraction to calculate monthly car payments.

You are shopping for a sports car and have put aside a certain amount of money each month for a car payment. Your instructor will assign this amount to you. After looking through a variety of resources, you have narrowed your choices to the cars listed in the table.

| *Year/Make/Model* | *Retail Price* | *Fuel Tank Size (in gallons)* | *Miles per Gallon (city)* | *Miles per Gallon (highway)* |
|---|---|---|---|---|
| 2002 Mitsubishi Eclipse | $24,197 | 16.4 | 21 | 28 |
| 2002 Chevy Camaro | 18,415 | 16.8 | 19 | 31 |
| 2002 Ford Mustang | 18,080 | 15.7 | 20 | 29 |
| 2002 Pontiac Firebird | 20,050 | 16.8 | 19 | 31 |
| 2002 BMW-Z3 | 37,700 | 13.5 | 21 | 29 |
| 2002 Toyota Celica GTS | 21,555 | 14.5 | 23 | 32 |

*Source:* http://aolsvc.carguides.aol.com

As a group, work through the following steps to determine which car you can afford to buy.

**A.** Decide which cars you think are within your budget.

**B.** Select one of the cars you identified in part A. Have each member of the group calculate the monthly payment for this car using a different financing option. Use the formula given below, where $P$ is principal, $r$ is interest rate, and $m$ is the number of monthly payments, along with the financing options table.

*Financing Options*

| *Time (in years)* | *Interest Rate* |
|---|---|
| 4 | 7.0% |
| 5 | 8.5% |
| 6 | 10.0% |

$$\text{Monthly Payment} = \frac{\dfrac{Pr}{12}}{1 - \left(\dfrac{12}{12 + r}\right)^m}$$

**C.** Have each group member determine the amount of money paid in interest over the duration of the loan for their financing option.

**D.** Consider fuel expenses.

1. Assume you will travel an average of 75 mi in the city and 400 mi on the highway each week. How many gallons of gas will you need to buy each month?
2. Using typical prices for gas in your area at this time, how much money will you need to have available for buying gas?

**E.** Repeat parts B–D as necessary until your group can reach a consensus on the car you will buy and the financing option you will use. Write a paragraph to explain your choices.

# CHAPTER 6 SUMMARY

## KEY TERMS

**6.1** rational expression
lowest terms

**6.3** least common denominator (LCD)

**6.5** complex fraction

## TEST YOUR WORD POWER

*See how well you have learned the vocabulary in this chapter. Answers, with examples, follow the Quick Review.*

1. A **rational expression** is
   **A.** an algebraic expression made up of a term or the sum of a finite number of terms with real coefficients and whole number exponents
   **B.** a polynomial equation of degree 2
   **C.** an expression with one or more fractions in the numerator, denominator, or both
   **D.** the quotient of two polynomials with denominator not 0.

2. A **complex fraction** is
   **A.** an algebraic expression made up of a term or the sum of a finite number of terms with real coefficients and whole number exponents
   **B.** a polynomial equation of degree 2
   **C.** a rational expression with one or more fractions in the numerator, denominator, or both
   **D.** the quotient of two polynomials with denominator not 0.

## QUICK REVIEW

| CONCEPTS | EXAMPLES |
|---|---|

### 6.1 THE FUNDAMENTAL PROPERTY OF RATIONAL EXPRESSIONS

To find the value(s) for which a rational expression is undefined, set the denominator equal to 0 and solve the equation.

Find the values for which the expression $\frac{x-4}{x^2-16}$ is undefined.

$$x^2 - 16 = 0$$
$$(x-4)(x+4) = 0$$
$$x - 4 = 0 \quad \text{or} \quad x + 4 = 0$$
$$x = 4 \quad \text{or} \quad x = -4$$

The rational expression is undefined for 4 and $-4$, so $x \neq 4$ and $x \neq -4$.

**Writing a Rational Expression in Lowest Terms**

*Step 1* Factor the numerator and denominator.

*Step 2* Use the fundamental property to divide out common factors.

Write $\frac{x^2-1}{(x-1)^2}$ in lowest terms.

$$\frac{x^2-1}{(x-1)^2} = \frac{(x-1)(x+1)}{(x-1)(x-1)} = \frac{x+1}{x-1}$$

There are often several different equivalent forms of a rational expression.

Give four equivalent forms of $-\frac{x-1}{x+2}$.

Distribute the $-$ sign in the numerator to get $\frac{-(x-1)}{x+2}$ or $\frac{-x+1}{x+2}$; do so in the denominator to get $\frac{x-1}{-(x+2)}$ or $\frac{x-1}{-x-2}$.

| CONCEPTS | EXAMPLES |
|---|---|

## 6.2 MULTIPLYING AND DIVIDING RATIONAL EXPRESSIONS

**Multiplying Rational Expressions**

*Step 1* Multiply numerators and multiply denominators.

*Step 2* Factor.

*Step 3* Write in lowest terms.

Multiply. $\frac{3x+9}{x-5} \cdot \frac{x^2-3x-10}{x^2-9}$

$$= \frac{(3x+9)(x^2-3x-10)}{(x-5)(x^2-9)}$$

$$= \frac{3(x+3)(x-5)(x+2)}{(x-5)(x+3)(x-3)}$$

$$= \frac{3(x+2)}{x-3}$$

**Dividing Rational Expressions**

*Step 1* Multiply the first rational expression by the reciprocal of the second.

*Step 2* Multiply numerators and multiply denominators.

*Step 3* Factor.

*Step 4* Write in lowest terms.

Divide. $\frac{2x+1}{x+5} \div \frac{6x^2-x-2}{x^2-25}$

$$= \frac{2x+1}{x+5} \cdot \frac{x^2-25}{6x^2-x-2}$$

$$= \frac{(2x+1)(x^2-25)}{(x+5)(6x^2-x-2)}$$

$$= \frac{(2x+1)(x+5)(x-5)}{(x+5)(2x+1)(3x-2)}$$

$$= \frac{x-5}{3x-2}$$

## 6.3 LEAST COMMON DENOMINATORS

**Finding the LCD**

*Step 1* Factor each denominator into prime factors.

*Step 2* List each different factor the greatest number of times it appears.

*Step 3* Multiply the factors from Step 2 to get the LCD.

Find the LCD for $\frac{3}{k^2-8k+16}$ and $\frac{1}{4k^2-16k}$.

$$k^2-8k+16 = (k-4)^2$$

$$4k^2-16k = 4k(k-4)$$

$$\text{LCD} = (k-4)^2 \cdot 4 \cdot k$$

$$= 4k(k-4)^2$$

**Writing a Rational Expression with a Specified Denominator**

*Step 1* Factor both denominators.

*Step 2* Decide what factors the denominator must be multiplied by to equal the specified denominator.

*Step 3* Multiply the rational expression by that factor divided by itself (multiply by 1).

Find the numerator: $\frac{5}{2z^2-6z} = \frac{\quad}{4z^3-12z^2}$.

$$\frac{5}{2z(z-3)} = \frac{\quad}{4z^2(z-3)}$$

$2z(z-3)$ must be multiplied by $2z$.

$$\frac{5}{2z(z-3)} \cdot \frac{2z}{2z} = \frac{10z}{4z^2(z-3)} = \frac{10z}{4z^3-12z^2}$$

(continued)

| CONCEPTS | EXAMPLES |
|---|---|
| **6.4 ADDING AND SUBTRACTING RATIONAL EXPRESSIONS** | |
| **Adding Rational Expressions** | Add. $\frac{2}{3m+6}+\frac{m}{m^2-4}$ |
| *Step 1* Find the LCD. | $3m+6=3(m+2)$ <br> $m^2-4=(m+2)(m-2)$ <br> The LCD is $3(m+2)(m-2)$. |
| *Step 2* Rewrite each rational expression with the LCD as denominator. | $=\frac{2(m-2)}{3(m+2)(m-2)}+\frac{3m}{3(m+2)(m-2)}$ |
| *Step 3* Add the numerators to get the numerator of the sum. The LCD is the denominator of the sum. | $=\frac{2m-4+3m}{3(m+2)(m-2)}$ |
| *Step 4* Write in lowest terms. | $=\frac{5m-4}{3(m+2)(m-2)}$ |
| **Subtracting Rational Expressions** <br> Follow the same steps as for addition, but subtract in Step 3. <br> Be careful with signs when subtracting the numerators. | Subtract. $\frac{6}{k+4}-\frac{2}{k}$ <br> The LCD is $k(k+4)$. <br> $\frac{6k}{(k+4)k}-\frac{2(k+4)}{k(k+4)}=\frac{6k-2(k+4)}{k(k+4)}$ <br> $=\frac{6k-2k-8}{k(k+4)}$ <br> $=\frac{4k-8}{k(k+4)}$ or $\frac{4(k-2)}{k(k+4)}$ |
| **6.5 COMPLEX FRACTIONS** | |
| **Simplifying Complex Fractions** | Simplify. |
| *Method 1* Simplify the numerator and denominator separately. Then divide the simplified numerator by the simplified denominator. | *Method 1* $\frac{\frac{1}{a}-a}{1-a}=\frac{\frac{1}{a}-\frac{a^2}{a}}{1-a}=\frac{\frac{1-a^2}{a}}{1-a}$ <br> $=\frac{1-a^2}{a}\div(1-a)$ <br> $=\frac{1-a^2}{a}\cdot\frac{1}{1-a}$ <br> $=\frac{(1-a)(1+a)}{a(1-a)}=\frac{1+a}{a}$ |
| *Method 2* Multiply the numerator and denominator of the complex fraction by the LCD of all the denominators in the complex fraction. Write in lowest terms. | *Method 2* $\frac{\frac{1}{a}-a}{1-a}=\frac{\frac{1}{a}-a}{1-a}\cdot\frac{a}{a}=\frac{\frac{a}{a}-a^2}{(1-a)a}$ <br> $=\frac{1-a^2}{(1-a)a}=\frac{(1+a)(1-a)}{(1-a)a}$ <br> $=\frac{1+a}{a}$ |

| CONCEPTS | EXAMPLES |
|---|---|

## 6.6 SOLVING EQUATIONS WITH RATIONAL EXPRESSIONS

**Solving Equations with Rational Expressions**

*Step 1* Find the LCD of all denominators in the equation.

*Step 2* Multiply each side of the equation by the LCD.

*Step 3* Solve the resulting equation, which should have no fractions.

*Step 4* Check each proposed solution.

Solve $\frac{2}{x-1} + \frac{3}{4} = \frac{5}{x-1}$.

The LCD is $4(x-1)$. Note that 1 cannot be a solution.

$$4(x-1)\left(\frac{2}{x-1} + \frac{3}{4}\right) = 4(x-1)\left(\frac{5}{x-1}\right)$$
$$4(x-1)\left(\frac{2}{x-1}\right) + 4(x-1)\left(\frac{3}{4}\right) = 4(x-1)\left(\frac{5}{x-1}\right)$$
$$8 + 3(x-1) = 20$$
$$8 + 3x - 3 = 20$$
$$3x = 15$$
$$x = 5$$

The proposed solution, 5, checks. The solution set is $\{5\}$.

## 6.7 APPLICATIONS OF RATIONAL EXPRESSIONS

**Solving Problems about Distance, Rate, Time**
Use the six-step method.

*Step 1* **Read** the problem carefully.

*Step 2* **Assign a variable.** Use a table to identify distance, rate, and time. Solve $d = rt$ for the unknown quantity in the table.

*Step 3* **Write an equation.** From the wording in the problem, decide the relationship between the quantities. Use those expressions to write an equation.

*Step 4* **Solve** the equation.

*Step 5* **State the answer.**

*Step 6* **Check** the solution.

On a trip from Sacramento to Monterey, Marge traveled at an average speed of 60 mph. The return trip, at an average speed of 64 mph, took $\frac{1}{4}$ hr less. How far did she travel between the two cities?

Let $x$ = the unknown distance.

| | $d$ | $r$ | $t = \frac{d}{r}$ |
|---|---|---|---|
| *Going* | $x$ | 60 | $\frac{x}{60}$ |
| *Returning* | $x$ | 64 | $\frac{x}{64}$ |

Since the time for the return trip was $\frac{1}{4}$ hr less, the time going equals the time returning plus $\frac{1}{4}$.

$$\frac{x}{60} = \frac{x}{64} + \frac{1}{4}$$
$$16x = 15x + 240 \quad \text{Multiply by 960.}$$
$$x = 240 \quad \text{Subtract } 15x.$$

She traveled 240 mi.

The trip there took $\frac{240}{60} = 4$ hr, while the return trip took $\frac{240}{64} = 3\frac{3}{4}$ hr, which is $\frac{1}{4}$ hr less time. The solution checks.

(continued)

| CONCEPTS | EXAMPLES |
|---|---|
| **Solving Problems about Work** | |
| *Step 1* **Read** the problem carefully. | It takes the regular mail carrier 6 hr to cover her route. A substitute takes 8 hr to cover the same route. How long would it take them to cover the route together? |
| *Step 2* **Assign a variable.** State what the variable represents. Put the information from the problem in a table. If a job is done in $t$ units of time, the rate is $\frac{1}{t}$. | Let $x$ = the number of hours to cover the route together. The rate of the regular carrier is $\frac{1}{6}$ job per hr; the rate of the substitute is $\frac{1}{8}$ job per hr. Multiply rate by time to get the fractional part of the job done. |
| *Step 3* **Write an equation.** The sum of the fractional parts should equal 1 (whole job). | The equation is $\frac{1}{6}x + \frac{1}{8}x = 1$. |
| *Step 4* **Solve** the equation. | The solution of the equation is $\frac{24}{7}$. |
| *Steps 5 and 6* **State the answer** and **check** the solution. | It would take them $\frac{24}{7}$ or $3\frac{3}{7}$ hr to cover the route together. The solution checks because $\frac{1}{6}\left(\frac{24}{7}\right) + \frac{1}{8}\left(\frac{24}{7}\right) = 1$ is true. |

| | *Rate* | *Time* | *Part of the Job Done* |
|---|---|---|---|
| *Regular* | $\frac{1}{6}$ | $x$ | $\frac{1}{6}x$ |
| *Substitute* | $\frac{1}{8}$ | $x$ | $\frac{1}{8}x$ |

**Answers to Test Your Word Power**

**1.** D; *Examples:* $-\frac{3}{4y}, \frac{5x^3}{x+2}, \frac{a+3}{a^2-4a-5}$ **2.** C; *Examples:* $\frac{\frac{2}{3}}{\frac{4}{7}}, \frac{x-\frac{1}{y}}{x+\frac{1}{y}}, \frac{\frac{2}{a+1}}{a^2-1}$

# CHAPTER 6 REVIEW EXERCISES

**[6.1]** *Find the numerical value of each rational expression when **(a)** $x = -2$ and **(b)** $x = 4$.*

**1.** $\frac{4x-3}{5x+2}$

**2.** $\frac{3x}{x^2-4}$

*Find the value(s) of the variable for which each rational expression is undefined.*

**3.** $\frac{4}{x-3}$

**4.** $\frac{y+3}{2y}$

**5.** $\frac{2k+1}{3k^2+17k+10}$

**6.** How would you determine the values of the variable for which a rational expression is undefined?

*Write each rational expression in lowest terms.*

7. $\dfrac{5a^3b^3}{15a^4b^2}$

8. $\dfrac{m-4}{4-m}$

9. $\dfrac{4x^2-9}{6-4x}$

10. $\dfrac{4p^2+8pq-5q^2}{10p^2-3pq-q^2}$

*Write four equivalent forms for each rational expression.*

11. $-\dfrac{4x-9}{2x+3}$

12. $\dfrac{8-3x}{3+6x}$

**[6.2]** *Multiply or divide, and write each answer in lowest terms.*

13. $\dfrac{18p^3}{6}\cdot\dfrac{24}{p^4}$

14. $\dfrac{8x^2}{12x^5}\cdot\dfrac{6x^4}{2x}$

15. $\dfrac{x-3}{4}\cdot\dfrac{5}{2x-6}$

16. $\dfrac{2r+3}{r-4}\cdot\dfrac{r^2-16}{6r+9}$

17. $\dfrac{6a^2+7a-3}{2a^2-a-6}\div\dfrac{a+5}{a-2}$

18. $\dfrac{y^2-6y+8}{y^2+3y-18}\div\dfrac{y-4}{y+6}$

19. $\dfrac{2p^2+13p+20}{p^2+p-12}\cdot\dfrac{p^2+2p-15}{2p^2+7p+5}$

20. $\dfrac{3z^2+5z-2}{9z^2-1}\cdot\dfrac{9z^2+6z+1}{z^2+5z+6}$

**[6.3]** *Find the least common denominator for each list of fractions.*

21. $\dfrac{4}{9y}, \dfrac{7}{12y^2}, \dfrac{5}{27y^4}$

22. $\dfrac{3}{x^2+4x+3}, \dfrac{5}{x^2+5x+4}$

*Rewrite each rational expression with the given denominator.*

23. $\dfrac{3}{2a^3}=\dfrac{}{10a^4}$

24. $\dfrac{9}{x-3}=\dfrac{}{18-6x}$

25. $\dfrac{-3y}{2y-10}=\dfrac{}{50-10y}$

26. $\dfrac{4b}{b^2+2b-3}=\dfrac{}{(b+3)(b-1)(b+2)}$

**[6.4]** *Add or subtract, and write each answer in lowest terms.*

27. $\dfrac{10}{x}+\dfrac{5}{x}$

28. $\dfrac{6}{3p}-\dfrac{12}{3p}$

29. $\dfrac{9}{k}-\dfrac{5}{k-5}$

30. $\dfrac{4}{y}+\dfrac{7}{7+y}$

31. $\dfrac{m}{3}-\dfrac{2+5m}{6}$

32. $\dfrac{12}{x^2}-\dfrac{3}{4x}$

33. $\dfrac{5}{a-2b}+\dfrac{2}{a+2b}$

34. $\dfrac{4}{k^2-9}-\dfrac{k+3}{3k-9}$

35. $\dfrac{8}{z^2+6z}-\dfrac{3}{z^2+4z-12}$

36. $\dfrac{11}{2p-p^2}-\dfrac{2}{p^2-5p+6}$

[6.5]

**37.** Simplify the complex fraction $\dfrac{\frac{a^4}{b^2}}{\frac{a^3}{b}}$ by

**(a)** Method 1 as described in Section 6.5.

**(b)** Method 2 as described in Section 6.5.

**(c)** Explain which method you prefer, and why.

*Simplify each complex fraction.*

**38.** $\dfrac{\frac{2}{3}-\frac{1}{6}}{\frac{1}{4}+\frac{2}{5}}$

**39.** $\dfrac{\frac{y-3}{y}}{\frac{y+3}{4y}}$

**40.** $\dfrac{\frac{1}{p}-\frac{1}{q}}{\frac{1}{q-p}}$

**41.** $\dfrac{x+\frac{1}{w}}{x-\frac{1}{w}}$

**42.** $\dfrac{\frac{1}{r+t}-1}{\frac{1}{r+t}+1}$

[6.6]

**43.** Before even beginning the solution process, how do you know that 2 cannot be a solution to the equation found in Exercise 46?

*Solve each equation, and check your solutions.*

**44.** $\dfrac{4-z}{z}+\dfrac{3}{2}=\dfrac{-4}{z}$

**45.** $\dfrac{3x-1}{x-2}=\dfrac{5}{x-2}+1$

**46.** $\dfrac{3}{m-2}+\dfrac{1}{m-1}=\dfrac{7}{m^2-3m+2}$

*Solve each formula for the specified variable.*

**47.** $m=\dfrac{Ry}{t}$ for $t$

**48.** $x=\dfrac{3y-5}{4}$ for $y$

**49.** $p^2=\dfrac{4}{3m-q}$ for $m$

[6.7] *Solve each problem.*

**50.** In a certain fraction, the denominator is 4 less than the numerator. If 3 is added to both the numerator and the denominator, the resulting fraction is equal to $\frac{3}{2}$. Find the original fraction.

**51.** The denominator of a certain fraction is 3 times the numerator. If 2 is added to the numerator and subtracted from the denominator, the resulting fraction is equal to 1. Find the original fraction.

**52.** A plane flies 350 mi with the wind in the same time that it can fly 310 mi against the wind. The plane has a still-air speed of 165 mph. Find the speed of the wind.

**53.** A man can plant his garden in 5 hr, working alone. His daughter can do the same job in 8 hr. How long would it take them if they worked together?

**54.** The head gardener can mow the lawns in the city park twice as fast as his assistant. Working together, they can complete the job in $1\frac{1}{3}$ hr. How long would it take the head gardener working alone?

## MIXED REVIEW EXERCISES

*Perform each indicated operation.*

55. $\frac{4}{m-1} - \frac{3}{m+1}$

56. $\frac{8p^5}{5} \div \frac{2p^3}{10}$

57. $\frac{r-3}{8} \div \frac{3r-9}{4}$

58. $\frac{\frac{5}{x} - 1}{\frac{5-x}{3x}}$

59. $\frac{4}{z^2 - 2z + 1} - \frac{3}{z^2 - 1}$

*Solve.*

60. $a = \frac{v - w}{t}$ for $v$

61. $\frac{2}{z} - \frac{z}{z+3} = \frac{1}{z+3}$

62. $\frac{5+m}{m} + \frac{3}{4} = -\frac{2}{m}$

63. "If Joe can paint a house in 3 hours, and Sam can paint the same house in 5 hours, how long does it take for them to do it together?" (From the movie *Little Big League*)

64. Anne Kelly flew her plane 400 km with the wind in the same time it took her to go 200 km against the wind. The speed of the wind is 50 km per hr. Find the speed of the plane in still air.

## RELATING CONCEPTS (EXERCISES 65–74)

**For Individual or Group Work**

*In these exercises, we summarize the various concepts involving rational expressions.* ***Work Exercises 65–74 in order.***

*Let P, Q, and R be rational expressions defined as follows.*

$$P = \frac{6}{x+3} \qquad Q = \frac{5}{x+1} \qquad R = \frac{4x}{x^2 + 4x + 3}$$

65. Find the value or values for which the expression is undefined.

**(a)** $P$ **(b)** $Q$ **(c)** $R$

66. Find and express in lowest terms: $(P \cdot Q) \div R$.

67. Why is $(P \cdot Q) \div R$ not defined if $x = 0$?

68. Find the LCD for $P$, $Q$, and $R$.

69. Perform the operations and express in lowest terms: $P + Q - R$.

70. Simplify the complex fraction $\frac{P+Q}{R}$.

(continued)

**71.** Solve the equation $P + Q = R$.

**72.** How does your answer to Exercise 65 help you work Exercise 71?

**73.** Suppose that a car travels 6 mi in $x + 3$ min. Explain why $P$ represents the rate of the car (in miles per minute).

**74.** For what value or values of $x$ is $R = \frac{40}{77}$?

# CHAPTER 6 TEST

**1.** Find the numerical value of $\dfrac{6r + 1}{2r^2 - 3r - 20}$ when

**(a)** $r = -2$ **(b)** $r = 4$.

**2.** Find any values for which $\dfrac{3x - 1}{x^2 - 2x - 8}$ is undefined.

**3.** Write four rational expressions equivalent to $-\dfrac{6x - 5}{2x + 3}$.

*Write each rational expression in lowest terms.*

**4.** $\dfrac{-15x^6y^4}{5x^4y}$

**5.** $\dfrac{6a^2 + a - 2}{2a^2 - 3a + 1}$

*Multiply or divide. Write each answer in lowest terms.*

**6.** $\dfrac{5(d - 2)}{9} \div \dfrac{3(d - 2)}{5}$

**7.** $\dfrac{6k^2 - k - 2}{8k^2 + 10k + 3} \cdot \dfrac{4k^2 + 7k + 3}{3k^2 + 5k + 2}$

**8.** $\dfrac{4a^2 + 9a + 2}{3a^2 + 11a + 10} \div \dfrac{4a^2 + 17a + 4}{3a^2 + 2a - 5}$

*Find the least common denominator for each list of fractions.*

**9.** $\dfrac{-3}{10p^2}, \dfrac{21}{25p^3}, \dfrac{-7}{30p^5}$

**10.** $\dfrac{r + 1}{2r^2 + 7r + 6}, \dfrac{-2r + 1}{2r^2 - 7r - 15}$

*Rewrite each rational expression with the given denominator.*

**11.** $\dfrac{15}{4p} = \dfrac{\quad}{64p^3}$

**12.** $\dfrac{3}{6m - 12} = \dfrac{\quad}{42m - 84}$

*Add or subtract. Write each answer in lowest terms.*

**13.** $\dfrac{4x + 2}{x + 5} + \dfrac{-2x + 8}{x + 5}$

**14.** $\dfrac{-4}{y + 2} + \dfrac{6}{5y + 10}$

**15.** $\dfrac{x + 1}{3 - x} + \dfrac{x^2}{x - 3}$

**16.** $\dfrac{3}{2m^2 - 9m - 5} - \dfrac{m + 1}{2m^2 - m - 1}$

*Simplify each complex fraction.*

**17.** $\dfrac{\frac{2p}{k^2}}{\frac{3p^2}{k^3}}$

**18.** $\dfrac{\frac{1}{x+3} - 1}{1 + \frac{1}{x+3}}$

*Solve.*

**19.** $\dfrac{2x}{x-3} + \dfrac{1}{x+3} = \dfrac{-6}{x^2-9}$

**20.** $F = \dfrac{k}{d-D}$ for $D$

*Solve each problem.*

**21.** A boat goes 7 mph in still water. It takes as long to go 20 mi upstream as 50 mi downstream. Find the speed of the current.

**22.** A man can paint a room in his house, working alone, in 5 hr. His wife can do the job in 4 hr. How long will it take them to paint the room if they work together?

## CUMULATIVE REVIEW EXERCISES CHAPTERS 1–6

**1.** Use the order of operations to evaluate $3 + 4\left(\frac{1}{2} - \frac{3}{4}\right)$.

*Solve.*

**2.** $3(2y - 5) = 2 + 5y$

**3.** $A = \dfrac{1}{2}bh$ for $b$

**4.** $\dfrac{2+m}{2-m} = \dfrac{3}{4}$

**5.** $5y \leq 6y + 8$

**6.** $5m - 9 > 2m + 3$

**7.** For the graph of $4x + 3y = -12$,

**(a)** what is the $x$-intercept? **(b)** what is the $y$-intercept?

*Sketch each graph.*

**8.** $y = -3x + 2$

**9.** $y = -x^2 + 1$

*Find the slope of each line described in Exercises 10 and 11.*

**10.** Through $(-5, 8)$ and $(-1, 2)$

**11.** Perpendicular to $4x - 3y = 12$

*Simplify each expression. Write with only positive exponents.*

**12.** $\dfrac{(2x^3)^{-1} \cdot x}{2^3x^5}$

**13.** $\dfrac{(m^{-2})^3m}{m^5m^{-4}}$

**14.** $\dfrac{2p^3q^4}{8p^5q^3}$

*Perform each indicated operation.*

**15.** $(2k^2 + 3k) - (k^2 + k - 1)$

**16.** $8x^2y^2(9x^4y^5)$

**17.** $(2a - b)^2$

**18.** $(y^2 + 3y + 5)(3y - 1)$

**19.** $\dfrac{12p^3 + 2p^2 - 12p + 4}{2p - 2}$

**20.** A computer can do one operation in $1.4 \times 10^{-7}$ sec. How long would it take for the computer to do one trillion ($10^{12}$) operations?

*Factor completely.*

**21.** $8t^2 + 10tv + 3v^2$

**22.** $8r^2 - 9rs + 12s^2$

**23.** $16x^4 - 1$

*Solve each equation.*

**24.** $r^2 = 2r + 15$

**25.** $(r - 5)(2r + 1)(3r - 2) = 0$

*Solve each problem.*

**26.** One number is 4 more than another. The product of the numbers is 2 less than the smaller number. Find the smaller number.

**27.** The length of a rectangle is 2 m less than twice the width. The area is 60 m². Find the width of the rectangle.

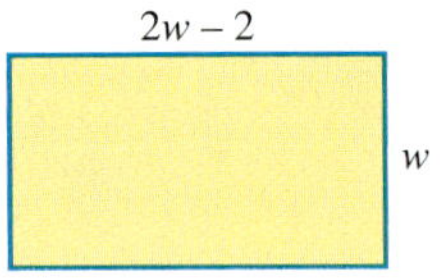


**28.** One of the following is equal to 1 for *all* real numbers. Which one is it?

**A.** $\dfrac{k^2 + 2}{k^2 + 2}$ **B.** $\dfrac{4 - m}{4 - m}$ **C.** $\dfrac{2x + 9}{2x + 9}$ **D.** $\dfrac{x^2 - 1}{x^2 - 1}$

**29.** Which one of the following rational expressions is *not* equivalent to $\dfrac{4 - 3x}{7}$?

**A.** $-\dfrac{-4 + 3x}{7}$ **B.** $-\dfrac{4 - 3x}{-7}$ **C.** $\dfrac{-4 + 3x}{-7}$ **D.** $\dfrac{-(3x + 4)}{7}$

*Perform each operation and write the answer in lowest terms.*

**30.** $\dfrac{5}{q} - \dfrac{1}{q}$

**31.** $\dfrac{3}{7} + \dfrac{4}{r}$

**32.** $\dfrac{4}{5q - 20} - \dfrac{1}{3q - 12}$

**33.** $\dfrac{2}{k^2 + k} - \dfrac{3}{k^2 - k}$

**34.** $\dfrac{7z^2 + 49z + 70}{16z^2 + 72z - 40} \div \dfrac{3z + 6}{4z^2 - 1}$

**35.** $\dfrac{\dfrac{4}{a} + \dfrac{5}{2a}}{\dfrac{7}{6a} - \dfrac{1}{5a}}$

**36.** What values of $x$ cannot possibly be solutions of the equation $\dfrac{1}{x - 4} = \dfrac{3}{2x}$?

*Solve each equation. Check your solutions.*

**37.** $\dfrac{r + 2}{5} = \dfrac{r - 3}{3}$

**38.** $\dfrac{1}{x} = \dfrac{1}{x + 1} + \dfrac{1}{2}$

*Solve each problem.*

**39.** Juanita can weed the yard in 3 hr. Benito can weed the yard in 2 hr. How long would it take them if they worked together?

**40.** A canal has a current of 2 mph. Find the speed of Amy's boat in still water if it goes 11 mi downstream in the same time that it goes 8 mi upstream.

# Equations of Lines; Functions

# 7

Graphs are widely used in the media because they present a lot of information in an easy-to-understand form. As the saying goes, "A picture is worth a thousand words." It is important to be able to read graphs correctly and understand how to use the data they provide. In Section 7.1, Example 10, we use a graph to find the average rate of change each year from 1997 to 2001 in the number of U.S. households owning more than one personal computer.

# 7.1 Review of Graphs and Slopes of Lines

**OBJECTIVES**

1. Plot ordered pairs.
2. Graph lines and find intercepts.
3. Recognize equations of vertical and horizontal lines.
4. Find the slope of a line.
5. Graph a line given its slope and a point on the line.
6. Use slopes to determine whether two lines are parallel, perpendicular, or neither.
7. Solve problems involving average rate of change.

This section and the next review some of the main topics of linear equations in two variables, first introduced in Chapter 3.

**OBJECTIVE 1 Plot ordered pairs.** Each of the pairs of numbers (3, 1), (−5, 6), and (4, −1) is an example of an **ordered pair;** that is, a pair of numbers written within parentheses in which the order of the numbers is important. We graph an ordered pair using two perpendicular number lines that intersect at their 0 points, as shown in Figure 1. The common 0 point is called the **origin.** The position of any point in this plane is determined by referring to the horizontal number line, the ***x*-axis,** and the vertical number line, the ***y*-axis.** The first number in the ordered pair indicates the position relative to the *x*-axis, and the second number indicates the position relative to the *y*-axis. The *x*-axis and the *y*-axis make up a **rectangular** (or **Cartesian,** for Descartes) **coordinate system.**

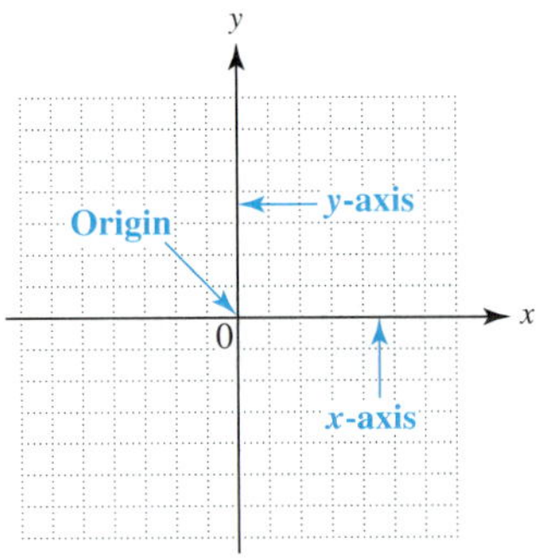


FIGURE 1

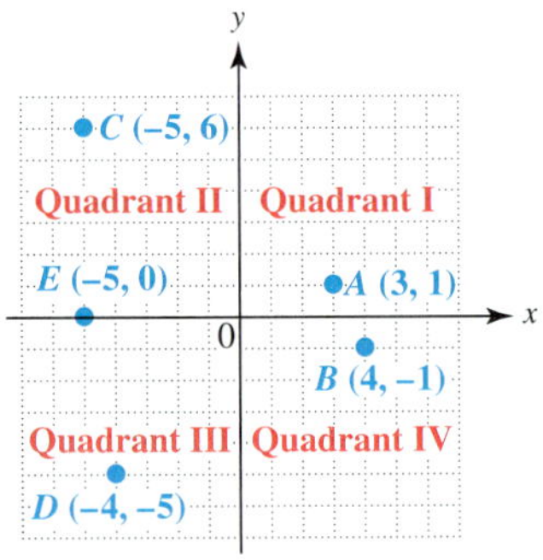


FIGURE 2

To locate, or **plot,** the point on the graph that corresponds to the ordered pair (3, 1), we move three units from 0 to the right along the *x*-axis, and then one unit up parallel to the *y*-axis. The point corresponding to the ordered pair (3, 1) is labeled *A* in Figure 2. Additional points are labeled *B*–*E*. The phrase "the point corresponding to the ordered pair (3, 1)" is often abbreviated as "the point (3, 1)." The numbers in an ordered pair are called the **coordinates** of the corresponding point.

The four regions of the graph, shown in Figure 2, are called **quadrants I, II, III,** and **IV,** reading counterclockwise from the upper right quadrant. The points on the *x*-axis and *y*-axis do not belong to any quadrant. For example, point *E* in Figure 2 belongs to no quadrant.

**Now Try Exercises 3, 5, 7, and 9.**

**OBJECTIVE 2 Graph lines and find intercepts.** Each solution to an equation with two variables, such as $2x + 3y = 6$, includes two numbers, one for each variable. To keep track of which number goes with which variable, we write the solutions as ordered pairs. (If *x* and *y* are used as the variables, the *x*-value is given first.) For example, we can show that (6, −2) is a solution of $2x + 3y = 6$ by substitution.

$$2x + 3y = 6$$
$$2(6) + 3(-2) \stackrel{?}{=} 6 \qquad \text{Let } x = 6,\ y = -2.$$
$$12 - 6 \stackrel{?}{=} 6$$
$$6 = 6 \qquad \text{True}$$

Because the ordered pair $(6, -2)$ makes the equation true, it is a solution. On the other hand, $(5, 1)$ is *not* a solution of the equation $2x + 3y = 6$ because

$$2(5) + 3(1) = 10 + 3 = 13 \neq 6.$$

To find ordered pairs that satisfy an equation, select any number for one of the variables, substitute it into the equation for that variable, and then solve for the other variable. Two other ordered pairs satisfying $2x + 3y = 6$ are $(0, 2)$ and $(3, 0)$. Since any real number could be selected for one variable and would lead to a real number for the other variable, linear equations in two variables have an infinite number of ordered-pair solutions.

How might we express the solution set of an equation like $2x + 3y = 6$? The graph of an equation is the set of points corresponding to *all* ordered pairs that satisfy the equation. It gives a "picture" of the equation. The graph of the equation $2x + 3y = 6$ is shown in Figure 3 along with a table of ordered pairs.

| $x$ | $y$ |
|---|---|
| −3 | 4 |
| 0 | 2 |
| 3 | 0 |
| 6 | −2 |
| 9 | −4 |

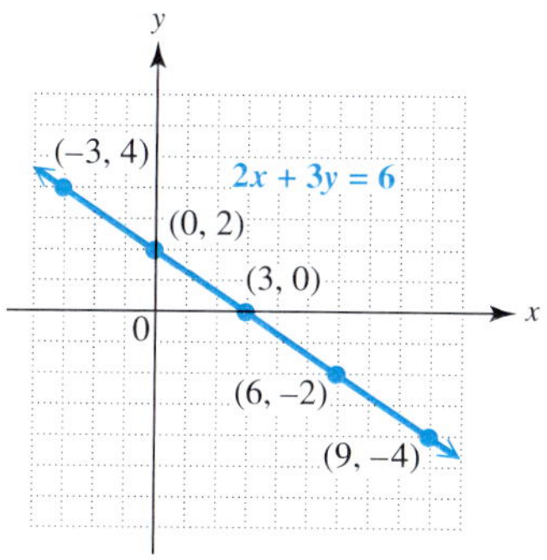


FIGURE 3

**Now Try Exercise 11.**

The equation $2x + 3y = 6$ is called a **first-degree equation** because it has no term with a variable to a power greater than 1.

**The graph of any first-degree equation in two variables is a straight line.**

Since first-degree equations with two variables have straight-line graphs, they are called *linear equations in two variables.*

**Linear Equation in Two Variables**

A **linear equation in two variables** can be written in the form

$$Ax + By = C,$$

where $A$, $B$, and $C$ are real numbers ($A$ and $B$ not both 0).

Thus, retail spending on prescription drugs increased by about \$19 billion per year. Using this slope, one of the points, say (7, 79), and the point-slope form, we obtain

$$y - y_1 = m(x - x_1) \qquad \text{Point-slope form}$$
$$y - 79 = 19(x - 7) \qquad (x_1, y_1) = (7, 79);\ m = 19$$
$$y - 79 = 19x - 133 \qquad \text{Distributive property}$$
$$y = 19x - 54. \qquad \text{Slope-intercept form}$$

Thus, retail spending $y$ (in billions of dollars) on prescription drugs in the United States in year $x$ can be approximated by the equation $y = 19x - 54$.

**(b)** Use the equation from part (a) to predict retail spending on prescription drugs in the United States in 2004. (Assume a constant rate of change.)

Since $x = 7$ represents 1997 and 2004 is 7 yr after 1997, $x = 14$ represents 2004. We substitute 14 for $x$ in the equation.

$$y = 19x - 54 = 19(14) - 54 = 212$$

According to the model, \$212 billion will be spent on prescription drugs in 2004.

**Now Try Exercise 81.**

## CONNECTIONS

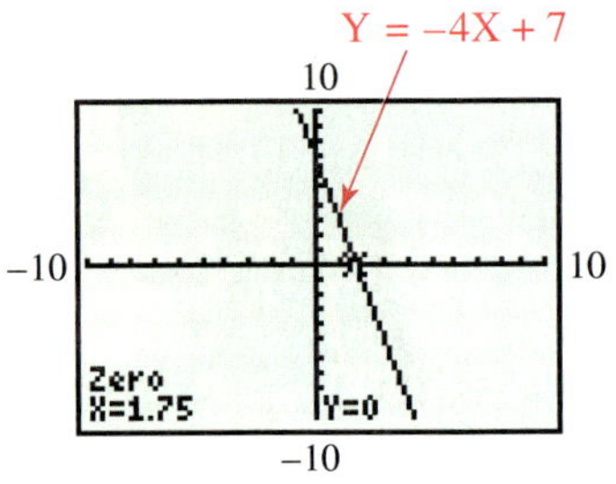


In the Connections box in Section 7.1, we graphed linear equations in two variables and located their intercepts using a graphing calculator. The top screen in the margin shows the graph of

$$y = -4x + 7.$$

From the values at the bottom of the screen, we see that when $x = 1.75$, $y = 0$. This means that $x = 1.75$ satisfies the equation

$$-4x + 7 = 0,$$

a linear equation in *one* variable. Therefore, the solution set of $-4x + 7 = 0$ is $\{1.75\}$. We can verify this algebraically by substitution. (Recall that the word "Zero" indicates that the $x$-intercept has been located.)

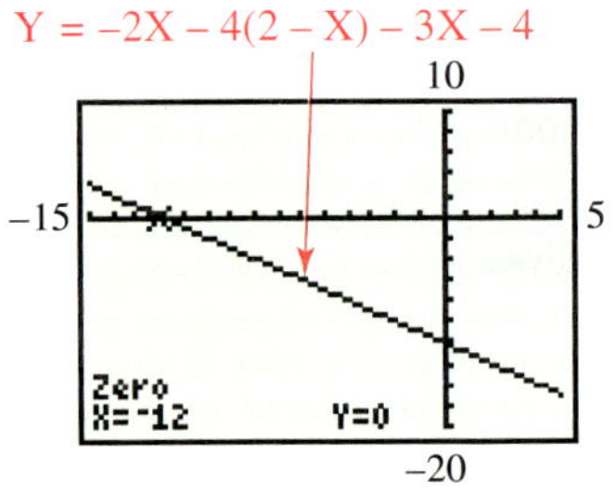


To solve $-2x - 4(2 - x) = 3x + 4$ using a graphing calculator, we must write the equation as an equivalent equation with 0 on one side.

$$-2x - 4(2 - x) - 3x - 4 = 0 \qquad \text{Subtract } 3x \text{ and } 4.$$

Then we graph

$$y = -2x - 4(2 - x) - 3x - 4$$

to find the $x$-intercept. The standard viewing window cannot be used because the $x$-intercept does not lie in the interval $[-10, 10]$. As seen in the bottom screen in the margin, the $x$-intercept of the graph is $(-12, 0)$, and thus the solution (or zero) of the equation is $-12$. The solution set is $\{-12\}$. Check this algebraically by substitution.

### For Discussion or Writing

Use a graphing calculator to solve each linear equation in one variable. Check solutions algebraically by substitution.

**1.** $2x + 7 - x = 4x - 2$

**2.** $7x - 2x + 4 - 5 = 3x + 1$

**3.** $4(x - 3) - x = x - 6$

**4.** $3(2x + 1) - 2(x - 2) = 5$

## 7.2 EXERCISES

**For Extra Help**

- Student's Solutions Manual
- MyMathLab
- InterAct Math Tutorial Software
- AW Math Tutor Center
- MathXL
- Digital Video Tutor CD 11/Videotape 11

**1.** The following equations all represent the same line. Which one is in standard form as defined in the text?

**A.** $3x - 2y = 5$ **B.** $2y = 3x - 5$ **C.** $\frac{3}{5}x - \frac{2}{5}y = 1$ **D.** $3x = 2y + 5$

**2.** Which equation is in point-slope form?

**A.** $y = 6x + 2$ **B.** $4x + y = 9$ **C.** $y - 3 = 2(x - 1)$ **D.** $2y = 3x - 7$

**3.** Which equation in Exercise 2 is in slope-intercept form?

**4.** Write the equation $y + 2 = -3(x - 4)$ in slope-intercept form.

**5.** Write the equation from Exercise 4 in standard form.

**6.** Write the equation $10x - 7y = 70$ in slope-intercept form.

*Match each equation with the graph that it most closely resembles. (Hint: Determine the signs of m and b to help you make your decision.)*

**7.** $y = 2x + 3$

**8.** $y = -2x + 3$

**9.** $y = -2x - 3$

**10.** $y = 2x - 3$

**11.** $y = 2x$

**12.** $y = -2x$

**13.** $y = 3$

**14.** $y = -3$

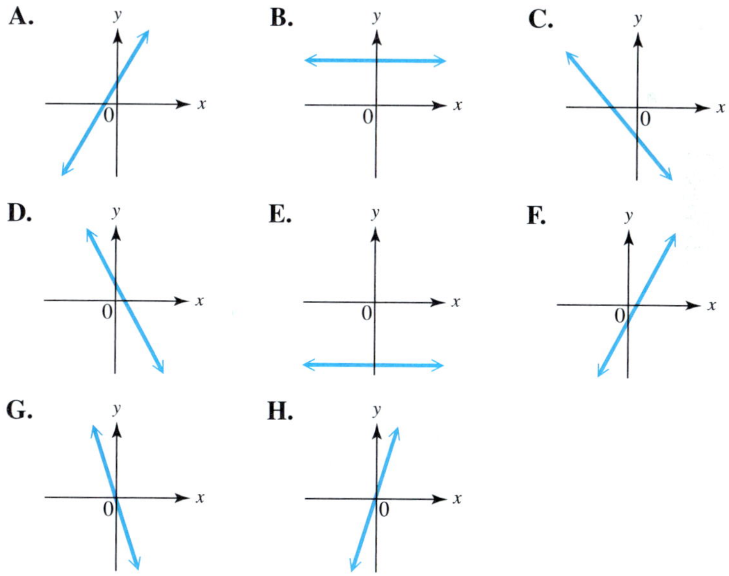


*Find the equation in slope-intercept form of the line satisfying the given conditions. See Example 1.*

**15.** $m = 5$; $b = 15$

**16.** $m = -2$; $b = 12$

**17.** $m = -\frac{2}{3}$; $b = \frac{4}{5}$

**18.** $m = -\frac{5}{8}$; $b = -\frac{1}{3}$

**19.** Slope $\frac{2}{5}$; $y$-intercept $(0, 5)$

**20.** Slope $-\frac{3}{4}$; $y$-intercept $(0, 7)$

*Write an equation in slope-intercept form of the line shown in each graph. (Hint: Use the indicated points to find the slope.)*

**21.**

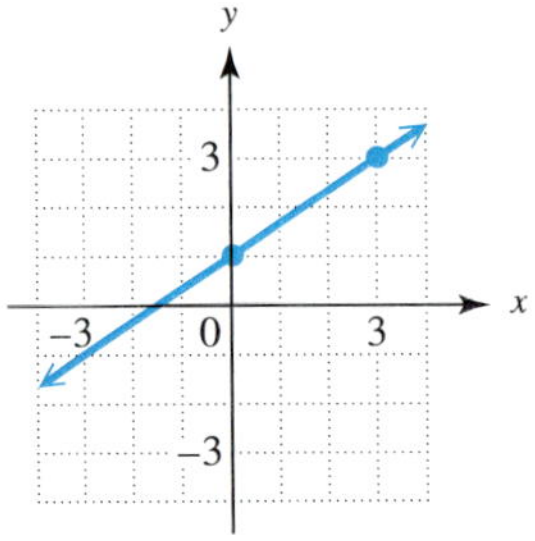


**22.**

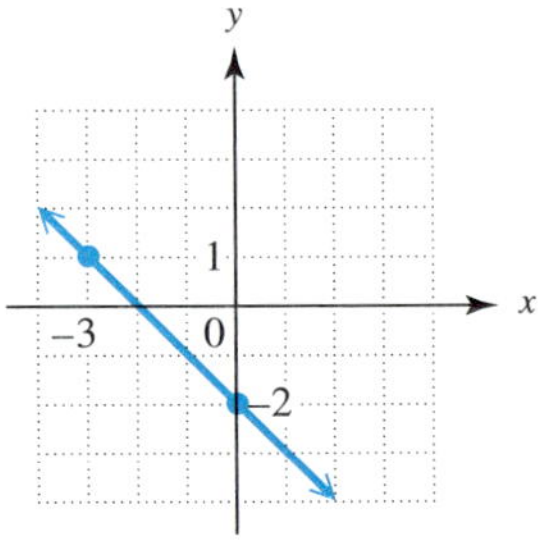


*For each equation, **(a)** write it in slope-intercept form, **(b)** give the slope of the line, **(c)** give the y-intercept, and **(d)** graph the line. See Example 2.*

**23.** $-x + y = 4$

**24.** $-x + y = 6$

**25.** $6x + 5y = 30$

**26.** $3x + 4y = 12$

**27.** $4x - 5y = 20$

**28.** $7x - 3y = 3$

**29.** $x + 2y = -4$

**30.** $x + 3y = -9$

*Find an equation of the line that satisfies the given conditions. Write the equation in standard form. See Example 3.*

**31.** Through $(-2, 4)$; slope $-\frac{3}{4}$

**32.** Through $(-1, 6)$; slope $-\frac{5}{6}$

**33.** Through $(5, 8)$; slope $-2$

**34.** Through $(12, 10)$; slope 1

**35.** Through $(-5, 4)$; slope $\frac{1}{2}$

**36.** Through $(7, -2)$; slope $\frac{1}{4}$

**37.** $x$-intercept $(3, 0)$; slope 4

**38.** $x$-intercept $(-2, 0)$; slope $-5$

**39.** In your own words, list all the forms of linear equations in two variables and describe when each form should be used.

**40.** Explain why the point-slope form of an equation cannot be used to find the equation of a vertical line.

*Write an equation of the line that satisfies the given conditions.*

**41.** Through $(9, 5)$; slope 0

**42.** Through $(-4, -2)$; slope 0

**43.** Through $(9, 10)$; undefined slope

**44.** Through $(-2, 8)$; undefined slope

**45.** Through $(.5, .2)$; vertical

**46.** Through $\left(\frac{5}{8}, \frac{2}{9}\right)$; vertical

**47.** Through $(-7, 8)$; horizontal

**48.** Through $(2, 7)$; horizontal

*Find an equation of the line passing through the given points. Write the equation in standard form. See Example 4.*

**49.** $(3, 4)$ and $(5, 8)$

**50.** $(5, -2)$ and $(-3, 14)$

**51.** $(6, 1)$ and $(-2, 5)$

**52.** $(-2, 5)$ and $(-8, 1)$

**53.** $\left(-\frac{2}{5}, \frac{2}{5}\right)$ and $\left(\frac{4}{3}, \frac{2}{3}\right)$

**54.** $\left(\frac{3}{4}, \frac{8}{3}\right)$ and $\left(\frac{2}{5}, \frac{2}{3}\right)$

**55.** $(2, 5)$ and $(1, 5)$

**56.** $(-2, 2)$ and $(4, 2)$

**57.** (7, 6) and (7, −8)

**58.** (13, 5) and (13, −1)

**59.** (1, −3) and (−1, −3)

**60.** (−4, −6) and (5, −6)

*Find an equation of the line satisfying the given conditions. Write the equation in slope-intercept form. See Example 5.*

**61.** Through (7, 2); parallel to $3x - y = 8$

**62.** Through (4, 1); parallel to $2x + 5y = 10$

**63.** Through (−2, −2); parallel to $-x + 2y = 10$

**64.** Through (−1, 3); parallel to $-x + 3y = 12$

**65.** Through (8, 5); perpendicular to $2x - y = 7$

**66.** Through (2, −7); perpendicular to $5x + 2y = 18$

**67.** Through (−2, 7); perpendicular to $x = 9$

**68.** Through (8, 4); perpendicular to $x = -3$

*Write an equation in the form $y = mx$ for each situation. Then give the three ordered pairs associated with the equation for x-values 0, 5, and 10. See Example 6(a).*

**69.** $x$ represents the number of hours traveling at 45 mph, and $y$ represents the distance traveled (in miles).

**70.** $x$ represents the number of compact discs sold at $16 each, and $y$ represents the total cost of the discs (in dollars).

**71.** $x$ represents the number of gallons of gas sold at $1.50 per gal, and $y$ represents the total cost of the gasoline (in dollars).

**72.** $x$ represents the number of days a videocassette is rented at $3.50 per day, and $y$ represents the total charge for the rental (in dollars).

*For each situation, **(a)** write an equation in the form $y = mx + b$; **(b)** find and interpret the ordered pair associated with the equation for $x = 5$; and **(c)** answer the question. See Examples 6(b) and 6(c).*

**73.** A membership to the Midwest Athletic Club costs $99 plus $39 per month. (*Source:* Midwest Athletic Club.) Let $x$ represent the number of months selected. How much does the first year's membership cost?

**74.** For a family membership, the athletic club in Exercise 73 charges a membership fee of $159 plus $60 for each additional family member after the first. Let $x$ represent the number of additional family members. What is the membership fee for a four-person family?

**75.** A cell phone plan includes 900 anytime minutes for $50 per month, plus a one-time activation fee of $25. A Nokia 5165 cell phone is included at no additional charge. (*Source:* U.S. Cellular.) Let $x$ represent the number of months of service. If you sign a 2-yr contract, how much will this cell phone plan cost? (Assume that you never use more than the allotted number of minutes.)

**76.** Another cell phone plan includes 450 anytime minutes for \$35 per month, plus \$19.95 for a Nokia 5165 cell phone and \$25 for a one-time activation fee. (*Source:* U.S. Cellular.) Let $x$ represent the number of months of service. If you sign a 1-yr contract, how much will this cell phone package cost? (Assume that you never use more than the allotted number of minutes.)

**77.** A rental car costs \$50 plus \$.20 per mile. Let $x$ represent the number of miles driven, and $y$ represent the total charge to the renter. How many miles was the car driven if the renter paid \$84.60?

**78.** There is a \$30 fee to rent a chain saw, plus \$6 per day. Let $x$ represent the number of days the saw is rented and $y$ represent the charge to the user in dollars. If the total charge is \$138, for how many days is the saw rented?

*Solve each problem. In part (a), give equations in slope-intercept form. See Examples 7 and 8.* (*Source for Exercises 79 and 80:* Jupiter Media Metrix.)

**79.** The percent of households that access the Internet by high-speed broadband is shown in the graph, where the year 2000 corresponds to $x = 0$.

**(a)** Use the ordered pairs from the graph to write an equation that models the data. What does the slope tell us in the context of this problem?

**(b)** Use the equation from part (a) to predict the percent of U.S. households that will access the Internet by broadband in 2006. Round your answer to the nearest percent.

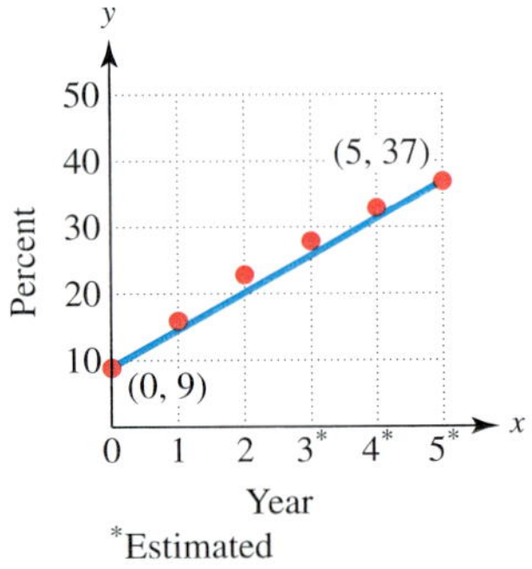


**80.** The percent of U.S. households that access the Internet by dial-up is shown in the graph, where the year 2000 corresponds to $x = 0$.

**(a)** Use the ordered pairs from the graph to write an equation that models the data. What does the slope tell us in the context of this problem?

**(b)** Use the equation from part (a) to predict the percent of U.S. households that will access the Internet by dial-up in 2006. Round your answer to the nearest percent.

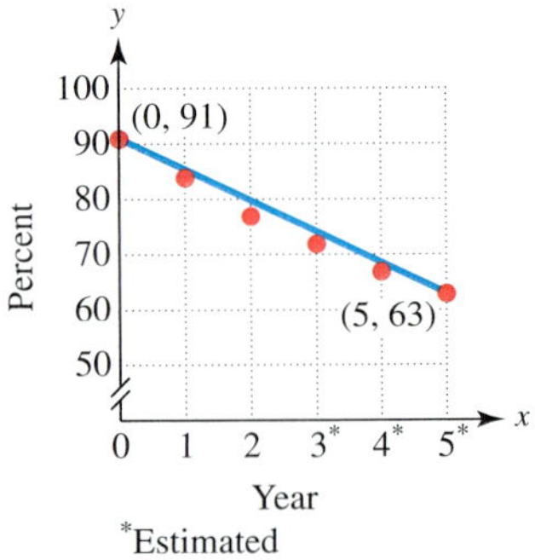


**81.** The number of post offices in the United States is shown in the bar graph.

**(a)** Use the information given for the years 1995 and 2000, letting $x = 5$ represent 1995, $x = 10$ represent 2000, and $y$ represent the number of post offices, to write an equation that models the data.

**(b)** Use the equation to approximate the number of post offices in 1998. How does this result compare to the actual value, 27,952?

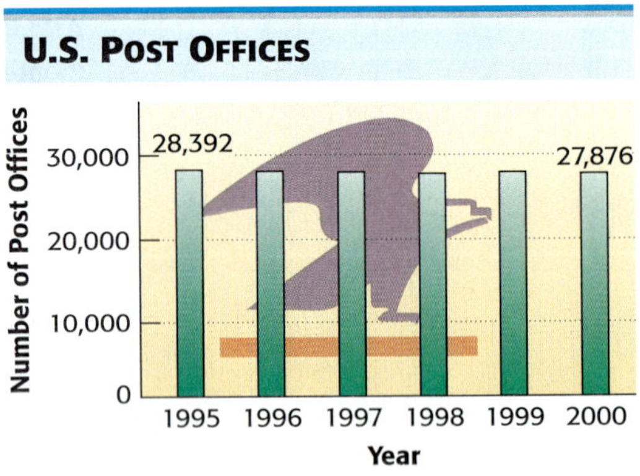


*Source:* U.S. Postal Service, *Annual Report of the Postmaster General.*

**82.** Median household income of African-Americans is shown in the bar graph.

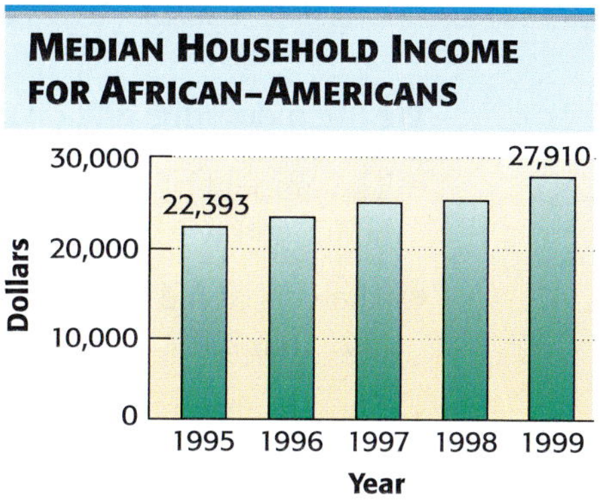


*Source:* U.S. Bureau of the Census.

**(a)** Use the information given for the years 1995 and 1999, letting $x = 5$ represent 1995, $x = 9$ represent 1999, and $y$ represent the median income, to write an equation that models median household income.

**(b)** Use the equation to approximate the median income for 1997. How does your result compare to the actual value, \$25,050?

## RELATING CONCEPTS (EXERCISES 83–88)

**For Individual or Group Work**

*In Section 2.5 we learned how formulas can be applied to problem solving.* ***Work Exercises 83–88 in order,*** *to see how the formula that relates Celsius and Fahrenheit temperatures is derived.*

**83.** There is a linear relationship between Celsius and Fahrenheit temperatures. When $C = 0°$, $F =$ _____°, and when $C = 100°$, $F =$ _____°.

**84.** Think of ordered pairs of temperatures $(C, F)$, where $C$ and $F$ represent corresponding Celsius and Fahrenheit temperatures. The equation that relates the two scales has a straight-line graph that contains the two points determined in Exercise 83. What are these two points?

**85.** Find the slope of the line described in Exercise 84.

**86.** Now think of the point-slope form of the equation in terms of $C$ and $F$, where $C$ replaces $x$ and $F$ replaces $y$. Use the slope you found in Exercise 85 and one of the two points determined earlier, and find the equation that gives $F$ in terms of $C$.

**87.** To obtain another form of the formula, use the equation you found in Exercise 86 and solve for $C$ in terms of $F$.

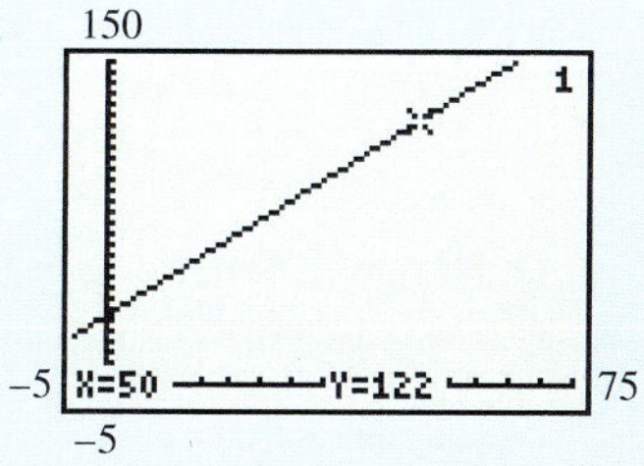


**88.** The equation found in Exercise 86 is graphed on the graphing calculator screen shown here. Interpret the display at the bottom, in the context of this group of exercises.

# 7.3 Functions

**OBJECTIVES**

1 Define and identify relations and functions.
2 Find domain and range.
3 Identify functions defined by graphs and equations.
4 Use function notation.
5 Identify linear functions.

We often describe one quantity in terms of another. Consider the following.

- The amount of your paycheck if you are paid hourly depends on the number of hours you worked.
- The cost at the gas station depends on the number of gallons of gas you pumped into your car.
- The distance traveled by a car moving at a constant speed depends on the time traveled.

We can use ordered pairs to represent these corresponding quantities. For example, we indicate the relationship between the amount of your paycheck and hours worked by writing ordered pairs in which the first number represents hours worked and the second number represents paycheck amount in dollars. Then the ordered pair (5, 40) indicates that when you work 5 hr, your paycheck is \$40. Similarly, the ordered pairs (10, 80) and (20, 160) show that working 10 hr results in an \$80 paycheck and working 20 hr results in a \$160 paycheck. In this example, what would the ordered pair (40, 320) indicate?

Since the amount of your paycheck *depends* on the number of hours worked, your paycheck amount is called the *dependent variable,* and the number of hours worked is called the *independent variable.* Generalizing, if the value of the variable $y$ depends on the value of the variable $x$, then $y$ is the **dependent variable** and $x$ is the **independent variable.**

Independent variable ↓ ↓ Dependent variable

$$(x, y)$$

**OBJECTIVE 1 Define and identify relations and functions.** Since we can write related quantities using ordered pairs, a set of ordered pairs such as

$$\{(5, 40), (10, 80), (20, 160), (40, 320)\}$$

is called a *relation.*

**Relation**

A **relation** is a set of ordered pairs.

A special kind of relation, called a *function,* is very important in mathematics and its applications.

**Function**

A **function** is a relation in which, for each value of the first component of the ordered pairs, there is *exactly one value* of the second component.

## EXAMPLE 1 Determining Whether Relations Are Functions

Tell whether each relation defines a function.

$$F = \{(1, 2), (-2, 4), (3, -1)\}$$
$$G = \{(-2, -1), (-1, 0), (0, 1), (1, 2), (2, 2)\}$$
$$H = \{(-4, 1), (-2, 1), (-2, 0)\}$$

Relations $F$ and $G$ are functions, because for each different $x$-value there is exactly one $y$-value. Notice that in $G$, the last two ordered pairs have the same $y$-value (1 is paired with 2, and 2 is paired with 2). This does not violate the definition of function, since the first components ($x$-values) are different and each is paired with only one second component ($y$-value).

In relation $H$, however, the last two ordered pairs have the *same* $x$-value paired with *two different* $y$-values ($-2$ is paired with both 1 and 0), so $H$ is a relation but not a function. ***In a function, no two ordered pairs can have the same first component and different second components.***

Different $y$-values

$$H = \{(-4, 1), (-2, 1), (-2, 0)\}$$ Not a function

Same $x$-value

**Now Try Exercises 5 and 7.**

In a function, there is *exactly one* value of the dependent variable, the second component, for each value of the independent variable, the first component. This is what makes functions so important in applications.

> **NOTE** The relation from the beginning of this section representing hours worked and corresponding paycheck amount is a function since each $x$-value is paired with exactly one $y$-value. You would not be happy, for example, if you and a coworker each worked 20 hr at the same hourly rate and your paycheck was \$160 while his was \$200.

Relations and functions can also be expressed as a correspondence or *mapping* from one set to another, as shown in Figure 25 for function $F$ and relation $H$ from Example 1. The arrow from 1 to 2 indicates that the ordered pair (1, 2) belongs to $F$—each first component is paired with exactly one second component. In the mapping for set $H$, which is not a function, the first component $-2$ is paired with two different second components, 1 and 0.

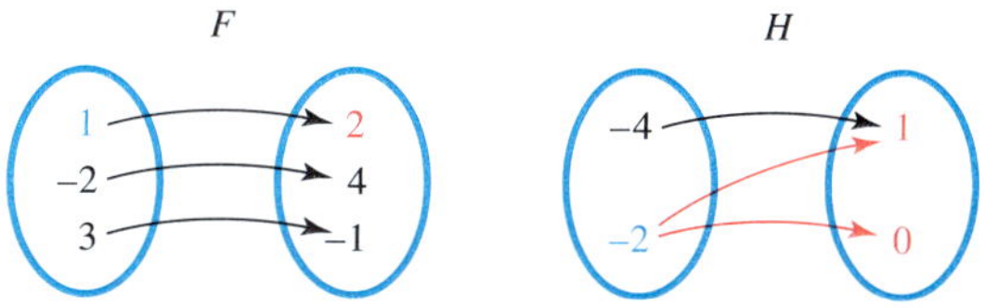


*F* is a function. *H* is not a function.

**FIGURE 25**

Since relations and functions are sets of ordered pairs, we can represent them using tables and graphs. A table and graph for function $F$ from Example 1 is shown in Figure 26.

| $x$ | $y$ |
|---|---|
| 1 | 2 |
| −2 | 4 |
| 3 | −1 |

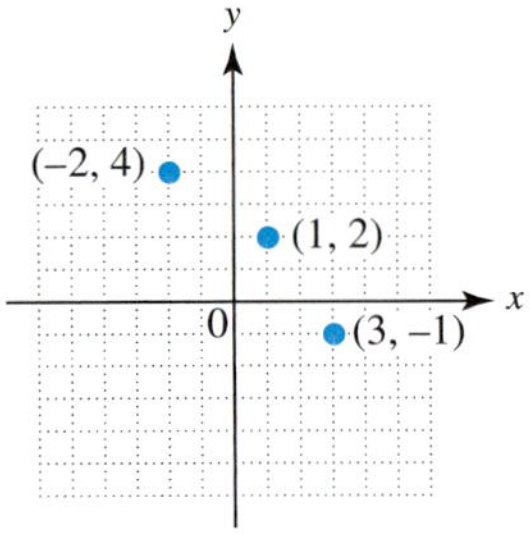


Graph of $F$

**FIGURE 26**

Finally, we can describe a relation or function using a rule that tells how to determine the dependent variable for a specific value of the independent variable. The rule may be given in words: the dependent variable is twice the independent variable. Usually the rule is an equation:

$$y = 2x.$$

$y$ ↑ Dependent variable; $x$ ↑ Independent variable

This is the most efficient way to define a relation or function.

**NOTE** Another way to think of a function relationship is to think of the independent variable as an input and the dependent variable as an output. This is illustrated by the input-output (function) machine for the function defined by $y = 2x$.

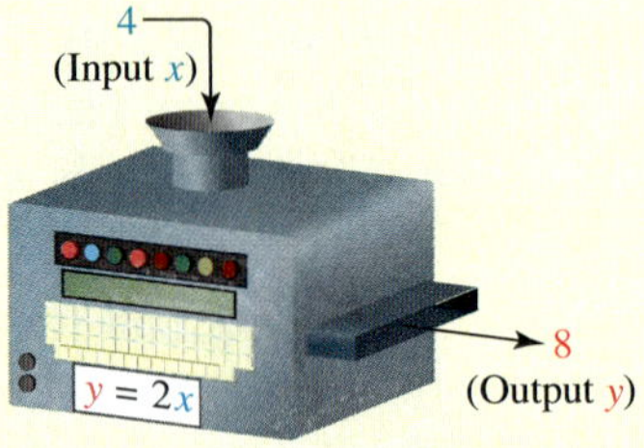


Function machine

**OBJECTIVE 2 Find domain and range.**

**Domain and Range**

In a relation, the set of all values of the independent variable ($x$) is the **domain;** the set of all values of the dependent variable ($y$) is the **range.**

## EXAMPLE 2 Finding Domains and Ranges of Relations

Give the domain and range of each relation. Tell whether the relation defines a function.

**(a)** $\{(3, -1), (4, 2), (4, 5), (6, 8)\}$

The domain, the set of $x$-values, is $\{3, 4, 6\}$; the range, the set of $y$-values, is $\{-1, 2, 5, 8\}$. This relation is not a function because the same $x$-value 4 is paired with two different $y$-values, 2 and 5.

**(b)**

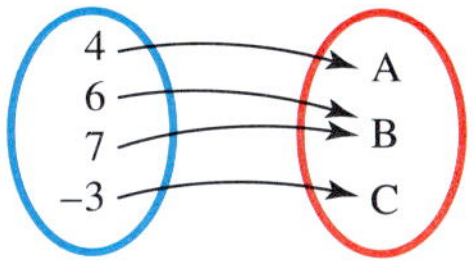


The domain of this relation is

$$\{4, 6, 7, -3\};$$

the range is

$$\{A, B, C\}.$$

This mapping defines a function—each $x$-value corresponds to exactly one $y$-value.

**(c)**

| $x$ | $y$ |
|---|---|
| $-5$ | 2 |
| 0 | 2 |
| 5 | 2 |

This is a table of ordered pairs, so the domain is the set of $x$-values $\{-5, 0, 5\}$ and the range is the set of $y$-values $\{2\}$. The table defines a function because each different $x$-value corresponds to exactly one $y$-value (even though it is the same $y$-value).

**Now Try Exercises 11, 13, and 15.**

As mentioned previously, the graph of a relation is the graph of its ordered pairs. The graph gives a picture of the relation, which can be used to determine its domain and range.

## EXAMPLE 3 Finding Domains and Ranges from Graphs

Give the domain and range of each relation.

**(a)**

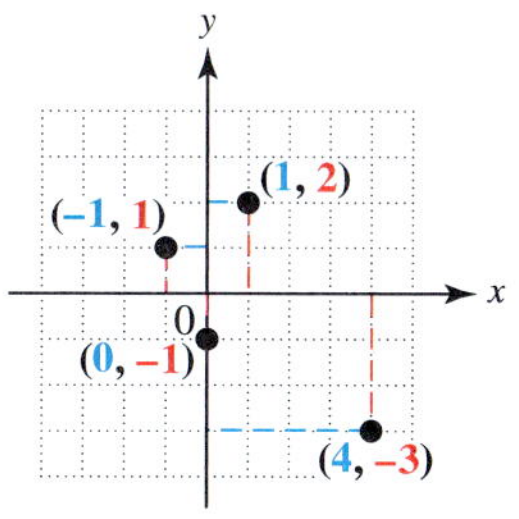


The domain is the set of $x$-values,

$$\{-1, 0, 1, 4\}.$$

The range is the set of $y$-values,

$$\{-3, -1, 1, 2\}.$$

**(b)**

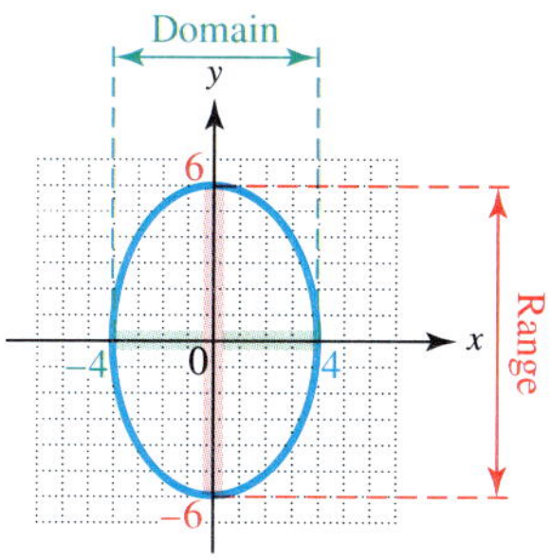


The $x$-values of the points on the graph include all numbers between $-4$ and 4, inclusive. The $y$-values include all numbers between $-6$ and 6, inclusive. Using interval notation,

the domain is $[-4, 4]$;
the range is $[-6, 6]$.

**(c)**

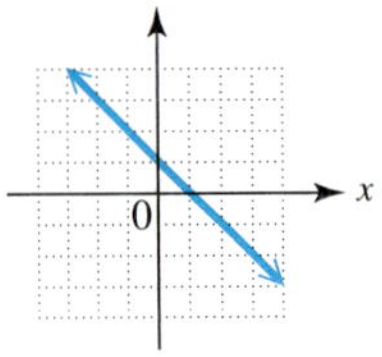


The arrowheads indicate that the line extends indefinitely left and right, as well as up and down. Therefore, both the domain and the range include all real numbers, written $(-\infty, \infty)$.

**(d)**

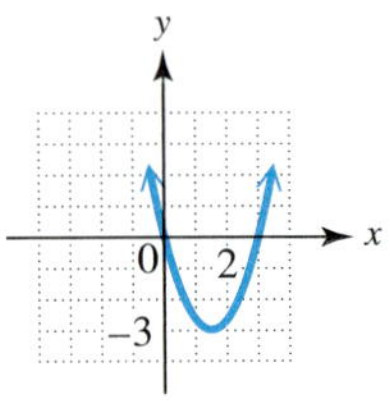


The arrowheads indicate that the graph extends indefinitely left and right, as well as upward. The domain is $(-\infty, \infty)$. Because there is a least $y$-value, $-3$, the range includes all numbers greater than or equal to $-3$, written $[-3, \infty)$.

**Now Try Exercises 17 and 19.**

Since relations are often defined by equations, such as $y = 2x + 3$ and $y^2 = x$, we must sometimes determine the domain of a relation from its equation. In this book, we assume the following agreement on the domain of a relation.

**Agreement on Domain**

Unless specified otherwise, the domain of a relation is assumed to be all real numbers that produce real numbers when substituted for the independent variable.

To illustrate this agreement, since any real number can be used as a replacement for $x$ in $y = 2x + 3$, the domain of this function is the set of all real numbers. As another example, the function defined by $y = \frac{1}{x}$ has all real numbers except 0 as domain, since $y$ is undefined if $x = 0$. In general, the domain of a function defined by an algebraic expression is all real numbers, except those numbers that lead to division by 0 or an even root of a negative number.

**OBJECTIVE 3 Identify functions defined by graphs and equations.** Most of the relations we have seen in the examples are functions—that is, each $x$-value corresponds to exactly one $y$-value. Since each value of $x$ leads to only one value of $y$ in a function, any vertical line drawn through the graph of a function must intersect the graph in at most one point. This is the *vertical line test* for a function.

**Vertical Line Test**

If every vertical line intersects the graph of a relation in no more than one point, then the relation represents a function.

For example, the graph shown in Figure 27(a) is not the graph of a function since a vertical line intersects the graph in more than one point. The graph in Figure 27(b) does represent a function.

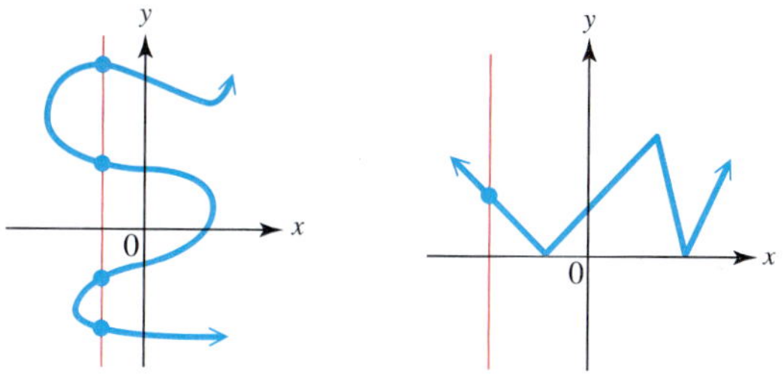


**Not a function**–the same $x$-value corresponds to four different $y$-values.

**(a)**

**Function**–each $x$-value corresponds to only one $y$-value.

**(b)**

**FIGURE 27**

## EXAMPLE 4 Using the Vertical Line Test

Use the vertical line test to determine whether each relation graphed in Example 3 is a function.

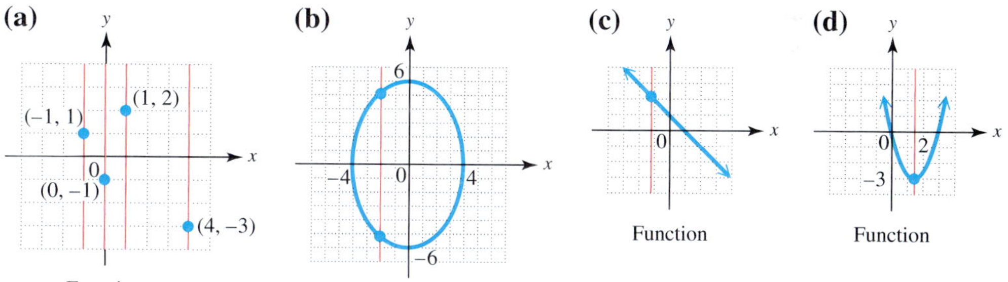


The graphs in (a), (c), and (d) represent functions. The graph of the relation in (b) fails the vertical line test, since the same $x$-value corresponds to two different $y$-values; therefore, it is not the graph of a function.

**Now Try Exercise 21.**

**NOTE** Graphs that do not represent functions are still relations. Remember that all equations and graphs represent relations and that all relations have a domain and range.

The vertical line test is a simple method for identifying a function defined by a graph. It is more difficult to decide whether a relation defined by an equation is a function. The next example gives some hints that may help.

## EXAMPLE 5 Identifying Functions from Their Equations

Decide whether each relation defines a function and give the domain.

**(a)** $y = x + 4$

In the defining equation (or rule), $y = x + 4$, $y$ is always found by adding 4 to $x$. Thus, each value of $x$ corresponds to just one value of $y$ and the relation defines a function; $x$ can be any real number, so the domain is $\{x \mid x \text{ is a real number}\}$ or $(-\infty, \infty)$.

**(b)** $y^2 = x$

The ordered pairs (16, 4) and (16, −4) both satisfy this equation. Since one value of $x$, 16, corresponds to two values of $y$, 4 and −4, this equation does not define a function. Because $x$ is equal to the square of $y$, the values of $x$ must always be nonnegative. The domain of the relation is $[0, \infty)$.

**(c)** $y = \dfrac{5}{x - 1}$

Given any value of $x$ in the domain, we find $y$ by subtracting 1, then dividing the result into 5. This process produces exactly one value of $y$ for each value in the domain, so this equation defines a function. The domain includes all real numbers except those that make the denominator 0. We find these numbers by setting the denominator equal to 0 and solving for $x$.

$$x - 1 = 0$$
$$x = 1$$

Thus, the domain includes all real numbers except 1. In interval notation this is written as $(-\infty, 1) \cup (1, \infty)$.

**Now Try Exercises 25, 29, and 33.**

In summary, three variations of the definition of function are given here.

**Variations of the Definition of Function**

1. A **function** is a relation in which, for each value of the first component of the ordered pairs, there is exactly one value of the second component.
2. A **function** is a set of ordered pairs in which no first component is repeated.
3. A **function** is a rule or correspondence that assigns exactly one range value to each domain value.

**OBJECTIVE 4 Use function notation.** When a function $f$ is defined with a rule or an equation using $x$ and $y$ for the independent and dependent variables, we say "$y$ is a function of $x$" to emphasize that $y$ *depends on* $x$. We use the notation

$$y = f(x),$$

called **function notation,** to express this and read $f(x)$ as "$f$ of $x$." (In this special notation the parentheses do not indicate multiplication.) The letter $f$ stands for *function.* For example, if $y = 9x - 5$, we can name this function $f$ and write

$$f(x) = 9x - 5.$$

Note that ***$f(x)$ is just another name for the dependent variable y.*** For example, if $y = f(x) = 9x - 5$ and $x = 2$, then we find $y$, or $f(2)$, by replacing $x$ with 2.

$$\begin{aligned} y &= f(2) \\ &= 9 \cdot 2 - 5 \\ &= 18 - 5 \\ &= 13. \end{aligned}$$

The statement "if $x = 2$, then $y = 13$" represents the ordered pair (2, 13) and is abbreviated with function notation as

$$f(2) = 13.$$

Read $f(2)$ as "$f$ of 2" or "$f$ at 2." Also,

$$f(0) = 9 \cdot 0 - 5 = -5 \quad \text{and} \quad f(-3) = 9(-3) - 5 = -32.$$

These ideas and the symbols used to represent them can be illustrated as follows.

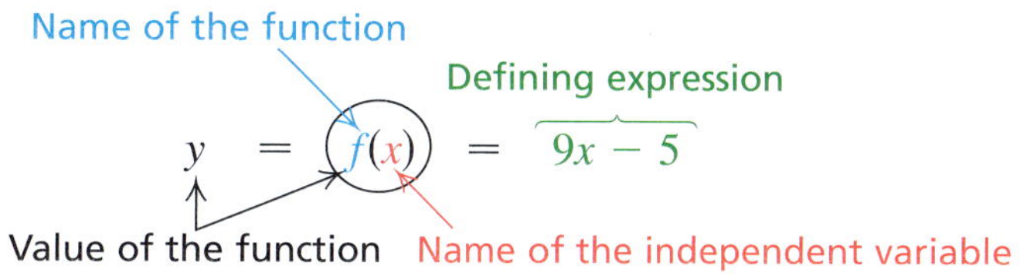


**CAUTION** The symbol $f(x)$ *does not* indicate "$f$ times $x$," but represents the $y$-value for the indicated $x$-value. As just shown, $f(2)$ is the $y$-value that corresponds to the $x$-value 2.

### EXAMPLE 6 Using Function Notation

Let $f(x) = -x^2 + 5x - 3$. Find the following.

**(a)** $f(2)$

Replace $x$ with 2.

$$\begin{aligned} f(x) &= -x^2 + 5x - 3 \\ f(2) &= -2^2 + 5 \cdot 2 - 3 \\ &= -4 + 10 - 3 \\ &= 3 \end{aligned}$$

Thus, $f(2) = 3$; the ordered pair (2, 3) belongs to $f$.

**(b)** $f(q)$

$$\begin{aligned} f(x) &= -x^2 + 5x - 3 \\ f(q) &= -q^2 + 5q - 3 \qquad \text{Replace } x \text{ with } q. \end{aligned}$$

The replacement of one variable with another is important in later courses.

**Now Try Exercises 41 and 45.**

Sometimes letters other than $f$, such as $g$, $h$, or capital letters $F$, $G$, and $H$ are used to name functions.

### EXAMPLE 7 Using Function Notation

Let $g(x) = 2x + 3$. Find and simplify $g(a + 1)$.

$$\begin{aligned} g(x) &= 2x + 3 \\ g(a + 1) &= 2(a + 1) + 3 \qquad \text{Replace } x \text{ with } a + 1. \\ &= 2a + 2 + 3 \\ &= 2a + 5 \end{aligned}$$

**Now Try Exercise 49.**

Functions can be evaluated in a variety of ways, as shown in Example 8.

## EXAMPLE 8 Using Function Notation

For each function, find $f(3)$.

**(a)** $f(x) = 3x - 7$

$f(3) = 3(3) - 7$ Replace $x$ with 3.

$f(3) = 2$

**(b)** $f = \{(-3, 5), (0, 3), (3, 1), (6, -1)\}$

We want $f(3)$, the $y$-value of the ordered pair where $x = 3$. As indicated by the ordered pair $(3, 1)$, when $x = 3$, $y = 1$, so $f(3) = 1$.

**(c)**

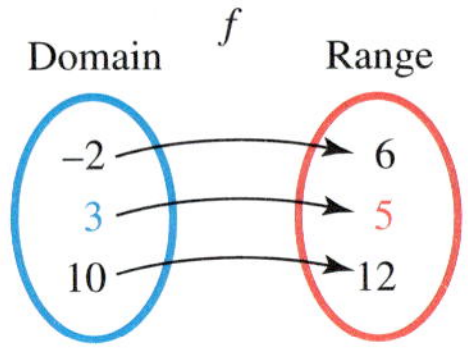


The domain element 3 is paired with 5 in the range, so $f(3) = 5$.

**(d)**

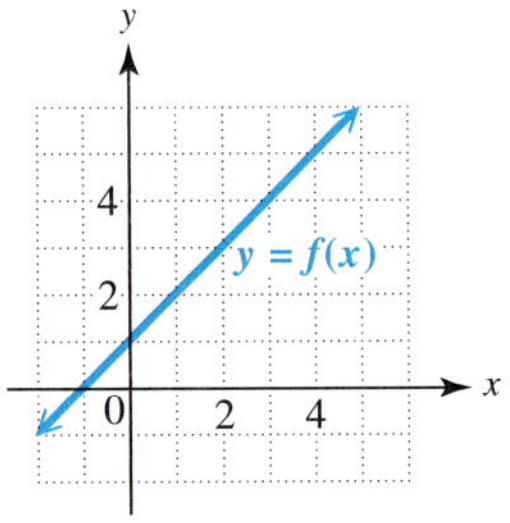


To evaluate $f(3)$, find 3 on the $x$-axis. See Figure 28. Then move up until the graph of $f$ is reached. Moving horizontally to the $y$-axis gives 4 for the corresponding $y$-value. Thus, $f(3) = 4$.

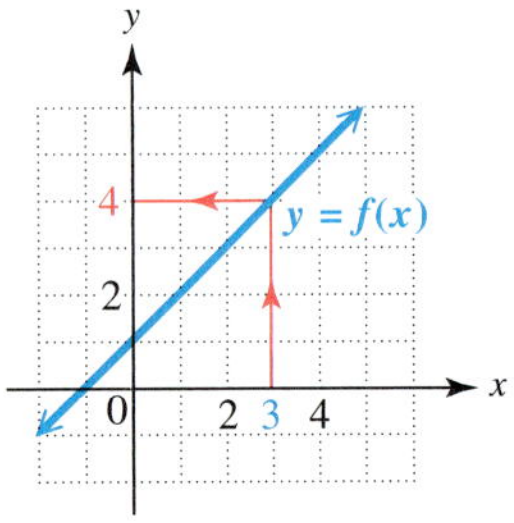


**FIGURE 28**

**Now Try Exercises 53, 55, and 57.**

If a function $f$ is defined by an equation with $x$ and $y$, not with function notation, use the following steps to find $f(x)$.

**Finding an Expression for $f(x)$**

*Step 1* Solve the equation for $y$.

*Step 2* Replace $y$ with $f(x)$.

**EXAMPLE 9 Writing Equations Using Function Notation**

Rewrite each equation using function notation. Then find $f(-2)$ and $f(a)$.

**(a)** $y = x^2 + 1$

This equation is already solved for $y$. Since $y = f(x)$,

$$f(x) = x^2 + 1.$$

To find $f(-2)$, let $x = -2$.

$$\begin{aligned} f(-2) &= (-2)^2 + 1 \\ &= 4 + 1 \\ &= 5 \end{aligned}$$

Find $f(a)$ by letting $x = a$: $\quad f(a) = a^2 + 1.$

**(b)** $x - 4y = 5$

First solve $x - 4y = 5$ for $y$. Then replace $y$ with $f(x)$.

$$\begin{aligned} x - 4y &= 5 \\ x - 5 &= 4y \\ y &= \frac{x-5}{4} \quad \text{so} \quad f(x) = \frac{1}{4}x - \frac{5}{4} \end{aligned}$$

Now find $f(-2)$ and $f(a)$.

$$f(-2) = \frac{1}{4}(-2) - \frac{5}{4} = -\frac{7}{4} \qquad \text{Let } x = -2.$$

$$f(a) = \frac{1}{4}a - \frac{5}{4} \qquad \text{Let } x = a.$$

**Now Try Exercise 59.**

**OBJECTIVE 5 Identify linear functions.** Our first two-dimensional graphing was of straight lines. Linear equations (except for vertical lines with equations $x = a$) define *linear functions.*

**Linear Function**

A function that can be defined by

$$f(x) = mx + b$$

for real numbers $m$ and $b$ is a **linear function.**

Recall that $m$ is the slope of the line and $(0, b)$ is the $y$-intercept. In Example 9(b), we wrote the equation $x - 4y = 5$ as the linear function defined by

$$f(x) = \frac{1}{4}x - \frac{5}{4}.$$

Slope ↑ $\frac{1}{4}$; ↑ $y$-intercept is $(0, -\frac{5}{4})$.

To graph this function, plot the $y$-intercept and use the definition of slope as $\frac{\text{rise}}{\text{run}}$ to find a second point on the line. Draw the straight line through the points to obtain the graph. See Figure 29.

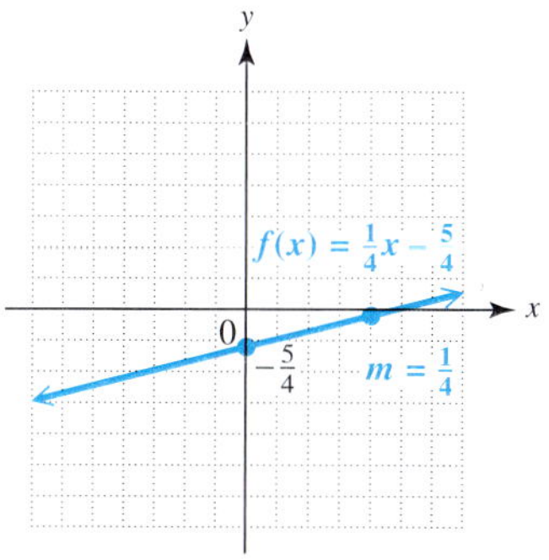


FIGURE 29

**86.** Refer to the graph to answer the questions.

**(a)** What numbers are possible values of the independent variable? the dependent variable?

**(b)** For how long is the water level increasing? decreasing?

**(c)** How many gallons of water are in the pool after 90 hr?

**(d)** Call this function $f$. What is $f(0)$? What does it mean?

**(e)** What is $f(25)$? What does it mean?

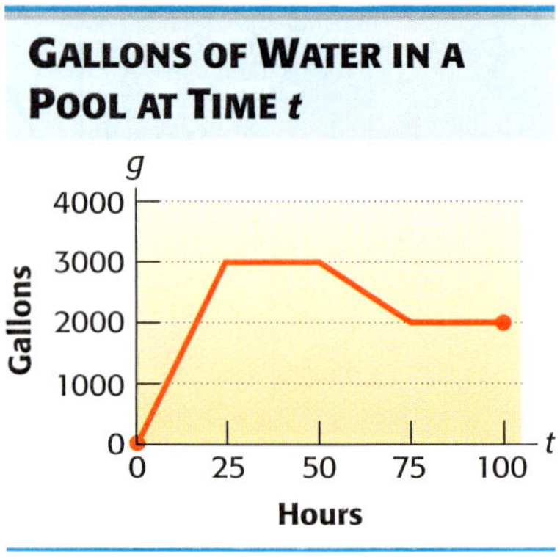


**TECHNOLOGY INSIGHTS** (EXERCISES 87 AND 88)

**87.** The calculator screen shows the graph of a linear function $y = f(x)$, along with the display of coordinates of a point on the graph. Use function notation to write what the display indicates.

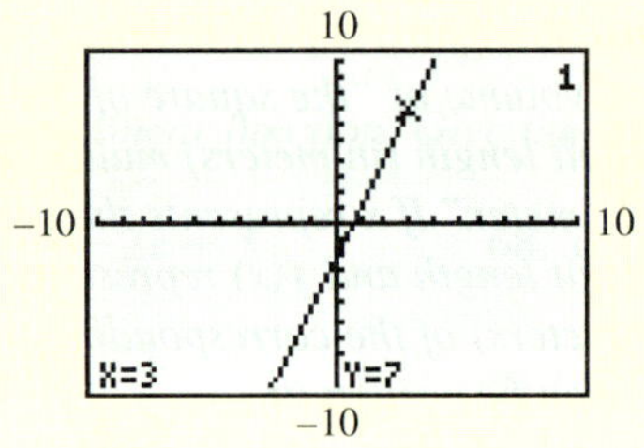


**88.** The table was generated by a graphing calculator for a linear function $Y_1 = f(X)$. Use the table to work parts (a)–(e).

**(a)** What is $f(2)$?

**(b)** If $f(X) = -3.7$, what is the value of X?

**(c)** What is the slope of the line?

**(d)** What is the $y$-intercept of the line?

**(e)** Find the expression for $f(X)$.

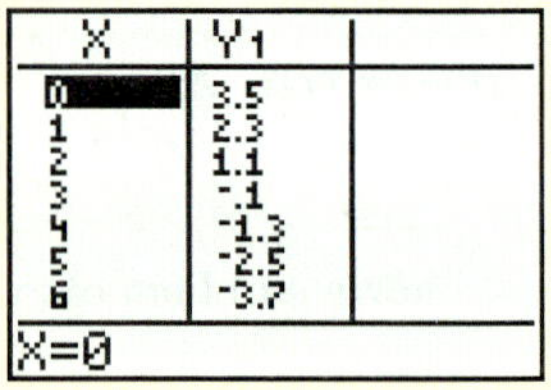

| X | Y1 |
|---|---|
| 0 | 3.5 |
| 1 | 2.3 |
| 2 | 1.1 |
| 3 | -.1 |
| 4 | -1.3 |
| 5 | -2.5 |
| 6 | -3.7 |

X=0

## 7.4 Variation

**OBJECTIVES**

1. Write an equation expressing direct variation.
2. Find the constant of variation, and solve direct variation problems.

Certain types of functions are very common, especially in business and the physical sciences. These are functions where $y$ depends on a multiple of $x$, or $y$ depends on a number divided by $x$. In such situations, $y$ is said to *vary directly as x* (in the first case) or *vary inversely as x* (in the second case). For example, by the distance formula, the distance traveled varies directly as the rate (or speed) and the time. Formulas for area and volume are other familiar examples of *direct variation.*

On the other hand, the force required to keep a car from skidding on a curve varies inversely as the radius of the curve. Another example of *inverse variation* is how travel time is inversely proportional to rate or speed.

**3** Solve inverse variation problems.

**4** Solve joint variation problems.

**5** Solve combined variation problems.

**OBJECTIVE 1 Write an equation expressing direct variation.** The circumference of a circle is given by the formula $C = 2\pi r$, where $r$ is the radius of the circle. See the figure. Circumference is always a constant multiple of the radius. ($C$ is always found by multiplying $r$ by the constant $2\pi$.) Thus,

As the *radius increases,* the *circumference increases.*

The reverse is also true.

As the *radius decreases,* the *circumference decreases.*

Because of this, the circumference is said to *vary directly* as the radius.

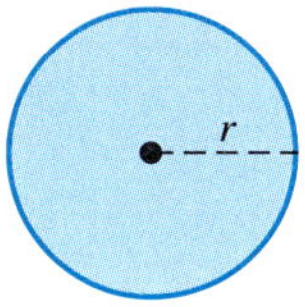


$C = 2\pi r$

**Direct Variation**

***y* varies directly as *x*** if there exists a real number $k$ such that

$$y = kx.$$

Also, $y$ is said to be **proportional to** $x$. The number $k$ is called the **constant of variation.** In direct variation, for $k > 0$, as the value of $x$ increases, the value of $y$ also increases. Similarly, as $x$ decreases, $y$ decreases.

**OBJECTIVE 2 Find the constant of variation, and solve direct variation problems.** The direct variation equation $y = kx$ defines a linear function, where the constant of variation $k$ is the slope of the line. For example, we wrote the equation

$$y = 1.60x$$

to describe the cost $y$ to buy $x$ gal of gas in Example 6 of Section 7.2. The cost varies directly as, or is proportional to, the number of gallons of gas purchased. That is, as the number of gallons of gas increases, cost increases; also, as the number of gallons of gas decreases, cost decreases. The constant of variation $k$ is 1.60, the cost of 1 gal of gas.

### EXAMPLE 1 Finding the Constant of Variation and the Variation Equation

Steven Pusztai is paid an hourly wage. One week he worked 43 hr and was paid \$795.50. How much does he earn per hour?

Let $h$ represent the number of hours he works and $P$ represent his corresponding pay. Then, $P$ varies directly as $h$, so

$$P = kh.$$

Here, $k$ represents Steven's hourly wage. Since $P = 795.50$ when $h = 43$,

$$795.50 = 43k$$

$$k = 18.50. \quad \text{Use a calculator.}$$

His hourly wage is \$18.50, and $P$ and $h$ are related by

$$P = 18.50h.$$

**Now Try Exercise 31.**

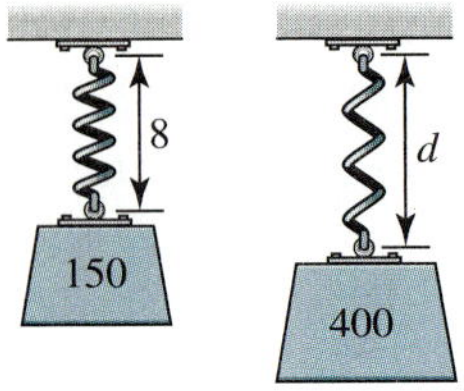


FIGURE 30

## EXAMPLE 2 Solving a Direct Variation Problem

Hooke's law for an elastic spring states that the distance a spring stretches is proportional to the force applied. If a force of 150 newtons* stretches a certain spring 8 cm, how much will a force of 400 newtons stretch the spring? See Figure 30.

If $d$ is the distance the spring stretches and $f$ is the force applied, then $d = kf$ for some constant $k$. Since a force of 150 newtons stretches the spring 8 cm, use these values to find $k$.

$$d = kf \qquad \text{Variation equation}$$
$$8 = k \cdot 150 \qquad \text{Let } d = 8 \text{ and } f = 150.$$
$$k = \frac{8}{150} \qquad \text{Find } k.$$
$$k = \frac{4}{75}$$

Substitute $\frac{4}{75}$ for $k$ in the variation equation $d = kf$ to get

$$d = \frac{4}{75}f.$$

For a force of 400 newtons,

$$d = \frac{4}{75}(400) \qquad \text{Let } f = 400.$$
$$= \frac{64}{3}.$$

The spring will stretch $\frac{64}{3}$ cm if a force of 400 newtons is applied.

**Now Try Exercise 35.**

In summary, use the following steps to solve a variation problem.

**Solving a Variation Problem**

*Step 1* Write the variation equation.

*Step 2* Substitute the initial values and solve for $k$.

*Step 3* Rewrite the variation equation with the value of $k$ from Step 2.

*Step 4* Substitute the remaining values, solve for the unknown, and find the required answer.

The direct variation equation $y = kx$ is a linear equation. However, other kinds of variation involve other types of equations. For example, one variable can be proportional to a power of another variable.

*A newton is a unit of measure of force used in physics.

**Direct Variation as a Power**

***y* varies directly as the *n*th power of *x*** if there exists a real number $k$ such that

$$y = kx^n.$$

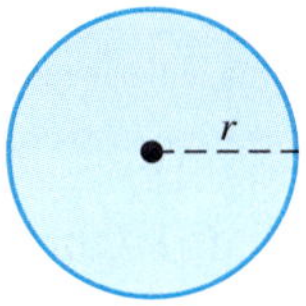


$A = \pi r^2$

An example of direct variation as a power is the formula for the area of a circle, $A = \pi r^2$. Here, $\pi$ is the constant of variation, and the area varies directly as the *square* of the radius.

## EXAMPLE 3 Solving a Direct Variation Problem

The distance a body falls from rest varies directly as the square of the time it falls (disregarding air resistance). If a skydiver falls 64 ft in 2 sec, how far will she fall in 8 sec?

*Step 1* If $d$ represents the distance the skydiver falls and $t$ the time it takes to fall, then $d$ is a function of $t$, and, for some constant $k$,

$$d = kt^2.$$

*Step 2* To find the value of $k$, use the fact that the skydiver falls 64 ft in 2 sec.

$$d = kt^2 \quad \text{Variation equation}$$
$$64 = k(2)^2 \quad \text{Let } d = 64 \text{ and } t = 2.$$
$$k = 16 \quad \text{Find } k.$$

*Step 3* Using 16 for $k$, the variation equation becomes

$$d = 16t^2.$$

*Step 4* Now let $t = 8$ to find the number of feet the skydiver will fall in 8 sec.

$$d = 16(8)^2 \quad \text{Let } t = 8.$$
$$= 1024$$

The skydiver will fall 1024 ft in 8 sec.

**Now Try Exercise 37.**

**OBJECTIVE 3 Solve inverse variation problems.** In direct variation, where $k > 0$, as $x$ increases, $y$ increases. Similarly, as $x$ decreases, $y$ decreases. Another type of variation is *inverse variation.* With inverse variation, where $k > 0$, as one variable increases, the other variable decreases. For example, in a closed space, volume decreases as pressure increases, as illustrated by a trash compactor. See Figure 31. As the compactor presses down, the pressure on the trash increases; in turn, the trash occupies a smaller space.

As pressure on trash increases, volume of trash decreases.

**FIGURE 31**

**Inverse Variation**

***y* varies inversely as *x*** if there exists a real number $k$ such that

$$y = \frac{k}{x}.$$

(continued)

Also, **$y$ varies inversely as the $n$th power of $x$** if there exists a real number $k$ such that

$$y = \frac{k}{x^n}.$$

The inverse variation equation also defines a function. Since $x$ is in the denominator, these functions are *rational functions.* (See Chapter 6 and Section 11.1.) Another example of inverse variation comes from the distance formula. In its usual form, the formula is

$$d = rt.$$

Dividing each side by $r$ gives

$$t = \frac{d}{r}.$$

Here, $t$ (time) varies inversely as $r$ (rate or speed), with $d$ (distance) serving as the constant of variation. For example, if the distance between Chicago and Des Moines is 300 mi, then

$$t = \frac{300}{r}$$

and the values of $r$ and $t$ might be any of the following.

| | | | |
|---|---|---|---|
| $r = 50, t = 6$ | As $r$ increases, $t$ decreases. | $r = 30, t = 10$ | As $r$ decreases, $t$ increases. |
| $r = 60, t = 5$ | | $r = 25, t = 12$ | |
| $r = 75, t = 4$ | | $r = 20, t = 15$ | |

If we *increase* the rate (speed) we drive, time *decreases.* If we *decrease* the rate (speed) we drive, what happens to time?

## EXAMPLE 4 Solving an Inverse Variation Problem

In the manufacturing of a certain medical syringe, the cost of producing the syringe varies inversely as the number produced. If 10,000 syringes are produced, the cost is $2 per syringe. Find the cost per syringe to produce 25,000 syringes.

Let $x =$ the number of syringes produced,
and $c =$ the cost per syringe.

Here, as production increases, cost decreases and as production decreases, cost increases. Since $c$ varies inversely as $x$, there is a constant $k$ such that

$$c = \frac{k}{x}.$$

Find $k$ by replacing $c$ with 2 and $x$ with 10,000.

$$2 = \frac{k}{10{,}000}$$

$$20{,}000 = k \quad \text{Multiply by 10,000.}$$

Since $c = \frac{k}{x}$,

$$c = \frac{20{,}000}{25{,}000} = .80. \qquad \text{Let } k = 20{,}000 \text{ and } x = 25{,}000.$$

The cost per syringe to make 25,000 syringes is \$.80.

**Now Try Exercise 39.**

## EXAMPLE 5 Solving an Inverse Variation Problem

The weight of an object above Earth varies inversely as the square of its distance from the center of Earth. A space shuttle in an elliptical orbit has a maximum distance from the center of Earth (*apogee*) of 6700 mi. Its minimum distance from the center of Earth (*perigee*) is 4090 mi. See Figure 32. If an astronaut in the shuttle weighs 57 lb at its apogee, what does the astronaut weigh at its perigee?

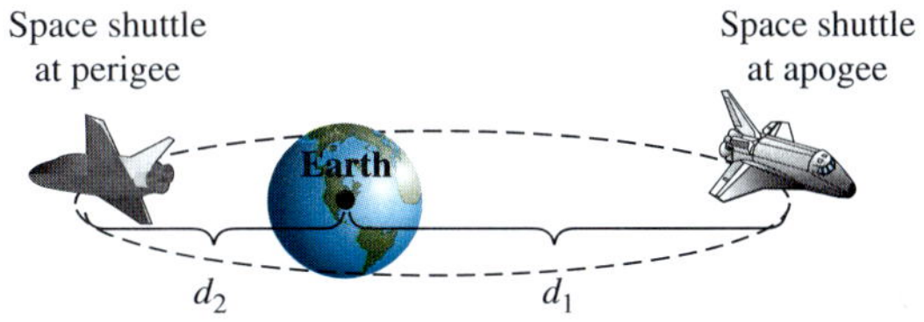


**FIGURE 32**

If $w$ is the weight and $d$ is the distance from the center of Earth, then

$$w = \frac{k}{d^2}$$

for some constant $k$. At the apogee the astronaut weighs 57 lb, and the distance from the center of Earth is 6700 mi. Use these values to find $k$.

$$57 = \frac{k}{(6700)^2} \qquad \text{Let } w = 57 \text{ and } d = 6700.$$

$$k = 57(6700)^2$$

Then the weight at the perigee with $d = 4090$ mi is

$$w = \frac{57(6700)^2}{(4090)^2} \approx 153 \text{ lb.} \qquad \text{Use a calculator.}$$

**Now Try Exercise 43.**

**OBJECTIVE 4 Solve joint variation problems.** It is common for one variable to depend on several others. If one variable varies directly as the *product* of several other variables (perhaps raised to powers), the first variable is said to *vary jointly* as the others.

**Joint Variation**

***y* varies jointly as *x* and *z*** if there exists a real number $k$ such that

$$y = kxz.$$

**CAUTION** Note that *and* in the expression "$y$ varies directly as *x and z*" translates as the product $y = kxz$. The word *and* does not indicate addition here.

### EXAMPLE 6 Solving a Joint Variation Problem

The interest on a loan or an investment is given by the formula $I = prt$. Here, for a given principal $p$, the interest earned $I$ varies jointly as the interest rate $r$ and the time $t$ the principal is left at interest. If an investment earns \$100 interest at 5% for 2 yr, how much interest will the same principal earn at 4.5% for 3 yr?

We use the formula $I = prt$, where $p$ is the constant of variation because it is the same for both investments. For the first investment, $I = 100$, $r = .05$, and $t = 2$, so

$$I = prt$$

$$100 = p(.05)(2) \quad \text{Let } I = 100,\ r = .05, \text{ and } t = 2.$$

$$100 = .1p$$

$$\frac{100}{.1} = p$$

$$p = 1000.$$

Now we find $I$ when $p = 1000$, $r = .045$, and $t = 3$.

$$I = 1000(.045)(3) \quad \text{Let } p = 1000,\ r = .045, \text{ and } t = 3.$$

$$I = 135$$

The interest will be \$135.

**Now Try Exercise 45.**

**OBJECTIVE 5 Solve combined variation problems.** There are many combinations of direct and inverse variation. Example 7 shows a typical **combined variation** problem.

### EXAMPLE 7 Solving a Combined Variation Problem

Body mass index, or BMI, is used by physicians to assess a person's level of fatness. A BMI from 19 through 25 is considered desirable. BMI varies directly as an individual's weight in pounds and inversely as the square of the individual's height in inches. A person who weighs 118 lb and is 64 in. tall has a BMI of 20. (The BMI is rounded to the nearest whole number.) Find the BMI of a person who weighs 165 lb with a height of 70 in. (*Source: Washington Post.*)

Let $B$ represent the BMI, $w$ the weight, and $h$ the height. Then

$$B = \frac{kw}{h^2}.$$

← BMI varies directly as the weight.
← BMI varies inversely as the square of the height.

To find $k$, let $B = 20$, $w = 118$, and $h = 64$.

$$20 = \frac{k(118)}{64^2}$$

$$k = \frac{20(64^2)}{118} \quad \text{Multiply by } 64^2\text{; divide by 118.}$$

$$k \approx 694 \quad \text{Use a calculator.}$$

Now find $B$ when $k = 694$, $w = 165$, and $h = 70$.

$$B = \frac{694(165)}{70^2} \approx 23 \quad \text{Nearest whole number}$$

The person's BMI is 23.

**Now Try Exercise 47.**

## 7.4 EXERCISES

**For Extra Help**

- Student's Solutions Manual
- MyMathLab
- InterAct Math Tutorial Software
- AW Math Tutor Center
- MathXL
- Digital Video Tutor CD 11/Videotape 11

*Use personal experience or intuition to determine whether the situation suggests* direct *or* inverse *variation.*

**1.** The number of different lottery tickets you buy and your probability of winning that lottery

**2.** The rate and the distance traveled by a pickup truck in 3 hr

**3.** The amount of pressure put on the accelerator of a car and the speed of the car

**4.** The number of days from now until December 25 and the magnitude of the frenzy of Christmas shopping

**5.** Your age and the probability that you believe in Santa Claus

**6.** The surface area of a balloon and its diameter

**7.** The number of days until the end of the baseball season and the number of home runs that Barry Bonds has

**8.** The amount of gasoline you pump and the amount you pay

*Determine whether each equation represents* direct, inverse, joint, *or* combined *variation.*

**9.** $y = \frac{3}{x}$ **10.** $y = \frac{8}{x}$ **11.** $y = 10x^2$ **12.** $y = 2x^3$

**13.** $y = 3xz^4$ **14.** $y = 6x^3z^2$ **15.** $y = \frac{4x}{wz}$ **16.** $y = \frac{6x}{st}$

**17.** For $k > 0$, if $y$ varies directly as $x$, when $x$ increases, $y$ ________________, and when $x$ decreases, $y$ ________________.

**18.** For $k > 0$, if $y$ varies inversely as $x$, when $x$ increases, $y$ ________________, and when $x$ decreases, $y$ ________________.

*Solve each problem.*

**19.** If $x$ varies directly as $y$, and $x = 9$ when $y = 3$, find $x$ when $y = 12$.

**20.** If $x$ varies directly as $y$, and $x = 10$ when $y = 7$, find $y$ when $x = 50$.

**21.** If $a$ varies directly as the square of $b$, and $a = 4$ when $b = 3$, find $a$ when $b = 2$.

**22.** If $h$ varies directly as the square of $m$, and $h = 15$ when $m = 5$, find $h$ when $m = 7$.

**23.** If $z$ varies inversely as $w$, and $z = 10$ when $w = .5$, find $z$ when $w = 8$.

**24.** If $t$ varies inversely as $s$, and $t = 3$ when $s = 5$, find $s$ when $t = 5$.

**25.** If $m$ varies inversely as $p^2$, and $m = 20$ when $p = 2$, find $m$ when $p = 5$.

**26.** If $a$ varies inversely as $b^2$, and $a = 48$ when $b = 4$, find $a$ when $b = 7$.

**27.** $p$ varies jointly as $q$ and $r^2$, and $p = 200$ when $q = 2$ and $r = 3$. Find $p$ when $q = 5$ and $r = 2$.

**28.** $f$ varies jointly as $g^2$ and $h$, and $f = 50$ when $g = 4$ and $h = 2$. Find $f$ when $g = 3$ and $h = 6$.

**29.** Explain the difference between inverse variation and direct variation.

**30.** What is meant by the constant of variation in a direct variation problem? If you were to graph the linear equation $y = kx$ for some nonnegative constant $k$, what role would the value of $k$ play in the graph?

*Solve each problem involving variation. See Examples 1–7.*

**31.** Todd bought 8 gal of gasoline and paid $13.59. To the nearest tenth of a cent, what is the price of gasoline per gallon?

**32.** Melissa gives horseback rides at Shadow Mountain Ranch. A 2.5-hr ride costs $50.00. What is the price per hour?

**33.** The volume of a can of tomatoes is proportional to the height of the can. If the volume of the can is 300 $cm^3$ when its height is 10.62 cm, find the volume of a can with height 15.92 cm.

**34.** The force required to compress a spring is proportional to the change in length of the spring. If a force of 20 newtons is required to compress a certain spring 2 cm, how much force is required to compress the spring from 20 cm to 8 cm?

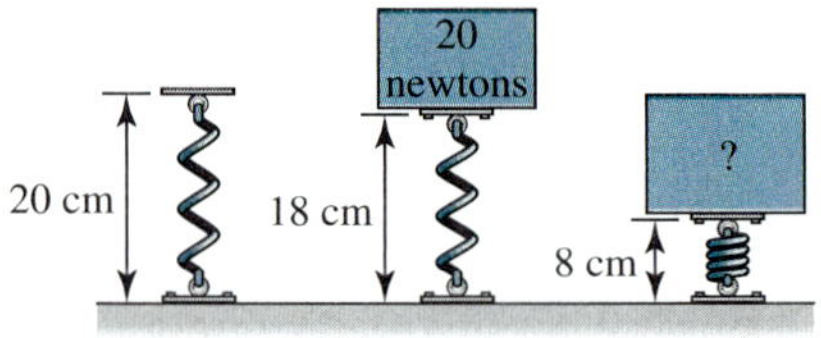


**35.** The weight of an object on Earth is directly proportional to the weight of that same object on the moon. A 200-lb astronaut would weigh 32 lb on the moon. How much would a 50-lb dog weigh on the moon?

**36.** The pressure exerted by a certain liquid at a given point varies directly as the depth of the point beneath the surface of the liquid. The pressure at 30 m is 80 newtons per $m^2$. What pressure is exerted at 50 m?

**37.** For a body falling freely from rest (disregarding air resistance), the distance the body falls varies directly as the square of the time. If an object is dropped from the top of a tower 576 ft high and hits the ground in 6 sec, how far did it fall in the first 4 sec?

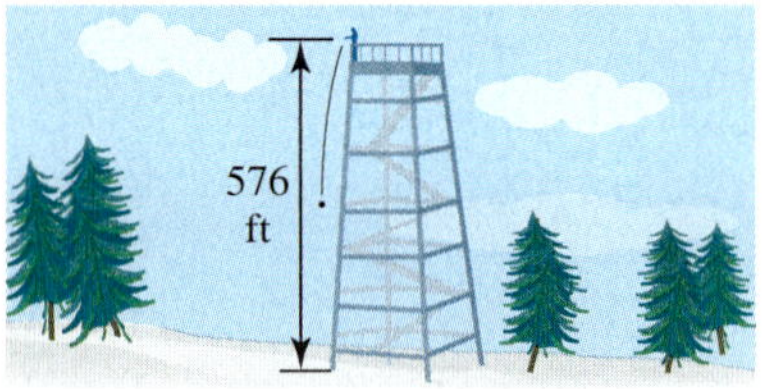


**38.** The amount of water emptied by a pipe varies directly as the square of the diameter of the pipe. For a certain constant water flow, a pipe emptying into a canal will allow 200 gal of water to escape in an hour. The diameter of the pipe is 6 in. How much water would a 12-in. pipe empty into the canal in an hour, assuming the same water flow?

**39.** Over a specified distance, speed varies inversely with time. If a Dodge Viper on a test track goes a certain distance in one-half minute at 160 mph, what speed is needed to go the same distance in three-fourths minute?

**40.** For a constant area, the length of a rectangle varies inversely as the width. The length of a rectangle is 27 ft when the width is 10 ft. Find the width of a rectangle with the same area if the length is 18 ft.

**41.** The frequency of a vibrating string varies inversely as its length. That is, a longer string vibrates fewer times in a second than a shorter string. Suppose a piano string 2 ft long vibrates 250 cycles per sec. What frequency would a string 5 ft long have?

**42.** The current in a simple electrical circuit varies inversely as the resistance. If the current is 20 amps when the resistance is 5 ohms, find the current when the resistance is 7.5 ohms.

**43.** The amount of light (measured in foot-candles) produced by a light source varies inversely as the square of the distance from the source. If the illumination produced 1 m from a light source is 768 foot-candles, find the illumination produced 6 m from the same source.

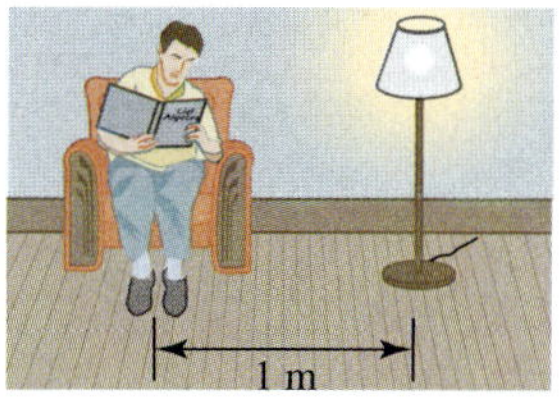


**44.** The force with which Earth attracts an object above Earth's surface varies inversely with the square of the distance of the object from the center of Earth. If an object 4000 mi from the center of Earth is attracted with a force of 160 lb, find the force of attraction if the object were 6000 mi from the center of Earth.

**45.** For a given interest rate, simple interest varies jointly as principal and time. If \$2000 left in an account for 4 yr earned interest of \$280, how much interest would be earned in 6 yr?

**46.** The collision impact of an automobile varies jointly as its mass and the square of its speed. Suppose a 2000-lb car traveling at 55 mph has a collision impact of 6.1. What is the collision impact of the same car at 65 mph?

**47.** The force needed to keep a car from skidding on a curve varies inversely as the radius of the curve and jointly as the weight of the car and the square of the speed. If 242 lb of force keep a 2000-lb car from skidding on a curve of radius 500 ft at 30 mph, what force would keep the same car from skidding on a curve of radius 750 ft at 50 mph?

**48.** The maximum load that a cylindrical column with a circular cross section can hold varies directly as the fourth power of the diameter of the cross section and inversely as the square of the height. A 9-m column 1 m in diameter will support 8 metric tons. How many metric tons can be supported by a column 12 m high and $\frac{2}{3}$ m in diameter?

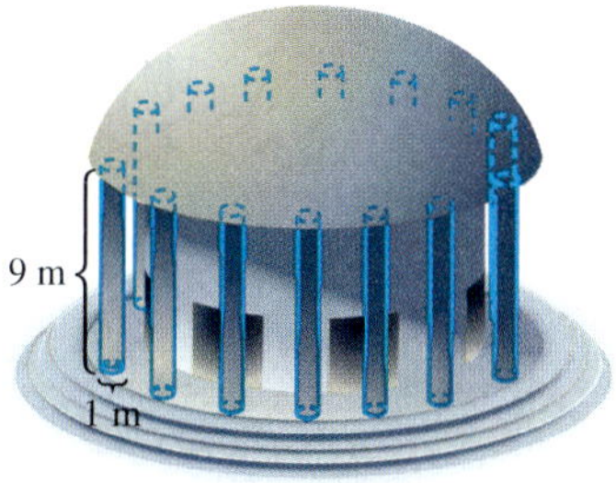


Load = 8 metric tons

**49.** The number of long-distance phone calls between two cities in a certain time period varies jointly as the populations of the cities, $p_1$ and $p_2$, and inversely as the distance between them. If 80,000 calls are made between two cities 400 mi apart, with populations of 70,000 and 100,000, how many calls are made between cities with populations of 50,000 and 75,000 that are 250 mi apart?

**50.** A body mass index from 27 through 29 carries a slight risk of weight-related health problems, while one of 30 or more indicates a great increase in risk. Use your own height and weight and the information in Example 7 to determine your BMI and whether you are at risk.

**51.** Natural gas provides 35.8% of U.S. energy. (*Source:* U.S. Energy Department.) The volume of gas varies inversely as the pressure and directly as the temperature. (Temperature must be measured in *Kelvin* (K), a unit of measurement used in physics.) If a certain gas occupies a volume of 1.3 L at 300 K and a pressure of 18 newtons per $cm^2$, find the volume at 340 K and a pressure of 24 newtons per $cm^2$.

**52.** The maximum load of a horizontal beam that is supported at both ends varies directly as the width and the square of the height and inversely as the length between the supports. A beam 6 m long, .1 m wide, and .06 m high supports a load of 360 kg. What is the maximum load supported by a beam 16 m long, .2 m wide, and .08 m high?

*Exercises 53 and 54 describe weight-estimation formulas that fishermen have used over the years.* Girth *is the distance around the body of the fish. (Source: Sacramento Bee,* November 9, 2000.)

**53.** The weight of a bass varies jointly as its girth and the square of its length. A prize-winning bass weighed in at 22.7 lb and measured 36 in. long with 21 in. girth. How much would a bass 28 in. long with 18 in. girth weigh?

**54.** The weight of a trout varies jointly as its length and the square of its girth. One angler caught a trout that weighed 10.5 lb and measured 26 in. long with 18 in. girth. Find the weight of a trout that is 22 in. long with 15 in. girth.

## RELATING CONCEPTS (EXERCISES 55–62)

**For Individual or Group Work**

*A routine activity such as pumping gasoline can be related to many of the concepts studied in this chapter. Suppose that premium unleaded costs* \$1.25 *per gallon.* ***Work Exercises 55–62 in order.***

**55.** Zero gallons of gasoline cost \$0.00, while 1 gallon costs \$1.25. Represent these two pieces of information as ordered pairs of the form (gallons, price).

**56.** Use the information from Exercise 55 to find the slope of the line on which the two points lie.

**57.** Write the slope-intercept form of the equation of the line on which the two points lie.

**58.** Using function notation, if $f(x) = ax + b$ represents the line from Exercise 57, what are the values of $a$ and $b$?

**59.** How does the value of $a$ from Exercise 58 relate to gasoline in this situation? With relationship to the line, what do we call this number?

**60.** Why does the equation from Exercise 57 satisfy the conditions for direct variation? In the context of variation, what do we call the value of $a$?

**61.** The graph of the equation from Exercise 57 is shown in the calculator screen. How is the display at the bottom of the screen interpreted in the context of these exercises?

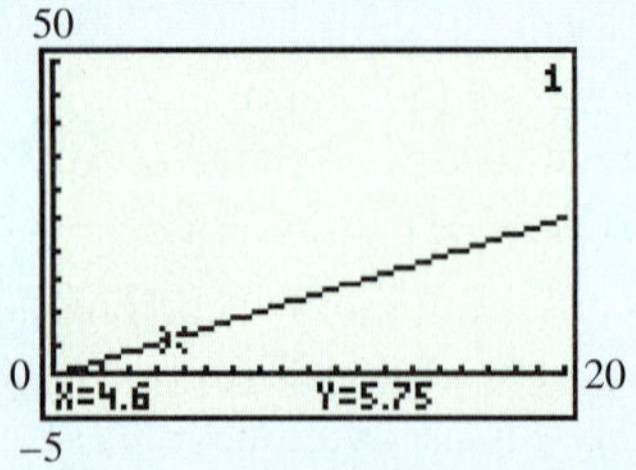


**62.** The table was generated by a graphing calculator, with $Y_1$ entered as the equation from Exercise 57. Interpret the entry for X = 12 in the context of these exercises.

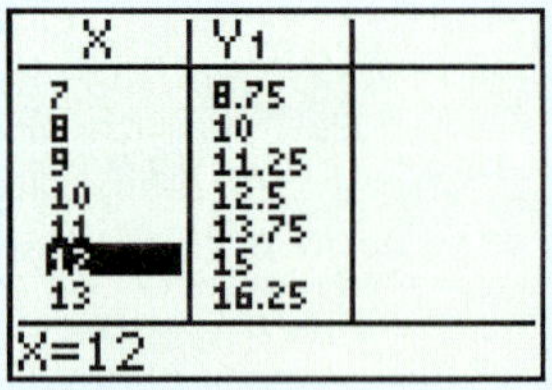

## Chapter 7 Group Activity

### Choosing an Energy Source (or How to Get Hot Water)

**OBJECTIVE** Write and graph linear functions that model given data.

There are many different ways to heat water. In this activity you will look at three different energy sources that may be used to provide heat for a 40-gal home water tank.

Have each student in your group choose one of the three types of water heaters listed in the table.

| *Type of Hot Water Heater* | *Size (in gallons)* | *Price* | *Operating Cost per Month (manufacturer's estimate)* | *Hot Water Temperature* |
|---|---|---|---|---|
| Kenmore Economizer 6 —Electric | 40 | \$139.99 | \$35.00 | 120°–130° |
| Kenmore Economizer 6 —Natural Gas | 40 | \$139.99 | \$13.25 | 120°–130° |
| Sunbather Water Heater —Solar | 40 | \$950.00 | \$0.00 | * |

*Source:* Jade Mountain 1999.

**A.** Using data for the water heater you selected, find a linear equation that represents total cost $y$ of heating water with respect to time. Let $x$ represent number of months. Write the equation in slope-intercept form.

**B.** Graph your equation using domain [0, 60] and range [0, 1000].

**C.** As a group, compare the graphs of your equations.

**1.** What are the $y$-intercepts?

**2.** How do the slopes compare? Which is the steepest? Which has 0 slope?

**D.** Discuss other factors to consider when choosing each type of water heater. Which of these water heaters would you choose to heat your home?

*No temperature listed but the ad says "Best for warm climates, preheating water, summer only in cold places, or when hot water needed only in afternoons and evenings."

# CHAPTER 7 SUMMARY

## KEY TERMS

**7.1** ordered pair
origin
$x$-axis
$y$-axis
rectangular (Cartesian) coordinate system
plot
coordinate
quadrant
graph of an equation
first-degree equation
linear equation in two variables
$x$-intercept
$y$-intercept
rise
run
slope

**7.2** slope-intercept form
point-slope form
standard form

**7.3** dependent variable
independent variable
relation
function
domain
range
function notation
linear function
constant function

**7.4** vary directly
proportional
constant of variation
vary inversely
vary jointly
combined variation

## NEW SYMBOLS

$(a, b)$ ordered pair

$x_1$ a specific value of the variable $x$ (read "$x$-sub-one")

$m$ slope

$f(x)$ function of $x$ (read "$f$ of $x$")

## TEST YOUR WORD POWER

*See how well you have learned the vocabulary in this chapter. Answers, with examples, follow the Quick Review.*

**1.** A **linear equation in two variables** is an equation that can be written in the form
**A.** $Ax + By < C$
**B.** $ax = b$
**C.** $y = x^2$
**D.** $Ax + By = C$.

**2.** The **slope** of a line is
**A.** the measure of the run over the rise of the line
**B.** the distance between two points on the line
**C.** the ratio of the change in $y$ to the change in $x$ along the line
**D.** the horizontal change compared to the vertical change between two points on the line.

**3.** In a relationship between two variables $x$ and $y$, the **independent variable** is
**A.** $x$, if $x$ depends on $y$
**B.** $x$, if $y$ depends on $x$
**C.** either $x$ or $y$
**D.** the larger of $x$ and $y$.

**4.** In a relationship between two variables $x$ and $y$, the **dependent variable** is
**A.** $y$, if $y$ depends on $x$
**B.** $y$, if $x$ depends on $y$
**C.** either $x$ or $y$
**D.** the smaller of $x$ and $y$.

**5.** A **relation** is
**A.** a set of ordered pairs
**B.** the ratio of the change in $y$ to the change in $x$ along a line
**C.** the set of all possible values of the independent variable
**D.** all the second components of a set of ordered pairs.

**6.** A **function** is
**A.** the numbers in an ordered pair
**B.** a set of ordered pairs in which each $x$-value corresponds to exactly one $y$-value
**C.** a pair of numbers written between parentheses in which order matters
**D.** the set of all ordered pairs that satisfy an equation.

**7.** The **domain** of a function is
**A.** the set of all possible values of the dependent variable $y$
**B.** a set of ordered pairs
**C.** the difference between the $x$-values
**D.** the set of all possible values of the independent variable $x$.

**8.** The **range** of a function is
**A.** the set of all possible values of the dependent variable $y$
**B.** a set of ordered pairs
**C.** the difference between the $y$-values
**D.** the set of all possible values of the independent variable $x$.

## QUICK REVIEW

| CONCEPTS | EXAMPLES |
|---|---|

### 7.1 REVIEW OF GRAPHS AND SLOPES OF LINES

**Finding Intercepts**

To find the $x$-intercept, let $y = 0$.

To find the $y$-intercept, let $x = 0$.

The graph of $2x + 3y = 12$ has

$x$-intercept (6, 0)

and $y$-intercept (0, 4).

**Finding Slope**

If $x_2 \neq x_1$, then

$$m = \frac{\text{rise}}{\text{run}} = \frac{\text{change in } y}{\text{change in } x} = \frac{y_2 - y_1}{x_2 - x_1}.$$

For $2x + 3y = 12$,

$$m = \frac{4 - 0}{0 - 6} = -\frac{2}{3}.$$

A vertical line has undefined slope.

$x = 3$ has undefined slope.

A horizontal line has 0 slope.

$y = -5$ has $m = 0$.

Parallel lines have equal slopes.

| | |
|---|---|
| $y = 2x + 3$<br>$m = 2$ | $4x - 2y = 6$<br>$-2y = -4x + 6$<br>$y = 2x - 3$<br>$m = 2$ |

These lines are parallel.

The slopes of perpendicular lines are negative reciprocals with a product of $-1$.

| | |
|---|---|
| $y = 3x - 1$<br>$m = 3$ | $x + 3y = 4$<br>$3y = -x + 4$<br>$y = -\frac{1}{3}x + \frac{4}{3}$<br>$m = -\frac{1}{3}$ |

These lines are perpendicular.

### 7.2 REVIEW OF EQUATIONS OF LINES; LINEAR MODELS

**Slope-Intercept Form**

$y = mx + b$

$y = 2x + 3$ $m = 2$, $y$-intercept is (0, 3).

**Point-Slope Form**

$y - y_1 = m(x - x_1)$

$y - 3 = 4(x - 5)$ (5, 3) is on the line, $m = 4$.

**Standard Form**

$Ax + By = C$ ($A$, $B$, $C$ integers, $A > 0$, $B \neq 0$)

$2x - 5y = 8$

**Horizontal Line**

$y = b$

$y = 4$

**Vertical Line**

$x = a$

$x = -1$

(continued)

| CONCEPTS | EXAMPLES |
|---|---|
| **7.3 FUNCTIONS** | |
| A **function** is a set of ordered pairs such that for each first component there is one and only one second component. The set of first components is called the **domain,** and the set of second components is called the **range.** | $y = f(x) = x^2$ defines a function $f$, with domain $(-\infty, \infty)$ and range $[0, \infty)$. |
| To evaluate a function using function notation (that is, $f(x)$ notation) for a given value of $x$, substitute the value wherever $x$ appears. | If $f(x) = x^2 - 7x + 12$, then $f(1) = 1^2 - 7(1) + 12 = 6.$ |
| To write an equation that defines a function in function notation, | Write $2x + 3y = 12$ using function notation. |
| | $3y = -2x + 12$ Subtract $2x$. |
| *Step 1* Solve the equation for $y$. | $y = -\frac{2}{3}x + 4$ Divide by 3. |
| *Step 2* Replace $y$ with $f(x)$. | $f(x) = -\frac{2}{3}x + 4$ |
| **7.4 VARIATION** | |
| If there is some constant $k$ such that: | The area of a circle varies directly as the square of the radius. |
| $y = kx^n$, then $y$ varies directly as $x^n$. | $A = kr^2$ Here, $k = \pi$. |
| $y = \frac{k}{x^n}$, then $y$ varies inversely as $x^n$. | Pressure varies inversely as volume. $p = \frac{k}{V}$ |
| $y = kxz$, then $y$ varies jointly as $x$ and $z$. | For a given principal, interest varies jointly as interest rate and time. $I = krt$ $k$ is the given principal. |

**Answers to Test Your Word Power**

**1.** D; *Examples:* $3x + 2y = 6$, $x = y - 7$, $4x = y$ **2.** C; *Example:* The line through (3, 6) and (5, 4) has slope $\frac{4-6}{5-3} = \frac{-2}{2} = -1$. **3.** B; *Example:* See Answer 4, which follows. **4.** A; *Example:* When borrowing money, the amount you borrow (independent variable) determines the size of your payments (dependent variable). **5.** A; *Example:* The set $\{(2, 0), (4, 3), (6, 6), (8, 9)\}$ defines a relation. **6.** B; The relation given in Answer 5 is a function since the $x$-value of each ordered pair corresponds to exactly one $y$-value. **7.** D; *Example:* In the function in Answer 5, the domain is the set of $x$-values, $\{2, 4, 6, 8\}$. **8.** A; *Example:* In the function in Answer 5, the range is the set of $y$-values, $\{0, 3, 6, 9\}$.

# CHAPTER 7 REVIEW EXERCISES

[7.1] *Complete the table of ordered pairs for each equation. Then graph the equation.*

1. $3x + 2y = 10$

| $x$ | $y$ |
|---|---|
| 0 | |
| | 0 |
| 2 | |
| | $-2$ |

2. $x - y = 8$

| $x$ | $y$ |
|---|---|
| 2 | |
| | $-3$ |
| 3 | |
| | $-2$ |

*Find the x- and y-intercepts and then graph each equation.*

3. $4x - 3y = 12$
4. $5x + 7y = 28$
5. $2x + 5y = 20$
6. $x - 4y = 8$
7. Explain how the signs of the $x$- and $y$-coordinates of a point determine the quadrant in which the point lies.

*Find the slope of each line.*

8. Through $(-1, 2)$ and $(4, -5)$
9. Through $(0, 3)$ and $(-2, 4)$
10. $y = 2x + 3$
11. $3x - 4y = 5$
12. $x = 5$
13. Parallel to $3y = 2x + 5$
14. Perpendicular to $3x - y = 4$
15. Through $(-1, 5)$ and $(-1, -4)$
16. $y + 6 = 0$
17.

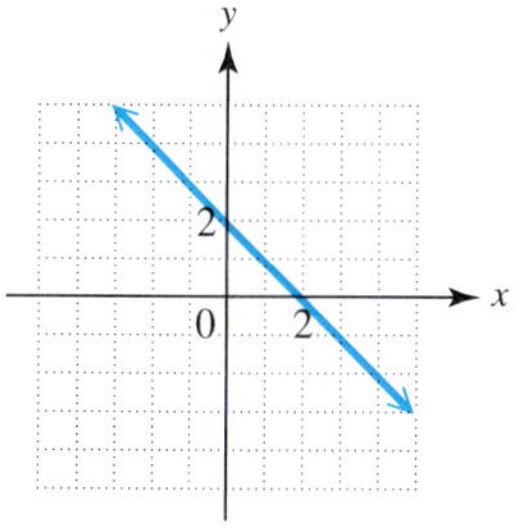


*Tell whether each line has* positive, negative, 0, *or* undefined *slope.*

18.

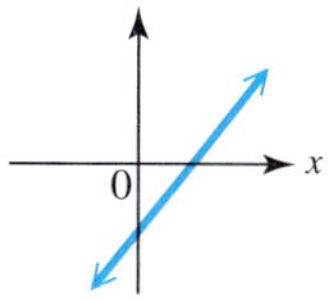


19.

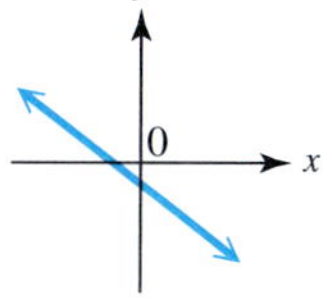


20.

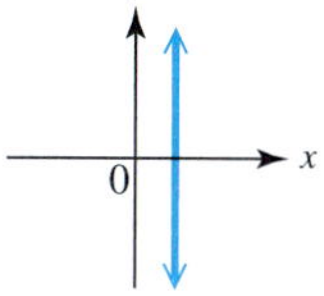


21.

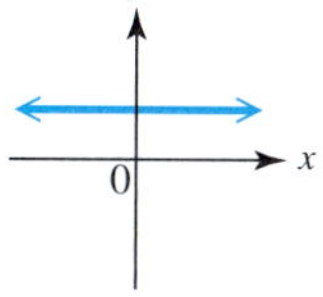


22. If a walkway rises 2 ft for every 10 ft on the horizontal, which of the following express its slope (or grade)? (There are several correct choices.)

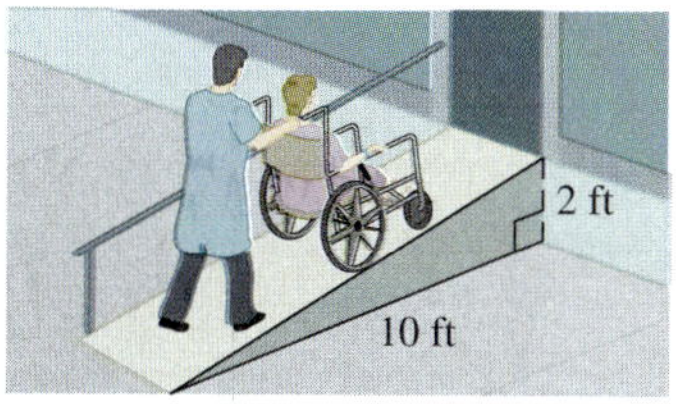


**A.** .2 **B.** $\frac{2}{10}$ **C.** $\frac{1}{5}$ **D.** 20%

**E.** 5 **F.** $\frac{20}{100}$ **G.** 500% **H.** $\frac{10}{2}$

**23.** If the pitch of a roof is $\frac{1}{4}$, how many feet in the horizontal direction correspond to a rise of 3 ft?

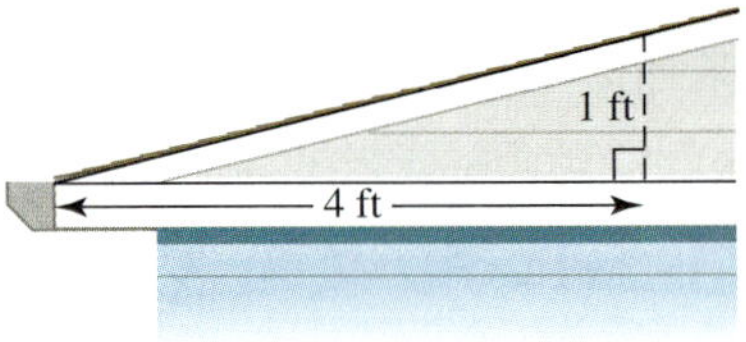


**24.** Family income in the United States has steadily increased for many years (primarily due to inflation). In 1970 the median family income was about \$10,000 a year. In 1999 it was about \$49,000 a year. Find the average rate of change of median family income to the nearest dollar over that period. (*Source:* U.S. Bureau of the Census.)

[7.2] *Find an equation for each line, if possible. Write the equation in slope-intercept form.*

**25.** Slope $-\dfrac{1}{3}$; $y$-intercept $(0, -1)$

**26.** Slope 0; $y$-intercept $(0, -2)$

**27.** Slope $-\dfrac{4}{3}$; through $(2, 7)$

**28.** Slope 3; through $(-1, 4)$

**29.** Vertical; through $(2, 5)$

**30.** Through $(2, -5)$ and $(1, 4)$

**31.** Through $(-3, -1)$ and $(2, 6)$

**32.** The line pictured in Exercise 17

**33.** Parallel to $4x - y = 3$ and through $(7, -1)$

**34.** Perpendicular to $2x - 5y = 7$ and through $(4, 3)$

**35.** The Midwest Athletic Club (Section 7.2, Exercises 73 and 74) offers two special membership plans. (*Source:* Midwest Athletic Club.) For each plan, write a linear equation in slope-intercept form and give the cost $y$ in dollars of a 1-yr membership. Let $x$ represent the number of months.

**(a)** Executive VIP/Gold membership: \$159 fee plus \$57 per month

**(b)** Executive Regular/Silver membership: \$159 fee plus \$47 per month

**36.** The percent of tax returns filed electronically for the years 1996–2001 is shown in the graph.

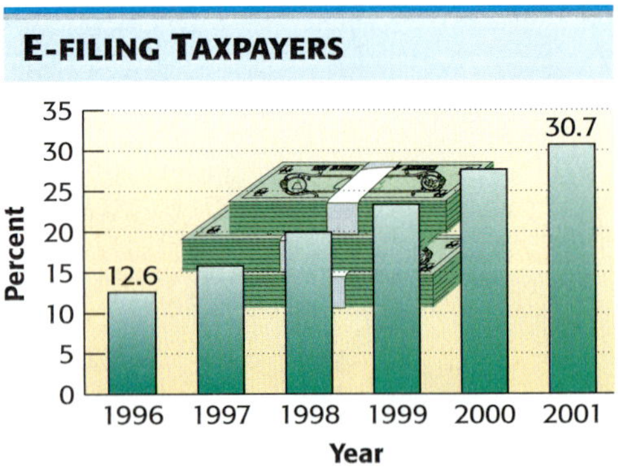


*Source:* Internal Revenue Service.

**(a)** Use the information given for the years 1996 and 2001, letting $x = 6$ represent 1996, $x = 11$ represent 2001, and $y$ represent the percent of returns filed electronically to find a linear equation that models the data. Write the equation in slope-intercept form. Interpret the slope of this equation.

**(b)** Use your equation from part (a) to predict the percent of tax returns that will be filed electronically in 2005. (Assume a constant rate of change.)

[7.3] *In Exercises 37–40, give the domain and range of each relation. Identify any functions.*

**37.** $\{(-4, 2), (-4, -2), (1, 5), (1, -5)\}$

**38.**

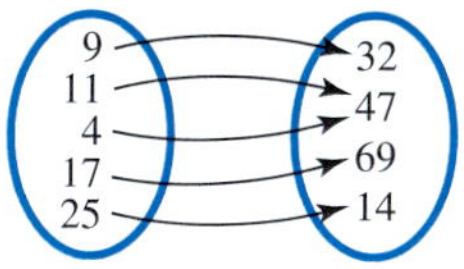


**39.**

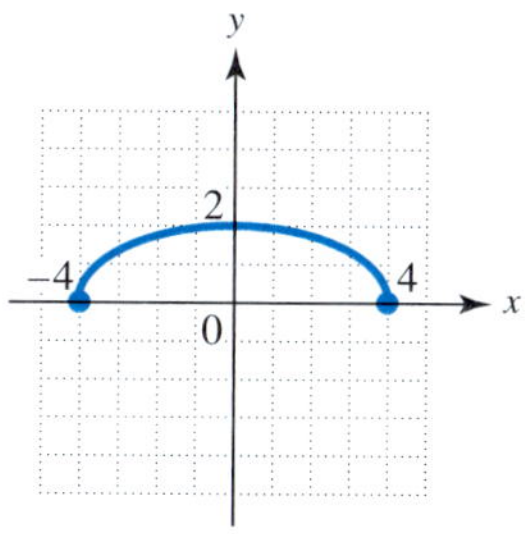


**40.**

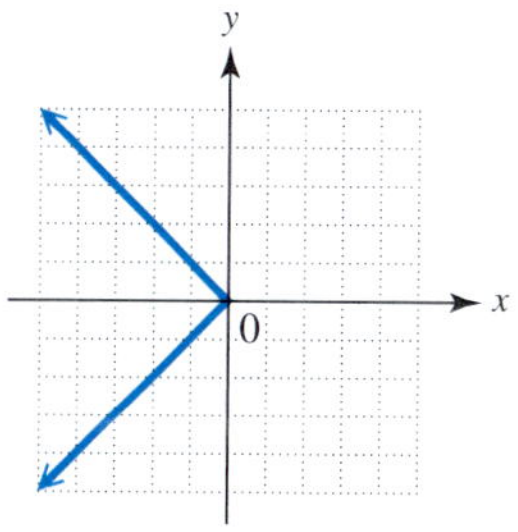


*Determine whether each equation defines y as a function of x. Give the domain in each case. Identify any linear functions.*

**41.** $y = 3x - 3$

**42.** $x = y^2$

**43.** $y = \dfrac{7}{x - 6}$

**44.** Explain the test that allows us to determine whether a graph is that of a function.

*Given* $f(x) = -2x^2 + 3x - 6$, *find each function value or expression.*

**45.** $f(0)$

**46.** $f(2.1)$

**47.** $f\left(-\dfrac{1}{2}\right)$

**48.** $f(k)$

**49.** The table shows profits for cable television station CNBC from 1994 through 1999.

(a) Does the table define a function?
(b) What are the domain and range?
(c) Call this function $f$. Give two ordered pairs that belong to $f$.
(d) Find $f(1994)$. What does it mean?
(e) If $f(x) = 180$, what does $x$ equal?

| Year | Profit (in millions of dollars) |
|---|---|
| 1994 | 40 |
| 1995 | 60 |
| 1996 | 80 |
| 1997 | 130 |
| 1998 | 180 |
| 1999 | 200 |

*Source: Fortune,* May 24, 1999, p. 142.

**50.** The equation $2x^2 - y = 0$ defines $y$ as a function of $x$. Rewrite it using $f(x)$ notation, and find $f(3)$.

**51.** Suppose that $2x - 5y = 7$ defines a function. If $y = f(x)$, which one of the following defines the same function?

**A.** $f(x) = \dfrac{7 - 2x}{5}$ **B.** $f(x) = \dfrac{-7 - 2x}{5}$

**C.** $f(x) = \dfrac{-7 + 2x}{5}$ **D.** $f(x) = \dfrac{7 + 2x}{5}$

**52.** Can the graph of a linear function have undefined slope? Explain.

## RELATING CONCEPTS (EXERCISES 53–64)

**For Individual or Group Work**

*Refer to the straight-line graph and* ***work Exercises 53–64 in order.***

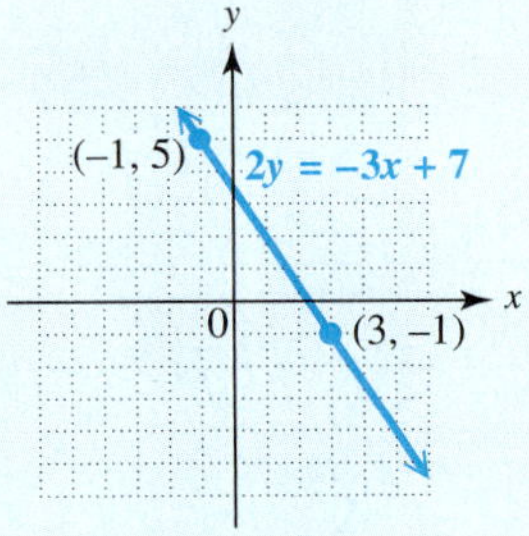


**53.** By just looking at the graph, how can you tell whether the slope is positive, negative, 0, or undefined?

**54.** Use the slope formula to find the slope of the line.

**55.** What is the slope of any line parallel to the line shown? perpendicular to the line shown?

**56.** Find the $x$-intercept of the graph.

**57.** Find the $y$-intercept of the graph.

**58.** Use function notation to write the equation of the line. Use $f$ to designate the function.

**59.** Find $f(8)$.

**60.** If $f(x) = -8$, what is the value of $x$?

**61.** Graph the solution set of $f(x) \geq 0$.

**62.** What is the solution set of $f(x) = 0$?

**63.** What is the solution set of $f(x) < 0$? (Use the graph and the result of Exercise 62.)

**64.** What is the solution set of $f(x) > 0$? (Use the graph and the result of Exercise 62.)

[7.4]

**65.** In which one of the following does $y$ vary inversely as $x$?

**A.** $y = 2x$ **B.** $y = \frac{x}{3}$ **C.** $y = \frac{3}{x}$ **D.** $y = x^2$

*Solve each problem.*

**66.** For the subject in a photograph to appear in the same perspective in the photograph as in real life, the viewing distance must be properly related to the amount of enlargement. For a particular camera, the viewing distance varies directly as the amount of enlargement. A picture taken with this camera that is enlarged 5 times should be viewed from a distance of 250 mm. Suppose a print 8.6 times the size of the negative is made. From what distance should it be viewed?

**67.** The frequency (number of vibrations per second) of a vibrating guitar string varies inversely as its length. That is, a longer string vibrates fewer times in a second than a shorter string. Suppose a guitar string .65 m long vibrates 4.3 times per sec. What frequency would a string .5 m long have?

**68.** The volume of a rectangular box of a given height is proportional to its width and length. A box with width 2 ft and length 4 ft has volume 12 $\text{ft}^3$. Find the volume of a box with the same height that is 3 ft wide and 5 ft long.

# CHAPTER 7 TEST

1. Complete the table of ordered pairs for the equation $2x - 3y = 12$.

| $x$ | $y$ |
|---|---|
| 1 | |
| 3 | |
| | $-4$ |

*Find the x- and y-intercepts, and graph each equation.*

2. $3x - 2y = 20$

3. $y = 5$

4. $x = 2$

5. Find the slope of the line through the points $(6, 4)$ and $(-4, -1)$.

6. Describe how the graph of a line with undefined slope is situated in a rectangular coordinate system.

*Determine whether each pair of lines is* parallel, perpendicular, *or* neither.

7. $5x - y = 8$ and $5y = -x + 3$

8. $2y = 3x + 12$ and $3y = 2x - 5$

9. In 1980, there were 119,000 farms in Iowa. As of 2001, there were 93,500. Find and interpret the average rate of change in the number of farms per year. Round your answer to the nearest whole number. (*Source:* Iowa Agricultural Statistics Service.)

*Find an equation of each line, and write it in slope-intercept form.*

10. Through $(4, -1)$; $m = -5$

11. Through $(-3, 14)$; horizontal

12. Through $(-7, 2)$ and parallel to $3x + 5y = 6$

13. Through $(-7, 2)$ and perpendicular to $y = 2x$

14. Through $(-2, 3)$ and $(6, -1)$

15. Which one of the following has positive slope and negative $y$-coordinate for its $y$-intercept?

**A.**
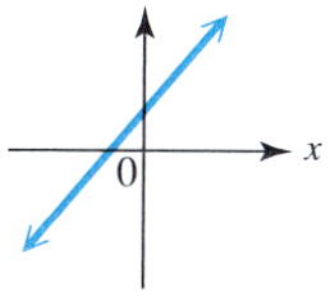


**B.**
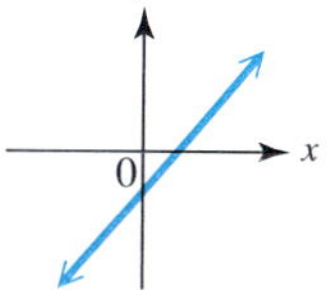


**C.**
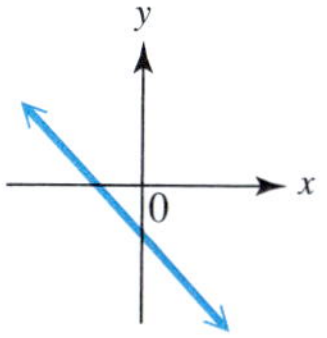


**D.**
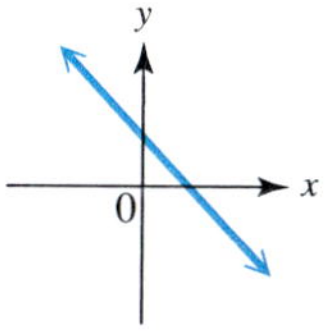

**16.** The bar graph shows median household income for Hispanics.

**(a)** Use the information for the years 1995 and 1999 to find an equation that models the data. Let $x = 5$ represent 1995, $x = 9$ represent 1999, and $y$ represent the median income. Write the equation in slope-intercept form.

**(b)** Use the equation from part (a) to approximate median household income for 1997 to the nearest dollar. How does your result compare to the actual value, \$26,628?

*Source:* U.S. Bureau of the Census.

**17.** Which one of the following is the graph of a function?

**A.**

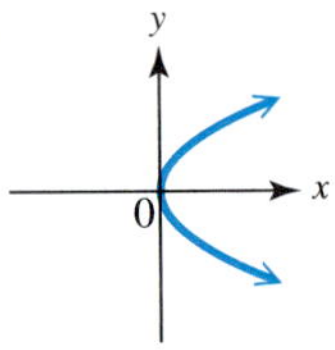


**B.**

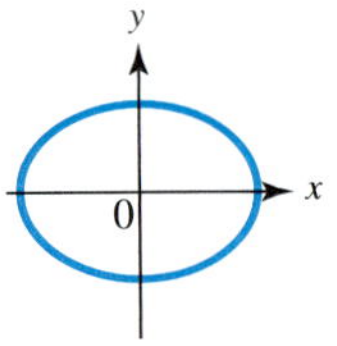


**C.**

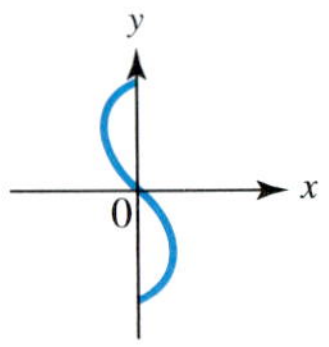


**D.**

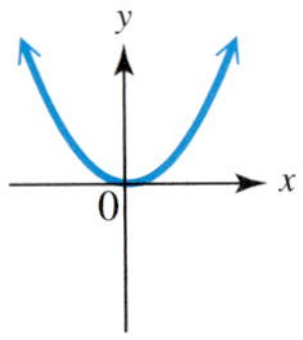


**18.** Which of the following does not define a function?

**A.** $\{(0, 1), (-2, 3), (4, 8)\}$ **B.** $y = 2x - 6$ **C.** $y = \dfrac{1}{x + 2}$ **D.**

| $x$ | $y$ |
|---|---|
| 0 | 1 |
| 3 | 2 |
| 0 | 2 |
| 6 | 3 |

**19.** Give the domain and range of the relation shown in **(a)** choice A of Problem 17 and **(b)** choice A of Problem 18.

**20.** If $f(x) = -x^2 + 2x - 1$, find $f(1)$ and $f(a)$.

**21.** Graph the linear function defined by $f(x) = \frac{2}{3}x - 1$. What is its domain and range?

*Solve each problem.*

**22.** The current in a simple electrical circuit is inversely proportional to the resistance. If the current is 80 amps when the resistance is 30 ohms, find the current when the resistance is 12 ohms.

**23.** The force of the wind blowing on a vertical surface varies jointly as the area of the surface and the square of the velocity. If a wind blowing at 40 mph exerts a force of 50 lb on a surface of 500 ft$^2$, how much force will a wind of 80 mph place on a surface of 2 ft$^2$?

# CUMULATIVE REVIEW EXERCISES CHAPTERS 1–7

*Decide whether each statement is* always true, sometimes true, *or* never true. *If the statement is* sometimes true, *give examples where it is true and where it is false.*

**1.** The absolute value of a negative number equals the additive inverse of the number.

**2.** The quotient of two integers with nonzero denominator is a rational number.

**3.** The sum of two negative numbers is positive.

**4.** The sum of a positive number and a negative number is 0.

*Simplify.*

**5.** $-|-2| - 4 + |-3| + 7$

**6.** $3x^2 - 4x + 4 + 9x - x^2$

*Evaluate each expression if $p = -4$, $q = -2$, and $r = 5$.*

**7.** $-3(2q - 3p)$

**8.** $\dfrac{r}{-p + 2q}$

*Solve.*

**9.** $2z - 5 + 3z = 4 - (z + 2)$

**10.** $\dfrac{3a - 1}{5} + \dfrac{a + 2}{2} = -\dfrac{3}{10}$

**11.** $V = \dfrac{1}{3}\pi r^2 h$ for $h$

*Solve each problem.*

**12.** If each side of a square were increased by 4 in., the perimeter would be 8 in. less than twice the perimeter of the original square. Find the length of a side of the original square.

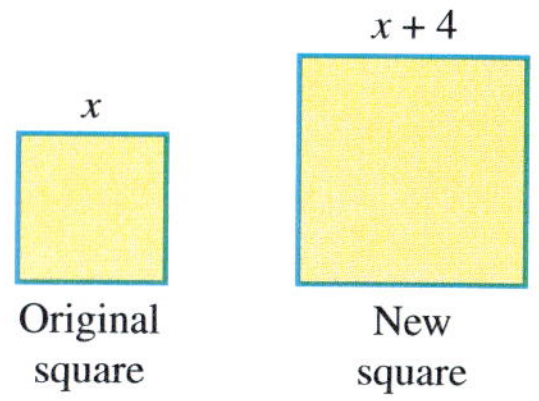


**13.** Two planes leave the Dallas–Fort Worth airport at the same time. One travels east at 550 mph, and the other travels west at 500 mph. Assuming no wind, how long will it take for the planes to be 2100 mi apart?

*Solve. Write each solution set in interval notation and graph it.*

**14.** $-4 < 3 - 2k < 9$

**15.** $-.3x + 2.1(x - 4) \leq -6.6$

**16.** Find the $x$- and $y$-intercepts of the line with equation $3x + 5y = 12$ and graph the line.

**17.** Consider the points $A(-2, 1)$ and $B(3, -5)$.

**(a)** Find the slope of the line $AB$.

**(b)** Find the slope of a line perpendicular to line $AB$.

*Write an equation for each line. Express the equation in slope-intercept form.*

**18.** Slope $-\frac{3}{4}$; $y$-intercept $(0, -1)$

**19.** Horizontal; through $(2, -2)$

**20.** Through $(4, -3)$ and $(1, 1)$

*Perform the indicated operations. In Exercises 21 and 22, assume that variables represent nonzero real numbers.*

**21.** $(3x^2y^{-1})^{-2}(2x^{-3}y)^{-1}$

**22.** $\frac{5m^{-2}y^3}{3m^{-3}y^{-1}}$

**23.** $(3x^3 + 4x^2 - 7) - (2x^3 - 8x^2 + 3x)$

**24.** $(7x + 3y)^2$

**25.** $(2p + 3)(5p^2 - 4p - 8)$

**26.** $\frac{m^3 - 3m^2 + 5m - 3}{m - 1}$

*Factor.*

**27.** $16w^2 + 50wz - 21z^2$

**28.** $4x^2 - 4x + 1 - y^2$

**29.** $4y^2 - 36y + 81$

**30.** $100x^4 - 81$

**31.** $8p^3 + 27$

*Solve.*

**32.** $(p - 1)(2p + 3)(p + 4) = 0$

**33.** $9q^2 = 6q - 1$

**34.** A sign is to have the shape of a triangle with a height 3 ft greater than the length of the base. How long should the base be if the area is to be 14 ft$^2$?

**35.** A game board has the shape of a rectangle. The longer sides are each 2 in. longer than the distance between them. The area of the board is 288 in.$^2$. Find the length of the longer sides and the distance between them.

*Perform each indicated operation. Write the answer in lowest terms.*

**36.** $\frac{8}{x + 1} - \frac{2}{x + 3}$

**37.** $\frac{x^2 - 25}{3x + 6} \cdot \frac{4x + 8}{x^2 + 10x + 25}$

**38.** $\frac{x^2 + 5x + 6}{3x} \div \frac{x^2 - 4}{x^2 + x - 6}$

**39.** $\dfrac{\frac{12}{x + 6}}{\frac{4}{2x + 12}}$

**40.** Solve $\frac{2}{x - 1} = \frac{5}{x - 1} - \frac{3}{4}$.

**41.** Give the domain and range of the relation. Does it define a function? Explain.

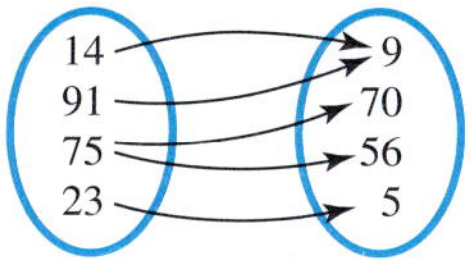


**42.** For the function defined by

$$f(x) = -4x + 10,$$

**(a)** find the domain and range.
**(b)** what is $f(-3)$?

**43.** Use the information in the graph to find and interpret the average rate of change in millions of U.S. cell phone subscribers per year from 1992 to 2000.

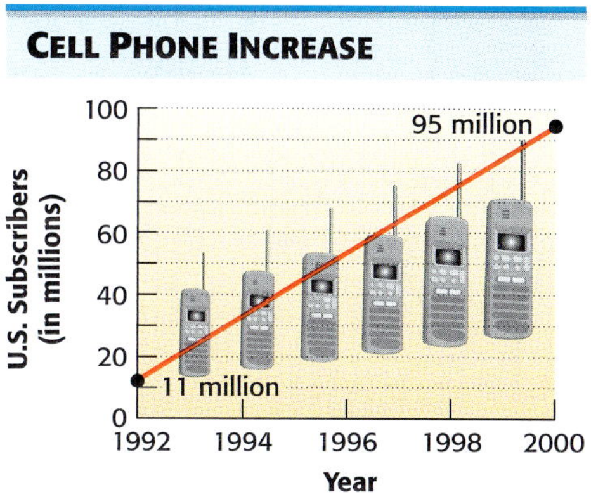


*Source:* Cellular Telecommunications Industry Association, Intel Corp.

**44.** The cost of a pizza varies directly as the square of its radius. If a pizza with a 7-in. radius costs \$6.00, how much should a pizza with a 9-in. radius cost?

# Systems of Linear Equations

# 8

Americans continue to flock to movie theaters in record numbers. Box office grosses reached $8.14 billion in 2001, the first time they exceeded $8 billion, as more than 1.4 billion movie tickets were sold. Surprisingly, the top three movies of the year were family films—*Harry Potter and the Sorcerer's Stone, Shrek,* and *Monsters, Inc.* not only attracted scores of adults with kids, but also unaccompanied adults, perhaps wishing to get away from it all for a few hours. (*Source:* ACNielsen EDI.) In Exercise 29 of the Chapter 8 Review Exercises, we use systems of linear equations to find out just how much money these top films earned.

# 8.1 Solving Systems of Linear Equations by Graphing

**OBJECTIVES**

1. Decide whether a given ordered pair is a solution of a system.
2. Solve linear systems by graphing.
3. Solve special systems by graphing.
4. Identify special systems without graphing.
5. Use a graphing calculator to solve a linear system.

A **system of linear equations,** often called a **linear system,** consists of two or more linear equations with the same variables. Examples of systems of two linear equations include

$$\begin{aligned} 2x + 3y &= 4 \\ 3x - y &= -5 \end{aligned} \qquad \begin{aligned} x + 3y &= 1 \\ -y &= 4 - 2x \end{aligned} \qquad \begin{aligned} x - y &= 1 \\ y &= 3. \end{aligned}$$

In the system on the right, think of $y = 3$ as an equation in two variables by writing it as $0x + y = 3$.

**OBJECTIVE 1 Decide whether a given ordered pair is a solution of a system.** A **solution of a system** of linear equations is an ordered pair that makes both equations true at the same time. A solution of an equation is said to *satisfy* the equation.

**EXAMPLE 1 Determining Whether an Ordered Pair Is a Solution**

Is $(4, -3)$ a solution of each system?

**(a)** $x + 4y = -8$
$3x + 2y = 6$

To decide whether $(4, -3)$ is a solution of the system, substitute 4 for $x$ and $-3$ for $y$ in each equation.

| | | | |
|---|---|---|---|
| $x + 4y = -8$ | | $3x + 2y = 6$ | |
| $4 + 4(-3) = -8$ ? | | $3(4) + 2(-3) = 6$ ? | |
| $4 + (-12) = -8$ ? | Multiply. | $12 + (-6) = 6$ ? | Multiply. |
| $-8 = -8$ | True | $6 = 6$ | True |

Because $(4, -3)$ satisfies both equations, it is a solution of the system.

**(b)** $2x + 5y = -7$
$3x + 4y = 2$

Again, substitute 4 for $x$ and $-3$ for $y$ in both equations.

| | | | |
|---|---|---|---|
| $2x + 5y = -7$ | | $3x + 4y = 2$ | |
| $2(4) + 5(-3) = -7$ ? | | $3(4) + 4(-3) = 2$ ? | |
| $8 + (-15) = -7$ ? | Multiply. | $12 + (-12) = 2$ ? | Multiply. |
| $-7 = -7$ | True | $0 = 2$ | False |

The ordered pair $(4, -3)$ is not a solution of this system because it does not satisfy the second equation.

**Now Try Exercises 3 and 5.**

We discuss three methods of solving a system of two linear equations in two variables in this chapter.

**OBJECTIVE 2 Solve linear systems by graphing.** The set of all ordered pairs that are solutions of a system is its **solution set.** One way to find the solution set of a system of two linear equations is to graph both equations on the same axes. The graph of each line shows points whose coordinates satisfy the equation of that line. Any intersection point would be on both lines and would therefore be a solution of both equations. Thus, the coordinates of any point where the lines intersect give a solution of the system. Because two *different* straight lines can intersect at no more than one point, there can never be more than one solution for such a system. The graph in Figure 1 shows that the solution of the system in Example 1(a) is the intersection point (4, −3).

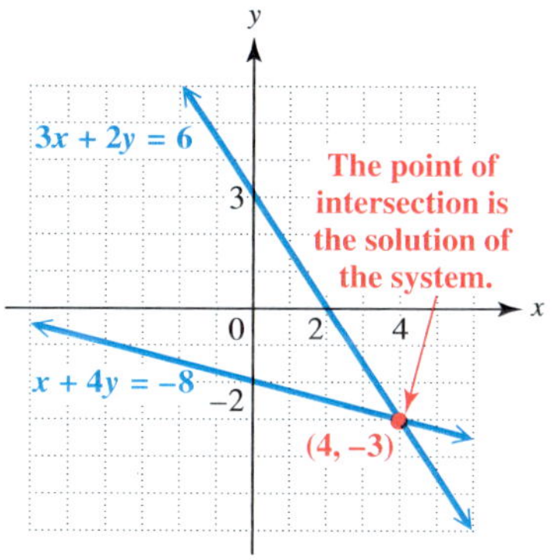


**FIGURE 1**

## EXAMPLE 2 Solving a System by Graphing

Solve the system of equations by graphing both equations on the same axes.

$$2x + 3y = 4$$
$$3x - y = -5$$

To graph these two equations, we choose *any* number for either $x$ or $y$ to get an ordered pair. The intercepts are often convenient choices. It is a good idea to use a third ordered pair as a check.

$2x + 3y = 4$

| $x$ | $y$ |
|---|---|
| 0 | $\frac{4}{3}$ |
| 2 | 0 |
| −2 | $\frac{8}{3}$ |

$3x - y = -5$

| $x$ | $y$ |
|---|---|
| 0 | 5 |
| $-\frac{5}{3}$ | 0 |
| −2 | −1 |

The lines in Figure 2 suggest that the graphs intersect at the point (−1, 2). We check this by substituting −1 for $x$ and 2 for $y$ in both equations.

$$2x + 3y = 4$$
$$2(-1) + 3(2) = 4 \quad ?$$
$$4 = 4 \quad \text{True}$$

$$3x - y = -5$$
$$3(-1) - 2 = -5 \quad ?$$
$$-5 = -5 \quad \text{True}$$

Because (−1, 2) satisfies both equations, the solution set of this system is {(−1, 2)}.

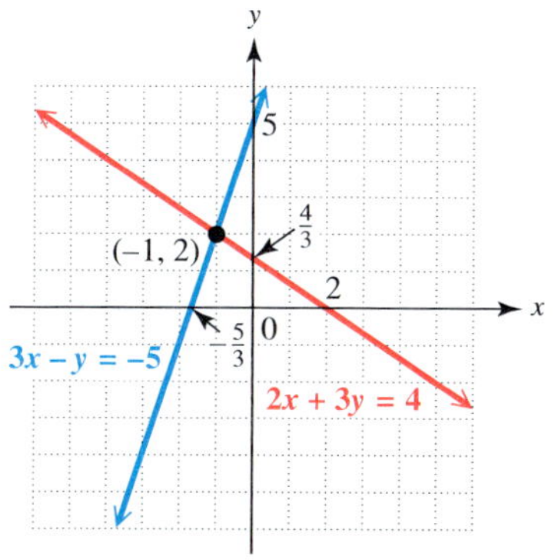


**FIGURE 2**

**Now Try Exercise 13.**

To solve a system by graphing, follow these steps.

**Solving a Linear System by Graphing**

*Step 1* **Graph each equation** of the system on the same coordinate axes.

*Step 2* **Find the coordinates of the point of intersection** of the graphs, if possible. This is the solution of the system.

*Step 3* **Check** the solution in both of the original equations. Then write the solution set.

**CAUTION** A difficulty with the graphing method of solution is that it may not be possible to determine from the graph the exact coordinates of the point that represents the solution, particularly if these coordinates are not integers. The graphing method does, however, show geometrically how solutions are found and is useful when approximate answers will do.

**OBJECTIVE 3 Solve special systems by graphing.** Sometimes the graphs of the two equations in a system either do not intersect at all or are the same line.

### EXAMPLE 3 Solving Special Systems by Graphing

Solve each system by graphing.

**(a)** $2x + y = 2$
$2x + y = 8$

The graphs of these lines are shown in Figure 3. The two lines are parallel and have no points in common. For such a system, there is no solution; we write the solution set as $\emptyset$.

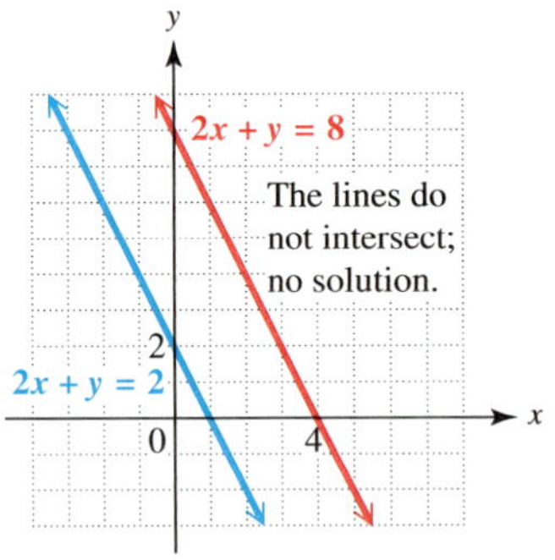


FIGURE 3

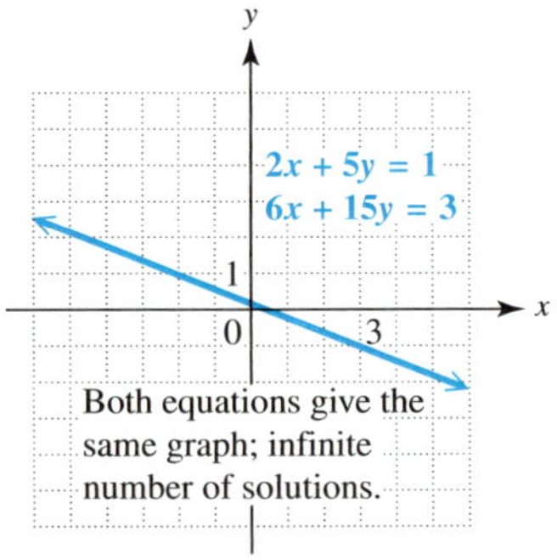


FIGURE 4

**(b)** $2x + 5y = 1$
$6x + 15y = 3$

The graphs of these two equations are the same line. See Figure 4. We can obtain the second equation by multiplying each side of the first equation by 3. In this case, every point on the line is a solution of the system, and the solution set contains an infinite number of ordered pairs. We write the solution set using set-builder notation as $\{(x, y) \mid 2x + 5y = 1\}$.

**NOTE** In Example 3(b), either equation of the system could be used to write the solution set. We prefer to use the equation (in standard form) with coefficients that are integers ($A > 0$) having no common factor (except 1). Other texts may express such solutions differently.

The system in Example 2 has exactly one solution. A system with at least one solution is called a **consistent system.** A system with no solution, such as the one in Example 3(a), is called an **inconsistent system.** The equations in Example 2 are **independent equations** with different graphs. The equations of the system in Example 3(b) have the same graph and are equivalent. Because they are different forms of the same equation, these equations are called **dependent equations.** Examples 2 and 3 show the three cases that may occur when solving a system of equations with two variables.

**Possible Types of Solutions**

1. The graphs intersect at exactly one point, which gives the (single) ordered pair solution of the system. The **system is consistent** and the **equations are independent.** See Figure 5(a).
2. The graphs are parallel lines, so there is no solution and the solution set is $\emptyset$. The **system is inconsistent.** See Figure 5(b).
3. The graphs are the same line. There are an infinite number of solutions. The **equations are dependent.** See Figure 5(c).

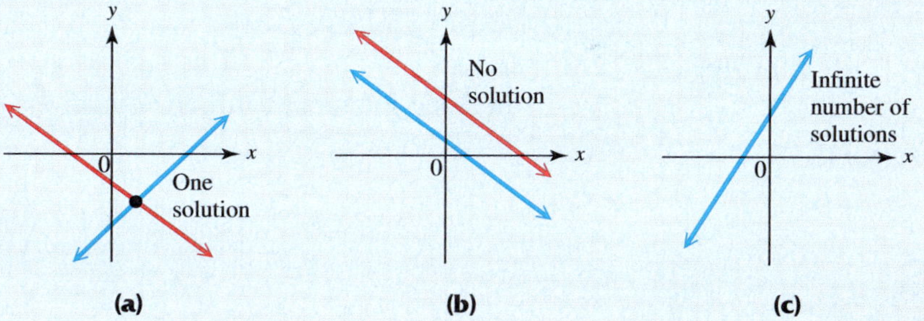


**FIGURE 5**

**Now Try Exercises 23 and 25.**

**OBJECTIVE 4 Identify special systems without graphing.** Example 3 showed that the graphs of an inconsistent system are parallel lines and the graphs of a system of dependent equations are the same line. We can recognize these special kinds of systems without graphing by using slopes.

## EXAMPLE 4 Identifying the Three Cases Using Slopes

Describe each system without graphing.

**(a)** $3x + 2y = 6$
$-2y = 3x - 5$

Write each equation in slope-intercept form by solving for $y$.

| | |
|---|---|
| $3x + 2y = 6$ | $-2y = 3x - 5$ |
| $2y = -3x + 6$ | $y = -\frac{3}{2}x + \frac{5}{2}$ |
| $y = -\frac{3}{2}x + 3$ | |

Both equations have slope $-\frac{3}{2}$ but they have different $y$-intercepts, 3 and $\frac{5}{2}$. In Chapter 3 we found that lines with the same slope are parallel, so these equations have graphs that are parallel lines. The system has no solution.

**(b)** $2x - y = 4$

$x = \frac{y}{2} + 2$

Again, write the equations in slope-intercept form.

| | |
|---|---|
| $2x - y = 4$ | $x = \frac{y}{2} + 2$ |
| $-y = -2x + 4$ | $\frac{y}{2} + 2 = x$ |
| $y = 2x - 4$ | $\frac{y}{2} = x - 2$ |
| | $y = 2x - 4$ |

The equations are exactly the same; their graphs are the same line. The system has an infinite number of solutions.

**(c)** $x - 3y = 5$

$2x + y = 8$

In slope-intercept form, the equations are as follows.

| | |
|---|---|
| $x - 3y = 5$ | $2x + y = 8$ |
| $-3y = -x + 5$ | $y = -2x + 8$ |
| $y = \frac{1}{3}x - \frac{5}{3}$ | |

The graphs of these equations are neither parallel nor the same line since the slopes are different. This system has exactly one solution. ■

**Now Try Exercises 31, 33, and 35.**

**OBJECTIVE 5 Use a graphing calculator to solve a linear system.** In the next example we use a graphing calculator to find the solution of the system in Example 1(a).

## EXAMPLE 5 Finding the Solution Set of a System with a Graphing Calculator

Solve the system by graphing with a calculator.

$$x + 4y = -8$$
$$3x + 2y = 6$$

To enter the equations in a graphing calculator, first solve each equation for $y$.

| | |
|---|---|
| $x + 4y = -8$ | $3x + 2y = 6$ |
| $4y = -x - 8$ | $2y = -3x + 6$ |
| $y = -\frac{1}{4}x - 2$ | $y = -\frac{3}{2}x + 3$ |

The calculator allows us to enter several equations to be graphed at the same time. We designate the first one $Y_1$ and the second one $Y_2$. See Figure 6(a). Notice the careful use of parentheses with the fractions. We graph the two equations using a standard window and then use the capability of the calculator to find the coordinates of the point of intersection of the graphs. The display at the bottom of Figure 6(b) indicates that the solution set is $\{(4, -3)\}$.

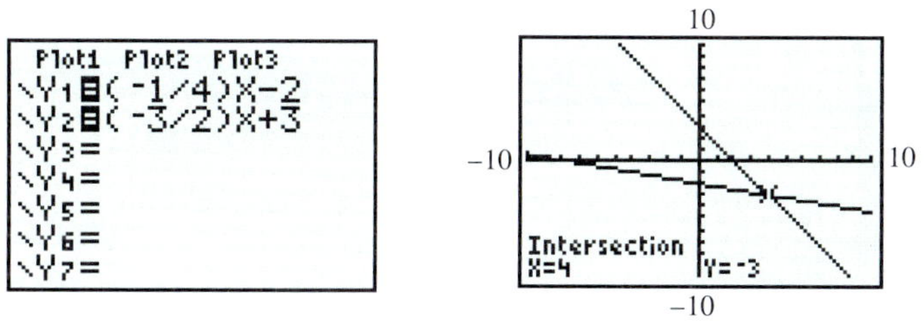


**(a)** **(b)**

**FIGURE 6**

**Now Try Exercise 51.**

## 8.1 EXERCISES

**For Extra Help**

Student's Solutions Manual

MyMathLab

InterAct Math Tutorial Software

AW Math Tutor Center

MathXL

Digital Video Tutor CD 12/Videotape 12

**1.** Each ordered pair in (a)–(d) is a solution of one of the systems graphed in A–D. Because of the location of the point of intersection, you should be able to determine the correct system for each solution. Match each system from A–D with its solution from (a)–(d).

**(a)** (3, 4)

**(b)** (−2, 3)

**(c)** (−4, −1)

**(d)** (5, −2)

**A.**

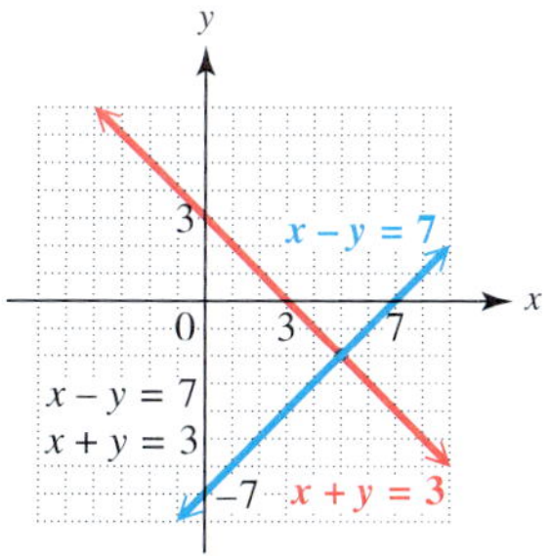


**B.**

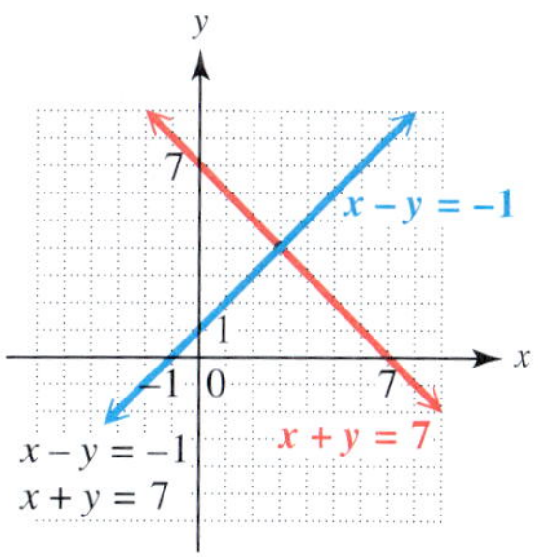


**C.**

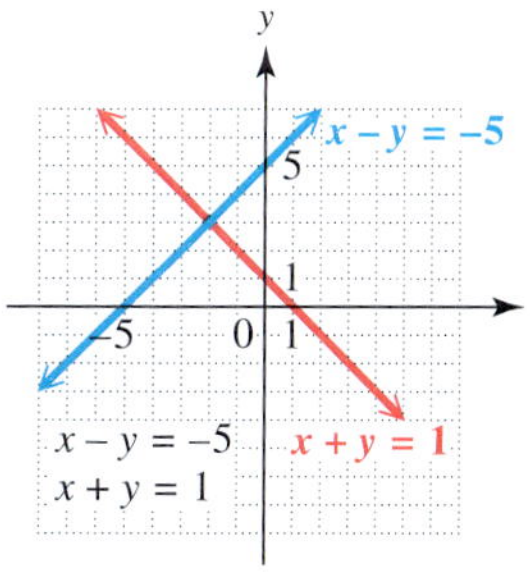


**D.**

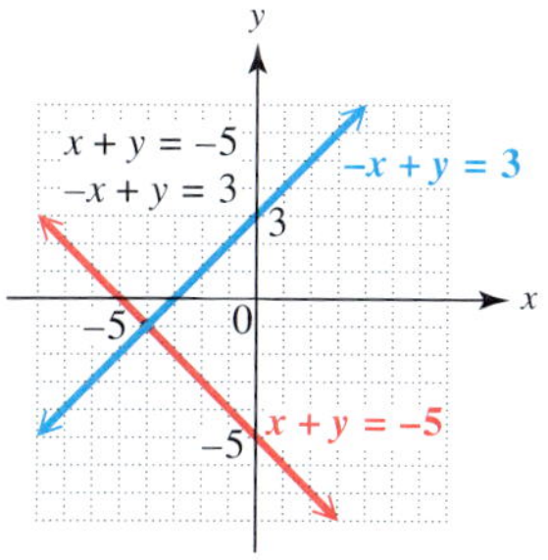

**2.** When a student was asked to determine whether the ordered pair $(1, -2)$ is a solution of the system

$$x + y = -1$$
$$2x + y = 4,$$

he answered "yes." His reasoning was that the ordered pair satisfies the equation $x + y = -1$ since $1 + (-2) = -1$ is true. Why is the student's answer wrong?

*Decide whether the given ordered pair is a solution of the given system. See Example 1.*

**3.** $(2, -3)$
$x + y = -1$
$2x + 5y = 19$

**4.** $(4, 3)$
$x + 2y = 10$
$3x + 5y = 3$

**5.** $(-1, -3)$
$3x + 5y = -18$
$4x + 2y = -10$

**6.** $(-9, -2)$
$2x - 5y = -8$
$3x + 6y = -39$

**7.** $(7, -2)$
$4x = 26 - y$
$3x = 29 + 4y$

**8.** $(9, 1)$
$2x = 23 - 5y$
$3x = 24 + 3y$

**9.** $(6, -8)$
$-2y = x + 10$
$3y = 2x + 30$

**10.** $(-5, 2)$
$5y = 3x + 20$
$3y = -2x - 4$

**11.** Which ordered pair could not possibly be a solution of the system graphed? Why is it the only valid choice?

**A.** $(-4, -4)$ **B.** $(-2, 2)$
**C.** $(-4, 4)$ **D.** $(-3, 3)$

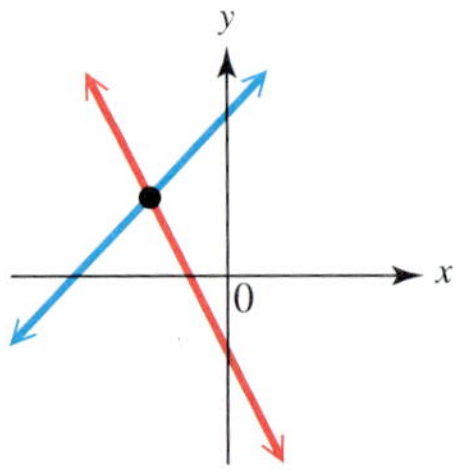


**12.** Which ordered pair could possibly be a solution of the system graphed? Why is it the only valid choice?

**A.** $(2, 0)$ **B.** $(0, 2)$
**C.** $(-2, 0)$ **D.** $(0, -2)$

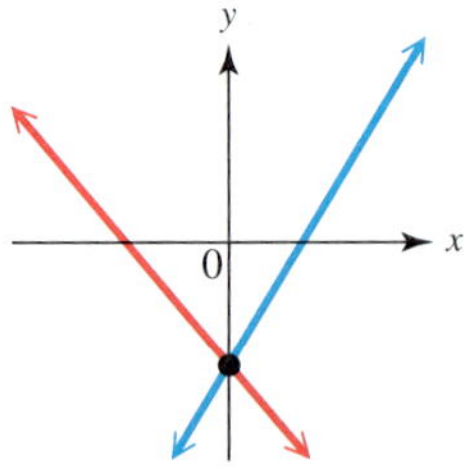


*Solve each system of equations by graphing. If the system is inconsistent or the equations are dependent, say so. See Examples 2 and 3.*

**13.** $x - y = 2$
$x + y = 6$

**14.** $x - y = 3$
$x + y = -1$

**15.** $x + y = 4$
$y - x = 4$

**16.** $x + y = -5$
$x - y = 5$

**17.** $x - 2y = 6$
$x + 2y = 2$

**18.** $2x - y = 4$
$4x + y = 2$

**19.** $3x - 2y = -3$
$-3x - y = -6$

**20.** $2x - y = 4$
$2x + 3y = 12$

**21.** $2x - 3y = -6$
$y = -3x + 2$

**22.** $x + 2y = 4$
$2x + 4y = 12$

**23.** $2x - y = 6$
$4x - 2y = 8$

**24.** $2x - y = 4$
$4x = 2y + 8$

**25.** $3x = 5 - y$
$6x + 2y = 10$

**26.** $-3x + y = -3$
$y = x - 3$

**27.** $3x - 4y = 24$
$y = -\frac{3}{2}x + 3$

**28.** Solve the system

$$\begin{aligned} 2x + 3y &= 6 \\ x - 3y &= 5 \end{aligned}$$

by graphing. Can you check your solution? Why or why not?

**29.** Explain one of the drawbacks of solving a system of equations graphically.

**30.** Explain the three situations that may occur (regarding the number of solutions) when solving a system of two linear equations in two variables by graphing.

*Without graphing, answer the following questions for each linear system. See Example 4.*

**(a)** *Is the system inconsistent, are the equations dependent, or neither?*
**(b)** *Is the graph a pair of intersecting lines, a pair of parallel lines, or one line?*
**(c)** *Does the system have one solution, no solution, or an infinite number of solutions?*

**31.** $y - x = -5$
$x + y = 1$

**32.** $2x + y = 6$
$x - 3y = -4$

**33.** $x + 2y = 0$
$4y = -2x$

**34.** $y = 3x$
$y + 3 = 3x$

**35.** $5x + 4y = 7$
$10x + 8y = 4$

**36.** $2x + 3y = 12$
$2x - y = 4$

**37.** $x - 3y = 5$
$2x + y = 8$

**38.** $2x - y = 4$
$x = \frac{1}{2}y + 2$

*An application of mathematics in economics deals with* **supply and demand.** *Typically, as the price of an item increases, the demand for the item decreases, while the supply increases. (There are exceptions to this, however.) If supply and demand can be described by straight-line equations, the point at which the lines intersect determines the* **equilibrium supply** *and* **equilibrium demand.**

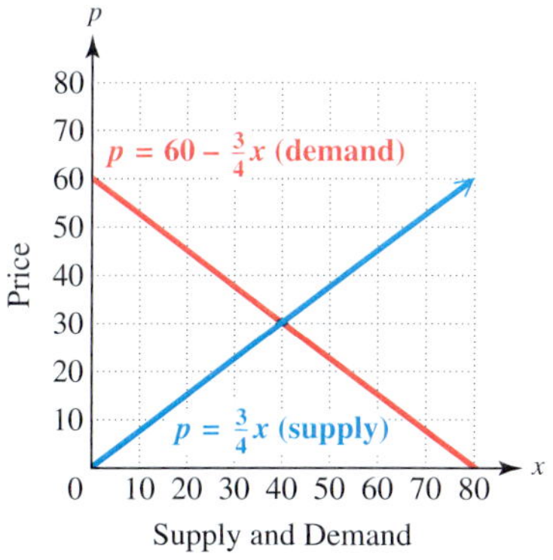


*Suppose that an economist has studied the supply and demand for aluminum siding and has concluded that the price per unit, p, and the demand, x, are related by the demand equation* $p = 60 - \frac{3}{4}x$*, while the supply is given by the equation* $p = \frac{3}{4}x$*. The graphs of these two equations are shown in the figure.*

*Use the graph to answer Exercises 39–42.*

**39.** At what value of $x$ does supply equal demand?

**40.** At what value of $p$ does supply equal demand?

**41.** What are the coordinates of the point of intersection of the two lines?

**42.** When $x > 40$, does demand exceed supply or does supply exceed demand?

*The personal computer market share for two manufacturers is shown in the graph.*

**43.** For which years was Hewlett-Packard's share less than Packard Bell, NEC's share?

**44.** Estimate the year in which market share for Hewlett-Packard and Packard Bell, NEC was the same. About what was this share?

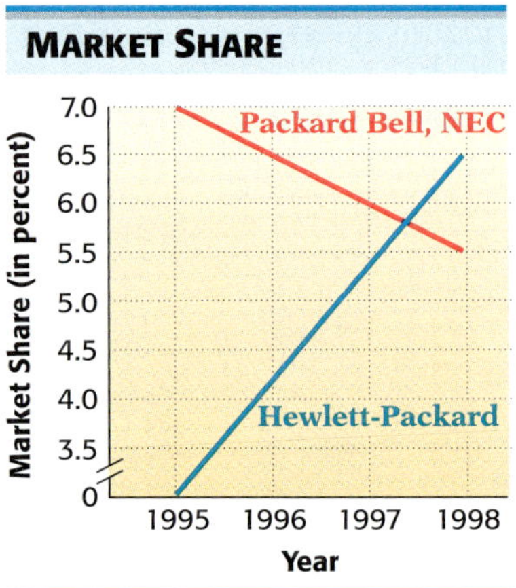


***Source:*** Intelliquest; IDC.

**45.** The graph shows network share (the percentage of TV sets in use) for the early evening news programs for the three major broadcast networks from 1986 through 2000.

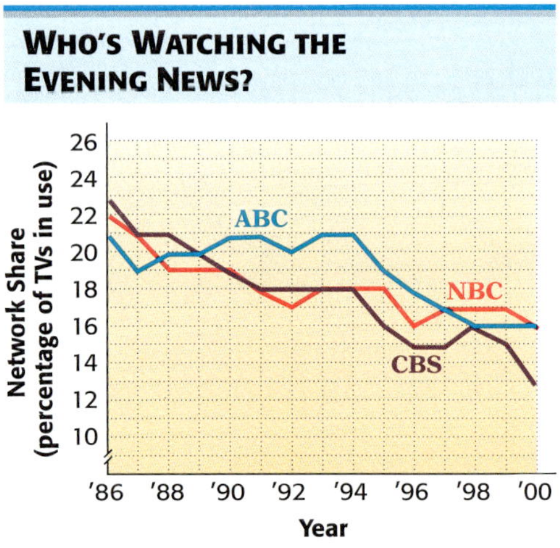


***Source:*** Nielsen Media Research.

**(a)** Between what years did the ABC early evening news dominate?

**(b)** During what year did ABC's dominance end? Which network equaled ABC's share that year? What was that share?

**(c)** During what years did ABC and CBS have equal network share? What was the share for each of these years?

**(d)** Which networks most recently had equal share? Write their share as an ordered pair of the form (year, share).

**(e)** Describe the general trend in viewership for the three major networks during these years.

**46.** The graph shows how the production of vinyl LPs, audiocassettes, and compact discs (CDs) changed over the years from 1986 through 1998.

**(a)** In what year did cassette production and CD production reach equal levels? What was that level?

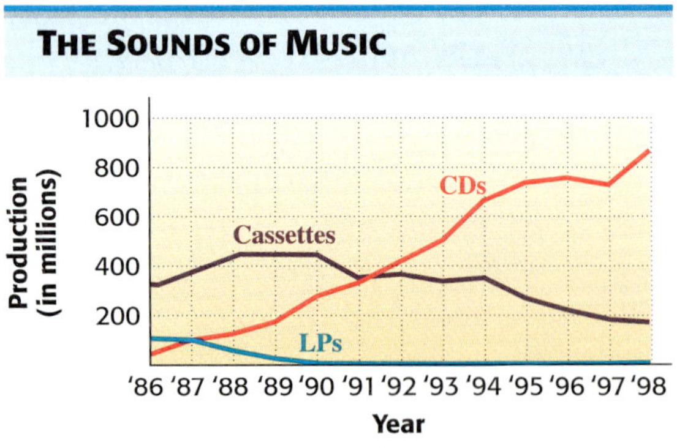


***Source:*** Recording Industry Association of America.

**(b)** Express the point of intersection of the graphs of LP production and CD production as an ordered pair of the form (year, production level).

**(c)** Between what years did cassette production first stabilize and remain fairly constant?

**(d)** Describe the trend in CD production from 1986 through 1998. If a straight line were used to approximate its graph, would the line have positive, negative, or 0 slope?

**(e)** If a straight line were used to approximate the graph of cassette production from 1990 through 1998, would the line have positive, negative, or 0 slope? Explain.

## TECHNOLOGY INSIGHTS (EXERCISES 47–50)

*Match the graphing calculator screens in choices A–D with the appropriate system in Exercises 47–50. See Example 5.*

**A.**

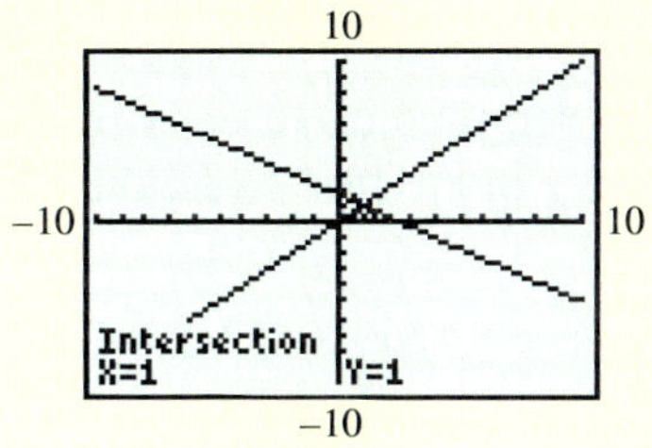


**B.**

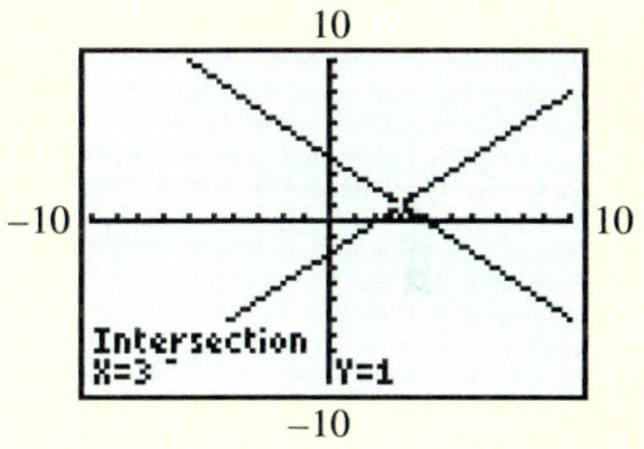


**C.**

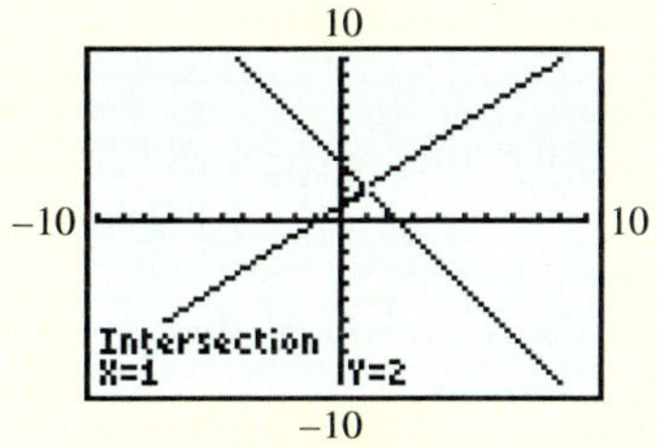


**D.**

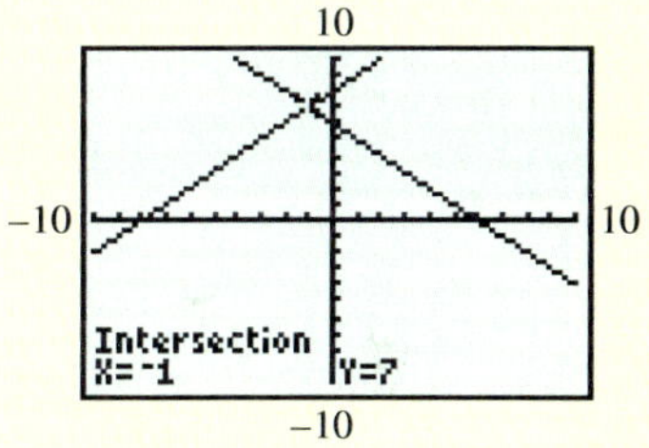


**47.** $x + y = 4$
$x - y = 2$

**48.** $x + y = 6$
$x - y = -8$

**49.** $2x + 3y = 5$
$x - y = 0$

**50.** $3x + 2y = 7$
$-x + y = 1$

*Use a graphing calculator to solve each system. See Example 5.*

**51.** $3x + y = 2$
$2x - y = -7$

**52.** $x + 2y = -2$
$2x - y = 11$

**53.** $8x + 4y = 0$
$4x - 2y = 2$

**54.** $3x + 3y = 0$
$4x + 2y = 3$

# 8.2 Solving Systems of Linear Equations by Substitution

**OBJECTIVES**

1 Solve linear systems by substitution.
2 Solve special systems by substitution.
3 Solve linear systems with fractions.

**OBJECTIVE 1 Solve linear systems by substitution.** Graphing to solve a system of equations has a serious drawback: It is difficult to accurately find a solution such as $\left(\frac{1}{3}, -\frac{5}{6}\right)$ from a graph. One algebraic method for solving a system of equations is the **substitution method.** This method is particularly useful for solving systems where one equation is already solved, or can be solved quickly, for one of the variables.

### EXAMPLE 1 Using the Substitution Method

Solve the system by the substitution method.

$$3x + 5y = 26$$
$$y = 2x$$

The second equation is already solved for $y$. This equation says that $y = 2x$. Substituting $2x$ for $y$ in the first equation gives

$$\begin{aligned} 3x + 5y &= 26 \\ 3x + 5(2x) &= 26 \quad \text{Let } y = 2x. \\ 3x + 10x &= 26 \quad \text{Multiply.} \\ 13x &= 26 \quad \text{Combine like terms.} \\ x &= 2. \quad \text{Divide by 13.} \end{aligned}$$

Because $x = 2$, we find $y$ from the equation $y = 2x$ by substituting 2 for $x$.

$$y = 2(2) = 4 \quad \text{Let } x = 2.$$

We check that the solution of the given system is (2, 4) by substituting 2 for $x$ and 4 for $y$ in *both* equations.

*Check:*

$$\begin{aligned} 3x + 5y &= 26 \\ 3(2) + 5(4) &= 26 \quad ? \\ 6 + 20 &= 26 \quad ? \\ 26 &= 26 \quad \text{True} \end{aligned} \qquad \begin{aligned} y &= 2x \\ 4 &= 2(2) \quad ? \\ 4 &= 4 \quad \text{True} \end{aligned}$$

Since (2, 4) satisfies both equations, the solution set is $\{(2, 4)\}$.

**Now Try Exercise 3.**

### EXAMPLE 2 Using the Substitution Method

Solve the system

$$2x + 5y = 7$$
$$x = -1 - y.$$

The second equation gives $x$ in terms of $y$. Substitute $-1 - y$ for $x$ in the first equation.

$$\begin{aligned} 2x + 5y &= 7 \\ 2(-1 - y) + 5y &= 7 \quad \text{Let } x = -1 - y. \\ -2 - 2y + 5y &= 7 \quad \text{Distributive property} \end{aligned}$$

$$-2 + 3y = 7 \quad \text{Combine like terms.}$$
$$3y = 9 \quad \text{Add 2.}$$
$$y = 3 \quad \text{Divide by 3.}$$

To find $x$, substitute 3 for $y$ in the equation $x = -1 - y$ to get

$$x = -1 - 3 = -4.$$

Check that the solution set of the given system is $\{(-4, 3)\}$.

**Now Try Exercise 5.**

**CAUTION** Be careful when you write the ordered pair solution of a system. Even though we found $y$ first in Example 2, the $x$-coordinate is *always* written first in the ordered pair.

To solve a system by substitution, follow these steps.

**Solving a Linear System by Substitution**

*Step 1* **Solve one equation for either variable.** If one of the variables has coefficient 1 or $-1$, choose it, since the substitution method is usually easier this way.

*Step 2* **Substitute** for that variable in the other equation. The result should be an equation with just one variable.

*Step 3* **Solve** the equation from Step 2.

*Step 4* **Substitute** the result from Step 3 into the equation from Step 1 to find the value of the other variable.

*Step 5* **Check** the solution in both of the original equations. Then write the solution set.

### EXAMPLE 3 Using the Substitution Method

Use substitution to solve the system

$$2x = 4 - y \quad (1)$$
$$5x + 3y = 10. \quad (2)$$

*Step 1* We must solve one of the equations for either $x$ or $y$. Because the coefficient of $y$ in equation (1) is $-1$, we avoid fractions by solving this equation for $y$.

$$2x = 4 - y \quad (1)$$
$$2x - 4 = -y \quad \text{Subtract 4.}$$
$$-2x + 4 = y \quad \text{Multiply by } -1.$$

*Step 2* Now substitute $-2x + 4$ for $y$ in equation (2).

$$5x + 3y = 10 \quad (2)$$
$$5x + 3(-2x + 4) = 10 \quad \text{Let } y = -2x + 4.$$

*Step 3* Solve the equation from Step 2.

$$5x - 6x + 12 = 10 \quad \text{Distributive property}$$
$$-x + 12 = 10 \quad \text{Combine like terms.}$$
$$-x = -2 \quad \text{Subtract 12.}$$
$$x = 2 \quad \text{Multiply by } -1.$$

*Step 4* Since $y = -2x + 4$ and $x = 2$,

$$y = -2(2) + 4 = 0,$$

and the solution is (2, 0).

*Step 5* *Check:*

| | | | |
|---|---|---|---|
| $2x = 4 - y$ | (1) | $5x + 3y = 10$ | (2) |
| $2(2) = 4 - 0$ ? | | $5(2) + 3(0) = 10$ ? | |
| $4 = 4$ | True | $10 = 10$ | True |

Since both results are true, the solution set of the system is $\{(2, 0)\}$.

**Now Try Exercise 7.**

**OBJECTIVE 2 Solve special systems by substitution.** In the previous section we solved inconsistent systems with graphs that are parallel lines and systems of dependent equations with graphs that are the same line. We can also solve these special systems with the substitution method.

**EXAMPLE 4 Solving an Inconsistent System by Substitution**

Use substitution to solve the system

$$x = 5 - 2y \quad (1)$$
$$2x + 4y = 6. \quad (2)$$

Substitute $5 - 2y$ for $x$ in equation (2).

$$2x + 4y = 6 \quad (2)$$
$$2(5 - 2y) + 4y = 6 \quad \text{Let } x = 5 - 2y.$$
$$10 - 4y + 4y = 6 \quad \text{Distributive property}$$
$$10 = 6 \quad \text{False}$$

This false result means that the equations in the system have graphs that are parallel lines. The system is inconsistent and has no solution, so the solution set is $\emptyset$. See Figure 7.

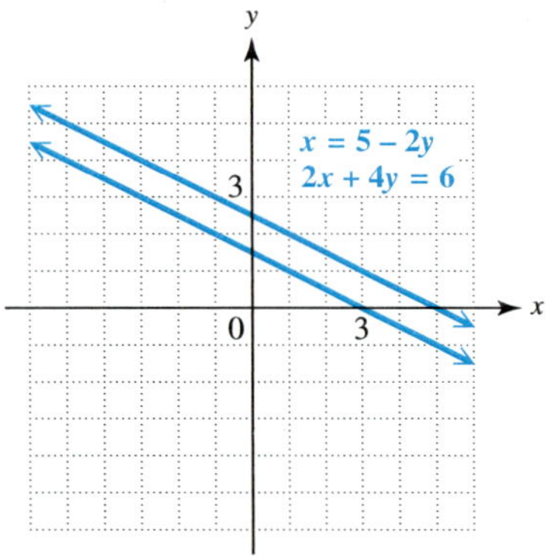


FIGURE 7

**Now Try Exercise 17.**

**CAUTION** It is a common error to give "false" as the solution of an inconsistent system. The correct response is $\emptyset$.

**EXAMPLE 5 Solving a System with Dependent Equations by Substitution**

Solve the system by the substitution method.

$$3x - y = 4 \qquad (1)$$
$$-9x + 3y = -12 \qquad (2)$$

Begin by solving equation (1) for $y$ to get $y = 3x - 4$. Substitute $3x - 4$ for $y$ in equation (2) and solve the resulting equation.

$$-9x + 3y = -12 \qquad (2)$$
$$-9x + 3(3x - 4) = -12 \qquad \text{Let } y = 3x - 4.$$
$$-9x + 9x - 12 = -12 \qquad \text{Distributive property}$$
$$0 = 0 \qquad \text{Add 12; combine like terms.}$$

This true result means that every solution of one equation is also a solution of the other, so the system has an infinite number of solutions: all the ordered pairs corresponding to points that lie on the common graph. The solution set is $\{(x, y) \mid 3x - y = 4\}$. A graph of the equations of this system is shown in Figure 8.

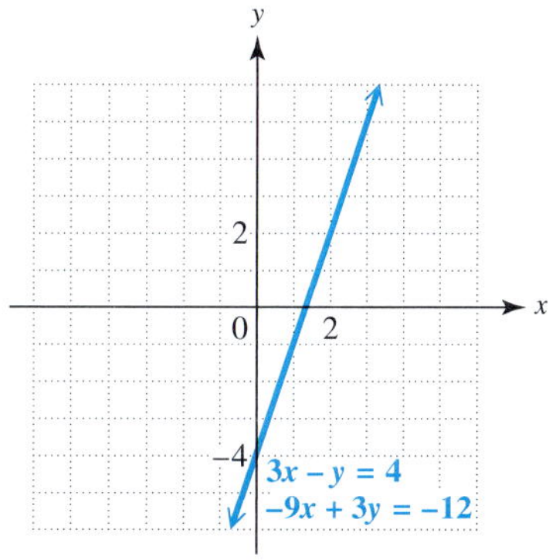


**FIGURE 8**

**Now Try Exercise 19.**

**CAUTION** It is a common error to give "true" as the solution of a system of dependent equations. The correct response is "infinite number of solutions."

**OBJECTIVE 3 Solve linear systems with fractions.** When a system includes an equation with fractions as coefficients, eliminate the fractions by multiplying each side of the equation by a common denominator. Then solve the resulting system.

**EXAMPLE 6 Using the Substitution Method with Fractions as Coefficients**

Solve the system by the substitution method.

$$3x + \frac{1}{4}y = 2 \qquad (1)$$

$$\frac{1}{2}x + \frac{3}{4}y = -\frac{5}{2} \qquad (2)$$

Clear equation (1) of fractions by multiplying each side by 4.

$$4\left(3x + \frac{1}{4}y\right) = 4(2) \qquad \text{Multiply by 4.}$$

$$4(3x) + 4\left(\frac{1}{4}y\right) = 4(2) \qquad \text{Distributive property}$$

$$12x + y = 8 \qquad (3)$$

Now clear equation (2) of fractions by multiplying each side by the common denominator 4.

$$4\left(\frac{1}{2}x + \frac{3}{4}y\right) = 4\left(-\frac{5}{2}\right) \qquad \text{Multiply by 4.}$$

$$4\left(\frac{1}{2}x\right) + 4\left(\frac{3}{4}y\right) = 4\left(-\frac{5}{2}\right) \qquad \text{Distributive property}$$

$$2x + 3y = -10 \qquad (4)$$

The given system of equations has been simplified to the equivalent system

$$12x + y = 8 \qquad (3)$$

$$2x + 3y = -10. \qquad (4)$$

Solve this system by the substitution method. Equation (3) can be solved for $y$ by subtracting $12x$ from each side.

$$12x + y = 8 \qquad (3)$$

$$y = -12x + 8 \qquad \text{Subtract } 12x.$$

Now substitute this result for $y$ in equation (4).

$$2x + 3y = -10 \qquad (4)$$

$$2x + 3(-12x + 8) = -10 \qquad \text{Let } y = -12x + 8.$$

$$2x - 36x + 24 = -10 \qquad \text{Distributive property}$$

$$-34x = -34 \qquad \text{Combine like terms; subtract 24.}$$

$$x = 1 \qquad \text{Divide by } -34.$$

Substitute 1 for $x$ in $y = -12x + 8$ to get

$$y = -12(1) + 8 = -4.$$

Check by substituting 1 for $x$ and $-4$ for $y$ in both of the original equations. The solution set is $\{(1, -4)\}$.

**Now Try Exercise 25.**

# 8.2 EXERCISES

**For Extra Help**

Student's Solutions Manual

MyMathLab

InterAct Math Tutorial Software

AW Math Tutor Center

MathXL

Digital Video Tutor CD 12/Videotape 12

**1.** A student solves the system

$$\begin{aligned} 5x - y &= 15 \\ 7x + y &= 21 \end{aligned}$$

and finds that $x = 3$, which is the correct value for $x$. The student gives the solution set as $\{3\}$. Is this correct? Explain.

**2.** A student solves the system

$$\begin{aligned} x + y &= 4 \\ 2x + 2y &= 8 \end{aligned}$$

and obtains the equation $0 = 0$. The student gives the solution set as $\{(0, 0)\}$. Is this correct? Explain.

*Solve each system by the substitution method. Check each solution. See Examples 1–5.*

**3.** $x + y = 12$; $y = 3x$

**4.** $x + 3y = -28$; $y = -5x$

**5.** $3x + 2y = 27$; $x = y + 4$

**6.** $4x + 3y = -5$; $x = y - 3$

**7.** $3x + 5y = 25$; $x - 2y = -10$

**8.** $5x + 2y = -15$; $2x - y = -6$

**9.** $3x + 4 = -y$; $2x + y = 0$

**10.** $2x - 5 = -y$; $x + 3y = 0$

**11.** $7x + 4y = 13$; $x + y = 1$

**12.** $3x - 2y = 19$; $x + y = 8$

**13.** $3x - y = 5$; $y = 3x - 5$

**14.** $4x - y = -3$; $y = 4x + 3$

**15.** $2x + y = 0$; $4x - 2y = 2$

**16.** $x + y = 0$; $4x + 2y = 3$

**17.** $2x + 8y = 3$; $x = 8 - 4y$

**18.** $2x + 10y = 3$; $x = 1 - 5y$

**19.** $2x = -12 + y$; $2y = 4x + 24$

**20.** $3x = 7 - y$; $2y = 14 - 6x$

**21.** When you use the substitution method, how can you tell that a system has

**(a)** no solution?

**(b)** an infinite number of solutions?

**22.** Solve each system.

**(a)** $5x - 4y = 7$; $x = 3$ **(b)** $5x - 4y = 7$; $y = -3$

Why are these systems easier to solve than the examples in this section?

*Solve each system by the substitution method. First clear all fractions. Check each solution. See Example 6.*

**23.** $\frac{1}{2}x + \frac{1}{3}y = 3$; $y = 3x$

**24.** $\frac{1}{4}x - \frac{1}{5}y = 9$; $y = 5x$

**25.** $\frac{1}{2}x + \frac{1}{3}y = -\frac{1}{3}$; $\frac{1}{2}x + 2y = -7$

**26.** $\frac{1}{6}x + \frac{1}{6}y = 1$; $-\frac{1}{2}x - \frac{1}{3}y = -5$

**27.** $\frac{x}{5} + 2y = \frac{8}{5}$; $\frac{3x}{5} + \frac{y}{2} = -\frac{7}{10}$

**28.** $\frac{x}{2} + \frac{y}{3} = \frac{7}{6}$; $\frac{x}{4} - \frac{3y}{2} = \frac{9}{4}$

**29.** $\frac{x}{5} + y = \frac{6}{5}$
$\frac{x}{10} + \frac{y}{3} = \frac{5}{6}$

**30.** $\frac{1}{2}x - \frac{1}{8}y = -\frac{1}{4}$
$-4x + y = 2$

**31.** $\frac{1}{6}x + \frac{1}{3}y = 8$
$\frac{1}{4}x + \frac{1}{2}y = 12$

**32.** One student solved the system

$$\frac{1}{3}x - \frac{1}{2}y = 7$$
$$\frac{1}{6}x + \frac{1}{3}y = 0$$

and wrote as his answer "$x = 12$," while another solved it and wrote as her answer "$y = -6$." Who, if either, was correct? Why?

## RELATING CONCEPTS (EXERCISES 33–36)

**For Individual or Group Work**

*A system of linear equations can be used to model the cost and the revenue of a business.* ***Work Exercises 33–36 in order.***

**33.** Suppose that you start a business manufacturing and selling bicycles, and it costs you \$5000 to get started. You determine that each bicycle will cost \$400 to manufacture. Explain why the linear equation $y_1 = 400x + 5000$ gives your *total* cost to manufacture $x$ bicycles ($y_1$ in dollars).

**34.** You decide to sell each bike for \$600. What expression in $x$ represents the revenue you will take in if you sell $x$ bikes? Write an equation using $y_2$ to express your revenue when you sell $x$ bikes ($y_2$ in dollars).

**35.** Form a system from the two equations in Exercises 33 and 34 and then solve the system.

**36.** The value of $x$ from Exercise 35 is the number of bikes it takes to *break even.* Fill in the blanks: When _____ bikes are sold, the break-even point is reached. At that point, you have spent _____ dollars and taken in _____ dollars.

*Solve each system by substitution. Then graph both lines in the standard viewing window of a graphing calculator and use the intersection feature to support your answer. See Example 5 in Section 8.1. (In Exercises 41 and 42, you will need to solve each equation for y first before graphing.)*

**37.** $y = 6 - x$
$y = 2x$

**38.** $y = 4x - 4$
$y = -3x - 11$

**39.** $y = -\frac{4}{3}x + \frac{19}{3}$
$y = \frac{15}{2}x - \frac{5}{2}$

**40.** $y = -\frac{15}{2}x + 10$
$y = \frac{25}{3}x - \frac{65}{3}$

**41.** $4x + 5y = 5$
$2x + 3y = 1$

**42.** $6x + 5y = 13$
$3x + 3y = 4$

**43.** If the point of intersection does not appear on your screen when solving a linear system using a graphing calculator, how can you find the point of intersection?

**44.** Suppose that you were asked to solve the system

$$y = 1.73x + 5.28$$
$$y = -2.94x - 3.85.$$

Why would it probably be easier to solve this system using a graphing calculator than using the substitution method?

# 8.3 Solving Systems of Linear Equations by Elimination

**OBJECTIVES**

1. Solve linear systems by elimination.
2. Multiply when using the elimination method.
3. Use an alternative method to find the second value in a solution.
4. Use the elimination method to solve special systems.

**OBJECTIVE 1 Solve linear systems by elimination.** An algebraic method that depends on the addition property of equality can also be used to solve systems. As mentioned earlier, adding the same quantity to each side of an equation results in equal sums.

$$\text{If} \quad A = B, \qquad \text{then} \qquad A + C = B + C.$$

We can take this addition a step further. Adding *equal* quantities, rather than the *same* quantity, to each side of an equation also results in equal sums.

$$\text{If} \quad A = B \quad \text{and} \quad C = D, \qquad \text{then} \qquad A + C = B + D.$$

Using the addition property to solve systems is called the **elimination method.** When using this method, the idea is to *eliminate* one of the variables. To do this, one pair of variable terms in the two equations must have coefficients that are opposites.

**EXAMPLE 1 Using the Elimination Method**

Use the elimination method to solve the system

$$x + y = 5$$
$$x - y = 3.$$

Each equation in this system is a statement of equality, so the sum of the right sides equals the sum of the left sides. Adding in this way gives

$$(x + y) + (x - y) = 5 + 3.$$

Combine terms and simplify to get

$$2x = 8$$
$$x = 4. \qquad \text{Divide by 2.}$$

Notice that $y$ has been eliminated. The result, $x = 4$, gives the $x$-value of the solution of the given system. To find the $y$-value of the solution, substitute 4 for $x$ in either of the two equations of the system. Choosing the first equation, $x + y = 5$, gives

$$x + y = 5$$
$$4 + y = 5 \qquad \text{Let } x = 4.$$
$$y = 1. \qquad \text{Subtract 4.}$$

Check the solution, (4, 1), by substituting 4 for $x$ and 1 for $y$ in both equations of the given system.

| | |
|---|---|
| $x + y = 5$ | $x - y = 3$ |
| $4 + 1 = 5$ ? | $4 - 1 = 3$ ? |
| $5 = 5$ True | $3 = 3$ True |

Since both results are true, the solution set of the system is $\{(4, 1)\}$.

**Now Try Exercise 5.**

**CAUTION** A system is not completely solved until values for *both* $x$ and $y$ are found. Do not stop after finding the value of only one variable. Remember to write the solution set as a set containing an ordered pair.

In general, use the following steps to solve a linear system of equations by the elimination method.

**Solving a Linear System by Elimination**

*Step 1* **Write both equations in standard form** $Ax + By = C$.

*Step 2* **Transform so that the coefficients of one pair of variable terms are opposites.** Multiply one or both equations by appropriate numbers so that the sum of the coefficients of either the $x$- or $y$-terms is 0.

*Step 3* **Add** the new equations to eliminate a variable. The sum should be an equation with just one variable.

*Step 4* **Solve** the equation from Step 3 for the remaining variable.

*Step 5* **Substitute** the result from Step 4 into either of the original equations and solve for the other variable.

*Step 6* **Check** the solution in both of the original equations. Then write the solution set.

It does not matter which variable is eliminated first. Usually we choose the one that is more convenient to work with.

**EXAMPLE 2 Using the Elimination Method**

Solve the system

$$y + 11 = 2x$$
$$5x = y + 26.$$

*Step 1* Rewrite both equations in standard form $Ax + By = C$ to get the system

$$-2x + y = -11 \quad \text{Subtract } 2x \text{ and } 11.$$
$$5x - y = 26. \quad \text{Subtract } y.$$

*Step 2* Because the coefficients of $y$ are 1 and $-1$, adding will eliminate $y$. It is not necessary to multiply either equation by a number.

*Step 3* Add the two equations. This time we use vertical addition.

$$\begin{aligned} -2x + y &= -11 \\ 5x - y &= \phantom{-}26 \\ \hline 3x \phantom{- y} &= \phantom{-}15 \end{aligned}$$ Add in columns.

*Step 4* Solve the equation.

$$3x = 15$$
$$x = 5 \qquad \text{Divide by 3.}$$

*Step 5* Find the value of $y$ by substituting 5 for $x$ in either of the original equations. Choosing the first gives

$$\begin{aligned} y + 11 &= 2x \\ y + 11 &= 2(5) \\ y + 11 &= 10 && \text{Let } x = 5. \\ y &= -1. && \text{Subtract 11.} \end{aligned}$$

*Step 6* Check by substituting $x = 5$ and $y = -1$ in both of the original equations.

*Check:*

| | |
|---|---|
| $y + 11 = 2x$ | $5x = y + 26$ |
| $(-1) + 11 = 2(5)$ ? | $5(5) = -1 + 26$ ? |
| $10 = 10$ True | $25 = 25$ True |

Since $(5, -1)$ is a solution of *both* equations, the solution set is $\{(5, -1)\}$.

**Now Try Exercise 9.**

**OBJECTIVE 2 Multiply when using the elimination method.** In both of the preceding examples, a variable was eliminated by adding the equations. Sometimes we need to multiply each side of one or both equations in a system by some number before adding will eliminate a variable.

## EXAMPLE 3 Multiplying Both Equations When Using the Elimination Method

Solve the system

$$2x + 3y = -15 \qquad (1)$$
$$5x + 2y = 1. \qquad (2)$$

Adding the two equations gives $7x + 5y = -14$, which does not eliminate either variable. However, we can multiply each equation by a suitable number so that the coefficients of one of the two variables are opposites. For example, to eliminate $x$, we multiply each side of equation (1) by 5, and each side of equation (2) by $-2$.

$$\begin{aligned} 10x + 15y &= -75 && \text{Multiply equation (1) by 5.} \\ -10x - \phantom{1}4y &= \phantom{-}{-2} && \text{Multiply equation (2) by } -2. \\ \hline 11y &= -77 && \text{Add.} \\ y &= -7 && \text{Divide by 11.} \end{aligned}$$

Substituting $-7$ for $y$ in either equation (1) or (2) gives $x = 3$. Check that the solution set of the system is $\{(3, -7)\}$.

**Now Try Exercise 19.**

**CAUTION** When using the elimination method, remember to *multiply both sides* of an equation by the same nonzero number.

**OBJECTIVE 3 Use an alternative method to find the second value in a solution.** Sometimes it is easier to find the value of the second variable in a solution by using the elimination method twice. The next example shows this approach.

### EXAMPLE 4 Finding the Second Value Using an Alternative Method

Solve the system

$$4x = 9 - 3y \quad (1)$$
$$5x - 2y = 8. \quad (2)$$

Rearrange the terms in equation (1) so that like terms are aligned in columns. Add $3y$ to each side to get the system

$$4x + 3y = 9 \quad (3)$$
$$5x - 2y = 8. \quad (2)$$

One way to proceed is to eliminate $y$ by multiplying each side of equation (3) by 2 and each side of equation (2) by 3, and then adding.

$$8x + 6y = 18 \quad \text{Multiply equation (3) by 2.}$$
$$15x - 6y = 24 \quad \text{Multiply equation (2) by 3.}$$
$$23x \quad = 42 \quad \text{Add.}$$
$$x = \frac{42}{23} \quad \text{Divide by 23.}$$

Substituting $\frac{42}{23}$ for $x$ in one of the given equations would give $y$, but the arithmetic involved would be messy. Instead, solve for $y$ by starting again with the original equations and eliminating $x$. Multiply each side of equation (3) by 5 and each side of equation (2) by $-4$, and then add.

$$20x + 15y = 45 \quad \text{Multiply equation (3) by 5.}$$
$$-20x + 8y = -32 \quad \text{Multiply equation (2) by } -4.$$
$$23y = 13 \quad \text{Add.}$$
$$y = \frac{13}{23} \quad \text{Divide by 23.}$$

Check that the solution set is $\left\{\left(\frac{42}{23}, \frac{13}{23}\right)\right\}$.

**Now Try Exercise 27.**

**NOTE** When the value of the first variable is a fraction, the method used in Example 4 helps avoid arithmetic errors. Of course, this method could be used to solve any system of equations.

**OBJECTIVE 4 Use the elimination method to solve special systems.** The next example shows the elimination method when a system is inconsistent or the equations of the system are dependent. To contrast the elimination method with the substitution method, in part (b) we use the same system solved in Example 5 of the previous section.

**EXAMPLE 5 Using the Elimination Method for an Inconsistent System or Dependent Equations**

Solve each system by the elimination method.

**(a)** $2x + 4y = 5$
$4x + 8y = -9$

Multiply each side of $2x + 4y = 5$ by $-2$; then add to $4x + 8y = -9$.

$$\begin{aligned} -4x - 8y &= -10 \\ 4x + 8y &= -9 \\ \hline 0 &= -19 \quad \text{False} \end{aligned}$$

The false statement $0 = -19$ indicates that the given system has solution set $\emptyset$.

**(b)** $3x - y = 4$
$-9x + 3y = -12$

Multiply each side of the first equation by 3; then add the two equations.

$$\begin{aligned} 9x - 3y &= 12 \\ -9x + 3y &= -12 \\ \hline 0 &= 0 \quad \text{True} \end{aligned}$$

A true statement occurs when the equations are equivalent. As before, this result indicates that every solution of one equation is also a solution of the other; there are an infinite number of solutions. The solution set is $\{(x, y) \mid 3x - y = 4\}$.

**Now Try Exercises 35 and 37.**

## 8.3 EXERCISES

**For Extra Help**

- Student's Solutions Manual
- MyMathLab
- InterAct Math Tutorial Software
- AW Math Tutor Center
- MathXL
- Digital Video Tutor CD 12/Videotape 12

*Answer* true *or* false *for each statement. If false, tell why.*

**1.** To eliminate the $y$-terms in the system

$$\begin{aligned} 2x + 12y &= 7 \\ 3x + 4y &= 1, \end{aligned}$$

we should multiply the bottom equation by 3 and then add.

**2.** The ordered pair (0, 0) *must* be a solution of a system of the form

$$\begin{aligned} Ax + By &= 0 \\ Cx + Dy &= 0. \end{aligned}$$

**3.** The system

$$\begin{aligned} x + y &= 1 \\ x + y &= 2 \end{aligned}$$

has $\emptyset$ as its solution set.

4. The ordered pair $(4, -5)$ cannot be a solution of a system that contains the equation $5x - 4y = 0$.

*Solve each system by the elimination method. Check each solution. See Examples 1 and 2.*

5. $x - y = -2$
$x + y = 10$

6. $x + y = 10$
$x - y = -6$

7. $2x + y = -5$
$x - y = 2$

8. $2x + y = -15$
$-x - y = 10$

9. $2y = -3x$
$-3x - y = 3$

10. $5x = y + 5$
$-5x + 2y = 0$

11. $6x - y = -1$
$5y = 17 + 6x$

12. $y = 9 - 6x$
$-6x + 3y = 15$

*Solve each system by the elimination method.* (Hint: *In Exercises 33 and 34, first clear all fractions.*) *Check each solution. See Examples 3–5.**

13. $2x - y = 12$
$3x + 2y = -3$

14. $x + y = 3$
$-3x + 2y = -19$

15. $x + 4y = 16$
$3x + 5y = 20$

16. $2x + y = 8$
$5x - 2y = -16$

17. $2x - 8y = 0$
$4x + 5y = 0$

18. $3x - 15y = 0$
$6x + 10y = 0$

19. $3x + 3y = 33$
$5x - 2y = 27$

20. $4x - 3y = -19$
$3x + 2y = 24$

21. $5x + 4y = 12$
$3x + 5y = 15$

22. $2x + 3y = 21$
$5x - 2y = -14$

23. $5x - 4y = 15$
$-3x + 6y = -9$

24. $4x + 5y = -16$
$5x - 6y = -20$

25. $6x - 2y = -21$
$-3x + 4y = 36$

26. $6x - 2y = -21$
$3x + 4y = 36$

27. $3x - 7y = 1$
$-5x + 4y = 4$

28. $-4x + 3y = 2$
$5x - 2y = -3$

29. $2x + 3y = 0$
$4x + 12 = 9y$

30. $-4x + 3y = 2$
$5x + 3 = -2y$

31. $24x + 12y = -7$
$16x - 17 = 18y$

32. $9x + 4y = -3$
$6x + 7 = -6y$

33. $3x = 3 + 2y$
$-\frac{4}{3}x + y = \frac{1}{3}$

34. $3x = 27 + 2y$
$x - \frac{7}{2}y = -25$

35. $-x + 3y = 4$
$-2x + 6y = 8$

36. $6x - 2y = 24$
$-3x + y = -12$

37. $5x - 2y = 3$
$10x - 4y = 5$

38. $3x - 5y = 1$
$6x - 10y = 4$

## RELATING CONCEPTS (EXERCISES 39–44)

**For Individual or Group Work**

*The graph on the next page shows movie attendance from 1991 through 1999. In 1991, attendance was* 1141 *million, as represented by the point P*(1991, 1141). *In 1999, attendance was* 1465 *million, as represented by the point Q*(1999, 1465). *We can find an equation of line segment PQ using a system of equations and then use the equation to approximate the attendance in any of the years between 1991 and 1999.* ***Work Exercises 39–44 in order.***

*The authors thank Mitchel Levy of Broward Community College for his suggestions for this group of exercises.

**MOVIE BOX OFFICE ATTENDANCE/ ADMISSIONS**

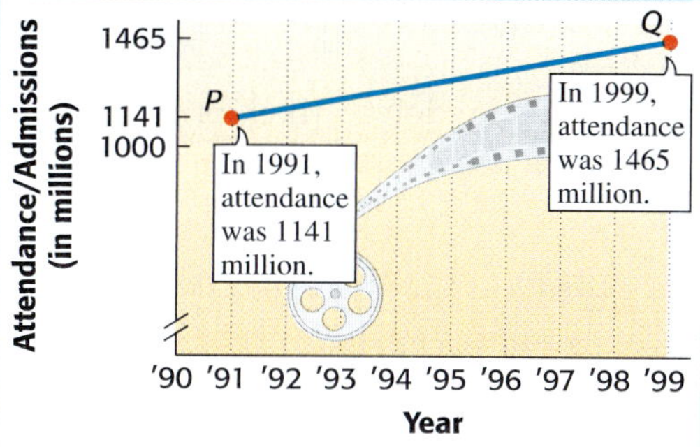


***Source:*** Motion Picture Association of America.

**39.** The line segment has an equation that can be written in the form $y = ax + b$. Using the coordinates of point $P$ with $x = 1991$ and $y = 1141$, write an equation in the variables $a$ and $b$.

**40.** Using the coordinates of point $Q$ with $x = 1999$ and $y = 1465$, write a second equation in the variables $a$ and $b$.

**41.** Write the system of equations formed from the two equations in Exercises 39 and 40, and solve the system using the elimination method.

**42.** What is the equation of the segment $PQ$?

**43.** Let $x = 1998$ in the equation of Exercise 42, and solve for $y$. How does the result compare with the actual figure of 1481 million?

**44.** The actual data points for the years 1991 through 1999 do not lie in a perfectly straight line. Explain the pitfalls of relying too heavily on using the equation in Exercise 42 to approximate attendance.

## SUMMARY EXERCISES ON SOLVING SYSTEMS OF LINEAR EQUATIONS

The exercises in this summary include a variety of problems on solving systems of linear equations. Since we do not usually specify the method of solution, use the following guidelines to help you decide whether to use substitution or elimination.

### Choosing a Method When Solving a System of Linear Equations

**1.** If one of the equations of the system is already solved for one of the variables, as in the systems

$$\begin{aligned} 3x + 4y &= 9 \\ y &= 2x - 6 \end{aligned} \qquad \text{or} \qquad \begin{aligned} -5x + 3y &= 9 \\ x &= 3y - 7, \end{aligned}$$

the substitution method is the better choice.

**2.** If both equations are in standard $Ax + By = C$ form, as in

$$\begin{aligned} 4x - 11y &= 3 \\ -2x + 3y &= 4, \end{aligned}$$

and none of the variables has coefficient $-1$ or $1$, the elimination method is the better choice.

(continued)

**3.** If one or both of the equations are in standard form and the coefficient of one of the variables is $-1$ or 1, as in the systems

$$\begin{aligned} 3x + y &= -2 \\ -5x + 2y &= 4 \end{aligned} \qquad \text{or} \qquad \begin{aligned} -x + 3y &= -4 \\ 3x - 2y &= 8, \end{aligned}$$

use the elimination method, or solve for the variable with coefficient $-1$ or 1 and then use the substitution method.

*Use the information in the preceding box to solve each problem.*

**1.** Assuming you want to minimize the amount of work required, tell whether you would use the substitution or elimination method to solve each system. Explain your answers. *Do not actually solve.*

**(a)** $\begin{aligned} 3x + 5y &= 69 \\ y &= 4x \end{aligned}$ **(b)** $\begin{aligned} 3x + y &= -7 \\ x - y &= -5 \end{aligned}$ **(c)** $\begin{aligned} 3x - 2y &= 0 \\ 9x + 8y &= 7 \end{aligned}$

**2.** Which one of the following systems would be easier to solve using the substitution method? Why?

$$\begin{aligned} 5x - 3y &= 7 \\ 2x + 8y &= 3 \end{aligned} \qquad \begin{aligned} 7x + 2y &= 4 \\ y &= -3x + 1 \end{aligned}$$

*In Exercises 3 and 4, **(a)** solve the system by the elimination method, **(b)** solve the system by the substitution method, and **(c)** tell which method you prefer for that particular system and why.*

**3.** $\begin{aligned} 4x - 3y &= -8 \\ x + 3y &= 13 \end{aligned}$

**4.** $\begin{aligned} 2x + 5y &= 0 \\ x &= -3y + 1 \end{aligned}$

*Solve each system by the method of your choice. (For Exercises 5–7, see your answers for Exercise 1.)*

**5.** $\begin{aligned} 3x + 5y &= 69 \\ y &= 4x \end{aligned}$

**6.** $\begin{aligned} 3x + y &= -7 \\ x - y &= -5 \end{aligned}$

**7.** $\begin{aligned} 3x - 2y &= 0 \\ 9x + 8y &= 7 \end{aligned}$

**8.** $\begin{aligned} x + y &= 7 \\ x &= -3 - y \end{aligned}$

**9.** $\begin{aligned} 6x + 7y &= 4 \\ 5x + 8y &= -1 \end{aligned}$

**10.** $\begin{aligned} 6x - y &= 5 \\ y &= 11x \end{aligned}$

**11.** $\begin{aligned} 4x - 6y &= 10 \\ -10x + 15y &= -25 \end{aligned}$

**12.** $\begin{aligned} 3x - 5y &= 7 \\ 2x + 3y &= 30 \end{aligned}$

**13.** $\begin{aligned} 5x &= 7 + 2y \\ 5y &= 5 - 3x \end{aligned}$

**14.** $\begin{aligned} 4x + 3y &= 1 \\ 3x + 2y &= 2 \end{aligned}$

**15.** $\begin{aligned} 2x - 3y &= 7 \\ -4x + 6y &= 14 \end{aligned}$

**16.** $\begin{aligned} 2x + 3y &= 10 \\ -3x + y &= 18 \end{aligned}$

**17.** $\begin{aligned} 2x + 5y &= 4 \\ x + y &= -1 \end{aligned}$

**18.** $\begin{aligned} x - 3y &= 7 \\ 4x + y &= 5 \end{aligned}$

*Solve each system by any method. First clear all fractions.*

**19.** $\begin{aligned} \frac{1}{3}x - \frac{1}{2}y &= \frac{1}{6} \\ 3x - 2y &= 9 \end{aligned}$

**20.** $\begin{aligned} \frac{1}{5}x + \frac{2}{3}y &= -\frac{8}{5} \\ 3x - y &= 9 \end{aligned}$

**21.** $\begin{aligned} \frac{1}{6}x + \frac{1}{6}y &= 2 \\ -\frac{1}{2}x - \frac{1}{3}y &= -8 \end{aligned}$

**22.** $\begin{aligned} \frac{x}{2} - \frac{y}{3} &= 9 \\ \frac{x}{5} - \frac{y}{4} &= 5 \end{aligned}$

**23.** $\begin{aligned} \frac{x}{3} - \frac{3y}{4} &= -\frac{1}{2} \\ \frac{x}{6} + \frac{y}{8} &= \frac{3}{4} \end{aligned}$

**24.** $\begin{aligned} \frac{x}{5} + 2y &= \frac{16}{5} \\ \frac{3x}{5} + \frac{y}{2} &= -\frac{7}{5} \end{aligned}$

# 8.4 Systems of Linear Equations in Three Variables

**OBJECTIVES**

1. Understand the geometry of systems of three equations in three variables.
2. Solve linear systems (with three equations and three variables) by elimination.
3. Solve linear systems (with three equations and three variables) where some of the equations have missing terms.
4. Solve special systems (with three equations and three variables).

A solution of an equation in three variables, such as

$$2x + 3y - z = 4,$$

is called an **ordered triple** and is written $(x, y, z)$. For example, the ordered triple $(0, 1, -1)$ is a solution of the equation, because

$$2(0) + 3(1) - (-1) = 0 + 3 + 1 = 4.$$

Verify that another solution of this equation is $(10, -3, 7)$.

The term *linear equation* can be extended to equations of the form

$$Ax + By + Cz + \cdots + Dw = K,$$

where not all the coefficients $A, B, C, \ldots, D$ equal 0. For example,

$$2x + 3y - 5z = 7 \qquad \text{and} \qquad x - 2y - z + 3u - 2w = 8$$

are linear equations, the first with three variables and the second with five variables.

**OBJECTIVE 1 Understand the geometry of systems of three equations in three variables.** Consider the solution of a system of linear equations in three variables, such as

$$\begin{aligned} 4x + 8y + z &= 2 \\ x + 7y - 3z &= -14 \\ 2x - 3y + 2z &= 3. \end{aligned}$$

Theoretically, a system of this type can be solved by graphing. However, the graph of a linear equation with three variables is a *plane,* not a line. Since the graph of each equation of the system is a plane, which requires three-dimensional graphing, this method is not practical. However, it does illustrate the number of solutions possible for such systems, as shown in Figure 9.

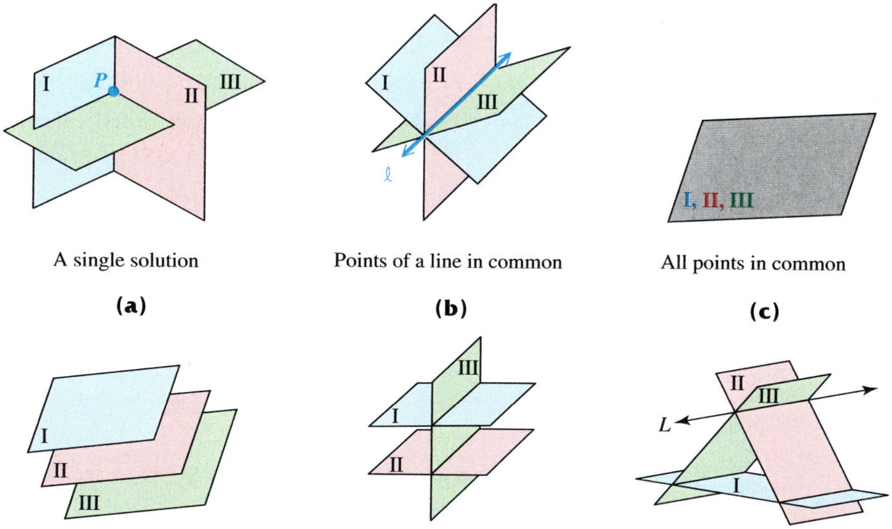


A single solution
**(a)**

Points of a line in common
**(b)**

All points in common
**(c)**

No points in common
**(d)**

No points in common
**(e)**

No points in common
**(f)**

**FIGURE 9**

Figure 9 illustrates the following cases.

**Graphs of Linear Systems in Three Variables**

1. The three planes may meet at a single, common point that is the solution of the system. See Figure 9(a).
2. The three planes may have the points of a line in common so that the infinite set of points that satisfy the equation of the line is the solution of the system. See Figure 9(b).
3. The three planes may coincide so that the solution of the system is the set of all points on a plane. See Figure 9(c).
4. The planes may have no points common to all three so that there is no solution of the system. See Figures 9(d), (e), and (f).

**OBJECTIVE 2 Solve linear systems (with three equations and three variables) by elimination.** Since graphing to find the solution set of a system of three equations in three variables is impractical, these systems are solved with an extension of the elimination method, summarized as follows.

**Solving a Linear System in Three Variables**

*Step 1* **Eliminate a variable.** Use the elimination method to eliminate any variable from any two of the original equations. The result is an equation in two variables.

*Step 2* **Eliminate the same variable again.** Eliminate the *same* variable from any *other* two equations. The result is an equation in the same two variables as in Step 1.

*Step 3* **Eliminate a different variable and solve.** Use the elimination method to eliminate a second variable from the two equations in two variables that result from Steps 1 and 2. The result is an equation in one variable that gives the value of that variable.

*Step 4* **Find a second value.** Substitute the value of the variable found in Step 3 into either of the equations in two variables to find the value of the second variable.

*Step 5* **Find a third value.** Use the values of the two variables from Steps 3 and 4 to find the value of the third variable by substituting into an appropriate equation.

*Step 6* **Check** the solution in all of the original equations. Then write the solution set.

**EXAMPLE 1 Solving a System in Three Variables**

Solve the system

$$4x + 8y + z = 2 \qquad (1)$$
$$x + 7y - 3z = -14 \qquad (2)$$
$$2x - 3y + 2z = 3. \qquad (3)$$

*Step 1* As before, the elimination method involves eliminating a variable from the sum of two equations. The choice of which variable to eliminate is arbitrary. Suppose we decide to begin by eliminating $z$. We multiply equation (1) by 3 and then add the result to equation (2).

$$\begin{array}{rl} 12x + 24y + 3z = 6 & \text{Multiply each side of (1) by 3.} \\ x + 7y - 3z = -14 & \text{(2)} \\ \hline 13x + 31y = -8 & \text{Add. (4)} \end{array}$$

*Step 2* Equation (4) has only two variables. To get another equation without $z$, we multiply equation (1) by $-2$ and add the result to equation (3). It is essential at this point to *eliminate the same variable, z*.

$$\begin{array}{rl} -8x - 16y - 2z = -4 & \text{Multiply each side of (1) by } -2. \\ 2x - 3y + 2z = 3 & \text{(3)} \\ \hline -6x - 19y = -1 & \text{Add. (5)} \end{array}$$

*Step 3* Now we solve the system of equations (4) and (5) for $x$ and $y$. This step is possible only if the *same* variable is eliminated in Steps 1 and 2.

$$\begin{array}{rl} 78x + 186y = -48 & \text{Multiply each side of (4) by 6.} \\ -78x - 247y = -13 & \text{Multiply each side of (5) by 13.} \\ \hline -61y = -61 & \text{Add.} \\ y = 1 & \end{array}$$

*Step 4* Now we substitute 1 for $y$ in either equation (4) or (5). Choosing (5) gives

$$\begin{array}{rl} -6x - 19y = -1 & \text{(5)} \\ -6x - 19(1) = -1 & \text{Let } y = 1. \\ -6x - 19 = -1 & \\ -6x = 18 & \\ x = -3. & \end{array}$$

*Step 5* We substitute $-3$ for $x$ and 1 for $y$ in any one of the three original equations to find $z$. Choosing (1) gives

$$\begin{array}{rl} 4x + 8y + z = 2 & \text{(1)} \\ 4(-3) + 8(1) + z = 2 & \text{Let } x = -3 \text{ and } y = 1. \\ -4 + z = 2 & \\ z = 6. & \end{array}$$

*Step 6* It appears that the ordered triple $(-3, 1, 6)$ is the only solution of the system. We must check that the solution satisfies all three equations of the system. For equation (1),

$$\begin{array}{rl} 4x + 8y + z = 2 & \text{(1)} \\ 4(-3) + 8(1) + 6 = 2 & ? \\ -12 + 8 + 6 = 2 & ? \\ 2 = 2. & \text{True} \end{array}$$

Because $(-3, 1, 6)$ also satisfies equations (2) and (3), the solution set is $\{(-3, 1, 6)\}$.

**Now Try Exercise 3.**

**OBJECTIVE 3 Solve linear systems (with three equations and three variables) where some of the equations have missing terms.** When this happens, one elimination step can be omitted.

**EXAMPLE 2 Solving a System of Equations with Missing Terms**

Solve the system

$$\begin{aligned} 6x - 12y &= -5 \quad (1) \\ 8y + z &= 0 \quad (2) \\ 9x - z &= 12. \quad (3) \end{aligned}$$

Since equation (3) is missing the variable $y$, eliminate $y$ using equations (1) and (2).

$$\begin{array}{rl} 12x - 24y \phantom{+ 3z} = -10 & \text{Multiply each side of (1) by 2.} \\ 24y + 3z = \phantom{-1}0 & \text{Multiply each side of (2) by 3.} \\ \hline 12x \phantom{- 24y} + 3z = -10 & \text{Add. (4)} \end{array}$$

Use this result, together with equation (3), to eliminate $z$. Multiply equation (3) by 3.

$$\begin{array}{rl} 27x - 3z = \phantom{-}36 & \text{Multiply each side of (3) by 3.} \\ 12x + 3z = -10 & \text{(4)} \\ \hline 39x \phantom{+ 3z} = \phantom{-}26 & \text{Add.} \end{array}$$

$$x = \frac{26}{39} = \frac{2}{3}$$

Substituting into equation (3) gives

$$\begin{aligned} 9x - z &= 12 \quad (3) \\ 9\left(\frac{2}{3}\right) - z &= 12 \quad \text{Let } x = \tfrac{2}{3}. \\ 6 - z &= 12 \\ z &= -6. \end{aligned}$$

Substituting $-6$ for $z$ in equation (2) gives

$$\begin{aligned} 8y + z &= 0 \quad (2) \\ 8y - 6 &= 0 \quad \text{Let } z = -6. \\ 8y &= 6 \\ y &= \frac{3}{4}. \end{aligned}$$

Check in each of the original equations of the system to verify that the solution set of the system is $\left\{\left(\frac{2}{3}, \frac{3}{4}, -6\right)\right\}$.

**Now Try Exercise 21.**

**OBJECTIVE 4 Solve special systems (with three equations and three variables).** Linear systems with three variables may be inconsistent or may include dependent equations. The next examples illustrate these cases.

**EXAMPLE 3 Solving an Inconsistent System with Three Variables**

Solve the system

$$2x - 4y + 6z = 5 \quad (1)$$
$$-x + 3y - 2z = -1 \quad (2)$$
$$x - 2y + 3z = 1. \quad (3)$$

Eliminate $x$ by adding equations (2) and (3) to get the equation

$$y + z = 0.$$

Now, *eliminate x again,* using equations (1) and (3).

$$-2x + 4y - 6z = -2 \quad \text{Multiply each side of (3) by } -2.$$
$$2x - 4y + 6z = 5 \quad (1)$$
$$0 = 3 \quad \text{False}$$

The resulting false statement indicates that equations (1) and (3) have no common solution. Thus, the system is inconsistent and the solution set is $\emptyset$. The graph of this system would show these two planes parallel to one another.

**Now Try Exercise 29.**

**NOTE** If you get a false statement when adding as in Example 3, you do not need to go any further with the solution. Since two of the three planes are parallel, it is not possible for the three planes to have any common points.

**EXAMPLE 4 Solving a System of Dependent Equations with Three Variables**

Solve the system

$$2x - 3y + 4z = 8 \quad (1)$$
$$-x + \frac{3}{2}y - 2z = -4 \quad (2)$$
$$6x - 9y + 12z = 24. \quad (3)$$

Multiplying each side of equation (1) by 3 gives equation (3). Multiplying each side of equation (2) by $-6$ also gives equation (3). Because of this, the equations are dependent. All three equations have the same graph, as illustrated in Figure 9(c). The solution set is written

$$\{(x, y, z) \mid 2x - 3y + 4z = 8\}.$$

Although any one of the three equations could be used to write the solution set, we use the equation in standard form with coefficients that are integers with no common factor (except 1), as we did in Section 8.1.

**Now Try Exercise 33.**

We can extend the method discussed in this section to solve larger systems. For example, to solve a system of four equations in four variables, eliminate a variable from three pairs of equations to get a system of three equations in three unknowns. Then proceed as shown above.

## 8.4 EXERCISES

**For Extra Help**

- Student's Solutions Manual
- MyMathLab
- InterAct Math Tutorial Software
- AW Math Tutor Center
- MathXL
- Digital Video Tutor CD 12/Videotape 12

**1.** Explain what the following statement means: The solution set of the system

$$\begin{aligned} 2x + y + z &= 3 \\ 3x - y + z &= -2 \\ 4x - y + 2z &= 0 \end{aligned}$$

is $\{(-1, 2, 3)\}$.

**2.** The two equations

$$\begin{aligned} x + y + z &= 6 \\ 2x - y + z &= 3 \end{aligned}$$

have a common solution of $(1, 2, 3)$. Which equation would complete a system of three linear equations in three variables having solution set $\{(1, 2, 3)\}$?

**A.** $3x + 2y - z = 1$ **B.** $3x + 2y - z = 4$
**C.** $3x + 2y - z = 5$ **D.** $3x + 2y - z = 6$

*Solve each system of equations. See Example 1.*

**3.** $$\begin{aligned} 2x - 5y + 3z &= -1 \\ x + 4y - 2z &= 9 \\ x - 2y - 4z &= -5 \end{aligned}$$

**4.** $$\begin{aligned} x + 3y - 6z &= 7 \\ 2x - y + z &= 1 \\ x + 2y + 2z &= -1 \end{aligned}$$

**5.** $$\begin{aligned} 3x + 2y + z &= 8 \\ 2x - 3y + 2z &= -16 \\ x + 4y - z &= 20 \end{aligned}$$

**6.** $$\begin{aligned} -3x + y - z &= -10 \\ -4x + 2y + 3z &= -1 \\ 2x + 3y - 2z &= -5 \end{aligned}$$

**7.** $$\begin{aligned} 2x + 5y + 2z &= 0 \\ 4x - 7y - 3z &= 1 \\ 3x - 8y - 2z &= -6 \end{aligned}$$

**8.** $$\begin{aligned} 5x - 2y + 3z &= -9 \\ 4x + 3y + 5z &= 4 \\ 2x + 4y - 2z &= 14 \end{aligned}$$

**9.** $$\begin{aligned} x + 2y + z &= 4 \\ 2x + y - z &= -1 \\ x - y - z &= -2 \end{aligned}$$

**10.** $$\begin{aligned} x - 2y + 5z &= -7 \\ -2x - 3y + 4z &= -14 \\ -3x + 5y - z &= -7 \end{aligned}$$

**11.** $$\begin{aligned} \frac{1}{3}x + \frac{1}{6}y - \frac{2}{3}z &= -1 \\ -\frac{3}{4}x - \frac{1}{3}y - \frac{1}{4}z &= 3 \\ \frac{1}{2}x + \frac{3}{2}y + \frac{3}{4}z &= 21 \end{aligned}$$

**12.** $$\begin{aligned} \frac{2}{3}x - \frac{1}{4}y + \frac{5}{8}z &= 0 \\ \frac{1}{5}x + \frac{2}{3}y - \frac{1}{4}z &= -7 \\ -\frac{3}{5}x + \frac{4}{3}y - \frac{7}{8}z &= -5 \end{aligned}$$

**13.** $$\begin{aligned} -x + 2y + 6z &= 2 \\ 3x + 2y + 6z &= 6 \\ x + 4y - 3z &= 1 \end{aligned}$$

**14.** $$\begin{aligned} 2x + y + 2z &= 1 \\ x + 2y + z &= 2 \\ x - y - z &= 0 \end{aligned}$$

**15.** $$\begin{aligned} x + y - z &= -2 \\ 2x - y + z &= -5 \\ -x + 2y - 3z &= -4 \end{aligned}$$

**16.** $$\begin{aligned} x + 2y + 3z &= 1 \\ -x - y + 3z &= 2 \\ -6x + y + z &= -2 \end{aligned}$$

*Solve each system of equations. See Example 2.*

**17.** $$\begin{aligned} 2x - 3y + 2z &= -1 \\ x + 2y + z &= 17 \\ 2y - z &= 7 \end{aligned}$$

**18.** $$\begin{aligned} 2x - y + 3z &= 6 \\ x + 2y - z &= 8 \\ 2y + z &= 1 \end{aligned}$$

**19.** $\begin{aligned} 4x + 2y - 3z &= 6 \\ x - 4y + z &= -4 \\ -x + 2z &= 2 \end{aligned}$

**20.** $\begin{aligned} 2x + 3y - 4z &= 4 \\ x - 6y + z &= -16 \\ -x + 3z &= 8 \end{aligned}$

**21.** $\begin{aligned} 2x + y &= 6 \\ 3y - 2z &= -4 \\ 3x - 5z &= -7 \end{aligned}$

**22.** $\begin{aligned} 4x - 8y &= -7 \\ 4y + z &= 7 \\ -8x + z &= -4 \end{aligned}$

**23.** $\begin{aligned} -5x + 2y + z &= 5 \\ -3x - 2y - z &= 3 \\ -x + 6y &= 1 \end{aligned}$

**24.** $\begin{aligned} x + y - z &= 0 \\ 2y - z &= 1 \\ 2x + 3y - 4z &= -4 \end{aligned}$

**25.** $\begin{aligned} 4x - z &= -6 \\ \frac{3}{5}y + \frac{1}{2}z &= 0 \\ \frac{1}{3}x + \frac{2}{3}z &= -5 \end{aligned}$

**26.** $\begin{aligned} 5x - 2z &= 8 \\ 4y + 3z &= -9 \\ \frac{1}{2}x + \frac{2}{3}y &= -1 \end{aligned}$

**27.** Using your immediate surroundings, give an example of three planes that

**(a)** intersect in a single point;
**(b)** do not intersect;
**(c)** intersect in infinitely many points.

**28.** Suppose that a system has infinitely many ordered triple solutions of the form $(x, y, z)$ such that

$$x + y + 2z = 1.$$

Give three specific ordered triples that are solutions of the system.

*Solve each system of equations. If the system is inconsistent or has dependent equations, say so. See Examples 1, 3, and 4.*

**29.** $\begin{aligned} 2x + 2y - 6z &= 5 \\ -3x + y - z &= -2 \\ -x - y + 3z &= 4 \end{aligned}$

**30.** $\begin{aligned} -2x + 5y + z &= -3 \\ 5x + 14y - z &= -11 \\ 7x + 9y - 2z &= -5 \end{aligned}$

**31.** $\begin{aligned} -5x + 5y - 20z &= -40 \\ x - y + 4z &= 8 \\ 3x - 3y + 12z &= 24 \end{aligned}$

**32.** $\begin{aligned} x + 4y - z &= 3 \\ -2x - 8y + 2z &= -6 \\ 3x + 12y - 3z &= 9 \end{aligned}$

**33.** $\begin{aligned} 2x + y - z &= 6 \\ 4x + 2y - 2z &= 12 \\ -x - \frac{1}{2}y + \frac{1}{2}z &= -3 \end{aligned}$

**34.** $\begin{aligned} 2x - 8y + 2z &= -10 \\ -x + 4y - z &= 5 \\ \frac{1}{8}x - \frac{1}{2}y + \frac{1}{8}z &= -\frac{5}{8} \end{aligned}$

**35.** $\begin{aligned} x + y - 2z &= 0 \\ 3x - y + z &= 0 \\ 4x + 2y - z &= 0 \end{aligned}$

**36.** $\begin{aligned} 2x + 3y - z &= 0 \\ x - 4y + 2z &= 0 \\ 3x - 5y - z &= 0 \end{aligned}$

*Extend the method of this section to solve each system. Express the solution in the form $(x, y, z, w)$.*

**37.** $\begin{aligned} x + y + z - w &= 5 \\ 2x + y - z + w &= 3 \\ x - 2y + 3z + w &= 18 \\ -x - y + z + 2w &= 8 \end{aligned}$

**38.** $\begin{aligned} 3x + y - z + 2w &= 9 \\ x + y + 2z - w &= 10 \\ x - y - z + 3w &= -2 \\ -x + y - z + w &= -6 \end{aligned}$

**39.**
$$\begin{aligned} 3x + y - z + w &= -3 \\ 2x + 4y + z - w &= -7 \\ -2x + 3y - 5z + w &= 3 \\ 5x + 4y - 5z + 2w &= -7 \end{aligned}$$

**40.**
$$\begin{aligned} x - 3y + 7z + w &= 11 \\ 2x + 4y + 6z - 3w &= -3 \\ 3x + 2y + z + 2w &= 19 \\ 4x + y - 3z + w &= 22 \end{aligned}$$

## RELATING CONCEPTS (EXERCISES 41–50)

**For Individual or Group Work**

*Suppose that on a distant planet a function of the form*

$$f(x) = ax^2 + bx + c \quad (a \neq 0)$$

*describes the height in feet of a projectile x sec after it has been projected upward.* ***Work Exercises 41–50 in order,*** *to see how this can be related to a system of three equations in three variables a, b, and c.*

**41.** After 1 sec, the height of a certain projectile is 128 ft. Thus, $f(1) = 128$. Use this information to find one equation in the variables $a$, $b$, and $c$. (*Hint:* Substitute 1 for $x$ and 128 for $f(x)$.)

**42.** After 1.5 sec, the height is 140 ft. Find a second equation in $a$, $b$, and $c$.

**43.** After 3 sec, the height is 80 ft. Find a third equation in $a$, $b$, and $c$.

**44.** Write a system of three equations in $a$, $b$, and $c$, based on your answers in Exercises 41–43. Solve the system.

**45.** What is the function $f$ for this particular projectile?

**46.** In the function $f$ written in Exercise 45, the _____ of the projectile is a function of the _____ elapsed since it was projected.

**47.** What was the initial height of the projectile? (*Hint:* Find $f(0)$.)

**48.** The projectile reaches its maximum height in 1.625 sec. Find its maximum height.

**49.** In Chapter 11 we discuss graphs of functions of the form $f(x) = ax^2 + bx + c$ $(a \neq 0)$. Use a system of equations to find the values of $a$, $b$, and $c$ for the function of this form that satisfies $f(1) = 2$, $f(-1) = 0$, and $f(-2) = 8$. Then write the expression for $f(x)$.

**50.** The accompanying table was generated by a graphing calculator for a function $Y_1 = aX^2 + bX + c$. Use any three points shown to find the values of $a$, $b$, and $c$. Then write the expression for $Y_1$.

| X | Y1 | |
|---|---|---|
| 1 | 8 | |
| 2 | 15 | |
| 3 | 24 | |
| 4 | 35 | |
| 5 | 48 | |
| 6 | 63 | |
| 7 | 80 | |

X=1

**51.** Discuss why it is necessary to eliminate the same variable in the first two steps of the elimination method with three equations and three variables.

**52.** In Step 3 of the elimination method for solving systems in three variables, does it matter which variable is eliminated? Explain.

# 8.5 Applications of Systems of Linear Equations

**OBJECTIVES**

1 Solve geometry problems using two variables.

2 Solve money problems using two variables.

3 Solve mixture problems using two variables.

4 Solve distance-rate-time problems using two variables.

5 Solve problems with three variables using a system of three equations.

Many applied problems involve more than one unknown quantity. Although some problems with two unknowns can be solved using just one variable, it is often easier to use two variables. To solve a problem with two unknowns, we must write two equations that relate the unknown quantities. The system formed by the pair of equations can then be solved using the methods of this chapter.

The following steps, based on the six-step problem-solving method first introduced in Chapter 2, give a strategy for solving applied problems using more than one variable.

**Solving an Applied Problem by Writing a System of Equations**

*Step 1* **Read** the problem carefully until you understand what is given and what is to be found.

*Step 2* **Assign variables** to represent the unknown values, using diagrams or tables as needed. *Write down* what each variable represents.

*Step 3* **Write a system of equations** that relates the unknowns.

*Step 4* **Solve** the system of equations.

*Step 5* **State the answer** to the problem. Does it seem reasonable?

*Step 6* **Check** the answer in the words of the original problem.

**OBJECTIVE 1 Solve geometry problems using two variables.** Problems about the perimeter of a geometric figure often involve two unknowns and can be solved using systems of equations.

**EXAMPLE 1 Finding the Dimensions of a Soccer Field**

Unlike football, where the dimensions of a playing field cannot vary, a rectangular soccer field may have a width between 50 and 100 yd and a length between 50 and 100 yd. Suppose that one particular field has a perimeter of 320 yd. Its length measures 40 yd more than its width. What are the dimensions of this field? (*Source: Microsoft Encarta Encyclopedia 2000.*)

*Step 1* **Read** the problem again. We are asked to find the dimensions of the field.

*Step 2* **Assign variables.** Let $L =$ the length and $W =$ the width. Figure 10 shows a soccer field with the length labeled $L$ and the width labeled $W$.

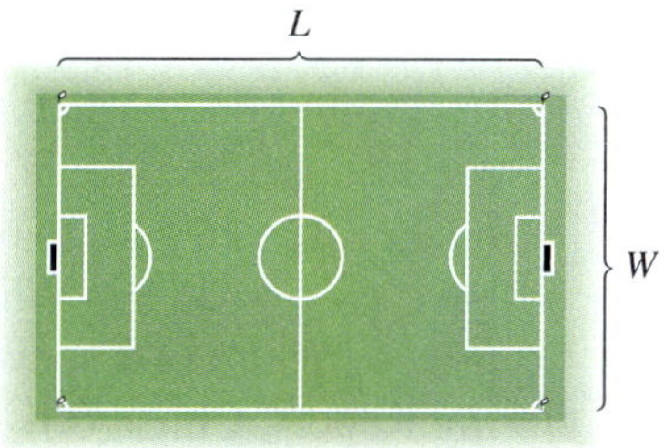


**FIGURE 10**

*Step 3* **Write a system of equations.** Because the perimeter is 320 yd, we find one equation by using the perimeter formula:

$$2L + 2W = 320.$$

Because the length is 40 yd more than the width, we have

$$L = W + 40.$$

The system is

$$\begin{aligned} 2L + 2W &= 320 && (1) \\ L &= W + 40. && (2) \end{aligned}$$

*Step 4* **Solve** the system of equations. Since equation (2) is solved for $L$, we can use the substitution method. We substitute $W + 40$ for $L$ in equation (1), and solve for $W$.

$$\begin{aligned} 2L + 2W &= 320 && (1) \\ 2(W + 40) + 2W &= 320 && \text{Let } L = W + 40. \\ 2W + 80 + 2W &= 320 && \text{Distributive property} \\ 4W + 80 &= 320 && \text{Combine terms.} \\ 4W &= 240 && \text{Subtract 80.} \\ W &= 60 && \text{Divide by 4.} \end{aligned}$$

Let $W = 60$ in the equation $L = W + 40$ to find $L$.

$$L = 60 + 40 = 100$$

*Step 5* **State the answer.** The length is 100 yd, and the width is 60 yd. The answer is reasonable, since both dimensions are within the ranges given in the problem.

*Step 6* **Check.** The perimeter of this soccer field is

$$2(100) + 2(60) = 320 \text{ yd},$$

and the length, 100 yd, is 40 yd more than the width, since

$$100 - 40 = 60.$$

The answer is correct.

**Now Try Exercise 3.**

**OBJECTIVE 2 Solve money problems using two variables.** Professional sport ticket prices increase annually. Average per-ticket prices in three of the four major sports (football, basketball, and hockey) now exceed $30.00.

### EXAMPLE 2 Solving a Problem about Ticket Prices

During recent National Hockey League and National Basketball Association seasons, two hockey tickets and one basketball ticket purchased at their average prices would have cost $110.40. One hockey ticket and two basketball tickets would have cost $106.32. What were the average ticket prices for the two sports? (*Source:* Team Marketing Report, Chicago.)

*Step 1* **Read** the problem again. There are two unknowns.

*Step 2* **Assign variables.** Let $h$ represent the average price for a hockey ticket and $b$ represent the average price for a basketball ticket.

*Step 3* **Write a system of equations.** Because two hockey tickets and one basketball ticket cost a total of \$110.40, one equation for the system is

$$2h + b = 110.40.$$

By similar reasoning, the second equation is

$$h + 2b = 106.32.$$

Therefore, the system is

$$2h + b = 110.40 \quad (1)$$
$$h + 2b = 106.32. \quad (2)$$

*Step 4* **Solve** the system of equations. To eliminate $h$, multiply equation (2) by $-2$ and add.

$$2h + b = 110.40 \quad (1)$$
$$-2h - 4b = -212.64 \quad \text{Multiply each side of (2) by } -2.$$
$$-3b = -102.24 \quad \text{Add.}$$
$$b = 34.08 \quad \text{Divide by } -3.$$

To find the value of $h$, let $b = 34.08$ in equation (2).

$$h + 2b = 106.32 \quad (2)$$
$$h + 2(34.08) = 106.32 \quad \text{Let } b = 34.08.$$
$$h + 68.16 = 106.32 \quad \text{Multiply.}$$
$$h = 38.16 \quad \text{Subtract 68.16.}$$

*Step 5* **State the answer.** The average price for one basketball ticket was \$34.08. For one hockey ticket, the average price was \$38.16.

*Step 6* **Check** that these values satisfy the conditions stated in the problem.

**Now Try Exercise 11.**

**OBJECTIVE 3 Solve mixture problems using two variables.** We solved mixture problems earlier using one variable. For many mixture problems it seems more natural to use more than one variable and a system of equations.

### EXAMPLE 3 Solving a Mixture Problem

How many ounces each of 5% hydrochloric acid and 20% hydrochloric acid must be combined to get 10 oz of solution that is 12.5% hydrochloric acid?

*Step 1* **Read** the problem. Two solutions of different strengths are being mixed together to get a specific amount of a solution with an "in-between" strength.

*Step 2* **Assign variables.** Let $x$ represent the number of ounces of 5% solution and $y$ represent the number of ounces of 20% solution. Use a table to summarize the information from the problem.

| Ounces of Solution | Percent (as a decimal) | Ounces of Pure Acid |
|---|---|---|
| $x$ | 5% = .05 | $.05x$ |
| $y$ | 20% = .20 | $.20y$ |
| 10 | 12.5% = .125 | $(.125)10$ |

Figure 11 illustrates what is happening in the problem.

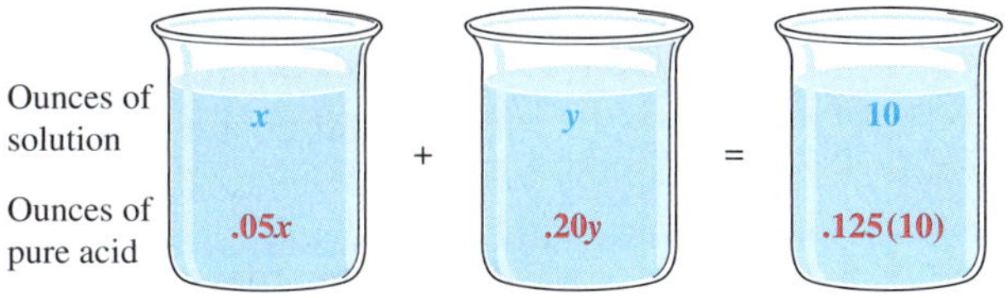


FIGURE 11

*Step 3* **Write a system of equations.** When the $x$ oz of 5% solution and the $y$ oz of 20% solution are combined, the total number of ounces is 10, so

$$x + y = 10. \qquad (1)$$

The ounces of acid in the 5% solution ($.05x$) plus the ounces of acid in the 20% solution ($.20y$) should equal the total ounces of acid in the mixture, which is $(.125)10$, or 1.25. That is,

$$.05x + .20y = 1.25. \qquad (2)$$

Notice that these equations can be quickly determined by reading down in the table or using the labels in Figure 9.

*Step 4* **Solve** the system of equations (1) and (2). Eliminate $x$ by first multiplying equation (2) by 100 to clear it of decimals and then multiplying equation (1) by $-5$.

$$\begin{aligned} 5x + 20y &= 125 && \text{Multiply each side of (2) by 100.} \\ -5x - 5y &= -50 && \text{Multiply each side of (1) by } -5. \\ \hline 15y &= 75 && \text{Add.} \\ y &= 5 \end{aligned}$$

Because $y = 5$ and $x + y = 10$, $x$ is also 5.

*Step 5* **State the answer.** The desired mixture will require 5 oz of the 5% solution and 5 oz of the 20% solution.

*Step 6* **Check** that these values satisfy both equations of the system.

**Now Try Exercise 17.**

## CONNECTIONS

Problems that can be solved by writing a system of equations have been of interest historically. The following problem appeared in a Hindu work that dates back to about 850 A.D.

> The mixed price of 9 citrons (a lemonlike fruit shown in the photo) and 7 fragrant wood apples is 107; again, the mixed price of 7 citrons and 9 fragrant wood apples is 101. O you arithmetician, tell me quickly the price of a citron and the price of a wood apple here, having distinctly separated those prices well.

### For Discussion or Writing

What do you think is meant by "the mixed price" in the problem quoted above? Write a system of equations for this problem. (You will be asked to solve it in Exercise 35.)

**OBJECTIVE 4 Solve distance-rate-time problems using two variables.** Motion problems require the distance formula, $d = rt$, where $d$ is distance, $r$ is rate (or speed), and $t$ is time. These applications often lead to systems of equations, as in the next example.

## EXAMPLE 4 Solving a Motion Problem

A car travels 250 km in the same time that a truck travels 225 km. If the speed of the car is 8 km per hr faster than the speed of the truck, find both speeds.

*Step 1* **Read** the problem again. Given the distances traveled, we need to find the speed of each vehicle.

*Step 2* **Assign variables.**

Let $x =$ the speed of the car,

and $y =$ the speed of the truck.

As in Example 3, a table helps organize the information. Fill in the given information for each vehicle (in this case, distance) and use the assigned variables for the unknown speeds (rates).

| | *d* | *r* | *t* |
|---|---|---|---|
| *Car* | 250 | *x* | |
| *Truck* | 225 | *y* | |

The table shows nothing about time. To get an expression for time, solve the distance formula, $d = rt$, for $t$.

$$\frac{d}{r} = t$$

The two times can be written as $\frac{250}{x}$ and $\frac{225}{y}$.

*Step 3* **Write a system of equations.** The problem states that the car travels 8 km per hr faster than the truck. Since the two speeds are $x$ and $y$,

$$x = y + 8. \qquad (1)$$

Both vehicles travel for the same time, so from the table,

$$\frac{250}{x} = \frac{225}{y}.$$

This is not a linear equation. However, multiplying each side by $xy$ gives

$$250y = 225x,$$

which is linear. The system is

$$x = y + 8$$
$$250y = 225x. \qquad (2)$$

*Step 4* **Solve** the system of equations by substitution. Replace $x$ with $y + 8$ in equation (2).

$$250y = 225x \qquad (2)$$
$$250y = 225(y + 8) \qquad \text{Let } x = y + 8.$$
$$250y = 225y + 1800 \qquad \text{Distributive property}$$
$$25y = 1800 \qquad \text{Subtract } 225y.$$
$$y = 72 \qquad \text{Divide by 25.}$$

Because $x = y + 8$, the value of $x$ is $72 + 8 = 80$.

*Step 5* **State the answer.** The car's speed is 80 km per hr, and the truck's speed is 72 km per hr.

*Step 6* **Check.** This is especially important since one of the equations had variable denominators.

$$\text{Car: } t = \frac{d}{r} = \frac{250}{80} = 3.125$$

$$\text{Truck: } t = \frac{d}{r} = \frac{225}{72} = 3.125$$

Times are equal.

Since $80 - 72 = 8$, the conditions of the problem are satisfied.

**Now Try Exercise 27.**

**OBJECTIVE 5 Solve problems with three variables using a system of three equations.** To solve such problems, we extend the method used for two unknowns. Since three variables are used, three equations are necessary to find a solution.

## EXAMPLE 5 Solving a Problem Involving Prices

At Panera Bread, a loaf of honey wheat bread costs \$2.40, a loaf of pumpernickel bread costs \$3.35, and a loaf of French bread costs \$2.10. On a recent day, three times as many loaves of honey wheat were sold as pumpernickel. The number of loaves of French bread sold was 5 less than the number of loaves of honey wheat sold. Total receipts for these breads were \$56.90. How many loaves of each type of bread were sold? (*Source:* Panera Bread menu.)

*Step 1* **Read** the problem again. There are three unknowns in this problem.

*Step 2* **Assign variables** to represent the three unknowns.

Let $x$ = the number of loaves of honey wheat,
$y$ = the number of loaves of pumpernickel,
and $z$ = the number of loaves of French bread.

Step 3 **Write a system of three equations** using the information in the problem. Since three times as many loaves of honey wheat were sold as pumpernickel,

$$x = 3y, \quad \text{or} \quad x - 3y = 0. \qquad (1)$$

Also,

| Number of loaves of French bread | equals | 5 less than the number of loaves of honey wheat. |
|---|---|---|
| ↓ | ↓ | ↓ |
| $z$ | $=$ | $x - 5$, |

so $x - z = 5.$ (2)

Multiplying the cost of a loaf of each kind of bread by the number of loaves of that kind sold and adding gives the total receipts.

$$2.40x + 3.35y + 2.10z = 56.90$$

Multiply each side of this equation by 100 to clear it of decimals.

$$240x + 335y + 210z = 5690 \qquad (3)$$

*Step 4* **Solve** the system of three equations using the method shown in Section 8.4. Solving the system

$$x - 3y = 0 \quad (1)$$
$$x - z = 5 \quad (2)$$
$$240x + 335y + 210z = 5690 \quad (3)$$

leads to

$$x = 12, \quad y = 4, \quad \text{and} \quad z = 7.$$

*Step 5* **State the answer.** The solution is (12, 4, 7), so 12 loaves of honey wheat, 4 loaves of pumpernickel, and 7 loaves of French bread were sold.

*Step 6* **Check.** Since $12 = 3 \cdot 4$, the number of loaves of honey wheat is three times the number of loaves of pumpernickel. Also, $12 - 7 = 5$, so the number of loaves of French bread is 5 less than the number of loaves of honey wheat. Multiply the appropriate cost per loaf by the number of loaves sold and add the results to check that total receipts were \$56.90.

**Now Try Exercise 45.**

### EXAMPLE 6 Solving a Business Production Problem

A company produces three color television sets, models X, Y, and Z. Each model X set requires 2 hr of electronics work, 2 hr of assembly time, and 1 hr of finishing time. Each model Y requires 1, 3, and 1 hr of electronics, assembly, and finishing time, respectively. Each model Z requires 3, 2, and 2 hr of the same work, respectively. There are 100 hr available for electronics, 100 hr available for assembly, and 65 hr available for finishing per week. How many of each model should be produced each week if all available time must be used?

*Step 1* **Read** the problem again. There are three unknowns.

*Step 2* **Assign variables.**

Let $x =$ the number of model X produced per week,
$y =$ the number of model Y produced per week,
and $z =$ the number of model Z produced per week.

We organize the information in a table.

| | *Each Model X* | *Each Model Y* | *Each Model Z* | *Totals* |
|---|---|---|---|---|
| *Hours of Electronics Work* | 2 | 1 | 3 | 100 |
| *Hours of Assembly Time* | 2 | 3 | 2 | 100 |
| *Hours of Finishing Time* | 1 | 1 | 2 | 65 |

*Step 3* **Write a system of three equations.** The $x$ model X sets require $2x$ hr of electronics, the $y$ model Y sets require $1y$ (or $y$) hr of electronics, and the

$z$ model Z sets require $3z$ hr of electronics. Since 100 hr are available for electronics,

$$2x + y + 3z = 100. \qquad (1)$$

Similarly, from the fact that 100 hr are available for assembly,

$$2x + 3y + 2z = 100, \qquad (2)$$

and the fact that 65 hr are available for finishing leads to the equation

$$x + y + 2z = 65. \qquad (3)$$

Again, notice the advantage of setting up a table. By reading across, we can easily determine the coefficients and constants in the equations of the system.

*Step 4* **Solve** the system

$$\begin{aligned} 2x + \phantom{3}y + 3z &= 100 \\ 2x + 3y + 2z &= 100 \\ x + \phantom{3}y + 2z &= 65 \end{aligned}$$

to find $x = 15$, $y = 10$, and $z = 20$.

*Step 5* **State the answer.** The company should produce 15 model X, 10 model Y, and 20 model Z sets per week.

*Step 6* **Check** that these values satisfy the conditions of the problem.

**Now Try Exercise 47.**

## 8.5 EXERCISES

**For Extra Help**

- Student's Solutions Manual
- MyMathLab
- InterAct Math Tutorial Software
- AW Math Tutor Center
- MathXL
- Digital Video Tutor CD 12/Videotape 12

*Solve each problem. See Example 1.*

**1.** During the 2000 Major League Baseball season, the St. Louis Cardinals played 162 games. They won 28 more games than they lost. What was their win–loss record that year?

**2.** Refer to Exercise 1. During the same 162-game season, the Chicago Cubs lost 32 more games than they won. What was the team's win–loss record?

**2000 MLB FINAL STANDINGS NATIONAL LEAGUE CENTRAL**

| Team | W | L |
|---|---|---|
| St. Louis | ___ | ___ |
| Cincinnati | 85 | 77 |
| Milwaukee | 73 | 89 |
| Houston | 72 | 90 |
| Pittsburgh | 69 | 93 |
| Chicago | ___ | ___ |

*Source:* www.mlb.com

**3.** Venus and Serena measured a tennis court and found that it was 42 ft longer than it was wide and had a perimeter of 228 ft. What were the length and the width of the tennis court?

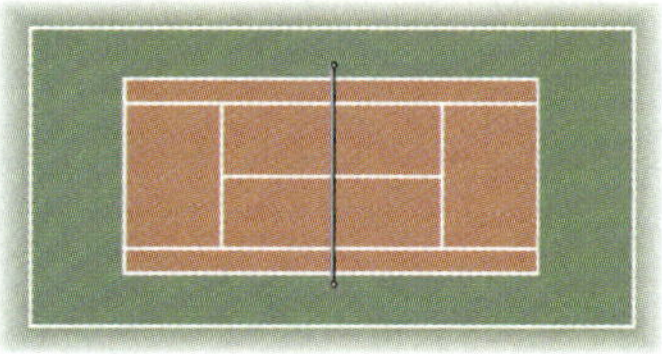

**4.** Shaq and Kobe found that the width of their basketball court was 44 ft less than the length. If the perimeter was 288 ft, what were the length and the width of their court?

**5.** The two biggest U.S. companies in terms of revenue in 2000 were ExxonMobil and General Motors. ExxonMobil's revenue was \$29 billion more than that of General Motors. Total revenue for the two companies was \$399 billion. What was the revenue for each company? (*Source:* Bridge News, MarketGuide.com)

**6.** The top two U.S. trading partners during the first four months of 2000 were Canada and Mexico. Exports and imports with Mexico were \$57 billion less than those with Canada. Total exports and imports involving these two countries were \$211 billion. How much were U.S. exports and imports with each country? (*Source:* U.S. Bureau of the Census.)

*In Exercises 7 and 8, find the measures of the angles marked x and y. Remember that* (1) *the sum of the measures of the angles of a triangle is* 180°, (2) *supplementary angles have a sum of* 180°, *and* (3) *vertical angles have equal measures.*

**7.**

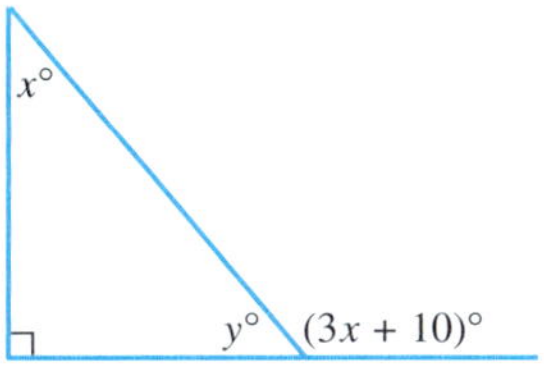


**8.**

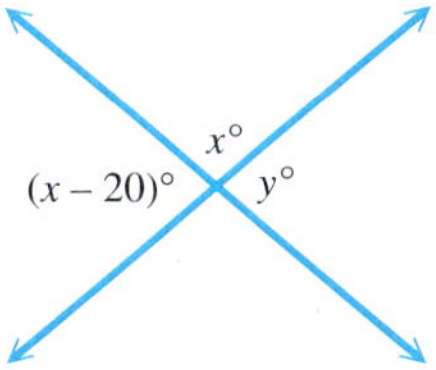


*The Fan Cost Index (FCI) represents the cost of four average-price tickets, four small soft drinks, two small beers, four hot dogs, parking for one car, two game programs, and two souvenir caps to a sporting event. For example, in a recent year, the FCI for Major League Baseball was* \$105.63. *This was by far the least for the four major professional sports.* (*Source:* Team Marketing Report, Chicago.)

*Use the concept of FCI in Exercises 9 and 10. See Example 2.*

**9.** The FCI prices for the National Hockey League and the National Basketball Association totaled \$423.12. The hockey FCI was \$16.36 more than that of basketball. What were the FCIs for these sports?

**10.** The FCI prices for Major League Baseball and the National Football League totaled \$311.03. The football FCI was \$105.87 more than that of baseball. What were the FCIs for these sports?

*Solve each problem. See Example 2.*

**11.** Andrew McGinnis works at Wendy's Old Fashioned Hamburgers. During one particular lunch hour, he sold 15 single hamburgers and 10 double hamburgers, totaling \$63.25. Another lunch hour, he sold 30 singles and 5 doubles, totaling \$78.65. How much did each type of burger cost? (*Source:* Wendy's Old Fashioned Hamburgers menu.)

**12.** Tokyo and New York are among the most expensive cities worldwide for business travelers. Using average costs per day for each city (which includes room, meals, laundry, and two taxi fares), 2 days in Tokyo and 3 days in New York cost $2015. Four days in Tokyo and 2 days in New York cost $2490. What is the average cost per day for each city? (*Source:* ECA International.)

*The formulas $p = br$ (percentage = base × rate) and $I = prt$ (simple interest = principal × rate × time) are used in the applications in Exercises 17–24. In general, we are using*

$$\text{portion} = \text{whole} \times \text{percent}.$$

*To prepare to use these formulas, answer the questions in Exercises 13 and 14.*

**13.** If a container of liquid contains 60 oz of solution, what is the number of ounces of pure acid if the given solution contains the following acid concentrations?

**(a)** 10% **(b)** 25% **(c)** 40% **(d)** 50%

**14.** If $5000 is invested in an account paying simple annual interest, how much interest will be earned during the first year at the following rates?

**(a)** 2% **(b)** 3% **(c)** 4% **(d)** 3.5%

**15.** If a pound of turkey costs $.99, how much will $x$ pounds cost?

**16.** If a ticket to the movie *Eight Legged Freaks* costs $8 and $y$ tickets are sold, how much is collected from the sale?

*Solve each problem. See Example 3.*

**17.** How many gallons each of 25% alcohol and 35% alcohol should be mixed to get 20 gal of 32% alcohol?

| Gallons of Solution | Percent (as a decimal) | Gallons of Pure Alcohol |
|---|---|---|
| $x$ | 25% = .25 | |
| $y$ | 35% = .35 | |
| 20 | 32% = | |

**18.** How many liters each of 15% acid and 33% acid should be mixed to get 120 L of 21% acid?

| Liters of Solution | Percent (as a decimal) | Liters of Pure Acid |
|---|---|---|
| $x$ | 15% = .15 | |
| $y$ | 33% = | |
| 120 | 21% = | |

**19.** Pure acid is to be added to a 10% acid solution to obtain 54 L of a 20% acid solution. What amounts of each should be used?

**20.** A truck radiator holds 36 L of fluid. How much pure antifreeze must be added to a mixture that is 4% antifreeze to fill the radiator with a mixture that is 20% antifreeze?

**21.** A party mix is made by adding nuts that sell for \$2.50 per kg to a cereal mixture that sells for \$1 per kg. How much of each should be added to get 30 kg of a mix that will sell for \$1.70 per kg?

| | *Number of Kilograms* | *Price per Kilogram* | *Value* |
|---|---|---|---|
| *Nuts* | $x$ | 2.50 | |
| *Cereal* | $y$ | 1.00 | |
| *Mixture* | | 1.70 | |

**22.** A popular fruit drink is made by mixing fruit juices. Such a drink with 50% juice is to be mixed with another drink that is 30% juice to get 200 L of a drink that is 45% juice. How much of each should be used?

| | *Liters of Drink* | *Percent (as a decimal)* | *Liters of Pure Juice* |
|---|---|---|---|
| *50% Juice* | $x$ | .50 | |
| *30% Juice* | $y$ | .30 | |
| *Mixture* | | .45 | |

**23.** A total of \$3000 is invested, part at 2% simple interest and part at 4%. If the total annual return from the two investments is \$100, how much is invested at each rate?

| *Principal* | *Rate (as a decimal)* | *Interest* |
|---|---|---|
| $x$ | .02 | $.02x$ |
| $y$ | .04 | $.04y$ |
| 3000 | | 100 |

**24.** An investor will invest a total of \$15,000 in two accounts, one paying 4% annual simple interest, and the other 3%. If he wants to earn \$550 annual interest, how much should he invest at each rate?

| *Principal* | *Rate (as a decimal)* | *Interest* |
|---|---|---|
| $x$ | .04 | |
| $y$ | .03 | |
| 15,000 | | |

*The formula $d = rt$ (distance = rate × time) is used in the applications in Exercises 27–30. To prepare to use this formula, answer the questions in Exercises 25 and 26.*

**25.** If the speed of a killer whale is 25 mph and the whale swims for $y$ hr, how many miles does the whale travel?

**26.** If the speed of a boat in still water is 10 mph, and the speed of the current of a river is $x$ mph, what is the speed of the boat

**(a)** going upstream (that is, against the current, which slows the boat down);

**(b)** going downstream (that is, with the current, which speeds the boat up)?

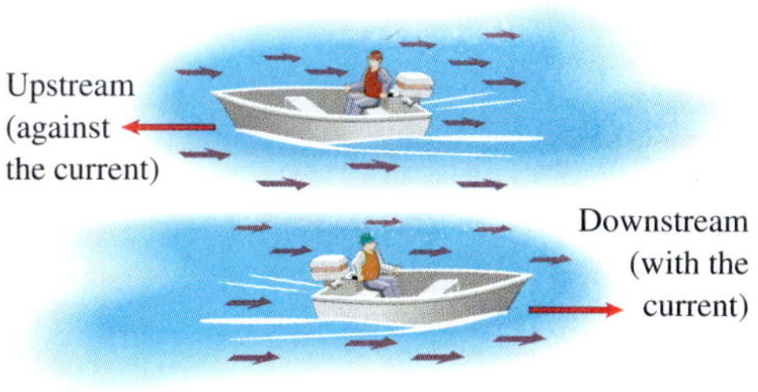


*Solve each problem. See Example 4.*

**27.** A train travels 150 km in the same time that a plane covers 400 km. If the speed of the plane is 20 km per hr less than 3 times the speed of the train, find both speeds.

| | *r* | *t* | *d* |
|---|---|---|---|
| *Train* | $x$ | | 150 |
| *Plane* | $y$ | | 400 |

**28.** A freight train and an express train leave towns 390 km apart, traveling toward one another. The freight train travels 30 km per hr slower than the express train. They pass one another 3 hr later. What are their speeds?

| | *r* | *t* | *d* |
|---|---|---|---|
| *Freight Train* | $x$ | 3 | |
| *Express Train* | $y$ | 3 | |

**29.** In his motorboat, Bill Ruhberg travels upstream at top speed to his favorite fishing spot, a distance of 36 mi, in 2 hr. Returning, he finds that the trip downstream, still at top speed, takes only 1.5 hr. Find the speed of Bill's boat and the speed of the current.

| | $r$ | $t$ | $d$ |
|---|---|---|---|
| Upstream | $x - y$ | 2 | |
| Downstream | $x + y$ | | |

**30.** Traveling for 3 hr into a steady headwind, a plane flies 1650 mi. The pilot determines that flying *with* the same wind for 2 hr, he could make a trip of 1300 mi. Find the speed of the plane and the speed of the wind.

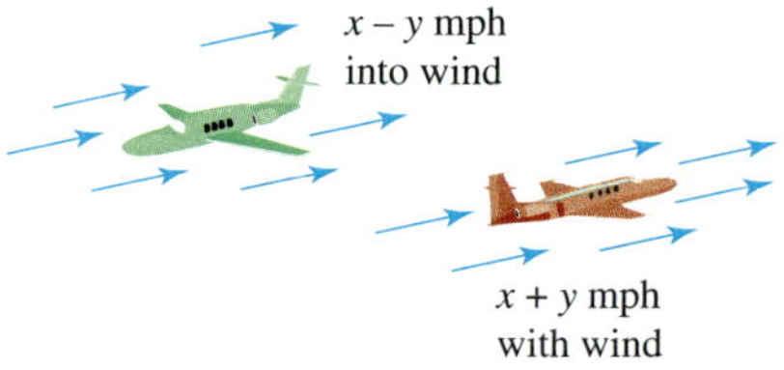


*Use the problem-solving techniques of this section to solve each problem with two variables. See Examples 1–4.*

**31.** At age 61, rock icon Tina Turner generated the most revenue on the concert circuit in 2000. Turner and second-place 'N Sync together took in \$157 million from ticket sales. If 'N Sync took in \$3.8 million less than Turner, how much did each generate? (*Source:* Pollstar.)

**32.** Carol Britz plans to mix pecan clusters that sell for \$3.60 per lb with chocolate truffles that sell for \$7.20 per lb to get a mixture that she can sell in Valentine boxes for \$4.95 per lb. How much of the \$3.60 clusters and the \$7.20 truffles should she use to create 80 lb of the mix?

| | Number of Pounds | Price per Pound | Value |
|---|---|---|---|
| Pecan Clusters | $x$ | | |
| Chocolate Truffles | $y$ | | |
| Valentine Mixture | 80 | | |

**33.** Tickets to a production of *King Lear* at the College of DuPage cost \$5 for general admission or \$4 with a student ID. If 184 people paid to see a performance and \$812 was collected, how many of each type of ticket were sold?

**34.** At a business meeting at Panera Bread, the bill for two cappuccinos and three house lattes was \$10.95. At another table, the bill for one cappuccino and two house lattes was \$6.65. How much did each type of beverage cost? (*Source:* Panera Bread menu.)

**35.** The mixed price of 9 citrons and 7 fragrant wood apples is 107; again, the mixed price of 7 citrons and 9 fragrant wood apples is 101. O you arithmetician, tell me quickly the price of a citron and the price of a wood apple here, having distinctly separated those prices well. (*Source:* Hindu work, A.D. 850.)

**36.** Braving blizzard conditions on the planet Hoth, Luke Skywalker sets out at top speed in his snow speeder for a rebel base 4800 mi away. He travels into a steady headwind and makes the trip in 3 hr. Returning, he finds that the trip back, still at top speed but now with a tailwind, takes only 2 hr. Find the top speed of Luke's snow speeder and the speed of the wind.

| | $r$ | $t$ | $d$ |
|---|---|---|---|
| Into Headwind | | | |
| With Tailwind | | | |

*Solve each problem involving three variables. See Examples 5 and 6. (In Exercises 37–40, remember that the sum of the measures of the angles of a triangle is* 180°.*)*

**37.** In the figure, $z = x + 10$ and $x + y = 100$. Determine a third equation involving $x$, $y$, and $z$, and then find the measures of the three angles.

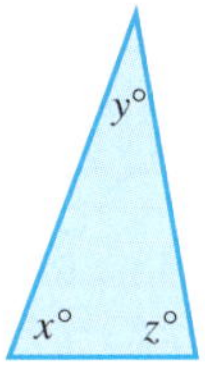


**38.** In the figure, $x$ is 10 less than $y$ and 20 less than $z$. Write a system of equations and find the measures of the three angles.

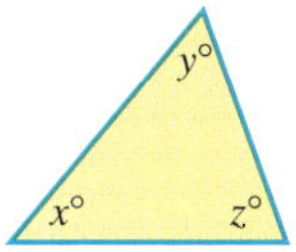


**39.** In a certain triangle, the measure of the second angle is 10° more than three times the first. The third angle measure is equal to the sum of the measures of the other two. Find the measures of the three angles.

**40.** The measure of the largest angle of a triangle is 12° less than the sum of the measures of the other two. The smallest angle measures 58° less than the largest. Find the measures of the angles.

**41.** The perimeter of a triangle is 70 cm. The longest side is 4 cm less than the sum of the other two sides. Twice the shortest side is 9 cm less than the longest side. Find the length of each side of the triangle.

**42.** The perimeter of a triangle is 56 in. The longest side measures 4 in. less than the sum of the other two sides. Three times the shortest side is 4 in. more than the longest side. Find the lengths of the three sides.

**43.** In a random sample of 100 Americans of voting age, 10 more Americans identify themselves as Independents than Republicans. Six fewer Americans identify themselves as Republicans than Democrats. Assuming that all of those sampled are Republican, Democrat, or Independent, how many of those in the sample identify themselves with each political affiliation? (*Source:* The Gallup Organization.)

**44.** In the 2000 Summer Olympics in Sydney, Australia, the United States earned 14 more gold medals than silver. The number of bronze medals earned was 17 less than twice the number of silver medals. The United States earned a total of 97 medals. How many of each kind of medal did the United States earn? (*Source: The Gazette,* October 2, 2000.)

**45.** Tickets for one show on the Harlem Globetrotters' 75th Anniversary Tour cost \$10, \$18, or, for VIP seats, \$30. So far, five times as many \$18 tickets have been sold as VIP tickets. The number of \$10 tickets equals the number of \$18 tickets plus twice the number of VIP tickets. Sales of these tickets total \$9500. How many of each kind of ticket have been sold? (*Source:* www.ticketmaster.com)

**46.** Three kinds of tickets are available for a *Prosthetic Forehead* concert: "up close," "in the middle," and "far out." "Up close" tickets cost \$10 more than "in the middle" tickets, while "in the middle" tickets cost \$10 more than "far out" tickets. Twice the cost of an "up close" ticket is \$20 more than 3 times the cost of a "far out" ticket. Find the price of each kind of ticket.

**47.** A hardware supplier manufactures three kinds of clamps, types A, B, and C. Production restrictions require it to make 10 units more type C clamps than the total of the other types and twice as many type B clamps as type A. The shop must produce a total of 490 units of clamps per day. How many units of each type can be made per day?

**48.** A Mardi Gras trinket manufacturer supplies three wholesalers, A, B, and C. The output from a day's production is 320 cases of trinkets. She must send wholesaler A three times as many cases as she sends B, and she must send wholesaler C 160 cases less than she provides A and B together. How many cases should she send to each wholesaler to distribute the entire day's production to them?

**49.** A plant food is to be made from three chemicals. The mix must include 60% of the first and second chemicals. The second and third chemicals must be in a ratio of 4 to 3 by weight. How much of each chemical is needed to make 750 kg of the plant food?

**50.** How many ounces of 5% hydrochloric acid, 20% hydrochloric acid, and water must be combined to get 10 oz of solution that is 8.5% hydrochloric acid, if the amount of water used must equal the total amount of the other two solutions?

**51.** During a recent National Hockey League regular season, the Dallas Stars played 82 games. Together, their wins and losses totaled 74. They tied 18 fewer games than they lost. How many wins, losses, and ties did they have that year?

| Team | GP | W | L | T | GF | GA | Pts |
|---|---|---|---|---|---|---|---|
| Dallas | 82 | ___ | ___ | ___ | 252 | 198 | 104 |
| Detroit | 82 | 38 | 26 | 18 | 253 | 197 | 94 |
| Phoenix | 82 | 38 | 37 | 7 | 240 | 243 | 83 |
| St. Louis | 82 | 36 | 35 | 11 | 236 | 239 | 83 |
| Chicago | 82 | 34 | 35 | 13 | 223 | 210 | 81 |
| Toronto | 82 | 30 | 44 | 8 | 230 | 273 | 68 |

***Source:*** *Sports Illustrated Sports Almanac.*

**52.** During a recent National Hockey League season, the Boston Bruins played 82 games. Their losses and ties totaled 56, and they had 21 fewer wins than losses. How many wins, losses, and ties did they have that year?

| Team | GP | W | L | T | GF | GA | Pts |
|---|---|---|---|---|---|---|---|
| Buffalo | 82 | 40 | 30 | 12 | 237 | 208 | 92 |
| Pittsburgh | 82 | 38 | 36 | 8 | 285 | 280 | 84 |
| Ottawa | 82 | 31 | 36 | 15 | 226 | 234 | 77 |
| Montreal | 82 | 31 | 36 | 15 | 249 | 276 | 77 |
| Hartford | 82 | 32 | 39 | 11 | 226 | 256 | 75 |
| Boston | 82 | ___ | ___ | ___ | 234 | 300 | 61 |

***Source:*** *Sports Illustrated Sports Almanac.*

# 8.6 Solving Systems of Linear Equations by Matrix Methods

**OBJECTIVES**

1. Define a matrix.
2. Write the augmented matrix for a system.
3. Use row operations to solve a system with two equations.
4. Use row operations to solve a system with three equations.
5. Use row operations to solve special systems.

**OBJECTIVE 1 Define a matrix.** An ordered array of numbers such as

Columns

$$\text{Rows}\quad \begin{bmatrix} 2 & 3 & 5 \\ 7 & 1 & 2 \end{bmatrix}$$

is called a **matrix.** The numbers are called **elements** of the matrix. Matrices (the plural of *matrix*) are named according to the number of **rows** and **columns** they contain. The rows are read horizontally, and the columns are read vertically. For example, the first row in the preceding matrix is 2 3 5 and the first column is $\begin{matrix} 2 \\ 7 \end{matrix}$. This matrix is a $2 \times 3$ (read "two by three") matrix because it has 2 rows and 3 columns. The number of rows is given first, and then the number of columns. Two other examples follow.

$$\begin{bmatrix} -1 & 0 \\ 1 & -2 \end{bmatrix} \quad 2 \times 2 \text{ matrix} \qquad \begin{bmatrix} 8 & -1 & -3 \\ 2 & 1 & 6 \\ 0 & 5 & -3 \\ 5 & 9 & 7 \end{bmatrix} \quad 4 \times 3 \text{ matrix}$$

A **square matrix** is one that has the same number of rows as columns. The $2 \times 2$ matrix is a square matrix.

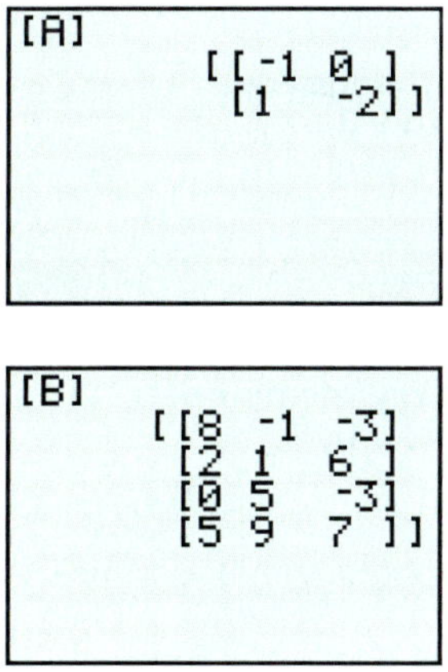


FIGURE 12

Figure 12 shows how a graphing calculator displays the preceding two matrices. Work with matrices is made much easier by using technology when available. Consult your owner's manual for details.

In this section, we discuss a matrix method of solving linear systems that is really just a very structured way of using the elimination method. The advantage of this new method is that it can be done by a graphing calculator or a computer, allowing large systems of equations to be solved easily.

**OBJECTIVE 2 Write the augmented matrix for a system.** To begin, we write an *augmented matrix* for the system. An **augmented matrix** has a vertical bar that separates the columns of the matrix into two groups. For example, to solve the system

$$\begin{aligned} x - 3y &= 1 \\ 2x + \ \ y &= -5, \end{aligned}$$

start with the augmented matrix

$$\left[\begin{array}{cc|c} 1 & -3 & 1 \\ 2 & 1 & -5 \end{array}\right].$$

Place the coefficients of the variables to the left of the bar, and the constants to the right. The bar separates the coefficients from the constants. The matrix is just a shorthand way of writing the system of equations, so the rows of the augmented matrix can be treated the same as the equations of a system of equations.

We know that exchanging the position of two equations in a system does not change the system. Also, multiplying any equation in a system by a nonzero number does not change the system. Comparable changes to the augmented matrix of a system of equations produce new matrices that correspond to systems with the same solutions as the original system.

The following **row operations** produce new matrices that lead to systems having the same solutions as the original system.

**Matrix Row Operations**

1. Any two rows of the matrix may be interchanged.
2. The elements in any row may be multiplied by any nonzero real number.
3. Any row may be changed by adding to the elements of the row the product of a real number and the corresponding elements of another row.

Examples of these row operations follow.

Row operation 1:

$$\begin{bmatrix} 2 & 3 & 9 \\ 4 & 8 & -3 \\ 1 & 0 & 7 \end{bmatrix} \text{ becomes } \begin{bmatrix} 1 & 0 & 7 \\ 4 & 8 & -3 \\ 2 & 3 & 9 \end{bmatrix}.$$

Interchange row 1 and row 3.

Row operation 2:

$$\begin{bmatrix} 2 & 3 & 9 \\ 4 & 8 & -3 \\ 1 & 0 & 7 \end{bmatrix} \text{ becomes } \begin{bmatrix} 6 & 9 & 27 \\ 4 & 8 & -3 \\ 1 & 0 & 7 \end{bmatrix}.$$

Multiply the numbers in row 1 by 3.

Row operation 3:

$$\begin{bmatrix} 2 & 3 & 9 \\ 4 & 8 & -3 \\ 1 & 0 & 7 \end{bmatrix} \text{ becomes } \begin{bmatrix} 0 & 3 & -5 \\ 4 & 8 & -3 \\ 1 & 0 & 7 \end{bmatrix}.$$

Multiply the numbers in row 3 by −2; add them to the corresponding numbers in row 1.

The third row operation corresponds to the way we eliminated a variable from a pair of equations in the previous sections.

**OBJECTIVE 3 Use row operations to solve a system with two equations.** Row operations can be used to rewrite a matrix until it is the matrix of a system where the solution is easy to find. The goal is a matrix in the form

$$\left[\begin{array}{cc|c} 1 & a & b \\ 0 & 1 & c \end{array}\right] \quad \text{or} \quad \left[\begin{array}{ccc|c} 1 & a & b & c \\ 0 & 1 & d & e \\ 0 & 0 & 1 & f \end{array}\right]$$

for systems with two or three equations, respectively. Notice that there are 1s down the diagonal from upper left to lower right and 0s below the 1s. A matrix written this way is said to be in **row echelon form.** When these matrices are rewritten as systems of equations, the value of one variable is known, and the rest can be found by substitution. The following examples illustrate this method.

### EXAMPLE 1 Using Row Operations to Solve a System with Two Variables

Use row operations to solve the system

$$\begin{aligned} x - 3y &= 1 \\ 2x + y &= -5. \end{aligned}$$

We start with the augmented matrix of the system.

$$\left[\begin{array}{cc|c} 1 & -3 & 1 \\ 2 & 1 & -5 \end{array}\right]$$

Now we use the various row operations to change this matrix into one that leads to a system that is easier to solve.

It is best to work by columns. We start with the first column and make sure that there is a 1 in the first row, first column position. There is already a 1 in this position. Next, we get 0 in every position below the first. To get a 0 in row two, column one, we use the third row operation and add to the numbers in row two the result of multiplying each number in row one by $-2$. (We abbreviate this as $-2R_1 + R_2$.) Row one remains unchanged.

$$\left[\begin{array}{cc|c} 1 & -3 & 1 \\ 2 + 1(-2) & 1 + -3(-2) & -5 + 1(-2) \end{array}\right]$$

Original number from row two ↑    ↑ $-2$ times number from row one

$$\left[\begin{array}{cc|c} 1 & -3 & 1 \\ 0 & 7 & -7 \end{array}\right] \quad -2R_1 + R_2$$

The matrix now has a 1 in the first position of column one, with 0 in every position below the first.

Now we go to column two. A 1 is needed in row two, column two. We get this 1 by using the second row operation, multiplying each number of row two by $\frac{1}{7}$.

$$\left[\begin{array}{cc|c} 1 & -3 & 1 \\ 0 & 1 & -1 \end{array}\right] \quad \tfrac{1}{7}R_2$$

This augmented matrix leads to the system of equations

$$\begin{aligned} 1x - 3y &= 1 \\ 0x + 1y &= -1 \end{aligned} \quad \text{or} \quad \begin{aligned} x - 3y &= 1 \\ y &= -1. \end{aligned}$$

From the second equation, $y = -1$. We substitute $-1$ for $y$ in the first equation to get

$$\begin{aligned} x - 3y &= 1 \\ x - 3(-1) &= 1 \\ x + 3 &= 1 \\ x &= -2. \end{aligned}$$

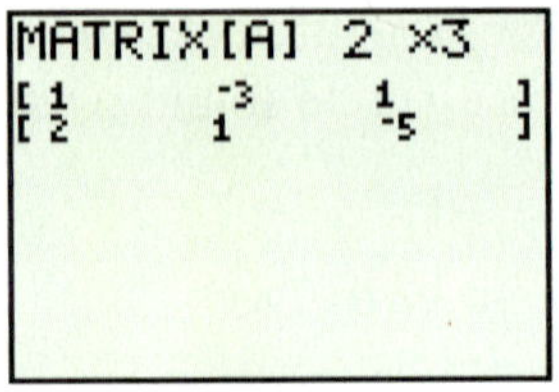


(a)

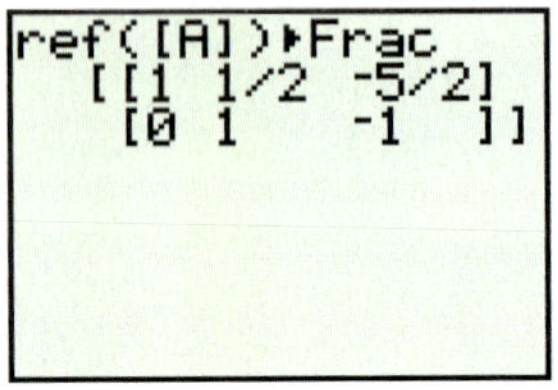


(b)

**FIGURE 13**

The solution set of the system is $\{(-2, -1)\}$. Check this solution by substitution in both equations of the system.

**Now Try Exercise 3.**

If the augmented matrix of the system in Example 1 is entered as matrix [A] in a graphing calculator (Figure 13(a)) and the row echelon form of the matrix is found (Figure 13(b)), then the system becomes

$$\begin{aligned} x + \frac{1}{2}y &= -\frac{5}{2} \\ y &= -1. \end{aligned}$$

While this system looks different from the one we obtained in Example 1, it is equivalent, since its solution set is also $\{(-2, -1)\}$.

**OBJECTIVE 4 Use row operations to solve a system with three equations.** A linear system with three equations is solved in a similar way. We use row operations to get 1s down the diagonal from left to right and all 0s below each 1.

### EXAMPLE 2 Using Row Operations to Solve a System with Three Variables

Use row operations to solve the system

$$\begin{aligned} x - y + 5z &= -6 \\ 3x + 3y - z &= 10 \\ x + 3y + 2z &= 5. \end{aligned}$$

Start by writing the augmented matrix of the system.

$$\left[\begin{array}{rrr|r} 1 & -1 & 5 & -6 \\ 3 & 3 & -1 & 10 \\ 1 & 3 & 2 & 5 \end{array}\right]$$

This matrix already has 1 in row one, column one. Next get 0s in the rest of column one. First, add to row two the results of multiplying each number of row one by $-3$. This gives the matrix

$$\left[\begin{array}{rrr|r} 1 & -1 & 5 & -6 \\ 0 & 6 & -16 & 28 \\ 1 & 3 & 2 & 5 \end{array}\right]. \quad -3R_1 + R_2$$

Now add to the numbers in row three the results of multiplying each number of row one by $-1$.

$$\left[\begin{array}{rrr|r} 1 & -1 & 5 & -6 \\ 0 & 6 & -16 & 28 \\ 0 & 4 & -3 & 11 \end{array}\right] \quad -1R_1 + R_3$$

Introduce 1 in row two, column two by multiplying each number in row two by $\frac{1}{6}$.

$$\left[\begin{array}{ccc|c} 1 & -1 & 5 & -6 \\ 0 & 1 & -\frac{8}{3} & \frac{14}{3} \\ 0 & 4 & -3 & 11 \end{array}\right] \quad \frac{1}{6}R_2$$

To obtain 0 in row three, column two, add to row three the results of multiplying each number in row two by $-4$.

$$\left[\begin{array}{ccc|c} 1 & -1 & 5 & -6 \\ 0 & 1 & -\frac{8}{3} & \frac{14}{3} \\ 0 & 0 & \frac{23}{3} & -\frac{23}{3} \end{array}\right] \quad -4R_2 + R_3$$

Finally, obtain 1 in row three, column three by multiplying each number in row three by $\frac{3}{23}$.

$$\left[\begin{array}{ccc|c} 1 & -1 & 5 & -6 \\ 0 & 1 & -\frac{8}{3} & \frac{14}{3} \\ 0 & 0 & 1 & -1 \end{array}\right] \quad \frac{3}{23}R_3$$

This final matrix gives the system of equations

$$\begin{aligned} x - y + 5z &= -6 \\ y - \frac{8}{3}z &= \frac{14}{3} \\ z &= -1. \end{aligned}$$

Substitute $-1$ for $z$ in the second equation, $y - \frac{8}{3}z = \frac{14}{3}$, to find that $y = 2$. Finally, substitute 2 for $y$ and $-1$ for $z$ in the first equation, $x - y + 5z = -6$, to determine that $x = 1$. The solution set of the original system is $\{(1, 2, -1)\}$. Check by substitution.

**Now Try Exercise 15.**

**OBJECTIVE 5 Use row operations to solve special systems.** In the final example we show how to recognize inconsistent systems or systems with dependent equations when solving these systems with row operations.

### EXAMPLE 3 Recognizing Inconsistent Systems or Dependent Equations

Use row operations to solve each system.

**(a)** $$\begin{aligned} 2x - 3y &= 8 \\ -6x + 9y &= 4 \end{aligned}$$

$$\left[\begin{array}{cc|c} 2 & -3 & 8 \\ -6 & 9 & 4 \end{array}\right] \quad \text{Write the augmented matrix.}$$

$$\left[\begin{array}{cc|c} 1 & -\frac{3}{2} & 4 \\ -6 & 9 & 4 \end{array}\right] \quad \frac{1}{2}R_1$$

$$\left[\begin{array}{cc|c} 1 & -\frac{3}{2} & 4 \\ 0 & 0 & 28 \end{array}\right] \quad 6R_1 + R_2$$

The corresponding system of equations is

$$x - \frac{3}{2}y = 4$$
$$0 = 28, \qquad \text{False}$$

which has no solution and is inconsistent. The solution set is $\emptyset$.

**(b)** $-10x + 12y = 30$
$5x - 6y = -15$

$$\left[\begin{array}{rr|r} -10 & 12 & 30 \\ 5 & -6 & -15 \end{array}\right] \quad \text{Write the augmented matrix.}$$

$$\left[\begin{array}{rr|r} 1 & -\frac{6}{5} & -3 \\ 5 & -6 & -15 \end{array}\right] \quad -\tfrac{1}{10}R_1$$

$$\left[\begin{array}{rr|r} 1 & -\frac{6}{5} & -3 \\ 0 & 0 & 0 \end{array}\right] \quad -5R_1 + R_2$$

The corresponding system is

$$x - \frac{6}{5}y = -3$$
$$0 = 0, \qquad \text{True}$$

which has dependent equations. Using the second equation of the original system, we write the solution set as

$$\{(x, y) \,|\, 5x - 6y = -15\}.$$

**Now Try Exercises 11 and 13.**

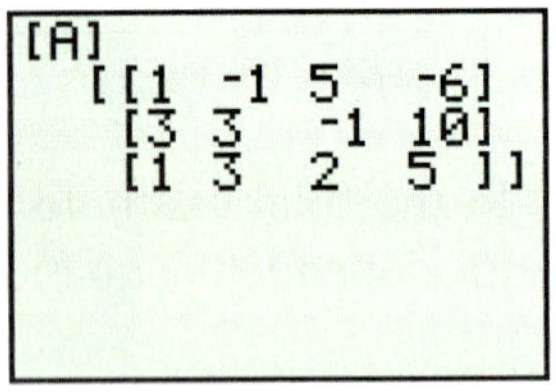


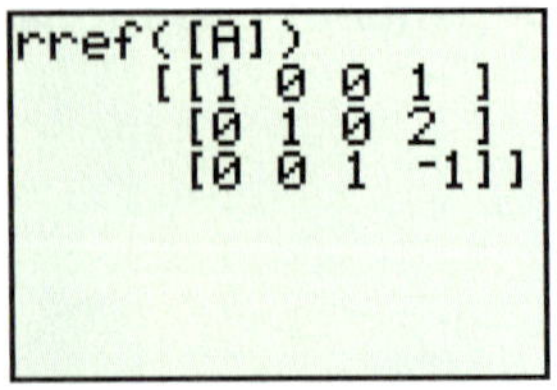


FIGURE 14

## CONNECTIONS

An extension of the matrix method described in this section involves transforming an augmented matrix into **reduced row echelon form.** This form has 1s down the main diagonal and 0s above and below this diagonal. For example, the matrix for the system in Example 2 could be transformed into the matrix

$$\left[\begin{array}{rrr|r} 1 & 0 & 0 & 1 \\ 0 & 1 & 0 & 2 \\ 0 & 0 & 1 & -1 \end{array}\right] \quad \text{which gives the equivalent system} \quad \begin{aligned} x &= 1 \\ y &= 2 \\ z &= -1. \end{aligned}$$

The calculator screens in Figure 14 indicate how easily this transformation can be obtained using technology.

### For Discussion or Writing

1. Write the reduced row echelon form for the matrix of the system in Example 1.
2. If transforming to reduced row echelon form leads to all 0s in the final row, what kind of system is represented?

## 8.6 EXERCISES

**For Extra Help**

- Student's Solutions Manual
- MyMathLab
- InterAct Math Tutorial Software
- AW Math Tutor Center
- MathXL
- Digital Video Tutor CD 12/Videotape 12

**1.** Consider the matrix $\begin{bmatrix} -2 & 3 & 1 \\ 0 & 5 & -3 \\ 1 & 4 & 8 \end{bmatrix}$ and answer the following.

**(a)** What are the elements of the second row?
**(b)** What are the elements of the third column?
**(c)** Is this a square matrix? Explain why or why not.
**(d)** Give the matrix obtained by interchanging the first and third rows.
**(e)** Give the matrix obtained by multiplying the first row by $-\frac{1}{2}$.
**(f)** Give the matrix obtained by multiplying the third row by 3 and adding to the first row.

**2.** Give the dimensions of each matrix.

**(a)** $\begin{bmatrix} 3 & -7 \\ 4 & 5 \\ -1 & 0 \end{bmatrix}$ **(b)** $\begin{bmatrix} 4 & 9 & 0 \\ -1 & 2 & -4 \end{bmatrix}$

**(c)**
```
[A]
      [[6  3 ]
       [-2 5 ]
       [4  10]
       [1  -1]]
```

**(d)**
```
[B]
   [[-1 7  1]
    [0  2  1]
    [-3 5  4]
    [8  -9 1]]
```

*Complete the steps in the matrix solution of each system by filling in the boxes. Give the final system and the solution set. See Example 1.*

**3.** $4x + 8y = 44$
$2x - y = -3$

$\left[\begin{array}{cc|c} 4 & 8 & 44 \\ 2 & -1 & -3 \end{array}\right]$

$\left[\begin{array}{cc|c} 1 & \square & \square \\ 2 & -1 & -3 \end{array}\right]$ $\quad \frac{1}{4}R_1$

$\left[\begin{array}{cc|c} 1 & 2 & 11 \\ 0 & \square & \square \end{array}\right]$ $\quad -2R_1 + R_2$

$\left[\begin{array}{cc|c} 1 & 2 & 11 \\ 0 & 1 & \square \end{array}\right]$ $\quad -\frac{1}{5}R_2$

**4.** $2x - 5y = -1$
$3x + y = 7$

$\left[\begin{array}{cc|c} 2 & -5 & -1 \\ 3 & 1 & 7 \end{array}\right]$

$\left[\begin{array}{cc|c} 1 & -\frac{5}{2} & \square \\ 3 & 1 & 7 \end{array}\right]$ $\quad \frac{1}{2}R_1$

$\left[\begin{array}{cc|c} 1 & -\frac{5}{2} & -\frac{1}{2} \\ 0 & \square & \square \end{array}\right]$ $\quad -3R_1 + R_2$

$\left[\begin{array}{cc|c} 1 & -\frac{5}{2} & -\frac{1}{2} \\ 0 & 1 & \square \end{array}\right]$ $\quad \frac{2}{17}R_2$

*Use row operations to solve each system. See Examples 1 and 3.*

**5.** $x + y = 5$
$x - y = 3$

**6.** $x + 2y = 7$
$x - y = -2$

**7.** $2x + 4y = 6$
$3x - y = 2$

**8.** $4x + 5y = -7$
$x - y = 5$

**9.** $3x + 4y = 13$
$2x - 3y = -14$

**10.** $5x + 2y = 8$
$3x - y = 7$

**11.** $-4x + 12y = 36$
$x - 3y = 9$

**12.** $2x - 4y = 8$
$-3x + 6y = 5$

**13.** $2x + y = 4$
$4x + 2y = 8$

**14.** $-3x - 4y = 1$
$6x + 8y = -2$

*Complete the steps in the matrix solution of each system by filling in the boxes. Give the final system and the solution set. See Example 2.*

**15.** $x + y - z = -3$
$2x + y + z = 4$
$5x - y + 2z = 23$

$$\left[\begin{array}{ccc|c} 1 & 1 & -1 & -3 \\ 2 & 1 & 1 & 4 \\ 5 & -1 & 2 & 23 \end{array}\right]$$

$$\left[\begin{array}{ccc|c} 1 & 1 & -1 & -3 \\ 0 & \square & \square & \square \\ 0 & \square & \square & \square \end{array}\right] \quad \begin{array}{l} \\ -2R_1 + R_2 \\ -5R_1 + R_3 \end{array}$$

$$\left[\begin{array}{ccc|c} 1 & 1 & -1 & -3 \\ 0 & 1 & \square & \square \\ 0 & -6 & 7 & 38 \end{array}\right] \quad -1R_2$$

$$\left[\begin{array}{ccc|c} 1 & 1 & -1 & -3 \\ 0 & 1 & -3 & -10 \\ 0 & 0 & \square & \square \end{array}\right] \quad 6R_2 + R_3$$

$$\left[\begin{array}{ccc|c} 1 & 1 & -1 & -3 \\ 0 & 1 & -3 & -10 \\ 0 & 0 & 1 & \square \end{array}\right] \quad -\tfrac{1}{11}R_3$$

**16.** $2x + y + 2z = 11$
$2x - y - z = -3$
$3x + 2y + z = 9$

$$\left[\begin{array}{ccc|c} 2 & 1 & 2 & 11 \\ 2 & -1 & -1 & -3 \\ 3 & 2 & 1 & 9 \end{array}\right]$$

$$\left[\begin{array}{ccc|c} 1 & \square & \square & \square \\ 2 & -1 & -1 & -3 \\ 3 & 2 & 1 & 9 \end{array}\right] \quad \tfrac{1}{2}R_1$$

$$\left[\begin{array}{ccc|c} 1 & \frac{1}{2} & 1 & \frac{11}{2} \\ 0 & \square & \square & \square \\ 0 & \square & \square & \square \end{array}\right] \quad \begin{array}{l} \\ -2R_1 + R_2 \\ -3R_1 + R_3 \end{array}$$

$$\left[\begin{array}{ccc|c} 1 & \frac{1}{2} & 1 & \frac{11}{2} \\ 0 & 1 & \square & \square \\ 0 & \frac{1}{2} & -2 & -\frac{15}{2} \end{array}\right] \quad -\tfrac{1}{2}R_2$$

$$\left[\begin{array}{ccc|c} 1 & \frac{1}{2} & 1 & \frac{11}{2} \\ 0 & 1 & \frac{3}{2} & 7 \\ 0 & 0 & \square & \square \end{array}\right] \quad -\tfrac{1}{2}R_2 + R_3$$

$$\left[\begin{array}{ccc|c} 1 & \frac{1}{2} & 1 & \frac{11}{2} \\ 0 & 1 & \frac{3}{2} & 7 \\ 0 & 0 & 1 & \square \end{array}\right] \quad -\tfrac{4}{11}R_3$$

*Use row operations to solve each system. See Examples 2 and 3.*

**17.** $x + y - 3z = 1$
$2x - y + z = 9$
$3x + y - 4z = 8$

**18.** $2x + 4y - 3z = -18$
$3x + y - z = -5$
$x - 2y + 4z = 14$

**19.** $x + y - z = 6$
$2x - y + z = -9$
$x - 2y + 3z = 1$

**20.** $x + 3y - 6z = 7$
$2x - y + 2z = 0$
$x + y + 2z = -1$

**21.**
$$\begin{aligned} x - y &= 1 \\ y - z &= 6 \\ x + z &= -1 \end{aligned}$$

**22.**
$$\begin{aligned} x + \phantom{2}y \phantom{+ 2z} &= 1 \\ 2x - \phantom{y +2} z &= 0 \\ y + 2z &= -2 \end{aligned}$$

**23.**
$$\begin{aligned} x - 2y + z &= 4 \\ 3x - 6y + 3z &= 12 \\ -2x + 4y - 2z &= -8 \end{aligned}$$

**24.**
$$\begin{aligned} 4x + 8y + 4z &= 9 \\ x + 3y + 4z &= 10 \\ 5x + 10y + 5z &= 12 \end{aligned}$$

**25.**
$$\begin{aligned} x + 2y + 3z &= -2 \\ 2x + 4y + 6z &= -5 \\ x - y + 2z &= 6 \end{aligned}$$

**26.**
$$\begin{aligned} x + 3y + z &= 1 \\ 2x + 6y + 2z &= 2 \\ 3x + 9y + 3z &= 3 \end{aligned}$$

**27.** Write a short explanation of each term. Include examples.

**(a)** Matrix
**(b)** Row of a matrix
**(c)** Column of a matrix
**(d)** Square matrix
**(e)** Augmented matrix
**(f)** Row operations on a matrix

**28.** Compare the use of the third row operation on a matrix and the elimination method of solving a system of linear equations. Give examples.

*The augmented matrix for the system in Exercise 3 is shown in the graphing calculator screen on the left as matrix* [A]. *The screen in the middle shows the row echelon form for* [A]. *Compare it to the matrix shown in the answer section for Exercise 3. The screen on the right shows the reduced row echelon form, and from this it can be determined by inspection that the solution set of the system is* $\{(1, 5)\}$.

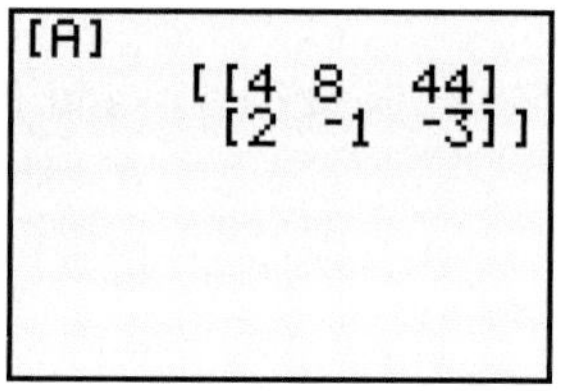


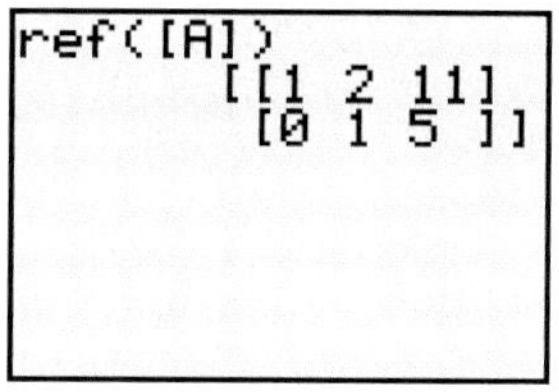


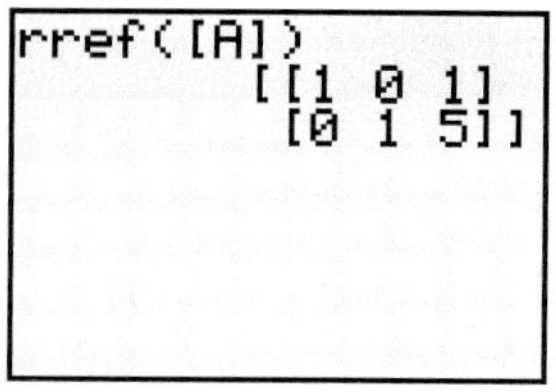


*Use a graphing calculator and either one of the two matrix methods illustrated to solve each system.*

**29.**
$$\begin{aligned} 4x + y &= 5 \\ 2x + y &= 3 \end{aligned}$$

**30.**
$$\begin{aligned} 5x + 3y &= 7 \\ 7x - 3y &= -19 \end{aligned}$$

**31.**
$$\begin{aligned} 5x + y - 3z &= -6 \\ 2x + 3y + z &= 5 \\ -3x - 2y + 4z &= 3 \end{aligned}$$

**32.**
$$\begin{aligned} x + y + z &= 3 \\ 3x - 3y - 4z &= -1 \\ x + y + 3z &= 11 \end{aligned}$$

**33.**
$$\begin{aligned} x + z &= -3 \\ y + z &= 3 \\ x + y &= 8 \end{aligned}$$

**34.**
$$\begin{aligned} x - y &= -1 \\ -y + z &= -2 \\ x + z &= -2 \end{aligned}$$

## Chapter 8 Group Activity

### Top Concert Tours

**OBJECTIVE** Use systems of equations to analyze data.

This activity will use systems of equations to analyze the top-grossing concert tours of 1997 and 2000.

- In 1997 the top-grossing North American concert tours were The Rolling Stones and U2. Together they grossed \$169.2 million. The Rolling Stones grossed \$9.4 million more than U2. The two groups performed a total of 79 shows, and U2 performed 13 more shows than The Rolling Stones. (*Source:* Pollstar.)
- In 2000 the top-grossing North American concert tours were Tina Turner and 'N Sync. Together they grossed \$156.6 million. Tina Turner grossed \$3.8 million more than 'N Sync. The two groups performed 181 shows, and 'N Sync performed 9 less shows than Tina Turner. (*Source:* Pollstar.)

Using the given information, complete the table. One student should complete the information for 1997 and the other student for 2000.

**A.** Write a system of equations to find the total gross for each group.

**B.** Write a system of equations to find the number of shows per year for each group.

**C.** Use the information from Exercises A and B to find the gross per show.

| *Year* | *Group* | *Total Gross* | *Shows per Year* | *Gross per Show* |
|---|---|---|---|---|
| 1997 | Rolling Stones | | | |
| 1997 | U2 | | | |
| 2000 | Tina Turner | | | |
| 2000 | 'N Sync | | | |

**D.** Once you have completed the table, compare the results. Answer these questions.

1. What differences do you notice between 1997 and 2000?
2. Which group had the highest total gross?
3. Which group had the highest gross per show?

# CHAPTER 8 SUMMARY

## KEY TERMS

**8.1** system of linear equations (linear system)
solution of a system
solution set of a system
consistent system
inconsistent system
independent equations
dependent equations

**8.4** ordered triple

**8.6** matrix
element of a matrix
row
column
square matrix
augmented matrix
row operations
row echelon form
reduced row echelon form

## NEW SYMBOLS

$(x, y, z)$ ordered triple

$\begin{bmatrix} a & b & c \\ d & e & f \end{bmatrix}$ matrix with two rows, three columns

## TEST YOUR WORD POWER

*See how well you have learned the vocabulary in this chapter. Answers, with examples, follow the Quick Review.*

**1.** A **system of linear equations** consists of
**A.** at least two linear equations with different variables
**B.** two or more linear equations that have an infinite number of solutions
**C.** two or more linear equations with the same variables
**D.** two or more linear inequalities.

**2.** A **solution of a system** of linear equations is
**A.** an ordered pair that makes one equation of the system true
**B.** an ordered pair that makes all the equations of the system true at the same time
**C.** any ordered pair that makes one or the other or both equations of the system true
**D.** the set of values that make all the equations of the system false.

**3.** A **consistent system** is a system of equations
**A.** with one solution
**B.** with no solution
**C.** with an infinite number of solutions
**D.** that have the same graph.

**4.** An **inconsistent system** is a system of equations
**A.** with one solution
**B.** with no solution
**C.** with an infinite number of solutions
**D.** that have the same graph.

**5.** **Dependent equations**
**A.** have different graphs
**B.** have no solution
**C.** have one solution
**D.** are different forms of the same equation.

**6.** A **matrix** is
**A.** an ordered pair of numbers
**B.** an array of numbers with the same number of rows and columns
**C.** a pair of numbers written between brackets
**D.** a rectangular array of numbers.

## QUICK REVIEW

**CONCEPTS** | **EXAMPLES**

### 8.1 SOLVING SYSTEMS OF LINEAR EQUATIONS BY GRAPHING

An ordered pair is a solution of a system if it makes all equations of the system true at the same time.

Is $(4, -1)$ a solution of the system $\begin{aligned} x + y &= 3 \\ 2x - y &= 9? \end{aligned}$

Yes, because $4 + (-1) = 3$, and $2(4) - (-1) = 9$ are both true.

To solve a linear system by graphing,

*Step 1* Graph each equation of the system on the same axes.

*Step 2* Find the coordinates of the point of intersection.

*Step 3* Check. Write the solution set.

Solve the system by graphing: $\begin{aligned} x + y &= 5 \\ 2x - y &= 4. \end{aligned}$

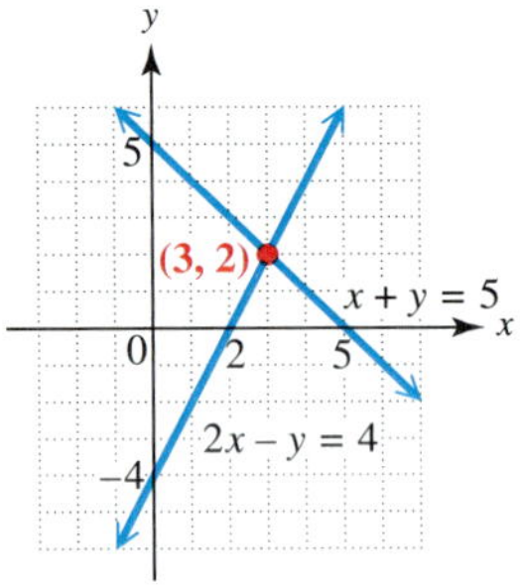


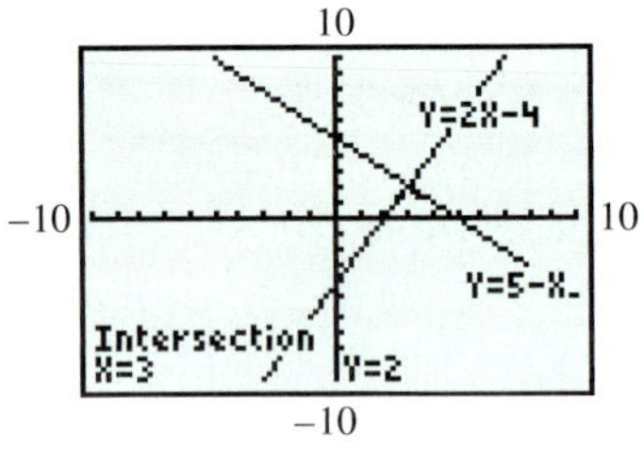


A graphing calculator can find the solution of a system by locating the point of intersection of the graphs.

The solution $(3, 2)$ checks, so $\{(3, 2)\}$ is the solution set.

### 8.2 SOLVING SYSTEMS OF LINEAR EQUATIONS BY SUBSTITUTION

*Step 1* Solve one equation for either variable.

Solve by substitution.

$$x + 2y = -5 \qquad (1)$$
$$y = -2x - 1 \qquad (2)$$

Equation (2) is already solved for $y$.

*Step 2* Substitute for that variable in the other equation to get an equation in one variable.

Substitute $-2x - 1$ for $y$ in equation (1).

$$x + 2(-2x - 1) = -5$$

*Step 3* Solve the equation from Step 2.

Solve to get $x = 1$.

*Step 4* Substitute the result into the equation from Step 1 to get the value of the other variable.

To find $y$, let $x = 1$ in equation (2).

$$y = -2(1) - 1 = -3$$

*Step 5* Check. Write the solution set.

The solution, $(1, -3)$, checks so $\{(1, -3)\}$ is the solution set.

| CONCEPTS | EXAMPLES |
|---|---|

## 8.3 SOLVING SYSTEMS OF LINEAR EQUATIONS BY ELIMINATION

| | |
|---|---|
| *Step 1* Write both equations in standard form $Ax + By = C$. | Solve by elimination. $$\begin{aligned} x + 3y &= 7 \quad (1)\\ 3x - y &= 1 \quad (2)\end{aligned}$$ |
| *Step 2* Multiply to make the coefficients of one pair of variable terms opposites. | Multiply equation (1) by $-3$ to eliminate the $x$-terms. |
| *Step 3* Add the equations to get an equation with only one variable. | $$\begin{aligned} -3x - 9y &= -21\\ 3x - y &= 1\\ \hline -10y &= -20 \quad \text{Add.}\end{aligned}$$ |
| *Step 4* Solve the equation from Step 3. | $$y = 2 \quad \text{Divide by } -10.$$ |
| *Step 5* Substitute the solution from Step 4 into either of the original equations to find the value of the remaining variable. | Substitute to get the value of $x$. $$\begin{aligned} x + 3y &= 7 \quad (1)\\ x + 3(2) &= 7 \quad \text{Let } y = 2.\\ x + 6 &= 7 \quad \text{Multiply.}\\ x &= 1 \quad \text{Subtract 6.}\end{aligned}$$ |
| *Step 6* Check. Write the solution set. | Since $1 + 3(2) = 7$ and $3(1) - 2 = 1$, the solution set is $\{(1, 2)\}$. |
| If the result of the addition step (Step 3) is a false statement, such as $0 = 4$, the graphs are parallel lines and *the solution set is* $\emptyset$. | $$\begin{aligned} x - 2y &= 6\\ -x + 2y &= -2\\ \hline 0 &= 4 \quad \text{Solution set: } \emptyset\end{aligned}$$ |
| If the result is a true statement, such as $0 = 0$, the graphs are the same line, and an *infinite number of ordered pairs are solutions.* | $$\begin{aligned} x - 2y &= 6\\ -x + 2y &= -6\\ \hline 0 &= 0 \quad \text{Solution set: } \{(x, y) \mid x - 2y = 6\}\end{aligned}$$ |

## 8.4 SYSTEMS OF LINEAR EQUATIONS IN THREE VARIABLES

| | |
|---|---|
| **Solving a Linear System in Three Variables** | Solve the system |
| *Step 1* Use the elimination method to eliminate any variable from any two of the original equations. | $$\begin{aligned} x + 2y - z &= 6 \quad (1)\\ x + y + z &= 6 \quad (2)\\ 2x + y - z &= 7. \quad (3)\end{aligned}$$ Add equations (1) and (2); $z$ is eliminated and the result is $2x + 3y = 12$. |
| *Step 2* Eliminate the *same* variable from any *other* two equations. | Eliminate $z$ again by adding equations (2) and (3) to get $3x + 2y = 13$. Now solve the system $$\begin{aligned} 2x + 3y &= 12 \quad (4)\\ 3x + 2y &= 13. \quad (5)\end{aligned}$$ |

(continued)

| CONCEPTS | EXAMPLES |
|---|---|
| *Step 3* Eliminate a second variable from the two equations in two variables that result from Steps 1 and 2. The result is an equation in one variable that gives the value of that variable. | To eliminate $x$, multiply equation (4) by $-3$ and equation (5) by 2. $$\begin{aligned} -6x - 9y &= -36 \\ 6x + 4y &= \phantom{-}26 \\ \hline -5y &= -10 \\ y &= 2 \end{aligned}$$ |
| *Step 4* Substitute the value of the variable found in Step 3 into either of the equations in two variables to find the value of the second variable. | Let $y = 2$ in equation (4). $$\begin{aligned} 2x + 3(2) &= 12 \\ 2x + 6 &= 12 \\ 2x &= 6 \\ x &= 3 \end{aligned}$$ |
| *Step 5* Use the values of the two variables from Steps 3 and 4 to find the value of the third variable by substituting into an appropriate equation. | Let $y = 2$ and $x = 3$ in any of the original equations to find $z = 1$. |
| *Step 6* Check the solution in all of the original equations. Then write the solution set. | Check. The solution set is $\{(3, 2, 1)\}$. |

**8.5 APPLICATIONS OF SYSTEMS OF LINEAR EQUATIONS**

Use the six-step problem-solving method.

| | |
|---|---|
| *Step 1* Read the problem carefully. | The perimeter of a rectangle is 18 ft. The length is 3 ft more than twice the width. What are the dimensions of the rectangle? |
| *Step 2* Assign variables. | Let $x$ represent the length and $y$ represent the width. 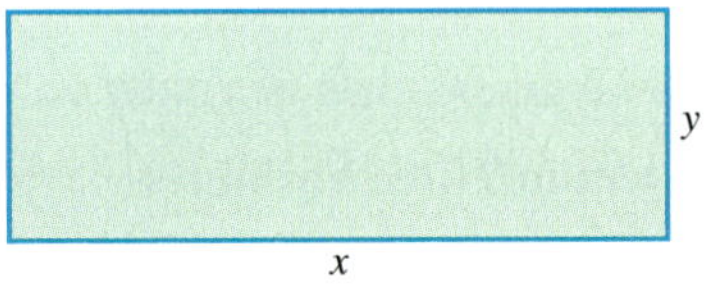  |
| *Step 3* Write a system of equations that relates the unknowns. | From the perimeter formula, one equation is $2x + 2y = 18$. From the problem, another equation is $x = 3 + 2y$. |
| *Step 4* Solve the system. | Solve the system $$\begin{aligned} 2x + 2y &= 18 \\ x &= 3 + 2y \end{aligned}$$ to find that $x = 7$ and $y = 2$. |
| *Steps 5 and 6* State the answer and check. | The length is 7 ft, and the width is 2 ft. Since the perimeter is $$2(7) + 2(2) = 18, \quad \text{and} \quad 3 + 2(2) = 7,$$ the answer checks. |

| CONCEPTS | EXAMPLES |
|---|---|

## 8.6 SOLVING SYSTEMS OF LINEAR EQUATIONS BY MATRIX METHODS

**Matrix Row Operations**

1. Any two rows of the matrix may be interchanged.

$$\begin{bmatrix} 1 & 5 & 7 \\ 3 & 9 & -2 \\ 0 & 6 & 4 \end{bmatrix} \text{ becomes } \begin{bmatrix} 3 & 9 & -2 \\ 1 & 5 & 7 \\ 0 & 6 & 4 \end{bmatrix}$$ Interchange $R_1$ and $R_2$.

2. The elements in any row may be multiplied by any nonzero real number.

$$\begin{bmatrix} 1 & 5 & 7 \\ 3 & 9 & -2 \\ 0 & 6 & 4 \end{bmatrix} \text{ becomes } \begin{bmatrix} 1 & 5 & 7 \\ 1 & 3 & -\frac{2}{3} \\ 0 & 6 & 4 \end{bmatrix}$$ $\frac{1}{3}R_2$

3. Any row may be changed by adding to the elements of the row the product of a real number and the elements of another row.

$$\begin{bmatrix} 1 & 5 & 7 \\ 3 & 9 & -2 \\ 0 & 6 & 4 \end{bmatrix} \text{ becomes } \begin{bmatrix} 1 & 5 & 7 \\ 0 & -6 & -23 \\ 0 & 6 & 4 \end{bmatrix}$$ $-3R_1 + R_2$

A system can be solved by matrix methods. Write the augmented matrix and use row operations to obtain a matrix in row echelon form.

Solve using row operations:
$$\begin{aligned} x + 3y &= 7 \\ 2x + y &= 4. \end{aligned}$$

$$\left[\begin{array}{cc|c} 1 & 3 & 7 \\ 2 & 1 & 4 \end{array}\right]$$ Augmented matrix

$$\left[\begin{array}{cc|c} 1 & 3 & 7 \\ 0 & -5 & -10 \end{array}\right]$$ $-2R_1 + R_2$

$$\left[\begin{array}{cc|c} 1 & 3 & 7 \\ 0 & 1 & 2 \end{array}\right]$$ $-\frac{1}{5}R_2$

$$\begin{aligned} x + 3y &= 7 \\ y &= 2 \end{aligned}$$

When $y = 2$, $x + 3(2) = 7$, so $x = 1$. The solution set is $\{(1, 2)\}$.

**Answers to Test Your Word Power**

**1.** C; *Example:* $2x + y = 7$, $3x - y = 3$ **2.** B; *Example:* The ordered pair (2, 3) satisfies both equations of the system in Answer 1, so it is a solution of the system. **3.** A; *Example:* The system in Answer 1 is consistent. The graphs of the equations intersect at exactly one point, in this case the solution (2, 3). **4.** B; *Example:* The equations of two parallel lines make up an inconsistent system; their graphs never intersect, so there is no solution to the system. **5.** D; *Example:* The equations $4x - y = 8$ and $8x - 2y = 16$ are dependent because their graphs are the same line. **6.** D; *Examples:* $\begin{bmatrix} 3 & -1 & 0 \\ 4 & 2 & 1 \end{bmatrix}$, $\begin{bmatrix} 1 & 2 \\ 4 & 3 \end{bmatrix}$

# CHAPTER 8 REVIEW EXERCISES

[8.1] *Decide whether the given ordered pair is a solution of the given system.*

**1.** (3, 4)
$4x - 2y = 4$
$5x + y = 19$

**2.** (−5, 2)
$x - 4y = -13$
$2x + 3y = 4$

*Solve each system by graphing.*

**3.** $x + y = 4$
$2x - y = 5$

**4.** $x - 2y = 4$
$2x + y = -2$

**5.** $2x + 4 = 2y$
$y - x = -3$

**6.** The graph shows the trends during the years 1974 through 1996 relating to bachelor's degrees awarded in the United States.

**(a)** Between what years shown on the horizontal axis did the number of degrees for men and women reach equal numbers?

**(b)** When the number of degrees for men and women reached equal numbers, what was that number (approximately)?

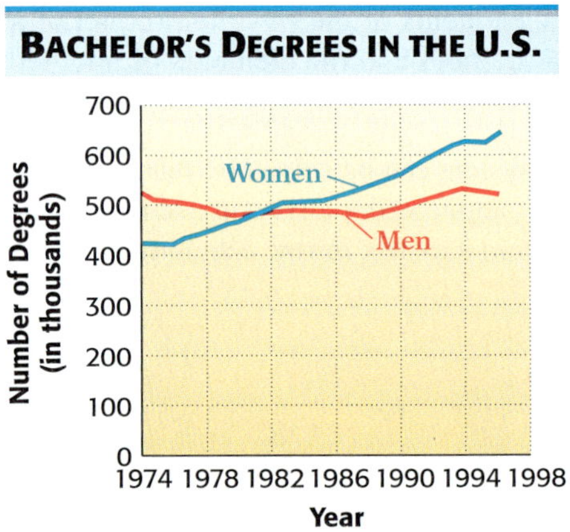


*Source:* U.S. National Center for Education Statistics, *Digest of Education Statistics*, annual.

[8.2]

**7.** A student solves the system $\begin{aligned} 2x + y &= 6 \\ -2x - y &= 4 \end{aligned}$ and gets the equation $0 = 10$. The student gives the solution set as {(0, 10)}. Is this correct? Explain.

*Solve each system by the substitution method.*

**8.** $3x + y = 7$
$x = 2y$

**9.** $2x - 5y = -19$
$y = x + 2$

**10.** $5x + 15y = 30$
$x + 3y = 6$

**11.** After solving a system of linear equations by the substitution method, a student obtained the equation "$0 = 0$." He gave the solution set of the system as {(0, 0)}. Was his answer correct? Why or why not?

**12.** Suppose that you were asked to solve the system $\begin{aligned} 5x - 3y &= 7 \\ -x + 2y &= 4 \end{aligned}$ by substitution. Which variable in which equation would be easiest to solve for in your first step? Why?

[8.3]

**13.** Only one of the following systems does not require that we multiply one or both equations by a constant to solve the system by the elimination method. Which one is it?

**A.** $-4x + 3y = 7$
$3x - 4y = 4$

**B.** $5x + 8y = 13$
$12x + 24y = 36$

**C.** $2x + 3y = 5$
$x - 3y = 12$

**D.** $x + 2y = 9$
$3x - y = 6$

**14.** For the system

$$2x + 12y = 7$$
$$3x + 4y = 1,$$

if we were to multiply the first equation by $-3$, by what number would we have to multiply the second equation to

**(a)** eliminate the $x$-terms when solving by the elimination method?
**(b)** eliminate the $y$-terms when solving by the elimination method?

*Solve each system by the elimination method.*

**15.** $2x - y = 13$; $x + y = 8$

**16.** $-4x + 3y = 25$; $6x - 5y = -39$

**17.** $3x - 4y = 9$; $6x - 8y = 18$

**18.** $2x + y = 3$; $-4x - 2y = 6$

[8.1–8.3] *Solve each system by any method.*

**19.** $2x + 3y = -5$; $3x + 4y = -8$

**20.** $6x - 9y = 0$; $2x - 3y = 0$

**21.** $x - 2y = 5$; $y = x - 7$

**22.** $\frac{x}{2} + \frac{y}{3} = 7$; $\frac{x}{4} + \frac{2y}{3} = 8$

**23.** List the methods of solving systems discussed in this chapter. Choose one and discuss its advantages and disadvantages.

**24.** Why would it be easier to solve system B by the substitution method than system A?

**A:** $-5x + 6y = 7$; $2x + 5y = -5$

**B:** $2x + 9y = 13$; $y = 3x - 2$

[8.4] *Solve each system. If a system is inconsistent or has dependent equations, say so.*

**25.** $2x + 3y - z = -16$; $x + 2y + 2z = -3$; $-3x + y + z = -5$

**26.** $4x - y = 2$; $3y + z = 9$; $x + 2z = 7$

**27.** $3x - y - z = -8$; $4x + 2y + 3z = 15$; $-6x + 2y + 2z = 10$

[8.5] *Solve each problem using a system of equations.*

**28.** A regulation National Hockey League ice rink has perimeter 570 ft. The length is 30 ft longer than twice the width. What are the dimensions of an NHL ice rink? (*Source: Microsoft Encarta Encyclopedia 2000.*)

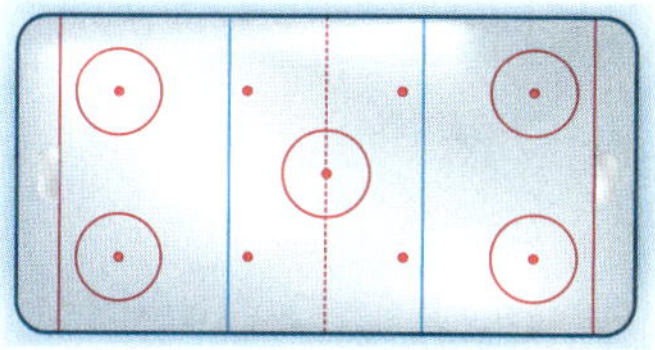

**29.** The two top-grossing movies of 2001 were *Harry Potter and the Sorcerer's Stone* and *Shrek*. *Shrek* grossed \$26 million less than *Harry Potter and the Sorcerer's Stone,* and together the two films took in \$562 million. (*Source:* ACNielsen EDI.)

**(a)** How much did each of these movies earn?
**(b)** If *Shrek* earned \$27 million more than *Monsters, Inc.*, how much did *Monsters, Inc.* earn? (*Hint:* Use your answer from part (a).)
**(c)** What is the total amount these top three films earned?

**30.** A plane flies 560 mi in 1.75 hr traveling with the wind. The return trip later against the same wind takes the plane 2 hr. Find the speed of the plane and the speed of the wind.

| | *r* | *t* | *d* |
|---|---|---|---|
| *With Wind* | $x + y$ | 1.75 | |
| *Against Wind* | | 2 | |

**31.** Sweet's Candy Store is offering a special mix for Valentine's Day. Ms. Sweet will mix some \$2-per-lb nuts with some \$1-per-lb chocolate candy to get 100 lb of mix, which she will sell at \$1.30 per lb. How many pounds of each should she use?

| | *Number of Pounds* | *Price per Pound* | *Value* |
|---|---|---|---|
| *Nuts* | $x$ | | |
| *Chocolate* | $y$ | | |
| *Mixture* | 100 | | |

**32.** The sum of the measures of the angles of a triangle is 180°. The largest angle measures 10° less than the sum of the other two. The measure of the middle-sized angle is the average of the other two. Find the measures of the three angles.

**33.** Maria Gonzales sells real estate. On three recent sales, she made 10% commission, 6% commission, and 5% commission. Her total commissions on these sales were \$17,000, and she sold property worth \$280,000. If the 5% sale amounted to the sum of the other two, what were the three sales prices?

**34.** How many liters each of 8%, 10%, and 20% hydrogen peroxide should be mixed together to get 8 L of 12.5% solution, if the amount of 8% solution used must be 2 L more than the amount of 20% solution used?

**35.** In the great baseball year of 1961, Yankee teammates Mickey Mantle, Roger Maris, and John Blanchard combined for 136 home runs. Mantle hit 7 fewer than Maris. Maris hit 40 more than Blanchard. What were the home run totals for each player? (*Source:* Neft, David S. and Richard M. Cohen, *The Sports Encyclopedia: Baseball 1997.*)

**[8.6]** *Solve each system of equations using row operations.*

**36.** $2x + 5y = -4$
$4x - y = 14$

**37.** $6x + 3y = 9$
$-7x + 2y = 17$

**38.** $x + 2y - z = 1$
$3x + 4y + 2z = -2$
$-2x - y + z = -1$

**39.** $x + 3y = 7$
$3x + z = 2$
$y - 2z = 4$

## MIXED REVIEW EXERCISES

*Solve by any method.*

**40.** $\frac{2}{3}x + \frac{1}{6}y = \frac{19}{2}$
$\frac{1}{3}x - \frac{2}{9}y = 2$

**41.** $2x + 5y - z = 12$
$-x + y - 4z = -10$
$-8x - 20y + 4z = 31$

**42.** $x = 7y + 10$
$2x + 3y = 3$

**43.** $x + 4y = 17$
$-3x + 2y = -9$

**44.** $-7x + 3y = 12$
$5x + 2y = 8$

**45.** $2x - 5y = 8$
$3x + 4y = 10$

**46.** To make a 10% acid solution for chemistry class, Xavier wants to mix some 5% solution with 10 L of 20% solution. How many liters of 5% solution should he use?

| *Liters of Solution* | *Percent (as a decimal)* | *Liters of Pure Acid* |
|---|---|---|
| | | |
| | | |
| | | |

**47.** At the end of 2001, Subway topped McDonald's as the largest restaurant chain in the United States. Subway operated 148 more restaurants than McDonald's, and together the two chains had 26,346 restaurants. How many restaurants did each company operate? (*Source: USA Today,* February 4, 2002.)

**48.** Candy that sells for \$1.30 per lb is to be mixed with candy selling for \$.90 per lb to get 100 lb of a mix that will sell for \$1 per lb. How much of each type should be used?

**49.** In the 2000 Summer Olympics in Sydney, Australia, the top medal-winning countries were the United States, Russia, and China, with a combined total of 244 medals. The United States won 9 more medals than Russia, while China won 29 fewer medals than Russia. How many medals did each country win? (*Source: The Gazette,* October 2, 2000.)

## RELATING CONCEPTS (EXERCISES 50–54)

**For Individual or Group Work**

*Thus far in this text we have studied only* linear *equations. In later chapters we will study the graphs of other kinds of equations. One such graph is a* circle, *which has an equation of the form*

$$x^2 + y^2 + ax + by + c = 0.$$

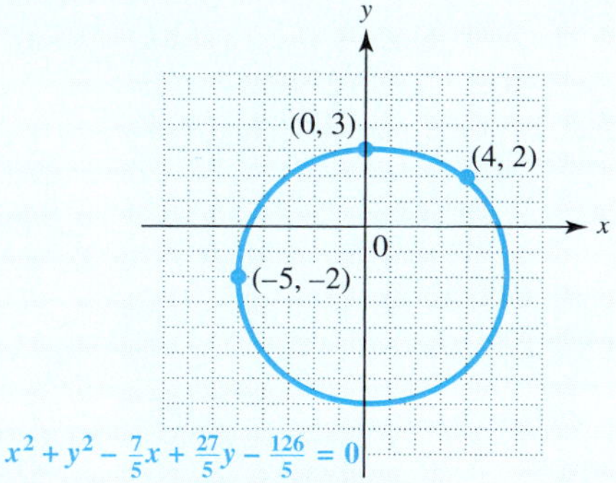


*It is a fact from geometry that given three* noncollinear *points (that is, points that do not all lie on the same straight line), there will be a circle that contains them. For example, the points* (4, 2), (−5, −2), *and* (0, 3) *lie on the circle whose equation is shown in the figure.* ***Work Exercises 50–54 in order,*** *to find an equation of the circle passing through the points* (2, 1), (−1, 0), *and* (3, 3).

**50.** Let $x = 2$ and $y = 1$ in the equation $x^2 + y^2 + ax + by + c = 0$ to find an equation in $a$, $b$, and $c$.

(continued)

**51.** Let $x = -1$ and $y = 0$ to find a second equation in $a$, $b$, and $c$.

**52.** Let $x = 3$ and $y = 3$ to find a third equation in $a$, $b$, and $c$.

**53.** Solve the system of equations formed by your answers in Exercises 50–52 to find the values of $a$, $b$, and $c$. What is the equation of the circle?

**54.** Explain why the relation whose graph is a circle is not a function.

# CHAPTER 8 TEST

*The graph shows a company's costs to produce computer parts and the revenue from the sale of those parts.*

**1.** At what production level does the cost equal the revenue?

**2.** What is the revenue at that point?

**COMPUTER PARTS**

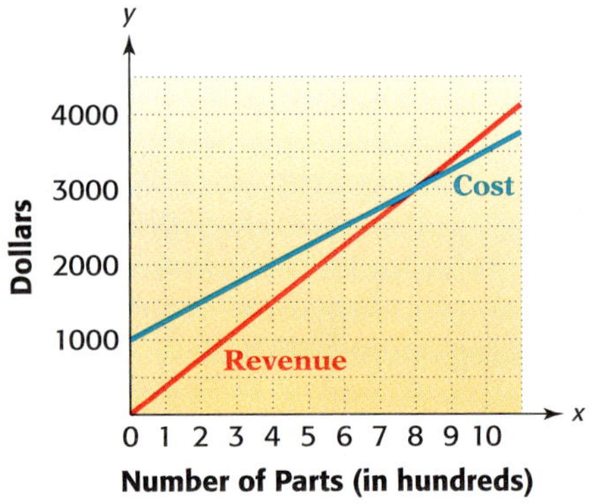


**3.** Decide whether each ordered pair is a solution of the system

$$\begin{aligned} 2x + y &= -3 \\ x - y &= -9. \end{aligned}$$

**(a)** $(1, -5)$ **(b)** $(1, 10)$ **(c)** $(-4, 5)$

**4.** Use a graph to solve the system

$$\begin{aligned} x + y &= 7 \\ x - y &= 5. \end{aligned}$$

*Solve each system by substitution. If a system is inconsistent or has dependent equations, say so.*

**5.** $2x - 3y = 24$
$y = -\frac{2}{3}x$

**6.** $12x - 5y = 8$
$3x = \frac{5}{4}y + 2$

*Solve each system by elimination. If a system is inconsistent or has dependent equations, say so.*

**7.** $3x + y = 12$
$2x - y = 3$

**8.** $-5x + 2y = -4$
$6x + 3y = -6$

**9.** $3x + 4y = 8$
$8y = 7 - 6x$

**10.** $\frac{6}{5}x - \frac{1}{3}y = -20$
$-\frac{2}{3}x + \frac{1}{6}y = 11$

**11.** $3x + 5y + 3z = 2$
$6x + 5y + z = 0$
$3x + 10y - 2z = 6$

**12.** $4x + y + z = 11$
$x - y - z = 4$
$y + 2z = 0$

*Solve each problem using a system of equations.*

**13.** Julia Roberts is one of the biggest box office stars in Hollywood. As of July 2001, her two top-grossing domestic films, *Pretty Woman* and *Runaway Bride,* together earned \$330.7 million. If *Runaway Bride* grossed \$26.1 million less than *Pretty Woman,* how much did each film gross? (*Source:* ACNielsen EDI.)

**14.** Two cars start from points 420 mi apart and travel toward each other. They meet after 3.5 hr. Find the average speed of each car if one travels 30 mph slower than the other.

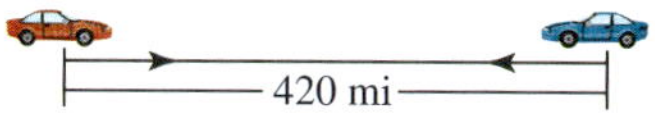


**15.** A chemist needs 12 L of a 40% alcohol solution. She must mix a 20% solution and a 50% solution. How many liters of each will be required to obtain what she needs?

**16.** A local electronics store will sell 7 AC adaptors and 2 rechargeable flashlights for \$86, or 3 AC adaptors and 4 rechargeable flashlights for \$84. What is the price of a single AC adaptor and a single rechargeable flashlight?

**17.** The owner of a tea shop wants to mix three kinds of tea to make 100 oz of a mixture that will sell for \$.83 per oz. He uses Orange Pekoe, which sells for \$.80 per oz, Irish Breakfast, for \$.85 per oz, and Earl Grey, for \$.95 per oz. If he wants to use twice as much Orange Pekoe as Irish Breakfast, how much of each kind of tea should he use?

*Solve each system using row operations.*

**18.** $$\begin{aligned} 3x + 2y &= 4 \\ 5x + 5y &= 9 \end{aligned}$$

**19.** $$\begin{aligned} x + 3y + 2z &= 11 \\ 3x + 7y + 4z &= 23 \\ 5x + 3y - 5z &= -14 \end{aligned}$$

**20.** Use any method described in this chapter to solve the system
$$\begin{aligned} 4x - 2y &= -8 \\ 3y - 5z &= 14 \\ 2x + z &= -10. \end{aligned}$$

## CUMULATIVE REVIEW EXERCISES CHAPTERS 1–8

**1.** Find the value of the expression $\dfrac{3x^2 + 2y^2}{10y + 3}$ if $x = 1$ and $y = 5$.

*Name the property that justifies each statement.*

**2.** $5 + (-4) = (-4) + 5$

**3.** $r(s - k) = rs - rk$

**4.** Evaluate $-2 + 6[3 - (4 - 9)]$.

*Solve each linear equation.*

**5.** $2 - 3(6x + 2) = 4(x + 1) + 18$

**6.** $\dfrac{3}{2}\left(\dfrac{1}{3}x + 4\right) = 6\left(\dfrac{1}{4} + x\right)$

**7.** Solve the formula $P = \dfrac{kT}{V}$ for $T$.

*Solve each linear inequality.*

**8.** $-\frac{5}{6}x < 15$

**9.** $-8 < 2x + 3$

**10.** A recent survey measured public recognition of the most popular contemporary advertising slogans. Complete the results shown in the table if 2500 people were surveyed.

| *Slogan (product or company)* | *Percent Recognition (nearest tenth of a percent)* | *Actual Number Who Recognized Slogan (nearest whole number)* |
|---|---|---|
| Please Don't Squeeze the . . . (Charmin) | 80.4% | |
| The Breakfast of Champions (Wheaties) | 72.5% | |
| The King of Beers (Budweiser) | | 1570 |
| Like a Good Neighbor (State Farm) | | 1430 |

(Other slogans included "You're in Good Hands" (Allstate), "Snap, Crackle, Pop" (Rice Krispies), and "The Un-Cola" (7-Up).)
*Source:* Department of Integrated Marketing Communications, Northwestern University.

*Solve each problem.*

**11.** On February 12, 1999, the U.S. Senate voted to acquit William Jefferson Clinton on both counts of impeachment (perjury and obstruction of justice). Of the 200 "guilty" or "not guilty" votes cast that day, there were 10 more "not guilty" votes than "guilty" votes. How many of each vote were there? (*Source:* MSNBC Web site, February 13, 1999.)

**12.** Two angles of a triangle have the same measure. The measure of the third angle is 4° less than twice the measure of each of the equal angles. Find the measures of the three angles.

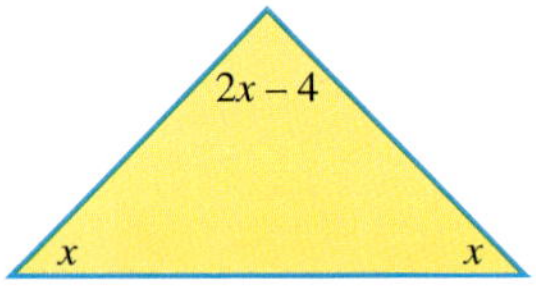

Measures are in degrees.

**13.** No baseball fan should be without a copy of *The Sports Encyclopedia: Baseball 2000* by David S. Neft and Richard M. Cohen. The 20th edition provides exhaustive statistics for professional baseball dating back to 1876. This book has a perimeter of 38 in., and its width measures 2.5 in. less than its length. What are the dimensions of the book?

*Graph each linear equation.*

**14.** $x - y = 4$

**15.** $3x + y = 6$

*Find the slope of each line.*

**16.** Through $(-5, 6)$ and $(1, -2)$

**17.** Perpendicular to the line $y = 4x - 3$

*Find an equation for each line. Write it in slope-intercept form.*

**18.** Through $(-4, 1)$ with slope $\frac{1}{2}$

**19.** Through the points $(1, 3)$ and $(-2, -3)$

**20.** **(a)** Write an equation of the vertical line through $(9, -2)$.
**(b)** Write an equation of the horizontal line through $(4, -1)$.

*Simplify. Write each answer with only positive exponents. Assume all variables represent nonzero real numbers.*

**21.** $\left(\dfrac{a^{-3}b^4}{a^2b^{-1}}\right)^{-2}$

**22.** $\left(\dfrac{m^{-4}n^2}{m^2n^{-3}}\right) \cdot \left(\dfrac{m^5n^{-1}}{m^{-2}n^5}\right)$

*Perform the indicated operations.*

**23.** $(3y^2 - 2y + 6) - (-y^2 + 5y + 12)$

**24.** $(4f + 3)(3f - 1)$

**25.** $\left(\dfrac{1}{4}x + 5\right)^2$

**26.** $(3x^3 + 13x^2 - 17x - 7) \div (3x + 1)$

*Factor each polynomial completely.*

**27.** $2x^2 - 13x - 45$

**28.** $100t^4 - 25$

**29.** $8p^3 + 125$

**30.** Solve the equation $3x^2 + 4x = 7$.

**31.** Write $\dfrac{y^2 - 16}{y^2 - 8y + 16}$ in lowest terms.

*Perform the indicated operations. Express the answer in lowest terms.*

**32.** $\dfrac{2a^2}{a + b} \cdot \dfrac{a - b}{4a}$

**33.** $\dfrac{2x}{2x - 1} + \dfrac{4}{2x + 1} + \dfrac{8}{4x^2 - 1}$

**34.** Solve the equation $\dfrac{-3x}{x + 1} + \dfrac{4x + 1}{x} = \dfrac{-3}{x^2 + x}$.

*Decide whether each relation defined in Exercises 35 and 36 is a function, and give its domain and range.*

**35. Average Hourly Wages in Mexico**

| Year | Wage (in dollars) |
|---|---|
| 1990 | 1.25 |
| 1992 | 1.61 |
| 1994 | 1.80 |
| 1996 | 1.21 |
| 1998 | 1.94 |
| 2000 | 2.26 |

*Source:* John Christman, CIEMEX-WEFA.

**36.**

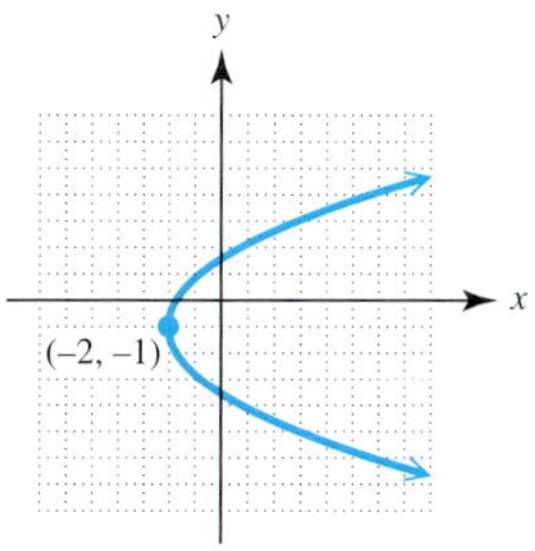


**37.** Given the equation $5x - 3y = 8$,

**(a)** write it with function notation $f(x)$;
**(b)** find $f(1)$.

**38.** If $y$ varies directly as $x$ and $y = 5$ when $x = 12$, find $y$ when $x = 42$.

*Solve by any method.*

**39.** $-2x + 3y = -15$
$4x - y = 15$

**40.** $x - 3y = 7$
$2x - 6y = 14$

**41.** $x + y + z = 10$
$x - y - z = 0$
$-x + y - z = -4$

*Solve each problem using a system of equations.*

**42.** Two of the best-selling toys in a recent year were Tickle Me Elmo and Snacktime Kid. Based on their average retail prices, Elmo cost \$8.63 less than Kid, and together they cost \$63.89. What was the average retail price for each toy? (*Source:* NPD Group, Inc.)

**43.** At the Chalmette Nut Shop, 6 lb of peanuts and 12 lb of cashews cost \$60, while 3 lb of peanuts and 4 lb of cashews cost \$22. Find the cost of each type of nut.

**44.** Eboni Perkins compared the monthly payments she would incur for two types of mortgages: fixed-rate and variable-rate. Her observations led to the graph at the right. Use it to answer the following questions.

**(a)** For which years would the monthly payment be more for the fixed-rate mortgage than for the variable-rate mortgage?

**(b)** In what year would the payments be the same, and what would those payments be?

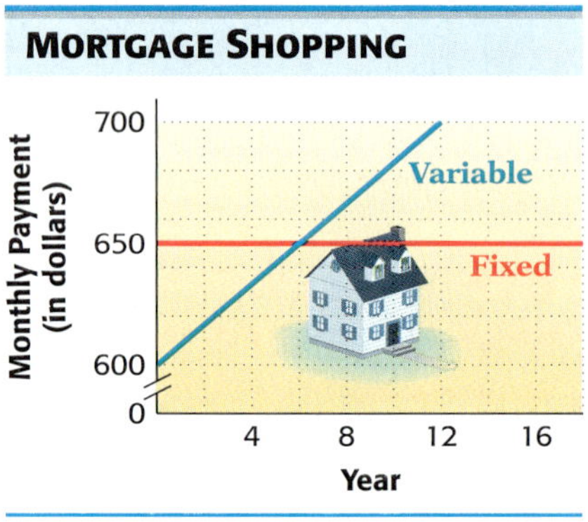

# Inequalities and Absolute Value

# 9



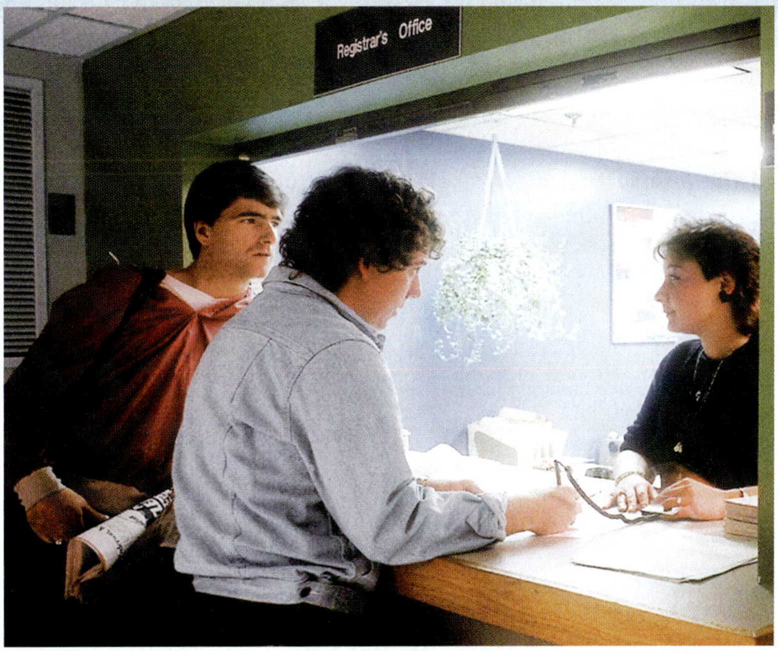


The cost of a college education has risen rapidly in the last decade. Average higher education tuition and fees increased by 78.8% from the 1990–1991 school year to the 2000–2001 school year. (*Source:* National Center for Education Statistics, U.S. Department of Education.) In Exercises 69–72 of Section 9.1, we apply the concepts of this chapter to college student expenses.

# 9.1 Set Operations and Compound Inequalities

**OBJECTIVES**

1. Find the intersection of two sets.
2. Solve compound inequalities with the word *and*.
3. Find the union of two sets.
4. Solve compound inequalities with the word *or*.

The table shows symptoms of an underactive thyroid and an overactive thyroid.

| *Underactive Thyroid* | *Overactive Thyroid* |
|---|---|
| Sleepiness, $s$ | Insomnia, $i$ |
| Dry hands, $d$ | Moist hands, $m$ |
| Intolerance of cold, $c$ | Intolerance of heat, $h$ |
| Goiter, $g$ | Goiter, $g$ |

*Source: The Merck Manual of Diagnosis and Therapy,* 16th Edition, Merck Research Laboratories, 1992.

Let $N$ be the set of symptoms for an underactive thyroid, and let $O$ be the set of symptoms for an overactive thyroid. Suppose we are interested in the set of symptoms that are found in *both* sets $N$ *and* $O$. In this section we discuss the use of the words *and* and *or* as they relate to sets and inequalities.

**OBJECTIVE 1 Find the intersection of two sets.** The intersection of two sets is defined using the word *and*.

**Intersection of Sets**

For any two sets $A$ and $B$, the **intersection** of $A$ and $B$, symbolized $A \cap B$, is defined as follows:

$$A \cap B = \{x \mid x \text{ is an element of } A \textbf{ and } x \text{ is an element of } B\}.$$

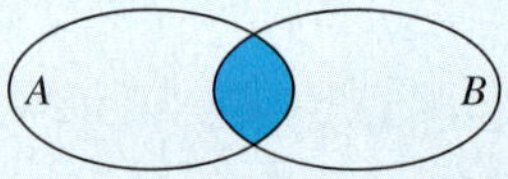


**EXAMPLE 1 Finding the Intersection of Two Sets**

Let $A = \{1, 2, 3, 4\}$ and $B = \{2, 4, 6\}$. Find $A \cap B$.

The set $A \cap B$ contains those elements that belong to both $A$ *and* $B$: the numbers 2 and 4. Therefore,

$$A \cap B = \{1, 2, 3, 4\} \cap \{2, 4, 6\}$$
$$= \{2, 4\}.$$

**Now Try Exercise 7.**

A **compound inequality** consists of two inequalities linked by a connective word such as *and* or *or*. Examples of compound inequalities are

$$x + 1 \leq 9 \quad \text{and} \quad x - 2 \geq 3$$

and

$$2x > 4 \quad \text{or} \quad 3x - 6 < 5.$$

**OBJECTIVE 2 Solve compound inequalities with the word *and*.** Use the following steps.

**Solving a Compound Inequality with *and***

*Step 1* Solve each inequality in the compound inequality individually.

*Step 2* Since the inequalities are joined with *and,* the solution set of the compound inequality will include all numbers that satisfy both inequalities in Step 1 (the intersection of the solution sets).

### EXAMPLE 2 Solving a Compound Inequality with *and*

Solve the compound inequality

$$x + 1 \leq 9 \quad \text{and} \quad x - 2 \geq 3.$$

*Step 1* Solve each inequality in the compound inequality individually.

$$\begin{aligned} x + 1 &\leq 9 & &\text{and} & x - 2 &\geq 3 \\ x + 1 - 1 &\leq 9 - 1 & &\text{and} & x - 2 + 2 &\geq 3 + 2 \\ x &\leq 8 & &\text{and} & x &\geq 5 \end{aligned}$$

*Step 2* Because the inequalities are joined with the word *and,* the solution set will include all numbers that satisfy both inequalities in Step 1 at the same time. Thus, the compound inequality is true whenever $x \leq 8$ and $x \geq 5$ are both true. The top graph in Figure 1 shows $x \leq 8$, and the bottom graph shows $x \geq 5$.

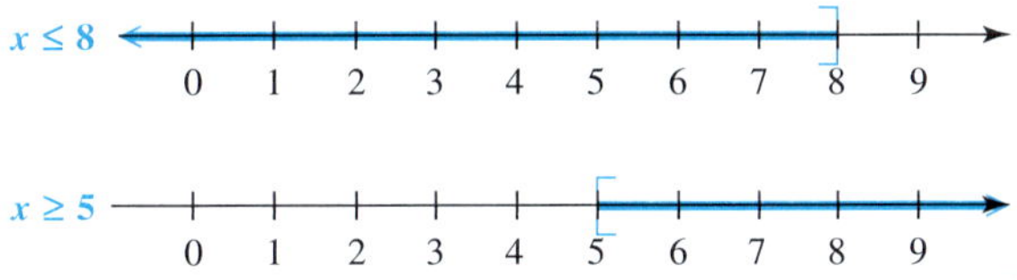


**FIGURE 1**

Find the intersection of the two graphs in Figure 1 to get the solution set of the compound inequality. The intersection of the two graphs in Figure 2 shows that the solution set in interval notation is [5, 8].

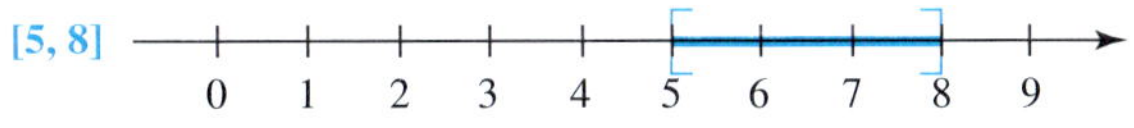


**FIGURE 2**

**Now Try Exercise 27.**

### EXAMPLE 3 Solving a Compound Inequality with *and*

Solve the compound inequality

$$-3x - 2 > 5 \quad \text{and} \quad 5x - 1 \leq -21.$$

*Step 1* Solve each inequality separately.

$$\begin{aligned} -3x - 2 &> 5 & &\text{and} & 5x - 1 &\leq -21 \\ -3x &> 7 & &\text{and} & 5x &\leq -20 \\ x &< -\frac{7}{3} & &\text{and} & x &\leq -4 \end{aligned}$$

The graphs of $x < -\frac{7}{3}$ and $x \leq -4$ are shown in Figure 3.

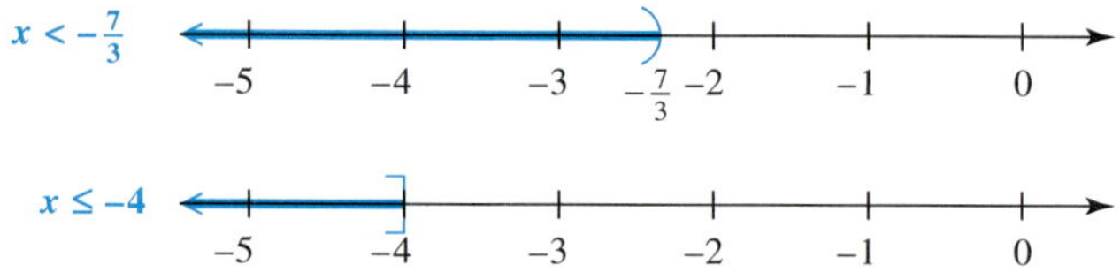


**FIGURE 3**

*Step 2* Now find all values of $x$ that satisfy both conditions; that is, the real numbers that are less than $-\frac{7}{3}$ and also less than or equal to $-4$. As shown by the graph in Figure 4, the solution set is $(-\infty, -4]$.

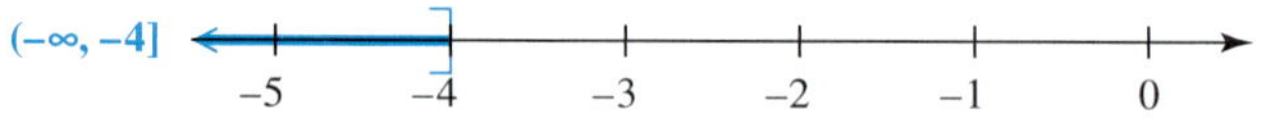


**FIGURE 4**

**Now Try Exercise 31.**

## EXAMPLE 4 Solving a Compound Inequality with *and*

Solve $x + 2 < 5$ and $x - 10 > 2$.

First solve each inequality separately.

$$x + 2 < 5 \quad \text{and} \quad x - 10 > 2$$
$$x < 3 \quad \text{and} \quad x > 12$$

The graphs of $x < 3$ and $x > 12$ are shown in Figure 5.

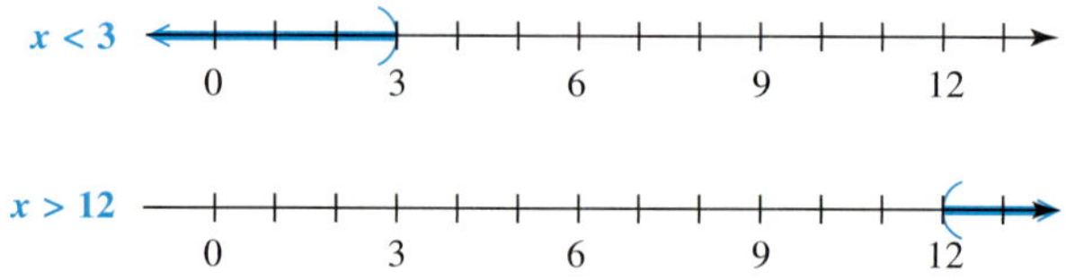


**FIGURE 5**

There is no number that is both less than 3 *and* greater than 12, so the given compound inequality has no solution. The solution set is $\emptyset$. See Figure 6.

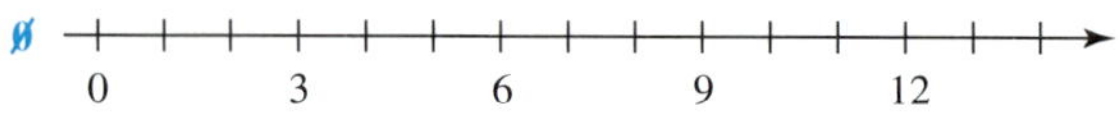


**FIGURE 6**

**Now Try Exercise 25.**

**OBJECTIVE 3 Find the union of two sets.** The union of two sets is defined using the word *or*.

**Union of Sets**

For any two sets $A$ and $B$, the **union** of $A$ and $B$, symbolized $A \cup B$, is defined as follows:

$$A \cup B = \{x \mid x \text{ is an element of } A \textbf{ or } x \text{ is an element of } B\}.$$

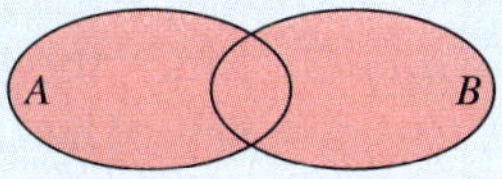


### EXAMPLE 5 Finding the Union of Two Sets

Let $A = \{1, 2, 3, 4\}$ and $B = \{2, 4, 6\}$. Find $A \cup B$.

Begin by listing all the elements of set $A$: 1, 2, 3, 4. Then list any additional elements from set $B$. In this case the elements 2 and 4 are already listed, so the only additional element is 6. Therefore,

$$\begin{aligned} A \cup B &= \{1, 2, 3, 4\} \cup \{2, 4, 6\} \\ &= \{1, 2, 3, 4, 6\}. \end{aligned}$$

The union consists of all elements in either $A$ *or* $B$ (or both).

**Now Try Exercise 13.**

In Example 5, notice that although the elements 2 and 4 appeared in both sets $A$ and $B$, they are written only once in $A \cup B$.

**OBJECTIVE 4 Solve compound inequalities with the word *or*.** Use the following steps.

**Solving a Compound Inequality with *or***

*Step 1* Solve each inequality in the compound inequality individually.

*Step 2* Since the inequalities are joined with *or*, the solution set of the compound inequality includes all numbers that satisfy either one of the two inequalities in Step 1 (the union of the solution sets).

### EXAMPLE 6 Solving a Compound Inequality with *or*

Solve $6x - 4 < 2x$ or $-3x \leq -9$.

*Step 1* Solve each inequality separately.

$$\begin{aligned} 6x - 4 &< 2x \quad \text{or} \quad -3x \leq -9 \\ 4x &< 4 \\ x &< 1 \quad \text{or} \quad x \geq 3 \end{aligned}$$

The graphs of these two inequalities are shown in Figure 7 on the next page.

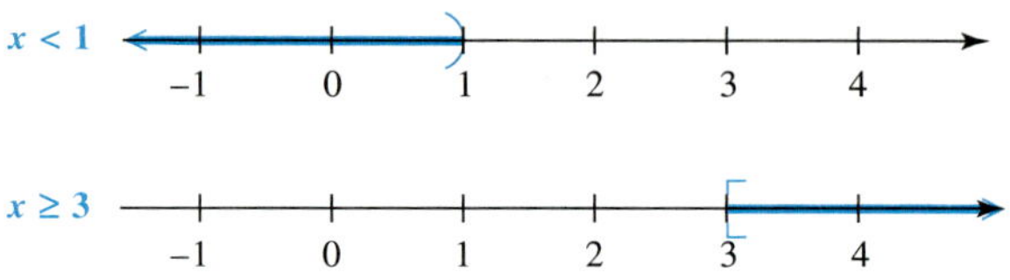


FIGURE 7

*Step 2* Since the inequalities are joined with *or*, find the union of the two solution sets. The union is shown in Figure 8 and is written

$$(-\infty, 1) \cup [3, \infty).$$

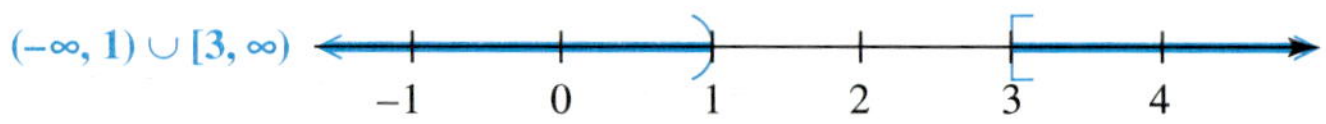


FIGURE 8

**Now Try Exercise 43.**

**CAUTION** When inequalities are used to write the solution set in Example 6, it *should* be written as

$$x < 1 \quad \text{or} \quad x \geq 3,$$

which keeps the numbers 1 and 3 in their order on the number line. Writing $3 \leq x < 1$ would imply that $3 \leq 1$, which is ***FALSE.*** There is no other way to write the solution set of such a union.

## EXAMPLE 7 Solving a Compound Inequality with *or*

Solve $-4x + 1 \geq 9$ or $5x + 3 \leq -12$.

First we solve each inequality separately.

$$\begin{aligned} -4x + 1 &\geq 9 &\quad \text{or} \quad 5x + 3 &\leq -12 \\ -4x &\geq 8 &\quad \text{or} \quad 5x &\leq -15 \\ x &\leq -2 &\quad \text{or} \quad x &\leq -3 \end{aligned}$$

The graphs of these two inequalities are shown in Figure 9.

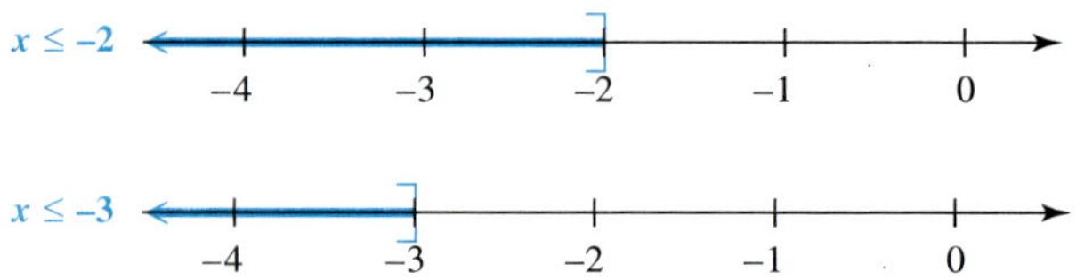


FIGURE 9

By taking the union, we obtain the interval $(-\infty, -2]$. It is graphed in Figure 10.

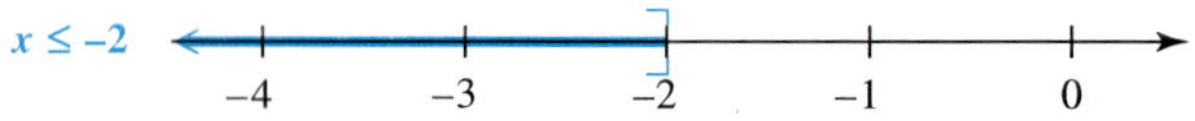


FIGURE 10

**Now Try Exercise 37.**

### EXAMPLE 8 Applying Intersection and Union

The five highest domestic grossing films (adjusted for inflation) are listed in the table.

**Five All-Time Highest Grossing Films**

| Film | Admissions | Gross Income |
|---|---|---|
| *Gone with the Wind* | 200,605,313 | \$972,900,000 |
| *Star Wars* | 178,119,595 | \$863,900,000 |
| *The Sound of Music* | 142,415,376 | \$690,700,000 |
| *E.T.* | 135,987,938 | \$659,500,000 |
| *The Ten Commandments* | 131,000,000 | \$635,400,000 |

*Source: New York Times Almanac*, 2001.

List the elements of the following sets.

**(a)** The set of top five films with admissions greater than 180,000,000 *and* gross greater than \$800,000,000

The only film that satisfies both conditions is *Gone with the Wind,* so the set is

{*Gone with the Wind*}.

**(b)** The set of top five films with admissions less than 170,000,000 *or* gross greater than \$700,000,000

Here, a film that satisfies at least one of the conditions is in the set. This set includes all five films:

{*Gone with the Wind, Star Wars, The Sound of Music, E.T., The Ten Commandments*}.

**Now Try Exercises 69 and 71.**

## 9.1 EXERCISES

**For Extra Help**

- Student's Solutions Manual
- MyMathLab
- InterAct Math Tutorial Software
- AW Math Tutor Center
- MathXL
- Digital Video Tutor CD 13/Videotape 13

*Decide whether each statement is* true *or* false. *If it is false, explain why.*

**1.** The union of the solution sets of $x + 1 = 5$, $x + 1 < 5$, and $x + 1 > 5$ is $(-\infty, \infty)$.

**2.** The intersection of the sets $\{x \mid x \geq 7\}$ and $\{x \mid x \leq 7\}$ is $\emptyset$.

**3.** The union of the sets $(-\infty, 8)$ and $(8, \infty)$ is $\{8\}$.

**4.** The intersection of the sets $(-\infty, 8]$ and $[8, \infty)$ is $\{8\}$.

**5.** The intersection of the set of rational numbers and the set of irrational numbers is $\{0\}$.

**6.** The union of the set of rational numbers and the set of irrational numbers is the set of real numbers.

*Let* $A = \{1, 2, 3, 4, 5, 6\}$, $B = \{1, 3, 5\}$, $C = \{1, 6\}$, *and* $D = \{4\}$. *Specify each set. See Examples 1 and 5.*

**7.** $B \cap A$ **8.** $A \cap B$ **9.** $A \cap D$ **10.** $B \cap C$

**11.** $B \cap \emptyset$ **12.** $A \cap \emptyset$ **13.** $A \cup B$ **14.** $B \cup D$

15. Give an example of intersection applied to a real-life situation.

16. A compound inequality uses one of the words *and* or *or*. Explain how you will determine whether to use *intersection* or *union* when graphing the solution set.

*Two sets are specified by graphs. Graph the intersection of the two sets.*

17.
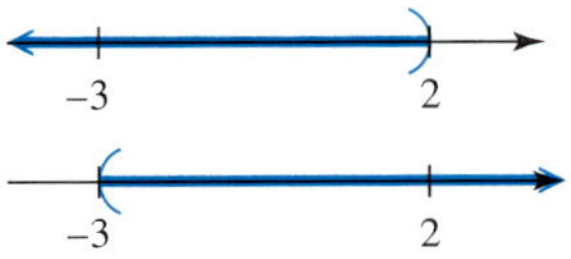


18.

19.
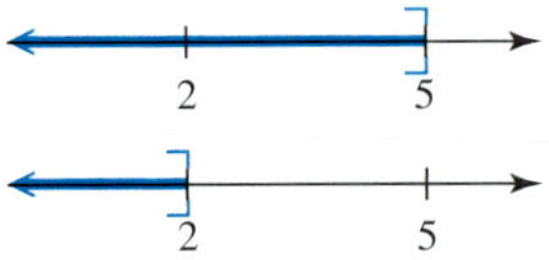


20.
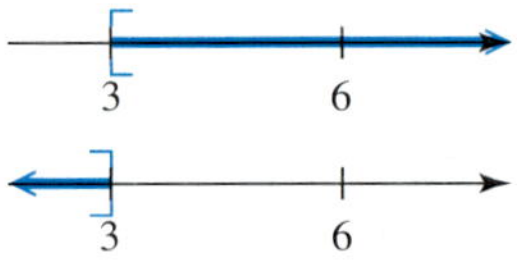


*For each compound inequality, give the solution set in both interval and graph forms. See Examples 2–4.*

21. $x < 2$ and $x > -3$

22. $x < 5$ and $x > 0$

23. $x \leq 2$ and $x \leq 5$

24. $x \geq 3$ and $x \geq 6$

25. $x \leq 3$ and $x \geq 6$

26. $x \leq -1$ and $x \geq 3$

27. $x - 3 \leq 6$ and $x + 2 \geq 7$

28. $x + 5 \leq 11$ and $x - 3 \geq -1$

29. $-3x > 3$ and $x + 3 > 0$

30. $-3x < 3$ and $x + 2 < 6$

31. $3x - 4 \leq 8$ and $-4x + 1 \geq -15$

32. $7x + 6 \leq 48$ and $-4x \geq -24$

*Two sets are specified by graphs. Graph the union of the two sets.*

33.
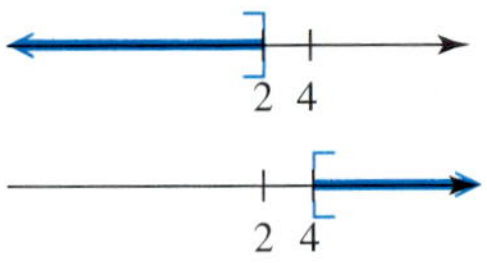


34.
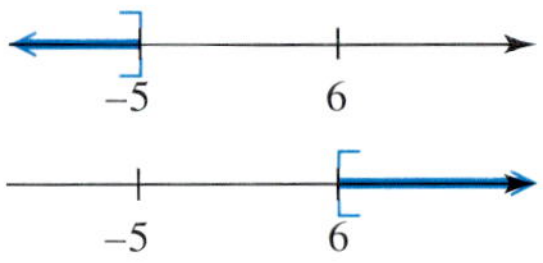


35.
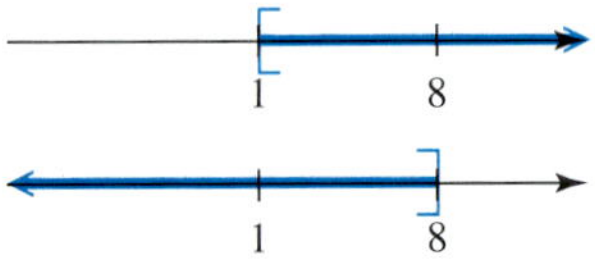


36.
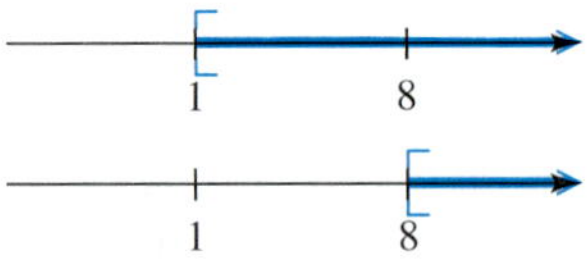


*For each compound inequality, give the solution set in both interval and graph forms. See Examples 6 and 7.*

37. $x \leq 1$ or $x \leq 8$

38. $x \geq 1$ or $x \geq 8$

39. $x \geq -2$ or $x \geq 5$

40. $x \leq -2$ or $x \leq 6$

41. $x \geq -2$ or $x \leq 4$

42. $x \geq 5$ or $x \leq 7$

43. $x + 2 > 7$ or $1 - x > 6$

44. $x + 1 > 3$ or $x + 4 < 2$

45. $x + 1 > 3$ or $-4x + 1 > 5$

46. $3x < x + 12$ or $x + 1 > 10$

*Express each set in the simplest interval form. (Hint: Graph each set and look for the intersection or union.)*

47. $(-\infty, -1] \cap [-4, \infty)$

48. $[-1, \infty) \cap (-\infty, 9]$

49. $(-\infty, -6] \cap [-9, \infty)$

50. $(5, 11] \cap [6, \infty)$

51. $(-\infty, 3) \cup (-\infty, -2)$

52. $[-9, 1] \cup (-\infty, -3)$

53. $[3, 6] \cup (4, 9)$

54. $[-1, 2] \cup (0, 5)$

*For each compound inequality, decide whether* intersection *or* union *should be used. Then give the solution set in both interval and graph forms. See Examples 2, 3, 4, 6, and 7.*

**55.** $x < -1$ and $x > -5$

**56.** $x > -1$ and $x < 7$

**57.** $x < 4$ or $x < -2$

**58.** $x < 5$ or $x < -3$

**59.** $-3x \leq -6$ or $-3x \geq 0$

**60.** $2x - 6 \leq -18$ and $2x \geq -18$

**61.** $x + 1 \geq 5$ and $x - 2 \leq 10$

**62.** $-8x \leq -24$ or $-5x \geq 15$

## RELATING CONCEPTS (EXERCISES 63–68)

**For Individual or Group Work**

*The figures represent the backyards of neighbors Luigi, Maria, Than, and Joe. Find the area and the perimeter of each yard. Suppose that each resident has* 150 *ft of fencing and enough sod to cover* 1400 *ft*$^2$ *of lawn. Give the name or names of the residents whose yards satisfy each description.* ***Work Exercises 63–68 in order.***

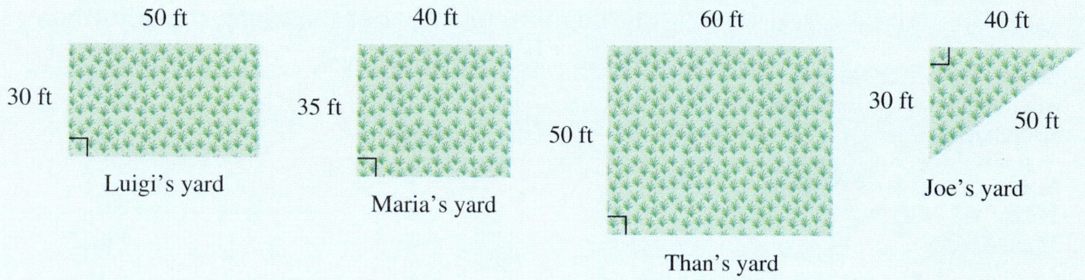


**63.** The yard can be fenced *and* the yard can be sodded.

**64.** The yard can be fenced *and* the yard cannot be sodded.

**65.** The yard cannot be fenced *and* the yard can be sodded.

**66.** The yard cannot be fenced *and* the yard cannot be sodded.

**67.** The yard can be fenced *or* the yard can be sodded.

**68.** The yard cannot be fenced *or* the yard can be sodded.

*Average expenses for full-time college students during the 2000–2001 academic year are shown in the table.*

**College Expenses in 2000–2001 (in dollars)**

| *Type of Expense* | *Public Schools* | *Private Schools* |
|---|---|---|
| Tuition and fees | 2600 | 14,690 |
| Board rates | 2454 | 2989 |
| Dormitory charges | 2566 | 3370 |

*Source:* U.S. National Center for Education Statistics, U.S. Department of Education.

*Use the table to list the elements of each set.*

**69.** The set of expenses less than \$2700 for public schools *and* greater than \$3500 for private schools

**70.** The set of expenses less than \$2600 for public schools *and* less than \$3500 for private schools

**71.** The set of expenses less than \$2500 for public schools *or* greater than \$3500 for private schools

**72.** The set of expenses greater than \$10,000 for private schools *or* less than \$2000 for public schools

# 9.2 Absolute Value Equations and Inequalities

**OBJECTIVES**

1. Use the distance definition of absolute value.
2. Solve equations of the form $|ax + b| = k$, for $k > 0$.
3. Solve inequalities of the form $|ax + b| < k$ and of the form $|ax + b| > k$, for $k > 0$.
4. Solve absolute value equations that involve rewriting.
5. Solve equations of the form $|ax + b| = |cx + d|$.
6. Solve special cases of absolute value equations and inequalities.

In a production line, quality is controlled by randomly choosing items from the line and checking to see how selected measurements vary from the optimum measure. These differences are sometimes positive and sometimes negative, so they are expressed with absolute value. For example, a machine that fills quart milk cartons might be set to release 1 qt (32 oz) plus or minus 2 oz per carton. Then the number of ounces in each carton should satisfy the *absolute value inequality* $|x - 32| \leq 2$, where $x$ is the number of ounces.

**OBJECTIVE 1 Use the distance definition of absolute value.** In Chapter 1 we saw that the absolute value of a number $x$, written $|x|$, represents the distance from $x$ to 0 on the number line. For example, the solutions of $|x| = 4$ are 4 and $-4$, as shown in Figure 11.

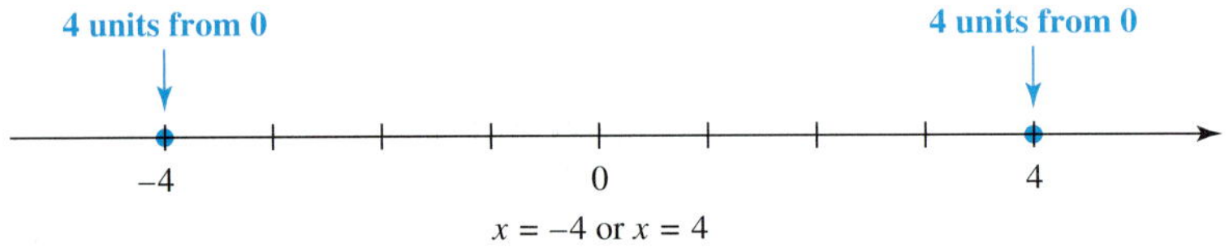


**FIGURE 11**

Because absolute value represents distance from 0, it is reasonable to interpret the solutions of $|x| > 4$ to be all numbers that are *more* than 4 units from 0. The set $(-\infty, -4) \cup (4, \infty)$ fits this description. Figure 12 shows the graph of the solution set of $|x| > 4$. Because the graph consists of two separate intervals, the solution set is described using *or* as

$$x < -4 \quad \text{or} \quad x > 4.$$

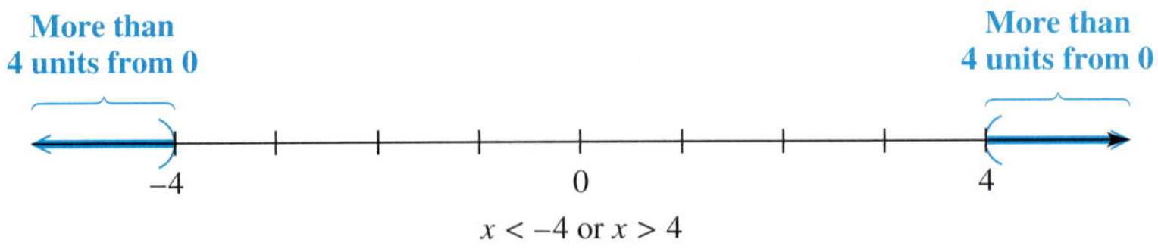


**FIGURE 12**

The solution set of $|x| < 4$ consists of all numbers that are *less* than 4 units from 0 on the number line. Another way of thinking of this is to think of all numbers *between* $-4$ and 4. This set of numbers is given by $(-4, 4)$, as shown in Figure 13. Here, the graph shows that $-4 < x < 4$, which means $x > -4$ *and* $x < 4$.

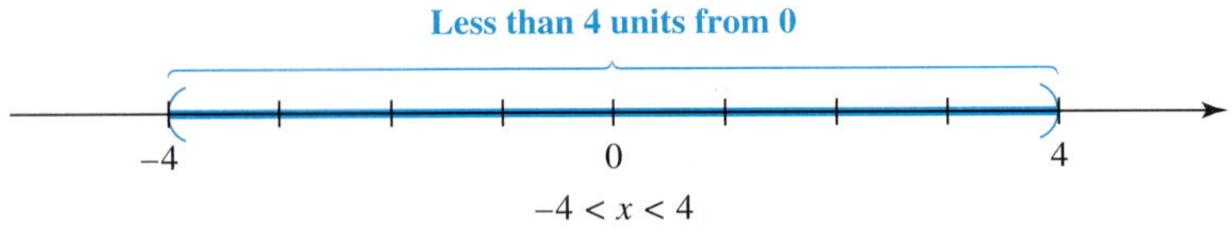


**FIGURE 13**

The equation and inequalities just described are examples of **absolute value equations and inequalities.** They involve the absolute value of a variable expression and generally take the form

$$|ax + b| = k, \qquad |ax + b| > k, \qquad \text{or} \qquad |ax + b| < k,$$

where $k$ is a positive number. From Figures 11–13, we see that

$|x| = 4$ has the same solution set as $x = -4$ or $x = 4$,
$|x| > 4$ has the same solution set as $x < -4$ or $x > 4$,
$|x| < 4$ has the same solution set as $x > -4$ and $x < 4$.

Thus, we solve an absolute value equation or inequality by first rewriting it as an equivalent statement without absolute value bars. Notice that, except for special cases, the solution set of an absolute value *equation* includes exactly *two points.* However, the solution set of an absolute value *inequality* includes one or more *intervals.* We summarize these facts in the next box.

**Solving Absolute Value Equations and Inequalities**

Let $k$ be a positive real number, and $p$ and $q$ be real numbers.

**1.** To solve $|ax + b| = k$, solve the compound equation

$$ax + b = k \quad \textbf{or} \quad ax + b = -k.$$

The solution set is usually of the form $\{p, q\}$, which includes two numbers.

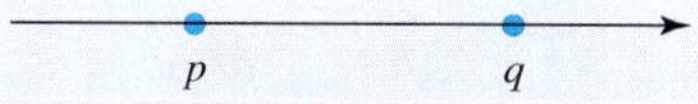


**2.** To solve $|ax + b| > k$, solve the compound inequality

$$ax + b > k \quad \textbf{or} \quad ax + b < -k.$$

The solution set is of the form $(-\infty, p) \cup (q, \infty)$, which consists of two separate intervals.

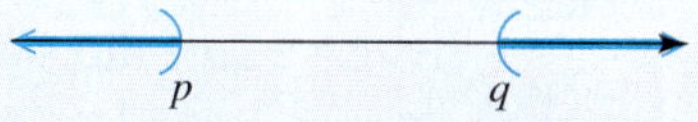


**3.** To solve $|ax + b| < k$, solve the compound inequality

$$-k < ax + b < k.$$

The solution set is of the form $(p, q)$, a single interval.

**NOTE** Some people prefer to write the compound statements in parts 1 and 2 of the preceding summary as

$$ax + b = k \quad \text{or} \quad -(ax + b) = k$$

and

$$ax + b > k \quad \text{or} \quad -(ax + b) > k.$$

These forms are equivalent to those we give in the summary and produce the same results.

**OBJECTIVE 2 Solve equations of the form $|ax + b| = k$, for $k > 0$.** The next example shows how we use a compound equation to solve a typical absolute value equation. Remember that because absolute value refers to distance from the origin, each absolute value equation will have two parts.

### EXAMPLE 1 Solving an Absolute Value Equation

Solve $|2x + 1| = 7$.

For $|2x + 1|$ to equal 7, $2x + 1$ must be 7 units from 0 on the number line. This can happen only when $2x + 1 = 7$ or $2x + 1 = -7$. This is the first case in the preceding summary. Solve this compound equation as follows.

$$2x + 1 = 7 \quad \text{or} \quad 2x + 1 = -7$$
$$2x = 6 \quad \text{or} \quad 2x = -8$$
$$x = 3 \quad \text{or} \quad x = -4$$

Check by substitution in the original absolute value equation to verify that the solution set is $\{-4, 3\}$. The graph is shown in Figure 14.

**FIGURE 14**

**Now Try Exercise 11.**

**OBJECTIVE 3 Solve inequalities of the form $|ax + b| < k$ and of the form $|ax + b| > k$, for $k > 0$.**

### EXAMPLE 2 Solving an Absolute Value Inequality with >

Solve $|2x + 1| > 7$.

By part 2 of the summary, this absolute value inequality is rewritten as

$$2x + 1 > 7 \quad \text{or} \quad 2x + 1 < -7,$$

because $2x + 1$ must represent a number that is *more* than 7 units from 0 on either side of the number line. Now, solve the compound inequality.

$$2x + 1 > 7 \quad \text{or} \quad 2x + 1 < -7$$
$$2x > 6 \quad \text{or} \quad 2x < -8$$
$$x > 3 \quad \text{or} \quad x < -4$$

Check these solutions. The solution set is $(-\infty, -4) \cup (3, \infty)$. See Figure 15. Notice that the graph consists of two intervals.

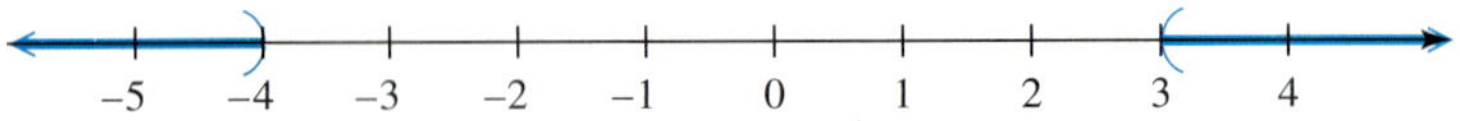


**FIGURE 15**

**Now Try Exercise 25.**

## EXAMPLE 3 Solving an Absolute Value Inequality with <

Solve $|2x + 1| < 7$.

The expression $2x + 1$ must represent a number that is less than 7 units from 0 on either side of the number line. Another way of thinking of this is to realize that $2x + 1$ must be between $-7$ and 7. As part 3 of the summary shows, this is written as the three-part inequality

$$-7 < 2x + 1 < 7.$$

We solved such inequalities in Section 2.8 by working with all three parts at the same time.

$$-7 < 2x + 1 < 7$$

$$-8 < 2x < 6 \qquad \text{Subtract 1 from each part.}$$

$$-4 < x < 3 \qquad \text{Divide each part by 2.}$$

Check that the solution set is $(-4, 3)$, so the graph consists of the single interval shown in Figure 16.

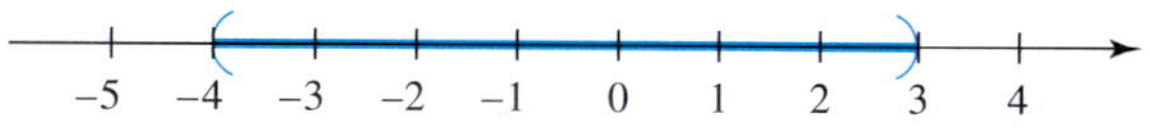


**FIGURE 16**

**Now Try Exercise 39.**

Look back at Figures 14, 15, and 16, with the graphs of $|2x + 1| = 7$, $|2x + 1| > 7$, and $|2x + 1| < 7$. If we find the union of the three sets, we get the set of all real numbers. This is because for any value of $x$, $|2x + 1|$ will satisfy one and only one of the following: it is equal to 7, greater than 7, or less than 7.

**CAUTION** When solving absolute value equations and inequalities of the types in Examples 1, 2, and 3, remember the following.

1. The methods described apply when the constant is alone on one side of the equation or inequality and is *positive.*
2. Absolute value equations and absolute value inequalities in the form $|ax + b| > k$ translate into "or" compound statements.
3. Absolute value inequalities in the form $|ax + b| < k$ translate into "and" compound statements, which may be written as three-part inequalities.
4. An "or" statement *cannot* be written in three parts. It would be incorrect to write

$$-7 > 2x + 1 > 7$$

in Example 2, because this would imply that $-7 > 7$, which is *false.*

**OBJECTIVE 4 Solve absolute value equations that involve rewriting.** Sometimes an absolute value equation or inequality requires some rewriting before it can be set up as a compound statement, as shown in the next example.

### EXAMPLE 4 Solving an Absolute Value Equation That Requires Rewriting

Solve $|x + 3| + 5 = 12$.

First get the absolute value alone on one side of the equals sign by subtracting 5 from each side.

$$|x + 3| + 5 - 5 = 12 - 5 \quad \text{Subtract 5.}$$
$$|x + 3| = 7$$

Now use the method shown in Example 1.

$$x + 3 = 7 \quad \text{or} \quad x + 3 = -7$$
$$x = 4 \quad \text{or} \quad x = -10$$

Check that the solution set is $\{4, -10\}$ by substituting into the original equation.

**Now Try Exercise 63.**

We use a similar method to solve an absolute value *inequality* that requires rewriting.

**OBJECTIVE 5 Solve equations of the form $|ax + b| = |cx + d|$.** By definition, for two expressions to have the same absolute value, they must either be equal or be negatives of each other.

**Solving $|ax + b| = |cx + d|$**

To solve an absolute value equation of the form

$$|ax + b| = |cx + d|,$$

solve the compound equation

$$ax + b = cx + d \quad \text{or} \quad ax + b = -(cx + d).$$

### EXAMPLE 5 Solving an Equation with Two Absolute Values

Solve $|z + 6| = |2z - 3|$.

This equation is satisfied either if $z + 6$ and $2z - 3$ are equal to each other or if $z + 6$ and $2z - 3$ are negatives of each other.

$$z + 6 = 2z - 3 \quad \text{or} \quad z + 6 = -(2z - 3)$$

Solve each equation.

$$z + 6 = 2z - 3 \quad \text{or} \quad z + 6 = -2z + 3$$
$$6 + 3 = 2z - z \qquad\qquad 3z = -3$$
$$9 = z \quad \text{or} \quad z = -1$$

The solution set is $\{9, -1\}$.

**Now Try Exercise 71.**

**OBJECTIVE 6 Solve special cases of absolute value equations and inequalities.** When a typical absolute value equation or inequality involves a *negative constant or 0* alone on one side, use the properties of absolute value to solve. Keep in mind the following.

1. The absolute value of an expression can never be negative: $|a| \geq 0$ for all real numbers $a$.
2. The absolute value of an expression equals 0 only when the expression is equal to 0.

The next two examples illustrate these special cases.

### EXAMPLE 6 Solving Special Cases of Absolute Value Equations

Solve each equation.

**(a)** $|5r - 3| = -4$

Since the absolute value of an expression can never be negative, there are no solutions for this equation. The solution set is $\emptyset$.

**(b)** $|7x - 3| = 0$

The expression $7x - 3$ will equal 0 *only* if

$$7x - 3 = 0.$$

The solution of this equation is $\frac{3}{7}$. The solution set is $\{\frac{3}{7}\}$. It consists of only one element that checks by substitution in the original equation.

**Now Try Exercises 79 and 81.**

### EXAMPLE 7 Solving Special Cases of Absolute Value Inequalities

Solve each inequality.

**(a)** $|x| \geq -4$

The absolute value of a number is never negative. For this reason, $|x| \geq -4$ is true for *all* real numbers. The solution set is $(-\infty, \infty)$.

**(b)** $|k + 6| - 3 < -5$

Add 3 to both sides to get the absolute value expression alone on one side.

$$|k + 6| < -2$$

There is no number whose absolute value is less than $-2$, so this inequality has no solution. The solution set is $\emptyset$.

**(c)** $|m - 7| + 4 \leq 4$

Adding $-4$ to both sides gives

$$|m - 7| \leq 0.$$

The value of $|m - 7|$ will never be less than 0. However, $|m - 7|$ will equal 0 when $m = 7$. Therefore, the solution set is $\{7\}$.

**Now Try Exercises 85 and 91.**

## CONNECTIONS

Absolute value is used to find the relative error of a measurement in science, engineering, manufacturing, and other fields. If $x_t$ represents the expected value of a measurement and $x$ represents the actual measurement, then the *relative error in x* equals the absolute value of the difference between $x_t$ and $x$ divided by $x_t$. That is,

$$\text{relative error in } x = \left|\frac{x_t - x}{x_t}\right|.$$

In many situations in the work world, the relative error must be less than some predetermined amount. For example, suppose a machine filling *quart* milk cartons is set for a relative error no greater than .05. Here $x_t = 32$ oz, the relative error $= .05$ oz, and we must find $x$, given

$$\left|\frac{32 - x}{32}\right| = \left|1 - \frac{x}{32}\right| \leq .05.$$

**For Discussion or Writing**

With this tolerance level, how many ounces may a carton contain?

## 9.2 EXERCISES

**For Extra Help**

- Student's Solutions Manual
- MyMathLab
- InterAct Math Tutorial Software
- AW Math Tutor Center
- MathXL
- Digital Video Tutor CD 13/Videotape 13

*Match each absolute value equation or inequality in Column I with the graph of its solution set in Column II.*

**1.**

| I | II |
|---|---|
| $\lvert x\rvert = 5$ | **A.** 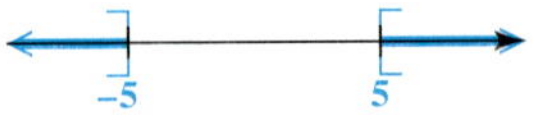 |
| $\lvert x\rvert < 5$ | **B.**  |
| $\lvert x\rvert > 5$ | **C.**  |
| $\lvert x\rvert \leq 5$ | **D.**  |
| $\lvert x\rvert \geq 5$ | **E.**  |

**2.**

| I | II |
|---|---|
| $\lvert x\rvert = 9$ | **A.**  |
| $\lvert x\rvert > 9$ | **B.**  |
| $\lvert x\rvert \geq 9$ | **C.**  |
| $\lvert x\rvert < 9$ | **D.**  |
| $\lvert x\rvert \leq 9$ | **E.** 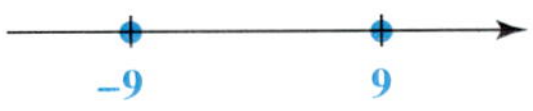 |

**3.** Explain when to use *and* and when to use *or* if you are solving an absolute value equation or inequality of the form $|ax + b| = k$, $|ax + b| < k$, or $|ax + b| > k$, where $k$ is a positive number.

**4.** How many solutions will $|ax + b| = k$ have if

**(a)** $k = 0$; **(b)** $k > 0$; **(c)** $k < 0$?

*Solve each equation. See Example 1.*

**5.** $|x| = 12$ **6.** $|k| = 14$ **7.** $|4x| = 20$ **8.** $|5x| = 30$

**9.** $|y - 3| = 9$ **10.** $|p - 5| = 13$ **11.** $|2x - 1| = 11$ **12.** $|2y + 3| = 19$

**13.** $|4r - 5| = 17$ **14.** $|5t - 1| = 21$ **15.** $|2y + 5| = 14$ **16.** $|2x - 9| = 18$

**17.** $\left|\frac{1}{2}x + 3\right| = 2$ **18.** $\left|\frac{2}{3}q - 1\right| = 5$

**19.** $\left|1 + \frac{3}{4}k\right| = 7$ **20.** $\left|2 - \frac{5}{2}m\right| = 14$

*Solve each inequality and graph the solution set. See Example 2.*

**21.** $|x| > 3$ **22.** $|y| > 5$ **23.** $|k| \geq 4$ **24.** $|r| \geq 6$

**25.** $|r + 5| \geq 20$ **26.** $|3x - 1| \geq 8$ **27.** $|t + 2| > 10$ **28.** $|4x + 1| \geq 21$

**29.** $|3 - x| > 5$ **30.** $|5 - x| > 3$ **31.** $|-5x + 3| \geq 12$ **32.** $|-2x - 4| \geq 5$

**33.** The graph of the solution set of $|2x + 1| = 9$ is given here.

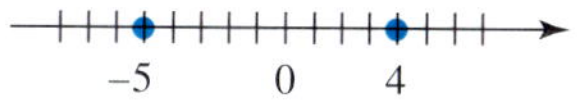


Without actually doing the algebraic work, graph the solution set of each inequality, referring to the graph above.

**(a)** $|2x + 1| < 9$ **(b)** $|2x + 1| > 9$

**34.** The graph of the solution set of $|3x - 4| < 5$ is given here.

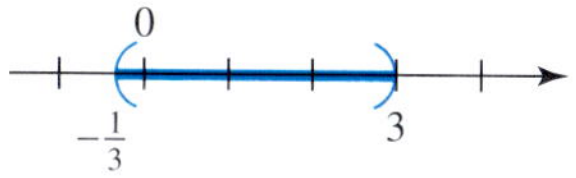


Without actually doing the algebraic work, graph the solution set of the following, referring to the graph above.

**(a)** $|3x - 4| = 5$ **(b)** $|3x - 4| > 5$

*Solve each inequality and graph the solution set. See Example 3. (Hint: Compare your answers to those in Exercises 21–32.)*

**35.** $|x| \leq 3$ **36.** $|y| \leq 5$ **37.** $|k| < 4$ **38.** $|r| < 6$

**39.** $|r + 5| \leq 20$ **40.** $|3x - 1| < 8$ **41.** $|t + 2| \leq 10$ **42.** $|4x + 1| < 21$

**43.** $|3 - x| \leq 5$ **44.** $|5 - x| \leq 3$ **45.** $|-5x + 3| \leq 12$ **46.** $|-2x - 4| \leq 5$

*Decide which method you should use to solve each absolute value equation or inequality. Find the solution set and in Exercises 47–58, graph the solution set. See Examples 1–3.*

**47.** $|-4 + k| > 9$ **48.** $|-3 + t| > 8$ **49.** $|r + 5| > 20$ **50.** $|2x - 1| < 7$

**51.** $|7 + 2z| = 5$ **52.** $|9 - 3p| = 3$ **53.** $|3r - 1| \leq 11$ **54.** $|2s - 6| \leq 6$

**55.** $|-6x - 6| \leq 1$ **56.** $|-2x - 6| \leq 5$ **57.** $|2x - 1| \geq 7$ **58.** $|-4 + k| \leq 9$

**59.** $|x| - 1 = 4$ **60.** $|x + 3| = 10$ **61.** $|x + 2| = 3$ **62.** $|x - 4| = 1$

# CUMULATIVE REVIEW EXERCISES CHAPTERS 1–9

**1.** Match each number in Column I with the choice or choices of sets of numbers in Column II to which the number belongs.

| I | II |
|---|---|
| **(a)** 34 | **A.** Natural numbers |
| **(b)** 0 | **B.** Whole numbers |
| **(c)** 2.16 | **C.** Integers |
| **(d)** $-\pi$ | **D.** Rational numbers |
| **(e)** $\sqrt{3}$ | **E.** Irrational numbers |
| **(f)** $-\frac{4}{5}$ | **F.** Real numbers |

*Evaluate.*

**2.** $9 \cdot 4 - 16 \div 4$

**3.** $-|8 - 13| - |-4| + |-9|$

*Solve.*

**4.** $-5(8 - 2z) + 4(7 - z) = 7(8 + z) - 3$

**5.** $3(x + 2) - 5(x + 2) = -2x - 4$

**6.** $A = p + prt$ for $t$

**7.** $2(m + 5) - 3m + 1 > 5$

**8.** A recent survey polled teens about the most important inventions of the twentieth century. Complete the results shown in the table if 1500 teens were surveyed.

| *Most Important Invention* | *Percent* | *Actual Number* |
|---|---|---|
| Personal computer | | 480 |
| Pacemaker | 26% | |
| Wireless communication | 18% | |
| Television | | 150 |

*Source:* Lemelson-MIT Program.

**9.** Find the measure of each angle of the triangle. (*Hint:* The sum of the measures of any triangle is 180°.)

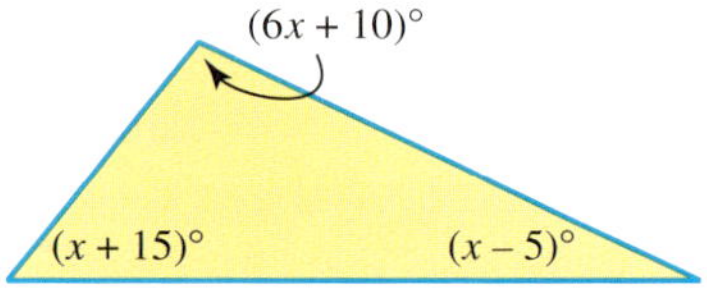


*Find the slope of each line described.*

**10.** Through $(-4, 5)$ and $(2, -3)$

**11.** Horizontal, through $(4, 5)$

*Find an equation of each line. Write the equation in slope-intercept form.*

**12.** Through $(4, -1)$, $m = -4$

**13.** Through $(0, 0)$ and $(1, 4)$

**14.** Graph $-3x + 4y = 12$.

*Simplify. Write answers with only positive exponents. Assume all variables represent positive real numbers.*

**15.** $\left(\dfrac{2m^3n}{p^2}\right)^3$

**16.** $\dfrac{x^{-6}y^3z^{-1}}{x^7y^{-4}z}$

*Perform the indicated operations.*

**17.** $2(3x^2 - 8x + 1) - 4(x^2 - 3x - 9)$

**18.** $(3x + 2y)(5x - y)$

**19.** $(8m + 5n)(8m - 5n)$

**20.** $(x + 2y)(x^2 - 2xy + 4y^2)$

**21.** $\dfrac{m^3 - 3m^2 + 5m - 3}{m - 1}$

*Factor each polynomial completely.*

**22.** $m^2 + 12m + 32$

**23.** $25t^4 - 36$

**24.** $81z^2 + 72z + 16$

**25.** Solve the equation $(x + 4)(x - 1) = -6$.

**26.** For what real number(s) is the expression $\dfrac{3}{x^2 + 5x - 14}$ undefined?

*Perform each indicated operation. Express answers in lowest terms.*

**27.** $\dfrac{x^2 - 3x - 4}{x^2 + 3x} \cdot \dfrac{x^2 + 2x - 3}{x^2 - 5x + 4}$

**28.** $\dfrac{t^2 + 4t - 5}{t + 5} \div \dfrac{t - 1}{t^2 + 8t + 15}$

**29.** $\dfrac{2}{x + 3} - \dfrac{4}{x - 1}$

**30.** $\dfrac{\dfrac{2}{3} + \dfrac{1}{2}}{\dfrac{1}{9} - \dfrac{1}{6}}$

**31.** Solve the equation $\dfrac{x}{x + 8} - \dfrac{3}{x - 8} = \dfrac{128}{x^2 - 64}$.

**32.** The graph shows the annual number of twin births in the United States for selected years.

(a) Use the information given on the graph to find and interpret the average rate of change in the number of twin births per year.

(b) If $x = 0$ represents 1990, use your answer from part (a) to write an equation of the line in slope-intercept form that models the annual number of twin births for the years 1990 through 2000.

(c) Use the equation from part (b) to approximate the number of twin births in 2002.

**SEEING DOUBLE**

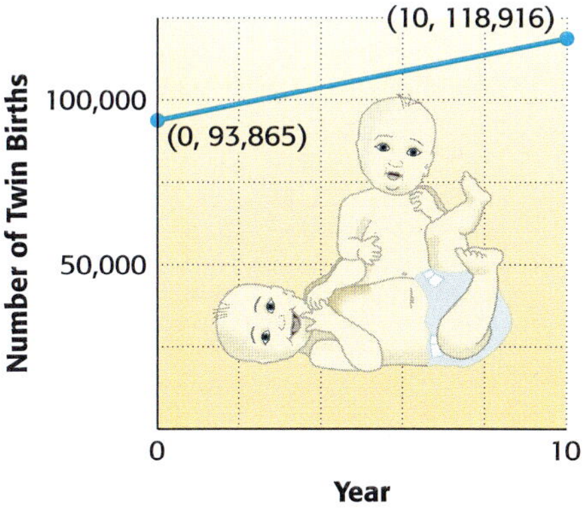


*Source:* National Center for Health Statistics.

**33.** Give the domain and range of the relation $\{(-4, -2), (-1, 0), (2, 0), (5, 2)\}$. Does this relation define a function?

**34.** One source of renewable energy is wind, although as of 2000, it provided less than 5 trillion BTUs in the United States. (*Source:* U.S. Energy Information Administration, *Annual Energy Review.*) The force of the wind blowing on a vertical surface varies jointly as the area of the surface and the square of the velocity. If a wind of 40 mph exerts a force of 50 lb on a surface of $\frac{1}{2}$ ft$^2$, how much force will a wind of 80 mph place on a surface of 2 ft$^2$?

*Solve each system.*

**35.** $3x - 4y = 1$
$2x + 3y = 12$

**36.** $3x - 2y = 4$
$-6x + 4y = 7$

**37.** $x + 3y - 6z = 7$
$2x - y + z = 1$
$x + 2y + 2z = -1$

*Solve each problem using a system of equations.*

**38.** The Star-Spangled Banner that flew over Fort McHenry during the War of 1812 had a perimeter of 144 ft. Its length measured 12 ft more than its width. Find the dimensions of this flag, which is displayed in the Smithsonian Institution's Museum of American History in Washington, D.C. (*Source:* National Park Service brochure.)

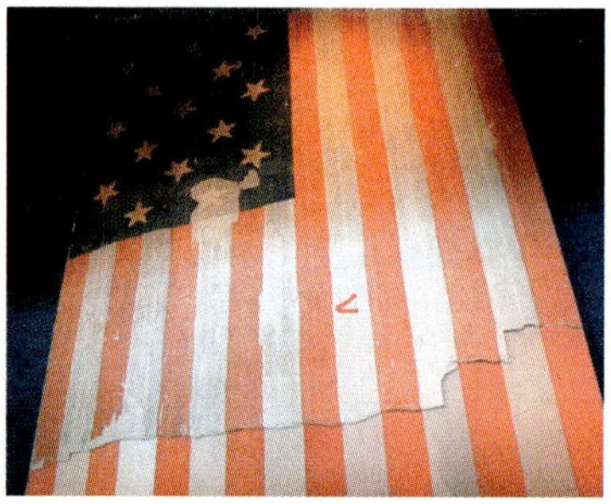

**39.** A chemist needs 9 L of a 20% solution of alcohol. She has a 15% solution on hand, as well as a 30% solution. How many liters of the 15% solution and the 30% solution should she mix to get the 20% solution she needs?

*Solve each equation or inequality.*

**40.** $x > -4$ and $x < 4$

**41.** $2x + 1 > 5$ or $2 - x \geq 2$

**42.** $|3x - 1| = 2$

**43.** $|3z + 1| \geq 7$

*Graph the solution set of each inequality or compound inequality.*

**44.** $y \leq 2x - 6$

**45.** $3x + 2y < 0$

**46.** $x - y \geq 3$ and $3x + 4y \leq 12$

# Roots, Radicals, and Root Functions

# 10

Tom Skilling is the chief meteorologist for the *Chicago Tribune.* He writes a column titled "Ask Tom Why," where readers question him on a variety of topics. In the Saturday, August 17, 2002 issue, reader Ted Fleischaker wrote: "I cannot remember the formula to calculate the distance to the horizon. I have a stunning view from my 14th floor condo, 150 feet above the ground. How far can I see?" (See Exercise 118 in Section 10.3.)

In Skilling's answer, he explained the formula for finding the distance $d$ to the horizon in miles,

$$d = 1.224\sqrt{h},$$

where $h$ is the height in feet. Square roots such as this one are often found in formulas. This chapter deals with roots and radicals.

# 10.1 Radical Expressions and Graphs

**OBJECTIVES**

1. Find square roots.
2. Decide whether a given root is rational, irrational, or not a real number.
3. Find higher roots.
4. Graph functions defined by radical expressions.
5. Find *n*th roots of *n*th powers.
6. Use a calculator to find roots.

In Section 1.2, we discussed the idea of the *square* of a number. Recall that squaring a number means multiplying the number by itself.

$$\text{If} \quad a = 8, \qquad \text{then} \quad a^2 = 8 \cdot 8 = 64.$$

$$\text{If} \quad a = -4, \qquad \text{then} \quad a^2 = (-4)(-4) = 16.$$

$$\text{If} \quad a = -\frac{1}{2}, \qquad \text{then} \quad a^2 = \left(-\frac{1}{2}\right)\left(-\frac{1}{2}\right) = \frac{1}{4}.$$

In this chapter, we consider the opposite process.

$$\text{If} \quad a^2 = 49, \qquad \text{then} \quad a = ?.$$

$$\text{If} \quad a^2 = 100, \qquad \text{then} \quad a = ?.$$

$$\text{If} \quad a^2 = 25, \qquad \text{then} \quad a = ?.$$

**OBJECTIVE 1 Find square roots.** To find $a$ in the three preceding statements, we must find a number that when multiplied by itself results in the given number. The number $a$ is called a **square root** of the number $a^2$.

### EXAMPLE 1 Finding All Square Roots of a Number

Find all square roots of 49.

To find a square root of 49, think of a number that when multiplied by itself gives 49. One square root is 7 because $7 \cdot 7 = 49$. Another square root of 49 is $-7$ because $(-7)(-7) = 49$. The number 49 has two square roots, 7 and $-7$; one is positive, and one is negative.

**Now Try Exercise 7.**

The **positive** or **principal square root** of a number is written with the symbol $\sqrt{\phantom{x}}$. For example, the positive square root of 121 is 11, written

$$\sqrt{121} = 11.$$

The symbol $-\sqrt{\phantom{x}}$ is used for the **negative square root** of a number. For example, the negative square root of 121 is $-11$, written

$$-\sqrt{121} = -11.$$

The symbol $\sqrt{\phantom{x}}$ called a **radical sign,** always represents the positive square root (except that $\sqrt{0} = 0$). The number inside the radical sign is called the **radicand,** and the entire expression—radical sign and radicand—is called a **radical.**

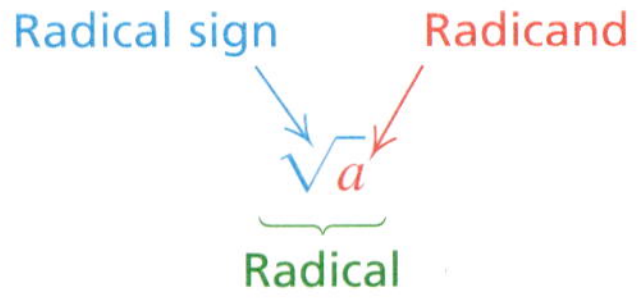

Early radical symbol

Radicals have a long mathematical history. The radical sign $\sqrt{\phantom{x}}$ has been used since sixteenth-century Germany and was probably derived from the letter *R*. The radical symbol in the margin on the preceding page comes from the Latin word for root, *radix*. It was first used by Leonardo of Pisa (Fibonacci) in 1220.

We summarize our discussion of square roots as follows.

**Square Roots of *a***

If $a$ is a positive real number, then

$\sqrt{a}$ is the positive or principal square root of $a$,

and $-\sqrt{a}$ is the negative square root of $a$.

For nonnegative $a$,

$$\sqrt{a} \cdot \sqrt{a} = \left(\sqrt{a}\right)^2 = a \quad \text{and} \quad -\sqrt{a} \cdot \left(-\sqrt{a}\right) = \left(-\sqrt{a}\right)^2 = a.$$

Also, $\sqrt{0} = 0$.

**EXAMPLE 2 Finding Square Roots**

Find each square root.

**(a)** $\sqrt{144}$

The radical $\sqrt{144}$ represents the positive or principal square root of 144. Think of a positive number whose square is 144.

$$12^2 = 144, \quad \text{so} \quad \sqrt{144} = 12.$$

**(b)** $-\sqrt{1024}$

This symbol represents the negative square root of 1024. A calculator with a square root key can be used to find $\sqrt{1024} = 32$. Then, $-\sqrt{1024} = -32$.

**(c)** $\sqrt{\frac{4}{9}} = \frac{2}{3}$ **(d)** $-\sqrt{\frac{16}{49}} = -\frac{4}{7}$

**Now Try Exercises 19, 21, and 23.**

As shown in the preceding definition, when the square root of a positive real number is squared, the result is that positive real number. (Also, $\left(\sqrt{0}\right)^2 = 0$.)

**EXAMPLE 3 Squaring Radical Expressions**

Find the *square* of each radical expression.

**(a)** $\sqrt{13}$

$\left(\sqrt{13}\right)^2 = 13$ Definition of square root

**(b)** $-\sqrt{29}$

$\left(-\sqrt{29}\right)^2 = 29$ The square of a *negative* number is positive.

**(c)** $\sqrt{p^2 + 1}$

$\left(\sqrt{p^2 + 1}\right)^2 = p^2 + 1$

**Now Try Exercises 27, 29, and 33.**

**OBJECTIVE 2 Decide whether a given root is rational, irrational, or not a real number.** All numbers with square roots that are rational are called **perfect squares.** For example, 144 and $\frac{4}{9}$ are perfect squares since their respective square roots, 12 and $\frac{2}{3}$, are rational numbers.

A number that is not a perfect square has a square root that is not a rational number. For example, $\sqrt{5}$ is not a rational number because it cannot be written as the ratio of two integers. Its decimal neither terminates nor repeats. However, $\sqrt{5}$ is a real number and corresponds to a point on the number line. As mentioned in Chapter 1, a real number that is not rational is called an *irrational number*. The number $\sqrt{5}$ is irrational. Many square roots of integers are irrational.

If $a$ is a positive real number that is not a perfect square, then $\sqrt{a}$ is irrational.

Not every number has a *real number* square root. For example, there is no real number that can be squared to get $-36$. (The square of a real number can never be negative.) Because of this, $\sqrt{-36}$ is not a real number.

If $a$ is a negative real number, then $\sqrt{a}$ is not a real number.

**CAUTION** Be careful not to confuse $\sqrt{-36}$ and $-\sqrt{36}$. $\sqrt{-36}$ is not a real number since there is no real number that can be squared to get $-36$. However, $-\sqrt{36}$ is the negative square root of 36, which is $-6$.

### EXAMPLE 4 Identifying Types of Square Roots

Tell whether each square root is *rational, irrational,* or *not a real number.*

**(a)** $\sqrt{17}$

Because 17 is not a perfect square, $\sqrt{17}$ is irrational.

**(b)** $\sqrt{64}$

The number 64 is a perfect square, $8^2$, so $\sqrt{64} = 8$, a rational number.

**(c)** $\sqrt{-25}$

There is no real number whose square is $-25$. Therefore, $\sqrt{-25}$ is not a real number.

**Now Try Exercises 39, 41, and 47.**

**NOTE** Not all irrational numbers are square roots of integers. For example, $\pi$ (approximately 3.14159) is an irrational number that is not a square root of any integer.

**OBJECTIVE 3 Find higher roots.** Finding the square root of a number is the inverse (reverse) of squaring a number. In a similar way, there are inverses to finding the cube of a number, or finding the fourth or higher power of a number. These inverses are the **cube root,** written $\sqrt[3]{a}$, and the **fourth root,** written $\sqrt[4]{a}$. Similar symbols are used for higher roots. In general, we have the following.

**$\sqrt[n]{a}$**

The $n$th root of $a$, written $\sqrt[n]{a}$, is a number whose $n$th power equals $a$. That is,

$$\sqrt[n]{a} = b \quad \text{means} \quad b^n = a.$$

In $\sqrt[n]{a}$, the number $n$ is the **index** or **order** of the radical. It is possible to write $\sqrt[2]{a}$ instead of $\sqrt{a}$, but the simpler symbol $\sqrt{a}$ is customary since the square root is the most commonly used root.

When working with cube roots or fourth roots, it is helpful to memorize the first few *perfect cubes* ($2^3 = 8$, $3^3 = 27$, and so on) and the first few perfect fourth powers ($2^4 = 16$, $3^4 = 81$, and so on).

### EXAMPLE 5 Finding Cube Roots

Find each cube root.

**(a)** $\sqrt[3]{8}$

Look for a number that can be cubed to give 8. Because $2^3 = 8$, $\sqrt[3]{8} = 2$.

**(b)** $\sqrt[3]{-8} = -2$ because $(-2)^3 = -8$.

**(c)** $-\sqrt[3]{216} = -6$ because $\sqrt[3]{216} = 6$, and thus $-\sqrt[3]{216} = -6$.

**Now Try Exercises 65, 67, and 69.**

Notice in Example 5(b) that we can find the cube root of a negative number. (Contrast this with the square root of a negative number, which is not real.) In fact, the cube root of a positive number is positive, and the cube root of a negative number is negative. *There is only one real number cube root for each real number.*

When the index of the radical is even (square root, fourth root, and so on), *the radicand must be nonnegative* to get a real number root. Also, for even indexes, the symbols $\sqrt{\phantom{x}}$, $\sqrt[4]{\phantom{x}}$, $\sqrt[6]{\phantom{x}}$, and so on are used for the positive (principal) roots. The symbols $-\sqrt{\phantom{x}}$, $-\sqrt[4]{\phantom{x}}$, $-\sqrt[6]{\phantom{x}}$, and so on are used for the negative roots.

### EXAMPLE 6 Finding Higher Roots

Find each root.

**(a)** $\sqrt[4]{16} = 2$ because 2 is positive and $2^4 = 16$.

**(b)** $-\sqrt[4]{16}$

From part (a), $\sqrt[4]{16} = 2$, so the negative root $-\sqrt[4]{16} = -2$.

**(c)** $\sqrt[4]{-16}$

For a real number fourth root, the radicand must be nonnegative. There is no real number that equals $\sqrt[4]{-16}$.

**(d)** $-\sqrt[5]{32}$

First find $\sqrt[5]{32}$. Because 2 is the number whose fifth power is 32, $\sqrt[5]{32} = 2$. If $\sqrt[5]{32} = 2$, then

$$-\sqrt[5]{32} = -2.$$

**(e)** $\sqrt[5]{-32} = -2$, because $(-2)^5 = -32$.

**Now Try Exercises 71, 73, 75, and 77.**

**120.** The Vietnam Veterans' Memorial in Washington, D.C., is in the shape of an unenclosed isosceles triangle with equal sides of length 246.75 ft. If the triangle were enclosed, the third side would have length 438.14 ft. Use Heron's formula from the previous exercise to find the area of this enclosure to the nearest hundred square feet. (*Source:* Information pamphlet obtained at the Vietnam Veterans' Memorial.)

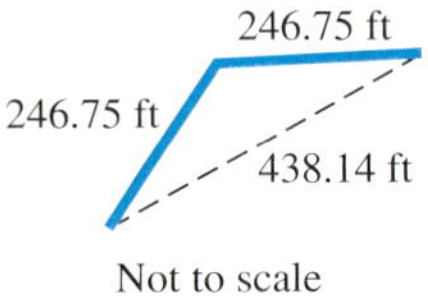


**121.** The formula

$$I = \sqrt{\frac{2P}{L}}$$

relates the coefficient of self-induction $L$ (in henrys), the energy $P$ stored in an electronic circuit (in joules), and the current $I$ (in amps). Find $I$ if $P = 120$ and $L = 80$.

**122.** When the wind blows, the air feels much colder than the actual temperature. The **windchill factor** measures the cooling effect that the wind has on one's skin. Until recently, the formula that the National Weather Service used to compute windchill was

$$T_{wc} = .0817(3.71\sqrt{V} + 5.81 - .25V)(T - 91.4) + 91.4,$$

where $T_{wc}$ is windchill, $V$ is wind speed in miles per hour (mph), and $T$ is air temperature in degrees Fahrenheit. The windchill for various wind speeds and temperatures is shown in the table.

**Windchill Factor**

| Wind Speed (mph) | Air Temperature (°Fahrenheit) | | | | | | | | | | | | | | |
|---|---|---|---|---|---|---|---|---|---|---|---|---|---|---|---|
| | **35** | **30** | **25** | **20** | **15** | **10** | **5** | **0** | **−5** | **−10** | **−15** | **−20** | **−25** | **−30** | **−35** |
| **4** | 35 | 30 | 25 | 20 | 15 | 10 | 5 | 0 | −5 | −10 | −15 | −20 | −25 | −30 | −35 |
| **5** | 32 | 27 | 22 | 16 | 11 | 6 | 0 | −5 | −10 | −15 | −21 | −26 | −31 | −36 | −42 |
| **10** | 22 | 16 | 10 | 3 | −3 | −9 | −15 | −22 | −27 | −34 | −40 | −46 | −52 | −58 | −64 |
| **15** | 16 | 9 | 2 | −5 | −11 | −18 | −25 | −31 | −38 | −45 | −51 | −58 | −65 | −72 | −78 |
| **20** | 12 | 4 | −3 | −10 | −17 | −24 | −31 | −39 | −46 | −53 | −60 | −67 | −74 | −81 | −88 |
| **25** | 8 | 1 | −7 | −15 | −22 | −29 | −36 | −44 | −51 | −59 | −66 | −74 | −81 | −88 | −96 |
| **30** | 6 | −2 | −10 | −18 | −25 | −33 | −41 | −49 | −56 | −64 | −71 | −79 | −86 | −93 | −101 |
| **35** | 4 | −4 | −12 | −20 | −27 | −35 | −43 | −52 | −58 | −67 | −74 | −82 | −89 | −97 | −105 |
| **40** | 3 | −5 | −13 | −21 | −29 | −37 | −45 | −53 | −60 | −69 | −76 | −84 | −92 | −100 | −107 |
| **45** | 2 | −6 | −14 | −22 | −30 | −38 | −46 | −54 | −62 | −70 | −78 | −85 | −93 | −102 | −109 |

*Source: USA Today.*

Choose a temperature of 10°F. Use the formula to calculate the windchill for wind speeds of 4, 10, 25, and 40 mph. Round the results to the nearest degree. Do your results match those in the tables?

# 10.2 Rational Exponents

**OBJECTIVES**

1. Use exponential notation for $n$th roots.
2. Define and use expressions of the form $a^{m/n}$.
3. Convert between radicals and rational exponents.
4. Use the rules for exponents with rational exponents.

**OBJECTIVE 1 Use exponential notation for $n$th roots.** In mathematics we often formulate definitions so that previous rules remain valid. In Chapter 4 we defined 0 as an exponent in such a way that the rules for products, quotients, and powers would still be valid. Now we look at exponents that are rational numbers of the form $\frac{1}{n}$, where $n$ is a natural number.

For the rules of exponents to remain valid, the product $(3^{1/2})^2 = 3^{1/2} \cdot 3^{1/2}$ should be found by adding exponents.

$$\begin{aligned}(3^{1/2})^2 &= 3^{1/2} \cdot 3^{1/2} \\ &= 3^{1/2+1/2} \\ &= 3^1 \\ &= 3\end{aligned}$$

However, by definition $(\sqrt{3})^2 = \sqrt{3} \cdot \sqrt{3} = 3$. Since both $(3^{1/2})^2$ and $(\sqrt{3})^2$ are equal to 3, it is reasonable to have

$$3^{1/2} = \sqrt{3}.$$

This suggests the following generalization.

> **$a^{1/n}$**
>
> If $\sqrt[n]{a}$ is a real number, then $\qquad \boldsymbol{a^{1/n} = \sqrt[n]{a}.}$

**EXAMPLE 1 Evaluating Exponentials of the Form $a^{1/n}$**

Evaluate each expression.

**(a)** $64^{1/3} = \sqrt[3]{64} = 4$

**(b)** $100^{1/2} = \sqrt{100} = 10$

**(c)** $-256^{1/4} = -\sqrt[4]{256} = -4$

**(d)** $(-256)^{1/4} = \sqrt[4]{-256}$ is not a real number because the radicand, $-256$, is negative and the index is even.

**(e)** $(-32)^{1/5} = \sqrt[5]{-32} = -2$

**(f)** $\left(\frac{1}{8}\right)^{1/3} = \sqrt[3]{\frac{1}{8}} = \frac{1}{2}$

**Now Try Exercises 11, 13, 19, and 25.**

**CAUTION** Notice the difference between parts (c) and (d) in Example 1. The radical in part (c) is the *negative fourth root* of a positive number, while the radical in part (d) is the *principal fourth root of a negative number,* which is not a real number.

**OBJECTIVE 2 Define and use expressions of the form $a^{m/n}$.** We know that $8^{1/3} = \sqrt[3]{8}$. How should we define a number like $8^{2/3}$? For past rules of exponents to be valid,

$$8^{2/3} = 8^{(1/3)2} = (8^{1/3})^2.$$

Since $8^{1/3} = \sqrt[3]{8}$,

$$8^{2/3} = \left(\sqrt[3]{8}\right)^2 = 2^2 = 4.$$

Generalizing from this example, we define $a^{m/n}$ as follows.

**$a^{m/n}$**

If $m$ and $n$ are positive integers with $m/n$ in lowest terms, then

$$a^{m/n} = (a^{1/n})^m,$$

provided that $a^{1/n}$ is a real number. If $a^{1/n}$ is not a real number, then $a^{m/n}$ is not a real number.

## EXAMPLE 2 Evaluating Exponentials of the Form $a^{m/n}$

Evaluate each exponential.

**(a)** $36^{3/2} = (36^{1/2})^3 = 6^3 = 216$

**(b)** $125^{2/3} = (125^{1/3})^2 = 5^2 = 25$

**(c)** $-4^{5/2} = -(4^{5/2}) = -(4^{1/2})^5 = -(2)^5 = -32$

**(d)** $(-27)^{2/3} = [(-27)^{1/3}]^2 = (-3)^2 = 9$

Notice how the $-$ sign is used in parts (c) and (d). In part (c), we first evaluate the exponential and then find its negative. In part (d), the $-$ sign is part of the base, $-27$.

**(e)** $(-100)^{3/2}$ is not a real number since $(-100)^{1/2}$ is not a real number.

**Now Try Exercises 21 and 23.**

When a rational exponent is negative, the earlier interpretation of negative exponents is applied.

**$a^{-m/n}$**

If $a^{m/n}$ is a real number, then

$$a^{-m/n} = \frac{1}{a^{m/n}} \quad (a \neq 0).$$

## EXAMPLE 3 Evaluating Exponentials with Negative Rational Exponents

Evaluate each exponential.

**(a)** $16^{-3/4}$

By the definition of a negative exponent,

$$16^{-3/4} = \frac{1}{16^{3/4}}.$$

Since $16^{3/4} = \left(\sqrt[4]{16}\right)^3 = 2^3 = 8$,

$$16^{-3/4} = \frac{1}{16^{3/4}} = \frac{1}{8}.$$

**(b)** $25^{-3/2} = \dfrac{1}{25^{3/2}} = \dfrac{1}{(\sqrt{25})^3} = \dfrac{1}{5^3} = \dfrac{1}{125}$

**(c)** $\left(\dfrac{8}{27}\right)^{-2/3} = \dfrac{1}{\left(\dfrac{8}{27}\right)^{2/3}} = \dfrac{1}{\left(\sqrt[3]{\dfrac{8}{27}}\right)^2} = \dfrac{1}{\left(\dfrac{2}{3}\right)^2} = \dfrac{1}{\dfrac{4}{9}} = \dfrac{9}{4}$

We could also use the rule $\left(\dfrac{b}{a}\right)^{-m} = \left(\dfrac{a}{b}\right)^m$ here, as follows.

$$\left(\frac{8}{27}\right)^{-2/3} = \left(\frac{27}{8}\right)^{2/3} = \left(\sqrt[3]{\frac{27}{8}}\right)^2 = \left(\frac{3}{2}\right)^2 = \frac{9}{4}$$

**Now Try Exercises 27 and 29.**

**CAUTION** When using the rule in Example 3(c), we take the reciprocal only of the base, *not* the exponent. Also, be careful to distinguish between exponential expressions like $-16^{1/4}$, $16^{-1/4}$, and $-16^{-1/4}$.

$$-16^{1/4} = -2, \qquad 16^{-1/4} = \frac{1}{2}, \qquad \text{and} \qquad -16^{-1/4} = -\frac{1}{2}.$$

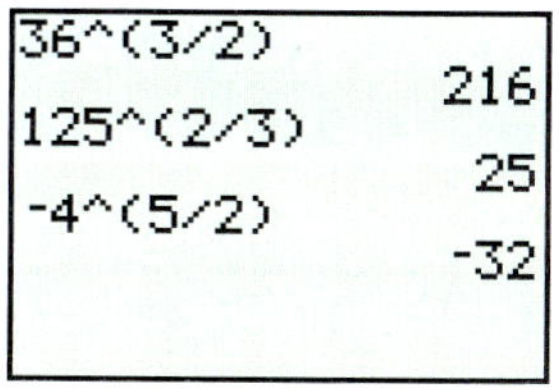


**FIGURE 6**

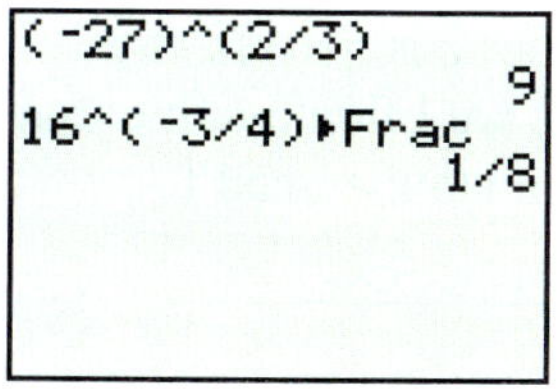


**FIGURE 7**

The screens in Figures 6 and 7 illustrate how a graphing calculator performs some of the evaluations seen in Examples 2 and 3. (All results on the screens are rational numbers.)

We obtain an alternative definition of $a^{m/n}$ by using the power rule for exponents a little differently than in the earlier definition. If all indicated roots are real numbers, then

$$a^{m/n} = a^{m(1/n)} = (a^m)^{1/n},$$

so

$$a^{m/n} = (a^m)^{1/n}.$$

### $a^{m/n}$

If all indicated roots are real numbers, then

$$a^{m/n} = (a^{1/n})^m = (a^m)^{1/n}.$$

We can now evaluate an expression such as $27^{2/3}$ in two ways:

$$27^{2/3} = (27^{1/3})^2 = 3^2 = 9$$

or

$$27^{2/3} = (27^2)^{1/3} = 729^{1/3} = 9.$$

In most cases, it is easier to use $(a^{1/n})^m$.

This rule can also be expressed with radicals as follows.

### Radical Form of $a^{m/n}$

If all indicated roots are real numbers, then

$$a^{m/n} = \sqrt[n]{a^m} = \left(\sqrt[n]{a}\right)^m.$$

That is, raise $a$ to the power and then take the root, or take the root and then raise $a$ to the power.

For example,

$$8^{2/3} = \sqrt[3]{8^2} = \sqrt[3]{64} = 4, \quad \text{and} \quad 8^{2/3} = \left(\sqrt[3]{8}\right)^2 = 2^2 = 4,$$

so

$$8^{2/3} = \sqrt[3]{8^2} = \left(\sqrt[3]{8}\right)^2.$$

**OBJECTIVE 3 Convert between radicals and rational exponents.** Using the definition of rational exponents, we can simplify many problems involving radicals by converting the radicals to numbers with rational exponents. After simplifying, we convert the answer back to radical form.

**EXAMPLE 4 Converting between Rational Exponents and Radicals**

Write each exponential as a radical. Assume all variables represent positive real numbers. Use the definition that takes the root first.

**(a)** $13^{1/2} = \sqrt{13}$

**(b)** $6^{3/4} = \left(\sqrt[4]{6}\right)^3$

**(c)** $9m^{5/8} = 9\left(\sqrt[8]{m}\right)^5$

**(d)** $6x^{2/3} - (4x)^{3/5} = 6\left(\sqrt[3]{x}\right)^2 - \left(\sqrt[5]{4x}\right)^3$

**(e)** $r^{-2/3} = \dfrac{1}{r^{2/3}} = \dfrac{1}{\left(\sqrt[3]{r}\right)^2}$

**(f)** $(a^2 + b^2)^{1/2} = \sqrt{a^2 + b^2}$ Note that $\sqrt{a^2 + b^2} \neq a + b$.

In (g)–(i), write each radical as an exponential. Simplify. Assume all variables represent positive real numbers.

**(g)** $\sqrt{10} = 10^{1/2}$

**(h)** $\sqrt[4]{3^8} = 3^{8/4} = 3^2 = 9$

**(i)** $\sqrt[6]{z^6} = z$ since $z$ is positive.

**Now Try Exercises 33, 35, 37, 49, and 51.**

**NOTE** In Example 4(i), it was not necessary to use absolute value bars since the directions specifically stated that the variable represents a positive real number. Since the absolute value of the positive real number $z$ is $z$ itself, the answer is simply $z$. When working exercises with radicals, we often assume that variables represent positive real numbers, which will eliminate the need for absolute value.

**OBJECTIVE 4 Use the rules for exponents with rational exponents.** The definition of rational exponents allows us to apply the rules for exponents first introduced in Chapter 4.

**Rules for Rational Exponents**

Let $r$ and $s$ be rational numbers. For all real numbers $a$ and $b$ for which the indicated expressions exist:

$$a^r \cdot a^s = a^{r+s} \qquad a^{-r} = \frac{1}{a^r} \qquad \frac{a^r}{a^s} = a^{r-s} \qquad \left(\frac{a}{b}\right)^{-r} = \frac{b^r}{a^r}$$

$$(a^r)^s = a^{rs} \qquad (ab)^r = a^r b^r \qquad \left(\frac{a}{b}\right)^r = \frac{a^r}{b^r} \qquad a^{-r} = \left(\frac{1}{a}\right)^r.$$

## EXAMPLE 5 Applying Rules for Rational Exponents

Write with only positive exponents. Assume all variables represent positive real numbers.

**(a)** $2^{1/2} \cdot 2^{1/4} = 2^{1/2+1/4} = 2^{3/4}$ Product rule

**(b)** $\dfrac{5^{2/3}}{5^{7/3}} = 5^{2/3-7/3} = 5^{-5/3} = \dfrac{1}{5^{5/3}}$ Quotient rule

**(c)**
$$\frac{(x^{1/2}y^{2/3})^4}{y} = \frac{(x^{1/2})^4(y^{2/3})^4}{y} \quad \text{Power rule}$$
$$= \frac{x^2y^{8/3}}{y^1} \quad \text{Power rule}$$
$$= x^2y^{8/3-1} \quad \text{Quotient rule}$$
$$= x^2y^{5/3}$$

**(d)**
$$\left(\frac{x^4y^{-6}}{x^{-2}y^{1/3}}\right)^{-2/3} = \frac{(x^4)^{-2/3}(y^{-6})^{-2/3}}{(x^{-2})^{-2/3}(y^{1/3})^{-2/3}}$$
$$= \frac{x^{-8/3}y^4}{x^{4/3}y^{-2/9}} \quad \text{Power rule}$$
$$= x^{-8/3-4/3}y^{4-(-2/9)} \quad \text{Quotient rule}$$
$$= x^{-4}y^{38/9}$$
$$= \frac{y^{38/9}}{x^4} \quad \text{Definition of negative exponent}$$

The same result is obtained if we simplify within the parentheses first, leading to

$$(x^6y^{-19/3})^{-2/3}.$$

Then, apply the power rule. (Show that the result is the same.)

**(e)**
$$m^{3/4}(m^{5/4} - m^{1/4}) = m^{3/4} \cdot m^{5/4} - m^{3/4} \cdot m^{1/4} \quad \text{Distributive property}$$
$$= m^{3/4+5/4} - m^{3/4+1/4} \quad \text{Product rule}$$
$$= m^{8/4} - m^{4/4}$$
$$= m^2 - m$$

Do not make the common mistake of multiplying exponents in the first step.

**Now Try Exercises 57, 59, 65, 75, and 77.**

**CAUTION** Use the rules of exponents in problems like those in Example 5. Do not convert the expressions to radical form.

## EXAMPLE 6 Applying Rules for Rational Exponents

Rewrite all radicals as exponentials, and then apply the rules for rational exponents. Leave answers in exponential form. Assume all variables represent positive real numbers.

**(a)** $\sqrt[3]{x^2} \cdot \sqrt[4]{x} = x^{2/3} \cdot x^{1/4}$ Convert to rational exponents.

$= x^{2/3+1/4}$ Product rule

$= x^{8/12+3/12}$ Write exponents with a common denominator.

$= x^{11/12}$

**(b)** $\dfrac{\sqrt{x^3}}{\sqrt[3]{x^2}} = \dfrac{x^{3/2}}{x^{2/3}} = x^{3/2-2/3} = x^{5/6}$

**(c)** $\sqrt{\sqrt[4]{z}} = \sqrt{z^{1/4}} = (z^{1/4})^{1/2} = z^{1/8}$

**Now Try Exercises 83, 85, and 89.**

# 10.2 EXERCISES

**For Extra Help**

- Student's Solutions Manual
- MyMathLab
- InterAct Math Tutorial Software
- AW Math Tutor Center
- MathXL
- Digital Video Tutor CD 14/Videotape 14

*Match each expression from Column I with the equivalent choice from Column II.*

| I | II |
|---|---|
| **1.** $2^{1/2}$ | **A.** $-4$ |
| **2.** $(-27)^{1/3}$ | **B.** $8$ |
| **3.** $-16^{1/2}$ | **C.** $\sqrt{2}$ |
| **4.** $(-16)^{1/2}$ | **D.** $-\sqrt{6}$ |
| **5.** $(-32)^{1/5}$ | **E.** $-3$ |
| **6.** $(-32)^{2/5}$ | **F.** $\sqrt{6}$ |
| **7.** $4^{3/2}$ | **G.** $4$ |
| **8.** $6^{2/4}$ | **H.** $-2$ |
| **9.** $-6^{2/4}$ | **I.** $6$ |
| **10.** $36^{.5}$ | **J.** Not a real number |

*Simplify each expression involving rational exponents. See Examples 1–3.*

**11.** $169^{1/2}$ **12.** $121^{1/2}$ **13.** $729^{1/3}$ **14.** $512^{1/3}$

**15.** $16^{1/4}$ **16.** $625^{1/4}$ **17.** $\left(\dfrac{64}{81}\right)^{1/2}$ **18.** $\left(\dfrac{8}{27}\right)^{1/3}$

**19.** $(-27)^{1/3}$ **20.** $(-32)^{1/5}$ **21.** $100^{3/2}$ **22.** $64^{3/2}$

**23.** $-16^{5/2}$ **24.** $-32^{3/5}$ **25.** $(-144)^{1/2}$ **26.** $(-36)^{1/2}$

**27.** $64^{-3/2}$ **28.** $81^{-3/2}$ **29.** $\left(-\dfrac{8}{27}\right)^{-2/3}$ **30.** $\left(-\dfrac{64}{125}\right)^{-2/3}$

**31.** Explain why $(-64)^{1/2}$ is not a real number, while $-64^{1/2}$ is a real number.

**32.** Explain why $a^{1/n}$ is defined to be equal to $\sqrt[n]{a}$ when $\sqrt[n]{a}$ is real.

*Write with radicals. Assume all variables represent positive real numbers. See Example 4.*

**33.** $12^{1/2}$ **34.** $3^{1/2}$ **35.** $8^{3/4}$

**36.** $7^{2/3}$ **37.** $(9q)^{5/8} - (2x)^{2/3}$ **38.** $(3p)^{3/4} + (4x)^{1/3}$

39. $(2m)^{-3/2}$
40. $(5y)^{-3/5}$
41. $(2y + x)^{2/3}$
42. $(r + 2z)^{3/2}$
43. $(3m^4 + 2k^2)^{-2/3}$
44. $(5x^2 + 3z^3)^{-5/6}$

45. Show that, in general, $\sqrt{a^2 + b^2} \neq a + b$ by replacing $a$ with 3 and $b$ with 4.

46. Suppose someone claims that $\sqrt[n]{a^n + b^n}$ must equal $a + b$, since when $a = 1$ and $b = 0$, a true statement results:

$$\sqrt[n]{a^n + b^n} = \sqrt[n]{1^n + 0^n} = \sqrt[n]{1^n} = 1 = 1 + 0 = a + b.$$

Explain why this is faulty reasoning.

*Simplify by first converting to rational exponents. Assume all variables represent positive real numbers. See Example 4.*

47. $\sqrt{2^{12}}$
48. $\sqrt{5^{10}}$
49. $\sqrt[3]{4^9}$
50. $\sqrt[4]{6^8}$
51. $\sqrt{x^{20}}$
52. $\sqrt{r^{50}}$
53. $\sqrt[3]{x} \cdot \sqrt{x}$
54. $\sqrt[4]{y} \cdot \sqrt[5]{y^2}$
55. $\dfrac{\sqrt[3]{t^4}}{\sqrt[5]{t^4}}$
56. $\dfrac{\sqrt[4]{w^3}}{\sqrt[6]{w}}$

*Use the rules of exponents to simplify each expression. Write all answers with positive exponents. Assume all variables represent positive real numbers. See Example 5.*

57. $3^{1/2} \cdot 3^{3/2}$
58. $6^{4/3} \cdot 6^{2/3}$
59. $\dfrac{64^{5/3}}{64^{4/3}}$
60. $\dfrac{125^{7/3}}{125^{5/3}}$
61. $y^{7/3} \cdot y^{-4/3}$
62. $r^{-8/9} \cdot r^{17/9}$
63. $\dfrac{k^{1/3}}{k^{2/3} \cdot k^{-1}}$
64. $\dfrac{z^{3/4}}{z^{5/4} \cdot z^{-2}}$
65. $\dfrac{(x^{1/4}y^{2/5})^{20}}{x^2}$
66. $\dfrac{(r^{1/5}s^{2/3})^{15}}{r^2}$
67. $\dfrac{(x^{2/3})^2}{(x^2)^{7/3}}$
68. $\dfrac{(p^3)^{1/4}}{(p^{5/4})^2}$
69. $\dfrac{m^{3/4}n^{-1/4}}{(m^2n)^{1/2}}$
70. $\dfrac{(a^2b^5)^{-1/4}}{(a^{-3}b^2)^{1/6}}$
71. $\dfrac{p^{1/5}p^{7/10}p^{1/2}}{(p^3)^{-1/5}}$
72. $\dfrac{z^{1/3}z^{-2/3}z^{1/6}}{(z^{-1/6})^3}$
73. $\left(\dfrac{b^{-3/2}}{c^{-5/3}}\right)^2(b^{-1/4}c^{-1/3})^{-1}$
74. $\left(\dfrac{m^{-2/3}}{a^{-3/4}}\right)^4(m^{-3/8}a^{1/4})^{-2}$
75. $\left(\dfrac{p^{-1/4}q^{-3/2}}{3^{-1}p^{-2}q^{-2/3}}\right)^{-2}$
76. $\left(\dfrac{2^{-2}w^{-3/4}x^{-5/8}}{w^{3/4}x^{-1/2}}\right)^{-3}$
77. $p^{2/3}(p^{1/3} + 2p^{4/3})$
78. $z^{5/8}(3z^{5/8} + 5z^{11/8})$
79. $k^{1/4}(k^{3/2} - k^{1/2})$
80. $r^{3/5}(r^{1/2} + r^{3/4})$
81. $6a^{7/4}(a^{-7/4} + 3a^{-3/4})$
82. $4m^{5/3}(m^{-2/3} - 4m^{-5/3})$

*Write with rational exponents, and then apply the properties of exponents. Assume all radicands represent positive real numbers. Give answers in exponential form. See Example 6.*

83. $\sqrt[5]{x^3} \cdot \sqrt[4]{x}$
84. $\sqrt[6]{y^5} \cdot \sqrt[3]{y^2}$
85. $\dfrac{\sqrt{x^5}}{\sqrt{x^8}}$
86. $\dfrac{\sqrt[3]{k^5}}{\sqrt[3]{k^7}}$
87. $\sqrt{y} \cdot \sqrt[3]{yz}$
88. $\sqrt[3]{xz} \cdot \sqrt{z}$
89. $\sqrt[4]{\sqrt[3]{m}}$
90. $\sqrt[3]{\sqrt{k}}$
91. $\sqrt{\sqrt[3]{\sqrt[4]{x}}}$
92. $\sqrt[3]{\sqrt[5]{\sqrt{y}}}$

*Solve each problem.*

**93.** Meteorologists can determine the duration of a storm by using the function defined by

$$T(D) = .07D^{3/2},$$

where $D$ is the diameter of the storm in miles and $T$ is the time in hours. Find the duration of a storm with a diameter of 16 mi. Round your answer to the nearest tenth of an hour.

**94.** The threshold weight $T$, in pounds, for a person is the weight above which the risk of death increases greatly. The threshold weight in pounds for men aged 40–49 is related to height in inches by the function defined by

$$h(T) = (1860.867T)^{1/3}.$$

What height corresponds to a threshold weight of 200 lb for a 46-yr-old man? Round your answer to the nearest inch, and then to the nearest tenth of a foot.

## RELATING CONCEPTS (EXERCISES 95–102)

**For Individual or Group Work**

*Earlier, we factored expressions like $x^4 - x^5$ by factoring out the greatest common factor to get*

$$x^4 - x^5 = x^4(1 - x).$$

*We can adapt this approach to factor expressions with rational exponents. When one or more of the exponents is negative or a fraction, we use order on the number line discussed in Chapter 1 to decide on the common factor. In this type of factoring, we want the binomial factor to have only positive exponents, so we always factor out the variable with the* least *exponent. A positive exponent is greater than a negative exponent, so in $7z^{5/8} + z^{-3/4}$, we factor out $z^{-3/4}$, because $-\frac{3}{4}$ is less than $\frac{5}{8}$.*

*Factor out the given common factor from each expression. Assume all variables represent positive real numbers.*

**95.** $3x^{-1/2} - 4x^{1/2}; \quad x^{-1/2}$

**96.** $m^3 - 3m^{5/2}; \quad m^{5/2}$

**97.** $4t^{-1/2} + 7t^{3/2}; \quad t^{-1/2}$

**98.** $8x^{2/3} + 5x^{-1/3}; \quad x^{-1/3}$

**99.** $4p - p^{3/4}; \quad p^{3/4}$

**100.** $2m^{1/8} - m^{5/8}; \quad m^{1/8}$

**101.** $9k^{-3/4} - 2k^{-1/4}; \quad k^{-3/4}$

**102.** $7z^{-5/8} - z^{-3/4}; \quad z^{-3/4}$

# 10.3 Simplifying Radical Expressions

**OBJECTIVES**

1. Use the product rule for radicals.
2. Use the quotient rule for radicals.
3. Simplify radicals.
4. Simplify products and quotients of radicals with different indexes.
5. Use the Pythagorean formula.
6. Use the distance formula.

**OBJECTIVE 1 Use the product rule for radicals.** We now develop rules for multiplying and dividing radicals that have the same index. For example, is the product of two $n$th-root radicals equal to the $n$th root of the product of the radicands? Are the expressions $\sqrt{36 \cdot 4}$ and $\sqrt{36} \cdot \sqrt{4}$ equal? To find out, we do the computations:

$$\sqrt{36 \cdot 4} = \sqrt{144} = 12$$
$$\sqrt{36} \cdot \sqrt{4} = 6 \cdot 2 = 12.$$

Notice that in both cases the result is the same. This is an example of the **product rule for radicals.**

**Product Rule for Radicals**

If $\sqrt[n]{a}$ and $\sqrt[n]{b}$ are real numbers and $n$ is a natural number, then

$$\sqrt[n]{a} \cdot \sqrt[n]{b} = \sqrt[n]{ab}.$$

That is, the product of two radicals is the radical of the product.

We justify the product rule using the rules for rational exponents. Since $\sqrt[n]{a} = a^{1/n}$ and $\sqrt[n]{b} = b^{1/n}$,

$$\sqrt[n]{a} \cdot \sqrt[n]{b} = a^{1/n} \cdot b^{1/n} = (ab)^{1/n} = \sqrt[n]{ab}.$$

**CAUTION** Use the product rule only when the radicals have the *same* index.

**EXAMPLE 1 Using the Product Rule**

Multiply. Assume all variables represent positive real numbers.

(a) $\sqrt{5} \cdot \sqrt{7} = \sqrt{5 \cdot 7} = \sqrt{35}$

(b) $\sqrt{2} \cdot \sqrt{19} = \sqrt{2 \cdot 19} = \sqrt{38}$

(c) $\sqrt{11} \cdot \sqrt{p} = \sqrt{11p}$

(d) $\sqrt{7} \cdot \sqrt{11xyz} = \sqrt{77xyz}$

**Now Try Exercises 7, 9, and 11.**

**EXAMPLE 2 Using the Product Rule**

Multiply. Assume all variables represent positive real numbers.

(a) $\sqrt[3]{3} \cdot \sqrt[3]{12} = \sqrt[3]{3 \cdot 12} = \sqrt[3]{36}$

(b) $\sqrt[4]{8y} \cdot \sqrt[4]{3r^2} = \sqrt[4]{24yr^2}$

(c) $\sqrt[6]{10m^4} \cdot \sqrt[6]{5m} = \sqrt[6]{50m^5}$

(d) $\sqrt[4]{2} \cdot \sqrt[5]{2}$ cannot be simplified using the product rule for radicals because the indexes (4 and 5) are different.

**Now Try Exercises 13, 15, 17, and 19.**

**OBJECTIVE 2 Use the quotient rule for radicals.** The **quotient rule for radicals** is similar to the product rule.

**Quotient Rule for Radicals**

If $\sqrt[n]{a}$ and $\sqrt[n]{b}$ are real numbers, $b \neq 0$, and $n$ is a natural number, then

$$\sqrt[n]{\frac{a}{b}} = \frac{\sqrt[n]{a}}{\sqrt[n]{b}}.$$

That is, the radical of a quotient is the quotient of the radicals.

**EXAMPLE 3 Using the Quotient Rule**

Simplify. Assume all variables represent positive real numbers.

(a) $\sqrt{\frac{16}{25}} = \frac{\sqrt{16}}{\sqrt{25}} = \frac{4}{5}$

(b) $\sqrt{\frac{7}{36}} = \frac{\sqrt{7}}{\sqrt{36}} = \frac{\sqrt{7}}{6}$

(c) $\sqrt[3]{-\frac{8}{125}} = \sqrt[3]{\frac{-8}{125}} = \frac{\sqrt[3]{-8}}{\sqrt[3]{125}} = \frac{-2}{5} = -\frac{2}{5}$

(d) $\sqrt[3]{\frac{7}{216}} = \frac{\sqrt[3]{7}}{\sqrt[3]{216}} = \frac{\sqrt[3]{7}}{6}$

(e) $\sqrt[5]{\frac{x}{32}} = \frac{\sqrt[5]{x}}{\sqrt[5]{32}} = \frac{\sqrt[5]{x}}{2}$

(f) $-\sqrt[3]{\frac{m^6}{125}} = -\frac{\sqrt[3]{m^6}}{\sqrt[3]{125}} = -\frac{m^2}{5}$

**Now Try Exercises 23, 25, 31, 33, and 35.**

**OBJECTIVE 3 Simplify radicals.** We use the product and quotient rules to simplify radicals. A radical is **simplified** if the following four conditions are met.

**Simplified Radical**

1. The radicand has no factor raised to a power greater than or equal to the index.
2. The radicand has no fractions.
3. No denominator contains a radical.
4. Exponents in the radicand and the index of the radical have no common factor (except 1).

**EXAMPLE 4 Simplifying Roots of Numbers**

Simplify.

(a) $\sqrt{24}$

Check to see whether 24 is divisible by a perfect square (the square of a natural number) such as 4, 9, . . . . Choose the largest perfect square that divides into 24. The largest such number is 4. Write 24 as the product of 4 and 6, and then use the product rule.

$$\sqrt{24} = \sqrt{4 \cdot 6} = \sqrt{4} \cdot \sqrt{6} = 2\sqrt{6}$$

**(b)** $\sqrt{108}$

The number 108 is divisible by the perfect square 36: $\sqrt{108} = \sqrt{36 \cdot 3}$. If this is not obvious, try factoring 108 into its prime factors.

$$\begin{aligned}\sqrt{108} &= \sqrt{2^2 \cdot 3^3} \\ &= \sqrt{2^2 \cdot 3^2 \cdot 3} \\ &= 2 \cdot 3 \cdot \sqrt{3} && \text{Product rule} \\ &= 6\sqrt{3}\end{aligned}$$

**(c)** $\sqrt{10}$

No perfect square (other than 1) divides into 10, so $\sqrt{10}$ cannot be simplified further.

**(d)** $\sqrt[3]{16}$

Look for the largest perfect *cube* that divides into 16. The number 8 satisfies this condition, so write 16 as $8 \cdot 2$ (or factor 16 into prime factors).

$$\sqrt[3]{16} = \sqrt[3]{8 \cdot 2} = \sqrt[3]{8} \cdot \sqrt[3]{2} = 2\sqrt[3]{2}$$

**(e)**
$$\begin{aligned}-\sqrt[4]{162} &= -\sqrt[4]{81 \cdot 2} && \text{81 is a perfect 4th power.} \\ &= -\sqrt[4]{81} \cdot \sqrt[4]{2} && \text{Product rule} \\ &= -3\sqrt[4]{2}\end{aligned}$$

**Now Try Exercises 39, 41, 49, and 55.**

**CAUTION** In simplifying an expression like that in Example 4(b), be careful with which factors belong *outside* the radical sign and which belong *inside*. Note how $2 \cdot 3$ is written outside because $\sqrt{2^2} = 2$ and $\sqrt{3^2} = 3$, while the remaining 3 is left inside the radical.

## EXAMPLE 5 Simplifying Radicals Involving Variables

Simplify. Assume all variables represent positive real numbers.

**(a)**
$$\begin{aligned}\sqrt{16m^3} &= \sqrt{16m^2 \cdot m} \\ &= \sqrt{16m^2} \cdot \sqrt{m} \\ &= 4m\sqrt{m}\end{aligned}$$

No absolute value bars are needed around the $m$ in color because of the assumption that all the variables represent *positive* real numbers.

**(b)**
$$\begin{aligned}\sqrt{200k^7q^8} &= \sqrt{10^2 \cdot 2 \cdot (k^3)^2 \cdot k \cdot (q^4)^2} && \text{Factor.} \\ &= 10k^3q^4\sqrt{2k} && \text{Remove perfect square factors.}\end{aligned}$$

**(c)**
$$\begin{aligned}\sqrt[3]{8x^4y^5} &= \sqrt[3]{(8x^3y^3)(xy^2)} && 8x^3y^3 \text{ is the largest perfect cube that divides } 8x^4y^5. \\ &= \sqrt[3]{8x^3y^3} \cdot \sqrt[3]{xy^2} \\ &= 2xy\sqrt[3]{xy^2}\end{aligned}$$

**(d)** $-\sqrt[4]{32y^9} = -\sqrt[4]{(16y^8)(2y)}$  $16y^8$ is the largest 4th power that divides $32y^9$.

$= -\sqrt[4]{16y^8} \cdot \sqrt[4]{2y}$

$= -2y^2\sqrt[4]{2y}$

**Now Try Exercises 75, 79, 83, and 87.**

**NOTE** From Example 5 we see that if a variable is raised to a power with an exponent divisible by 2, it is a perfect square. If it is raised to a power with an exponent divisible by 3, it is a perfect cube. In general, if it is raised to a power with an exponent divisible by $n$, it is a perfect $n$th power.

The conditions for a simplified radical given earlier state that an exponent in the radicand and the index of the radical should have no common factor (except 1). The next example shows how to simplify radicals with such common factors.

### EXAMPLE 6 Simplifying Radicals by Using Smaller Indexes

Simplify. Assume all variables represent positive real numbers.

**(a)** $\sqrt[9]{5^6}$

We can write this radical using rational exponents and then write the exponent in lowest terms. We then express the answer as a radical.

$$\sqrt[9]{5^6} = 5^{6/9} = 5^{2/3} = \sqrt[3]{5^2} \quad \text{or} \quad \sqrt[3]{25}$$

**(b)** $\sqrt[4]{p^2} = p^{2/4} = p^{1/2} = \sqrt{p}$  (Recall the assumption that $p > 0$.)

**Now Try Exercises 93 and 97.**

These examples suggest the following rule.

If $m$ is an integer, $n$ and $k$ are natural numbers, and all indicated roots exist, then

$$\sqrt[kn]{a^{km}} = \sqrt[n]{a^m}.$$

**OBJECTIVE 4 Simplify products and quotients of radicals with different indexes.** Since the product and quotient rules for radicals apply only when they have the same index, we multiply and divide radicals with different indexes by using rational exponents.

### EXAMPLE 7 Multiplying Radicals with Different Indexes

Simplify $\sqrt{7} \cdot \sqrt[3]{2}$.

Because the different indexes, 2 and 3, have a least common index of 6, use rational exponents to write each radical as a sixth root.

$$\sqrt{7} = 7^{1/2} = 7^{3/6} = \sqrt[6]{7^3} = \sqrt[6]{343}$$

$$\sqrt[3]{2} = 2^{1/3} = 2^{2/6} = \sqrt[6]{2^2} = \sqrt[6]{4}$$

Therefore,

$$\sqrt{7} \cdot \sqrt[3]{2} = \sqrt[6]{343} \cdot \sqrt[6]{4} = \sqrt[6]{1372}. \quad \text{Product rule}$$

**Now Try Exercise 99.**

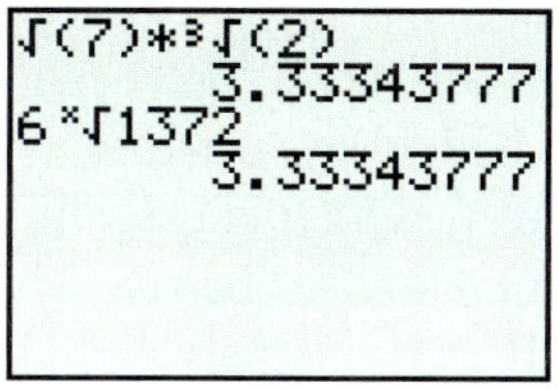


FIGURE 8

Results such as the one in Example 7 can be supported using a calculator, as shown in Figure 8. Notice that the calculator gives the same approximation for the initial product and the final radical that we obtained.

**CAUTION** The computation in Figure 8 is not *proof* that the two expressions are equal. The algebra in Example 7, however, is valid proof of their equality.

**OBJECTIVE 5 Use the Pythagorean formula.** Recall from Section 5.6 that the **Pythagorean formula** relates the lengths of the three sides of a right triangle.

### Pythagorean Formula

If $c$ is the length of the longest side of a right triangle and $a$ and $b$ are the lengths of the shorter sides, then

$$c^2 = a^2 + b^2.$$

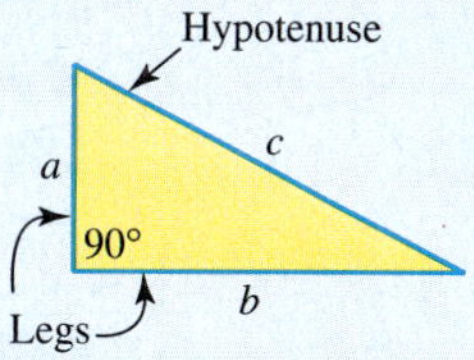


The longest side is the **hypotenuse** and the two shorter sides are the **legs** of the triangle. The hypotenuse is the side opposite the right angle.

### EXAMPLE 8 Using the Pythagorean Formula

Use the Pythagorean formula to find the length of the hypotenuse in the triangle in Figure 9.

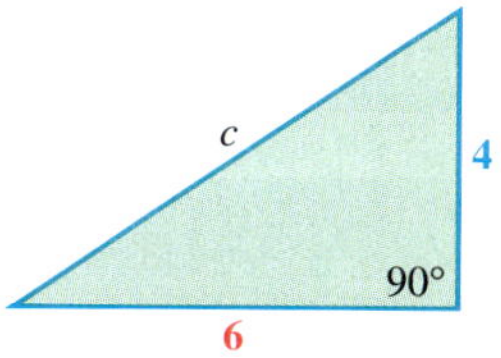


FIGURE 9

To find the length of the hypotenuse $c$, let $a = 4$ and $b = 6$. Then use the formula.

$$c^2 = a^2 + b^2$$
$$c^2 = 4^2 + 6^2 \quad \text{Let } a = 4 \text{ and } b = 6.$$
$$c^2 = 52$$
$$c = \sqrt{52} \quad \text{Choose the principal root.}$$
$$c = \sqrt{4 \cdot 13} \quad \text{Factor.}$$
$$c = \sqrt{4} \cdot \sqrt{13} \quad \text{Product rule}$$
$$c = 2\sqrt{13}$$

The length of the hypotenuse is $2\sqrt{13}$.

**Now Try Exercise 109.**

**CAUTION** When using the equation $c^2 = a^2 + b^2$, be sure that the length of the hypotenuse is substituted for $c$, and that the lengths of the legs are substituted for $a$ and $b$. Errors often occur because values are substituted incorrectly.

## CONNECTIONS

The Pythagorean formula is undoubtedly one of the most widely used and oldest formulas we have. It is very important in trigonometry, which is used in surveying, drafting, engineering, navigation, and many other fields. There is evidence that the Babylonians knew the concept quite well. Although attributed to Pythagoras, it was known to every surveyor from Egypt to China for a thousand years before Pythagoras. In the 1939 movie *The Wizard of Oz,* the Scarecrow asks the Wizard for a brain. When the Wizard presents him with a diploma granting him a Th.D. (Doctor of Thinkology), the Scarecrow recites the following:

> The sum of the square roots of any two sides of an isosceles triangle is equal to the square root of the remaining side. . . .
> Oh joy! Rapture! I've got a brain.

**For Discussion or Writing**

Did the Scarecrow recite the Pythagorean formula? (An *isosceles triangle* is a triangle with two equal sides.) Is his statement true? Explain.

### EXAMPLE 9 Using a Formula from Electronics

The impedance $Z$ of an alternating series circuit is given by the formula

$$Z = \sqrt{R^2 + X^2},$$

where $R$ is the resistance and $X$ is the reactance, both in ohms. Find the value of the impedance if $R = 40$ ohms and $X = 30$ ohms.

Substitute 40 for $R$ and 30 for $X$ in the formula.

$$\begin{aligned} Z &= \sqrt{R^2 + X^2} && \text{Given formula} \\ &= \sqrt{40^2 + 30^2} && \text{Let } R = 40 \text{ and } X = 30. \\ &= \sqrt{1600 + 900} \\ &= \sqrt{2500} \\ &= 50 \end{aligned}$$

The impedance is 50 ohms.

**Now Try Exercise 113.**

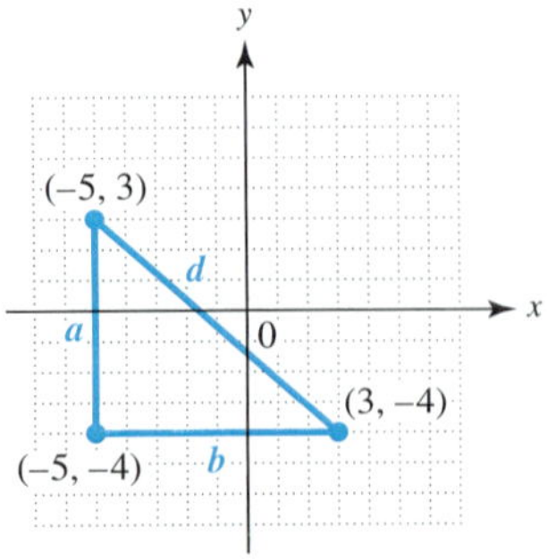


FIGURE 10

**OBJECTIVE 6 Use the distance formula.** An important result in algebra is derived by using the Pythagorean formula. The *distance formula* allows us to find the distance between two points in the coordinate plane, or the length of the line segment joining those two points. Figure 10 shows the points $(3, -4)$ and $(-5, 3)$. The vertical line through $(-5, 3)$ and the horizontal line through $(3, -4)$ intersect at the point $(-5, -4)$. Thus, the point $(-5, -4)$ becomes the vertex of the right angle in a right triangle. By the Pythagorean formula, the square of the length of the hypotenuse, $d$, of the right triangle in Figure 10 is equal to the sum of the squares of the lengths of the two legs $a$ and $b$:

$$d^2 = a^2 + b^2.$$

The length $a$ is the difference between the $y$-coordinates of the endpoints. Since the $x$-coordinate of both points in Figure 10 is $-5$, the side is vertical, and we can find $a$ by finding the difference between the $y$-coordinates. We subtract $-4$ from 3 to get a positive value for $a$.

$$a = 3 - (-4) = 7$$

Similarly, we find $b$ by subtracting $-5$ from 3.

$$b = 3 - (-5) = 8$$

Substituting these values into the formula, we obtain

$$d^2 = a^2 + b^2$$
$$d^2 = 7^2 + 8^2 \qquad \text{Let } a = 7 \text{ and } b = 8.$$
$$d^2 = 49 + 64$$
$$d^2 = 113$$
$$d = \sqrt{113}.$$

We choose the principal root since distance cannot be negative. Therefore, the distance between $(-5, 3)$ and $(3, -4)$ is $\sqrt{113}$.

**NOTE** It is customary to leave the distance in radical form. Do not use a calculator to get an approximation unless you are specifically directed to do so.

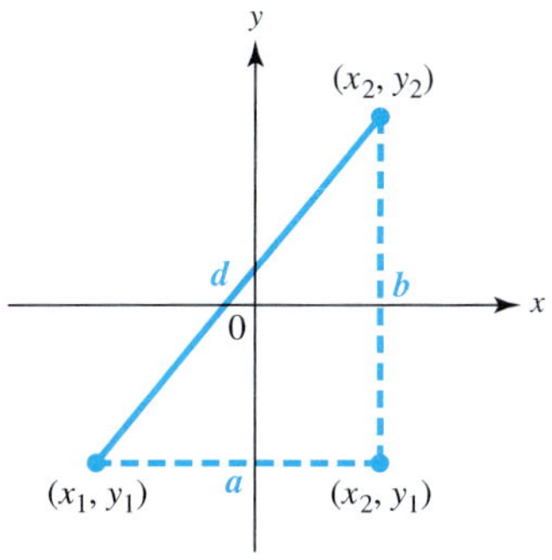

**FIGURE 11**

This result can be generalized. Figure 11 shows the two points $(x_1, y_1)$ and $(x_2, y_2)$. To find a formula for the distance $d$ between these two points, notice that the distance between $(x_1, y_1)$ and $(x_2, y_1)$ is given by

$$a = |x_2 - x_1|,$$

and the distance between $(x_2, y_2)$ and $(x_2, y_1)$ is given by

$$b = |y_2 - y_1|.$$

From the Pythagorean formula,

$$d^2 = a^2 + b^2$$
$$d^2 = (x_2 - x_1)^2 + (y_2 - y_1)^2. \qquad |p - q|^2 = (p - q)^2$$

Choosing the principal square root gives the **distance formula.**

**Distance Formula**

The distance between the points $(x_1, y_1)$ and $(x_2, y_2)$ is

$$d = \sqrt{(x_2 - x_1)^2 + (y_2 - y_1)^2}.$$

## EXAMPLE 10 Using the Distance Formula

Find the distance between the points $(-3, 5)$ and $(6, 4)$.

When using the distance formula to find the distance between two points, designating the points as $(x_1, y_1)$ and $(x_2, y_2)$ is arbitrary. We choose $(x_1, y_1) = (-3, 5)$ and $(x_2, y_2) = (6, 4)$.

$$
\begin{aligned}
d &= \sqrt{(x_2 - x_1)^2 + (y_2 - y_1)^2} \\
&= \sqrt{[6 - (-3)]^2 + (4 - 5)^2} \qquad x_2 = 6,\ y_2 = 4,\ x_1 = -3,\ y_1 = 5 \\
&= \sqrt{9^2 + (-1)^2} \\
&= \sqrt{82}
\end{aligned}
$$

**Now Try Exercise 121.**

## 10.3 EXERCISES

**For Extra Help**

- Student's Solutions Manual
- MyMathLab
- InterAct Math Tutorial Software
- AW Math Tutor Center
- MathXL
- Digital Video Tutor CD 14/Videotape 14

*Decide whether each statement is* true *or* false *by using the product rule explained in this section. Then support your answer by finding a calculator approximation for each expression.*

**1.** $2\sqrt{12} = \sqrt{48}$ **2.** $\sqrt{72} = 2\sqrt{18}$

**3.** $3\sqrt{8} = 2\sqrt{18}$ **4.** $5\sqrt{72} = 6\sqrt{50}$

**5.** Which one of the following is *not* equal to $\sqrt{\frac{1}{2}}$? (Do not use calculator approximations.)

**A.** $\sqrt{.5}$ **B.** $\sqrt{\frac{2}{4}}$ **C.** $\sqrt{\frac{3}{6}}$ **D.** $\frac{\sqrt{4}}{\sqrt{16}}$

**6.** Use the $\pi$ key on your calculator to get a value for $\pi$. Now find an approximation for $\sqrt[4]{\frac{2143}{22}}$. Does the result mean that $\pi$ is actually equal to $\sqrt[4]{\frac{2143}{22}}$? Why or why not?

*Multiply using the product rule. Assume all variables represent positive real numbers. See Examples 1 and 2.*

**7.** $\sqrt{5} \cdot \sqrt{6}$ **8.** $\sqrt{10} \cdot \sqrt{3}$ **9.** $\sqrt{14} \cdot \sqrt{x}$ **10.** $\sqrt{23} \cdot \sqrt{t}$

**11.** $\sqrt{14} \cdot \sqrt{3pqr}$ **12.** $\sqrt{7} \cdot \sqrt{5xt}$ **13.** $\sqrt[3]{7x} \cdot \sqrt[3]{2y}$ **14.** $\sqrt[3]{9x} \cdot \sqrt[3]{4y}$

**15.** $\sqrt[4]{11} \cdot \sqrt[4]{3}$ **16.** $\sqrt[4]{6} \cdot \sqrt[4]{9}$ **17.** $\sqrt[4]{2x} \cdot \sqrt[4]{3y^2}$ **18.** $\sqrt[4]{3y^2} \cdot \sqrt[4]{6yz}$

**19.** $\sqrt[3]{7} \cdot \sqrt[4]{3}$ **20.** $\sqrt[5]{8} \cdot \sqrt[6]{12}$

**21.** Explain the product rule for radicals in your own words. Give examples.

**22.** Explain the quotient rule for radicals in your own words. Give examples.

*Simplify each radical. Assume all variables represent positive real numbers. See Example 3.*

**23.** $\sqrt{\frac{64}{121}}$ **24.** $\sqrt{\frac{16}{49}}$ **25.** $\sqrt{\frac{3}{25}}$ **26.** $\sqrt{\frac{13}{49}}$

**27.** $\sqrt{\frac{x}{25}}$ **28.** $\sqrt{\frac{k}{100}}$ **29.** $\sqrt{\frac{p^6}{81}}$ **30.** $\sqrt{\frac{w^{10}}{36}}$

**31.** $\sqrt[3]{\frac{27}{64}}$ **32.** $\sqrt[3]{\frac{216}{125}}$ **33.** $\sqrt[3]{-\frac{r^2}{8}}$ **34.** $\sqrt[3]{-\frac{t}{125}}$

**35.** $-\sqrt[4]{\frac{81}{x^4}}$ **36.** $-\sqrt[4]{\frac{625}{y^4}}$ **37.** $\sqrt[5]{\frac{1}{x^{15}}}$ **38.** $\sqrt[5]{\frac{32}{y^{20}}}$

*Express each radical in simplified form. See Example 4.*

39. $\sqrt{12}$
40. $\sqrt{18}$
41. $\sqrt{288}$
42. $\sqrt{72}$
43. $-\sqrt{32}$
44. $-\sqrt{48}$
45. $-\sqrt{28}$
46. $-\sqrt{24}$
47. $\sqrt{-300}$
48. $\sqrt{-150}$
49. $\sqrt[3]{128}$
50. $\sqrt[3]{24}$
51. $\sqrt[3]{-16}$
52. $\sqrt[3]{-250}$
53. $\sqrt[3]{40}$
54. $\sqrt[3]{375}$
55. $-\sqrt[4]{512}$
56. $-\sqrt[4]{1250}$
57. $\sqrt[5]{64}$
58. $\sqrt[5]{128}$

59. A student claimed that $\sqrt[3]{14}$ is not in simplified form, since $14 = 8 + 6$, and 8 is a perfect cube. Was his reasoning correct? Why or why not?

60. Explain in your own words why $\sqrt[3]{k^4}$ is not a simplified radical.

*Express each radical in simplified form. Assume all variables represent positive real numbers. See Example 5.*

61. $\sqrt{72k^2}$
62. $\sqrt{18m^2}$
63. $\sqrt{144x^3y^9}$
64. $\sqrt{169s^5t^{10}}$
65. $\sqrt{121x^6}$
66. $\sqrt{256z^{12}}$
67. $-\sqrt[3]{27t^{12}}$
68. $-\sqrt[3]{64y^{18}}$
69. $-\sqrt{100m^8z^4}$
70. $-\sqrt{25t^6s^{20}}$
71. $-\sqrt[3]{-125a^6b^9c^{12}}$
72. $-\sqrt[3]{-216y^{15}x^6z^3}$
73. $\sqrt[4]{\frac{1}{16}r^8t^{20}}$
74. $\sqrt[4]{\frac{81}{256}t^{12}u^8}$
75. $\sqrt{50x^3}$
76. $\sqrt{300z^3}$
77. $-\sqrt{500r^{11}}$
78. $-\sqrt{200p^{13}}$
79. $\sqrt{13x^7y^8}$
80. $\sqrt{23k^9p^{14}}$
81. $\sqrt[3]{8z^6w^9}$
82. $\sqrt[3]{64a^{15}b^{12}}$
83. $\sqrt[3]{-16z^5t^7}$
84. $\sqrt[3]{-81m^4n^{10}}$
85. $\sqrt[4]{81x^{12}y^{16}}$
86. $\sqrt[4]{81t^8u^{28}}$
87. $-\sqrt[4]{162r^{15}s^{10}}$
88. $-\sqrt[4]{32k^5m^{10}}$
89. $\sqrt{\frac{y^{11}}{36}}$
90. $\sqrt{\frac{v^{13}}{49}}$
91. $\sqrt[3]{\frac{x^{16}}{27}}$
92. $\sqrt[3]{\frac{y^{17}}{125}}$

*Simplify each radical. Assume that $x \geq 0$. See Example 6.*

93. $\sqrt[4]{48^2}$
94. $\sqrt[4]{50^2}$
95. $\sqrt[4]{25}$
96. $\sqrt[6]{8}$
97. $\sqrt[10]{x^{25}}$
98. $\sqrt[12]{x^{44}}$

*Simplify by first writing the radicals as radicals with the same index. Then multiply. Assume all variables represent positive real numbers. See Example 7.*

99. $\sqrt[3]{4} \cdot \sqrt{3}$
100. $\sqrt[3]{5} \cdot \sqrt{6}$
101. $\sqrt[4]{3} \cdot \sqrt[3]{4}$
102. $\sqrt[5]{7} \cdot \sqrt[7]{5}$
103. $\sqrt{x} \cdot \sqrt[3]{x}$
104. $\sqrt[3]{y} \cdot \sqrt[4]{y}$

## TECHNOLOGY INSIGHTS (EXERCISES 105–108)

*A graphing calculator can be used to test whether two quantities are equal. In the screen shown here, the first two lines of entries both represent true statements, and thus the calculator returns a* 1 *to indicate* true. *The third entry is* false, *and the calculator returns a* 0. *These can be verified algebraically using the rules for radicals found in this section.*

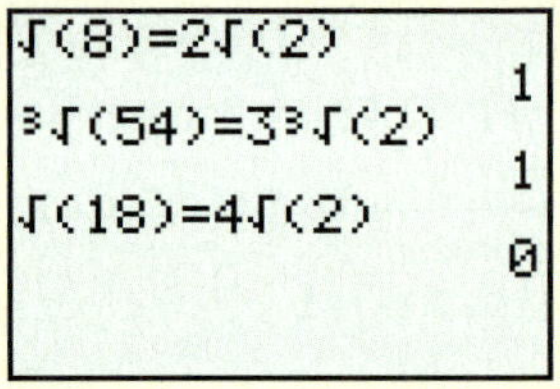


*Determine whether the calculator should return a* 1 *or a* 0 *for each screen.*

**105.**

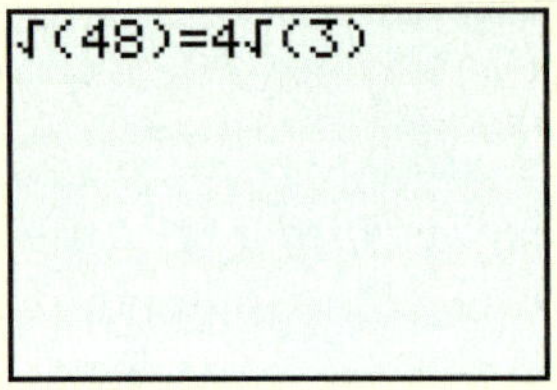


**106.**

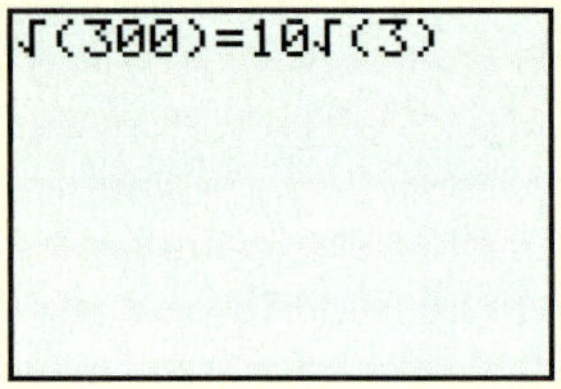


**107.**

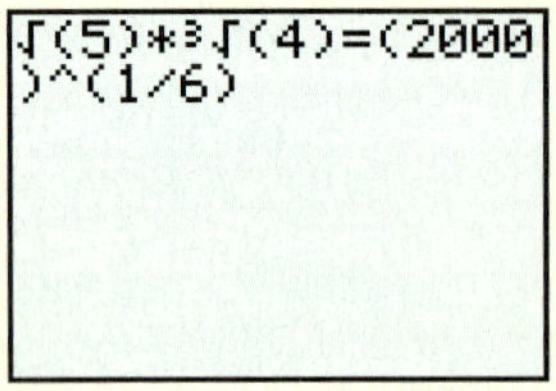


**108.**

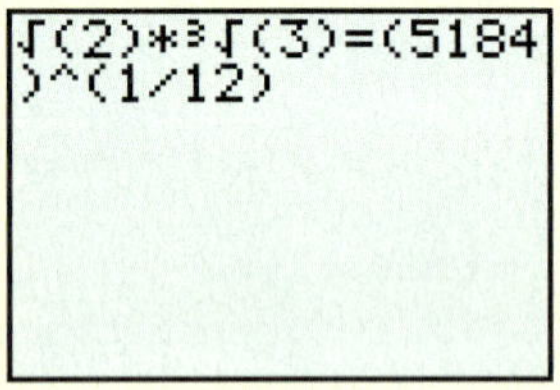


*Find the unknown length in each right triangle. Simplify the answer if necessary. See Example 8.*

**109.**

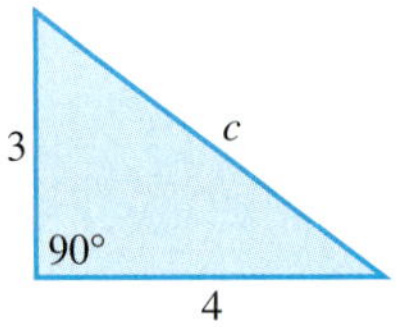


**110.**

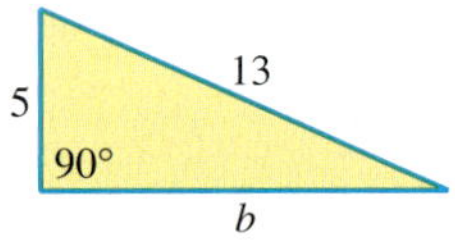


**111.**

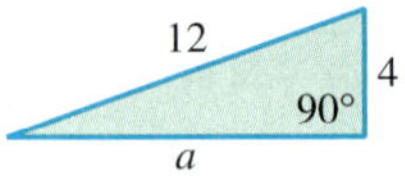


**112.**

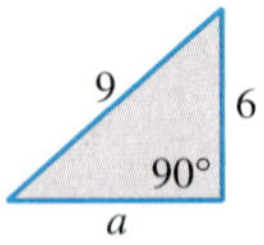


*Solve each problem. See Example 9.*

**113.** The length of the diagonal of a box is given by

$$D = \sqrt{L^2 + W^2 + H^2},$$

where $L$, $W$, and $H$ are the length, width, and height of the box. Find the length of the diagonal, $D$, of a box that is 4 ft long, 3 ft high, and 2 ft wide. Give the exact value, then round to the nearest tenth of a foot.

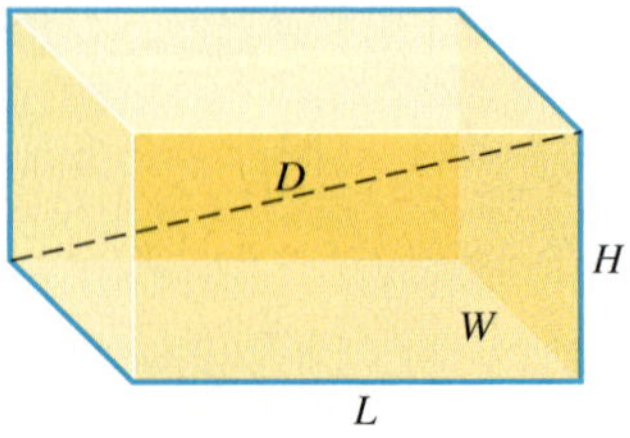

**114.** A Sanyo color television, model AVM-2755, has a rectangular screen with a 21.7-in. width. Its height is 16 in. What is the diagonal of the screen to the nearest tenth of an inch? (*Source:* Actual measurements of the author's television.)

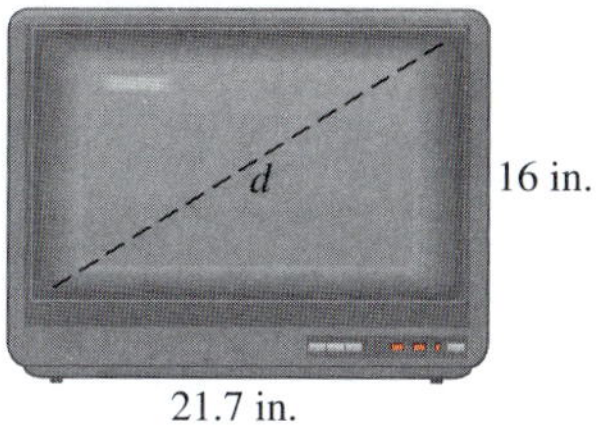


**115.** A formula from electronics dealing with impedance of parallel resonant circuits is

$$I = \frac{E}{\sqrt{R^2 + \omega^2 L^2}},$$

where the variables are in appropriate units. Find $I$ if $E = 282$, $R = 100$, $L = 264$, and $\omega = 120\pi$. Give your answer to the nearest thousandth.

**116.** In the study of sound, one version of the law of tensions is

$$f_1 = f_2\sqrt{\frac{F_1}{F_2}}.$$

If $F_1 = 300$, $F_2 = 60$, and $f_2 = 260$, find $f_1$ to the nearest unit.

**117.** The illumination $I$, in foot-candles, produced by a light source is related to the distance $d$, in feet, from the light source by the equation

$$d = \sqrt{\frac{k}{I}},$$

where $k$ is a constant. If $k = 640$, how far from the light source will the illumination be 2 foot-candles? Give the exact value, and then round to the nearest tenth of a foot.

**118.** The following letter appeared in the column "Ask Tom Why," written by Tom Skilling of the *Chicago Tribune.*

*Dear Tom,*

*I cannot remember the formula to calculate the distance to the horizon. I have a stunning view from my 14th floor condo, 150 feet above the ground. How far can I see?*

*Ted Fleischaker; Indianapolis, Ind.*

Skilling's answer was as follows.

To find the distance to the horizon in miles, take the square root of the height of your view in feet and multiply that result by 1.224. Your answer will be the number of miles to the horizon. (*Source: Chicago Tribune,* August 17, 2002.)

Assuming Ted's eyes are 6 ft above the ground, the total height from the ground is $150 + 6 = 156$ ft. To the nearest tenth of a mile, how far can he see to the horizon?

*Find the distance between each pair of points. See Example 10.*

**119.** $(6, 13)$ and $(1, 1)$

**120.** $(8, 13)$ and $(2, 5)$

**121.** $(-6, 5)$ and $(3, -4)$

**122.** $(-1, 5)$ and $(-7, 7)$

**123.** $(-8, 2)$ and $(-4, 1)$

**124.** $(-1, 2)$ and $(5, 3)$

**125.** (4.7, 2.3) and (1.7, −1.7)

**126.** (−2.9, 18.2) and (2.1, 6.2)

**127.** $(\sqrt{2}, \sqrt{6})$ and $(-2\sqrt{2}, 4\sqrt{6})$

**128.** $(\sqrt{7}, 9\sqrt{3})$ and $(-\sqrt{7}, 4\sqrt{3})$

**129.** $(x + y, y)$ and $(x - y, x)$

**130.** $(c, c - d)$ and $(d, c + d)$

**131.** As given in the text, the distance formula is expressed with a radical. Write the distance formula using rational exponents.

**132.** An alternative form of the distance formula is

$$d = \sqrt{(x_1 - x_2)^2 + (y_1 - y_2)^2}.$$

Compare this to the form given in this section, and explain why the two forms are equivalent.

*Find the perimeter of each triangle.* (Hint: *For Exercise 133,* $\sqrt{k} + \sqrt{k} = 2\sqrt{k}$.)

**133.**

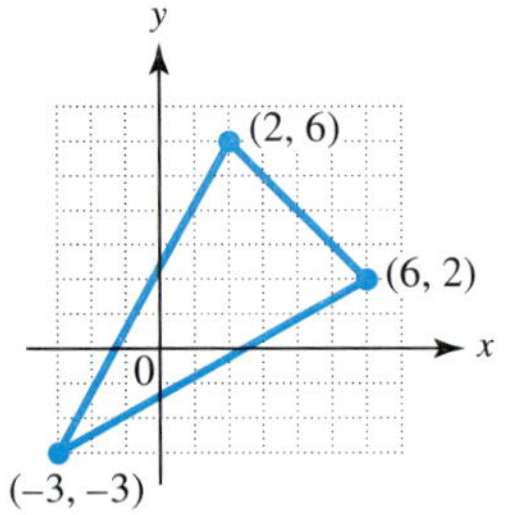


**134.**

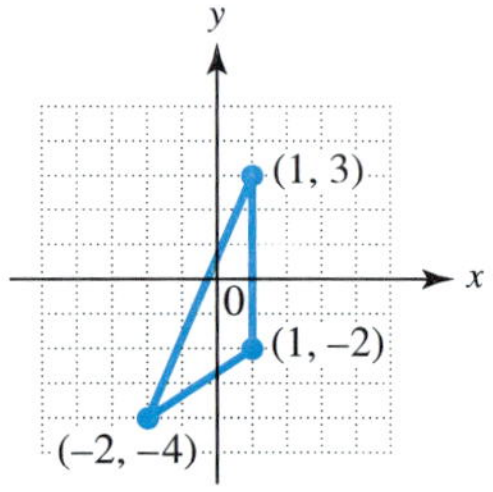


# 10.4 Adding and Subtracting Radical Expressions

**OBJECTIVES**

1. Define a radical expression.
2. Simplify radical expressions involving addition and subtraction.

**OBJECTIVE 1 Define a radical expression.** Recall from Section 10.1 that a **radical expression** is an algebraic expression that contains radicals. For example,

$$\sqrt[4]{3} + \sqrt{6}, \qquad \sqrt{x + 2y} - 1, \qquad \text{and} \qquad \sqrt{8} - \sqrt{2r}$$

are radical expressions. The examples in the previous section discussed simplifying radical expressions that involve multiplication and division. Now we show how to simplify radical expressions that involve addition and subtraction.

**OBJECTIVE 2 Simplify radical expressions involving addition and subtraction.** An expression such as $4\sqrt{2} + 3\sqrt{2}$ can be simplified by using the distributive property.

$$4\sqrt{2} + 3\sqrt{2} = (4 + 3)\sqrt{2} = 7\sqrt{2}$$

As another example, $2\sqrt{3} - 5\sqrt{3} = (2 - 5)\sqrt{3} = -3\sqrt{3}$. This is similar to simplifying $4x + 3x$ to $7x$ or $2y - 5y$ to $-3y$.

**CAUTION** Only radical expressions with the *same index* and the *same radicand* may be combined. Expressions such as $5\sqrt{3} + 2\sqrt{2}$ or $3\sqrt{3} + 2\sqrt[3]{3}$ cannot be simplified by combining terms.

## EXAMPLE 1 Adding and Subtracting Radicals

Add or subtract to simplify each radical expression.

**(a)** $3\sqrt{24} + \sqrt{54}$

Begin by simplifying each radical; then use the distributive property to combine terms.

$$\begin{aligned} 3\sqrt{24} + \sqrt{54} &= 3\sqrt{4} \cdot \sqrt{6} + \sqrt{9} \cdot \sqrt{6} && \text{Product rule} \\ &= 3 \cdot 2\sqrt{6} + 3\sqrt{6} \\ &= 6\sqrt{6} + 3\sqrt{6} \\ &= 9\sqrt{6} && \text{Combine terms.} \end{aligned}$$

**(b)** $$\begin{aligned} 2\sqrt{20x} - \sqrt{45x} &= 2\sqrt{4} \cdot \sqrt{5x} - \sqrt{9} \cdot \sqrt{5x} && \text{Product rule} \\ &= 2 \cdot 2\sqrt{5x} - 3\sqrt{5x} \\ &= 4\sqrt{5x} - 3\sqrt{5x} \\ &= \sqrt{5x}, \quad x \geq 0 && \text{Combine terms.} \end{aligned}$$

**(c)** $2\sqrt{3} - 4\sqrt{5}$

Here the radicands differ and are already simplified, so $2\sqrt{3} - 4\sqrt{5}$ cannot be simplified further.

**Now Try Exercises 7, 15, and 19.**

**CAUTION** Do not confuse the product rule with combining like terms. The root of a sum ***does not equal*** the sum of the roots. For example,

$$\sqrt{9 + 16} \neq \sqrt{9} + \sqrt{16}$$

since $\sqrt{9 + 16} = \sqrt{25} = 5$, but $\sqrt{9} + \sqrt{16} = 3 + 4 = 7$.

## EXAMPLE 2 Adding and Subtracting Radicals with Higher Indexes

Add or subtract to simplify each radical expression. Assume all variables represent positive real numbers.

**(a)** $$\begin{aligned} 2\sqrt[3]{16} - 5\sqrt[3]{54} &= 2\sqrt[3]{8 \cdot 2} - 5\sqrt[3]{27 \cdot 2} && \text{Factor.} \\ &= 2\sqrt[3]{8} \cdot \sqrt[3]{2} - 5\sqrt[3]{27} \cdot \sqrt[3]{2} && \text{Product rule} \\ &= 2 \cdot 2 \cdot \sqrt[3]{2} - 5 \cdot 3 \cdot \sqrt[3]{2} \\ &= 4\sqrt[3]{2} - 15\sqrt[3]{2} \\ &= -11\sqrt[3]{2} && \text{Combine terms.} \end{aligned}$$

**(b)** $$\begin{aligned} 2\sqrt[3]{x^2y} + \sqrt[3]{8x^5y^4} &= 2\sqrt[3]{x^2y} + \sqrt[3]{(8x^3y^3)x^2y} && \text{Factor.} \\ &= 2\sqrt[3]{x^2y} + 2xy\sqrt[3]{x^2y} && \text{Product rule} \\ &= (2 + 2xy)\sqrt[3]{x^2y} && \text{Distributive property} \end{aligned}$$

**Now Try Exercises 23 and 29.**

**CAUTION** Remember to write the index when working with cube roots, fourth roots, and so on.

## EXAMPLE 3 Adding and Subtracting Radicals with Fractions

Perform the indicated operations. Assume all variables represent positive real numbers.

**(a)** $2\sqrt{\frac{75}{16}} + 4\frac{\sqrt{8}}{\sqrt{32}} = 2\frac{\sqrt{25 \cdot 3}}{\sqrt{16}} + 4\frac{\sqrt{4 \cdot 2}}{\sqrt{16 \cdot 2}}$ Quotient rule

$= 2\left(\frac{5\sqrt{3}}{4}\right) + 4\left(\frac{2\sqrt{2}}{4\sqrt{2}}\right)$ Product rule

$= \frac{5\sqrt{3}}{2} + 2$ Multiply; $\frac{\sqrt{2}}{\sqrt{2}} = 1$.

$= \frac{5\sqrt{3}}{2} + \frac{4}{2}$ Write with a common denominator.

$= \frac{5\sqrt{3} + 4}{2}$

**(b)** $10\sqrt[3]{\frac{5}{x^6}} - 3\sqrt[3]{\frac{4}{x^9}} = 10\frac{\sqrt[3]{5}}{\sqrt[3]{x^6}} - 3\frac{\sqrt[3]{4}}{\sqrt[3]{x^9}}$ Quotient rule

$= \frac{10\sqrt[3]{5}}{x^2} - \frac{3\sqrt[3]{4}}{x^3}$

$= \frac{10x\sqrt[3]{5}}{x^3} - \frac{3\sqrt[3]{4}}{x^3}$ Write with a common denominator.

$= \frac{10x\sqrt[3]{5} - 3\sqrt[3]{4}}{x^3}$

**Now Try Exercises 47 and 53.**

A calculator can support some of the results obtained in the examples of this section. In Example 1(a), we simplified $3\sqrt{24} + \sqrt{54}$ to obtain $9\sqrt{6}$. The screen in Figure 12(a) shows that the approximations are the same, suggesting that our simplification was correct. Figure 12(b) shows support for the result of Example 2(a): $2\sqrt[3]{16} - 5\sqrt[3]{54} = -11\sqrt[3]{2}$. Figure 12(c) supports the result of Example 3(a).

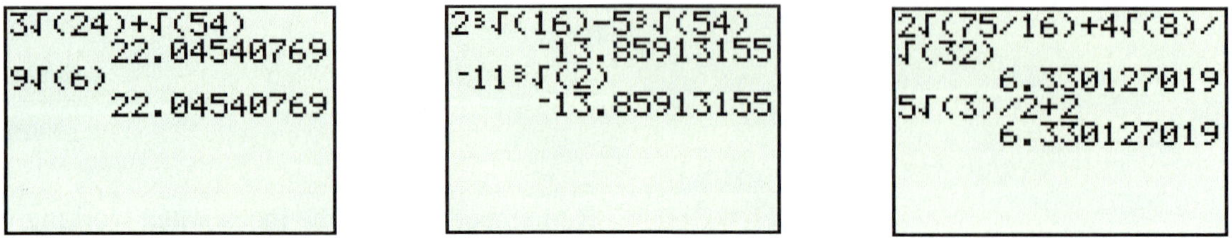


(a) (b) (c)

FIGURE 12

## CONNECTIONS

A triangle that has whole number measures for the lengths of two sides may have an irrational number as the measure of the third side. For example, a right triangle with the two shorter sides measuring 1 and 2 units will have a longest side measuring $\sqrt{5}$ units. The ratio of the dimensions of the *golden rectangle,* considered to have the most pleasing dimensions of any rectangle, is irrational. To sketch a golden rectangle, begin with the square *ONRS.* Divide it into two equal parts by segment *MK,* as shown in the figure. Let *M* be the center of a circle with radius *MN*. Sketch the rectangle *PQRS.* This is a golden rectangle, with the property that if the original square is taken away, *PQNO* is still a golden rectangle. If the square with side *OP* is taken away, another golden rectangle results, and so on.

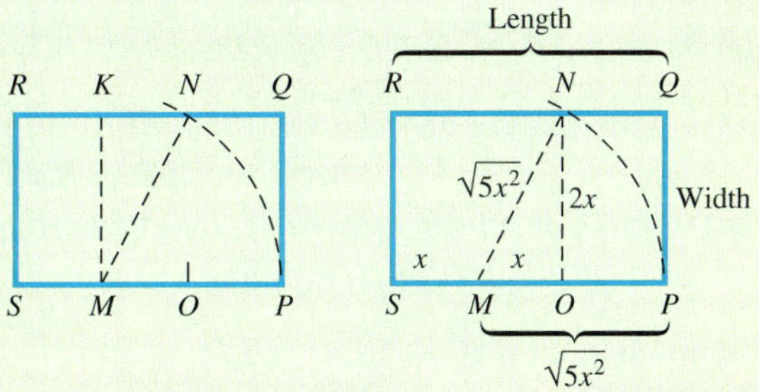


If the sides of the generating square have measure $2x$, then by the Pythagorean formula,

$$MN = \sqrt{x^2 + (2x)^2} = \sqrt{x^2 + 4x^2} = \sqrt{5x^2}.$$

Since *NP* is an arc of the circle with radius *MN*,

$$MP = MN = \sqrt{5x^2}.$$

The ratio of length to width is

$$\frac{\text{length}}{\text{width}} = \frac{x + \sqrt{5x^2}}{2x} = \frac{x + x\sqrt{5}}{2x} = \frac{x\left(1 + \sqrt{5}\right)}{2x} = \frac{1 + \sqrt{5}}{2},$$

which is an irrational number.

### For Discussion or Writing

1. The golden rectangle has been widely used in art and architecture. See whether you can find some examples of its use. Use a calculator to approximate the ratio found above, called the *golden ratio*.
2. The sequence 1, 1, 2, 3, 5, 8, 13, 21, 34, 55, . . . is called the *Fibonacci sequence.* After the first two terms, both 1, every term is found by adding the two preceding terms. Form a sequence of ratios of the successive terms:

$$\frac{1}{1}, \frac{2}{1}, \frac{3}{2}, \frac{5}{3}, \frac{8}{5}, \frac{13}{8}, \frac{21}{13}, \frac{34}{21}, \frac{55}{34}, \ldots.$$

Now use a calculator to find approximations of these ratios. What seems to be happening?

# 10.4 EXERCISES

**For Extra Help**

- Student's Solutions Manual
- MyMathLab
- InterAct Math Tutorial Software
- AW Math Tutor Center
- MathXL
- Digital Video Tutor CD 14/Videotape 14

**1.** Which one of the following sums could be simplified without first simplifying the individual radical expressions?

**A.** $\sqrt{50} + \sqrt{32}$ **B.** $3\sqrt{6} + 9\sqrt{6}$ **C.** $\sqrt[3]{32} - \sqrt[3]{108}$ **D.** $\sqrt[5]{6} - \sqrt[5]{192}$

**2.** Let $a = 1$ and let $b = 64$.

**(a)** Evaluate $\sqrt{a} + \sqrt{b}$. Then find $\sqrt{a + b}$. Are they equal?
**(b)** Evaluate $\sqrt[3]{a} + \sqrt[3]{b}$. Then find $\sqrt[3]{a + b}$. Are they equal?
**(c)** Complete the following: In general, $\sqrt[n]{a} + \sqrt[n]{b} \neq$ ______________________, based on the observations in parts (a) and (b) of this exercise.

**3.** Even though the root indexes of the terms are not equal, the sum $\sqrt{64} + \sqrt[3]{125} + \sqrt[4]{16}$ can be simplified quite easily. What is this sum? Why can we add these terms so easily?

**4.** Explain why $28 - 4\sqrt{2}$ is not equal to $24\sqrt{2}$. (This is a common error among algebra students.)

*Simplify. Assume all variables represent positive real numbers. See Examples 1 and 2.*

**5.** $\sqrt{36} - \sqrt{100}$
**6.** $\sqrt{25} - \sqrt{81}$
**7.** $-2\sqrt{48} + 3\sqrt{75}$
**8.** $4\sqrt{32} - 2\sqrt{8}$
**9.** $\sqrt[3]{16} + 4\sqrt[3]{54}$
**10.** $3\sqrt[3]{24} - 2\sqrt[3]{192}$
**11.** $\sqrt[4]{32} + 3\sqrt[4]{2}$
**12.** $\sqrt[4]{405} - 2\sqrt[4]{5}$
**13.** $6\sqrt{18} - \sqrt{32} + 2\sqrt{50}$
**14.** $5\sqrt{8} + 3\sqrt{72} - 3\sqrt{50}$
**15.** $5\sqrt{6} + 2\sqrt{10}$
**16.** $3\sqrt{11} - 5\sqrt{13}$
**17.** $2\sqrt{5} + 3\sqrt{20} + 4\sqrt{45}$
**18.** $5\sqrt{54} - 2\sqrt{24} - 2\sqrt{96}$
**19.** $8\sqrt{2x} - \sqrt{8x} + \sqrt{72x}$
**20.** $4\sqrt{18k} - \sqrt{72k} + \sqrt{50k}$
**21.** $3\sqrt{72m^2} - 5\sqrt{32m^2} - 3\sqrt{18m^2}$
**22.** $9\sqrt{27p^2} - 14\sqrt{108p^2} + 2\sqrt{48p^2}$
**23.** $-\sqrt[3]{54} + 2\sqrt[3]{16}$
**24.** $15\sqrt[3]{81} - 4\sqrt[3]{24}$
**25.** $2\sqrt[3]{27x} - 2\sqrt[3]{8x}$
**26.** $6\sqrt[3]{128m} + 3\sqrt[3]{16m}$
**27.** $\sqrt[3]{x^2y} - \sqrt[3]{8x^2y}$
**28.** $3\sqrt[3]{x^2y^2} - 2\sqrt[3]{64x^2y^2}$
**29.** $3x\sqrt[3]{xy^2} - 2\sqrt[3]{8x^4y^2}$
**30.** $6q^2\sqrt[3]{5q} - 2q\sqrt[3]{40q^4}$
**31.** $5\sqrt[4]{32} + 3\sqrt[4]{162}$
**32.** $2\sqrt[4]{512} + 4\sqrt[4]{32}$
**33.** $3\sqrt[4]{x^5y} - 2x\sqrt[4]{xy}$
**34.** $2\sqrt[4]{m^9p^6} - 3m^2p\sqrt[4]{mp^2}$
**35.** $2\sqrt[4]{32a^3} + 5\sqrt[4]{2a^3}$
**36.** $-\sqrt[4]{16r} + 5\sqrt[4]{r}$
**37.** $\sqrt[3]{64xy^2} + \sqrt[3]{27x^4y^5}$
**38.** $\sqrt[4]{625s^3t} - \sqrt[4]{81s^7t^5}$

*Simplify. Assume all variables represent positive real numbers. See Example 3.*

**39.** $\sqrt{8} - \dfrac{\sqrt{64}}{\sqrt{16}}$
**40.** $\sqrt{48} - \dfrac{\sqrt{81}}{\sqrt{9}}$
**41.** $\dfrac{2\sqrt{5}}{3} + \dfrac{\sqrt{5}}{6}$

**42.** $\frac{4\sqrt{3}}{3} + \frac{2\sqrt{3}}{9}$

**43.** $\sqrt{\frac{8}{9}} + \sqrt{\frac{18}{36}}$

**44.** $\sqrt{\frac{12}{16}} + \sqrt{\frac{48}{64}}$

**45.** $\frac{\sqrt{32}}{3} + \frac{2\sqrt{2}}{3} - \frac{\sqrt{2}}{\sqrt{9}}$

**46.** $\frac{\sqrt{27}}{2} - \frac{3\sqrt{3}}{2} + \frac{\sqrt{3}}{\sqrt{4}}$

**47.** $3\sqrt{\frac{50}{9}} + 8\frac{\sqrt{2}}{\sqrt{8}}$

**48.** $9\sqrt{\frac{48}{25}} - 2\frac{\sqrt{2}}{\sqrt{98}}$

**49.** $\sqrt{\frac{25}{x^8}} - \sqrt{\frac{9}{x^6}}$

**50.** $\sqrt{\frac{100}{y^4}} + \sqrt{\frac{81}{y^{10}}}$

**51.** $3\sqrt[3]{\frac{m^5}{27}} - 2m\sqrt[3]{\frac{m^2}{64}}$

**52.** $2a\sqrt[4]{\frac{a}{16}} - 5a\sqrt[4]{\frac{a}{81}}$

**53.** $3\sqrt[3]{\frac{2}{x^6}} - 4\sqrt[3]{\frac{5}{x^9}}$

**54.** $-4\sqrt[3]{\frac{4}{t^9}} + 3\sqrt[3]{\frac{9}{t^{12}}}$

*In Example 1(a) we showed that $3\sqrt{24} + \sqrt{54} = 9\sqrt{6}$. To support this result, we can find a calculator approximation of $3\sqrt{24}$, then find a calculator approximation of $\sqrt{54}$, and add these two approximations. Then, we find a calculator approximation of $9\sqrt{6}$. It should correspond to the sum that we just found. (For this example, both approximations are 22.04540769. Due to rounding procedures, there may be a discrepancy in the final digit if you try to duplicate this work.) Follow this procedure to support the statements in Exercises 55 and 56.*

**55.** $3\sqrt{32} - 2\sqrt{8} = 8\sqrt{2}$

**56.** $2\sqrt{40} + 6\sqrt{90} - 3\sqrt{160} = 10\sqrt{10}$

**57.** A rectangular yard has a length of $\sqrt{192}$ m and a width of $\sqrt{48}$ m. Choose the best estimate of its dimensions. Then estimate the perimeter.

**A.** 14 m by 7 m **B.** 5 m by 7 m **C.** 14 m by 8 m **D.** 15 m by 8 m

**58.** If the sides of a triangle are $\sqrt{65}$ in., $\sqrt{35}$ in., and $\sqrt{26}$ in., which one of the following is the best estimate of its perimeter?

**A.** 20 in. **B.** 26 in. **C.** 19 in. **D.** 24 in.

*Solve each problem. Give answers as simplified radical expressions.*

**59.** Find the perimeter of the triangle.

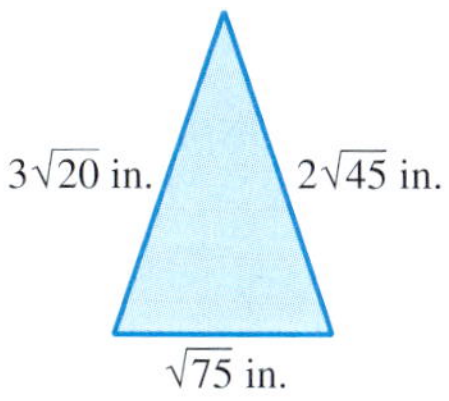


**60.** Find the perimeter of the rectangle.

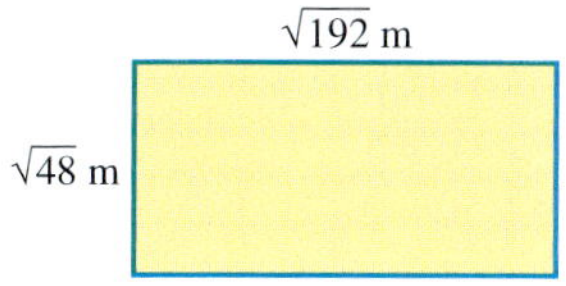


**61.** What is the perimeter of the computer graphic?

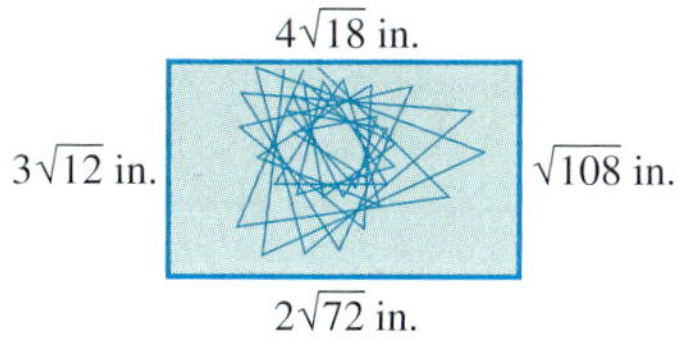


**62.** Find the area of the trapezoid.

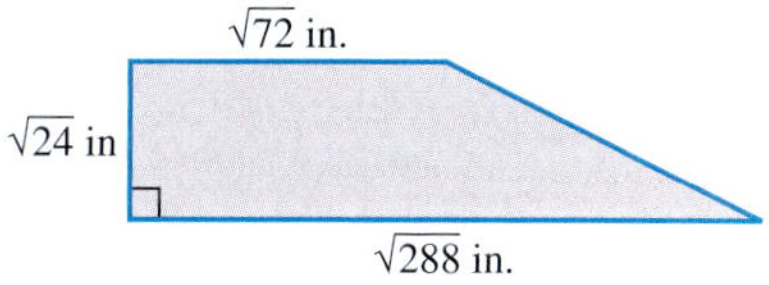

# 10.5 Multiplying and Dividing Radical Expressions

**OBJECTIVES**

1. Multiply radical expressions.
2. Rationalize denominators with one radical term.
3. Rationalize denominators with binomials involving radicals.
4. Write radical quotients in lowest terms.

**OBJECTIVE 1 Multiply radical expressions.** We multiply binomial expressions involving radicals by using the FOIL (First, Outer, Inner, Last) method. For example, we find the product of the binomials $\sqrt{5} + 3$ and $\sqrt{6} + 1$ as follows.

$$\begin{aligned}(\sqrt{5}+3)(\sqrt{6}+1) &= \underbrace{\sqrt{5}\cdot\sqrt{6}}_{\text{First}} + \underbrace{\sqrt{5}\cdot 1}_{\text{Outer}} + \underbrace{3\cdot\sqrt{6}}_{\text{Inner}} + \underbrace{3\cdot 1}_{\text{Last}} \\ &= \sqrt{30} + \sqrt{5} + 3\sqrt{6} + 3\end{aligned}$$

This result cannot be simplified further.

**EXAMPLE 1 Multiplying Binomials Involving Radical Expressions**

Multiply using FOIL.

**(a)** $$\begin{aligned}(7-\sqrt{3})(\sqrt{5}+\sqrt{2}) &= \underbrace{7\sqrt{5}}_{\text{F}} + \underbrace{7\sqrt{2}}_{\text{O}} - \underbrace{\sqrt{3}\cdot\sqrt{5}}_{\text{I}} - \underbrace{\sqrt{3}\cdot\sqrt{2}}_{\text{L}} \\ &= 7\sqrt{5} + 7\sqrt{2} - \sqrt{15} - \sqrt{6}\end{aligned}$$

**(b)** $$\begin{aligned}&(\sqrt{10}+\sqrt{3})(\sqrt{10}-\sqrt{3}) \\ &= \sqrt{10}\cdot\sqrt{10} - \sqrt{10}\cdot\sqrt{3} + \sqrt{10}\cdot\sqrt{3} - \sqrt{3}\cdot\sqrt{3} \\ &= 10 - 3 \\ &= 7\end{aligned}$$

Notice that this is the kind of product that results in the difference of squares:

$$(x+y)(x-y) = x^2 - y^2.$$

Here, $x = \sqrt{10}$ and $y = \sqrt{3}$.

**(c)** $$\begin{aligned}(\sqrt{7}-3)^2 &= (\sqrt{7}-3)(\sqrt{7}-3) \\ &= \sqrt{7}\cdot\sqrt{7} - 3\sqrt{7} - 3\sqrt{7} + 3\cdot 3 \\ &= 7 - 6\sqrt{7} + 9 \\ &= 16 - 6\sqrt{7}\end{aligned}$$

**(d)** $$\begin{aligned}(5-\sqrt[3]{3})(5+\sqrt[3]{3}) &= 5\cdot 5 + 5\sqrt[3]{3} - 5\sqrt[3]{3} - \sqrt[3]{3}\cdot\sqrt[3]{3} \\ &= 25 - \sqrt[3]{3^2} \\ &= 25 - \sqrt[3]{9}\end{aligned}$$

**(e)** $$\begin{aligned}(\sqrt{k}+\sqrt{y})(\sqrt{k}-\sqrt{y}) &= (\sqrt{k})^2 - (\sqrt{y})^2 \\ &= k - y, \quad k \geq 0 \text{ and } y \geq 0\end{aligned}$$

**Now Try Exercises 13, 17, 23, 27, and 39.**

**NOTE** In Example 1(c) we could have used the formula for the square of a binomial,

$$(x - y)^2 = x^2 - 2xy + y^2,$$

to obtain the same result:

$$\begin{aligned}(\sqrt{7} - 3)^2 &= (\sqrt{7})^2 - 2(\sqrt{7})(3) + 3^2 \\ &= 7 - 6\sqrt{7} + 9 \\ &= 16 - 6\sqrt{7}.\end{aligned}$$

**OBJECTIVE 2 Rationalize denominators with one radical term.** As defined earlier, a simplified radical expression will have no radical in the denominator. The origin of this agreement no doubt occurred before the days of high-speed calculation, when computation was a tedious process performed by hand. To see this, consider the radical expression $\frac{1}{\sqrt{2}}$. To find a decimal approximation by hand, it would be necessary to divide 1 by a decimal approximation for $\sqrt{2}$, such as 1.414. It would be much easier if the divisor were a whole number. This can be accomplished by multiplying $\frac{1}{\sqrt{2}}$ by 1 in the form $\frac{\sqrt{2}}{\sqrt{2}}$:

$$\frac{1}{\sqrt{2}} \cdot \frac{\sqrt{2}}{\sqrt{2}} = \frac{\sqrt{2}}{2}.$$

Now the computation would require dividing 1.414 by 2 to obtain .707, a much easier task.

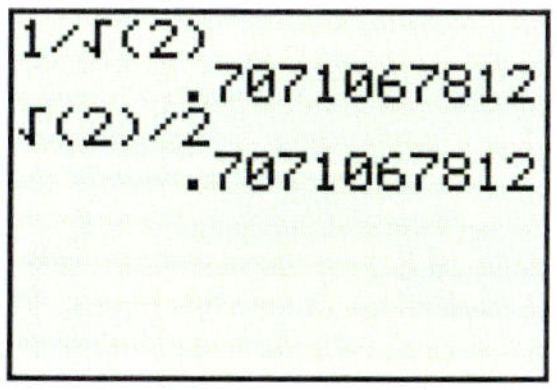


FIGURE 13

With current technology, either form of this fraction can be approximated with the same number of keystrokes. See Figure 13, which shows how a calculator gives the same approximation for both forms of the expression.

A common way of "standardizing" the form of a radical expression is to have the denominator contain no radicals. The process of removing radicals from a denominator so that the denominator contains only rational numbers is called **rationalizing the denominator.**

### EXAMPLE 2 Rationalizing Denominators with Square Roots

Rationalize each denominator.

**(a)** $\frac{3}{\sqrt{7}}$

Multiply the numerator and denominator by $\sqrt{7}$. This is, in effect, multiplying by 1.

$$\frac{3}{\sqrt{7}} = \frac{3 \cdot \sqrt{7}}{\sqrt{7} \cdot \sqrt{7}}$$

In the denominator, since $\sqrt{7} \cdot \sqrt{7} = \sqrt{7 \cdot 7} = \sqrt{49} = 7$,

$$\frac{3}{\sqrt{7}} = \frac{3\sqrt{7}}{7}.$$

The denominator is now a rational number.

**(b)** $\dfrac{5\sqrt{2}}{\sqrt{5}} = \dfrac{5\sqrt{2} \cdot \sqrt{5}}{\sqrt{5} \cdot \sqrt{5}} = \dfrac{5\sqrt{10}}{5} = \sqrt{10}$

**(c)** $\dfrac{6}{\sqrt{12}}$

Less work is involved if we simplify the radical in the denominator first.

$$\frac{6}{\sqrt{12}} = \frac{6}{\sqrt{4 \cdot 3}} = \frac{6}{2\sqrt{3}} = \frac{3}{\sqrt{3}}$$

Now we rationalize the denominator by multiplying the numerator and denominator by $\sqrt{3}$.

$$\frac{3 \cdot \sqrt{3}}{\sqrt{3} \cdot \sqrt{3}} = \frac{3\sqrt{3}}{3} = \sqrt{3}$$

**Now Try Exercises 43, 49, and 51.**

### EXAMPLE 3 Rationalizing Denominators in Roots of Fractions

Simplify each radical. In part (b), $p > 0$.

**(a)**
$$\sqrt{\frac{18}{125}} = \frac{\sqrt{18}}{\sqrt{125}} \quad \text{Quotient rule}$$
$$= \frac{\sqrt{9 \cdot 2}}{\sqrt{25 \cdot 5}} \quad \text{Factor.}$$
$$= \frac{3\sqrt{2}}{5\sqrt{5}} \quad \text{Product rule}$$
$$= \frac{3\sqrt{2} \cdot \sqrt{5}}{5\sqrt{5} \cdot \sqrt{5}} \quad \text{Multiply by } \frac{\sqrt{5}}{\sqrt{5}}.$$
$$= \frac{3\sqrt{10}}{5 \cdot 5} \quad \text{Product rule}$$
$$= \frac{3\sqrt{10}}{25}$$

**(b)**
$$\sqrt{\frac{50m^4}{p^5}} = \frac{\sqrt{50m^4}}{\sqrt{p^5}} \quad \text{Quotient rule}$$
$$= \frac{5m^2\sqrt{2}}{p^2\sqrt{p}} \quad \text{Product rule}$$
$$= \frac{5m^2\sqrt{2} \cdot \sqrt{p}}{p^2\sqrt{p} \cdot \sqrt{p}} \quad \text{Multiply by } \frac{\sqrt{p}}{\sqrt{p}}.$$
$$= \frac{5m^2\sqrt{2p}}{p^2 \cdot p} \quad \text{Product rule}$$
$$= \frac{5m^2\sqrt{2p}}{p^3}$$

**Now Try Exercises 55 and 63.**

**EXAMPLE 4 Rationalizing Denominators with Cube and Fourth Roots**

Simplify.

**(a)** $\sqrt[3]{\dfrac{27}{16}}$

Use the quotient rule and simplify the numerator and denominator.

$$\sqrt[3]{\frac{27}{16}} = \frac{\sqrt[3]{27}}{\sqrt[3]{16}} = \frac{3}{\sqrt[3]{8} \cdot \sqrt[3]{2}} = \frac{3}{2\sqrt[3]{2}}$$

To get a rational denominator, multiply the numerator and denominator by a number that will result in a perfect cube in the radicand in the denominator. Since $2 \cdot 4 = 8$, a perfect cube, multiply the numerator and denominator by $\sqrt[3]{4}$.

$$\sqrt[3]{\frac{27}{16}} = \frac{3}{2\sqrt[3]{2}} = \frac{3 \cdot \sqrt[3]{4}}{2\sqrt[3]{2} \cdot \sqrt[3]{4}} = \frac{3\sqrt[3]{4}}{2\sqrt[3]{8}} = \frac{3\sqrt[3]{4}}{2 \cdot 2} = \frac{3\sqrt[3]{4}}{4}$$

**(b)** $\sqrt[4]{\dfrac{5x}{z}} = \dfrac{\sqrt[4]{5x}}{\sqrt[4]{z}} \cdot \dfrac{\sqrt[4]{z^3}}{\sqrt[4]{z^3}} = \dfrac{\sqrt[4]{5xz^3}}{\sqrt[4]{z^4}} = \dfrac{\sqrt[4]{5xz^3}}{z}, \quad x \geq 0, z > 0$

**Now Try Exercises 71 and 81.**

**CAUTION** It is easy to make mistakes in problems like the one in Example 4(a). A typical error is to multiply the numerator and denominator by $\sqrt[3]{2}$, forgetting that

$$\sqrt[3]{2} \cdot \sqrt[3]{2} \neq 2.$$

You need *three* factors of 2 to obtain $2^3$ under the radical. As implied in Example 4(a),

$$\sqrt[3]{2} \cdot \sqrt[3]{2} \cdot \sqrt[3]{2} = 2.$$

**OBJECTIVE 3 Rationalize denominators with binomials involving radicals.** Recall the special product

$$(x + y)(x - y) = x^2 - y^2.$$

To rationalize a denominator that contains a binomial expression (one that contains exactly two terms) involving radicals, such as

$$\frac{3}{1 + \sqrt{2}},$$

we must use *conjugates.* The conjugate of $1 + \sqrt{2}$ is $1 - \sqrt{2}$. In general, $x + y$ and $x - y$ are **conjugates.**

**Rationalizing a Binomial Denominator**

Whenever a radical expression has a sum or difference with square root radicals in the denominator, rationalize the denominator by multiplying both the numerator and denominator by the conjugate of the denominator.

For the expression $\frac{3}{1+\sqrt{2}}$, we rationalize the denominator by multiplying both the numerator and denominator by $1 - \sqrt{2}$, the conjugate of the denominator.

$$\frac{3}{1+\sqrt{2}} = \frac{3(1-\sqrt{2})}{(1+\sqrt{2})(1-\sqrt{2})}$$

Then $(1+\sqrt{2})(1-\sqrt{2}) = 1^2 - (\sqrt{2})^2 = 1 - 2 = -1$. Placing $-1$ in the denominator gives

$$= \frac{3(1-\sqrt{2})}{-1}$$

$$= \frac{3}{-1}(1-\sqrt{2})$$

$$= -3(1-\sqrt{2}) \quad \text{or} \quad -3 + 3\sqrt{2}.$$

## EXAMPLE 5 Rationalizing Binomial Denominators

Rationalize each denominator.

**(a)** $\frac{5}{4-\sqrt{3}}$

To rationalize the denominator, multiply both the numerator and denominator by the conjugate of the denominator, $4 + \sqrt{3}$.

$$\frac{5}{4-\sqrt{3}} = \frac{5(4+\sqrt{3})}{(4-\sqrt{3})(4+\sqrt{3})}$$

$$= \frac{5(4+\sqrt{3})}{16-3}$$

$$= \frac{5(4+\sqrt{3})}{13}$$

Notice that the numerator is left in factored form. This makes it easier to determine whether the expression is written in lowest terms.

**(b)** $\frac{\sqrt{2}-\sqrt{3}}{\sqrt{5}+\sqrt{3}}$

Multiply the numerator and denominator by $\sqrt{5} - \sqrt{3}$ to rationalize the denominator.

$$\frac{\sqrt{2}-\sqrt{3}}{\sqrt{5}+\sqrt{3}} = \frac{(\sqrt{2}-\sqrt{3})(\sqrt{5}-\sqrt{3})}{(\sqrt{5}+\sqrt{3})(\sqrt{5}-\sqrt{3})}$$

$$= \frac{\sqrt{10}-\sqrt{6}-\sqrt{15}+3}{5-3}$$

$$= \frac{\sqrt{10}-\sqrt{6}-\sqrt{15}+3}{2}$$

(c) $$\frac{3}{\sqrt{5m}-\sqrt{p}} = \frac{3(\sqrt{5m}+\sqrt{p})}{(\sqrt{5m}-\sqrt{p})(\sqrt{5m}+\sqrt{p})}$$

$$= \frac{3(\sqrt{5m}+\sqrt{p})}{5m-p}, \quad 5m \neq p, m > 0, p > 0$$

**Now Try Exercises 85, 91, and 95.**

**OBJECTIVE 4 Write radical quotients in lowest terms.**

## EXAMPLE 6 Writing Radical Quotients in Lowest Terms

Write each quotient in lowest terms.

(a) $\frac{6+2\sqrt{5}}{4}$

Factor the numerator and denominator, then write in lowest terms.

$$\frac{6+2\sqrt{5}}{4} = \frac{2(3+\sqrt{5})}{2 \cdot 2} = \frac{3+\sqrt{5}}{2}$$

Here is an alternative method for writing this expression in lowest terms.

$$\frac{6+2\sqrt{5}}{4} = \frac{6}{4} + \frac{2\sqrt{5}}{4} = \frac{3}{2} + \frac{\sqrt{5}}{2} = \frac{3+\sqrt{5}}{2}$$

(b) $$\frac{5y-\sqrt{8y^2}}{6y} = \frac{5y-2y\sqrt{2}}{6y}, \quad y > 0 \qquad \text{Product rule}$$

$$= \frac{y(5-2\sqrt{2})}{6y} \qquad \text{Factor the numerator.}$$

$$= \frac{5-2\sqrt{2}}{6}$$

Note that the final fraction cannot be simplified further because there is no common factor of 2 in the numerator.

**Now Try Exercises 107 and 109.**

**CAUTION** Be careful to factor *before* writing a quotient in lowest terms.

**CONNECTIONS**

In calculus, it is sometimes desirable to rationalize the *numerator* in an expression. The procedure is similar to rationalizing the denominator. For example, to rationalize the numerator of

$$\frac{6-\sqrt{2}}{4},$$

we multiply the numerator and the denominator by the conjugate of the numerator.

(continued)

*Step 2* Now square both sides.

$$\left(\sqrt{5q-1}\right)^2 = (-3)^2$$

*Step 3*

$$5q - 1 = 9$$
$$5q = 10$$
$$q = 2$$

*Step 4* Check the potential solution, 2, by substituting it in the original equation.

*Check:*

$$\sqrt{5q-1} + 3 = 0$$
$$\sqrt{5 \cdot 2 - 1} + 3 \stackrel{?}{=} 0 \qquad \text{Let } q = 2.$$
$$3 + 3 = 0 \qquad \text{False}$$

This false result shows that 2 is *not* a solution of the original equation; it is extraneous. The solution set is $\emptyset$.

**Now Try Exercise 11.**

**NOTE** We could have determined after Step 1 that the equation in Example 2 has no solution because the expression on the left cannot be negative.

**OBJECTIVE 2 Solve radical equations that require additional steps.** The next examples involve finding the square of a binomial. Recall that

$$(x + y)^2 = x^2 + 2xy + y^2.$$

### EXAMPLE 3 Using the Power Rule; Squaring a Binomial

Solve $\sqrt{4-x} = x + 2$.

*Step 1* The radical is alone on the left side of the equation.

*Step 2* Square both sides; the square of $x + 2$ is $(x + 2)^2 = x^2 + 4x + 4$.

$$\left(\sqrt{4-x}\right)^2 = (x+2)^2$$
$$4 - x = x^2 + 4x + 4$$

Twice the product of 2 and $x$

*Step 3* The new equation is quadratic, so get 0 on one side.

$$0 = x^2 + 5x \qquad \text{Subtract 4 and add } x.$$
$$0 = x(x+5) \qquad \text{Factor.}$$
$$x = 0 \quad \text{or} \quad x + 5 = 0 \qquad \text{Zero-factor property}$$
$$x = -5$$

*Step 4* Check each potential solution in the original equation.

*Check:*

| If $x = 0$, then | | If $x = -5$, then | |
|---|---|---|---|
| $\sqrt{4-x} = x + 2$ | | $\sqrt{4-x} = x + 2$ | |
| $\sqrt{4-0} \stackrel{?}{=} 0 + 2$ | | $\sqrt{4-(-5)} \stackrel{?}{=} -5 + 2$ | |
| $\sqrt{4} \stackrel{?}{=} 2$ | | $\sqrt{9} \stackrel{?}{=} -3$ | |
| $2 = 2.$ | True | $3 = -3.$ | False |

The solution set is $\{0\}$. The other potential solution, $-5$, is extraneous.

**Now Try Exercise 27.**

**CAUTION** When a radical equation requires squaring a binomial as in Example 3, remember to include the middle term.

| $(x + 2)^2 \neq x^2 + 4$ | $(x + 2)^2 = x^2 + 4x + 4$ |
|---|---|
| **INCORRECT** | **CORRECT** |

### EXAMPLE 4 Using the Power Rule; Squaring a Binomial

Solve $\sqrt{x^2 - 4x + 9} = x - 1$.

Squaring both sides gives $(x - 1)^2 = x^2 - 2(x)(1) + 1^2$ on the right.

$$\left(\sqrt{x^2 - 4x + 9}\right)^2 = (x - 1)^2$$

$$x^2 - 4x + 9 = x^2 - 2x + 1$$

Twice the product of $x$ and $-1$

Subtract $x^2$ and 1 from each side; then add $4x$ to each side to obtain

$$8 = 2x$$

$$4 = x.$$

Check this potential solution in the original equation.

*Check:*

$$\sqrt{x^2 - 4x + 9} = x - 1$$

$$\sqrt{4^2 - 4 \cdot 4 + 9} \stackrel{?}{=} 4 - 1 \quad \text{Let } x = 4.$$

$$3 = 3 \quad \text{True}$$

The solution set of the original equation is $\{4\}$.

**Now Try Exercise 29.**

### EXAMPLE 5 Using the Power Rule; Squaring Twice

Solve $\sqrt{5x + 6} + \sqrt{3x + 4} = 2$.

Start by getting one radical alone on one side of the equation by subtracting $\sqrt{3x + 4}$ from each side.

$$\sqrt{5x + 6} = 2 - \sqrt{3x + 4}$$

$$\left(\sqrt{5x + 6}\right)^2 = \left(2 - \sqrt{3x + 4}\right)^2 \quad \text{Square both sides.}$$

$$5x + 6 = 4 - 4\sqrt{3x + 4} + (3x + 4)$$

Twice the product of 2 and $-\sqrt{3x + 4}$

This equation still contains a radical, so square both sides again. Before doing this, isolate the radical term on the right.

$$5x + 6 = 8 + 3x - 4\sqrt{3x + 4}$$

$$2x - 2 = -4\sqrt{3x + 4} \quad \text{Subtract 8 and } 3x.$$

$$x - 1 = -2\sqrt{3x + 4} \quad \text{Divide by 2.}$$

$$(x - 1)^2 = \left(-2\sqrt{3x + 4}\right)^2 \quad \text{Square both sides again.}$$

$$x^2 - 2x + 1 = (-2)^2\left(\sqrt{3x + 4}\right)^2 \quad (ab)^2 = a^2b^2$$

$$x^2 - 2x + 1 = 4(3x + 4)$$

$$x^2 - 2x + 1 = 12x + 16 \quad \text{Distributive property}$$

$$x^2 - 14x - 15 = 0 \qquad \text{Standard form}$$
$$(x - 15)(x + 1) = 0 \qquad \text{Factor.}$$
$$x - 15 = 0 \quad \text{or} \quad x + 1 = 0 \qquad \text{Zero-factor property}$$
$$x = 15 \quad \text{or} \quad x = -1$$

Check each of these potential solutions in the original equation. Only $-1$ checks, so the solution set, $\{-1\}$, has only one element.

**Now Try Exercise 51.**

**OBJECTIVE 3 Solve radical equations with indexes greater than 2.** The power rule also works for powers greater than 2.

**EXAMPLE 6 Using the Power Rule for a Power Greater than 2**

Solve $\sqrt[3]{z + 5} = \sqrt[3]{2z - 6}$.

Raise both sides to the third power.

$$\left(\sqrt[3]{z + 5}\right)^3 = \left(\sqrt[3]{2z - 6}\right)^3$$
$$z + 5 = 2z - 6$$
$$11 = z$$

*Check:*
$$\sqrt[3]{z + 5} = \sqrt[3]{2z - 6} \qquad \text{Original equation}$$
$$\sqrt[3]{11 + 5} = \sqrt[3]{2 \cdot 11 - 6} \quad ? \qquad \text{Let } z = 11.$$
$$\sqrt[3]{16} = \sqrt[3]{16} \qquad \text{True}$$

The solution set is $\{11\}$.

**Now Try Exercise 37.**

**OBJECTIVE 4 Solve radical equations using a graphing calculator.** In Example 4 we solved the equation $\sqrt{x^2 - 4x + 9} = x - 1$ using algebraic methods. If we write this equation with one side equal to 0, we get

$$\sqrt{x^2 - 4x + 9} - x + 1 = 0.$$

Using a graphing calculator to graph the function defined by

$$f(x) = \sqrt{x^2 - 4x + 9} - x + 1,$$

we obtain the graph shown in Figure 14. Notice that its zero ($x$-value of the $x$-intercept) is 4, which is the solution we found in Example 4.

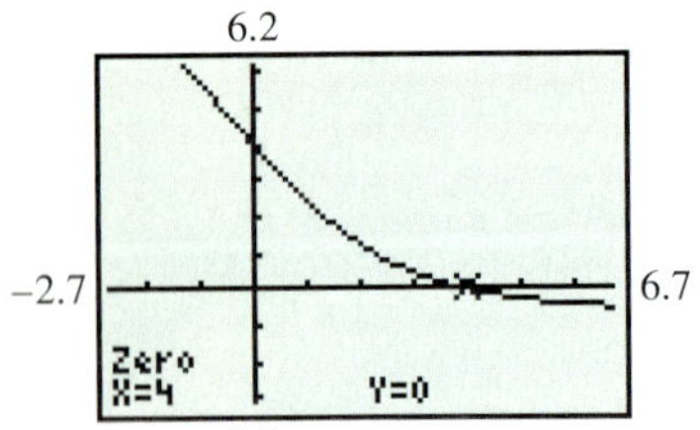


FIGURE 14

In Example 3, we found that the single solution of $\sqrt{4 - x} = x + 2$ is 0, with an extraneous value of $-5$. If we graph $f(x) = \sqrt{4 - x}$ and $g(x) = x + 2$ in the same window, we find that the $x$-coordinate of the point of intersection of the two graphs is 0, which is the solution of the equation. See Figure 15.

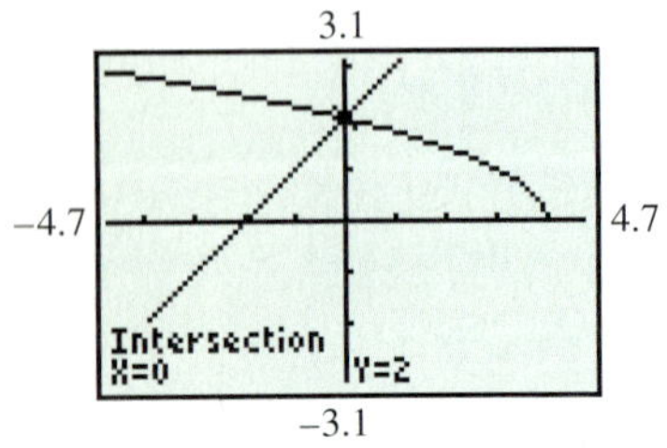


FIGURE 15

We solved the equation in Example 3 by squaring both sides, obtaining $4 - x = x^2 + 4x + 4$. In Figure 16 on the next page, we show that the two functions defined by $f(x) = 4 - x$ and $g(x) = x^2 + 4x + 4$ have two points of intersection. The extraneous value $-5$ that we found in Example 3 shows up as an $x$-value of one of these points of intersection. However, our check showed that $-5$ was not a solution of the *original* equation (before the squaring step). Here we see a graphical interpretation of the extraneous value.

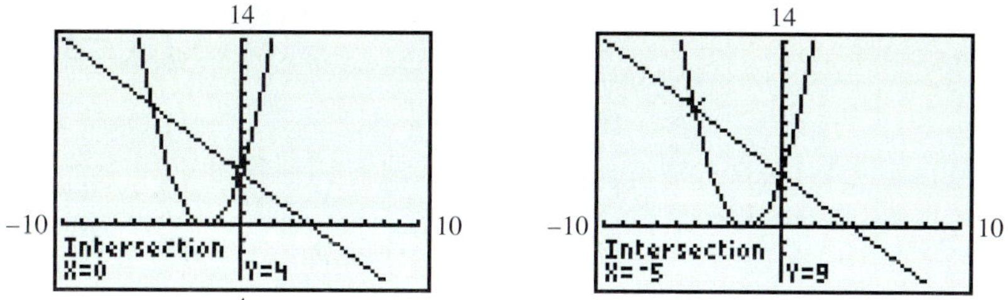


0 is a solution of the original equation.

−5 is *not* a solution of the original equation.

**FIGURE 16**

**OBJECTIVE 5** **Use the power rule to solve a formula for a specified variable.**

### EXAMPLE 7 Solving a Formula from Electronics for a Variable

An important property of a radio frequency transmission line is its *characteristic impedance,* represented by $Z$ and measured in ohms. If $L$ and $C$ are the inductance and capacitance, respectively, per unit of length of the line, then these quantities are related by the formula $Z = \sqrt{\frac{L}{C}}$. Solve this formula for $C$.

$$Z = \sqrt{\frac{L}{C}} \qquad \text{Given formula}$$

$$Z^2 = \frac{L}{C} \qquad \text{Square both sides.}$$

$$CZ^2 = L \qquad \text{Multiply by } C.$$

$$C = \frac{L}{Z^2} \qquad \text{Divide by } Z^2.$$

**Now Try Exercise 67.**

## 10.6 EXERCISES

**For Extra Help**

- Student's Solutions Manual
- MyMathLab
- InterAct Math Tutorial Software
- AW Math Tutor Center
- MathXL
- Digital Video Tutor CD 15/Videotape 15

*Check each equation to see if the given value for x is a solution.*

**1.** $\sqrt{3x + 18} = x$

**(a)** 6 **(b)** −3

**2.** $\sqrt{3x - 3} = x - 1$

**(a)** 1 **(b)** 4

**3.** $\sqrt{x + 2} = \sqrt{9x - 2} - 2\sqrt{x - 1}$

**(a)** 2 **(b)** 7

**4.** $\sqrt{8x - 3} = 2x$

**(a)** $\frac{3}{2}$ **(b)** $\frac{1}{2}$

**5.** Is 9 a solution of the equation $\sqrt{x} = -3$? If not, what is the solution of this equation? Explain.

**6.** Before even attempting to solve $\sqrt{3x + 18} = x$, how can you be sure that the equation cannot have a negative solution?

*Solve each equation. See Examples 1–4.*

**7.** $\sqrt{r-2}=3$

**8.** $\sqrt{q+1}=7$

**9.** $\sqrt{6k-1}=1$

**10.** $\sqrt{7m-3}=5$

**11.** $\sqrt{4r+3}+1=0$

**12.** $\sqrt{5k-3}+2=0$

**13.** $\sqrt{3k+1}-4=0$

**14.** $\sqrt{5z+1}-11=0$

**15.** $4-\sqrt{x-2}=0$

**16.** $9-\sqrt{4k+1}=0$

**17.** $\sqrt{9a-4}=\sqrt{8a+1}$

**18.** $\sqrt{4p-2}=\sqrt{3p+5}$

**19.** $2\sqrt{x}=\sqrt{3x+4}$

**20.** $2\sqrt{m}=\sqrt{5m-16}$

**21.** $3\sqrt{z-1}=2\sqrt{2z+2}$

**22.** $5\sqrt{4a+1}=3\sqrt{10a+25}$

**23.** $k=\sqrt{k^2+4k-20}$

**24.** $p=\sqrt{p^2-3p+18}$

**25.** $a=\sqrt{a^2+3a+9}$

**26.** $z=\sqrt{z^2-4z-8}$

**27.** $\sqrt{9-x}=x+3$

**28.** $\sqrt{5-x}=x+1$

**29.** $\sqrt{k^2+2k+9}=k+3$

**30.** $\sqrt{a^2-3a+3}=a-1$

**31.** $\sqrt{r^2+9r+3}=-r$

**32.** $\sqrt{p^2-15p+15}=p-5$

**33.** $\sqrt{z^2+12z-4}+4-z=0$

**34.** $\sqrt{m^2+3m+12}-m-2=0$

**35.** What is *wrong* with this first step in the solution process for $\sqrt{3x+4}=8-x$. Solve it correctly.

$$3x+4=64+x^2$$

**36.** Explain what is *wrong* with this first step in the solution process for the equation $\sqrt{5x+6}-\sqrt{x+3}=3$. Then solve it correctly.

$$(5x+6)+(x+3)=9$$

*Solve each equation. See Examples 5 and 6.*

**37.** $\sqrt[3]{2x+5}=\sqrt[3]{6x+1}$

**38.** $\sqrt[3]{p-1}=2$

**39.** $\sqrt[3]{a^2+5a+1}=\sqrt[3]{a^2+4a}$

**40.** $\sqrt[3]{r^2+2r+8}=\sqrt[3]{r^2}$

**41.** $\sqrt[3]{2m-1}=\sqrt[3]{m+13}$

**42.** $\sqrt[3]{2k-11}-\sqrt[3]{5k+1}=0$

**43.** $\sqrt[4]{a+8}=\sqrt[4]{2a}$

**44.** $\sqrt[4]{z+11}=\sqrt[4]{2z+6}$

**45.** $\sqrt[3]{x-8}+2=0$

**46.** $\sqrt[3]{r+1}+1=0$

**47.** $\sqrt[4]{2k-5}+4=0$

**48.** $\sqrt[4]{8z-3}+2=0$

**49.** $\sqrt{k+2}-\sqrt{k-3}=1$

**50.** $\sqrt{r+6}-\sqrt{r-2}=2$

**51.** $\sqrt{2r+11}-\sqrt{5r+1}=-1$

**52.** $\sqrt{3x-2}-\sqrt{x+3}=1$

**53.** $\sqrt{3p+4}-\sqrt{2p-4}=2$

**54.** $\sqrt{4x+5}-\sqrt{2x+2}=1$

**55.** $\sqrt{3-3p}-3=\sqrt{3p+2}$

**56.** $\sqrt{4x+7}-4=\sqrt{4x-1}$

**57.** $\sqrt{2\sqrt{x+11}}=\sqrt{4x+2}$

**58.** $\sqrt{1+\sqrt{24-10x}}=\sqrt{3x+5}$

**59.** What is the smallest power to which you can raise both sides of the radical equation $\sqrt{x+3}=\sqrt[3]{5+4x}$ so that the radicals are eliminated?

**60.** What is the smallest power to which you can raise both sides of the radical equation $\sqrt[4]{x+3}=\sqrt[3]{10x+14}$ so that the radicals are eliminated?

61. Use a graphing calculator to solve $\sqrt{3-3x} = 3 + \sqrt{3x+2}$. What is the domain of $y = \sqrt{3-3x} - 3 - \sqrt{3x+2}$?

62. Use a graphing calculator with a window of $[-1, 4]$ by $[-1, 3]$ to solve $\sqrt{2\sqrt{7x+2}} = \sqrt{3x+2}$. What is the domain of $f(x) = \sqrt{2\sqrt{7x+2}} - \sqrt{3x+2}$?

*For each equation, rewrite the expressions with rational exponents as radical expressions, and then solve using the procedures explained in this section.*

63. $(2x-9)^{1/2} = 2 + (x-8)^{1/2}$

64. $(3w+7)^{1/2} = 1 + (w+2)^{1/2}$

65. $(2w-1)^{2/3} - w^{1/3} = 0$

66. $(x^2-2x)^{1/3} - x^{1/3} = 0$

*Solve each formula from electricity and radio for the indicated variable. See Example 7. (*Source: *Cooke, Nelson M., and Joseph B. Orleans,* Mathematics Essential to Electricity and Radio, *McGraw-Hill, 1943.)*

67. $V = \sqrt{\dfrac{2K}{m}}$ for $K$

68. $V = \sqrt{\dfrac{2K}{m}}$ for $m$

69. $f = \dfrac{1}{2\pi\sqrt{LC}}$ for $L$

70. $r = \sqrt{\dfrac{Mm}{F}}$ for $F$

71. A number of useful formulas involve radicals or radical expressions. Many occur in the mathematics needed for working with objects in space. The formula

$$N = \frac{1}{2\pi}\sqrt{\frac{a}{r}}$$

is used to find the rotational rate $N$ of a space station. Here $a$ is the acceleration and $r$ represents the radius of the space station in meters. To find the value of $r$ that will make $N$ simulate the effect of gravity on Earth, the equation must be solved for $r$, using the required value of $N$. (*Source:* Kastner, Bernice, *Space Mathematics,* NASA, 1972.)

**(a)** Solve the equation for $r$.

**(b)** Find the value of $r$ that makes $N = .063$ rotation per sec, if $a = 9.8$ m per sec$^2$.

**(c)** Find the value of $r$ that makes $N = .04$ rotation per sec, if $a = 9.8$ m per sec$^2$.

*If x is the number of years since 1900, the equation $y = x^{.7}$ approximates the timber grown in the United States in billions of cubic feet. Let $x = 20$ represent 1920, $x = 52$ represent 1952, and so on.*

72. Replace $x$ in the equation for each year shown in the graph and use a calculator to find the value of $y$. (Round answers to the nearest billion.)

73. Use the graph to estimate the amount of timber grown for each year shown.

74. Compare the values found from the equation with your estimates from the graph. Does the equation give a good approximation of the data from the graph? In which year is the approximation best?

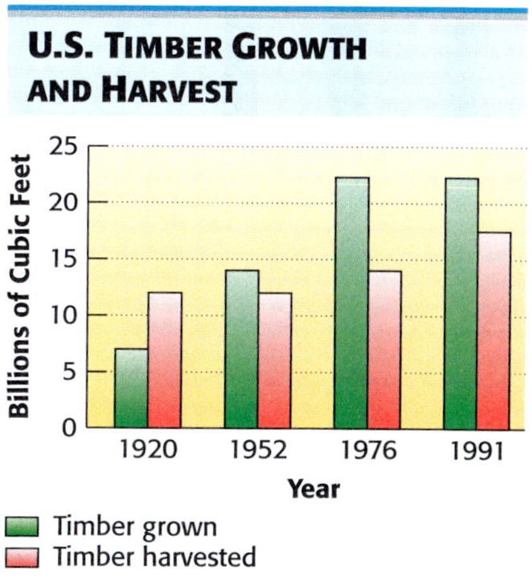


***Source:*** Figures from U.S. Forest Service. Adapted from *"The Truth about America's Forests,"* Evergreen, 4025 Crater Lake Hwy., Medford, Ore. 97504.

**75.** From the graph, estimate the amount of timber harvested in each year shown.

**76.** Use the equation $y = x^{.62}$ and a calculator to approximate the amount of timber harvested in each of the given years. (Round answers to the nearest billion.)

**77.** Compare your answers from Exercises 75 and 76. Does the equation give a good approximation? For which year is it poorest?

# 10.7 Complex Numbers

**OBJECTIVES**

1. Simplify numbers of the form $\sqrt{-b}$, where $b > 0$.
2. Recognize imaginary complex numbers.
3. Add and subtract complex numbers.
4. Multiply complex numbers.
5. Divide complex numbers.
6. Find powers of $i$.

As we saw in Chapter 1, the set of real numbers includes many other number sets (the rational numbers, integers, and natural numbers, for example). In this section a new set of numbers is introduced that includes the set of real numbers, as well as numbers that are even roots of negative numbers, like $\sqrt{-2}$.

**OBJECTIVE 1 Simplify numbers of the form $\sqrt{-b}$, where $b > 0$.** The equation $x^2 + 1 = 0$ has no real number solution since any solution must be a number whose square is $-1$. In the set of real numbers, all squares are nonnegative numbers because the product of two positive numbers or two negative numbers is positive and $0^2 = 0$. To provide a solution for the equation $x^2 + 1 = 0$, a new number $i$ is defined so that

$$i^2 = -1.$$

That is, $i$ is a number whose square is $-1$, so $i = \sqrt{-1}$. This definition of $i$ makes it possible to define any square root of a negative real number as follows.

**$\sqrt{-b}$**

For any positive real number $b$,

$$\sqrt{-b} = i\sqrt{b}.$$

**EXAMPLE 1 Simplifying Square Roots of Negative Numbers**

Write each number as a product of a real number and $i$.

**(a)** $\sqrt{-100} = i\sqrt{100} = 10i$ **(b)** $-\sqrt{-36} = -i\sqrt{36} = -6i$

**(c)** $\sqrt{-2} = i\sqrt{2}$

**Now Try Exercises 7, 9, and 11.**

**CAUTION** It is easy to mistake $\sqrt{2}i$ for $\sqrt{2i}$, with the $i$ under the radical. For this reason, we usually write $\sqrt{2}i$ as $i\sqrt{2}$, as in the definition of $\sqrt{-b}$.

When finding a product such as $\sqrt{-4} \cdot \sqrt{-9}$, we cannot use the product rule for radicals because it applies only to nonnegative radicands. For this reason, we change $\sqrt{-b}$ to the form $i\sqrt{b}$ before performing any multiplications or divisions. For example,

$$\begin{aligned}\sqrt{-4} \cdot \sqrt{-9} &= i\sqrt{4} \cdot i\sqrt{9} \\ &= i \cdot 2 \cdot i \cdot 3 \\ &= 6i^2 \\ &= 6(-1) \qquad \text{Substitute: } i^2 = -1. \\ &= -6.\end{aligned}$$

**CAUTION** Using the product rule for radicals *before* using the definition of $\sqrt{-b}$ gives a *wrong* answer. The preceding example shows that

$$\sqrt{-4} \cdot \sqrt{-9} = -6,$$

but

$$\sqrt{-4(-9)} = \sqrt{36} = 6,$$

so

$$\sqrt{-4} \cdot \sqrt{-9} \neq \sqrt{-4(-9)}.$$

**EXAMPLE 2 Multiplying Square Roots of Negative Numbers**

Multiply.

**(a)** $$\begin{aligned}\sqrt{-3} \cdot \sqrt{-7} &= i\sqrt{3} \cdot i\sqrt{7} \\ &= i^2\sqrt{3 \cdot 7} \\ &= (-1)\sqrt{21} \qquad \text{Substitute: } i^2 = -1. \\ &= -\sqrt{21}\end{aligned}$$

**(b)** $$\begin{aligned}\sqrt{-2} \cdot \sqrt{-8} &= i\sqrt{2} \cdot i\sqrt{8} \\ &= i^2\sqrt{2 \cdot 8} \\ &= (-1)\sqrt{16} \\ &= (-1)4 \\ &= -4\end{aligned}$$

**(c)** $\sqrt{-5} \cdot \sqrt{6} = i\sqrt{5} \cdot \sqrt{6} = i\sqrt{30}$

**Now Try Exercises 15, 17, and 19.**

The methods used to find products also apply to quotients.

**EXAMPLE 3 Dividing Square Roots of Negative Numbers**

Divide.

**(a)** $\dfrac{\sqrt{-75}}{\sqrt{-3}} = \dfrac{i\sqrt{75}}{i\sqrt{3}} = \sqrt{\dfrac{75}{3}} = \sqrt{25} = 5$

**(b)** $\dfrac{\sqrt{-32}}{\sqrt{8}} = \dfrac{i\sqrt{32}}{\sqrt{8}} = i\sqrt{\dfrac{32}{8}} = i\sqrt{4} = 2i$

**Now Try Exercises 21 and 23.**

**OBJECTIVE 2 Recognize imaginary complex numbers.** With the imaginary number $i$ and the real numbers, a new set of numbers can be formed that includes the real numbers as a subset. The *complex numbers* are defined as follows.

**Complex Number**

If $a$ and $b$ are real numbers, then any number of the form $\boldsymbol{a + bi}$ is called a **complex number.**

In the complex number $a + bi$, the number $a$ is called the **real part** and $b$ is called the **imaginary part.** When $b = 0$, $a + bi$ is a real number, so the real numbers are a subset of the complex numbers. Complex numbers with $b \neq 0$ are called **imaginary numbers.*** In spite of their name, imaginary numbers are very useful in applications, particularly in work with electricity.

The relationships among the various sets of numbers discussed in this book are shown in Figure 17.

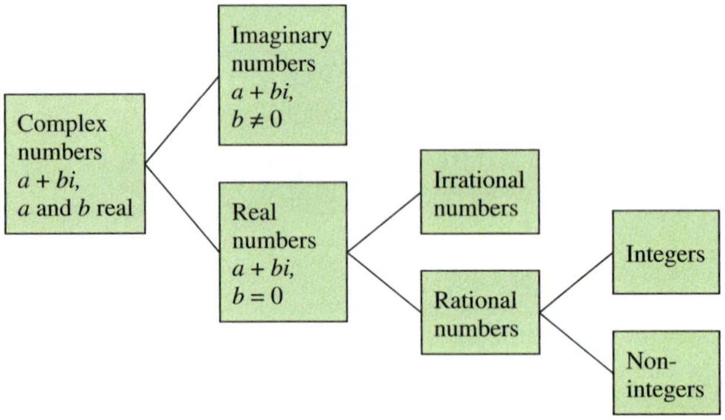


FIGURE 17

**OBJECTIVE 3 Add and subtract complex numbers.** The commutative, associative, and distributive properties for real numbers are also valid for complex numbers. Thus, to add complex numbers, we add their real parts and add their imaginary parts.

### EXAMPLE 4 Adding Complex Numbers

Add.

**(a)** $(2 + 3i) + (6 + 4i)$

$= (2 + 6) + (3 + 4)i$ Commutative, associative, and distributive properties

$= 8 + 7i$

**(b)** $(4 + 2i) + (3 - i) + (-6 + 3i) = [4 + 3 + (-6)] + [2 + (-1) + 3]i$

$= 1 + 4i$

**Now Try Exercises 27 and 35.**

*Some texts define $bi$ as the imaginary part of the complex number $a + bi$. Also, imaginary numbers are sometimes defined as complex numbers with $a = 0$ and $b \neq 0$.

We subtract complex numbers by subtracting their real parts and subtracting their imaginary parts.

### EXAMPLE 5 Subtracting Complex Numbers

Subtract.

**(a)** $(6 + 5i) - (3 + 2i) = (6 - 3) + (5 - 2)i$
$= 3 + 3i$

**(b)** $(7 - 3i) - (8 - 6i) = (7 - 8) + [-3 - (-6)]i$
$= -1 + 3i$

**(c)** $(-9 + 4i) - (-9 + 8i) = (-9 + 9) + (4 - 8)i$
$= 0 - 4i$
$= -4i$

**Now Try Exercises 31 and 33.**

In Example 5(c), the answer was written as $0 - 4i$ and then as just $-4i$. A complex number written in the form $a + bi$, like $0 - 4i$, is in **standard form.** In this section, most answers will be given in standard form, but if $a$ or $b$ is 0, we consider answers such as $a$ or $bi$ to be in standard form.

**OBJECTIVE 4 Multiply complex numbers.** We multiply complex numbers as we multiply polynomials. Complex numbers of the form $a + bi$ have the same form as binomials, so we multiply two complex numbers in standard form by using the FOIL method for multiplying binomials. (Recall that FOIL stands for *First, Outer, Inner, Last.*)

### EXAMPLE 6 Multiplying Complex Numbers

Multiply.

**(a)** $4i(2 + 3i) = 4i(2) + 4i(3i)$ Distributive property
$= 8i + 12i^2$
$= 8i + 12(-1)$ Substitute: $i^2 = -1$.
$= -12 + 8i$

**(b)** $(3 + 5i)(4 - 2i)$

Use the FOIL method.

$$(3 + 5i)(4 - 2i) = \underbrace{3(4)}_{\text{First}} + \underbrace{3(-2i)}_{\text{Outer}} + \underbrace{5i(4)}_{\text{Inner}} + \underbrace{5i(-2i)}_{\text{Last}}$$

$= 12 - 6i + 20i - 10i^2$
$= 12 + 14i - 10(-1)$ Substitute: $i^2 = -1$.
$= 12 + 14i + 10$
$= 22 + 14i$

**(c)** $(2 + 3i)(1 - 5i) = 2(1) + 2(-5i) + 3i(1) + 3i(-5i)$    FOIL

$= 2 - 10i + 3i - 15i^2$

$= 2 - 7i - 15(-1)$

$= 2 - 7i + 15$

$= 17 - 7i$

**Now Try Exercises 45 and 47.**

The two complex numbers $a + bi$ and $a - bi$ are called **complex conjugates,** or simply *conjugates,* of each other. The product of a complex number and its conjugate is always a real number, as shown here.

$$(a + bi)(a - bi) = a^2 - abi + abi - b^2i^2$$
$$= a^2 - b^2(-1)$$
$$(a + bi)(a - bi) = a^2 + b^2$$

For example, $(3 + 7i)(3 - 7i) = 3^2 + 7^2 = 9 + 49 = 58$.

**OBJECTIVE 5 Divide complex numbers.** The quotient of two complex numbers should be a complex number. To write the quotient as a complex number, we need to eliminate $i$ in the denominator. We use conjugates to do this.

## EXAMPLE 7 Dividing Complex Numbers

Find each quotient.

**(a)** $\frac{8 + 9i}{5 + 2i}$

Multiply both the numerator and denominator by the conjugate of the denominator. The conjugate of $5 + 2i$ is $5 - 2i$.

$$\frac{8 + 9i}{5 + 2i} = \frac{(8 + 9i)(5 - 2i)}{(5 + 2i)(5 - 2i)} \qquad \frac{5 - 2i}{5 - 2i} = 1$$

$$= \frac{40 - 16i + 45i - 18i^2}{5^2 + 2^2}$$

$$= \frac{58 + 29i}{29} \qquad \text{Substitute: } i^2 = -1; \text{ combine terms.}$$

$$= \frac{29(2 + i)}{29} \qquad \text{Factor the numerator.}$$

$$= 2 + i \qquad \text{Lowest terms}$$

Notice that this is just like rationalizing a denominator. The final result is in standard form.

**(b)** $\dfrac{1+i}{i}$

The conjugate of $i$ is $-i$. Multiply both the numerator and denominator by $-i$.

$$\begin{aligned}\frac{1+i}{i} &= \frac{(1+i)(-i)}{i(-i)}\\ &= \frac{-i-i^2}{-i^2}\\ &= \frac{-i-(-1)}{-(-1)} \qquad \text{Substitute: } i^2=-1.\\ &= \frac{-i+1}{1}\\ &= 1-i\end{aligned}$$

**Now Try Exercises 61 and 67.**

In Examples 4–7, we showed how complex numbers can be added, subtracted, multiplied, and divided using algebraic methods. Many current models of graphing calculators can perform these operations. Figure 18 shows how the computations in parts of Examples 4–7 are carried out by a TI-83 Plus calculator. It is important to use parentheses as shown.

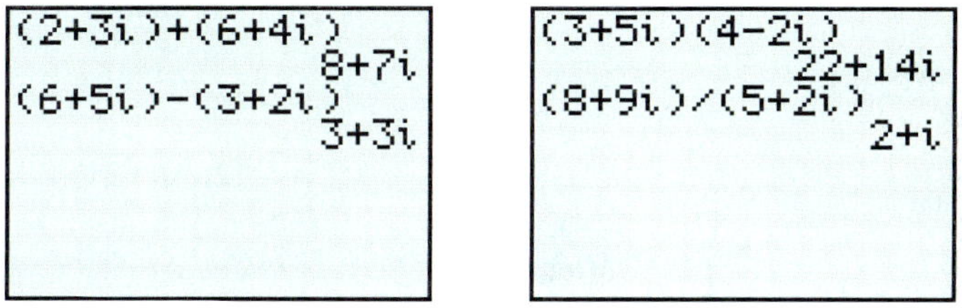


FIGURE 18

**OBJECTIVE 6 Find powers of *i*.** Because $i^2$ is defined to be $-1$, we can find higher powers of $i$ as shown in the following examples.

$$\begin{aligned}i^3 &= i\cdot i^2 = i(-1) = -i & i^6 &= i^2\cdot i^4 = (-1)\cdot 1 = -1\\ i^4 &= i^2\cdot i^2 = (-1)(-1) = 1 & i^7 &= i^3\cdot i^4 = (-i)\cdot 1 = -i\\ i^5 &= i\cdot i^4 = i\cdot 1 = i & i^8 &= i^4\cdot i^4 = 1\cdot 1 = 1\end{aligned}$$

As these examples suggest, the powers of $i$ rotate through the four numbers $i$, $-1$, $-i$, and 1. Larger powers of $i$ can be simplified by using the fact that $i^4 = 1$. For example,

$$i^{75} = (i^4)^{18}\cdot i^3 = 1^{18}\cdot i^3 = 1\cdot i^3 = i^3 = -i.$$

This example suggests a quick method for simplifying larger powers of $i$.

## EXAMPLE 8 Simplifying Powers of *i*

Find each power of $i$.

**(a)** $i^{12} = (i^4)^3 = 1^3 = 1$

**(b)** $i^{39} = i^{36}\cdot i^3 = (i^4)^9\cdot i^3 = 1^9\cdot(-i) = -i$

**(c)** $i^{-2} = \frac{1}{i^2} = \frac{1}{-1} = -1$

**(d)** $i^{-1} = \frac{1}{i}$

To simplify this quotient, multiply both the numerator and denominator by $-i$, the conjugate of $i$.

$$\frac{1}{i} = \frac{1(-i)}{i(-i)} = \frac{-i}{-i^2} = \frac{-i}{-(-1)} = \frac{-i}{1} = -i$$

**Now Try Exercises 73 and 81.**

## 10.7 EXERCISES

**For Extra Help**

- Student's Solutions Manual
- MyMathLab
- InterAct Math Tutorial Software
- AW Math Tutor Center
- MathXL
- Digital Video Tutor CD 15/Videotape 15

*Decide whether each expression is equal to 1, −1, i, or −i.*

**1.** $\sqrt{-1}$ **2.** $-\sqrt{-1}$ **3.** $i^2$ **4.** $-i^2$ **5.** $\frac{1}{i}$ **6.** $(-i)^2$

*Write each number as a product of a real number and i. Simplify all radical expressions. See Example 1.*

**7.** $\sqrt{-169}$ **8.** $\sqrt{-225}$ **9.** $-\sqrt{-144}$ **10.** $-\sqrt{-196}$

**11.** $\sqrt{-5}$ **12.** $\sqrt{-21}$ **13.** $\sqrt{-48}$ **14.** $\sqrt{-96}$

*Multiply or divide as indicated. See Examples 2 and 3.*

**15.** $\sqrt{-7} \cdot \sqrt{-15}$ **16.** $\sqrt{-3} \cdot \sqrt{-19}$ **17.** $\sqrt{-4} \cdot \sqrt{-25}$ **18.** $\sqrt{-9} \cdot \sqrt{-81}$

**19.** $\sqrt{-3} \cdot \sqrt{11}$ **20.** $\sqrt{-10} \cdot \sqrt{2}$ **21.** $\frac{\sqrt{-300}}{\sqrt{-100}}$ **22.** $\frac{\sqrt{-40}}{\sqrt{-10}}$

**23.** $\frac{\sqrt{-75}}{\sqrt{3}}$ **24.** $\frac{\sqrt{-160}}{\sqrt{10}}$

**25. (a)** Every real number is a complex number. Explain why this is so.
**(b)** Not every complex number is a real number. Give an example of this and explain why this statement is true.

**26.** Explain how to add, subtract, multiply, and divide complex numbers. Give examples.

*Add or subtract as indicated. Write your answers in the form a + bi. See Examples 4 and 5.*

**27.** $(3 + 2i) + (-4 + 5i)$ **28.** $(7 + 15i) + (-11 + 14i)$

**29.** $(5 - i) + (-5 + i)$ **30.** $(-2 + 6i) + (2 - 6i)$

**31.** $(4 + i) - (-3 - 2i)$ **32.** $(9 + i) - (3 + 2i)$

**33.** $(-3 - 4i) - (-1 - 4i)$ **34.** $(-2 - 3i) - (-5 - 3i)$

**35.** $(-4 + 11i) + (-2 - 4i) + (7 + 6i)$ **36.** $(-1 + i) + (2 + 5i) + (3 + 2i)$

**37.** $[(7 + 3i) - (4 - 2i)] + (3 + i)$ **38.** $[(7 + 2i) + (-4 - i)] - (2 + 5i)$

**39.** Fill in the blank with the correct response:

Because $(4 + 2i) - (3 + i) = 1 + i$, using the definition of subtraction, we can check this to find that $(1 + i) + (3 + i) =$ ______________.

**40.** Fill in the blank with the correct response:

Because $\dfrac{-5}{2 - i} = -2 - i$, using the definition of division, we can check this to find that $(-2 - i)(2 - i) =$ ______________.

*Multiply. See Example 6.*

| | | |
|---|---|---|
| **41.** $(3i)(27i)$ | **42.** $(5i)(125i)$ | **43.** $(-8i)(-2i)$ |
| **44.** $(-32i)(-2i)$ | **45.** $5i(-6 + 2i)$ | **46.** $3i(4 + 9i)$ |
| **47.** $(4 + 3i)(1 - 2i)$ | **48.** $(7 - 2i)(3 + i)$ | **49.** $(4 + 5i)^2$ |
| **50.** $(3 + 2i)^2$ | **51.** $2i(-4 - i)^2$ | **52.** $3i(-3 - i)^2$ |
| **53.** $(12 + 3i)(12 - 3i)$ | **54.** $(6 + 7i)(6 - 7i)$ | **55.** $(4 + 9i)(4 - 9i)$ |

**56.** $(7 + 2i)(7 - 2i)$

**57.** What is the conjugate of $a + bi$?

**58.** If we multiply $a + bi$ by its conjugate, we get ______________, which is always a real number.

*Write each expression in standard form $a + bi$. See Example 7.*

| | | | |
|---|---|---|---|
| **59.** $\dfrac{2}{1 - i}$ | **60.** $\dfrac{29}{5 + 2i}$ | **61.** $\dfrac{-7 + 4i}{3 + 2i}$ | **62.** $\dfrac{-38 - 8i}{7 + 3i}$ |
| **63.** $\dfrac{8i}{2 + 2i}$ | **64.** $\dfrac{-8i}{1 + i}$ | **65.** $\dfrac{2 - 3i}{2 + 3i}$ | **66.** $\dfrac{-1 + 5i}{3 + 2i}$ |
| **67.** $\dfrac{3 + i}{i}$ | **68.** $\dfrac{5 - i}{-i}$ | | |

## TECHNOLOGY INSIGHTS (EXERCISES 69–70)

*Predict the answer that the calculator screen will provide for the given complex number operation entry.*

**69.**
```
(8-5i)+2(11+6i)-
(1+i)²
```

**70.**
```
(4+3i)(2-6i)-2i(
5+7i)
```

**71.** Recall that if $a \neq 0$, then $\frac{1}{a}$ is called the reciprocal of $a$. Use this definition to express the reciprocal of $5 - 4i$ in the form $a + bi$.

**72.** Recall that if $a \neq 0$, then $a^{-1}$ is defined to be $\frac{1}{a}$. Use this definition to express $(4 - 3i)^{-1}$ in the form $a + bi$.

*Find each power of i. See Example 8.*

**73.** $i^{18}$ **74.** $i^{26}$ **75.** $i^{89}$ **76.** $i^{48}$ **77.** $i^{38}$

**78.** $i^{102}$ **79.** $i^{43}$ **80.** $i^{83}$ **81.** $i^{-5}$ **82.** $i^{-17}$

**83.** A student simplified $i^{-18}$ as follows:

$$i^{-18} = i^{-18} \cdot i^{20} = i^{-18+20} = i^2 = -1.$$

Explain the mathematical justification for this correct work.

**84.** Explain why

$$(46 + 25i)(3 - 6i) \quad \text{and} \quad (46 + 25i)(3 - 6i)i^{12}$$

must be equal. (Do not actually perform the computation.)

*Ohm's law for the current I in a circuit with voltage E, resistance R, capacitance reactance $X_c$, and inductive reactance $X_L$ is*

$$I = \frac{E}{R + (X_L - X_c)i}.$$

*Use this law to work Exercises 85 and 86.*

**85.** Find $I$ if $E = 2 + 3i$, $R = 5$, $X_L = 4$, and $X_c = 3$.

**86.** Find $E$ if $I = 1 - i$, $R = 2$, $X_L = 3$, and $X_c = 1$.

*Complex numbers will appear again in this book in Chapter 11, when we study quadratic equations. The following exercises examine how a complex number can be a solution of a quadratic equation.*

**87.** Show that $1 + 5i$ is a solution of $x^2 - 2x + 26 = 0$. Then show that its conjugate is also a solution.

**88.** Show that $3 + 2i$ is a solution of $x^2 - 6x + 13 = 0$. Then show that its conjugate is also a solution.

## RELATING CONCEPTS (EXERCISES 89–94)

**For Individual or Group Work**

*Consider the following expressions:*

| Binomials | Complex Numbers |
|---|---|
| $x + 2, \quad 3x - 1$ | $1 + 2i, \quad 3 - i$ |

*When we add, subtract, or multiply complex numbers in standard form, the rules are the same as those for the corresponding operations on binomials. That is, we add or subtract like terms, and we use FOIL to multiply. Division, however, is comparable to division by the sum or difference of radicals, where we multiply by the conjugate to get a rational denominator. To express the quotient of two complex numbers in standard form, we also multiply by the conjugate of the denominator.* ***Work Exercises 89–94 in order,*** *to better understand these ideas.*

**89. (a)** Add the two binomials. **(b)** Add the two complex numbers.

**90.** **(a)** Subtract the second binomial from the first.
**(b)** Subtract the second complex number from the first.

**91.** **(a)** Multiply the two binomials.
**(b)** Multiply the two complex numbers.

**92.** **(a)** Rationalize the denominator: $\frac{\sqrt{3}-1}{1+\sqrt{2}}$.
**(b)** Write in standard form: $\frac{3-i}{1+2i}$.

**93.** Explain why the answers for (a) and (b) in Exercise 91 do not correspond as the answers in Exercises 89–90 do.

**94.** Explain why the answers for (a) and (b) in Exercise 92 do not correspond as the answers in Exercises 89–90 do.

*Perform the indicated operations. Give answers in standard form.*

**95.** $\frac{3}{2-i}+\frac{5}{1+i}$

**96.** $\frac{2}{3+4i}+\frac{4}{1-i}$

**97.** $\left(\frac{2+i}{2-i}+\frac{i}{1+i}\right)i$

**98.** $\left(\frac{4-i}{1+i}-\frac{2i}{2+i}\right)4i$

## Chapter 10 Group Activity

### Solar Electricity

**OBJECTIVE** Apply the Pythagorean formula.

In this activity you will determine the sizes of frames needed to support solar electric panels on a flat roof.

**A.** The following table gives three different solar modules by Solarex. Have each member of the group choose one of the solar panels.

| *Model* | *Watts* | *Volts* | *Amps* | *Size (in inches)* | *Cost (in dollars)* |
|---|---|---|---|---|---|
| MSX-77 | 77 | 16.9 | 4.56 | $44 \times 26$ | 475 |
| MSX-83 | 83 | 17.1 | 4.85 | $44 \times 24$ | 490 |
| MSX-60 | 60 | 17.1 | 3.5 | $44 \times 20$ | 382 |

*Source:* Solarex table in *Jade Mountain* catalog.

**B.** To use your solar panel, you must make a wooden frame to support it. The sides of this frame will form a right triangle. The hypotenuse of the triangle will be the width of the solar panel you chose. Make a sketch and use the Pythagorean formula to find the dimensions of the legs for each frame given the following conditions. Round answers to the nearest tenth.

1. The legs have equal length.
2. One leg is twice the length of the other.
3. One leg is 3 times the length of the other.

**C.** Compare the different frame sizes for each panel. What factors might determine which of the triangles you would use in your frame?

# CHAPTER 10 SUMMARY

## KEY TERMS

**10.1** square root, principal square root, radicand, radical, perfect square, cube root, fourth root, index (order), radical expression, square root function, cube root function

**10.3** simplified radical

**10.5** rationalizing the denominator, conjugates

**10.6** radical equation, extraneous solution

**10.7** complex number, real part, imaginary part, imaginary numbers, standard form, complex conjugates

## NEW SYMBOLS

$\sqrt{\phantom{a}}$ radical sign

$\sqrt[3]{a}$ cube root of $a$

$\sqrt[4]{a}$ principal fourth root of $a$

$\sqrt[n]{a}$ principal $n$th root of $a$

$\approx$ is approximately equal to

$a^{1/n}$ $a$ to the power $\frac{1}{n}$

$a^{m/n}$ $a$ to the power $\frac{m}{n}$

$i$ a number whose square is $-1$

## TEST YOUR WORD POWER

*See how well you have learned the vocabulary in this chapter. Answers, with examples, follow the Quick Review.*

**1.** A **radicand** is
**A.** the index of a radical
**B.** the number or expression under the radical sign
**C.** the positive root of a number
**D.** the radical sign.

**2.** The **Pythagorean formula** states that, in a right triangle,
**A.** the sum of the measures of the angles is 180°
**B.** the sum of the lengths of the two shorter sides equals the length of the longest side
**C.** the longest side is opposite the right angle
**D.** the square of the length of the longest side equals the sum of the squares of the lengths of the two shorter sides.

**3.** A **hypotenuse** is
**A.** either of the two shorter sides of a triangle
**B.** the shortest side of a triangle
**C.** the side opposite the right angle in a right triangle
**D.** the longest side in any triangle.

**4.** **Rationalizing the denominator** is the process of
**A.** eliminating fractions from a radical expression
**B.** changing the denominator of a fraction from a radical to a rational number
**C.** clearing a radical expression of radicals
**D.** multiplying radical expressions.

**5.** An **extraneous solution** is a solution
**A.** that does not satisfy the original equation
**B.** that makes an equation true
**C.** that makes an expression equal 0
**D.** that checks in the original equation.

**6.** A **complex number** is
**A.** a real number that includes a complex fraction
**B.** a zero multiple of $i$
**C.** a number of the form $a + bi$, where $a$ and $b$ are real numbers
**D.** the square root of $-1$.

## QUICK REVIEW

| CONCEPTS | EXAMPLES |
|---|---|
| **10.1 RADICAL EXPRESSIONS AND GRAPHS** | |
| $\sqrt[n]{a} = b$ means $b^n = a$.<br>$\sqrt[n]{a}$ is the principal $n$th root of $a$.<br>$\sqrt[n]{a^n} = \lvert a \rvert$ if $n$ is even.<br>$\sqrt[n]{a^n} = a$ if $n$ is odd. | The two square roots of 64 are $\sqrt{64} = 8$, the principal square root, and $-\sqrt{64} = -8$.<br>$\sqrt[4]{(-2)^4} = \lvert -2 \rvert = 2$<br>$\sqrt[3]{-27} = -3$ |

(continued)

| CONCEPTS | EXAMPLES |
|---|---|

**Functions Defined by Radical Expressions**
The square root function with $f(x) = \sqrt{x}$ and the cube root function with $f(x) = \sqrt[3]{x}$ are two important functions defined by radical expressions.

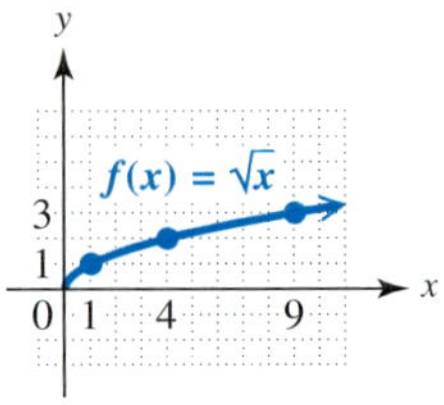


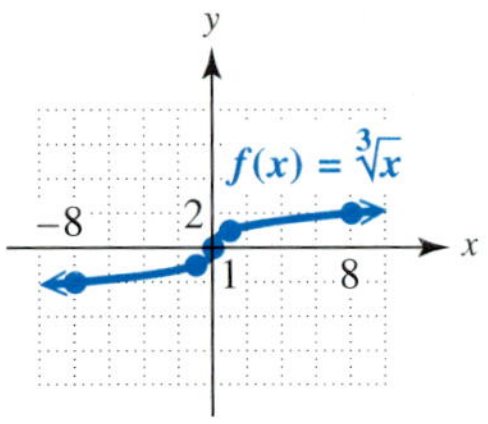


## 10.2 RATIONAL EXPONENTS

$a^{1/n} = \sqrt[n]{a}$ whenever $\sqrt[n]{a}$ exists.

If $m$ and $n$ are positive integers with $\frac{m}{n}$ in lowest terms, then $a^{m/n} = (a^{1/n})^m$, provided that $a^{1/n}$ is a real number.

All of the usual definitions and rules for exponents are valid for rational exponents.

$$81^{1/2} = \sqrt{81} = 9 \qquad -64^{1/3} = -\sqrt[3]{64} = -4$$

$$8^{5/3} = (8^{1/3})^5 = 2^5 = 32$$

$$5^{-1/2} \cdot 5^{1/4} = 5^{-1/2+1/4} = 5^{-1/4} = \frac{1}{5^{1/4}} \qquad (y^{2/5})^{10} = y^4$$

$$\frac{x^{-1/3}}{x^{-1/2}} = x^{-1/3-(-1/2)} = x^{-1/3+1/2} = x^{1/6}, \quad x > 0$$

## 10.3 SIMPLIFYING RADICAL EXPRESSIONS

**Product and Quotient Rules for Radicals**
If $\sqrt[n]{a}$ and $\sqrt[n]{b}$ are real numbers and $n$ is a natural number, then

$$\sqrt[n]{a} \cdot \sqrt[n]{b} = \sqrt[n]{ab}$$

and

$$\sqrt[n]{\frac{a}{b}} = \frac{\sqrt[n]{a}}{\sqrt[n]{b}}, \qquad b \neq 0.$$

$$\sqrt{3} \cdot \sqrt{7} = \sqrt{21}$$

$$\sqrt[5]{x^3y} \cdot \sqrt[5]{xy^2} = \sqrt[5]{x^4y^3}$$

$$\frac{\sqrt{x^5}}{\sqrt{x^4}} = \sqrt{\frac{x^5}{x^4}} = \sqrt{x}, \quad x > 0$$

**Simplified Radical**
1. The radicand has no factor raised to a power greater than or equal to the index.
2. The radicand has no fractions.
3. No denominator contains a radical.
4. Exponents in the radicand and the index of the radical have no common factor (except 1).

$$\sqrt{18} = \sqrt{9 \cdot 2} = 3\sqrt{2}$$

$$\sqrt[3]{54x^5y^3} = \sqrt[3]{27x^3y^3 \cdot 2x^2} = 3xy\sqrt[3]{2x^2}$$

$$\sqrt{\frac{7}{4}} = \frac{\sqrt{7}}{\sqrt{4}} = \frac{\sqrt{7}}{2}$$

$$\sqrt[9]{x^3} = x^{3/9} = x^{1/3} \quad \text{or} \quad \sqrt[3]{x}$$

**Pythagorean Formula**
If $c$ is the length of the longest side of a right triangle and $a$ and $b$ are the lengths of the shorter sides, then $c^2 = a^2 + b^2$. The longest side is the hypotenuse and the two shorter sides are the legs of the triangle. The hypotenuse is opposite the right angle.

Find $b$ for the triangle in the figure.

$$10^2 + b^2 = (2\sqrt{61})^2$$

$$b^2 = 4(61) - 100$$

$$b^2 = 144$$

$$b = 12$$

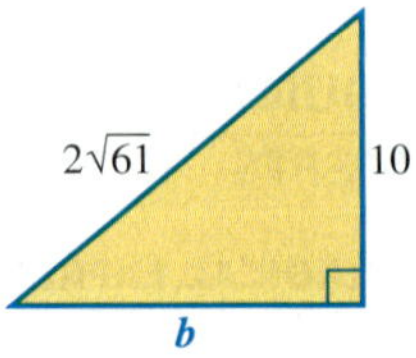


**Distance Formula**
The distance between $(x_1, y_1)$ and $(x_2, y_2)$ is

$$d = \sqrt{(x_2 - x_1)^2 + (y_2 - y_1)^2}.$$

The distance between $(3, -2)$ and $(-1, 1)$ is

$$\sqrt{(-1-3)^2 + [1-(-2)]^2} = \sqrt{(-4)^2 + 3^2}$$

$$= \sqrt{16 + 9} = \sqrt{25} = 5.$$

| CONCEPTS | EXAMPLES |
|---|---|

**10.4 ADDING AND SUBTRACTING RADICAL EXPRESSIONS**

Only radical expressions with the same index and the same radicand may be combined.

$$3\sqrt{17} + 2\sqrt{17} - 8\sqrt{17} = (3 + 2 - 8)\sqrt{17} = -3\sqrt{17}$$

$$\sqrt[3]{2} - \sqrt[3]{250} = \sqrt[3]{2} - 5\sqrt[3]{2} = -4\sqrt[3]{2}$$

$$\left.\begin{matrix}\sqrt{15} + \sqrt{30} \\ \sqrt{3} + \sqrt[3]{9}\end{matrix}\right\} \text{cannot be simplified further}$$

**10.5 MULTIPLYING AND DIVIDING RADICAL EXPRESSIONS**

Multiply binomial radical expressions by using the FOIL method. Special products from Section 4.6 may apply.

$$(\sqrt{2} + \sqrt{7})(\sqrt{3} - \sqrt{6}) = \sqrt{6} - 2\sqrt{3} + \sqrt{21} - \sqrt{42} \qquad \sqrt{12} = 2\sqrt{3}$$

$$(\sqrt{5} - \sqrt{10})(\sqrt{5} + \sqrt{10}) = 5 - 10 = -5$$

$$(\sqrt{3} - \sqrt{2})^2 = 3 - 2\sqrt{3} \cdot \sqrt{2} + 2 = 5 - 2\sqrt{6}$$

Rationalize the denominator by multiplying both the numerator and denominator by the same expression.

$$\frac{\sqrt{7}}{\sqrt{5}} = \frac{\sqrt{7}}{\sqrt{5}} \cdot \frac{\sqrt{5}}{\sqrt{5}} = \frac{\sqrt{35}}{5}$$

$$\frac{\sqrt[3]{2}}{\sqrt[3]{4}} = \frac{\sqrt[3]{2}}{\sqrt[3]{4}} \cdot \frac{\sqrt[3]{2}}{\sqrt[3]{2}} = \frac{\sqrt[3]{4}}{\sqrt[3]{8}} = \frac{\sqrt[3]{4}}{2}$$

$$\frac{4}{\sqrt{5} - \sqrt{2}} = \frac{4}{\sqrt{5} - \sqrt{2}} \cdot \frac{\sqrt{5} + \sqrt{2}}{\sqrt{5} + \sqrt{2}} = \frac{4(\sqrt{5} + \sqrt{2})}{5 - 2} = \frac{4(\sqrt{5} + \sqrt{2})}{3}$$

$$\frac{5 + 15\sqrt{6}}{10} = \frac{5(1 + 3\sqrt{6})}{5 \cdot 2} = \frac{1 + 3\sqrt{6}}{2}$$

**10.6 SOLVING EQUATIONS WITH RADICALS**

**Solving an Equation with Radicals**

*Step 1* Isolate one radical on one side of the equation.

*Step 2* Raise both sides of the equation to a power that is the same as the index of the radical.

*Step 3* Solve the resulting equation; if it still contains a radical, repeat Steps 1 and 2.

*Step 4* Check all potential solutions in the *original* equation.

Potential solutions that do not check are extraneous; they are not part of the solution set.

Solve $\sqrt{2x + 3} - x = 0$.

$$\sqrt{2x + 3} = x$$

$$(\sqrt{2x + 3})^2 = x^2$$

$$2x + 3 = x^2$$

$$x^2 - 2x - 3 = 0$$

$$(x - 3)(x + 1) = 0$$

$$x - 3 = 0 \quad \text{or} \quad x + 1 = 0$$

$$x = 3 \quad \text{or} \quad x = -1$$

A check shows that 3 is a solution, but $-1$ is extraneous. The solution set is $\{3\}$.

(continued)

| CONCEPTS | EXAMPLES |
|---|---|
| **10.7 COMPLEX NUMBERS** | |
| $i^2 = -1$, so $i = \sqrt{-1}$. For any positive number $b$, $\sqrt{-b} = i\sqrt{b}$. To multiply radicals with negative radicands, first change each factor to the form $i\sqrt{b}$, then multiply. The same procedure applies to quotients. | $\sqrt{-25} = i\sqrt{25} = 5i$ <br> $\sqrt{-3} \cdot \sqrt{-27} = i\sqrt{3} \cdot i\sqrt{27} = i^2\sqrt{81} = -1 \cdot 9 = -9$ <br> $\dfrac{\sqrt{-18}}{\sqrt{-2}} = \dfrac{i\sqrt{18}}{i\sqrt{2}} = \sqrt{\dfrac{18}{2}} = \sqrt{9} = 3$ |
| **Adding and Subtracting Complex Numbers** Add (or subtract) the real parts and add (or subtract) the imaginary parts. | $(5 + 3i) + (8 - 7i) = 13 - 4i$ <br> $(5 + 3i) - (8 - 7i) = -3 + 10i$ |
| **Multiplying and Dividing Complex Numbers** Multiply complex numbers by using the FOIL method. | $(2 + i)(5 - 3i) = 10 - 6i + 5i - 3i^2 = 10 - i - 3(-1) = 10 - i + 3 = 13 - i$ |
| Divide complex numbers by multiplying the numerator and the denominator by the conjugate of the denominator. | $\dfrac{20}{3 + i} = \dfrac{20(3 - i)}{(3 + i)(3 - i)} = \dfrac{20(3 - i)}{9 - i^2} = \dfrac{20(3 - i)}{10} = 2(3 - i) = 6 - 2i$ |

**Answers to Test Your Word Power**

**1.** B; *Example:* In $\sqrt{3xy}$, $3xy$ is the radicand. **2.** D; *Example:* In a right triangle where $a = 6$, $b = 8$, and $c = 10$, $6^2 + 8^2 = 10^2$. **3.** C; *Example:* In a right triangle where the sides measure 9, 12, and 15 units, the hypotenuse is the side opposite the right angle, with measure 15 units. **4.** B; *Example:* To rationalize the denominator of $\frac{5}{\sqrt{3} + 1}$, multiply both the numerator and denominator by $\sqrt{3} - 1$ to get $\frac{5(\sqrt{3} - 1)}{2}$. **5.** A; *Example:* The potential solution 2 is extraneous in $\sqrt{5q - 1} + 3 = 0$. **6.** C; *Examples:* $-5$ (or $-5 + 0i$), $7i$ (or $0 + 7i$), $\sqrt{2} - 4i$

# CHAPTER 10 REVIEW EXERCISES

[10.1] *Find each root.*

**1.** $\sqrt{1764}$ **2.** $-\sqrt{289}$ **3.** $\sqrt[3]{216}$

**4.** $\sqrt[3]{-125}$ **5.** $-\sqrt[3]{27}$ **6.** $\sqrt[5]{-32}$

**7.** Under what conditions is $\sqrt[n]{a}$ not a real number?

**8.** Simplify each radical so that no radicals appear. Assume $x$ represents any real number.

**(a)** $\sqrt{x^2}$ **(b)** $-\sqrt{x^2}$ **(c)** $\sqrt[3]{x^3}$

*Use a calculator to find a decimal approximation for each number. Give the answer to the nearest thousandth.*

**9.** $-\sqrt{47}$ **10.** $\sqrt[3]{-129}$ **11.** $\sqrt[4]{605}$

**12.** $500^{-3/4}$ **13.** $-500^{4/3}$ **14.** $-28^{-1/2}$

*Graph each function. Give the domain and the range.*

**15.** $f(x) = \sqrt{x-1}$ **16.** $f(x) = \sqrt[3]{x} + 4$

**17.** What is the best estimate of the area of the triangle shown here?

**A.** 3600 **B.** 30 **C.** 60 **D.** 360

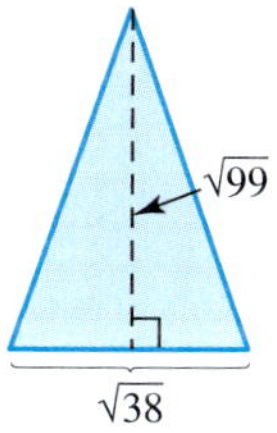


**[10.2]**

**18.** Fill in the blanks with the correct responses: One way to evaluate $8^{2/3}$ is to first find the _____ root of _____, which is _____. Then raise that result to the _____ power, to get an answer of _____. Therefore, $8^{2/3} =$ _____.

**19.** Which one of the following is a positive number?

**A.** $(-27)^{2/3}$ **B.** $(-64)^{5/3}$ **C.** $(-100)^{1/2}$ **D.** $(-32)^{1/5}$

**20.** If $a$ is a negative number and $n$ is odd, then what must be true about $m$ for $a^{m/n}$ to be

**(a)** positive **(b)** negative?

**21.** If $a$ is negative and $n$ is even, then what can be said about $a^{1/n}$?

*Simplify. If the expression does not represent a real number, say so.*

**22.** $49^{1/2}$ **23.** $-121^{1/2}$ **24.** $16^{5/4}$ **25.** $-8^{2/3}$

**26.** $-\left(\frac{36}{25}\right)^{3/2}$ **27.** $\left(-\frac{1}{8}\right)^{-5/3}$ **28.** $\left(\frac{81}{10{,}000}\right)^{-3/4}$ **29.** $(-16)^{3/4}$

**30.** Solve the Pythagorean formula $a^2 + b^2 = c^2$ for $b$, where $b > 0$.

**31.** Explain the relationship between the expressions $a^{m/n}$ and $\sqrt[n]{a^m}$. Give an example.

*Write each expression as a radical.*

**32.** $(m + 3n)^{1/2}$ **33.** $(3a + b)^{-5/3}$

*Write each expression with a rational exponent.*

**34.** $\sqrt{7^9}$ **35.** $\sqrt[5]{p^4}$

*Use the rules for exponents to simplify each expression. Write the answer with only positive exponents. Assume all variables represent positive real numbers.*

**36.** $5^{1/4} \cdot 5^{7/4}$ **37.** $\frac{96^{2/3}}{96^{-1/3}}$ **38.** $\frac{(a^{1/3})^4}{a^{2/3}}$

**39.** $\frac{y^{-1/3} \cdot y^{5/6}}{y}$ **40.** $\left(\frac{z^{-1}x^{-3/5}}{2^{-2}z^{-1/2}x}\right)^{-1}$ **41.** $r^{-1/2}(r + r^{3/2})$

*Simplify by first writing each radical in exponential form. Leave the answer in exponential form. Assume all variables represent positive real numbers.*

**42.** $\sqrt[8]{s^4}$ **43.** $\sqrt[6]{r^9}$ **44.** $\dfrac{\sqrt{p^5}}{p^2}$

**45.** $\sqrt[4]{k^3} \cdot \sqrt{k^3}$ **46.** $\sqrt[3]{m^5} \cdot \sqrt[3]{m^8}$ **47.** $\sqrt[4]{\sqrt[3]{z}}$

**48.** $\sqrt{\sqrt{\sqrt{x}}}$ **49.** $\sqrt[3]{\sqrt[5]{x}}$ **50.** $\sqrt{\sqrt[6]{\sqrt[3]{x}}}$

**51.** By the product rule for exponents, we know that $2^{1/4} \cdot 2^{1/5} = 2^{9/20}$. However, there is no exponent rule to simplify $3^{1/4} \cdot 2^{1/5}$. Why?

**[10.3]** *Simplify each radical. Assume all variables represent positive real numbers.*

**52.** $\sqrt{6} \cdot \sqrt{11}$ **53.** $\sqrt{5} \cdot \sqrt{r}$ **54.** $\sqrt[3]{6} \cdot \sqrt[3]{5}$ **55.** $\sqrt[4]{7} \cdot \sqrt[4]{3}$

**56.** $\sqrt{20}$ **57.** $\sqrt{75}$ **58.** $-\sqrt{125}$ **59.** $\sqrt[3]{-108}$

**60.** $\sqrt{100y^7}$ **61.** $\sqrt[3]{64p^4q^6}$ **62.** $\sqrt[3]{108a^8b^5}$ **63.** $\sqrt[3]{632r^8t^4}$

**64.** $\sqrt{\dfrac{y^3}{144}}$ **65.** $\sqrt[3]{\dfrac{m^{15}}{27}}$ **66.** $\sqrt[3]{\dfrac{r^2}{8}}$ **67.** $\sqrt[4]{\dfrac{a^9}{81}}$

*Simplify each radical expression.*

**68.** $\sqrt[6]{15^3}$ **69.** $\sqrt[4]{p^6}$ **70.** $\sqrt[3]{2} \cdot \sqrt[4]{5}$ **71.** $\sqrt{x} \cdot \sqrt[5]{x}$

**72.** Find the missing length in the right triangle. Simplify the answer if applicable.

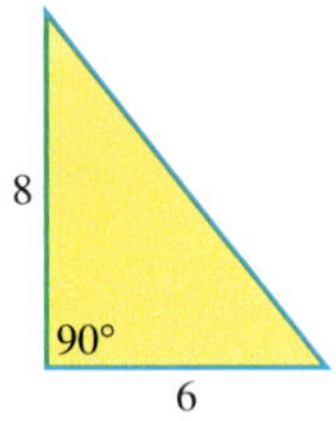


**73.** Find the distance between the points $(-4, 7)$ and $(10, 6)$.

**[10.4]** *Perform the indicated operations. Assume all variables represent positive real numbers.*

**74.** $2\sqrt{8} - 3\sqrt{50}$ **75.** $8\sqrt{80} - 3\sqrt{45}$ **76.** $-\sqrt{27y} + 2\sqrt{75y}$

**77.** $2\sqrt{54m^3} + 5\sqrt{96m^3}$ **78.** $3\sqrt[3]{54} + 5\sqrt[3]{16}$ **79.** $-6\sqrt[4]{32} + \sqrt[4]{512}$

**80.** $\dfrac{3}{\sqrt{16}} - \dfrac{\sqrt{5}}{2}$ **81.** $\dfrac{4}{\sqrt{25}} + \dfrac{\sqrt{5}}{4}$

*In Exercises 82 and 83, leave answers as simplified radicals.*

**82.** Find the perimeter of a rectangular electronic billboard having sides of lengths shown in the figure.

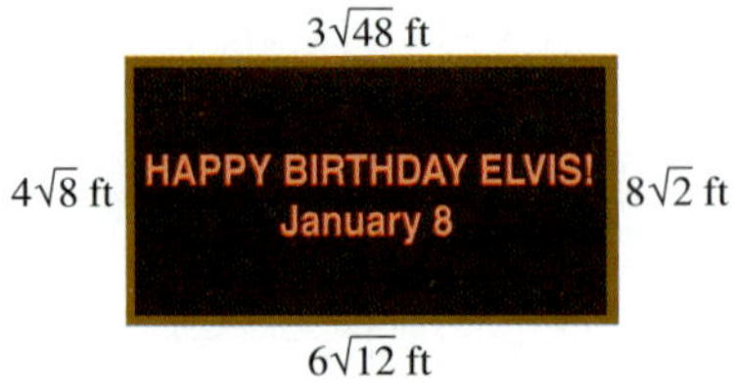


**83.** Find the perimeter of a triangular electronic highway road sign having the dimensions shown in the figure.

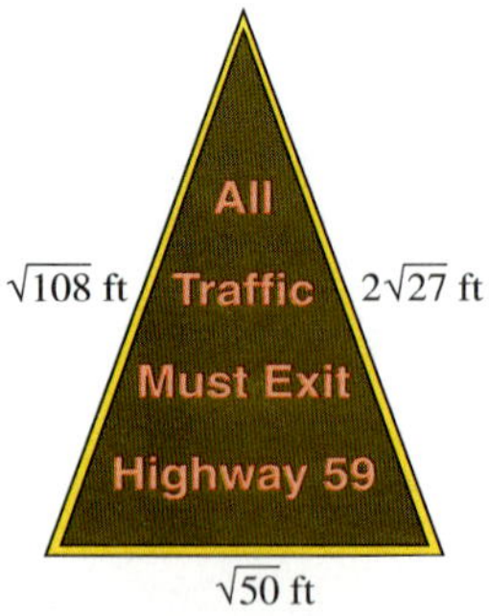

[10.5] *Multiply.*

**84.** $(\sqrt{3}+1)(\sqrt{3}-2)$

**85.** $(\sqrt{7}+\sqrt{5})(\sqrt{7}-\sqrt{5})$

**86.** $(3\sqrt{2}+1)(2\sqrt{2}-3)$

**87.** $(\sqrt{13}-\sqrt{2})^2$

**88.** $(\sqrt[3]{2}+3)(\sqrt[3]{4}-3\sqrt[3]{2}+9)$

**89.** $(\sqrt[3]{4y}-1)(\sqrt[3]{4y}+3)$

**90.** Use a calculator to show that the answer to Exercise 87, $15-2\sqrt{26}$, is not equal to $13\sqrt{26}$.

**91.** A friend wants to rationalize the denominator of the fraction $\frac{5}{\sqrt[3]{6}}$, and she decides to multiply the numerator and denominator by $\sqrt[3]{6}$. Why will her plan *not* work?

*Rationalize each denominator. Assume all variables represent positive real numbers.*

**92.** $\frac{\sqrt{6}}{\sqrt{5}}$

**93.** $\frac{-6\sqrt{3}}{\sqrt{2}}$

**94.** $\frac{3\sqrt{7p}}{\sqrt{y}}$

**95.** $\sqrt{\frac{11}{8}}$

**96.** $-\sqrt[3]{\frac{9}{25}}$

**97.** $\sqrt[3]{\frac{108m^3}{n^5}}$

**98.** $\frac{1}{\sqrt{2}+\sqrt{7}}$

**99.** $\frac{-5}{\sqrt{6}-3}$

*Write in lowest terms.*

**100.** $\frac{2-2\sqrt{5}}{8}$

**101.** $\frac{4-8\sqrt{8}}{12}$

**102.** $\frac{-18+\sqrt{27}}{6}$

[10.6] *Solve each equation.*

**103.** $\sqrt{8x+9}=5$

**104.** $\sqrt{2z-3}-3=0$

**105.** $\sqrt{3m+1}-2=-3$

**106.** $\sqrt{7z+1}=z+1$

**107.** $3\sqrt{m}=\sqrt{10m-9}$

**108.** $\sqrt{p^2+3p+7}=p+2$

**109.** $\sqrt{a+2}-\sqrt{a-3}=1$

**110.** $\sqrt[3]{5m-1}=\sqrt[3]{3m-2}$

**111.** $\sqrt[3]{2x^2+3x-7}=\sqrt[3]{2x^2+4x+6}$

**112.** $\sqrt[3]{3y^2-4y+6}=\sqrt[3]{3y^2-2y+8}$

**113.** $\sqrt[3]{1-2k}-\sqrt[3]{-k-13}=0$

**114.** $\sqrt[3]{11-2t}-\sqrt[3]{-1-5t}=0$

**115.** $\sqrt[4]{x-1}+2=0$

**116.** $\sqrt[4]{2k+3}+1=0$

**117.** $\sqrt[4]{x+7}=\sqrt[4]{2x}$

**118.** $\sqrt[4]{x+8}=\sqrt[4]{3x}$

## RELATING CONCEPTS (EXERCISES 119–125)

**For Individual or Group Work**

***Solve the equations in Exercises 119–124 in order,*** *and then use a generalization to fill in the blanks in Exercise 125.*

**119.** $x=3$

**120.** $x=-3$

**121.** $x^2=9$

**122.** $x^3=27$

**123.** $x^4=81$

**124.** $x^5=-243$

**125.** Suppose both sides of $x=k$ are raised to the $n$th power.

**(a)** If $n$ is even, the number of solutions of the new equation is ______________ the number of solutions of the original
(more than/the same as/fewer than)
equation.

**(b)** If $n$ is odd, the number of solutions of the new equation is ______________ the number of solutions of the original
(more than/the same as/fewer than)
equation.

**126.** Carpenters stabilize wall frames with a diagonal brace, as shown in the figure. The length of the brace is given by $L = \sqrt{H^2 + W^2}$.

**(a)** Solve this formula for $H$.

**(b)** If the bottom of the brace is attached 9 ft from the corner and the brace is 12 ft long, how far up the corner post should it be nailed? Give your answer to the nearest tenth of a foot.

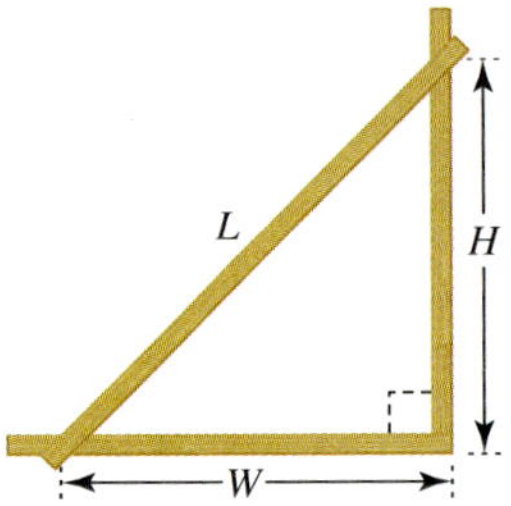


[10.7] *Write each expression as a product of a real number and i.*

**127.** $\sqrt{-25}$

**128.** $\sqrt{-200}$

**129.** If $a$ is a positive real number, is $-\sqrt{-a}$ a real number?

*Perform the indicated operations. Write each imaginary number answer in standard form $a + bi$.*

**130.** $(-2 + 5i) + (-8 - 7i)$

**131.** $(5 + 4i) - (-9 - 3i)$

**132.** $\sqrt{-5} \cdot \sqrt{-7}$

**133.** $\sqrt{-25} \cdot \sqrt{-81}$

**134.** $\dfrac{\sqrt{-72}}{\sqrt{-8}}$

**135.** $(2 + 3i)(1 - i)$

**136.** $(6 - 2i)^2$

**137.** $\dfrac{3 - i}{2 + i}$

**138.** $\dfrac{5 + 14i}{2 + 3i}$

*Find each power of i.*

**139.** $i^{11}$

**140.** $i^{36}$

**141.** $i^{-10}$

**142.** $i^{-8}$

## MIXED REVIEW EXERCISES

*Simplify. Assume all variables represent positive real numbers.*

**143.** $-\sqrt[4]{256}$

**144.** $1000^{-2/3}$

**145.** $\dfrac{z^{-1/5} \cdot z^{3/10}}{z^{7/10}}$

**146.** $\sqrt[4]{k^{24}}$

**147.** $\sqrt[3]{54z^9t^8}$

**148.** $-5\sqrt{18} + 12\sqrt{72}$

**149.** $8\sqrt[3]{x^3y^2} - 2x\sqrt[3]{y^2}$

**150.** $(\sqrt{5} - \sqrt{3})(\sqrt{7} + \sqrt{3})$

**151.** $\dfrac{-1}{\sqrt{12}}$

**152.** $\sqrt[3]{\dfrac{12}{25}}$

**153.** $i^{-1000}$

**154.** $\sqrt{-49}$

**155.** $(4 - 9i) + (-1 + 2i)$

**156.** $\dfrac{\sqrt{50}}{\sqrt{-2}}$

**157.** $\dfrac{3 + \sqrt{54}}{6}$

**158.** $(3 + 2i)^2$

*Solve each equation.*

**159.** $\sqrt{x + 4} = x - 2$

**160.** $\sqrt[3]{2x - 9} = \sqrt[3]{5x + 3}$

**161.** $\sqrt{6 + 2x} - 1 = \sqrt{7 - 2x}$

**162.** $\sqrt{7x + 11} - 5 = 0$

**163.** $\sqrt{6x + 2} - \sqrt{5x + 3} = 0$

**164.** $\sqrt{3 + 5x} - \sqrt{x + 11} = 0$

**165.** $3\sqrt{x} = \sqrt{8x + 9}$

**166.** $6\sqrt{p} = \sqrt{30p + 24}$

**167.** $\sqrt{11 + 2x} + 1 = \sqrt{5x + 1}$

**168.** $\sqrt{5x + 6} - \sqrt{x + 3} = 3$

# CHAPTER 10 TEST

*Evaluate.*

1. $-\sqrt{841}$
2. $\sqrt[3]{-512}$
3. $125^{1/3}$
4. For $\sqrt{146.25}$, which choice gives the best estimate?
   **A.** 10 **B.** 11 **C.** 12 **D.** 13

*Use a calculator to approximate each root to the nearest thousandth.*

5. $\sqrt{478}$
6. $\sqrt[3]{-832}$
7. Graph the function defined by $f(x) = \sqrt{x + 6}$, and give the domain and range.

*Simplify each expression. Assume all variables represent positive real numbers.*

8. $\left(\dfrac{16}{25}\right)^{-3/2}$
9. $(-64)^{-4/3}$
10. $\dfrac{3^{2/5}x^{-1/4}y^{2/5}}{3^{-8/5}x^{7/4}y^{1/10}}$
11. $\left(\dfrac{x^{-4}y^{-6}}{x^{-2}y^{3}}\right)^{-2/3}$
12. $7^{3/4} \cdot 7^{-1/4}$
13. $\sqrt[3]{a^4} \cdot \sqrt[3]{a^7}$
14. Use the Pythagorean formula to find the exact length of side $b$ in the figure.

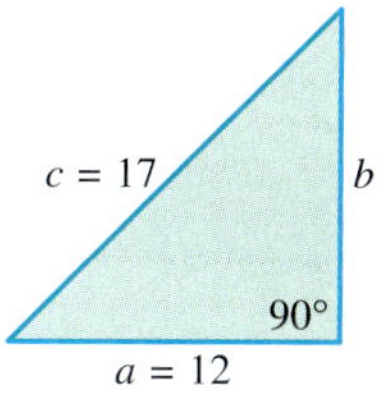


15. Find the distance between the points $(-4, 2)$ and $(2, 10)$.

*Simplify each expression. Assume all variables represent positive real numbers.*

16. $\sqrt{54x^5y^6}$
17. $\sqrt[4]{32a^7b^{13}}$
18. $\sqrt{2} \cdot \sqrt[3]{5}$ (Express as a radical.)
19. $3\sqrt{20} - 5\sqrt{80} + 4\sqrt{500}$
20. $\sqrt[3]{16t^3s^5} - \sqrt[3]{54t^6s^2}$
21. $(7\sqrt{5} + 4)(2\sqrt{5} - 1)$
22. $(\sqrt{3} - 2\sqrt{5})^2$
23. $\dfrac{-5}{\sqrt{40}}$
24. $\dfrac{2}{\sqrt[3]{5}}$
25. $\dfrac{-4}{\sqrt{7} + \sqrt{5}}$
26. Write $\dfrac{6 + \sqrt{24}}{2}$ in lowest terms.
27. The following formula is used in physics, relating the velocity of sound $V$ to the temperature $T$.

$$V = \frac{V_0}{\sqrt{1 - kT}}$$

   **(a)** Find an approximation of $V$ to the nearest tenth if $V_0 = 50$, $k = .01$, and $T = 30$. Use a calculator.

   **(b)** Solve the formula for $T$.

*Solve each equation.*

**28.** $\sqrt[3]{5x} = \sqrt[3]{2x - 3}$

**29.** $x + \sqrt{x + 6} = 9 - x$

**30.** $\sqrt{x + 4} - \sqrt{1 - x} = -1$

*Perform the indicated operations. Express the answers in standard form $a + bi$.*

**31.** $(-2 + 5i) - (3 + 6i) - 7i$

**32.** $(1 + 5i)(3 + i)$

**33.** $\dfrac{7 + i}{1 - i}$

**34.** Simplify $i^{37}$.

**35.** Answer *true* or *false* to each of the following.

**(a)** $i^2 = -1$ **(b)** $i = \sqrt{-1}$ **(c)** $i = -1$ **(d)** $\sqrt{-3} = i\sqrt{3}$

## CUMULATIVE REVIEW EXERCISES CHAPTERS 1–10

*Evaluate each expression if $a = -3$, $b = 5$, and $c = -4$.*

**1.** $|2a^2 - 3b + c|$

**2.** $\dfrac{(a + b)(a + c)}{3b - 6}$

*Solve each equation or inequality.*

**3.** $3(x + 2) - 4(2x + 3) = -3x + 2$

**4.** $\dfrac{1}{3}x + \dfrac{1}{4}(x + 8) = x + 7$

**5.** $.04x + .06(100 - x) = 5.88$

**6.** $-5 - 3(m - 2) < 11 - 2(m + 2)$

**7.** $2k + 4 < 10$ and $3k - 1 > 5$

**8.** $2k + 4 > 10$ or $3k - 1 < 5$

**9.** $|6x + 7| = 13$

**10.** $|-2x + 4| = |-2x - 3|$

**11.** $|2p - 5| \geq 9$

**12.** Find the measures of the marked angles.

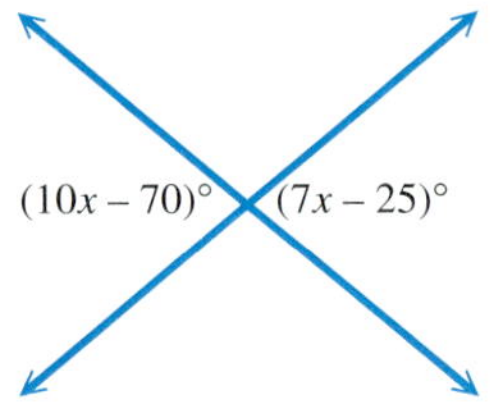


*Solve each problem.*

**13.** A piggy bank has 50 coins, all of which are nickels and quarters. The total value of the money is $8.90. How many of each denomination are there in the bank?

**14.** How many liters of pure alcohol must be mixed with 40 L of 18% alcohol to obtain a 22% alcohol solution?

**15.** Graph the equation $4x - 3y = 12$.

**16.** Find the slope of the line passing through the points $(-4, 6)$ and $(2, -3)$. Then find the equation of the line and write it in the form $y = mx + b$.

*Perform the indicated operations.*

**17.** $(3k^3 - 5k^2 + 8k - 2) - (4k^3 + 11k + 7) + (2k^2 - 5k)$

**18.** $(8x - 7)(x + 3)$

**19.** $\dfrac{8z^3 - 16z^2 + 24z}{8z^2}$

**20.** $\dfrac{6y^4 - 3y^3 + 5y^2 + 6y - 9}{2y + 1}$

*Factor each polynomial completely.*

**21.** $2p^2 - 5pq + 3q^2$

**22.** $3k^4 + k^2 - 4$

**23.** $x^3 + 512$

*Solve by factoring.*

**24.** $2x^2 + 11x + 15 = 0$

**25.** $5t(t - 1) = 2(1 - t)$

*Perform each operation and express the answer in lowest terms.*

**26.** $\dfrac{y^2 + y - 12}{y^3 + 9y^2 + 20y} \div \dfrac{y^2 - 9}{y^3 + 3y^2}$

**27.** $\dfrac{1}{x + y} + \dfrac{3}{x - y}$

*Simplify each complex fraction.*

**28.** $\dfrac{\dfrac{-6}{x - 2}}{\dfrac{8}{3x - 6}}$

**29.** $\dfrac{\dfrac{1}{a} - \dfrac{1}{b}}{\dfrac{a}{b} - \dfrac{b}{a}}$

**30.** Natalie can ride her bike 4 mph faster than her husband, Chuck. If Natalie can ride 48 mi in the same time that Chuck can ride 24 mi, what are their speeds?

**31.** Solve the equation $\dfrac{p + 1}{p - 3} = \dfrac{4}{p - 3} + 6$.

**32.** If $f(x) = 3x - 7$, find $f(-10)$.

**33.** Solve the system by elimination or substitution.

$$\begin{aligned} 3x - y &= 23 \\ 2x + 3y &= 8 \end{aligned}$$

**34.** Solve the system by matrix methods.

$$\begin{aligned} x + y + z &= 1 \\ x - y - z &= -3 \\ x + y - z &= -1 \end{aligned}$$

*Solve the problem by using a system of equations.*

**35.** In 1997, if you had sent five 2-oz letters and three 3-oz letters using first-class mail, it would have cost you \$5.09. Sending three 2-oz letters and five 3-oz letters would have cost \$5.55. What was the 1997 postage rate for one 2-oz letter and for one 3-oz letter? (*Source:* U.S. Postal Service.)

*Write each expression in simplest form, using only positive exponents. Assume all variables represent positive real numbers.*

**36.** $27^{-2/3}$

**37.** $\sqrt{200x^4}$

**38.** $\sqrt[3]{16x^2y} \cdot \sqrt[3]{3x^3y}$

**39.** $\sqrt{50} + \sqrt{8}$

**40.** $\dfrac{1}{\sqrt{10} - \sqrt{8}}$

**41.** $(2\sqrt{x} + \sqrt{y})(-3\sqrt{x} - 4\sqrt{y})$

**42.** Find the distance between the points $(-4, 4)$ and $(-2, 9)$.

**43.** Solve the equation $\sqrt{3r - 8} = r - 2$.

**44.** The *fall speed*, in miles per hour, of a vehicle running off the road into a ditch is given by

$$S = \frac{2.74D}{\sqrt{h}},$$

where $D$ is the horizontal distance traveled from the level surface to the bottom of the ditch and $h$ is the height (or depth) of the ditch. What is the fall speed of a vehicle that traveled 32 ft horizontally into a 5-ft-deep ditch?

*Write in standard form $a + bi$.*

**45.** $(5 + 7i) - (3 - 2i)$

**46.** $\dfrac{6 - 2i}{1 - i}$

# 11 Quadratic Equations, Inequalities, and Functions

In recent years, the number of publicly traded U.S. companies filing for bankruptcy has been at its highest level since the recession of the early 1990s. One casualty of this trend was retailer Montgomery Ward. Started in 1872 as a mail-order catalog business, the company grew to include 250 stores in 30 states. After filing for Chapter 11 bankruptcy protection, the retailer closed for good in 2001. (*Source: USA Today,* December 29, 2000.)

Since then, other high-profile U.S. companies including Enron Corp., Adelphia Communications Corp., Kmart Corp., WorldCom Inc., US Airways, and United Airlines have become victims of this trend. In Sections 11.5 and 11.6, we use *quadratic functions* to model the number of company bankruptcy filings.

# 11.1 Solving Quadratic Equations by the Square Root Property

**OBJECTIVES**

1 Review the zero-factor property.

2 Solve equations of the form $x^2 = k$, where $k > 0$.

3 Solve equations of the form $(ax + b)^2 = k$, where $k > 0$.

4 Solve quadratic equations with solutions that are not real numbers.

We introduced quadratic equations in Section 5.5. Recall that a *quadratic equation* is defined as follows.

**Quadratic Equation**

An equation that can be written in the form

$$ax^2 + bx + c = 0,$$

where $a$, $b$, and $c$ are real numbers, with $a \neq 0$, is a **quadratic equation.** The given form is called **standard form.**

A quadratic equation is a *second-degree equation,* that is, an equation with a squared term and no terms of higher degree. For example,

$$4m^2 + 4m - 5 = 0 \quad \text{and} \quad 3x^2 = 4x - 8$$

are quadratic equations, with the first equation in standard form.

**OBJECTIVE 1 Review the zero-factor property.** In Section 5.5 we used factoring and the zero-factor property to solve quadratic equations.

**Zero-Factor Property**

If two numbers have a product of 0, then at least one of the numbers must be 0. That is, if $ab = 0$, then $a = 0$ or $b = 0$.

We solved a quadratic equation such as $3x^2 - 5x - 28 = 0$ using the zero-factor property as follows.

$$3x^2 - 5x - 28 = 0$$

$$(3x + 7)(x - 4) = 0 \quad \text{Factor.}$$

$$3x + 7 = 0 \quad \text{or} \quad x - 4 = 0 \quad \text{Zero-factor property}$$

$$3x = -7 \quad \text{or} \quad x = 4 \quad \text{Solve each equation.}$$

$$x = -\frac{7}{3}$$

The solution set is $\left\{-\frac{7}{3}, 4\right\}$.

Although factoring is the simplest way to solve quadratic equations, not every quadratic equation can be solved easily by factoring. In this section and Sections 11.2 and 11.3, we develop three other methods of solving quadratic equations.

**OBJECTIVE 2 Solve equations of the form $x^2 = k$, where $k > 0$.** We can solve a quadratic equation such as $x^2 = 9$ by factoring as follows.

$$x^2 = 9$$

$$x^2 - 9 = 0 \qquad \text{Subtract 9.}$$

$$(x + 3)(x - 3) = 0 \qquad \text{Factor.}$$

$$x + 3 = 0 \quad \text{or} \quad x - 3 = 0 \qquad \text{Zero-factor property}$$

$$x = -3 \quad \text{or} \quad x = 3$$

The solution set is $\{-3, 3\}$.

We might also have solved $x^2 = 9$ by noticing that $x$ must be a number whose square is 9. Thus, $x = \sqrt{9} = 3$ or $x = -\sqrt{9} = -3$. This is generalized as the **square root property of equations.**

**Square Root Property of Equations**

If $k$ is a positive number and if $x^2 = k$, then

$$\boldsymbol{x = \sqrt{k}} \quad \text{or} \quad \boldsymbol{x = -\sqrt{k},}$$

and the solution set is $\{-\sqrt{k}, \sqrt{k}\}$.

**NOTE** When we solve an equation, we want to find *all* values of the variable that satisfy the equation. Therefore, we want both the positive and negative square roots of $k$.

**EXAMPLE 1 Solving Quadratic Equations of the Form $x^2 = k$**

Solve each equation. Write radicals in simplified form.

**(a)** $x^2 = 16$

By the square root property, since $x^2 = 16$,

$$x = \sqrt{16} = 4 \quad \text{or} \quad x = -\sqrt{16} = -4.$$

An abbreviation for "$x = 4$ or $x = -4$" is written $x = \pm 4$ (read "positive or negative 4"). Check each solution by substituting it for $x$ in the original equation. The solution set is $\{-4, 4\}$.

**(b)** $z^2 = 5$

By the square root property, the solutions are $z = \sqrt{5}$ or $z = -\sqrt{5}$. The solution set is $\{\sqrt{5}, -\sqrt{5}\}$, which may be written $\{\pm\sqrt{5}\}$.

**(c)**

$$m^2 - 8 = 0$$

$$m^2 = 8 \qquad \text{Add 8.}$$

$$m = \sqrt{8} \quad \text{or} \quad m = -\sqrt{8} \qquad \text{Square root property}$$

$$m = 2\sqrt{2} \quad \text{or} \quad m = -2\sqrt{2} \qquad \text{Simplify } \sqrt{8}.$$

The solution set is $\{2\sqrt{2}, -2\sqrt{2}\}$.

**(d)** $4x^2 - 48 = 0$

Solve for $x^2$.

$$4x^2 - 48 = 0$$

$$4x^2 = 48 \quad \text{Add 48.}$$

$$x^2 = 12 \quad \text{Divide by 4.}$$

$$x = \sqrt{12} \quad \text{or} \quad x = -\sqrt{12} \quad \text{Square root property}$$

$$x = 2\sqrt{3} \quad \text{or} \quad x = -2\sqrt{3} \quad \sqrt{12} = \sqrt{4} \cdot \sqrt{3} = 2\sqrt{3}$$

The solutions are $2\sqrt{3}$ and $-2\sqrt{3}$. Check each in the original equation.

*Check:* $4x^2 - 48 = 0$ Original equation

| | |
|---|---|
| $4(2\sqrt{3})^2 - 48 = 0$ ? | $4(-2\sqrt{3})^2 - 48 = 0$ ? |
| $4(12) - 48 = 0$ ? | $4(12) - 48 = 0$ ? |
| $48 - 48 = 0$ ? | $48 - 48 = 0$ ? |
| $0 = 0$ True | $0 = 0$ True |

The solution set is $\{2\sqrt{3}, -2\sqrt{3}\}$.

**(e)** $3x^2 + 5 = 11$

Solve for $x^2$.

$$3x^2 + 5 = 11$$

$$3x^2 = 6 \quad \text{Subtract 5.}$$

$$x^2 = 2 \quad \text{Divide by 3.}$$

$$x = \sqrt{2} \quad \text{or} \quad x = -\sqrt{2} \quad \text{Square root property}$$

Check each solution as shown in part (d). The solution set is $\{\sqrt{2}, -\sqrt{2}\}$.

**Now Try Exercises 5, 7, 11, 15, and 19.**

### EXAMPLE 2 Using the Square Root Property in an Application

Galileo Galilei (1564–1642) developed a formula for freely falling objects described by

$$d = 16t^2,$$

where $d$ is the distance in feet that an object falls (disregarding air resistance) in $t$ sec, regardless of weight. Galileo dropped objects from the Leaning Tower of Pisa to develop this formula. If the Leaning Tower is about 180 ft tall, use Galileo's formula to determine how long it would take an object dropped from the tower to fall to the ground. (*Source: Microsoft Encarta Encyclopedia 2002.*)

We substitute 180 for $d$ in Galileo's formula.

$$d = 16t^2$$

$$180 = 16t^2 \quad \text{Let } d = 180.$$

$$11.25 = t^2 \quad \text{Divide by 16.}$$

$$t = \sqrt{11.25} \quad \text{or} \quad t = -\sqrt{11.25} \quad \text{Square root property}$$

Since time cannot be negative, we discard the negative solution. In applied problems, we usually prefer approximations to exact values. Using a calculator, $\sqrt{11.25} \approx 3.4$ so $t \approx 3.4$. The object would fall to the ground in about 3.4 sec.

**Now Try Exercise 59.**

**OBJECTIVE 3 Solve equations of the form $(ax + b)^2 = k$, where $k > 0$.** In each equation in Example 1, the exponent 2 appeared with a single variable as its base. We can extend the square root property to solve equations where the base is a binomial, as shown in the next examples.

### EXAMPLE 3 Solving Quadratic Equations of the Form $(x + b)^2 = k$

Solve each equation.

**(a)** $(x - 3)^2 = 16$

Apply the square root property, using $x - 3$ as the base.

$$(x - 3)^2 = 16$$

$$x - 3 = \sqrt{16} \quad \text{or} \quad x - 3 = -\sqrt{16}$$

$$x - 3 = 4 \quad \text{or} \quad x - 3 = -4 \qquad \sqrt{16} = 4$$

$$x = 7 \quad \text{or} \quad x = -1 \qquad \text{Add 3.}$$

Check each solution in the original equation.

$$(x - 3)^2 = 16$$
$$(7 - 3)^2 = 16 \quad ? \quad \text{Let } x = 7.$$
$$4^2 = 16 \quad ?$$
$$16 = 16 \quad \text{True}$$

$$(x - 3)^2 = 16$$
$$(-1 - 3)^2 = 16 \quad ? \quad \text{Let } x = -1.$$
$$(-4)^2 = 16 \quad ?$$
$$16 = 16 \quad \text{True}$$

The solution set is $\{7, -1\}$.

**(b)** $(x + 1)^2 = 6$

By the square root property,

$$x + 1 = \sqrt{6} \quad \text{or} \quad x + 1 = -\sqrt{6}$$

$$x = -1 + \sqrt{6} \quad \text{or} \quad x = -1 - \sqrt{6}.$$

*Check:*

$$\left(-1 + \sqrt{6} + 1\right)^2 = \left(\sqrt{6}\right)^2 = 6;$$
$$\left(-1 - \sqrt{6} + 1\right)^2 = \left(-\sqrt{6}\right)^2 = 6.$$

The solution set is $\left\{-1 + \sqrt{6}, -1 - \sqrt{6}\right\}$.

**Now Try Exercises 21 and 27.**

**NOTE** The solutions in Example 3(b) may be written in abbreviated form as $-1 \pm \sqrt{6}$. If they are written this way, keep in mind that *two* solutions are indicated, one with the + sign and the other with the − sign.

### EXAMPLE 4 Solving a Quadratic Equation of the Form $(ax + b)^2 = k$

Solve $(3r - 2)^2 = 27$.

$$3r - 2 = \sqrt{27} \quad \text{or} \quad 3r - 2 = -\sqrt{27} \qquad \text{Square root property}$$

$$3r - 2 = 3\sqrt{3} \quad \text{or} \quad 3r - 2 = -3\sqrt{3} \qquad \sqrt{27} = \sqrt{9 \cdot 3} = 3\sqrt{3}$$

$$3r = 2 + 3\sqrt{3} \quad \text{or} \quad 3r = 2 - 3\sqrt{3} \qquad \text{Add 2.}$$

$$r = \frac{2 + 3\sqrt{3}}{3} \quad \text{or} \quad r = \frac{2 - 3\sqrt{3}}{3} \qquad \text{Divide by 3.}$$

We show the check for the first solution. The check for the second solution is similar.

*Check:*

$$(3r - 2)^2 = 27 \qquad \text{Original equation}$$

$$\left[3\left(\frac{2 + 3\sqrt{3}}{3}\right) - 2\right]^2 = 27 \quad ?$$

$$\left(2 + 3\sqrt{3} - 2\right)^2 = 27 \quad ?$$

$$\left(3\sqrt{3}\right)^2 = 27 \quad ?$$

$$27 = 27 \qquad \text{True}$$

The solution set is $\left\{\frac{2 + 3\sqrt{3}}{3}, \frac{2 - 3\sqrt{3}}{3}\right\}$.

**Now Try Exercise 31.**

**CAUTION** The solutions in Example 4 are fractions that cannot be simplified, since 3 is *not* a common factor in the numerator.

**OBJECTIVE 4 Solve quadratic equations with solutions that are not real numbers.** So far, all the equations we have solved using the square root property have had two real solutions. In the equation $x^2 = k$, if $k < 0$, there will be two imaginary solutions.

### EXAMPLE 5 Solving Quadratic Equations with Imaginary Solutions

Solve each equation.

**(a)** $x^2 = -15$

$$x = \sqrt{-15} \quad \text{or} \quad x = -\sqrt{-15} \qquad \text{Square root property}$$

$$x = i\sqrt{15} \quad \text{or} \quad x = -i\sqrt{15} \qquad \sqrt{-1} = i$$

The solution set is $\{i\sqrt{15}, -i\sqrt{15}\}$. Check by substitution.

**(b)** $(t + 2)^2 = -16$

$$t + 2 = \sqrt{-16} \quad \text{or} \quad t + 2 = -\sqrt{-16} \qquad \text{Square root property}$$

$$t + 2 = 4i \quad \text{or} \quad t + 2 = -4i \qquad \sqrt{-16} = 4i$$

$$t = -2 + 4i \quad \text{or} \quad t = -2 - 4i$$

*Check:*

$$(t + 2)^2 = -16 \qquad \text{Original equation}$$

$$(-2 + 4i + 2)^2 = -16 \quad ?$$

$$(4i)^2 = -16 \quad ?$$

$$-16 = -16 \qquad i^2 = -1$$

The check for the other solution is similar. The solution set is

$$\{-2 + 4i, -2 - 4i\}.$$

**Now Try Exercises 49 and 51.**

# 11.1 EXERCISES

**For Extra Help**

Student's Solutions Manual

MyMathLab

InterAct Math Tutorial Software

AW Math Tutor Center

MathXL

Digital Video Tutor CD 16/Videotape 16

**1.** A student was asked to solve the quadratic equation $x^2 = 16$ and did not get full credit for the solution set $\{4\}$. Why?

**2.** Why can't the zero-factor property be used to solve every quadratic equation?

**3.** Give a one-sentence description or explanation of each phrase.

**(a)** Quadratic equation in standard form **(b)** Zero-factor property
**(c)** Square root property

**4.** What is wrong with the following "solution"?

$$x^2 - x - 2 = 5$$
$$(x - 2)(x + 1) = 5$$
$$x - 2 = 5 \quad \text{or} \quad x + 1 = 5 \qquad \text{Zero-factor property}$$
$$x = 7 \quad \text{or} \quad x = 4$$

*Use the square root property to solve each equation. See Example 1.*

**5.** $x^2 = 81$ **6.** $z^2 = 225$ **7.** $t^2 = 17$ **8.** $x^2 = 19$

**9.** $m^2 = 32$ **10.** $x^2 = 54$ **11.** $r^2 - 21 = 0$ **12.** $x^2 - 14 = 0$

**13.** $t^2 - 20 = 0$ **14.** $p^2 - 50 = 0$ **15.** $3n^2 - 72 = 0$ **16.** $5z^2 - 200 = 0$

**17.** $5a^2 + 4 = 8$ **18.** $4p^2 - 3 = 7$ **19.** $2t^2 + 7 = 61$ **20.** $3x^2 - 8 = 64$

*Solve each equation by using the square root property. See Examples 3 and 4.*

**21.** $(x - 3)^2 = 25$ **22.** $(k - 7)^2 = 16$ **23.** $(3k + 2)^2 = 49$

**24.** $(5t + 3)^2 = 36$ **25.** $(4x - 3)^2 = 9$ **26.** $(7z - 5)^2 = 25$

**27.** $(x - 4)^2 = 3$ **28.** $(x + 3)^2 = 11$ **29.** $(t + 5)^2 = 48$

**30.** $(m - 6)^2 = 27$ **31.** $(3x - 1)^2 = 7$ **32.** $(2x + 4)^2 = 10$

**33.** $(4p + 1)^2 = 24$ **34.** $(5t - 2)^2 = 12$ **35.** $(5 - 2x)^2 = 30$

**36.** $(3 - 2a)^2 = 70$ **37.** $(3k + 1)^2 = 18$ **38.** $(5z + 6)^2 = 75$

**39.** $\left(\frac{1}{2}x + 5\right)^2 = 12$ **40.** $\left(\frac{1}{3}m + 4\right)^2 = 27$ **41.** $(4k - 1)^2 - 48 = 0$

**42.** $(2s - 5)^2 - 180 = 0$

**43.** Johnny solved the equation in Exercise 35 and wrote his answer as $\left\{\frac{5 + \sqrt{30}}{2}, \frac{5 - \sqrt{30}}{2}\right\}$. Linda solved the same equation and wrote her answer as $\left\{\frac{-5 + \sqrt{30}}{-2}, \frac{-5 - \sqrt{30}}{-2}\right\}$. The teacher gave them both full credit. Explain why both students were correct, although their answers look different.

**44.** In the solutions found in Example 4 of this section, why is it not valid to simplify the answers by dividing out the 3s in the numerator and denominator?

*Use a calculator with a square root key to solve each equation. Round your answers to the nearest hundredth.*

**45.** $(k + 2.14)^2 = 5.46$

**46.** $(r - 3.91)^2 = 9.28$

**47.** $(2.11p + 3.42)^2 = 9.58$

**48.** $(1.71m - 6.20)^2 = 5.41$

*Find the imaginary solutions of each equation. See Example 5.*

**49.** $x^2 = -12$

**50.** $x^2 = -18$

**51.** $(r - 5)^2 = -3$

**52.** $(t + 6)^2 = -5$

**53.** $(6k - 1)^2 = -8$

**54.** $(4m - 7)^2 = -27$

## RELATING CONCEPTS (EXERCISES 55–58)

**For Individual or Group Work**

*In Section 5.4 we saw how certain trinomials can be factored as squares of binomials. Use this idea and* ***work Exercises 55–58 in order,*** *considering the equation*

$$x^2 + 6x + 9 = 100.$$

**55.** Factor the left side of the equation as the square of a binomial, and write the resulting equation.

**56.** Use the square root property to solve the equation from Exercise 55.

**57.** What is the solution set of the original equation?

**58.** Solve the equation $4k^2 - 12k + 9 = 81$ using the method described in Exercises 55–57.

*Solve Exercises 59 and 60 using Galileo's formula,* $d = 16t^2$*. Round answers to the nearest tenth. See Example 2.*

**59.** The Gateway Arch in St. Louis, Missouri, is 630 ft tall. How long would it take an object dropped from the top of it to fall to the ground? (*Source: Home & Away,* November/December 2000.)

**60.** Mount Rushmore National Memorial in South Dakota features a sculpture of four of America's favorite presidents carved into the rim of the mountain, 500 ft above the valley floor. How long would it take a rock dropped from the top of the sculpture to fall to the ground? (*Source: Microsoft Encarta Encyclopedia 2002.*)

*Solve each problem. See Example 2.*

**61.** The area $A$ of a circle with radius $r$ is given by the formula

$$A = \pi r^2.$$

If a circle has area $81\pi$ in.$^2$, what is its radius?

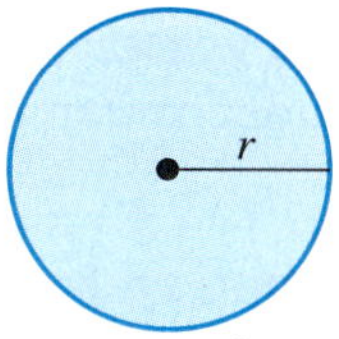


$A = \pi r^2$

**62.** The surface area $S$ of a sphere with radius $r$ is given by the formula

$$S = 4\pi r^2.$$

If a sphere has surface area $36\pi$ ft$^2$, what is its radius?

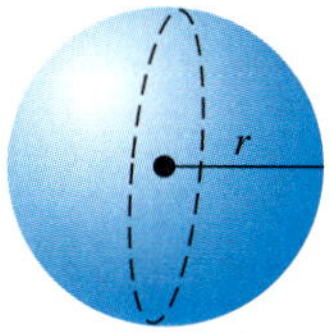


$S = 4\pi r^2$

*The amount A that P dollars invested at an annual rate of interest r will grow to in 2 yr is*

$$A = P(1 + r)^2.$$

**63.** At what interest rate will \$100 grow to \$110.25 in 2 yr?

**64.** At what interest rate will \$500 grow to \$572.45 in 2 yr?

# 11.2 Solving Quadratic Equations by Completing the Square

**OBJECTIVES**

1 Solve quadratic equations by completing the square when the coefficient of the squared term is 1.

2 Solve quadratic equations by completing the square when the coefficient of the squared term is not 1.

3 Simplify an equation before solving.

**OBJECTIVE 1 Solve quadratic equations by completing the square when the coefficient of the squared term is 1.** The methods we have studied so far are not enough to solve the equation

$$x^2 + 6x + 7 = 0.$$

If we could write the equation in the form $(x + b)^2 = k$, we could solve it with the square root property discussed in the previous section. To do that, we need to have a perfect square trinomial on the left side of the equation.

Recall from Section 5.4 that the perfect square trinomial

$$x^2 + 6x + 9$$

can be factored as $(x + 3)^2$. In the trinomial, the coefficient of $x$ (the first-degree term) is 6 and the constant term is 9. Notice that if we take half of 6 and square it, we get the constant term, 9.

Coefficient of $x$ Constant

$$\left[\frac{1}{2}(6)\right]^2 = 3^2 = 9$$

Similarly, in

$$x^2 + 12x + 36, \qquad \left[\frac{1}{2}(12)\right]^2 = 6^2 = 36,$$

and in

$$m^2 - 10m + 25, \qquad \left[\frac{1}{2}(-10)\right]^2 = (-5)^2 = 25.$$

This relationship is true in general and is the idea behind rewriting a quadratic equation so the square root property can be applied.

### EXAMPLE 1 Rewriting an Equation to Use the Square Root Property

Solve $x^2 + 6x + 7 = 0$.

This quadratic equation cannot be solved by factoring, and it is not in the correct form to solve using the square root property. To get this form, we need a perfect square trinomial on the left side of the equation. We first subtract 7 from each side.

$$x^2 + 6x + 7 = 0$$

$$x^2 + 6x = -7 \qquad \text{Subtract 7.}$$

We must add a constant to get a perfect square trinomial on the left.

$$\underbrace{x^2 + 6x + \underline{\ ?\ }}_{\text{Needs to be a perfect square trinomial}}$$

To find this constant, we apply the ideas preceding this example—we take half the coefficient of the first-degree term and square the result.

$$\left[\frac{1}{2}(6)\right]^2 = 3^2 = 9 \leftarrow \text{Desired constant}$$

We add this constant, 9, to *each* side of the equation. (Why?)

$$x^2 + 6x + 9 = -7 + 9$$

Now we factor the perfect square trinomial on the left and add on the right.

$$(x + 3)^2 = 2$$

We can solve this equation using the square root property.

$$x + 3 = \sqrt{2} \quad \text{or} \quad x + 3 = -\sqrt{2}$$

$$x = -3 + \sqrt{2} \quad \text{or} \quad x = -3 - \sqrt{2}$$

*Check:*

$$x^2 + 6x + 7 = 0 \qquad \text{Original equation}$$

$$(-3 + \sqrt{2})^2 + 6(-3 + \sqrt{2}) + 7 = 0 \quad ? \qquad \text{Let } x = -3 + \sqrt{2}.$$

$$9 - 6\sqrt{2} + 2 - 18 + 6\sqrt{2} + 7 = 0 \quad ?$$

$$0 = 0 \qquad \text{True}$$

The check for the second solution is similar. The solution set is

$$\{-3 + \sqrt{2}, -3 - \sqrt{2}\}.$$

**Now Try Exercise 11.**

The process of changing the form of the equation in Example 1 from

$$x^2 + 6x + 7 = 0 \qquad \text{to} \qquad (x + 3)^2 = 2$$

is called **completing the square.** Completing the square changes only the form of the equation. To see this, multiply out the left side of $(x + 3)^2 = 2$ and combine terms. Then subtract 2 from both sides to see that the result is $x^2 + 6x + 7 = 0$.

### EXAMPLE 2 Completing the Square to Solve a Quadratic Equation

Solve $x^2 - 8x = 5$.

To complete the square on $x^2 - 8x$, take half the coefficient of $x$ and square it.

$$\frac{1}{2}(-8) = -4 \qquad \text{and} \qquad (-4)^2 = 16$$

↑ Coefficient of $x$

Add the result, 16, to both sides of the equation.

$$x^2 - 8x = 5 \qquad \text{Given equation}$$

$$x^2 - 8x + 16 = 5 + 16 \qquad \text{Add 16.}$$

$$(x - 4)^2 = 21 \qquad \text{Factor the left side as the square of a binomial; add on the right.}$$

Now apply the square root property.

$$x - 4 = \sqrt{21} \quad \text{or} \quad x - 4 = -\sqrt{21} \qquad \text{Square root property}$$

$$x = 4 + \sqrt{21} \quad \text{or} \quad x = 4 - \sqrt{21} \qquad \text{Add 4.}$$

A check indicates that the solution set is

$$\{4 + \sqrt{21}, 4 - \sqrt{21}\}.$$

**Now Try Exercise 13.**

The steps to solve a quadratic equation $ax^2 + bx + c = 0$ $(a \neq 0)$ by completing the square are summarized here.

### Solving a Quadratic Equation by Completing the Square

*Step 1* **Be sure the squared term has coefficient 1.** If the coefficient of the squared term is 1, go to Step 2. If it is not 1 but some other nonzero number $a$, divide both sides of the equation by $a$.

*Step 2* **Write in correct form.** Make sure that all variable terms are on one side of the equation and that all constant terms are on the other.

*Step 3* **Complete the square.** Take half the coefficient of the first-degree term, and square it. Add the square to both sides of the equation. Factor the variable side and combine terms on the other side.

*Step 4* **Solve.** Use the square root property to solve the equation.

**EXAMPLE 3 Solving a Quadratic Equation by Completing the Square ($a = 1$)**

Solve $k^2 + 5k - 1 = 0$.

Follow the steps in the box. Since the coefficient of the squared term is 1, begin with Step 2. Write the equation in correct form.

*Step 2* $$k^2 + 5k = 1$$ Add 1 to each side.

*Step 3* Complete the square. Take half the coefficient of the first-degree term and square the result.

$$\left[\frac{1}{2}(5)\right]^2 = \left(\frac{5}{2}\right)^2 = \frac{25}{4}$$

Add the square to each side of the equation to get

$$k^2 + 5k + \frac{25}{4} = 1 + \frac{25}{4}.$$

$$\left(k + \frac{5}{2}\right)^2 = \frac{29}{4}$$ Factor on the left; add on the right.

*Step 4* Use the square root property and solve for $k$.

$$k + \frac{5}{2} = \sqrt{\frac{29}{4}} \quad \text{or} \quad k + \frac{5}{2} = -\sqrt{\frac{29}{4}}$$

$$k + \frac{5}{2} = \frac{\sqrt{29}}{2} \quad \text{or} \quad k + \frac{5}{2} = -\frac{\sqrt{29}}{2}$$

$$k = -\frac{5}{2} + \frac{\sqrt{29}}{2} \quad \text{or} \quad k = -\frac{5}{2} - \frac{\sqrt{29}}{2}$$

$$k = \frac{-5 + \sqrt{29}}{2} \quad \text{or} \quad k = \frac{-5 - \sqrt{29}}{2}$$

Check that the solution set is $\left\{\frac{-5 + \sqrt{29}}{2}, \frac{-5 - \sqrt{29}}{2}\right\}$.

**Now Try Exercise 15.**

**OBJECTIVE 2 Solve quadratic equations by completing the square when the coefficient of the squared term is not 1.** To complete the square, the coefficient of the squared term must be 1 (Step 1). The next examples show what to do when this coefficient is not 1.

**EXAMPLE 4 Solving a Quadratic Equation by Completing the Square ($a \neq 1$)**

Solve $4z^2 + 16z - 9 = 0$.

*Step 1* Divide each side by 4 so that the coefficient of $z^2$ is 1.

$$4z^2 + 16z - 9 = 0$$

$$z^2 + 4z - \frac{9}{4} = 0 \qquad \text{Divide by 4.}$$

*Step 2* Add $\frac{9}{4}$ to each side to get the variable terms on the left and the constant on the right.

$$z^2 + 4z = \frac{9}{4}$$

*Step 3* To complete the square, take half the coefficient of $z$ and square the result: $\left[\frac{1}{2}(4)\right]^2 = 2^2 = 4$. Add 4 to both sides of the equation.

$$z^2 + 4z + 4 = \frac{9}{4} + 4 \qquad \text{Add 4.}$$

$$(z + 2)^2 = \frac{25}{4} \qquad \text{Factor on the left; add on the right.}$$

*Step 4* Use the square root property and solve for $z$.

$$z + 2 = \sqrt{\frac{25}{4}} \quad \text{or} \quad z + 2 = -\sqrt{\frac{25}{4}}$$

$$z + 2 = \frac{5}{2} \quad \text{or} \quad z + 2 = -\frac{5}{2} \qquad \text{Square root property}$$

$$z = -2 + \frac{5}{2} \quad \text{or} \quad z = -2 - \frac{5}{2} \qquad \text{Subtract 2.}$$

$$z = \frac{1}{2} \quad \text{or} \quad z = -\frac{9}{2}$$

Check by substituting each solution into the original equation. The solution set is $\left\{\frac{1}{2}, -\frac{9}{2}\right\}$.

**Now Try Exercise 17.**

### EXAMPLE 5 Solving a Quadratic Equation by Completing the Square ($a \neq 1$)

Solve $2x^2 - 4x - 5 = 0$.

First divide each side of the equation by 2 to get 1 as the coefficient of the squared term.

$$x^2 - 2x - \frac{5}{2} = 0 \qquad \text{Step 1}$$

$$x^2 - 2x = \frac{5}{2} \qquad \text{Step 2}$$

$$\left[\frac{1}{2}(-2)\right]^2 = (-1)^2 = 1 \qquad \text{Step 3}$$

$$x^2 - 2x + 1 = \frac{5}{2} + 1 \qquad \text{Add 1.}$$

$$(x - 1)^2 = \frac{7}{2} \qquad \text{Factor; add.}$$

$$x - 1 = \sqrt{\frac{7}{2}} \quad \text{or} \quad x - 1 = -\sqrt{\frac{7}{2}} \qquad \text{Step 4}$$

$$x = 1 + \sqrt{\frac{7}{2}} \quad \text{or} \quad x = 1 - \sqrt{\frac{7}{2}}$$

$$x = 1 + \frac{\sqrt{14}}{2} \quad \text{or} \quad x = 1 - \frac{\sqrt{14}}{2} \qquad \text{Rationalize denominators.}$$

Add the two terms in each solution as follows.

$$1 + \frac{\sqrt{14}}{2} = \frac{2}{2} + \frac{\sqrt{14}}{2} = \frac{2 + \sqrt{14}}{2}$$

$$1 - \frac{\sqrt{14}}{2} = \frac{2}{2} - \frac{\sqrt{14}}{2} = \frac{2 - \sqrt{14}}{2}.$$

Check that the solution set is $\left\{\frac{2 + \sqrt{14}}{2}, \frac{2 - \sqrt{14}}{2}\right\}$.

**Now Try Exercise 21.**

**EXAMPLE 6 Solving a Quadratic Equation by Completing the Square (Imaginary Solutions)**

Solve $4p^2 + 8p + 5 = 0$.

$$p^2 + 2p + \frac{5}{4} = 0 \qquad \text{Divide by 4.}$$

$$p^2 + 2p = -\frac{5}{4} \qquad \text{Subtract } \tfrac{5}{4}.$$

The coefficient of $p$ is 2. Take half of 2, square the result, and add it to both sides: $\left[\frac{1}{2}(2)\right]^2 = 1^2 = 1$. The left side can then be factored as a perfect square.

$$p^2 + 2p + 1 = -\frac{5}{4} + 1 \qquad \text{Add 1.}$$

$$(p + 1)^2 = -\frac{1}{4} \qquad \text{Factor; add.}$$

Because the constant on the right side of the equation is negative, this equation will have imaginary solutions.

$$p + 1 = \sqrt{-\frac{1}{4}} \quad \text{or} \quad p + 1 = -\sqrt{-\frac{1}{4}} \qquad \text{Square root property}$$

$$p + 1 = \frac{1}{2}i \quad \text{or} \quad p + 1 = -\frac{1}{2}i \qquad \sqrt{-\tfrac{1}{4}} = \tfrac{1}{2}i$$

$$p = -1 + \frac{1}{2}i \quad \text{or} \quad p = -1 - \frac{1}{2}i \qquad \text{Subtract 1.}$$

*Check:*

$$4p^2 + 8p + 5 = 0 \qquad \text{Original equation}$$

$$4\left(-1 + \frac{1}{2}i\right)^2 + 8\left(-1 + \frac{1}{2}i\right) + 5 = 0 \qquad ?$$

$$4\left[1 + 2(-1)\left(\frac{1}{2}i\right) - \frac{1}{4}\right] - 8 + 4i + 5 = 0 \qquad ?$$

$$4\left[\frac{3}{4} - i\right] - 3 + 4i = 0 \qquad ?$$

$$3 - 4i - 3 + 4i = 0 \qquad ?$$

$$0 = 0 \qquad \text{True}$$

The check of the other solution is similar. The solution set is

$$\left\{-1 + \frac{1}{2}i, -1 - \frac{1}{2}i\right\}.$$

**Now Try Exercise 39.**

**OBJECTIVE 3 Simplify an equation before solving.** Sometimes an equation must be simplified before completing the square. The next example illustrates this.

### EXAMPLE 7 Simplifying an Equation Before Completing the Square

Solve $(x + 3)(x - 1) = 2$.

$$(x + 3)(x - 1) = 2$$

$$x^2 + 2x - 3 = 2 \qquad \text{Use FOIL.}$$

$$x^2 + 2x = 5 \qquad \text{Add 3.}$$

$$x^2 + 2x + 1 = 5 + 1 \qquad \text{Add } \left[\tfrac{1}{2}(2)\right]^2 = 1.$$

$$(x + 1)^2 = 6 \qquad \text{Factor on the left; add on the right.}$$

$$x + 1 = \sqrt{6} \quad \text{or} \quad x + 1 = -\sqrt{6} \qquad \text{Square root property}$$

$$x = -1 + \sqrt{6} \quad \text{or} \quad x = -1 - \sqrt{6} \qquad \text{Subtract 1.}$$

The solution set is $\{-1 + \sqrt{6}, -1 - \sqrt{6}\}$.

**Now Try Exercise 25.**

**NOTE** The procedure for completing the square is also used in other areas of mathematics. For example, we use it in Section 11.7 when we graph quadratic equations and again in Chapter 13 when we work with circles.

## 11.2 EXERCISES

**For Extra Help**

Student's Solutions Manual

MyMathLab.com

InterAct Math Tutorial Software

AW Math Tutor Center

MathXL www.mathxl.com

Digital Video Tutor CD 16/Videotape 16

**1.** Decide what number must be added to make each expression a perfect square trinomial.

**(a)** $x^2 + 6x +$ ____ **(b)** $x^2 + 14x +$ ____ **(c)** $p^2 - 12p +$ ____

**(d)** $x^2 + 3x +$ ____ **(e)** $q^2 - 9q +$ ____ **(f)** $t^2 - \frac{1}{2}t +$ ____

**2.** Which one of the following steps is an appropriate way to begin solving the quadratic equation

$$2x^2 - 4x = 9$$

by completing the square?

**A.** Add 4 to both sides of the equation. **B.** Factor the left side as $2x(x - 2)$.
**C.** Factor the left side as $x(2x - 4)$. **D.** Divide both sides by 2.

*Determine the number that will complete the square to solve each equation after the constant term has been written on the right side. Do not actually solve. See Examples 1–7.*

**3.** $x^2 + 4x - 2 = 0$ **4.** $t^2 + 2t - 1 = 0$ **5.** $x^2 + 10x + 18 = 0$

**6.** $x^2 + 8x + 11 = 0$ **7.** $3w^2 - w - 24 = 0$ **8.** $4z^2 - z - 39 = 0$

*Solve each equation by completing the square. Use the results of Exercises 3–6 to solve Exercises 11–14. See Examples 1–3.*

**9.** $x^2 - 2x - 24 = 0$ **10.** $m^2 - 4m - 32 = 0$ **11.** $x^2 + 4x - 2 = 0$

**12.** $t^2 + 2t - 1 = 0$ **13.** $x^2 + 10x + 18 = 0$ **14.** $x^2 + 8x + 11 = 0$

**15.** $x^2 + 3x - 2 = 0$ **16.** $p^2 + 5p + 5 = 0$

*Solve each equation by completing the square. Use the results of Exercises 7 and 8 to solve Exercises 17 and 18. See Examples 4, 5, and 7.*

**17.** $3w^2 - w = 24$ **18.** $4z^2 - z = 39$ **19.** $2k^2 + 5k - 2 = 0$

**20.** $3r^2 + 2r - 2 = 0$ **21.** $5x^2 - 10x + 2 = 0$ **22.** $2x^2 - 16x + 25 = 0$

**23.** $9x^2 - 24x = -13$ **24.** $25n^2 - 20n = 1$ **25.** $(x + 2)(x + 1) = 10$

**26.** $z(z + 6) + 4 = 0$ **27.** $z^2 - \frac{4}{3}z = -\frac{1}{9}$ **28.** $p^2 - \frac{8}{3}p = -1$

**29.** $.1x^2 - .2x - .1 = 0$ **30.** $.1p^2 - .4p + .1 = 0$

*Solve each equation by completing the square. Give **(a)** exact solutions and **(b)** solutions rounded to the nearest thousandth.*

**31.** $3r^2 - 2 = 6r + 3$ **32.** $4p + 3 = 2p^2 + 2p$

**33.** $(x + 1)(x + 3) = 2$ **34.** $(x - 3)(x + 1) = 1$

**35.** In using the method of completing the square to solve $2x^2 - 10x = -8$, a student began by adding the square of half the coefficient of $x$ $\left(\text{that is, } \left[\frac{1}{2}(-10)\right]^2 = 25\right)$ to both sides of the equation. He then encountered difficulty in his later steps. What was his error? Explain the steps needed to solve the problem, and give the solution set.

**36.** The equation $x^4 - 2x^2 = 8$ can be solved by completing the square, even though it is not a quadratic equation. What number should be added to both sides of the equation so that the left side can be factored as the square of a binomial? Solve the equation for its real solutions.

*Find the imaginary solutions of each equation. See Example 6.*

**37.** $m^2 + 4m + 13 = 0$ **38.** $t^2 + 6t + 10 = 0$ **39.** $3r^2 + 4r + 4 = 0$

**40.** $4x^2 + 5x + 5 = 0$ **41.** $-m^2 - 6m - 12 = 0$ **42.** $-k^2 - 5k - 10 = 0$

## RELATING CONCEPTS (EXERCISES 43–48)

**For Individual or Group Work**

*The Greeks had a method of completing the square geometrically in which they literally changed a figure into a square. For example, to complete the square for $x^2 + 6x$, we begin with a square of side $x$, as in the figure on the left. We add three rectangles of width 1 to the right side and the bottom to get a region with area $x^2 + 6x$. To fill in the corner (complete the square), we must add 9 1-by-1 squares as shown in the figure on the right.*

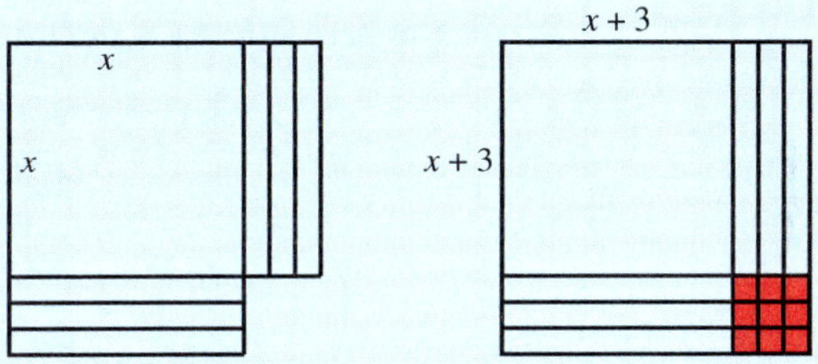


***Work Exercises 43–48 in order.***

**43.** What is the area of the original square?

**44.** What is the area of each strip?

**45.** What is the total area of the six strips?

**46.** What is the area of each small square in the corner of the second figure?

**47.** What is the total area of the small squares?

**48.** What is the area of the new, larger square?

# 11.3 Solving Quadratic Equations by the Quadratic Formula

**OBJECTIVES**

1. Derive the quadratic formula.
2. Solve quadratic equations using the quadratic formula.
3. Use the discriminant to determine the number and type of solutions.

The examples in the previous section showed that any quadratic equation can be solved by completing the square; however, completing the square can be tedious and time consuming. In this section, we complete the square to solve the general quadratic equation

$$ax^2 + bx + c = 0,$$

where $a$, $b$, and $c$ are complex numbers and $a \neq 0$. The solution of this general equation gives a formula for finding the solution of any specific quadratic equation.

**OBJECTIVE 1 Derive the quadratic formula.** To solve $ax^2 + bx + c = 0$ by completing the square (assuming $a > 0$), we follow the steps given in Section 11.2.

$$ax^2 + bx + c = 0$$

$$x^2 + \frac{b}{a}x + \frac{c}{a} = 0 \qquad \text{Divide by } a. \text{ (Step 1)}$$

$$x^2 + \frac{b}{a}x = -\frac{c}{a} \qquad \text{Subtract } \tfrac{c}{a}. \text{ (Step 2)}$$

$$\left[\frac{1}{2}\left(\frac{b}{a}\right)\right]^2 = \left(\frac{b}{2a}\right)^2 = \frac{b^2}{4a^2} \qquad \text{(Step 3)}$$

$$x^2 + \frac{b}{a}x + \frac{b^2}{4a^2} = -\frac{c}{a} + \frac{b^2}{4a^2} \qquad \text{Add } \tfrac{b^2}{4a^2} \text{ to each side.}$$

Write the left side as a perfect square and rearrange the right side.

$$\left(x + \frac{b}{2a}\right)^2 = \frac{b^2}{4a^2} + \frac{-c}{a}$$

$$\left(x + \frac{b}{2a}\right)^2 = \frac{b^2}{4a^2} + \frac{-4ac}{4a^2} \qquad \text{Write with a common denominator.}$$

$$\left(x + \frac{b}{2a}\right)^2 = \frac{b^2 - 4ac}{4a^2} \qquad \text{Add fractions.}$$

$$x + \frac{b}{2a} = \sqrt{\frac{b^2 - 4ac}{4a^2}} \quad \text{or} \quad x + \frac{b}{2a} = -\sqrt{\frac{b^2 - 4ac}{4a^2}} \qquad \text{Square root property (Step 4)}$$

Since
$$\sqrt{\frac{b^2 - 4ac}{4a^2}} = \frac{\sqrt{b^2 - 4ac}}{\sqrt{4a^2}} = \frac{\sqrt{b^2 - 4ac}}{2a},$$

the right side of each equation can be expressed as

$$x + \frac{b}{2a} = \frac{\sqrt{b^2 - 4ac}}{2a} \quad \text{or} \quad x + \frac{b}{2a} = \frac{-\sqrt{b^2 - 4ac}}{2a}$$

$$x = \frac{-b}{2a} + \frac{\sqrt{b^2 - 4ac}}{2a} \quad \text{or} \quad x = \frac{-b}{2a} - \frac{\sqrt{b^2 - 4ac}}{2a}$$

$$x = \frac{-b + \sqrt{b^2 - 4ac}}{2a} \quad \text{or} \quad x = \frac{-b - \sqrt{b^2 - 4ac}}{2a}.$$

If $a < 0$, the same two solutions are obtained. The result is the **quadratic formula,** which is abbreviated as follows.

**Quadratic Formula**

The solutions of $ax^2 + bx + c = 0$ $(a \neq 0)$ are given by

$$x = \frac{-b \pm \sqrt{b^2 - 4ac}}{2a}.$$

**CAUTION** In the quadratic formula, $x = \dfrac{-b \pm \sqrt{b^2 - 4ac}}{2a}$, the square root is added to or subtracted from the value of $-b$ *before* dividing by $2a$.

**OBJECTIVE 2 Solve quadratic equations using the quadratic formula.** To use the quadratic formula, first write the given equation in standard form $ax^2 + bx + c = 0$; then identify the values of $a$, $b$, and $c$ and substitute them into the formula.

### EXAMPLE 1 Using the Quadratic Formula (Rational Solutions)

Solve $6x^2 - 5x - 4 = 0$.

First, identify the values of $a$, $b$, and $c$ of the general quadratic equation, $ax^2 + bx + c = 0$. Here $a$, the coefficient of the second-degree term, is 6, while $b$, the coefficient of the first-degree term, is $-5$, and the constant $c$ is $-4$. Substitute these values into the quadratic formula.

$$x = \frac{-b \pm \sqrt{b^2 - 4ac}}{2a}$$

$$x = \frac{-(-5) \pm \sqrt{(-5)^2 - 4(6)(-4)}}{2(6)} \qquad a = 6,\ b = -5,\ c = -4$$

$$x = \frac{5 \pm \sqrt{25 + 96}}{12}$$

$$x = \frac{5 \pm \sqrt{121}}{12}$$

$$x = \frac{5 \pm 11}{12}$$

This last statement leads to two solutions, one from + and one from −.

$$x = \frac{5 + 11}{12} = \frac{16}{12} = \frac{4}{3} \quad \text{or} \quad x = \frac{5 - 11}{12} = \frac{-6}{12} = -\frac{1}{2}$$

Check each solution by substituting it in the original equation. The solution set is $\left\{-\frac{1}{2}, \frac{4}{3}\right\}$.

**Now Try Exercise 5.**

We could have used factoring to solve the equation in Example 1.

$$6x^2 - 5x - 4 = 0$$

$$(3x - 4)(2x + 1) = 0 \qquad \text{Factor.}$$

$$3x - 4 = 0 \quad \text{or} \quad 2x + 1 = 0 \qquad \text{Zero-factor property}$$

$$3x = 4 \quad \text{or} \quad 2x = -1 \qquad \text{Solve each equation.}$$

$$x = \frac{4}{3} \quad \text{or} \quad x = -\frac{1}{2} \qquad \text{Same solutions as in Example 1}$$

When solving quadratic equations, it is a good idea to try factoring first. If the equation cannot be factored or if factoring is difficult, then use the quadratic formula.

Later in this section, we will show a way to determine whether factoring can be used to solve a quadratic equation.

**EXAMPLE 2 Using the Quadratic Formula (Irrational Solutions)**

Solve $4r^2 = 8r - 1$.

Write the equation in standard form as

$$4r^2 - 8r + 1 = 0,$$

and identify $a = 4$, $b = -8$, and $c = 1$. Now use the quadratic formula.

$$r = \frac{-b \pm \sqrt{b^2 - 4ac}}{2a}$$

$$r = \frac{-(-8) \pm \sqrt{(-8)^2 - 4(4)(1)}}{2(4)} \qquad a = 4,\ b = -8,\ c = 1$$

$$= \frac{8 \pm \sqrt{64 - 16}}{8}$$

$$= \frac{8 \pm \sqrt{48}}{8}$$

$$= \frac{8 \pm 4\sqrt{3}}{8} \qquad \sqrt{48} = \sqrt{16} \cdot \sqrt{3} = 4\sqrt{3}$$

$$= \frac{4(2 \pm \sqrt{3})}{4(2)} \qquad \text{Factor.}$$

$$= \frac{2 \pm \sqrt{3}}{2} \qquad \text{Lowest terms}$$

The solution set is $\left\{\frac{2 + \sqrt{3}}{2}, \frac{2 - \sqrt{3}}{2}\right\}$.

**Now Try Exercise 9.**

**CAUTION** *Every* quadratic equation must be expressed in standard form $ax^2 + bx + c = 0$ before we begin to solve it, whether we use factoring or the quadratic formula. Also, when writing solutions in lowest terms, be sure to *factor first;* then divide out the common factor, as shown in the last two steps in Example 2.

**EXAMPLE 3 Using the Quadratic Formula (Imaginary Solutions)**

Solve $(9q + 3)(q - 1) = -8$.

To write this equation in standard form, we first multiply and collect all nonzero terms on the left.

$$(9q + 3)(q - 1) = -8$$

$$9q^2 - 6q - 3 = -8$$

$$9q^2 - 6q + 5 = 0 \qquad \text{Standard form}$$

From the equation $9q^2 - 6q + 5 = 0$, we identify $a = 9$, $b = -6$, and $c = 5$, and use the quadratic formula.

$$q = \frac{-(-6) \pm \sqrt{(-6)^2 - 4(9)(5)}}{2(9)}$$

$$= \frac{6 \pm \sqrt{-144}}{18}$$

$$= \frac{6 \pm 12i}{18} \qquad \sqrt{-144} = 12i$$

$$= \frac{6(1 \pm 2i)}{6(3)} \qquad \text{Factor.}$$

$$= \frac{1 \pm 2i}{3} \qquad \text{Lowest terms}$$

The solution set, written in standard form $a + bi$ for complex numbers, is $\left\{\frac{1}{3} + \frac{2}{3}i, \frac{1}{3} - \frac{2}{3}i\right\}$.

**Now Try Exercise 33.**

**OBJECTIVE 3 Use the discriminant to determine the number and type of solutions.** The solutions of the quadratic equation $ax^2 + bx + c = 0$ are given by

$$x = \frac{-b \pm \sqrt{b^2 - 4ac}}{2a}. \quad \leftarrow \text{Discriminant}$$

If $a$, $b$, and $c$ are integers, the type of solutions of a quadratic equation—that is, rational, irrational, or imaginary—is determined by the expression under the radical sign, $b^2 - 4ac$. Because it distinguishes among the three types of solutions, $b^2 - 4ac$ is called the *discriminant.* By calculating the discriminant before solving a quadratic equation, we can predict whether the solutions will be rational numbers, irrational numbers, or imaginary numbers. (This can be useful in an applied problem, for example, where irrational or imaginary solutions are not acceptable.)

**Discriminant**

The **discriminant** of $ax^2 + bx + c = 0$ is $\mathbf{b^2 - 4ac}$. If $a$, $b$, and $c$ are integers, then the number and type of solutions are determined as follows.

| *Discriminant* | *Number and Type of Solutions* |
|---|---|
| Positive, and the square of an integer | Two rational solutions |
| Positive, but not the square of an integer | Two irrational solutions |
| Zero | One rational solution |
| Negative | Two imaginary solutions |

Calculating the discriminant can also help you decide whether to solve a quadratic equation by factoring or by using the quadratic formula. If the discriminant is a perfect square (including 0), then the equation can be solved by factoring. Otherwise, the quadratic formula should be used.

**EXAMPLE 4 Using the Discriminant**

Find the discriminant. Use it to predict the number and type of solutions for each equation. Tell whether the equation can be solved by factoring or whether the quadratic formula should be used.

**(a)** $6x^2 - x - 15 = 0$

We find the discriminant by evaluating $b^2 - 4ac$.

$$\begin{aligned} b^2 - 4ac &= (-1)^2 - 4(6)(-15) \qquad a = 6, b = -1, c = -15 \\ &= 1 + 360 \\ &= 361 \end{aligned}$$

A calculator shows that $361 = 19^2$, a perfect square. Since $a$, $b$, and $c$ are integers and the discriminant is a perfect square, there will be two rational solutions and the equation can be solved by factoring.

**(b)** $3m^2 - 4m = 5$

Write the equation in standard form as $3m^2 - 4m - 5 = 0$ to find that $a = 3$, $b = -4$, and $c = -5$.

$$\begin{aligned} b^2 - 4ac &= (-4)^2 - 4(3)(-5) \\ &= 16 + 60 \\ &= 76 \end{aligned}$$

Because 76 is positive but not the square of an integer and $a$, $b$, and $c$ are integers, the equation will have two irrational solutions and is best solved using the quadratic formula.

**(c)** $4x^2 + x + 1 = 0$

Since $a = 4$, $b = 1$, and $c = 1$, the discriminant is

$$1^2 - 4(4)(1) = -15.$$

Since the discriminant is negative and $a$, $b$, and $c$ are integers, this quadratic equation will have two imaginary solutions. The quadratic formula should be used to solve it.

**(d)** $4t^2 + 9 = 12t$

Write the equation as $4t^2 - 12t + 9 = 0$ to find $a = 4$, $b = -12$, and $c = 9$. The discriminant is

$$\begin{aligned} b^2 - 4ac &= (-12)^2 - 4(4)(9) \\ &= 144 - 144 \\ &= 0. \end{aligned}$$

Because the discriminant is 0, the quantity under the radical in the quadratic formula is 0, and there is only one rational solution. Again, the equation can be solved by factoring.

**Now Try Exercises 37 and 39.**

**EXAMPLE 5 Using the Discriminant**

Find $k$ so that $9x^2 + kx + 4 = 0$ will have only one rational solution.

The equation will have only one rational solution if the discriminant is 0. Since $a = 9$, $b = k$, and $c = 4$, the discriminant is

$$b^2 - 4ac = k^2 - 4(9)(4) = k^2 - 144.$$

Set the discriminant equal to 0 and solve for $k$.

$$k^2 - 144 = 0$$
$$k^2 = 144 \quad \text{Subtract 144.}$$
$$k = 12 \quad \text{or} \quad k = -12 \quad \text{Square root property}$$

The equation will have only one rational solution if $k = 12$ or $k = -12$.

**Now Try Exercise 53.**

## 11.3 EXERCISES

**For Extra Help**

- Student's Solutions Manual
- MyMathLab
- InterAct Math Tutorial Software
- AW Math Tutor Center
- MathXL
- Digital Video Tutor CD 16/Videotape 16

*Answer each question in Exercises 1–4.*

**1.** An early version of Microsoft *Word* for Windows included the 1.0 edition of *Equation Editor.* The documentation used the following for the quadratic formula.

$$x = -b \pm \frac{\sqrt{b^2 - 4ac}}{2a}$$

Was this correct? Explain.

**2.** The Cadillac Bar in Houston, Texas, encourages patrons to write (tasteful) messages on the walls. One person attempted to write the quadratic formula, as shown here.

$$x = \frac{-b\sqrt{b^2 - 4ac}}{2a}$$

Was this correct? Explain.

**3.** What is wrong with the following "solution" of $5x^2 - 5x + 1 = 0$?

$$x = \frac{5 \pm \sqrt{25 - 4(5)(1)}}{2(5)} \qquad a = 5,\ b = -5,\ c = 1$$

$$x = \frac{5 \pm \sqrt{5}}{10}$$

$$x = \frac{1}{2} \pm \sqrt{5}$$

**4.** A student claimed that the equation $2x^2 - 5 = 0$ cannot be solved using the quadratic formula because there is no first-degree $x$-term. Was the student correct? Explain.

*Use the quadratic formula to solve each equation. (All solutions for these equations are real numbers.) See Examples 1 and 2.*

**5.** $m^2 - 8m + 15 = 0$

**6.** $x^2 + 3x - 28 = 0$

**7.** $2k^2 + 4k + 1 = 0$

**8.** $2w^2 + 3w - 1 = 0$

**9.** $2x^2 - 2x = 1$

**10.** $9t^2 + 6t = 1$

**11.** $x^2 + 18 = 10x$

**12.** $x^2 - 4 = 2x$

**13.** $4k^2 + 4k - 1 = 0$

**14.** $4r^2 - 4r - 19 = 0$

**15.** $2 - 2x = 3x^2$

**16.** $26r - 2 = 3r^2$

**17.** $\frac{x^2}{4} - \frac{x}{2} = 1$

**18.** $p^2 + \frac{p}{3} = \frac{1}{6}$

**19.** $-2t(t + 2) = -3$

**20.** $-3x(x + 2) = -4$

**21.** $(r - 3)(r + 5) = 2$

**22.** $(k + 1)(k - 7) = 1$

**23.** $(g + 2)(g - 3) = 1$

**24.** $(x - 5)(x + 2) = 6$

**25.** $p = \frac{5(5 - p)}{3(p + 1)}$

**26.** $k = \frac{k + 15}{3(k - 1)}$

*Use the quadratic formula to solve each equation. (All solutions for these equations are imaginary numbers.) See Example 3.*

**27.** $x^2 - 3x + 6 = 0$

**28.** $x^2 - 5x + 20 = 0$

**29.** $r^2 - 6r + 14 = 0$

**30.** $t^2 + 4t + 11 = 0$

**31.** $4x^2 - 4x = -7$

**32.** $9x^2 - 6x = -7$

**33.** $x(3x + 4) = -2$

**34.** $z(2z + 3) = -2$

**35.** $(x + 5)(x - 6) = (2x - 1)(x - 4)$

**36.** $(3x - 4)(x + 2) = (2x - 5)(x + 5)$

*Use the discriminant to determine whether the solutions for each equation are*

**A.** *two rational numbers;* **B.** *one rational number;*
**C.** *two irrational numbers;* **D.** *two imaginary numbers.*

*Do not actually solve. See Example 4.*

**37.** $25x^2 + 70x + 49 = 0$

**38.** $4k^2 - 28k + 49 = 0$

**39.** $x^2 + 4x + 2 = 0$

**40.** $9x^2 - 12x - 1 = 0$

**41.** $3x^2 = 5x + 2$

**42.** $4x^2 = 4x + 3$

**43.** $3m^2 - 10m + 15 = 0$

**44.** $18x^2 + 60x + 82 = 0$

**45.** Using the discriminant, which equations in Exercises 37–44 can be solved by factoring?

*Based on your answer in Exercise 45, solve the equation given in each exercise.*

**46.** Exercise 37

**47.** Exercise 38

**48.** Exercise 41

**49.** Exercise 42

**50.** Find the discriminant for each quadratic equation. Use it to tell whether the equation can be solved by factoring or whether the quadratic formula should be used. Then solve each equation.

**(a)** $3k^2 + 13k = -12$ **(b)** $2x^2 + 19 = 14x$

**51.** Is it possible for the solution of a quadratic equation with integer coefficients to include just one irrational number? Why or why not?

**52.** Can the solution of a quadratic equation with integer coefficients include one real and one imaginary number? Why or why not?

*Find the value of a, b, or c so that each equation will have exactly one rational solution. See Example 5.*

**53.** $p^2 + bp + 25 = 0$

**54.** $r^2 - br + 49 = 0$

**55.** $am^2 + 8m + 1 = 0$

**56.** $at^2 + 24t + 16 = 0$

**57.** $9x^2 - 30x + c = 0$

**58.** $4m^2 + 12m + c = 0$

**59.** One solution of $4x^2 + bx - 3 = 0$ is $-\frac{5}{2}$. Find $b$ and the other solution.

**60.** One solution of $3x^2 - 7x + c = 0$ is $\frac{1}{3}$. Find $c$ and the other solution.

# 11.4 Equations Quadratic in Form

**OBJECTIVES**

1 Solve an equation with fractions by writing it in quadratic form.

2 Use quadratic equations to solve applied problems.

3 Solve an equation with radicals by writing it in quadratic form.

4 Solve an equation that is quadratic in form by substitution.

We have introduced four methods for solving quadratic equations written in standard form $ax^2 + bx + c = 0$. The following table lists some advantages and disadvantages of each method.

**Methods for Solving Quadratic Equations**

| *Method* | *Advantages* | *Disadvantages* |
|---|---|---|
| **Factoring** | This is usually the fastest method. | Not all polynomials are factorable; some factorable polynomials are hard to factor. |
| **Square root property** | This is the simplest method for solving equations of the form $(ax + b)^2 = c$. | Few equations are given in this form. |
| **Completing the square** | This method can always be used, although most people prefer the quadratic formula. | It requires more steps than other methods. |
| **Quadratic formula** | This method can always be used. | It is more difficult than factoring because of the square root, although calculators can simplify its use. |

**OBJECTIVE 1 Solve an equation with fractions by writing it in quadratic form.** A variety of nonquadratic equations can be written in the form of a quadratic equation and solved by using one of the methods in the table. As you solve the equations in this section, try to decide which method is best for each equation.

**EXAMPLE 1 Solving an Equation with Fractions that Leads to a Quadratic Equation**

Solve $\frac{1}{x} + \frac{1}{x-1} = \frac{7}{12}$.

Clear fractions by multiplying each term by the least common denominator, $12x(x - 1)$. (Note that the domain must be restricted to $x \neq 0$ and $x \neq 1$.)

$$12x(x-1)\frac{1}{x} + 12x(x-1)\frac{1}{x-1} = 12x(x-1)\frac{7}{12}$$

$$12(x-1) + 12x = 7x(x-1)$$

$$12x - 12 + 12x = 7x^2 - 7x \quad \text{Distributive property}$$

$$24x - 12 = 7x^2 - 7x \quad \text{Combine terms.}$$

Recall that a quadratic equation must be in standard form before it can be solved by factoring or the quadratic formula. Combine and rearrange terms so that one side

$$u - 4 = 0 \quad \text{or} \quad u - 9 = 0 \qquad \text{Zero-factor property}$$

$$u = 4 \quad \text{or} \quad u = 9 \qquad \text{Solve.}$$

To find $x$, we substitute $x^2$ for $u$.

$$x^2 = 4 \quad \text{or} \quad x^2 = 9$$

$$x = \pm 2 \quad \text{or} \quad x = \pm 3 \qquad \text{Square root property}$$

The equation $x^4 - 13x^2 + 36 = 0$, a fourth-degree equation, has four solutions.* The solution set is $\{-3, -2, 2, 3\}$, which can be verified by substituting into the original equation for $x$.

**(b)** $4x^6 + 1 = 5x^3$

This equation is quadratic in form with $u = x^3$ and $u^2 = x^6$.

$$4x^6 + 1 = 5x^3$$

$$4(x^3)^2 + 1 = 5x^3$$

$$4u^2 + 1 = 5u \qquad \text{Let } u = x^3.$$

$$4u^2 - 5u + 1 = 0 \qquad \text{Standard form}$$

$$(4u - 1)(u - 1) = 0 \qquad \text{Factor.}$$

$$4u - 1 = 0 \quad \text{or} \quad u - 1 = 0 \qquad \text{Zero-factor property}$$

$$u = \frac{1}{4} \quad \text{or} \quad u = 1 \qquad \text{Solve.}$$

$$x^3 = \frac{1}{4} \quad \text{or} \quad x^3 = 1 \qquad u = x^3$$

From these equations,

$$x = \sqrt[3]{\frac{1}{4}} = \frac{\sqrt[3]{1}}{\sqrt[3]{4}} = \frac{1}{\sqrt[3]{4}} \cdot \frac{\sqrt[3]{2}}{\sqrt[3]{2}} = \frac{\sqrt[3]{2}}{2} \qquad \text{or} \qquad x = \sqrt[3]{1} = 1.$$

There are other complex solutions for this equation, but finding them involves trigonometry. The real number solution set of $4x^6 + 1 = 5x^3$ is $\left\{\frac{\sqrt[3]{2}}{2}, 1\right\}$.

**(c)** $x^4 = 6x^2 - 3$

First write the equation as

$$x^4 - 6x^2 + 3 = 0 \qquad \text{or} \qquad (x^2)^2 - 6(x^2) + 3 = 0,$$

which is quadratic in form with $u = x^2$. Substitute $u$ for $x^2$ and $u^2$ for $x^4$ to get

$$u^2 - 6u + 3 = 0.$$

*In general, an equation in which an $n$th-degree polynomial equals 0 has $n$ complex solutions, although some of them may be repeated.

Since this equation cannot be solved by factoring, use the quadratic formula.

$$u = \frac{6 \pm \sqrt{36 - 12}}{2} \qquad a = 1, b = -6, c = 3$$

$$u = \frac{6 \pm \sqrt{24}}{2}$$

$$u = \frac{6 \pm 2\sqrt{6}}{2} \qquad \sqrt{24} = \sqrt{4} \cdot \sqrt{6} = 2\sqrt{6}$$

$$u = \frac{2(3 \pm \sqrt{6})}{2} \qquad \text{Factor.}$$

$$u = 3 \pm \sqrt{6} \qquad \text{Lowest terms}$$

Since $u = x^2$, find $x$ by using the square root property.

$$x^2 = 3 + \sqrt{6} \quad \text{or} \quad x^2 = 3 - \sqrt{6}$$

$$x = \pm\sqrt{3 + \sqrt{6}} \quad \text{or} \quad x = \pm\sqrt{3 - \sqrt{6}}$$

The solution set contains four numbers:

$$\left\{\sqrt{3 + \sqrt{6}}, -\sqrt{3 + \sqrt{6}}, \sqrt{3 - \sqrt{6}}, -\sqrt{3 - \sqrt{6}}\right\}.$$

**Now Try Exercises 55, 79, and 83.**

**NOTE** Some students prefer to solve equations like those in Examples 5(a) and (b) by factoring directly. For example,

$$x^4 - 13x^2 + 36 = 0 \qquad \text{Example 5(a) equation}$$

$$(x^2 - 9)(x^2 - 4) = 0 \qquad \text{Factor.}$$

$$(x + 3)(x - 3)(x + 2)(x - 2) = 0. \qquad \text{Factor again.}$$

Using the zero-factor property gives the same solutions obtained in Example 5(a). Equations that cannot be solved by factoring, like that in Example 5(c), must be solved using the method of substitution and the quadratic formula.

## EXAMPLE 6 Solving Equations That Are Quadratic in Form

Solve each equation.

**(a)** $2(4m - 3)^2 + 7(4m - 3) + 5 = 0$

Because of the repeated quantity $4m - 3$, this equation is quadratic in form with $u = 4m - 3$.

$$2(4m - 3)^2 + 7(4m - 3) + 5 = 0$$

$$2u^2 + 7u + 5 = 0 \qquad \text{Let } 4m - 3 = u.$$

$$(2u + 5)(u + 1) = 0 \qquad \text{Factor.}$$

$$2u + 5 = 0 \quad \text{or} \quad u + 1 = 0 \qquad \text{Zero-factor property}$$

$$u = -\frac{5}{2} \quad \text{or} \quad u = -1$$

$$4m - 3 = -\frac{5}{2} \quad \text{or} \quad 4m - 3 = -1 \qquad \text{Substitute } 4m - 3 \text{ for } u.$$

$$4m = \frac{1}{2} \quad \text{or} \quad 4m = 2 \qquad \text{Solve for } m.$$

$$m = \frac{1}{8} \quad \text{or} \quad m = \frac{1}{2}$$

Check that the solution set of the original equation is $\{\frac{1}{8}, \frac{1}{2}\}$.

**(b)** $2a^{2/3} - 11a^{1/3} + 12 = 0$

Let $a^{1/3} = u$; then $a^{2/3} = (a^{1/3})^2 = u^2$. Substitute into the given equation.

$$2u^2 - 11u + 12 = 0 \qquad \text{Let } a^{1/3} = u;\ a^{2/3} = u^2.$$

$$(2u - 3)(u - 4) = 0 \qquad \text{Factor.}$$

$$2u - 3 = 0 \quad \text{or} \quad u - 4 = 0 \qquad \text{Zero-factor property}$$

$$u = \frac{3}{2} \quad \text{or} \quad u = 4$$

$$a^{1/3} = \frac{3}{2} \quad \text{or} \quad a^{1/3} = 4 \qquad u = a^{1/3}$$

$$(a^{1/3})^3 = \left(\frac{3}{2}\right)^3 \quad \text{or} \quad (a^{1/3})^3 = 4^3 \qquad \text{Cube each side.}$$

$$a = \frac{27}{8} \quad \text{or} \quad a = 64$$

Check that the solution set is $\{\frac{27}{8}, 64\}$.

**Now Try Exercises 59 and 65.**

**CAUTION** A common error when solving problems like those in Examples 5 and 6 is to stop too soon. Once you have solved for $u$, remember to substitute and solve for the values of the *original* variable.

## 11.4 EXERCISES

**For Extra Help**

- Student's Solutions Manual
- MyMathLab
- InterAct Math Tutorial Software
- AW Math Tutor Center
- MathXL
- Digital Video Tutor CD 16/Videotape 16

*Refer to the box at the beginning of this section. Decide whether* factoring, the square root property, *or* the quadratic formula *is most appropriate for solving each quadratic equation. Do not actually solve the equations.*

**1.** $(2x + 3)^2 = 4$ **2.** $4x^2 - 3x = 1$ **3.** $z^2 + 5z - 8 = 0$

**4.** $2k^2 + 3k = 1$ **5.** $3m^2 = 2 - 5m$ **6.** $p^2 = 5$

*Write a sentence describing the first step you would take to solve each equation. Do not actually solve.*

**7.** $\frac{14}{x} = x - 5$ **8.** $\sqrt{1 + x} + x = 5$

**9.** $(r^2 + r)^2 - 8(r^2 + r) + 12 = 0$ **10.** $3t = \sqrt{16 - 10t}$

**11.** What is wrong with the following "solution"?

$$x = \sqrt{3x + 4}$$
$$x^2 = 3x + 4 \quad \text{Square both sides.}$$
$$x^2 - 3x - 4 = 0$$
$$(x - 4)(x + 1) = 0$$
$$x - 4 = 0 \quad \text{or} \quad x + 1 = 0$$
$$x = 4 \quad \text{or} \quad x = -1$$

Solution set: $\{4, -1\}$

**12.** What is wrong with the following "solution"?

$$2(m - 1)^2 - 3(m - 1) + 1 = 0$$
$$2u^2 - 3u + 1 = 0 \quad \text{Let } u = m - 1.$$
$$(2u - 1)(u - 1) = 0$$
$$2u - 1 = 0 \quad \text{or} \quad u - 1 = 0$$
$$u = \frac{1}{2} \quad \text{or} \quad u = 1$$

Solution set: $\left\{\frac{1}{2}, 1\right\}$

*Solve each equation. Check your solutions. See Example 1.*

**13.** $\frac{14}{x} = x - 5$

**14.** $\frac{-12}{x} = x + 8$

**15.** $1 - \frac{3}{x} - \frac{28}{x^2} = 0$

**16.** $4 - \frac{7}{r} - \frac{2}{r^2} = 0$

**17.** $3 - \frac{1}{t} = \frac{2}{t^2}$

**18.** $1 + \frac{2}{k} = \frac{3}{k^2}$

**19.** $\frac{1}{x} + \frac{2}{x + 2} = \frac{17}{35}$

**20.** $\frac{2}{m} + \frac{3}{m + 9} = \frac{11}{4}$

**21.** $\frac{2}{x + 1} + \frac{3}{x + 2} = \frac{7}{2}$

**22.** $\frac{4}{3 - p} + \frac{2}{5 - p} = \frac{26}{15}$

**23.** $\frac{3}{2x} - \frac{1}{2(x + 2)} = 1$

**24.** $\frac{4}{3x} - \frac{1}{2(x + 1)} = 1$

**25.** $3 = \frac{1}{t + 2} + \frac{2}{(t + 2)^2}$

**26.** $1 + \frac{2}{3z + 2} = \frac{15}{(3z + 2)^2}$

**27.** $\frac{6}{p} = 2 + \frac{p}{p + 1}$

**28.** $\frac{k}{2 - k} + \frac{2}{k} = 5$

*Use the concepts of this section to answer each question.*

**29.** A boat goes 20 mph in still water, and the rate of the current is $t$ mph.

**(a)** What is the rate of the boat when it travels upstream?

**(b)** What is the rate of the boat when it travels downstream?

**30.** If it takes $m$ hr to grade a set of papers, what is the grader's rate (in job per hour)?

*Solve each problem. See Examples 2 and 3.*

**31.** On a windy day Yoshiaki found that he could go 16 mi downstream and then 4 mi back upstream at top speed in a total of 48 min. What was the top speed of Yoshiaki's boat if the current was 15 mph?

| | $d$ | $r$ | $t$ |
|---|---|---|---|
| *Upstream* | 4 | $x - 15$ | |
| *Downstream* | 16 | | |

**32.** Lekesha flew her plane for 6 hr at a constant speed. She traveled 810 mi with the wind, then turned around and traveled 720 mi against the wind. The wind speed was a constant 15 mph. Find the speed of the plane.

| | $d$ | $r$ | $t$ |
|---|---|---|---|
| *With Wind* | 810 | | |
| *Against Wind* | 720 | | |

**33.** In Canada, Medicine Hat and Cranbrook are 300 km apart. Harry rides his Honda 20 km per hr faster than Yoshi rides his Yamaha. Find Harry's average speed if he travels from Cranbrook to Medicine Hat in $1\frac{1}{4}$ hr less time than Yoshi. (*Source: State Farm Road Atlas.*)

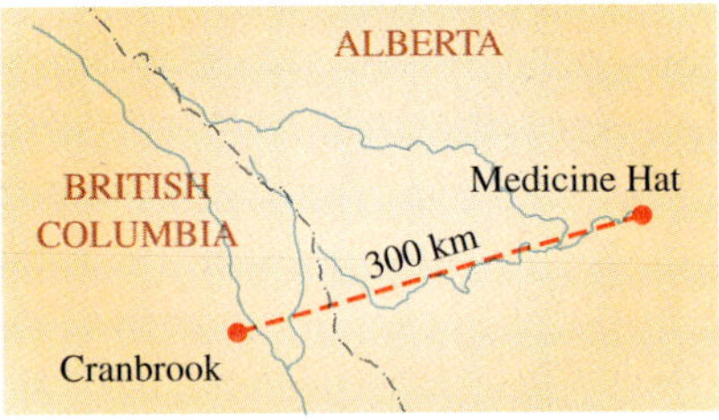


**34.** In California, the distance from Jackson to Lodi is about 40 mi, as is the distance from Lodi to Manteca. Rico drove from Jackson to Lodi during the rush hour, stopped in Lodi for a root beer, and then drove on to Manteca at 10 mph faster. Driving time for the entire trip was 88 min. Find his speed from Jackson to Lodi. (*Source: State Farm Road Atlas.*)

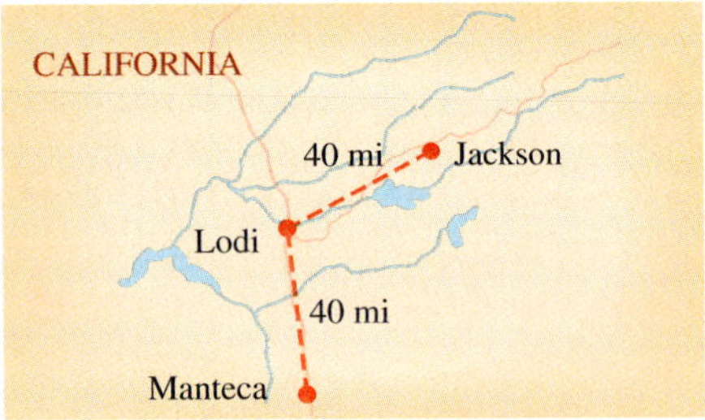


**35.** Working together, two people can cut a large lawn in 2 hr. One person can do the job alone in 1 hr less time than the other. How long (to the nearest tenth) would it take the faster person to do the job? (*Hint: x* is the time of the faster person.)

| | Rate | Time Working Together | Fractional Part of the Job Done |
|---|---|---|---|
| Faster Worker | $\frac{1}{x}$ | 2 | |
| Slower Worker | | 2 | |

**36.** A janitorial service provides two people to clean an office building. Working together, the two can clean the building in 5 hr. One person is new to the job and would take 2 hr longer than the other person to clean the building alone. How long (to the nearest tenth) would it take the new worker to clean the building alone?

| | Rate | Time Working Together | Fractional Part of the Job Done |
|---|---|---|---|
| Faster Worker | | | |
| Slower Worker | | | |

**37.** Rusty and Nancy Brauner are planting flats of spring flowers. Working alone, Rusty would take 2 hr longer than Nancy to plant the flowers. Working together, they do the job in 12 hr. How long would it have taken each person working alone?

**38.** Jay Beckenstein can work through a stack of invoices in 1 hr less time than Colleen Manley Jones can. Working together they take $1\frac{1}{2}$ hr. How long would it take each person working alone?

**39.** A washing machine can be filled in 6 min if both the hot and cold water taps are fully opened. Filling the washer with hot water alone takes 9 min longer than filling it with cold water alone. How long does it take to fill the washer with cold water?

**40.** Two pipes together can fill a large tank in 2 hr. One of the pipes, used alone, takes 3 hr longer than the other to fill the tank. How long would each pipe take to fill the tank alone?

*Solve each equation. Check your solutions. See Example 4.*

**41.** $x = \sqrt{7x - 10}$
**42.** $z = \sqrt{5z - 4}$
**43.** $2x = \sqrt{11x + 3}$
**44.** $4x = \sqrt{6x + 1}$
**45.** $3x = \sqrt{16 - 10x}$
**46.** $4t = \sqrt{8t + 3}$
**47.** $p - 2\sqrt{p} = 8$
**48.** $k + \sqrt{k} = 12$
**49.** $m = \sqrt{\dfrac{6 - 13m}{5}}$
**50.** $r = \sqrt{\dfrac{20 - 19r}{6}}$

*Solve each equation. Check your solutions. See Examples 5 and 6.*

**51.** $t^4 - 18t^2 + 81 = 0$
**52.** $x^4 - 8x^2 + 16 = 0$
**53.** $4k^4 - 13k^2 + 9 = 0$
**54.** $9x^4 - 25x^2 + 16 = 0$
**55.** $x^4 + 48 = 16x^2$
**56.** $z^4 = 17z^2 - 72$
**57.** $(x + 3)^2 + 5(x + 3) + 6 = 0$
**58.** $(k - 4)^2 + (k - 4) - 20 = 0$
**59.** $3(m + 4)^2 - 8 = 2(m + 4)$
**60.** $(t + 5)^2 + 6 = 7(t + 5)$
**61.** $2 + \dfrac{5}{3k - 1} = \dfrac{-2}{(3k - 1)^2}$
**62.** $3 - \dfrac{7}{2p + 2} = \dfrac{6}{(2p + 2)^2}$
**63.** $2 - 6(m - 1)^{-2} = (m - 1)^{-1}$
**64.** $3 - 2(x - 1)^{-1} = (x - 1)^{-2}$
**65.** $x^{2/3} + x^{1/3} - 2 = 0$
**66.** $x^{2/3} - 2x^{1/3} - 3 = 0$
**67.** $r^{2/3} + r^{1/3} - 12 = 0$
**68.** $3x^{2/3} - x^{1/3} - 24 = 0$
**69.** $4k^{4/3} - 13k^{2/3} + 9 = 0$
**70.** $9m^{2/5} = 16 - 10m^{1/5}$
**71.** $2(1 + \sqrt{r})^2 = 13(1 + \sqrt{r}) - 6$
**72.** $(k^2 + k)^2 + 12 = 8(k^2 + k)$
**73.** $2x^4 + x^2 - 3 = 0$
**74.** $4k^4 + 5k^2 + 1 = 0$

*The equations in Exercises 75–84 are not grouped by type. Decide which method of solution applies, and then solve each equation. Give only real solutions. See Examples 1 and 4–6.*

**75.** $12x^4 - 11x^2 + 2 = 0$
**76.** $\left(x - \dfrac{1}{2}\right)^2 + 5\left(x - \dfrac{1}{2}\right) - 4 = 0$
**77.** $\sqrt{2x + 3} = 2 + \sqrt{x - 2}$
**78.** $\sqrt{m + 1} = -1 + \sqrt{2m}$
**79.** $2m^6 + 11m^3 + 5 = 0$
**80.** $8x^6 + 513x^3 + 64 = 0$
**81.** $6 = 7(2w - 3)^{-1} + 3(2w - 3)^{-2}$
**82.** $m^6 - 10m^3 = -9$
**83.** $2x^4 - 9x^2 = -2$
**84.** $8x^4 + 1 = 11x^2$

## RELATING CONCEPTS (EXERCISES 85–90)

**For Individual or Group Work**

*Consider the following equation, and* ***work Exercises 85–90 in order.***

$$\frac{x^2}{(x - 3)^2} + \frac{3x}{x - 3} - 4 = 0.$$

**85.** Why must 3 be excluded from the domain of this equation?

**86.** Multiply both sides of the equation by the LCD, $(x - 3)^2$, and solve. There is only one solution—what is it?

**87.** Write the equation so that it is quadratic in form using the rational expression $\frac{x}{x-3}$.

**88.** Explain why the expression $\frac{x}{x-3}$ cannot equal 1.

**89.** Solve the equation from Exercise 87 by making the substitution $t = \frac{x}{x-3}$. You should get two values for $t$. Why is one of them impossible for this equation?

**90.** Solve the equation $x^2(x-3)^{-2} + 3x(x-3)^{-1} - 4 = 0$ by letting $s = (x-3)^{-1}$. You should get two values for $s$. Why is this impossible for this equation?

## SUMMARY EXERCISES ON SOLVING QUADRATIC EQUATIONS

*Exercises marked * require knowledge of imaginary numbers.*

*We have introduced four algebraic methods for solving quadratic equations written in the form $ax^2 + bx + c = 0$:* ***factoring, the square root property, completing the square, and the quadratic formula.*** *Refer to the summary box at the beginning of Section 11.4 to review some of the advantages and disadvantages of each method. Then solve each quadratic equation by the method of your choice.*

**1.** $p^2 = 7$

**2.** $6x^2 - x - 15 = 0$

**3.** $n^2 + 6n + 4 = 0$

**4.** $(x-4)^2 = 25$

**5.** $\frac{5}{m} + \frac{12}{m^2} = 2$

**6.** $3m^2 = 3 - 8m$

**7.** $2r^2 - 4r + 1 = 0$

***8.** $x^2 = -12$

**9.** $x\sqrt{2} = \sqrt{5x - 2}$

**10.** $m^4 - 10m^2 + 9 = 0$

**11.** $(2k+3)^2 = 8$

**12.** $\frac{2}{x} + \frac{1}{x-2} = \frac{5}{3}$

**13.** $t^4 + 14 = 9t^2$

**14.** $8x^2 - 4x = 2$

***15.** $z^2 + z + 1 = 0$

**16.** $5x^6 + 2x^3 - 7 = 0$

**17.** $4t^2 - 12t + 9 = 0$

**18.** $x\sqrt{3} = \sqrt{2 - x}$

**19.** $r^2 - 72 = 0$

**20.** $-3x^2 + 4x = -4$

**21.** $x^2 - 5x - 36 = 0$

**22.** $w^2 = 169$

***23.** $3p^2 = 6p - 4$

**24.** $z = \sqrt{\frac{5z+3}{2}}$

**25.** $2(3k-1)^2 + 5(3k-1) = -2$

***26.** $\frac{4}{r^2} + 3 = \frac{1}{r}$

# 11.5 Formulas and Further Applications

**OBJECTIVES**

1. Solve formulas for variables involving squares and square roots.
2. Solve applied problems using the Pythagorean formula.
3. Solve applied problems using area formulas.
4. Solve applied problems using quadratic functions as models.

**OBJECTIVE 1 Solve formulas for variables involving squares and square roots.** The methods presented earlier can be used to solve such formulas.

**EXAMPLE 1 Solving for Variables Involving Squares or Square Roots**

Solve each formula for the given variable.

**(a)** $w = \frac{kFr}{v^2}$ for $v$

$$w = \frac{kFr}{v^2} \quad \text{Get } v \text{ alone on one side.}$$

$$v^2 w = kFr \qquad \text{Multiply by } v^2.$$

$$v^2 = \frac{kFr}{w} \qquad \text{Divide by } w.$$

$$v = \pm\sqrt{\frac{kFr}{w}} \qquad \text{Square root property}$$

$$v = \frac{\pm\sqrt{kFr}}{\sqrt{w}} \cdot \frac{\sqrt{w}}{\sqrt{w}} = \frac{\pm\sqrt{kFrw}}{w} \qquad \text{Rationalize the denominator.}$$

**(b)** $d = \sqrt{\frac{4A}{\pi}}$ for $A$

$$d = \sqrt{\frac{4A}{\pi}} \quad \text{Get } A \text{ alone on one side.}$$

$$d^2 = \frac{4A}{\pi} \qquad \text{Square both sides.}$$

$$\pi d^2 = 4A \qquad \text{Multiply by } \pi.$$

$$\frac{\pi d^2}{4} = A \qquad \text{Divide by 4.}$$

**Now Try Exercises 9 and 19.**

**NOTE** In many formulas like $v = \frac{\pm\sqrt{kFrw}}{w}$ in Example 1(a), we choose the positive value. In our work here, we will include both positive and negative values.

**EXAMPLE 2 Solving for a Variable that Appears in First- and Second-Degree Terms**

Solve $s = 2t^2 + kt$ for $t$.

Since the given equation has terms with $t^2$ and $t$, write it in standard form $ax^2 + bx + c = 0$, with $t$ as the variable instead of $x$.

$$2t^2 + kt - s = 0$$

*Solve each problem. When appropriate, round answers to the nearest tenth. See Example 5.*

**43.** An object is projected directly upward from the ground. After $t$ sec its distance in feet above the ground is

$$s(t) = 144t - 16t^2.$$

After how many seconds will the object be 128 ft above the ground? (*Hint:* Look for a common factor before solving the equation.)

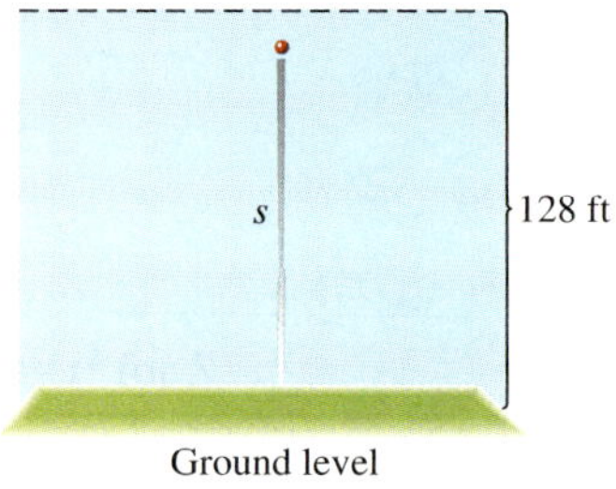


**44.** When does the object in Exercise 43 strike the ground?

**45.** A ball is projected upward from the ground. Its distance in feet from the ground in $t$ sec is given by

$$s(t) = -16t^2 + 128t.$$

At what times will the ball be 213 ft from the ground?

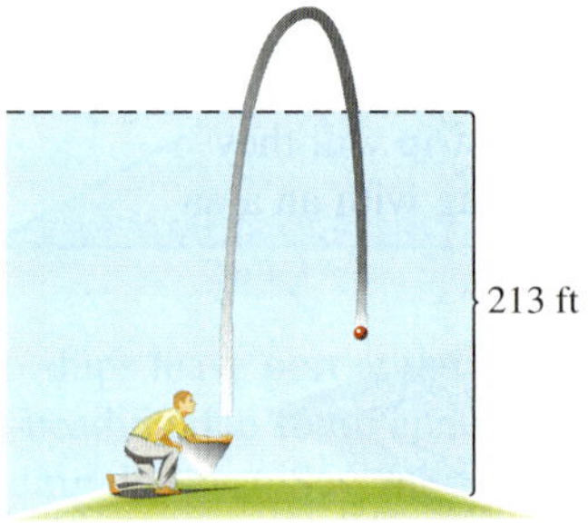


**46.** A toy rocket is launched from ground level. Its distance in feet from the ground in $t$ sec is given by

$$s(t) = -16t^2 + 208t.$$

At what times will the rocket be 550 ft from the ground?

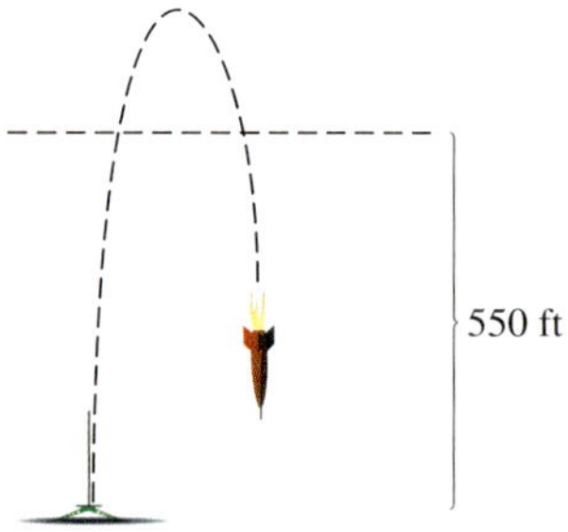


**47.** The function defined by

$$D(t) = 13t^2 - 100t$$

gives the distance in feet a car going approximately 68 mph will skid in $t$ sec. Find the time it would take for the car to skid 180 ft.

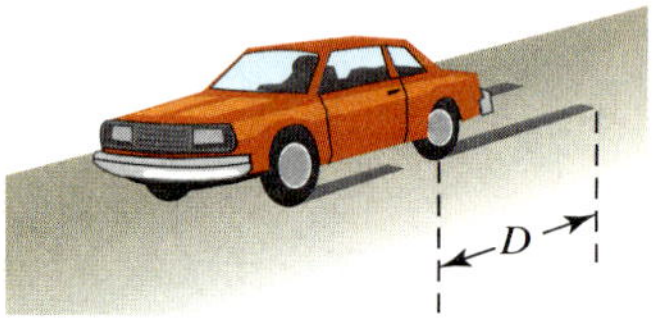


**48.** The function given in Exercise 47 becomes

$$D(t) = 13t^2 - 73t$$

for a car going 50 mph. Find the time for this car to skid 218 ft.

*A rock is projected upward from ground level, and its distance in feet from the ground in t sec is given by* $s(t) = -16t^2 + 160t$. *Use algebra and a short explanation to answer Exercises 49 and 50.*

**49.** After how many seconds does it reach a height of 400 ft? How would you describe in words its position at this height?

**50.** After how many seconds does it reach a height of 425 ft? How would you interpret the mathematical result here?

*Solve each problem using a quadratic equation.*

**51.** A certain bakery has found that the daily demand for bran muffins is $\frac{3200}{p}$, where $p$ is the price of a muffin in cents. The daily supply is $3p - 200$. Find the price at which supply and demand are equal.

**52.** In one area the demand for compact discs is $\frac{700}{P}$ per day, where $P$ is the price in dollars per disc. The supply is $5P - 1$ per day. At what price does supply equal demand?

**53.** The formula $A = P(1 + r)^2$ gives the amount $A$ in dollars that $P$ dollars will grow to in 2 yr at interest rate $r$ (where $r$ is given as a decimal), using compound interest. What interest rate will cause \$2000 to grow to \$2142.25 in 2 yr?

**54.** If a square piece of cardboard has 3-in. squares cut from its corners and then has the flaps folded up to form an open-top box, the volume of the box is given by the formula $V = 3(x - 6)^2$, where $x$ is the length of each side of the original piece of cardboard in inches. What original length would yield a box with volume 432 in.$^3$?

*Sales of SUVs (sport utility vehicles) in the United States (in millions) for the years 1990 through 1999 are shown in the bar graph and can be modeled by the quadratic function defined by*

$$f(x) = .016x^2 + .124x + .787.$$

*Here, $x = 0$ represents 1990, $x = 1$ represents 1991, and so on. Use the graph and the model to work Exercises 55–58. See Example 6.*

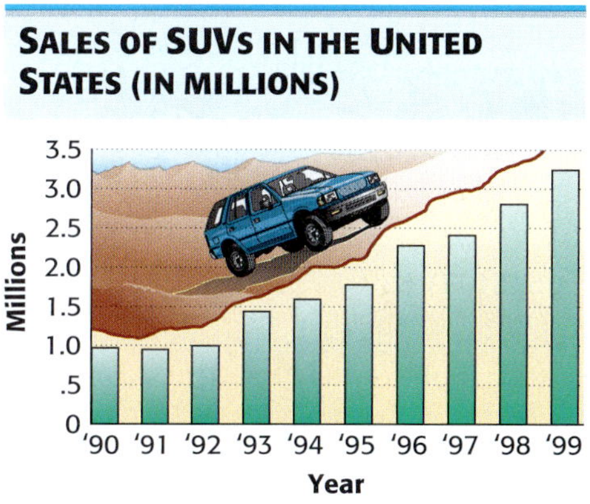


***Source:*** CNW Marketing Research of Bandon, OR, based on automakers' reported sales.

**55. (a)** Use the graph to estimate sales in 1997 to the nearest tenth.

**(b)** Use the model to approximate sales in 1997 to the nearest tenth. How does this result compare to your estimate from part (a)?

**56. (a)** Use the model to estimate sales in 2000 to the nearest tenth.

**(b)** Sales through October 2000 were about 2.9 million. Based on this, is the sales estimate for 2000 from part (a) reasonable? Explain.

**57.** Based on the model, in what year did sales reach 2 million? (Round down to the nearest year.) How does this result compare to the sales shown in the graph?

**58.** Based on the model, in what year did sales reach 3 million? (Round down to the nearest year.) How does this result compare to the sales shown in the graph?

*William Froude was a 19th century naval architect who used the expression*

$$\frac{v^2}{g\ell}$$

*in shipbuilding. This expression, known as the Froude number, was also used by R. McNeill Alexander in his research on dinosaurs.* (Source: *"How Dinosaurs Ran,"* Scientific American, *April 1991.) In Exercises 59 and 60, find the value of v (in meters per second), given g* $= 9.8$ m per sec$^2$.

**59.** Rhinoceros: $\ell = 1.2$; Froude number $= 2.57$

**60.** Triceratops: $\ell = 2.8$; Froude number $= .16$

*Recall that corresponding sides of similar triangles are proportional. Use this fact to find the lengths of the indicated sides of each pair of similar triangles. Check all possible solutions in both triangles. Sides of a triangle cannot be negative (and are not drawn to scale here).*

**61.** Side *AC*

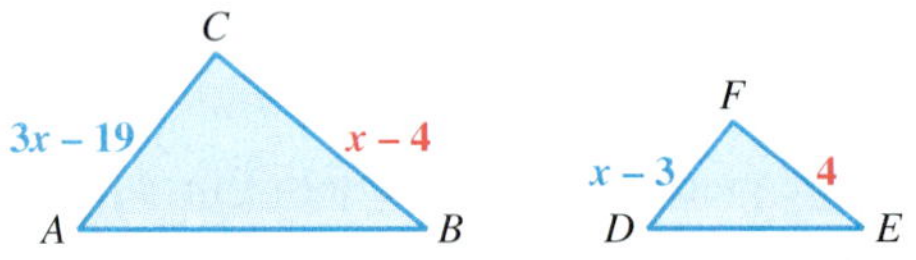


**62.** Side *RQ*

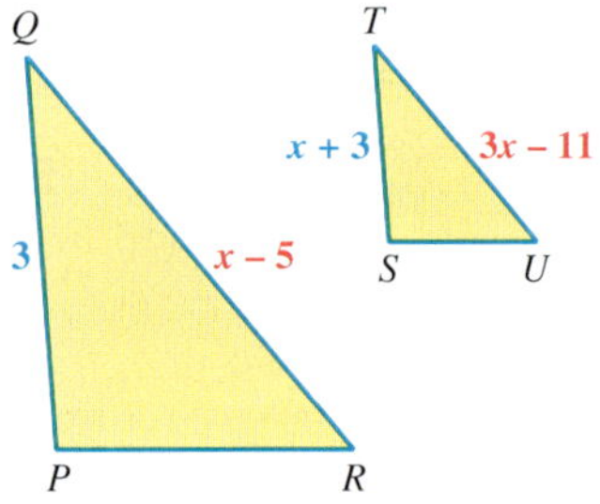


## 11.6 Graphs of Quadratic Functions

**OBJECTIVES**

1 Graph a quadratic function.
2 Graph parabolas with horizontal and vertical shifts.
3 Predict the shape and direction of a parabola from the coefficient of $x^2$.
4 Find a quadratic function to model data.

**OBJECTIVE 1 Graph a quadratic function.** In Chapter 4, we graphed a few simple second-degree polynomial equations by point-plotting. In Figure 4, we repeat a table of ordered pairs for the simplest quadratic equation, $y = x^2$, and the resulting graph.

| x | y |
|---|---|
| −2 | 4 |
| −1 | 1 |
| 0 | 0 |
| 1 | 1 |
| 2 | 4 |

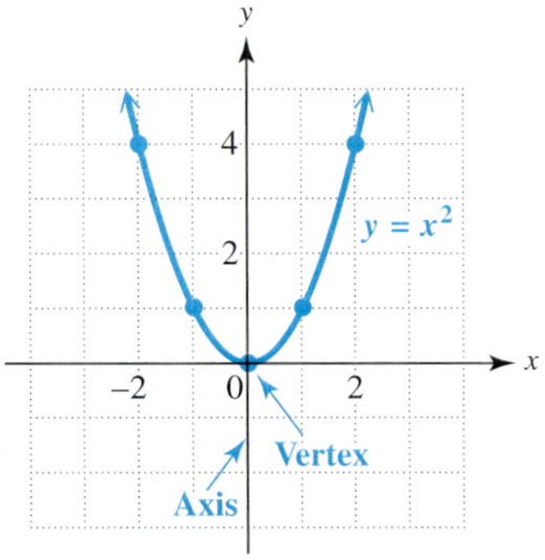


**FIGURE 4**

As mentioned in Chapter 4, this graph is called a **parabola.** The point (0, 0), the lowest point on the curve, is the **vertex** of this parabola. The vertical line through the vertex is the **axis** of the parabola, here $x = 0$. A parabola is **symmetric about its axis;**

that is, if the graph were folded along the axis, the two portions of the curve would coincide. As Figure 4 suggests, the graph is that of a function. Because $x$ can be any real number, the domain of the function defined by $y = x^2$ is $(-\infty, \infty)$. Since $y$ is always nonnegative, the range is $[0, \infty)$.

In Section 11.5, we solved applications modeled by *quadratic functions.* In this section and the next, we consider graphs of more general quadratic functions as defined here.

**Quadratic Function**

A function that can be written in the form

$$f(x) = ax^2 + bx + c$$

for real numbers $a$, $b$, and $c$, with $a \neq 0$, is a **quadratic function.**

The graph of any quadratic function is a parabola with a vertical axis.

**NOTE** We use the variable $y$ and function notation $f(x)$ interchangeably when discussing parabolas. Although we use the letter $f$ most often to name quadratic functions, other letters can be used. We use the capital letter $F$ to distinguish between different parabolas graphed on the same coordinate axes.

Parabolas, which are a type of *conic section* (Chapter 13), have many applications. The large dishes seen on the sidelines of televised football games, which are used by television crews to pick up the shouted signals of players on the field, have cross sections that are parabolas. Cross sections of satellite dishes and automobile headlights also form parabolas. The cables that are used to support suspension bridges are shaped like parabolas.

**OBJECTIVE 2 Graph parabolas with horizontal and vertical shifts.** Parabolas need not have their vertices at the origin, as does the graph of $f(x) = x^2$. For example, to graph a parabola of the form $F(x) = x^2 + k$, start by selecting sample values of $x$ like those that were used to graph $f(x) = x^2$. The corresponding values of $F(x)$ in $F(x) = x^2 + k$ differ by $k$ from those of $f(x) = x^2$. For this reason, the graph of $F(x) = x^2 + k$ is *shifted,* or *translated,* $k$ units vertically compared with that of $f(x) = x^2$.

## EXAMPLE 1 Graphing a Parabola with a Vertical Shift

Graph $F(x) = x^2 - 2$.

This graph has the same shape as that of $f(x) = x^2$, but since $k$ here is $-2$, the graph is shifted 2 units down, with vertex $(0, -2)$. Every function value is 2 less than the corresponding function value of $f(x) = x^2$. Plotting points on both sides of the vertex gives the graph in Figure 5 on the next page. Notice that since the parabola is symmetric about its axis $x = 0$, the plotted points are "mirror images" of each other. Since $x$ can be any real number, the domain is still $(-\infty, \infty)$; the value of $y$ (or $F(x)$) is always greater than or equal to $-2$, so the range is $[-2, \infty)$. The graph of $f(x) = x^2$ is shown in Figure 5 for comparison.

| $x$ | $f(x) = x^2$ | $F(x) = x^2 - 2$ |
|---|---|---|
| $-2$ | 4 | 2 |
| $-1$ | 1 | $-1$ |
| 0 | 0 | $-2$ |
| 1 | 1 | $-1$ |
| 2 | 4 | 2 |

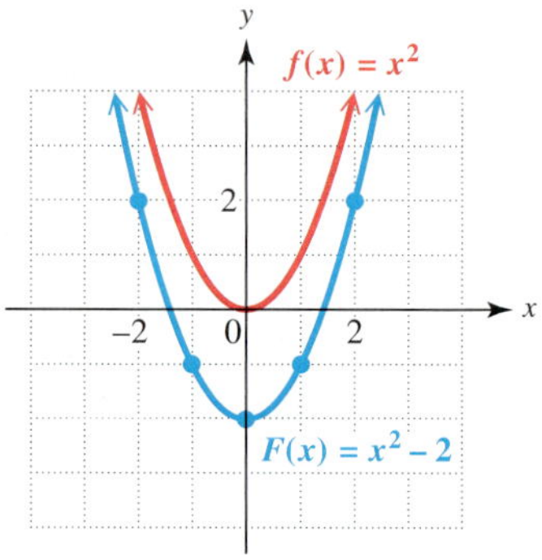


**FIGURE 5**

**Now Try Exercise 23.**

### Vertical Shift

The graph of $F(x) = x^2 + k$ is a parabola with the same shape as the graph of $f(x) = x^2$. The parabola is shifted $k$ units up if $k > 0$, and $|k|$ units down if $k < 0$. The vertex is $(0, k)$.

The graph of $F(x) = (x - h)^2$ is also a parabola with the same shape as that of $f(x) = x^2$. Because $(x - h)^2 \geq 0$ for all $x$, the vertex of $F(x) = (x - h)^2$ is the lowest point on the parabola. The lowest point occurs here when $F(x)$ is 0. To get $F(x)$ equal to 0, let $x = h$ so the vertex of $F(x) = (x - h)^2$ is $(h, 0)$. Based on this, the graph of $F(x) = (x - h)^2$ is shifted $h$ units horizontally compared with that of $f(x) = x^2$.

### EXAMPLE 2 Graphing a Parabola with a Horizontal Shift

Graph $F(x) = (x - 2)^2$.

If $x = 2$, then $F(x) = 0$, giving the vertex $(2, 0)$. The graph of $F(x) = (x - 2)^2$ has the same shape as that of $f(x) = x^2$ but is shifted 2 units to the right. Plotting several points on one side of the vertex and using symmetry about the axis $x = 2$ to find corresponding points on the other side of the vertex gives the graph in Figure 6. Again, the domain is $(-\infty, \infty)$; the range is $[0, \infty)$.

| $x$ | $F(x) = (x - 2)^2$ |
|---|---|
| 0 | 4 |
| 1 | 1 |
| 2 | 0 |
| 3 | 1 |
| 4 | 4 |

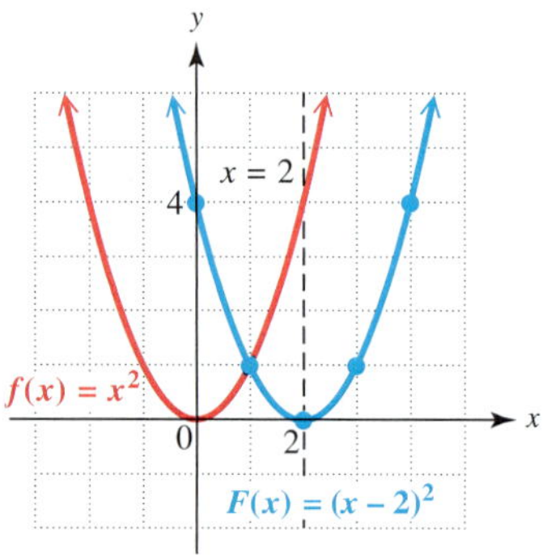


**FIGURE 6**

**Now Try Exercise 27.**

**Horizontal Shift**

The graph of $\boldsymbol{F(x) = (x - h)^2}$ is a parabola with the same shape as the graph of $f(x) = x^2$. The parabola is shifted $h$ units horizontally: $h$ units to the right if $h > 0$, and $|h|$ units to the left if $h < 0$. The vertex is $(h, 0)$.

**CAUTION** Errors frequently occur when horizontal shifts are involved. To determine the direction and magnitude of a horizontal shift, find the value that would cause the expression $x - h$ to equal 0. For example, the graph of $F(x) = (x - 5)^2$ would be shifted 5 units to the *right,* because $+5$ would cause $x - 5$ to equal 0. On the other hand, the graph of $F(x) = (x + 5)^2$ would be shifted 5 units to the *left,* because $-5$ would cause $x + 5$ to equal 0.

A parabola can have both horizontal and vertical shifts.

## EXAMPLE 3 Graphing a Parabola with Horizontal and Vertical Shifts

Graph $F(x) = (x + 3)^2 - 2$.

This graph has the same shape as that of $f(x) = x^2$, but is shifted 3 units to the left (since $x + 3 = 0$ if $x = -3$) and 2 units down (because of the $-2$). As shown in Figure 7, the vertex is $(-3, -2)$, with axis $x = -3$. This function has domain $(-\infty, \infty)$ and range $[-2, \infty)$.

| $x$ | $F(x)$ |
|---|---|
| $-5$ | 2 |
| $-4$ | $-1$ |
| $-3$ | $-2$ |
| $-2$ | $-1$ |
| $-1$ | 2 |

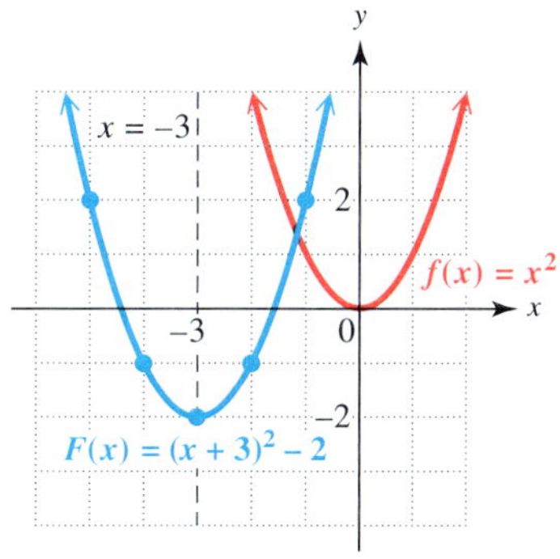


FIGURE 7

**Now Try Exercise 29.**

The characteristics of the graph of a parabola of the form $F(x) = (x - h)^2 + k$ are summarized as follows.

**Vertex and Axis of a Parabola**

The graph of $\boldsymbol{F(x) = (x - h)^2 + k}$ is a parabola with the same shape as the graph of $f(x) = x^2$, but with vertex $(h, k)$. The axis is the vertical line $x = h$.

**OBJECTIVE 3 Predict the shape and direction of a parabola from the coefficient of $x^2$.** Not all parabolas open up, and not all parabolas have the same shape as the graph of $f(x) = x^2$.

## EXAMPLE 4 Graphing a Parabola That Opens Down

Graph $f(x) = -\frac{1}{2}x^2$.

This parabola is shown in Figure 8. The coefficient $-\frac{1}{2}$ affects the shape of the graph; the $\frac{1}{2}$ makes the parabola wider (since the values of $\frac{1}{2}x^2$ increase more slowly than those of $x^2$), and the negative sign makes the parabola open down. The graph is not shifted in any direction; the vertex is still (0, 0) and the axis is $x = 0$. Unlike the parabolas graphed in Examples 1–3, the vertex here has the *largest* function value of any point on the graph. The domain is $(-\infty, \infty)$; the range is $(-\infty, 0]$.

| $x$ | $f(x)$ |
|---|---|
| $-2$ | $-2$ |
| $-1$ | $-\frac{1}{2}$ |
| $0$ | $0$ |
| $1$ | $-\frac{1}{2}$ |
| $2$ | $-2$ |

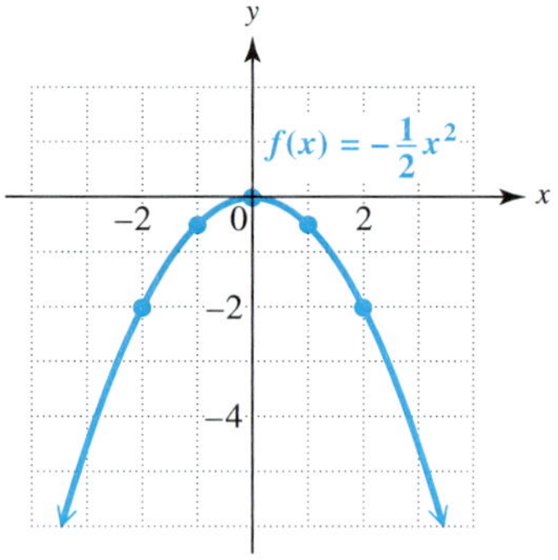


FIGURE 8

**Now Try Exercise 21.**

Some general principles concerning the graph of $F(x) = a(x - h)^2 + k$ are summarized as follows.

**General Principles**

1. The graph of the quadratic function defined by
$$F(x) = a(x - h)^2 + k, \quad a \neq 0$$
is a parabola with vertex $(h, k)$ and the vertical line $x = h$ as axis.
2. The graph opens up if $a$ is positive and down if $a$ is negative.
3. The graph is wider than that of $f(x) = x^2$ if $0 < |a| < 1$. The graph is narrower than that of $f(x) = x^2$ if $|a| > 1$.

## EXAMPLE 5 Using the General Principles to Graph a Parabola

Graph $F(x) = -2(x + 3)^2 + 4$.

The parabola opens down (because $a < 0$) and is narrower than the graph of $f(x) = x^2$, since $|-2| = 2 > 1$, causing values of $F(x)$ to decrease more quickly than those of $f(x) = -x^2$. This parabola has vertex $(-3, 4)$, as shown in Figure 9. To complete the graph, we plotted the ordered pairs $(-4, 2)$ and, by symmetry, $(-2, 2)$. Symmetry can be used to find additional ordered pairs that satisfy the equation, if desired.

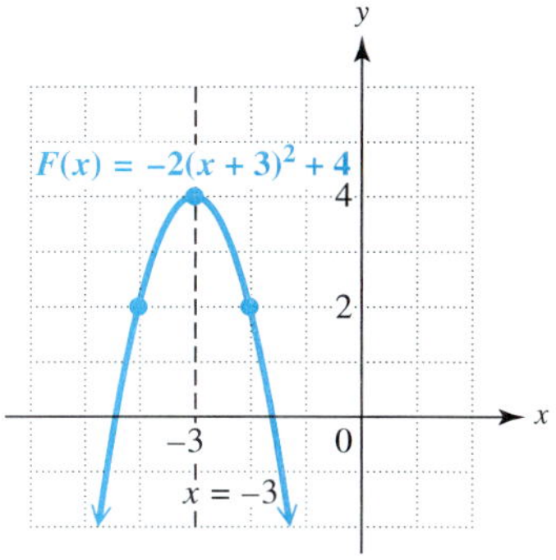


**FIGURE 9**

**Now Try Exercise 33.**

**OBJECTIVE 4 Find a quadratic function to model data.**

## EXAMPLE 6 Finding a Quadratic Function to Model the Rise in Multiple Births

The number of higher-order multiple births in the United States is rising. Let $x$ represent the number of years since 1970 and $y$ represent the rate of higher-order multiples born per 100,000 births since 1971. The data are shown in the following table.

| Year | $x$ | $y$ |
|---|---|---|
| 1971 | 1 | 29.1 |
| 1976 | 6 | 35.0 |
| 1981 | 11 | 40.0 |
| 1986 | 16 | 47.0 |
| 1991 | 21 | 100.0 |
| 1996 | 26 | 152.6 |

*Source:* National Center for Health Statistics.

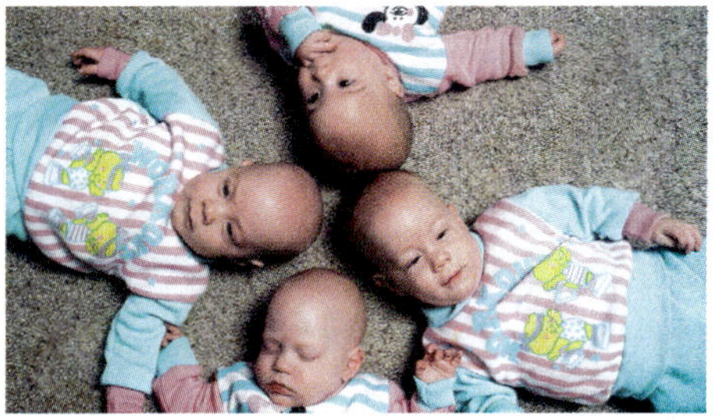

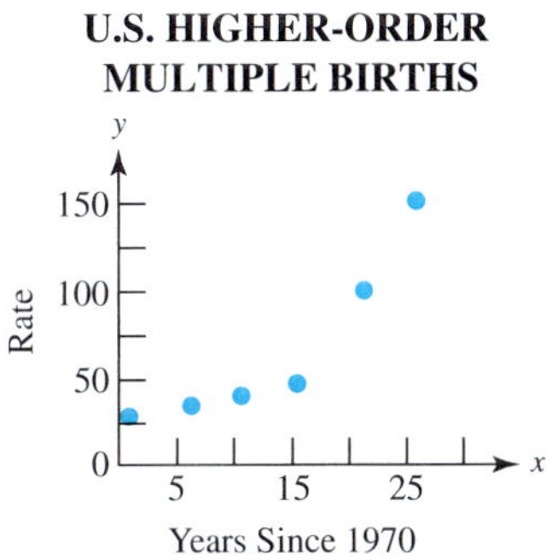


**FIGURE 10**

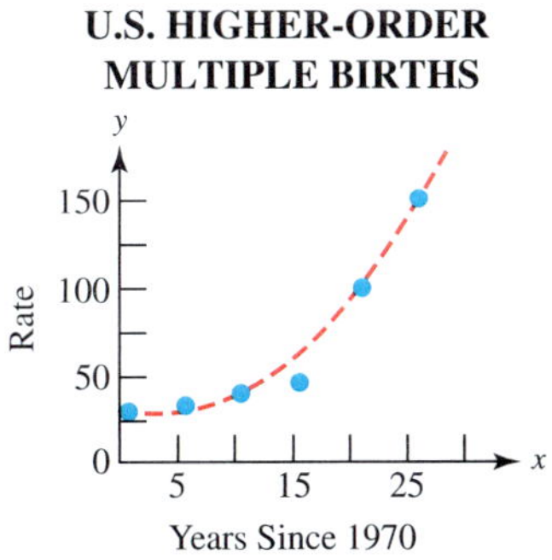


**FIGURE 11**

Find a quadratic function that models the data.

A scatter diagram of the ordered pairs $(x, y)$ is shown in Figure 10. Notice that the graphed points do not follow a linear pattern, so a linear function would not model the data very well. Instead, the general shape suggested by the scatter diagram indicates that a parabola should approximate these points, as shown by the dashed curve in Figure 11. The equation for such a parabola would have a positive coefficient for $x^2$ since the graph opens up. To find a quadratic function of the form

$$y = ax^2 + bx + c$$

that models, or *fits,* these data, we choose three representative ordered pairs and use them to write a system of three equations. Using (1, 29.1), (11, 40), and (21, 100), we substitute the $x$- and $y$-values from the ordered pairs into the quadratic form $y = ax^2 + bx + c$ to get the following three equations.

$$a(1)^2 + b(1) + c = 29.1 \quad \text{or} \quad a + b + c = 29.1 \quad (1)$$
$$a(11)^2 + b(11) + c = 40 \quad \text{or} \quad 121a + 11b + c = 40 \quad (2)$$
$$a(21)^2 + b(21) + c = 100 \quad \text{or} \quad 441a + 21b + c = 100 \quad (3)$$

We can find the values of $a$, $b$, and $c$ by solving this system of three equations in three variables using the methods of Section 8.4. Multiplying equation (1) by $-1$ and adding the result to equation (2) gives

$$120a + 10b = 10.9. \qquad (4)$$

Multiplying equation (2) by $-1$ and adding the result to equation (3) gives

$$320a + 10b = 60. \qquad (5)$$

We eliminate $b$ from this system of two equations in two variables by multiplying equation (4) by $-1$ and adding the result to equation (5) to obtain

$$200a = 49.1$$

$$a = .2455. \qquad \text{Use a calculator.}$$

We substitute .2455 for $a$ in equation (4) or (5) to find that $b = -1.856$. Substituting the values of $a$ and $b$ into equation (1) gives $c = 30.7105$. Using these values of $a$, $b$, and $c$, our model is defined by

$$y = .2455x^2 - 1.856x + 30.7105.$$

**Now Try Exercise 49.**

**NOTE** In Example 6, if we had chosen three different ordered pairs of data, a slightly different model would result. The *quadratic regression* feature on a graphing calculator can also be used to generate the quadratic model that best fits given data. See your owner's manual for details.

## 11.6 EXERCISES

**For Extra Help**

- Student's Solutions Manual
- MyMathLab
- InterAct Math Tutorial Software
- AW Math Tutor Center
- MathXL
- Digital Video Tutor CD 17/Videotape 17

**1.** Match each quadratic function with its graph from choices A–D.

**(a)** $f(x) = (x + 2)^2 - 1$

**(b)** $f(x) = (x + 2)^2 + 1$

**(c)** $f(x) = (x - 2)^2 - 1$

**(d)** $f(x) = (x - 2)^2 + 1$

**A.**

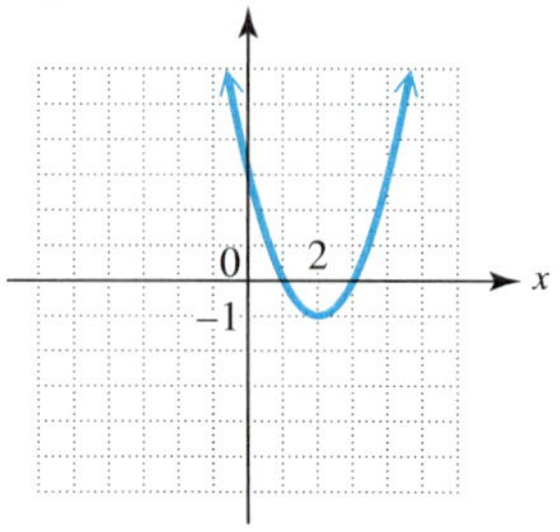


**B.**

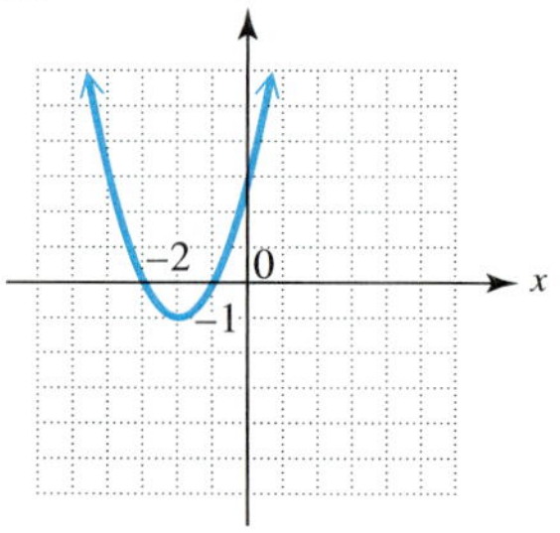


**C.**

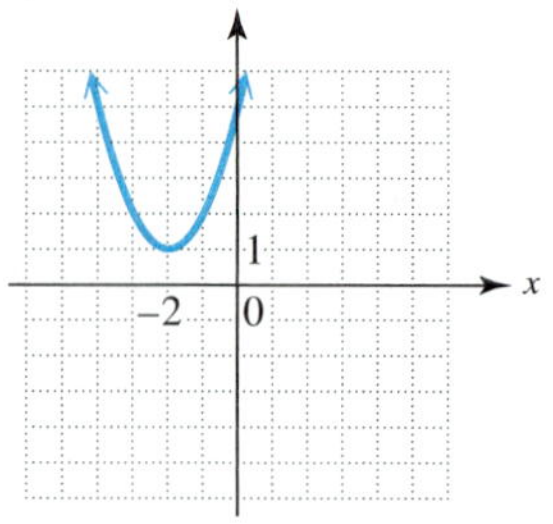


**D.**

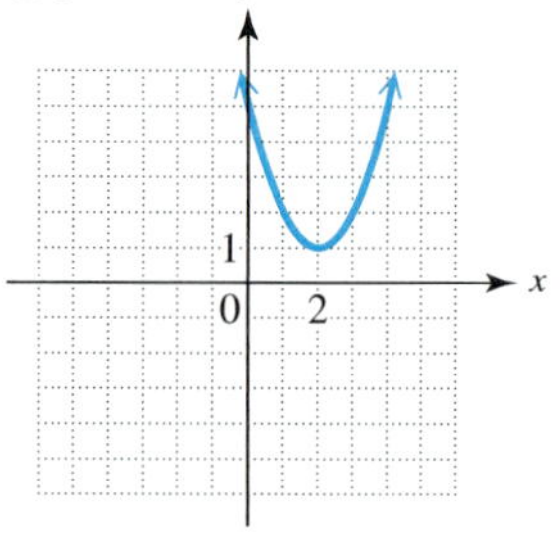

**2.** Match each quadratic function with its graph from choices A–D.

**(a)** $f(x) = -x^2 + 2$

**(b)** $f(x) = -x^2 - 2$

**(c)** $f(x) = -(x + 2)^2$

**(d)** $f(x) = -(x - 2)^2$

**A.**

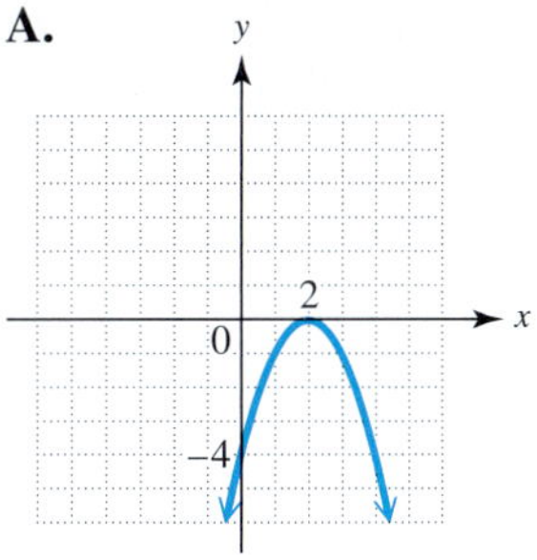


**B.**

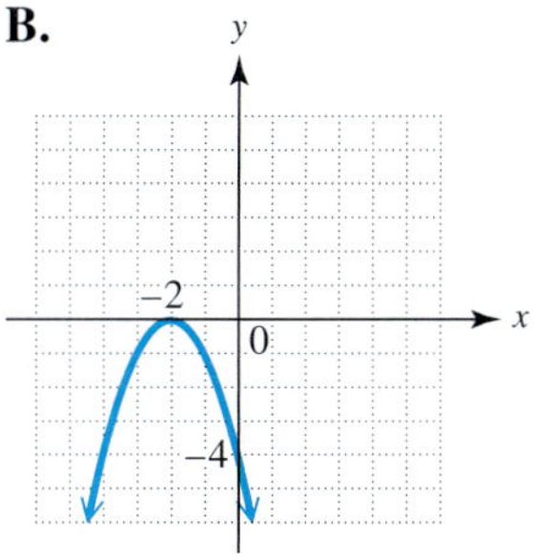


**C.**

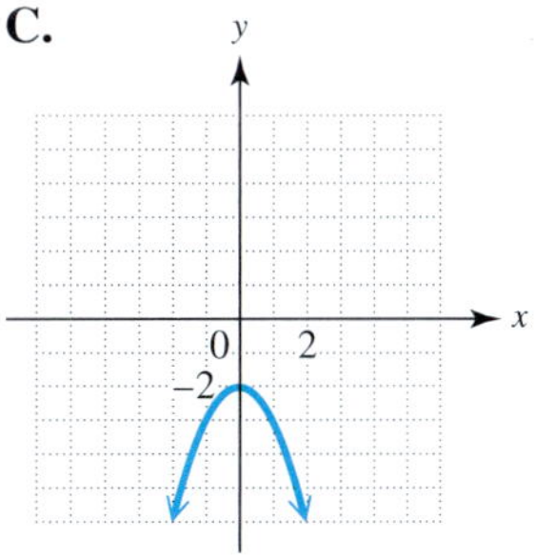


**D.**

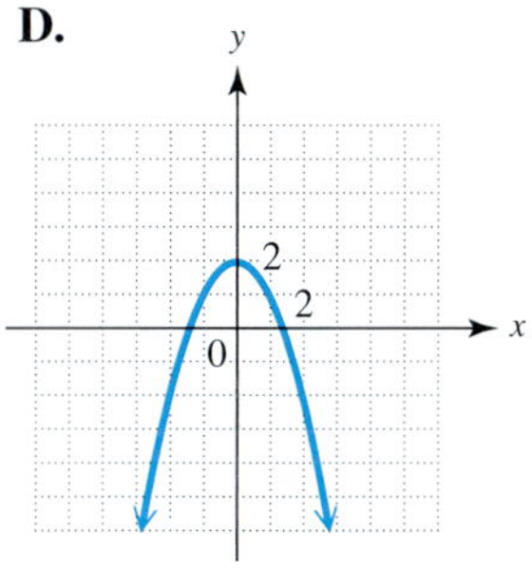


*Identify the vertex of each parabola. See Examples 1–4.*

**3.** $f(x) = -3x^2$ **4.** $f(x) = \frac{1}{2}x^2$ **5.** $f(x) = x^2 + 4$

**6.** $f(x) = x^2 - 4$ **7.** $f(x) = (x - 1)^2$ **8.** $f(x) = (x + 3)^2$

**9.** $f(x) = (x + 3)^2 - 4$ **10.** $f(x) = (x - 5)^2 - 8$

**11.** Describe how each of the parabolas in Exercises 9 and 10 is shifted compared to the graph of $f(x) = x^2$.

**12.** What does the value of $a$ in $F(x) = a(x - h)^2 + k$ tell you about the graph of the function compared to the graph of $f(x) = x^2$?

*For each quadratic function, tell whether the graph opens up or down and whether the graph is wider, narrower, or the same shape as the graph of $f(x) = x^2$. See Examples 4 and 5.*

**13.** $f(x) = -\frac{2}{5}x^2$ **14.** $f(x) = -2x^2$

**15.** $f(x) = 3x^2 + 1$ **16.** $f(x) = \frac{2}{3}x^2 - 4$

**17.** For $f(x) = a(x - h)^2 + k$, in what quadrant is the vertex if

**(a)** $h > 0, k > 0$; **(b)** $h > 0, k < 0$; **(c)** $h < 0, k > 0$; **(d)** $h < 0, k < 0$?

**18.** **(a)** What is the value of $h$ if the graph of $f(x) = a(x - h)^2 + k$ has vertex on the $y$-axis?

**(b)** What is the value of $k$ if the graph of $f(x) = a(x - h)^2 + k$ has vertex on the $x$-axis?

**19.** Match each quadratic function with the description of the parabola that is its graph.

| | |
|---|---|
| **(a)** $f(x) = (x - 4)^2 - 2$ | **A.** Vertex $(2, -4)$, opens down |
| **(b)** $f(x) = (x - 2)^2 - 4$ | **B.** Vertex $(2, -4)$, opens up |
| **(c)** $f(x) = -(x - 4)^2 - 2$ | **C.** Vertex $(4, -2)$, opens down |
| **(d)** $f(x) = -(x - 2)^2 - 4$ | **D.** Vertex $(4, -2)$, opens up |

**20.** Explain in your own words the meaning of each term.

**(a)** Vertex of a parabola **(b)** Axis of a parabola

*Graph each parabola. Plot at least two points in addition to the vertex. Give the vertex, axis, domain, and range in Exercises 27–36. See Examples 1–5.*

**21.** $f(x) = -2x^2$ **22.** $f(x) = \frac{1}{3}x^2$ **23.** $f(x) = x^2 - 1$

**24.** $f(x) = x^2 + 3$ **25.** $f(x) = -x^2 + 2$ **26.** $f(x) = 2x^2 - 2$

**27.** $f(x) = (x - 4)^2$ **28.** $f(x) = -2(x + 1)^2$ **29.** $f(x) = (x + 2)^2 - 1$

**30.** $f(x) = (x - 1)^2 + 2$ **31.** $f(x) = 2(x - 2)^2 - 4$ **32.** $f(x) = -3(x - 2)^2 + 1$

**33.** $f(x) = -\frac{1}{2}(x + 1)^2 + 2$ **34.** $f(x) = -\frac{2}{3}(x + 2)^2 + 1$

**35.** $f(x) = 2(x - 2)^2 - 3$ **36.** $f(x) = \frac{4}{3}(x - 3)^2 - 2$

## RELATING CONCEPTS (EXERCISES 37–42)

**For Individual or Group Work**

*The procedures that allow the graph of $y = x^2$ to be shifted vertically and horizontally apply to other types of functions. In Section 7.3 we introduced linear functions of the form $g(x) = ax + b$. Consider the graph of the simplest linear function defined by $g(x) = x$, shown here.* ***Work Exercises 37–42 in order.***

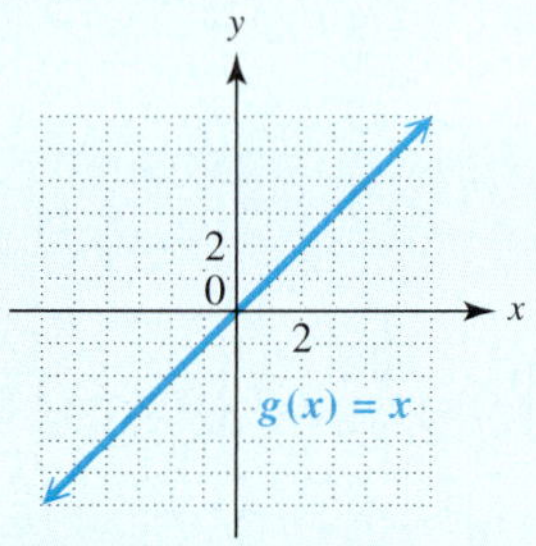


**37.** Based on the concepts of this section, how does the graph of $F(x) = x^2 + 6$ compare to the graph of $f(x) = x^2$ if a *vertical* shift is considered?

**38.** Graph the linear function defined by $G(x) = x + 6$.

**39.** Based on the concepts of Chapter 7, how does the graph of $G(x) = x + 6$ compare to the graph of $g(x) = x$ if a vertical shift is considered? (*Hint:* Look at the $y$-intercept.)

**40.** Based on the concepts of this section, how does the graph of $F(x) = (x - 6)^2$ compare to the graph of $f(x) = x^2$ if a *horizontal* shift is considered?

**41.** Graph the linear function $G(x) = x - 6$.

**42.** Based on the concepts of Chapter 7, how does the graph of $G(x) = x - 6$ compare to the graph of $g(x) = x$ if a horizontal shift is considered? (*Hint:* Look at the $x$-intercept.)

*In Exercises 43–48, tell whether a linear or quadratic function would be a more appropriate model for each set of graphed data. If linear, tell whether the slope should be positive or negative. If quadratic, tell whether the coefficient a of $x^2$ should be positive or negative. See Example 6.*

**43.**

**U.S. TRADE DEFICIT**

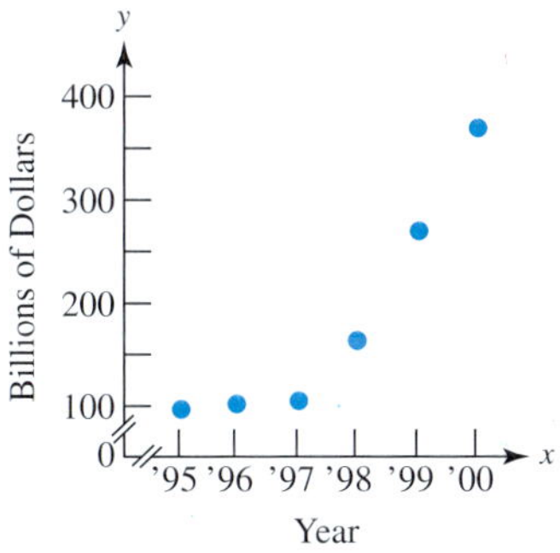


***Source:*** U.S. Department of Commerce.

**44. AVERAGE DAILY VOLUME OF FIRST-CLASS MAIL***

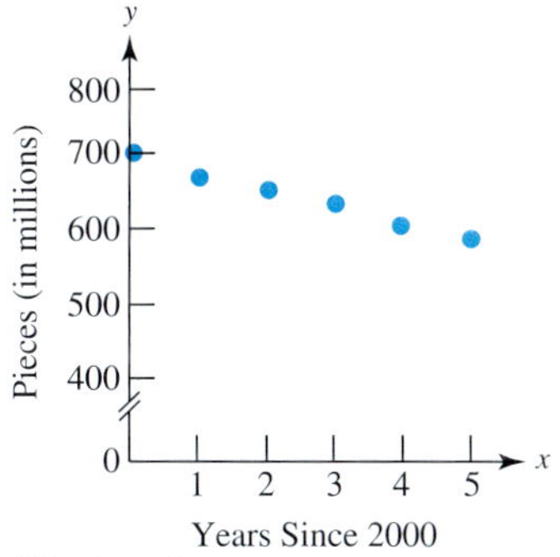


*Projected

***Source:*** General Accounting Office.

**45.**

**SOCIAL SECURITY ASSETS***

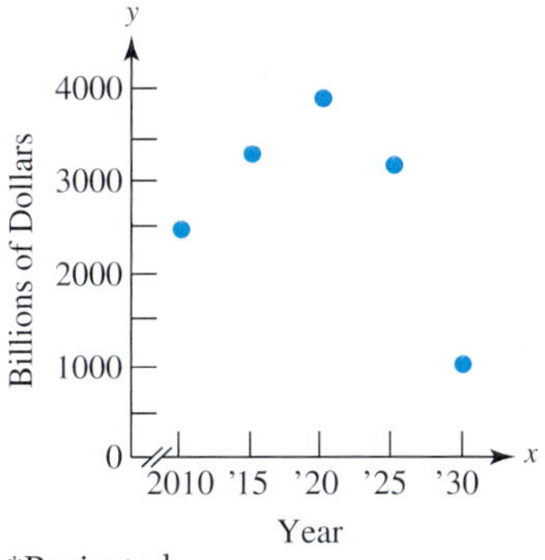


*Projected

***Source:*** Social Security Administration.

**46. CEDAR RAPIDS SCHOOLS— GENERAL RESERVE FUND**

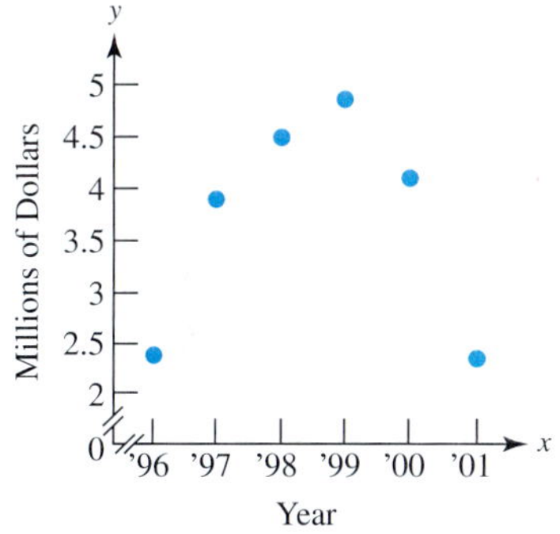


***Source:*** Cedar Rapids School District.

**47. CONSUMER DEMAND FOR ELECTRICITY**

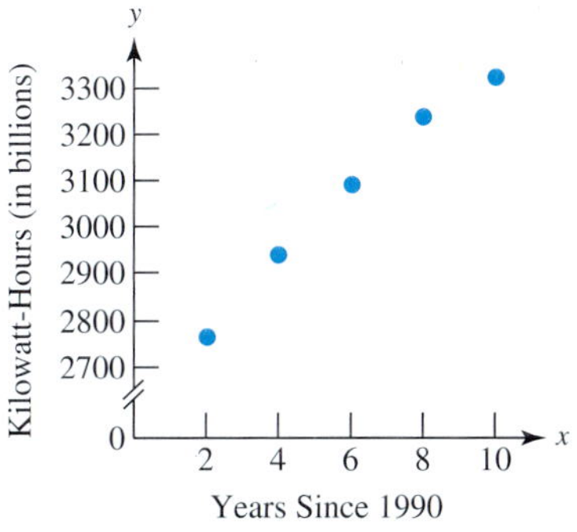


***Source:*** U.S. Department of Energy.

**48. U.S. COMMERCIAL BANK FAILURES**

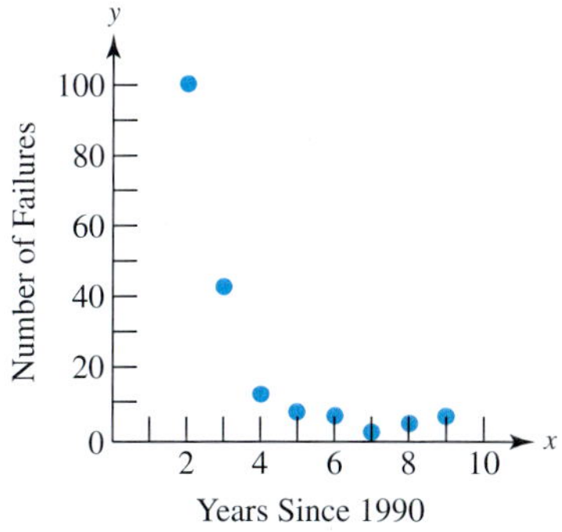


***Source:*** www.ABA.com

*Solve each problem. See Example 6.*

**49.** The number of publicly traded companies filing for bankruptcy for selected years between 1990 and 2000 are shown in the table. In the year column, 0 represents 1990, 2 represents 1992, and so on.

| Year | Number of Bankruptcies |
|---|---|
| 0 | 115 |
| 2 | 91 |
| 4 | 70 |
| 6 | 84 |
| 8 | 120 |
| 10 | 176 |

*Source:* www.BankruptcyData.com

**(a)** Use the ordered pairs (year, number of bankruptcies) to make a scatter diagram of the data.

**(b)** Use the scatter diagram to decide whether a linear or quadratic function would better model the data. If quadratic, should the coefficient $a$ of $x^2$ be positive or negative?

**(c)** Use the ordered pairs (0, 115), (4, 70), and (8, 120) to find a quadratic function that models the data. Round the values of $a$, $b$, and $c$ in your model to three decimal places, as necessary.

**(d)** Use your model from part (c) to approximate the number of company bankruptcy filings in 2002. Round your answer to the nearest whole number.

**(e)** The number of company bankruptcy filings through August 16, 2002 was 129. Based on this, is your estimate from part (d) reasonable? Explain.

**50.** In a study, the number of new AIDS patients who survived the first year for the years from 1991 through 1997 are shown in the table. In the year column, 1 represents 1991, 2 represents 1992, and so on.

| Year | Number of Patients |
|---|---|
| 1 | 55 |
| 2 | 130 |
| 3 | 155 |
| 4 | 160 |
| 5 | 155 |
| 6 | 150 |
| 7 | 115 |

*Source:* HIV Health Services Planning Council.

**(a)** Use the ordered pairs (year, number of patients) to make a scatter diagram of the data.

**(b)** Would a linear or quadratic function better model the data?

**(c)** Should the coefficient $a$ of $x^2$ in a quadratic model be positive or negative?

**(d)** Use the ordered pairs (2, 130), (3, 155), and (7, 115) to find a quadratic function that models the data.

**(e)** Use your model from part (d) to approximate the number of AIDS patients who survived the first year in 1994 and 1996. How well does the model approximate the actual data from the table?

**51.** In Example 6, we determined that the quadratic function defined by

$$y = .2455x^2 - 1.856x + 30.7105$$

modeled the rate of higher-order multiple births, where $x$ represents the number of years since 1970.

**(a)** Use this model to approximate the rate of higher-order births in 1999 to the nearest tenth.

**(b)** The actual rate of higher-order births in 1999 was 184.9. (*Source:* National Center for Health Statistics.) How does the approximation using the model compare to the actual rate for 1999?

## TECHNOLOGY INSIGHTS (EXERCISES 52–56)

*Recall from Chapters 3 and 7 that the x-value of the x-intercept of the graph of the line $y = mx + b$ is the solution of the linear equation $mx + b = 0$. In the same way, the x-values of the x-intercepts of the graph of the parabola $y = ax^2 + bx + c$ are the real solutions of the quadratic equation $ax^2 + bx + c = 0$.*

*In Exercises 52–55, the calculator graphs show the x-values of the x-intercepts of the graph of the polynomial in the equation. Use the graphs to solve each equation.*

**52.** $x^2 - x - 20 = 0$

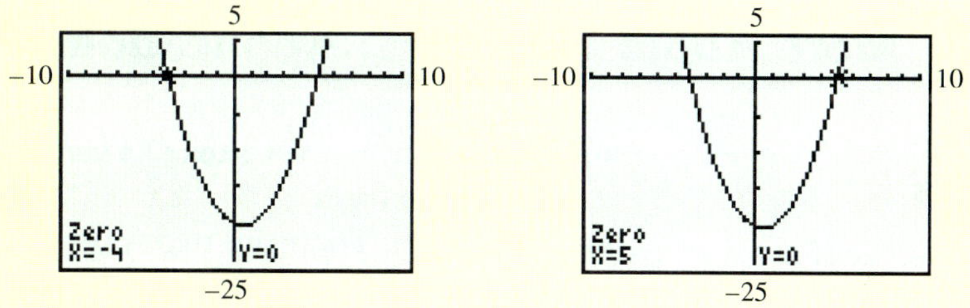


**53.** $x^2 + 9x + 14 = 0$

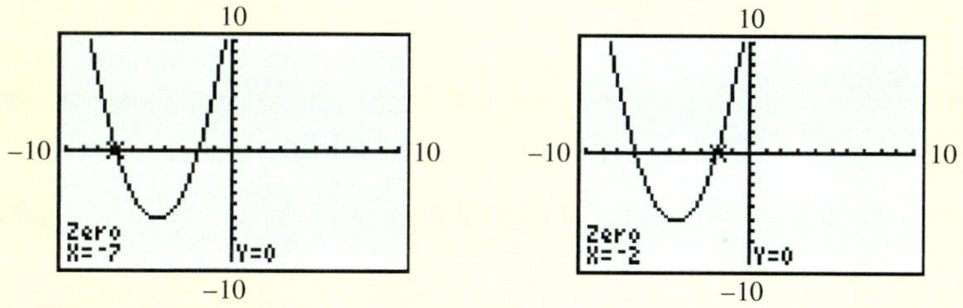


**54.** $-2x^2 + 5x + 3 = 0$

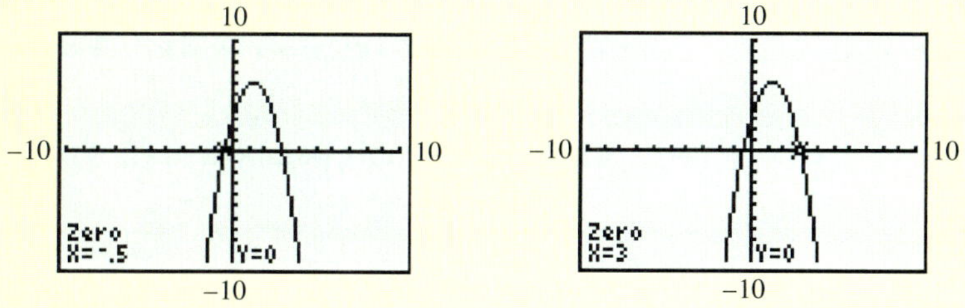


**55.** $-8x^2 + 6x + 5 = 0$

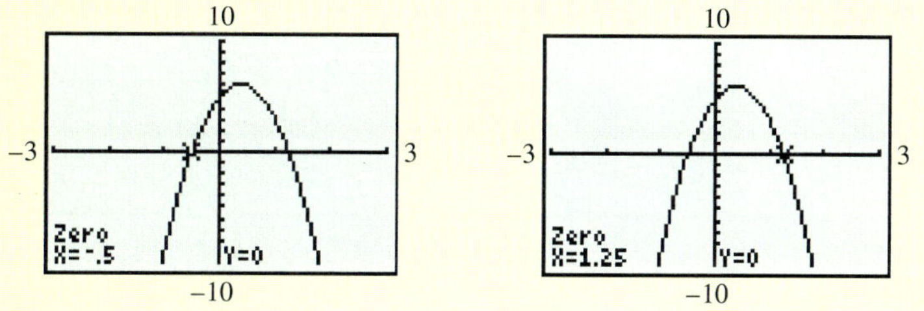

**56.** The graph of a quadratic function defined by $y = f(x)$ is shown in the standard viewing window, without $x$-axis tick marks. Which one of the following choices would be the only possible solution set for the equation $f(x) = 0$?

**A.** $\{-4, 1\}$ **B.** $\{1, 4\}$ **C.** $\{-1, -4\}$ **D.** $\{4, -1\}$

Explain your answer.

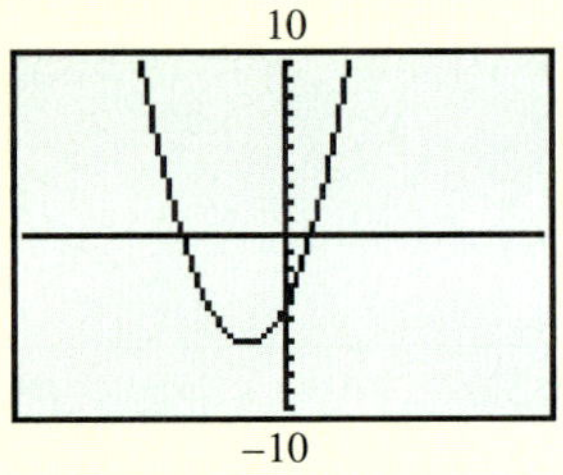


# 11.7 More About Parabolas; Applications

**OBJECTIVES**

1. Find the vertex of a vertical parabola.
2. Graph a quadratic function.
3. Use the discriminant to find the number of $x$-intercepts of a vertical parabola.
4. Use quadratic functions to solve problems involving maximum or minimum value.
5. Graph horizontal parabolas.

**OBJECTIVE 1 Find the vertex of a vertical parabola.** When the equation of a parabola is given in the form $f(x) = ax^2 + bx + c$, we need to locate the vertex to sketch an accurate graph. There are two ways to do this: complete the square as shown in Examples 1 and 2, or use a formula derived by completing the square.

## EXAMPLE 1 Completing the Square to Find the Vertex

Find the vertex of the graph of $f(x) = x^2 - 4x + 5$.

To find the vertex, we need to express $x^2 - 4x + 5$ in the form $(x - h)^2 + k$. We do this by completing the square on $x^2 - 4x$, as in Section 11.2. The process is slightly different here because we want to keep $f(x)$ alone on one side of the equation. Instead of adding the appropriate number to each side, we *add and subtract* it on the right. This is equivalent to adding 0.

$$f(x) = x^2 - 4x + 5$$

$$= (x^2 - 4x \qquad) + 5 \qquad \text{Group the variable terms.}$$

$$\left[\frac{1}{2}(-4)\right]^2 = (-2)^2 = 4$$

$$= (x^2 - 4x + 4 - 4) + 5 \qquad \text{Add and subtract 4.}$$

$$= (x^2 - 4x + 4) - 4 + 5 \qquad \text{Bring } -4 \text{ outside the parentheses.}$$

$$f(x) = (x - 2)^2 + 1 \qquad \text{Factor; combine terms.}$$

The vertex of this parabola is $(2, 1)$.

**Now Try Exercise 5.**

## EXAMPLE 2 Completing the Square to Find the Vertex When $a \neq 1$

Find the vertex of the graph of $f(x) = -3x^2 + 6x - 1$.

We must complete the square on $-3x^2 + 6x$. Because the $x^2$-term has a coefficient other than 1, we factor that coefficient out of the first two terms and then proceed as in Example 1.

$$f(x) = -3x^2 + 6x - 1$$

$$= -3(x^2 - 2x) - 1 \qquad \text{Factor out } -3.$$

$$\left[\frac{1}{2}(-2)\right]^2 = (-1)^2 = 1$$

$$= -3(x^2 - 2x + 1 - 1) - 1 \qquad \text{Add and subtract 1 within the parentheses.}$$

Bring $-1$ outside the parentheses; be sure to multiply it by $-3$.

$$= -3(x^2 - 2x + 1) + (-3)(-1) - 1 \quad \text{Distributive property}$$
$$= -3(x^2 - 2x + 1) + 3 - 1$$
$$f(x) = -3(x - 1)^2 + 2 \quad \text{Factor; combine terms.}$$

The vertex is $(1, 2)$.

**Now Try Exercise 7.**

To derive a formula for the vertex of the graph of the quadratic function defined by $f(x) = ax^2 + bx + c$ $(a \neq 0)$, complete the square.

$$f(x) = ax^2 + bx + c \quad \text{Standard form}$$
$$= a\left(x^2 + \frac{b}{a}x\right) + c \quad \text{Factor } a \text{ from the first two terms.}$$
$$\left[\frac{1}{2}\left(\frac{b}{a}\right)\right]^2 = \left(\frac{b}{2a}\right)^2 = \frac{b^2}{4a^2}$$
$$= a\left(x^2 + \frac{b}{a}x + \frac{b^2}{4a^2} - \frac{b^2}{4a^2}\right) + c \quad \text{Add and subtract } \frac{b^2}{4a^2}.$$
$$= a\left(x^2 + \frac{b}{a}x + \frac{b^2}{4a^2}\right) + a\left(-\frac{b^2}{4a^2}\right) + c \quad \text{Distributive property}$$
$$= a\left(x^2 + \frac{b}{a}x + \frac{b^2}{4a^2}\right) - \frac{b^2}{4a} + c$$
$$= a\left(x + \frac{b}{2a}\right)^2 + \frac{4ac - b^2}{4a} \quad \text{Factor; combine terms.}$$
$$f(x) = a\left[x - \underbrace{\left(\frac{-b}{2a}\right)}_{h}\right]^2 + \underbrace{\frac{4ac - b^2}{4a}}_{k} \quad f(x) = (x - h)^2 + k$$

Thus, the vertex $(h, k)$ can be expressed in terms of $a$, $b$, and $c$. However, it is not necessary to remember this expression for $k$, since it can be found by replacing $x$ with $\frac{-b}{2a}$. Using function notation, if $y = f(x)$, then the $y$-value of the vertex is $f\left(\frac{-b}{2a}\right)$.

**Vertex Formula**

The graph of the quadratic function defined by $f(x) = ax^2 + bx + c$ $(a \neq 0)$ has vertex

$$\left(\frac{-b}{2a}, f\left(\frac{-b}{2a}\right)\right),$$

and the axis of the parabola is the line

$$x = \frac{-b}{2a}.$$

## EXAMPLE 3 Using the Formula to Find the Vertex

Use the vertex formula to find the vertex of the graph of

$$f(x) = x^2 - x - 6.$$

For this function, $a = 1$, $b = -1$, and $c = -6$. The $x$-coordinate of the vertex of the parabola is given by

$$\frac{-b}{2a} = \frac{-(-1)}{2(1)} = \frac{1}{2}.$$

The $y$-coordinate is $f\left(\frac{-b}{2a}\right) = f\left(\frac{1}{2}\right)$.

$$f\left(\frac{1}{2}\right) = \left(\frac{1}{2}\right)^2 - \frac{1}{2} - 6 = \frac{1}{4} - \frac{1}{2} - 6 = -\frac{25}{4}$$

The vertex is $\left(\frac{1}{2}, -\frac{25}{4}\right)$.

**Now Try Exercise 9.**

**OBJECTIVE 2 Graph a quadratic function.** We give a general approach for graphing any quadratic function here.

### Graphing a Quadratic Function *f*

*Step 1* **Determine whether the graph opens up or down.** If $a > 0$, the parabola opens up; if $a < 0$, it opens down.

*Step 2* **Find the vertex.** Use either the vertex formula or completing the square.

*Step 3* **Find any intercepts.** To find the $x$-intercepts (if any), solve $f(x) = 0$. To find the $y$-intercept, evaluate $f(0)$.

*Step 4* **Complete the graph.** Plot the points found so far. Find and plot additional points as needed, using symmetry about the axis.

## EXAMPLE 4 Using the Steps to Graph a Quadratic Function

Graph the quadratic function defined by

$$f(x) = x^2 - x - 6.$$

*Step 1* From the equation, $a = 1$, so the graph of the function opens up.

*Step 2* The vertex, $\left(\frac{1}{2}, -\frac{25}{4}\right)$, was found in Example 3 using the vertex formula.

*Step 3* Find any intercepts. Since the vertex, $\left(\frac{1}{2}, -\frac{25}{4}\right)$, is in quadrant IV and the graph opens up, there will be two $x$-intercepts. To find them, let $f(x) = 0$ and solve.

$$f(x) = x^2 - x - 6$$

$$0 = x^2 - x - 6 \qquad \text{Let } f(x) = 0.$$

$$0 = (x - 3)(x + 2) \qquad \text{Factor.}$$

$$x - 3 = 0 \quad \text{or} \quad x + 2 = 0 \qquad \text{Zero-factor property}$$

$$x = 3 \quad \text{or} \quad x = -2$$

The $x$-intercepts are $(3, 0)$ and $(-2, 0)$.

To find the $y$-intercept, evaluate $f(0)$.

$$f(x) = x^2 - x - 6$$
$$f(0) = 0^2 - 0 - 6 \quad \text{Let } x = 0.$$
$$f(0) = -6$$

The $y$-intercept is $(0, -6)$.

*Step 4* Plot the points found so far and additional points as needed using symmetry about the axis, $x = \frac{1}{2}$. The graph is shown in Figure 12. The domain is $(-\infty, \infty)$, and the range is $\left[-\frac{25}{4}, \infty\right)$.

| $x$ | $y$ |
|---|---|
| $-2$ | $0$ |
| $-1$ | $-4$ |
| $0$ | $-6$ |
| $\frac{1}{2}$ | $-\frac{25}{4}$ |
| $2$ | $-4$ |
| $3$ | $0$ |

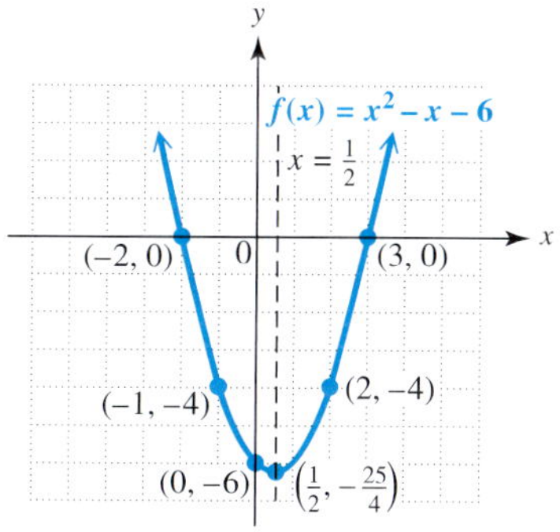


**FIGURE 12**

**Now Try Exercise 17.**

**OBJECTIVE 3 Use the discriminant to find the number of $x$-intercepts of a vertical parabola.** The graph of a quadratic function may have two $x$-intercepts, one $x$-intercept, or no $x$-intercepts, as shown in Figure 13. Recall from Section 11.3 that $b^2 - 4ac$ is called the *discriminant* of the quadratic equation $ax^2 + bx + c = 0$ and that we can use it to determine the number of real solutions of a quadratic equation.

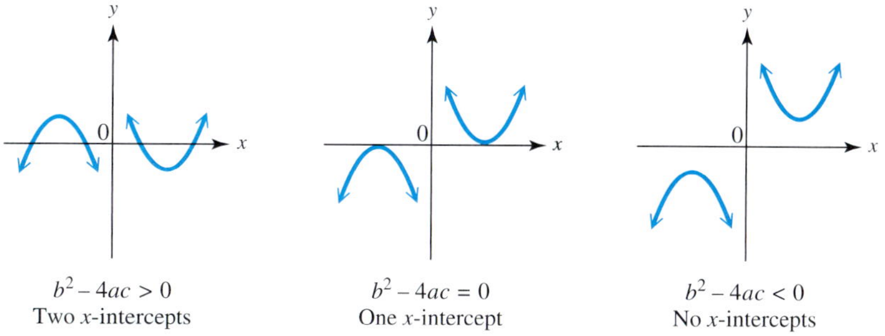


**FIGURE 13**

In a similar way, we can use the discriminant of a quadratic *function* to determine the number of $x$-intercepts of its graph. If the discriminant is positive, the parabola will have two $x$-intercepts. If the discriminant is 0, there will be only one $x$-intercept, and it will be the vertex of the parabola. If the discriminant is negative, the graph will have no $x$-intercepts.

**EXAMPLE 3 Solving Special Cases**

Solve $(2t - 3)^2 > -1$. Then solve $(2t - 3)^2 < -1$.

Because $(2t - 3)^2$ is never negative, it is always greater than $-1$. Thus, the solution for $(2t - 3)^2 > -1$ is the set of all real numbers, $(-\infty, \infty)$. In the same way, there is no solution for $(2t - 3)^2 < -1$ and the solution set is $\emptyset$.

**Now Try Exercises 25 and 27.**

**OBJECTIVE 2 Solve polynomial inequalities of degree 3 or more.** Higher-degree polynomial inequalities that can be factored are solved in the same way as quadratic inequalities.

**EXAMPLE 4 Solving a Third-Degree Polynomial Inequality**

Solve $(x - 1)(x + 2)(x - 4) \le 0$.

This is a *cubic* (third-degree) inequality rather than a quadratic inequality, but it can be solved using the method shown in the box by extending the zero-factor property to more than two factors. Begin by setting the factored polynomial *equal* to 0 and solving the equation. (Step 1)

$$(x - 1)(x + 2)(x - 4) = 0$$

$$x - 1 = 0 \quad \text{or} \quad x + 2 = 0 \quad \text{or} \quad x - 4 = 0$$

$$x = 1 \quad \text{or} \quad x = -2 \quad \text{or} \quad x = 4$$

Locate the numbers $-2$, 1, and 4 on a number line, as in Figure 20, to determine the Intervals A, B, C, and D. (Step 2)

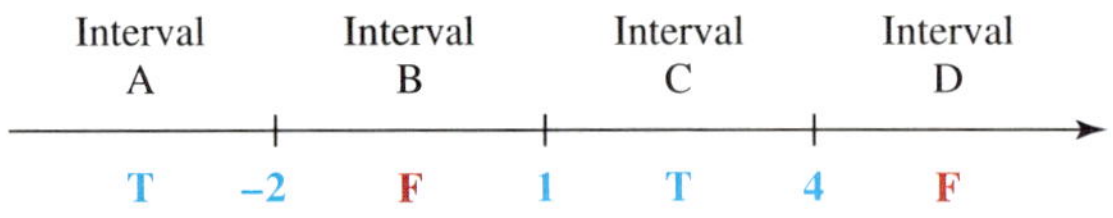


FIGURE 20

Substitute a test number from each interval in the *original* inequality to determine which intervals satisfy the inequality. (Step 3) Use a table to organize this information.

| *Interval* | *Test Number* | *Test of Inequality* | *True or False?* |
|---|---|---|---|
| A | $-3$ | $-28 \le 0$ | T |
| B | 0 | $8 \le 0$ | F |
| C | 2 | $-8 \le 0$ | T |
| D | 5 | $28 \le 0$ | F |

Verify the information given in the table and graphed in Figure 21 on the next page. The numbers in Intervals A and C are in the solution set, which is written in interval notation as

$$(-\infty, -2] \cup [1, 4].$$

The three endpoints are included since the inequality symbol is $\le$. (Step 4)

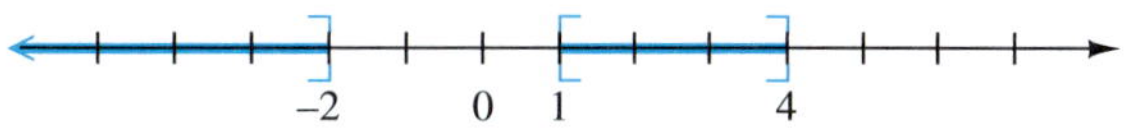

FIGURE 21

**Now Try Exercise 29.**

**OBJECTIVE 3 Solve rational inequalities.** Inequalities that involve rational expressions, called **rational inequalities,** are solved similarly using the following steps.

**Solving a Rational Inequality**

*Step 1* **Write the inequality** so that 0 is on one side and there is a single fraction on the other side.

*Step 2* **Determine the numbers that make the numerator or denominator equal to 0.**

*Step 3* **Divide a number line into intervals.** Use the numbers from Step 2.

*Step 4* **Find the intervals that satisfy the inequality.** Test a number from each interval by substituting it into the *original* inequality.

*Step 5* **Consider the endpoints separately.** Exclude any values that make the denominator 0.

**CAUTION** As indicated in Step 5, any number that makes the denominator 0 *must* be excluded from the solution set.

### EXAMPLE 5 Solving a Rational Inequality

Solve $\frac{-1}{p-3} > 1$.

Write the inequality so that 0 is on one side. (Step 1)

$$\frac{-1}{p-3} - 1 > 0 \qquad \text{Subtract 1.}$$

$$\frac{-1}{p-3} - \frac{p-3}{p-3} > 0 \qquad \text{Use } p-3 \text{ as the common denominator.}$$

$$\frac{-1-p+3}{p-3} > 0 \qquad \text{Write the left side as a single fraction; be careful with signs in the numerator.}$$

$$\frac{-p+2}{p-3} > 0 \qquad \text{Combine terms.}$$

The sign of the rational expression $\frac{-p+2}{p-3}$ will change from positive to negative or negative to positive only at those numbers that make the numerator or denominator 0. The number 2 makes the numerator 0, and 3 makes the denominator 0. (Step 2) These two numbers, 2 and 3, divide a number line into three intervals. See Figure 22. (Step 3)

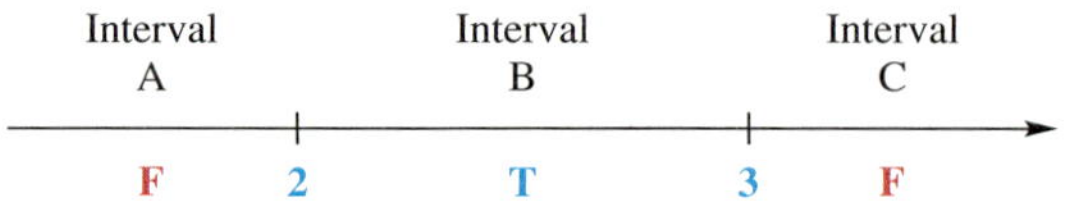


**FIGURE 22**

Testing a number from each interval in the *original* inequality, $\frac{-1}{p-3} > 1$, gives the results shown in the table. (Step 4)

| *Interval* | *Test Number* | *Test of Inequality* | *True or False?* |
|---|---|---|---|
| A | 0 | $\frac{1}{3} > 1$ | F |
| B | 2.5 | $2 > 1$ | T |
| C | 4 | $-1 > 1$ | F |

The solution set is the interval (2, 3). This interval does not include 3 since it would make the denominator of the original equality 0; 2 is not included either since the inequality symbol is >. (Step 5) A graph of the solution set is given in Figure 23.

**FIGURE 23**

**Now Try Exercise 37.**

## EXAMPLE 6 Solving a Rational Inequality

Solve $\frac{m-2}{m+2} \leq 2$.

Write the inequality so that 0 is on one side. (Step 1)

$$\frac{m-2}{m+2} - 2 \leq 0 \qquad \text{Subtract 2.}$$

$$\frac{m-2}{m+2} - \frac{2(m+2)}{m+2} \leq 0 \qquad \text{Use } m + 2 \text{ as the common denominator.}$$

$$\frac{m - 2 - 2m - 4}{m+2} \leq 0 \qquad \text{Write as a single fraction.}$$

$$\frac{-m-6}{m+2} \leq 0 \qquad \text{Combine terms.}$$

The number $-6$ makes the numerator 0, and $-2$ makes the denominator 0. (Step 2) These two numbers determine three intervals. (Step 3) Test one number from each interval (Step 4) to see that the solution set is

$$(-\infty, -6] \cup (-2, \infty).$$

The number $-6$ satisfies the original inequality, but $-2$ cannot be used as a solution since it makes the denominator 0. (Step 5) A graph of the solution set is shown in Figure 24.

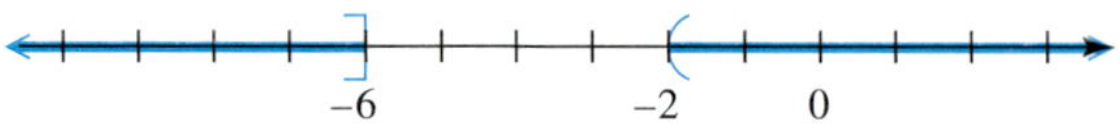


**FIGURE 24**

**Now Try Exercise 41.**

# 11.8 EXERCISES

**For Extra Help**

- Student's Solutions Manual
- MyMathLab
- InterAct Math Tutorial Software
- AW Math Tutor Center
- MathXL
- Digital Video Tutor CD 17/Videotape 17

In Example 1, we determined the solution sets of the quadratic inequalities $x^2 - x - 12 > 0$ and $x^2 - x - 12 < 0$ by graphing $f(x) = x^2 - x - 12$. The $x$-intercepts of this graph indicated the solutions of the equation $x^2 - x - 12 = 0$. The $x$-values of the points on the graph that were above the $x$-axis formed the solution set of $x^2 - x - 12 > 0$, and the $x$-values of the points on the graph that were below the $x$-axis formed the solution set of $x^2 - x - 12 < 0$.

*In Exercises 1–4, the graph of a quadratic function f is given. Use the graph to find the solution set of each equation or inequality. See Example 1.*

**1. (a)** $x^2 - 4x + 3 = 0$
**(b)** $x^2 - 4x + 3 > 0$
**(c)** $x^2 - 4x + 3 < 0$

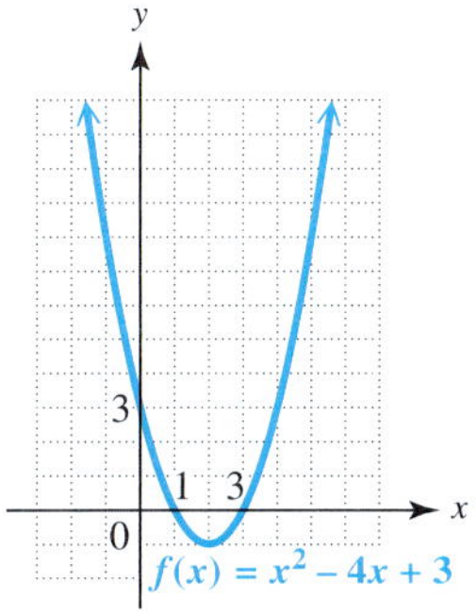


**2. (a)** $3x^2 + 10x - 8 = 0$
**(b)** $3x^2 + 10x - 8 \geq 0$
**(c)** $3x^2 + 10x - 8 < 0$

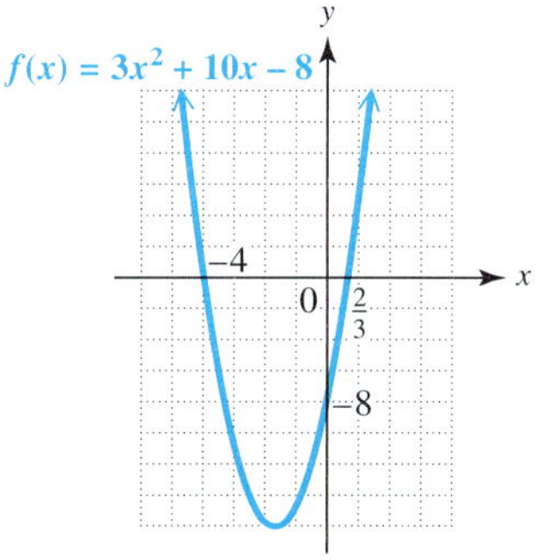


**3. (a)** $-2x^2 - x + 15 = 0$
**(b)** $-2x^2 - x + 15 \geq 0$
**(c)** $-2x^2 - x + 15 \leq 0$

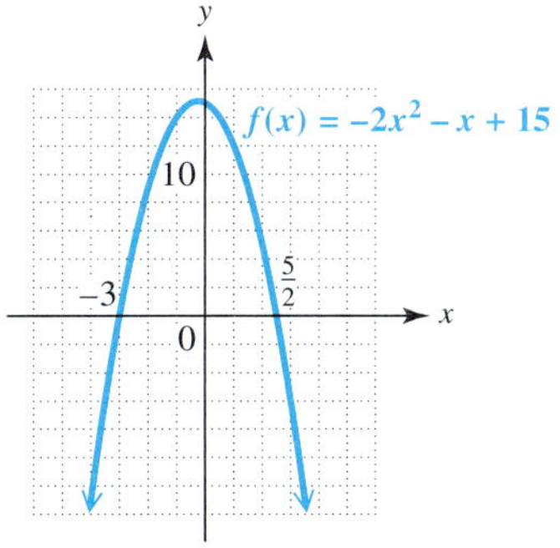


**4. (a)** $-x^2 + 3x + 10 = 0$
**(b)** $-x^2 + 3x + 10 \geq 0$
**(c)** $-x^2 + 3x + 10 \leq 0$

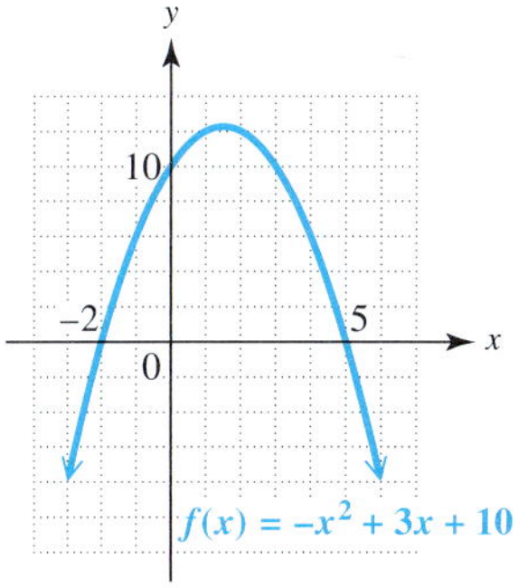


**5.** Explain how to determine whether to include or exclude endpoints when solving a quadratic or higher-degree inequality.

**6.** The solution set of the inequality $x^2 + x - 12 < 0$ is the interval $(-4, 3)$. Without actually performing any work, give the solution set of the inequality $x^2 + x - 12 \geq 0$.

*Solve each inequality and graph the solution set. See Example 2. (Hint: In Exercises 23 and 24, use the quadratic formula.)*

**7.** $(x + 1)(x - 5) > 0$
**8.** $(m + 6)(m - 2) > 0$
**9.** $(r + 4)(r - 6) < 0$
**10.** $(x + 4)(x - 8) < 0$
**11.** $x^2 - 4x + 3 \geq 0$
**12.** $m^2 - 3m - 10 \geq 0$

**13.** $10t^2 + 9t \geq 9$ **14.** $3r^2 + 10r \geq 8$ **15.** $4x^2 - 9 \leq 0$

**16.** $9x^2 - 25 \leq 0$ **17.** $6x^2 + x \geq 1$ **18.** $4p^2 + 7p \geq -3$

**19.** $z^2 - 4z \geq 0$ **20.** $x^2 + 2x < 0$ **21.** $3k^2 - 5k \leq 0$

**22.** $2z^2 + 3z > 0$ **23.** $x^2 - 6x + 6 \geq 0$ **24.** $3k^2 - 6k + 2 \leq 0$

*Solve each inequality. See Example 3.*

**25.** $(4 - 3x)^2 \geq -2$ **26.** $(6z + 7)^2 \geq -1$

**27.** $(3x + 5)^2 \leq -4$ **28.** $(8t + 5)^2 \leq -5$

*Solve each inequality and graph the solution set. See Example 4.*

**29.** $(p - 1)(p - 2)(p - 4) < 0$ **30.** $(2r + 1)(3r - 2)(4r + 7) < 0$

**31.** $(x - 4)(2x + 3)(3x - 1) \geq 0$ **32.** $(z + 2)(4z - 3)(2z + 7) \geq 0$

*Solve each inequality and graph the solution set. See Examples 5 and 6.*

**33.** $\frac{x - 1}{x - 4} > 0$ **34.** $\frac{x + 1}{x - 5} > 0$ **35.** $\frac{2n + 3}{n - 5} \leq 0$

**36.** $\frac{3t + 7}{t - 3} \leq 0$ **37.** $\frac{8}{x - 2} \geq 2$ **38.** $\frac{20}{x - 1} \geq 1$

**39.** $\frac{3}{2t - 1} < 2$ **40.** $\frac{6}{m - 1} < 1$ **41.** $\frac{g - 3}{g + 2} \geq 2$

**42.** $\frac{m + 4}{m + 5} \geq 2$ **43.** $\frac{x - 8}{x - 4} < 3$ **44.** $\frac{2t - 3}{t + 1} > 4$

**45.** $\frac{4k}{2k - 1} < k$ **46.** $\frac{r}{r + 2} < 2r$ **47.** $\frac{2x - 3}{x^2 + 1} \geq 0$

**48.** $\frac{9x - 8}{4x^2 + 25} < 0$ **49.** $\frac{(3x - 5)^2}{x + 2} > 0$ **50.** $\frac{(5x - 3)^2}{2x + 1} \leq 0$

## RELATING CONCEPTS (EXERCISES 51–54)

**For Individual or Group Work**

*A rock is projected vertically upward from the ground. Its distance s in feet above the ground after t sec is given by the quadratic function defined by*

$$s(t) = -16t^2 + 256t.$$

***Work Exercises 51–54 in order,*** *to see how quadratic equations and inequalities are related.*

**51.** At what times will the rock be 624 ft above the ground? (*Hint:* Let $s(t) = 624$ and solve the quadratic *equation.*)

**52.** At what times will the rock be more than 624 ft above the ground? (*Hint:* Set $s(t) > 624$ and solve the quadratic *inequality.*)

**53.** At what times will the rock be at ground level? (*Hint:* Let $s(t) = 0$ and solve the quadratic *equation.*)

**54.** At what times will the rock be less than 624 ft above the ground? (*Hint:* Set $s(t) < 624$, solve the quadratic *inequality,* and observe the solutions in Exercises 52 and 53 to determine the smallest and largest possible values of $t$.)

## Chapter 11 Group Activity

### Finding the Path of a Comet

**OBJECTIVE** Find and graph an equation of a parabola with a given focus.

The orbit that a comet takes as it approaches the sun depends on its velocity, as well as other factors. If the velocity of a comet equals escape velocity—that is, it is going just fast enough to get away from the sun—then its orbit will be parabolic. (Other possible orbits are *hyperbolic* or *elliptical.* These figures are discussed in Chapter 13.) The sun is at the focus of the parabola. The vertex is the point where the comet is closest to the sun.

**A.** Refer to Section 11.7, Exercise 51, and make a sketch of the comet and the sun.

   **1.** Place the vertex of the parabola at the origin, with the focus on the $y$-axis and a horizontal directrix.

   **2.** Assume that the comet is .75 astronomical unit* from the sun at its closest point.

**B.** What are the coordinates of the focus?

**C.** What is the equation of the directrix?

**D.** Using the information from Section 11.7, Exercise 51, find an equation of the parabola.

**E.** Graph the parabola (as a vertical parabola). Include the focus and directrix on your graph.

*One astronomical unit (AU) is the average distance between Earth and the sun.

# CHAPTER 11 SUMMARY

## KEY TERMS

**11.1** quadratic equation
**11.3** quadratic formula
discriminant
**11.4** quadratic in form
**11.6** parabola
vertex
axis
quadratic function
**11.8** quadratic inequality
rational inequality

## TEST YOUR WORD POWER

*See how well you have learned the vocabulary in this chapter. Answers, with examples, follow the Quick Review.*

**1.** The **quadratic formula** is
**A.** a formula to find the number of solutions of a quadratic equation
**B.** a formula to find the type of solutions of a quadratic equation
**C.** the standard form of a quadratic equation
**D.** a general formula for solving any quadratic equation.

**2.** A **quadratic function** is a function that can be written in the form
**A.** $f(x) = mx + b$ for real numbers $m$ and $b$
**B.** $f(x) = \frac{P(x)}{Q(x)}$, where $Q(x) \neq 0$
**C.** $f(x) = ax^2 + bx + c$ for real numbers $a$, $b$, and $c$ $(a \neq 0)$
**D.** $f(x) = \sqrt{x}$ for $x \geq 0$.

**3.** A **parabola** is the graph of
**A.** any equation in two variables
**B.** a linear equation
**C.** an equation of degree 3
**D.** a quadratic equation in two variables.

**4.** The **vertex** of a parabola is
**A.** the point where the graph intersects the $y$-axis
**B.** the point where the graph intersects the $x$-axis
**C.** the lowest point on a parabola that opens up or the highest point on a parabola that opens down
**D.** the origin.

**5.** The **axis** of a parabola is
**A.** either the $x$-axis or the $y$-axis
**B.** the vertical line (of a vertical parabola) or the horizontal line (of a horizontal parabola) through the vertex
**C.** the lowest or highest point on the graph of a parabola
**D.** a line through the origin.

**6.** A parabola is **symmetric about its axis** since
**A.** its graph is near the axis
**B.** its graph is identical on each side of the axis
**C.** its graph looks different on each side of the axis
**D.** its graph intersects the axis.

## QUICK REVIEW

| CONCEPTS | EXAMPLES |
|---|---|
| **11.1 SOLVING QUADRATIC EQUATIONS BY THE SQUARE ROOT PROPERTY** | |
| **Square Root Property**<br>If $x$ and $k$ are complex numbers and $x^2 = k$, then<br>$x = \sqrt{k}$ or $x = -\sqrt{k}$. | Solve $(x - 1)^2 = 8$.<br>$x - 1 = \sqrt{8}$ or $x - 1 = -\sqrt{8}$<br>$x = 1 + 2\sqrt{2}$ or $x = 1 - 2\sqrt{2}$<br>Solution set: $\{1 + 2\sqrt{2}, 1 - 2\sqrt{2}\}$ |
| **11.2 SOLVING QUADRATIC EQUATIONS BY COMPLETING THE SQUARE** | |
| **Completing the Square**<br>To solve $ax^2 + bx + c = 0$ $(a \neq 0)$:<br>*Step 1* If $a \neq 1$, divide each side by $a$.<br>*Step 2* Write the equation with the variable terms on one side and the constant on the other. | Solve $2x^2 - 4x - 18 = 0$.<br>$x^2 - 2x - 9 = 0$ Divide by 2.<br>$x^2 - 2x = 9$ Add 9. |

| CONCEPTS | EXAMPLES |
|---|---|
| *Step 3* Take half the coefficient of $x$ and square it. Add the square to each side. Factor the perfect square trinomial, and write it as the square of a binomial. Simplify the other side. | $\left[\frac{1}{2}(-2)\right]^2 = (-1)^2 = 1$<br>$x^2 - 2x + 1 = 9 + 1$<br>$(x - 1)^2 = 10$ |
| *Step 4* Use the square root property to complete the solution. | $x - 1 = \sqrt{10}$ or $x - 1 = -\sqrt{10}$<br>$x = 1 + \sqrt{10}$ or $x = 1 - \sqrt{10}$<br>Solution set: $\{1 + \sqrt{10}, 1 - \sqrt{10}\}$ |

### 11.3 SOLVING QUADRATIC EQUATIONS BY THE QUADRATIC FORMULA

**Quadratic Formula**

The solutions of $ax^2 + bx + c = 0$ $(a \neq 0)$ are given by

$$x = \frac{-b \pm \sqrt{b^2 - 4ac}}{2a}.$$

Solve $3x^2 + 5x + 2 = 0$.

$$x = \frac{-5 \pm \sqrt{5^2 - 4(3)(2)}}{2(3)} = \frac{-5 \pm 1}{6}$$

$$x = -1 \quad \text{or} \quad x = -\frac{2}{3}$$

Solution set: $\{-1, -\frac{2}{3}\}$

**The Discriminant**

If $a$, $b$, and $c$ are integers, then the discriminant, $b^2 - 4ac$, of $ax^2 + bx + c = 0$ determines the number and type of solutions as follows.

| Discriminant | Number and Type of Solutions |
|---|---|
| Positive, the square of an integer | Two rational solutions |
| Positive, not the square of an integer | Two irrational solutions |
| Zero | One rational solution |
| Negative | Two imaginary solutions |

For $x^2 + 3x - 10 = 0$, the discriminant is

$3^2 - 4(1)(-10) = 49$. Two rational solutions

For $4x^2 + x + 1 = 0$, the discriminant is

$1^2 - 4(4)(1) = -15$. Two imaginary solutions

### 11.4 EQUATIONS QUADRATIC IN FORM

A nonquadratic equation that can be written in the form

$$au^2 + bu + c = 0,$$

for $a \neq 0$ and an algebraic expression $u$, is called quadratic in form. Substitute $u$ for the expression, solve for $u$, and then solve for the variable in the expression.

Solve $3(x + 5)^2 + 7(x + 5) + 2 = 0$.

$3u^2 + 7u + 2 = 0$ Let $u = x + 5$.

$(3u + 1)(u + 2) = 0$

$u = -\frac{1}{3}$ or $u = -2$

$x + 5 = -\frac{1}{3}$ or $x + 5 = -2$ $x + 5 = u$

$x = -\frac{16}{3}$ or $x = -7$

Solution set: $\{-7, -\frac{16}{3}\}$

(continued)

| CONCEPTS | EXAMPLES |
|---|---|

## 11.5 FORMULAS AND FURTHER APPLICATIONS

To solve a formula for a squared variable, proceed as follows.

**(a) If the variable appears only to the second power:** Isolate the squared variable on one side of the equation, and then use the square root property.

Solve $A = \frac{2mp}{r^2}$ for $r$.

$r^2 A = 2mp$ — Multiply by $r^2$.

$r^2 = \frac{2mp}{A}$ — Divide by $A$.

$r = \pm\sqrt{\frac{2mp}{A}}$ — Square root property

$r = \frac{\pm\sqrt{2mpA}}{A}$ — Rationalize the denominator.

**(b) If the variable appears to the first and second powers:** Write the equation in standard form, and then use the quadratic formula.

Solve $m^2 + rm = t$ for $m$.

$m^2 + rm - t = 0$ — Standard form

$m = \frac{-r \pm \sqrt{r^2 - 4(1)(-t)}}{2(1)}$ — $a = 1, b = r, c = -t$

$m = \frac{-r \pm \sqrt{r^2 + 4t}}{2}$

## 11.6 GRAPHS OF QUADRATIC FUNCTIONS

1. The graph of the quadratic function defined by $F(x) = a(x - h)^2 + k$, $a \neq 0$, is a parabola with vertex at $(h, k)$ and the vertical line $x = h$ as axis.
2. The graph opens up if $a$ is positive and down if $a$ is negative.
3. The graph is wider than the graph of $f(x) = x^2$ if $0 < |a| < 1$ and narrower if $|a| > 1$.

Graph $f(x) = -(x + 3)^2 + 1$.

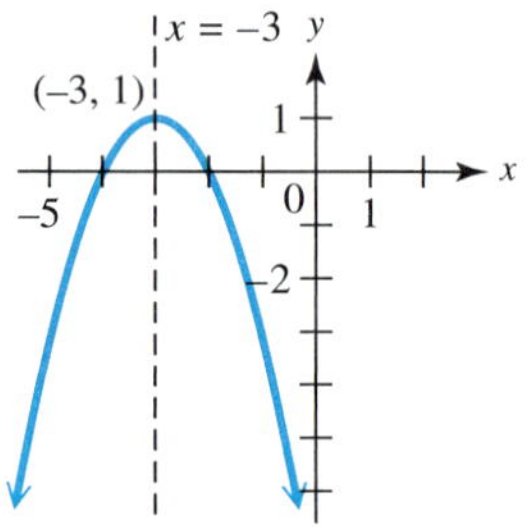


The graph opens down since $a < 0$. It is shifted 3 units left and 1 unit up, so the vertex is $(-3, 1)$, with axis $x = -3$. The domain is $(-\infty, \infty)$; the range is $(-\infty, 1]$.

## 11.7 MORE ABOUT PARABOLAS; APPLICATIONS

The vertex of the graph of $f(x) = ax^2 + bx + c$, $a \neq 0$, may be found by completing the square. The vertex has coordinates

$$\left(\frac{-b}{2a}, f\left(\frac{-b}{2a}\right)\right).$$

**Graphing a Quadratic Function**

*Step 1* Determine whether the graph opens up or down.

*Step 2* Find the vertex.

*Step 3* Find the $x$-intercepts (if any). Find the $y$-intercept.

*Step 4* Find and plot additional points as needed.

Graph $f(x) = x^2 + 4x + 3$.

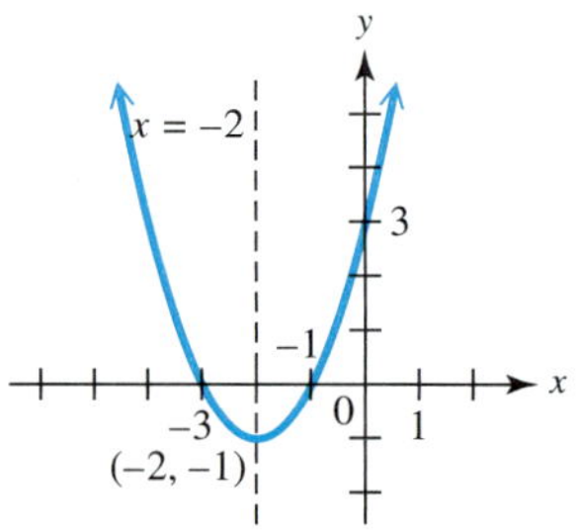


The graph opens up since $a > 0$. The vertex is $(-2, -1)$. The solutions of $x^2 + 4x + 3 = 0$ are $-1$ and $-3$, so the $x$-intercepts are $(-1, 0)$ and $(-3, 0)$. Since $f(0) = 3$, the $y$-intercept is $(0, 3)$. The domain is $(-\infty, \infty)$; the range is $[-1, \infty)$.

| CONCEPTS | EXAMPLES |
| --- | --- |

**Horizontal Parabolas**

The graph of $x = ay^2 + by + c$ is a horizontal parabola, opening to the right if $a > 0$ or to the left if $a < 0$. Horizontal parabolas do not represent functions.

Graph $x = 2y^2 + 6y + 5$.

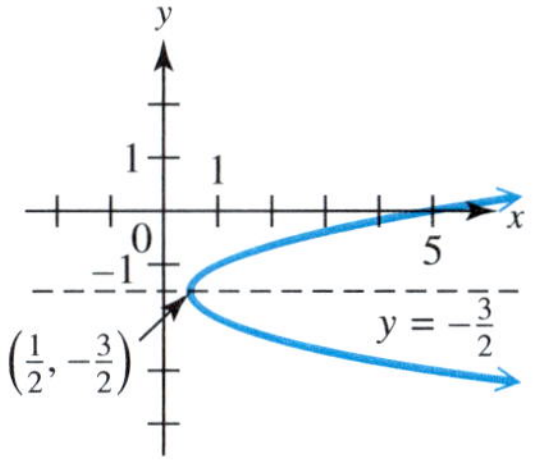


The graph opens to the right since $a > 0$. The vertex is $\left(\frac{1}{2}, -\frac{3}{2}\right)$. The axis is $y = -\frac{3}{2}$. The domain is $\left[\frac{1}{2}, \infty\right)$; the range is $(-\infty, \infty)$.

## 11.8 QUADRATIC AND RATIONAL INEQUALITIES

**Solving a Quadratic (or Higher-Degree Polynomial) Inequality**

Solve $2x^2 + 5x + 2 < 0$.

*Step 1* Write the inequality as an equation and solve.

$$2x^2 + 5x + 2 = 0$$

$$x = -\frac{1}{2} \quad \text{or} \quad x = -2$$

*Step 2* Use the numbers found in Step 1 to divide a number line into intervals.

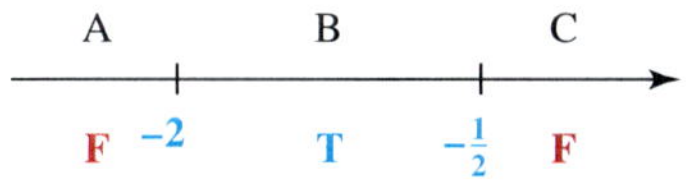


*Step 3* Substitute a test number from each interval into the original inequality to determine the intervals that belong to the solution set.

$x = -3$ makes the original inequality false; $x = -1$ makes it true; $x = 0$ makes it false.

*Step 4* Consider the endpoints separately.

The solution set is the interval $\left(-2, -\frac{1}{2}\right)$.

**Solving a Rational Inequality**

Solve $\dfrac{x}{x+2} \geq 4$.

*Step 1* Write the inequality so that 0 is on one side and there is a single fraction on the other side.

$$\frac{x}{x+2} - 4 \geq 0$$

$$\frac{x}{x+2} - \frac{4(x+2)}{x+2} \geq 0$$

$$\frac{-3x - 8}{x+2} \geq 0$$

*Step 2* Determine the numbers that make the numerator or denominator 0.

$-\frac{8}{3}$ makes the numerator 0; $-2$ makes the denominator 0.

*Step 3* Use the numbers from Step 2 to divide a number line into intervals.

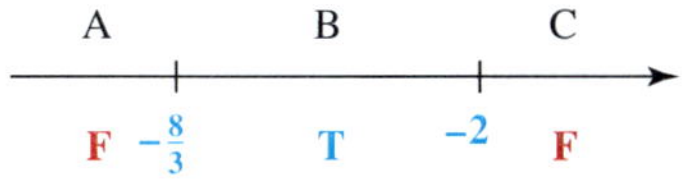


*Step 4* Substitute a test number from each interval into the original inequality to determine the intervals that belong to the solution set.

$-4$ makes the original inequality false; $-\frac{7}{3}$ makes it true; 0 makes it false.

*Step 5* Consider the endpoints separately.

The solution set is the interval $\left[-\frac{8}{3}, -2\right)$. The endpoint $-2$ is not included since it makes the denominator 0.

(continued)

**Answers to Test Your Word Power**

**1.** D; *Example:* The solutions of $ax^2 + bx + c = 0$ $(a \neq 0)$ are given by $x = \frac{-b \pm \sqrt{b^2 - 4ac}}{2a}$. **2.** C; *Examples:* $f(x) = x^2 - 2$, $f(x) = (x + 4)^2 + 1$, $f(x) = x^2 - 4x + 5$ **3.** D; *Examples:* See the figures in the Quick Review for Sections 11.6 and 11.7. **4.** C; *Example:* The graph of $y = (x + 3)^2$ has vertex $(-3, 0)$, which is the lowest point on the graph. **5.** B; *Example:* The axis of $y = (x + 3)^2$ is the vertical line $x = -3$. **6.** B; *Example:* Since the graph of $y = (x + 3)^2$ is symmetric about its axis $x = -3$, the points $(-2, 1)$ and $(-4, 1)$ are on the graph.

# CHAPTER 11 REVIEW EXERCISES

*Exercises marked * require knowledge of imaginary numbers.*

**[11.1]** *Solve each equation by using the square root property.*

**1.** $t^2 = 121$ **2.** $p^2 = 3$ **3.** $m^2 - 128 = 0$

**4.** $(r - 3)^2 = 10$ **5.** $(2x + 5)^2 = 100$ ***6.** $(3k - 2)^2 = -25$

**7.** A student gave the following "solution" to the equation $x^2 = 12$.

$$x^2 = 12$$
$$x = \sqrt{12} \quad \text{Square root property}$$
$$x = 2\sqrt{3}$$

What is wrong with this solution?

**8.** Navy Pier Center in Chicago, Illinois, features a 150-ft tall Ferris wheel. Use Galileo's formula $d = 16t^2$ to find how long it would take a wallet dropped from the top of the Ferris wheel to fall to the ground. Round your answer to the nearest tenth of a second. (*Source: Microsoft Encarta Encyclopedia 2002.*)

**[11.2]** *Solve each equation by completing the square.*

**9.** $m^2 + 6m + 5 = 0$ **10.** $p^2 + 4p = 7$ **11.** $-x^2 + 5 = 2x$

**12.** $2z^2 - 3 = -8z$ **13.** $5k^2 - 3k - 2 = 0$ ***14.** $(4a + 1)(a - 1) = -7$

**[11.3]** *Solve each equation by using the quadratic formula.*

**15.** $2x^2 + x - 21 = 0$ **16.** $k^2 + 5k = 7$ **17.** $(t + 3)(t - 4) = -2$

***18.** $2x^2 + 3x + 4 = 0$ ***19.** $3p^2 = 2(2p - 1)$ **20.** $m(2m - 7) = 3m^2 + 3$

*Use the discriminant to predict whether the solutions to the equations in Exercises 21–24 are*

**A.** *two rational numbers;* **B.** *one rational number;*
**C.** *two irrational numbers;* **D.** *two imaginary numbers.*

**21.** $x^2 + 5x + 2 = 0$

**22.** $4t^2 = 3 - 4t$

**23.** $4x^2 = 6x - 8$

**24.** $9z^2 + 30z + 25 = 0$

[11.4] *Solve each equation.*

**25.** $\frac{15}{x} = 2x - 1$

**26.** $\frac{1}{n} + \frac{2}{n + 1} = 2$

**27.** $-2r = \sqrt{\frac{48 - 20r}{2}}$

**28.** $8(3x + 5)^2 + 2(3x + 5) - 1 = 0$

**29.** $2x^{2/3} - x^{1/3} - 28 = 0$

**30.** $p^4 - 10p^2 + 9 = 0$

*Solve each problem. Round answers to the nearest tenth, as necessary.*

**31.** Phong paddled his canoe 20 mi upstream, then paddled back. If the speed of the current was 3 mph and the total trip took 7 hr, what was Phong's speed?

**32.** Maureen O'Connor drove 8 mi to pick up her friend Laurie, and then drove 11 mi to a mall at a speed 15 mph faster. If Maureen's total travel time was 24 min, what was her speed on the trip to pick up Laurie?

**33.** An old machine processes a batch of checks in 1 hr more time than a new one. How long would it take the old machine to process a batch of checks that the two machines together process in 2 hr?

**34.** Greg Tobin can process a stack of invoices 1 hr faster than Carter Fenton can. Working together, they take 1.5 hr. How long would it take each person working alone?

[11.5] *Solve each formula for the indicated variable. (Give answers with ±.)*

**35.** $k = \frac{rF}{wv^2}$ for $v$

**36.** $p = \sqrt{\frac{yz}{6}}$ for $y$

**37.** $mt^2 = 3mt + 6$ for $t$

*Solve each problem. Round answers to the nearest tenth, as necessary.*

**38.** A large machine requires a part in the shape of a right triangle with a hypotenuse 9 ft less than twice the length of the longer leg. The shorter leg must be $\frac{3}{4}$ the length of the longer leg. Find the lengths of the three sides of the part.

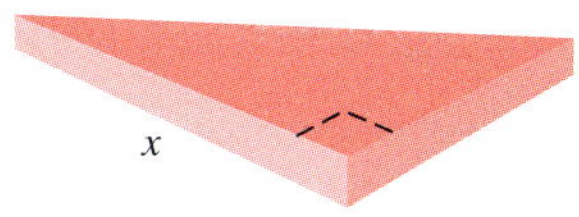


**39.** A square has an area of 256 cm². If the same amount is removed from one dimension and added to the other, the resulting rectangle has an area 16 cm² less. Find the dimensions of the rectangle.

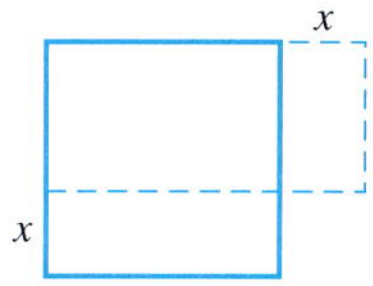


**40.** Nancy wants to buy a mat for a photograph that measures 14 in. by 20 in. She wants to have an even border around the picture when it is mounted on the mat. If the area of the mat she chooses is 352 in.², how wide will the border be?

**41.** A searchlight moves horizontally back and forth along a wall with the distance of the light from a starting point at $t$ min given by the quadratic function defined by

$$f(t) = 100t^2 - 300t.$$

How long will it take before the light returns to the starting point?

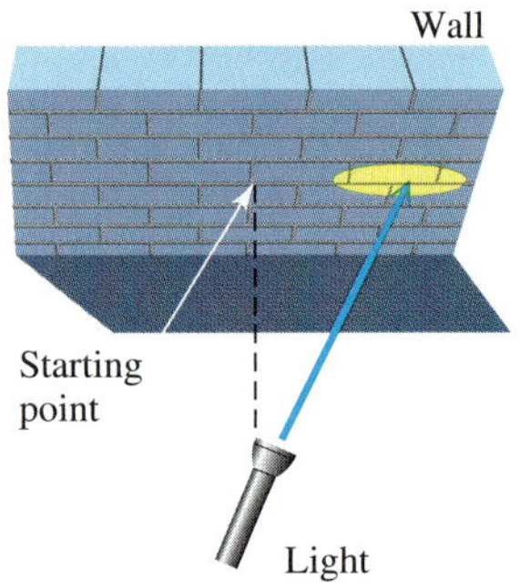


**42.** The Mart Hotel in Dallas, Texas, is 400 ft high. Suppose that a ball is projected upward from the top of the Mart, and its position in feet above the ground is given by the quadratic function defined by

$$f(t) = -16t^2 + 45t + 400,$$

where $t$ is the number of seconds elapsed. How long will it take for the ball to reach a height of 200 ft above the ground? (*Source: World Almanac and Book of Facts,* 2002.)

**43.** The Toronto Dominion Center in Winnipeg, Manitoba, is 407 ft high. Suppose that a ball is projected upward from the top of the center, and its position in feet above the ground is given by the quadratic function defined by

$$s(t) = -16t^2 + 75t + 407,$$

where $t$ is the number of seconds elapsed. How long will it take for the ball to reach a height of 450 ft above the ground? (*Source: World Almanac and Book of Facts,* 2002.)

**44.** The manager of a restaurant has determined that the demand for frozen yogurt is $\frac{25}{p}$ units per day, where $p$ is the price (in dollars) per unit. The supply is $70p + 15$ units per day. Find the price at which supply and demand are equal.

**45.** Use the formula $A = P(1 + r)^2$ to find the interest rate $r$ at which a principal $P$ of \$10,000 will increase to \$10,920.25 in 2 yr.

**46.** The number of e-mail boxes in North America (in millions) for the years 1995 through 2001 are shown in the graph and can be modeled by the quadratic function defined by

$$f(x) = 3.29x^2 - 10.4x + 21.6.$$

In the model, $x = 5$ represents 1995, $x = 10$ represents 2000, and so on.

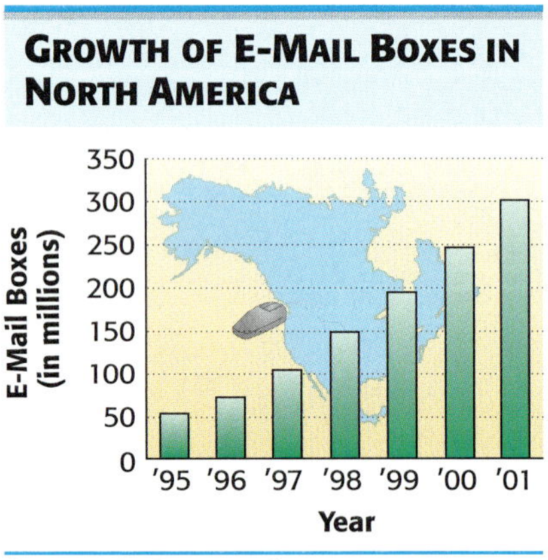

*Source:* IDC research.

**(a)** Use the model to approximate the number of e-mail boxes in 2001 to the nearest whole number. How does this result compare to the number shown in the graph?

**(b)** Based on the model, in what year did the number of e-mail boxes reach 200 million? (Round down to the nearest year.) How does this result compare to the number shown in the graph?

[11.6–11.7] *Identify the vertex of each parabola.*

**47.** $y = 6 - 2x^2$ **48.** $f(x) = -(x - 1)^2$ **49.** $f(x) = (x - 3)^2 + 7$

**50.** $y = -3x^2 + 4x - 2$ **51.** $x = (y - 3)^2 - 4$

**52.** If the discriminant of a quadratic function is negative, what do you know about the graph of the function?

*Graph each parabola. Give the vertex, axis, domain, and range.*

**53.** $y = 2(x - 2)^2 - 3$ **54.** $f(x) = -2x^2 + 8x - 5$

**55.** $x = 2(y + 3)^2 - 4$ **56.** $x = -\frac{1}{2}y^2 + 6y - 14$

*Solve each problem.*

**57.** Consumer spending for home video games in dollars per person per year is given in the table. Let $x = 0$ represent 1990, $x = 2$ represent 1992, and so on.

**CONSUMER SPENDING FOR HOME VIDEO GAMES**

| Year | Dollars |
|---|---|
| 1990 | 12.39 |
| 1992 | 13.08 |
| 1994 | 15.78 |
| 1996 | 19.43 |
| 1997 | 22.71 |
| 1998 | 24.14 |
| 1999 | 25.08 |

*Source: Statistical Abstract of the United States.*

**(a)** Use the data for 1990, 1994, and 1997 in the quadratic form $ax^2 + bx + c = y$ to write a system of three equations.

**(b)** Solve the system from part (a) to find a quadratic function $f$ that models the data.

**(c)** Use the model found in part (b) to approximate consumer spending for home video games in 1998 to the nearest cent. How does your answer compare to the actual data from the table?

**58.** The height (in feet) of a projectile $t$ sec after being fired from Earth into the air is given by

$$f(t) = -16t^2 + 160t.$$

Find the number of seconds required for the projectile to reach maximum height. What is the maximum height?

**59.** Find the length and width of a rectangle having a perimeter of 200 m if the area is to be a maximum.

[11.8] *Solve each inequality and graph the solution set.*

**60.** $(x - 4)(2x + 3) > 0$ **61.** $x^2 + x \leq 12$

**62.** $(x + 2)(x - 3)(x + 5) \leq 0$ **63.** $(4m + 3)^2 \leq -4$

**64.** $\frac{6}{2z - 1} < 2$ **65.** $\frac{3t + 4}{t - 2} \leq 1$

## MIXED REVIEW EXERCISES

*Solve.*

**66.** $V = r^2 + R^2h$ for $R$

***67.** $3t^2 - 6t = -4$

***68.** $x^4 - 1 = 0$

**69.** $(x^2 - 2x)^2 = 11(x^2 - 2x) - 24$

**70.** $(r - 1)(2r + 3)(r + 6) < 0$

**71.** $2x - \sqrt{x} = 6$

**72.** $(3k + 11)^2 = 7$

**73.** $S = \dfrac{Id^2}{k}$ for $d$

**74.** $(8k - 7)^2 \geq -1$

**75.** $6 + \dfrac{15}{s^2} = -\dfrac{19}{s}$

**76.** $x^4 - 8x^2 = -1$

**77.** $\dfrac{-2}{x + 5} \leq -5$

*Match each equation with the figure that most closely resembles its graph.*

**78.** $g(x) = x^2 - 5$

**79.** $h(x) = -x^2 + 4$

**80.** $F(x) = (x - 1)^2$

**81.** $G(x) = (x + 1)^2$

**82.** $H(x) = (x - 1)^2 + 1$

**83.** $K(x) = (x + 1)^2 + 1$

**A.**

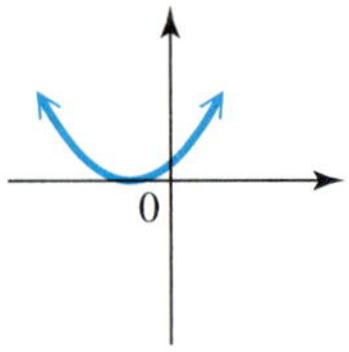


**B.**

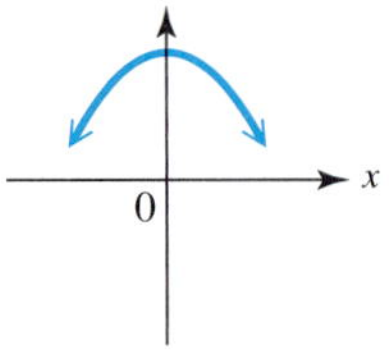


**C.**

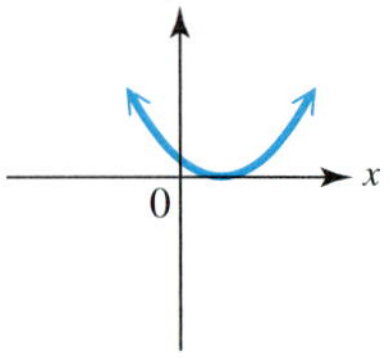


**D.**

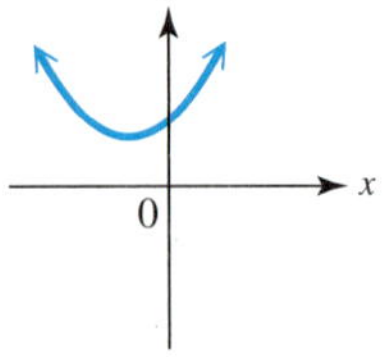


**E.**

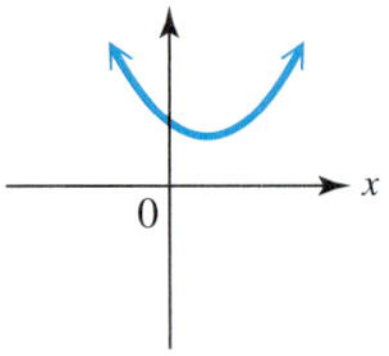


**F.**

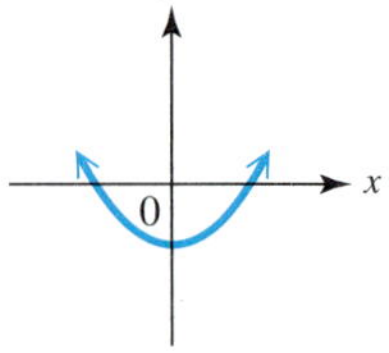


**84.** Graph $f(x) = 4x^2 + 4x - 2$. Give the vertex, axis, domain, and range.

*Solve each problem.*

**85.** Natural gas use in the United States in trillions of cubic feet ($ft^3$) from 1970 through 1999 can be modeled by the quadratic function defined by

$$f(x) = .014x^2 - .396x + 21.2,$$

where $x = 0$ represents 1970, $x = 5$ represents 1975, and so on. (*Source:* Energy Information Administration.)

**(a)** Use the model to approximate natural gas use in 2000.

**(b)** Based on the model, in what year will natural gas use reach 25 trillion $ft^3$? (Round down to the nearest year.)

**86.** In 4 hr, Kerrie can go 15 mi upriver and come back. The speed of the current is 5 mph. Find the speed of the boat in still water.

**87.** Refer to Exercise 42. Suppose that a wire is attached to the top of the Mart and pulled tight. It is attached to the ground 100 ft from the base of the building. How long is the wire?

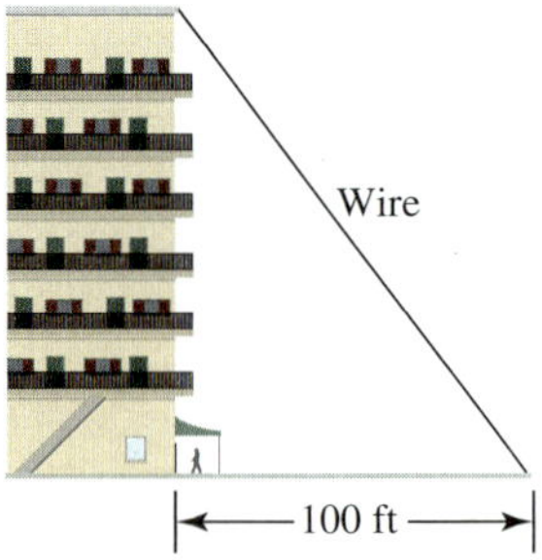

## RELATING CONCEPTS (EXERCISES 88–92)

**For Individual or Group Work**

***Work Exercises 88–92 in order,*** *to see the connections between equations and inequalities.*

**88.** Use the methods of Chapter 2 to solve the equation or inequality, and graph the solution set.

**(a)** $3x - (4x + 2) = 0$ **(b)** $3x - (4x + 2) > 0$ **(c)** $3x - (4x + 2) < 0$

**89.** Use the methods of this chapter to solve the equation or inequality, and graph the solution set.

**(a)** $x^2 - 6x + 5 = 0$ **(b)** $x^2 - 6x + 5 > 0$ **(c)** $x^2 - 6x + 5 < 0$

**90.** Use the methods of Sections 6.6 and 11.8 to solve the equation or inequality, and graph the solution set.

**(a)** $\dfrac{-5x + 20}{x - 2} = 0$ **(b)** $\dfrac{-5x + 20}{x - 2} > 0$ **(c)** $\dfrac{-5x + 20}{x - 2} < 0$

**91.** Fill in the blanks in the following statement: If we solve a linear, quadratic, or rational equation and the two inequalities associated with it, the union of the three solution sets will be ________; the only exception will be in the case of the rational equation and inequalities, where the number or numbers that cause the ________ to be 0 will be excluded.

**92.** Suppose that the solution set of a quadratic equation is $\{-5, 3\}$ and the solution set of one of the associated inequalities is $(-\infty, -5) \cup (3, \infty)$. What is the solution set of the other associated inequality?

# CHAPTER 11 TEST

*Problems marked * require knowledge of imaginary numbers.*

*Solve by using the square root property.*

**1.** $t^2 = 54$

**2.** $(7x + 3)^2 = 25$

**3.** Solve $2x^2 + 4x = 8$ by completing the square.

*Solve by using the quadratic formula.*

**4.** $2x^2 - 3x - 1 = 0$

***5.** $3t^2 - 4t = -5$

***6.** If $k$ is a negative number, then which one of the following equations will have two imaginary solutions?

**A.** $x^2 = 4k$ **B.** $x^2 = -4k$ **C.** $(x + 2)^2 = -k$ **D.** $x^2 + k = 0$

**7.** What is the discriminant for $2x^2 - 8x - 3 = 0$? How many and what type of solutions does this equation have? (Do not actually solve.)

*Solve by any method.*

**8.** $3x = \sqrt{\frac{9x + 2}{2}}$ **9.** $3 - \frac{16}{x} - \frac{12}{x^2} = 0$ **10.** $4x^2 + 7x - 3 = 0$

**11.** $9x^4 + 4 = 37x^2$ **12.** $12 = (2n + 1)^2 + (2n + 1)$

**13.** Solve $S = 4\pi r^2$ for $r$. (Leave $\pm$ in your answer.)

*Solve each problem.*

**14.** Andrew and Kent do word processing. For a certain prospectus, Kent can prepare it 2 hr faster than Andrew can. If they work together, they can do the entire prospectus in 5 hr. How long will it take each of them working alone to prepare the prospectus? Round your answers to the nearest tenth of an hour.

**15.** Abby Tanenbaum paddled her canoe 10 mi upstream and then paddled back to her starting point. If the rate of the current was 3 mph and the entire trip took $3\frac{1}{2}$ hr, what was Abby's rate?

**16.** Tyler McGinnis has a pool 24 ft long and 10 ft wide. He wants to construct a concrete walk around the pool. If he plans for the walk to be of uniform width and cover 152 $ft^2$, what will the width of the walk be?

**17.** At a point 30 m from the base of a tower, the distance to the top of the tower is 2 m more than twice the height of the tower. Find the height of the tower.

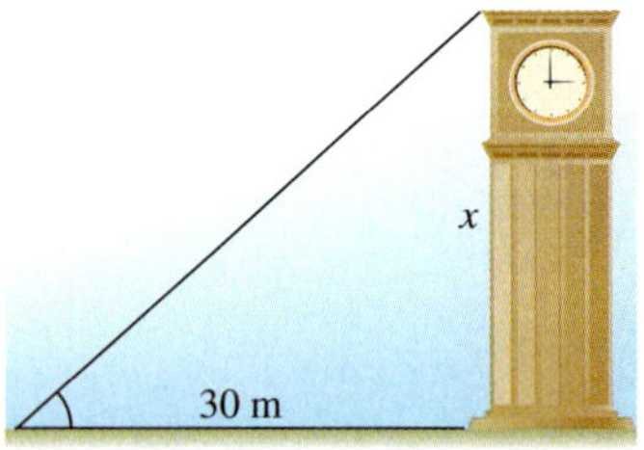


**18.** Which one of the following most closely resembles the graph of $f(x) = a(x - h)^2 + k$ if $a < 0$, $h > 0$, and $k < 0$?

**A.**
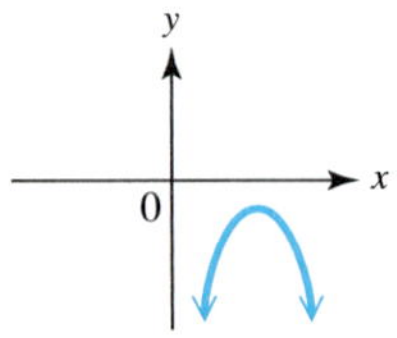


**B.**
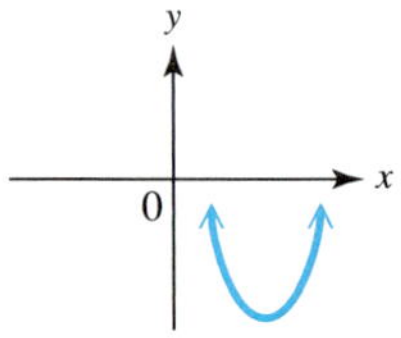


**C.**
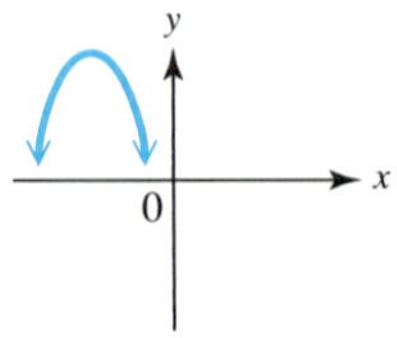


**D.**
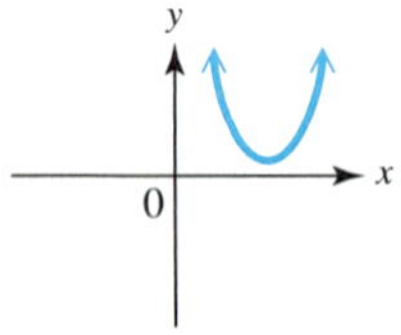


*Graph each parabola. Identify the vertex, axis, domain, and range.*

**19.** $f(x) = \frac{1}{2}x^2 - 2$ **20.** $f(x) = -x^2 + 4x - 1$ **21.** $x = -(y - 2)^2 + 2$

**22.** The percent increase for in-state tuition at Iowa public universities during the years 1992 through 2002 can be modeled by the quadratic function defined by

$$f(x) = .156x^2 - 2.05x + 10.2,$$

where $x = 2$ represents 1992, $x = 3$ represents 1993, and so on. (*Source:* Iowa Board of Regents.)

**(a)** Based on this model, by what percent (to the nearest tenth) did tuition increase in 2001?

**(b)** In what year was the minimum tuition increase? (Round down to the nearest year.) To the nearest tenth, by what percent did tuition increase that year?

**23.** Palo Alto College is planning to construct a rectangular parking lot on land bordered on one side by a highway. The plan is to use 640 ft of fencing to fence off the other three sides. What should the dimensions of the lot be if the enclosed area is to be a maximum?

*Solve, and graph each solution set.*

**24.** $2x^2 + 7x > 15$

**25.** $\dfrac{5}{t - 4} \le 1$

## CUMULATIVE REVIEW EXERCISES CHAPTERS 1–11

**1.** Let $S = \left\{-\frac{7}{3}, -2, -\sqrt{3}, 0, .7, \sqrt{12}, \sqrt{-8}, 7, \frac{32}{3}\right\}$. List the elements of $S$ that are elements of each set.

**(a)** Integers **(b)** Rational numbers **(c)** Real numbers **(d)** Complex numbers

*Simplify each expression.*

**2.** $|-3| + 8 - |-9| - (-7 + 3)$

**3.** $2(-3)^2 + (-8)(-5) + (-17)$

*Solve each equation.*

**4.** $7 - (4 + 3t) + 2t = -6(t - 2) - 5$

**5.** $|6x - 9| = |-4x + 2|$

**6.** $2x = \sqrt{\dfrac{5x + 2}{3}}$

**7.** $\dfrac{3}{x - 3} - \dfrac{2}{x - 2} = \dfrac{3}{x^2 - 5x + 6}$

**8.** $(r - 5)(2r + 3) = 1$

**9.** $x^4 - 5x^2 + 4 = 0$

*Solve each inequality.*

**10.** $-2x + 4 \le -x + 3$

**11.** $|3x - 7| \le 1$

**12.** $x^2 - 4x + 3 < 0$

**13.** $\dfrac{3}{p + 2} > 1$

*Graph each relation. Tell whether or not each is a function, and if it is, give its domain and range.*

**14.** $4x - 5y = 15$

**15.** $4x - 5y < 15$

**16.** $f(x) = -2(x - 1)^2 + 3$

**17.** Find the slope and intercepts of the line with equation $-2x + 7y = 16$.

**18.** Write an equation for the specified line. Express each equation in slope-intercept form.

**(a)** Through $(2, -3)$ and parallel to the line with equation $5x + 2y = 6$
**(b)** Through $(-4, 1)$ and perpendicular to the line with equation $5x + 2y = 6$

*Write with positive exponents only. Assume variables represent positive real numbers.*

**19.** $\left(\dfrac{x^{-3}y^2}{x^5y^{-2}}\right)^{-1}$

**20.** $\dfrac{(4x^{-2})^2(2y^3)}{8x^{-3}y^5}$

*Perform the indicated operations.*

**21.** $(7x + 4)(2x - 3)$

**22.** $\left(\dfrac{2}{3}t + 9\right)^2$

**23.** $(3t^3 + 5t^2 - 8t + 7) - (6t^3 + 4t - 8)$

**24.** Divide $4x^3 + 2x^2 - x + 26$ by $x + 2$.

*Factor completely.*

**25.** $16x - x^3$

**26.** $24m^2 + 2m - 15$

**27.** $8x^3 + 27y^3$

**28.** $9x^2 - 30xy + 25y^2$

*Perform the indicated operations and express each answer in lowest terms. Assume denominators are nonzero.*

**29.** $\dfrac{x^2 - 3x - 10}{x^2 + 3x + 2} \cdot \dfrac{x^2 - 2x - 3}{x^2 + 2x - 15}$

**30.** $\dfrac{3}{2 - k} - \dfrac{5}{k} + \dfrac{6}{k^2 - 2k}$

**31.** $\dfrac{\dfrac{r}{s} - \dfrac{s}{r}}{\dfrac{r}{s} + 1}$

**32.** The record track-qualifying speeds at North Carolina Motor Speedway since Richard Petty captured the first pole in 1965 are given in the table. Let $x = 0$ represent 1965, $x = 10$ represent 1975, and so on.

**(a)** Use the ordered pairs (year, speed) to make a scatter diagram of the data.
**(b)** A linear equation can be used to model the data. Will its slope be positive or negative?
**(c)** Use the ordered pairs $(0, 116.26)$ and $(20, 141.85)$ to write a linear equation that models the data.
**(d)** Use your model to approximate the record speed for 1998 to the nearest hundredth. How does it compare to the actual value from the table?

**QUALIFYING RECORDS**

| *Year* | *Speed (in mph)* |
|---|---|
| 1965 | 116.26 |
| 1975 | 132.02 |
| 1985 | 141.85 |
| 1995 | 155.38 |
| 1998 | 156.36 |

*Source:* NASCAR.

**33.** Does the relation $x = 5$ define a function? Explain.

**34.** For the function defined by $f(x) = 2(x - 1)^2 - 5$, find

**(a)** $f(-2)$; **(b)** the domain and range.

*Solve each system of equations.*

**35.**
$$\begin{aligned} 2x - 4y &= 10 \\ 9x + 3y &= 3 \end{aligned}$$

**36.**
$$\begin{aligned} x + y + 2z &= 3 \\ -x + y + z &= -5 \\ 2x + 3y - z &= -8 \end{aligned}$$

**37.** The merger in 2000 of America Online and Time Warner was the largest in U.S. history. The two companies had combined sales of \$34.2 billion. Sales for AOL were \$.3 billion less than 4 times the sales for Time Warner. What were sales for each company? (*Source:* Company reports.)

**(a)** Write a system of equations to solve the problem.
**(b)** Solve the problem.

*Simplify each radical expression.*

**38.** $\sqrt[3]{\frac{27}{16}}$

**39.** $\frac{2}{\sqrt{7}-\sqrt{5}}$

*Solve each problem.*

**40.** Tri rode his bicycle for 12 mi and then walked an additional 8 mi. The total time for the trip was 5 hr. If his rate while walking was 10 mph less than his rate while riding, what was each rate?

**41.** Two cars left an intersection at the same time, one heading due south and the other due east. Later they were exactly 95 mi apart. The car heading east had gone 38 mi less than twice as far as the car heading south. How far had each car traveled?

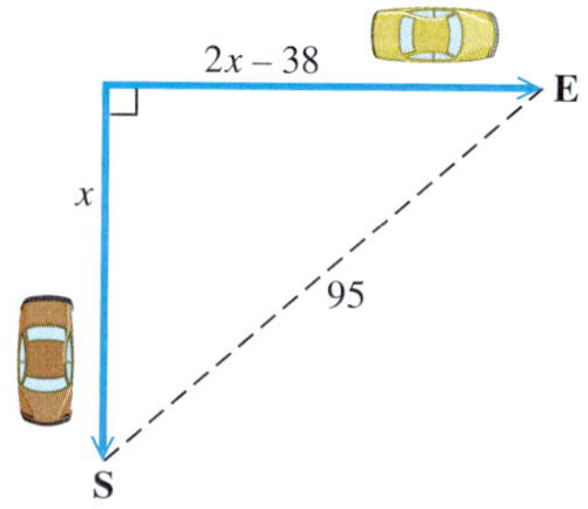


*In this day of Automated Teller Machines (ATMs), people often find themselves doing what they have done for years when faced with a soft drink machine that won't respond: They talk to it. According to one report, the following are percentages of people in the United States, the United Kingdom (UK), and Germany who talk to ATMs and what they say.*

| | *United States* | *UK* | *Germany* |
|---|---|---|---|
| *Thanking the ATM* | 22% | 24% | 14% |
| *Cursing the ATM* | 31% | 41% | 53% |
| *Telling the ATM to Hurry Up* | 47% | 36% | 33% |

*Source:* BMRB International for NCR.

*In a random sample of* 3000 *people, how many would there be in each category?*

**42.** People in the United States who curse the ATM

**43.** People in the UK who thank the ATM

**44.** People in Germany who tell the ATM to hurry up

**45.** How many more German cursers would there be than United States thankers?

# Appendix A
# An Introduction to Calculators

There is little doubt that the appearance of handheld calculators three decades ago and the later development of scientific and graphing calculators have changed the methods of learning and studying mathematics forever. For example, computations with tables of logarithms and slide rules made up an important part of mathematics courses prior to 1970. Today, with the widespread availability of calculators, these topics are studied only for their historical significance.

Calculators come in a large array of different types, sizes, and prices. *For the course for which this textbook is intended, the most appropriate type is the scientific calculator,* which costs \$10–\$20.

In this introduction, we explain some of the features of scientific and graphing calculators. However, remember that calculators vary among manufacturers and models, and that while the methods explained here apply to many of them, they may not apply to your specific calculator. *This introduction is only a guide and is not intended to take the place of your owner's manual.* Always refer to the manual in the event you need an explanation of how to perform a particular operation.

## Scientific Calculators

Scientific calculators are capable of much more than the typical four-function calculator that you might use for balancing your checkbook. Most scientific calculators use *algebraic logic.* (Models sold by Texas Instruments, Sharp, Casio, and Radio Shack, for example, use algebraic logic.) A notable exception is Hewlett-Packard, a company whose calculators use *Reverse Polish Notation* (RPN). In this introduction, we explain the use of calculators with algebraic logic.

**Arithmetic Operations** To perform an operation of arithmetic, simply enter the first number, press the operation key, (+), (−), (×), or (÷), enter the second number, and then press the (=) key. For example, to add 4 and 3, use the following keystrokes.

(4) (+) (3) (=) [ 7 ]

**Change Sign Key** The key marked (+/−) allows you to change the sign of a display. This is particularly useful when you wish to enter a negative number. For example, to enter $-3$, use the following keystrokes.

(3) (+/−) ( −3)

**Memory Key** Scientific calculators can hold a number in memory for later use. The label of the memory key varies among models; two of these are (M) and (STO). The (M+) and (M−) keys allow you to add to or subtract from the value currently in memory. The memory recall key, labeled (MR), (RM), or (RCL), allows you to retrieve the value stored in memory.

Suppose that you wish to store the number 5 in memory. Enter 5, then press the key for memory. You can then perform other calculations. When you need to retrieve the 5, press the key for memory recall.

If a calculator has a constant memory feature, the value in memory will be retained even after the power is turned off. Some advanced calculators have more than one memory. Read the owner's manual for your model to see exactly how memory is activated.

**Clearing/Clear Entry Keys** The keys (C) or (CE) allow you to clear the display or clear the last entry entered into the display. In some models, pressing the (C) key once will clear the last entry, while pressing it twice will clear the entire operation in progress.

**Second Function Key** This key, usually marked (2nd), is used in conjunction with another key to activate a function that is printed *above* an operation key (and not on the key itself). For example, suppose you wish to find the square of a number, and the squaring function (explained in more detail later) is printed above another key. You would need to press (2nd) before the desired squaring function can be activated.

**Square Root Key** Pressing ($\sqrt{\ }$) or ($\sqrt{x}$) will give the square root (or an approximation of the square root) of the number in the display. On many newer scientific calculators, the square root key is pressed *before* entering the number, while other calculators use the opposite order. Experiment with your calculator to see which method it uses. For example, to find the square root of 36, use the following keystrokes.

($\sqrt{\ }$) (3) (6) ( 6) or (3) (6) ($\sqrt{\ }$) ( 6)

The square root of 2 is an example of an irrational number (Chapter 10). The calculator will give an approximation of its value, since the decimal for $\sqrt{2}$ never terminates and never repeats. The number of digits shown will vary among models. To find an approximation for $\sqrt{2}$, use the following keystrokes.

($\sqrt{\ }$) (2) (1.4142136) or (2) ($\sqrt{\ }$) (1.4142136)

An approximation for $\sqrt{2}$

**Squaring Key** The ($x^2$) key allows you to square the entry in the display. For example, to square 35.7, use the following keystrokes.

(3) (5) (.) (7) ($x^2$) (1274.49)

The squaring key and the square root key are often found on the same key, with one of them being a second function (that is, activated by the second function key previously described).

**Reciprocal Key** The key marked (1/x) is the reciprocal key. (When two numbers have a product of 1, they are called *reciprocals.* See Chapter 1.) Suppose that you wish to find the reciprocal of 5. Use the following keystrokes.

(5) (1/x) (0.2)

**Inverse Key** Some calculators have an inverse key, marked (INV). Inverse operations are operations that "undo" each other. For example, the operations of squaring and taking the square root are inverse operations. The use of the (INV) key varies among different models of calculators, so read your owner's manual carefully.

**Exponential Key** The key marked ($x^y$) or ($y^x$) allows you to raise a number to a power. For example, if you wish to raise 4 to the fifth power (that is, find $4^5$, as explained in Chapter 1), use the following keystrokes.

(4) ($x^y$) (5) (=) (1024)

**Root Key** Some calculators have a key specifically marked ($\sqrt[x]{x}$) or ($\sqrt[x]{y}$); with others, the operation of taking roots is accomplished by using the inverse key in conjunction with the exponential key. Suppose, for example, your calculator is of the latter type and you wish to find the fifth root of 1024. Use the following keystrokes.

(1) (0) (2) (4) (INV) ($x^y$) (5) (=) (4)

Notice how this "undoes" the operation explained in the exponential key discussion.

**Pi Key** The number $\pi$ is an important number in mathematics. It occurs, for example, in the area and circumference formulas for a circle. By pressing the ($\pi$) key, you can display the first few digits of $\pi$. (Because $\pi$ is irrational, the display shows only an approximation.) One popular model gives the following display when the ($\pi$) key is pressed.

(3.1415927) An approximation for $\pi$

**Methods of Display** When decimal approximations are shown on scientific calculators, they are either *truncated* or *rounded.* To see how a particular model is programmed, evaluate 1/18 as an example. If the display shows .0555555 (last digit 5), it truncates the display. If it shows .0555556 (last digit 6), it rounds the display.

When very large or very small numbers are obtained as answers, scientific calculators often express these numbers in scientific notation (Chapter 4). For example, if you multiply 6,265,804 by 8,980,591, the display might look like this:

(5.6270623 13)

The 13 at the far right means that the number on the left is multiplied by $10^{13}$. This means that the decimal point must be moved 13 places to the right if the answer is to be expressed in its usual form. Even then, the value obtained will only be an approximation: 56,270,623,000,000.

## Graphing Calculators

While you are not expected to have a graphing calculator to study from this book, we include the following as background information and reference should your course or future courses require the use of graphing calculators.

**Basic Features** In addition to the typical keys found on scientific calculators, graphing calculators have keys that can be used to create graphs, make tables, analyze data, and change settings. One of the major differences between graphing and scientific calculators is that a graphing calculator has a larger viewing screen with graphing capabilities. The screens below illustrate the graphs of Y = X and Y = $X^2$.

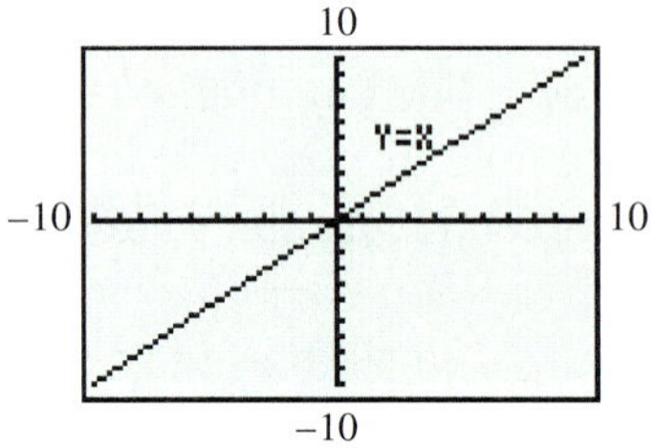


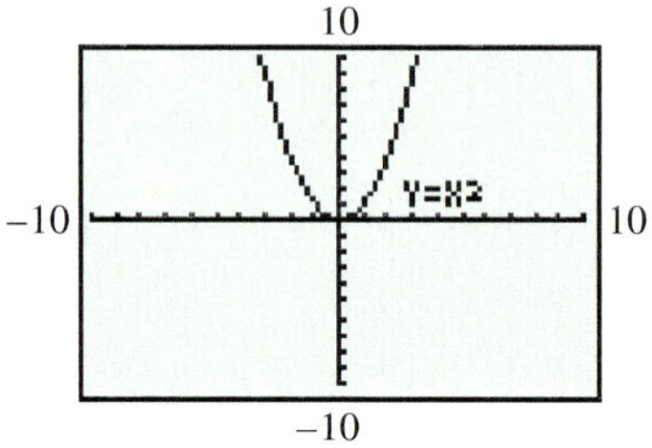


If you look closely at the screens, you will see that the graphs appear to be jagged rather than smooth, as they should be. The reason for this is that graphing calculators have much lower resolution than computer screens. Because of this, graphs generated by graphing calculators must be interpreted carefully.

**Editing Input** The screen of a graphing calculator can display several lines of text at a time. This feature allows you to view both previous and current expressions. If an incorrect expression is entered, an error message is displayed. The erroneous expression can be viewed and corrected by using various editing keys, much like a word-processing program. You do not need to enter the entire expression again. Many graphing calculators can also recall past expressions for editing or updating. The screen on the left shows how two expressions are evaluated. The final line is entered incorrectly, and the resulting error message is shown in the screen on the right.

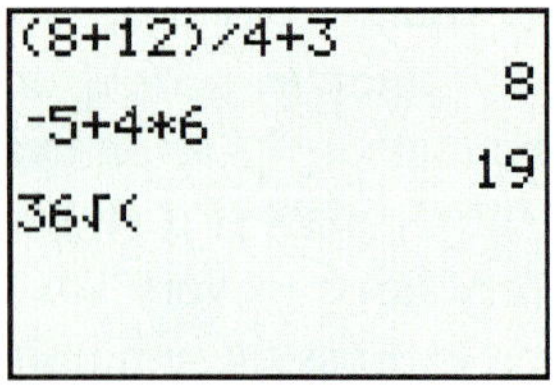


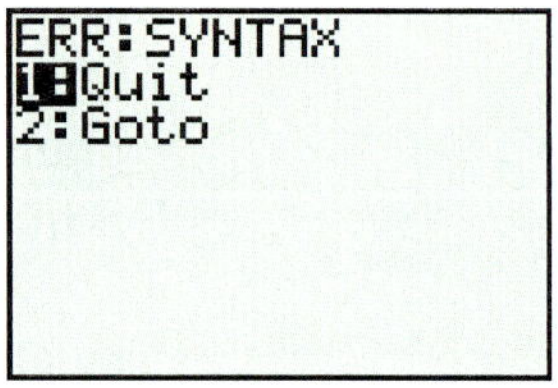


**Order of Operations** Arithmetic operations on graphing calculators are usually entered as they are written in mathematical expressions. For example, to evaluate $\sqrt{36}$ you would first press the square root key, and then enter 36. See the left screen at the top of the next page. The order of operations on a graphing calculator is also important, and current models insert parentheses when typical errors might occur. The open parenthesis that follows the square root symbol is automatically entered by the calculator so that an expression such as $\sqrt{2 \times 8}$ will not be calculated incorrectly as $\sqrt{2} \times 8$. Compare the two entries and their results in the screen on the right.

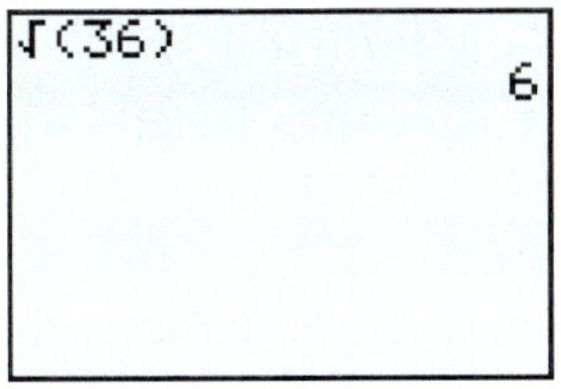


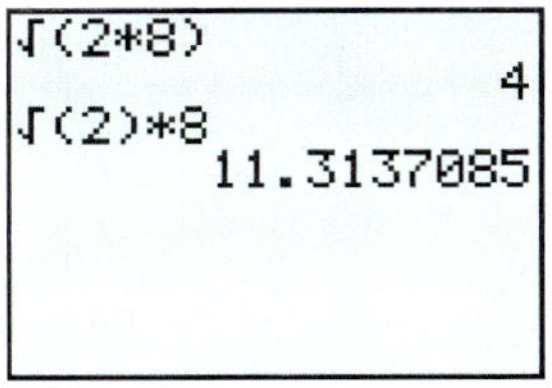


**Viewing Windows** The viewing window for a graphing calculator is similar to the viewfinder in a camera. A camera usually cannot take a photograph of an entire view of a scene. The camera must be centered on some object and can capture only a portion of the available scenery. A camera with a zoom lens can photograph different views of the same scene by zooming in and out. Graphing calculators have similar capabilities. The *xy*-coordinate plane is infinite. The calculator screen can only show a finite, rectangular region in the plane, and it must be specified before the graph can be drawn. This is done by setting both minimum and maximum values for the *x*- and *y*-axes. The scale (distance between tick marks) is usually specified as well. Determining an appropriate viewing window for a graph is often a challenge, and many times it will take a few attempts before a satisfactory window is found.

The screen on the left shows a standard viewing window, and the graph of $Y = 2X + 1$ is shown on the right. Using a different window would give a different view of the line.

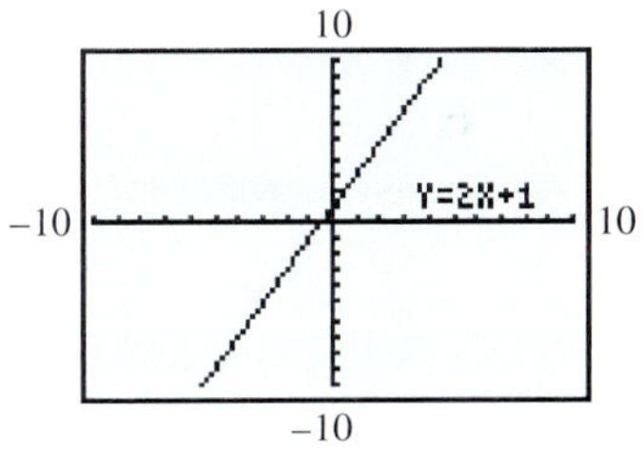


**Locating Points on a Graph: Tracing and Tables** Graphing calculators allow you to trace along the graph of an equation and display the coordinates of points on the graph. See the screen on the left below, which indicates that the point (2, 5) lies on the graph of $Y = 2X + 1$. Tables for equations can also be displayed. The screen on the right shows a partial table for this same equation. Note the middle of the screen, which indicates that when $X = 2$, $Y = 5$.

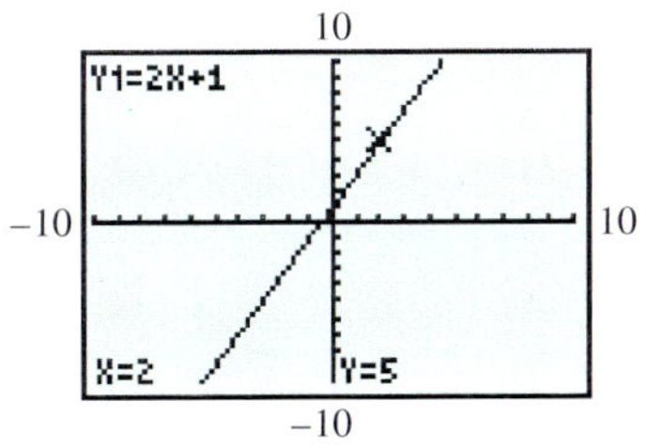


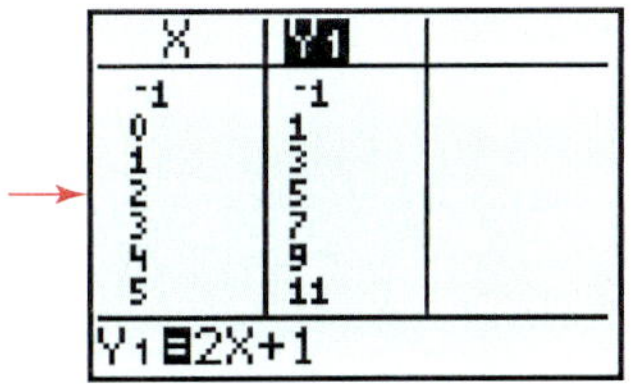

**Additional Features** There are many features of graphing calculators that go far beyond the scope of this book. These calculators can be programmed, much like computers. Many of them can solve equations at the stroke of a key, analyze statistical data, and perform symbolic algebraic manipulations. Calculators also provide the opportunity to ask "What if . . . ?" more easily. Values in algebraic expressions can be altered and conjectures tested quickly.

**Final Comments** Despite the power of today's calculators, they cannot replace human thought. ***In the entire problem-solving process, your brain is the most important component.*** Calculators are only tools and, like any tool, they must be used appropriately in order to enhance our ability to understand mathematics. Mathematical insight may often be the quickest and easiest way to solve a problem; a calculator may neither be needed nor appropriate. By applying mathematical concepts, you can make the decision whether or not to use a calculator.

# Appendix B Review of Decimals and Percents

**OBJECTIVES**

1. Add and subtract decimals.
2. Multiply and divide decimals.
3. Convert percents to decimals and decimals to percents.
4. Find percentages by multiplying.

**OBJECTIVE 1 Add and subtract decimals.** A **decimal** is a number written with a decimal point, such as 4.2. The operations on decimals—addition, subtraction, multiplication, and division—are reviewed in the next examples.

**EXAMPLE 1 Adding and Subtracting Decimals**

Add or subtract as indicated.

**(a)** $6.92 + 14.8 + 3.217$

Place the numbers in a column, with decimal points lined up, then add. If you like, attach 0s to make all the numbers the same length; this is a good way to avoid errors.

Decimal points lined up

$$\begin{array}{r} 6.92 \\ 14.8 \\ +\ 3.217 \\ \hline 24.937 \end{array} \quad \text{or} \quad \begin{array}{r} 6.920 \\ 14.800 \\ +\ 3.217 \\ \hline 24.937 \end{array}$$

**(b)** $47.6 - 32.509$

Write the numbers in a column, attaching 0s to 47.6.

$$\begin{array}{r} 47.6 \\ -32.509 \\ \hline \end{array} \quad \text{becomes} \quad \begin{array}{r} 47.600 \\ -32.509 \\ \hline 15.091 \end{array}$$

**(c)** $3 - .253$

$$\begin{array}{r} 3.000 \\ -\ \ .253 \\ \hline 2.747 \end{array}$$

**Now Try Exercises 1 and 5.**

**OBJECTIVE 2 Multiply and divide decimals.** Multiplication and division of decimals are similar to the same operations with whole numbers.

**EXAMPLE 2 Multiplying Decimals**

Multiply.

**(a)** $29.3 \times 4.52$

Multiply as if the numbers were whole numbers. To find the number of decimal places in the answer, add the numbers of decimal places in the factors.

$$\begin{array}{rl}
29.3 & \text{1 decimal place in first factor} \\
\times\ 4.52 & \text{2 decimal places in second factor} \\
\hline
5\ 86 & \\
14\ 6\ 5\ \ & 1 + 2 = 3 \\
117\ 2\ \ \ \ & \\
\hline
132.4\ 36 & \text{3 decimal places in answer}
\end{array}$$

**(b)** $7.003 \times 55.8$

$$\begin{array}{rl}
7.003 & \text{3 decimal places} \\
\times\ 55.8 & \text{1 decimal place} \\
\hline
5\ 602\ 4 & \\
35\ 015\ \ & 3 + 1 = 4 \\
350\ 15\ \ \ \ & \\
\hline
390.767\ 4 & \text{4 decimal places in answer}
\end{array}$$

**Now Try Exercises 11 and 15.**

### EXAMPLE 3 Dividing Decimals

Divide: $279.45 \div 24.3$.

Move the decimal point in 24.3 one place to the right, to get 243. Move the decimal point the same number of places in 279.45. By doing this, 24.3 is converted into the whole number 243.

$$243.\overline{)2794.5}$$

Bring the decimal point straight up and divide as with whole numbers.

$$\begin{array}{r}
11.5 \\
243.\overline{)2794.5} \\
\underline{243}\ \ \ \ \\
364\ \ \\
\underline{243}\ \ \\
121\ 5 \\
\underline{121\ 5} \\
0
\end{array}$$

**Now Try Exercise 19.**

**OBJECTIVE 3 Convert percents to decimals and decimals to percents.** One of the main uses of decimals is in percent problems. The word **percent** means "per one hundred." Percent is written with the sign %. One percent means "one per one hundred."

**Percent**

$$\mathbf{1\% = .01} \qquad \text{or} \qquad \mathbf{1\% = \frac{1}{100}}$$

### EXAMPLE 4 Converting between Decimals and Percents

Convert.

**(a)** 75% to a decimal

Since $1\% = .01$,

$$75\% = 75 \cdot 1\% = 75 \cdot .01 = .75.$$

The fraction form $1\% = \frac{1}{100}$ can also be used to convert 75% to a decimal.

$$75\% = 75 \cdot 1\% = 75 \cdot \frac{1}{100} = .75$$

**(b)** 2.63 to a percent

$$2.63 = 263 \cdot .01 = 263 \cdot 1\% = 263\%$$

**Now Try Exercises 23 and 33.**

**OBJECTIVE 4 Find percentages by multiplying.** A part of a whole is called a **percentage.** For example, since 50% represents $\frac{50}{100} = \frac{1}{2}$ of a whole, 50% of 800 is half of 800, or 400. Multiply to find percentages, as in the next example.

### EXAMPLE 5 Finding Percentages

Find each percentage.

**(a)** 15% of 600

The word *of* indicates multiplication here. For this reason, 15% of 600 is found by multiplying.

$$15\% \cdot 600 = .15 \cdot 600 = 90$$

**(b)** 125% of 80

$$125\% \cdot 80 = 1.25 \cdot 80 = 100$$

**(c)** What percent of 52 is 7.8?

We can translate this sentence to symbols word by word.

| What percent | of | 52 | is | 7.8? |
|---|---|---|---|---|
| ↓ | ↓ | ↓ | ↓ | ↓ |
| $p$ | $\cdot$ | 52 | = | 7.8 |

$$52p = 7.8$$

$$p = .15 \quad \text{Divide both sides by 52.}$$

$$p = 15\% \quad \text{Change to percent.}$$

**Now Try Exercises 39 and 43.**

### EXAMPLE 6 Using Percent in a Consumer Problem

A DVD movie with a regular price of \$18 is on sale at 22% off. Find the amount of the discount.

The discount is 22% of \$18. Change 22% to .22 and use the fact that *of* indicates multiplication.

$$22\% \cdot 18 = .22 \cdot 18 = 3.96$$

The discount is \$3.96.

**Now Try Exercise 55.**

## APPENDIX B EXERCISES

*Perform each indicated operation. See Examples 1–3.*

1. $14.23 + 9.81 + 74.63 + 18.715$
2. $89.416 + 21.32 + 478.91 + 298.213$
3. $19.74 - 6.53$
4. $27.96 - 8.39$
5. $219 - 68.51$
6. $283 - 12.42$
7. $\begin{array}{r} 48.96 \\ 37.421 \\ +\ 9.72 \\ \hline \end{array}$
8. $\begin{array}{r} 9.71 \\ 4.8 \\ 3.6 \\ 5.2 \\ +8.17 \\ \hline \end{array}$
9. $\begin{array}{r} 8.6 \\ -3.751 \\ \hline \end{array}$
10. $\begin{array}{r} 27.8 \\ -13.582 \\ \hline \end{array}$
11. $39.6 \times 4.2$
12. $18.7 \times 2.3$
13. $42.1 \times 3.9$
14. $19.63 \times 4.08$
15. $.042 \times 32$
16. $571 \times 2.9$
17. $24.84 \div 6$
18. $32.84 \div 4$
19. $7.6266 \div 3.42$
20. $14.9202 \div 2.43$
21. $2496 \div .52$
22. $.56984 \div .034$

*Convert each percent to a decimal. See Example 4(a).*

23. 53%
24. 38%
25. 129%
26. 174%
27. 96%
28. 11%
29. .9%
30. .1%

*Convert each decimal to a percent. See Example 4(b).*

31. .80
32. .75
33. .007
34. 1.4
35. .67
36. .003
37. .125
38. .983

*Respond to each statement or question. Round your answer to the nearest hundredth if appropriate. See Example 5.*

39. What is 14% of 780?
40. Find 12% of 350.
41. Find 22% of 1086.
42. What is 20% of 1500?
43. 4 is what percent of 80?
44. 1300 is what percent of 2000?
45. What percent of 5820 is 6402?
46. What percent of 75 is 90?
47. 121 is what percent of 484?
48. What percent of 3200 is 64?
49. Find 118% of 125.8.
50. Find 3% of 128.
51. What is 91.72% of 8546.95?
52. Find 12.741% of 58.902.
53. What percent of 198.72 is 14.68?
54. 586.3 is what percent of 765.4?

*Solve each problem. See Example 6.*

55. A retailer has $23,000 invested in her business. She finds that she is earning 12% per year on this investment. How much money is she earning per year?
56. Harley Dabler recently bought a duplex for $144,000. He expects to earn 16% per year on the purchase price. How many dollars per year will he earn?
57. For a recent tour of the eastern United States, a travel agent figured that the trip totaled 2300 mi, with 35% of the trip by air. How many miles of the trip were by air?
58. Capitol Savings Bank pays 3.2% interest per year. What is the annual interest on an account of $3000?

**59.** An ad for steel-belted radial tires promises 15% better mileage when the tires are used. Alexandria's Escort now goes 420 mi on a tank of gas. If she switched to the new tires, how many extra miles could she drive on a tank of gas?

**60.** A home worth $77,000 is located in an area where home prices are increasing at a rate of 12% per year. By how much would the value of this home increase in 1 year?

**61.** A family of four with a monthly income of $2000 spends 90% of its earnings and saves the rest. Find the *annual* savings of this family.

# Appendix C
# Sets

**OBJECTIVES**

1. List the elements of a set.
2. Learn the vocabulary and symbols used to discuss sets.
3. Decide whether a set is finite or infinite.
4. Decide whether a given set is a subset of another set.
5. Find the complement of a set.
6. Find the union and the intersection of two sets.

**OBJECTIVE 1 List the elements of a set.** A **set** is a collection of things. The objects in a set are called the **elements** of the set. A set is represented by listing its elements between **set braces,** $\{\ \ \}$. The order in which the elements of a set are listed is unimportant.

### EXAMPLE 1 Listing the Elements of Sets

Represent each set by listing the elements.

**(a)** The set of states in the United States that border on the Pacific Ocean is

{California, Oregon, Washington, Hawaii, Alaska}.

**(b)** The set of all counting numbers less than $6 = \{1, 2, 3, 4, 5\}$.

**Now Try Exercises 1 and 3.**

**OBJECTIVE 2 Learn the vocabulary and symbols used to discuss sets.** Capital letters are used to name sets. To state that 5 is an element of

$$S = \{1, 2, 3, 4, 5\},$$

write $5 \in S$. The statement $6 \notin S$ means that 6 is not an element of $S$.

A set with no elements is called the **empty set,** or the **null set.** The symbols $\emptyset$ or $\{\ \ \}$ are used for the empty set. If we let $A$ be the set of all cats that fly, then $A$ is the empty set.

$$A = \emptyset \qquad \text{or} \qquad A = \{\ \ \}$$

**CAUTION** Do not make the common error of writing the empty set as $\{\emptyset\}$.

In any discussion of sets, there is some set that includes all the elements under consideration. This set is called the **universal set** for that situation. For example, if the discussion is about presidents of the United States, then the set of all presidents of the United States is the universal set. The universal set is denoted $U$.

**OBJECTIVE 3 Decide whether a set is finite or infinite.** In Example 1, there are five elements in the set in part (a), and five in part (b). If the number of elements in a set is either 0 or a counting number, then the set is **finite.** On the other hand, the set of natural numbers, for example, is an **infinite** set, because there is no final natural number. We can list the elements of the set of natural numbers as

$$N = \{1, 2, 3, 4, \ldots\},$$

where the three dots indicate that the set continues indefinitely. Not all infinite sets can be listed in this way. For example, there is no way to list the elements in the set of all real numbers between 1 and 2.

### EXAMPLE 2 Distinguishing between Finite and Infinite Sets

List the elements of each set, if possible. Decide whether each set is finite or infinite.

**(a)** The set of all integers
One way to list the elements is $\{\ldots, -2, -1, 0, 1, 2, \ldots\}$. The set is infinite.

**(b)** The set of all natural numbers between 0 and 5
$\{1, 2, 3, 4\}$ The set is finite.

**(c)** The set of all irrational numbers
This is an infinite set whose elements cannot be listed.

**Now Try Exercise 11.**

Two sets are equal if they have exactly the same elements. Thus, the set of natural numbers and the set of positive integers are equal sets. Also, the sets

$$\{1, 2, 4, 7\} \qquad \text{and} \qquad \{4, 2, 7, 1\}$$

are equal. The order of the elements does not make a difference.

**OBJECTIVE 4 Decide whether a given set is a subset of another set.** If all elements of a set $A$ are also elements of another set $B$, then we say $A$ is a **subset** of $B$, written $A \subseteq B$. We use the symbol $A \not\subseteq B$ to mean that $A$ is not a subset of $B$.

### EXAMPLE 3 Using Subset Notation

Let $A = \{1, 2, 3, 4\}$, $B = \{1, 4\}$, and $C = \{1\}$. Then

$$B \subseteq A, \qquad C \subseteq A, \qquad \text{and} \qquad C \subseteq B,$$

but

$$A \not\subseteq B, \qquad A \not\subseteq C, \qquad \text{and} \qquad B \not\subseteq C.$$

**Now Try Exercises 21 and 25.**

The set $M = \{a, b\}$ has four subsets: $\{a, b\}$, $\{a\}$, $\{b\}$, and $\emptyset$. The empty set is defined to be a subset of any set. How many subsets does $N = \{a, b, c\}$ have? There is one subset with 3 elements: $\{a, b, c\}$. There are three subsets with 2 elements:

$$\{a, b\}, \qquad \{a, c\}, \qquad \text{and} \qquad \{b, c\}.$$

There are three subsets with 1 element:

$$\{a\}, \qquad \{b\}, \qquad \text{and} \qquad \{c\}.$$

There is one subset with 0 elements: $\emptyset$. Thus, set $N$ has eight subsets.

The following generalization can be made.

**Number of Subsets of a Set**

A set with $n$ elements has $2^n$ subsets.

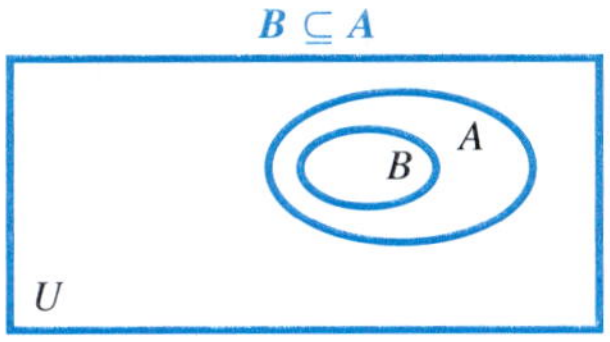


FIGURE 1

To illustrate the relationships between sets, **Venn diagrams** are often used. A rectangle represents the universal set, $U$. The sets under discussion are represented by regions within the rectangle. The Venn diagram in Figure 1 shows that $B \subseteq A$.

**OBJECTIVE 5 Find the complement of a set.** For every set $A$, there is a set $A'$, the **complement** of $A$, that contains all the elements of $U$ that are not in $A$. The shaded region in the Venn diagram in Figure 2 represents $A'$.

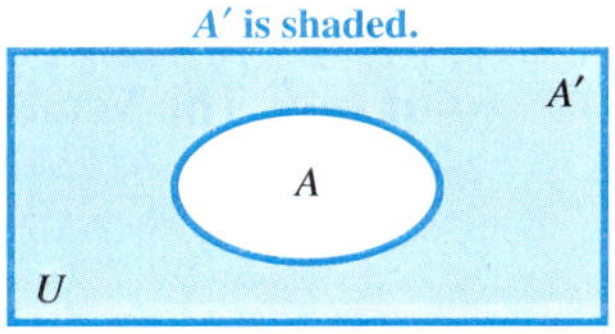


FIGURE 2

### EXAMPLE 4 Determining Complements of a Set

Given $U = \{a, b, c, d, e, f, g\}$, $A = \{a, b, c\}$, $B = \{a, d, f, g\}$, and $C = \{d, e\}$, then

$$A' = \{d, e, f, g\}, \quad B' = \{b, c, e\}, \quad \text{and} \quad C' = \{a, b, c, f, g\}.$$

**Now Try Exercises 45 and 47.**

**OBJECTIVE 6 Find the union and the intersection of two sets.** The **union** of two sets $A$ and $B$, written $A \cup B$, is the set of all elements of $A$ together with all elements of $B$. Thus, for the sets in Example 4,

$$A \cup B = \{a, b, c, d, f, g\}$$

and

$$A \cup C = \{a, b, c, d, e\}.$$

In Figure 3 the shaded region is the union of sets $A$ and $B$.

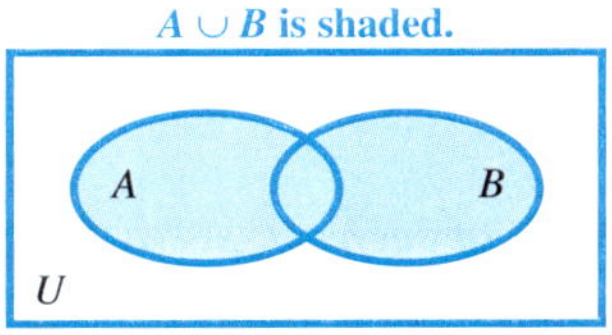


FIGURE 3

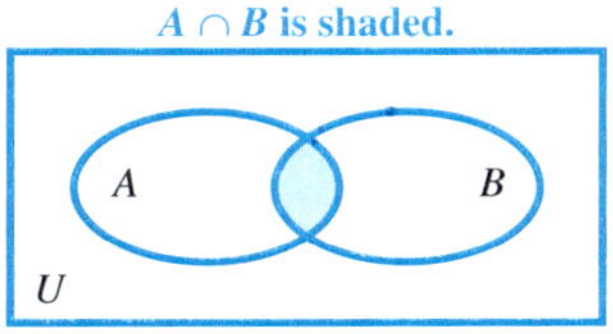


FIGURE 4

### EXAMPLE 5 Finding the Union of Two Sets

If $M = \{2, 5, 7\}$ and $N = \{1, 2, 3, 4, 5\}$, then

$$M \cup N = \{1, 2, 3, 4, 5, 7\}.$$

**Now Try Exercise 55.**

The **intersection** of two sets $A$ and $B$, written $A \cap B$, is the set of all elements that belong to both $A$ and $B$. For example if,

$$A = \{\text{Jose, Ellen, Marge, Kevin}\}$$

and

$$B = \{\text{Jose, Patrick, Ellen, Sue}\},$$

then

$$A \cap B = \{\text{Jose, Ellen}\}.$$

The shaded region in Figure 4 represents the intersection of the two sets $A$ and $B$.

### EXAMPLE 6 Finding the Intersection of Two Sets

Suppose that $P = \{3, 9, 27\}$, $Q = \{2, 3, 10, 18, 27, 28\}$, and $R = \{2, 10, 28\}$. List the elements in each set.

**(a)** $P \cap Q = \{3, 27\}$ **(b)** $Q \cap R = \{2, 10, 28\} = R$ **(c)** $P \cap R = \emptyset$

**Now Try Exercises 49 and 51.**

Sets like $P$ and $R$ in Example 6 that have no elements in common are called **disjoint sets.** The Venn diagram in Figure 5 shows a pair of disjoint sets.

Disjoint sets; $A \cap B = \emptyset$

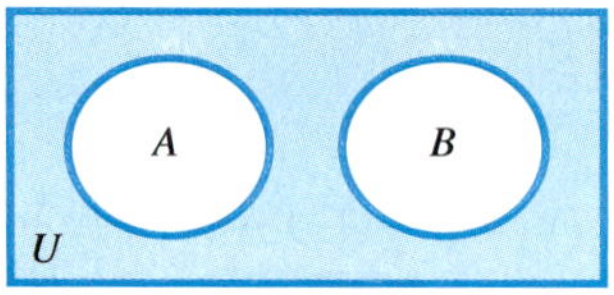


FIGURE 5

### EXAMPLE 7 Using Set Operations

Let $U = \{2, 5, 7, 10, 14, 20\}$, $A = \{2, 10, 14, 20\}$, $B = \{5, 7\}$, and $C = \{2, 5, 7\}$. Find each set.

**(a)** $A \cup B = \{2, 5, 7, 10, 14, 20\} = U$ **(b)** $A \cap B = \emptyset$

**(c)** $B \cup C = \{2, 5, 7\} = C$ **(d)** $B \cap C = \{5, 7\} = B$

**(e)** $A' = \{5, 7\} = B$

**Now Try Exercises 53 and 57.**

## APPENDIX C EXERCISES

*List the elements of each set. See Examples 1 and 2.*

**1.** The set of all natural numbers less than 8

**2.** The set of all integers between 4 and 10

**3.** The set of seasons

**4.** The set of months of the year

**5.** The set of women presidents of the United States

**6.** The set of all living humans who are more than 200 years old

**7.** The set of letters of the alphabet between K and M

**8.** The set of letters of the alphabet between D and H

**9.** The set of positive even integers

**10.** The set of all multiples of 5

**11.** Which of the sets described in Exercises 1–10 are infinite sets?

**12.** Which of the sets described in Exercises 1–10 are finite sets?

*Tell whether each statement is* true *or* false.

**13.** $5 \in \{1, 2, 5, 8\}$

**14.** $6 \in \{1, 2, 3, 4, 5\}$

**15.** $2 \in \{1, 3, 5, 7, 9\}$

**16.** $1 \in \{6, 2, 5, 1\}$

**17.** $7 \notin \{2, 4, 6, 8\}$

**18.** $7 \notin \{1, 3, 5, 7\}$

**19.** $\{2, 4, 9, 12, 13\} = \{13, 12, 9, 4, 2\}$

**20.** $\{7, 11, 4\} = \{7, 11, 4, 0\}$

*Let*

$$A = \{1, 3, 4, 5, 7, 8\}, \quad B = \{2, 4, 6, 8\},$$
$$C = \{1, 3, 5, 7\}, \quad D = \{1, 2, 3\},$$
$$E = \{3, 7\}, \quad \text{and} \quad U = \{1, 2, 3, 4, 5, 6, 7, 8, 9, 10\}.$$

*Tell whether each statement is* true *or* false. *See Examples 3, 5, 6, and 7.*

**21.** $A \subseteq U$

**22.** $D \subseteq A$

**23.** $\emptyset \subseteq A$

**24.** $\{1, 2\} \subseteq D$

**25.** $C \subseteq A$

**26.** $A \subseteq C$

**27.** $D \subseteq B$

**28.** $E \subseteq C$

**29.** $D \not\subseteq E$

**30.** $E \not\subseteq A$

**31.** There are exactly 4 subsets of $E$.

**32.** There are exactly 8 subsets of $D$.

**33.** There are exactly 12 subsets of $C$.

**34.** There are exactly 16 subsets of $B$.

**35.** $\{4, 6, 8, 12\} \cap \{6, 8, 14, 17\} = \{6, 8\}$

**36.** $\{2, 5, 9\} \cap \{1, 2, 3, 4, 5\} = \{2, 5\}$

**37.** $\{3, 1, 0\} \cap \{0, 2, 4\} = \{0\}$

**38.** $\{4, 2, 1\} \cap \{1, 2, 3, 4\} = \{1, 2, 3\}$

**39.** $\{3, 9, 12\} \cap \emptyset = \{3, 9, 12\}$

**40.** $\{3, 9, 12\} \cup \emptyset = \emptyset$

**41.** $\{4, 9, 11, 7, 3\} \cup \{1, 2, 3, 4, 5\} = \{1, 2, 3, 4, 5, 7, 9, 11\}$

**42.** $\{1, 2, 3\} \cup \{1, 2, 3\} = \{1, 2, 3\}$

**43.** $\{3, 5, 7, 9\} \cup \{4, 6, 8\} = \emptyset$

**44.** $\{5, 10, 15, 20\} \cup \{5, 15, 30\} = \{5, 15\}$

*Let*

$$U = \{\text{a, b, c, d, e, f, g, h}\}, \quad A = \{\text{a, b, c, d, e, f}\},$$
$$B = \{\text{a, c, e}\}, \quad C = \{\text{a, f}\},$$

and $$D = \{\text{d}\}.$$

*List the elements in each set. See Examples 4–7.*

**45.** $A'$

**46.** $B'$

**47.** $C'$

**48.** $D'$

**49.** $A \cap B$

**50.** $B \cap A$

**51.** $A \cap D$

**52.** $B \cap D$

**53.** $B \cap C$

**54.** $A \cup B$

**55.** $B \cup D$

**56.** $B \cup C$

**57.** $C \cup B$

**58.** $C \cup D$

**59.** $A \cap \emptyset$

**60.** $B \cup \emptyset$

**61.** Name every pair of disjoint sets among sets $A$–$D$ above.

# Appendix D
# Mean, Median, and Mode

**OBJECTIVES**

1. Find the mean of a list of numbers.
2. Find a weighted mean.
3. Find the median.
4. Find the mode.

**OBJECTIVE 1 Find the mean of a list of numbers.** Making sense of a long list of numbers can be difficult. So when we analyze data, one of the first things to look for is a *measure of central tendency*—a single number that we can use to represent the entire list of numbers. One such measure is the *average* or **mean.** The mean can be found with the following formula.

**Finding the Mean (Average)**

$$\textbf{mean} = \frac{\textbf{sum of all values}}{\textbf{number of values}}$$

### EXAMPLE 1 Finding the Mean (Average)

David had test scores of 84, 90, 95, 98, and 88. Find his mean (average) score.

Use the formula for finding the mean. Add up all the test scores and then divide the sum by the number of tests.

$$\text{mean} = \frac{84 + 90 + 95 + 98 + 88}{5}$$ ← Sum of test scores ← Number of tests

$$\text{mean} = \frac{455}{5}$$

$$\text{mean} = 91$$ Divide.

David has a mean (average) score of 91.

**Now Try Exercise 1.**

### EXAMPLE 2 Applying the Mean (Average)

The sales of photo albums at Sarah's Card Shop for each day last week were \$86, \$103, \$118, \$117, \$126, \$158, and \$149. Find the mean daily sales of photo albums.

To find the mean, add all the daily sales amounts and then divide the sum by the number of days (7).

$$\text{mean} = \frac{\$86 + \$103 + \$118 + \$117 + \$126 + \$158 + \$149}{7}$$ ← Sum of sales ← Number of days

$$\text{mean} = \frac{\$857}{7}$$

$$\text{mean} \approx \$122.43$$ Nearest cent

**Now Try Exercise 7.**

**OBJECTIVE 2 Find a weighted mean.** Some items in a list of data might appear more than once. In this case, we find a **weighted mean,** in which each value is "weighted" by multiplying it by the number of times it occurs.

### EXAMPLE 3 Finding a Weighted Mean

The table shows the amount of contribution and the number of times the amount was given (frequency) to a food pantry. Find the weighted mean.

| *Contribution Value* | *Frequency* |
|---|---|
| \$ 3 | 4 ← 4 people each contributed \$3. |
| \$ 5 | 2 |
| \$ 7 | 1 |
| \$ 8 | 5 |
| \$ 9 | 3 |
| \$10 | 2 |
| \$12 | 1 |
| \$13 | 2 |

In most cases, the same amount was given by more than one person: for example, \$3 was given by four people, and \$8 was given by five people. Other amounts, such as \$12, were given by only one person.

To find the mean, multiply each contribution value by its frequency. Then add the products. Next, add the numbers in the *frequency* column to find the total number of values, that is, the total number of people who contributed money.

| *Value* | *Frequency* | *Product* |
|---|---|---|
| \$ 3 | 4 | $(\$3 \cdot 4) = \$12$ |
| \$ 5 | 2 | $(\$5 \cdot 2) = \$10$ |
| \$ 7 | 1 | $(\$7 \cdot 1) = \$\ 7$ |
| \$ 8 | 5 | $(\$8 \cdot 5) = \$40$ |
| \$ 9 | 3 | $(\$9 \cdot 3) = \$27$ |
| \$10 | 2 | $(\$10 \cdot 2) = \$20$ |
| \$12 | 1 | $(\$12 \cdot 1) = \$12$ |
| \$13 | 2 | $(\$13 \cdot 2) = \$26$ |
| Totals | 20 | \$154 |

Finally, divide the totals.

$$\text{mean} = \frac{\$154}{20} = \$7.70$$

The mean contribution to the food pantry was \$7.70.

**Now Try Exercise 11.**

A common use of the weighted mean is to find a student's *grade point average (GPA),* as shown in the next example.

## EXAMPLE 4 Applying the Weighted Mean

Find the GPA (grade point average) for a student who earned the following grades last semester. Assume A = 4, B = 3, C = 2, D = 1, and F = 0. The number of credits determines how many times the grade is counted (the frequency).

| Course | Credits | Grade | Credits · Grade |
|---|---|---|---|
| Mathematics | 4 | A (= 4) | $4 \cdot 4 = 16$ |
| Speech | 3 | C (= 2) | $3 \cdot 2 = 6$ |
| English | 3 | B (= 3) | $3 \cdot 3 = 9$ |
| Computer Science | 2 | A (= 4) | $2 \cdot 4 = 8$ |
| Theater | 2 | D (= 1) | $2 \cdot 1 = 2$ |
| Totals | 14 | | 41 |

It is common to round grade point averages to the nearest hundredth, so

$$\text{GPA} = \frac{41}{14} \approx 2.93.$$

**Now Try Exercise 15.**

**OBJECTIVE 3 Find the median.** Because it can be affected by extremely high or low numbers, the mean is often a poor indicator of central tendency for a list of numbers. In cases like this, another measure of central tendency, called the *median,* can be used. The **median** divides a group of numbers in half; half the numbers lie above the median, and half lie below the median.

Find the median by listing the numbers *in order* from *smallest* to *largest.* If the list contains an *odd* number of items, the median is the *middle number.*

## EXAMPLE 5 Finding the Median

Find the median for this list of prices.

$7, $23, $15, $6, $18, $12, $24

First arrange the numbers in numerical order from smallest to largest.

Smallest → 6, 7, 12, 15, 18, 23, 24 ← Largest

Next, find the middle number in the list.

6, 7, 12, 15, 18, 23, 24

Three are below. ↓ Three are above.
Middle number

The median price is $15.

**Now Try Exercise 19.**

If a list contains an *even* number of items, there is no single middle number. In this case, the median is defined as the mean (average) of the *middle two* numbers.

### EXAMPLE 6 Finding the Median

Find the median for this list of ages, in years.

74, 7, 15, 13, 25, 28, 47, 59, 32, 68

First arrange the numbers in numerical order from smallest to largest. Then, because the list has an even number of ages, find the middle *two* numbers.

Smallest → 7, 13, 15, 25, 28, 32, 47, 59, 68, 74 ← Largest

(28, 32: Middle two numbers)

The median age is the mean of the two middle numbers.

$$\text{median} = \frac{28 + 32}{2} = \frac{60}{2} = 30 \text{ yr}$$

**Now Try Exercise 21.**

**OBJECTIVE 4 Find the mode.** Another statistical measure is the **mode,** which is the number that occurs *most often* in a list of numbers. For example, if the test scores for 10 students were

74, 81, 39, 74, 82, 80, 100, 92, 74, and 85,

then the mode is 74. Three students earned a score of 74, so 74 appears more times on the list than any other score. It is *not* necessary to place the numbers in numerical order when looking for the mode, although that may help you find it more easily.

A list can have two modes; such a list is sometimes called **bimodal.** If no number occurs more frequently than any other number in a list, the list has *no mode.*

### EXAMPLE 7 Finding the Mode

Find the mode for each list of numbers.

**(a)** 51, 32, 49, 51, 49, 90, 49, 60, 17, 60

The number 49 occurs three times, which is more often than any other number. Therefore, 49 is the mode.

**(b)** 482, 485, 483, 485, 487, 487, 489, 486

Because both 485 and 487 occur twice, each is a mode. This list is binomial.

**(c)** 10,708; 11,519; 10,972; 12,546; 13,905; 12,182

No number occurs more than once. This list has no mode.

**Now Try Exercises 29, 31, and 33.**

**Measures of Central Tendency**

The **mean** is the sum of all the values divided by the number of values. It is the mathematical *average.*

The **median** is the middle number in a group of values that are listed from smallest to largest. It divides a group of numbers in half.

The **mode** is the value that occurs most often in a group of values.

## APPENDIX D EXERCISES

*Find the mean for each list of numbers. Round answers to the nearest tenth when necessary. See Example 1.*

1. Ages of infants at the child care center (in months): 4, 9, 6, 4, 7, 10, 9
2. Monthly electric bills: \$53, \$77, \$38, \$29, \$49, \$48
3. Final exam scores: 92, 51, 59, 86, 68, 73, 49, 80
4. Quiz scores: 18, 25, 21, 8, 16, 13, 23, 19
5. Annual salaries: \$31,900; \$32,850; \$34,930; \$39,712; \$38,340; \$60,000
6. Numbers of people attending baseball games: 27,500; 18,250; 17,357; 14,298; 33,110

*Solve each problem. See Examples 2 and 3.*

7. The Athletic Shoe Store sold shoes at the following prices: \$75.52, \$36.15, \$58.24, \$21.86, \$47.68, \$106.57, \$82.72, \$52.14, \$28.60, \$72.92. Find the mean shoe sales amount.
8. In one evening, a waitress collected the following checks from her dinner customers: \$30.10, \$42.80, \$91.60, \$51.20, \$88.30, \$21.90, \$43.70, \$51.20. Find the mean dinner check amount.
9. The table shows the face value (policy amount) of life insurance policies sold and the number of policies sold for each amount by the New World Life Company during one week. Find the weighted mean amount for the policies sold.

| Policy Amount | Number of Policies Sold |
|---|---|
| \$ 10,000 | 6 |
| \$ 20,000 | 24 |
| \$ 25,000 | 12 |
| \$ 30,000 | 8 |
| \$ 50,000 | 5 |
| \$100,000 | 3 |
| \$250,000 | 2 |

10. Detroit Metro-Sales Company prepared the following table showing the gasoline mileage obtained by each of the cars in their automobile fleet. Find the weighted mean to determine the miles per gallon for the fleet of cars.

| Miles per Gallon | Number of Autos |
|---|---|
| 15 | 5 |
| 20 | 6 |
| 24 | 10 |
| 30 | 14 |
| 32 | 5 |
| 35 | 6 |
| 40 | 4 |

*Find each weighted mean. Round answers to the nearest tenth when necessary. See Example 3.*

11.

| Quiz Scores | Frequency |
|---|---|
| 3 | 4 |
| 5 | 2 |
| 6 | 5 |
| 8 | 5 |
| 9 | 2 |

12.

| Credits per Student | Frequency |
|---|---|
| 9 | 3 |
| 12 | 5 |
| 13 | 2 |
| 15 | 6 |
| 18 | 1 |

**13.**

| Hours Worked | Frequency |
|---|---|
| 12 | 4 |
| 13 | 2 |
| 15 | 5 |
| 19 | 3 |
| 22 | 1 |
| 23 | 5 |

**14.**

| Students per Class | Frequency |
|---|---|
| 25 | 1 |
| 26 | 2 |
| 29 | 5 |
| 30 | 4 |
| 32 | 3 |
| 33 | 5 |

*Find the GPA (grade point average) for students earning the following grades. Assume* A = 4, B = 3, C = 2, D = 1, *and* F = 0. *Round answers to the nearest hundredth. See Example 4.*

**15.**

| Course | Credits | Grade |
|---|---|---|
| Biology | 4 | B |
| Biology Lab | 2 | A |
| Mathematics | 5 | C |
| Health | 1 | F |
| Psychology | 3 | B |

**16.**

| Course | Credits | Grade |
|---|---|---|
| Chemistry | 3 | A |
| English | 3 | B |
| Mathematics | 4 | B |
| Theater | 2 | C |
| Astronomy | 3 | C |

**17.** Look again at the grades in Exercise 15. Find the student's GPA in each of these situations.

**(a)** The student earned a B instead of an F in the 1-credit class.
**(b)** The student earned a B instead of a C in the 5-credit class.
**(c)** Both (a) and (b) happened.

**18.** List the credits for the courses you're taking at this time. List the lowest grades you think you will earn in each class and find your GPA. Then list the highest grades you think you will earn and find your GPA.

*Find the median for each list of numbers. See Examples 5 and 6.*

**19.** Number of e-mail messages received: 9, 12, 14, 15, 23, 24, 28

**20.** Deliveries by a newspaper distributor: 99, 108, 109, 123, 126, 129, 146, 168, 170

**21.** Students enrolled in algebra each semester: 328, 549, 420, 592, 715, 483

**22.** Number of cars in the parking lot each day: 520, 523, 513, 1283, 338, 509, 290, 420

**23.** Number of computer service calls taken each day: 51, 48, 96, 40, 47, 23, 95, 56, 34, 48

**24.** Number of gallons of paint sold per week: 1072, 1068, 1093, 1042, 1056, 205, 1009, 1081

*The table lists the cruising speed and distance flown without refueling for several types of larger airplanes used to carry passengers. Use the table to answer Exercises 25–28.*

| Type of Airplane | Cruising Speed (miles per hour) | Distance without Refueling (miles) |
|---|---|---|
| 747-400 | 565 | 7650 |
| 747-200 | 558 | 6450 |
| DC-9 | 505 | 1100 |
| DC-10 | 550 | 5225 |
| 727 | 530 | 1550 |
| 757 | 530 | 2875 |

*Source:* Northwest Airlines *WorldTraveler.*

**25.** What is the mean distance flown without refueling, to the nearest mile?

**26.** Find the mean cruising speed.

**27. (a)** Find the median distance.
**(b)** Is the median distance similar to the mean distance from Exercise 25? Explain why or why not.

**28. (a)** Find the median cruising speed.
**(b)** Is the median speed similar to the mean speed from Exercise 26? Explain why or why not.

*Find the mode(s) for each list of numbers. Indicate whether a list is bimodal or has no mode. See Example 7.*

**29.** Number of samples taken each hour: 3, 8, 5, 1, 7, 6, 8, 4, 5, 8

**30.** Monthly water bills: \$21, \$32, \$46, \$32, \$49, \$32, \$49, \$25, \$32

**31.** Ages of retirees (in years): 74, 68, 68, 68, 75, 75, 74, 74, 70, 77

**32.** Patients admitted to the hospital each week: 30, 19, 25, 78, 36, 20, 45, 85, 38

**33.** The number of boxes of candy sold by each child: 5, 9, 17, 3, 2, 8, 19, 1, 4, 20, 10, 6

**34.** The weights of soccer players (in pounds): 158, 161, 165, 162, 165, 157, 163, 162

# Appendix E
# The Metric System and Conversions

**OBJECTIVES**

1. Learn the basic metric units of length.
2. Use unit fractions to convert among units.
3. Move the decimal point to convert among units.
4. Learn the basic metric units of capacity.
5. Convert among metric capacity units.
6. Learn the basic metric units of weight (mass).
7. Convert among metric weight (mass) units.

**OBJECTIVE 1 Learn the basic metric units of length.** Around 1790, a group of French scientists developed the metric system of measurement. It is an organized system based on multiples of 10, like our number system and our money. The basic unit of length in the metric system is the **meter** (also spelled *metre*). Use the symbol **m** for meter; do not put a period after it. Look at a yardstick—a meter is just a little longer. A yard is 36 inches long; a meter is about 39 inches long.

To make longer or shorter length units in the metric system, **prefixes** are written in front of the word *meter*. For example, the prefix *kilo* means 1000, so a *kilo*meter is 1000 meters. The table below shows how to use the prefixes for length measurements. It is helpful to memorize the prefixes because they are also used with weight and capacity measurements. The blue boxes indicate the units we use most often in daily life.

| *Prefix* | *kilo*-meter | *hecto*-meter | *deka*-meter | meter | *deci*-meter | *centi*-meter | *milli*-meter |
|---|---|---|---|---|---|---|---|
| *Meaning* | 1000 meters | 100 meters | 10 meters | 1 meter | $\frac{1}{10}$ of a meter | $\frac{1}{100}$ of a meter | $\frac{1}{1000}$ of a meter |
| *Symbol* | km | hm | dam | m | dm | cm | mm |

Units that are used most often: km, m, cm, mm

Here are some comparisons that involve the most commonly used length units: km, m, cm, mm.

*Kilo*meters are used instead of miles. A kilometer is 1000 meters. It is about .6 mile (a little more than half a mile) or about 5 to 6 city blocks. If you participate in a 10 km run, you'll run about 6 miles.

A meter is divided into 100 smaller pieces called *centi*meters. Each centimeter is $\frac{1}{100}$ of a meter. A centimeter is a little shorter than $\frac{1}{2}$ inch. A nickel is about 2 cm across. The width of your little finger is probably about 1 cm.

A meter is divided into 1000 smaller pieces called *milli*meters. Each millimeter is $\frac{1}{1000}$ of a meter. It takes 10 mm to equal 1 cm, so it is a very small length. The thickness of a dime is about 1 mm.

### EXAMPLE 1 Using Metric Length Units

Write the most reasonable metric unit in each blank. Choose from km, m, cm, and mm.

**(a)** The distance from home to work is 20 ______.

Choose 20 km because kilometers are used instead of miles; 20 km is about 12 miles.

**(b)** The wedding ring is 4 ______ wide.

Choose 4 mm because the width of a ring is very small.

**(c)** The newborn baby is 50 ______ long.

Choose 50 cm, which is half of a meter; a meter is about 39 inches, so half a meter is around 20 inches.

**Now Try Exercises 5, 7, 9, and 11.**

**OBJECTIVE 2 Use unit fractions to convert among units.** We can convert among metric length units using unit fractions. Keep these relationships in mind when setting up the unit fractions.

**Metric Length Relationships**

| | |
|---|---|
| **1 km = 1000 m,** so the unit fractions are: $\frac{1\text{ km}}{1000\text{ m}}$ or $\frac{1000\text{ m}}{1\text{ km}}$ | **1 m = 1000 mm,** so the unit fractions are: $\frac{1\text{ m}}{1000\text{ mm}}$ or $\frac{1000\text{ mm}}{1\text{ m}}$ |
| **1 m = 100 cm,** so the unit fractions are: $\frac{1\text{ m}}{100\text{ cm}}$ or $\frac{100\text{ cm}}{1\text{ m}}$ | **1 cm = 10 mm,** so the unit fractions are: $\frac{1\text{ cm}}{10\text{ mm}}$ or $\frac{10\text{ mm}}{1\text{ cm}}$ |

### EXAMPLE 2 Using Unit Fractions to Convert Length Measurements

Convert the following.

**(a)** 5 km to m

Put the unit for the answer (meters) in the numerator of the unit fraction; put the unit you want to change (km) in the denominator.

Unit fraction equivalent to 1 $\left\{ \frac{1000\text{ m}}{1\text{ km}} \right.$ ← Unit for answer; ← Unit being changed

Multiply. Divide out common units where possible.

$$5\text{ km} \cdot \frac{1000\text{ m}}{1\text{ km}} = \frac{5\text{ }\cancel{\text{km}}}{1} \cdot \frac{1000\text{ m}}{1\text{ }\cancel{\text{km}}} = \frac{5 \cdot 1000\text{ m}}{1} = 5000\text{ m}$$

These units should match.

The answer makes sense because a kilometer is much longer than a meter, so 5 km will contain many meters.

**(b)** 18.6 cm to m

Multiply by a unit fraction that allows you to divide out centimeters.

$$\frac{18.6\ \cancel{\text{cm}}}{1} \cdot \underbrace{\frac{1\text{ m}}{100\ \cancel{\text{cm}}}}_{\text{Unit fraction}} = \frac{18.6}{100}\text{ m} = .186\text{ m}$$

There are 100 cm in a meter, so 18.6 cm will be a small part of a meter. The answer makes sense.

**Now Try Exercises 23 and 25.**

**OBJECTIVE 3 Move the decimal point to convert among units.** By now you have probably noticed that conversions among metric units are made by multiplying or dividing by 10, 100, or 1000. A quick way to *multiply* by 10 is to move the decimal point one place to the *right.* Move it two places to the right to multiply by 100, three places to multiply by 1000. *Dividing* is done by moving the decimal point to the *left* in the same manner.

An alternate conversion method to unit fractions is moving the decimal point using this **metric conversion line.**

| 1000 | 100 | 10 | 1 | $\frac{1}{10}$ | $\frac{1}{100}$ | $\frac{1}{1000}$ |
|---|---|---|---|---|---|---|
| km | hm | dam | m | dm | cm | mm |

Here are the steps for using the conversion line.

**Using the Metric Conversion Line**

*Step 1* Find the given unit on the metric conversion line.

*Step 2* Count the number of places to get from the given unit to the unit needed in the answer.

*Step 3* Move the decimal point the ***same number of places*** and in the ***same direction*** as on the conversion line.

## EXAMPLE 3 Using the Metric Conversion Line

Use the metric conversion line to make the following conversions.

**(a)** 5.702 km to m

Find **km** on the metric conversion line. To get to **m** , move *three places* to the *right.* So move the decimal point in 5.702 *three places* to the *right.*

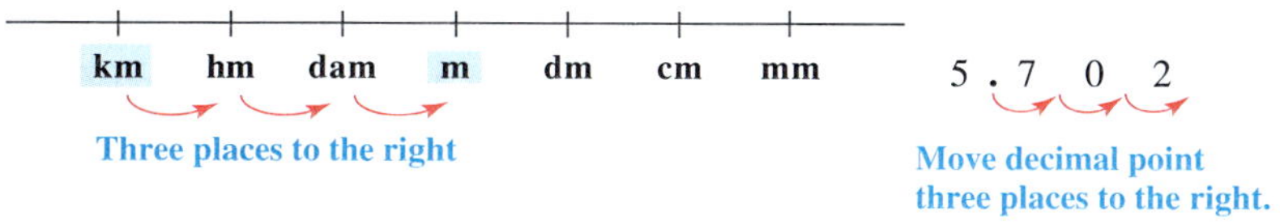


5.702 km = 5702 m

**(b)** 69.5 cm to m

Find **cm** on the conversion line. To get to **m**, move *two places* to the *left.*

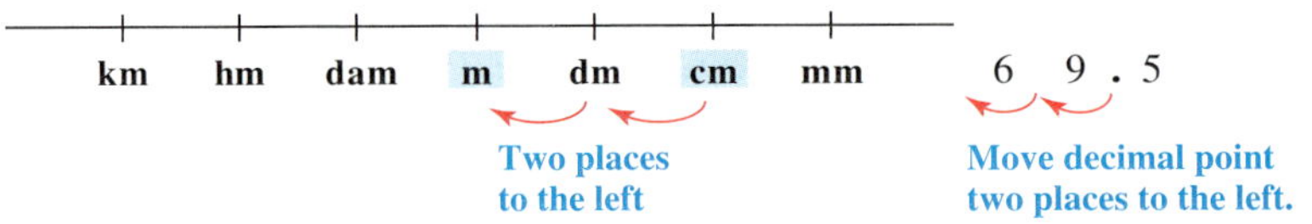


69.5 cm = .695 m

**(c)** 8.1 cm to mm

From **cm** to **mm** is *one place* to the *right.*

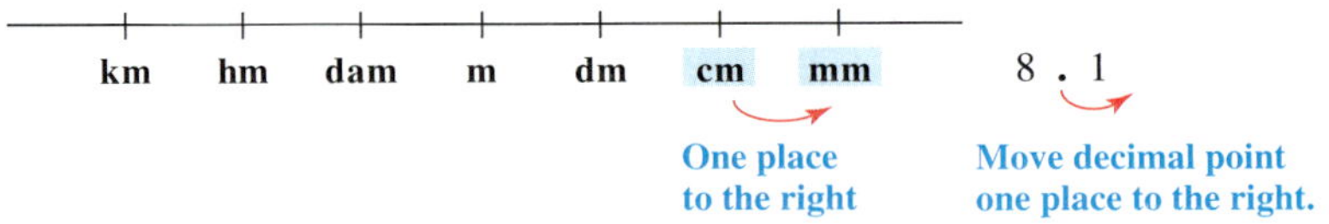


8.1 cm = 81 mm

**Now Try Exercises 19, 21, and 27.**

**OBJECTIVE 4 Learn the basic metric units of capacity.** The basic metric unit for capacity is the **liter** (also spelled *litre*). The capital letter **L** is the symbol for liter, to avoid confusion with the numeral 1. The liter is related to metric length in this way: A box that measures 10 cm on every side holds exactly one liter. (The volume of the box is 1000 cubic centimeters.) A liter is just a little more than 1 quart.

To make larger or smaller capacity units, we use the same prefixes as we did with length units. For example, *kilo* means 1000 so a *kilo*meter is 1000 meters. In the same way, a *kilo*liter is 1000 liters.

| *Prefix* | *kilo-* liter | *hecto-* liter | *deka-* liter | liter | *deci-* liter | *centi-* liter | *milli-* liter |
|---|---|---|---|---|---|---|---|
| *Meaning* | 1000 liters | 100 liters | 10 liters | 1 liter | $\frac{1}{10}$ of a liter | $\frac{1}{100}$ of a liter | $\frac{1}{1000}$ of a liter |
| *Symbol* | kL | hL | daL | L | dL | cL | mL |

Used most often (liter and milliliter)

The capacity units we use most often in daily life are liters (L) and *milli*liters (mL). A tiny box that measures 1 cm on every side holds exactly one milliliter. (In medicine, this small amount is also called 1 cubic centimeter, or 1 cc for short.) It takes 1000 mL to make 1 L.

### EXAMPLE 4 Using Metric Capacity Units

Write the most reasonable metric unit in each blank. Choose from L and mL.

**(a)** The bottle of shampoo held 500 ______.

Choose 500 mL because 500 L would be about 500 quarts, which is too much.

**(b)** I bought a 2 ______ carton of orange juice.

Choose 2 L because 2 mL would be less than a teaspoon.

**Now Try Exercises 37 and 39.**

**OBJECTIVE 5 Convert among metric capacity units.** Just as with length units, we can convert between milliliters and liters using unit fractions.

**Metric Capacity Relationships**

**1 L = 1000 mL,** so the unit fractions are:

$$\frac{\mathbf{1\ L}}{\mathbf{1000\ mL}} \quad \text{or} \quad \frac{\mathbf{1000\ mL}}{\mathbf{1\ L}}$$

We can also use a metric conversion line to decide how to move the decimal point.

| 1000 | 100 | 10 | 1 | $\frac{1}{10}$ | $\frac{1}{100}$ | $\frac{1}{1000}$ |
|---|---|---|---|---|---|---|
| kL | hL | daL | L | dL | cL | mL |

## EXAMPLE 5 Converting among Metric Capacity Units

Convert using the metric conversion line or unit fractions.

**(a)** 2.5 L to mL

Using the metric conversion line:

From **L** to **mL** is *three places* to the *right.*

2.500 Write two zeros as placeholders.

Using unit fractions:

Multiply by a unit fraction that allows you to divide out liters.

$$\frac{2.5\ \cancel{L}}{1} \cdot \frac{1000\ mL}{1\ \cancel{L}} = 2500\ mL$$

2.5 L = 2500 mL

**(b)** 80 mL to L

Using the metric conversion line:

From **mL** to **L** is *three places* to the *left.*

80. — Decimal point starts here.

080. — Move three places left.

Using unit fractions:

Multiply by a unit fraction that allows you to divide out mL.

$$\frac{80\ \cancel{mL}}{1} \cdot \frac{1\ L}{1000\ \cancel{mL}} = \frac{80}{1000}\ L = .08\ L$$

80 mL = .080 L or .08 L

**Now Try Exercises 59 and 61.**

**OBJECTIVE 6 Learn the basic metric units of weight (mass).** The **gram** is the basic metric unit for *mass*. Although we often call it "weight," there is a difference. Weight is a measure of the pull of gravity; the farther you are from the center of the earth, the less you weigh. In outer space you become weightless, but your mass, the amount of matter in your body, stays the same regardless of where you are. For everyday purposes, we will use the word *weight.*

The gram is related to metric length in this way: The weight of the water in a box measuring 1 cm on every side is 1 gram. This is a very tiny amount of water (1 mL) and a very small weight. One gram is also the weight of a dollar bill or a single raisin. A nickel weighs 5 grams. A regular hamburger weighs from 175 to 200 grams.

To make larger or smaller weight units, we use the same prefixes as we did with length and capacity units. For example, *kilo* means 1000 so a *kilo*meter is 1000 meters, a *kilo*liter is 1000 liters, and a *kilo*gram is 1000 grams.

| *Prefix* | *kilo-* gram | *hecto-* gram | *deka-* gram | gram | *deci-* gram | *centi-* gram | *milli-* gram |
|---|---|---|---|---|---|---|---|
| *Meaning* | 1000 grams | 100 grams | 10 grams | 1 gram | $\frac{1}{10}$ of a gram | $\frac{1}{100}$ of a gram | $\frac{1}{1000}$ of a gram |
| *Symbol* | kg | hg | dag | g | dg | cg | mg |

Units that are used most often (kilogram, gram, milligram)

The units we use most often in daily life are kilograms (kg), grams (g), and milligrams (mg). *Kilo*grams are used instead of pounds. A kilogram is 1000 grams. It is about 2.2 pounds. Two one-pound packages of butter plus one stick of butter weigh about 1 kg. An average newborn baby weighs 3 to 4 kg; a college football player might weigh 100 to 130 kg.

Extremely small weights are measured in *milli*grams. It takes 1000 mg to make 1 g. Recall that a dollar bill weighs about 1 g. Imagine cutting it into 1000 pieces; the weight of one tiny piece would be 1 mg. Dosages of medicine and vitamins are given in milligrams.

### EXAMPLE 6 Using Metric Weight Units

Write the most reasonable metric unit in each blank. Choose from kg, g, and mg.

**(a)** Ramon's suitcase weighed 20 ______.

Choose 20 kg because kilograms are used instead of pounds; 20 kg is about 44 pounds.

**(b)** LeTia took a 350 ______ aspirin tablet.

Choose 350 mg because 350 g would be more than the weight of a hamburger, which is too much.

**(c)** Jenny mailed a letter that weighed 30 ______.

Choose 30 g because 30 kg would be much too heavy and 30 mg is less than the weight of a dollar bill.

**Now Try Exercises 41, 43, and 47.**

**OBJECTIVE 7 Convert among metric weight (mass) units.** As with length and capacity, we can convert among metric weight units by using unit fractions. The unit fractions we need are as follows.

**Metric Weight (Mass) Relationships**

| | |
|---|---|
| **1 kg = 1000 g,** so the unit fractions are: $\dfrac{\mathbf{1\ kg}}{\mathbf{1000\ g}}$ or $\dfrac{\mathbf{1000\ g}}{\mathbf{1\ kg}}$ | **1 g = 1000 mg,** so the unit fractions are: $\dfrac{\mathbf{1\ g}}{\mathbf{1000\ mg}}$ or $\dfrac{\mathbf{1000\ mg}}{\mathbf{1\ g}}$ |

Or we can use a metric conversion line to decide how to move the decimal point.

| 1000 | 100 | 10 | 1 | $\frac{1}{10}$ | $\frac{1}{100}$ | $\frac{1}{1000}$ |
|---|---|---|---|---|---|---|
| kg | hg | dag | g | dg | cg | mg |

**EXAMPLE 7 Converting among Metric Weight Units**

Convert using the metric conversion line or unit fractions.

**(a)** 7 mg to g

Using the metric conversion line:

From **mg** to **g** is *three places* to the *left.*

7. ← Decimal point starts here.

007. ← Move three places left.

Using unit fractions:

Multiply by a unit fraction that allows you to divide out mg.

$$\frac{7\ \cancel{mg}}{1} \cdot \frac{1\ g}{1000\ \cancel{mg}} = \frac{7}{1000}\ g = .007\ g$$

7 mg = .007 mg

**(b)** 13.72 kg to g

Using the metric conversion line:

From **kg** to **g** is *three* places to the *right.*

13.720 ← Decimal point moves three places to the right.

13.72 kg = 13,720 g

↑ A comma (not a decimal point)

Using unit fractions:

Multiply by a unit fraction that allows you to divide out kg.

$$\frac{13.72\ \cancel{kg}}{1} \cdot \frac{1000\ g}{1\ \cancel{kg}} = 13{,}720\ g$$

↑ A comma (not a decimal point)

13.72 kg = 13,720 g

**Now Try Exercises 71 and 75.**

## APPENDIX E EXERCISES

*Use this ruler to measure your thumb and hand. Determine each measurement.*

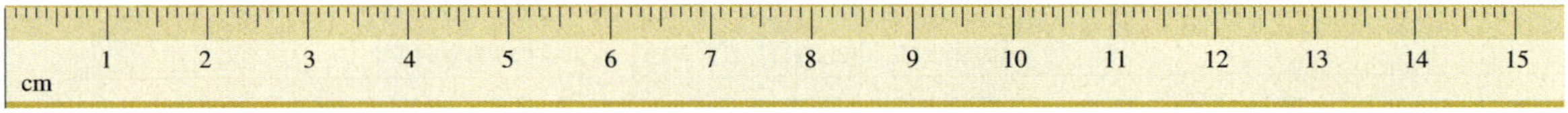


**1.** The width of your hand in centimeters

**2.** The width of your hand in millimeters

**3.** The width of your thumb in millimeters

**4.** The width of your thumb in centimeters

*Determine the most reasonable metric length unit for each blank. Choose from km, m, cm, and mm. See Example 1.*

**5.** The child was 91 _____ tall.

**6.** The cardboard was 3 _____ thick.

**7.** Ming-Na swam in the 200 _____ backstroke race.

**8.** The bookcase is 75 _____ wide.

**9.** Adriana drove 400 _____ on her vacation.

**10.** The door is 2 _____ high.

**11.** An aspirin tablet is 10 _____ across.

**12.** Lamard jogs 4 _____ every morning.

**13.** A paper clip is about 3 _____ long.

**14.** My pen is 145 _____ long.

**15.** Dave's truck is 5 _____ long.

**16.** Wheelchairs need doorways that are at least 80 _____ wide.

**17.** Describe at least three examples of metric length units that you have come across in your daily life.

**18.** Explain one reason why the metric system would be easier for a child to learn than the English system.

*Convert each measurement. Use unit fractions or the metric conversion line. See Examples 2 and 3.*

**19.** 7 m to cm

**20.** 18 m to cm

**21.** 40 mm to m

**22.** 6 mm to m

**23.** 9.4 km to m

**24.** .7 km to m

**25.** 509 cm to m

**26.** 30 cm to m

**27.** 400 mm to cm

**28.** 25 mm to cm

**29.** .91 m to mm

**30.** 4 m to mm

*Solve each problem.*

**31.** Many cameras use film that is 35 mm wide. Film for movie theaters may be 70 mm wide. Using the ruler given at the top of this page, draw a line segment that is 35 mm long and a line segment 70 mm long. Then convert each measurement to centimeters.

**32.** Gold wedding bands may be very narrow or quite wide. Common widths are 3 mm, 5 mm, and 10 mm. Using the ruler given at the top of this page, draw line segments that are 3 mm, 5 mm, and 10 mm long. Then convert each measurement to centimeters.

5 mm

**33.** The Roe River near Great Falls, Montana, is the shortest river in the world, with a north fork that is just under 18 m long. (*Source: Guinness Book of Amazing Nature.*) How many kilometers long is the north fork of the river?

**34.** There are 60,000 km of blood vessels in the human body. (*Source: Big Book of Knowledge.*) How many meters of blood vessels are in the body?

**35.** Use two unit fractions to convert 5.6 mm to km.

**36.** Use two unit fractions to convert 16.5 km to mm.

*Determine the most reasonable metric unit for each blank. Choose from L, mL, kg, g, and mg. See Examples 4 and 6.*

**37.** The glass held 250 _____ of water.

**38.** Hiromi used 20 _____ of water to wash the kitchen floor.

**39.** Dolores can make 10 _____ of soup in that pot.

**40.** Jay gave 2 _____ of vitamin drops to the baby.

**41.** Our yellow Labrador dog grew up to weigh 40 _____ .

**42.** A small safety pin weighs 750 _____ .

**43.** Lori caught a small sunfish weighing 150 _____ .

**44.** One dime weighs 2 _____ .

**45.** Andre donated 500 _____ of blood today.

**46.** Barbara bought the 2 _____ bottle of cola.

**47.** The patient received a 250 _____ tablet of medication each hour.

**48.** The 8 people on the elevator weighed a total of 500 _____ .

**49.** The gas can for the lawn mower holds 4 _____ .

**50.** Kevin poured 10 _____ of vanilla into the mixing bowl.

*Today, medical measurements are usually given in the metric system. Since we convert among metric units of measure by moving the decimal point, it is possible that mistakes can be made. Examine the following dosages and indicate whether they are reasonable or unreasonable. If a dose is unreasonable, indicate whether it is too much or too little.*

**51.** Drink 4.1 L of Kaopectate after each meal.

**52.** Drop 1 mL of solution into the eye twice a day.

**53.** Soak your feet in 5 kg of Epsom salts per liter of water.

**54.** Inject .5 L of insulin each morning.

**55.** Take 15 mL of cough syrup every four hours.

**56.** Take 200 mg of vitamin C each day.

**57.** Describe at least two examples of metric capacity units and two examples of metric weight units that you have come across in your daily life.

**58.** Explain in your own words how the meter, liter, and gram are related.

*Convert each measurement. Use unit fractions or the metric conversion line. See Examples 5 and 7.*

**59.** 15 L to mL

**60.** 6 L to mL

**61.** 3000 mL to L

**62.** 18,000 mL to L

**63.** 925 mL to L

**64.** 200 mL to L

**65.** 8 mL to L

**66.** 25 mL to L

**67.** 4.15 L to mL

**68.** 11.7 L to mL

**69.** 8000 g to kg

**70.** 25,000 g to kg

**71.** 5.2 kg to g

**72.** 12.42 kg to g

**73.** .85 g to mg

**74.** .2 g to mg

**75.** 30,000 mg to g

**76.** 7500 mg to g

*Solve each application. (Source for Exercises 77–82:* Top 10 of Everything, 2000.*)*

**77.** Human skin has about 3 million sweat glands, which release an average of 300 mL of sweat per day. How many liters of sweat are released each day?

**78.** In hot climates, the sweat glands in a person's skin may release up to 3.5 L of sweat in one day. How many milliliters is that?

**79.** The average weight of a human brain is 1.34 kg. How many grams is that?

**80.** In the Victorian era, people believed that heavier brains meant greater intelligence. They were impressed that Otto von Bismarck's brain weighted 1907 g, which is how many kilograms?

**81.** A healthy human heart pumps about 70 mL of blood per beat. How many liters of blood does it pump per beat?

**82.** On average, we breath in and out roughly 900 mL of air every 10 seconds. How many liters of air is that?

**83.** A small adult cat weighs from 3000 g to 4000 g. How many kilograms is that? (*Source:* Lyndale Animal Hospital.)

**84.** If the letter you are mailing weighs 29 g, you must put additional postage on it. (*Source:* U.S. Postal Service.) How many kilograms does the letter weigh?

**85.** One nickel weighs 5 g. How many nickels are in 1 kg of nickels?

**86.** Seawater contains about 3.5 g of salt per 1000 mL of water. How many grams of salt would be in 1 L of seawater?

# Appendix F
# Review of Exponents, Polynomials, and Factoring

**(Transition from Beginning to Intermediate Algebra)**

**OBJECTIVES**

1. Review the basic rules for exponents.
2. Review addition, subtraction, and multiplication of polynomials.
3. Review factoring techniques.

**OBJECTIVE 1 Review the basic rules for exponents.** In Sections 4.1 and 4.2 we introduced the following definitions and rules for working with exponents.

**Definitions and Rules for Exponents**

If no denominators are 0, for any integers $m$ and $n$:

| | | Examples |
|---|---|---|
| **Product rule** | $a^m \cdot a^n = a^{m+n}$ | $7^4 \cdot 7^5 = 7^9$ |
| **Zero exponent** | $a^0 = 1$ | $(-3)^0 = 1$ |
| **Negative exponent** | $a^{-n} = \dfrac{1}{a^n}$ | $5^{-3} = \dfrac{1}{5^3}$ |
| **Quotient rule** | $\dfrac{a^m}{a^n} = a^{m-n}$ | $\dfrac{2^2}{2^5} = 2^{-3} = \dfrac{1}{2^3}$ |
| **Power rules (a)** | $(a^m)^n = a^{mn}$ | $(4^2)^3 = 4^6$ |
| **(b)** | $(ab)^m = a^m b^m$ | $(3k)^4 = 3^4k^4$ |
| **(c)** | $\left(\dfrac{a}{b}\right)^m = \dfrac{a^m}{b^m}$ | $\left(\dfrac{2}{3}\right)^2 = \dfrac{2^2}{3^2}$ |
| **Negative to positive rules** | $\dfrac{a^{-m}}{b^{-n}} = \dfrac{b^n}{a^m}$ | $\dfrac{2^{-4}}{5^{-3}} = \dfrac{5^3}{2^4}$ |
| | $\left(\dfrac{a}{b}\right)^{-m} = \left(\dfrac{b}{a}\right)^m$ | $\left(\dfrac{4}{7}\right)^{-2} = \left(\dfrac{7}{4}\right)^2$ |

**EXAMPLE 1 Applying Definitions and Rules for Exponents**

Simplify each expression. Write answers using only positive exponents. Assume all variables represent nonzero real numbers.

**(a)** 
$$\begin{aligned}(x^2y^{-3})(x^{-5}y^7) &= (x^{2+(-5)})(y^{-3+7})\\ &= x^{-3}y^4\\ &= \frac{1}{x^3}y^4\\ &= \frac{y^4}{x^3}\end{aligned}$$

**(b)** $(-5)^0 + (-5^0) = 1 + (-1)$ $\quad -5^0 = -1 \cdot 5^0 = -1 \cdot 1 = -1$

$= 0$

**(c)** $\dfrac{(t^5s^{-4})^2}{(t^{-3}s^5)^3} = \dfrac{t^{10}s^{-8}}{t^{-9}s^{15}} = \dfrac{t^{10}t^9}{s^{15}s^8} = \dfrac{t^{19}}{s^{23}}$

**(d)** $\left(\dfrac{-3x^{-4}y}{x^5y^{-4}}\right)^{-2} = \left(\dfrac{x^5y^{-4}}{-3x^{-4}y}\right)^2 = \dfrac{x^{10}y^{-8}}{9x^{-8}y^2} = \dfrac{x^{18}}{9y^{10}}$

**(e)** $(2x^2y^3z)^2(x^4y^2)^3 = (4x^4y^6z^2)(x^{12}y^6) = 4x^{16}y^{12}z^2$

**Now Try Exercises 1, 5, 7, 9, and 11.**

**OBJECTIVE 2 Review addition, subtraction, and multiplication of polynomials.** These arithmetic operations with polynomials were covered in Sections 4.4–4.6.

**Adding and Subtracting Polynomials**

To add polynomials, add like terms. To subtract polynomials, change all signs on the second polynomial and add the result to the first polynomial.

**EXAMPLE 2 Adding and Subtracting Polynomials**

Add or subtract as indicated.

**(a)** $(-4x^3 + 3x^2 - 8x + 2) + (5x^3 - 8x^2 + 12x - 3)$

$= (-4 + 5)x^3 + (3 - 8)x^2 + (-8 + 12)x + (2 - 3)$

$= x^3 - 5x^2 + 4x - 1$

**(b)** $-4(x^2 + 3x - 6) - (2x^2 - 3x + 7)$

$= -4x^2 - 12x + 24 - 2x^2 + 3x - 7$

$= -6x^2 - 9x + 17$

**(c)** Subtract.

$$\begin{array}{r} 2t^2 - 3t - 4 \\ \underline{-8t^2 + 4t - 1} \end{array}$$

Change the sign of each term in $-8t^2 + 4t - 1$, and add.

$$\begin{array}{rl} 2t^2 - 3t - 4 & \\ \underline{8t^2 - 4t + 1} & \quad \text{Change signs.} \\ 10t^2 - 7t - 3 & \quad \text{Add.} \end{array}$$

**Now Try Exercises 13, 19, and 21.**

**Multiplying Polynomials**

To multiply two polynomials, multiply each term of the second polynomial by each term of the first polynomial and add the products. In particular, when multiplying two binomials, use the FOIL method. (See Section 4.5.)

There are also several special product rules that are useful when multiplying binomials.

**Special Product Rules**

$$(x + y)^2 = x^2 + 2xy + y^2$$
$$(x - y)^2 = x^2 - 2xy + y^2$$
$$(x + y)(x - y) = x^2 - y^2$$

**EXAMPLE 3 Multiplying Polynomials**

Find each product.

**(a)** $(4y - 1)(3y + 2) = 4y(3y) + 4y(2) - 1(3y) - 1(2)$ FOIL
$= 12y^2 + 8y - 3y - 2$
$= 12y^2 + 5y - 2$

**(b)** $(3x + 5y)(3x - 5y) = (3x)^2 - (5y)^2$ $(x + y)(x - y) = x^2 - y^2$
$= 9x^2 - 25y^2$

**(c)** $(2t + 3)^2 = (2t)^2 + 2(2t)(3) + 3^2$ $(x + y)^2 = x^2 + 2xy + y^2$
$= 4t^2 + 12t + 9$

**(d)** $(5x - 1)^2 = (5x)^2 - 2(5x)(1) + 1^2$ $(x - y)^2 = x^2 - 2xy + y^2$
$= 25x^2 - 10x + 1$

**(e)** $(3x + 2)(9x^2 - 6x + 4)$
Multiply vertically.

$$\begin{array}{rl} 9x^2 - 6x + 4 & \\ 3x + 2 & \\ \hline 18x^2 - 12x + 8 & \leftarrow 2(9x^2 - 6x + 4) \\ 27x^3 - 18x^2 + 12x \phantom{+ 8} & \leftarrow 3x(9x^2 - 6x + 4) \\ \hline 27x^3 \phantom{- 18x^2 + 12x} + 8 & \text{Add like terms.} \end{array}$$

The product is the sum of cubes, $27x^3 + 8$.

**Now Try Exercises 23, 27, 31, and 35.**

**OBJECTIVE 3 Review factoring techniques.** Factoring, which involves writing a polynomial as a product, was covered in Chapter 5. Here are some general guidelines to use when factoring.

**Factoring a Polynomial**

1. Is there a common factor? If so, factor it out.
2. How many terms are in the polynomial?

   *Two terms:* Check to see whether it is a difference of squares or the sum or difference of cubes. If so, factor as in Section 5.4.

(continued)

*Three terms:* Is it a perfect square trinomial? If the trinomial is not a perfect square, check to see whether the coefficient of the squared term is 1. If so, use the method of Section 5.2. If the coefficient of the squared term of the trinomial is not 1, use the general factoring methods of Section 5.3.

*Four terms:* Try to factor the polynomial by grouping using the method of Section 5.1.

**3.** Can any factors be factored further? If so, factor them.

### EXAMPLE 4 Factoring Polynomials

Factor each polynomial completely.

**(a)** $6x^2y^3 - 12x^3y^2 = 6x^2y^2(y - 2x)$ $6x^2y^2$ is the greatest common factor.

**(b)** $3x^2 - x - 2$

To find the factors, find two terms that multiply to give $3x^2$ ($3x$ and $x$) and two terms that multiply to give $-2$ ($+2$ and $-1$). Make sure that the sum of the outer and inner products in the factored form is $-x$.

$$3x^2 - x - 2 = (3x + 2)(x - 1)$$

To check, multiply the factors using the FOIL method.

**(c)** $100t^2 - 81 = (10t)^2 - 9^2$ Difference of squares

$= (10t + 9)(10t - 9)$ $x^2 - y^2 = (x + y)(x - y)$

**(d)** $4x^2 + 20xy + 25y^2$

The terms $4x^2$ and $25y^2$ are both perfect squares, so factor as a perfect square trinomial.

$$4x^2 + 20xy + 25y^2 = (2x + 5y)^2$$

To check, take twice the product of the two terms in the squared binomial.

$$2(2x)(5y) = 20xy$$

Twice — First term — Last term

Since $20xy$ is the middle term of the trinomial, the trinomial is a perfect square and can be factored as $(2x + 5y)^2$.

**(e)** $1000x^3 - 27 = (10x)^3 - 3^3$ Difference of cubes

$= (10x - 3)[(10x)^2 + 10x(3) + 3^2]$ $x^3 - y^3 =$

$= (10x - 3)(100x^2 + 30x + 9)$ $(x - y)(x^2 + xy + y^2)$

**(f)** $6xy - 3x + 4y - 2$

Since there are four terms, try factoring by grouping.

$6xy - 3x + 4y - 2 = (6xy - 3x) + (4y - 2)$ Group terms.

$= 3x(2y - 1) + 2(2y - 1)$ Factor each group.

$= (2y - 1)(3x + 2)$ Factor out $2y - 1$.

In the final step, factor out the greatest common factor, the binomial $2y - 1$.

**Now Try Exercises 41, 45, 47, 49, and 59.**

## APPENDIX F EXERCISES

*Simplify each expression. Write the answers using only positive exponents. Assume all variables represent positive real numbers. See Example 1.*

**1.** $(a^4b^{-3})(a^{-6}b^2)$

**2.** $(t^{-3}s^{-5})(t^8s^{-2})$

**3.** $(5x^{-2}y)^2(2xy^4)^2$

**4.** $(7x^{-3}y^4)^3(2x^{-1}y^{-4})^2$

**5.** $-6^0 + (-6)^0$

**6.** $(-12)^0 - 12^0$

**7.** $\dfrac{(2w^{-1}x^2y^{-1})^3}{(4w^5x^{-2}y)^2}$

**8.** $\dfrac{(5p^{-3}q^2r^{-4})^2}{(10p^4q^{-1}r^5)^{-1}}$

**9.** $\left(\dfrac{-4a^{-2}b^4}{a^3b^{-1}}\right)^{-3}$

**10.** $\left(\dfrac{r^{-3}s^{-8}}{-6r^2s^{-4}}\right)^{-2}$

**11.** $(7x^{-4}y^2z^{-2})^{-2}(7x^4y^{-1}z^3)^2$

**12.** $(3m^{-5}n^2p^{-4})^3(3m^4n^{-3}p^5)^{-2}$

*Add or subtract as indicated. See Example 2.*

**13.** $(2a^4 + 3a^3 - 6a^2 + 5a - 12) + (-8a^4 + 8a^3 - 14a^2 + 21a - 3)$

**14.** $(-6r^4 - 3r^3 + 12r^2 - 9r + 9) + (8r^4 - 13r^3 - 14r^2 - 10r - 3)$

**15.** $(6x^3 - 12x^2 + 3x - 4) - (-2x^3 + 6x^2 - 3x + 12)$

**16.** $(10y^3 - 4y^2 + 8y + 7) - (7y^3 + 5y^2 - 2y - 13)$

**17.** Add.

$$\begin{array}{r} 5x^2y + 2xy^2 + \phantom{5}y^3 \\ \underline{-4x^2y - 3xy^2 + 5y^3} \end{array}$$

**18.** Add.

$$\begin{array}{r} 6ab^3 - \phantom{1}2a^2b^2 + 3b^5 \\ \underline{8ab^3 + 12a^2b^2 - 8b^5} \end{array}$$

**19.** $3(5x^2 - 12x + 4) - 2(9x^2 + 13x - 10)$

**20.** $-4(2t^3 - 3t^2 + 4t - 1) - 3(-8t^3 + 3t^2 - 2t + 9)$

**21.** Subtract.

$$\begin{array}{r} 6x^3 - 2x^2 + 3x - 1 \\ \underline{-4x^3 + 2x^2 - 6x + 3} \end{array}$$

**22.** Subtract.

$$\begin{array}{r} -9y^3 - 2y^2 + 3y - 8 \\ \underline{-8y^3 + 4y^2 + 3y + 1} \end{array}$$

*Find each product. See Example 3.*

**23.** $(3x + 1)(2x - 7)$

**24.** $(5z + 3)(2z - 3)$

**25.** $(4x - 1)(x - 2)$

**26.** $(7t - 3)(t - 4)$

**27.** $(4t + 3)(4t - 3)$

**28.** $(6x + 1)(6x - 1)$

**29.** $(2y^2 + 4)(2y^2 - 4)$

**30.** $(3b^3 + 2t)(3b^3 - 2t)$

**31.** $(4x - 3)^2$

**32.** $(9t + 2)^2$

**33.** $(6r + 5y)^2$

**34.** $(8m - 3n)^2$

**35.** $(c + 2d)(c^2 - 2cd + 4d^2)$

**36.** $(f + 3g)(f^2 - 3fg + 9g^2)$

**37.** $(4x - 1)(16x^2 + 4x + 1)$

**38.** $(5r - 2)(25r^2 + 10r + 4)$

**39.** $(7t + 5s)(2t^2 + 5st - s^2)$

**40.** $(8p + 3q)(2p^2 - 4pq + q^2)$

*Factor each polynomial completely. See Example 4.*

**41.** $8x^3y^4 + 12x^2y^3 + 36xy^4$

**42.** $10m^5n + 4m^2n^3 + 18m^3n^2$

**43.** $x^2 - 2x - 15$

**44.** $x^2 + x - 12$

**45.** $2x^2 - 9x - 18$

**46.** $3x^2 + 2x - 8$

**47.** $36t^2 - 25$

**48.** $49r^2 - 9$

**49.** $16t^2 + 24t + 9$

**50.** $25t^2 + 90t + 81$

**51.** $4m^2p - 12mnp + 9n^2p$

**52.** $16p^2r - 40pqr + 25q^2r$

**53.** $x^3 + 1$

**54.** $x^3 + 27$

**55.** $8t^3 + 125$

**56.** $27s^3 + 64$

**57.** $t^6 - 125$

**58.** $w^6 - 27$

**59.** $5xt + 15xr + 2yt + 6yr$

**60.** $3am + 18mb + 2an + 12nb$

**61.** $6ar + 12br - 5as - 10bs$

**62.** $7mt + 35ms - 2nt - 10ns$

**63.** $t^4 - 1$

**64.** $r^4 - 81$

**65.** $4x^2 + 12xy + 9y^2 - 1$

**66.** $81t^2 + 36ty + 4y^2 - 9$

# Appendix G
# Synthetic Division

**OBJECTIVES**

1. Use synthetic division to divide by a polynomial of the form $x - k$.
2. Use the remainder theorem to evaluate a polynomial.
3. Decide whether a given number is a solution of an equation.

**OBJECTIVE 1 Use synthetic division to divide by a polynomial of the form $x - k$.** Often, when one polynomial is divided by a second, the second polynomial is of the form $x - k$, where the coefficient of the $x$ term is 1. To see how a shortcut for these divisions works, look first below left, where the division of $3x^3 - 2x + 5$ by $x - 3$ is shown. Notice that 0 was inserted for the missing $x^2$-term.

$$\begin{array}{r} 3x^2 + 9x + 25 \\ x-3\overline{)3x^3+0x^2-2x+5} \\ \underline{3x^3-9x^2}\qquad\qquad\quad \\ 9x^2-2x\qquad\; \\ \underline{9x^2-27x}\qquad\; \\ 25x+5 \\ \underline{25x-75} \\ 80 \end{array} \qquad\qquad \begin{array}{rrrrr} & & 3 & 9 & 25 \\ \cline{2-5} 1-3) & 3 & 0 & -2 & 5 \\ & \underline{3} & \underline{-9} & & \\ & & 9 & -2 & \\ & & \underline{9} & \underline{-27} & \\ & & & 25 & 5 \\ & & & \underline{25} & \underline{-75} \\ & & & & 80 \end{array}$$

On the right, exactly the same division is shown written without the variables. This is why it is *essential* to use 0 as a placeholder in synthetic division. All the numbers in color on the right are repetitions of the numbers directly above them, so we omit them, as shown on the left below.

$$\begin{array}{rrrrr} & & 3 & 9 & 25 \\ \cline{2-5} 1-3) & 3 & 0 & -2 & 5 \\ & & \underline{-9} & & \\ & & 9 & -2 & \\ & & & \underline{-27} & \\ & & & 25 & 5 \\ & & & & \underline{-75} \\ & & & & 80 \end{array} \qquad\qquad \begin{array}{rrrrr} & & 3 & 9 & 25 \\ \cline{2-5} 1-3) & 3 & 0 & -2 & 5 \\ & & \underline{-9} & & \\ & & 9 & & \\ & & & \underline{-27} & \\ & & & 25 & \\ & & & & \underline{-75} \\ & & & & 80 \end{array}$$

The numbers in color on the left are again repetitions of the numbers directly above them; they too are omitted, as shown on the right above.

Now we can condense the problem. If we bring the 3 in the dividend down to the beginning of the bottom row, the top row can be omitted, since it duplicates the bottom row.

$$\begin{array}{rrrrr} & & & & \\ \cline{2-5} 1-3) & 3 & 0 & -2 & 5 \\ & & -9 & -27 & -75 \\ \cline{2-5} & 3 & 9 & 25 & 80 \end{array}$$

Finally, we omit the 1 at the upper left. Also, to simplify the arithmetic, we replace subtraction in the second row by addition. To compensate for this, we change the $-3$ at the upper left to its additive inverse, 3, as shown on the next page.

Additive inverse → 

$$\begin{array}{r|rrrr} 3 & 3 & 0 & -2 & 5 \\ & & 9 & 27 & 75 \\ \hline & 3 & 9 & 25 & 80 \end{array}$$

← Signs changed (75); ← Remainder (80)

The quotient is read from the bottom row.

$$3x^2 + 9x + 25 + \frac{80}{x-3}$$

The first three numbers in the bottom row are the coefficients of the quotient polynomial with degree 1 less than the degree of the dividend. The last number gives the remainder.

**Synthetic Division**

This shortcut procedure is called **synthetic division.** It is used only when dividing a polynomial by a binomial of the form $x - k$.

## EXAMPLE 1 Using Synthetic Division

Use synthetic division to divide $5x^2 + 16x + 15$ by $x + 2$.

As mentioned, we use synthetic division only when dividing by a polynomial of the form $x - k$. We change $x + 2$ into this form by writing it as

$$x + 2 = x - (-2),$$

where $k = -2$. Now write the coefficients of $5x^2 + 16x + 15$, placing $-2$ to the left.

$x + 2$ leads to $-2$. → $\;-2\overline{)5 \quad 16 \quad 15}$ ← Coefficients

Bring down the 5, and multiply: $-2 \cdot 5 = -10$.

$$\begin{array}{r|rrr} -2 & 5 & 16 & 15 \\ & & -10 & \\ \hline & 5 & & \end{array}$$

Add 16 and $-10$, getting 6, and multiply 6 and $-2$ to get $-12$.

$$\begin{array}{r|rrr} -2 & 5 & 16 & 15 \\ & & -10 & -12 \\ \hline & 5 & 6 & \end{array}$$

Add 15 and $-12$, getting 3.

$$\begin{array}{r|rrr} -2 & 5 & 16 & 15 \\ & & -10 & -12 \\ \hline & 5 & 6 & 3 \end{array}$$

← Remainder (3)

The result is read from the bottom row.

$$\frac{5x^2 + 16x + 15}{x + 2} = 5x + 6 + \frac{3}{x + 2}$$

**Now Try Exercise 7.**

### EXAMPLE 2 Using Synthetic Division with a Missing Term

Use synthetic division to find $(-4x^5 + x^4 + 6x^3 + 2x^2 + 50) \div (x - 2)$.

Use the steps given above, inserting a 0 for the missing $x$-term.

$$\begin{array}{r|rrrrrr} 2 & -4 & 1 & 6 & 2 & 0 & 50 \\ & & -8 & -14 & -16 & -28 & -56 \\ \hline & -4 & -7 & -8 & -14 & -28 & -6 \end{array}$$

Read the result from the bottom row.

$$\frac{-4x^5 + x^4 + 6x^3 + 2x^2 + 50}{x - 2} = -4x^4 - 7x^3 - 8x^2 - 14x - 28 + \frac{-6}{x - 2}$$

**Now Try Exercise 13.**

**OBJECTIVE 2 Use the remainder theorem to evaluate a polynomial.** We can use synthetic division to evaluate polynomials. For example, in the synthetic division of Example 2, where the polynomial was divided by $x - 2$, the remainder was $-6$.

Replacing $x$ in the polynomial with 2 gives

$$\begin{aligned} -4x^5 + x^4 + 6x^3 + 2x^2 + 50 &= -4 \cdot 2^5 + 2^4 + 6 \cdot 2^3 + 2 \cdot 2^2 + 50 \\ &= -4 \cdot 32 + 16 + 6 \cdot 8 + 2 \cdot 4 + 50 \\ &= -128 + 16 + 48 + 8 + 50 \\ &= -6, \end{aligned}$$

the same number as the remainder; that is, dividing by $x - 2$ produced a remainder equal to the result when $x$ is replaced with 2. This always happens, as the following **remainder theorem** states.

**Remainder Theorem**

If the polynomial $P(x)$ is divided by $x - k$, then the remainder is equal to $P(k)$.

This result is proved in more advanced courses.

### EXAMPLE 3 Using the Remainder Theorem

Let $P(x) = 2x^3 - 5x^2 - 3x + 11$. Find $P(-2)$.

Use the remainder theorem; divide $P(x)$ by $x - (-2)$.

$$\text{Value of } x \rightarrow \begin{array}{r|rrrr} -2 & 2 & -5 & -3 & 11 \\ & & -4 & 18 & -30 \\ \hline & 2 & -9 & 15 & -19 \end{array} \leftarrow \text{Remainder}$$

By this result, $P(-2) = -19$.

**Now Try Exercise 21.**

**OBJECTIVE 3 Decide whether a given number is a solution of an equation.** We can also use the remainder theorem to show that a given number is a solution of an equation.

### EXAMPLE 4 Using the Remainder Theorem

Show that $-5$ is a solution of the equation

$$2x^4 + 12x^3 + 6x^2 - 5x + 75 = 0.$$

One way to show that $-5$ is a solution is by substituting $-5$ for $x$ in the equation. However, an easier way is to use synthetic division and the remainder theorem.

| | | | | | | |
|---|---|---|---|---|---|---|
| Proposed solution → | $-5)$ | 2 | 12 | 6 | $-5$ | 75 |
| | | | $-10$ | $-10$ | 20 | $-75$ |
| | | 2 | 2 | $-4$ | 15 | 0 ← Remainder |

Since the remainder is 0, the polynomial has value 0 when $k = -5$, and so $-5$ is a solution of the given equation.

**Now Try Exercises 29 and 31.**

The synthetic division in Example 4 shows that $x - (-5)$ divides the polynomial with 0 remainder. Thus $x - (-5) = x + 5$ is a *factor* of the polynomial and

$$2x^4 + 12x^3 + 6x^2 - 5x + 75 = (x + 5)(2x^3 + 2x^2 - 4x + 15).$$

The second factor is the quotient polynomial found in the last row of the synthetic division.

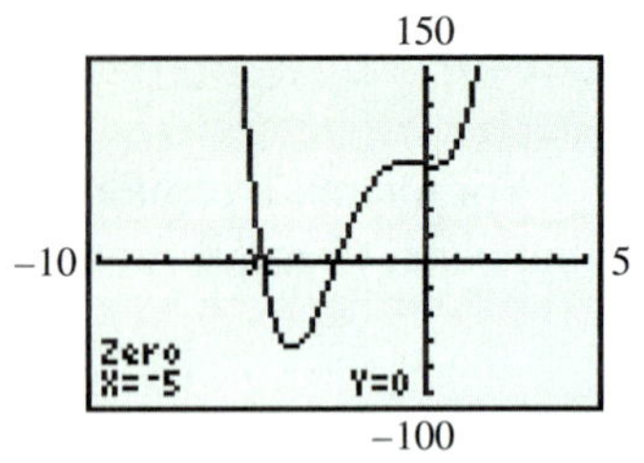


#### CONNECTIONS

The procedure in Example 4 is exactly how we use a graphing calculator to find real solutions of an equation by determining the $x$-intercepts of the graph. The screen shows the graph of $P(x) = 2x^4 + 12x^3 + 6x^2 - 5x + 75$ and shows that one value of $x$ that makes $P(x) = 0$ is $-5$. This agrees with our result in Example 4.

**For Discussion or Writing**

Estimate the other $x$-intercept to the nearest tenth. Verify your answer using synthetic division.

## APPENDIX G EXERCISES

**1.** What is the purpose of synthetic division?

**2.** What type of polynomial divisors may be used with synthetic division?

*Use synthetic division to find each quotient. See Examples 1 and 2.*

**3.** $\dfrac{x^2 - 6x + 5}{x - 1}$ **4.** $\dfrac{x^2 - 4x - 21}{x + 3}$ **5.** $\dfrac{4m^2 + 19m - 5}{m + 5}$

**6.** $\dfrac{3k^2 - 5k - 12}{k - 3}$ **7.** $\dfrac{2a^2 + 8a + 13}{a + 2}$ **8.** $\dfrac{4y^2 - 5y - 20}{y - 4}$

**9.** $(p^2 - 3p + 5) \div (p + 1)$ **10.** $(z^2 + 4z - 6) \div (z - 5)$

**11.** $\dfrac{4a^3 - 3a^2 + 2a - 3}{a - 1}$ **12.** $\dfrac{5p^3 - 6p^2 + 3p + 14}{p + 1}$

13. $(x^5 - 2x^3 + 3x^2 - 4x - 2) \div (x - 2)$

14. $(2y^5 - 5y^4 - 3y^2 - 6y - 23) \div (y - 3)$

15. $(-4r^6 - 3r^5 - 3r^4 + 5r^3 - 6r^2 + 3r + 3) \div (r - 1)$

16. $(2t^6 - 3t^5 + 2t^4 - 5t^3 + 6t^2 - 3t - 2) \div (t - 2)$

17. $(-3y^5 + 2y^4 - 5y^3 - 6y^2 - 1) \div (y + 2)$

18. $(m^6 + 2m^4 - 5m + 11) \div (m - 2)$

19. $\dfrac{y^3 + 1}{y - 1}$

20. $\dfrac{z^4 + 81}{z - 3}$

*Use the remainder theorem to find $P(k)$. See Example 3.*

21. $P(x) = 2x^3 - 4x^2 + 5x - 3$; $k = 2$

22. $P(y) = y^3 + 3y^2 - y + 5$; $k = -1$

23. $P(r) = -r^3 - 5r^2 - 4r - 2$; $k = -4$

24. $P(z) = -z^3 + 5z^2 - 3z + 4$; $k = 3$

25. $P(y) = 2y^3 - 4y^2 + 5y - 33$; $k = 3$

26. $P(x) = x^3 - 3x^2 + 4x - 4$; $k = 2$

27. Explain why a 0 remainder in synthetic division of $P(x)$ by $x - k$ indicates that $k$ is a solution of the equation $P(x) = 0$.

28. Explain why it is important to insert 0s as placeholders for missing terms before performing synthetic division.

*Use synthetic division to decide whether the given number is a solution of the equation. See Example 4.*

29. $x^3 - 2x^2 - 3x + 10 = 0$; $x = -2$

30. $x^3 - 3x^2 - x + 10 = 0$; $x = -2$

31. $m^4 + 2m^3 - 3m^2 + 8m - 8 = 0$; $m = -2$

32. $r^4 - r^3 - 6r^2 + 5r + 10 = 0$; $r = -2$

33. $3a^3 + 2a^2 - 2a + 11 = 0$; $a = -2$

34. $3z^3 + 10z^2 + 3z - 9 = 0$; $z = -2$

35. $2x^3 - x^2 - 13x + 24 = 0$; $x = -3$

36. $5p^3 + 22p^2 + p - 28 = 0$; $p = -4$

## RELATING CONCEPTS (EXERCISES 37–41)

**For Individual or Group Work**

*We can show a connection between dividing one polynomial by another and factoring the first polynomial. Let $P(x) = 2x^2 + 5x - 12$.* ***Work Exercises 37–41 in order.***

37. Factor $P(x)$.

38. Solve $P(x) = 0$.

39. Find $P(-4)$ and $P(\frac{3}{2})$.

40. Complete the following sentence. If $P(a) = 0$, then $x -$ _____ is a factor of $P(x)$.

41. Use the conclusion in Exercise 40 to decide whether $x - 3$ is a factor of $Q(x) = 3x^3 - 4x^2 - 17x + 6$. Factor $Q(x)$ completely.

## TECHNOLOGY INSIGHTS (EXERCISES 42–45)

*Use the graph to determine a solution of each equation.*

**42.** $2x^3 + 12x^2 + 24x + 16 = 0$

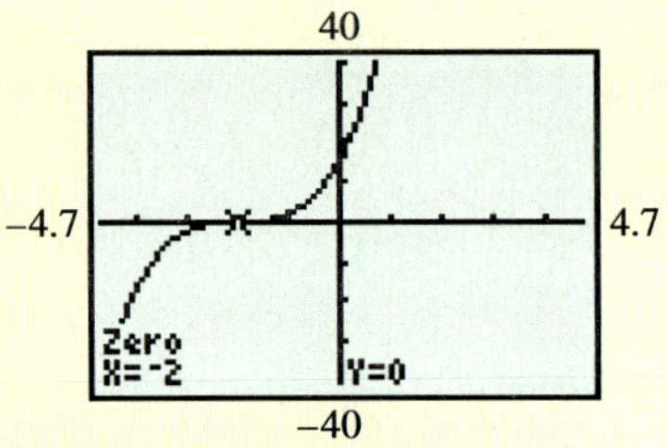


**43.** $x^3 - x^2 - 21x + 45 = 0$

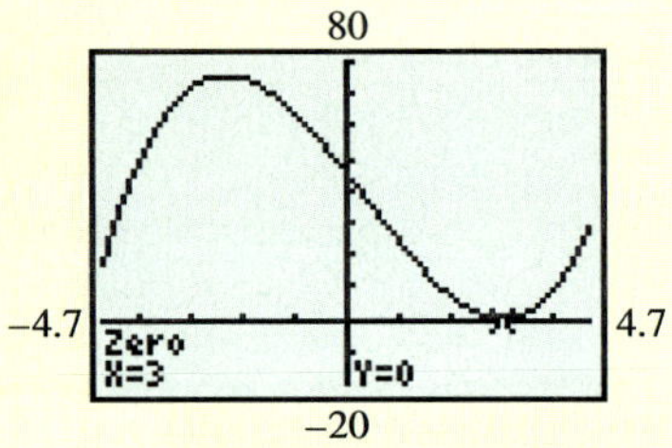


**44.** $x^3 + 3x^2 - 10x - 24 = 0$

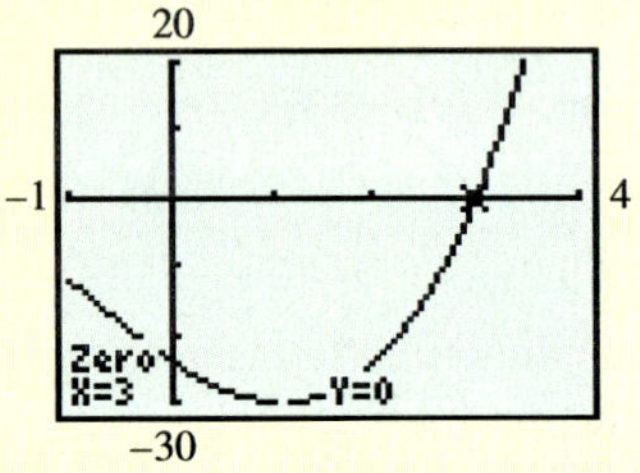


**45.** $x^3 + 3x^2 - 13x - 15 = 0$

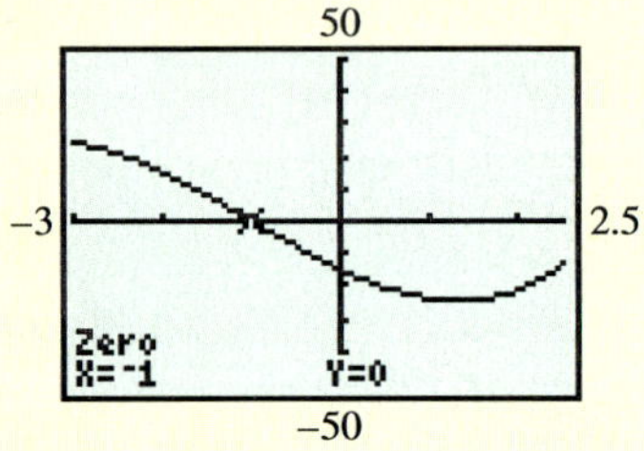

# Appendix H Determinants and Cramer's Rule

**OBJECTIVES**

1. Evaluate $2 \times 2$ determinants.
2. Use expansion by minors to evaluate $3 \times 3$ determinants.
3. Use a graphing calculator to evaluate determinants.
4. Understand the derivation of Cramer's rule.
5. Apply Cramer's rule to solve linear systems.

Recall from Section 8.6 that an ordered array of numbers within square brackets is called a *matrix* (plural *matrices*). Matrices are named according to the number of rows and columns they contain. A *square matrix* has the same number of rows and columns.

Columns

Rows $\begin{bmatrix} 2 & 3 & 5 \\ 7 & 1 & 2 \end{bmatrix}$ $2 \times 3$ matrix $\qquad$ $\begin{bmatrix} -1 & 0 \\ 1 & -2 \end{bmatrix}$ $2 \times 2$ square matrix

Associated with every *square matrix* is a real number called the **determinant** of the matrix. A determinant is symbolized by the entries of the matrix placed between two vertical lines, such as

$$\begin{vmatrix} 2 & 3 \\ 7 & 1 \end{vmatrix} \quad 2 \times 2 \text{ determinant} \qquad \begin{vmatrix} 7 & 4 & 3 \\ 0 & 1 & 5 \\ 6 & 0 & 1 \end{vmatrix}. \quad 3 \times 3 \text{ determinant}$$

Like matrices, determinants are named according to the number of rows and columns they contain.

**OBJECTIVE 1 Evaluate 2 × 2 determinants.** As mentioned above, the value of a determinant is a *real number.* The value of the $2 \times 2$ determinant

$$\begin{vmatrix} a & b \\ c & d \end{vmatrix}$$

is defined as follows.

**Value of a 2 × 2 Determinant**

$$\begin{vmatrix} \boldsymbol{a} & \boldsymbol{b} \\ \boldsymbol{c} & \boldsymbol{d} \end{vmatrix} = \boldsymbol{ad - bc}$$

**EXAMPLE 1 Evaluating a 2 × 2 Determinant**

Evaluate the determinant.

$$\begin{vmatrix} -1 & -3 \\ 4 & -2 \end{vmatrix}$$

Here $a = -1$, $b = -3$, $c = 4$, and $d = -2$, so

$$\begin{vmatrix} -1 & -3 \\ 4 & -2 \end{vmatrix} = -1(-2) - (-3)4 = 2 + 12 = 14.$$

**Now Try Exercise 3.**

A $3 \times 3$ determinant can be evaluated in a similar way.

**Value of a 3 × 3 Determinant**

$$\begin{vmatrix} a_1 & b_1 & c_1 \\ a_2 & b_2 & c_2 \\ a_3 & b_3 & c_3 \end{vmatrix} = (a_1b_2c_3 + b_1c_2a_3 + c_1a_2b_3) - (a_3b_2c_1 + b_3c_2a_1 + c_3a_2b_1)$$

This rule for evaluating a $3 \times 3$ determinant is hard to remember. A method for calculating a $3 \times 3$ determinant that is easier to use is based on the rule. Rearranging terms and using the distributive property gives

$$\begin{vmatrix} a_1 & b_1 & c_1 \\ a_2 & b_2 & c_2 \\ a_3 & b_3 & c_3 \end{vmatrix} = a_1(b_2c_3 - b_3c_2) - a_2(b_1c_3 - b_3c_1) + a_3(b_1c_2 - b_2c_1). \qquad (1)$$

Each of the quantities in parentheses represents a $2 \times 2$ determinant that is the part of the $3 \times 3$ determinant remaining when the row and column of the multiplier are eliminated, as shown below.

$$a_1(b_2c_3 - b_3c_2) \qquad \begin{vmatrix} a_1 & b_1 & c_1 \\ a_2 & b_2 & c_2 \\ a_3 & b_3 & c_3 \end{vmatrix}$$

$$a_2(b_1c_3 - b_3c_1) \qquad \begin{vmatrix} a_1 & b_1 & c_1 \\ a_2 & b_2 & c_2 \\ a_3 & b_3 & c_3 \end{vmatrix}$$

$$a_3(b_1c_2 - b_2c_1) \qquad \begin{vmatrix} a_1 & b_1 & c_1 \\ a_2 & b_2 & c_2 \\ a_3 & b_3 & c_3 \end{vmatrix}$$

These $2 \times 2$ determinants are called **minors** of the elements in the $3 \times 3$ determinant. In the determinant above, the minors of $a_1$, $a_2$, and $a_3$ are, respectively,

$$\begin{vmatrix} b_2 & c_2 \\ b_3 & c_3 \end{vmatrix}, \quad \begin{vmatrix} b_1 & c_1 \\ b_3 & c_3 \end{vmatrix}, \quad \text{and} \quad \begin{vmatrix} b_1 & c_1 \\ b_2 & c_2 \end{vmatrix}.$$

**OBJECTIVE 2 Use expansion by minors to evaluate 3 × 3 determinants.** A $3 \times 3$ determinant can be evaluated by multiplying each element in the first column by its minor and combining the products as indicated in equation (1). This is called **expansion of the determinant by minors** about the first column.

### EXAMPLE 2 Evaluating a 3 × 3 Determinant

Evaluate the determinant using expansion by minors about the first column.

$$\begin{vmatrix} 1 & 3 & -2 \\ -1 & -2 & -3 \\ 1 & 1 & 2 \end{vmatrix}$$

In this determinant, $a_1 = 1$, $a_2 = -1$, and $a_3 = 1$. Multiply each of these numbers by its minor, and combine the three terms using the definition. Notice that the second term in the definition is *subtracted.*

$$\begin{vmatrix} 1 & 3 & -2 \\ -1 & -2 & -3 \\ 1 & 1 & 2 \end{vmatrix} = 1\begin{vmatrix} -2 & -3 \\ 1 & 2 \end{vmatrix} - (-1)\begin{vmatrix} 3 & -2 \\ 1 & 2 \end{vmatrix} + 1\begin{vmatrix} 3 & -2 \\ -2 & -3 \end{vmatrix}$$

$$= 1[-2(2) - (-3)1] + 1[3(2) - (-2)1]$$
$$+ 1[3(-3) - (-2)(-2)]$$
$$= 1(-1) + 1(8) + 1(-13)$$
$$= -1 + 8 - 13$$
$$= -6$$

**Now Try Exercise 9.**

To obtain equation (1), we could have rearranged terms in the definition of the determinant and used the distributive property to factor out the three elements of the second or third column or of any of the three rows. Therefore, expanding by minors about any row or any column results in the same value for a 3 × 3 determinant. To determine the correct signs for the terms of other expansions, the following **array of signs** is helpful.

**Array of Signs for a 3 × 3 Determinant**

$$\begin{matrix} + & - & + \\ - & + & - \\ + & - & + \end{matrix}$$

The signs alternate for each row and column beginning with a + in the first row, first column position. For example, if the expansion is to be about the second column, the first term would have a minus sign associated with it, the second term a plus sign, and the third term a minus sign.

### EXAMPLE 3 Evaluating a 3 × 3 Determinant

Evaluate the determinant of Example 2 using expansion by minors about the second column.

$$\begin{vmatrix} 1 & 3 & -2 \\ -1 & -2 & -3 \\ 1 & 1 & 2 \end{vmatrix} = -3\begin{vmatrix} -1 & -3 \\ 1 & 2 \end{vmatrix} + (-2)\begin{vmatrix} 1 & -2 \\ 1 & 2 \end{vmatrix} - 1\begin{vmatrix} 1 & -2 \\ -1 & -3 \end{vmatrix}$$

$$= -3(1) - 2(4) - 1(-5)$$

$$= -3 - 8 + 5$$

$$= -6$$

As expected, the result is the same as in Example 2.

**Now Try Exercise 15.**

**OBJECTIVE 3 Use a graphing calculator to evaluate determinants.** The graphing calculator function det(A) assigns to each square matrix [A] one and only one real number, the determinant of A.

### EXAMPLE 4 Evaluating Determinants Using a Graphing Calculator

Evaluate the determinants in Examples 1 and 2 using a graphing calculator.

Figure 1 shows how a graphing calculator displays the correct value for the determinant in Example 1. Similarly, Figure 2 supports the result of Example 2.

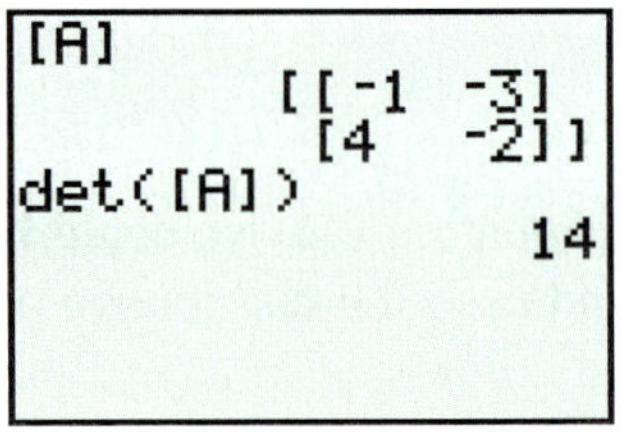


FIGURE 1

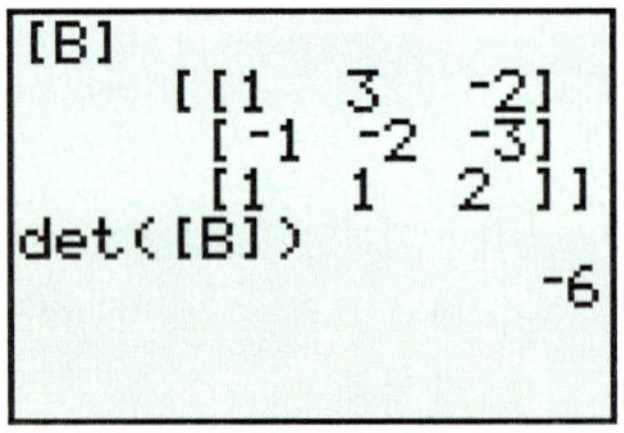


FIGURE 2

**Now Try Exercise 23.**

**CONNECTIONS**

Determinants of larger dimensions (such as 4 × 4) can be evaluated by extending the concepts presented thus far. However, because of the tedious calculations and chance for error, they are usually evaluated by computer or graphing calculator. For example, the determinant

$$\begin{vmatrix} -1 & -2 & 3 & 2 \\ 0 & 1 & 4 & -2 \\ 3 & -1 & 4 & 0 \\ 2 & 1 & 0 & 3 \end{vmatrix}$$

is equal to −185, as shown in the graphing calculator screen in Figure 3.

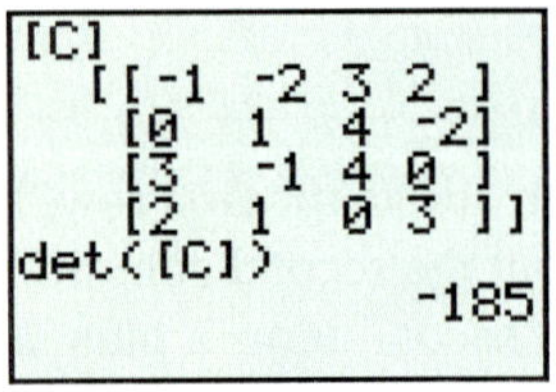


FIGURE 3

**For Discussion or Writing**

**1.** Use the array of signs

$$\begin{matrix} + & - & + & - \\ - & + & - & + \\ + & - & + & - \\ - & + & - & + \end{matrix}$$

to evaluate the preceding determinant by hand, expanding about the fourth row.

**2.** Explain how finding a determinant illustrates the function concept.

**OBJECTIVE 4 Understand the derivation of Cramer's rule.** Determinants can be used to solve a system of the form

$$a_1x + b_1y = c_1 \quad (1)$$
$$a_2x + b_2y = c_2. \quad (2)$$

The result will be a formula that can be used to solve any system of two equations with two variables. To get this general solution, we eliminate $y$ and solve for $x$ by first multiplying each side of equation (1) by $b_2$ and each side of equation (2) by $-b_1$. Then we add these results and solve for $x$.

$$\begin{aligned} a_1b_2x + b_1b_2y &= c_1b_2 && \text{Multiply equation (1) by } b_2. \\ -a_2b_1x - b_1b_2y &= -c_2b_1 && \text{Multiply equation (2) by } -b_1. \\ \hline (a_1b_2 - a_2b_1)x &= c_1b_2 - c_2b_1 \\ x &= \frac{c_1b_2 - c_2b_1}{a_1b_2 - a_2b_1} && (\text{if } a_1b_2 - a_2b_1 \neq 0) \end{aligned}$$

To solve for $y$, we multiply each side of equation (1) by $-a_2$ and each side of equation (2) by $a_1$ and add.

$$\begin{aligned} -a_1a_2x - a_2b_1y &= -a_2c_1 && \text{Multiply equation (1) by } -a_2. \\ a_1a_2x + a_1b_2y &= a_1c_2 && \text{Multiply equation (2) by } a_1. \\ \hline (a_1b_2 - a_2b_1)y &= a_1c_2 - a_2c_1 \\ y &= \frac{a_1c_2 - a_2c_1}{a_1b_2 - a_2b_1} && (\text{if } a_1b_2 - a_2b_1 \neq 0) \end{aligned}$$

Both numerators and the common denominator of these values for $x$ and $y$ can be written as determinants because

$$a_1c_2 - a_2c_1 = \begin{vmatrix} a_1 & c_1 \\ a_2 & c_2 \end{vmatrix},$$

$$c_1b_2 - c_2b_1 = \begin{vmatrix} c_1 & b_1 \\ c_2 & b_2 \end{vmatrix},$$

and

$$a_1b_2 - a_2b_1 = \begin{vmatrix} a_1 & b_1 \\ a_2 & b_2 \end{vmatrix}.$$

Using these results, the solutions for $x$ and $y$ become

$$x = \frac{\begin{vmatrix} c_1 & b_1 \\ c_2 & b_2 \end{vmatrix}}{\begin{vmatrix} a_1 & b_1 \\ a_2 & b_2 \end{vmatrix}} \quad \text{and} \quad y = \frac{\begin{vmatrix} a_1 & c_1 \\ a_2 & c_2 \end{vmatrix}}{\begin{vmatrix} a_1 & b_1 \\ a_2 & b_2 \end{vmatrix}}, \quad \begin{vmatrix} a_1 & b_1 \\ a_2 & b_2 \end{vmatrix} \neq 0.$$

For convenience, we denote the three determinants in the solution as

$$\begin{vmatrix} a_1 & b_1 \\ a_2 & b_2 \end{vmatrix} = D, \quad \begin{vmatrix} c_1 & b_1 \\ c_2 & b_2 \end{vmatrix} = D_x, \quad \text{and} \quad \begin{vmatrix} a_1 & c_1 \\ a_2 & c_2 \end{vmatrix} = D_y.$$

Notice that the elements of $D$ are the four coefficients of the variables in the given system; the elements of $D_x$ are obtained by replacing the coefficients of $x$ by the respective constants; the elements of $D_y$ are obtained by replacing the coefficients of $y$ by the respective constants.

These results are summarized as **Cramer's rule.**

**Cramer's Rule for 2 × 2 Systems**

Given the system

$$a_1x + b_1y = c_1$$
$$a_2x + b_2y = c_2 \quad \text{with } a_1b_2 - a_2b_1 = D \neq 0,$$

then

$$\boldsymbol{x} = \frac{\begin{vmatrix} c_1 & b_1 \\ c_2 & b_2 \end{vmatrix}}{\begin{vmatrix} a_1 & b_1 \\ a_2 & b_2 \end{vmatrix}} = \frac{D_x}{D} \quad \text{and} \quad \boldsymbol{y} = \frac{\begin{vmatrix} a_1 & c_1 \\ a_2 & c_2 \end{vmatrix}}{\begin{vmatrix} a_1 & b_1 \\ a_2 & b_2 \end{vmatrix}} = \frac{D_y}{D}.$$

**OBJECTIVE 5 Apply Cramer's rule to solve linear systems.** To use Cramer's rule to solve a system of equations, find the three determinants, $D$, $D_x$, and $D_y$, and then write the necessary quotients for $x$ and $y$.

**CAUTION** As indicated in the box, Cramer's rule does not apply if $D = a_1b_2 - a_2b_1 = 0$. When $D = 0$, the system is inconsistent or has dependent equations. For this reason, it is a good idea to evaluate $D$ first.

### EXAMPLE 5 Using Cramer's Rule to Solve a 2 × 2 System

Use Cramer's rule to solve the system

$$5x + 7y = -1$$
$$6x + 8y = 1.$$

By Cramer's rule, $x = \frac{D_x}{D}$ and $y = \frac{D_y}{D}$. As previously mentioned, it is a good idea to find $D$ first since if $D = 0$, Cramer's rule does not apply. If $D \neq 0$, then find $D_x$ and $D_y$.

$$D = \begin{vmatrix} 5 & 7 \\ 6 & 8 \end{vmatrix} = 5(8) - 7(6) = -2$$

$$D_x = \begin{vmatrix} -1 & 7 \\ 1 & 8 \end{vmatrix} = -1(8) - 7(1) = -15$$

$$D_y = \begin{vmatrix} 5 & -1 \\ 6 & 1 \end{vmatrix} = 5(1) - (-1)6 = 11$$

From Cramer's rule,

$$x = \frac{D_x}{D} = \frac{-15}{-2} = \frac{15}{2} \quad \text{and} \quad y = \frac{D_y}{D} = \frac{11}{-2} = -\frac{11}{2}.$$

The solution set is $\left\{\left(\frac{15}{2}, -\frac{11}{2}\right)\right\}$, as can be verified by checking in the given system.

**Now Try Exercise 27.**

Because graphing calculators can evaluate determinants, Cramer's rule can be applied using them. Figure 4 shows how the work of Example 5 is accomplished, with $D$ the determinant of matrix [A], $D_x$ the determinant of matrix [B], and $D_y$ the determinant of matrix [C].

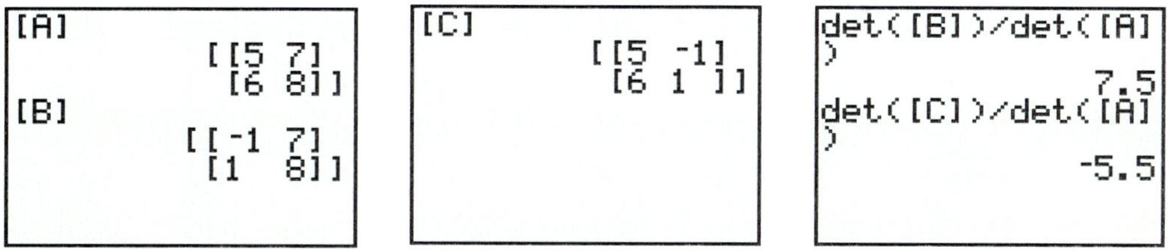


**FIGURE 4**

In a similar manner, Cramer's rule can be applied to systems of three equations with three variables.

**Cramer's Rule for 3 × 3 Systems**

Given the system

$$a_1x + b_1y + c_1z = d_1$$
$$a_2x + b_2y + c_2z = d_2$$
$$a_3x + b_3y + c_3z = d_3$$

(continued)

with

$$D_x = \begin{vmatrix} d_1 & b_1 & c_1 \\ d_2 & b_2 & c_2 \\ d_3 & b_3 & c_3 \end{vmatrix}, \quad D_y = \begin{vmatrix} a_1 & d_1 & c_1 \\ a_2 & d_2 & c_2 \\ a_3 & d_3 & c_3 \end{vmatrix},$$

$$D_z = \begin{vmatrix} a_1 & b_1 & d_1 \\ a_2 & b_2 & d_2 \\ a_3 & b_3 & d_3 \end{vmatrix}, \quad D = \begin{vmatrix} a_1 & b_1 & c_1 \\ a_2 & b_2 & c_2 \\ a_3 & b_3 & c_3 \end{vmatrix} \neq 0,$$

then

$$\boldsymbol{x = \frac{D_x}{D}}, \quad \boldsymbol{y = \frac{D_y}{D}}, \quad \text{and} \quad \boldsymbol{z = \frac{D_z}{D}}.$$

## EXAMPLE 6 Using Cramer's Rule to Solve a 3 × 3 System

Use Cramer's rule to solve the system

$$\begin{aligned} x + y - z + 2 &= 0 \\ 2x - y + z + 5 &= 0 \\ x - 2y + 3z - 4 &= 0. \end{aligned}$$

To use Cramer's rule, first rewrite the system in the form

$$\begin{aligned} x + y - z &= -2 \\ 2x - y + z &= -5 \\ x - 2y + 3z &= 4. \end{aligned}$$

Expand by minors about row 1 to find $D$.

$$\begin{aligned} D &= \begin{vmatrix} 1 & 1 & -1 \\ 2 & -1 & 1 \\ 1 & -2 & 3 \end{vmatrix} \\ &= 1\begin{vmatrix} -1 & 1 \\ -2 & 3 \end{vmatrix} - 1\begin{vmatrix} 2 & 1 \\ 1 & 3 \end{vmatrix} + (-1)\begin{vmatrix} 2 & -1 \\ 1 & -2 \end{vmatrix} \\ &= 1(-1) - 1(5) - 1(-3) \\ &= -3 \end{aligned}$$

Expanding $D_x$ by minors about row 1 gives

$$\begin{aligned} D_x &= \begin{vmatrix} -2 & 1 & -1 \\ -5 & -1 & 1 \\ 4 & -2 & 3 \end{vmatrix} \\ &= -2\begin{vmatrix} -1 & 1 \\ -2 & 3 \end{vmatrix} - 1\begin{vmatrix} -5 & 1 \\ 4 & 3 \end{vmatrix} + (-1)\begin{vmatrix} -5 & -1 \\ 4 & -2 \end{vmatrix} \\ &= -2(-1) - 1(-19) - 1(14) \\ &= 7. \end{aligned}$$

In the same way, $D_y = -22$ and $D_z = -21$, so that

$$x = \frac{D_x}{D} = \frac{7}{-3} = -\frac{7}{3}, \qquad y = \frac{D_y}{D} = \frac{-22}{-3} = \frac{22}{3}, \qquad z = \frac{D_z}{D} = \frac{-21}{-3} = 7.$$

Check that the solution set is $\{(-\frac{7}{3}, \frac{22}{3}, 7)\}$.

**Now Try Exercise 33.**

As mentioned earlier, Cramer's rule does not apply when $D = 0$. The next example illustrates this case.

### EXAMPLE 7 Determining When Cramer's Rule Does Not Apply

Use Cramer's rule to solve the system

$$\begin{aligned} 2x - 3y + 4z &= 8 \\ 6x - 9y + 12z &= 24 \\ x + 2y - 3z &= 5. \end{aligned}$$

First, find $D$.

$$\begin{aligned} D &= \begin{vmatrix} 2 & -3 & 4 \\ 6 & -9 & 12 \\ 1 & 2 & -3 \end{vmatrix} \\ &= 2\begin{vmatrix} -9 & 12 \\ 2 & -3 \end{vmatrix} - 6\begin{vmatrix} -3 & 4 \\ 2 & -3 \end{vmatrix} + 1\begin{vmatrix} -3 & 4 \\ -9 & 12 \end{vmatrix} \\ &= 2(3) - 6(1) + 1(0) \\ &= 0 \end{aligned}$$

Since $D = 0$ here, Cramer's rule does not apply and we must use another method to solve the system. Multiplying each side of the first equation by 3 shows that the first two equations have the same solution set, so this system has dependent equations and an infinite solution set.

**Now Try Exercise 35.**

Cramer's rule can be extended to $4 \times 4$ or larger systems. See a standard college algebra text for details.

## APPENDIX H EXERCISES

**1.** Which one of the following is the expression for the determinant $\begin{vmatrix} -2 & -3 \\ 4 & -6 \end{vmatrix}$?

**A.** $-2(-6) + (-3)4$ **B.** $-2(-6) - 3(4)$
**C.** $-3(4) - (-2)(-6)$ **D.** $-2(-6) - (-3)4$

**2.** Evaluate $\begin{vmatrix} 0 & 0 \\ 3 & -4 \end{vmatrix}$ and $\begin{vmatrix} 0 & 1 & 2 \\ 0 & -3 & 4 \\ 0 & 2 & 6 \end{vmatrix}$ and make a conjecture (educated guess) about the value of a determinant that has all 0s in a row or a column.

*Evaluate each determinant. See Example 1.*

**3.** $\begin{vmatrix} -2 & 5 \\ -1 & 4 \end{vmatrix}$ **4.** $\begin{vmatrix} 3 & -6 \\ 2 & -2 \end{vmatrix}$ **5.** $\begin{vmatrix} 1 & -2 \\ 7 & 0 \end{vmatrix}$

**6.** $\begin{vmatrix} -5 & -1 \\ 1 & 0 \end{vmatrix}$ **7.** $\begin{vmatrix} 0 & 4 \\ 0 & 4 \end{vmatrix}$ **8.** $\begin{vmatrix} 8 & -3 \\ 0 & 0 \end{vmatrix}$

*Evaluate each determinant by expansion by minors about the first column. See Example 2.*

**9.** $\begin{vmatrix} -1 & 2 & 4 \\ -3 & -2 & -3 \\ 2 & -1 & 5 \end{vmatrix}$ **10.** $\begin{vmatrix} 2 & -3 & -5 \\ 1 & 2 & 2 \\ 5 & 3 & -1 \end{vmatrix}$

**11.** $\begin{vmatrix} 1 & 0 & -2 \\ 0 & 2 & 3 \\ 1 & 0 & 5 \end{vmatrix}$ **12.** $\begin{vmatrix} 2 & -1 & 0 \\ 0 & -1 & 1 \\ 1 & 2 & 0 \end{vmatrix}$

**13.** Explain in your own words how to evaluate a $2 \times 2$ determinant. Illustrate with an example.

**14.** Explain in your own words how to evaluate a $3 \times 3$ determinant. Illustrate with an example.

*Evaluate each determinant by expansion by minors about any row or column. (Hint: The work is easier if you choose a row or a column with 0s.) See Example 3.*

**15.** $\begin{vmatrix} 4 & 4 & 2 \\ 1 & -1 & -2 \\ 1 & 0 & 2 \end{vmatrix}$ **16.** $\begin{vmatrix} 3 & -1 & 2 \\ 1 & 5 & -2 \\ 0 & 2 & 0 \end{vmatrix}$

**17.** $\begin{vmatrix} 3 & 5 & -2 \\ 1 & -4 & 1 \\ 3 & 1 & -2 \end{vmatrix}$ **18.** $\begin{vmatrix} 0 & 0 & 3 \\ 4 & 0 & -2 \\ 2 & -1 & 3 \end{vmatrix}$

**19.** $\begin{vmatrix} 3 & 0 & -2 \\ 1 & -4 & 1 \\ 3 & 1 & -2 \end{vmatrix}$ **20.** $\begin{vmatrix} 1 & 1 & 2 \\ 5 & 5 & 7 \\ 3 & 3 & 1 \end{vmatrix}$

**21.** Explain why a determinant with a row or column of 0s has a value of 0.

*Use a graphing calculator with matrix capabilities to find each determinant. See Example 4.*

**22.** $\begin{vmatrix} .68 & .94 \\ .31 & -.56 \end{vmatrix}$ **23.** $\begin{vmatrix} 1.5 & 2.6 & 9.3 \\ 5.2 & -1.4 & 8.6 \\ 0 & .7 & 1.2 \end{vmatrix}$

**24.** $\begin{vmatrix} \sqrt{5} & \sqrt{2} & -\sqrt{3} \\ \sqrt{7} & -\sqrt{6} & \sqrt{10} \\ -\sqrt{5} & -\sqrt{2} & \sqrt{17} \end{vmatrix}$ (To as many places as the calculator shows)

**25.** Consider the system
$$\begin{aligned} 4x + 3y - 2z &= 1 \\ 7x - 4y + 3z &= 2 \\ -2x + y - 8z &= 0. \end{aligned}$$

Match each determinant in parts (a)–(d) with its correct representation from choices A–D.

**(a)** $D$

**(b)** $D_x$

**(c)** $D_y$

**(d)** $D_z$

**A.** $\begin{vmatrix} 1 & 3 & -2 \\ 2 & -4 & 3 \\ 0 & 1 & -8 \end{vmatrix}$ **B.** $\begin{vmatrix} 4 & 3 & 1 \\ 7 & -4 & 2 \\ -2 & 1 & 0 \end{vmatrix}$

**C.** $\begin{vmatrix} 4 & 1 & -2 \\ 7 & 2 & 3 \\ -2 & 0 & -8 \end{vmatrix}$ **D.** $\begin{vmatrix} 4 & 3 & -2 \\ 7 & -4 & 3 \\ -2 & 1 & -8 \end{vmatrix}$

**26.** For the system
$$\begin{aligned} x + 3y - 6z &= 7 \\ 2x - y + z &= 1 \\ x + 2y + 2z &= -1, \end{aligned}$$
$D = -43$, $D_x = -43$, $D_y = 0$, and $D_z = 43$. What is the solution set of the system?

*Use Cramer's rule to solve each linear system in two variables. See Example 5.*

**27.** $\begin{aligned} 3x + 5y &= -5 \\ -2x + 3y &= 16 \end{aligned}$

**28.** $\begin{aligned} 5x + 2y &= -3 \\ 4x - 3y &= -30 \end{aligned}$

**29.** $\begin{aligned} 8x + 3y &= 1 \\ 6x - 5y &= 2 \end{aligned}$

**30.** $\begin{aligned} 3x - y &= 9 \\ 2x + 5y &= 8 \end{aligned}$

**31.** $\begin{aligned} 2x + 3y &= 4 \\ 5x + 6y &= 7 \end{aligned}$

**32.** $\begin{aligned} 4x + 5y &= 6 \\ 7x + 8y &= 9 \end{aligned}$

*Use Cramer's rule where applicable to solve each linear system in three variables. See Examples 6 and 7.*

**33.** $\begin{aligned} 2x + 3y + 2z &= 15 \\ x - y + 2z &= 5 \\ x + 2y - 6z &= -26 \end{aligned}$

**34.** $\begin{aligned} x - y + 6z &= 19 \\ 3x + 3y - z &= 1 \\ x + 9y + 2z &= -19 \end{aligned}$

**35.** $\begin{aligned} 2x - 3y + 4z &= 8 \\ 6x - 9y + 12z &= 24 \\ -4x + 6y - 8z &= -16 \end{aligned}$

**36.** $\begin{aligned} 7x + y - z &= 4 \\ 2x - 3y + z &= 2 \\ -6x + 9y - 3z &= -6 \end{aligned}$

**37.** $\begin{aligned} 3x + 5z &= 0 \\ 2x + 3y &= 1 \\ -y + 2z &= -11 \end{aligned}$

**38.** $\begin{aligned} -x + 2y &= 4 \\ 3x + y &= -5 \\ 2x + z &= -1 \end{aligned}$

**39.** $\begin{aligned} x - 3y &= 13 \\ 2y + z &= 5 \\ -x + z &= -7 \end{aligned}$

**40.** $\begin{aligned} -5x - y &= -10 \\ 3x + 2y + z &= -3 \\ -y - 2z &= -13 \end{aligned}$

*Use a graphing calculator and the approach described with Figure 4 to solve each system using Cramer's rule.*

**41.** $\begin{aligned} x + 2y + z &= 10 \\ 2x - y - 3z &= -20 \\ -x + 4y + z &= 18 \end{aligned}$

**42.** $\begin{aligned} 2x + y + 3z &= 1 \\ x - 2y + z &= -3 \\ -3x + y - 2z &= -4 \end{aligned}$

**43.** $$\begin{aligned} -8w + 4x - 2y + z &= -28 \\ -w + x - y + z &= -10 \\ w + x + y + z &= -4 \\ 27w + 9x + 3y + z &= 2 \end{aligned}$$

**44.** $$\begin{aligned} 5w + 2x - 3y + z &= 4.7 \\ -2w + x + 2y - z &= -3.2 \\ w + 3x - y + 2z &= 2.1 \\ 2w + x - 5y + 3z &= 3.4 \end{aligned}$$

There is another method for evaluating a $3 \times 3$ determinant. Refer to Example 2, and copy the first two columns to the right of the original determinant to obtain

$$\begin{vmatrix} 1 & 3 & -2 & 1 & 3 \\ -1 & -2 & -3 & -1 & -2 \\ 1 & 1 & 2 & 1 & 1 \end{vmatrix}.$$

Multiply along the diagonals as shown, placing the product at the end of the arrow.

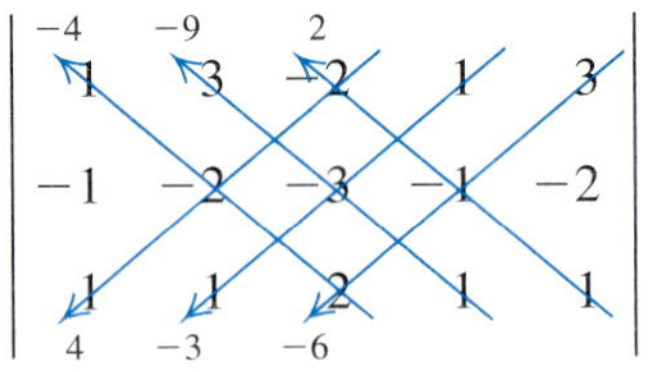


Add the top numbers: $-4 - 9 + 2 = -11.$

Add the bottom numbers: $4 - 3 - 6 = -5.$

Find the *difference* between these sums to obtain the final answer:

$$-11 - (-5) = -6.$$

*Use this method to find each determinant in the indicated exercise.*

**45.** Exercise 15 **46.** Exercise 16 **47.** Exercise 17

**48.** Exercise 18 **49.** Exercise 19 **50.** Exercise 20

*Solve each equation by finding an expression for the determinant on the left, and then solving using the methods of Chapter 2.*

**51.** $\begin{vmatrix} 4 & x \\ 2 & 3 \end{vmatrix} = 8$ **52.** $\begin{vmatrix} 5 & 3 \\ x & x \end{vmatrix} = 20$ **53.** $\begin{vmatrix} x & 4 \\ x & -3 \end{vmatrix} = 0$

**54.** Look at the coefficients and constants in the systems in Exercises 31 and 32. Notice that in both cases, the six numbers are consecutive integers. Make up a system having this same pattern for its coefficients and constants, and solve it using Cramer's rule. Compare the solutions in Exercises 31, 32, and here. What do you notice?

**55.** Use Cramer's rule to prove that the following system has solution set $\{(-1, 2)\}$.

$$\begin{aligned} ax + (a + 1)y &= a + 2 \\ (a + 3)x + (a + 4)y &= a + 5, \quad \text{where } D \neq 0. \end{aligned}$$

**56.** Under what conditions can a system *not* be solved using Cramer's rule?

## RELATING CONCEPTS (EXERCISES 57–64)

**For Individual or Group Work**

*In this section we have seen how determinants can be used to solve systems of equations. There are other applications of determinants. Here, we show how a determinant can be used to find the area of a triangle if we know the coordinates of its vertices.*

*Suppose that $A(x_1, y_1)$, $B(x_2, y_2)$, and $C(x_3, y_3)$ are the coordinates of the vertices of triangle ABC in the coordinate plane. Then it can be shown that the area of the triangle is given by the absolute value of*

$$\frac{1}{2}\begin{vmatrix} x_1 & y_1 & 1 \\ x_2 & y_2 & 1 \\ x_3 & y_3 & 1 \end{vmatrix}.$$

***Work Exercises 57–60 in order.***

**57.** Sketch triangle $ABC$ in the coordinate plane, given that the coordinates of $A$ are $(0, 0)$, of $B$ are $(-3, -4)$, and of $C$ are $(2, -2)$.

**58.** Write the determinant expression described above that gives the area of triangle $ABC$ described in Exercise 57.

**59.** Evaluate the absolute value of the determinant expression in Exercise 58 to find the area.

**60.** Use the determinant expression described above to find the area of the triangle with vertices at $(3, 8)$, $(-1, 4)$, and $(0, 1)$.

*Here is yet another application of determinants. Recall the formula for slope and the point-slope form of the equation of a line from Chapters 3 and 7. Use these formulas to* ***work Exercises 61–64 in order*** *and see how a determinant can be used in writing the equation of a line.*

**61.** Write the expression for the slope of a line passing through the points $(x_1, y_1)$ and $(x_2, y_2)$.

**62.** Using the expression from Exercise 61 as $m$, and the point $(x_1, y_1)$, write the point-slope form of the equation of the line.

**63.** Using the equation obtained in Exercise 62, multiply both sides by $x_2 - x_1$, and write the equation so that 0 is on the right side.

**64.** Consider the *determinant equation*

$$\begin{vmatrix} x & y & 1 \\ x_1 & y_1 & 1 \\ x_2 & y_2 & 1 \end{vmatrix} = 0.$$

Expand by minors on the left and show that this determinant equation yields the same result that you obtained in Exercise 63.

# Answers to Selected Exercises

In this section we provide the answers that we think most students will obtain when they work the exercises using the methods explained in the text. If your answer does not look exactly like the one given here, it is not necessarily wrong. In many cases there are equivalent forms of the answer that are correct. For example, if the answer section shows $\frac{3}{4}$ and your answer is .75, you have obtained the right answer but written it in a different (yet equivalent) form. Unless the directions specify otherwise, .75 is just as valid an answer as $\frac{3}{4}$.

In general, if your answer does not agree with the one given in the text, see whether it can be transformed into the other form. If it can, then it is the correct answer. If you still have doubts, talk with your instructor. You might also want to obtain a copy of the *Student's Solutions Manual* that goes with this book. Your college bookstore either has this manual or can order it for you.

## CHAPTER 1 THE REAL NUMBER SYSTEM

### Section 1.1 (page 10)

**Exercises** **1.** true **3.** false; The fraction $\frac{17}{51}$ is written in lowest terms as $\frac{1}{3}$. **5.** false; *Product* refers to multiplication, so the product of 8 and 2 is 16. **7.** prime **9.** composite; $2 \cdot 2 \cdot 2 \cdot 2 \cdot 2 \cdot 2$ **11.** composite; $2 \cdot 7 \cdot 13 \cdot 19$ **13.** neither **15.** composite; $2 \cdot 3 \cdot 5$ **17.** composite; $2 \cdot 2 \cdot 5 \cdot 5 \cdot 5$ **19.** composite; $2 \cdot 2 \cdot 31$ **21.** prime **23.** $\frac{1}{2}$ **25.** $\frac{5}{6}$ **27.** $\frac{3}{10}$ **29.** $\frac{6}{5}$ **31.** C **33.** $\frac{24}{35}$ **35.** $\frac{6}{25}$ **37.** $\frac{6}{5}$ or $1\frac{1}{5}$ **39.** $\frac{232}{15}$ or $15\frac{7}{15}$ **41.** $\frac{10}{3}$ or $3\frac{1}{3}$ **43.** 12 **45.** $\frac{1}{16}$ **47.** $\frac{84}{47}$ or $1\frac{37}{47}$ **49.** To multiply two fractions, multiply their numerators to get the numerator of the product and multiply their denominators to get the denominator of the product. For example, $\frac{2}{3} \cdot \frac{8}{5} = \frac{2 \cdot 8}{3 \cdot 5} = \frac{16}{15}$. To divide two fractions, replace the divisor with its reciprocal and then multiply. For example, $\frac{2}{5} \div \frac{7}{9} = \frac{2}{5} \cdot \frac{9}{7} = \frac{2 \cdot 9}{5 \cdot 7} = \frac{18}{35}$. **51.** $\frac{2}{3}$ **53.** $\frac{8}{9}$ **55.** $\frac{43}{8}$ or $5\frac{3}{8}$ **57.** $\frac{2}{3}$ **59.** $\frac{17}{36}$ **61.** $\frac{11}{12}$ **63.** 6 cups **65.** 34 dollars **67.** $\frac{9}{16}$ in. **69.** $618\frac{3}{4}$ ft **71.** $5\frac{5}{24}$ in. **73.** $\frac{1}{3}$ cup **75.** $\frac{1}{20}$ **77.** more than $1\frac{1}{25}$ million **79. (a)** $\frac{1}{2}$ **(b)** $\frac{1}{4}$ **(c)** $\frac{1}{3}$ **(d)** $\frac{1}{6}$

### Section 1.2 (page 20)

**Exercises** **1.** false; $4 + 3(8 - 2) = 4 + 3 \cdot 6 = 4 + 18 = 22$. The common error leading to 42 is adding 4 to 3 and then multiplying by 6. One must follow the order of operations. **3.** false; The correct interpretation is $4 = 16 - 12$. **5.** 49 **7.** 144 **9.** 64 **11.** 1000 **13.** 81 **15.** 1024 **17.** $\frac{16}{81}$ **19.** .000064 **21.** Write the base as a factor the number of times indicated by the exponent. For example, $6^3 = 6 \cdot 6 \cdot 6 = 216$. **23.** 32 **25.** $\frac{49}{30}$ or $1\frac{19}{30}$ **27.** 12 **29.** 42 **31.** 95 **33.** 90 **35.** 14 **37.** 9

**39.** Begin by squaring 2. Then subtract 1, to get a result of $4 - 1 = 3$ within the parentheses. Next, raise 3 to the third power to get $3^3 = 27$. Multiply this result by 3 to obtain 81. Finally, add this result to 4 to get the final answer, 85. **41.** $16 \le 16$; true **43.** $61 \le 60$; false **45.** $0 \ge 0$; true **47.** $45 \ge 46$; false **49.** $66 > 72$; false **51.** $2 \ge 3$; false **53.** $3 \ge 3$; true **55.** $15 = 5 + 10$ **57.** $9 > 5 - 4$ **59.** $16 \ne 19$ **61.** $2 \le 3$ **63.** Seven is less that nineteen; true **65.** Three is not equal to six; true **67.** Eight is greater than or equal to eleven; false **69.** Answers will vary. One example is $5 + 3 \ge 2 \cdot 2$. **71.** $30 > 5$ **73.** $3 \le 12$ **75.** is younger than **77.** The inequality symbol $\ge$ implies a true statement if 12 equals 12 *or* if 12 is greater than 12. **79.** 1998, 1999 **81.** **(a)** \$16.96 **(b)** approximately 31.5% **83.** $3 \cdot (6 + 4) \cdot 2 = 60$ **85.** $10 - (7 - 3) = 6$ **87.** $(8 + 2)^2 = 100$

## Section 1.3 (page 27)

**Exercises** **1.** 10 **3.** $12 + x$; 21 **5.** no **7.** $2x^3 = 2 \cdot x \cdot x \cdot x$, while $2x \cdot 2x \cdot 2x = (2x)^3$. **9.** The exponent 2 applies only to its base, which is $x$. (The expression $(4x)^2$ would require multiplying 4 by $x = 3$ first.) **11.** Answers will vary. Two such pairs are $x = 0$, $y = 6$ and $x = 1$, $y = 4$. To determine them, choose a value for $x$, substitute it into the expression $2x + y$, and then subtract the value of $2x$ from 6. **13.** **(a)** 13 **(b)** 15 **15.** **(a)** 20 **(b)** 30 **17.** **(a)** 64 **(b)** 144 **19.** **(a)** $\frac{5}{3}$ **(b)** $\frac{7}{3}$ **21.** **(a)** $\frac{7}{8}$ **(b)** $\frac{13}{12}$ **23.** **(a)** 52 **(b)** 114 **25.** **(a)** 25.836 **(b)** 38.754 **27.** **(a)** 24 **(b)** 28 **29.** **(a)** 12 **(b)** 33 **31.** **(a)** 6 **(b)** $\frac{9}{5}$ **33.** **(a)** $\frac{4}{3}$ **(b)** $\frac{13}{6}$ **35.** **(a)** $\frac{2}{7}$ **(b)** $\frac{16}{27}$ **37.** **(a)** 12 **(b)** 55 **39.** **(a)** 1 **(b)** $\frac{28}{17}$ **41.** **(a)** 3.684 **(b)** 8.841 **43.** $12x$ **45.** $x + 7$ **47.** $x - 2$ **49.** $7 - x$ **51.** $x - 6$ **53.** $\frac{12}{x}$ **55.** $6(x - 4)$ **57.** No, it is a connective word that joins the two factors: the number and 6. **59.** yes **61.** no **63.** yes **65.** yes **67.** yes **69.** $x + 8 = 18$; 10 **71.** $16 - \frac{3}{4}x = 13$; 4 **73.** $2x + 1 = 5$; 2 **75.** $3x = 2x + 8$; 8 **77.** expression **79.** equation **81.** equation **83.** 128.02 ft **85.** 187.08 ft

## Section 1.4 (page 36)

**Exercises** **1.** 1,198,000 **3.** 925 **5.** $-5074$ **7.** $-11.35$ **9.** 4 **11.** 0 **13.** One example is $\sqrt{12}$. There are others. **15.** true **17.** true **19.** **(a)** 3, 7 **(b)** 0, 3, 7 **(c)** $-9$, 0, 3, 7 **(d)** $-9$, $-1\frac{1}{4}$, $-\frac{3}{5}$, 0, 3, 5.9, 7 **(e)** $-\sqrt{7}$, $\sqrt{5}$ **(f)** All are real numbers. **21.** The *natural numbers* are the numbers with which we count. An example is 1. The *whole numbers* are the natural numbers with 0 also included. An example is 0. The *integers* are the whole numbers and their negatives. An example is $-1$. The *rational numbers* are the numbers that can be represented by a quotient of integers, such as $\frac{1}{2}$. The *irrational numbers,* such as $\sqrt{2}$, cannot be represented as a quotient of integers. The *real numbers* include all positive numbers, negative numbers, and zero. All the numbers discussed are real. **23.**

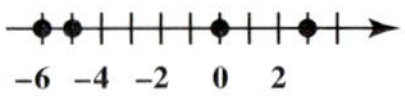


**25.**

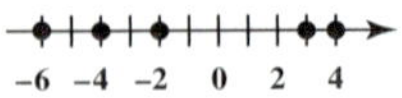


**27.**

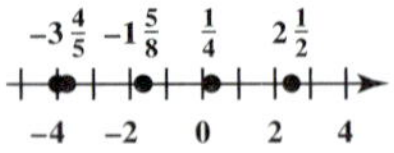


**29.** **(a)** A **(b)** A **(c)** B **(d)** B **31.** **(a)** 2 **(b)** 2 **33.** **(a)** $-6$ **(b)** 6 **35.** 6 **37.** $-12$ **39.** 3 **41.** $-12$ **43.** $-8$ **45.** 3 **47.** $|-3|$ or 3 **49.** $-|-6|$ or $-6$ **51.** $|5 - 3|$ or 2 **53.** true **55.** true **57.** true **59.** false **61.** true **63.** false **65.** softwood plywood, 1998 to 1999 **67.** paving mixtures and blocks, 1998 to 1999

In Exercises 69–73, answers will vary.

**69.** $\frac{1}{2}, \frac{5}{8}, 1\frac{3}{4}$ **71.** $-3\frac{1}{2}, -\frac{2}{3}, \frac{3}{7}$ **73.** $\sqrt{5}, \pi, -\sqrt{3}$ **75.** This is not true. The absolute value of 0 is 0, and 0 is not positive. A more accurate way of describing absolute value is to say that *absolute value is never negative*, or *absolute value is always nonnegative.*

## Section 1.5 (page 46)

**Exercises** **1.** negative

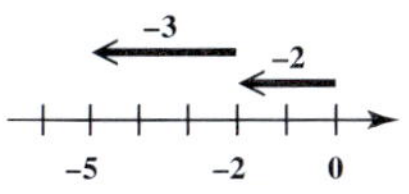


**3.** negative

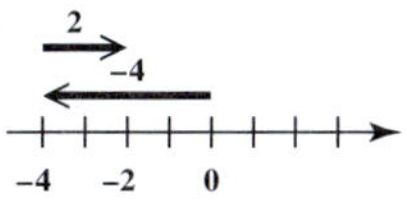


**5.** To add two numbers with the same sign, add their absolute values and keep the same sign for the sum. For example, $3 + 4 = 7$ and $-3 + (-4) = -7$. To add two numbers with different signs, subtract the smaller absolute value from the larger absolute value, and use the sign of the number with the larger absolute value. For example, $6 + (-4) = 2$ and $(-6) + 4 = -2$. **7.** $-8$ **9.** $-12$ **11.** 2 **13.** $-2$ **15.** 4 **17.** 12 **19.** 5 **21.** 2 **23.** $-9$ **25.** 0 **27.** $\frac{1}{2}$ **29.** $-\frac{19}{24}$ **31.** $-\frac{3}{4}$ **33.** $-7.7$ **35.** $-8$ **37.** 0 **39.** $-20$ **41.** $-3$ **43.** $-4$ **45.** $-8$ **47.** $-14$ **49.** 9 **51.** $-4$ **53.** 4 **55.** $\frac{3}{4}$ **57.** $-\frac{11}{8}$ or $-1\frac{3}{8}$ **59.** $\frac{15}{8}$ or $1\frac{7}{8}$ **61.** 11.6 **63.** $-9.9$ **65.** 10 **67.** $-5$ **69.** 11 **71.** $-10$ **73.** 22 **75.** $-2$ **77.** $-6$ **79.** $-12$ **81.** $-5.90617$ **83.** $-5 + 12 + 6$; 13 **85.** $[-19 + (-4)] + 14$; $-9$ **87.** $[-4 + (-10)] + 12$; $-2$ **89.** $[8 + (-18)] + 4$; $-6$ **91.** $4 - (-8)$; 12 **93.** $-2 - 8$; $-10$ **95.** $[9 + (-4)] - 7$; $-2$ **97.** $[8 - (-5)] - 12$; 1 **99.** $-3.4$ (billion dollars) **101.** $-2.7$ (billion dollars) **103.** 50,395 ft **105.** 1345 ft **107.** 136 ft **109.** $-12$ **111.** 45°F **113.** $-58$°F **115.** 27 ft **117.** +31,900 ft **119.** $-\$107$

## Section 1.6 (page 60)

**Exercises** **1.** greater than 0 **3.** greater than 0 **5.** less than 0 **7.** greater than 0 **9.** equal to 0 **11.** 12 **13.** $-12$ **15.** 120 **17.** $-33$ **19.** $-165$ **21.** $\frac{5}{12}$ **23.** $-\frac{1}{6}$ **25.** 6 **27.** $-32, -16, -8, -4, -2, -1, 1, 2, 4, 8, 16, 32$ **29.** $-40, -20, -10, -8, -5, -4, -2, -1, 1, 2, 4, 5, 8, 10, 20, 40$ **31.** $-31, -1, 1, 31$ **33.** 3 **35.** $-5$ **37.** 7 **39.** $-6$ **41.** $\frac{32}{3}$ or $10\frac{2}{3}$ **43.** $-4$ **45.** 0 **47.** undefined **49.** $-11$ **51.** $-2$ **53.** 35 **55.** 6 **57.** $-18$ **59.** 67 **61.** $-8$ **63.** 3 **65.** 7 **67.** 4 **69.** $-3$ **71.** 10 **73.** negative **75.** 47 **77.** 72 **79.** $-\frac{78}{25}$ **81.** 0 **83.** $-23$ **85.** 2 **87.** $9 + (-9)(2)$; $-9$ **89.** $-4 - 2(-1)(6)$; 8 **91.** $(1.5)(-3.2) - 9$; $-13.8$ **93.** $12[9 - (-8)]$; 204 **95.** $\frac{-12}{-5 + (-1)}$; 2 **97.** $\frac{15 + (-3)}{4(-3)}$; $-1$ **99.** $2(8 + 9)$; 34 **101.** $.20(-5 \cdot 6)$; $-6$ **103.** $\frac{x}{3} = -3$; $-9$ **105.** $x - 6 = 4$; 10 **107.** $x + 5 = -5$; $-10$ **109.** $8\frac{2}{5}$ **111.** 2 **113.** 0 **115. (a)** 6 is divisible by 2. **(b)** 9 is not divisible by 2. **117. (a)** 64 is divisible by 4. **(b)** 35 is not divisible by 4. **119. (a)** 2 is divisible by 2 and $1 + 5 + 2 + 4 + 8 + 2 + 2 = 24$ is divisible by 3. **(b)** While 0 is divisible by 2, $2 + 8 + 7 + 3 + 5 + 9 + 0 = 34$ is not divisible by 3. **121. (a)** $4 + 1 + 1 + 4 + 1 + 0 + 7 = 18$ is divisible by 9. **(b)** $2 + 2 + 8 + 7 + 3 + 2 + 1 = 25$ is not divisible by 9.

## Summary Exercises on Operations with Real Numbers (page 63)

**1.** $-16$ **2.** 4 **3.** 0 **4.** $-24$ **5.** $-17$ **6.** 76 **7.** $-18$ **8.** 90 **9.** 38 **10.** 4 **11.** $-5$ **12.** 5 **13.** $-\frac{7}{2}$ or $-3\frac{1}{2}$ **14.** 4 **15.** 13 **16.** $\frac{5}{4}$ or $1\frac{1}{4}$ **17.** 9 **18.** $\frac{37}{10}$ or $3\frac{7}{10}$ **19.** 0 **20.** 25 **21.** 14 **22.** 0 **23.** $-4$ **24.** $\frac{6}{5}$ or $1\frac{1}{5}$ **25.** $-1$ **26.** $\frac{52}{37}$ or $1\frac{15}{37}$ **27.** $\frac{17}{16}$ or $1\frac{1}{16}$ **28.** $-\frac{2}{3}$ **29.** 3.33 **30.** 1.02 **31.** $-13$ **32.** 0 **33.** 24 **34.** $-7$ **35.** 37 **36.** $-3$ **37.** $-1$ **38.** $\frac{1}{2}$ **39.** $-\frac{5}{13}$ **40.** 5

## Section 1.7 (page 70)

**Exercises** **1.** $-12$; commutative property **3.** 3; commutative property **5.** 7; associative property **7.** 8; associative property **9.** **(a)** B **(b)** F **(c)** C **(d)** I **(e)** B **(f)** D, F **(g)** B **(h)** A **(i)** G **(j)** H **11.** commutative property **13.** associative property **15.** associative property **17.** inverse property **19.** inverse property **21.** identity property **23.** commutative property **25.** distributive property **27.** identity property **29.** distributive property **31.** identity property **33.** The identity properties allow us to perform an operation so that the result is the number we started with. The inverse properties allow us to perform an operation that gives an identity element as a result. **35.** 150 **37.** 2010 **39.** 400 **41.** 1400 **43.** 11 **45.** 0 **47.** $-.38$ **49.** 1 **51.** Subtraction is not associative. **53.** The expression following the first equals sign should be $-3(4) - 3(-6)$. The student forgot that 6 should be preceded by a $-$ sign. The correct work is $-3(4 - 6) = -3(4) - 3(-6) = -12 + 18 = 6$. **55.** 85 **57.** $4t + 12$ **59.** $-8r - 24$ **61.** $-5y + 20$ **63.** $-16y - 20z$ **65.** $8(z + w)$ **67.** $7(2v + 5r)$ **69.** $24r + 32s - 40y$ **71.** $-24x - 9y - 12z$ **73.** $5(x + 3)$ **75.** $-4t - 3m$ **77.** $5c + 4d$ **79.** $3q - 5r + 8s$ **81.** Answers will vary; for example, "putting on your socks" and "putting on your shoes." **83.** 0 **84.** $-3(5) + (-3)(-5)$ **85.** $-15$ **86.** We must interpret $(-3)(-5)$ as 15, since it is the additive inverse of $-15$. **87.** **(a)** no **(b)** distributive property

## Section 1.8 (page 76)

**Exercises** **1.** $4r + 11$ **3.** $5 + 2x - 6y$ **5.** $-7 + 3p$ **7.** $2 - 3x$ **9.** $-12$ **11.** 5 **13.** 1 **15.** $-1$ **17.** 74 **19.** like **21.** unlike **23.** like **25.** unlike **27.** $17y$ **29.** $-6a$ **31.** $13b$ **33.** $7k + 15$ **35.** $-4y$ **37.** $2x + 6$ **39.** $14 - 7m$ **41.** $-17 + x$ **43.** $23x$ **45.** $9y^2$ **47.** $-14p^3 + 5p^2$ **49.** $8x + 15$ **51.** $5x + 15$ **53.** $-4y + 22$ **55.** $-16y + 63$ **57.** $4r + 15$ **59.** $12k - 5$ **61.** $-2k - 3$ **63.** $4k - 7$ **65.** $-23.7y - 12.6$ **67.** $(x + 3) + 5x$; $6x + 3$ **69.** $(13 + 6x) - (-7x)$; $13 + 13x$ **71.** $2(3x + 4) - (-4 + 6x)$; 12 **73.** 2, 3, 4, 5 **74.** 1 **75.** **(a)** 1, 2, 3, 4 **(b)** 3, 4, 5, 6 **(c)** 4, 5, 6, 7 **76.** The value of $x + b$ also increases by 1 unit. **77.** **(a)** 2, 4, 6, 8 **(b)** 2, 5, 8, 11 **(c)** 2, 6, 10, 14 **78.** $m$ **79.** **(a)** 7, 9, 11, 13 **(b)** 5, 8, 11, 14 **(c)** 1, 5, 9, 13; In comparison, we see that while the values themselves are different, the number of units of increase is the same as the corresponding parts of Exercise 77. **80.** $m$ **81.** Apples and oranges are examples of unlike fruits, just like $x$ and $y$ are unlike terms. We cannot add $x$ and $y$ to get an expression any simpler than $x + y$; we cannot add, for example, 2 apples and 3 oranges to obtain 5 fruits that are all alike. **83.** Wording will vary. One example is "the difference between 9 times a number and the sum of the number and 2."

## Chapter 1 Review Exercises (page 84)

**1.** $\frac{3}{4}$ **3.** $\frac{9}{40}$ **5.** 625 **7.** .0000000032 **9.** 27 **11.** 39 **13.** true **15.** false **17.** $5 + 2 \neq 10$ **19.** 30 **21.** 14 **23.** $x + 6$ **25.** $6x - 9$ **27.** yes **29.** $2x - 6 = 10$; 8 **31.**

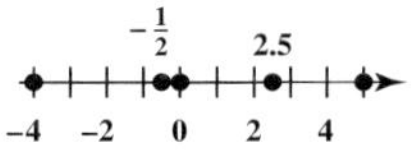


**33.** rational numbers, real numbers **35.** $-10$ **37.** $-\frac{3}{4}$ **39.** true **41.** true **43.** **(a)** 9 **(b)** 9 **45.** **(a)** $-6$ **(b)** 6 **47.** 12 **49.** $-19$ **51.** $-6$ **53.** $-17$ **55.** $-21.8$ **57.** $-10$ **59.** $-11$ **61.** 7 **63.** 10.31 **65.** 2 **67.** $(-31 + 12) + 19$; 0 **69.** $-4 - (-6)$; 2 **71.** $-2$ **73.** \$26.25 **75.** $-\$29$ **77.** It gained 4 yd. **79.** 36 **81.** $\frac{1}{2}$ **83.** $-20$ **85.** $-24$ **87.** 4 **89.** $-\frac{3}{4}$ **91.** $-1$ **93.** 1 **95.** $-18$ **97.** 125 **99.** $-4(5) - 9$; $-29$ **101.** $\frac{12}{8 + (-4)}$; 3 **103.** $8x = -24$; $-3$ **105.** 32 **107.** identity property **109.** inverse property **111.** associative property **113.** distributive property **115.** $7(y + 2)$ **117.** $3(2s + 5y)$ **119.** $25 - (5 - 2) = 22$ and $(25 - 5) - 2 = 18$. Because different groupings lead to different results, we conclude that in general subtraction is not associative. **121.** $11m$ **123.** $16p^2 + 2p$ **125.** $-2m + 29$ **127.** C **129.** A **131.** 16 **133.** $\frac{8}{3}$ or $2\frac{2}{3}$ **135.** 2 **137.** $-\frac{3}{2}$ or $-1\frac{1}{2}$ **139.** $-\frac{28}{15}$ or $-1\frac{13}{15}$ **141.** $8x^2 - 21y^2$ **143.** When dividing 0 *by* a nonzero number, the quotient will be 0. However, dividing a number *by* 0 is undefined. **145.** $5(x + 7)$; $5x + 35$

## Chapter 1 Test (page 89)

[1.1] **1.** $\frac{7}{11}$ **2.** $\frac{241}{120}$ or $2\frac{1}{120}$ **3.** $\frac{19}{18}$ or $1\frac{1}{18}$ **4. (a)** 492 million **(b)** 861 million [1.2] **5.** true [1.4] **6.** 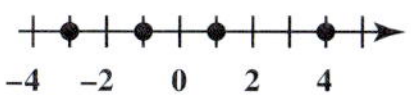
**7.** rational numbers, real numbers **8.** If $-8$ and $-1$ are both graphed on a number line, we see that the point for $-8$ is to the *left* of the point for $-1$. This indicates $-8 < -1$. [1.6] **9.** $\frac{-6}{2 + (-8)}$; 1 [1.1, 1.4–1.6] **10.** 4 **11.** $-\frac{17}{6}$ or $-2\frac{5}{6}$ **12.** 2 **13.** 6 **14.** 108 **15.** 3 **16.** $\frac{30}{7}$ or $4\frac{2}{7}$ [1.3, 1.5, 1.6] **17.** 6 **18.** 4 [1.4–1.6] **19.** $-70$ **20.** 3 **21.** 7000 m **22.** 15 **23. (a)** \$25.1 billion **(b)** $-$\$11.3 billion **(c)** \$5.0 billion **(d)** $-$\$2.2 billion [1.7] **24.** B **25.** D **26.** E **27.** A **28.** C **29.** distributive property **30. (a)** $-18$ **(b)** $-18$ **(c)** The distributive property assures us that the answers must be the same, because $a(b + c) = ab + ac$ for all $a$, $b$, $c$. [1.8] **31.** $21x$ **32.** $15x - 3$

# CHAPTER 2 LINEAR EQUATIONS AND INEQUALITIES IN ONE VARIABLE; APPLICATIONS

## Section 2.1 (page 98)

Exercises **1.** A and C **3.** The addition property of equality says that the same number (or expression) added to each side of an equation results in an equivalent equation. Example: $-x$ can be added to each side of $2x + 3 = x - 5$ to get the equivalent equation $x + 3 = -5$. **5.** $\{12\}$ **7.** $\{31\}$ **9.** $\{-3\}$ **11.** $\{4\}$ **13.** $\{-9\}$ **15.** $\{-10\}$ **17.** $\{-13\}$ **19.** $\{10\}$ **21.** $\{6.3\}$ **23.** $\{-16.9\}$ **25.** $\{-6\}$ **27.** $\{-2\}$ **29.** $\{4\}$ **31.** $\{0\}$ **33.** $\{-2\}$ **35.** $\{-7\}$ **37.** A and B; A sample answer might be, "A linear equation in one variable is an equation that can be written using only one variable term with the variable to the first power." **39.** $\{13\}$ **41.** $\{-4\}$ **43.** $\{0\}$ **45.** $\left\{\frac{7}{15}\right\}$ **47.** $\{7\}$ **49.** $\{-4\}$ **51.** $\{13\}$ **53.** $\{29\}$ **55.** $\{18\}$ **57.** $\{12\}$ **59.** Answers will vary. One example is $x - 6 = -8$. **61.** $3x = 2x + 17$; $\{17\}$ **63.** $7x - 6x = -9$; $\{-9\}$

## Section 2.2 (page 104)

Exercises **1.** The multiplication property of equality says that the same nonzero number (or expression) multiplied on each side of the equation results in an equivalent equation. Example: Multiplying each side of $7x = 4$ by $\frac{1}{7}$ gives the equivalent equation $x = \frac{4}{7}$. **3.** C **5.** To get $x$ alone on the left side, divide each side by 4, the coefficient of $x$. **7.** $\frac{3}{2}$ **9.** 10 **11.** $-\frac{2}{9}$ **13.** $-1$ **15.** 6 **17.** $-4$ **19.** .12 **21.** $-1$ **23.** $\{6\}$ **25.** $\left\{\frac{15}{2}\right\}$ **27.** $\{-5\}$ **29.** $\{-4\}$ **31.** $\left\{-\frac{18}{5}\right\}$ **33.** $\{12\}$ **35.** $\{0\}$ **37.** $\{40\}$ **39.** $\{-12.2\}$ **41.** $\{-48\}$ **43.** $\{72\}$ **45.** $\{-35\}$ **47.** $\{14\}$ **49.** $\{18\}$ **51.** $\left\{-\frac{27}{35}\right\}$ **53.** $\{-12\}$ **55.** $\left\{\frac{3}{4}\right\}$ **57.** $\{3\}$ **59.** $\{-5\}$ **61.** $\{7\}$ **63.** $\{0\}$ **65.** $\left\{-\frac{3}{5}\right\}$ **67.** Answers will vary. One example is $\frac{3}{2}x = -6$. **69.** $4x = 6$; $\left\{\frac{3}{2}\right\}$ **71.** $\frac{x}{-5} = 2$; $\{-10\}$

## Section 2.3 (page 112)

Exercises **1.** *Step 1:* Clear parentheses and combine like terms, as needed. *Step 2:* Use the addition property to get all variable terms on one side of the equation and all numbers on the other. Then combine like terms. *Step 3:* Use the multiplication property to get the equation in the form $x =$ a number. *Step 4:* Check the solution. Examples will vary. **3.** D **5.** $\{-1\}$ **7.** $\{5\}$ **9.** $\left\{-\frac{5}{3}\right\}$ **11.** $\left\{\frac{4}{3}\right\}$ **13.** $\{5\}$

**15.** $\emptyset$ **17.** {all real numbers} **19.** {1} **21.** $\emptyset$ **23.** {5} **25.** {0} **27.** $\left\{-\frac{7}{5}\right\}$ **29.** {120} **31.** {6} **33.** {15,000} **35.** {8} **37.** {0} **39.** {4} **41.** {20} **43.** {all real numbers} **45.** $\emptyset$ **47.** $11 - q$ **49.** $x + 7$ **51.** $a + 12$; $a - 5$ **53.** $\frac{t}{5}$

## Summary Exercises on Solving Linear Equations (page 114)

**1.** {−5} **2.** {4} **3.** {−5.1} **4.** {12} **5.** {−25} **6.** {−6} **7.** {−3} **8.** {−16} **9.** {7} **10.** $\left\{-\frac{96}{5}\right\}$ **11.** {5} **12.** {23.7} **13.** {all real numbers} **14.** {1} **15.** {−6} **16.** $\emptyset$ **17.** {6} **18.** {3} **19.** $\emptyset$ **20.** $\left\{\frac{7}{3}\right\}$ **21.** {25} **22.** {−10.8} **23.** {3} **24.** {7} **25.** {2} **26.** {all real numbers} **27.** {−2} **28.** {70} **29.** $\left\{\frac{14}{17}\right\}$ **30.** $\left\{-\frac{5}{2}\right\}$

## Section 2.4 (page 122)

**Connections** **(page 122)** Polya's Step 1 corresponds to our Steps 1 and 2. Polya's Step 2 corresponds to our Step 3. Polya's Step 3 corresponds to our Steps 4 and 5. Polya's Step 4 corresponds to our Step 6. Trial and error or guessing and checking fit into Polya's Step 2, devising a plan.

**Exercises** **1.** The procedure should include the following steps: read the problem carefully; assign a variable to represent the unknown to be found, and write down variable expressions for any other unknown quantities; translate into an equation; solve the equation; state the answer; check your solution. **3.** D; there cannot be a fractional number of cars. **5.** 3 **7.** 6 **9.** −3 **11.** California: 59 screens; New York: 48 screens **13.** Democrats: 45; Republicans: 55 **15.** U2: \$109.7 million; 'N Sync: \$86.8 million **17.** wins: 61; losses: 21 **19.** 1950 Denver nickel: \$14.00; 1945 Philadelphia nickel: \$12.00 **21.** ice cream: 44,687.9 lb; topping: 537.1 lb **23.** 18 prescriptions **25.** peanuts: $22\frac{1}{2}$ oz; cashews: $4\frac{1}{2}$ oz **27.** Airborne Express: 3; Federal Express: 9; United Parcel Service: 1 **29.** gold: 10; silver: 13; bronze: 11 **31.** 36 million mi **33.** $A$ and $B$: 40°; $C$: 100° **35.** $k - m$ **37.** no **39.** $x - 1$ **41.** 18° **43.** 39° **45.** 50° **47.** 68, 69 **49.** 10, 12 **51.** 101, 102 **53.** 10, 11 **55.** 18 **57.** 15, 17, 19 **59.** \$2.78 billion; \$3.33 billion; \$3.53 billion

## Section 2.5 (page 133)

**Exercises** **1. (a)** The perimeter of a plane geometric figure is the distance around the figure. **(b)** The area of a plane geometric figure is the measure of the surface covered or enclosed by the figure. **3.** four **5.** area **7.** perimeter **9.** area **11.** area **13.** $P = 26$ **15.** $A = 64$ **17.** $b = 4$ **19.** $t = 5.6$ **21.** $I = 1575$ **23.** $B = 14$ **25.** $r = 2.6$ **27.** $A = 50.24$ **29.** $V = 150$ **31.** $V = 52$ **33.** $V = 7234.56$ **35.** about 154,000 $\text{ft}^2$ **37.** perimeter: 13 in., area: 10.5 $\text{in.}^2$ **39.** 132.665 $\text{ft}^2$ **41.** 23,800.10 $\text{ft}^2$ **43.** length: 36 in.; volume: 11,664 $\text{in.}^3$ **45.** 48°, 132° **47.** 51°, 51° **49.** 105°, 105° **51.** $t = \frac{d}{r}$ **53.** $b = \frac{A}{h}$ **55.** $d = \frac{C}{\pi}$ **57.** $H = \frac{V}{LW}$ **59.** $r = \frac{I}{pt}$ **61.** $h = \frac{2A}{b}$ **63.** $h = \frac{3V}{\pi r^2}$ **65.** $b = P - a - c$ **67.** $W = \frac{P - 2L}{2}$ or $W = \frac{P}{2} - L$ **69.** $m = \frac{y - b}{x}$ **71.** $y = \frac{C - Ax}{B}$ **73.** $r = \frac{M - C}{C}$

## Section 2.6 (page 141)

**Exercises** **1. (a)** C **(b)** D **(c)** B **(d)** A **3.** $\frac{6}{7}$ **5.** $\frac{18}{55}$ **7.** $\frac{5}{16}$ **9.** $\frac{4}{15}$ **11.** $\frac{3}{1}$ **13.** 17-oz size **15.** 64-oz can **17.** 500-count **19.** 28-oz size **21.** A ratio is a comparison, while a proportion is a statement that two ratios are equal. For example, $\frac{2}{3}$ is a ratio and $\frac{2}{3} = \frac{8}{12}$ is a proportion. **23.** true **25.** false **27.** true **29.** {35} **31.** {7} **33.** $\left\{\frac{45}{2}\right\}$ **35.** {2} **37.** {−1} **39.** {5}

**41.** $\left\{-\frac{31}{5}\right\}$ **43.** \$67.50 **45.** \$28.35 **47.** 4 ft **49.** 6.875 fluid oz **51.** \$670.48 **53.** 50,000 fish **55.** **(a)** $\frac{26}{100} = \frac{x}{350}$; \$91 million **(b)** \$112 million; \$11.2 million **(c)** \$119 million **57.** 4 **59.** 1 **61.** **(a)**

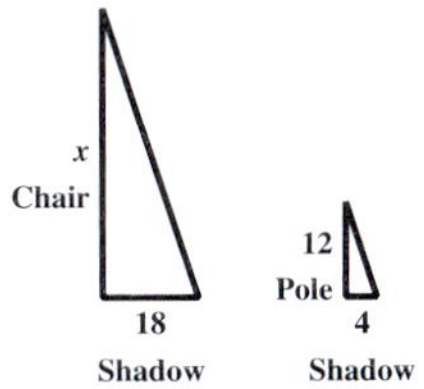


**(b)** 54 ft **63.** \$270

**65.** \$287 **67.** 30 **68.** **(a)** $5x = 12$ **(b)** $\left\{\frac{12}{5}\right\}$ **69.** $\left\{\frac{12}{5}\right\}$ **70.** Both methods give the same solution set.

## Section 2.7 (page 152)

**Exercises** **1.** 35 mL **3.** \$350 **5.** \$14.15 **7.** C **9.** **(a)** 375,000 **(b)** 575,000 **(c)** 275,000 **11.** 4.5% **13.** D **15.** 160 gal **17.** $53\frac{1}{3}$ kg **19.** 4 L **21.** $13\frac{1}{3}$ L **23.** 25 mL **25.** \$5000 at 3%; \$1000 at 5% **27.** \$40,000 at 3%; \$110,000 at 4% **29.** 25 fives **31.** fives: 84; tens: 42 **33.** 20 lb **35.** A **37.** 530 mi **39.** 4.059 hr **41.** 7.91 m per sec **43.** 8.42 m per sec **45.** 10 hr **47.** $2\frac{1}{2}$ hr **49.** 5 hr **51.** northbound: 60 mph; southbound: 80 mph **53.** 50 km per hr; 65 km per hr

## Section 2.8 (page 168)

**Connections** **(page 168)** The revenue is represented by $5x - 100$. The production cost is $125 + 4x$. The profit is represented by $R - C = (5x - 100) - (125 + 4x) = x - 225$. The solution of $x - 225 > 0$ is $x > 225$. To make a profit, more than 225 cassettes must be produced and sold.

**Exercises** **1.** Use a parenthesis if the symbol is $<$ or $>$. Use a square bracket if the symbol is $\leq$ or $\geq$. **3.** $x > -4$ **5.** $x \leq 4$

**7.** $(-\infty, 4]$ 

**9.** $(-\infty, -3)$ 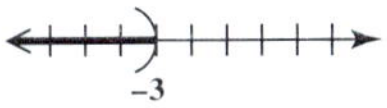


**11.** $(4, \infty)$ 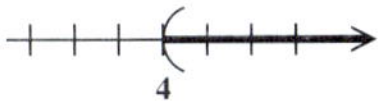


**13.** $[8, 10]$ 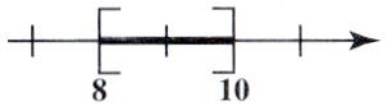


**15.** $(0, 10]$ 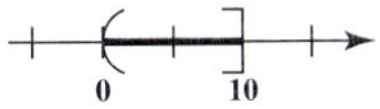


**17.** It would imply that $3 < -2$, a false statement.

**19.** $[1, \infty)$ 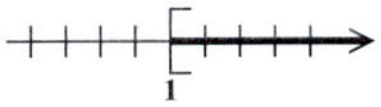


**21.** $[5, \infty)$ 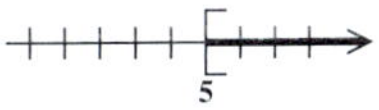


**23.** $(-\infty, -11)$ 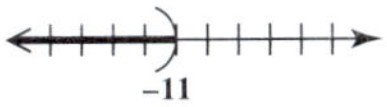


**25.** It must be reversed when multiplying or dividing by a negative number.

**27.** $(-\infty, 6)$ 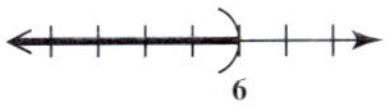


**29.** $[-10, \infty)$ 

**31.** $(-\infty, -3)$ 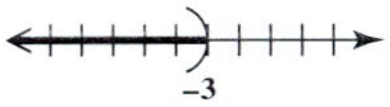


**33.** $(-\infty, 0]$ 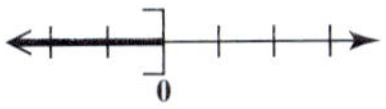


**35.** $(20, \infty)$ 

**37.** $[-3, \infty)$ 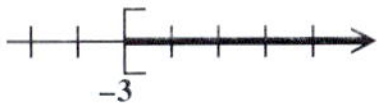


**39.** $[-5, \infty)$ 

**41.** $(-\infty, 1)$ 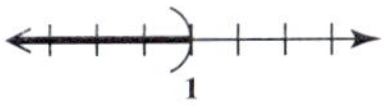


**43.** $(-\infty, 0]$ 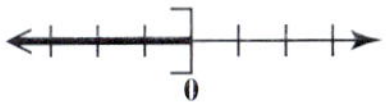


**45.** $[4, \infty)$ 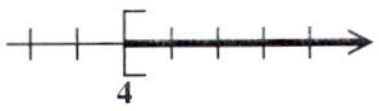


**47.** $(-\infty, 32)$ 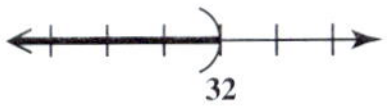


**49.** $\left(-\infty, \frac{76}{11}\right)$ 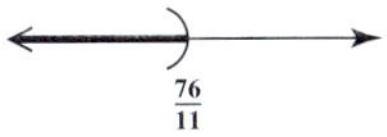


**51.** $\left[\frac{5}{12}, \infty\right)$ 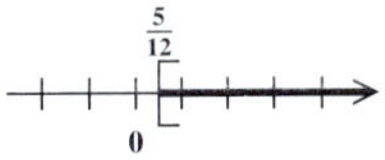


**53.** $(-21, \infty)$ 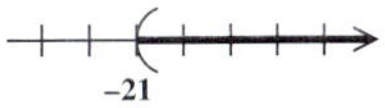


**55.** $-1 < x < 2$ **57.** $-1 < x \leq 2$

**59.** $[-1, 6]$ 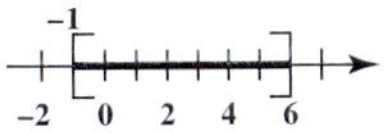


**61.** $\left(-\frac{11}{6}, -\frac{2}{3}\right)$ 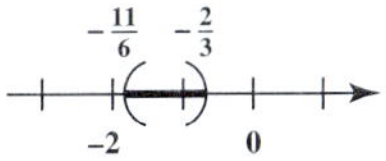


**63.** $(1, 3)$ 

**65.** $[-26, 6]$ 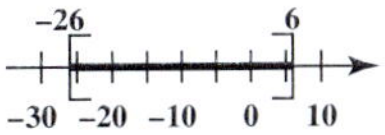


**67.** $[-3, 6]$ 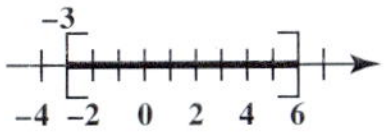


**69.** $\left[-\frac{24}{5}, 0\right]$ 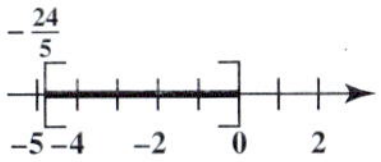

**71.** {4} 

**72.** $(4, \infty)$ 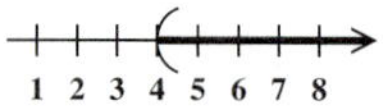

**73.** $(-\infty, 4)$ 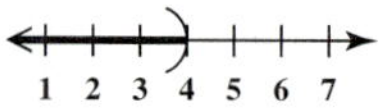

**74.** The graph would be all real numbers. **75.** The graph would be all real numbers. **76.** If a point on the number line satisfies an equation, the points on one side of that point will satisfy the corresponding less-than inequality, and the points on the other side will satisfy the corresponding greater-than inequality. **77.** 83 or greater **79.** all numbers greater than 16 **81.** It is never more than 86° Fahrenheit. **83.** 32 or greater **85.** 15 min **87.** 9.5 gal **89.** $x \geq 500$ **91.** **(a)** 140 to 184 lb **(b)** Answers will vary. **93.** from about 2:30 P.M. to 6:00 P.M. **95.** about 84°F–91°F

## Chapter 2 Review Exercises (page 178)

**1.** {6} **3.** {7} **5.** {11} **7.** {5} **9.** {5} **11.** $\left\{\frac{64}{5}\right\}$ **13.** {all real numbers} **15.** {all real numbers} **17.** $\emptyset$
**19.** Democrats: 75; Republicans: 45 **21.** Seven Falls: 300 ft; Twin Falls: 120 ft **23.** 11, 13 **25.** $A = 28$ **27.** $V = 904.32$
**29.** $h = \frac{2A}{b + B}$ **31.** 100°; 100° **33.** diameter: approximately 19.9 ft; radius: approximately 9.95 ft; area: approximately 311 $ft^2$
**35.** Not enough information is given. We also need the value of $B$. **37.** $\frac{5}{14}$ **39.** $\frac{1}{12}$ **41.** $\left\{-\frac{8}{3}\right\}$ **43.** $6\frac{2}{3}$ lb **45.** 375 km
**47.** 25.5-oz size **49.** 3.75 L **51.** 8.2 mph **53.** $2\frac{1}{2}$ hr **55.** $(-\infty, 7)$ 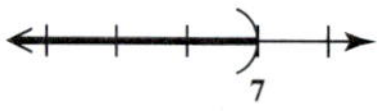

**57.** $[-3, \infty)$ 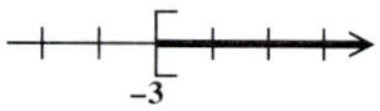

**59.** $[3, \infty)$ 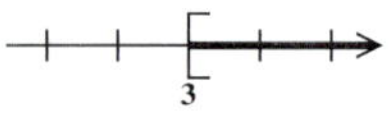

**61.** $(-\infty, -5)$ 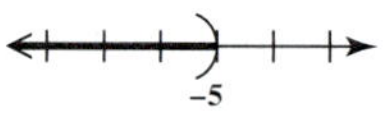

**63.** $\left[-2, \frac{3}{2}\right]$ 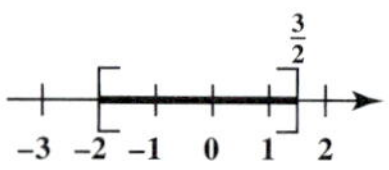

**65.** 88 or more
**67.** {7} **69.** $[-3, 3]$ **71.** {70} **73.** $\emptyset$ **75.** Since $-(8 + 4x) = -1(8 + 4x)$, the first step is to distribute $-1$ over *both* terms in the parentheses, to get $3 - 8 - 4x$ on the left side of the equation. The student got $3 - 8 + 4x$ instead. The correct solution set is $\{-2\}$.
**77.** Golden Gate Bridge: 4200 ft; Brooklyn Bridge: 1595 ft **79.** 8 qt **81.** 44 m

## Chapter 2 Test (page 182)

[2.1–2.3] **1.** $\{-6\}$ **2.** {21} **3.** $\emptyset$ **4.** {30} **5.** {all real numbers} [2.4] **6.** wins: 116; losses: 46 **7.** Hawaii: 4021 $mi^2$; Maui: 728 $mi^2$; Kauai: 551 $mi^2$ **8.** 50° [2.5] **9.** **(a)** $W = \frac{P - 2L}{2}$ or $W = \frac{P}{2} - L$ **(b)** 18 **10.** 75°, 75° [2.6] **11.** {6} **12.** $\{-29\}$
**13.** 8 slices for \$2.19 **14.** 2300 mi [2.7] **15.** \$8000 at 3%; \$14,000 at 4.5% **16.** 4 hr [2.8] **17.** $(-\infty, 4]$ 

**18.** $(-2, 6]$ 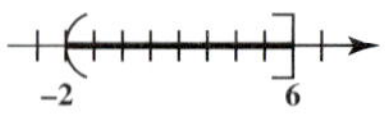

**19.** 83 or more **20.** When an inequality is multiplied or divided by a negative number, the direction of the inequality symbol must be reversed.

## Cumulative Review Exercises Chapters 1–2 (page 184)

[1.1] **1.** $\frac{3}{4}$ **2.** $\frac{37}{60}$ **3.** $\frac{48}{5}$ [1.2] **4.** $\frac{1}{2}x - 18$ **5.** $\frac{6}{x + 12} = 2$ [1.4] **6.** true [1.5–1.6] **7.** 11 **8.** $-8$ **9.** 28
[1.3] **10.** $-\frac{19}{3}$ [1.7] **11.** distributive property **12.** commutative property [1.8] **13.** $2k - 11$ [2.1–2.3] **14.** $\{-1\}$ **15.** $\{-1\}$
**16.** $\{-12\}$ [2.6] **17.** {26} [2.5] **18.** $y = \frac{24 - 3x}{4}$ **19.** $n = \frac{A - P}{iP}$ [2.8] **20.** $(-\infty, 1]$ 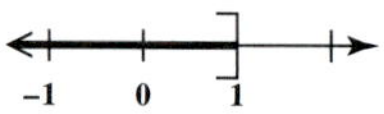

**21.** $(-1, 2]$ 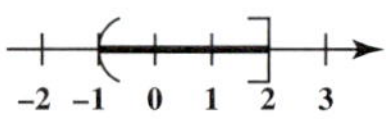

[2.4] **22.** 4 cm; 9 cm; 27 cm [2.5] **23.** 12.42 cm [2.6] **24.** $\frac{25}{6}$ or $4\frac{1}{6}$ cups
[2.7] **25.** 40 mph; 60 mph

# CHAPTER 3 LINEAR EQUATIONS IN TWO VARIABLES

## Section 3.1 (page 196)

**Exercises** **1.** Ohio (OH): about 680 million eggs; Iowa (IA): about 550 million eggs **3.** Indiana (IN) and Pennsylvania (PA); about 490 million eggs each **5.** from 1975 to 1980; about \$.75 **7.** The price of a gallon of gas was decreasing. **9.** does; do not **11.** II **13.** 3 **15.** A linear equation in one variable can be written in the form $Ax + B = C$, where $A \neq 0$. Examples are $2x + 5 = 0$, $3x + 6 = 2$, and $x = -5$. A linear equation in two variables can be written in the form $Ax + By = C$, where $A$ and $B$ cannot both equal 0. Examples are $2x + 3y = 8$, $3x = 5y$, and $x - y = 0$. **17.** yes **19.** yes **21.** no **23.** yes **25.** yes **27.** no **29.** No, the ordered pair $(3, 4)$ represents the point 3 units to the right of the origin and 4 units up from the $x$-axis. The ordered pair $(4, 3)$ represents the point 4 units to the right of the origin and 3 units up from the $x$-axis. **31.** 17 **33.** $-5$ **35.** $-1$ **37.** $-7$ **39.** $b$ **41.** 8; 6; 3 **43.** $-9$; 4; 9 **45.** 12; 12; 12 **47.** $-10$; $-10$; $-10$ **49.–55.**

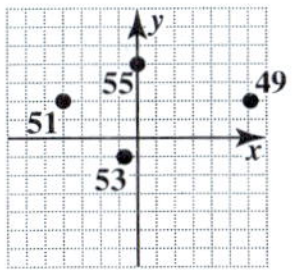


**57.** negative; negative **59.** positive; negative

**61.** $-3$; 6; $-2$; 4

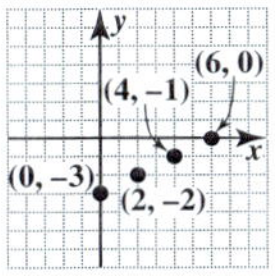


**63.** $-3$; 4; $-6$; $-\frac{4}{3}$

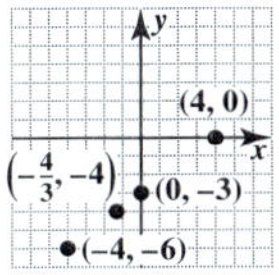


**65.** $-4$; $-4$; $-4$; $-4$

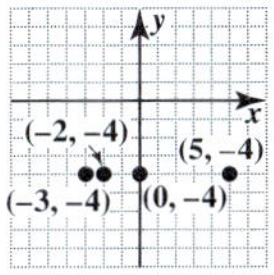


**67.** The points in each graph appear to lie on a straight line. **69.** **(a)** (5, 45) **(b)** (6, 50) **71.** **(a)** (1996, 53.3), (1997, 52.8), (1998, 52.1), (1999, 51.6) **(b)** (1995, 54.0) means that in 1995, the graduation rate for 4-yr college students within 5 yr was 54.0%. **(c)**

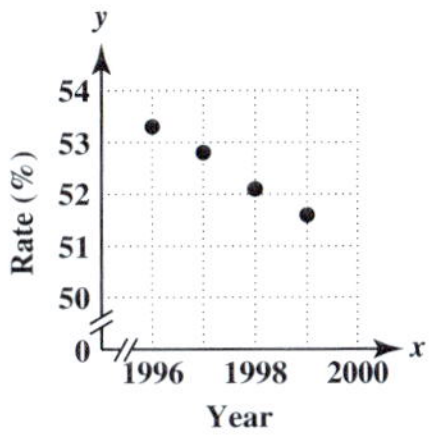


**(d)** The points appear to lie on a straight line. Graduation rates for 4-yr college students within 5 yr are decreasing. **73.** **(a)** 170; 154; 138; 122 **(b)** (20, 170), (40, 154), (60, 138), (80, 122) **(c)**

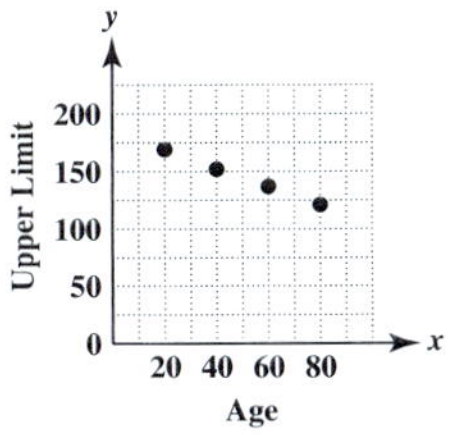


The points lie in a linear pattern.

## Section 3.2 (page 208)

**Connections** **(page 206)** **1.** $3x + 4 - 2x - 7 - 4x - 3 = 0$ **2.** $5x - 15 - 3(x - 2) = 0$

**Exercises** **1.** 5; 5; 3

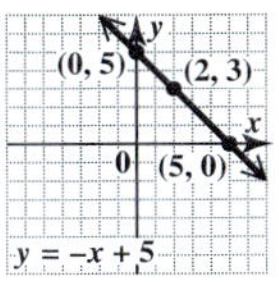


**3.** 1; 3; $-1$

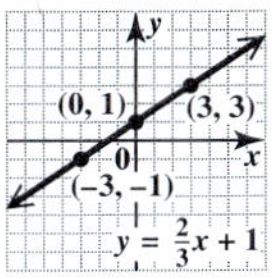


**5.** $-6$; $-2$; $-5$

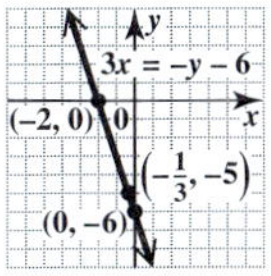


**7.** C **9.** D **11.** B

**13.** (12, 0); (0, −8) **15.** (0, 0); (0, 0) **17.** (4, 0); (0, −10) **19.** (4, 0); none **21.** $y = 0$; $x = 0$ **23.**

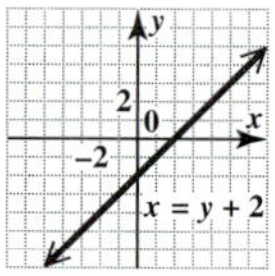


**25.**

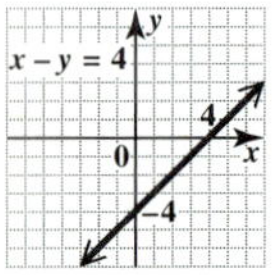


**27.**

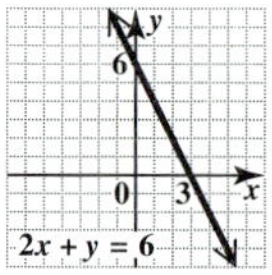


**29.**

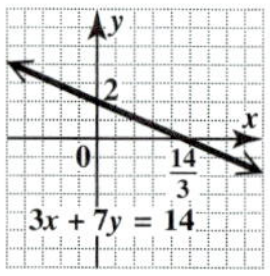


**31.**

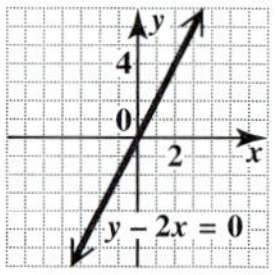


**33.**

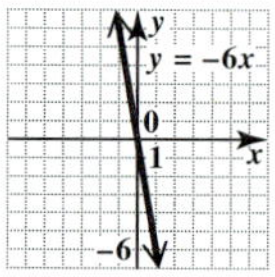


**35.**

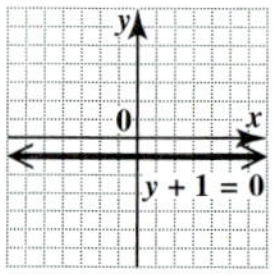


**37.**

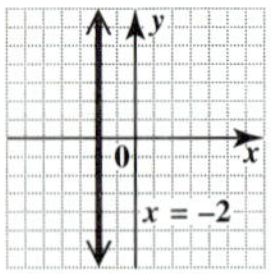


**39.** Find two ordered pairs that satisfy the equation. Plot the corresponding points on a coordinate system. Draw a straight line through the two points. As a check, find a third ordered pair and verify that it lies on the line you drew. **41.** 8; yes **43.** 7; yes **45.** **(a)** 151.5 cm, 174.9 cm, 159.3 cm **(b)**

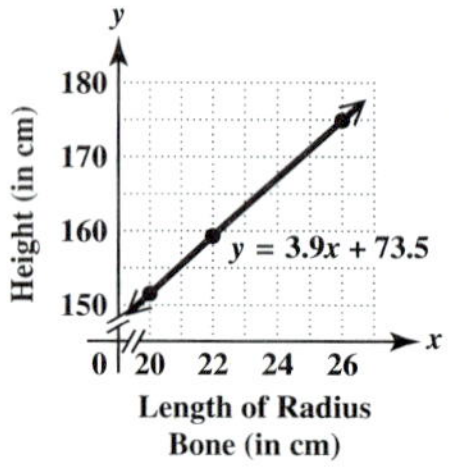


**(c)** 24 cm; 24 cm **47.** **(a)** 130 **(b)** 133 **(c)** They are quite close.

**49.** between 133 and 162 **51.** **(a)** 1993: 49.8 gal; 1995: 51.4 gal; 1997: 53 gal **(b)** 1993: 50.1 gal; 1995: 51.6 gal; 1997: 53 gal **(c)** Corresponding values are quite close. **53.** **(a)** \$30,000 **(b)** \$15,000 **(c)** \$5000 **(d)** After 5 yr, the SUV has a value of \$5000.

## Section 3.3 (page 219)

**Exercises** **1.** Rise is the vertical change between two different points on a line. Run is the horizontal change between two different points on a line. **3.** 4 **5.** $-\frac{1}{2}$ **7.** 0 **9.** Yes; it doesn't matter which point you start with. Both differences will be the negatives of the differences in Exercise 8, and the quotient will be the same. **In Exercises 11 and 13, sketches will vary.** **11.** The line must rise from left to right. **13.** The line must be horizontal. **15.** Because he found the difference $3 - 5 = -2$ in the numerator, he should have subtracted in the same order in the denominator to get $-1 - 2 = -3$. The correct slope is $\frac{-2}{-3} = \frac{2}{3}$. **17.** $\frac{5}{4}$ **19.** $\frac{3}{2}$ **21.** −3 **23.** 0 **25.** undefined **27.** $-\frac{1}{2}$ **29.** 5 **31.** $\frac{1}{4}$ **33.** $\frac{3}{2}$ **35.** $\frac{3}{2}$ **37.** 0 **39.** undefined **41.** $-3$; $\frac{1}{3}$ **43.** **(a)** negative **(b)** zero **45.** **(a)** positive **(b)** negative **47.** **(a)** zero **(b)** negative **49.** A **51.** $-\frac{2}{5}$; $-\frac{2}{5}$; parallel **53.** $\frac{8}{9}$; $-\frac{4}{3}$; neither **55.** $\frac{3}{2}$; $-\frac{2}{3}$; perpendicular **57.** 5; $\frac{1}{5}$; neither **59.** $\frac{3}{10}$ **61.** 232 thousand or 232,000 **62.** positive; increased **63.** 232,000 students **64.** −1.66 **65.** negative; decreased **66.** 1.66 students per computer **67.** The change for each year is .1 billion (or 100,000,000) ft$^2$, so the graph is a straight line. **69.** $\frac{2}{5}$ **71.** (0, 4)

## Section 3.4 (page 228)

**Exercises** **1.** D **3.** B **5.** The slope $m$ of a vertical line is undefined, so it is not possible to write the equation of the line in the form $y = mx + b$. **7.** $y = 3x - 3$ **9.** $y = -x + 3$ **11.** $y = 4x - 3$ **13.** $y = 3$ **15.** $x = 0$ **17.**

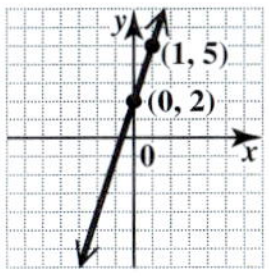


**19.**

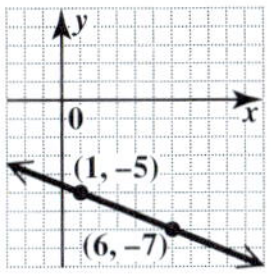


**21.**

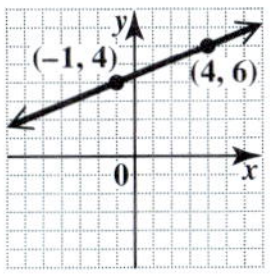


**23.**

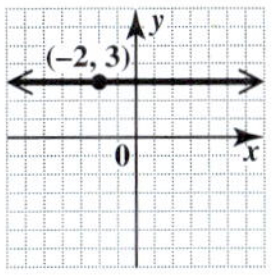


**25.**

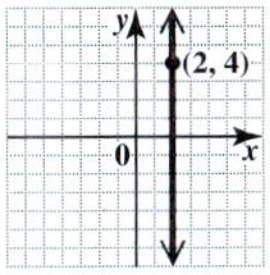


**27.** the $x$-axis **29.** $y = -4x - 1$

**31.** $y = \frac{2}{3}x + \frac{19}{3}$ **33.** $y = \frac{3}{4}x + 4$ **35.** $y = \frac{5}{2}x - 4$ **37.** $y = x$ (There are other forms as well.) **39.** $y = x + 6$

**41.** $y = -\frac{3}{5}x - \frac{11}{5}$ **43.** $y = \frac{1}{2}x + 2$ **45.** $y = -\frac{1}{3}x + \frac{22}{9}$ **47.** (0, 32), (100, 212) **48.** $\frac{9}{5}$ **49.** $F - 32 = \frac{9}{5}(C - 0)$

**50.** $F = \frac{9}{5}C + 32$ **51.** $C = \frac{5}{9}(F - 32)$ **52.** 86° **53.** 10° **54.** −40° **55.** $y = \frac{3}{4}x - \frac{9}{2}$ **57.** $y = -2x - 3$

**59.** **(a)** \$400 **(b)** \$.25 **(c)** $y = .25x + 400$ **(d)** \$425 **(e)** 1500 **61.** $y = -3x + 6$ **63.** $Y_1 = \frac{3}{4}X + 1$

## Chapter 3 Review Exercises (page 236)

**1.** **(a)** \$1.05 per gal; \$1.75 per gal **(b)** \$.70 per gal; about 67% **(c)** between April and June 2000; about \$.40 per gal **(d)** August–October 1999 and February–April 2000 **3.** 2; $\frac{3}{2}$; $\frac{14}{3}$ **5.** 7; 7; 7 **7.** no **9.** I Exercises 9 and 11

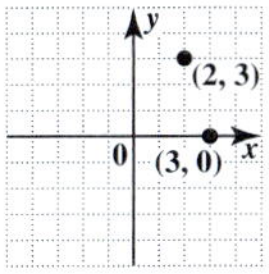


**11.** none **13.** I or III

**15.** $\left(\frac{8}{3}, 0\right)$; (0, 4)

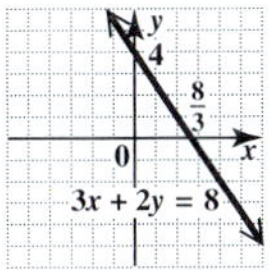


**17.** $-\frac{1}{2}$ **19.** 3 **21.** $\frac{3}{2}$ **23.** $\frac{3}{2}$ **25.** parallel **27.** neither **29.** $y = -\frac{1}{2}x + 4$

**31.** $y = \frac{2}{3}x + \frac{14}{3}$ **33.** $y = -\frac{1}{4}x + \frac{3}{2}$ **35.** $x = \frac{1}{3}$; It is not possible to express this equation in the form $y = mx + b$.

**37.** C, D **39.** D **41.** B **43.** (0, 0); (0, 0); $-\frac{1}{3}$

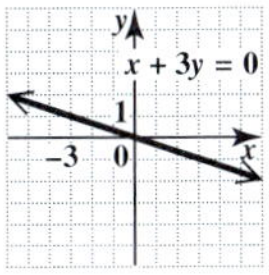


**45.** $y = -\frac{1}{4}x - \frac{5}{4}$ **47.** $y = -\frac{4}{7}x - \frac{23}{7}$

**48.** about \$1.5 billion **49.** It will have negative slope since the total spent on video rentals is decreasing over these years. **50.** (1996, 11.1), (2000, 9.6); −.375 **51.** −.375; Yes, the slope is negative as stated in the answer to Exercise 49 and the same as the slope calculated in Exercise 50. **52.** 10.7; 10.4; 10.0 **53.** The actual amounts are fairly close to those given by the equation. **54.** \$8.9 billion

## Chapter 3 Test (page 239)

[3.1] **1.** $-6, -10, -5$ **2.** no [3.2] **3.** To find the $x$-intercept, let $y = 0$, and to find the $y$-intercept, let $x = 0$. **4.** $x$-intercept: $(2, 0)$; $y$-intercept: $(0, 6)$

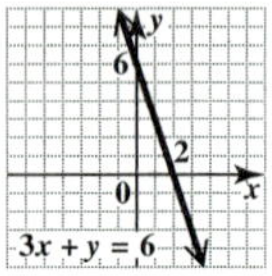


**5.** $x$-intercept: $(0, 0)$; $y$-intercept: $(0, 0)$

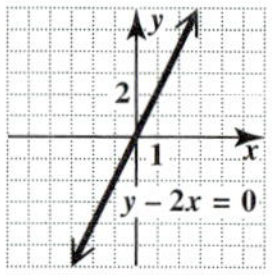


**6.** $x$-intercept: $(-3, 0)$; $y$-intercept: none

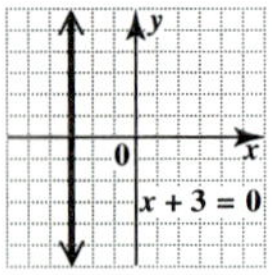


**7.** $x$-intercept: none; $y$-intercept: $(0, 1)$

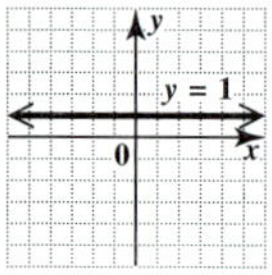


**8.** $x$-intercept: $(4, 0)$; $y$-intercept: $(0, -4)$

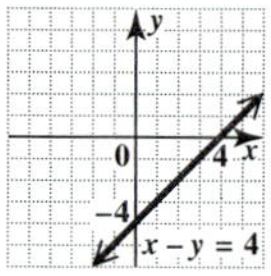


[3.3] **9.** $-\frac{8}{3}$ **10.** $-2$ **11.** undefined **12.** $\frac{5}{2}$ **13.** 0 [3.4] **14.** $y = 2x + 6$ **15.** $y = \frac{5}{2}x - 4$ **16.** $y = -9x + 12$

[3.1–3.3] **17.** The slope is positive since food and drink sales are increasing. **18.** $(0, 43)$, $(30, 376)$; 11.1 **19.** 1990: \$265 billion; 1995: \$320.5 billion **20.** In 2000, food and drink sales were \$376 billion.

## Cumulative Review Exercises Chapters 1–3 (page 240)

[1.1] **1.** $\frac{301}{40}$ or $7\frac{21}{40}$ **2.** 6 [1.5] **3.** 7 [1.6] **4.** $\frac{73}{18}$ or $4\frac{1}{18}$ [1.2–1.6] **5.** true **6.** $-43$ [1.7] **7.** distributive property

[1.8] **8.** $-p + 2$ [2.5] **9.** $h = \frac{3V}{\pi r^2}$ [2.3] **10.** $\{-1\}$ **11.** $\{2\}$ [2.6] **12.** $\{-13\}$ [2.8] **13.** $(-2.6, \infty)$

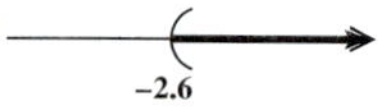


**14.** $(0, \infty)$

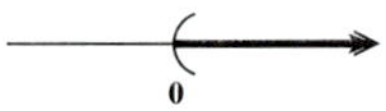


**15.** $(-\infty, -4]$

[2.4, 2.5] **16.** high school diploma: \$22,895; bachelor's degree: \$40,478 **17.** 13 mi [3.1] **18.** **(a)** 89.45; 81.95; 78.20 **(b)** In 1980, the winning time was 85.7 sec.

**19.** **(a)** \$7000 **(b)** \$10,000 **(c)** about \$30,000 [3.2] **20.** $(-4, 0)$; $(0, 3)$ [3.3] **21.** $\frac{3}{4}$ [3.2] **22.**

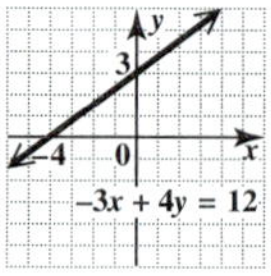


[3.3] **23.** perpendicular [3.4] **24.** $y = 3x - 11$ **25.** $y = 4$

# CHAPTER 4 EXPONENTS AND POLYNOMIALS

## Section 4.1 (page 248)

**Exercises** **1.** false **3.** false **5.** $w^6$ **7.** $\frac{1}{4^4}$ **9.** $(-7x)^4$ **11.** $\left(\frac{1}{2}\right)^6$ **13.** In $(-3)^4$, $-3$ is the base, while in $-3^4$, 3 is the base. $(-3)^4 = 81$, while $-3^4 = -81$. **15.** base: 3; exponent: 5; 243 **17.** base: $-3$; exponent: 5; $-243$ **19.** base: $-6x$; exponent: 4 **21.** base: $x$; exponent: 4 **23.** $5^2 + 5^3$ is a sum, not a product. $5^2 + 5^3 = 25 + 125 = 150$. **25.** $5^8$ **27.** $4^{12}$ **29.** $(-7)^9$ **31.** $t^{24}$ **33.** $-56r^7$ **35.** $42p^{10}$ **37.** $-30x^9$ **39.** The product rule does not apply. **41.** The product rule does not apply. **43.** $4^6$ **45.** $t^{20}$ **47.** $7^3r^3$ **49.** $5^5x^5y^5$ **51.** $5^{12}$ **53.** $-8^{15}$ **55.** $8q^3r^3$ **57.** $\frac{1}{2^3}$ **59.** $\frac{a^3}{b^3}$ **61.** $\frac{9^8}{5^8}$ **63.** $\frac{5^5}{2^5}$ **65.** $\frac{9^5}{8^3}$ **67.** $2^{12}x^{12}$ **69.** $-6^5p^5$ **71.** $6^5x^{10}y^{15}$ **73.** $x^{21}$ **75.** $2^2w^4x^{26}y^7$ **77.** $-r^{18}s^{17}$ **79.** $\frac{5^3a^6b^{15}}{c^{18}}$ **81.** This is incorrect. Using the product rule, it is simplified as follows: $(10^2)^3 = 10^{2\cdot3} = 10^6 = 1{,}000{,}000$. **83.** $12x^5$ **85.** $6p^7$ **87.** $125x^6$ **89.** $-a^4, -a^3, -(-a)^3, (-a)^4$; One way is to choose a positive number greater than 1 and substitute it for $a$ in each expression. Then arrange the terms from smallest to largest. **91.** \$304.16 **93.** \$1843.88

## Section 4.2 (page 257)

**Exercises** **1.** 1 **3.** 1 **5.** $-1$ **7.** 0 **9.** 0 **11.** C **13.** F **15.** E **17.** 2 **19.** $\frac{1}{64}$ **21.** 16 **23.** $\frac{49}{36}$ **25.** $\frac{1}{81}$ **27.** $\frac{8}{15}$ **29.** $5^3$ **31.** $\frac{5^3}{3^2}$ or $\frac{125}{9}$ **33.** $5^2$ **35.** $x^{15}$ **37.** $6^3$ **39.** $2r^4$ **41.** $\frac{5^2}{4^3}$ **43.** $\frac{p^5}{q^8}$ **45.** $r^9$ **47.** $\frac{yz^2}{4x^3}$ **49.** $a+b$ **51.** $(x+2y)^2$ **53.** 1 **54.** $\frac{5^2}{5^2}$ **55.** $5^0$ **56.** $1 = 5^0$; This supports the definition of 0 as an exponent. **57.** $7^3$ or 343 **59.** $\frac{1}{x^2}$ **61.** $\frac{64x}{9}$ **63.** $\frac{x^2z^4}{y^2}$ **65.** $6x$ **67.** $\frac{1}{m^{10}n^5}$ **69.** $\frac{1}{xyz}$ **71.** $x^3y^9$ **73.** $\frac{a^{11}}{2b^5}$ **75.** $\frac{108}{y^5z^3}$ **77.** $\frac{9z^2}{400x^3}$ **79.** The student attempted to use the quotient rule with unequal bases. The correct way to simplify this expression is $\frac{16^3}{2^2} = \frac{(2^4)^3}{2^2} = \frac{2^{12}}{2^2} = 2^{10} = 1024$.

## Summary Exercises on the Rules for Exponents (page 259)

**1.** $\frac{6^{12}x^{24}}{5^{12}}$ **2.** $\frac{r^6s^{12}}{729t^6}$ **3.** $10^5x^7y^{14}$ **4.** $-128a^{10}b^{15}c^4$ **5.** $\frac{729w^3x^9}{y^{12}}$ **6.** $\frac{x^4y^6}{16}$ **7.** $c^{22}$ **8.** $\frac{1}{k^4t^{12}}$ **9.** $\frac{11}{30}$ **10.** $y^{12}z^3$ **11.** $\frac{x^6}{y^5}$ **12.** 0 **13.** $\frac{1}{z^2}$ **14.** $\frac{9}{r^2s^2t^{10}}$ **15.** $\frac{300x^3}{y^3}$ **16.** $\frac{3}{5x^6}$ **17.** $x^8$ **18.** $\frac{y^{11}}{x^{11}}$ **19.** $\frac{a^6}{b^4}$ **20.** $6ab$ **21.** $\frac{61}{900}$ **22.** 1 **23.** $\frac{343a^6b^9}{8}$ **24.** 1 **25.** $-1$ **26.** 0 **27.** $\frac{27y^{18}}{4x^8}$ **28.** $\frac{1}{a^8b^{12}c^{16}}$ **29.** $\frac{x^{15}}{216z^9}$ **30.** $\frac{q}{8p^6r^3}$ **31.** $x^6y^6$ **32.** 0 **33.** $\frac{343}{x^{15}}$ **34.** $\frac{9}{x^6}$ **35.** $5p^{10}q^9$ **36.** $\frac{7}{24}$ **37.** $\frac{r^{14}t}{2s^2}$ **38.** 1 **39.** $8p^{10}q$ **40.** $\frac{1}{mn^3p^3}$ **41.** $-1$ **42.** $\frac{3}{40}$

## Section 4.3 (page 264)

**Connections** **(page 263)** **1.** The Erzincan earthquake was ten times as powerful as the Sumatra earthquake. **2.** The Adana earthquake had one hundredth the power of the San Francisco earthquake. **3.** The Erzincan earthquake was about 50.12 times as powerful as the Adana earthquake. **4.** "+3" corresponds to a factor of 1000 times stronger; "−1" corresponds to a factor of one-tenth.

**Exercises** **1.** A **3.** C **5.** in scientific notation **7.** not in scientific notation; $5.6 \times 10^6$ **9.** not in scientific notation; $8 \times 10^1$ **11.** not in scientific notation; $4 \times 10^{-3}$ **13.** It is written as the product of a power of 10 and a number whose absolute value is between 1 and 10 (inclusive of 1). Some examples are $2.3 \times 10^{-4}$ and $6.02 \times 10^{23}$. **15.** $5.876 \times 10^9$ **17.** $8.235 \times 10^4$ **19.** $7 \times 10^{-6}$ **21.** $2.03 \times 10^{-3}$ **23.** 750,000 **25.** 5,677,000,000,000 **27.** 6.21 **29.** .00078 **31.** .000000005134 **33.** 600,000,000,000 **35.** 15,000,000 **37.** 60,000 **39.** .0003 **41.** 40 **43.** .000013 **45.** 4.7E−7 **47.** 2E7 **49.** 1E1 **51.** $1 \times 10^{10}$ **53.** 2,000,000,000 **55.** $\$6.03 \times 10^8$ **57.** $\$5.2466 \times 10^{10}$ **59.** about 15,300 sec **61.** about $9.2 \times 10^8$ acres **63.** about 3.3 **65.** about .276 lb

## Section 4.4 (page 274)

**Connections** **(page 274)** The polynomial gives the following: for 1996, 5730 million (greater than actual); for 1997, 5594 million (less than actual); for 1999, 5260 million (less than actual); for 2000, 5062 million (less than actual); for 2001, 4844 million (greater than actual).

**Exercises** **1.** 7; 5 **3.** 8 **5.** 26 **7.** 0 **9.** 1; 6 **11.** 1; 1 **13.** 2; $-19$, $-1$ **15.** 3; 1, 8, 5 **17.** $2m^5$ **19.** $-r^5$ **21.** cannot be simplified **23.** $-5x^5$ **25.** $5p^9 + 4p^7$ **27.** $-2xy^2$ **29.** already simplified; 4; binomial **31.** $11m^4 - 7m^3 - 3m^2$; 4; trinomial **33.** $x^4$; 4; monomial **35.** 7; 0; monomial **37.** **(a)** 36 **(b)** $-12$ **39.** **(a)** 14 **(b)** $-19$ **41.** **(a)** $-3$ **(b)** 0 **43.** $5x^2 - 2x$

**45.** $5m^2 + 3m + 2$ **47.** $\frac{7}{6}x^2 - \frac{2}{15}x + \frac{5}{6}$ **49.** $6m^3 + m^2 + 12m - 14$ **51.** $3y^3 - 11y^2$ **53.** $4x^4 - 4x^2 + 4x$

**55.** $15m^3 - 13m^2 + 8m + 11$ **57.** Answers will vary. **59.** $5m^2 - 14m + 6$ **61.** $4x^3 + 2x^2 + 5x$ **63.** $-11y^4 + 8y^2 + y$ **65.** $a^4 - a^2 + 1$ **67.** $5m^2 + 8m - 10$ **69.** $-6x^2 - 12x + 12$ **71.** $-10$ **73.** $4b - 5c$ **75.** $6x - xy - 7$ **77.** $-3x^2y - 15xy - 3xy^2$ **79.** $8x^2 + 8x + 6$ **81. (a)** $23y + 5t$ **(b)** approximately 37.26°, 79.52°, and 63.22° **83.** $-6x^2 + 6x - 7$ **85.** $-7x - 1$ **87.** 0, −3, −4, −3, 0

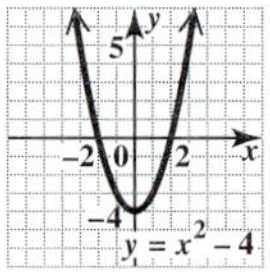


**89.** 7, 1, −1, 1, 7

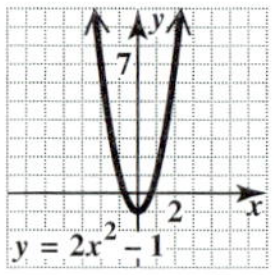


**91.** 0, 3, 4, 3, 0

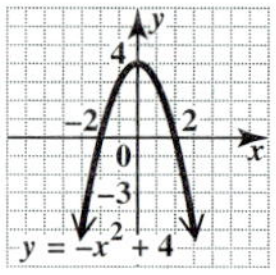


**93.** 4, 1, 0, 1, 4

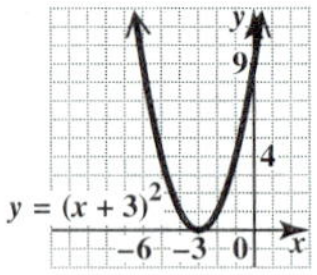


**95.** 175 ft **97.** 4; \$5.00 **98.** 6; \$27 **99.** 2.5; 130 **100.** 459.87 billion

## Section 4.5 (page 283)

**Exercises** **1. (a)** B **(b)** D **(c)** A **(d)** C **3.** $15y^{11}$ **5.** $30a^9$ **7.** $15pq^2$ **9.** $-18m^3n^2$ **11.** $6m^2 + 4m$ **13.** $-6p^4 + 12p^3$ **15.** $-16z^2 - 24z^3 - 24z^4$ **17.** $6y^3 + 4y^4 + 10y^7$ **19.** $28r^5 - 32r^4 + 36r^3$ **21.** $6a^4 - 12a^3b + 15a^2b^2$ **23.** $21m^5n^2 + 14m^4n^3 - 7m^3n^5$ **25.** $12x^3 + 26x^2 + 10x + 1$ **27.** $81a^3 + 27a^2 + 11a + 2$ **29.** $20m^4 - m^3 - 8m^2 - 17m - 15$ **31.** $6x^6 - 3x^5 - 4x^4 + 4x^3 - 5x^2 + 8x - 3$ **33.** $5x^4 - 13x^3 + 20x^2 + 7x + 5$ **35.** $3x^5 + 18x^4 - 2x^3 - 8x^2 + 24x$ **37.** $m^2 + 12m + 35$ **39.** $x^2 - 25$ **41.** $12x^2 + 10x - 12$ **43.** $9x^2 - 12x + 4$ **45.** $10a^2 + 37a + 7$ **47.** $12 + 8m - 15m^2$ **49.** $20 - 7x - 3x^2$ **51.** $8xy - 4x + 6y - 3$ **53.** $15x^2 + xy - 6y^2$ **55.** $6y^5 - 21y^4 - 45y^3$ **57.** $-200r^7 + 32r^3$ **59.** $3y^2 + 10y + 7$ **61.** $6p^2 - \frac{5}{2}pq - \frac{25}{12}q^2$ **63.** $x^2 + 14x + 49$ **65.** $a^2 - 16$ **67.** $4p^2 - 20p + 25$ **69.** $25k^2 + 30kq + 9q^2$ **71.** $m^3 - 15m^2 + 75m - 125$ **73.** $8a^3 + 12a^2 + 6a + 1$ **75.** $56m^2 - 14m - 21$ **77.** $-9a^3 + 33a^2 + 12a$ **79.** $81r^4 - 216r^3s + 216r^2s^2 - 96rs^3 + 16s^4$ **81.** $6p^8 + 15p^7 + 12p^6 + 36p^5 + 15p^4$ **83.** $-24x^8 - 28x^7 + 32x^6 + 20x^5$ **85.** $14x + 49$ **87.** $\pi x^2 - 9$ **89.** $30x + 60$ **90.** $30x + 60 = 600$ **91.** 18 **92.** 10 yd by 60 yd **93.** \$2100 **94.** 140 yd **95.** \$1260 **96. (a)** $30kx + 60k$ (dollars) **(b)** $6rx + 32r$ **97.** Suppose that we want to multiply $(x + 3)(2x + 5)$. The letter F stands for *first.* We multiply the two first terms to get $2x^2$. The letter O represents *outer.* We next multiply the two outer terms to get $5x$. The letter I represents *inner.* The product of the two inner terms is $6x$. L stands for *last.* The product of the two last terms is 15. Very often the outer and inner products are like terms, as they are in this case. So we simplify $2x^2 + 5x + 6x + 15$ to get the final product, $2x^2 + 11x + 15$.

## Section 4.6 (page 289)

**Exercises** **1. (a)** $4x^2$ **(b)** $12x$ **(c)** 9 **(d)** $4x^2 + 12x + 9$ **3.** $m^2 + 4m + 4$ **5.** $r^2 - 6r + 9$ **7.** $x^2 + 4xy + 4y^2$ **9.** $25p^2 + 20pq + 4q^2$ **11.** $16a^2 + 40ab + 25b^2$ **13.** $49t^2 + 14ts + s^2$ **15.** $36m^2 - \frac{48}{5}mn + \frac{16}{25}n^2$ **17.** $9t^3 - 6t^2 + t$ **19.** $-9y^2 + 48y - 64$ **21.** To find the product of the sum and the difference of two terms, we find the difference of the squares of the two terms. For example, $(5x + 7y)(5x - 7y) = (5x)^2 - (7y)^2 = 25x^2 - 49y^2$. **23.** $a^2 - 64$ **25.** $4 - p^2$ **27.** $4m^2 - 25$ **29.** $9x^2 - 16y^2$ **31.** $25y^2 - 9x^2$ **33.** $169r^2 - 4z^2$ **35.** $81y^4 - 4$ **37.** $81y^2 - \frac{4}{9}$ **39.** $25q^3 - q$ **41.** $x^3 + 3x^2 + 3x + 1$ **43.** $t^3 - 9t^2 + 27t - 27$ **45.** $r^3 + 15r^2 + 75r + 125$ **47.** $8a^3 + 12a^2 + 6a + 1$ **49.** $81r^4 - 216r^3t + 216r^2t^2 - 96rt^3 + 16t^4$ **51.** $(a + b)^2$ **52.** $a^2$ **53.** $2ab$ **54.** $b^2$ **55.** $a^2 + 2ab + b^2$ **56.** They both represent the area of the entire large square. **57.** 1225 **58.** $30^2 + 2(30)(5) + 5^2$ **59.** 1225 **60.** They are equal. **61.** 9999 **63.** 39,999 **65.** $399\frac{3}{4}$ **67.** $\frac{1}{2}m^2 - 2n^2$ **69.** $9a^2 - 4$ **71.** $\pi x^2 + 4\pi x + 4\pi$ **73.** $x^3 + 6x^2 + 12x + 8$

## Section 4.7 (page 298)

**Connections** (page 298) **1.** $-104$ **2.** $-104$ **3.** They are both $-104$. **4.** The answers should agree.

**Exercises** **1.** $6x^2 + 8$; 2; $3x^2 + 4$ **3.** $3x^2 + 4$; 2 (These may be reversed.); $6x^2 + 8$ **5.** The first is a polynomial divided by a monomial, covered in Objective 1. This section does not cover dividing a monomial by a polynomial of several terms. **7.** $30x^3 - 10x + 5$ **9.** $4m^3 - 2m^2 + 1$ **11.** $4t^4 - 2t^2 + 2t$ **13.** $a^4 - a + \frac{2}{a}$ **15.** $4x^3 - 3x^2 + 2x$ **17.** $1 + 5x - 9x^2$ **19.** $\frac{12}{x} + 8 + 2x$ **21.** $\frac{4x^2}{3} + x + \frac{2}{3x}$ **23.** $9r^3 - 12r^2 + 2r + \frac{26}{3} - \frac{2}{3r}$ **25.** $-m^2 + 3m - \frac{4}{m}$ **27.** $4 - 3a + \frac{5}{a}$ **29.** $\frac{12}{x} - \frac{6}{x^2} + \frac{14}{x^3} - \frac{10}{x^4}$ **31.** $6x^4y^2 - 4xy + 2xy^2 - x^4y$ **33.** $5x^3 + 4x^2 - 3x + 1$ **35.** $-63m^4 - 21m^3 - 35m^2 + 14m$ **37.** 1423 **38.** $(1 \times 10^3) + (4 \times 10^2) + (2 \times 10^1) + (3 \times 10^0)$ **39.** $x^3 + 4x^2 + 2x + 3$ **40.** They are similar in that the coefficients of powers of ten are equal to the coefficients of the powers of $x$. They are different in that one is a constant while the other is a polynomial. They are equal if $x = 10$ (the base of our decimal system). **41.** $x + 2$ **43.** $2y - 5$ **45.** $p - 4 + \frac{44}{p + 6}$ **47.** $r - 5$ **49.** $6m - 1$ **51.** $2a - 14 + \frac{74}{2a + 3}$ **53.** $4x^2 - 7x + 3$ **55.** $4k^3 - k + 2$ **57.** $5y^3 + 2y - 3$ **59.** $3k^2 + 2k - 2 + \frac{6}{k - 2}$ **61.** $2p^3 - 6p^2 + 7p - 4 + \frac{14}{3p + 1}$ **63.** $r^2 - 1 + \frac{4}{r^2 - 1}$ **65.** $y^2 - y + 1$ **67.** $a^2 + 1$ **69.** $x^2 - 4x + 2 + \frac{9x - 4}{x^2 + 3}$ **71.** $x^3 + 3x^2 - x + 5$ **73.** $\frac{3}{2}a - 10 + \frac{77}{2a + 6}$ **75.** The process stops when the degree of the remainder is less than the degree of the divisor. **77.** $x^2 + x - 3$ units **79.** $5x^2 - 11x + 14$ hr **81.** A is correct; B is incorrect. **82.** A is incorrect; B is correct. **83.** A is correct; B is incorrect. **84.** Because any power of 1 is 1, to evaluate a polynomial for 1 we simply add the coefficients. If the divisor is $x - 1$, this method does not apply, since $1 - 1 = 0$ and division by 0 is undefined.

## Chapter 4 Review Exercises (page 307)

**1.** $4^{11}$ **3.** $-72x^7$ **5.** $19^5x^5$ **7.** $5p^4t^4$ **9.** $6^2x^{16}y^4z^{16}$ **11.** The expression is a *sum* of powers of 7, not a *product*. **13.** 0 **15.** $-\frac{1}{49}$ **17.** $5^8$ **19.** $\frac{3}{4}$ **21.** $x^2$ **23.** $\frac{r^8}{81}$ **25.** $\frac{1}{a^3b^5}$ **27.** $4.8 \times 10^7$ **29.** $8.24 \times 10^{-8}$ **31.** 78,300,000 **33.** 800 **35.** .025 **37.** .0000000000016 **39.** $9.7 \times 10^4$; $5 \times 10^3$ **41.** **(a)** $1 \times 10^3$ **(b)** $2 \times 10^3$ **(c)** $5 \times 10^4$ **(d)** $1 \times 10^5$ **43.** $p^3 - p^2 + 4p + 2$; degree 3; none of these **45.** $-8y^5 - 7y^4 + 9y$; degree 5; trinomial **47.** $13x^3y^2 - 5xy^5 + 21x^2$ **49.** $y^2 - 10y + 9$ **51.** 1, 4, 5, 4, 1

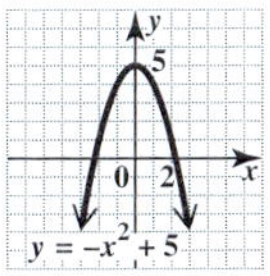


**53.** $a^3 - 2a^2 - 7a + 2$ **55.** $5p^5 - 2p^4 - 3p^3 + 25p^2 + 15p$ **57.** $6k^2 - 9k - 6$ **59.** $12k^2 - 32kq - 35q^2$ **61.** $2x^2 + x - 6$ **63.** $a^2 + 8a + 16$ **65.** $36m^2 - 25$ **67.** $r^3 + 6r^2 + 12r + 8$ **69.** **(a)** Answers will vary. For example, let $x = 1$ and $y = 2$. $(1 + 2)^2 \neq 1^2 + 2^2$, because $9 \neq 5$. **(b)** Answers will vary. For example, let $x = 1$ and $y = 2$. $(1 + 2)^3 \neq 1^3 + 2^3$, because $27 \neq 9$. **71.** In both cases, $x = 0$ and $y = 1$ lead to 1 on each side of the inequality. This would not be sufficient to show that *in general* the inequality is true. It would be necessary to choose other values of $x$ and $y$. **73.** $\frac{4}{3}\pi(x + 1)^3$ or $\frac{4}{3}\pi x^3 + 4\pi x^2 + 4\pi x + \frac{4}{3}\pi$ in.$^3$ **75.** $y^3 - 2y + 3$ **77.** $2mn + 3m^4n^2 - 4n$ **79.** $2r + 7$ **81.** $x^2 + 3x - 4$ **83.** $4x - 5$ **85.** $y^2 + 2y + 4$ **87.** $2y^2 - 5y + 4 + \frac{-5}{3y^2 + 1}$ **89.** 2 **91.** $144a^2 - 1$ **93.** $\frac{1}{8^{12}}$ **95.** $\frac{2}{3m^3}$ **97.** $r^{13}$ **99.** $-y^2 - 4y + 4$ **101.** $y^2 + 5y + 1$ **103.** $10p^2 - 3p - 5$ **105.** $49 - 28k + 4k^2$

## Chapter 4 Test (page 311)

[4.1, 4.2] **1.** $\frac{1}{625}$ **2.** 2 **3.** $\frac{7}{12}$ **4.** $9x^3y^5$ **5.** $8^5$ **6.** $x^2y^6$ **7.** **(a)** positive **(b)** positive **(c)** negative **(d)** positive **(e)** zero **(f)** negative [4.3] **8.** **(a)** $4.5 \times 10^{10}$ **(b)** .0000036 **(c)** .00019 **9.** **(a)** $1 \times 10^3$; $5.89 \times 10^{12}$ **(b)** $5.89 \times 10^{15}$ mi [4.4] **10.** $-7x^2 + 8x$; 2; binomial **11.** $4n^4 + 13n^3 - 10n^2$; 4; trinomial **12.** 4, −2, −4, −2, 4

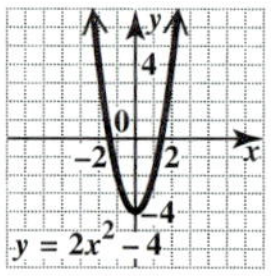


**13.** $-2y^2 - 9y + 17$ **14.** $-21a^3b^2 + 7ab^5 - 5a^2b^2$ **15.** $-12t^2 + 5t + 8$ [4.5] **16.** $-27x^5 + 18x^4 - 6x^3 + 3x^2$ **17.** $t^2 - 5t - 24$ **18.** $8x^2 + 2xy - 3y^2$ [4.6] **19.** $25x^2 - 20xy + 4y^2$ **20.** $100v^2 - 9w^2$ [4.5] **21.** $2r^3 + r^2 - 16r + 15$ [4.6] **22.** $9x^2 + 54x + 81$ [4.7] **23.** $4y^2 - 3y + 2 + \frac{5}{y}$ **24.** $-3xy^2 + 2x^3y^2 + 4y^2$ **25.** $3x^2 + 6x + 11 + \frac{26}{x-2}$

## Cumulative Review Exercises Chapters 1–4 (page 312)

[1.1] **1.** $\frac{7}{4}$ **2.** 5 **3.** $\frac{19}{24}$ **4.** $-\frac{1}{20}$ **5.** $31\frac{1}{4}$ yd$^3$ [1.6] **6.** \$1836 **7.** 1, 3, 5, 9, 15, 45 **8.** −8 **9.** $\frac{1}{2}$ **10.** −4 [1.7] **11.** associative property **12.** distributive property [1.8] **13.** $-10x^2 + 21x - 29$ [2.1–2.3] **14.** $\left\{\frac{13}{4}\right\}$ **15.** $\emptyset$ [2.5] **16.** $r = \frac{d}{t}$ [2.6] **17.** {−5} [2.1–2.3] **18.** {−12} **19.** {20} **20.** {all real numbers} [2.4] **21.** mouse: 160; elephant: 10 **22.** 4 [2.8] **23.** 11 ft and 22 ft [2.8] **24.** $[10, \infty)$ **25.** $\left(-\infty, -\frac{14}{5}\right)$ **26.** $[-4, 2)$ [3.2] **27.**

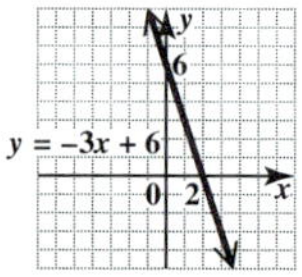


[3.3, 3.4] **28.** **(a)** 1 **(b)** $y = x + 6$ [4.1, 4.2] **29.** $\frac{5}{4}$ **30.** 2 **31.** 1 **32.** $\frac{2b}{a^{10}}$ [4.3] **33.** $3.45 \times 10^4$ **34.** .000000536 **35.** about 10,800,000 km [4.4] **36.**

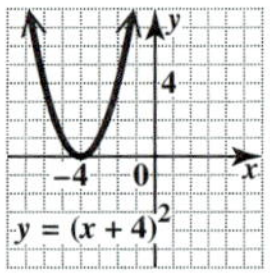


**37.** $11x^3 - 14x^2 - x + 14$ [4.5] **38.** $18x^7 - 54x^6 + 60x^5$ **39.** $63x^2 + 57x + 12$ [4.6] **40.** $25x^2 + 80x + 64$ [4.7] **41.** $2x^2 - 3x + 1$ **42.** $y^2 - 2y + 6$

# CHAPTER 5 FACTORING AND APPLICATIONS

## Section 5.1 (page 322)

**Exercises** **1.** 4 **3.** 6 **5.** First, verify that you have factored completely. Then multiply the factors. The product should be the original polynomial. **7.** 8 **9.** $10x^3$ **11.** $6m^3n^2$ **13.** $xy^2$ **15.** 6 **17.** factored **19.** not factored **21.** yes; $x^3y^2$ **23.** $3m^2$ **25.** $2z^4$ **27.** $2mn^4$ **29.** $y + 2$ **31.** $a - 2$ **33.** $2 + 3xy$ **35.** $9m(3m^2 - 1)$ **37.** $8z^2(2z^2 + 3)$ **39.** $\frac{1}{4}d(d - 3)$ **41.** $6x^2(2x + 1)$

**43.** $5y^6(13y^4 + 7)$ **45.** no common factor (except 1) **47.** $8m^2n^2(n + 3)$ **49.** $13y^2(y^6 + 2y^2 - 3)$ **51.** $9qp^3(5q^3p^2 + 4p^3 + 9q)$ **53.** $a^3(a^2 + 2b^2 - 3a^2b^2 + 4ab^3)$ **55.** $(x + 2)(c - d)$ **57.** $(m + 2n)(m + n)$ **59.** not in factored form; $(7t + 4)(8 + x)$ **61.** in factored form **63.** not in factored form; $(y + 4)(18x^2 + 7)$ **65.** The quantities in parentheses are not the same, so there is no common factor in the two terms $12k^3(s - 3)$ and $7(s + 3)$. **67.** $(p + 4)(p + q)$ **69.** $(a - 2)(a + b)$ **71.** $(z + 2)(7z - a)$ **73.** $(3r + 2y)(6r - x)$ **75.** $(a^2 + b^2)(3a + 2b)$ **77.** $(1 - a)(1 - b)$ **79.** $(4m - p^2)(4m^2 - p)$ **81.** $(5 - 2p)(m + 3)$ **83.** $(3r + 2y)(6r - t)$ **85.** $(a^5 - 3)(1 + 2b)$ **87.** commutative property **88.** $2x(y - 1) - 3(y - 1)$ **89.** No, because it is not a product. It is the difference between $2x(y - 1)$ and $3(y - 1)$. **90.** $(2x - 3)(y - 1)$; yes **91.** **(a)** yes **(b)** When either one is multiplied out, the product is $1 - a + ab - b$.

## Section 5.2 (page 328)

**Exercises** **1.** 1 and 48, −1 and −48, 2 and 24, −2 and −24, 3 and 16, −3 and −16, 4 and 12, −4 and −12, 6 and 8, −6 and −8; the pair with a sum of −19 is −3 and −16. **3.** 1 and −24, −1 and 24, 2 and −12, −2 and 12, 3 and −8, −3 and 8, 4 and −6, −4 and 6; the pair with a sum of −5 is 3 and −8. **5.** $a$ and $b$ must have different signs, one positive and one negative. **7.** A prime polynomial is one that cannot be factored using only integers in the factors. **9.** C **11.** $p + 6$ **13.** $x + 11$ **15.** $x - 8$ **17.** $y - 5$ **19.** $x + 11$ **21.** $y - 9$ **23.** $(y + 8)(y + 1)$ **25.** $(b + 3)(b + 5)$ **27.** $(m + 5)(m - 4)$ **29.** $(y - 5)(y - 3)$ **31.** prime **33.** $(z - 7)(z - 8)$ **35.** $(r - 6)(r + 5)$ **37.** $(a + 4)(a - 12)$ **39.** prime **41.** Factor $8 + 6x + x^2$ directly to get $(2 + x)(4 + x)$. Alternatively, use the commutative property to write the trinomial as $x^2 + 6x + 8$ and factor to get $(x + 2)(x + 4)$, an equivalent answer. **43.** $(r + 2a)(r + a)$ **45.** $(t + 2z)(t - 3z)$ **47.** $(x + y)(x + 3y)$ **49.** $(v - 5w)(v - 6w)$ **51.** $4(x + 5)(x - 2)$ **53.** $2t(t + 1)(t + 3)$ **55.** $2x^4(x - 3)(x + 7)$ **57.** $5m^2(m^3 + 5m^2 - 8)$ **59.** $mn(m - 6n)(m - 4n)$ **61.** $(2x + 4)(x - 3)$ is incorrect because $2x + 4$ has a common factor of 2, which must be factored out. **63.** $a^3(a + 4b)(a - b)$ **65.** $yz(y + 3z)(y - 2z)$ **67.** $z^8(z - 7y)(z + 3y)$ **69.** $(a + b)(x + 4)(x - 3)$ **71.** $(2p + q)(r - 9)(r - 3)$ **73.** $a^2 + 13a + 36$

## Section 5.3 (page 335)

**Exercises** **1.** $(2t + 1)(5t + 2)$ **3.** $(3z - 2)(5z - 3)$ **5.** $(2s - t)(4s + 3t)$ **7.** B **9.** **(a)** 2, 12, 24, 11 **(b)** 3, 8 (Order is irrelevant.) **(c)** $3m$, $8m$ **(d)** $2m^2 + 3m + 8m + 12$ **(e)** $(2m + 3)(m + 4)$ **(f)** $(2m + 3)(m + 4) = 2m^2 + 11m + 12$ **11.** B **13.** A **15.** $2a + 5b$ **17.** $x^2 + 3x - 4$; $x + 4$, $x - 1$ or $x - 1$, $x + 4$ **19.** The binomial $2x - 6$ cannot be a factor because it has a common factor of 2, which the polynomial does not have. **21.** $(3a + 7)(a + 1)$ **23.** $(2y + 3)(y + 2)$ **25.** $(3m - 1)(5m + 2)$ **27.** $(3s - 1)(4s + 5)$ **29.** $(5m - 4)(2m - 3)$ **31.** $(4w - 1)(2w - 3)$ **33.** $(4y + 1)(5y - 11)$ **35.** prime **37.** $2(5x + 3)(2x + 1)$ **39.** $3(4x - 1)(2x - 3)$ **41.** $q(5m + 2)(8m - 3)$ **43.** $3n^2(5n - 3)(n - 2)$ **45.** $y^2(5x - 4)(3x + 1)$ **47.** $(5a + 3b)(a - 2b)$ **49.** $(4s + 5t)(3s - t)$ **51.** $m^4n(3m + 2n)(2m + n)$ **53.** $(5 - x)(1 - x)$ **55.** $(4 + 3x)(4 + x)$ **57.** $-5x(2x + 7)(x - 4)$ **59.** The student stopped too soon. He needs to factor out the common factor $4x - 1$ to get $(4x - 1)(4x - 5)$ as the correct answer. **61.** $-1(x + 7)(x - 3)$ **63.** $-1(3x + 4)(x - 1)$ **65.** $-1(a + 2b)(2a + b)$ **67.** Yes, $(x + 7)(3 - x)$ is equivalent to $-1(x + 7)(x - 3)$ because $-1(x - 3) = -x + 3 = 3 - x$. **69.** $(m + 1)^3(5q - 2)(5q + 1)$ **71.** $(r + 3)^3(5x + 2y)(3x - 8y)$ **73.** −4, 4 **75.** −11, −7, 7, 11

## Section 5.4 (page 344)

**Exercises** **1.** 1; 4; 9; 16; 25; 36; 49; 64; 81; 100; 121; 144; 169; 196; 225; 256; 289; 324; 361; 400 **3.** 1; 8; 27; 64; 125; 216; 343; 512; 729; 1000 **5.** **(a)** both of these **(b)** a perfect cube **(c)** a perfect square **(d)** a perfect square **7.** $(y + 5)(y - 5)$ **9.** $\left(p + \frac{1}{3}\right)\left(p - \frac{1}{3}\right)$ **11.** prime **13.** $(3r + 2)(3r - 2)$ **15.** $\left(6m + \frac{4}{5}\right)\left(6m - \frac{4}{5}\right)$ **17.** $4(3x + 2)(3x - 2)$ **19.** $(14p + 15)(14p - 15)$ **21.** $(4r + 5a)(4r - 5a)$ **23.** prime **25.** $(p^2 + 7)(p^2 - 7)$ **27.** $(x^2 + 1)(x + 1)(x - 1)$ **29.** $(p^2 + 16)(p + 4)(p - 4)$ **31.** The teacher was justified. $x^2 - 9$ can be factored as $(x + 3)(x - 3)$. The complete factored form is $(x^2 + 9)(x + 3)(x - 3)$. **33.** $(w + 1)^2$ **35.** $(x - 4)^2$ **37.** $\left(t + \frac{1}{2}\right)^2$ **39.** $(x - .5)^2$ **41.** $2(x + 6)^2$ **43.** $(4x - 5)^2$

**45.** $(7x - 2y)^2$ **47.** $(8x + 3y)^2$ **49.** $-2(5h - 2y)^2$ **51.** $k(4k^2 - 4k + 9)$ **53.** $z^2(25z^2 + 5z + 1)$ **55.** 10 **57.** 9 **59.** $(a - 1)(a^2 + a + 1)$ **61.** $(m + 2)(m^2 - 2m + 4)$ **63.** $(3x - 4)(9x^2 + 12x + 16)$ **65.** $6(p + 1)(p^2 - p + 1)$ **67.** $5(x + 2)(x^2 - 2x + 4)$ **69.** $2(y - 2x)(y^2 + 2xy + 4x^2)$ **71.** $(2p + 9q)(4p^2 - 18pq + 81q^2)$ **73.** $(3a + 4b)(9a^2 - 12ab + 16b^2)$ **75.** $(5t + 2s)(25t^2 - 10ts + 4s^2)$ **77.** $x^2 - y^2 = (x + y)(x - y)$ Difference of Squares; See Exercises 7–10, 13–22, and 25–30. $x^2 + 2xy + y^2 = (x + y)^2$ Perfect Square Trinomial; See Exercises 33, 34, 37, 38, 41, 47, 48, 50. $x^2 - 2xy + y^2 = (x - y)^2$ Perfect Square Trinomial; See Exercises 35, 36, 39, 40, 42–46, 49. $x^3 - y^3 = (x - y)(x^2 + xy + y^2)$ Difference of Cubes; See Exercises 59, 60, 63, 64, 68–70, 74. $x^3 + y^3 = (x + y)(x^2 - xy + y^2)$ Sum of Cubes; See Exercises 61, 62, 65–67, 71–73, 75, 76. **78.** $(5x - 2)(2x + 3)$ **79.** $5x - 2$ **80.** Yes. If $10x^2 + 11x - 6$ factors as $(5x - 2)(2x + 3)$, then when $10x^2 + 11x - 6$ is divided by $2x + 3$, the quotient should be $5x - 2$. **81.** $x^2 + x + 1$; $(x - 1)(x^2 + x + 1)$ **83.** $-2b(3a^2 + b^2)$ **85.** $3(r - k)(1 + r + k)$

## Summary Exercises on Factoring (page 346)

**1.** $(a - 6)(a + 2)$ **2.** $(a + 8)(a + 9)$ **3.** $6(y - 2)(y + 1)$ **4.** $7y^4(y + 6)(y - 4)$ **5.** $6(a + 2b + 3c)$ **6.** $(m - 4n)(m + n)$ **7.** $(p - 11)(p - 6)$ **8.** $(z + 7)(z - 6)$ **9.** $(5z - 6)(2z + 1)$ **10.** $2(m - 8)(m + 3)$ **11.** $(m + n + 5)(m - n)$ **12.** $5(3y + 1)$ **13.** $8a^3(a - 3)(a + 2)$ **14.** $(4k + 1)(2k - 3)$ **15.** $(z - 5a)(z + 2a)$ **16.** $50(z^2 - 2)$ **17.** $(x - 5)(x - 4)$ **18.** $10nr(10nr + 3r^2 - 5n)$ **19.** $(3n - 2)(2n - 5)$ **20.** $(3y - 1)(3y + 5)$ **21.** $4(4x + 5)$ **22.** $(m + 5)(m - 3)$ **23.** $(3y - 4)(2y + 1)$ **24.** $(m + 9)(m - 9)$ **25.** $(6z + 1)(z + 5)$ **26.** $(5z + 2)(z + 5)$ **27.** $(2k - 3)^2$ **28.** $(8p - 1)(p + 3)$ **29.** $6(3m + 2z)(3m - 2z)$ **30.** $(4m - 3)(2m + 1)$ **31.** $(3k - 2)(k + 2)$ **32.** $15a^3b^2(3b^3 - 4a + 5a^3b^2)$ **33.** $7k(2k + 5)(k - 2)$ **34.** $(5 + r)(1 - s)$ **35.** $(y^2 + 4)(y + 2)(y - 2)$ **36.** $10y^4(2y - 3)$ **37.** $8m(1 - 2m)$ **38.** $(k + 4)(k - 4)$ **39.** $(z - 2)(z^2 + 2z + 4)$ **40.** $(y - 8)(y + 7)$ **41.** prime **42.** $9p^8(3p + 7)(p - 4)$ **43.** $8m^3(4m^6 + 2m^2 + 3)$ **44.** $(2m + 5)(4m^2 - 10m + 25)$ **45.** $(4r + 3m)^2$ **46.** $(z - 6)^2$ **47.** $(5h + 7g)(3h - 2g)$ **48.** $5z(z - 7)(z - 2)$ **49.** $(k - 5)(k - 6)$ **50.** $4(4p - 5m)(4p + 5m)$ **51.** $3k(k - 5)(k + 1)$ **52.** $(y - 6k)(y + 2k)$ **53.** $(10p + 3)(100p^2 - 30p + 9)$ **54.** $(4r - 7)(16r^2 + 28r + 49)$ **55.** $(2 + m)(3 + p)$ **56.** $(2m - 3n)(m + 5n)$ **57.** $(4z - 1)^2$ **58.** $5m^2(5m - 3n)(5m - 13n)$ **59.** $3(6m - 1)^2$ **60.** $(10a + 9y)(10a - 9y)$ **61.** prime **62.** $(2y + 5)(2y - 5)$ **63.** $8z(4z - 1)(z + 2)$ **64.** $5(2m - 3)(m + 4)$ **65.** $(4 + m)(5 + 3n)$ **66.** $(2 - q)(2 - 3p)$ **67.** $2(3a - 1)(a + 2)$ **68.** $6y^4(3y + 4)(2y - 5)$ **69.** $(a - b)(a^2 + ab + b^2 + 2)$ **70.** $4(2k - 3)^2$ **71.** $(8m - 5n)^2$ **72.** $12y^2(6yz^2 + 1 - 2y^2z^2)$ **73.** $(4k - 3h)(2k + h)$ **74.** $(2a + 5)(a - 6)$ **75.** $(m + 2)(m^2 + m + 1)$ **76.** $(2a - 3)(4a^2 + 6a + 9)$ **77.** $(5y - 6z)(2y + z)$ **78.** $(m - 2)^2$ **79.** $(8a - b)(a + 3b)$ **80.** $(a^2 + 25)(a + 5)(a - 5)$ **81.** $(x^3 - 1)(x^3 + 1)$ **82.** $(x - 1)(x^2 + x + 1)(x + 1)(x^2 - x + 1)$ **83.** $(x^2 - 1)(x^4 + x^2 + 1)$ **84.** $(x - 1)(x + 1)(x^4 + x^2 + 1)$ **85.** The result in Exercise 82 is completely factored. **86.** Show that $x^4 + x^2 + 1 = (x^2 + x + 1)(x^2 - x + 1)$. **87.** difference of squares **88.** $(x - 3)(x^2 + 3x + 9)(x + 3)(x^2 - 3x + 9)$

## Section 5.5 (page 353)

**Exercises** **1.** $\{-5, 2\}$ **3.** $\left\{\frac{7}{2}, 3\right\}$ **5.** $\left\{-\frac{5}{6}, 0\right\}$ **7.** $\left\{0, \frac{4}{3}\right\}$ **9.** $\left\{-\frac{1}{2}, \frac{1}{6}\right\}$ **11.** $\{-.8, 2\}$ **13.** $\{9\}$ **15.** Set each *variable* factor equal to 0, to get $2x = 0$ or $3x - 4 = 0$. The solution set is $\left\{0, \frac{4}{3}\right\}$. **17.** $\{-2, -1\}$ **19.** $\{1, 2\}$ **21.** $\{-8, 3\}$ **23.** $\{-1, 3\}$ **25.** $\{-2, -1\}$ **27.** $\{-4\}$ **29.** $\left\{-2, \frac{1}{3}\right\}$ **31.** $\left\{-\frac{4}{3}, \frac{1}{2}\right\}$ **33.** $\left\{-\frac{2}{3}\right\}$ **35.** $\{-3, 3\}$ **37.** $\left\{-\frac{7}{4}, \frac{7}{4}\right\}$ **39.** $\{-11, 11\}$ **41.** $\{0, 7\}$ **43.** $\left\{0, \frac{1}{2}\right\}$ **45.** $\{2, 5\}$ **47.** $\left\{-4, \frac{1}{2}\right\}$ **49.** $\left\{-12, \frac{11}{2}\right\}$ **51.** $\left\{-\frac{5}{2}, \frac{1}{3}, 5\right\}$ **53.** $\left\{-\frac{7}{2}, -3, 1\right\}$ **55.** $\left\{-\frac{7}{3}, 0, \frac{7}{3}\right\}$ **57.** $\{-2, 0, 4\}$ **59.** $\{-5, 0, 4\}$ **61.** $\{-3, 0, 5\}$ **63.** $\{-1, 3\}$ **65.** $\{-1, 3\}$ **67.** $\{3\}$ **69.** $\left\{-\frac{4}{3}, -1, \frac{1}{2}\right\}$ **71.** $\left\{-\frac{2}{3}, 4\right\}$ **73.** **(a)** 64; 144; 4; 6 **(b)** No time has elapsed, so the object hasn't fallen (been released) yet. **(c)** Time cannot be negative. **75.** $\{1.5, 2.1\}$ **77.** $\{-2.75, 3\}$

## Section 5.6 (page 361)

**Exercises** **1.** Read; variable; equation; Solve; answer; Check, original **3.** *Step 3:* $45 = (2x + 1)(x + 1)$; *Step 4:* $x = 4$ or $x = -\frac{11}{2}$; *Step 5:* base: 9 units; height: 5 units; *Step 6:* $9 \cdot 5 = 45$ **5.** *Step 3:* $80 = (x + 8)(x - 8)$; *Step 4:* $x = 12$ or $x = -12$; *Step 5:* length: 20 units; width: 4 units; *Step 6:* $20 \cdot 4 = 80$ **7.** length: 7 in.; width: 4 in. **9.** length: 13 in.; width: 10 in. **11.** base: 12 in.; height: 5 in. **13.** height: 13 in.; width: 10 in. **15.** mirror: 7 ft; painting: 9 ft **17.** $-3, -2$ or $4, 5$ **19.** 7, 9, 11 **21.** $-2, 0, 2$ or 6, 8, 10 **23.** 12 cm **25.** 12 mi **27.** 8 ft **29.** **(a)** 1 sec **(b)** $\frac{1}{2}$ sec and $1\frac{1}{2}$ sec **(c)** 3 sec **(d)** The negative solution, $-1$, does not make sense since $t$ represents time, which cannot be negative. **31.** **(a)** 23.4 million; The result using the model is less than 24 million, the actual number for 1994. **(b)** 9 **(c)** 82.6 million; The result is less than 86 million, the actual number for 1999. **(d)** 137.6 million **32.** 107 billion dollars; 65% **33.** 1995: 66.9 billion dollars; 1997: 148.5 billion dollars; 1999: 230.1 billion dollars **34.** The answers using the linear equation are not at all close to the actual data. **35.** 1995: 104 billion dollars; 1997: 111.2 billion dollars; 1999: 266.4 billion dollars **36.** The answers in Exercise 35 are fairly close to the actual data. The quadratic equation models the data better. **37.** (0, 97.5), (1, 104.3), (2, 104.7), (3, 164.3), (4, 271.3) **38.** no

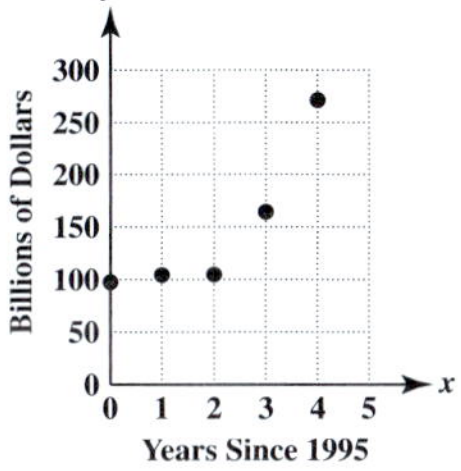


**39.** 399.5 billion dollars

**40.** **(a)** The actual deficit is about 30 billion dollars less than the prediction. **(b)** No, the equation is based on data for the years 1995–1999. Data for later years might not follow the same pattern.

## Chapter 5 Review Exercises (page 370)

**1.** $7(t + 2)$ **3.** $(x - 4)(2y + 3)$ **5.** $(x + 3)(x + 2)$ **7.** $(q + 9)(q - 3)$ **9.** $(r + 8s)(r - 12s)$ **11.** $8p(p + 2)(p - 5)$ **13.** $p^5(p - 2q)(p + q)$ **15.** $r$ and $6r$, $2r$ and $3r$ **17.** $(2k - 1)(k - 2)$ **19.** $(3r + 2)(2r - 3)$ **21.** $(v + 3)(8v - 7)$ **23.** $-3(x + 2)(2x - 5)$ **25.** B **27.** $(n + 7)(n - 7)$ **29.** $(7y + 5w)(7y - 5w)$ **31.** prime **33.** $(3t - 7)^2$ **35.** $(5k + 4x)(25k^2 - 20kx + 16x^2)$ **37.** $\left\{-\frac{3}{4}, 1\right\}$ **39.** $\left\{0, \frac{5}{2}\right\}$ **41.** $\{1, 4\}$ **43.** $\left\{-\frac{4}{3}, 5\right\}$ **45.** $\{0, 8\}$ **47.** $\{7\}$ **49.** $\left\{-\frac{2}{5}, -2, -1\right\}$ **51.** length: 10 ft; width: 4 ft **53.** length: 6 m; width: 2 m **55.** 6, 7 or $-5, -4$ **57.** 112 ft **59.** 256 ft **61.** **(a)** \$537 million **(b)** No, the prediction seems high. If eBay revenues in the last half of 2000 are comparable to those for the first half of the year, annual revenue in 2000 would be about \$366 million. **63.** The factor $2x + 8$ has a common factor of 2. The complete factored form is $2(x + 4)(3x - 4)$. **65.** $(3k + 5)(k + 2)$ **67.** $(y^2 + 25)(y + 5)(y - 5)$ **69.** $8abc(3b^2c - 7ac^2 + 9ab)$ **71.** $6xyz(2xz^2 + 2y - 5x^2yz^3)$ **73.** $(2r + 3q)(6r - 5)$ **75.** $(7t + 4)^2$ **77.** $\{-5, 2\}$ **79.** **(a)** 481,000 vehicles **(b)** The estimate may be unreliable because the conditions that prevailed in the years 1998–2001 may have changed, causing either a greater increase or a greater decrease in the numbers of alternative-fueled vehicles. **81.** length: 6 m; width: 4 m **83.** **(a)** 256 ft **(b)** 1024 ft **85.** 6 m

## Chapter 5 Test (page 374)

[5.1–5.4] **1.** D **2.** $6x(2x - 5)$ **3.** $m^2n(2mn + 3m - 5n)$ **4.** $(2x + y)(a - b)$ **5.** $(x + 3)(x - 8)$ **6.** $(2x + 3)(x - 1)$ **7.** $(5z - 1)(2z - 3)$ **8.** prime **9.** prime **10.** $(2 - a)(6 + b)$ **11.** $(3y + 8)(3y - 8)$ **12.** $(2x - 7y)^2$ **13.** $-2(x + 1)^2$ **14.** $3t^2(2t + 9)(t - 4)$ **15.** $(r - 5)(r^2 + 5r + 25)$ **16.** $8(k + 2)(k^2 - 2k + 4)$ **17.** $(x^2 + 9)(x + 3)(x - 3)$ **18.** The product $(p + 3)(p + 3) = p^2 + 6p + 9$, which does not equal $p^2 + 9$. The binomial $p^2 + 9$ is a prime polynomial. [5.5] **19.** $\left\{6, \frac{1}{2}\right\}$ **20.** $\left\{-\frac{2}{5}, \frac{2}{5}\right\}$ **21.** $\{10\}$ **22.** $\{-3, 0, 3\}$ [5.6] **23.** 6 ft by 9 ft **24.** $-2, -1$ **25.** 17 ft **26.** 181

## Cumulative Review Exercises Chapters 1–5 (page 375)

[2.1–2.3] **1.** {0} **2.** {.05} **3.** {6} [2.5] **4.** $P = \frac{A}{1 + rt}$ [2.7] **5.** 345; 210; 38%; 15% [2.4] **6.** gold: 12; silver: 9; bronze: 8 [2.7] **7.** 107 million [2.5] **8.** 110° and 70° [3.1] **9.** **(a)** negative, positive **(b)** negative, negative [3.2] **10.** $\left(-\frac{1}{4}, 0\right)$, (0, 3) [3.3] **11.** 12 [3.2] **12.**

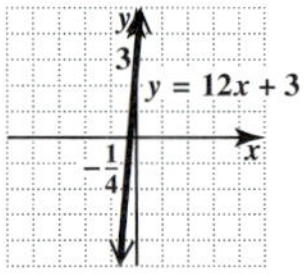


[3.3] **13.** 103; A slope of 103 means that the number of radio stations increased by about 103 stations per year. [3.4] **14.** $y = -\frac{1}{2}x + 5$ **15.** $y = -\frac{2}{3}$ [4.1, 4.2] **16.** 4 **17.** $\frac{16}{9}$ **18.** 256 **19.** $\frac{1}{p^2}$ **20.** $\frac{1}{m^6}$ [4.4] **21.** $-4k^2 - 4k + 8$ [4.5] **22.** $45x^2 + 3x - 18$ [4.6] **23.** $9p^2 + 12p + 4$ [4.7] **24.** $4x^3 + 6x^2 - 3x + 10$ [4.3] **25.** $5.5 \times 10^4$; $2.0 \times 10^6$ [5.2, 5.3] **26.** $(2a - 1)(a + 4)$ **27.** $(2m + 3)(5m + 2)$ **28.** $(4t + 3v)(2t + v)$ [5.4] **29.** $(2p - 3)^2$ **30.** $(5r + 9t)(5r - 9t)$ [5.3] **31.** $2pq(3p + 1)(p + 1)$ [5.5] **32.** $\left\{-\frac{2}{3}, \frac{1}{2}\right\}$ **33.** {0, 8} [5.6] **34.** 5 m, 12 m, 13 m

## CHAPTER 6 RATIONAL EXPRESSIONS AND APPLICATIONS

## Section 6.1 (page 385)

**Connections** **(page 385)** **1.** $3x^2 + 11x + 8$ cannot be factored, so this quotient cannot be simplified. By long division the quotient is $3x + 5 + \frac{-2}{x + 2}$. **2.** The numerator factors as $(x - 2)(x^2 + 2x + 4)$, so after simplifying the quotient is $x - 2$. Long division gives the same quotient.

**Exercises** **1.** A rational expression is a quotient of two polynomials, such as $\frac{x^2 + 3x - 6}{x + 4}$. One can think of this as an algebraic fraction. **3.** **(a)** $\frac{7}{10}$ **(b)** $\frac{8}{15}$ **5.** **(a)** 0 **(b)** $-1$ **7.** **(a)** $-\frac{64}{15}$ **(b)** undefined **9.** **(a)** undefined **(b)** $\frac{8}{25}$ **11.** Division by 0 is undefined, so if the denominator of a rational expression equals 0, the expression is undefined. **13.** 0 **15.** 6 **17.** $-\frac{5}{3}$ **19.** $-3, 2$ **21.** never undefined **23.** never undefined **25.** **(a)** numerator: $x^2$, $4x$; denominator: $x$, 4 **(b)** First factor the numerator, getting $x(x + 4)$, then divide the numerator and denominator by the common factor of $x + 4$ to get $\frac{x}{1}$ or $x$. **27.** $3r^2$ **29.** $\frac{2}{5}$ **31.** $\frac{x - 1}{x + 1}$ **33.** $\frac{7}{5}$ **35.** $\frac{6}{7}$ **37.** $m - n$ **39.** $\frac{2}{t - 3}$ **41.** $\frac{3(2m + 1)}{4}$ **43.** $\frac{3m}{5}$ **45.** $\frac{3r - 2s}{3}$ **47.** $k - 3$ **49.** $\frac{x - 3}{x + 1}$ **51.** $\frac{x + 1}{x - 1}$ **53.** $\frac{z - 3}{z + 5}$ **55.** $\frac{m + n}{2}$ **57.** $-\frac{b^2 + ba + a^2}{a + b}$ **59.** $\frac{z + 3}{z}$ **61.** B, D **63.** $-1$ **65.** $-(m + 1)$ **67.** $-1$ **69.** already in lowest terms **71.** B

**Answers may vary in Exercises 73, 75, and 77.** **73.** $\frac{-(x + 4)}{x - 3}, \frac{-x - 4}{x - 3}, \frac{x + 4}{-(x - 3)}, \frac{x + 4}{-x + 3}$ **75.** $\frac{-(2x - 3)}{x + 3}, \frac{-2x + 3}{x + 3}, \frac{2x - 3}{-(x + 3)}, \frac{2x - 3}{-x - 3}$ **77.** $-\frac{3x - 1}{5x - 6}, \frac{-(3x - 1)}{5x - 6}, -\frac{-3x + 1}{-5x + 6}, \frac{3x - 1}{-5x + 6}$ **79.** $x^2 + 3$ **81.** **(a)** 0 **(b)** 1.6 **(c)** 4.1 **(d)** The waiting time also increases.

## Section 6.2 (page 392)

**Exercises** **1.** **(a)** B **(b)** D **(c)** C **(d)** A **3.** $\frac{3a}{2}$ **5.** $-\frac{4x^4}{3}$ **7.** $\frac{2}{c+d}$ **9.** $4(x-y)$ **11.** $\frac{t^2}{2}$ **13.** $\frac{x+3}{2x}$ **15.** 5 **17.** $-\frac{3}{2t^4}$ **19.** $\frac{1}{4}$ **21.** $-\frac{35}{8}$ **23.** $\frac{2(x+2)}{x(x-1)}$ **25.** $\frac{x(x-3)}{6}$ **27.** Suppose I want to multiply $\frac{a^2-1}{6}\cdot\frac{9}{2a+2}$. I start by factoring where possible: $\frac{(a+1)(a-1)}{6}\cdot\frac{9}{2(a+1)}$. Next, I divide out common factors in the numerator and denominator to get $\frac{a-1}{2}\cdot\frac{3}{2}$. Finally, I multiply numerator times numerator and denominator times denominator to get the final product, $\frac{3(a-1)}{4}$. **29.** $\frac{10}{9}$ **31.** $-\frac{3}{4}$ **33.** $-\frac{9}{2}$ **35.** $\frac{p+4}{p+2}$ **37.** $-1$ **39.** $\frac{(2x-1)(x+2)}{x-1}$ **41.** $\frac{(k-1)^2}{(k+1)(2k-1)}$ **43.** $\frac{4k-1}{3k-2}$ **45.** $\frac{m+4p}{m+p}$ **47.** $\frac{m+6}{m+3}$ **49.** $\frac{y+3}{y+4}$ **51.** $\frac{m}{m+5}$ **53.** $\frac{r+6s}{r+s}$ **55.** $\frac{(q-3)^2(q+2)^2}{q+1}$ **57.** $\frac{x+10}{10}$ **59.** $\frac{3-a-b}{2a-b}$ **61.** $-\frac{(x+y)^2(x^2-xy+y^2)}{3y(y-x)(x-y)}$ or $\frac{(x+y)^2(x^2-xy+y^2)}{3y(x-y)^2}$ **63.** $\frac{5xy^2}{4q}$

## Section 6.3 (page 397)

**Exercises** **1.** C **3.** C **5.** 60 **7.** 1800 **9.** $x^5$ **11.** $30p$ **13.** $180y^4$ **15.** $84r^5$ **17.** $15a^5b^3$ **19.** $12p(p-2)$ **21.** $28m^2(3m-5)$ **23.** $30(b-2)$ **25.** $2^3\cdot 3\cdot 5$ **26.** $(t+4)^3(t-3)(t+8)$ **27.** The similarity is that 2 is replaced by $t+4$, 3 is replaced by $t-3$, and 5 is replaced by $t+8$. **28.** The procedure used is the same. The only difference is that for algebraic fractions, the factors may contain variables, while in common fractions, the factors are numbers. **29.** $18(r-2)$ **31.** $12p(p+5)^2$ **33.** $8(y+2)(y+1)$ **35.** $c-d$ or $d-c$ **37.** $m-3$ or $3-m$ **39.** $p-q$ or $q-p$ **41.** $k(k+5)(k-2)$ **43.** $a(a+6)(a-3)$ **45.** $(p+3)(p+5)(p-6)$ **47.** $(k+3)(k-5)(k+7)(k+8)$ **49.** Yes, because $(2x-5)^2=(5-2x)^2$. **51.** 7 **52.** 1 **53.** identity property of multiplication **54.** 7 **55.** 1 **56.** identity property of multiplication **57.** $\frac{20}{55}$ **59.** $\frac{-45}{9k}$ **61.** $\frac{60m^2k^3}{32k^4}$ **63.** $\frac{57z}{6z-18}$ **65.** $\frac{-4a}{18a-36}$ **67.** $\frac{6(k+1)}{k(k-4)(k+1)}$ **69.** $\frac{36r(r+1)}{(r-3)(r+2)(r+1)}$ **71.** $\frac{ab(a+2b)}{2a^3b+a^2b^2-ab^3}$ **73.** $\frac{(t-r)(4r-t)}{t^3-r^3}$ **75.** $\frac{2y(z-y)(y-z)}{y^4-z^3y}$ or $\frac{-2y(y-z)^2}{y^4-z^3y}$ **77.** *Step 1:* Factor each denominator into prime factors. *Step 2:* List each different denominator factor the greatest number of times it appears in any of the denominators. *Step 3:* Multiply the factors in the list to get the LCD. For example, the least common denominator for $\frac{1}{(x+y)^3}$ and $\frac{-2}{(x+y)^2(p+q)}$ is $(x+y)^3(p+q)$.

## Section 6.4 (page 406)

**Exercises** **1.** E **3.** C **5.** B **7.** G **9.** $\frac{11}{m}$ **11.** $\frac{4}{y+4}$ **13.** 1 **15.** $\frac{m-1}{m+1}$ **17.** $b$ **19.** $x$ **21.** $y-6$ **23.** To add or subtract rational expressions with the same denominators, combine the numerators and keep the same denominator. For example, $\frac{3x+2}{x-6}+\frac{-2x-8}{x-6}=\frac{x-6}{x-6}$. Then write in lowest terms. In this example, the sum simplifies to 1. **25.** $\frac{3z+5}{15}$ **27.** $\frac{10-7r}{14}$ **29.** $\frac{-3x-2}{4x}$ **31.** $\frac{57}{10x}$ **33.** $\frac{x+1}{2}$ **35.** $\frac{5x+9}{6x}$ **37.** $\frac{7-6p}{3p^2}$ **39.** $\frac{-k-8}{k(k+4)}$ **41.** $\frac{x+4}{x+2}$ **43.** $\frac{6m^2+23m-2}{(m+2)(m+1)(m+5)}$ **45.** $\frac{4y^2-y+5}{(y+1)^2(y-1)}$ **47.** $\frac{3}{t}$ **49.** $m-2$ or $2-m$ **51.** $\frac{-2}{x-5}$ or $\frac{2}{5-x}$ **53.** $-4$ **55.** $\frac{-5}{x-y^2}$ or $\frac{5}{y^2-x}$ **57.** $\frac{x+y}{5x-3y}$ or $\frac{-x-y}{3y-5x}$

**59.** $\frac{-6}{4p-5}$ or $\frac{6}{5-4p}$ **61.** $\frac{-(m+n)}{2(m-n)}$ **63.** $\frac{-x^2+6x+11}{(x+3)(x-3)(x+1)}$ **65.** $\frac{-5q^2-13q+7}{(3q-2)(q+4)(2q-3)}$ **67.** $\frac{9r+2}{r(r+2)(r-1)}$

**69.** $\frac{2(x^2+3xy+4y^2)}{(x+y)(x+y)(x+3y)}$ or $\frac{2(x^2+3xy+4y^2)}{(x+y)^2(x+3y)}$ **71.** $\frac{15r^2+10ry-y^2}{(3r+2y)(6r-y)(6r+y)}$ **73. (a)** $\frac{9k^2+6k+26}{5(3k+1)}$ **(b)** $\frac{1}{4}$ **75.** $\frac{10x}{49(101-x)}$

## Section 6.5 (page 414)

**Exercises** **1. (a)** $6; \frac{1}{6}$ **(b)** $12; \frac{3}{4}$ **(c)** $\frac{1}{6} \div \frac{3}{4}$ **(d)** $\frac{2}{9}$ **3.** Choice D is correct, because every sign has been changed in the fraction.

**5.** Method 1 indicates to write the complex fraction as a division problem, and then perform the division. For example, to simplify $\frac{\frac{1}{2}}{\frac{2}{3}}$, write $\frac{1}{2} \div \frac{2}{3}$. Then simplify as $\frac{1}{2} \cdot \frac{3}{2} = \frac{3}{4}$. **7.** $-6$ **9.** $\frac{1}{xy}$ **11.** $\frac{2a^2b}{3}$ **13.** $\frac{m(m+2)}{3(m-4)}$ **15.** $\frac{2}{x}$ **17.** $\frac{8}{x}$ **19.** $\frac{a^2-5}{a^2+1}$ **21.** $\frac{31}{50}$

**23.** $\frac{y^2+x^2}{xy(y-x)}$ **25.** $\frac{40-12p}{85p}$ **27.** $\frac{5y-2x}{3+4xy}$ **29.** $\frac{a-2}{2a}$ **31.** $\frac{z-5}{4}$ **33.** $\frac{-m}{m+2}$ **35.** $\frac{3m(m-3)}{(m-1)(m-8)}$ **37.** division

**39.** $\frac{\frac{3}{8}+\frac{5}{6}}{2}$ **40.** $\frac{29}{48}$ **41.** $\frac{29}{48}$ **42.** Answers will vary.

## Section 6.6 (page 423)

**Exercises** **1.** expression; $\frac{43}{40}x$ **3.** equation; $\left\{\frac{40}{43}\right\}$ **5.** expression; $-\frac{1}{10}x$ **7.** equation; $\{-10\}$ **9.** $-2, 0$ **11.** $-3, 4, -\frac{1}{2}$

**13.** $-9, 1, -2, 2$ **15.** When adding and subtracting, the LCD is retained as part of the answer. When solving an equation, the LCD is used as the multiplier in applying the multiplication property of equality. **17.** $\left\{\frac{1}{4}\right\}$ **19.** $\left\{-\frac{3}{4}\right\}$ **21.** $\{-15\}$ **23.** $\{7\}$ **25.** $\{-15\}$

**27.** $\{-5\}$ **29.** $\{-6\}$ **31.** $\emptyset$ **33.** $\{5\}$ **35.** $\{4\}$ **37.** $\{1\}$ **39.** $\{4\}$ **41.** $\{5\}$ **43.** $\{-4\}$ **45.** $\{-2, 12\}$ **47.** $\emptyset$ **49.** $\{3\}$

**51.** $\{3\}$ **53.** $\{-3\}$ **55.** $\left\{-\frac{1}{5}, 3\right\}$ **57.** $\left\{-\frac{1}{2}, 5\right\}$ **59.** $\{3\}$ **61.** $\left\{-\frac{1}{3}, 3\right\}$ **63.** $\{-1\}$ **65.** $\{-6\}$ **67.** $\left\{-6, \frac{1}{2}\right\}$ **69.** $\{6\}$

**71.** Transform the equation so that the terms with $k$ are on one side and the remaining term is on the other. **73.** $F = \frac{ma}{k}$ **75.** $a = \frac{kF}{m}$

**77.** $R = \frac{E-Ir}{I}$ or $R = \frac{E}{I} - r$ **79.** $A = \frac{h(B+b)}{2}$ **81.** $a = \frac{2S-ndL}{nd}$ or $a = \frac{2S}{nd} - L$ **83.** $y = \frac{xz}{x+z}$ **85.** $z = \frac{3y}{5-9xy}$ or $z = \frac{-3y}{9xy-5}$

## Summary Exercises on Operations and Equations with Rational Expressions (page 427)

**1.** operation; $\frac{10}{p}$ **2.** operation; $\frac{y^3}{x^3}$ **3.** operation; $\frac{1}{2x^2(x+2)}$ **4.** equation; $\{9\}$ **5.** operation; $\frac{y+2}{y-1}$ **6.** operation; $\frac{5k+8}{k(k-4)(k+4)}$

**7.** equation; $\{39\}$ **8.** operation; $\frac{t-5}{3(2t+1)}$ **9.** operation; $\frac{13}{3(p+2)}$ **10.** equation; $\left\{-1, \frac{12}{5}\right\}$ **11.** equation; $\left\{\frac{1}{7}, 2\right\}$

**12.** operation; $\frac{16}{3y}$ **13.** operation; $\frac{7}{12z}$ **14.** equation; $\{13\}$ **15.** operation; $\frac{3m+5}{(m+3)(m+2)(m+1)}$ **16.** operation; $\frac{k+3}{5(k-1)}$

**17.** equation; $\emptyset$ **18.** equation; $\emptyset$ **19.** operation; $\frac{t+2}{2(2t+1)}$ **20.** equation; $\{-7\}$

## Section 6.7 (page 432)

**Exercises** **1.** **(a)** the amount **(b)** $5 + x$ **(c)** $\frac{5+x}{6} = \frac{13}{3}$ **3.** $\frac{12}{18}$ **5.** $\frac{12}{3}$ **7.** 12 **9.** $\frac{1386}{97}$ **11.** 74.75 sec **13.** 367.197 m per min **15.** 3.090 hr **17.** $\frac{D}{R} = \frac{d}{r}$ **19.** $\frac{500}{x-10} = \frac{600}{x+10}$ **21.** 8 mph **23.** 165 mph **25.** 3 mph **27.** 18.5 mph **29.** $\frac{1}{10}$ job per hr **31.** $\frac{1}{8}x + \frac{1}{6}x = 1$ or $\frac{1}{8} + \frac{1}{6} = \frac{1}{x}$ **33.** $4\frac{4}{17}$ hr **35.** $5\frac{5}{11}$ hr **37.** 3 hr **39.** $2\frac{7}{10}$ hr **41.** $9\frac{1}{11}$ min

## Chapter 6 Review Exercises (page 442)

**1.** **(a)** $\frac{11}{8}$ **(b)** $\frac{13}{22}$ **3.** 3 **5.** $-5, -\frac{2}{3}$ **7.** $\frac{b}{3a}$ **9.** $\frac{-(2x+3)}{2}$ **11.** (Answers may vary.) $\frac{-(4x-9)}{2x+3}, \frac{-4x+9}{2x+3}, \frac{4x-9}{-(2x+3)}, \frac{4x-9}{-2x-3}$ **13.** $\frac{72}{p}$ **15.** $\frac{5}{8}$ **17.** $\frac{3a-1}{a+5}$ **19.** $\frac{p+5}{p+1}$ **21.** $108y^4$ **23.** $\frac{15a}{10a^4}$ **25.** $\frac{15y}{50-10y}$ **27.** $\frac{15}{x}$ **29.** $\frac{4k-45}{k(k-5)}$ **31.** $\frac{-2-3m}{6}$ **33.** $\frac{7a+6b}{(a-2b)(a+2b)}$ **35.** $\frac{5z-16}{z(z+6)(z-2)}$ **37.** **(a)** $\frac{a}{b}$ **(b)** $\frac{a}{b}$ **(c)** Answers will vary. **39.** $\frac{4(y-3)}{y+3}$ **41.** $\frac{xw+1}{xw-1}$ **43.** It would cause the first and third denominators to equal 0. **45.** $\emptyset$ **47.** $t = \frac{Ry}{m}$ **49.** $m = \frac{4+p^2q}{3p^2}$ **51.** $\frac{2}{6}$ **53.** $3\frac{1}{13}$ hr **55.** $\frac{m+7}{(m-1)(m+1)}$ **57.** $\frac{1}{6}$ **59.** $\frac{z+7}{(z+1)(z-1)^2}$ **61.** $\{-2, 3\}$ **63.** $1\frac{7}{8}$ hr **65.** **(a)** $-3$ **(b)** $-1$ **(c)** $-3, -1$ **66.** $\frac{15}{2x}$ **67.** If $x = 0$, the divisor $R$ is equal to 0, and division by 0 is undefined. **68.** $(x+3)(x+1)$ **69.** $\frac{7}{x+1}$ **70.** $\frac{11x+21}{4x}$ **71.** $\emptyset$ **72.** We know that $-3$ is not allowed because $P$ and $R$ are undefined for $x = -3$. **73.** Rate is equal to distance divided by time. Here, distance is 6 mi, and time is $x + 3$ min, so rate $= \frac{6}{x+3}$, which is the expression for $P$. **74.** $\frac{6}{5}, \frac{5}{2}$

## Chapter 6 Test (page 446)

[6.1] **1.** **(a)** $\frac{11}{6}$ **(b)** undefined **2.** $-2, 4$ **3.** (Answers may vary.) $\frac{-(6x-5)}{2x+3}, \frac{-6x+5}{2x+3}, \frac{6x-5}{-(2x+3)}, \frac{6x-5}{-2x-3}$ **4.** $-3x^2y^3$ **5.** $\frac{3a+2}{a-1}$ [6.2] **6.** $\frac{25}{27}$ **7.** $\frac{3k-2}{3k+2}$ **8.** $\frac{a-1}{a+4}$ [6.3] **9.** $150p^5$ **10.** $(2r+3)(r+2)(r-5)$ **11.** $\frac{240p^2}{64p^3}$ **12.** $\frac{21}{42m-84}$ [6.4] **13.** 2 **14.** $\frac{-14}{5(y+2)}$ **15.** $\frac{-x^2+x+1}{3-x}$ or $\frac{x^2-x-1}{x-3}$ **16.** $\frac{-m^2+7m+2}{(2m+1)(m-5)(m-1)}$ [6.5] **17.** $\frac{2k}{3p}$ **18.** $\frac{-2-x}{4+x}$ [6.6] **19.** $\left\{-\frac{1}{2}\right\}$ **20.** $D = \frac{dF-k}{F}$ or $D = \frac{k-dF}{-F}$ [6.7] **21.** 3 mph **22.** $2\frac{2}{9}$ hr

## Cumulative Review Exercises Chapters 1–6 (page 447)

[1.2, 1.5, 1.6] **1.** 2 [2.3] **2.** $\{17\}$ [2.5] **3.** $b = \frac{2A}{h}$ [2.6] **4.** $\left\{-\frac{2}{7}\right\}$ [2.8] **5.** $[-8, \infty)$ **6.** $(4, \infty)$ [3.1] **7.** **(a)** $(-3, 0)$ **(b)** $(0, -4)$ [3.2] **8.**

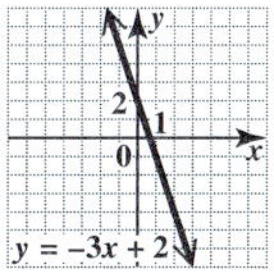


[4.4] **9.**

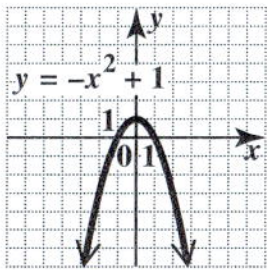


[3.3] **10.** $-\frac{3}{2}$ **11.** $-\frac{3}{4}$ [4.1, 4.2] **12.** $\frac{1}{2^4x^7}$ **13.** $\frac{1}{m^6}$ **14.** $\frac{q}{4p^2}$ [4.4] **15.** $k^2 + 2k + 1$ [4.1] **16.** $72x^6y^7$ [4.6] **17.** $4a^2 - 4ab + b^2$ [4.5] **18.** $3y^3 + 8y^2 + 12y - 5$

[4.7] **19.** $6p^2 + 7p + 1 + \frac{3}{p-1}$ [4.3] **20.** $1.4 \times 10^5$ sec [5.3] **21.** $(4t + 3v)(2t + v)$ **22.** prime [5.4] **23.** $(4x^2 + 1)(2x + 1)(2x - 1)$ [5.5] **24.** $\{-3, 5\}$ **25.** $\left\{5, -\frac{1}{2}, \frac{2}{3}\right\}$ [5.6] **26.** $-2$ or $-1$ **27.** 6 m [6.1] **28.** A **29.** D [6.4] **30.** $\frac{4}{q}$ **31.** $\frac{3r+28}{7r}$ **32.** $\frac{7}{15(q-4)}$ **33.** $\frac{-k-5}{k(k+1)(k-1)}$ [6.2] **34.** $\frac{7(2z+1)}{24}$ [6.5] **35.** $\frac{195}{29}$ [6.6] **36.** 4, 0 **37.** $\left\{\frac{21}{2}\right\}$ **38.** $\{-2, 1\}$ [6.7] **39.** $1\frac{1}{5}$ hr **40.** $12\frac{2}{3}$ mph

## CHAPTER 7 EQUATIONS OF LINES; FUNCTIONS

### Section 7.1 (page 461)

**Connections** (page 455) We give each equation solved for $y$, the $x$-intercept, and the $y$-intercept. **1.** $y = -2.5x - 5$; $(-2, 0)$; $(0, -5)$ **2.** $y = .75x + 1.5$; $(-2, 0)$; $(0, 1.5)$ **3.** $y = 3.2x + 5.8$; $(-1.8125, 0)$; $(0, 5.8)$ **4.** $y = -1.5x + 4.2$; $(2.8, 0)$; $(0, 4.2)$

**Exercises** **1.** **(a)** $x$ represents the year; $y$ represents the percent of women in math or computer science professions. **(b)** 1990–2000 **(c)** (1990, 36) **(d)** In 2000, the percent of women in math or computer science professions was 30%. **3.** **(a)** I **(b)** III **(c)** II **(d)** IV **(e)** none **(f)** none **5.–9.**

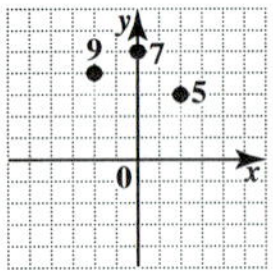

**11.** **(a)** $-3$; 3; 2; $-1$ **(b)**

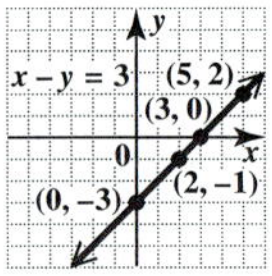


**13.** **(a)** $-4$; 5; $-\frac{12}{5}$; $\frac{5}{4}$ **(b)**

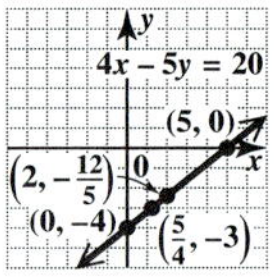


**15.** In quadrant III, both coordinates of the ordered pairs are negative. If $x + y = k$ and $k$ is positive, then either $x$ or $y$ must be positive because the sum of two negative numbers is negative. **17.** $(6, 0)$; $(0, 4)$

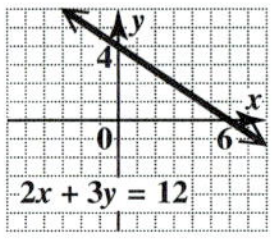


**19.** $(6, 0)$; $(0, -2)$

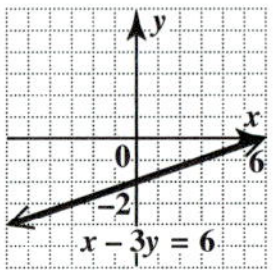


**21.** $\left(\frac{21}{2}, 0\right)$; $\left(0, -\frac{7}{3}\right)$

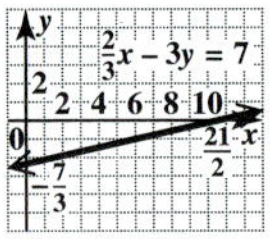

**23.** $(0, 0)$; $(0, 0)$

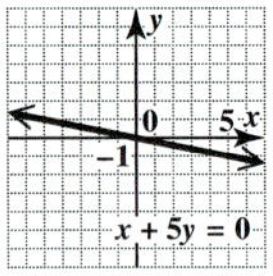


**25.** $(0, 0)$; $(0, 0)$

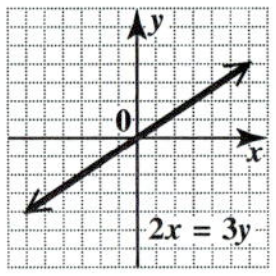


**27.** none; $(0, 5)$

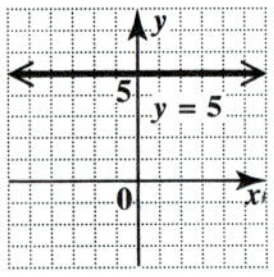


**29.** $(-4, 0)$; none

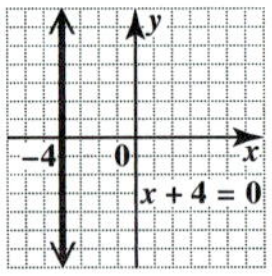


**31.** B **33.** The screen on the right is more useful because it shows the intercepts. **34.** $(6, -2)$ **35.** $(5, -2)$ **36.** $(6, 0)$ **37.** $(5, 0)$ **38.** 5; 0 **39.** The $x$-coordinate of $M$ is the average of the $x$-coordinates of $P$ and $Q$. The $y$-coordinate of $M$ is the average of the $y$-coordinates of $P$ and $Q$. **41.** $(2, 5)$ **43.** $\left(-\frac{3}{2}, \frac{5}{2}\right)$ **45.** $\left(\frac{3}{2}, 0\right)$ **47.** $(3.8, 2.6)$ **49.** A, B, and D **51.** **(a)** B **(b)** C **(c)** A **53.** 0 **55.** $-\frac{1}{3}$ **57.** B and D are correct. Choice A is wrong because the order of subtraction must be the same in the numerator and denominator. Choice C is wrong because slope is defined as the change in $y$ divided by the change in $x$. **59.** 1 **61.** $\frac{9}{8}$ **63.** 0 **65.** 2 **67.** $B$ **69.** $A$ **71.** $-\frac{1}{2}$

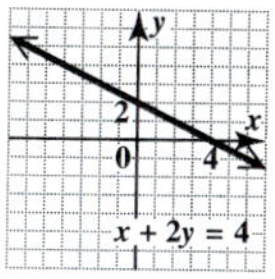


**73.** $\frac{5}{2}$

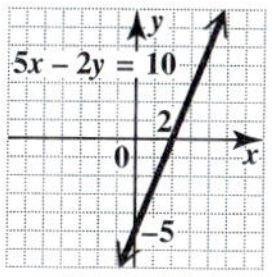


**75.** 4

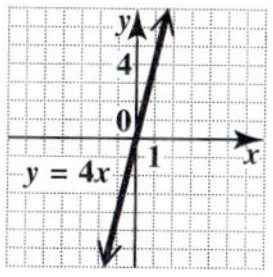


**77.** undefined

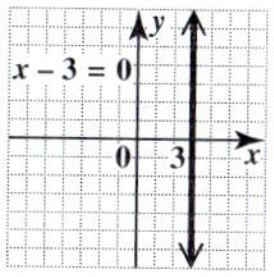


**79.** 0

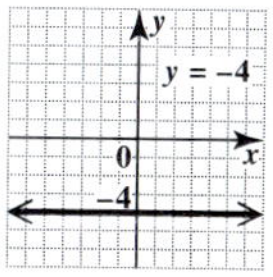


**81.**

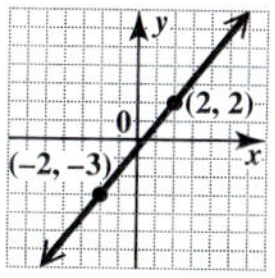


**83.**

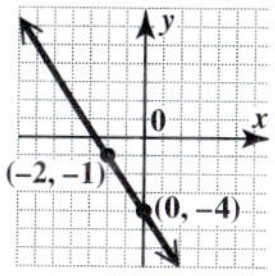


**85.**

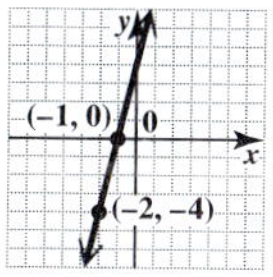


**87.**

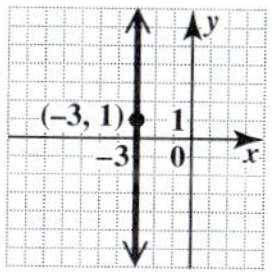


**89.** parallel **91.** perpendicular **93.** neither **95.** neither

**97.** perpendicular **99.** −$4000 per yr; The value of the machine is decreasing $4000 each year during these years. **101.** 0% per yr (or no change); The percent of pay raise is not changing—it is 3% each year during these years. **103.** **(a)** 1000 million per yr; 1000 million per yr; 1000 million per yr **(b)** The average rate of change is the same. When graphed, the data points lie on a straight line. **105.** **(a)** $200 million per yr **(b)** The positive slope means expenditures *increased* an average of $200 million each year. **107.** −$69 per yr; The price decreased an average of $69 each year from 1997 to 2002.

## Section 7.2 (page 477)

**Connections** **(page 476)** **1.** {3} **2.** {1} **3.** {3} **4.** {−.5}

**Exercises** **1.** A **3.** A **5.** $3x + y = 10$ **7.** A **9.** C **11.** H **13.** B **15.** $y = 5x + 15$ **17.** $y = -\frac{2}{3}x + \frac{4}{5}$

**19.** $y = \frac{2}{5}x + 5$ **21.** $y = \frac{2}{3}x + 1$ **23.** **(a)** $y = x + 4$ **(b)** 1 **(c)** 4 **(d)**

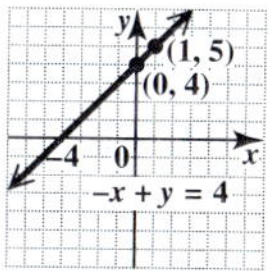


**25.** **(a)** $y = -\frac{6}{5}x + 6$ **(b)** $-\frac{6}{5}$ **(c)** 6

**(d)**

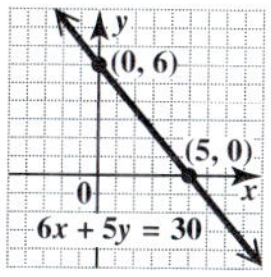


**27.** **(a)** $y = \frac{4}{5}x - 4$ **(b)** $\frac{4}{5}$ **(c)** −4 **(d)**

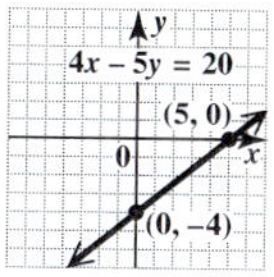


**29.** **(a)** $y = -\frac{1}{2}x - 2$ **(b)** $-\frac{1}{2}$ **(c)** −2 **(d)**

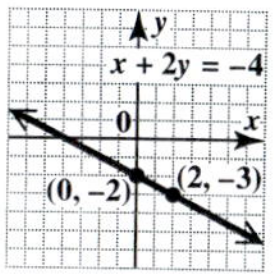


**31.** $3x + 4y = 10$ **33.** $2x + y = 18$ **35.** $x - 2y = -13$ **37.** $4x - y = 12$ **39.** The slope-intercept form, $y = mx + b$, is used when the slope and $y$-intercept are known. The point-slope form, $y - y_1 = m(x - x_1)$, is used when the slope and one point on a line or two points on a line are known. The standard form, $Ax + By = C$, is not useful for writing the equation, but the intercepts can be found quickly and used to graph the equation. The form $y = b$ is used for a horizontal line through the point $(a, b)$. The form $x = a$ is used for a vertical line through the point $(a, b)$. **41.** $y = 5$ **43.** $x = 9$ **45.** $x = .5$ **47.** $y = 8$ **49.** $2x - y = 2$ **51.** $x + 2y = 8$

**53.** $2x - 13y = -6$ **55.** $y = 5$ **57.** $x = 7$ **59.** $y = -3$ **61.** $y = 3x - 19$ **63.** $y = \frac{1}{2}x - 1$ **65.** $y = -\frac{1}{2}x + 9$

**67.** $y = 7$ **69.** $y = 45x$; (0, 0), (5, 225), (10, 450) **71.** $y = 1.50x$; (0, 0), (5, 7.50), (10, 15.00) **73.** **(a)** $y = 39x + 99$ **(b)** (5, 294); The cost of a 5-month membership is $294. **(c)** $567 **75.** **(a)** $y = 50x + 25$ **(b)** (5, 275); The cost of the plan for 5 months is $275. **(c)** $1225 **77.** **(a)** $y = .20x + 50$ **(b)** (5, 51); The charge for driving 5 mi is $51. **(c)** 173 mi **79.** **(a)** $y = 5.6x + 9$; The percent of households accessing the Internet by broadband is increasing 5.6% per year. **(b)** 43% **81.** **(a)** $y = -103.2x + 28{,}908$ **(b)** 28,082; The result using the model is a little high. **83.** 32; 212 **84.** (0, 32) and (100, 212) **85.** $\frac{9}{5}$ **86.** $F = \frac{9}{5}C + 32$ **87.** $C = \frac{5}{9}(F - 32)$ **88.** When the Celsius temperature is 50°, the Fahrenheit temperature is 122°.

**47. (a)**

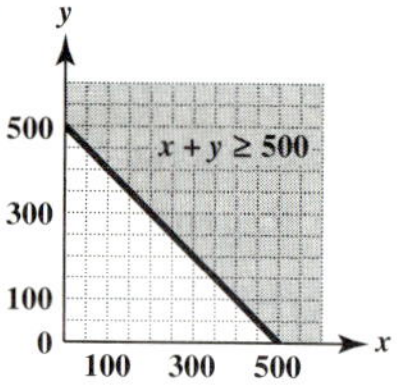


**(b)** (500, 0), (200, 400), (100, 450) Answers may vary.

**49. (a)**

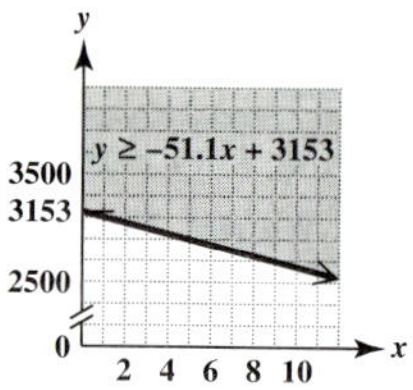


**(b)** (0, 3153), (2, 3050.8), (10, 2642) Answers may vary.

## Chapter 9 Review Exercises (page 623)

**1.** {a, c} **3.** {a, c, e, f, g} **5.** (6, 9)

**7.** $(-\infty, -3] \cup (5, \infty)$

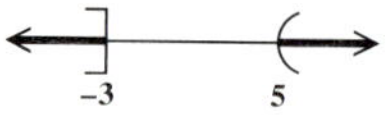


**9.** ∅ **11.** (−3, 4)

**13.** (4, ∞) **15.** {−7, 7} **17.** $\left\{-\frac{1}{3}, 5\right\}$ **19.** {0, 7} **21.** $\left\{-\frac{3}{4}, \frac{1}{2}\right\}$ **23.** (−14, 14) **25.** [−3, −2]

**27.** $\left(-\infty, -\frac{8}{5}\right) \cup (2, \infty)$ **29.** (−∞, ∞) **31.**

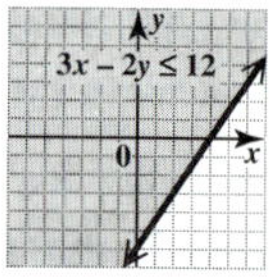


**33.**

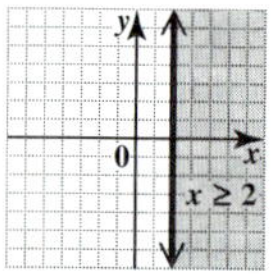


**35.**

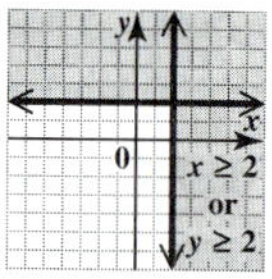


**37.** [−2, 3)

**39.** $\left\{-\frac{7}{3}, 1\right\}$ **41.** $\left\{-4, -\frac{2}{3}\right\}$ **43.** (−∞, ∞) **45.** [−4, −2] **47.**

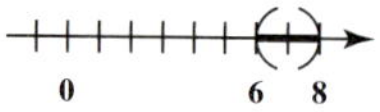


**49.**

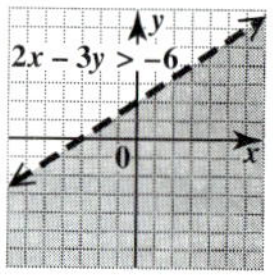


**51.** Managerial/Professional, Technical/Sales/Administrative Support, Service, Operators/Fabricators/Laborers

## Chapter 9 Test (page 625)

[9.1] **1.** none **2.** 1946, 1960, 1968, 1976, 1996, 1998, 1999 **3.** {1, 5} **4.** {1, 2, 5, 7, 9, 12}

**5.** [2, 9)

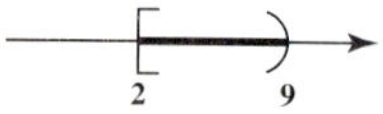


**6.** $(-\infty, 3) \cup [6, \infty)$

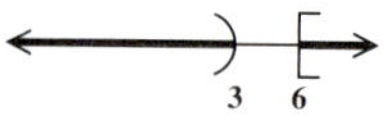


[9.2] **7.** $\left\{-1, \frac{5}{2}\right\}$ **8.** $\left(-\infty, -\frac{7}{6}\right) \cup \left(\frac{17}{6}, \infty\right)$

**9.** ∅ **10.** $\left\{-\frac{5}{7}, \frac{11}{3}\right\}$ **11.** $\left(\frac{1}{3}, \frac{7}{3}\right)$ **12.** (−∞, ∞) **13.** **(a)** ∅ **(b)** (−∞, ∞) **(c)** ∅

[9.3] **14.**

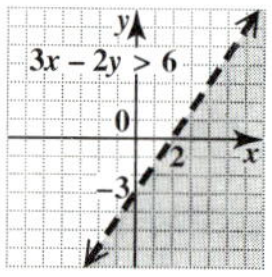


**15.**

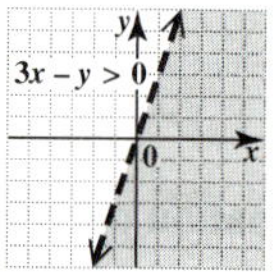


**16.**

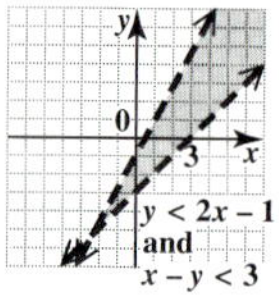


**17.**

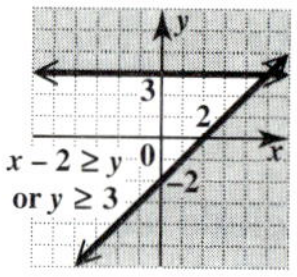


## Cumulative Review Exercises Chapters 1–9 (page 626)

[1.4] **1. (a)** A, B, C, D, F **(b)** B, C, D, F **(c)** D, F **(d)** E, F **(e)** E, F **(f)** D, F [1.2] **2.** 32 [1.4, 1.5] **3.** 0

[2.1–2.3] **4.** {−65} **5.** (−∞, ∞) [2.5] **6.** $t = \frac{A - p}{pr}$ [2.8] **7.** (−∞, 6) [2.7] **8.** 32%; 390; 270; 10% [2.4] **9.** 15°, 35°, 130°

[3.3, 7.1] **10.** $-\frac{4}{3}$ **11.** 0 [3.4, 7.2] **12.** $y = -4x + 15$ **13.** $y = 4x$ [3.2, 7.1] **14.**

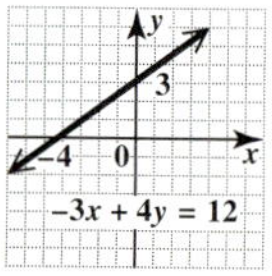


[4.1, 4.2] **15.** $\frac{8m^9n^3}{p^6}$ **16.** $\frac{y^7}{x^{13}z^2}$

[4.4] **17.** $2x^2 - 4x + 38$ [4.5] **18.** $15x^2 + 7xy - 2y^2$ [4.6] **19.** $64m^2 - 25n^2$ **20.** $x^3 + 8y^3$ [4.7] **21.** $m^2 - 2m + 3$ [5.2, 5.3] **22.** $(m + 8)(m + 4)$ [5.4] **23.** $(5t^2 + 6)(5t^2 - 6)$ **24.** $(9z + 4)^2$ [5.5] **25.** $\{-2, -1\}$ [6.1] **26.** $-7, 2$ [6.2] **27.** $\frac{x + 1}{x}$ **28.** $(t + 5)(t + 3)$ or $t^2 + 8t + 15$ [6.4] **29.** $\frac{-2x - 14}{(x + 3)(x - 1)}$ [6.5] **30.** $-21$ [6.6] **31.** $\{19\}$ [7.1, 7.2] **32.** **(a)** 2505.1 per yr; The number of twin births increased an average of 2505.1 per yr. **(b)** $y = 2505.1x + 93{,}865$ **(c)** about 123,926 [7.3] **33.** domain: $\{-4, -1, 2, 5\}$; range: $\{-2, 0, 2\}$; function [7.4] **34.** 800 lb [8.1–8.3, 8.6] **35.** $\{(3, 2)\}$ **36.** $\emptyset$ [8.4, 8.6] **37.** $\{(1, 0, -1)\}$ [8.5] **38.** length: 42 ft; width: 30 ft **39.** 15% solution: 6 L; 30% solution: 3 L [9.1] **40.** $(-4, 4)$ **41.** $(-\infty, 0] \cup (2, \infty)$ [9.2] **42.** $\left\{-\frac{1}{3}, 1\right\}$ **43.** $\left(-\infty, -\frac{8}{3}\right] \cup [2, \infty)$

[9.3] **44.**

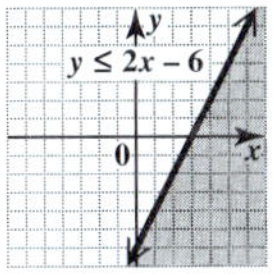


**45.**

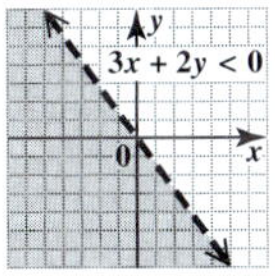


**46.**

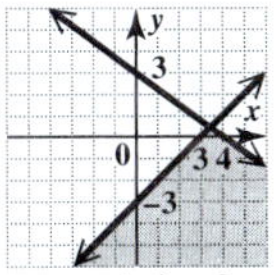


$x - y \geq 3$ and $3x + 4y \leq 12$

**CHAPTER 10 ROOTS, RADICALS, AND ROOT FUNCTIONS**

## Section 10.1 (page 637)

**Exercises** **1.** true **3.** false; Zero has only one square root. **5.** true **7.** $-3, 3$ **9.** $-8, 8$ **11.** $-12, 12$ **13.** $-\frac{5}{14}, \frac{5}{14}$ **15.** $-30, 30$ **17.** 1 **19.** 7 **21.** $-11$ **23.** $-\frac{12}{11}$ **25.** not a real number **27.** 19 **29.** 19 **31.** $\frac{2}{3}$ **33.** $3x^2 + 4$ **35.** $a$ must be positive. **37.** $a$ must be negative. **39.** rational; 5 **41.** irrational; 5.385 **43.** rational; $-8$ **45.** irrational; $-17.321$ **47.** not a real number **49.** irrational; 34.641 **51.** E **53.** D **55.** A **57.** C **59.** C **61.** **(a)** not a real number **(b)** negative **(c)** 0 **63.** 1 **65.** $-5$ **67.** $-4$ **69.** 8 **71.** 5 **73.** $-5$ **75.** not a real number **77.** 2 **79.** not a real number **81.** $\frac{4}{3}$ **83.** $\frac{2}{3}$ **85.** $\frac{1}{2}$ **In Exercises 87–93, we give the domain and then the range.** **87.** $[-3, \infty)$; $[0, \infty)$

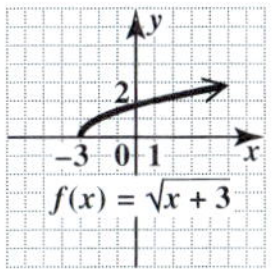


**89.** $[0, \infty)$; $[-2, \infty)$

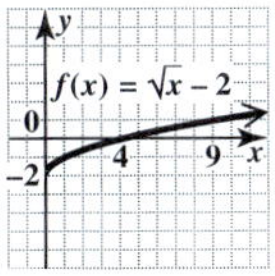


**91.** $(-\infty, \infty)$; $(-\infty, \infty)$

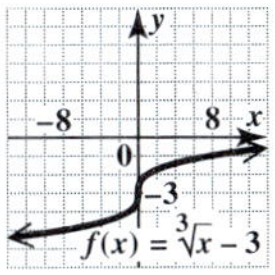


**93.** $(-\infty, \infty)$; $(-\infty, \infty)$

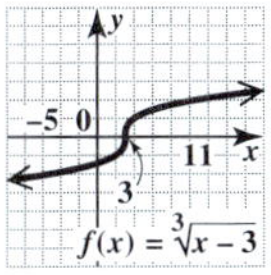


**95.** 12 **97.** 10 **99.** 2 **101.** $-9$ **103.** $|x|$ **105.** $x$ **107.** $x^5$ **109.** $-9.055$ **111.** 7.507 **113.** 3.162 **115.** 1.885 **117.** 1,183,000 cycles per sec **119.** 392,000 mi$^2$ **121.** 1.732 amps

## Section 10.2 (page 646)

**Exercises** **1.** C **3.** A **5.** H **7.** B **9.** D **11.** 13 **13.** 9 **15.** 2 **17.** $\frac{8}{9}$ **19.** $-3$ **21.** 1000 **23.** $-1024$ **25.** not a real number **27.** $\frac{1}{512}$ **29.** $\frac{9}{4}$ **31.** $(-64)^{1/2}$ is an even root of a negative number. No real number squared will give $-64$. On the other hand, $-64^{1/2} = -\sqrt{64} = -8$, which is a real number. ($-64^{1/2}$ is the opposite of $64^{1/2}$.) **33.** $\sqrt{12}$ **35.** $(\sqrt[4]{8})^3$ **37.** $(\sqrt[8]{9q})^5 - (\sqrt[3]{2x})^2$ **39.** $\frac{1}{(\sqrt{2m})^3}$ **41.** $(\sqrt[3]{2y + x})^2$ **43.** $\frac{1}{(\sqrt[3]{3m^4 + 2k^2})^2}$ **45.** $\sqrt{a^2 + b^2} = \sqrt{3^2 + 4^2} = 5$; $a + b = 3 + 4 = 7$; $5 \neq 7$ **47.** 64 **49.** 64 **51.** $x^{10}$ **53.** $\sqrt[6]{x^5}$ **55.** $\sqrt[15]{t^8}$ **57.** 9 **59.** 4 **61.** $y$ **63.** $k^{2/3}$ **65.** $x^3y^8$ **67.** $\frac{1}{x^{10/3}}$ **69.** $\frac{1}{m^{1/4}n^{3/4}}$ **71.** $p^2$ **73.** $\frac{c^{11/3}}{b^{11/4}}$ **75.** $\frac{q^{5/3}}{9p^{7/2}}$ **77.** $p + 2p^2$ **79.** $k^{7/4} - k^{3/4}$ **81.** $6 + 18a$ **83.** $x^{17/20}$ **85.** $\frac{1}{x^{3/2}}$ **87.** $y^{5/6}z^{1/3}$ **89.** $m^{1/12}$ **91.** $x^{1/24}$ **93.** 4.5 hr **95.** $x^{-1/2}(3 - 4x)$ **96.** $m^{5/2}(m^{1/2} - 3)$ **97.** $t^{-1/2}(4 + 7t^2)$ **98.** $x^{-1/3}(8x + 5)$ **99.** $p^{3/4}(4p^{1/4} - 1)$ **100.** $m^{1/8}(2 - m^{1/2})$ **101.** $k^{-3/4}(9 - 2k^{1/2})$ **102.** $z^{-3/4}(7z^{1/8} - 1)$

## Section 10.3 (page 656)

**Connections** **(page 654)** no; no; Answers will vary.

**Exercises** **1.** true; Both are equal to $4\sqrt{3}$ and approximately 6.92820323. **3.** true; Both are equal to $6\sqrt{2}$ and approximately 8.485281374. **5.** D **7.** $\sqrt{30}$ **9.** $\sqrt{14x}$ **11.** $\sqrt{42pqr}$ **13.** $\sqrt[3]{14xy}$ **15.** $\sqrt[4]{33}$ **17.** $\sqrt[4]{6xy^2}$ **19.** cannot be simplified using the product rule **21.** To multiply two radical expressions with the same index, multiply the radicands and keep the same index. For example, $\sqrt[3]{3} \cdot \sqrt[3]{5} = \sqrt[3]{15}$. **23.** $\frac{8}{11}$ **25.** $\frac{\sqrt{3}}{5}$ **27.** $\frac{\sqrt{x}}{5}$ **29.** $\frac{p^3}{9}$ **31.** $\frac{3}{4}$ **33.** $-\frac{\sqrt[3]{r^2}}{2}$ **35.** $-\frac{3}{x}$ **37.** $\frac{1}{x^3}$ **39.** $2\sqrt{3}$ **41.** $12\sqrt{2}$ **43.** $-4\sqrt{2}$ **45.** $-2\sqrt{7}$ **47.** not a real number **49.** $4\sqrt[3]{2}$ **51.** $-2\sqrt[3]{2}$ **53.** $2\sqrt[3]{5}$ **55.** $-4\sqrt[4]{2}$ **57.** $2\sqrt[5]{2}$ **59.** His reasoning was incorrect. Here 8 is a term, not a factor. **61.** $6k\sqrt{2}$ **63.** $12xy^4\sqrt{xy}$ **65.** $11x^3$ **67.** $-3t^4$ **69.** $-10m^4z^2$ **71.** $5a^2b^3c^4$ **73.** $\frac{1}{2}r^2t^5$ **75.** $5x\sqrt{2x}$ **77.** $-10r^5\sqrt{5r}$ **79.** $x^3y^4\sqrt{13x}$ **81.** $2z^2w^3$ **83.** $-2zt^2\sqrt[3]{2z^2t}$ **85.** $3x^3y^4$ **87.** $-3r^3s^2\sqrt[4]{2r^3s^2}$ **89.** $\frac{y^5\sqrt{y}}{6}$ **91.** $\frac{x^5\sqrt[3]{x}}{3}$ **93.** $4\sqrt{3}$ **95.** $\sqrt{5}$ **97.** $x^2\sqrt{x}$ **99.** $\sqrt[6]{432}$ **101.** $\sqrt[12]{6912}$ **103.** $\sqrt[6]{x^5}$ **105.** 1 **107.** 1 **109.** 5 **111.** $8\sqrt{2}$ **113.** $\sqrt{29}$ ft; 5.4 ft **115.** .003 **117.** $8\sqrt{5}$ ft; 17.9 ft **119.** 13 **121.** $9\sqrt{2}$ **123.** $\sqrt{17}$ **125.** 5 **127.** $6\sqrt{2}$ **129.** $\sqrt{5y^2 - 2xy + x^2}$ **131.** $d = [(x_2 - x_1)^2 + (y_2 - y_1)^2]^{1/2}$ **133.** $2\sqrt{106} + 4\sqrt{2}$

## Section 10.4 (page 664)

**Connections** **(page 663)** **1.** 1.618033989 **2.** As one goes farther and farther into the sequence, the successive ratios appear to become closer and closer to the golden ratio. This is indeed the case.

**Exercises** **1.** B **3.** 15; Each radical expression simplifies to a whole number. **5.** $-4$ **7.** $7\sqrt{3}$ **9.** $14\sqrt[3]{2}$ **11.** $5\sqrt[4]{2}$ **13.** $24\sqrt{2}$ **15.** cannot be simplified further **17.** $20\sqrt{5}$ **19.** $12\sqrt{2x}$ **21.** $-11m\sqrt{2}$ **23.** $\sqrt[3]{2}$ **25.** $2\sqrt[3]{x}$ **27.** $-\sqrt[3]{x^2y}$ **29.** $-x\sqrt[3]{xy^2}$ **31.** $19\sqrt[4]{2}$ **33.** $x\sqrt[4]{xy}$ **35.** $9\sqrt[4]{2a^3}$ **37.** $(4 + 3xy)\sqrt[3]{xy^2}$ **39.** $2\sqrt{2} - 2$ **41.** $\frac{5\sqrt{5}}{6}$ **43.** $\frac{7\sqrt{2}}{6}$ **45.** $\frac{5\sqrt{2}}{3}$ **47.** $5\sqrt{2} + 4$ **49.** $\frac{5 - 3x}{x^4}$ **51.** $\frac{m\sqrt[3]{m^2}}{2}$ **53.** $\frac{3x\sqrt[3]{2} - 4\sqrt[3]{5}}{x^3}$ **55.** Both are approximately 11.3137085. **57.** A; 42 m **59.** $(12\sqrt{5} + 5\sqrt{3})$ in. **61.** $(24\sqrt{2} + 12\sqrt{3})$ in.

## Section 10.5 (page 672)

**Connections** **(page 672)** **1.** $\frac{319}{6(8\sqrt{5} + 1)}$ **2.** $\frac{9a - b}{b(3\sqrt{a} - \sqrt{b})}$ **3.** $\frac{9a - b}{(\sqrt{b} - \sqrt{a})(3\sqrt{a} - \sqrt{b})}$ **4.** $\frac{(3\sqrt{a} + \sqrt{b})(\sqrt{b} + \sqrt{a})}{b - a}$; Instead of multiplying by the conjugate of the numerator, we use the conjugate of the denominator.

**Exercises** **1.** E **3.** A **5.** D **7.** $3\sqrt{6} + 2\sqrt{3}$ **9.** $20\sqrt{2}$ **11.** $-2$ **13.** $-1$ **15.** 6 **17.** $\sqrt{6} - \sqrt{2} + \sqrt{3} - 1$ **19.** $\sqrt{22} + \sqrt{55} - \sqrt{14} - \sqrt{35}$ **21.** $8 - \sqrt{15}$ **23.** $9 + 4\sqrt{5}$ **25.** $26 - 2\sqrt{105}$ **27.** $4 - \sqrt[3]{36}$ **29.** 10 **31.** $6x + 3\sqrt{x} - 2\sqrt{5x} - \sqrt{5}$ **33.** $9r - s$ **35.** $4\sqrt[3]{4y^2} - 19\sqrt[3]{2y} - 5$ **37.** $3x - 4$ **39.** $4x - y$ **41.** $2\sqrt{6} - 1$ **43.** $\sqrt{7}$

**45.** $5\sqrt{3}$ **47.** $\frac{\sqrt{6}}{2}$ **49.** $\frac{9\sqrt{15}}{5}$ **51.** $-\sqrt{2}$ **53.** $\frac{\sqrt{14}}{2}$ **55.** $-\frac{\sqrt{14}}{10}$ **57.** $\frac{2\sqrt{6x}}{x}$ **59.** $\frac{-8\sqrt{3k}}{k}$ **61.** $\frac{-5m^2\sqrt{6mn}}{n^2}$ **63.** $\frac{12x^3\sqrt{2xy}}{y^5}$ **65.** $\frac{5\sqrt{2my}}{y^2}$ **67.** $-\frac{4k\sqrt{3z}}{z}$ **69.** $\frac{\sqrt[3]{18}}{3}$ **71.** $\frac{\sqrt[3]{12}}{3}$ **73.** $\frac{\sqrt[3]{18}}{4}$ **75.** $-\frac{\sqrt[3]{2pr}}{r}$ **77.** $\frac{x^2\sqrt[3]{y^2}}{y}$ **79.** $\frac{2\sqrt[4]{x^3}}{x}$ **81.** $\frac{\sqrt[4]{2yz^3}}{z}$ **83.** Multiply both the numerator and the denominator by $4 - \sqrt{5}$. No, it would not. The new denominator would be $(4 + \sqrt{5})^2 = 21 + 8\sqrt{5}$, which is not rational. **85.** $\frac{3(4 - \sqrt{5})}{11}$ **87.** $\frac{6\sqrt{2} + 4}{7}$ **89.** $\frac{2(3\sqrt{5} - 2\sqrt{3})}{33}$ **91.** $2\sqrt{3} + \sqrt{10} - 3\sqrt{2} - \sqrt{15}$ **93.** $\sqrt{m} - 2$ **95.** $\frac{4\sqrt{x}(\sqrt{x} + 2\sqrt{y})}{x - 4y}$ **97.** $\frac{x - 2\sqrt{xy} + y}{x - y}$ **99.** $\frac{5\sqrt{k}(2\sqrt{k} - \sqrt{q})}{4k - q}$ **101.** Square both sides to show that each is equal to $\frac{2 - \sqrt{3}}{4}$. **103.** $3 - 2\sqrt{6}$ **105.** $1 - \sqrt{5}$ **107.** $\frac{4 - 2\sqrt{2}}{3}$ **109.** $\frac{6 + 2\sqrt{6p}}{3}$ **111.** $\frac{\sqrt{x + y}}{x + y}$ **113.** $\frac{p\sqrt{p + 2}}{p + 2}$ **115.** Each expression is approximately equal to .2588190451. **117.** $(\sqrt{x} + \sqrt{7})(\sqrt{x} - \sqrt{7})$ **118.** $(\sqrt[3]{x} - \sqrt[3]{7})(\sqrt[3]{x^2} + \sqrt[3]{7x} + \sqrt[3]{49})$ **119.** $(\sqrt[3]{x} + \sqrt[3]{7})(\sqrt[3]{x^2} - \sqrt[3]{7x} + \sqrt[3]{49})$ **120.** $\frac{(x + 3)(\sqrt{x} + \sqrt{7})}{x - 7}$ **121.** $\frac{(x + 3)(\sqrt[3]{x^2} + \sqrt[3]{7x} + \sqrt[3]{49})}{x - 7}$ **122.** $\frac{(x + 3)(\sqrt[3]{x} + \sqrt[3]{7})}{x + 7}$ **123.** $(\sqrt[3]{5} - \sqrt[3]{3})(\sqrt[3]{25} + \sqrt[3]{15} + \sqrt[3]{9})$ **124.** $\sqrt[3]{25} + \sqrt[3]{15} + \sqrt[3]{9}$ **125.** $\frac{17}{2(6 + \sqrt{2})}$ **127.** $\frac{9a - b}{b(3\sqrt{a} - \sqrt{b})}$

## Summary Exercises on Operations with Radicals (page 676)

**1.** $-6\sqrt{10}$ **2.** $7 - \sqrt{14}$ **3.** $2 + \sqrt{6} - 2\sqrt{3} - 3\sqrt{2}$ **4.** $4\sqrt{2}$ **5.** $73 + 12\sqrt{35}$ **6.** $\frac{-\sqrt{6}}{2}$ **7.** $4(\sqrt{7} - \sqrt{5})$ **8.** $3\sqrt[3]{2x^2}$ **9.** $-3 + 2\sqrt{2}$ **10.** $-2$ **11.** $-44$ **12.** $\frac{\sqrt{x} + \sqrt{5}}{x - 5}$ **13.** $2abc^3\sqrt[3]{b^2}$ **14.** $5\sqrt[3]{3}$ **15.** $3(\sqrt{5} - 2)$ **16.** $\frac{\sqrt{15x}}{5x}$ **17.** $\frac{8}{5}$ **18.** $\frac{\sqrt{2}}{8}$ **19.** $-\sqrt[3]{100}$ **20.** $11 + 2\sqrt{30}$ **21.** $-3\sqrt{3x}$ **22.** $52 - 30\sqrt{3}$ **23.** 1 **24.** $\frac{\sqrt[3]{117}}{9}$ **25.** $t^2\sqrt[4]{t}$

## Section 10.6 (page 681)

**Exercises** **1.** **(a)** yes **(b)** no **3.** **(a)** yes **(b)** no **5.** no; There is no solution. The radical expression, which is positive, cannot equal a negative number. **7.** $\{11\}$ **9.** $\left\{\frac{1}{3}\right\}$ **11.** $\emptyset$ **13.** $\{5\}$ **15.** $\{18\}$ **17.** $\{5\}$ **19.** $\{4\}$ **21.** $\{17\}$ **23.** $\{5\}$ **25.** $\emptyset$ **27.** $\{0\}$ **29.** $\{0\}$ **31.** $\left\{-\frac{1}{3}\right\}$ **33.** $\emptyset$ **35.** You cannot just square each term. The right side should be $(8 - x)^2 = 64 - 16x + x^2$. The correct first step is $3x + 4 = 64 - 16x + x^2$, and the solution set is $\{4\}$. **37.** $\{1\}$ **39.** $\{-1\}$ **41.** $\{14\}$ **43.** $\{8\}$ **45.** $\{0\}$ **47.** $\emptyset$ **49.** $\{7\}$ **51.** $\{7\}$ **53.** $\{4, 20\}$ **55.** $\emptyset$ **57.** $\left\{\frac{5}{4}\right\}$ **59.** 6 **61.** $\emptyset$; domain: $\left[-\frac{2}{3}, 1\right]$ **63.** $\{9, 17\}$ **65.** $\left\{\frac{1}{4}, 1\right\}$ **67.** $K = \frac{V^2m}{2}$ **69.** $L = \frac{1}{4\pi^2f^2C}$ **71.** **(a)** $r = \frac{a}{4\pi^2N^2}$ **(b)** 62.5 m **(c)** 155.1 m **73.** 7 billion $ft^3$; 14 billion $ft^3$; 22.5 billion $ft^3$; 22.5 billion $ft^3$ **75.** 12 billion $ft^3$; 12 billion $ft^3$; 14 billion $ft^3$; 17.5 billion $ft^3$ **77.** fairly good; 1920

## Section 10.7 (page 690)

**Exercises** **1.** $i$ **3.** $-1$ **5.** $-i$ **7.** $13i$ **9.** $-12i$ **11.** $i\sqrt{5}$ **13.** $4i\sqrt{3}$ **15.** $-\sqrt{105}$ **17.** $-10$ **19.** $i\sqrt{33}$ **21.** $\sqrt{3}$ **23.** $5i$ **25.** **(a)** Any real number $a$ can be written as $a + 0i$, and this is a complex number with imaginary part 0. **(b)** A complex number such as $2 + 3i$, with nonzero imaginary part, is not real. **27.** $-1 + 7i$ **29.** 0 **31.** $7 + 3i$ **33.** $-2$ **35.** $1 + 13i$ **37.** $6 + 6i$ **39.** $4 + 2i$ **41.** $-81$ **43.** $-16$ **45.** $-10 - 30i$ **47.** $10 - 5i$ **49.** $-9 + 40i$ **51.** $-16 + 30i$ **53.** 153 **55.** 97 **57.** $a - bi$ **59.** $1 + i$ **61.** $-1 + 2i$ **63.** $2 + 2i$ **65.** $-\frac{5}{13} - \frac{12}{13}i$ **67.** $1 - 3i$ **69.** $30 + 5i$ **71.** $\frac{5}{41} + \frac{4}{41}i$ **73.** $-1$ **75.** $i$ **77.** $-1$ **79.** $-i$ **81.** $-i$ **83.** Since $i^{20} = (i^4)^5 = 1^5 = 1$, the student multiplied by 1, which is justified by the identity property for multiplication. **85.** $\frac{1}{2} + \frac{1}{2}i$ **87.** Substitute both $1 + 5i$ and $1 - 5i$ for $x$ and show that the result is $0 = 0$ in each case. **89. (a)** $4x + 1$ **(b)** $4 + i$ **90. (a)** $-2x + 3$ **(b)** $-2 + 3i$ **91. (a)** $3x^2 + 5x - 2$ **(b)** $5 + 5i$ **92. (a)** $-\sqrt{3} + \sqrt{6} + 1 - \sqrt{2}$ **(b)** $\frac{1}{5} - \frac{7}{5}i$ **93.** In parts (a) and (b) of Exercises 89 and 90, real and imaginary parts are added, just like coefficients of similar terms in the binomials, and the answers correspond. In Exercise 91, introducing $i^2 = -1$ when a product is found leads to answers that do not correspond. **94.** In parts (a) and (b) of Exercises 89 and 90, real and imaginary parts are added, just like coefficients of similar terms in binomials, and the answers correspond. In Exercise 92, introducing $i^2 = -1$ when performing the division leads to answers that do not correspond. **95.** $\frac{37}{10} - \frac{19}{10}i$ **97.** $-\frac{13}{10} + \frac{11}{10}i$

## Chapter 10 Review Exercises (page 698)

**1.** 42 **3.** 6 **5.** $-3$ **7.** $\sqrt[n]{a}$ is not a real number if $n$ is even and $a$ is negative. **9.** $-6.856$ **11.** 4.960 **13.** $-3968.503$

**15.**

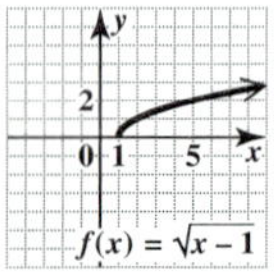


domain: $[1, \infty)$; range: $[0, \infty)$ **17.** B **19.** A **21.** It is not a real number. **23.** $-11$ **25.** $-4$ **27.** $-32$

**29.** It is not a real number. **31.** The radical $\sqrt[n]{a^m}$ is equivalent to $a^{m/n}$. For example, $\sqrt[3]{8^2} = \sqrt[3]{64} = 4$, and $8^{2/3} = (8^{1/3})^2 = 2^2 = 4$. **33.** $\frac{1}{(\sqrt[3]{3a + b})^5}$ or $\frac{1}{\sqrt[3]{(3a + b)^5}}$ **35.** $p^{4/5}$ **37.** 96 **39.** $\frac{1}{y^{1/2}}$ **41.** $r^{1/2} + r$ **43.** $r^{3/2}$ **45.** $k^{9/4}$ **47.** $z^{1/12}$ **49.** $x^{1/15}$ **51.** The product rule for exponents applies only if the bases are the same. **53.** $\sqrt{5r}$ **55.** $\sqrt[4]{21}$ **57.** $5\sqrt{3}$ **59.** $-3\sqrt[3]{4}$ **61.** $4pq^2\sqrt[3]{p}$ **63.** $2r^2t\sqrt[3]{79r^2t}$ **65.** $\frac{m^5}{3}$ **67.** $\frac{a^2\sqrt[4]{a}}{3}$ **69.** $p\sqrt{p}$ **71.** $\sqrt[10]{x^7}$ **73.** $\sqrt{197}$ **75.** $23\sqrt{5}$ **77.** $26m\sqrt{6m}$ **79.** $-8\sqrt[4]{2}$ **81.** $\frac{16 + 5\sqrt{5}}{20}$ **83.** $(12\sqrt{3} + 5\sqrt{2})$ ft **85.** 2 **87.** $15 - 2\sqrt{26}$ **89.** $2\sqrt[3]{2y^2} + 2\sqrt[3]{4y} - 3$ **91.** The denominator would become $\sqrt[3]{6^2} = \sqrt[3]{36}$, which is not rational. **93.** $-3\sqrt{6}$ **95.** $\frac{\sqrt{22}}{4}$ **97.** $\frac{3m\sqrt[3]{4n}}{n^2}$ **99.** $\frac{5(\sqrt{6} + 3)}{3}$ **101.** $\frac{1 - 4\sqrt{2}}{3}$ **103.** $\{2\}$ **105.** $\emptyset$ **107.** $\{9\}$ **109.** $\{7\}$ **111.** $\{-13\}$ **113.** $\{14\}$ **115.** $\emptyset$ **117.** $\{7\}$ **119.** $\{3\}$ **120.** $\{-3\}$ **121.** $\{\pm 3\}$ **122.** $\{3\}$ **123.** $\{\pm 3\}$ **124.** $\{-3\}$ **125. (a)** more than **(b)** the same as **127.** $5i$ **129.** no **131.** $14 + 7i$ **133.** $-45$ **135.** $5 + i$ **137.** $1 - i$ **139.** $-i$ **141.** $-1$ **143.** $-4$ **145.** $\frac{1}{z^{3/5}}$ **147.** $3z^3t^2\sqrt[3]{2t^2}$ **149.** $6x\sqrt[3]{y^2}$ **151.** $-\frac{\sqrt{3}}{6}$ **153.** 1 **155.** $3 - 7i$ **157.** $\frac{1 + \sqrt{6}}{2}$ **159.** $\{5\}$ **161.** $\left\{\frac{3}{2}\right\}$ **163.** $\{1\}$ **165.** $\{9\}$ **167.** $\{7\}$

## Chapter 10 Test (page 703)

[10.1] **1.** $-29$ **2.** $-8$ [10.2] **3.** 5 [10.1] **4.** C **5.** 21.863 **6.** $-9.405$ **7.** domain: $[-6, \infty)$; range: $[0, \infty)$

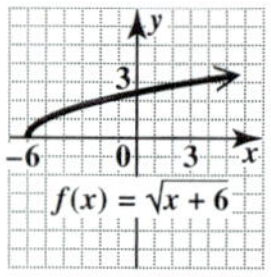

[10.2] **8.** $\frac{125}{64}$ **9.** $\frac{1}{256}$ **10.** $\frac{9y^{3/10}}{x^2}$ **11.** $x^{4/3}y^6$ **12.** $7^{1/2}$ or $\sqrt{7}$ [10.3] **13.** $a^3\sqrt[3]{a^2}$ or $a^{11/3}$ **14.** $\sqrt{145}$ **15.** 10 **16.** $3x^2y^3\sqrt{6x}$ **17.** $2ab^3\sqrt[4]{2a^3b}$ **18.** $\sqrt[6]{200}$ [10.4] **19.** $26\sqrt{5}$ **20.** $(2ts - 3t^2)\sqrt[3]{2s^2}$ [10.5] **21.** $66 + \sqrt{5}$ **22.** $23 - 4\sqrt{15}$ **23.** $-\frac{\sqrt{10}}{4}$ **24.** $\frac{2\sqrt[3]{25}}{5}$ **25.** $-2(\sqrt{7} - \sqrt{5})$ **26.** $3 + \sqrt{6}$ [10.6] **27.** **(a)** 59.8 **(b)** $T = \frac{V_0^2 - V^2}{-V^2k}$ or $T = \frac{V^2 - V_0^2}{V^2k}$ **28.** $\{-1\}$ **29.** $\{3\}$ **30.** $\{-3\}$ [10.7] **31.** $-5 - 8i$ **32.** $-2 + 16i$ **33.** $3 + 4i$ **34.** $i$ **35.** **(a)** true **(b)** true **(c)** false **(d)** true

## Cumulative Review Exercises Chapters 1–10 (page 704)

[1.4–1.6] **1.** 1 **2.** $-\frac{14}{9}$ [2.1–2.3] **3.** $\{-4\}$ **4.** $\{-12\}$ **5.** $\{6\}$ [2.8] **6.** $(-6, \infty)$ [9.1] **7.** $(2, 3)$ **8.** $(-\infty, 2) \cup (3, \infty)$ [9.2] **9.** $\left\{-\frac{10}{3}, 1\right\}$ **10.** $\left\{\frac{1}{4}\right\}$ **11.** $(-\infty, -2] \cup [7, \infty)$ [2.5] **12.** Both angles measure 80°. [2.7] **13.** 18 nickels; 32 quarters **14.** $2\frac{2}{39}$ L [3.2, 7.1] **15.**

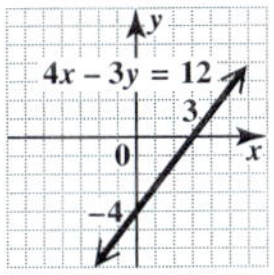


[3.3, 3.4, 7.1, 7.2] **16.** $-\frac{3}{2}$; $y = -\frac{3}{2}x$ [4.4] **17.** $-k^3 - 3k^2 - 8k - 9$ [4.5] **18.** $8x^2 + 17x - 21$ [4.7] **19.** $z - 2 + \frac{3}{z}$ **20.** $3y^3 - 3y^2 + 4y + 1 + \frac{-10}{2y + 1}$ [5.2, 5.3] **21.** $(2p - 3q)(p - q)$ [5.4] **22.** $(3k^2 + 4)(k - 1)(k + 1)$ **23.** $(x + 8)(x^2 - 8x + 64)$ [5.5] **24.** $\left\{-3, -\frac{5}{2}\right\}$ **25.** $\left\{-\frac{2}{5}, 1\right\}$ [6.2] **26.** $\frac{y}{y + 5}$ [6.4] **27.** $\frac{4x + 2y}{(x + y)(x - y)}$ [6.5] **28.** $-\frac{9}{4}$ **29.** $\frac{-1}{a + b}$ [6.7] **30.** Natalie: 8 mph; Chuck: 4 mph [6.6] **31.** $\emptyset$ [7.3] **32.** $-37$ [8.2, 8.3] **33.** $\{(7, -2)\}$ [8.6] **34.** $\{(-1, 1, 1)\}$ [8.5] **35.** 2-oz letter: \$.55; 3-oz letter: \$.78 [10.2] **36.** $\frac{1}{9}$ [10.3] **37.** $10x^2\sqrt{2}$ **38.** $2x\sqrt[3]{6x^2y^2}$ [10.4] **39.** $7\sqrt{2}$ [10.5] **40.** $\frac{\sqrt{10} + 2\sqrt{2}}{2}$ **41.** $-6x - 11\sqrt{xy} - 4y$ [10.3] **42.** $\sqrt{29}$ [10.6] **43.** $\{3, 4\}$ [10.1] **44.** 39.2 mph [10.7] **45.** $2 + 9i$ **46.** $4 + 2i$

# CHAPTER 11 QUADRATIC EQUATIONS, INEQUALITIES, AND FUNCTIONS

## Section 11.1 (page 713)

**Exercises** **1.** The equation is also true for $x = -4$. **3.** **(a)** A quadratic equation in standard form has a second-degree polynomial in decreasing powers equal to 0. **(b)** The zero-factor property states that if a product equals 0, then at least one of the factors equals 0. **(c)** The square root property states that if the square of a quantity equals a number, then the quantity equals the positive or negative square root of the number. **5.** $\{9, -9\}$ **7.** $\{\sqrt{17}, -\sqrt{17}\}$ **9.** $\{4\sqrt{2}, -4\sqrt{2}\}$ **11.** $\{\sqrt{21}, -\sqrt{21}\}$ **13.** $\{2\sqrt{5}, -2\sqrt{5}\}$ **15.** $\{2\sqrt{6}, -2\sqrt{6}\}$ **17.** $\left\{\frac{2\sqrt{5}}{5}, -\frac{2\sqrt{5}}{5}\right\}$ **19.** $\{3\sqrt{3}, -3\sqrt{3}\}$ **21.** $\{-2, 8\}$ **23.** $\left\{-3, \frac{5}{3}\right\}$ **25.** $\left\{0, \frac{3}{2}\right\}$ **27.** $\{4 + \sqrt{3}, 4 - \sqrt{3}\}$

**29.** $\{-5 + 4\sqrt{3}, -5 - 4\sqrt{3}\}$ **31.** $\left\{\frac{1 + \sqrt{7}}{3}, \frac{1 - \sqrt{7}}{3}\right\}$ **33.** $\left\{\frac{-1 + 2\sqrt{6}}{4}, \frac{-1 - 2\sqrt{6}}{4}\right\}$ **35.** $\left\{\frac{5 + \sqrt{30}}{2}, \frac{5 - \sqrt{30}}{2}\right\}$ **37.** $\left\{\frac{-1 + 3\sqrt{2}}{3}, \frac{-1 - 3\sqrt{2}}{3}\right\}$ **39.** $\{-10 + 4\sqrt{3}, -10 - 4\sqrt{3}\}$ **41.** $\left\{\frac{1 + 4\sqrt{3}}{4}, \frac{1 - 4\sqrt{3}}{4}\right\}$ **43.** Johnny's first solution, $\frac{5 + \sqrt{30}}{2}$, is equivalent to Linda's second solution, $\frac{-5 - \sqrt{30}}{-2}$. This can be verified by multiplying $\frac{5 + \sqrt{30}}{2}$ by 1 in the form $\frac{-1}{-1}$. Similarly, Johnny's second solution is equivalent to Linda's first one. **45.** $\{-4.48, .20\}$ **47.** $\{-3.09, -.15\}$ **49.** $\{2i\sqrt{3}, -2i\sqrt{3}\}$ **51.** $\{5 + i\sqrt{3}, 5 - i\sqrt{3}\}$ **53.** $\left\{\frac{1 + 2i\sqrt{2}}{6}, \frac{1 - 2i\sqrt{2}}{6}\right\}$ **55.** $(x + 3)^2 = 100$ **56.** $\{-13, 7\}$ **57.** $\{-13, 7\}$ **58.** $\{-3, 6\}$ **59.** 6.3 sec **61.** 9 in. **63.** 5%

## Section 11.2 (page 722)

**Exercises** **1.** **(a)** 9 **(b)** 49 **(c)** 36 **(d)** $\frac{9}{4}$ **(e)** $\frac{81}{4}$ **(f)** $\frac{1}{16}$ **3.** 4 **5.** 25 **7.** $\frac{1}{36}$ **9.** $\{-4, 6\}$ **11.** $\{-2 + \sqrt{6}, -2 - \sqrt{6}\}$ **13.** $\{-5 + \sqrt{7}, -5 - \sqrt{7}\}$ **15.** $\left\{\frac{-3 + \sqrt{17}}{2}, \frac{-3 - \sqrt{17}}{2}\right\}$ **17.** $\left\{-\frac{8}{3}, 3\right\}$ **19.** $\left\{\frac{-5 + \sqrt{41}}{4}, \frac{-5 - \sqrt{41}}{4}\right\}$ **21.** $\left\{\frac{5 + \sqrt{15}}{5}, \frac{5 - \sqrt{15}}{5}\right\}$ **23.** $\left\{\frac{4 + \sqrt{3}}{3}, \frac{4 - \sqrt{3}}{3}\right\}$ **25.** $\left\{\frac{-3 + \sqrt{41}}{2}, \frac{-3 - \sqrt{41}}{2}\right\}$ **27.** $\left\{\frac{2 + \sqrt{3}}{3}, \frac{2 - \sqrt{3}}{3}\right\}$ **29.** $\{1 + \sqrt{2}, 1 - \sqrt{2}\}$ **31.** **(a)** $\left\{\frac{3 + 2\sqrt{6}}{3}, \frac{3 - 2\sqrt{6}}{3}\right\}$ **(b)** $\{-.633, 2.633\}$ **33.** **(a)** $\{-2 + \sqrt{3}, -2 - \sqrt{3}\}$ **(b)** $\{-3.732, -.268\}$ **35.** The student should have divided both sides of the equation by 2 as his first step. The correct solution set is $\{1, 4\}$. **37.** $\{-2 + 3i, -2 - 3i\}$ **39.** $\left\{\frac{-2 + 2i\sqrt{2}}{3}, \frac{-2 - 2i\sqrt{2}}{3}\right\}$ **41.** $\{-3 + i\sqrt{3}, -3 - i\sqrt{3}\}$ **43.** $x^2$ **44.** $x$ **45.** $6x$ **46.** 1 **47.** 9 **48.** $(x + 3)^2$ or $x^2 + 6x + 9$

## Section 11.3 (page 729)

**Exercises** **1.** The documentation was incorrect, since the fraction bar should extend under the term $-b$. **3.** The last step is wrong. Because 5 is not a common factor in the numerator, the fraction cannot be simplified. The solutions are $\frac{5 \pm \sqrt{5}}{10}$. **5.** $\{3, 5\}$ **7.** $\left\{\frac{-2 + \sqrt{2}}{2}, \frac{-2 - \sqrt{2}}{2}\right\}$ **9.** $\left\{\frac{1 + \sqrt{3}}{2}, \frac{1 - \sqrt{3}}{2}\right\}$ **11.** $\{5 + \sqrt{7}, 5 - \sqrt{7}\}$ **13.** $\left\{\frac{-1 + \sqrt{2}}{2}, \frac{-1 - \sqrt{2}}{2}\right\}$ **15.** $\left\{\frac{-1 + \sqrt{7}}{3}, \frac{-1 - \sqrt{7}}{3}\right\}$ **17.** $\{1 + \sqrt{5}, 1 - \sqrt{5}\}$ **19.** $\left\{\frac{-2 + \sqrt{10}}{2}, \frac{-2 - \sqrt{10}}{2}\right\}$ **21.** $\{-1 + 3\sqrt{2}, -1 - 3\sqrt{2}\}$ **23.** $\left\{\frac{1 + \sqrt{29}}{2}, \frac{1 - \sqrt{29}}{2}\right\}$ **25.** $\left\{\frac{-4 + \sqrt{91}}{3}, \frac{-4 - \sqrt{91}}{3}\right\}$ **27.** $\left\{\frac{3}{2} + \frac{\sqrt{15}}{2}i, \frac{3}{2} - \frac{\sqrt{15}}{2}i\right\}$ **29.** $\{3 + i\sqrt{5}, 3 - i\sqrt{5}\}$ **31.** $\left\{\frac{1}{2} + \frac{\sqrt{6}}{2}i, \frac{1}{2} - \frac{\sqrt{6}}{2}i\right\}$ **33.** $\left\{-\frac{2}{3} + \frac{\sqrt{2}}{3}i, -\frac{2}{3} - \frac{\sqrt{2}}{3}i\right\}$ **35.** $\{4 + 3i\sqrt{2}, 4 - 3i\sqrt{2}\}$ **37.** B **39.** C **41.** A **43.** D **45.** The equations in Exercises 37, 38, 41, and 42 can be solved by factoring. **47.** $\left\{\frac{7}{2}\right\}$ **49.** $\left\{-\frac{1}{2}, \frac{3}{2}\right\}$ **51.** No, because an irrational solution occurs only if the discriminant is positive, but not the square of an integer. In that case, there will be two irrational solutions. **53.** $-10$ or 10 **55.** 16 **57.** 25 **59.** $b = \frac{44}{5}; \frac{3}{10}$

## Section 11.4 (page 738)

**Exercises** **1.** square root property **3.** quadratic formula **5.** factoring **7.** Multiply by the LCD, $x$. **9.** Substitute a variable for $r^2 + r$. **11.** The potential solution $-1$ does not check. The solution set is $\{4\}$. **13.** $\{-2, 7\}$ **15.** $\{-4, 7\}$ **17.** $\left\{-\frac{2}{3}, 1\right\}$ **19.** $\left\{-\frac{14}{17}, 5\right\}$ **21.** $\left\{-\frac{11}{7}, 0\right\}$ **23.** $\left\{\frac{-1+\sqrt{13}}{2}, \frac{-1-\sqrt{13}}{2}\right\}$ **25.** $\left\{-\frac{8}{3}, -1\right\}$ **27.** $\left\{\frac{2+\sqrt{22}}{3}, \frac{2-\sqrt{22}}{3}\right\}$ **29.** **(a)** $(20 - t)$ mph **(b)** $(20 + t)$ mph **31.** 25 mph **33.** 80 km per hr **35.** 3.6 hr **37.** Rusty: 25.0 hr; Nancy: 23.0 hr **39.** 9 min **41.** $\{2, 5\}$ **43.** $\{3\}$ **45.** $\left\{\frac{8}{9}\right\}$ **47.** $\{16\}$ **49.** $\left\{\frac{2}{5}\right\}$ **51.** $\{-3, 3\}$ **53.** $\left\{-\frac{3}{2}, -1, 1, \frac{3}{2}\right\}$ **55.** $\{-2\sqrt{3}, -2, 2, 2\sqrt{3}\}$ **57.** $\{-6, -5\}$ **59.** $\left\{-\frac{16}{3}, -2\right\}$ **61.** $\left\{-\frac{1}{3}, \frac{1}{6}\right\}$ **63.** $\left\{-\frac{1}{2}, 3\right\}$ **65.** $\{-8, 1\}$ **67.** $\{-64, 27\}$ **69.** $\left\{-\frac{27}{8}, -1, 1, \frac{27}{8}\right\}$ **71.** $\{25\}$ **73.** $\left\{-1, 1, -\frac{\sqrt{6}}{2}i, \frac{\sqrt{6}}{2}i\right\}$ **75.** $\left\{-\frac{\sqrt{6}}{3}, -\frac{1}{2}, \frac{1}{2}, \frac{\sqrt{6}}{3}\right\}$ **77.** $\{3, 11\}$ **79.** $\left\{-\sqrt[3]{5}, -\frac{\sqrt[3]{4}}{2}\right\}$ **81.** $\left\{\frac{4}{3}, \frac{9}{4}\right\}$ **83.** $\left\{\frac{\sqrt{9+\sqrt{65}}}{2}, -\frac{\sqrt{9+\sqrt{65}}}{2}, \frac{\sqrt{9-\sqrt{65}}}{2}, -\frac{\sqrt{9-\sqrt{65}}}{2}\right\}$ **85.** It would cause both denominators to be 0, and division by 0 is undefined. **86.** $\frac{12}{5}$ **87.** $\left(\frac{x}{x-3}\right)^2 + 3\left(\frac{x}{x-3}\right) - 4 = 0$ **88.** The numerator can never equal the denominator, since the denominator is 3 less than the numerator. **89.** $\left\{\frac{12}{5}\right\}$; The values for $t$ are $-4$ and 1. The value 1 is impossible because it leads to a contradiction $\left(\text{since } \frac{x}{x-3} \text{ is never equal to } 1\right)$. **90.** $\left\{\frac{12}{5}\right\}$; The values for $s$ are $\frac{1}{x}$ and $\frac{-4}{x}$. The value $\frac{1}{x}$ is impossible, since $\frac{1}{x} \neq \frac{1}{x-3}$ for all $x$.

## Summary Exercises on Solving Quadratic Equations (page 742)

**1.** $\{\sqrt{7}, -\sqrt{7}\}$ **2.** $\left\{-\frac{3}{2}, \frac{5}{3}\right\}$ **3.** $\{-3+\sqrt{5}, -3-\sqrt{5}\}$ **4.** $\{-1, 9\}$ **5.** $\left\{-\frac{3}{2}, 4\right\}$ **6.** $\left\{-3, \frac{1}{3}\right\}$ **7.** $\left\{\frac{2+\sqrt{2}}{2}, \frac{2-\sqrt{2}}{2}\right\}$ **8.** $\{2i\sqrt{3}, -2i\sqrt{3}\}$ **9.** $\left\{\frac{1}{2}, 2\right\}$ **10.** $\{-3, -1, 1, 3\}$ **11.** $\left\{\frac{-3+2\sqrt{2}}{2}, \frac{-3-2\sqrt{2}}{2}\right\}$ **12.** $\left\{\frac{4}{5}, 3\right\}$ **13.** $\{-\sqrt{7}, -\sqrt{2}, \sqrt{2}, \sqrt{7}\}$ **14.** $\left\{\frac{1+\sqrt{5}}{4}, \frac{1-\sqrt{5}}{4}\right\}$ **15.** $\left\{-\frac{1}{2}+\frac{\sqrt{3}}{2}i, -\frac{1}{2}-\frac{\sqrt{3}}{2}i\right\}$ **16.** $\left\{-\frac{\sqrt[3]{175}}{5}, 1\right\}$ **17.** $\left\{\frac{3}{2}\right\}$ **18.** $\left\{\frac{2}{3}\right\}$ **19.** $\{6\sqrt{2}, -6\sqrt{2}\}$ **20.** $\left\{-\frac{2}{3}, 2\right\}$ **21.** $\{-4, 9\}$ **22.** $\{13, -13\}$ **23.** $\left\{1+\frac{\sqrt{3}}{3}i, 1-\frac{\sqrt{3}}{3}i\right\}$ **24.** $\{3\}$ **25.** $\left\{-\frac{1}{3}, \frac{1}{6}\right\}$ **26.** $\left\{\frac{1}{6}+\frac{\sqrt{47}}{6}i, \frac{1}{6}-\frac{\sqrt{47}}{6}i\right\}$

## Section 11.5 (page 747)

**Exercises** **1.** Find a common denominator, and then multiply both sides by the common denominator. **3.** Write it in standard form (with 0 on one side, in decreasing powers of $w$). **5.** $m = \sqrt{p^2 - n^2}$ **7.** $t = \frac{\pm\sqrt{dk}}{k}$ **9.** $d = \frac{\pm\sqrt{skI}}{I}$ **11.** $v = \frac{\pm\sqrt{kAF}}{F}$ **13.** $r = \frac{\pm\sqrt{3\pi Vh}}{\pi h}$ **15.** $t = \frac{-B \pm \sqrt{B^2 - 4AC}}{2A}$ **17.** $h = \frac{D^2}{k}$ **19.** $\ell = \frac{p^2 g}{k}$ **21.** If $g$ is positive, the only way to have a real value for $p$ is to have $k\ell$ positive, since the quotient of two positive numbers is positive. If $k$ and $\ell$ have different signs, their product is negative, leading to a negative radicand. **23.** $R = \frac{E^2 - 2pr \pm E\sqrt{E^2 - 4pr}}{2p}$ **25.** $r = \frac{5pc}{4}$ or $r = -\frac{2pc}{3}$ **27.** $I = \frac{-cR \pm \sqrt{c^2R^2 - 4cL}}{2cL}$ **29.** 2.3, 5.3, 5.8

**31.** eastbound ship: 80 mi; southbound ship: 150 mi **33.** 5 cm, 12 cm, 13 cm **35.** length: 2 cm; width: 1.5 cm **37.** 1 ft **39.** length: 26 m; width: 16 m **41.** 20 in. by 12 in. **43.** 1 sec and 8 sec **45.** 2.4 sec and 5.6 sec **47.** 9.2 sec **49.** It reaches its *maximum* height at 5 sec because this is the only time it reaches 400 ft. **51.** \$.80 **53.** .035 or 3.5% **55.** **(a)** 2.4 million **(b)** 2.4 million; They are the same. **57.** 1995; The graph indicates that sales reached 2 million in 1996. **59.** 5.5 m per sec **61.** 5 or 14

## Section 11.6 (page 758)

**Exercises** **1.** **(a)** B **(b)** C **(c)** A **(d)** D **3.** (0, 0) **5.** (0, 4) **7.** (1, 0) **9.** (−3, −4) **11.** In Exercise 9, the parabola is shifted 3 units to the left and 4 units down. The parabola in Exercise 10 is shifted 5 units to the right and 8 units down. **13.** down; wider **15.** up; narrower **17.** **(a)** I **(b)** IV **(c)** II **(d)** III **19.** **(a)** D **(b)** B **(c)** C **(d)** A **21.**

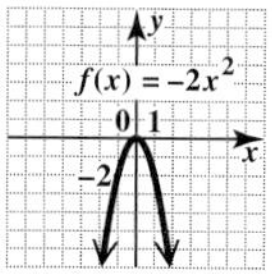


**23.**

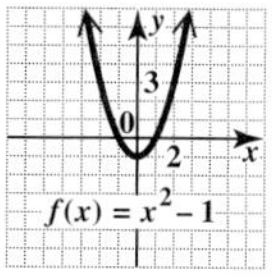


**25.**

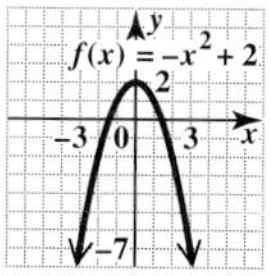


**27.** vertex: (4, 0); axis: $x = 4$; domain: $(-\infty, \infty)$; range: $[0, \infty)$

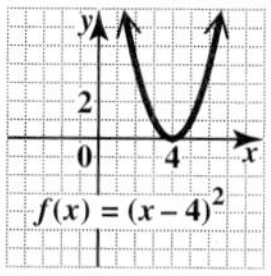


**29.** vertex: $(-2, -1)$; axis: $x = -2$; domain: $(-\infty, \infty)$; range: $[-1, \infty)$

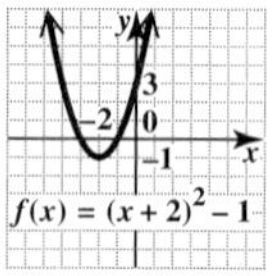


**31.** vertex: $(2, -4)$; axis: $x = 2$; domain: $(-\infty, \infty)$; range: $[-4, \infty)$

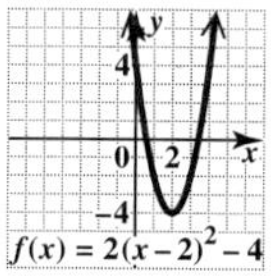


**33.** vertex: $(-1, 2)$; axis: $x = -1$; domain: $(-\infty, \infty)$; range: $(-\infty, 2]$

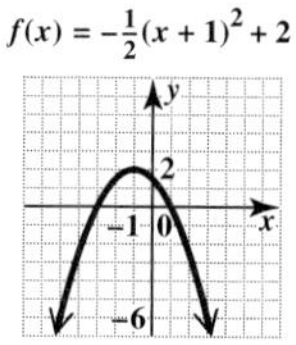


**35.** vertex: $(2, -3)$; axis: $x = 2$; domain: $(-\infty, \infty)$; range: $[-3, \infty)$

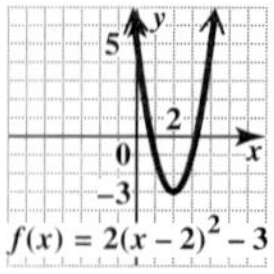


**37.** It is shifted 6 units up. **38.**

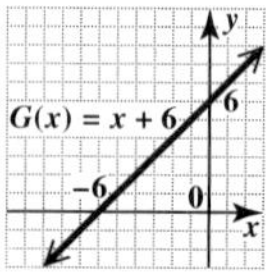


**39.** It is shifted 6 units up. **40.** It is shifted 6 units to the right.

**41.**

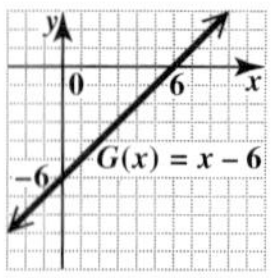


**42.** It is shifted 6 units to the right. **43.** quadratic; positive **45.** quadratic; negative **47.** linear; positive

**49.** **(a)**

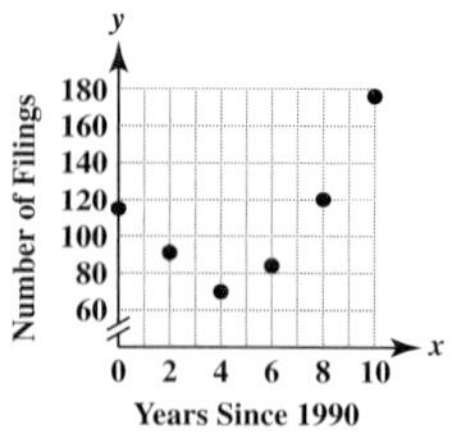


**(b)** quadratic; positive **(c)** $y = 2.969x^2 - 23.125x + 115$ **(d)** 265 **(e)** No. About 16 companies filed for bankruptcy each month, so at this rate, filings for 2002 would be about 192. The approximation from the model seems high. **51.** **(a)** 183.4 **(b)** The approximation using the model is quite close. **53.** $\{-7, -2\}$ **55.** $\{-.5, 1.25\}$

## Section 11.7 (page 772)

**Exercises** **1.** If $x$ is squared, it has a vertical axis; if $y$ is squared, it has a horizontal axis. **3.** Use the discriminant of the function. If it is positive, there are two $x$-intercepts. If it is 0, there is one $x$-intercept (at the vertex), and if it is negative, there is no $x$-intercept. **5.** $(-4, -6)$ **7.** $(1, -3)$ **9.** $(2, -1)$ **11.** $(-1, 3)$; up; narrower; no $x$-intercepts **13.** $\left(\frac{5}{2}, \frac{37}{4}\right)$; down; same; two $x$-intercepts **15.** $(-3, -9)$; to the right; wider **17.** F **19.** C **21.** D **23.** vertex: $(-4, -6)$; axis: $x = -4$; domain: $(-\infty, \infty)$; range: $[-6, \infty)$

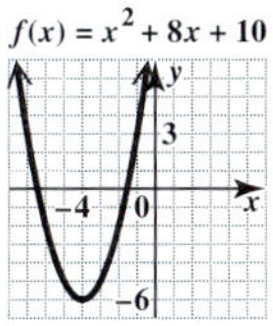


**25.** vertex: $(1, -3)$; axis: $x = 1$; domain: $(-\infty, \infty)$; range: $(-\infty, -3]$

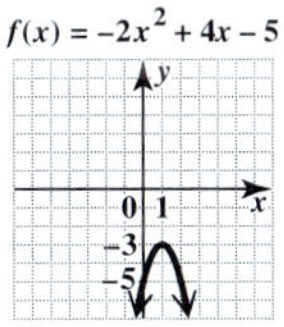


**27.** vertex: $(1, -2)$; axis: $y = -2$; domain: $[1, \infty)$; range: $(-\infty, \infty)$

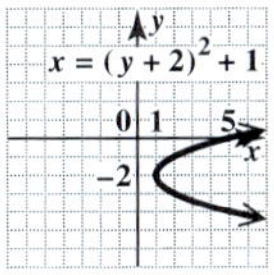


**29.** vertex: $(1, 5)$; axis: $y = 5$; domain: $(-\infty, 1]$; range: $(-\infty, \infty)$

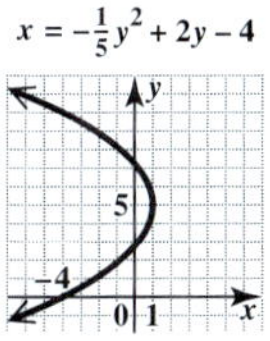


**31.** vertex: $(-7, -2)$; axis: $y = -2$; domain: $[-7, \infty)$; range: $(-\infty, \infty)$

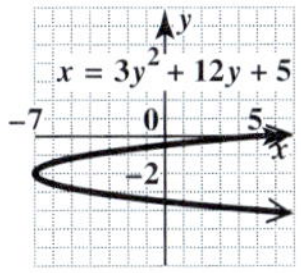


**33.** 30 and 30 **35.** 140 ft by 70 ft; 9800 ft$^2$ **37.** 16 ft; 2 sec

**39.** 2 sec; 67 ft **41.** **(a)** minimum **(b)** 1995; 1.7% **43.** **(a)** The coefficient of $x^2$ is negative because the parabola opens down. **(b)** (18.45, 3860) **(c)** In 2018 Social Security assets will reach their maximum value of \$3860 billion.

**45.** **(a)** $R(x) = (100 - x)(200 + 4x) = 20{,}000 + 200x - 4x^2$ **(b)**

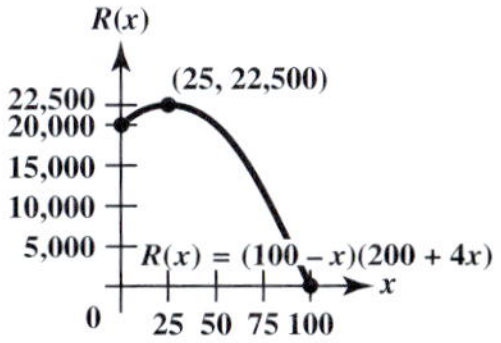


**(c)** 25 **(d)** \$22,500 **47.** B **49.** A

**51.** **(a)** $|x + p|$ **(b)** The distance from the focus to the origin should equal the distance from the directrix to the origin. **(c)** $\sqrt{(x - p)^2 + y^2}$ **(d)** $y^2 = 4px$

## Section 11.8 (page 783)

**Exercises** **1.** **(a)** $\{1, 3\}$ **(b)** $(-\infty, 1) \cup (3, \infty)$ **(c)** $(1, 3)$ **3.** **(a)** $\left\{-3, \frac{5}{2}\right\}$ **(b)** $\left[-3, \frac{5}{2}\right]$ **(c)** $(-\infty, -3] \cup \left[\frac{5}{2}, \infty\right)$

**5.** Include the endpoints if the symbol is $\geq$ or $\leq$. Exclude the endpoints if the symbol is $>$ or $<$.

**7.** $(-\infty, -1) \cup (5, \infty)$

**9.** $(-4, 6)$

**11.** $(-\infty, 1] \cup [3, \infty)$

**13.** $\left(-\infty, -\frac{3}{2}\right] \cup \left[\frac{3}{5}, \infty\right)$

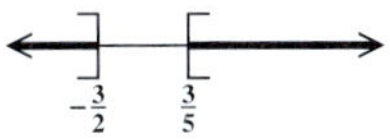


**15.** $\left[-\frac{3}{2}, \frac{3}{2}\right]$

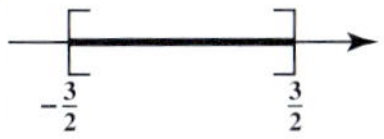


**17.** $\left(-\infty, -\frac{1}{2}\right] \cup \left[\frac{1}{3}, \infty\right)$

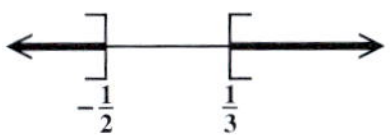


**19.** $(-\infty, 0] \cup [4, \infty)$

**21.** $\left[0, \frac{5}{3}\right]$

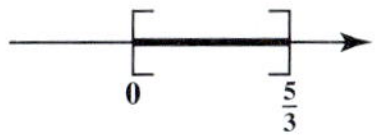

**23.** $(-\infty, 3-\sqrt{3}] \cup [3+\sqrt{3}, \infty)$ **25.** $(-\infty, \infty)$ **27.** $\emptyset$ **29.** $(-\infty, 1) \cup (2, 4)$ **31.** $\left[-\frac{3}{2}, \frac{1}{3}\right] \cup [4, \infty)$ **33.** $(-\infty, 1) \cup (4, \infty)$ **35.** $\left[-\frac{3}{2}, 5\right)$ **37.** $(2, 6]$ **39.** $\left(-\infty, \frac{1}{2}\right) \cup \left(\frac{5}{4}, \infty\right)$ **41.** $[-7, -2)$ **43.** $(-\infty, 2) \cup (4, \infty)$ **45.** $\left(0, \frac{1}{2}\right) \cup \left(\frac{5}{2}, \infty\right)$ **47.** $\left[\frac{3}{2}, \infty\right)$ **49.** $\left(-2, \frac{5}{3}\right) \cup \left(\frac{5}{3}, \infty\right)$ **51.** 3 sec and 13 sec **52.** between 3 sec and 13 sec **53.** at 0 sec (the time when it is initially projected) and at 16 sec (the time when it hits the ground) **54.** between 0 and 3 sec and between 13 and 16 sec

## Chapter 11 Review Exercises (page 790)

**1.** $\{11, -11\}$ **3.** $\{8\sqrt{2}, -8\sqrt{2}\}$ **5.** $\left\{-\frac{15}{2}, \frac{5}{2}\right\}$ **7.** By the square root property, the first step should be $x = \sqrt{12}$ or $x = -\sqrt{12}$. **9.** $\{-5, -1\}$ **11.** $\{-1+\sqrt{6}, -1-\sqrt{6}\}$ **13.** $\left\{-\frac{2}{5}, 1\right\}$ **15.** $\left\{-\frac{7}{2}, 3\right\}$ **17.** $\left\{\frac{1+\sqrt{41}}{2}, \frac{1-\sqrt{41}}{2}\right\}$ **19.** $\left\{\frac{2+i\sqrt{2}}{3}, \frac{2-i\sqrt{2}}{3}\right\}$ **21.** C **23.** D **25.** $\left\{-\frac{5}{2}, 3\right\}$ **27.** $\{-4\}$ **29.** $\left\{-\frac{343}{8}, 64\right\}$ **31.** 7 mph **33.** 4.6 hr **35.** $v = \frac{\pm\sqrt{rFkw}}{kw}$ **37.** $t = \frac{3m \pm \sqrt{9m^2 + 24m}}{2m}$ **39.** 12 cm by 20 cm **41.** 3 min **43.** .7 sec and 4.0 sec **45.** 4.5% **47.** $(0, 6)$ **49.** $(3, 7)$ **51.** $(-4, 3)$ **53.** vertex: $(2, -3)$; axis: $x = 2$; domain: $(-\infty, \infty)$; range: $[-3, \infty)$

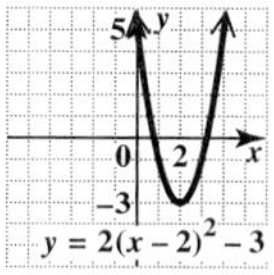


**55.** vertex: $(-4, -3)$; axis: $y = -3$; domain: $[-4, \infty)$; range: $(-\infty, \infty)$

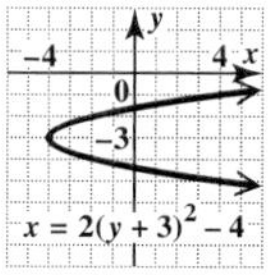


**57. (a)** $c = 12.39$, $16a + 4b + c = 15.78$, $49a + 7b + c = 22.71$ **(b)** $f(x) = .2089x^2 + .0118x + 12.39$ **(c)** \$25.85; The result using the model is a little high. **59.** length: 50 m; width: 50 m **61.** $[-4, 3]$ **63.** $\emptyset$ **65.** $[-3, 2)$ **67.** $\left\{\frac{3+i\sqrt{3}}{3}, \frac{3-i\sqrt{3}}{3}\right\}$ **69.** $\{-2, -1, 3, 4\}$ **71.** $\{4\}$ **73.** $d = \frac{\pm\sqrt{SkI}}{I}$ **75.** $\left\{-\frac{5}{3}, -\frac{3}{2}\right\}$ **77.** $\left(-5, -\frac{23}{5}\right]$ **79.** B **81.** A **83.** D **85. (a)** 21.92 trillion $ft^3$ **(b)** 2005 **87.** 412.3 ft **88. (a)** $\{-2\}$ **(b)** $(-\infty, -2)$ **(c)** $(-2, \infty)$ **89. (a)** $\{1, 5\}$ **(b)** $(-\infty, 1) \cup (5, \infty)$ **(c)** $(1, 5)$ **90. (a)** $\{4\}$ **(b)** $(2, 4)$ **(c)** $(-\infty, 2) \cup (4, \infty)$ **91.** $(-\infty, \infty)$; denominator **92.** $(-5, 3)$

## Chapter 11 Test (page 795)

[11.1] **1.** $\{3\sqrt{6}, -3\sqrt{6}\}$ **2.** $\left\{-\frac{8}{7}, \frac{2}{7}\right\}$ [11.2] **3.** $\{-1+\sqrt{5}, -1-\sqrt{5}\}$ [11.3] **4.** $\left\{\frac{3+\sqrt{17}}{4}, \frac{3-\sqrt{17}}{4}\right\}$

**5.** $\left\{\frac{2+i\sqrt{11}}{3}, \frac{2-i\sqrt{11}}{3}\right\}$ [11.1] **6.** A [11.3] **7.** discriminant: 88; There are two irrational solutions. [11.1–11.4] **8.** $\left\{\frac{2}{3}\right\}$

**9.** $\left\{-\frac{2}{3}, 6\right\}$ **10.** $\left\{\frac{-7+\sqrt{97}}{8}, \frac{-7-\sqrt{97}}{8}\right\}$ **11.** $\left\{-2, -\frac{1}{3}, \frac{1}{3}, 2\right\}$ **12.** $\left\{-\frac{5}{2}, 1\right\}$ [11.5] **13.** $r = \frac{\pm\sqrt{\pi S}}{2\pi}$

[11.4] **14.** Andrew: 11.1 hr; Kent: 9.1 hr **15.** 7 mph [11.5] **16.** 2 ft **17.** 16 m [11.6] **18.** A **19.** vertex: $(0, -2)$; axis: $x = 0$; domain: $(-\infty, \infty)$; range: $[-2, \infty)$

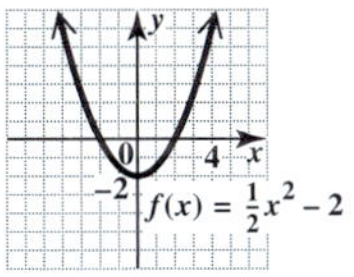


[11.7] **20.** vertex: $(2, 3)$; axis: $x = 2$; domain: $(-\infty, \infty)$; range: $(-\infty, 3]$

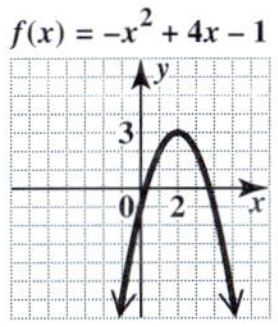


**21.** vertex: $(2, 2)$; axis: $y = 2$; domain: $(-\infty, 2]$; range: $(-\infty, \infty)$

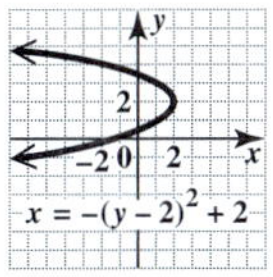


**22.** **(a)** 6.5% **(b)** 1996; 3.5% **23.** 160 ft by 320 ft [11.8] **24.** $(-\infty, -5) \cup \left(\frac{3}{2}, \infty\right)$

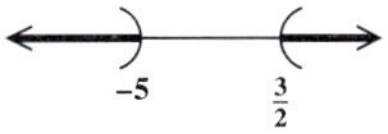


**25.** $(-\infty, 4) \cup [9, \infty)$

## Cumulative Review Exercises Chapters 1–11 (page 797)

[1.4, 10.7] **1.** **(a)** $-2, 0, 7$ **(b)** $-\frac{7}{3}, -2, 0, .7, 7, \frac{32}{3}$ **(c)** all except $\sqrt{-8}$ **(d)** All are complex numbers. [1.4–1.6] **2.** 6 **3.** 41

[2.3] **4.** $\left\{\frac{4}{5}\right\}$ [9.2] **5.** $\left\{\frac{11}{10}, \frac{7}{2}\right\}$ [10.6] **6.** $\left\{\frac{2}{3}\right\}$ [6.6] **7.** $\emptyset$ [11.2, 11.3] **8.** $\left\{\frac{7+\sqrt{177}}{4}, \frac{7-\sqrt{177}}{4}\right\}$ [11.4] **9.** $\{-2, -1, 1, 2\}$

[2.8] **10.** $[1, \infty)$ [9.2] **11.** $\left[2, \frac{8}{3}\right]$ [11.8] **12.** $(1, 3)$ **13.** $(-2, 1)$ [3.2, 7.1, 7.3] **14.** function; domain: $(-\infty, \infty)$; range: $(-\infty, \infty)$

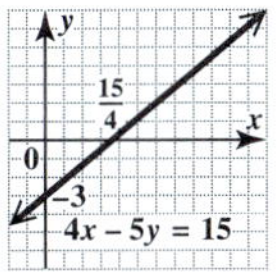


[7.3, 9.3] **15.** not a function

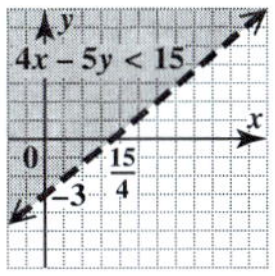


[11.6] **16.** function; domain: $(-\infty, \infty)$; range: $(-\infty, 3]$

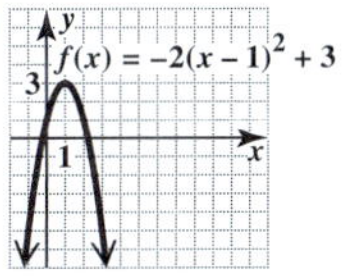


[3.3, 7.1] **17.** $m = \frac{2}{7}$; $x$-intercept: $(-8, 0)$; $y$-intercept: $\left(0, \frac{16}{7}\right)$ [3.4, 7.2] **18.** **(a)** $y = -\frac{5}{2}x + 2$

**(b)** $y = \frac{2}{5}x + \frac{13}{5}$ [4.1, 4.2] **19.** $\frac{x^8}{y^4}$ **20.** $\frac{4}{xy^2}$ [4.5] **21.** $14x^2 - 13x - 12$ [4.6] **22.** $\frac{4}{9}t^2 + 12t + 81$

[4.4] **23.** $-3t^3 + 5t^2 - 12t + 15$ [4.7] **24.** $4x^2 - 6x + 11 + \frac{4}{x+2}$ [5.1–5.4] **25.** $x(4+x)(4-x)$ **26.** $(4m-3)(6m+5)$

**27.** $(2x + 3y)(4x^2 - 6xy + 9y^2)$ **28.** $(3x - 5y)^2$ [6.2] **29.** $\frac{x-5}{x+5}$ [6.4] **30.** $-\frac{8}{k}$ [6.5] **31.** $\frac{r-s}{r}$ [7.2] **32. (a)**

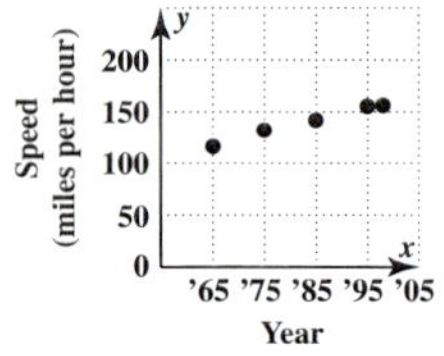


**(b)** positive **(c)** $y = 1.279x + 116.26$ **(d)** 158.47; It is a little too high. [7.3, 11.6] **33.** No, because the graph is a vertical line, which is not the graph of a function by the vertical line test. **34. (a)** 13 **(b)** domain: $(-\infty, \infty)$; range: $[-5, \infty)$ [8.1–8.3] **35.** $\{(1, -2)\}$ [8.4] **36.** $\{(3, -4, 2)\}$ [8.5] **37. (a)** $x + y = 34.2$; $x = 4y - .3$ **(b)** AOL: \$27.3 billion; Time Warner: \$6.9 billion [10.1] **38.** $\frac{3\sqrt[3]{4}}{4}$ [10.5] **39.** $\sqrt{7} + \sqrt{5}$ [6.7] **40.** biking: 12 mph; walking: 2 mph [11.5] **41.** southbound car: 57 mi; eastbound car: 76 mi [2.7] **42.** 930 **43.** 720 **44.** 990 **45.** 930

# CHAPTER 12 INVERSE, EXPONENTIAL, AND LOGARITHMIC FUNCTIONS

## Section 12.1 (page 807)

**Exercises** **1.** It is not one-to-one because both Illinois and Wisconsin are paired with the same range element, 40. **3.** Yes. By adding 1 to 1058 two distances would be the same, so the function would not be one-to-one. **5.** B **7.** A **9.** $\{(6, 3), (10, 2), (12, 5)\}$ **11.** not one-to-one **13.** $f^{-1}(x) = \frac{x-4}{2}$ **15.** $g^{-1}(x) = x^2 + 3,\ x \geq 0$ **17.** not one-to-one **19.** $f^{-1}(x) = \sqrt[3]{x+4}$ **21. (a)** 8 **(b)** 3 **23. (a)** 1 **(b)** 0 **25. (a)** one-to one **(b)**

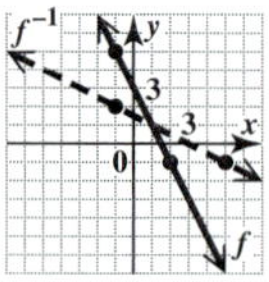


**27. (a)** not one-to one **29. (a)** one-to one **(b)**

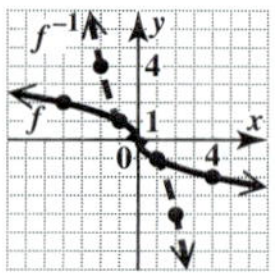


**31.**

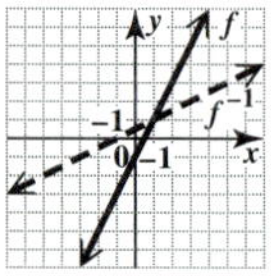


**33.**

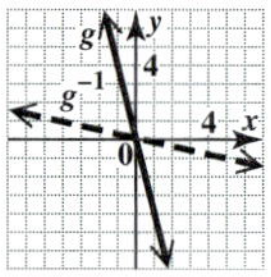


**35.**

| $x$ | $f(x)$ |
|---|---|
| 0 | 0 |
| 1 | 1 |
| 4 | 2 |

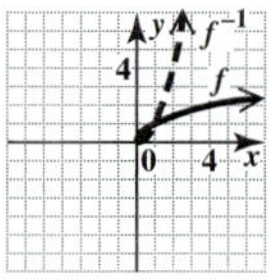


**37.**

| $x$ | $f(x)$ |
|---|---|
| $-1$ | $-3$ |
| 0 | $-2$ |
| 1 | $-1$ |
| 2 | 6 |

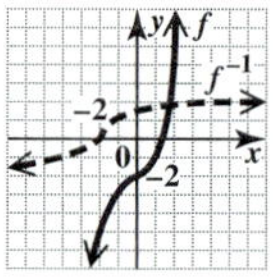


**39.** $f^{-1}(x) = \frac{x+5}{4}$

**40.** My graphing calculator is the greatest thing since sliced bread. **41.** If the function were not one-to-one, there would be ambiguity in some of the characters, as they could represent more than one letter. **42.** Answers will vary. For example, Jane Doe is 1004 5 2748 129 68 3379 129. **43.** $f^{-1}(x) = \frac{x+7}{2}$

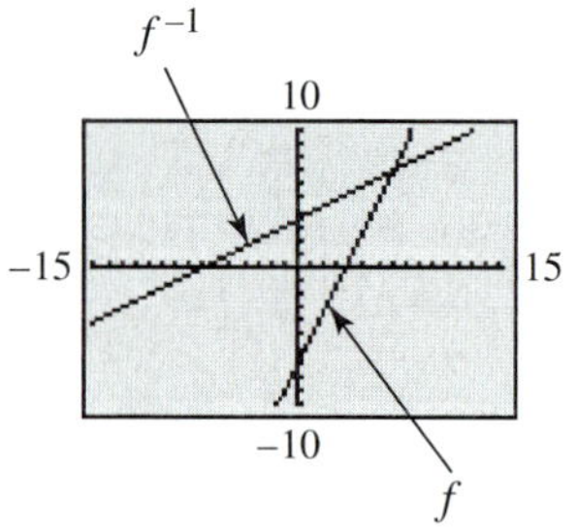

**45.** $f^{-1}(x) = \sqrt[3]{x-5}$

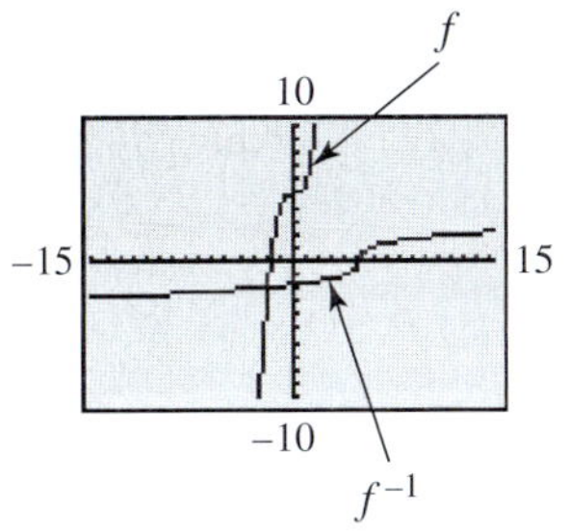


**47.**

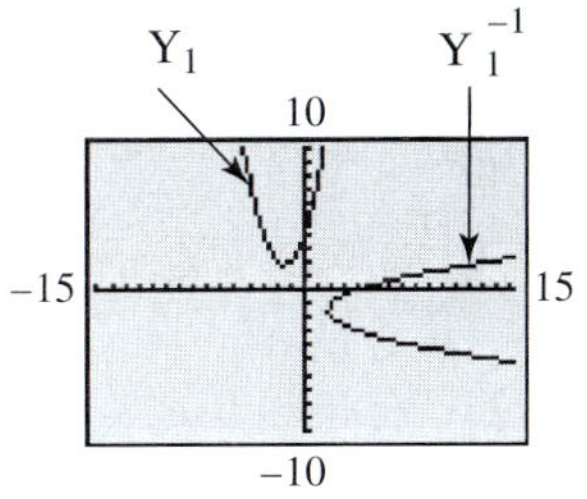


**49.** It is not a one-to-one function.

## Section 12.2 (page 818)

**Exercises** **1.** C **3.** A **5.**

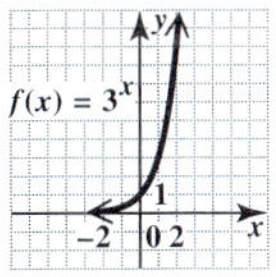


**7.**

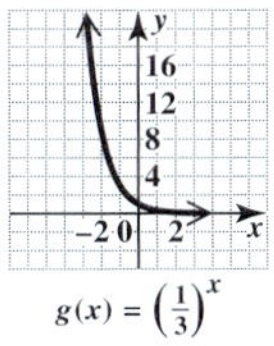


**9.**

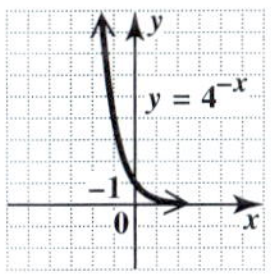


**11.**

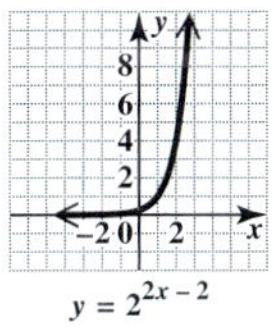


**13.** **(a)** rises; falls **(b)** It is one-to-one and thus has an inverse. **15.** $\{2\}$ **17.** $\left\{\frac{3}{2}\right\}$ **19.** $\{7\}$ **21.** $\{-3\}$ **23.** $\{-1\}$ **25.** $\{-3\}$ **27.** 639.545 **29.** .066 **31.** 12.179 **33.** In the definition of the exponential function defined by $F(x) = a^x$, $a$ must be a positive number. This corresponds to the peculiarity that some scientific calculators do not allow negative bases. **35.** **(a)** .5°C **(b)** .35°C **37.** **(a)** 1.6°C **(b)** .5°C **39.** **(a)** 132,359 thousand tons **(b)** 97,264 thousand tons **(c)** It is slightly less than what the model provides (93,733 thousand tons). **41.** **(a)** \$5000 **(b)** \$2973 **(c)** \$1768 **(d)**

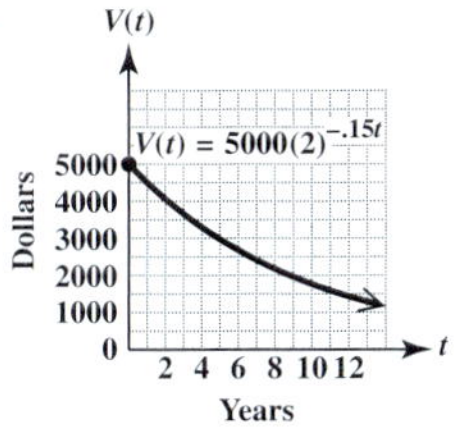


**43.** 6.67 yr after it was purchased **45.** $\left(\sqrt[4]{16}\right)^3$; 8 **46.** $\sqrt[4]{16^3}$; 8 **47.** 64; 8 **48.** Because $\sqrt{\sqrt{x}} = (x^{1/2})^{1/2} = x^{1/4} = \sqrt[4]{x}$, the fourth root of $16^3$ can be found by taking the square root twice. **49.** 8 **50.** $16^{75/100}$; $16^{3/4} = \left(\sqrt[4]{16}\right)^3 = 2^3 = 8$

## Section 12.3 (page 826)

**Connections** **(page 826)** **1.** almost 4 times as powerful **2.** about 300 times as powerful

**Exercises** **1.** **(a)** C **(b)** F **(c)** B **(d)** A **(e)** E **(f)** D **3.** $\log_4 1024 = 5$ **5.** $\log_{1/2} 8 = -3$ **7.** $\log_{10} .001 = -3$ **9.** $\log_{625} 5 = \frac{1}{4}$ **11.** $4^3 = 64$ **13.** $10^{-4} = \frac{1}{10{,}000}$ **15.** $6^0 = 1$ **17.** $9^{1/2} = 3$ **19.** By using the word "radically," the teacher meant for him to consider roots. Because 3 is the square (2nd) root of 9, $\log_9 3 = \frac{1}{2}$. **21.** $\left\{\frac{1}{3}\right\}$ **23.** $\{81\}$ **25.** $\left\{\frac{1}{5}\right\}$ **27.** $\{1\}$ **29.** $\{x \mid x > 0, x \neq 1\}$ **31.** $\{5\}$ **33.** $\left\{\frac{5}{3}\right\}$ **35.** $\{4\}$ **37.** $\left\{\frac{3}{2}\right\}$ **39.** $\{30\}$ **41.**

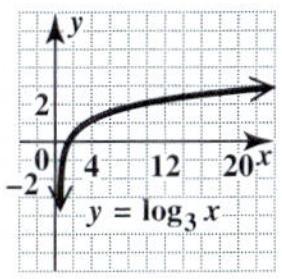


**43.**

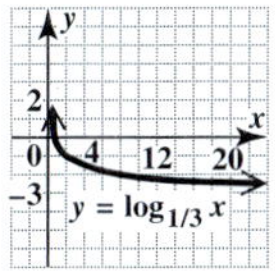


**45.** Every power of 1 is equal to 1,

and thus it cannot be used as a base. **47.** $(0, \infty)$; $(-\infty, \infty)$ **49.** 8 **51.** 24 **53.** **(a)** 645 sites **(b)** 962 sites **(c)** 1279 sites

**55.** **(a)** 130 thousand units **(b)** 190 thousand units **(c)**

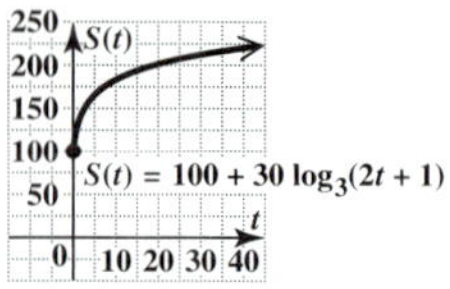


**57.** **(a)** 500 **(b)** 1000 **(c)** 1500 **(d)**

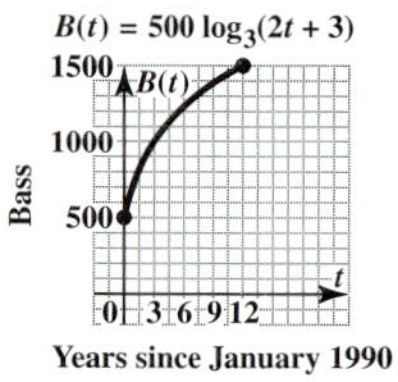


**59.**

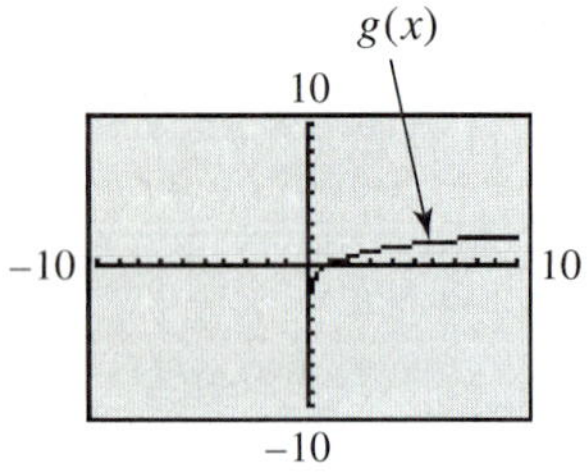


**61.**

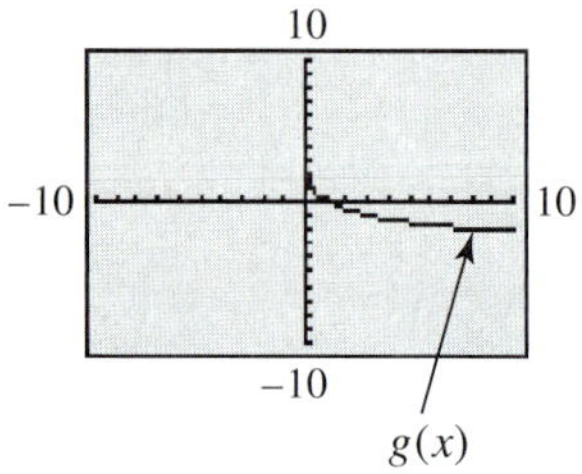


## Section 12.4 (page 836)

**Connections** (page 835) **1.**

$$\begin{array}{r} \log_{10} 458.3 \approx 2.661149857 \\ + \log_{10} 294.6 \approx 2.469232743 \\ \hline \approx 5.130382600 \\ 10^{5.130382600} \approx 135{,}015.18 \end{array}$$

A calculator gives $(458.3)(294.6) = 135{,}015.18$.

**2.** Answers will vary.

**Exercises** **1.** $\log_{10} 3 + \log_{10} 4$ **3.** 4 **5.** 4 **7.** $\log_7 4 + \log_7 5$ **9.** $\log_5 8 - \log_5 3$ **11.** $2 \log_4 6$ **13.** $\frac{1}{3} \log_3 4 - 2 \log_3 x - \log_3 y$ **15.** $\frac{1}{2} \log_3 x + \frac{1}{2} \log_3 y - \frac{1}{2} \log_3 5$ **17.** $\frac{1}{3} \log_2 x + \frac{1}{5} \log_2 y - 2 \log_2 r$ **19.** The distributive property tells us that the *product* $a(x + y)$ equals the sum $ax + ay$. In the notation $\log_a(x + y)$, the parentheses do not indicate multiplication. They indicate that $x + y$ is the result of raising $a$ to some power. **21.** $\log_b xy$ **23.** $\log_a \frac{m}{n}$ **25.** $\log_a \frac{rt^3}{s}$ **27.** $\log_a \frac{125}{81}$ **29.** $\log_{10}(x^2 - 9)$ **31.** $\log_p \frac{x^3 y^{1/2}}{z^{3/2} a^3}$ **33.** 1.2552 **35.** −.6532 **37.** 1.5562 **39.** .4771 **41.** .2386 **43.** 4.7710 **45.** false **47.** true **49.** true **51.** false **53.** The exponent of a quotient is the difference between the exponent of the numerator and the exponent of the denominator. **55.** $\log_2 8 - \log_2 4 = \log_2 \frac{8}{4} = \log_2 2 = 1$ **57.** 4 **58.** It is the exponent to which 3 must be raised to obtain 81. **59.** 81 **60.** It is the exponent to which 2 must be raised to obtain 19. **61.** 19 **62.** $m$

## Section 12.5 (page 843)

**Connections** (page 842) 2; 2.5; $2.\overline{6}$; $2.708\overline{3}$; $2.71\overline{6}$; The difference is .0016151618. It approaches $e$ fairly quickly.

**Exercises** **1.** C **3.** C **5.** 19.2 **7.** 1.6335 **9.** 2.5164 **11.** −1.4868 **13.** 9.6776 **15.** 2.0592 **17.** −2.8896 **19.** 5.9613 **21.** 4.1506 **23.** 2.3026 **25.** **(a)** 2.552424846 **(b)** 1.552424846 **(c)** .552424846 **(d)** The whole number parts will vary but the decimal parts are the same. **27.** An error message appears, because we cannot find the common logarithm of a negative number. **29.** bog **31.** 11.6 **33.** 4.3 **35.** $4.0 \times 10^{-8}$ **37.** $4.0 \times 10^{-6}$ **39.** **(a)** 800 yr **(b)** 5200 yr **(c)** 11,500 yr **41.** **(a)** 107 dB **(b)** 100 dB **(c)** 98 dB **43.** **(a)** 77% **(b)** 1989 **45.** **(a)** \$54 per ton **(b)** If $p = 0$, then $\ln(1 - p) = \ln 1 = 0$, so $T$ would be negative. If $p = 1$, then $\ln(1 - p) = \ln 0$, but the domain of $\ln x$ is $(0, \infty)$. **47.** 2.2619 **49.** .6826 **51.** .3155 **53.** .8736 **55.** 2.4849 **57.** Answers will vary. Suppose the name is Jeffery Cole, with $m = 7$ and $n = 4$. **(a)** $\log_7 4$ is the exponent to which 7 must be raised to obtain 4.

**(b)** .7124143742 **(c)** 4 **59.** 1333 sites **61.**

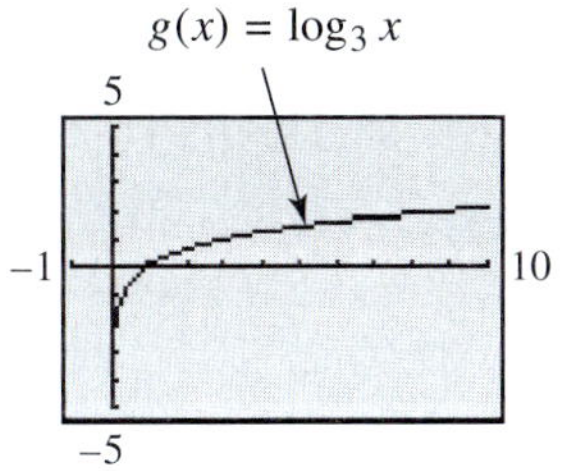


**63.**

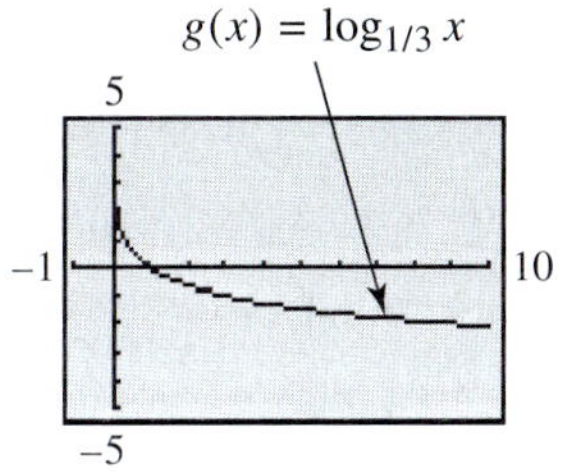


## Section 12.6 (page 853)

**Exercises** **1.** $\log 5^x = \log 125$ **2.** $x \log 5 = \log 125$ **3.** $x = \dfrac{\log 125}{\log 5}$ **4.** $\dfrac{\log 125}{\log 5} = 3$; $\{3\}$ **5.** $\{.827\}$ **7.** $\{.833\}$ **9.** $\{1.201\}$ **11.** $\{2.269\}$ **13.** $\{15.967\}$ **15.** $\{261.291\}$ **17.** $\{-10.718\}$ **19.** $\{3\}$ **21.** $\{5.879\}$ **23.** Natural logarithms are a better choice because $e$ is the base. **25.** $\left\{\dfrac{2}{3}\right\}$ **27.** $\left\{\dfrac{33}{2}\right\}$ **29.** $\{-1 + \sqrt[3]{49}\}$ **31.** 2 cannot be a solution because $\log(2 - 3) = \log(-1)$, and $-1$ is not in the domain of $\log x$. **33.** $\left\{\dfrac{1}{3}\right\}$ **35.** $\{2\}$ **37.** $\emptyset$ **39.** $\{8\}$ **41.** $\left\{\dfrac{4}{3}\right\}$ **43.** $\{8\}$ **45.** **(a)** \$2539.47 **(b)** 10.2 yr **47.** **(a)** \$4934.71 **(b)** 19.8 yr **49.** **(a)** \$11,260.96 **(b)** \$11,416.64 **(c)** \$11,497.99 **(d)** \$11,580.90 **(e)** \$11,581.83 **51.** \$137.41 **53.** **(a)** 78,840 million dollars **(b)** 92,316 million dollars **(c)** 108,095 million dollars **(d)** 70,967 million dollars **55.** 32,044 million dollars **57.** **(a)** 1.62 g **(b)** 1.18 g **(c)** .69 g **(d)** 2.00 g **59.** **(a)** 179.73 g **(b)** 21.66 yr **61.** 1997 **63.** 1.733 days **65.** It means that after 250 yr, approximately 2.9 g of the original sample remain. **67.** **(a)** 2.302585093 **(b)** .4342944819

## Chapter 12 Review Exercises (page 864)

**1.** not one-to-one **3.** This function is not one-to-one because two sodas in the list have 41 mg of caffeine. **5.** $f^{-1}(x) = \dfrac{x^3 + 4}{6}$

**7.**

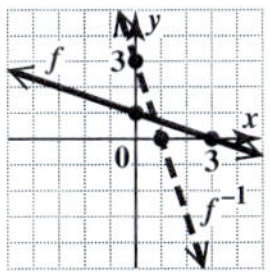


**9.**

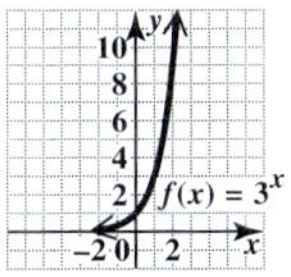


**11.**

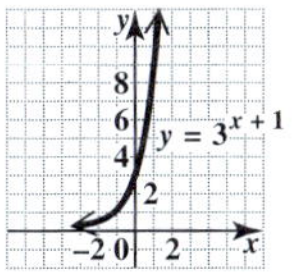


**13.** $\{4\}$ **15.** **(a)** 1.2 million tons **(b)** 3.8 million tons **(c)** 21.8 million tons

**17.**

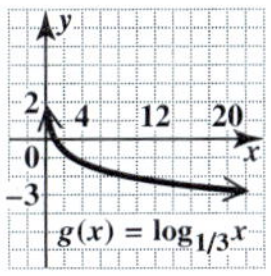


**19.** $\left\{\dfrac{3}{2}\right\}$ **21.** $\{8\}$ **23.** $\{b \mid b > 0, b \neq 1\}$ **25.** $a$ **27.** $\log_2 3 + \log_2 x + 2 \log_2 y$ **29.** $\log_b \dfrac{3x}{y^2}$ **31.** 1.4609

**33.** 3.3638 **35.** .9251 **37.** 6.4 **39.** $2.5 \times 10^{-5}$ **41.** **(a)** 18 yr **(b)** 12 yr **(c)** 7 yr **(d)** 6 yr **(e)** Each comparison shows approximately the same number. For example, in part (a) the doubling time is 18 yr (rounded) and $\dfrac{72}{4} = 18$. Thus, the formula $t = \dfrac{72}{100r}$ (called the *rule of 72*) is an excellent approximation of the doubling time formula. (It is used by bankers for that purpose.) **43.** $\{2.042\}$ **45.** $\{18.310\}$ **47.** $\{-6 + \sqrt[3]{25}\}$ **49.** $\left\{\dfrac{3}{8}\right\}$ **51.** $\{1\}$ **53.** \$28,295.56 **55.** Plan A is better, since it would pay \$2.92 more.

**57.** **(a)** about \$4267 **(b)** about 11% **59.** D **61.** $\dfrac{1}{4}, \dfrac{1}{2}, 1, 2, 4, 8$

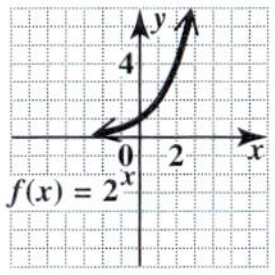


**62.** $-2, -1, 0, 1, 2, 3$

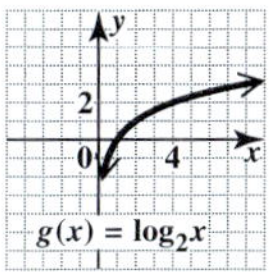

**63.** The roles of $x$ and $y$ are interchanged. They are inverses. **64.** horizontal; vertical **65.** 5; 2; 3; 5 **66.** 4; 8 (or 8; 4)

**67.** 3.700439718 (The number of displayed digits may vary.) **68.** $\log_2 13$ is the exponent to which 2 must be raised to obtain 13. **69.** 13

**70.** 13; The number in Exercise 67 is the exponent to which 2 must be raised to obtain 13. **71.** 3.700439718 **72.** $\left\{-\frac{8}{5}\right\}$ **73.** 7

**75.** 4 **77.** $-5$ **79.** $\{72\}$ **81.** $\left\{\frac{1}{9}\right\}$ **83.** $\{3\}$ **85.** $\left\{\frac{1}{8}\right\}$ **87.** $\{-2, -1\}$ **89.** 6.8 yr **91.** .325

## Chapter 12 Test (page 869)

[12.1] **1. (a)** not one-to-one **(b)** one-to-one **2.** $f^{-1}(x) = x^3 - 7$ **3.**

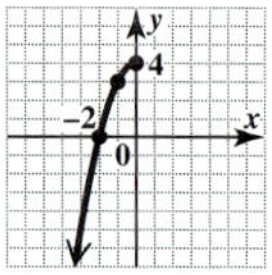


[12.2] **4.**

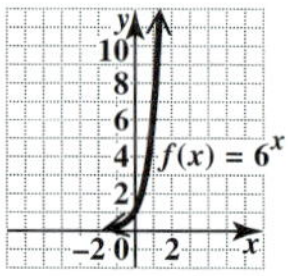


[12.3] **5.**

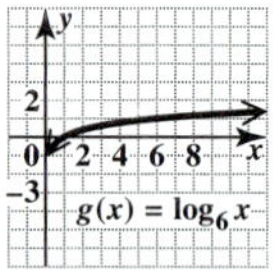


[12.1–12.3] **6.** Once the graph of $f(x) = 6^x$ is sketched, interchange the $x$- and $y$-values of its ordered pairs. The resulting points will be on the graph of $g(x) = \log_6 x$ since $f$ and $g$ are inverses. [12.2] **7.** $\{-4\}$ **8.** $\left\{-\frac{13}{3}\right\}$ [12.5] **9. (a)** 30.0 million **(b)** 37.7 million [12.3] **10.** $\log_4 .0625 = -2$ **11.** $7^2 = 49$ **12.** $\{32\}$ **13.** $\left\{\frac{1}{2}\right\}$ **14.** $\{2\}$ **15.** 5; 2; 5th; 32

[12.4] **16.** $2 \log_3 x + \log_3 y$ **17.** $\frac{1}{2}\log_5 x - \log_5 y - \log_5 z$ **18.** $\log_b \frac{s^3}{t}$ **19.** $\log_b \frac{r^{1/4}s^2}{t^{2/3}}$ [12.5] **20. (a)** 1.3636 **(b)** $-.1985$

**21. (a)** $\frac{\log 19}{\log 3}$ **(b)** $\frac{\ln 19}{\ln 3}$ **(c)** 2.6801 [12.6] **22.** $\{3.9656\}$ **23.** $\{3\}$ **24.** \$12,507.51 **25. (a)** \$19,260.38 **(b)** approximately 13.9 yr

## Cumulative Review Exercises Chapters 1–12 (page 871)

[1.4] **1.** $-2, 0, 6, \frac{30}{3}$ (or 10) **2.** $-\frac{9}{4}, -2, 0, .6, 6, \frac{30}{3}$ (or 10) **3.** $-\sqrt{2}, \sqrt{11}$ **4.** $-\frac{9}{4}, -2, -\sqrt{2}, 0, .6, \sqrt{11}, 6, \frac{30}{3}$ (or 10)

[1.4–1.6] **5.** 16 **6.** $-27$ **7.** $-39$ [2.3] **8.** $\left\{-\frac{2}{3}\right\}$ [2.8] **9.** $[1, \infty)$ [4.5] **10.** $6p^2 + 7p - 3$ [4.6] **11.** $16k^2 - 24k + 9$

[4.4] **12.** $-5m^3 + 2m^2 - 7m + 4$ [4.7] **13.** $2t^3 + 5t^2 - 3t + 4$ [5.1] **14.** $x(8 + x^2)$ [5.2, 5.3] **15.** $(3y - 2)(8y + 3)$

**16.** $z(5z + 1)(z - 4)$ [5.4] **17.** $(4a + 5b^2)(4a - 5b^2)$ **18.** $(2c + d)(4c^2 - 2cd + d^2)$ **19.** $(4r + 7q)^2$ [4.1, 4.2] **20.** $-\frac{1875p^{13}}{8}$

[6.2] **21.** $\frac{x + 5}{x + 4}$ [6.3, 6.4] **22.** $\frac{-3k - 19}{(k + 3)(k - 2)}$ **23.** $\frac{22 - p}{p(p - 4)(p + 2)}$ [3.2, 7.1] **24.**

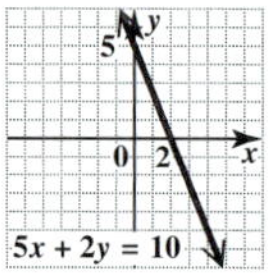


[9.3] **25.**

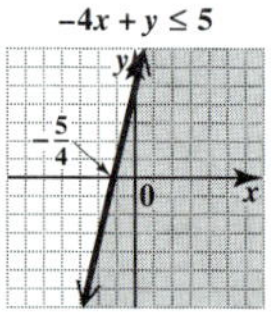


[3.3, 7.1, 7.3] **26. (a)** yes **(b)** approximately $-4000$; The number of acres harvested decreased by approximately 4000 acres per year during 1997–1999. [3.4, 7.2] **27.** $y = \frac{3}{4}x - \frac{19}{4}$ [8.1–8.3, 8.6] **28.** $\{(4, 2)\}$ **29.** $\emptyset$ [8.4, 8.6] **30.** $\{(1, -1, 4)\}$ [8.5] **31.** 6 lb

[9.2] **32.** $\{-2, 7\}$ **33.** $\left\{\pm\frac{16}{3}\right\}$ **34.** $\left[\frac{7}{3}, 3\right]$ **35.** $(-\infty, -3) \cup (2, \infty)$ [10.3] **36.** $12\sqrt{2}$ [10.4] **37.** $-27\sqrt{2}$ [10.6] **38.** $\{0, 4\}$

[10.7] **39.** 41 [11.1–11.3] **40.** $\left\{\frac{1 \pm \sqrt{13}}{6}\right\}$ [11.8] **41.** $(-\infty, -4) \cup (2, \infty)$ [11.4] **42.** $\{\pm 1, \pm 2\}$ [11.7] **43.** 150 and 150

[11.6] **44.**

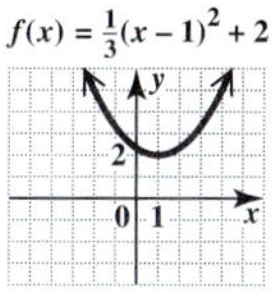


[12.2] **45.**

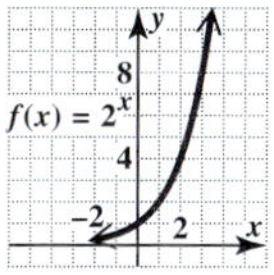


**46.** $\{-1\}$ [12.3] **47.**

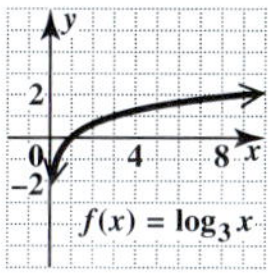


[12.4] **48.** 6.3398

**49.** $3 \log x + \frac{1}{2} \log y - \log z$ [12.6] **50.** **(a)** 25,000 **(b)** 30,500 **(c)** 37,300 **(d)** in about 3.5 hr, or at about 3:30 P.M.

# CHAPTER 13 NONLINEAR FUNCTIONS, CONIC SECTIONS, AND NONLINEAR SYSTEMS

## Section 13.1 (page 883)

**Exercises** **1.** 0 **3.** 0, 0 **5.** B **7.** A **9.** domain: $(-\infty, \infty)$; range: $[0, \infty)$

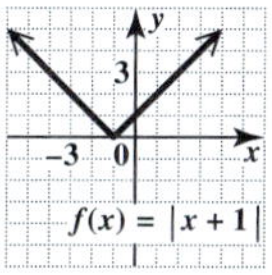


**11.** domain: $(-\infty, 0) \cup (0, \infty)$; range: $(-\infty, 1) \cup (1, \infty)$

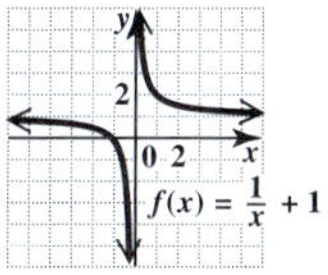


**13.** domain: $[2, \infty)$; range: $[0, \infty)$

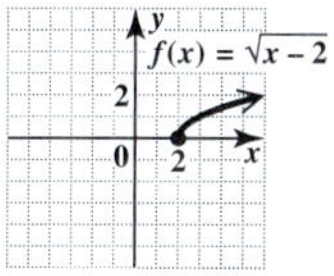


**15.** domain: $(-\infty, 2) \cup (2, \infty)$; range: $(-\infty, 0) \cup (0, \infty)$

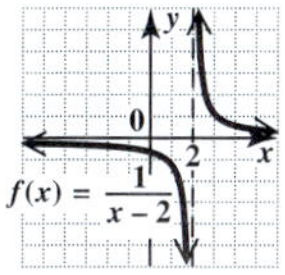


**17.** domain: $[-3, \infty)$; range: $[-3, \infty)$

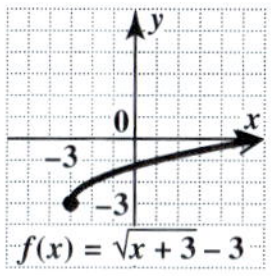


**19.**

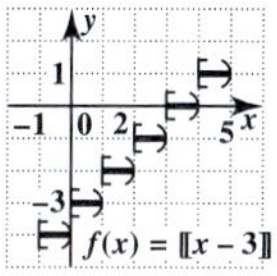


**21.**

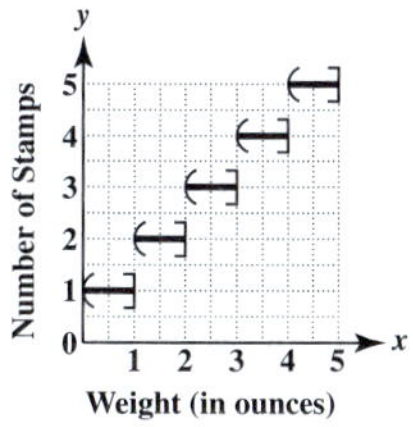


**23.** **(a)** $8x - 3$ **(b)** $2x - 17$ **25.** **(a)** $-x^2 + 12x - 12$ **(b)** $9x^2 + 4x + 6$ **27.** $10x^2 - 2x$

**29.** $2x^2 - x - 3$ **31.** $5x - 1; 0$ **33.** $2x - 3; -1$ **35.** $x^2 + 2x - 9$ **37.** 6 **39.** $x^2 - x - 6$ **41.** $2x^3 - 18x$ **43.** $-20$ **45.** $2x^2 - 6x$ **47.** $\frac{x^2 - 9}{2x}$, $x \neq 0$ **49.** $-\frac{5}{4}$ **51.** 0 **53.** For example, let $f(x) = 2x^3 + 3x^2 + x + 4$ and $g(x) = 2x^4 + 3x^3 - 9x^2 + 2x - 4$; $(f - g)(x) = -2x^4 - x^3 + 12x^2 - x + 8$, and $(g - f)(x) = 2x^4 + x^3 - 12x^2 + x - 8$. Because the two differences are not equal, subtraction of polynomial functions is not commutative. **55.** 16 **57.** 83 **59.** 13 **61.** $4x^2 + 12x + 13$ **63.** $x^2 + 10x + 29$ **65.** $2x + 8$ **67.** $(f \circ g)(x) = 63{,}360x$; It computes the number of inches in $x$ mi. **69.** $(A \circ r)(t) = 4\pi t^2$; This is the area of the circular layer as a function of time.

## Section 13.2 (page 891)

**Exercises** **1.** **(a)** (0, 0) **(b)** 5 **(c)**

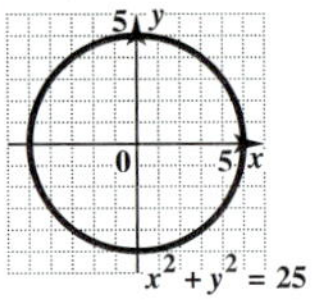


**3.** B **5.** D **7.** $(x + 4)^2 + (y - 3)^2 = 4$ **9.** $(x + 8)^2 + (y + 5)^2 = 5$

**11.** center: $(-2, -3)$; $r = 2$ **13.** center: $(-5, 7)$; $r = 9$ **15.** center: $(2, 4)$; $r = 4$ **17.** The thumbtack acts as the center and the length of the string acts as the radius. **19.**

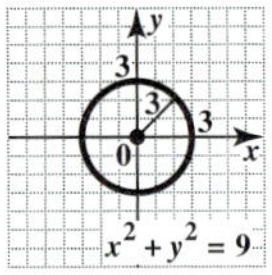


**21.**

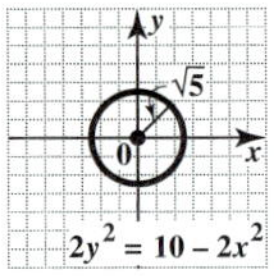


**23.** center: $(-3, 2)$

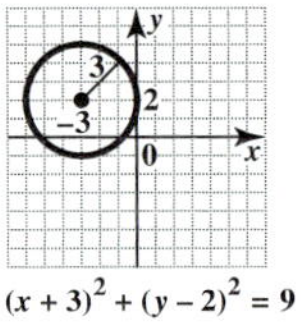


**25.** center: $(2, 3)$

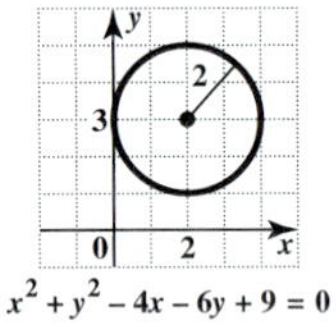


**27.**

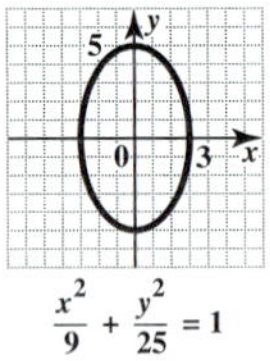


**29.**

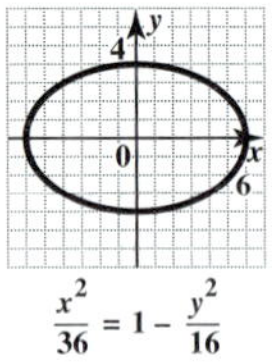


**31.**

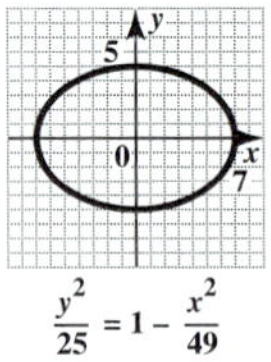


**33.**

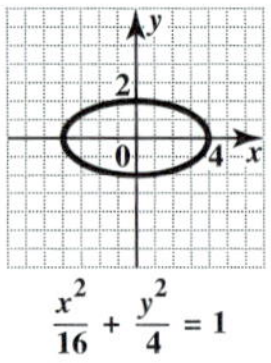


**35.**

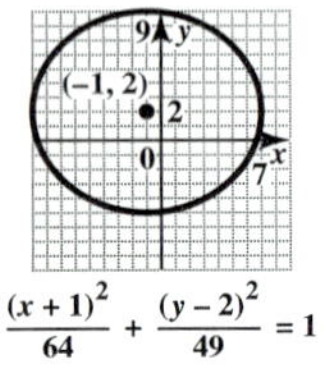


**37.**

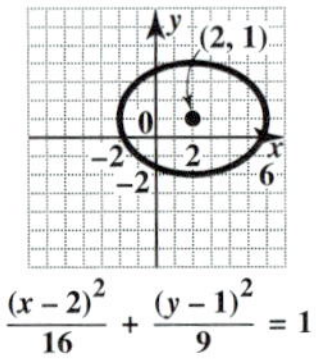


**39.** The fixed ends of the string are at the foci, and the constant length of the string represents the sum of the distances from any point on the curve to the foci. **41.** By the vertical line test the set is not a function, because a vertical line may intersect the graph of an ellipse in two points. **43.** $y_1 = 4 + \sqrt{16 - (x + 2)^2}$, $y_2 = 4 - \sqrt{16 - (x + 2)^2}$

**45.**

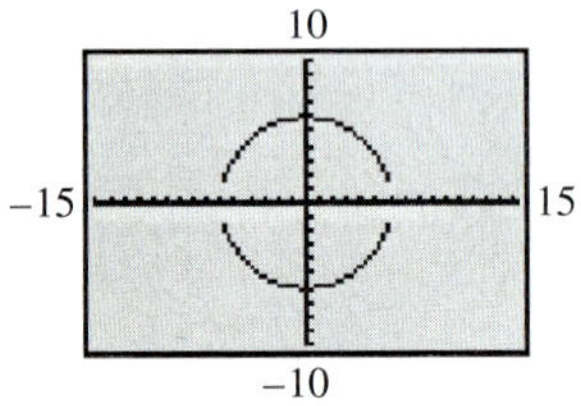


**47.**

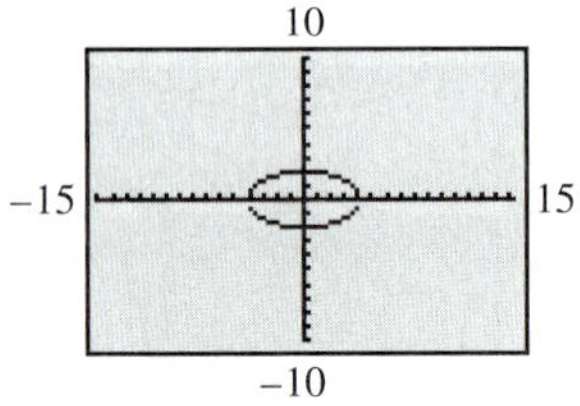


**49.** **(a)** 10 m **(b)** 36 m **51.** **(a)** 154.7 million mi **(b)** 128.7 million mi (Answers are rounded.) **53.** $3\sqrt{3}$ units **55.** Answers will vary.

## Section 13.3 (page 902)

**Connections** **(page 898)** Answers will vary.

**Exercises** **1.** C **3.** D **5.** When written in one of the forms given in the box titled "Equations of Hyperbolas" in this section, it will open up and down if the − sign precedes the $x^2$-term; it will open left and right if the − sign precedes the $y^2$-term. **7.**

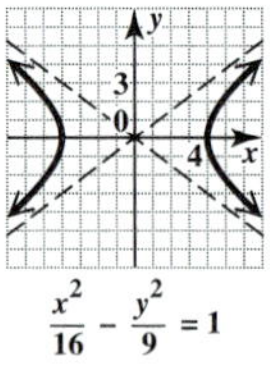

**9.** 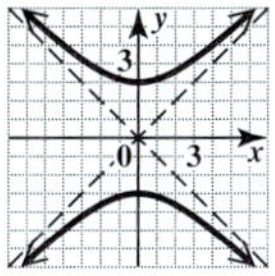


**11.** 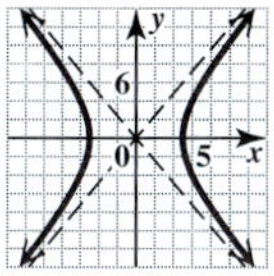


**13.** 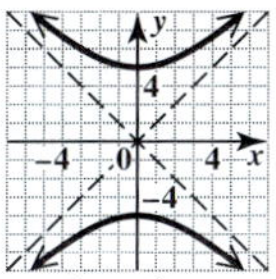


**15.** hyperbola 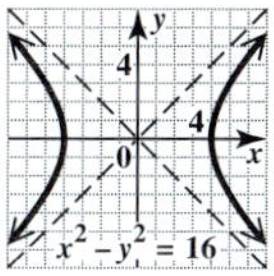


**17.** ellipse 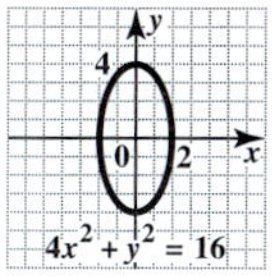


**19.** circle 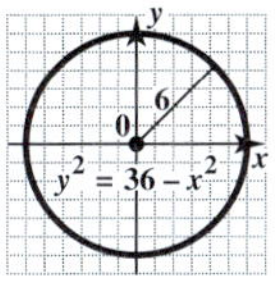


**21.** hyperbola 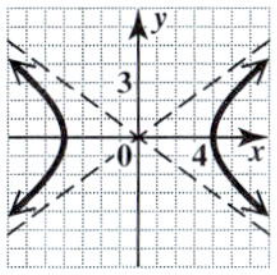


**23.** hyperbola 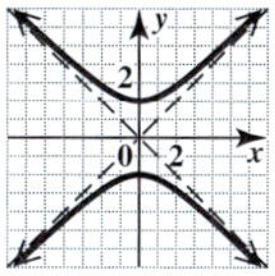


**25.** domain: $[-4, 4]$; range: $[0, 4]$ 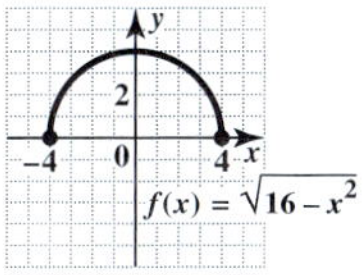


**27.** domain: $[-6, 6]$; range: $[-6, 0]$ 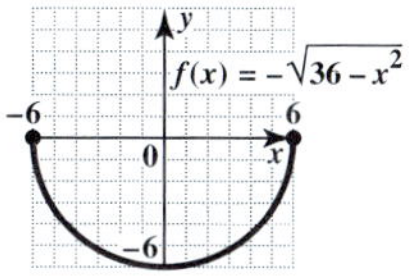


**29.** domain: $(-\infty, \infty)$; range: $[3, \infty)$ 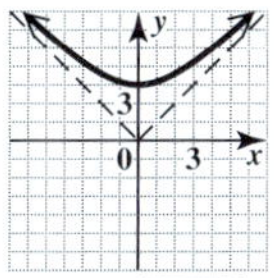


**31.** 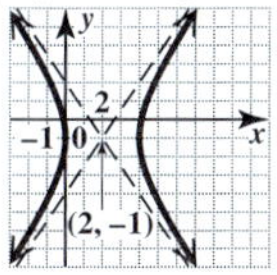


**33.** 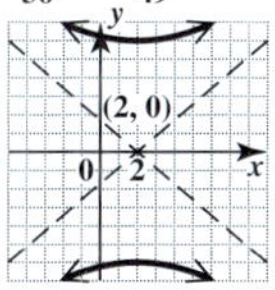


**35.** **(a)** 50 m **(b)** 69.3 m **37.** for $V$ greater than 4325.68 m per sec

**39.** $y_1 = \sqrt{\frac{x^2}{9} - 1}$, $y_2 = -\sqrt{\frac{x^2}{9} - 1}$ **41.** 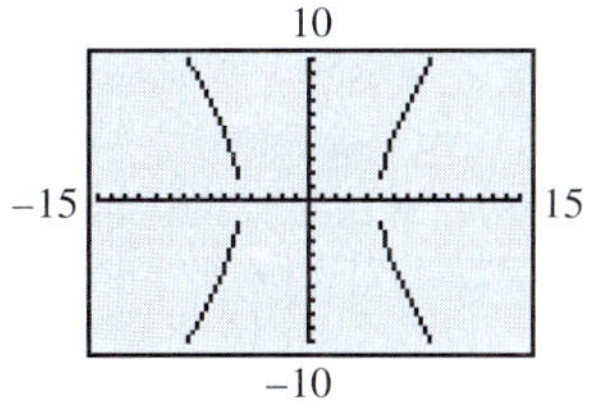


**43.** 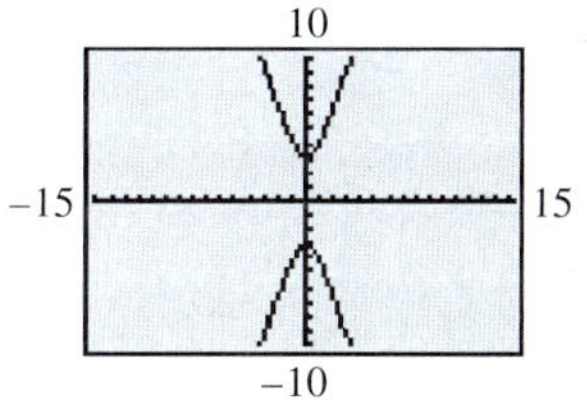


**45.** Answers will vary.

## Section 13.4 (page 910)

**Exercises** **1.** Substitute $x - 1$ for $y$ in the first equation. Then solve for $x$. Find the corresponding $y$-values by substituting back into $y = x - 1$. In the first equation, both variables are squared and in the second, both variables are to the first power, so the elimination method is not appropriate. **3.** one **5.** none **7.** 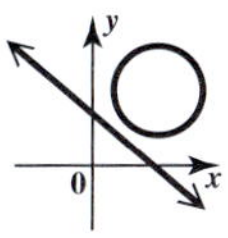 **9.** 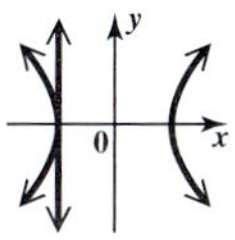 **11.** 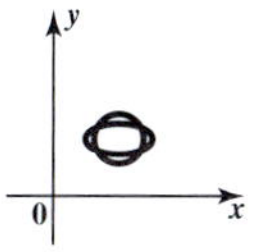 **13.** 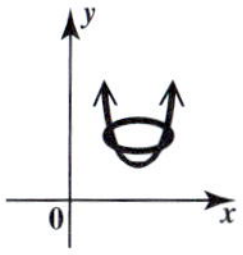 **15.** $\left\{(0, 0), \left(\frac{1}{2}, \frac{1}{2}\right)\right\}$

**17.** $\{(-6, 9), (-1, 4)\}$ **19.** $\left\{\left(-\frac{1}{5}, \frac{7}{5}\right), (1, -1)\right\}$ **21.** $\left\{(-2, -2), \left(-\frac{4}{3}, -3\right)\right\}$ **23.** $\{(-3, 1), (1, -3)\}$

**25.** $\left\{\left(-\frac{3}{2}, -\frac{9}{4}\right), (-2, 0)\right\}$ **27.** $\{(-\sqrt{3}, 0), (\sqrt{3}, 0), (-\sqrt{5}, 2), (\sqrt{5}, 2)\}$ **29.** $\left\{\left(\frac{i\sqrt{3}}{3}, \frac{-3 + i\sqrt{3}}{6}\right), \left(\frac{-i\sqrt{3}}{3}, \frac{-3 - i\sqrt{3}}{6}\right)\right\}$

**31.** $\{(-2, 0), (2, 0)\}$ **33.** $\{(\sqrt{3}, 0), (-\sqrt{3}, 0)\}$ **35.** $\{(1, 3), (1, -3), (-1, 3), (-1, -3)\}$ **37.** $\{(-2i\sqrt{2}, -2\sqrt{3}), (-2i\sqrt{2}, 2\sqrt{3}), (2i\sqrt{2}, -2\sqrt{3}), (2i\sqrt{2}, 2\sqrt{3})\}$ **39.** $\{(-\sqrt{5}, -\sqrt{5}), (\sqrt{5}, \sqrt{5})\}$ **41.** $\{(i, 2i), (-i, -2i), (2, -1), (-2, 1)\}$

**43.** $\{(2, -3), (-3, 2)\}$

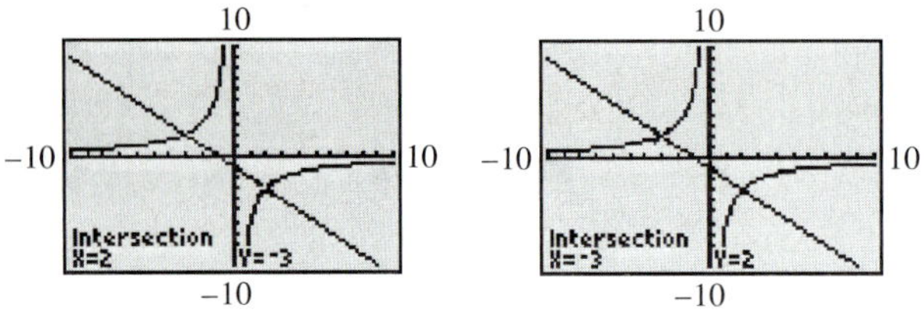


**45.** length: 12 ft; width: 7 ft **47.** \$20; $\frac{4}{5}$ thousand or 800 calculators **49.** 1981; 470 thousand

## Section 13.5 (page 916)

**Exercises** **1.** C **3.** Answers will vary. **5.** B **7.** A **9.**

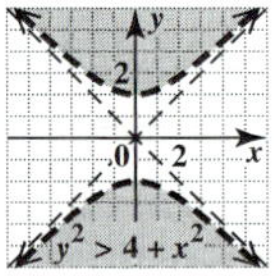


**11.**

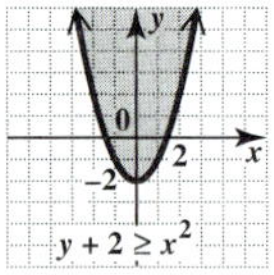


**13.**

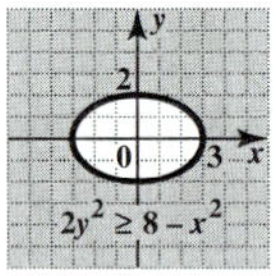


**15.**

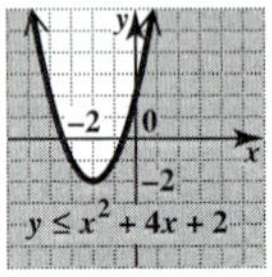


**17.**

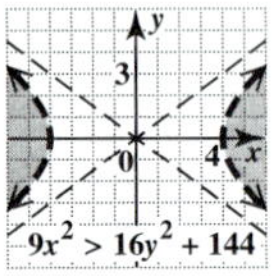


**19.**

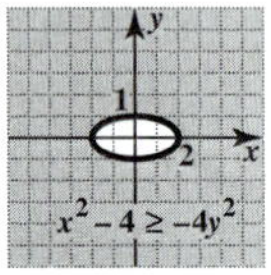


**21.**

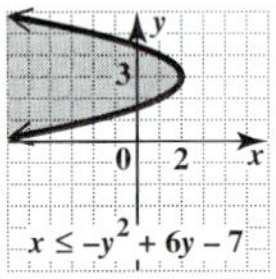


**23.**

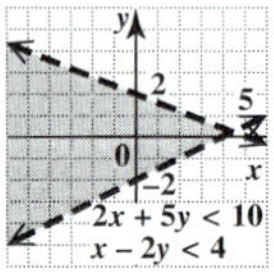


**25.**

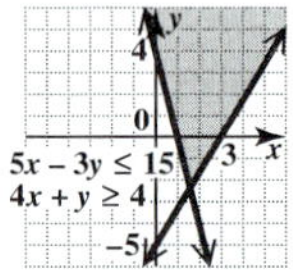


**27.**

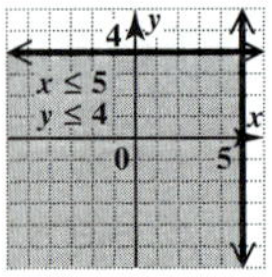


**29.**

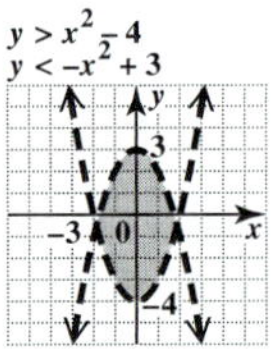


**31.**

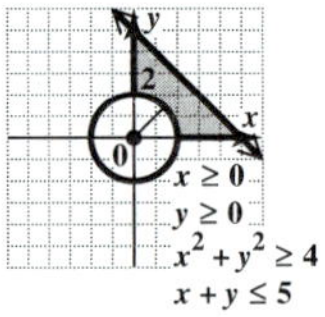


**33.**

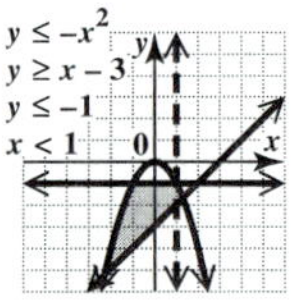


**35.**

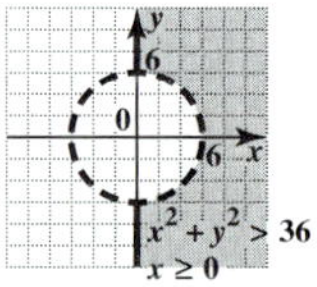


**37.**

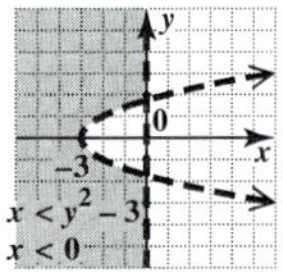


**39.**

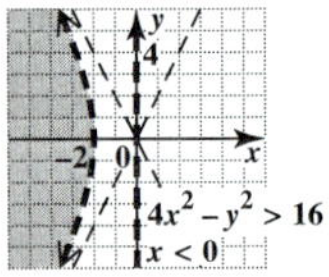


**41.**

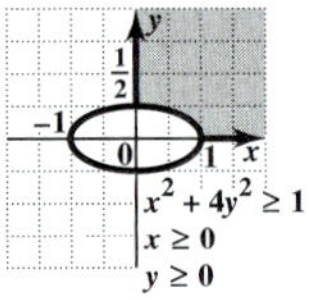


**43.**

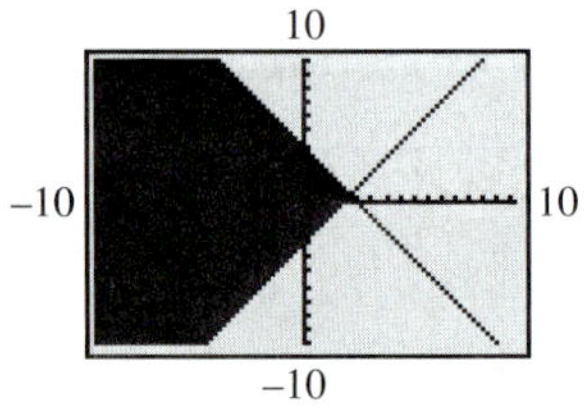

**45.**

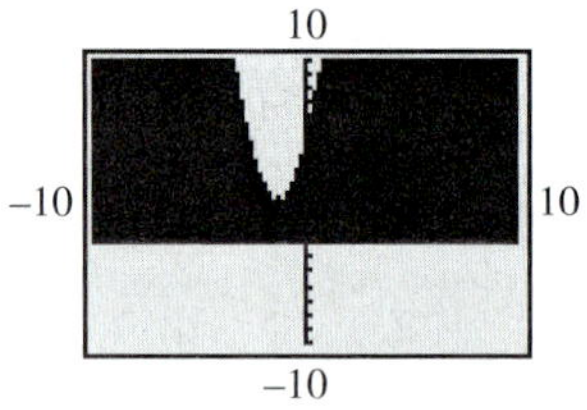

## Chapter 13 Review Exercises (page 923)

**1.**

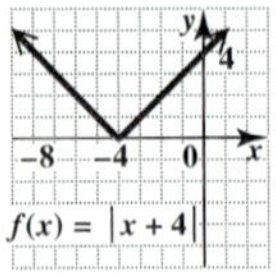


**3.**

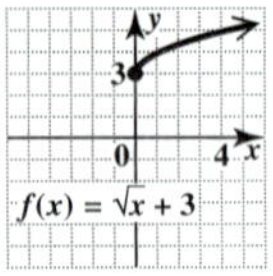


**5.** **(a)** $5x^2 - x + 5$ **(b)** $-5x^2 + 5x + 1$ **(c)** 11 **(d)** $-9$ **7.** 167 **9.** 20

**11.** $75x^2 + 220x + 160$ **13.** $(x + 2)^2 + (y - 4)^2 = 9$ **15.** $(x - 4)^2 + (y - 2)^2 = 36$ **17.** center: $(4, 1)$; $r = 2$

**19.** center: $(3, -2)$; $r = 5$ **21.**

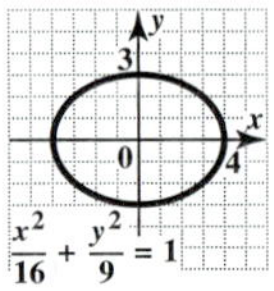


**23.** $\frac{x^2}{65{,}286{,}400} + \frac{y^2}{2{,}560{,}000} = 1$ **25.**

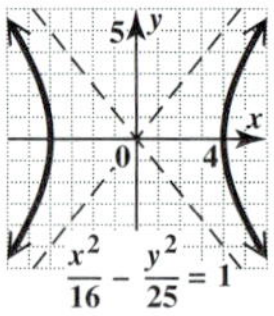


**27.**

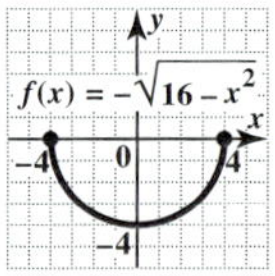

**29.** parabola **31.** ellipse **33.** hyperbola **35.** $\{(6, -9), (-2, -5)\}$ **37.** $\{(4, 2), (-1, -3)\}$

**39.** $\left\{\left(-\sqrt{2}, 2\right), \left(-\sqrt{2}, -2\right), \left(\sqrt{2}, -2\right), \left(\sqrt{2}, 2\right)\right\}$ **41.** 0, 1, or 2 **43.**

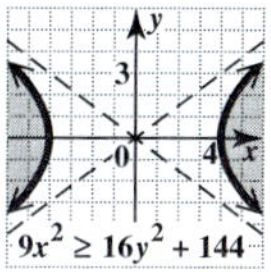


**45.**

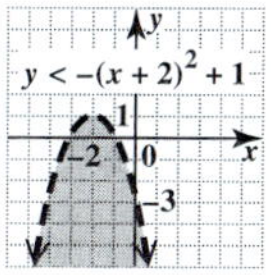


**47.**

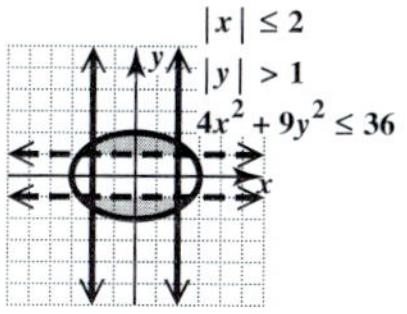


**49.** $2a + 4b + c = -20$ **50.** $5a + b + c = -26$ **51.** $-a + b + c = -2$ **52.** $\{(-4, -2, -4)\}$; $x^2 + y^2 - 4x - 2y - 4 = 0$

**53.** center: (2, 1); radius: 3 **55.**

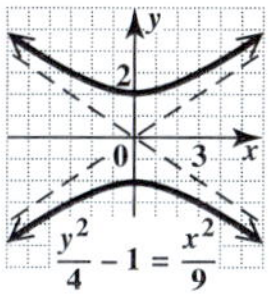


**57.**

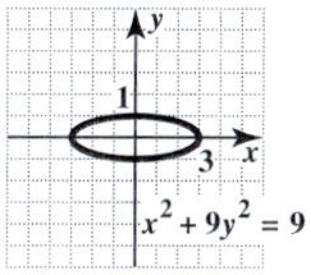


**59.**

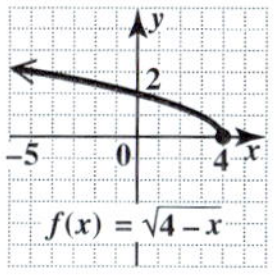


**61.**

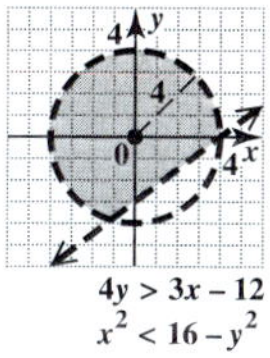


**63.** 69.8 million km

## Chapter 13 Test (page 925)

[13.1] **1.** C **2.** A **3.** D **4.** B **5.**

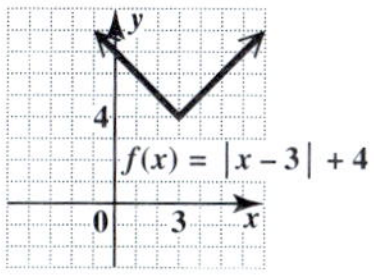


**6.** **(a)** $x^2 + 3x + 7$ **(b)** 5 **(c)** $3x^3 + 5x^2 + 6x + 10$ **(d)** 23 **(e)** $3x^2 + 11$ **(f)** $9x^2 + 30x + 27$ [13.2] **7.** center: (2, −3); radius: 4

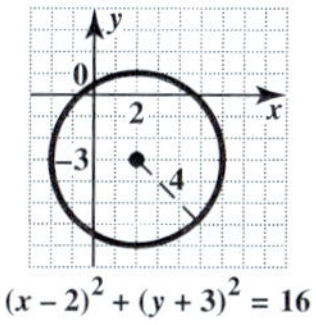


**8.** center: (−4, 1); radius: 5 [13.3] **9.**

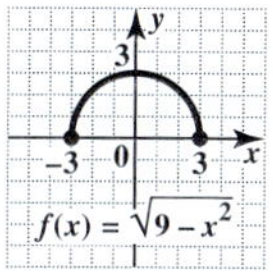


[13.2] **10.**

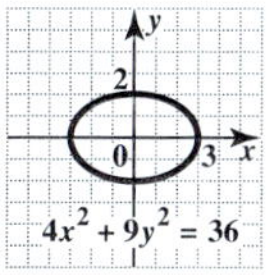


[13.3] **11.**

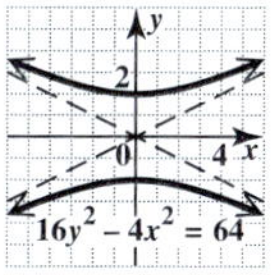


**12.**

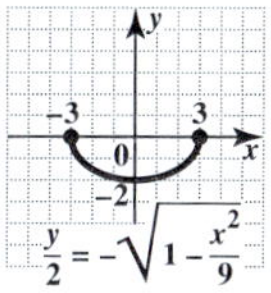


**13.** ellipse **14.** hyperbola **15.** parabola

[13.4] **16.** $\left\{\left(-\frac{1}{2}, -10\right), (5, 1)\right\}$ **17.** $\left\{(-2, -2), \left(\frac{14}{5}, -\frac{2}{5}\right)\right\}$ **18.** $\left\{\left(-\sqrt{22}, -\sqrt{3}\right), \left(-\sqrt{22}, \sqrt{3}\right), \left(\sqrt{22}, -\sqrt{3}\right), \left(\sqrt{22}, \sqrt{3}\right)\right\}$

[13.5] **19.**

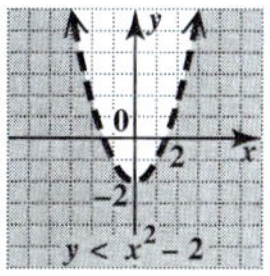


**20.**

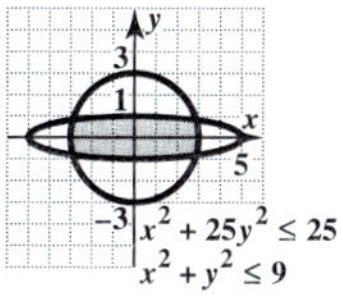


## Cumulative Review Exercises Chapters 1–13 (page 926)

[1.4, 1.5] **1.** −4 [2.3] **2.** $\left\{\frac{2}{3}\right\}$ [2.8] **3.** $\left(-\infty, \frac{3}{5}\right]$ [3.3, 7.1] **4.** $\frac{2}{3}$ [3.4, 7.2] **5.** $3x + 2y = -13$ [4.6] **6.** $25y^2 - 30y + 9$

[4.5] **7.** $12r^2 + 40r - 7$ [4.7] **8.** $4x^3 - 4x^2 + 3x + 5 + \frac{3}{2x + 1}$ [5.2, 5.3] **9.** $(3x + 2)(4x - 5)$ **10.** $(2y^2 - 1)(y^2 + 3)$

[5.4] **11.** $(z^2 + 1)(z + 1)(z - 1)$ **12.** $(a - 3b)(a^2 + 3ab + 9b^2)$ [6.2] **13.** $\frac{40}{9}$ **14.** $\frac{y - 1}{y(y - 3)}$ [6.4] **15.** $\frac{3c + 5}{(c + 5)(c + 3)}$ **16.** $\frac{1}{p}$

[6.7] **17.** $1\frac{1}{5}$ hr  [7.3] **18.** \$275  [8.1–8.3, 8.6] **19.** $\{(3, -3)\}$  [8.4, 8.6] **20.** $\{(4, 1, -2)\}$  [13.4] **21.** $\left\{(-1, 5), \left(\frac{5}{2}, -2\right)\right\}$

[8.5] **22.** 40 mph  [4.1, 4.2] **23.** $\frac{3}{4}$  **24.** $\frac{a^5}{4}$  [10.4] **25.** $2\sqrt[3]{2}$  [10.5] **26.** $\frac{3\sqrt{10}}{2}$  [10.7] **27.** $\frac{7}{5} + \frac{11}{5}i$  [9.2] **28.** $\{-4, 4\}$

**29.** $(-\infty, -5) \cup (10, \infty)$  [10.6] **30.** $\emptyset$  [5.5] **31.** $\left\{\frac{1}{5}, -\frac{3}{2}\right\}$  [11.1–11.3] **32.** $\left\{\frac{1 + 2\sqrt{2}}{4}, \frac{1 - 2\sqrt{2}}{4}\right\}$

**33.** $\left\{\frac{3 + \sqrt{33}}{6}, \frac{3 - \sqrt{33}}{6}\right\}$  [11.4] **34.** $\left\{-\frac{\sqrt{6}}{2}, \frac{\sqrt{6}}{2}, -\sqrt{7}, \sqrt{7}\right\}$  [11.5] **35.** $v = \frac{\pm\sqrt{rFkw}}{kw}$  [12.1] **36.** $f^{-1}(x) = \sqrt[3]{x - 4}$

[12.4] **37.** 4  [12.5] **38.** 7  [12.4] **39.** $\log \frac{(3x + 7)^2}{4}$  [12.6] **40.** $\{3\}$  **41.** **(a)** \$12,198.90 **(b)** \$12,214.03  **42.** \$16.9 billion

**43.** \$75.8 billion  [13.1] **44.** **(a)** $-1$ **(b)** $9x^2 + 18x + 4$  [7.3] **45.**

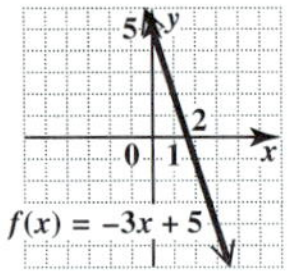


[11.6] **46.**

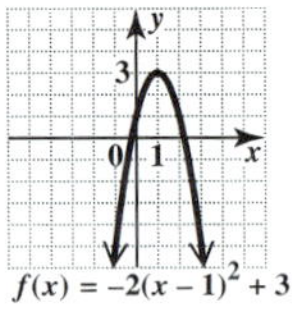


[13.5] **47.**

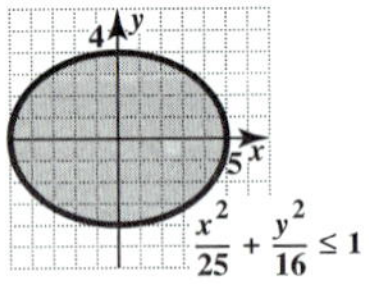


[13.1] **48.**

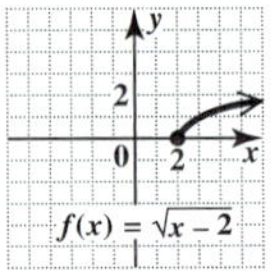


[13.3] **49.**

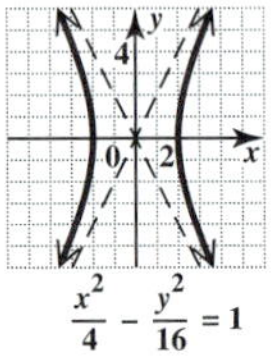


[12.2] **50.**

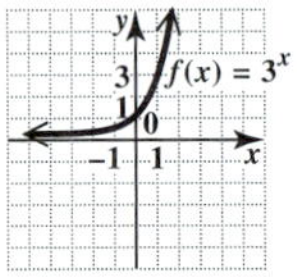


# CHAPTER 14 SEQUENCES AND SERIES

## Section 14.1 (page 935)

Exercises **1.** 2, 3, 4, 5, 6  **3.** $4, \frac{5}{2}, 2, \frac{7}{4}, \frac{8}{5}$  **5.** 3, 9, 27, 81, 243  **7.** $1, \frac{1}{4}, \frac{1}{9}, \frac{1}{16}, \frac{1}{25}$  **9.** $-1, 1, -1, 1, -1$  **11.** $-70$  **13.** $\frac{49}{23}$

**15.** 171  **17.** $4n$  **19.** $\frac{1}{3^n}$  **21.** \$110, \$109, \$108, \$107, \$106, \$105; \$400  **23.** \$6554  **25.** $4 + 5 + 6 + 7 + 8 = 30$

**27.** $3 + 6 + 11 = 20$  **29.** $-1 + 1 - 1 + 1 - 1 + 1 = 0$  **31.** $0 + 6 + 14 + 24 + 36 = 80$  **33.** $2x + 4x + 6x + 8x + 10x$

**35.** $x + 2x^2 + 3x^3 + 4x^4 + 5x^5$  **Answers may vary for Exercises 37–41.**  **37.** $\sum_{i=1}^{5}(i + 2)$  **39.** $\sum_{i=1}^{5}\frac{1}{i + 1}$  **41.** $\sum_{i=1}^{5} i^2$

**43.** The similarities are that both are defined by the same linear expression and that points satisfying both lie in a straight line. The difference is that the domain of $f$ consists of all real numbers, but the domain of the sequence is $\{1, 2, 3, \ldots\}$. An example of a similarity is $f(1) = 6$ and $a_1 = 6$. An example of a difference is $f\left(\frac{3}{2}\right) = 7$ but $a_{3/2}$ is not allowed.  **45.** A sequence is a list of terms in a specific order, while a series is the indicated sum of the terms of a sequence.  **47.** $\frac{59}{7}$  **49.** 5  **51.** 7243  **53.** $\sum_{i=1}^{6} i^2 + \sum_{i=1}^{6} 3i + \sum_{i=1}^{6} 5$  **54.** $3\sum_{i=1}^{6} i$  **55.** $6 \cdot 5 = 30$

**56.** $\sum_{i=1}^{n} i = \frac{n(n + 1)}{2}$  **57.** $\sum_{i=1}^{n} i^2 = \frac{n(n + 1)(2n + 1)}{6}$  **58.** $91 + 63 + 30 = 184$  **59.** 572  **60.** $-2620$

## Section 14.2 (page 943)

Exercises **1.** An arithmetic sequence is a sequence (list) of numbers in a specific order such that there is a common difference between any two successive terms. For example, the sequence 1, 5, 9, 13, . . . is arithmetic with difference $d = 5 - 1 = 9 - 5 = 13 - 9 = 4$. As another example, 2, −1, −4, −7, . . . is an arithmetic sequence with $d = -3$. **3.** $d = 1$ **5.** not arithmetic **7.** $d = 5$ **9.** 5, 9, 13, 17, 21 **11.** −2, −6, −10, −14, −18 **13.** $a_n = 5n - 3$ **15.** $a_n = \frac{3}{4}n + \frac{9}{4}$ **17.** $a_n = 3n - 6$ **19.** 76 **21.** 48 **23.** −1 **25.** 16 **27.** 6 **29.** $n$ represents the number of terms. **31.** 2$S$; 101; 100; 10,100 **32.** $2S = 10{,}100$ **33.** 5050 **34.** 20,100 **35.** 81 **37.** −3 **39.** 87 **41.** 390 **43.** 320 **45.** 31,375 **47.** \$465 **49.** \$2100 per month **51.** 68; 1100 **53.** no; 3; 9

## Section 14.3 (page 952)

**Exercises** **1.** A geometric sequence is an ordered list of numbers such that each term after the first is obtained by multiplying the previous term by a constant, $r$, called the common ratio. For example, if the first term is 3 and $r = 4$, then the sequence is 3, 12, 48, 192, . . . . If the first term is 2 and $r = -1$, the sequence is 2, −2, 2, −2, . . . . **3.** $r = 2$ **5.** not geometric **7.** $r = -3$ **9.** $r = -\frac{1}{2}$ **11.** $a_n = 5(2)^{n-1}$ **13.** $a_n = \frac{3^{n-1}}{9}$ **15.** $a_n = 10\left(-\frac{1}{5}\right)^{n-1}$ **17.** $2(5)^9 = 3{,}906{,}250$ **19.** $\frac{1}{2}\left(\frac{1}{3}\right)^{11}$ **21.** $2\left(\frac{1}{2}\right)^{24} = \frac{1}{2^{23}}$ **23.** 2, 6, 18, 54, 162 **25.** $5, -1, \frac{1}{5}, -\frac{1}{25}, \frac{1}{125}$ **27.** $\frac{121}{243}$ **29.** −1.997 **31.** 2.662 **33.** −2.982 **35.** \$66,988.91 **37.** \$130,159.72 **39.** 9 **41.** $\frac{10{,}000}{11}$ **43.** $-\frac{9}{20}$ **45.** does not exist **47.** $10\left(\frac{3}{5}\right)^4 \approx 1.3$ ft **49.** 3 days; $\frac{1}{4}$ g **51.** **(a)** $1.1(1.06)^5 \approx 1.5$ billion units **(b)** approximately 12 yr **53.** $\$50{,}000\left(\frac{3}{4}\right)^8 \approx \$5000$ **55.** .33333 . . . **56.** .66666 . . . **57.** .99999 . . .

**58.** $\frac{a_1}{1 - r} = \frac{.9}{1 - .1} = \frac{.9}{.9} = 1$; therefore, .99999 . . . = 1 **59.** B

**60.** $.49999\ldots = .4 + .09999\ldots = \frac{4}{10} + \frac{1}{10}(.9999\ldots) = \frac{4}{10} + \frac{1}{10}(1) = \frac{5}{10} = \frac{1}{2}$

## Section 14.4 (page 961)

**Connections** **(page 957)** **1.** 21 and 34 **2.** 15, 21, 28, 36, 45 **3.** Answers will vary.

**Exercises** **1.** 720 **3.** 15 **5.** 15 **7.** 1 **9.** 120 **11.** 78 **13.** $m^4 + 4m^3n + 6m^2n^2 + 4mn^3 + n^4$ **15.** $a^5 - 5a^4b + 10a^3b^2 - 10a^2b^3 + 5ab^4 - b^5$ **17.** $8x^3 + 36x^2 + 54x + 27$ **19.** $\frac{x^4}{16} - \frac{x^3y}{2} + \frac{3x^2y^2}{2} - 2xy^3 + y^4$ **21.** $m^3x^3 - 3m^2n^2x^2 + 3mn^4x - n^6$ **23.** $r^{12} + 24r^{11}s + 264r^{10}s^2 + 1760r^9s^3$ **25.** $3^{14}x^{14} - 14(3^{13})x^{13}y + 91(3^{12})x^{12}y^2 - 364(3^{11})x^{11}y^3$ **27.** $t^{20} + 10t^{18}u^2 + 45t^{16}u^4 + 120t^{14}u^6$ **29.** $120(2^7)m^7n^3$ **31.** $\frac{7x^2y^6}{16}$ **33.** $36k^7$ **35.** $-160x^6y^3$ **37.** $4320x^9y^4$

## Chapter 14 Review Exercises (page 966)

**1.** −1, 1, 3, 5 **3.** 1, 4, 9, 16 **5.** 0, 3, 8, 15 **7.** $2x + 3x^2 + 4x^3 + 5x^4 + 6x^5 + 7x^6$ **9.** 126 **11.** 1888.8 billion dollars **13.** arithmetic; $d = 4$ **15.** geometric; $r = -1$ **17.** geometric; $r = \frac{1}{2}$ **19.** 73 **21.** $a_n = -5n + 1$ **23.** 15 **25.** 152 **27.** $a_n = -1(4)^{n-1}$ **29.** $2(-3)^{10} = 118{,}098$ **31.** $\frac{341}{1024}$ **33.** 1 **35.** $32p^5 - 80p^4q + 80p^3q^2 - 40p^2q^3 + 10pq^4 - q^5$ **37.** $m^2 + 4m\sqrt{mn} + 6mn + 4n\sqrt{mn} + n^2$ **39.** $-18{,}400(3)^{22}k^3$ **41.** $a_{10} = 1536$; $S_{10} = 1023$ **43.** $a_{15} = 38$; $S_{10} = 95$ **45.** $a_n = 2(4)^{n-1}$ **47.** $a_n = -3n + 15$ **49.** \$24,898.41 **51.** $\frac{1}{128}$ **53.** No, the sum cannot be found because $r = 2$, and this value of $r$ does not satisfy $|r| < 1$.

## Chapter 14 Test (page 968)

[14.1] **1.** 0, 2, 0, 2, 0 [14.2] **2.** 4, 6, 8, 10, 12 [14.3] **3.** 48, 24, 12, 6, 3 [14.2] **4.** 0 [14.3] **5.** $\frac{64}{3}$ or $-\frac{64}{3}$ [14.2] **6.** 75 [14.3] **7.** 124 or 44 [14.1] **8.** 70,439 [14.3] **9.** \$137,925.91 **10.** It has a sum if $|r| < 1$. [14.2] **11.** 70 **12.** 33 **13.** 125,250 [14.3] **14.** 42 **15.** $\frac{1}{3}$ **16.** The sum does not exist. [14.4] **17.** 40,320 **18.** 1 **19.** 15 **20.** 66 **21.** $81k^4 - 540k^3 + 1350k^2 - 1500k + 625$ **22.** $\frac{14{,}080x^8y^4}{9}$ [14.1] **23.** \$324 [14.3] **24.** $20(3^{11}) = 3{,}542{,}940$

## Cumulative Review Exercises Chapters 1–14 (page 969)

[1.4] **1.** 10, 0, $\frac{45}{15}$ (or 3), $-3$ **2.** $-\frac{8}{3}$, 10, 0, $\frac{45}{15}$ (or 3), .82, $-3$ **3.** $\sqrt{13}, -\sqrt{3}$ **4.** all except $\sqrt{-7}$ [1.4–1.6] **5.** 8 **6.** $-35$ **7.** $-55$ [2.3] **8.** $\left\{\frac{1}{6}\right\}$ [2.8] **9.** $[10, \infty)$ [9.2] **10.** $\left\{-\frac{9}{2}, 6\right\}$ [2.3, 2.6] **11.** $\{9\}$ [9.1] **12.** $(-\infty, -3) \cup (4, \infty)$ [9.2] **13.** $(-\infty, -3] \cup [8, \infty)$ [4.5] **14.** $20p^2 - 2p - 6$ [4.6] **15.** $9k^2 - 42k + 49$ [4.4] **16.** $-5m^3 - 3m^2 + 3m + 8$ [4.7] **17.** $2t^3 + 3t^2 - 4t + 2 + \frac{3}{3t - 2}$ [5.1] **18.** $x(7 + x^2)$ [5.2, 5.3] **19.** $(7y - 4)(2y + 3)$ **20.** $z(3z + 4)(2z - 1)$ [5.4] **21.** $(7a^2 + 3b)(7a^2 - 3b)$ **22.** $(c + 3d)(c^2 - 3cd + 9d^2)$ **23.** $(8r + 3q)^2$ [4.1, 4.2] **24.** $\frac{9}{4}$ **25.** $-\frac{27p^2}{10}$ [6.1] **26.** $-9, 9$ [6.2] **27.** $\frac{x + 7}{x - 2}$ [6.4] **28.** $\frac{3p - 26}{p(p + 3)(p - 4)}$ [3.3, 7.1] **29.** $\frac{3}{4}$ [7.2] **30.** $3x + y = 4$ [3.2, 7.1] **31.**

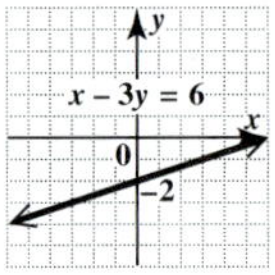


[9.3] **32.**

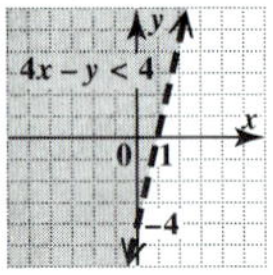


[7.3] **33.** **(a)** yes **(b)** $\{-3, -2, 0, 1, 2\}$ **(c)** $\{2, 6, 4\}$ [8.3] **34.** $\{(3, -5)\}$ [8.2] **35.** $\{(-1, -2)\}$ [8.6] **36.** $\{(2, 1, 4)\}$ [8.5] **37.** 2 lb [5.5] **38.** $\left\{-\frac{5}{2}, 2\right\}$ [11.8] **39.** $[-2, 3]$ [6.6] **40.** $\emptyset$ [11.3] **41.** $\left\{\frac{-5 + \sqrt{217}}{12}, \frac{-5 - \sqrt{217}}{12}\right\}$ [10.6] **42.** $\{1, 2\}$ [12.2] **43.** $\left\{\frac{5}{2}\right\}$ [12.6] **44.** $\{2\}$ [10.4] **45.** $10\sqrt{2}$ [10.7] **46.** 73 [12.1] **47.** $f^{-1}(x) = \frac{x - 5}{9}$ [11.6] **48.**

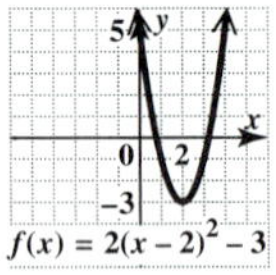


[12.2] **49.**

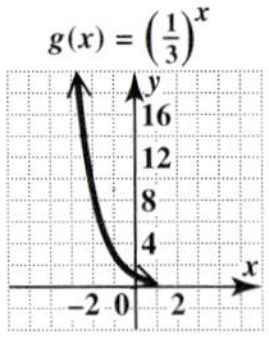


[12.3] **50.**

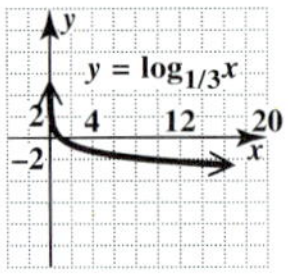


[13.2] **51.**

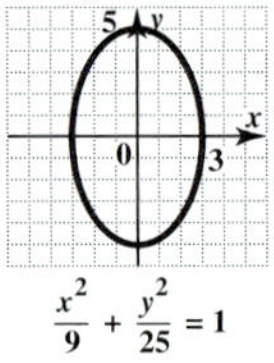


[13.3] **52.**

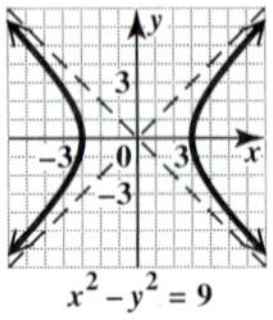


[13.4] **53.** $\left\{(-1, 5), \left(\frac{5}{2}, -2\right)\right\}$ [13.2] **54.** $(x + 5)^2 + (y - 12)^2 = 81$ [14.1] **55.** $-7, -2, 3, 8, 13$ [14.2, 14.3] **56.** **(a)** 78 **(b)** $\frac{75}{7}$ [14.2] **57.** 30 [14.4] **58.** 362,880 **59.** $32a^5 - 80a^4 + 80a^3 - 40a^2 + 10a - 1$ **60.** $-\frac{45x^8y^6}{4}$

**APPENDIXES**

## Appendix B (page 982)

**1.** 117.385 **3.** 13.21 **5.** 150.49 **7.** 96.101 **9.** 4.849 **11.** 166.32 **13.** 164.19 **15.** 1.344 **17.** 4.14 **19.** 2.23 **21.** 4800 **23.** .53 **25.** 1.29 **27.** .96 **29.** .009 **31.** 80% **33.** .7% **35.** 67% **37.** 12.5% **39.** 109.2 **41.** 238.92 **43.** 5% **45.** 110% **47.** 25% **49.** 148.44 **51.** 7839.26 **53.** 7.39% **55.** \$2760 **57.** 805 mi **59.** 63 mi **61.** \$2400

## Appendix C (page 988)

**1.** {1, 2, 3, 4, 5, 6, 7} **3.** {winter, spring, summer, fall} **5.** $\emptyset$ **7.** {L} **9.** {2, 4, 6, 8, 10, . . .} **11.** The sets in Exercises 9 and 10 are infinite sets. **13.** true **15.** false **17.** true **19.** true **21.** true **23.** true **25.** true **27.** false **29.** true **31.** true **33.** false **35.** true **37.** true **39.** false **41.** true **43.** false **45.** {g, h} **47.** {b, c, d, e, g, h} **49.** {a, c, e} $= B$ **51.** {d} $= D$ **53.** {a} **55.** {a, c, d, e} **57.** {a, c, e, f} **59.** $\emptyset$ **61.** *B* and *D*; *C* and *D*

## Appendix D (page 995)

**1.** 7 **3.** 69.8 (rounded) **5.** \$39,622 **7.** \$58.24 **9.** \$35,500 **11.** 6.1 **13.** 17.2 **15.** 2.60 **17.** **(a)** 2.80 **(b)** 2.93 (rounded) **(c)** 3.13 (rounded) **19.** 15 **21.** 516 **23.** 48 **25.** 4142 mi **27.** **(a)** 4050 mi **(b)** The median is somewhat different from the mean; the mean is more affected by the high and low numbers. **29.** 8 **31.** 68 and 74; bimodal **33.** no mode

## Appendix E (page 1006)

**1.** Answers will vary; about 9 to 11 cm. **3.** Answers will vary; about 20 to 25 mm. **5.** cm **7.** m **9.** km **11.** mm **13.** cm **15.** m **17.** Some possible answers are: 35 mm film for cameras, track and field events, metric auto parts, and lead refills for mechanical pencils. **19.** 700 cm **21.** .040 m or .04 m **23.** 9400 m **25.** 5.09 m **27.** 40 cm **29.** 910 mm

**31.** 35 mm = 3.5 cm

70 mm = 7 cm

**33.** .018 km **35.** .0000056 km **37.** mL **39.** L **41.** kg **43.** g **45.** mL

**47.** mg **49.** L **51.** unreasonable; too much **53.** unreasonable; too much **55.** reasonable **57.** Some capacity examples are 2 L bottles of soda and shampoo bottles marked in mL; weight examples are grams of fat listed on cereal boxes and vitamin doses in milligrams. **59.** 15,000 mL **61.** 3 L **63.** .925 L **65.** .008 L **67.** 4150 mL **69.** 8 kg **71.** 5200 g **73.** 850 mg **75.** 30 g **77.** .3 L **79.** 1340 g **81.** .07 L **83.** 3 kg to 4 kg **85.** 200 nickels

## Appendix F (page 1013)

**1.** $\frac{1}{a^2b}$ **3.** $\frac{100y^{10}}{x^2}$ **5.** 0 **7.** $\frac{x^{10}}{2w^{13}y^5}$ **9.** $\frac{a^{15}}{-64b^{15}}$ **11.** $\frac{x^{16}z^{10}}{y^6}$ **13.** $-6a^4 + 11a^3 - 20a^2 + 26a - 15$ **15.** $8x^3 - 18x^2 + 6x - 16$ **17.** $x^2y - xy^2 + 6y^3$ **19.** $-3x^2 - 62x + 32$ **21.** $10x^3 - 4x^2 + 9x - 4$ **23.** $6x^2 - 19x - 7$ **25.** $4x^2 - 9x + 2$ **27.** $16t^2 - 9$ **29.** $4y^4 - 16$ **31.** $16x^2 - 24x + 9$ **33.** $36r^2 + 60ry + 25y^2$ **35.** $c^3 + 8d^3$ **37.** $64x^3 - 1$ **39.** $14t^3 + 45st^2 + 18s^2t - 5s^3$ **41.** $4xy^3(2x^2y + 3x + 9y)$ **43.** $(x + 3)(x - 5)$ **45.** $(2x + 3)(x - 6)$ **47.** $(6t + 5)(6t - 5)$ **49.** $(4t + 3)^2$ **51.** $p(2m - 3n)^2$ **53.** $(x + 1)(x^2 - x + 1)$ **55.** $(2t + 5)(4t^2 - 10t + 25)$ **57.** $(t^2 - 5)(t^4 + 5t^2 + 25)$ **59.** $(5x + 2y)(t + 3r)$ **61.** $(6r - 5s)(a + 2b)$ **63.** $(t^2 + 1)(t + 1)(t - 1)$ **65.** $(2x + 3y - 1)(2x + 3y + 1)$

## Appendix G (page 1018)

**Connections** (page 1018) $-2.7$

**Exercises** **1.** Synthetic division provides a quick, easy way to divide a polynomial by a binomial of the form $x - k$. **3.** $x - 5$ **5.** $4m - 1$ **7.** $2a + 4 + \frac{5}{a + 2}$ **9.** $p - 4 + \frac{9}{p + 1}$ **11.** $4a^2 + a + 3$ **13.** $x^4 + 2x^3 + 2x^2 + 7x + 10 + \frac{18}{x - 2}$ **15.** $-4r^5 - 7r^4 - 10r^3 - 5r^2 - 11r - 8 + \frac{-5}{r - 1}$ **17.** $-3y^4 + 8y^3 - 21y^2 + 36y - 72 + \frac{143}{y + 2}$ **19.** $y^2 + y + 1 + \frac{2}{y - 1}$ **21.** 7 **23.** $-2$ **25.** 0 **27.** By the remainder theorem, a 0 remainder means that $P(k) = 0$; that is, $k$ is a number that makes $P(x) = 0$. **29.** yes **31.** no **33.** no **35.** yes **37.** $(2x - 3)(x + 4)$ **38.** $\left\{\frac{3}{2}, -4\right\}$ **39.** $P(-4) = 0, P\left(\frac{3}{2}\right) = 0$ **40.** $a$ **41.** Yes, $x - 3$ is a factor. $Q(x) = (x - 3)(3x - 1)(x + 2)$ **43.** 3 **45.** $-1$

## Appendix H (page 1029)

**Connections** (page 1025) **1.** $-185$ **2.** Every square matrix corresponds to *one and only one* real number, called its determinant.

**Exercises** **1.** D **3.** $-3$ **5.** 14 **7.** 0 **9.** 59 **11.** 14 **13.** Multiply the upper left and lower right entries. Then multiply the upper right and lower left entries. Subtract the second product from the first to obtain the determinant. For example, $\begin{vmatrix} 4 & 2 \\ 7 & 1 \end{vmatrix} = 4 \cdot 1 - 2 \cdot 7 = 4 - 14 = -10.$ **15.** $-22$ **17.** 20 **19.** $-5$ **21.** By choosing that row or column to expand about, all terms will have a factor of 0, and so the sum of all these terms will be 0. **23.** 6.078 **25.** **(a)** D **(b)** A **(c)** C **(d)** B **27.** $\{(-5, 2)\}$ **29.** $\left\{\left(\frac{11}{58}, -\frac{5}{29}\right)\right\}$ **31.** $\{(-1, 2)\}$ **33.** $\{(-2, 3, 5)\}$ **35.** Cramer's rule does not apply. **37.** $\{(20, -13, -12)\}$ **39.** $\left\{\left(\frac{62}{5}, -\frac{1}{5}, \frac{27}{5}\right)\right\}$ **41.** $\{(-1, 3, 5)\}$ **43.** $\{(1, -3, 2, -4)\}$ **45.** $-22$ **47.** 20 **49.** $-5$ **51.** $\{2\}$ **53.** $\{0\}$ **55.** Answers will vary. **57.**

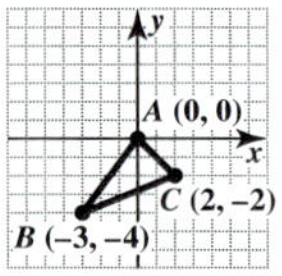


**58.** $\frac{1}{2}\begin{vmatrix} 0 & 0 & 1 \\ -3 & -4 & 1 \\ 2 & -2 & 1 \end{vmatrix}$ **59.** 7 **60.** 8 **61.** $\frac{y_2 - y_1}{x_2 - x_1}$ **62.** $y - y_1 = \frac{y_2 - y_1}{x_2 - x_1}(x - x_1)$ **63.** $x_2y - x_1y - x_2y_1 - xy_2 + x_1y_2 + xy_1 = 0$ **64.** The result is the same as in Exercise 63.

# Glossary

## A

**absolute value** The absolute value of a number is the distance between 0 and the number on a number line. (Section 1.4)

**absolute value equation** An absolute value equation is an equation that involves the absolute value of a variable expression. (Section 9.2)

**absolute value function** The function defined by $f(x) = |x|$ with a graph that includes portions of two lines is called the absolute value function. (Section 13.1)

**absolute value inequality** An absolute value inequality is an inequality that involves the absolute value of a variable expression. (Section 9.2)

**addition property of equality** The addition property of equality states that the same number can be added to (or subtracted from) both sides of an equation to obtain an equivalent equation. (Section 2.1)

**addition property of inequality** The addition property of inequality states that the same number can be added to (or subtracted from) both sides of an inequality without changing the solution set. (Section 2.8)

**additive inverse (negative, opposite)** Two numbers that are the same distance from 0 on a number line but on opposite sides of 0 are called additive inverses. (Section 1.4)

**algebraic expression** Any collection of numbers or variables joined by the basic operations of addition, subtraction, multiplication, or division (except by 0), or the operations of raising to powers or taking roots is called an algebraic expression. (Section 1.3)

**annuity** An annuity is a sequence of equal payments made at equal periods of time. (Section 14.3)

**area** Area is a measure of the surface covered by a two-dimensional (flat) figure. (Section 2.5)

**arithmetic mean** The arithmetic mean (average) of a group of numbers is the sum of all the numbers divided by the number of numbers. (Section 14.1, Appendix D)

**arithmetic sequence (arithmetic progression)** An arithmetic sequence is a sequence in which each term after the first differs from the preceding term by a constant amount. (Section 14.2)

**array of signs** An array of signs is used when evaluating a determinant using expansion by minors. The signs alternate for each row and column, beginning with + in the first row, first column position. (Appendix H)

**associative property of addition** The associative property of addition states that the way in which numbers being added are grouped does not change the sum. (Section 1.7)

**associative property of multiplication** The associative property of multiplication states that the way in which numbers being multiplied are grouped does not change the product. (Section 1.7)

**asymptote** A line that a graph more and more closely approaches as the graph gets farther away from the origin is called an asymptote of the graph. (Sections 12.2, 13.1)

**asymptotes of a hyperbola** The two intersecting straight lines that the branches of a hyperbola approach are called asymptotes of the hyperbola. (Section 13.3)

**augmented matrix** An augmented matrix is a matrix that has a vertical bar that separates the columns of the matrix into two groups. (Section 8.6)

**axis (axis of symmetry)** The axis of a parabola is the vertical or horizontal line through the vertex of the parabola. (Sections 4.4, 11.6)

## B

**base** The base is the number that is a repeated factor when written with an exponent. (Sections 1.2, 4.1)

**bimodal** A group of numbers with two modes is bimodal. (Appendix D)

**binomial** A binomial is a polynomial with exactly two terms. (Section 4.4)

**binomial theorem (general binomial expansion)** The binomial theorem is a formula used to expand a binomial raised to a power. (Section 14.4)

**boundary line** In the graph of a linear inequality, the boundary line separates the region that satisfies the inequality from the region that does not satisfy the inequality. (Sections 9.3, 13.5)

## C

**center of a circle** The fixed point that is a fixed distance from all the points that form a circle is the center of the circle. (Section 13.2)

**center of an ellipse** The center of an ellipse is the fixed point located exactly halfway between the two foci. (Section 13.2)

**center-radius form of the equation of a circle** The center-radius form of the equation of a circle with center $(h, k)$ and radius $r$ is $(x - h)^2 + (y - k)^2 = r^2$. (Section 13.2)

**circle** A circle is the set of all points in a plane that lie a fixed distance from a fixed point. (Section 13.2)

**coefficient (numerical coefficient)** A coefficient is the numerical factor of a term. (Sections 1.8, 4.2)

**column of a matrix** A column of a matrix is a group of elements that are read vertically. (Section 8.6)

**combined variation** If a problem involves a combination of direct and inverse variation, then it is called a combined variation problem. (Section 7.4)

**combining like terms** Combining like terms is a method of adding or subtracting like terms by using the properties of real numbers. (Section 1.8)

**common difference** The common difference $d$ is the difference between any two adjacent terms of an arithmetic sequence. (Section 14.2)

**common factor** An integer that is a factor of two or more integers is called a common factor of those integers. (Section 5.1)

**common logarithm** A common logarithm is a logarithm to base 10. (Section 12.5)

**common ratio** A common ratio $r$ is the constant multiplier between adjacent terms in a geometric sequence. (Section 14.3)

**commutative property of addition** The commutative property of addition states that the order of numbers in an addition problem can be changed without changing the sum. (Section 1.7)

**commutative property of multiplication** The commutative property of multiplication states that the product in a multiplication problem remains the same regardless of the order of the factors. (Section 1.7)

**complement of a set** The set of elements in the universal set that are not in a set $A$ is the complement of $A$, written $A'$. (Appendix C)

**complementary angles (complements)** Complementary angles are angles whose measures have a sum of 90°. (Section 2.4)

**completing the square** The process of adding to a binomial the number that makes it a perfect square trinomial is called completing the square. (Section 11.2)

**complex conjugates** The complex conjugate of $a + bi$ is $a - bi$. (Section 10.7)

**complex fraction** A complex fraction is an expression with one or more fractions in the numerator, denominator, or both. (Section 6.5)

**complex number** A complex number is any number that can be written in the form $a + bi$, where $a$ and $b$ are real numbers. (Section 10.7)

**composite function** A function in which some quantity depends on a variable that, in turn, depends on another variable is called a composite function. (Section 13.1)

**composite number** A composite number has at least one factor other than itself and 1. (Section 1.1)

**composition of functions** Replacing a variable with an algebraic expression is called composition of functions. (Section 13.1)

**compound inequality** A compound inequality consists of two inequalities linked by a connective word such as *and* or *or*. (Section 9.1)

**conditional equation** A conditional equation is true for some replacements of the variable and false for others. (Section 2.3)

**conic section** When a plane intersects an infinite cone at different angles, the figures formed by the intersections are called conic sections. (Section 13.2)

**conjugate** The conjugate of $a + b$ is $a - b$. (Section 10.5)

**consecutive integers** Two integers that differ by one are called consecutive integers. (Section 2.4)

**consistent system** A system of equations with a solution is called a consistent system. (Section 8.1)

**constant function** A linear function of the form $f(x) = b$, where $b$ is a constant, is called a constant function. (Section 7.3)

**constant of variation** In the variation equations $y = kx$, or $y = \frac{k}{x}$, or $y = kxz$, the number $k$ is called the constant of variation. (Section 7.4)

**contradiction** A contradiction is an equation that is never true. It has no solution. (Section 2.3)

**coordinate on a number line** Each number on a number line is called the coordinate of the point that it labels. (Section 1.4)

**coordinates of a point** The numbers in an ordered pair are called the coordinates of the corresponding point in the plane. (Sections 3.1, 7.1)

**Cramer's rule** Cramer's rule uses determinants to solve systems of linear equations. (Appendix H)

**cross products** The cross products in the proportion $\frac{a}{b} = \frac{c}{d}$ are $ad$ and $bc$. (Section 2.6)

**cube root** A number $b$ is the cube root of $a$ if $b^3 = a$. (Section 10.1)

**cube root function** The function defined by $f(x) = \sqrt[3]{x}$ is called the cube root function. (Section 10.1)

## D

**decimal** A decimal is a number written with a decimal point. (Appendix B)

**degree** A degree is a basic unit of measure for angles in which one degree (1°) is $\frac{1}{360}$ of a complete revolution. (Section 2.4)

**degree of a polynomial** The degree of a polynomial is the greatest degree of any of the terms in the polynomial. (Section 4.4)

**degree of a term** The degree of a term is the sum of the exponents on the variables in the term. (Section 4.4)

**denominator** The number below the fraction bar in a fraction is called the denominator. It shows the number of equal parts in a whole. (Section 1.1)

**dependent equations** Equations of a system that have the same graph (because they are different forms of the same equation) are called dependent equations. (Section 8.1)

**dependent variable** In an equation relating $x$ and $y$, if the value of the variable $y$ depends on the variable $x$, then $y$ is called the dependent variable. (Section 7.3)

**descending powers** A polynomial in one variable is written in descending powers of the variable if the degree of the terms of the polynomial decreases from left to right. (Section 4.4)

**determinant** Associated with every square matrix is a real number called the determinant of the matrix, symbolized by the entries of the matrix placed between two vertical lines. (Appendix H)

**difference** The answer to a subtraction problem is called the difference. (Section 1.1)

**difference of cubes** The difference of cubes, $x^3 - y^3$, can be factored as $x^3 - y^3 = (x - y)(x^2 + xy + y^2)$. (Section 5.4)

**difference of squares** The difference of squares, $x^2 - y^2$, can be factored as the product of the sum and difference of two terms, or $x^2 - y^2 = (x + y)(x - y)$. (Section 5.4)

**direct variation** $y$ varies directly as $x$ if there exists a real number $k$ such that $y = kx$. (Section 7.4)

**discriminant** The discriminant is the quantity under the radical, $b^2 - 4ac$, in the quadratic formula. (Section 11.3)

**disjoint sets** Sets that have no elements in common are disjoint sets. (Appendix C)

**distributive property** For any real numbers $a$, $b$, and $c$, the distributive property states that $a(b + c) = ab + ac$ and $(b + c)a = ba + ca$. (Section 1.7)

**domain** The set of all first components ($x$-values) in the ordered pairs of a relation is the domain. (Section 7.3)

## E

**element of a matrix** The numbers in a matrix are called the elements of the matrix. (Section 8.6)

**elements (members)** Elements are the objects that belong to a set. (Section 1.3, Appendix C)

**elimination method** The elimination method is an algebraic method used to solve a system of equations in which the equations of the system are combined so that one or more variables is eliminated. (Section 8.3)

**ellipse** An ellipse is the set of all points in a plane the sum of whose distances from two fixed points is constant. (Section 13.2)

**empty set (null set)** The empty set, denoted by { } or $\emptyset$, is the set containing no elements. (Section 2.3, Appendix C)

**equation** An equation is a statement that two algebraic expressions are equal. (Section 1.3)

**equivalent equations** Equivalent equations are equations that have the same solution set. (Section 2.1)

**equivalent inequalities** Equivalent inequalities are inequalities that have the same solution set. (Section 2.8)

**expansion by minors** A method of evaluating a $3 \times 3$ or larger determinant is called expansion by minors. (Appendix H)

**exponent (power)** An exponent is a number that indicates how many times a factor is repeated. (Sections 1.2, 4.1)

**exponential equation** An exponential equation is an equation that has a variable as an exponent. (Section 12.2)

**exponential expression** A number or letter (variable) written with an exponent is an exponential expression. (Sections 1.2, 4.1)

**exponential function** An exponential function is a function defined by an expression of the form $f(x) = a^x$, where $a > 0$ and $a \neq 1$ for all real numbers $x$. (Section 12.2)

**extraneous solution** A solution to a new equation that does not satisfy the original equation is called an extraneous solution. (Section 10.6)

**extremes of a proportion** In the proportion $\frac{a}{b} = \frac{c}{d}$, the $a$- and $d$-terms are called the extremes. (Section 2.6)

## F

**factor** A factor of a given number is any number that divides evenly (without remainder) into the given number. (Sections 1.1, 5.1)

**factored** A number is factored by writing it as the product of two or more numbers. (Section 1.1)

**factored form** An expression is in factored form when it is written as a product. (Section 5.1)

**factoring** Writing a polynomial as the product of two or more simpler polynomials is called factoring. (Section 5.1)

**factoring by grouping** Factoring by grouping is a method of grouping the terms of a polynomial in such a way that the polynomial can be factored even though its greatest common factor is 1. (Section 5.1)

**factoring out the greatest common factor** Factoring out the greatest common factor is the process of using the distributive property to write a polynomial as a product of the greatest common factor and a simpler polynomial. (Section 5.1)

**finite sequence** A finite sequence has a domain that includes only the first $n$ positive integers. (Section 14.1)

**first-degree equation** A first-degree (linear) equation has no term with the variable to a power greater than 1. (Section 7.1)

**foci** (singular, **focus**) Foci are fixed points used to determine the points that form a parabola, an ellipse, or a hyperbola. (Sections 13.2, 13.3)

**FOIL** FOIL is a method for multiplying two binomials $(A + B)(C + D)$. Multiply **F**irst terms $AC$, **O**uter terms $AD$, **I**nner terms $BC$, and **L**ast terms $BD$. Then combine like terms. (Section 4.5)

**formula** A formula is a mathematical expression in which letters are used to describe a relationship. (Section 2.5)

**fourth root** A number $b$ is a fourth root of $a$ if $b^4 = a$. (Section 10.1)

**function** A function is a set of ordered pairs (relation) in which each value of the first component $x$ corresponds to exactly one value of the second component $y$. (Section 7.3)

**function notation** Function notation $f(x)$ represents the value of the function at $x$, that is, the $y$-value that corresponds to $x$. (Section 7.3)

**fundamental rectangle** The asymptotes of a hyperbola are the extended diagonals of its fundamental rectangle, with corners at the points $(a, b)$, $(-a, b)$, $(-a, -b)$, and $(a, -b)$. (Section 13.3)

**future value of an annuity** The future value of an annuity is the sum of the compound amounts of all the payments, compounded to the end of the term. (Section 14.3)

## G

**general term of a sequence** The expression $a_n$, which defines a sequence, is called the general term of the sequence. (Section 14.1)

**geometric sequence (geometric progression)** A geometric sequence is a sequence in which each term after the first is a constant multiple of the preceding term. (Section 14.3)

**gram** The basic metric unit for mass is the gram, written using the symbol g. (Appendix E)

**graph of an equation** The graph of an equation is the set of all points that correspond to all of the ordered pairs that satisfy the equation. (Sections 3.2, 7.1)

**graph of a number** The point on a number line that corresponds to a number is its graph. (Section 1.4)

**graph of a relation** The graph of a relation is the graph of its ordered pairs. (Section 7.3)

**greatest common factor (GCF)** The greatest common factor of a list of integers is the largest common factor of those integers. The greatest common factor of a polynomial is the largest term that is a factor of all terms in the polynomial. (Sections 1.1, 5.1)

**greatest integer function** The function defined by $f(x) = [\![x]\!]$, where the symbol $[\![x]\!]$ is used to represent the greatest integer less than or equal to $x$, is called the greatest integer function. (Section 13.1)

**grouping symbols** Grouping symbols are parentheses ( ), square brackets [ ], or fraction bars. (Section 1.2)

## H

**horizontal line test** The horizontal line test states that a function is one-to-one if every horizontal line intersects the graph of the function at most once. (Section 12.1)

**hyperbola** A hyperbola is the set of all points in a plane such that the absolute value of the difference of the distances from two fixed points is constant. (Section 13.3)

**hypotenuse** The hypotenuse is the longest side in a right triangle. It is the side opposite the right angle. (Sections 5.6, 10.3)

## I

**identity** An identity is an equation that is true for all replacements of the variable. It has an infinite number of solutions. (Section 2.3)

**identity element for addition** Since adding 0 to a number does not change the number, 0 is called the identity element for addition. (Section 1.7)

**identity element for multiplication** Since multiplying a number by 1 does not change the number, 1 is called the identity element for multiplication. (Section 1.7)

**identity properties** The identity properties state that the sum of 0 and any number equals the number, and the product of 1 and any number equals the number. (Section 1.7)

**imaginary number** A complex number $a + bi$ with $b \neq 0$ is called an imaginary number. (Section 10.7)

**imaginary part** The imaginary part of the complex number $a + bi$ is $b$. (Section 10.7)

**inconsistent system** An inconsistent system of equations is a system with no solution. (Section 8.1)

**independent equations** Equations of a system that have different graphs are called independent equations. (Section 8.1)

**independent variable** In an equation relating $x$ and $y$, if the value of the variable $y$ depends on the variable $x$, then $x$ is called the independent variable. (Section 7.3)

**index (order)** In a radical of the form $\sqrt[n]{a}$, $n$ is called the index or order. (Section 10.1)

**index of summation** When using summation notation, $\sum_{i=1}^{n} f(i)$, the letter $i$ is called the index of summation. (Section 14.1)

**inequality** An inequality is a statement that two expressions are not equal. (Section 1.2)

**infinite sequence** An infinite sequence is a function with the set of positive integers as the domain. (Section 14.1)

**inner product** When using the FOIL method to multiply two binomials $(A + B)(C + D)$, the inner product is $BC$. (Section 4.5)

**integers** The set of integers is $\{\ldots, -3, -2, -1, 0, 1, 2, 3, \ldots\}$. (Section 1.4)

**intersection** The intersection of two sets $A$ and $B$, written $A \cap B$, is the set of elements that belong to both $A$ and $B$. (Section 9.1, Appendix C)

**interval** An interval is a portion of a number line. (Section 2.8)

**interval notation** Interval notation is a simplified notation that uses parentheses ( ) and/or brackets [ ] to describe an interval on a number line. (Section 2.8)

**inverse of a function *f*** If $f$ is a one-to-one function, then the inverse of $f$ is the set of all ordered pairs of the form $(y, x)$ where $(x, y)$ belongs to $f$. (Section 12.1)

**inverse properties** The inverse properties state that a number added to its opposite is 0, and a number multiplied by its reciprocal is 1. (Section 1.7)

**inverse variation** $y$ varies inversely as $x$ if there exists a real number $k$ such that $y = \frac{k}{x}$. (Section 7.4)

**irrational numbers** Irrational numbers cannot be written as the quotient of two integers but can be represented by points on the number line. (Section 1.4)

## J

**joint variation** $y$ varies jointly as $x$ and $z$ if there exists a real number $k$ such that $y = kxz$. (Section 7.4)

## L

**least common denominator (LCD)** Given several denominators, the smallest expression that is divisible by all the denominators is called the least common denominator. (Sections 1.1, 6.3)

**legs of a right triangle** The two shorter sides of a right triangle are called the legs. (Sections 5.6, 10.3)

**like terms** Terms with exactly the same variables raised to exactly the same powers are called like terms. (Sections 1.8, 4.4)

**linear equation in one variable** A linear equation in one variable can be written in the form $Ax + B = C$, where $A$, $B$, and $C$ are real numbers, with $A \neq 0$. (Section 2.1)

**linear equation in two variables** A linear equation in two variables is an equation that can be written in the form $Ax + By = C$, where $A$, $B$, and $C$ are real numbers and $A$ and $B$ are not both 0. (Sections 3.1, 7.1)

**linear function** A function defined by an equation of the form $f(x) = mx + b$, for real numbers $m$ and $b$, is a linear function. (Section 7.3)

**linear inequality in one variable** A linear inequality in one variable can be written in the form $Ax + B < C$ or $Ax + B > C$ (or with $\leq$ or $\geq$), where $A$, $B$, and $C$ are real numbers, with $A \neq 0$. (Section 2.8)

**linear inequality in two variables** A linear inequality in two variables can be written in the form $Ax + By < C$ or $Ax + By > C$ (or with $\leq$ or $\geq$), where $A$, $B$, and $C$ are real numbers and $A$ and $B$ are not both 0. (Section 9.3)

**linear system (system of linear equations)** Two or more linear equations form a linear system. (Section 8.1)

**line of symmetry** The axis of a parabola is a line of symmetry for the graph. It is a line that can be drawn through the graph in such a way that the part of the graph on one side of the line is an exact reflection of the part on the opposite side. (Sections 4.4, 11.6)

**liter** The basic metric unit for capacity is the liter, written using the symbol L. (Appendix E)

**logarithm** A logarithm is an exponent; $\log_a x$ is the exponent on the base $a$ that gives the number $x$. (Section 12.3)

**logarithmic equation** A logarithmic equation is an equation with a logarithm in at least one term. (Section 12.3)

**logarithmic function with base *a*** If $a$ and $x$ are positive numbers with $a \neq 1$, then $f(x) = \log_a x$ defines the logarithmic function with base $a$. (Section 12.3)

**lowest terms** A fraction is in lowest terms when there are no common factors in the numerator and denominator (except 1). (Sections 1.1, 6.1)

## M

**mathematical model** In a real-world problem, a mathematical model is one or more equations (or inequalities) that describe the situation. (Sections 1.3 Exercises, 3.1)

**matrix** (plural, **matrices**) A matrix is a rectangular array of numbers, consisting of horizontal rows and vertical columns. (Section 8.6, Appendix H)

**mean** The mean (average) of a group of numbers is the sum of the numbers divided by the number of numbers. (Section 14.1, Appendix D)

**means of a proportion** In the proportion $\frac{a}{b} = \frac{c}{d}$, the $b$- and $c$-terms are called the means. (Section 2.6)

**median** The median divides a group of numbers listed in order from smallest to largest in half; that is, half the numbers lie above the median and half lie below the median. (Appendix D)

**meter** The basic unit of length in the metric system is the meter, written using the symbol m. (Appendix E)

**minors** The minor of an element in a $3 \times 3$ determinant is the $2 \times 2$ determinant remaining when a row and a column of the $3 \times 3$ determinant are eliminated. (Appendix H)

**mixed number** A mixed number includes a whole number and a fraction written together and is understood to be the sum of the whole number and the fraction. (Section 1.1)

**mode** The mode is the number (or numbers) that occurs most often in a group of numbers. (Appendix D)

**monomial** A monomial is a polynomial with only one term. (Section 4.4)

**multiplication property of equality** The multiplication property of equality states that the same nonzero number can be multiplied by (or divided into) both sides of an equation to obtain an equivalent equation. (Section 2.2)

**multiplication property of inequality** The multiplication property of inequality states that both sides of an inequality may be multiplied (or divided) by a positive number without changing the direction of the inequality symbol. Multiplying (or dividing) by a negative number reverses the inequality symbol. (Section 2.8)

**multiplicative inverse (reciprocal)** The multiplicative inverse of a nonzero real number $a$ is $\frac{1}{a}$. (Section 1.6)

## N

**$n$-factorial ($n!$)** For any positive integer $n$, $n(n-1)(n-2)(n-3)\cdots(2)(1) = n!$. (Section 14.4)

**natural logarithm** A natural logarithm is a logarithm to base $e$. (Section 12.5)

**natural numbers (counting numbers)** The set of natural numbers includes the numbers used for counting: $\{1, 2, 3, 4, \ldots\}$. (Sections 1.1, 1.4)

**negative number** A negative number is located to the left of 0 on a number line. (Section 1.4)

**nonlinear equation** A nonlinear equation is an equation in which some terms have more than one variable or a variable of degree 2 or higher. (Section 13.4)

**nonlinear system of equations** A nonlinear system of equations is a system that includes at least one nonlinear equation. (Section 13.4)

**nonlinear system of inequalities** A nonlinear system of inequalities is two or more inequalities to be considered at the same time, at least one of which is nonlinear. (Section 13.5)

**number line** A number line is a line with a scale that is used to show how numbers relate to each other. (Section 1.4)

**numerator** The number above the fraction bar in a fraction is called the numerator. It shows how many of the equivalent parts are being considered. (Section 1.1)

**numerical coefficient** The numerical factor in a term is its numerical coefficient. (Sections 1.8, 4.4)

## O

**one-to-one function** A one-to-one function is a function in which each $x$-value corresponds to only one $y$-value and each $y$-value corresponds to just one $x$-value. (Section 12.1)

**ordered pair** An ordered pair is a pair of numbers written within parentheses in which the order of the numbers is important. (Sections 3.1, 7.1)

**ordered triple** A solution of an equation in three variables, written $(x, y, z)$, is called an ordered triple. (Section 8.4)

**ordinary annuity** An ordinary annuity is an annuity in which the payments are made at the end of each time period and the frequency of payments is the same as the frequency of compounding. (Section 14.3)

**origin** The point at which the $x$-axis and $y$-axis of a rectangular coordinate system intersect is called the origin. (Sections 3.1, 7.1)

**outer product** When using the FOIL method to multiply two binomials $(A + B)(C + D)$, the outer product is $AD$. (Section 4.5)

## P

**parabola** The graph of a second-degree (quadratic) equation in two variables is called a parabola. (Sections 4.4, 11.6)

**parallel lines** Parallel lines are two lines in the same plane that never intersect. (Sections 3.3, 7.1)

**Pascal's triangle** Pascal's triangle is a triangular array of numbers that is helpful in expanding binomials. (Section 14.4)

**payment period** In an annuity, the time between payments is called the payment period. (Section 14.3)

**percent** Percent, written with the sign %, means "per one hundred." (Appendix B)

**percentage** A percentage is a part of a whole. (Appendix B)

**perfect cube** A perfect cube is a number with a rational cube root. (Section 10.1)

**perfect square** A perfect square is a number with a rational square root. (Section 10.1)

**perfect square trinomial** A perfect square trinomial is a trinomial that can be factored as the square of a binomial. (Section 5.4)

**perimeter** The perimeter of a two-dimensional figure is a measure of the distance around the outside edges of the figure, or the sum of the lengths of its sides. (Section 2.5)

**perpendicular lines** Perpendicular lines are two lines that intersect to form a right (90°) angle. (Sections 3.3, 7.1)

**plot** To plot an ordered pair is to locate it on a rectangular coordinate system. (Sections 3.1, 7.1)

**point-slope form** A linear equation is written in point-slope form if it is in the form $y - y_1 = m(x - x_1)$, where $m$ is the slope of the line and $(x_1, y_1)$ is a point on the line. (Sections 3.4, 7.2)

**polynomial** A polynomial is a term or a finite sum of terms in which all coefficients are real, all variables have whole number exponents, and no variables appear in denominators. (Section 4.4)

**polynomial in $x$** A polynomial containing only the variable $x$ is called a polynomial in $x$. (Section 4.4)

**positive number** A positive number is located to the right of 0 on a number line. (Section 1.4)

**prime number** A natural number (except 1) is prime if it has only 1 and itself as factors. (Section 1.1)

**prime polynomial** A prime polynomial is a polynomial that cannot be factored using only integer coefficients. (Section 5.2)

**principal root (principal $n$th root)** For even indexes, the symbols $\sqrt{\ }$, $\sqrt[4]{\ }$, $\sqrt[6]{\ }$, . . . , $\sqrt[n]{\ }$, are used for nonnegative roots, which are called principal roots. (Section 10.1)

**product** The answer to a multiplication problem is called the product. (Section 1.1)

**product of the sum and difference of two terms** The product of the sum and difference of two terms is the difference of the squares of the terms: $(x + y)(x - y) = x^2 - y^2$. (Section 4.6)

**proportion** A proportion is a statement that two ratios are equal. (Section 2.6)

**proportional** If $y$ varies directly as $x$ and there exists some number (constant) $k$ such that $y = kx$, then $y$ is said to be proportional to $x$. (Section 7.4)

**Pythagorean formula** The Pythagorean formula states that the square of the length of the hypotenuse of a right triangle equals the sum of the squares of the lengths of the two legs. (Sections 5.6, 10.3)

## Q

**quadrant** A quadrant is one of the four regions in the plane determined by a rectangular coordinate system. (Sections 3.1, 7.1)

**quadratic equation** A quadratic equation is an equation that can be written in the form $ax^2 + bx + c = 0$, where $a$, $b$, and $c$ are real numbers, with $a \neq 0$. (Sections 5.5, 11.1)

**quadratic formula** The quadratic formula is a general formula used to solve any quadratic equation. (Section 11.3)

**quadratic function** A function defined by an equation of the form $f(x) = ax^2 + bx + c$, for real numbers $a$, $b$, and $c$, with $a \neq 0$, is a quadratic function. (Section 11.6)

**quadratic inequality** A quadratic inequality can be written in the form $ax^2 + bx + c < 0$ or $ax^2 + bx + c > 0$ (or with $\leq$ or $\geq$), where $a$, $b$, and $c$ are real numbers, with $a \neq 0$. (Section 11.8)

**quadratic in form** An equation that is written in the form $a[f(x)]^2 + b[f(x)] + c = 0$, for $a \neq 0$, is called quadratic in form. (Section 11.4)

**quotient** The answer to a division problem is called the quotient. (Section 1.1)

## R

**radical** A radical sign with a radicand is called a radical. (Section 10.1)

**radical equation** An equation that includes one or more radical expressions with a variable is called a radical equation. (Section 10.6)

**radical expression** A radical expression is an algebraic expression that contains radicals. (Section 10.1)

**radical sign** The symbol $\sqrt{\ }$ is called a radical sign. (Section 10.1)

**radicand** The number or expression under a radical sign is called the radicand. (Section 10.1)

**radius** The radius of a circle is the fixed distance between the center and any point on the circle. (Section 13.2)

**range** The set of all second components ($y$-values) in the ordered pairs of a relation is the range. (Section 7.3)

**ratio** A ratio is a comparison of two quantities with the same units. (Section 2.6)

**rational expression** The quotient of two polynomials with denominator not 0 is called a rational expression, or algebraic fraction. (Section 6.1)

**rational function** A function that is defined by a rational expression is called a rational function. (Section 13.1)

**rational inequality** An inequality that involves fractions is called a rational inequality. (Section 11.8)

**rational numbers** Rational numbers can be written as the quotient of two integers, with denominator not 0. (Section 1.4)

**rationalizing the denominator** The process of removing radicals from a denominator so that the denominator contains only rational numbers is called rationalizing the denominator. (Section 10.5)

**real numbers** Real numbers include all numbers that can be represented by points on the number line, that is, all rational and irrational numbers. (Section 1.4)

**real part** The real part of a complex number $a + bi$ is $a$. (Section 10.7)

**reciprocal** Pairs of numbers whose product is 1 are called reciprocals of each other. (Sections 1.1, 1.6)

**reciprocal function** The reciprocal function is defined by $f(x) = \frac{1}{x}$. (Section 13.1)

**rectangular (Cartesian) coordinate system** The $x$-axis and $y$-axis placed at a right angle at their zero points form a rectangular coordinate system, also called the Cartesian coordinate system. (Sections 3.1, 7.1)

**reduced row echelon form** The reduced row echelon form is an extension of row echelon form that has 0s above and below the diagonal of 1s. (Section 8.6)

**relation** A relation is a set of ordered pairs. (Section 7.3)

**right angle** A right angle measures 90°. (Section 2.4)

**rise** Rise is the vertical change between two points on a line, that is, the change in $y$-values. (Sections 3.3, 7.1)

**row echelon form** If a matrix is written with 1s on the diagonal from upper left to lower right and 0s below the 1s, it is said to be in row echelon form. (Section 8.6)

**row of a matrix** A row of a matrix is a group of elements that are read horizontally. (Section 8.6)

**row operations** Row operations are operations on a matrix that produce equivalent matrices leading to the solution of a system of equations. (Section 8.6)

**run** Run is the horizontal change between two points on a line, that is, the change in $x$-values. (Sections 3.3, 7.1)

## S

**scatter diagram** A scatter diagram is a graph of ordered pairs of data. (Section 3.1)

**scientific notation** A number is written in scientific notation when it is expressed in the form $a \times 10^n$, where $1 \leq |a| < 10$ and $n$ is an integer. (Section 4.3)

**second-degree inequality** A second-degree inequality is an inequality with at least one variable of degree 2 and no variable with degree greater than 2. (Section 13.5)

**sequence** A sequence is a function whose domain is the set of natural numbers. (Section 14.1)

**series** The indicated sum of the terms of a sequence is called a series. (Section 14.1)

**set** A set is a collection of objects. (Section 1.3, Appendix C)

**set-builder notation** Set-builder notation is used to describe a set of numbers without actually having to list all of the elements. (Section 1.4)

**signed numbers** Signed numbers are numbers that can be written with a positive or negative sign. (Section 1.4)

**simplified radical** A simplified radical meets four conditions:

1. The radicand has no factor raised to a power greater than or equal to the index.
2. The radicand has no fractions.
3. No denominator contains a radical.
4. Exponents in the radicand and the index of the radical have no common factor (except 1).

(Section 10.3)

**slope** The ratio of the change in $y$ to the change in $x$ along a line is called the slope of the line. (Sections 3.3, 7.1)

**slope-intercept form** A linear equation is written in slope-intercept form if it is in the form $y = mx + b$, where $m$ is the slope and $(0, b)$ is the $y$-intercept. (Sections 3.4, 7.2)

**solution of an equation** A solution of an equation is any replacement for the variable that makes the equation true. (Section 1.3)

**solution set** The solution set of an equation is the set of all solutions of the equation. (Section 2.1)

**solution set of a linear system** The solution set of a linear system of equations includes all ordered pairs that satisfy all the equations of the system at the same time. (Section 8.1)

**solution set of a system of linear inequalities** The solution set of a system of linear inequalities includes all ordered pairs that make all inequalities of the system true at the same time. (Section 13.5)

**square matrix** A square matrix is a matrix that has the same number of rows as columns. (Section 8.6, Appendix H)

**square of a binomial** The square of a binomial is the sum of the square of the first term, twice the product of the two terms, and the square of the last term: $(x + y)^2 = x^2 + 2xy + y^2$ or $(x - y)^2 = x^2 - 2xy + y^2$. (Section 4.6)

**square root** The opposite of squaring a number is called taking its square root; that is, a number $b$ is a square root of $a$ if $b^2 = a$. (Section 10.1)

**square root function** The function defined by $f(x) = \sqrt{x}$, with $x \geq 0$, is called the square root function. (Sections 10.1, 13.3)

**square root property** The square root property states that if $x^2 = k$, then $x = \sqrt{k}$ or $x = -\sqrt{k}$. (Section 11.1)

**standard form of a complex number** The standard form of a complex number is $a + bi$. (Section 10.7)

**standard form of a linear equation** A linear equation in two variables written in the form $Ax + By = C$, where $A$, $B$, and $C$ are integers with no common factor (except 1) and $A \geq 0$, is in standard form. (Sections 3.4, 7.2)

**standard form of a quadratic equation** A quadratic equation written in the form $ax^2 + bx + c = 0$, where $a$, $b$, and $c$ are real numbers with $a \neq 0$, is in standard form. (Sections 5.5, 11.1)

**step function** A function with a graph that looks like a series of steps is called a step function. (Section 13.1)

**straight angle** A straight angle measures 180°. (Section 2.4)

**subscript notation** Subscript notation is a way of indicating nonspecific values, such as $x_1$ and $x_2$. (Sections 3.3, 7.1)

**subset** If all elements of set $A$ are in set $B$, then $A$ is a subset of $B$, written $A \subseteq B$. (Appendix C)

**substitution method** The substitution method is an algebraic method for solving a system of equations in which one equation is solved for one of the variables and the result is substituted in the other equation. (Section 8.2)

**sum** The answer to an addition problem is called the sum. (Section 1.1)

**sum of cubes** The sum of cubes, $x^3 + y^3$, can be factored as $x^3 + y^3 = (x + y) \cdot (x^2 - xy + y^2)$. (Section 5.4)

**summation (sigma) notation** Summation notation is a compact way of writing a series using the general term of the corresponding sequence. (Section 14.1)

**supplementary angles (supplements)** Supplementary angles are angles whose measures have a sum of 180°. (Section 2.4)

**synthetic division** Synthetic division is a shortcut procedure for dividing a polynomial by a binomial of the form $x - k$. (Appendix G)

**system of inequalities** A system of inequalities consists of two or more inequalities to be solved at the same time. (Section 13.5)

**system of linear equations (linear system)** A system of linear equations consists of two or more linear equations to be solved at the same time. (Section 8.1)

## T

**table of values** A table of values is an organized way of displaying ordered pairs. (Section 3.1)

**term** A term is a number, a variable, or the product or quotient of a number and one or more variables raised to powers. (Section 1.8)

**term of an annuity** The time from the beginning of the first payment period to the end of the last period is called the term of an annuity. (Section 14.3)

**terms of a proportion** The terms of the proportion $\frac{a}{b} = \frac{c}{d}$ are $a$, $b$, $c$, and $d$. (Section 2.6)

**terms of a sequence** The function values written in order are called the terms of the sequence. (Section 14.1)

**three-part inequality** An inequality that says that one number is between two other numbers is called a three-part inequality. (Section 2.8)

**trinomial** A trinomial is a polynomial with exactly three terms. (Section 4.4)

## U

**union** The union of two sets $A$ and $B$, written $A \cup B$, is the set of elements that belong to either $A$ or $B$ (or both). (Section 9.1, Appendix C)

**universal set** The set that includes all elements under consideration is the universal set, written $U$. (Appendix C)

**unlike terms** Unlike terms are terms that do not have the same variable or the variables are not raised to the same powers. (Section 1.8)

## V

**variable** A variable is a symbol, usually a letter, used to represent an unknown number. (Section 1.3)

**vary directly (is proportional to)** $y$ varies directly as $x$ if there exists a real number (constant) $k$ such that $y = kx$. (Section 7.4)

**vary inversely** $y$ varies inversely as $x$ if there exists a real number (constant) $k$ such that $y = \frac{k}{x}$. (Section 7.4)

**vary jointly** If one variable varies as the product of several other variables (sometimes raised to powers), then the first variable is said to vary jointly as the others. (Section 7.4)

**Venn diagram** A Venn diagram represents the relationships between sets. (Appendix C)

**vertex** The point on a parabola that has the smallest $y$-value (if the parabola opens up) or the largest $y$-value (if the parabola opens down) is called the vertex of the parabola. (Sections 4.4, 11.6)

**vertical angles** When two intersecting lines are drawn, the angles that lie opposite each other have the same measure and are called vertical angles. (Section 2.5)

**vertical line test** The vertical line test states that any vertical line drawn through the graph of a function must intersect the graph in at most one point. (Section 7.3)

**volume** The volume of a three-dimensional figure is a measure of the space occupied by the figure. (Section 2.5)

## W

**weighted mean** A weighted mean is one in which each number is weighted by multiplying it by the number of times it occurs. (Appendix D)

**whole numbers** The set of whole numbers is {0, 1, 2, 3, 4, . . .}. (Section 1.4)

## X

**$x$-axis** The horizontal number line in a rectangular coordinate system is called the $x$-axis. (Sections 3.1, 7.1)

**$x$-intercept** A point where a graph intersects the $x$-axis is called an $x$-intercept. (Sections 3.1, 7.1)

## Y

**$y$-axis** The vertical number line in a rectangular coordinate system is called the $y$-axis. (Sections 3.1, 7.1)

**$y$-intercept** A point where a graph intersects the $y$-axis is called the $y$-intercept. (Sections 3.1, 7.1)

## Z

**zero-factor property** The zero-factor property states that if two numbers have a product of 0, then at least one of the numbers must be 0. (Sections 5.5, 11.1)

# Index

## D



## E

## F

## G



## H

## M

## Q



## R

## S

## T



## U

*Photo Credits:* **p. 1** © Alan Schein Photography/CORBIS; **p. 10** © Anthony Cooper; Ecoscene/CORBIS; **p. 12 top** © Alan Schein Photography/CORBIS; **p. 12 bottom** © 2002 PhotoDisc; **p. 22** © Keith Wood/Getty Images; **p. 28** © Pete Saloutos/CORBIS; **p. 36** © 2002 PhotoDisc; **p. 45** © Charles Gupton/CORBIS; **p. 48 top** © James Leynse/CORBIS SABA; **p. 48 bottom** © James L. Amos/CORBIS; **p. 87** Rob Tringali Jr./SportsChrome USA; **p. 90** AP/Wide World Photos; **p. 93** © Karl Weatherly/CORBIS; **p. 116** AP/Wide World Photos; **p. 123 left** AP/Wide World Photos; **p. 123 right** © White/Packert/Getty Images; **p. 124** AP/Wide World Photos; **p. 125** © Karl Weatherly/CORBIS; **p. 137** Courtesy of John Hornsby; **p. 140** Courtesy of John Hornsby; **p. 142** Jeff Greenberg/PhotoEdit; **p. 149** © 2002 Eyewire; **p. 150** © Reuters New Media Inc./CORBIS; **p. 154** © Paul Conklin/PhotoEdit; **p. 155 left** Greg Crisp/SportsChrome USA; **p. 155 right** Brian Spurlock/SportsChrome USA; **p. 156** © Karl Weatherly/CORBIS; **p. 167** © Gary Conner/PhotoEdit; **p. 173** © Karl Weatherly/CORBIS; **p. 180** © Douglas Peebles/CORBIS; **p. 182** © 2002 PhotoDisc; **p. 187** © 2002 PhotoDisc; **p. 194** © 2002 PhotoDisc; **p. 200** © 2002 PhotoDisc; **p. 207** © 2002 PhotoDisc; **p. 210** © 2002 PhotoDisc; **p. 223** © Alan Schein Photography/CORBIS; **p. 239** © Kate Powers/Getty Images; **p. 240** © 2002 PhotoDisc; **p. 243** Credit Line Henryk T. Kaiser/IndexStock; **p. 250** © Jim Cummins/Getty Images; **p. 262** Courtesy of NASA; **p. 263** © Roger Ressmeyer/CORBIS; **p. 265** Courtesy of NASA; **p. 266** © Digital Vision; **p. 267** THE KOBAL COLLECTION/20TH CENTURY FOX/ PARAMOUNT; **p. 274** Courtesy of NASA; **p. 277** © 2002 PhotoDisc; **p. 278** © Ron Levine/Getty Images; **p. 308** © Reuters New Media Inc./CORBIS; **p. 311** Courtesy of NASA; **p. 315** © 2002 PhotoDisc; **p. 348** © 2002 PhotoDisc; **p. 360** © 2002 PhotoDisc; **p. 365 top** © 2002 PhotoDisc; **p. 365 bottom** © Bettmann/CORBIS; **p. 372** © 2002 PhotoDisc; **p. 373** AP/Wide World Photos; **p. 377** © Neil Rabinowitz/CORBIS; **p. 409** © Neil Rabinowitz/CORBIS; **p. 433 left** AP/Wide World Photos; **p. 433 right** © Reuters New Media Inc./CORBIS; **p. 434** © Duomo/CORBIS; **p. 437** © Douglas Slone/CORBIS; **p. 445** © Joseph Sohm; ChromoSohm Inc./CORBIS; **p. 449** © PhotoDisc/Getty Images; **460** © PhotoDisc/Getty Images; **p. 473** © Digital Vision/Getty Images; **p. 474** © Digital Vision/Getty Images; **p. 479** © Lori Adamski Peek/Getty Images; **p. 482** © Steve Smith/Getty Images; **p. 494** © Walter Hodges/Getty Images; **p. 495** © Gail Shumway/Getty Images; **p. 500** © Robert Brenner/PhotoEdit; **p. 502** © Jonathan Nourok/PhotoEdit; **p. 505** © Randall/The Image Works; **p. 506** © Royalty-Free/CORBIS; **p. 512** © Lori Adamski Peek/Getty Images; **p. 515** © Eyewire/Getty Images; **p. 521** THE KOBAL COLLECTION/WARNER BROS/MOUNTAIN, PETER; **p. 538** © 2002 PhotoDisc; **p. 545** © Chuck Savage/CORBIS; **p. 585** THE KOBAL COLLECTION/WARNER BROS/MOUNTAIN, PETER; **p. 554** Courtesy of NASA; **p. 556** AP/Wide World Photos; **p. 558** © Nigel Cattlin, Holt Studios International/Photo Researchers; **p. 560** © Bill Aron/PhotoEdit; **p. 563** © Mark Richards/PhotoEdit; **p. 564** © Royalty-Free/CORBIS; **p. 566** © Mary Ellen Matthews/Photofest; **p. 567** © Charles Gupton/Stock Boston; **p. 578** © FORESTIER YVES/CORBIS SYGMA; **p. 586** © Daniel Bosler/Getty Images; **p. 587 top** © Tony Freeman/PhotoEdit; **p. 587 bottom** © AFP/CORBIS; **p. 589** Photofest; **p. 592** AP/Wide World Photos; **p. 593** Richard Pasley/Stock Boston; **p. 601** Richard Pasley/Stock Boston; **p. 610** © PhotoDisk/Getty Images; **p. 624** © Dana White/PhotoEdit; **p. 625** © Michael Zito/SportsChrome USA; **p. 628 top** © Eyewire/Getty Images; **p. 628 bottom** AP/Wide World Photos; **p. 629** © Alan Schein Photography/CORBIS; **p. 640 top** © James P. Blair/CORBIS; **p. 640 bottom** © David Pollack/CORBIS; **p. 648 left** © D. Jennings/The Image Works; **p. 648 right** © Cleve Bryant/PhotoEdit; **p. 654** THE KOBAL COLLECTION/MGM; **p. 659** © Alan Schein Photography/CORBIS; **p. 681** AP/Wide World Photos; **p. 684** © Dan Lamont/CORBIS; **p. 705** © Bob Collins III/The Image Works; **p. 706** AP/Wide World Photos; **p. 707** © Michael Newman/PhotoEdit; **p. 710** © PhotoDisc/Getty Images; **p. 714 left** © PhotoDisc/Getty Images; **p. 714 right** © PhotoDisc/Getty Images; **p. 715** © John Van Hasselt/CORBIS SYGMA; **p. 733** © Dick Blume/Syracuse Newspapers/The Image Works; **p. 746** © Michael Newman/PhotoEdit; **p. 751 left** © Steve Niedorf Photography/Getty Images; **p. 751 right** © Bonnie Kamin/PhotoEdit; **p. 753** © Gunter Marx Photography/CORBIS; **p. 757** © Doug Martin/Photo Researchers, Inc.; **p. 762** © Michael Newman/PhotoEdit; **p. 773** © Bonnie Kamin/Index Stock Imagery/PictureQuest; **p. 790** © Eric Neurath/Stock Boston; **p. 792** © Carl & Ann Purcell/CORBIS; **p. 793** © David Young-Wolff/PhotoEdit; **p. 797** Iowa State University; **p. 799** © David Young-Wolff/PhotoEdit; **p. 801** © Photofest; **p. 803** © John Coletti/Stock Boston; **p. 807** © Joseph Sohm; ChromoSohm Inc./CORBIS; **p. 810** Courtesy of National Security Agency; **p. 825 top** © Phyllis Picardi/Stock Boston; **p. 825 bottom** © Roger Ressmeyer/CORBIS; **p. 835** IBM Corporate Archives; **p. 839 top** © Kevin Fleming/CORBIS; **p. 839 bottom** © Bettmann/CORBIS; **p. 844** © Photofest; **p. 845** © Tony Freeman/ PhotoEdit; **p. 850** © Tony Freeman/PhotoEdit; **p. 852** Courtesy of NASA; **p. 858** Courtesy of NASA; **p. 864** © Michael Newman/PhotoEdit; **p. 865** © Spencer Grant/PhotoEdit; **p. 869** Courtesy of Panasonic; **p. 870** © Mark Richards/PhotoEdit; **p. 875** Courtesy of NASA; **p. 884** © Lee Snider; Lee Snider/CORBIS; **p. 886** © PhotoDisc/Getty Images; **p. 894** Courtesy of NASA; **p. 903** © Bob Daemmrich/Stock Boston; **p. 904** Courtesy of NASA; **p. 911** © David Young-Wolff/PhotoEdit; **p. 912 top** © A. Ramey/PhotoEdit; **p. 912 bottom** © PhotoDisc/Getty Images; **p. 923** © PhotoDisc/Getty Images; **p. 929** © David Seed Photography/Getty Images; **p. 930** © Paul Almasy/CORBIS; **p. 936** © V.C.L./Getty Images; **p. 953** © Michael Newman/PhotoEdit; **p. 956** © Stefano Bianchetti/CORBIS; **p. 968** © Laurence Monneret/Getty Images; **p. 969** © Tom McCarthy/PhotoEdit

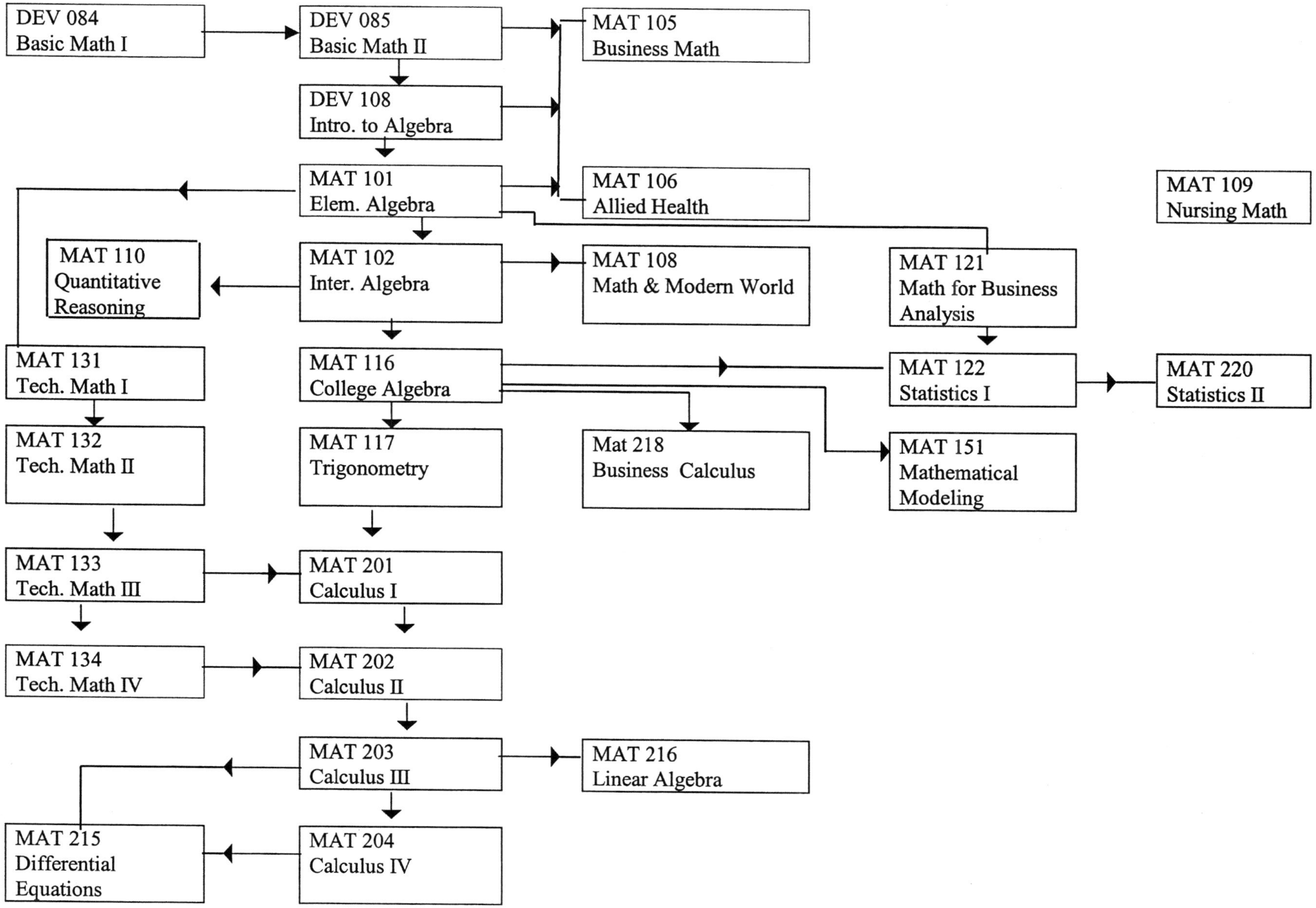


May 5, 2003

# SEQUENCE OF MATHEMATICS COURSES

## SINCLAIR POLICIES, PROCEDURES AND SERVICES FOR STUDENTS

The purpose of these policies, procedures, and services is to help you succeed as a student at Sinclair. Please carefully review and address any questions you might have to your instructor or the contacts provided below.

- **Educational Support Services**

  You may be eligible for free educational assistance if you are enrolled for credit. Contact Tutorial Services in Room 10-444 or call 512-2792. Tutoring is usually not provided for 200 level courses that have a prerequisite or developmental courses in which there are tutors. Assistance is available to eligible students through the Office of Disability Services. You are responsible for informing your instructor of any instructional accommodations and/or special learning needs at the beginning of the quarter. Tutoring information is available at http://www.sinclair.edu/departments/tutorial. English as a Second Language information and services are available in Room 10-421 or call 512-5113.

- **Important College Dates**

  Please check the listing of important college dates contained in the ***Quarterly Class Schedule*** including withdrawal dates, application date for graduation, etc.

- **Dropping a Course**

  If you drop a course during the fall, winter and spring quarters within the first eight calendar days (including Saturday and Sunday), you will receive a 100% refund and no record of a grade for that course. Deadlines for summer courses and short term, non-standard courses have different drop deadlines. Consult the ***Quarterly Class Schedule*** for these deadlines. If you drop a course after the first eight calendar days but before the end of the eighth week, you will receive a "W". You may not drop a course after the eighth week has ended. Consult the ***Quarterly Class Schedule*** for specific deadline dates. Drop forms are available at the Office of Registration & Student Records, second floor of Building 10. **PLEASE NOTE: It is your responsibility to obtain the drop form, complete the form, get a signature in Sinclair Central (10-242), and then turn that form into the Office of Registration & Student Records. Keep a signed and dated copy of the drop form for your records.** Failure to follow this procedure will result in "F" grades being recorded on your permanent record.

  You may also drop courses by calling the Office of Registration and Student Records (512-3000 or 1-800-315-3000) or accessing the telephone registration system (512-5454). If you call the Office of Registration and Student Records, make sure you ask that a copy of the drop form be mailed to you.

- **Withdrawing from College**

  If you decide to withdraw from all of your courses during the fall, winter and spring quarters within the first eight calendar days (including Saturday and Sunday), you will receive a 100% refund and no record of grades for the courses. Summer terms have different withdrawal deadlines. Consult the ***Quarterly Class Schedule*** for these deadlines. If you withdraw from your courses after the first eight calendar days but before the end of the eighth week, you will receive "Ws". You may not withdraw after the eighth week has ended. Consult the ***Quarterly Class Schedule*** for specific deadline dates. Withdrawal forms are available at the Office of Registration & Student Records, second floor of Building 10. **PLEASE NOTE: It is your responsibility to obtain the withdrawal form, complete the form, get a signature in Sinclair Central (10-242), and then turn that form into the Office of Registration &**

**Student Records. Keep a signed and dated copy of the drop form for your records.** Failure to follow this procedure will result in "F" grades being recorded on your permanent record.

You may also withdraw from all of your courses by calling the Office of Registration and Student Records (512-2736) or accessing the telephone registration system (512-5454). If you call the Office of Registration and Student Records, make sure you ask that a copy of the withdrawal form be mailed to you.

- **Late Registration & Change of Schedule**

Effective spring 2003, late and audit registration will be the week before classes begin. During this period, you may register for, or add, any class that is open. A late fee of $20 will be assessed if you register after the on-time registration ends. This late fee does not apply to audit courses. ***After the beginning of the quarter, you will not be permitted to register for any course that has already met for the first time.***

You may register to audit a course during the week before classes begin. **YOU MAY NOT CHANGE FROM AUDIT TO CREDIT, NOR CREDIT TO AUDIT, STATUS ANY TIME IN REGISTRATION PROCESS.** You will not receive credit for classes you audit, nor can you qualify for veterans benefits for classes you audit.

- **Attendance**

You are expected to be present at all class sessions. If you must be absent, it is your responsibility to contact your instructor and ask if missed assignments can be completed. Excessive absences and work not completed may affect your final grade, financial aid, and VA benefits.

- **Administrative Withdrawal**

You may be **administratively withdrawn** from a class by your instructor for nonattendance. He or she must advise you in writing **at the first class meeting** what attendance record would constitute cause for administrative withdrawal. If you do not attend the first class, it is your responsibility to obtain a copy of all materials which were distributed at the first class meeting.

You may also be administratively withdrawn from classes as a result of a **student conduct hearing** with either a hearing officer or the student conduct committee. Administrative withdrawals may be made when it has been determined that your presence on campus is potentially detrimental to the college, faculty, staff, students or yourself.

- **Grades**

A paper copy of your grades is no longer being sent out automatically at the end of each quarter. A paper copy of grades is sent to you only by request. To access your grades, call 1-800-613-9516 or log on to: http://www.sinclair.edu/stservices/rsr/index.cfm during the published dates listed in the ***Quarterly Class Schedule.***

Letter grades are issued at the end of each quarter:

| | | | |
|---|---|---|---|
| **A** | Excellent | **I** | Incomplete (unusual circumstances by contract only) |
| **B** | Good | **W** | Withdrawal |
| **C** | Average | **P** | Pass |
| **D** | Passing | **N** | Progress |
| **F** | Failure | **Y** | Proficiency Credit |
| **X** | Audit | **Z** | Non-attendance |

- **Student Behavior Guidelines**

Treat your classes as you would a desirable job. The instructor is a team leader and your fellow students are coworkers. Everyone must work together to complete learning objectives. These behaviors are expected of you:

1. Attend all classes on time.
2. Respect the rights of others to contribute by listening attentively. Show consideration for students, instructors and other college employees.
3. Participate appropriately and actively on topics presented in class.
4. Complete your assignments on time.
5. Ask for feedback from your instructors and peers to insure progress toward learning objectives.
6. Resolve problems by immediately discussing issues with your instructors and/or peers.

Your student rights and responsibilities are detailed in the ***Sinclair Community College Student Handbook*** available in Student Activities, Room 8025.

- **Safety and Security**

You may access the Campus Police website at http://www.sinclair.edu/departments/campuspolice for information on safety, crime reports/statistics, the department's annual report, and other services. For more information, contact the Campus Police Department in Building 7, Room 7112 or call 512-2534.

- **Academic Counseling**

You are encouraged to meet with an academic counselor early in your studies at Sinclair. Academic counselors are available to assist you in understanding your degree requirements, selecting courses, transferring to other institutions, etc. In addition, prior to taking any action that will impact your schedule, (e.g., dropping or adding a course) it is **strongly** recommended that you see your academic counselor.

| | | |
|---|---|---|
| Allied Health<br>Room 6120; 512-3029 | Developmental Studies<br>Room 6222; 512-2701 | Engineering & Industrial Technologies<br>Room 3142; 512-2282 |
| Business<br>Room 6131; 512-3054 | Fine & Performing Arts<br>Room 2222; 512-2544 | Liberal Arts & Sciences<br>Room 6121; 512-5134 |
| Extended Learning & Human Services<br>Room 9301; 512-2702 | | |

- **Other Counseling**

| | | |
|---|---|---|
| Counseling Services<br>Room 10324; 512-2752 | Sinclair Central<br>Room 10242; 512-2201 | Career Planning & Placement<br>Room 10315; 512-2772 |

- **Testing Center**

Exams are administered on a walk-in basis only at the Testing Center in Room 10445. You must present a Tartan ID Card, driver's license, or state issued photo ID at the time of testing. You may have to wait during peak periods (lunch time, after 4:00 p.m., midterms, and finals). You may call the Testing Center at 512-3076 to verify that your exam is on file and ready to be administered. Children are not permitted in the Center and may not be left unattended in the lobby.

# Math 102 Formulas

## Prerequisite Formulas

Formulas of special importance that students are expected to know upon entering this course:

- Geometric Formulas

| | Area | Perimeter |
|---|---|---|
| Square | $A = s^2$ | $P = 4s$ |
| Rectangle | $A = lw$ | $P = 2l + 2w$ |
| Triangle | $A = \frac{1}{2}(bh)$ | $P = a + b + c$ |

- Sum of the Angles in a Triangle $\quad A + B + C = 180°$

- Square of a Binomial:

$$(x+y)^2 = x^2 + 2xy + y^2$$
$$(x-y)^2 = x^2 - 2xy + y^2$$

- Pythagorean Theorem $\quad a^2 + b^2 = c^2$

- Absolute Value of $x$

$$|x| = \begin{cases} x & \text{if } x \geq 0 \\ -x & \text{if } x < 0 \end{cases}$$

## Course Formulas

Formulas that students are required to memorize in this course:

Chapter 5

- Difference of Squares $\quad x^2 - y^2 = (x+y)(x-y)$

- Sums and Differences of Cubes

$$x^3 + y^3 = (x+y)(x^2 - xy + y^2)$$
$$x^3 - y^3 = (x-y)(x^2 + xy + y^2)$$

Chapter 6

- Multiplying and Dividing Rational Expressions

$$\frac{P}{Q} \cdot \frac{R}{S} = \frac{PR}{QS} \quad (Q \neq 0, S \neq 0) \qquad \frac{P}{Q} \div \frac{R}{S} = \frac{P}{Q} \cdot \frac{S}{R} = \frac{PS}{QR} \quad (Q \neq 0, R \neq 0, S \neq 0)$$

- Adding and Subtracting Rational Expressions

$$\frac{P}{Q} + \frac{R}{Q} = \frac{P+R}{Q} \quad (Q \neq 0) \qquad \frac{P}{Q} - \frac{R}{Q} = \frac{P-R}{Q} \quad (Q \neq 0)$$

- Distance, Rate, Time $\quad d = rt, \ r = d/t, \ t = d/r$

- Rate of Work

If a job can be completed in $t$ units of time, then the rate of work is $\frac{1}{t}$ job per unit of time.

Chapter 7

- Slope of a Line $m = \frac{rise}{run} = \frac{change\ in\ y}{change\ in\ x} = \frac{y_2 - y_1}{x_2 - x_1} \quad (x_1 \neq x_2)$

- Forms of Linear Equations

| | |
|---|---|
| Point-slope form | $y - y_1 = m(x - x_1)$ |
| Slope-intercept form | $y = mx + b$ |
| General form | $Ax + By = C$ |
| Horizontal line | $y = b$ |
| Vertical line | $x = a$ |

- Midpoint Formula $\left(\frac{x_1 + x_2}{2}, \frac{y_1 + y_2}{2}\right)$

Chapter 9

- Absolute Value Equations and Inequalities

If $k$, $p$, and $q$ are real numbers with $k > 0$, then

$|ax + b| = k \iff ax + b = k$ or $ax + b = -k$

$|ax + b| > k \iff ax + b > k$ or $ax + b < -k$

$|ax + b| < k \iff -k < ax + b < k$

$|ax + b| = |cx + d| \iff ax + b = cx + d$ or $ax + b = -(cx + d)$

Chapter 10

- Principal Square Root of $a^2$ $\sqrt{a^2} = |a|$

If n is an even positive integer $\sqrt[n]{a^n} = |a|$

If n is an odd positive integer $\sqrt[n]{a^n} = a$

- Fractional Exponents

$$a^{\frac{1}{n}} = \sqrt[n]{a}$$

$$a^{\frac{m}{n}} = \left(a^{\frac{1}{n}}\right)^m = \left(a^m\right)^{\frac{1}{n}} = \sqrt[n]{a^m} = \left(\sqrt[n]{a}\right)^m$$

$$a^{-\frac{m}{n}} = \frac{1}{a^{\frac{m}{n}}} \quad (a \neq 0)$$

- Laws of Radicals

$$\sqrt[n]{a}\sqrt[n]{b} = \sqrt[n]{ab}$$

$$\frac{\sqrt[n]{a}}{\sqrt[n]{b}} = \sqrt[n]{\frac{a}{b}}$$

$$\sqrt[kn]{a^{km}} = \sqrt[n]{a^m}$$

- Distance Formula

$$d = \sqrt{(x_2 - x_1)^2 + (y_2 - y_1)^2}$$

- Imaginary Unit

$$i = \sqrt{-1} \qquad (i^2 = -1)$$

$$\sqrt{-b} = i\sqrt{b}$$

<u>Chapter 11</u>

- Square Root Property $\quad x^2 = k \quad \Leftrightarrow \quad x = \sqrt{k} \quad$ or $\quad x = -\sqrt{k}$

- Quadratic Formula

  If $\quad ax^2 + bx + c = 0,\;$ then $\quad x = \dfrac{-b \pm \sqrt{b^2 - 4ac}}{2a} \quad (a \neq 0)$

- Forms of Quadratic Functions

  General Form $\quad f(x) = ax^2 + bx + c$

  Standard Form $\quad F(x) = a(x-h)^2 + k,\;$ vertex $(h, k)$

- Line of Symmetry for a Parabola $\quad x = \dfrac{-b}{2a} \quad (a \neq 0)$

- Vertex Formula $\quad \left(\dfrac{-b}{2a}, f\left(\dfrac{-b}{2a}\right)\right) \quad (a \neq 0)$

# Formulas

| *Figure* | *Formulas* | *Illustration* |
|---|---|---|

**Square**

Perimeter: $P = 4s$
Area: $A = s^2$

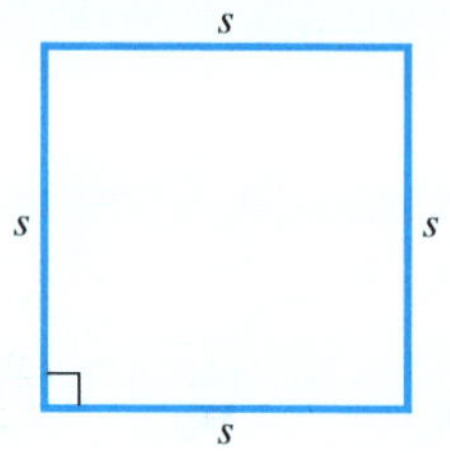


**Rectangle**

Perimeter: $P = 2L + 2W$
Area: $A = LW$

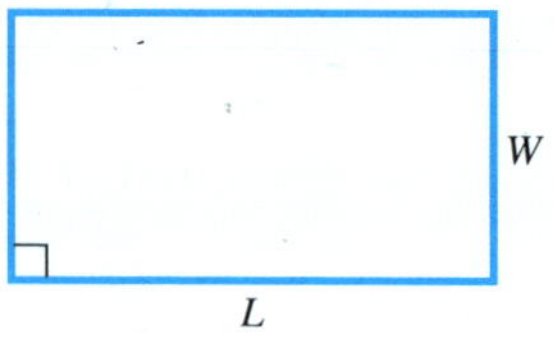


**Triangle**

Perimeter: $P = a + b + c$
Area: $A = \frac{1}{2}bh$

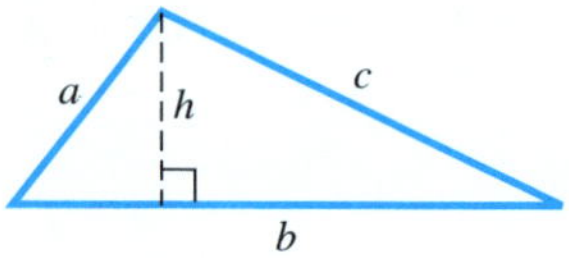


**Circle**

Diameter: $d = 2r$
Circumference: $C = 2\pi r$
$C = \pi d$

Area: $A = \pi r^2$

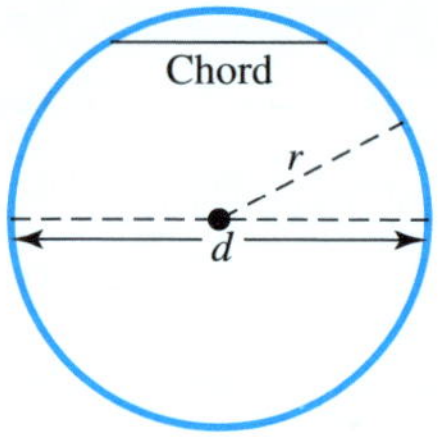


**Parallelogram**

Area: $A = bh$
Perimeter: $P = 2a + 2b$

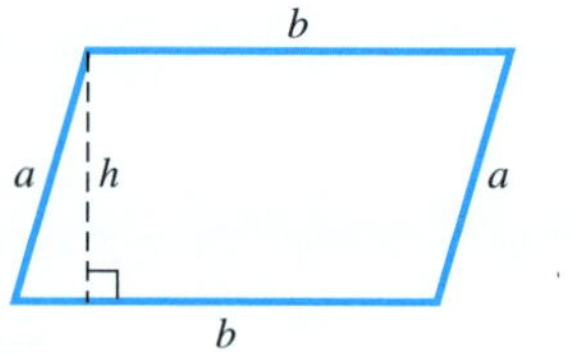


**Trapezoid**

Area: $A = \frac{1}{2}(B + b)h$

Perimeter: $P = a + b + c + B$

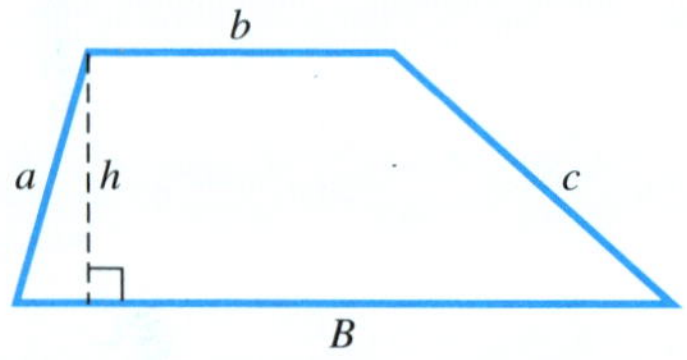